P9-CJM-863

ANNOTATED INSTRUCTOR'S EDITION

Developmental Mathematics

ARNOLD R. STEFFENSEN
L. MURPHY JOHNSON
Northern Arizona University

HarperCollins*Publishers*

Riverside Community College
Library
4800 Magnolia Avenue
Riverside, California 92506

MAY '95

To Barbara, Barbara, Becky, Cindy, and Pam

Sponsoring Editor:	Bill Poole
Project Editor:	Randee Wire
Art Direction:	Julie Anderson
Text and Cover Design:	Lucy Lesiak Design/Lucy Lesiak
Cover Photo:	Chicago Photographic Company ©
Photo Research:	Judy Ladendorf
Director of Production:	Jeanie Berke
Production Assistant:	Linda Murray
Compositor:	York Graphic Services
Printer and Binder:	Courier Corporation

Developmental Mathematics
Annotated Instructor's Edition

Copyright © 1991 by HarperCollins Publishers Inc.

All rights reserved. Printed in the United States of America. No part of this book
may be used or reproduced in any manner whatsoever without written permission,
except in the case of brief quotations embodied in critical articles and reviews. For
information address HarperCollins Publishers Inc. 10 East 53d Street, New York, NY
10022.

ISBN 0-673-53619-X

90 91 92 93 9 8 7 6 5 4 3 2 1

Developmental Mathematics is designed to give students a review of the basic skills of mathematics and to provide an introduction to the study of algebra. Some of the material in this text has been drawn from *Fundamentals of Mathematics, Third Edition,* and from *Introductory Algebra, Fourth Edition,* with special emphasis given to the TASP™ skill requirements, including graph reading and geometry.

Informal yet carefully worded explanations, detailed examples with accompanying practice exercises, pedagogical second color, abundant exercises, and comprehensive chapter reviews are hallmarks of the book. The text has been written for maximum instructor flexibility. Both core and peripheral topics can be selected to fit individual course needs. An annotated instructor's edition, testing manual, and test generator are provided for the instructor. Interactive tutorial software and a set of instructional videotapes are also available.

FEATURES

STUDENT GUIDEPOSTS Designed to help students locate important concepts as they study or review, student guideposts specify the major topics, rules, and procedures in each section. The guideposts are listed at the beginning of the section, then each is repeated as the corresponding material is discussed in the section.

EXAMPLES More than 1200 carefully selected examples include detailed, step-by-step solutions and side annotations in color. Each example is headed by a brief descriptive title to help students focus on the concept being developed and to aid in review.

PRACTICE EXERCISES These parallel each example and keep students involved with the presentation by allowing them to check immediately their understanding of ideas. Answers immediately follow these exercises.

CAUTIONS This feature calls students' attention to common mistakes and special problems to avoid.

HELPFUL HINTS These have been supplied throughout the early portions of the text to emphasize useful memory devices or identify more important mechanical procedures.

FIGURES AND ILLUSTRATIONS Abundant use has been made of figures and illustrations in examples and exercises, particularly applications.

COLOR Pedagogical color highlights important information throughout the book. Key definitions, rules, and procedures are set off in colored boxes for increased emphasis. Figures and graphs utilize color to clarify the concepts presented. Examples present important steps and helpful side comments in color.

EXERCISES As a key feature of the text, nearly 10,000 exercises, including about 1800 applied problems, are provided. Two parallel exercise sets (A and B) and a collection of extension exercises (C) follow each section and offer a wealth of practice for students and abundant flexibility for instructors. Exercises ranging from the routine to the more challenging, including application and calculator problems, are provided.

Exercises A This set of exercises includes space for working the problems, with answers immediately following the exercises. Some of these problems, identified by a colored circle ④ , have their solutions worked out at the back of the book. Many of these solutions are to exercises that students frequently have difficulty solving.

Exercises B This set matches the exercises in set A problem for problem but is presented without work space or answers.

Exercises C This set is designed to give students an extra challenge. These problems extend the concepts of the section or demand more thought than exercises in sets A and B. Answers or hints are given for selected exercises in this set.

REVIEW EXERCISES To provide ample opportunities for review, the text features a variety of review exercises.

For Review exercises are located at the end of most A and B exercise sets. They not only encourage continuous review of previously covered material, but also often provide special review preparation for topics covered in the upcoming section.

Chapter Review Exercises and a practice **Chapter Test** conclude each chapter. The Chapter Review Exercises are divided into two parts: the problems in Part I are ordered and marked by section. Those in Part II are not referenced to the source section, and are thoroughly mixed to help students prepare for examinations. Answers to all review and test exercises are provided in the text.

Final Review Exercises, referenced to each chapter and with answers supplied, are located at the back of the book.

CHAPTER REVIEWS In addition to the Chapter Review Exercises and Chapter Tests, comprehensive chapter reviews also include **Key Words** and **Key Concepts.** Key Words, listed by section, have brief definitions. Key Concepts summarize the major points of each section.

CALCULATORS The use of a calculator is discouraged until Section 5.8, where a brief introduction is presented. In working exercises that involve dividing decimals or approximating irrational numbers, a calculator can be an invaluable tool, and it is important for students to learn to judge for themselves when a calculator should or should not be used. Some exercises, however, do specifically refer to calculator use and are marked with this calculator symbol: ▤

INSTRUCTIONAL FLEXIBILITY

Developmental Mathematics offers proven flexibility for a variety of teaching situations such as individualized instruction, lab instruction, lecture classes, or a combination of methods.

Material in each section of the book is presented in a well-paced, easy-to-follow sequence. Students in a tutorial or lab instruction setting, aided by the student guideposts, can work through a section completely by reading the explanation, following the detailed steps in the examples, working the practice exercises, and then doing the exercises in set A.

The book can also serve as the basis for, or as a supplement to, classroom lectures. The straightforward presentation of material, numerous examples, practice exercises, and three sets of exercises offer the traditional lecture class an alternative approach within the convenient workbook format.

CONTENT HIGHLIGHTS

- Diagnostic Pretests for Chapters 1 and 2 help the instructor determine the extent of coverage necessary for a particular class.
- The proportion approach to percent is used with the percent equation in a side-by-side presentation in solving percent problems.
- Consistent emphasis is placed on real-world applications.
- A separate chapter on statistics (Chapter 7) provides extensive coverage of this important area.
- The development of operations with signed numbers has been given special consideration to provide a thorough discussion of the topic.
- An introduction to inductive and deductive reasoning has been placed at the beginning of the geometry chapter (Chapter 9) and referenced throughout.
- Special right triangles and the Pythagorean theorem are considered in detail in the chapter on radicals (Chapter 17).

SUPPLEMENTS

An extensive supplemental package is available for use with *Developmental Mathematics*.

The **ANNOTATED INSTRUCTOR'S EDITION** provides instructors with immediate access to the answers to every exercise in the text; each answer is printed in color next to the corresponding text exercise.

The **INSTRUCTOR'S TESTING MANUAL** contains a series of **ready-to-duplicate tests,** including a Placement Test, six different but equivalent tests for each chapter (four open-response and two multiple-choice), and two final exams, all with answers supplied. Section-by-section **teaching tips** provide suggestions for content implementation that an instructor, tutor, or teaching assistant might find helpful.

HARPERCOLLINS TEST GENERATOR FOR MATHEMATICS Available in Apple, IBM, and Macintosh versions, the test generator enables instructors to select questions by objective, section, or chapter, or to use a ready-made test for each chapter. Instructors may generate tests in multiple-choice or open-response formats, scramble the order of questions while printing, and produce multiple versions of each test (up to 9 with Apple, up to 25 with IBM and Macintosh). The system features printed graphics and accurate mathematics symbols. It also features a preview option that allows instructors to view questions before printing, to regenerate variables, and to replace or skip questions if desired. The IBM version includes an editor that allows instructors to add their own problems to existing data disks.

VIDEOTAPES A new videotape series, ALGEBRA CONNECTION: *The Developmental Mathematics Course,* has been developed to accompany *Developmental Mathematics*. Produced by an Emmy Award–winning team in consultation with a task force of academicians from both two-year and four-year colleges, the tapes cover all objectives, topics, and problem-solving techniques within the text. In addition, each lesson is preceded by motivational "launchers" that connect classroom activity to real-world applications.

INTERACTIVE TUTORIAL SOFTWARE This innovative package is also available in Apple, IBM, and Macintosh versions. It offers interactive modular units, specifically linked to the text, for reinforcement of selected topics. The tutorial is self-paced and provides unlimited opportunities to review lessons and to practice problem solving. When students give a wrong answer, they can request to see the problem worked out. The program is menu-driven for ease of use, and on-screen help can be obtained at any time with a single keystroke. Students' scores are automatically recorded and can be printed for a permanent record.

We extend our sincere gratitude to the students and instructors who used the previous editions of this book in other forms and offered many suggestions for improvement. Special thanks go to the instructors at Northern Arizona University and Yavapai Community College. In particular, the assistance given over the years by James Kirk and Michael Ratliff is most appreciated. Also, we sincerely appreciate the support and encouragement of the Northern Arizona University administration, especially President Eugene M. Hughes. It is a pleasure and privilege to serve on the faculty of a university that recognizes quality teaching as its primary role.

We also express our thanks to the following instructors who responded to questionnaires sent out by HarperCollins:

Tony Bower, *Saint Philip's College*

Jess Collins, *McLennan Community College*

Gregg Cox, *College of Boca Raton*

Robert Dunker, *Metropolitan Technical Community College*

Steve E. Green, *Tyler Junior College*

James Head, *Dalton Junior College*

Luther Henderson, *Broward Community College*

Kay Kriewold, *Laredo Junior College*

Virginia Licata, *Camden County College*

Nancy Long, *Trinity Valley Community College*

Vincent McGarry, *Austin Community College, Rio Grande Campus*

Susan L. Peterson, *Rock Valley College*

Douglas Proffer, *Collin County Community College*

J. Doug Richey, *Northeast Texas Community College*

Frances Rosamond, *National University*

Virginia Singer, *Northwest Community College*

Jim Swink, *Southwest Texas Junior College*

Sam Tinsley, *Eastfield College*

Andy Tomasulo, *Erie Community College, North Campus*

Glen Wallace, *Texas Southmost College*

Joseph Williams, *Essex County College*

Terry Wilson, *San Jacinto College, Central Campus*

We are also indebted to the following reviewers for their countless beneficial suggestions at various stages of the book's development:

Sharon Abramson, *Nassau Community College*

June A. Barrett, *California State University*

Janet Bickham, *Del Mar College*

LaVerne Blagmon-Earl, *University of the District of Columbia*

Douglas J. Campbell, *West Valley College*

John E. Chavez, *Skyline College*

Elizabeth J. Collins, *Glassboro State College*

Doris Cox, *Sauk Valley College*

Gladys Cummings, *Saint Petersburg Junior College*

Patricia Deamer, *Skyline College*

Sarah R. Evangelista, *Temple University*

Margaret H. Finster, *Erie Community College, South Campus*

Mary Lou Hart, *Brevard Community College*

Diana L. Hestwood, *Minneapolis Community College*

Winston Johnson, *Central Florida Community College*

Ginny Keen, *Western Michigan University*

Lucy Landesberg, *Nassau Community College*

Harry Lassiter, *Craven Community College*

Marion Littlepage, *Clark County Community College*

Lee H. Marsh, *Kalamazoo Valley Community College*

Kenneth R. Ohm, *Sheridan College*

John Pace, *Essex County College*

David Price, *Tarrant County Junior College*

Thomas J. Ribley, *Valencia Community College West*

C. Donald Rogers, *Erie Community College, North Campus*

David Sack, *Lincoln Land Community College*

Mark Serebransky, *Camden County College*

Virginia Singer, *Northwest Community College*

Gayle Smith, *Eastern Washington University*

Helen J. Smith, *South Mountain Community College*

Beverly Weatherwax, *Southwest Missouri State University*

Michael White, *Jefferson Community College*

Victoria J. Young, *Motlow State Community College*

We extend special appreciation to Joseph Mutter for the countless suggestions and support given over the years. Thanks also go to Diana Denlinger for typing this edition.

We thank our editors, Jack Pritchard, Bill Poole, Pam Carlson, Leslie Borns, Randee Wire, Sarah Joseph, and Ann-Marie Buesing, whose support has been most appreciated.

Finally, we are indebted to our families and in particular our wives, Barbara and Barbara, whose encouragement over the years cannot be measured.

Arnold R. Steffensen
L. Murphy Johnson

CONTENTS

17 RADICALS

18 QUADRATIC EQUATIONS

Students in this course traditionally have a variety of backgrounds. For some, the pace of Chapters 1 and 2 will be appropriate and provide a thorough review in preparation for the more complex topics that follow. On the other hand, some students will be able to review operations on whole numbers at a faster rate or omit it altogether, leaving more time for the remaining sections. As a result, we have provided the following two Diagnostic Pretests for Chapters 1 and 2 to help you determine the extent of coverage necessary for your particular class. Since the questions in each chapter are grouped by section, you may discover that some topics will need to be discussed, while others can be omitted. Answers to the Diagnostic Pretests are given in the Instructor's Guide that accompanies this text and in the Annotated Instructor's Edition. Also, multiple-choice versions of the Pretests have been included in the Instructor's Guide for those who wish to use this format of testing.

CHAPTER 1 DIAGNOSTIC PRETEST

1.1   **1.** Give the first five whole numbers.

1. _____ 0, 1, 2, 3, 4 _____

2. Consider the number line.

2. (a) _____ 8 _____

(a) What number corresponds to b?
(b) Place the appropriate symbol ($<$ or $>$) between b and 4.

(b) _____ > _____

$$b \underline{\ \ ?\ \ } 4$$

(c) _____ < _____

(c) Place the appropriate symbol ($<$ or $>$) between a and b.

$$a \underline{\ \ ?\ \ } b$$

3. Write $300,000 + 5000 + 200 + 6$ in standard notation.

3. _____ 305, 206 _____

4. In the number $62,451,903$, what digit represents the following places?

4. (a) _____ 2 _____

(a) millions   (b) ten thousands   (c) tens

(b) _____ 5 _____

(c) _____ 0 _____

5. Write $23,401$ in expanded notation.

5. _____ 20,000 + 3000 + 400 + 1 _____

1.2   **6.** What law of addition is illustrated by $6 + 10 = 10 + 6$?

6. _____ commutative law _____

7. What is the missing number? $2 + \underline{\ \ ?\ \ } = 2$

7. _____ 0 _____

1.3   *Find the following sums.*

8.	**9.**	**10.**	**11.**
21	648	2049	4235
+ 49	+ 37	+ 1167	217
			+ 6258

8. _____ 70 _____

9. _____ 685 _____

10. _____ 3216 _____

11. _____ 10,710 _____

12. What law of addition is illustrated by

$$(2 + 3) + 7 = 2 + (3 + 7)?$$

12. _____ associative law _____

1.4   **13.** Round $43,584$ to the nearest

13. (a) _____ 43,580 _____

(a) ten   (b) hundred   (c) thousand.

(b) _____ 43,600 _____

(c) _____ 44,000 _____

14. Estimate   2459 by rounding to the nearest
          +   375

   (a) ten   **(b)** hundred.

14. (a) _____ 2840 _____

     (b) _____ 2900 _____

15. During a two-game championship, 8176 people attended the first game, and 9857 attended the second. First estimate the total attendance, and then find the exact attendance.

15. _____ 18,000; 18,033 _____

1.5   **16.** What is the missing number? $7 - \underline{\quad?\quad} = 2$

16. _____ 5 _____

17. What is the missing number? $\underline{\quad?\quad} - 0 = 24$

17. _____ 24 _____

1.6   *Find the following differences.*

   18.   42   **19.**   643   **20.**   805   **21.**   9003
        − 35   − 97   − 486   − 157

18. _____ 7 _____

19. _____ 546 _____

20. _____ 319 _____

21. _____ 8846 _____

1.7   **22.** Professor Perko has three classes with 32, 45, and 13 students, respectively. How many students does Professor Perko have?

22. _____ 90 _____

23. On the first day of football practice, 137 players reported. After one week, 48 players had been cut or quit the team. How many players were still on the team at the end of the week?

23. _____ 89 _____

24. Cari received two checks for her graduation, one for $250 and the other for $325. She used the money to buy a sweater for $68, a skirt for $45, and a book for $7. She kept $50 in cash and deposited the rest of the money in her savings account. How much did she deposit?

24. _____ $405 _____

CHAPTER 2 DIAGNOSTIC PRETEST

2.1 **1.** Give the first three multiples of 5.

1. _____0, 5, 10_____

2. What law of multiplication is illustrated by

$$5 \times 9 = 9 \times 5?$$

2. _____commutative law_____

3. What is the missing number? $3 \times \underline{\ \ ?\ \ } = 3$

3. _____1_____

4. What is the missing number? $3 \times \underline{\ \ ?\ \ } = 0$

4. _____0_____

2.2 *Find the following products.*

5. 296	**6.** 43	**7.** 651	**8.** 4351
$\times\ \ 5$	$\times\ 15$	$\times\ 202$	$\times\ 600$

5. _____1480_____
6. _____645_____
7. _____131,502_____
8. _____2,610,600_____

2.3 **9.** What law of multiplication is illustrated by

$$(2 \times 3) \times 8 = 2 \times (3 \times 8)?$$

9. _____associative law_____

10. Estimate the product 897×603 by rounding to the nearest hundred.

10. _____540,000_____

11. Without actually finding the product, use estimation to determine whether the work shown appears to be correct.

```
  409
× 817
 2863
 409
3902
397,153
```

11. _____the estimated product is 320,000; appears to be incorrect_____

2.4 **12.** What is the missing number? $15 \div 3 = \underline{\ \ ?\ \ }$

12. _____5_____

13. What is the missing number? $12 \div \underline{\ \ ?\ \ } = 6$

13. _____2_____

14. What is the missing number? $0 \div 7 = \underline{\ \ ?\ \ }$

14. _____0_____

15. What number cannot be used as a divisor?

15. _____0_____

2.5 *Find the following quotients and remainders.*

16. $7\overline{)168}$ **17.** $21\overline{)745}$ **18.** $12\overline{)3100}$

16. _____Q: 24; R: 0_____

17. _____Q: 35; R: 10_____

18. _____Q: 258; R: 4_____

2.6 **19.** Dr. Besnette has 3 children and he wishes to divide equally an inheritance of $7695 among them. How much will each child receive?

19. _____ $2565

20. Nancy purchases 7 books at $16 each and 3 posters at $4 each. How much change will she receive if she pays with seven 20-dollar bills?

20. _____ $16

2.7 **21.** Is 17 a prime number? Explain.

21. _____ yes; 1 and 17 only divisors

22. What is the only even prime number?

22. _____ 2

23. Find the prime factorization of 150.

23. _____ $2 \cdot 3 \cdot 5 \cdot 5$

24. Is 1357 divisible by 3?

24. _____ no

25. What property do numbers that are divisible by 10 have?

25. _____ ones digit is 0

26. Numbers that are divisible by 2 or that are multiples of 2 have a special name. What is that name?

26. _____ even numbers

2.8 **27.** Is 16 a perfect square?

27. _____ yes

28. What is the square of 25?

28. _____ 625

29. What is the square root of 25?

29. _____ 5

30. Evaluate 2^4.

30. _____ 16

31. Write $3 \cdot 3 \cdot 3 \cdot 3 \cdot 3 \cdot 3$ using an exponent.

31. _____ 3^6

Perform the indicated operations.

32. $3 - 2 + 8 \div 4$

32. _____ 3

33. $(2 + 3)^2 - 2^2 + 3^2$

33. _____ 30

34. $5 \cdot 2^3$

34. _____ 40

35. $\sqrt{9} + 3$

35. _____ 1

During the past few years we have taught developmental mathematics to over 2000 students and have heard the following comments more than just a few times. ''I've always been afraid of math and have avoided it as much as possible.'' ''I don't like math, but it's required to graduate.'' ''I can't do percent problems!'' If you have ever made a similar statement, now is the time to think positively, stop making negative comments, and start down the path toward success in mathematics. Don't worry about this course as a whole. The material in the text is presented in a way that lets you take one small step at a time. Here are some general and specific guidelines that are necessary and helpful.

GENERAL GUIDELINES

1. Mastering mathematics requires motivation and dedication. A successful athlete does not become a champion without commitment to his or her goal. The same is true for the successful student of mathematics. Be prepared to work hard and spend time studying.

2. Mathematics is not learned simply by watching, listening, or reading; *it is learned by doing*. Use your pencil and practice. When your thoughts are organized and written in a neat and orderly fashion, you have taken a giant step toward success. Be complete and write out all details. The following are samples of two students' work on a word problem. Can you tell which one was the most successful in the course?

STUDENT A

Tax is 5% of total cost

$720 = 0.05 · □

□ = $\frac{720}{0.05}$ (related division Sentence)

= 14400

Thus, the total cost of the car was $14,400.

STUDENT F

tax = 5%

cost = $720

5% X 720 ~~3600%~~

.05 X 720 = 36

the cost of the car was $36

3. The use of a calculator is introduced in Section 5.8. Become familiar with the features of your calculator by consulting the owner's manual. When working problems that involve computing with decimals, a calculator can be a time-saving device. On the other hand, you should not be so dependent on a calculator that you use it to perform simple calculations that might be done mentally. For example, it would be ridiculous to use a calculator to find $8 \div 2$, while it would be helpful in finding $7343.84 \div 1.12$. It is important that you learn when to use and when not to use your calculator.

SPECIFIC GUIDELINES

1. As you begin to study each section, glance through the material and obtain a preview of what is coming.

2. Return to the beginning of the section and start reading slowly. The STUDENT GUIDEPOSTS will help you locate important concepts as you progress or as you review.

3. Read through each EXAMPLE and make sure that you understand each step. The side comments in color will help you if something is not quite clear.

4. After reading each example, work the parallel PRACTICE EXERCISE. This will reinforce what you have just read and start the process of practice.

5. Periodically you will encounter a CAUTION. These warn you of common mistakes and special problems that should be avoided.

6. The HELPFUL HINTS in the earlier portions of the text are designed to give you a better understanding of material or a more direct way of solving a problem.

7. After you have completed the material in the body of a section, you must test your mastery of the skills and practice, practice, practice! Begin with the exercises in set A. Answers to all of the problems are placed at the end of the set for easy reference. Some of the problems, identified by colored exercise numbers, have complete solutions at the back of the book. After trying these problems, refer to the step-by-step solutions if you have difficulty. You may want to practice more by doing the exercises in set B or to challenge yourself by trying the exercises in set C.

8. After you have completed all of the sections in the chapter, read the CHAPTER REVIEW that contains key words and key concepts. The review exercises provide additional practice before you take the CHAPTER TEST. Answers to the tests are at the back of the book.

9. To aid you in studying for your final examination, we have concluded the book with a comprehensive set of FINAL REVIEW EXERCISES.

10. To review the definitions of important terms covered previously, refer to the GLOSSARY that is also located at the back of the text.

If you follow these steps and work closely with your instructor, you will greatly improve your chance of success in mathematics.

Best of luck, and remember that you can do it!

Adding and Subtracting Whole Numbers

1.1 WHOLE NUMBERS

1 NUMBERS AND THE NUMBER LINE

The numbers that we learn about first are the **natural** or **counting numbers.** They are used to answer the question ''How many?'' The symbols or **numerals** representing these numbers are

$$1, 2, 3, 4, 5, 6, 7, 8, 9, 10, 11, \ldots .$$

The three dots are used to show that the pattern continues.

The **whole numbers** are formed by including 0 with the natural numbers. Thus,

$$0, 1, 2, 3, 4, 5, 6, 7, 8, 9, 10, 11$$

are the first twelve whole numbers.

Numbers are actually ideas that exist in our minds. It is helpful to ''picture'' these ideas using a **number line,** shown in Figure 1.1.

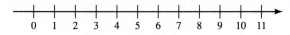

Figure 1.1

Every whole number corresponds to a point on a number line. The smallest whole number is 0. There is no largest whole number, since we can keep on extending the number line to the right. The arrowhead points to the direction in which the numbers get larger.

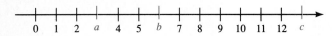

EXAMPLE 1   IDENTIFYING NUMBERS ON A NUMBER LINE

PRACTICE EXERCISE 1

What numbers are paired with *a, b,* and *c* in Figure 1.2?

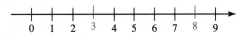

Figure 1.2

The point *a* is paired with the number 3, *b* with the number 6, and *c* with the number 13.

What numbers are paired with *x, y,* and *z* on the number line below?

Answer: *x* is paired with 5, *y* is paired with 1, and *z* is paired with 11.

② ORDER AND INEQUALITY SYMBOLS

The whole numbers occur in a natural **order.** For example, we know that 3 is less than 8, and that 8 is greater than 3. On a number line, 3 is to the left of 8, while 8 is to the right of 3.

Figure 1.3

The symbol $<$ means "is less than." We write "3 is less than 8" as

$$3 < 8.$$

Similarly, the symbol $>$ means "is greater than." Thus, we write "8 is greater than 3" as

$$8 > 3.$$

The symbols $<$ and $>$ are called **inequality symbols.**

Order on a Number Line
1. A number to the left of another number on a number line is *less* than the other number.
2. A number to the right of another number on a number line is *greater than* the other number.

HINT

When the inequality symbols $<$ and $>$ are used, remember that the symbol points to the smaller of the two numbers. Also, when referring to a number line, it may help to remember that "left" and "less" both begin with the same letter.

| EXAMPLE 2 ORDERING NUMBERS | PRACTICE EXERCISE 2 |

Place the correct symbol (< or >) between the numbers in the given pairs. Use part of a number line if necessary.

(a) 2____7

2 < 7. Notice 2 is to the left of 7 on the number line in Figure 1.4.

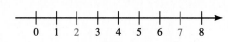

Figure 1.4

(b) 12____9

12 > 9. Notice that 12 is to the right of 9 on the number line in Figure 1.5.

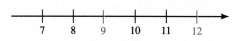

Figure 1.5

(c) 0____42

0 < 42. See Figure 1.6.

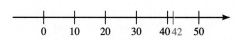

Figure 1.6

Place the correct symbol (< or >) between the numbers in the given pairs.

(a) 3____10

(b) 15____5

(c) 0____106

Answers: (a) <   (b) >   (c) <

H I N T

Notice that 13 > 6, read "13 is greater than 6," and 6 < 13, read "6 is less than 13," have the same meaning even though they are read differently.

❸ PLACE-VALUE NUMBER SYSTEM

The first ten whole numbers,

0, 1, 2, 3, 4, 5, 6, 7, 8, 9,

are called **digits** (from the Latin word *digitus* meaning finger). Any whole number can be written using a combination of these ten digits. For example,

245 is a three-digit number
3 is a one-digit number
75 is a two-digit number
4280 is a four-digit number.

The value of each digit in a number depends on its place in the number. In other words, our number system is a **place-value system.** For example, the four-digit number

$$2538,$$

written above in **standard notation,** is a short form of the **expanded notation**

$$2000 + 500 + 30 + 8,$$

as shown in the following diagram.

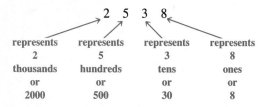

represents	represents	represents	represents
2	5	3	8
thousands	hundreds	tens	ones
or	or	or	or
2000	500	30	8

EXAMPLE 3 WRITING IN EXPANDED NOTATION	PRACTICE EXERCISE 3

Write each number in expanded notation.

(a) 532

Write each number in expanded notation.

(a) 103

$$5 \quad 3 \quad 2 = 500 + 30 + 2$$

represents	represents	represents
5	3	2
hundreds	tens	ones
or	or	or
500	30	2

(b) 1030

(b) 5032

$$5 \quad 0 \quad 3 \quad 2 = 5000 + 30 + 2$$

5	0	3	2
thousands	hundred	tens	ones
or	or	or	or
5000	0	30	2

Answers: **(a)** $100 + 3$
(b) $1000 + 30$

Notice the difference between 5032 and 532 in Examples 3 **(b)** and 3 **(a)**. In 5032, 0 represents 0 hundreds while the digit 5 represents 5 thousands. If 0 were not in the hundreds place in 5032, we would read the number as 532. It is clear from this example that 0 is an important **place-holder** in our system.

EXAMPLE 4 WRITING IN STANDARD NOTATION	PRACTICE EXERCISE 4

Write each number in standard notation.

(a) $300 + 20 + 8$ $300 + 20 + 8 = 328$

(b) $5000 + 400 + 30 + 9$ $5000 + 400 + 30 + 9 = 5439$

(c) $7000 + 20$

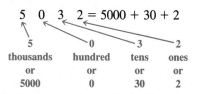

$7000 + 20 = 7020$ Zero in the hundreds place and ones place

(d) $90,000 + 300 + 1$ $90,000 + 300 + 1 = 90,301$

Write each number in standard notation.

(a) $900 + 20 + 7$

(b) $8000 + 200 + 70 + 1$

(c) $8000 + 70 + 1$

(d) $10,000 + 300 + 5$

Answers: **(a)** 927   **(b)** 8271
(c) 8071   **(d)** 10,305

④ WORD NAMES FOR NUMBERS

When a number contains more than four digits, commas are used to separate the digits into groups of three, as in the number

$$2,431,897,502,113.$$

Notice that the commas are placed starting at the right. The last group of digits on the left can have one, two, or three digits. Each group of three has a name, as does each digit in the group. These names are given in the following chart.

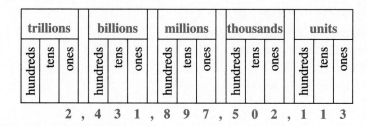

In words, the number above is

two trillion, four hundred thirty-one billion, eight hundred ninety-seven million, five hundred two thousand, one hundred thirteen.

The commas in the word name are in the same place as they are in the standard notation.

⫻⫻⫻⫻⫻⫻⫻⫻ C A U T I O N ⫻⫻⫻⫻⫻⫻⫻⫻

Do not use the word *and* when expressing whole numbers in words. For example, 423,123 is four hundred twenty-three thousand, one hundred twenty-three, *not* four hundred *and* twenty-three thousand, one hundred *and* twenty-three. The word *and* will be used later in Chapter 5 to refer to a decimal point in a number.

⫻⫻⫻⫻⫻⫻

EXAMPLE 5 WRITING WORD NAMES FROM STANDARD NOTATION

Write a word name for each number

(a) 39,205

$$\begin{array}{c} 39,205 \\ \downarrow \\ \text{thousand} \end{array}$$

The word name is thirty-nine thousand, two hundred five. The word *and* is not put between the words *hundred* and *five*.

(b) 27,532,141,269

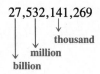

PRACTICE EXERCISE 5

Write a word name for each number.

(a) 436,109

(b) 84,000,227

(c) 7,643,000,058

The word name is twenty-seven billion, five hundred thirty-two million, one hundred forty-one thousand, two hundred sixty-nine.

(c) 904,000,102,425,000

The word name is nine hundred four trillion, one hundred two million, four hundred twenty-five thousand.

Answers: (a) four hundred thirty-six thousand, one hundred nine (b) eighty-four million, two hundred twenty-seven (c) seven billion, six hundred forty-three million, fifty-eight

HINT

Four-digit numbers are written without commas. For example, we write 4136 instead of 4,136. However, a number with five or more digits, such as 14,136, should be written with the comma.

EXAMPLE 6 WRITING STANDARD NOTATION FROM WORD NAMES

Write each number in standard notation.

(a) Four thousand, one hundred thirty-six.

4136

(b) Four hundred eighty-three billion, two hundred sixty million, five hundred twenty-one thousand, nine hundred ninety-six

billions	millions	thousands	units
4 8 3,	2 6 0,	5 2 1,	9 9 6

(c) Six trillion, three hundred twenty-four million, five hundred six

6,000,324,000,506

PRACTICE EXERCISE 6

Write each number in standard notation.

(a) Seven thousand, twenty-three

(b) Two hundred eighty-seven million, four hundred five thousand, nine hundred sixty-six

(c) Nine trillion, seven hundred fifty thousand, two hundred one

Answers: (a) 7023
(b) 287,405,966
(c) 9,000,000,750,201

The value of each digit in a number is determined by its place or position.

EXAMPLE 7 GIVING THE VALUE OF A DIGIT IN A NUMBER

Give the value of the digit 4 in each number.

(a) 304,152   4 thousands

(b) 23,241,309,117   4 ten millions

(c) 452,000,302,198   4 hundred billions

(d) 32,922,647   4 tens

PRACTICE EXERCISE 7

Give the value of the digit 7 in each number.

(a) 1,720,458

(b) 65,286,127,003

(c) 97,480,153

(d) 925,703

Answers: (a) 7 hundred thousands
(b) 7 thousands   (c) 7 millions
(d) 7 hundreds

EXAMPLE 8   IDENTIFYING DIGITS BY NAME

In the number 12,345,678 give the digit that represents the following places.

(a) Ten millions   1

(b) Hundred thousands   3

(c) Hundreds   6

(d) Ten thousands   4

(e) Ones   8

```
1 2 , 3 4 5 , 6 7 8
          ones
        hundreds
      ten thousands
    hundred thousands
  ten millions
```

PRACTICE EXERCISE 8

In the number 856,321,409 give the digit that represents the following places.

(a) Hundred millions

(b) Millions

(c) Ten thousands

(d) Thousands

(e) Tens

Answers: (a) **8**   (b) **6**   (c) **2**   (d) **1**   (e) **0**

Extremely large numbers, in the billions or larger, are difficult to imagine. For example, traveling day and night at a rate of 55 miles per hour, it would take over 2000 years to travel one billion miles and over 2,000,000 years to travel one trillion miles.

1.1   EXERCISES A

1. What are the numbers 1, 2, 3, . . . called?
   natural or counting numbers

2. How are numbers often "pictured" in mathematics?
   using a number line

3. How do we read the symbols 7 > 2?
   7 is greater than 2

4. What is the largest three-digit number?
   999

5. What is the smallest two-digit number?
   10

6. What is the largest digit?
   9

7. Is 10 a natural number?
   yes

8. Is 10 a digit?
   no

9. What is the smallest whole number?
   0

10. What is the largest whole number?
    There is no largest whole number.

11. On the given number line, what number is paired with each point?
    (a) a   (b) b   (c) c   (d) d
    2   5   10   13

```
+--+--+--+--+--+--+--+--+--+--+--+--+--+-->
0  1  a     b  7        c        d
```

Refer to the number line below to place the correct symbol (< or >) between the numbers in the following pairs.

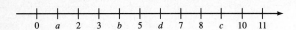

12. 2 <u><</u> 3

13. 8 <u>></u> 1

14. 0 <u><</u> 10

15. 7 <u>></u> 0

16. *a* <u><</u> 2

17. 7 <u>></u> *b*

18. 7 <u><</u> *c*

19. *d* <u>></u> *a*

20. *b* <u><</u> *d*

Write each number in expanded notation.

21. 2479

2000 + 400 + 70 + 9

22. 503

500 + 3

23. 207,519

200,000 + 7000 + 500 + 10 + 9

24. 4,127,982

4,000,000 + 100,000 + 20,000 + 7000 + 900 + 80 + 2

Write each number in standard notation.

25. 500 + 20 + 3

523

26. 10,000 + 600 + 20 + 1

10,621

27. 300 + 5

305

28. 400,000 + 30,000 + 9000 + 700 + 20 + 1

439,721

Write a word name for each number.

29. 4219

four thousand, two hundred nineteen

30. 107,586

one hundred seven thousand, five hundred eighty-six

31. 93,117

ninety-three thousand, one hundred seventeen

32. 13,219,475

thirteen million, two hundred nineteen thousand, four hundred seventy-five

Write each number in standard notation.

33. Six thousand, seven hundred five

6705

34. Twenty-four thousand, one hundred fifty-nine

24,159

35. Three billion, four hundred twenty-seven million, one hundred ninety-three thousand, two hundred

3,427,193,200

36. Eight hundred twenty-five million, one

825,000,001

Give the value of the digit 5 in each number.

37. 325,076   5 thousands

38. 51,032,299   5 ten millions

39. 3,279,005   5 ones

40. 5,361,291,447   5 billions

In the number 63,572,189, what digit represents each of the following places?

41. Millions   3

42. Ten thousands   7

43. Tens   8

44. Ten millions   6

45. Hundred thousands   5

46. Ones   9

In Exercises 47–48, write the number in the statement in standard notation.

47. The world population is approximately four billion, eight hundred forty-five million.   4,845,000,000

48. The U.S. Gross National Product in a recent year exceeded three trillion, six hundred twenty-five billion, one hundred fifty million dollars. $3,625,150,000,000

In Exercises 49–50, write a word name for the number in the statement.

49. One light-year, the distance light travels in one year, is approximately 5,879,195,000,000 miles.
five trillion, eight hundred seventy-nine billion, one hundred ninety-five million

50. It has been estimated that about 139,710,000 square kilometers of the earth's surface area is water.
one hundred thirty-nine million, seven hundred ten thousand

ANSWERS: 1. natural or counting numbers 2. using a number line 3. 7 is greater than 2 4. 999 5. 10 6. 9
7. yes 8. no 9. 0 10. There is no largest whole number. 11. (a) 2 (b) 5 (c) 10 (d) 13 12. 2 < 3 13. 8 > 1
14. 0 < 10 15. 7 > 0 16. $a < 2$ 17. 7 > b 18. 7 < c 19. $d > a$ 20. $b < d$ 21. 2000 + 400 + 70 + 9 22. 500 +
3 23. 200,000 + 7000 + 500 + 10 + 9 24. 4,000,000 + 100,000 + 20,000 + 7000 + 900 + 80 + 2 25. 523
26. 10,621 27. 305 28. 439,721 29. four thousand, two hundred nineteen 30. one hundred seven thousand, five
hundred eighty-six 31. ninety-three thousand, one hundred seventeen 32. thirteen million, two hundred nineteen
thousand, four hundred seventy-five 33. 6705 34. 24,159 35. 3,427,193,200 36. 825,000,001 37. 5 thousands
38. 5 ten millions 39. 5 ones 40. 5 billions 41. 3 42. 7 43. 8 44. 6 45. 5 46. 9 47. 4,845,000,000
48. $3,625,150,000,000 49. five trillion, eight hundred seventy-nine billion, one hundred ninety-five million 50. one
hundred thirty-nine million, seven hundred ten thousand

1.1 EXERCISES B

1. What numbers are formed by including 0 with the natural numbers?   **whole numbers**

2. What is the symbol for a number called?
numeral

3. How do we read the symbols 3 < 4?
3 is less than 4.

4. What is another name for the first ten whole numbers?   **digits**

5. What is the largest four-digit number?
9999

6. What is the smallest three-digit number?
100

7. Is 10 a counting number?
yes

8. Is 10 a whole number?
yes

9. What is the smallest natural number?
1

10. What is the largest natural number?
There is no largest natural number.

11. On the given number line, what number is paired with each point?   **(a)** a **(b)** b **(c)** c **(d)** d
3 **6** **9** **12**

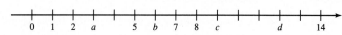

Refer to the number line below to place the correct symbol (< or >) between the numbers in the following pairs.

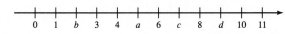

12. 4 _**<**_ 11

13. 9 _**>**_ 2

14. 5 _**>**_ 0

15. 0 _**<**_ 8

16. b _**<**_ 3

17. a _**<**_ 6

18. 2 _**<**_ c

19. c _**>**_ a

20. a _**<**_ d

Write each number in expanded notation.

21. 409
400 + 9

22. 21,155
**20,000 + 1000 +
100 + 50 + 5**

23. 603,114
**600,000 + 3000 +
100 + 10 + 4**

24. 7,123,321
**7,000,000 + 100,000 +
20,000 + 3000 +
300 + 20 + 1**

Write each number in standard notation.

25. 800 + 20 + 7
827

26. 5000 + 800 + 30 + 2
5832

27. 30,000 + 500 + 40 + 3
30,543

28. 70,000 + 300
70,300

Write a word name for each number.

29. 5143
five thousand, one
hundred forty-three

30. 203,195
two hundred three
thousand, one hundred
ninety-five

31. 65,728
sixty-five thousand,
seven hundred
twenty-eight

32. 21,205,414
twenty-one million,
two hundred five
thousand, four hundred
fourteen

Write each number in standard notation.

33. Eleven thousand, two hundred twenty-six
11,226

34. Ninety-five thousand, six hundred twenty-seven
95,627

35. Ten billion, two hundred five million, three hundred
forty-seven   10,205,000,347

36. Seven hundred six million, fifteen
706,000,015

Give the value of the digit 9 in each number.

37. 193,040
9 ten thousands

38. 923,147
9 hundred thousands

39. 27,925
9 hundreds

40. 8,145,329,500
9 thousands

In the number 520,193,678, *what digit represents each of the following places?*

41. Ten millions   2

42. Ten thousands   9

43. Tens   7

44. Ones   8

45. Hundred thousands   1

46. Millions   0

In Exercises 47–48, write the number in the statement in standard notation.

47. The population of the greater Los Angeles area exceeds three million, eight hundred thousand.
3,800,000

48. There are over one hundred sixty-five million, five hundred thousand television sets in the United States.   165,500,000

In Exercises 49–50, write a word name for the number in the statement.

49. It has been estimated that motorists in the United States waste over 250,700,000 gallons of gasoline annually just warming up their automobiles.
two hundred fifty million, seven hundred thousand

50. A variety store has over 1,570,300 items in its inventory
one million, five hundred seventy thousand, three hundred

1.1   EXERCISES C

Assume that a and b represent whole numbers in Exercises 1–4.

1. If $a < b$ is $2 + a < 2 + b$?
yes

2. If $a < 5$ and $5 < b$ is $a < b$?
yes

3. If $a < b$, could we also write $b > a$?   yes

4. If $a < b$ and $b < a$ are both false, what could we conclude?
a and b represent the same number.

1.2 BASIC ADDITION FACTS

1 ADDING TWO WHOLE NUMBERS

Addition of two whole numbers can be thought of as counting. The two numbers to be added are called **addends,** and the result is their **sum.** The symbol used in addition is the **plus sign, +.** As an example, suppose we add 2 coins and 5 coins. There are 2 coins in one stack and 5 coins in another stack, as in Figure 1.7. The addends are written

$$2 + 5.$$

The sum can be found by counting the total number of coins. When the two stacks are put together, the combined stack contains 7 coins, so

$$2 + 5 = 7.$$

Figure 1.7

Addition problems are written either horizontally or vertically.

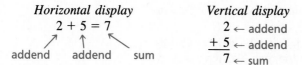

Horizontal display
$$2 + 5 = 7$$
addend   addend   sum

Vertical display
$$\begin{array}{r} 2 \leftarrow \text{addend} \\ + 5 \leftarrow \text{addend} \\ \hline 7 \leftarrow \text{sum} \end{array}$$

Addition can be shown using a number line. To add 2 and 5 on a number line, first move 2 units to the right of 0. From this point, move 5 more units to the right. The net result is a movement of 7 units from 0. Using the number line in Figure 1.8, we have shown that $2 + 5 = 7$.

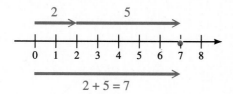

Figure 1.8

| EXAMPLE 1 ADDING TWO WHOLE NUMBERS | PRACTICE EXERCISE 1 |

Add.

Add.

(a) $4 + 2 = 6$ 4 objects + 2 objects results in 6 objects

(a) $6 + 3$

We could also show this using the number line in Figure 1.9.

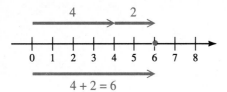

Figure 1.9

(b)   1
       $+\ 8$
       $\overline{\quad 9}$ 1 object + 8 objects results in 9 objects

(b)   9
       $+\ 5$

The number line in Figure 1.10 shows the same problem.

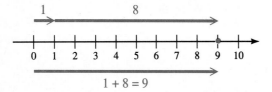

Figure 1.10

Answers: (a) 9   (b) 14

② BASIC ADDITION FACTS

Finding sums by counting or by using a number line takes time. However, by memorizing the **basic addition facts,** we can find the sum of any two one-digit numbers quickly. These facts, which serve as building blocks for larger sums, are summarized in the table below.

To find $7 + 3$ using the table, start at 7 in the left column and move straight across to the right until you are below the 3 in the top row. You have arrived at the number 10, which is $7 + 3$.

+	0	1	2	3	4	5	6	7	8	9
0	0	1	2	3	4	5	6	7	8	9
1	1	2	3	4	5	6	7	8	9	10
2	2	3	4	5	6	7	8	9	10	11
3	3	4	5	6	7	8	9	10	11	12
4	4	5	6	7	8	9	10	11	12	13
5	5	6	7	8	9	10	11	12	13	14
6	6	7	8	9	10	11	12	13	14	15
7	7	8	9	10	11	12	13	14	15	16
8	8	9	10	11	12	13	14	15	16	17
9	9	10	11	12	13	14	15	16	17	18

Addition Table

EXAMPLE 2   Using the Addition Table

Use the Addition Table to find the sums.

(a) 8 + 4   Start at 8 in the left column and move across to below 4. The number is 12. Thus 8 + 4 = 12.

(b) 4 + 8   Start at 4 in the left column and move across to below 8. Thus 4 + 8 = 12.

(c) 0 + 4   Start at 0 in the left column and move across to below 4. Thus 0 + 4 = 4.

(d) 4 + 0   Start at 4 in the left column and move across to below 0. Thus 4 + 0 = 4.

PRACTICE EXERCISE 2

Use the Addition Table to find the sums.

(a) 2 + 7

(b) 5 + 5

(c) 0 + 9

(d) 9 + 0

Answers: (a) 9   (b) 10   (c) 9 (d) 9

❸ SOME PROPERTIES OF ADDITION

Parts **(a)** and **(b)** of Example 2 show that changing the order of addition does not change the sum. We saw that

$$4 + 8 = 8 + 4.$$

This property of the number system is true in general.

The Commutative Law of Addition

The **commutative law of addition** tells us that the order in which two (or more) numbers are added does not change the sum.

Parts **(c)** and **(d)** of Example 2 show that

$$0 + 4 = 4 \qquad \text{and} \qquad 4 + 0 = 4.$$

In general, if zero is added to any number, the sum is the *identical* number. For this reason, zero is sometimes called the **additive identity.**

It will be necessary to find the sum of three one-digit numbers in the next section. We can use a number line as shown in the next example.

EXAMPLE 3   Adding Three One-Digit Numbers

Add 3 + 4 + 2 using a number line.

First move 3 units to the right of 0. From this point (corresponding to 3) move 4 more units to the right arriving at 7. Finally, move two additional units to reach the point that corresponds to 9, as shown in Figure 1.11.

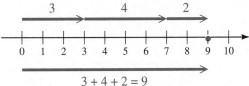

Figure 1.11

PRACTICE EXERCISE 3

Add 1 + 5 + 3 using a number line.

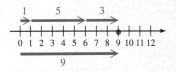

Answer: 9

4 APPLICATIONS OF ADDITION

Many problems are solved by adding whole numbers as shown in the next example.

| EXAMPLE 4 AN APPLICATION OF ADDITION | PRACTICE EXERCISE 4 |

Figure 1.12 shows a sketch of a trail map in a wilderness area. (Note: **mi** is the abbreviation for miles.)

Refer to Figure 1.12

(a) A scout troop hiked from the Twin Buttes to Trout Lake and on to the Hangman's Tree. How far did they hike?

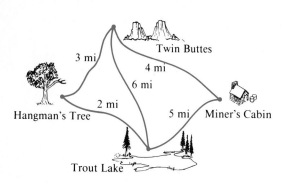

Figure 1.12

(a) If Joe hikes from the Miner's Cabin to Trout Lake and on to the Twin Buttes, how far has he hiked?

The distance hiked is the sum of 5 mi and 6 mi. Since

$$5 + 6 = 11,$$

Joe hiked 11 mi.

(b) Kelli hiked from Trout Lake to the Hangman's Tree, on to the Twin Buttes, and ended up at the Miner's Cabin. How far did she hike?

Since

$$2 + 3 + 4 = 9,$$

Kelli hiked a total of 9 mi.

(b) Ranger Dave hiked from the Miner's Cabin to the Twin Buttes, on to Trout Lake, and ended the hike at the Hangman's Tree. How far did he hike?

Answers: (a) 8 mi   (b) 12 mi

1.2 EXERCISES A

What addition problems are displayed below?

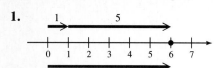

1.

$1 + 5 = 6$

2.

$5 + 0 = 5$

You should be able to complete Exercises 3–52 in 90 seconds or less.

3. 8 + 9 ‾17‾	**4.** 0 + 2 ‾2‾	**5.** 3 + 7 ‾10‾	**6.** 6 + 0 ‾6‾	**7.** 2 + 4 ‾6‾	**8.** 7 + 1 ‾8‾	**9.** 0 + 0 ‾0‾

3. 8
  $+\,9$
  $\overline{17}$

4. 0
  $+\,2$
  $\overline{2}$

5. 3
  $+\,7$
  $\overline{10}$

6. 6
  $+\,0$
  $\overline{6}$

7. 2
  $+\,4$
  $\overline{6}$

8. 7
  $+\,1$
  $\overline{8}$

9. 0
  $+\,0$
  $\overline{0}$

10. 2
  $+\,5$
  $\overline{7}$

11. 7
  $+\,7$
  $\overline{14}$

12. 8
  $+\,2$
  $\overline{10}$

13. 9
  $+\,0$
  $\overline{9}$

14. 6
  $+\,5$
  $\overline{11}$

15. 7
  $+\,8$
  $\overline{15}$

16. 3
  $+\,4$
  $\overline{7}$

17. 3
  $+\,0$
  $\overline{3}$

18. 2
  $+\,7$
  $\overline{9}$

19. 5
  $+\,9$
  $\overline{14}$

20. 1
  $+\,3$
  $\overline{4}$

21. 0
  $+\,4$
  $\overline{4}$

22. 5
  $+\,7$
  $\overline{12}$

23. 8
  $+\,1$
  $\overline{9}$

24. 1
  $+\,6$
  $\overline{7}$

25. 4
  $+\,0$
  $\overline{4}$

26. 6
  $+\,9$
  $\overline{15}$

27. 9
  $+\,2$
  $\overline{11}$

28. 8
  $+\,5$
  $\overline{13}$

29. 3
  $+\,5$
  $\overline{8}$

30. 7
  $+\,9$
  $\overline{16}$

31. 0
  $+\,3$
  $\overline{3}$

32. 7
  $+\,6$
  $\overline{13}$

33. 4
  $+\,7$
  $\overline{11}$

34. 9
  $+\,3$
  $\overline{12}$

35. 3
  $+\,1$
  $\overline{4}$

36. 8
  $+\,6$
  $\overline{14}$

37. 9
  $+\,9$
  $\overline{18}$

38. $2 + 8 = 10$ **39.** $6 + 3 = 9$ **40.** $5 + 5 = 10$ **41.** $2 + 6 = 8$ **42.** $0 + 9 = 9$

43. $9 + 1 = 10$ **44.** $8 + 0 = 8$ **45.** $4 + 3 = 7$ **46.** $2 + 3 = 5$ **47.** $5 + 2 = 7$

48. $9 + 4 = 13$ **49.** $6 + 1 = 7$ **50.** $4 + 5 = 9$ **51.** $5 + 6 = 11$ **52.** $7 + 3 = 10$

Add using a number line.

53. $2 + 3 + 1$

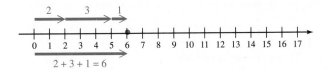

$2 + 3 + 1 = 6$

54. $4 + 6 + 3$

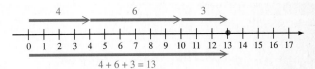

$4 + 6 + 3 = 13$

55. $4 + 4 + 7$

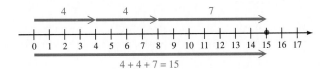

$4 + 4 + 7 = 15$

56. $1 + 5 + 1$

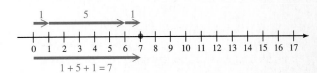

$1 + 5 + 1 = 7$

A hauling company has three trucks, one blue, one red, and one white. The blue truck can haul 2 tons of sand, the red one can haul 5 tons of sand, and the white one can haul 8 tons of sand. Use this information to solve Exercises 57–60.

57. How many tons can be hauled using the blue truck and the red truck?   **7 tons**

58. How many tons can be delivered using the white truck and the red truck?   **13 tons**

59. How many tons of sand can be hauled by all three trucks?   **15 tons**

60. If a contractor orders 10 tons of sand, which two trucks should be used for the delivery?
The blue truck and the white truck

FOR REVIEW

61. Write eight trillion, two hundred twenty-five billion, five hundred million, four hundred eighty-six in standard notation.   8,225,500,000,486

62. Write 63,247 in expanded notation.
60,000 + 3000 + 200 + 40 + 7

63. Write a word name for 8,275,111.
eight million, two hundred seventy-five thousand, one hundred eleven

64. In the number 2,458,769, what digit represents each of the following?
(a) ten thousands   **(b)** millions   **(c)** thousands   **(d)** tens
5   2   8   6

65. Refer to the given number line to place the correct symbol ($<$ or $>$) between the numbers in each pair.

(a) $2 \underline{<} a$ **(b)** $6 \underline{>} b$ **(c)** $a \underline{>} 5$ **(d)** $b \underline{<} a$

ANSWERS: 1. $1 + 5 = 6$ 2. $5 + 0 = 5$ 3. 17   4. 2   5. 10   6. 6   7. 6   8. 8   9. 0   10. 7   11. 14   12. 10   13. 9   14. 11   15. 15   16. 7   17. 3   18. 9   19. 14   20. 4   21. 4   22. 12   23. 9   24. 7   25. 4   26. 15   27. 11   28. 13   29. 8   30. 16   31. 3   32. 13   33. 11   34. 12   35. 4   36. 14   37. 18   38. 10   39. 9   40. 10   41. 8   42. 9   43. 10   44. 8   45. 7   46. 5   47. 7   48. 13   49. 7   50. 9   51. 11   52. 10

53.
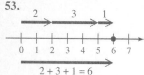
$2 + 3 + 1 = 6$

54.
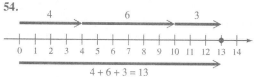
$4 + 6 + 3 = 13$

55.
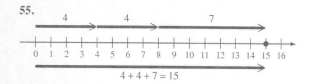
$4 + 4 + 7 = 15$

56.
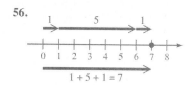
$1 + 5 + 1 = 7$

57. 7 tons   58. 13 tons   59. 15 tons   60. the blue truck and the white truck   61. 8,225,500,000,486   62. 60,000 + 3000 + 200 + 40 + 7   63. eight million, two hundred seventy-five thousand, one hundred eleven   64. (a) 5 (b) 2 (c) 8 (d) 6   65. (a) $2 < a$ (b) $6 > b$ (c) $a > 5$ (d) $b < a$

1.2 EXERCISES B

What addition problems are displayed below?

1.
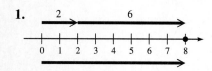
$2 + 6 = 8$

2.
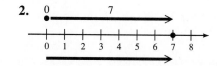
$0 + 7 = 7$

You should be able to complete Exercises 3–52 in 90 seconds or less.

| 3. | 4
+ 6
10 | 4. | 3
+ 4
7 | 5. | 5
+ 2
7 | 6. | 7
+ 4
11 | 7. | 9
+ 3
12 | 8. | 8
+ 7
15 | 9. | 3
+ 8
11 |

| 10. | 9
+ 5
14 | 11. | 8
+ 2
10 | 12. | 6
+ 1
7 | 13. | 3
+ 5
8 | 14. | 2
+ 1
3 | 15. | 0
+ 0
0 | 16. | 4
+ 3
7 |

| 17. | 1
+ 7
8 | 18. | 0
+ 7
7 | 19. | 2
+ 5
7 | 20. | 3
+ 9
12 | 21. | 1
+ 1
2 | 22. | 0
+ 3
3 | 23. | 6
+ 4
10 |

| 24. | 2
+ 7
9 | 25. | 1
+ 8
9 | 26. | 0
+ 1
1 | 27. | 6
+ 2
8 | 28. | 9
+ 7
16 | 29. | 8
+ 5
13 | 30. | 6
+ 6
12 |

| 31. | 4
+ 5
9 | 32. | 0
+ 6
6 | 33. | 9
+ 1
10 | 34. | 2
+ 4
6 | 35. | 1
+ 3
4 | 36. | 0
+ 4
4 | 37. | 5
+ 5
10 |

38. $2 + 9 = 11$ **39.** $1 + 3 = 4$ **40.** $8 + 1 = 9$ **41.** $9 + 2 = 11$ **42.** $7 + 1 = 8$

43. $3 + 6 = 9$ **44.** $4 + 7 = 11$ **45.** $3 + 7 = 10$ **46.** $0 + 5 = 5$ **47.** $9 + 4 = 13$

48. $1 + 4 = 5$ **49.** $4 + 9 = 13$ **50.** $5 + 8 = 13$ **51.** $6 + 7 = 13$ **52.** $5 + 6 = 11$

Add using a number line.

53. $4 + 5 + 3$

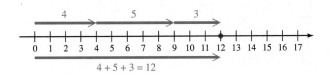

54. $5 + 3 + 8$

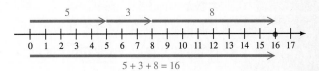

55. $5 + 5 + 6$

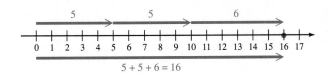

56. $2 + 9 + 2$

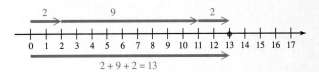

Kim Doane has three boxes for storing cassette tapes, a tan one that holds 5 tapes, a gray one that holds 6 tapes, and a red one that holds 8 tapes. Use this information to solve Exercises 57–60.

57. How many tapes can Kim store in the tan and gray cases?   **11 tapes**

58. How many tapes can be stored in the red case and the gray case?   **14 tapes**

59. How many tapes can Kim store in all three cases?   **19 tapes**

60. Kim plans to take 13 tapes along on a trip. Which two cases should she use?   **the tan case and the red case**

FOR REVIEW

61. Write three hundred twenty-one million, one hundred five thousand, six hundred twenty-two in standard notation.
321,105,622

62. Write 136,403 in expanded notation.
100,000 + 30,000 + 6000 + 400 + 3

63. Write a word name for 39,205,116.
thirty-nine million, two hundred five thousand, one hundred sixteen

64. In the number 6,258,403, what digit represents each of the following?
   (a) millions   **(b)** ones   **(c)** thousands   **(d)** hundred thousands
      6   3   8   2

65. Refer to the given number line to place the correct symbol ($<$ or $>$) between the numbers in each pair.

```
  +---+---+---+---+---+---+---+---+---+---+--->
  0   1   x   3           y   7   z   9
```

   (a) 4 $\underline{\ <\ }$ y   **(b)** 7 $\underline{\ >\ }$ x   **(c)** y $\underline{\ <\ }$ z   **(d)** 5 $\underline{\ <\ }$ z

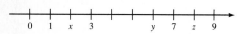

1.2   EXERCISES C

1. If a and b represent whole numbers, what law is illustrated by $a + b = b + a$?   commutative law of addition

2. If a is a whole number, what does $a + 0$ equal? a

1.3   ADDING WHOLE NUMBERS

═══════════════════ **STUDENT GUIDEPOSTS** ═══════════════════

1 Adding a One-Digit Number and a Two-Digit Number

2 Adding Two Two-Digit Numbers

3 Adding Two Many-Digit Numbers

4 Associative Law of Addition

5 Checking Addition (Reverse-Order Method)

1 ADDING A ONE-DIGIT NUMBER AND A TWO-DIGIT NUMBER

The last section gave the basic facts for adding two one-digit numbers. To add numbers with more digits, these facts are used together with knowledge of the place-value system. To illustrate, suppose we find the following sum.

$$\begin{array}{r} 45 \\ +\ \ 3 \\ \hline \end{array}$$

Write each addend in expanded notation.

$$\begin{array}{l} 45 = 40 + 5 = 4 \text{ tens} + 5 \text{ ones} \\ \underline{\ 3} = \underline{\ \ \ \ 3} = \underline{\ \ \ \ \ \ \ \ 3 \text{ ones}} \end{array}$$

Find the sum by adding the ones digits, 5 and 3, and placing the sum with the 4 tens.

$$\begin{array}{r} 4 \text{ tens} + 5 \text{ ones} \\ \underline{3 \text{ ones}} \\ 4 \text{ tens} + 8 \text{ ones} = 48 \end{array}$$

The sum, 48, can be found more directly as shown below using the circled steps.

$$
\begin{array}{r}
45 \\
+\ \ 3 \\
\hline
48
\end{array}
$$

② Bring down tens digit → 48 ← ① Add ones digits

H I N T

To add 45 and 3 as shown above, remember that you will write only the sum, 48. The other remarks shown in color are used for explanation.

❷ ADDING TWO TWO-DIGIT NUMBERS

Next we find the sum of two two-digit numbers such as the following.

$$
\begin{array}{r}
35 \\
+\ 24 \\
\hline
\end{array}
$$

Remember that

$$35 = 3 \text{ tens} + 5 \text{ ones} \quad \text{and} \quad 24 = 2 \text{ tens} + 4 \text{ ones}.$$

Find the number of ones in the sum by adding 5 and 4. Then find the number of tens by adding 3 and 2.

$$
\begin{array}{r}
35 = 3 \text{ tens} + 5 \text{ ones} \\
+\ 24 = 2 \text{ tens} + 4 \text{ ones} \\
\hline
5 \text{ tens} + 9 \text{ ones} = 59
\end{array}
$$

Obviously, the sum can be found more quickly by simply adding the ones digits and then adding the tens digits.

$$
\begin{array}{r}
35 \\
+\ 24 \\
\hline
59
\end{array}
$$

② Add tens digits → 59 ← ① Add ones digits

   Try to find the following sum in the same way. Adding the ones digits and placing the sum directly below the ones digits in the addends gives

$$
\begin{array}{r}
27 \\
+\ 65 \\
\hline
12.
\end{array}
$$
   7 + 5 = 12

If the tens digits are added, the resulting sum looks strange and is obviously incorrect.

$$
\begin{array}{r}
27 \\
+\ 65 \\
\hline
812
\end{array}
$$
 ← This is not correct

What have we done wrong? For one thing, most of us would recognize that the sum of 27 and 65 could not possibly be as large a number as 812! In fact, we are forgetting what we know about place value. When we add the 2 and 6, we are adding 2 *tens* and 6 *tens* so that the sum is 8 *tens,* not 8 hundreds, as in 812. This is easier to see if we write the addends in expanded notation.

$$
\begin{array}{r}
27 = 20 + 7 = 2 \text{ tens} + \ 7 \text{ ones} \\
+\ 65 = 60 + 5 = 6 \text{ tens} + \ 5 \text{ ones} \\
\hline
8 \text{ tens} + 12 \text{ ones}
\end{array}
$$

Since

$$12 \text{ ones} = 10 \text{ ones} + 2 \text{ ones}$$
$$= 1 \text{ ten} + 2 \text{ ones}, \quad \textbf{1 ten} = \textbf{10 ones}$$

the sum

$$8 \text{ tens} + 12 \text{ ones} = 8 \text{ tens} + 1 \text{ ten} + 2 \text{ ones}$$
$$= 8 \text{ tens} + 1 \text{ ten} + 2 \text{ ones}$$
$$= 9 \text{ tens} + 2 \text{ ones} = 92.$$

Thus,

$$
\begin{array}{r}
27 \\
+\ 65 \\
\hline
92.
\end{array}
$$

The steps shown above are usually shortened by using a process called **carrying.**

$$
\begin{array}{r}
27 \\
+\ 65
\end{array}
\quad \text{Add ones and think of}
\quad
\begin{array}{r}
7 \\
+\ 5 \\
\hline
12
\end{array}
$$

Write 2, the ones digit of this sum, below the ones digits of the addends. Write 1, the tens digit of the sum, above the tens digits of the addends.

$$
\begin{array}{r}
1 \quad \leftarrow \text{Write tens digit in the tens column}\\
27 \\
+\ 65 \\
\hline
2 \quad \leftarrow \text{Write ones digit here}
\end{array}
$$

The number 1 is called a **carry number.** Now add 1, 2, and 6 and write the sum, 9, below the tens digits.

$$
\begin{array}{r}
1 \\
27 \\
+\ 65 \\
\hline
92
\end{array}
$$

EXAMPLE 1   ADDING TWO TWO-DIGIT NUMBERS

Find each sum.

(a)
$$
\begin{array}{r}
34 \\
+\ 49
\end{array}
$$

The steps to follow are listed below.

① Add ones digits: $4 + 9 = 13$.
② Write 3 under ones digits and carry 1.
③ Add 1, 3, and 4 to obtain 8.

$$
\begin{array}{r}
1 \\
34 \\
+\ 49 \\
\hline
83
\end{array}
$$

(b)
$$
\begin{array}{r}
69 \\
+\ 85
\end{array}
$$

Begin as before.

$$
\begin{array}{r}
1 \\
69 \\
+\ 85 \\
\hline
4
\end{array}
\quad 9 + 5 = 14; \text{ write 4 and carry 1}
$$

PRACTICE EXERCISE 1

Find each sum.

(a)
$$
\begin{array}{r}
58 \\
+\ 25
\end{array}
$$

(b)
$$
\begin{array}{r}
87 \\
+\ 65
\end{array}
$$

At this point, you must add 1, 6, and 8. Notice that the sum, 15, is really

$$15 \text{ tens} = 10 \text{ tens} + 5 \text{ tens}$$
$$= 1 \text{ hundred} + 5 \text{ tens}. \qquad 10 \text{ tens} = 1 \text{ hundred}$$

Write the 5 under the tens digits of the addends, and the 1 in the hundreds place of the sum.

$$\begin{array}{r} 1 \\ 69 \\ + 85 \\ \hline 154 \end{array}$$

Answers: (a) 83   (b) 152

3 ADDING TWO MANY-DIGIT NUMBERS

When there are more digits in the addends, continue the process as shown in the next example.

EXAMPLE 2   ADDING TWO THREE-DIGIT NUMBERS

Find each sum.

(a) $\begin{array}{r} 875 \\ + 398 \end{array}$

$$\begin{array}{r} 11 \\ 875 \\ + 398 \\ \hline 1273 \end{array}$$

The completed addition is shown above. Remember to use the following steps.

① Add 5 and 8, write 3 and carry 1.

② Add 1, 7, and 9, write 7 and carry 1.

③ Add 1, 8, and 3, getting 12. Since

$$12 \text{ hundreds} = 10 \text{ hundreds} + 2 \text{ hundreds}$$
$$= 1 \text{ thousand} + 2 \text{ hundreds} \qquad 10 \text{ hundreds} = 1 \text{ thousand}$$

write the 2 hundreds in the hundreds place and the 1 in the thousands place.

(b) $\begin{array}{r} 11 \\ 5627 \\ + 7809 \\ \hline 13436 \end{array}$

Notice that there is no carry number for the hundreds digit, and that 13, the final sum of 1, 5, and 7, is placed in the thousands and ten thousands position.

PRACTICE EXERCISE 2

Find each sum.

(a) $\begin{array}{r} 963 \\ + 459 \end{array}$

(b) $\begin{array}{r} 8246 \\ + 6195 \end{array}$

Answers: (a) 1422   (b) 14,441

❹ ASSOCIATIVE LAW OF ADDITION

Addition is a **binary operation.** This means that only two numbers are added at a time (*bi* means two). Suppose we look again at the problem of adding three numbers. Find the following sum.

$$
\begin{array}{r}
3 \\
5 \\
+\ 2 \\
\hline
\end{array}
$$

Start by adding the top two numbers, 3 and 5. Then add their sum, 8, to 2, to obtain 10. The entire process is usually done mentally, but can be shown in the following way.

We could also calculate the sum by adding the bottom two numbers, 2 and 5, then adding their sum, 7, to 3 to obtain 10. Again, we can show the technique as follows.

Notice that both cases give the same result.

When the sum is written horizontally, as $3 + 5 + 2$, we use parentheses to show which numbers to add first. In the first case,

$$(3 + 5) + 2 = 8 + 2 = 10.$$

In the second case,

$$3 + (5 + 2) = 3 + 7 = 10.$$

Thus,

$$(3 + 5) + 2 = 3 + (5 + 2).$$

Regrouping the numbers does not change the sum. The same is true for any three whole numbers.

The Associative Law of Addition

The **associative law of addition** tells us that the sum of more than two numbers is not changed by grouping. No matter how many numbers are added, they are always added two at a time.

EXAMPLE 3 ADDING SEVERAL NUMBERS

Find the sum. $9369 + 387 + 18 + 538$

$$
\begin{array}{r}
123 \\
9369 \\
387 \\
18 \\
+\ \ \ 538 \\
\hline
10312
\end{array}
$$

PRACTICE EXERCISE 3

Find the sum.

$294 + 8451 + 12 + 1306$

① Add 9, 7, 8, and 8; write 2 and carry 3.

② Add 3, 6, 8, 1, and 3; write 1 and carry 2.

③ Add 2, 3, 3, and 5; write 3 and carry 1.

④ Add 1 and 9; write 0 in thousands column and 1 in ten thousands column.

Answer: 10,063

⑤ CHECKING ADDITION (REVERSE-ORDER METHOD)

We can check addition in one of two ways. One way is to repeat the addition. A second way, which avoids the risk of repeating any error, is to reverse the order of addition. That is, if we added from top to bottom the first time, we check by adding from bottom to top. This method, shown in Example 6, is called the **reverse-order check of addition**.

EXAMPLE 4   REVERSE-ORDER CHECK

Add and check.

(a) Add:
$$3 \leftarrow ① \ 3 + 7 = 10$$
$$7$$
$$+\ 8 \quad ② \ 10 + 8 = 18$$
$$18 \leftarrow$$

Check:
$$3 \quad ② \ 15 + 3 = 18$$
$$7$$
$$+\ 8 \leftarrow ① \ 8 + 7 = 15$$
$$18 \leftarrow$$

(b) Add, and check in reverse order.
$$21$$
$$35$$
$$+\ 76$$

The steps used to find the sum are numbered in order. Notice that both the ones column and the tens column are added from the top.

$$1$$
④ $1 + 2 = 3 \ \rightarrow 21 \leftarrow$ ① $1 + 5 = 6$
⑤ $3 + 3 = 6 \ \rightarrow 35$
⑥ $6 + 7 = 13 \rightarrow 76 \leftarrow$ ② $6 + 6 = 12$
⑦ Write 13 $\rightarrow 132 \leftarrow$ ③ Write 2 and carry 1

The steps used to check are numbered in order.

⑥ $12 + 1 = 13 \ \rightarrow 1$
⑤ $10 + 2 = 12 \rightarrow 21 \leftarrow$ ② $11 + 1 + 12$
④ $7 + 3 = 10 \ \rightarrow 35$
$76 \leftarrow$ ① $6 + 5 = 11$
⑦ Write 13 $\rightarrow 132 \leftarrow$ ③ Write 2 and carry 1

PRACTICE EXERCISE 4

Add and check.

(a) Add:
$$6$$
$$2$$
$$+\ 9$$

Check:
$$6$$
$$2$$
$$+\ 9$$

(b) Add:
$$36$$
$$51$$
$$+\ 89$$

Check:
$$36$$
$$51$$
$$+\ 89$$

Answers: (a) 17   (b) 176

HINT

Sometimes when adding several numbers it helps to look for pairs that add to 10, 20, 30, and so on. This is illustrated below.

Add.
$$11$$
$$7 \ \rightarrow 20$$
$$4 \quad \rightarrow 10 \rightarrow 30$$
$$9 \quad \ 4 \rightarrow \ 4$$
$$+\ 3 \qquad \qquad \overline{34}$$
$$\overline{34} \quad \text{Write only this sum}$$

EXAMPLE 5 AN APPLICATION OF ADDITION	PRACTICE EXERCISE 5

Farmer Adams owns 35 sheep and 48 cattle. How many sheep and cattle does he have on his farm?

Find the sum of 35 and 48.

$$\begin{array}{r} 35 \\ + 48 \\ \hline 83 \end{array}$$

Thus, Farmer Adams owns a total of 83 sheep and cattle.

Basketball player Andy Hurd scored 23 points in the first game of a tournament and 28 points in the second game. How many points did he score in the two games?

Answer: 51 points

1.3 EXERCISES A

Find the following sums and check using the reverse-order method.

1. $\begin{array}{r} 42 \\ + 5 \\ \hline 47 \end{array}$

2. $\begin{array}{r} 72 \\ + 6 \\ \hline 78 \end{array}$

3. $84 + 3$ 87

4. $\begin{array}{r} 69 \\ + 30 \\ \hline 99 \end{array}$

5. $\begin{array}{r} 33 \\ + 22 \\ \hline 55 \end{array}$

6. $\begin{array}{r} 49 \\ + 35 \\ \hline 84 \end{array}$

7. $\begin{array}{r} 59 \\ + 76 \\ \hline 135 \end{array}$

8. $43 + 79$ 122

9. $\begin{array}{r} 99 \\ + 56 \\ \hline 155 \end{array}$

10. $\begin{array}{r} 512 \\ + 65 \\ \hline 577 \end{array}$

11. $\begin{array}{r} 405 \\ + 26 \\ \hline 431 \end{array}$

12. $727 + 73$ 800

13. $173 + 469$ 642

14. $\begin{array}{r} 837 \\ + 175 \\ \hline 1012 \end{array}$

15. $\begin{array}{r} 5824 \\ + 1736 \\ \hline 7560 \end{array}$

16. $\begin{array}{r} 9999 \\ + 1111 \\ \hline 11,110 \end{array}$

17. $\begin{array}{r} 4002 \\ + 3009 \\ \hline 7011 \end{array}$

18. $3297 + 816$ 4113

19. $\begin{array}{r} 7 \\ 3 \\ + 6 \\ \hline 16 \end{array}$

20. $\begin{array}{r} 3 \\ 5 \\ + 8 \\ \hline 16 \end{array}$

21. $\begin{array}{r} 1 \\ 9 \\ + 7 \\ \hline 17 \end{array}$

22. $\begin{array}{r} 31 \\ 46 \\ + 72 \\ \hline 149 \end{array}$

23. $85 + 37 + 62$ 184

24. $\begin{array}{r} 456 \\ 767 \\ + 318 \\ \hline 1541 \end{array}$

25. $\begin{array}{r} 3 \\ 25 \\ 14 \\ + 2 \\ \hline 44 \end{array}$

26. $27 + 6 + 9 + 83$ 125

27. $\begin{array}{r} 3 \\ 79 \\ 5 \\ + 13 \\ \hline 100 \end{array}$

28. $\begin{array}{r} 615 \\ 203 \\ 970 \\ + 341 \\ \hline 2129 \end{array}$

29. $\begin{array}{r} 3846 \\ 1092 \\ 637 \\ + 4425 \\ \hline 10,000 \end{array}$

30. $\begin{array}{r} 5298 \\ 6137 \\ 8056 \\ + 4321 \\ \hline 23,812 \end{array}$

Solve.

31. Mary bought 8 pairs of shorts one day and 5 more pairs the next. How many pairs of shorts did she buy?   **13 pairs of shorts**

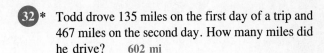

32.* Todd drove 135 miles on the first day of a trip and 467 miles on the second day. How many miles did he drive?   **602 mi**

33. Attendance at the Friday-night performance of a play was 324. The next night, 275 attended. What was the total attendance these two nights?   **599**

34. Janet had a balance of $365 in her checking account before making deposits of $418 and $279. How much did she deposit, and what was her new balance?   **$697; $1062**

ANSWERS: 1. 47 2. 78 3. 87 4. 99 5. 55 6. 84 7. 135 8. 122 9. 155 10. 577 11. 431 12. 800
13. 642 14. 1012 15. 7560 16. 11,110 17. 7011 18. 4113 19. 16 20. 16 21. 17 22. 149 23. 184
24. 1541 25. 44 26. 125 27. 100 28. 2129 29. 10,000 30. 23,812 31. 13 pairs of shorts 32. 602 mi 33. 599
34. $697; $1062

1.3 EXERCISES B

Find the following sums and check using the reverse-order method.

1. 32
  + 4
  ___
  36

2. 91
  + 8
  ___
  99

3. 52 + 7
  59

4. 25
  + 72
  ___
  97

5. 63
  + 18
  ___
  81

6. 27
  + 48
  ___
  75

7. 38
  + 83
  ___
  121

8. 27 + 87
  114

9. 213
  + 41
  ___
  254

10. 392
  + 12
  ___
  404

11. 519
  + 91
  ___
  610

12. 989 + 11
  1000

13. 385 + 297
  682

14. 487
  + 399
  ___
  886

15. 7904
  + 2311
  ___
  10,215

16. 6329
  + 4350
  ___
  10,679

17. 6195
  + 4805
  ___
  11,000

18. 4075 + 28
  4103

19. 8
  1
  + 5
  ___
  14

20. 4
  5
  + 4
  ___
  13

21. 7
  5
  + 2
  ___
  14

22. 52
  29
  + 14
  ___
  95

23. 47 + 38 + 91
  176

24. 806
  359
  + 461
  ___
  1626

25. 16
  5
  70
  + 19
  ___
  110

*The color indicates there is a complete solution in the back of the text. See *To the Student* for more details.

26. 4 + 27 + 9 + 41   **27.**   52   **28.**   225   **29.**   8371   **30.**   7896
        81   8   407   5   4327
                                       63   311   693   5131
                                     + 7   + 597   + 12   + 8454
                                      130   1540   9081   25,808

Solve.

31. The first floor of the mathematics building has 18 classrooms and the second floor has 14 classrooms. How many classrooms are on the two floors?
32 classrooms

32. A lab assistant poured 585 cubic centimeters of distilled water into a beaker followed by 428 cubic centimeters of alcohol. How much liquid was poured into the beaker?   **1013 cubic centimeters**

33. Suppose you spent $47 for groceries, $18 for gas, and $85 for a jacket. What was the total amount that you spent?   **$150**

34. The *odometer* on a car indicates how many miles the car has been driven. Prior to a trip, the odometer on Curt's car read 23,165 miles. He then drove 496 miles one day and 387 miles the next. How many miles did he drive those two days? What was the odometer reading at the end of the second day?   **883 mi; 24,048 mi**

1.3 EXERCISES C

Find the following sums.

1. 30,120
    1021
   40,112
   71,253

2. 99,876
   29,432
   67,132
   196,440

3. 46,001
   20,110
   31,010
      210
   97,331

4. 76,432
   49,612
   58,668
   24,688
   [Answer: 209,400]

5. Sam Passamonte is an insurance salesman. For income tax purposes he keeps track of the number of miles that he drives each year. Last year, the miles driven each month were as follows: 2643 mi, 1298 mi, 3045 mi, 1127 mi, 2213 mi, 1778 mi, 896 mi, 743 mi, 512 mi, 919 mi, 1037 mi, and 1428 mi. How many miles did he drive last year?   **17,639 mi**

6. If a, b, and c represent whole numbers, what law is illustrated by $(a + b) + c = a + (b + c)$?
associative law of addition

Find the following sums.

7.   6 ft   2 in
        + 5 ft   4 in
         11 ft   6 in

8.   6 ft   8 in
        + 5 ft   9 in
         12 ft   5 in

1.4 ROUNDING AND ESTIMATING WHOLE NUMBERS

═══════ STUDENT GUIDEPOSTS ═══════

1 Rounding a Whole Number   **2** Using Rounded Numbers to
                                           Estimate and Check Sums

1 ROUNDING A WHOLE NUMBER

In many situations we might use an approximate value of a whole number to
obtain an *estimated* answer. For example, suppose we know that one clerk in a
post office processed 12,780 letters while a second clerk processed 9125 letters
during one working day. We could **estimate** that the total processed by the two
was about 22,000 letters, if we *rounded* 12,780 to 13,000 and 9125 to 9000 and
mentally added these rounded numbers.

Numbers may be rounded to the nearest ten, hundred, thousand, ten thousand, and so on. To **round** a number to the nearest ten, use the nearest multiple
of ten (0, 10, 20, 30, 40, . . .) to approximate the given number.

For example, consider rounding 38 to the nearest ten. We may use part of a
number line showing the multiples of ten on either side of the given number. The
number 38 is between 30 and 40, as shown in Figure 1.13. Since 38 is closer to
40 than 30, we round up to 40. That is, 38 rounded to the nearest ten is 40.

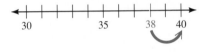

Figure 1.13

Figure 1.14

To round 53 to the nearest ten, consider Figure 1.14. The number 53 is
between 50 and 60. Since it is closer to 50, 53 rounded to the nearest ten is 50.
What if we wanted to round 25 to the nearest ten? Since 25 is halfway between
20 and 30, we would round down to 20 or round up to 30. When a number is
halfway between rounding numbers we will agree to round up. Thus, we will
round 25 up to 30 as shown in Figure 1.15.

Figure 1.15

The following rule allows us to round numbers without using a number line.

To Round a Whole Number to a Given Position

1. Identify the digit in that position.
2. If the first digit to the right of that position is 0, 1, 2, 3, or 4, do not
   change the digit.
3. If the first digit to the right of that position is 5, 6, 7, 8, or 9, increase
   the digit by 1.
4. *In both cases* replace all digits to the right of that position with zeros.

| EXAMPLE 1 ROUNDING TO THE NEAREST HUNDRED | PRACTICE EXERCISE 1 |

(a) Round 392 to the nearest hundred.

The digit 3 is in the hundreds position and 9 is the first digit to the right.

hundreds position
↓
392
↑
first digit to right

Increase 3 by 1 to 4, and replace 9 and 2 with zeros. Thus 400 is the desired rounded number.

(b) Round 6237 to the nearest hundred.

The digit 2 is in the hundreds position and 3 is the first digit to the right.

hundreds position
↓
6237
↑
first digit to right

Do not change the 2, and replace 3 and 7 with zeros. Thus 6200 is 6237 rounded to the nearest hundred.

(c) Round 2950 to the nearest hundred.

The 9 is in the hundreds position and 5 is the first digit to the right. Increasing 9 by 1 means that the thousands digit must be increased to 3 and the hundreds digit becomes 0. Thus 2950 to the nearest hundred is 3000.

(a) Round 541 to the nearest hundred.

(b) Round 1875 to the nearest hundred.

(c) Round 4950 to the nearest hundred.

Answers: (a) 500 (b) 1900 (c) 5000

| EXAMPLE 2 ROUNDING TO THE NEAREST THOUSAND | PRACTICE EXERCISE 2 |

(a) Round 7283 to the nearest thousand.

The 7 is in the thousands position. Since 2 is the first digit to the right, 7283 rounded to the nearest thousand is 7000.

(b) Round 26,501 to the nearest thousand.

The 6 is in the thousands position, and 5 is the first digit to the right. Increase 6 to 7 and the rounded number is 27,000.

(a) Round 4423 to the nearest thousand.

(b) Round 69,825 to the nearest thousand.

Answers: (a) 4000 (b) 70,000

| EXAMPLE 3 ROUNDING WHOLE NUMBERS | PRACTICE EXERCISE 3 |

(a) Round 4645 to the nearest ten.

There is a 4 in the tens position and 5 is the first digit to the right. Round 4645 to 4650.

(b) Round 4645 to the nearest hundred.

The 6 is in the hundreds position and 4 is the first digit to the right. Leave the 6 and write 4600 for the rounded number.

(a) Round 9745 to the nearest ten.

(b) Round 9745 to the nearest hundred.

Answers: (a) 9750 (b) 9700

///////////// **CAUTION** /////////////

In part **(b)** of Example 3, if 4645 had been rounded to tens before being rounded to hundreds, the final result would have been 4700 instead of 4600. This points out that this method of rounding *cannot* be done in steps; it must be done using the original number, not a rounded number.

/////////////

② USING ROUNDED NUMBERS TO ESTIMATE AND CHECK SUMS

Rounded numbers are used to make quick *estimates* of sums. These estimates may also be used for a quick *check* for large errors in addition problems.

EXAMPLE 4   ROUNDING TO THE NEAREST TEN

Estimate the sum by rounding to the nearest ten.

	Rounded to nearest ten
18	20
82	80
65	70
+ 21	+ 20
	190

The estimated sum is 190. The exact sum, which is 186, can be checked by comparing it to the estimate.

PRACTICE EXERCISE 4

Estimate the sum by rounding to the nearest ten.

```
  94
  48
  25
+ 12
```

Answer: The estimated sum is 180 (exact sum is 179).

EXAMPLE 5   ROUNDING TO THE NEAREST HUNDRED

Estimate the sum by rounding to the nearest hundred.

	Rounded to nearest hundred
425	400
607	600
587	600
+ 235	+ 200
	1800

The estimated sum is 1800. The exact sum, which is 1854, can be checked by comparing it to the estimate.

PRACTICE EXERCISE 5

Estimate the sum by rounding to the nearest hundred.

```
  355
  739
  462
+ 651
```

Answer: The estimated sum is 2300 (exact sum is 2207).

////////////// **CAUTION** //////////////

Using rounded numbers to estimate or check sums provides only a general technique that may not be totally accurate. It should be used with care. For example, in Practice Exercise 5, the estimated sum 2300 differs from the actual sum of 2207 by 93. In part **(a)** of Example 6 an error of 93 causes us to conclude that a mistake was made in addition. In one case, an error of 93 appears acceptable while in the other case it is not. As a result, remember that an *estimate* is an *approximation* not intended to replace actual calculations. Such approximations will help to provide us with some sense of accuracy in our work.

//////////

EXAMPLE 6   ESTIMATING SUMS BY ROUNDING

Without finding the exact sum, estimate the sum by rounding and use the estimate to check the work shown.

            Rounded to ten
(a)   382   380
            37   40
            409   410
         + 155   + 160
            983   990

With an estimated sum of 990, we would probably conclude that the given sum, 983, is correct.

            Rounded to hundred
(b)   625   600
            187   200
            362   400
         + 717   + 700
           1691   1900

With an estimated sum of 1900, we would probably conclude that the given sum, 1691, is incorrect. The correct sum is 1891. Can you find where the error was made?

PRACTICE EXERCISE 6

Estimate the sum by rounding and use your estimate to check the work shown.

(a) Check by rounding to the nearest ten.

            423
            165
             75
         + 349
           1012

(b) Check by rounding to the nearest hundred.

            835
            293
            441
         + 752
           2121

Answers: (a) The estimated sum is 1020; the sum appears to be correct. (b) The estimated sum is 2300; the sum appears to be incorrect.

Real-life situations often present problems similar to the one in the next example. By using rounded numbers, we can estimate or approximate a solution mentally.

EXAMPLE 7   ESTIMATING IN AN APPLIED PROBLEM

Harvey has $589 in one savings account and $807 in a second. Without finding the exact total in the two accounts, estimate this total using rounded numbers.

PRACTICE EXERCISE 7

There are 2985 acres on the Walter Ranch and 4120 acres on the Kirk Ranch. Without finding the exact number, estimate the total number of acres on the two ranches.

Harvey has $589 in one account. This is about $600, rounding to the nearest hundred. Similarly, he has about $800 in the second account. The total in the two accounts can be estimated by adding $600 and $800.

$$
\begin{array}{r}
600 \\
+\ \ 800 \\
\hline
1400
\end{array}
$$

Thus, Harvey has about $1400 in the two accounts.

Answer: 7000 acres (using 3000 and 4000 as the two estimates)

CAUTION

Do not estimate a sum by adding first and then rounding the answer. This would serve no useful purpose for checking or for getting a ''rough'' answer.

1.4 EXERCISES A

Round to the nearest ten.

1. 33   30   **2.** 68   70   **3.** 15   20   **4.** 625   630   **5.** 199   200

Round to the nearest hundred.

6. 327   300   **7.** 372   400   **8.** 68   100   **9.** 8550   8600   **10.** 4977   5000

Round to the nearest thousand.

11. 927   1000   **12.** 9499   9000   **13.** 9500   10,000   **14.** 23,450   23,000   **15.** 127,456   127,000

Estimate the sum by rounding to the nearest ten. Do not find the exact sum.

16	**17.**	**18.**	**19.**
23	65	602	1722
16	91	327	3345
54	19	455	2471
+ 79	+ 32	+ 281	+ 5128
170	210	1670	12,670

Estimate the sum by rounding to the nearest hundred. Do not find the exact sum.

20	**21.**	**22.**	**23.**
329	103	602	1722
556	75	327	3345
113	969	455	2471
+ 925	+ 473	+ 281	+ 5128
1900	1700	1700	12,600

Estimate the sum by rounding to the nearest thousand. Do not find the exact sum.

24.	4321	**25.**	6025	**26**	19,501	**27.**	1722
	1530		847		19,499		3345
	2905		9327		38,201		2471
	+ 3040		+ 1700		+ 21,005		+ 5128
	12,000		**18,000**		**98,000**		**12,000**

Without finding the exact sum, determine if the given sum seems correct or incorrect.

28.	72	**29.**	357	**30.**	245	**31.**	2193
	16		125		61		6407
	97		845		193		3556
	+ 33		+ 581		+ 702		+ 409
	218		2308		1201		12,565

220 (nearest 10); appears correct   1900 (nearest 100); appears incorrect   1200 (nearest 10); appears correct   12,600 (nearest 100); appears correct

Solve.

32. According to a recent census, the population of New York City is 7,908,421. Round this number to the nearest **(a)** million **(b)** hundred thousand.
(a) **8,000,000**   (b) **7,900,000**

33. During the first few weeks of distribution, the movie *Beverly Hills Cop* grossed over $68,495,240. Round this number to the nearest **(a)** million **(b)** ten million.
(a) **$68,000,000**   (b) **$70,000,000**

34 There are 7105 books on the first floor of a library, and 6898 books on the second floor. Without finding the exact total, estimate the number of books on the two floors.   **14,000 books**

35. Dr. Morgan purchased a new Buick for $14,958 and a new Honda for $11,141. Without finding the exact total, estimate the amount of money he spent buying these two automobiles.   **$26,000**

ANSWERS: 1. 30   2. 70   3. 20   4. 630   5. 200   6. 300   7. 400   8. 100   9. 8600   10. 5000   11. 1000   12. 9000   13. 10,000   14. 23,000   15. 127,000   16. 170   17. 210   18. 1670   19. 12,670   20. 1900   21. 1700   22. 1700   23. 12,600   24. 12,000   25. 18,000   26. 98,000   27. 12,000   28. Rounding to nearest ten, the estimated sum is 220; given sum appears correct.   29. Rounding to nearest hundred, the estimated sum is 1900; given sum appears incorrect.   30. Rounding to nearest ten, the estimated sum is 1200; given sum appears correct.   31. Rounding to the nearest hundred, the estimated sum is 12,600; given sum appears correct.   32. (a) 8,000,000 (b) 7,900,000   33. (a) $68,000,000 (b) $70,000,000   34. 14,000 books   35. $26,000

1.4 EXERCISES B

Round to the nearest ten.

1. 47   50   **2.** 22   20   **3.** 548   550   **4.** 195   200   **5.** 699   700

Round to the nearest hundred.

6. 538   500   **7.** 662   700   **8.** 59   100   **9.** 6850   6900   **10.** 2195
                                                                                                                              2200

Round to the nearest thousand.

11. 4501   5000   **12.** 4499   4000   **13.** 2799   3000   **14.** 8985   9000   **15.** 135,525
                                                                                                                                          136,000

Estimate the sum by rounding to the nearest ten. Do not find the exact sum.

16.	**17.**	**18.**	**19.**
34	75	503	2344
17	81	426	1665
65	18	545	2582
+ 88	+ 43	+ 397	+ 7138
210	220	1880	13,730

Estimate the sum by rounding to the nearest hundred. Do not find the exact sum.

20.	**21.**	**22.**	**23.**
239	102	427	1856
665	875	581	2345
113	53	326	2281
+ 849	+ 271	+ 108	+ 6195
1800	1400	1400	12,700

Estimate the sum by rounding to the nearest thousand. Do not find the exact sum.

24.	**25.**	**26.**	**27.**
6435	3175	39,501	4723
2505	955	39,499	2549
3916	7258	27,208	3449
+ 5050	+ 1900	+ 41,006	+ 6207
18,000	13,000	147,000	17,000

Without finding the exact sum, determine if the given sum seems correct or incorrect.

28.	**29.**	**30.**	**31.**
81	445	329	4293
19	165	83	5445
48	851	294	3795
+ 31	+ 558	+ 911	+ 603
179 180 (nearest 10); appears correct	1999 2000 (nearest 100); appears correct	1417 1610 (nearest 10); appears incorrect	14,136 14,100 (nearest 100); appears correct

Solve.

32. One year the profit made by the IBEX Corporation was $7,625,428. Round this number to the nearest **(a)** million **(b)** hundred thousand.
(a) $8,000,000   (b) $7,600,000

33. During the first few weeks of distribution, the movie *Raiders of the Lost Ark* grossed over $45,625,155. Round this number to the nearest **(a)** million **(b)** ten million.
(a) $46,000,000   (b) $50,000,000

34. Doug Christensen owns 5208 shares of one stock and 7112 shares of another. Without finding the exact total, estimate the number of shares of the two types of stock that he owns.   12,000 shares

35. A jet is flying at an altitude of 7895 feet and then increases its altitude by 5120 feet. Without finding the exact altitude, estimate its new altitude.
13,000 ft

1.4 EXERCISES C

The Farm Investment Company has four properties in the state of Nebraska. The values of these properties are $26,482,120, $5,842,580, $18,000,720, *and* $48,216,480. *Estimate the total value of these holdings by rounding to the indicated place.*

1. To the nearest thousand
[Answer: $98,542,000]

2. To the nearest hundred thousand
$98,500,000

3. To the nearest million   **$98,000,000**

4. To the nearest ten million
[Answer: $110,000,000]

5. Use your calculator to find the exact value of these four properties. Even with a calculator it is wise to know something about estimation, since errors can be made using a calculator by pressing the wrong button. **$98,541,900**

1.5 BASIC SUBTRACTION FACTS

1 Subtracting Whole Numbers

2 Related Addition and Subtraction Sentences

1 SUBTRACTING WHOLE NUMBERS

Subtraction of one whole number from another can be thought of as "removing" or "taking away." For example, suppose we subtract 4 coins from 7 coins. If there are 7 coins in a stack and 4 are removed, 3 coins remain. See Figure 1.16.

7 coins { }
Remove 4 coins
} 3 coins remain

Figure 1.16

Subtraction is written with a **minus sign,** **−,** and can be written either horizontally or vertically.

Horizontal display
7 − 4 = 3
minuend
subtrahend
difference

Vertical display
7 ← minuend
− 4 ← subtrahend
3 ← difference

The number being taken away is called the **subtrahend.** The number it is taken from is the **minuend.** The number remaining is the **difference.**

minuend − subtrahend = difference

Subtraction can be shown on a number line. Recall that to find the sum 5 + 2 using a number line, we move right 5 units from 0 to the point associated with 5 and then move 2 more units to the right to the point 7, as shown in Figure 1.17.

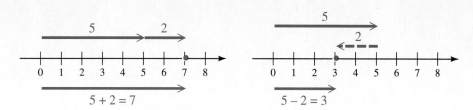

Figure 1.17	**Figure 1.18**

To find $5 - 2$ using a number line, we start the same way by moving 5 units to the right of 0 to the point 5. We then move 2 units to the *left* of 5 to the point 3. See Figure 1.18. Using a number line we have shown that $5 - 2 = 3$.

② RELATED ADDITION AND SUBTRACTION SENTENCES

Although we can subtract whole numbers using the ''take away'' method or the number line, it is better to relate subtraction to addition since we already know the basic addition facts. Each subtraction sentence is related to an addition sentence. For example, $4 + 5 = 9$ is the **related addition sentence** to $9 - 5 = 4$ and $9 - 5 = 4$ is the **related subtraction sentence** to $4 + 5 = 9$. Thus we can solve the subtraction problem,

$$9 - 5 = \square \quad \text{by solving} \quad \square + 5 = 9.$$

Since the addition sentence is true when 4 replaces $\square$, the subtraction sentence is true with 4 also.

EXAMPLE 1 USING THE RELATED ADDITION SENTENCE	**PRACTICE EXERCISE 1**
Subtract.	Subtract.
(a) $6 - 2 = 4$ 4 is correct since $4 + 2 = 6$	**(a)** $5 - 1$
(b) $8 - 6 = 2$ 2 is correct since $2 + 6 = 8$	**(b)** $7 - 7$
(c) $8 - 8 = 0$ 0 is correct since $0 + 8 = 8$	**(c)** $9 - 4$
(d) $9 - 1 = 8$ 8 is correct since $8 + 1 = 9$	**(d)** $4 - 2$
(e) $5 - 0 = 5$ 5 is correct since $5 + 0 = 5$	**(e)** $8 - 0$
	Answers: (a) **4** (b) **0** (c) **5** (d) **2** (e) **8**

With practice the basic subtraction facts, which come from the basic addition facts, should be easy to remember. Thus we should be able to do basic subtraction quickly. Example 2 gives more practice with the vertical display.

EXAMPLE 2 BASIC SUBTRACTION USING A VERTICAL DISPLAY	**PRACTICE EXERCISE 2**
Subtract.	Subtract.

Example 2:

	Subtraction	*Related addition*
(a)	9 $-\,7$ —— 2	2 $+\,7$ —— 9
(b)	5 $-\,1$ —— 4	4 $+\,1$ —— 5

Practice Exercise 2:

(a)   8
      $-\,3$

(b)   7
      $-\,4$

<table>
<tr><td>(c)</td><td>$\begin{array}{r} 4 \\ -4 \\ \hline 0 \end{array}$</td><td>$\begin{array}{r} 0 \\ +4 \\ \hline 4 \end{array}$</td></tr>
<tr><td>(d)</td><td>$\begin{array}{r} 3 \\ -0 \\ \hline 3 \end{array}$</td><td>$\begin{array}{r} 3 \\ +0 \\ \hline 3 \end{array}$</td></tr>
</table>

(c) $\begin{array}{r} 3 \\ -1 \\ \hline \end{array}$

(d) $\begin{array}{r} 9 \\ -9 \\ \hline \end{array}$

Answers: (a) 5   (b) 3   (c) 2
(d) 0

1.5 EXERCISES A

What subtraction problems are displayed below?

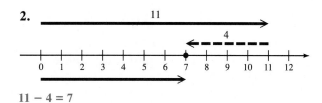

1. $9 - 8 = 1$

2. $11 - 4 = 7$

You should be able to complete Exercises 3–37 in two minutes or less.

3. $\begin{array}{r} 6 \\ -4 \\ \hline 2 \end{array}$ 4. $\begin{array}{r} 5 \\ -1 \\ \hline 4 \end{array}$ 5. $\begin{array}{r} 2 \\ -1 \\ \hline 1 \end{array}$ 6. $\begin{array}{r} 9 \\ -6 \\ \hline 3 \end{array}$ 7. $\begin{array}{r} 8 \\ -5 \\ \hline 3 \end{array}$ 8. $\begin{array}{r} 7 \\ -0 \\ \hline 7 \end{array}$ 9. $\begin{array}{r} 9 \\ -2 \\ \hline 7 \end{array}$

10. $\begin{array}{r} 9 \\ -4 \\ \hline 5 \end{array}$ 11. $\begin{array}{r} 8 \\ -2 \\ \hline 6 \end{array}$ 12. $\begin{array}{r} 7 \\ -1 \\ \hline 6 \end{array}$ 13. $\begin{array}{r} 7 \\ -4 \\ \hline 3 \end{array}$ 14. $\begin{array}{r} 5 \\ -4 \\ \hline 1 \end{array}$ 15. $\begin{array}{r} 6 \\ -2 \\ \hline 4 \end{array}$ 16. $\begin{array}{r} 9 \\ -8 \\ \hline 1 \end{array}$

17. $\begin{array}{r} 3 \\ -0 \\ \hline 3 \end{array}$ 18. $\begin{array}{r} 1 \\ -1 \\ \hline 0 \end{array}$ 19. $\begin{array}{r} 5 \\ -2 \\ \hline 3 \end{array}$ 20. $\begin{array}{r} 8 \\ -4 \\ \hline 4 \end{array}$ 21. $\begin{array}{r} 12 \\ -8 \\ \hline 4 \end{array}$ 22. $\begin{array}{r} 10 \\ -4 \\ \hline 6 \end{array}$ 23. $\begin{array}{r} 13 \\ -5 \\ \hline 8 \end{array}$

24. $\begin{array}{r} 15 \\ -8 \\ \hline 7 \end{array}$ 25. $\begin{array}{r} 10 \\ -2 \\ \hline 8 \end{array}$ 26. $\begin{array}{r} 17 \\ -9 \\ \hline 8 \end{array}$ 27. $\begin{array}{r} 11 \\ -5 \\ \hline 6 \end{array}$ 28. $\begin{array}{r} 14 \\ -7 \\ \hline 7 \end{array}$ 29. $\begin{array}{r} 10 \\ -1 \\ \hline 9 \end{array}$ 30. $\begin{array}{r} 12 \\ -4 \\ \hline 8 \end{array}$

31. $\begin{array}{r} 10 \\ -6 \\ \hline 4 \end{array}$ 32. $\begin{array}{r} 18 \\ -9 \\ \hline 9 \end{array}$ 33. $\begin{array}{r} 16 \\ -9 \\ \hline 7 \end{array}$ 34. $\begin{array}{r} 13 \\ -6 \\ \hline 7 \end{array}$ 35. $\begin{array}{r} 11 \\ -2 \\ \hline 9 \end{array}$ 36. $\begin{array}{r} 15 \\ -7 \\ \hline 8 \end{array}$ 37. $\begin{array}{r} 11 \\ -8 \\ \hline 3 \end{array}$

38. $14 - 6 = 8$ 39. $10 - 5 = 5$ 40. $12 - 7 = 5$ 41. $11 - 6 = 5$ 42. $15 - 9 = 6$

43. $10 - 8 = 2$ 44. $11 - 3 = 8$ 45. $13 - 9 = 4$ 46. $12 - 5 = 7$ 47. $16 - 8 = 8$

48. $12 - 9 = 3$ 49. $17 - 8 = 9$ 50. $14 - 9 = 5$ 51. $12 - 3 = 9$ 52. $13 - 4 = 9$

Solve.

53. Kathy had 9 kittens and gave away 4 of them. How many kittens did she have left?   **5 kittens**

54. Greg has $14 in his wallet. If he buys a book for $6, how much money will he have left?   **$8**

FOR REVIEW

55. Estimate the sum by rounding to the nearest   **(a)** ten   **(b)** hundred   **(c)** thousand.

$$\begin{array}{r} 2351 \\ 4932 \\ +\ 6145 \end{array}$$ (a) 13,430   (b) 13,400   (c) 13,000

Without finding the exact sum, determine if the given sum seems correct or incorrect.

56.	**57.**	**58.**
$$\begin{array}{r} 19 \\ 38 \\ 52 \\ +\ 87 \\ \hline 176 \end{array}$$	$$\begin{array}{r} 491 \\ 650 \\ 125 \\ +\ 749 \\ \hline 2015 \end{array}$$	$$\begin{array}{r} 2395 \\ 841 \\ 89 \\ +\ 153 \\ \hline 3478 \end{array}$$
200 (nearest 10); appears incorrect	2000 (nearest 100); appears correct	3500 (nearest 100); appears correct

---

ANSWERS: 1. $9-8=1$ 2. $11-4=7$ 3. 2 4. 4 5. 1 6. 3 7. 3 8. 7 9. 7 10. 5 11. 6 12. 6 13. 3 14. 1 15. 4 16. 1 17. 3 18. 0 19. 3 20. 4 21. 4 22. 6 23. 8 24. 7 25. 8 26. 8 27. 6 28. 7 29. 9 30. 8 31. 4 32. 9 33. 7 34. 7 35. 9 36. 8 37. 3 38. 8 39. 5 40. 5 41. 5 42. 6 43. 2 44. 8 45. 4 46. 7 47. 8 48. 3 49. 9 50. 5 51. 9 52. 9 53. 5 kittens 54. $8 55. (a) 13,430 (b) 13,400 (c) 13,000 56. Rounded to the nearest ten, the estimated sum is 200; given sum appears incorrect. 57. Rounded to the nearest hundred, the estimated sum is 2000; given sum appears correct. 58. Rounded to the nearest hundred, the estimated sum is 3500; given sum appears correct.

1.5   EXERCISES B

What subtraction problems are displayed below?

1.

$6-5=1$

2.

$9-9=0$

You should be able to complete Exercises 3–37 in two minutes or less.

3.	**4.**	**5.**	**6.**	**7.**	**8.**	**9.**
$$\begin{array}{r} 10 \\ -\ 1 \\ \hline 9 \end{array}$$	$$\begin{array}{r} 13 \\ -\ 4 \\ \hline 9 \end{array}$$	$$\begin{array}{r} 2 \\ -\ 0 \\ \hline 2 \end{array}$$	$$\begin{array}{r} 16 \\ -\ 7 \\ \hline 9 \end{array}$$	$$\begin{array}{r} 14 \\ -\ 9 \\ \hline 5 \end{array}$$	$$\begin{array}{r} 0 \\ -\ 0 \\ \hline 0 \end{array}$$	$$\begin{array}{r} 1 \\ -\ 0 \\ \hline 1 \end{array}$$

10.	**11.**	**12.**	**13.**	**14.**	**15.**	**16.**
$$\begin{array}{r} 16 \\ -\ 8 \\ \hline 8 \end{array}$$	$$\begin{array}{r} 6 \\ -\ 1 \\ \hline 5 \end{array}$$	$$\begin{array}{r} 6 \\ -\ 2 \\ \hline 4 \end{array}$$	$$\begin{array}{r} 4 \\ -\ 2 \\ \hline 2 \end{array}$$	$$\begin{array}{r} 9 \\ -\ 0 \\ \hline 9 \end{array}$$	$$\begin{array}{r} 8 \\ -\ 5 \\ \hline 3 \end{array}$$	$$\begin{array}{r} 12 \\ -\ 9 \\ \hline 3 \end{array}$$

17.	**18.**	**19.**	**20.**	**21.**	**22.**	**23.**
$$\begin{array}{r} 7 \\ -\ 3 \\ \hline 4 \end{array}$$	$$\begin{array}{r} 5 \\ -\ 2 \\ \hline 3 \end{array}$$	$$\begin{array}{r} 5 \\ -\ 3 \\ \hline 2 \end{array}$$	$$\begin{array}{r} 7 \\ -\ 4 \\ \hline 3 \end{array}$$	$$\begin{array}{r} 12 \\ -\ 6 \\ \hline 6 \end{array}$$	$$\begin{array}{r} 11 \\ -\ 5 \\ \hline 6 \end{array}$$	$$\begin{array}{r} 16 \\ -\ 9 \\ \hline 7 \end{array}$$

24. $\begin{array}{r} 11 \\ -\ 6 \\ \hline 5 \end{array}$ **25.** $\begin{array}{r} 5 \\ -\ 1 \\ \hline 4 \end{array}$ **26.** $\begin{array}{r} 9 \\ -\ 8 \\ \hline 1 \end{array}$ **27.** $\begin{array}{r} 10 \\ -\ 5 \\ \hline 5 \end{array}$ **28.** $\begin{array}{r} 14 \\ -\ 7 \\ \hline 7 \end{array}$ **29.** $\begin{array}{r} 4 \\ -\ 0 \\ \hline 4 \end{array}$ **30.** $\begin{array}{r} 10 \\ -\ 6 \\ \hline 4 \end{array}$

31. $\begin{array}{r} 6 \\ -\ 4 \\ \hline 2 \end{array}$ **32.** $\begin{array}{r} 6 \\ -\ 3 \\ \hline 3 \end{array}$ **33.** $\begin{array}{r} 9 \\ -\ 1 \\ \hline 8 \end{array}$ **34.** $\begin{array}{r} 17 \\ -\ 8 \\ \hline 9 \end{array}$ **35.** $\begin{array}{r} 11 \\ -\ 8 \\ \hline 3 \end{array}$ **36.** $\begin{array}{r} 13 \\ -\ 6 \\ \hline 7 \end{array}$ **37.** $\begin{array}{r} 10 \\ -\ 7 \\ \hline 3 \end{array}$

38. $9 - 4 = 5$ **39.** $7 - 6 = 1$ **40.** $15 - 8 = 7$ **41.** $12 - 3 = 9$ **42.** $17 - 9 = 8$

43. $3 - 1 = 2$ **44.** $7 - 7 = 0$ **45.** $6 - 6 = 0$ **46.** $8 - 3 = 5$ **47.** $7 - 0 = 7$

48. $14 - 5 = 9$ **49.** $12 - 4 = 8$ **50.** $2 - 2 = 0$ **51.** $18 - 9 = 9$ **52.** $13 - 9 = 4$

53. A delivery truck contains 17 cartons. At the first stop, 6 cartons are left off. How many cartons remain on the truck?   **11 cartons**

54. Wendy has 13 cassette tapes in a tan storage case and 7 cassette tapes in a blue case. How many more tapes are in the tan case than the blue case?
6 tapes

FOR REVIEW

55. Estimate the sum by rounding to the nearest **(a)** ten **(b)** hundred **(c)** thousand.
$\begin{array}{r} 6549 \\ 2387 \\ +\ 3501 \\ \hline \end{array}$ **(a) 12,440** **(b) 12,400** **(c) 13,000**

Without finding the exact sum, determine if the given sum seems correct or incorrect.

56. $\begin{array}{r} 38 \\ 45 \\ 23 \\ +\ 91 \\ \hline 197 \end{array}$

**200 (nearest 10);
appears correct**

57. $\begin{array}{r} 789 \\ 450 \\ 235 \\ +\ 549 \\ \hline 1823 \end{array}$

**2000 (nearest 100);
appears incorrect**

58. $\begin{array}{r} 4285 \\ 91 \\ 165 \\ +\ 2035 \\ \hline 6576 \end{array}$

**6600 (nearest 100);
appears correct**

1.5 EXERCISES C

Let n represent a whole number. Find n by writing the related addition or subtraction sentence.

1. $n - 5 = 9$
[Answer: 14]

2. $n + 5 = 9$ 4

3. $n - 1 = 6$ 7

4. $n + 2 = 7$
[Answer: 5]

5. $n - 0 = 3$ 3

6. $n + 0 = 6$ 6

7. $n - n = 0$
[Answer: *n* can be
any whole number]

8. $n + 0 = n$
n can be any
whole number.

1.6 SUBTRACTING WHOLE NUMBERS

STUDENT GUIDEPOSTS

1 Subtracting Numbers with More Digits

3 Checking Subtraction by Addition

2 Borrowing in Subtraction Problems

1 SUBTRACTING NUMBERS WITH MORE DIGITS

We now know basic facts of subtraction. To find differences between larger numbers, use the basic facts and the knowledge of place value. For example, find the following difference by writing the numbers in expanded form, then subtracting the ones digits, 8 and 5, and the tens digits, 3 and 2.

$$
\begin{aligned}
38 &= 30 + 8 = 3 \text{ tens} + 8 \text{ ones} \\
-\ 25 &= \underline{20 + 5} = \underline{2 \text{ tens} + 5 \text{ ones}} \\
& \qquad\qquad\quad\ 1 \text{ ten}\ + 3 \text{ ones} = 13
\end{aligned}
$$

The difference, 13, can be found more directly using the two steps shown below.

$$
\begin{aligned}
& \quad\ 38 \\
& -\ 25 \\
\text{② Subtract tens digits} \rightarrow & \quad\ 13\ \leftarrow \text{① Subtract ones digits}
\end{aligned}
$$

The same method works for numbers with more digits as shown in the next example.

EXAMPLE 1   SUBTRACTING LARGER NUMBERS

Find the difference.
$$
\begin{aligned}
& 896 \\
-\ & 572
\end{aligned}
$$

Subtract the ones digits, the tens digits, and the hundreds digits.

$$
\begin{aligned}
896 &= 8 \text{ hundreds} + 9 \text{ tens} + 6 \text{ ones} \\
-\ 572 &= \underline{5 \text{ hundreds} + 7 \text{ tens} + 2 \text{ ones}} \\
324 & \quad\ 3 \text{ hundreds} + 2 \text{ tens} + 4 \text{ ones} = 324
\end{aligned}
$$

PRACTICE EXERCISE 1

Find the difference.
$$
\begin{aligned}
& 785 \\
-\ & 643
\end{aligned}
$$

Answer: 142

---

2 BORROWING IN SUBTRACTION PROBLEMS

You probably noticed that in Example 1 only subtraction of one-digit numbers was needed. We now look at problems which require subtraction of a one-digit number from a two-digit number. The notion of place value plays an important part in this process. Consider the following problem.

$$
\begin{aligned}
& 63 \\
-\ & 48
\end{aligned}
$$

We quickly see that the ones digit 8 cannot be subtracted from the ones digit 3. The difficulty can be removed by ''borrowing'' ten ones from the tens column in 63. That is,

$$
\begin{aligned}
63 &= \textbf{6 tens} + 3 \text{ ones} \\
&= \textbf{5 tens} + \textbf{1 ten} + 3 \text{ ones} \qquad \text{6 tens} = \text{5 tens} + \text{1 ten} \\
&= 5 \text{ tens} + \boxed{1 \text{ ten} + 3 \text{ ones}} \\
&= 5 \text{ tens} + \boxed{10 \text{ ones} + 3 \text{ ones}} \qquad \text{1 ten} = \text{10 ones} \\
&= 5 \text{ tens} + 13 \text{ ones}.
\end{aligned}
$$

Thus,

$$
\begin{array}{rcl}
63 = 6 \text{ tens} + 3 \text{ ones} &=& 5 \text{ tens} + 13 \text{ ones} \\
- \ 48 = 4 \text{ tens} + 8 \text{ ones} &=& 4 \text{ tens} + \ \ 8 \text{ ones} \\
\hline
15 & & 1 \text{ ten} \ + \ 5 \text{ ones} = 15.
\end{array}
$$

We usually write the problem in shortened form as follows.

$$
\begin{array}{r}
^{5\,13} \\
\cancel{63} \\
- \ 48 \\
\hline
15
\end{array}
$$

Borrow 1 ten from 6 tens, change 6 to 5 and 3 to 13

Subtract corresponding columns

H I N T

Borrowing in a subtraction problem is the reverse of carrying in an addition problem.

EXAMPLE 2   BORROWING IN A SUBTRACTION PROBLEM

Find the difference.
$$
\begin{array}{r}
529 \\
- \ 365
\end{array}
$$

$$
\begin{array}{r}
^{4\,12} \\
\cancel{529} \\
- \ 365 \\
\hline
164
\end{array}
$$

Subtract 5 from 9 to obtain 4. Since $2 - 6$ is not a whole number, borrow 1 hundred from 5, change 5 to 4 and 2 to 12.

Subtract 6 from 12 and 3 from 4

PRACTICE EXERCISE 2

Find the difference.
$$
\begin{array}{r}
638 \\
- \ 491
\end{array}
$$

Answer: 147

Sometimes we need to borrow more than once in the same problem. This is illustrated in the next example.

EXAMPLE 3   REPEATED BORROWING

Find each difference.

(a)
$$
\begin{array}{r}
723 \\
- \ 475
\end{array}
$$

$$
\begin{array}{r}
^{1\,13} \\
7\,\cancel{2}\,\cancel{3} \\
- \ 4\,7\,5 \\
\hline
8
\end{array}
$$

Borrow 1 ten from 2, change 2 to 1 and 3 to 13

Subtract 5 from 13

PRACTICE EXERCISE 3

Find each difference.

(a)
$$
\begin{array}{r}
425 \\
- \ 187
\end{array}
$$

```
   11
  6 1 13
  7 2 3      Since 1 − 7 is not a whole number, borrow 1 hundred
 − 4 7 5       from 7, change 7 to 6 and 1 to 11
      4 8    Subtract 7 from 11
```

```
   11
  6 1 13
  7 2 3
 − 4 7 5
  2 4 8      Subtract 4 from 6
```

(b)   6312
         − 4795

```
       0 12
  6 3 1 2    Borrow 1 ten from 1, change 1 to 0 and 2 to 12
 − 4 7 9 5
        7    Subtract 5 from 12
```

```
     10
   2 0 12
  6 3 1 2    Since 0 − 9 is not a whole number, borrow 1 hundred
 − 4 7 9 5     from 3, change 3 to 2 and 0 to 10
      1 7    Subtract 9 from 10
```

```
  12 10
  5 2 0 12
  6 3 1 2    Since 2 − 7 is not a whole number, borrow 1 thousand
 − 4 7 9 5     from 6, change 6 to 5 and 2 to 12
  1 5 1 7    Subtract 7 from 12 and 4 from 5
```

(b)   7436
         − 2697

Answers: (a) **238**   (b) **4739**

When there is a zero in a position from which we must borrow, we borrow across that position. This is shown in the next example.

EXAMPLE 4   BORROWING ACROSS A ZERO

Find each difference.

(a)   605
         − 238

```
  5 10
  6 0 5      We must borrow 1 ten, but since 605 has no tens,
 − 2 3 8       we must first borrow 1 hundred (10 tens) from 6.
               Change 6 to 5 and 0 to 10.
```

```
      9
  5 10 15
  6 0 5      Now borrow 1 ten from 10, change 10 to 9 and 5 to 15
 − 2 3 8
  3 6 7      Subtract 8 from 15, 3 from 9, and 2 from 5
```

(b)   2003
         − 895

```
  1 10
  2 0 0 3    We must borrow 1 ten, but there are no tens, so we
 −   8 9 5     try to borrow 1 hundred. Since there are no
               hundreds, we must first borrow 1 thousand.
               Change 2 to 1 and 0 (the hundreds digit) to 10.
```

PRACTICE EXERCISE 4

Find each difference.

(a)   502
         − 267

(b)   6007
         − 2659

$$\begin{array}{r} 9 \\ 1\ \cancel{10}\cancel{0} \\ \cancel{2}\ \cancel{0}\ \cancel{0}\ 3 \\ -\quad 8\ 9\ 5 \end{array}$$ Next, change 10 (hundreds) to 9 and 0 (tens) to 10

$$\begin{array}{r} 9\ \ 9 \\ 1\ \cancel{10}\cancel{10}13 \\ \cancel{2}\ \cancel{0}\ \cancel{0}\ \cancel{3} \\ -\quad 8\ 9\ 5 \end{array}$$ Change 10 (tens) to 9 and 3 (ones) to 13

$$\begin{array}{r} 9\ \ 9 \\ 1\ \cancel{10}\cancel{10}13 \\ \cancel{2}\ \cancel{0}\ \cancel{0}\ \cancel{3} \\ -\quad 8\ 9\ 5 \\ \hline 1\ 1\ 0\ 8 \end{array}$$ Subtract 5 from 13, 9 from 9, 8 from 9, and bring down the 1

Answers: (a) 235 (b) 3348

/////////////// **CAUTION** ///////////////

From Section 1.2 we know that changing the order of addition does not change the sum. The same is *not* true for subtraction. For example,

$$8 - 5 = 3,$$

but $5 - 8$ is not a whole number.

Using the symbol $\neq$, which is read "is not equal to," we write

$$8 - 5 \neq 5 - 8.$$

Similarly, in Section 1.3, we discovered that regrouping when adding does not change the sum. Again the same is *not* true for subtraction. For example,

$$(7 - 3) - 2 = 4 - 2 = 2,$$

but

$$7 - (3 - 2) = 7 - 1 = 6,$$

so that

$$(7 - 3) - 2 \neq 7 - (3 - 2).$$

Thus, subtraction is neither commutative nor associative.

//////////

③ CHECKING SUBTRACTION BY ADDITION

Subtraction can be checked in one of two ways. If we check by repeating the subtraction step by step, chances are that any error made the first time will be repeated. Since it is better to check work using a different process, we use the related addition sentence.

$$7 - 3 = 4 \quad \text{is true since} \quad 4 + 3 = 7.$$

This example can be generalized.

To Check a Subtraction Problem

Add the difference to the subtrahend. If the sum is the minuend, the problem is correct.

The format to follow when checking a subtraction problem is illustrated in the following example.

EXAMPLE 5 CHECKING SUBTRACTION BY ADDITION

Check each subtraction problem.

(a)
```
   629
 − 342
   287
```

These must be equal
```
   629
 − 342     Add these and place sum below dashed line
   287
 − − − −
   629
```

The dashed line can be thought of as the line for the addition problem
```
   342
 + 287
   629.
```

(b)
```
   4030
 − 2376
   1664
```

```
   4030
 − 2376     Since these are not equal, the problem is
   1664         incorrect. The difference should be 1654.
 − − − −
   4040
```

EXAMPLE 6 AN APPLICATION OF SUBTRACTION

Peggy has $883 in her checking account and writes a check for $209. First estimate the number of dollars left in her account, and then find the exact amount.

Since $883 is about $900 and $209 is about $200 (rounding to the nearest 100), estimate the number of dollars left in the account by subtracting 200 from 900.

```
   900 ⎞
 − 200 ⎬ mental work
   700 ⎠
```

Peggy has about $700 left in the account. To find the exact amount, subtract 209 from 883.

```
   883
 − 209
   674
```

The exact amount left in her account is $674.

PRACTICE EXERCISE 5

Check each subtraction problem.

(a)
```
   847
 − 269
   578
```

(b)
```
   6701
 − 4985
   1726
```

Answers: (a) Since 269 + 578 = 847, the subtraction is correct. (b) Since 4985 + 1726 = 6711 ≠ 6701, the subtraction is incorrect.

PRACTICE EXERCISE 6

Phil Mortensen owned 3105 cattle and sold 1970 of them. First estimate the number of cattle he still owned, and then find the exact number.

Answer: Estimated number: 1000 cattle; exact number: 1135 cattle

//////// **CAUTION** ////////

Since subtraction is not commutative, we should be careful when solving a subtraction word problem. The numbers must be subtracted in the correct order.

1.6 EXERCISES A

Find the following differences.

1.	2.	3.	4.	5.	6.
63	97	57	983	738	634
− 21	− 62	− 36	− 365	− 429	− 519
42	35	21	618	309	115

7.	8.	9.	10.	11.	12.
205	603	101	300	400	800
− 173	− 381	− 90	− 215	− 260	− 473
32	222	11	85	140	327

13.	14.	15.	16.	17.	18.
6973	6281	3005	7002	4315	38,000
− 4325	− 3425	− 1624	− 1325	− 2549	− 23,999
2648	2856	1381	5677	1766	14,001

Check the following subtraction problems.

19.	20.	21.	22.	23.
96	425	325	4392	8002
− 23	− 132	− 279	− 1782	− 4293
73	293	156	2610	4709
correct	**correct**	**incorrect (should be 46)**	**correct**	**incorrect (should be 3709)**

Find the difference and check your work.

24.	25.	26.	27.	28.
423	700	302	7562	6006
− 105	− 268	− 78	− 3157	− 1991
318	432	224	4405	4015

Solve.

29 The Kerns make a house payment of $389 per month. They are considering buying a new house for which the payment will be $605 per month. First estimate the increase in payment, and then find the exact increase.

estimated increase: $200; actual increase: $216

30. A motor home will travel 402 miles on one 30-gallon tank of gasoline, while a Blazer can travel 595 miles on the same amount of gasoline. First estimate the difference in miles traveled, and then find the exact difference.

estimated difference: 200 mi; actual difference: 193 mi

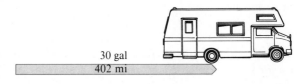

30 gal
402 mi

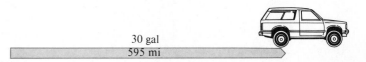

30 gal
595 mi

31. Suppose your bank balance was $614 when you wrote three checks for $105, $89, and $26.

   (a) What was the total of the checks written?   **$220**

   (b) What was your new bank balance?   **$394**

32. At the end of a three-day trip, the odometer on Walter's car read 36,297 miles. Suppose he drove 426 miles, 298 miles, and 512 miles on the three days.

   (a) What was the total number of miles driven on the three-day trip?   **1236 mi**

   (b) What was the odometer reading prior to the trip?   **35,061 mi**

ANSWERS: 1. 42 2. 35 3. 21 4. 618 5. 309 6. 115 7. 32 8. 222 9. 11 10. 85 11. 140 12. 327 13. 2648 14. 2856 15. 1381 16. 5677 17. 1766 18. 14,001 19. correct 20. correct 21. incorrect (should be 46) 22. correct 23. incorrect (should be 3709) 24. 318 25. 432 26. 224 27. 4405 28. 4015 29. estimated increase: $200; actual increase: $216 30. estimated difference: 200 mi; actual difference: 193 mi 31. (a) $220 (b) $394 32. (a) 1236 mi (b) 35,061 mi

1.6 EXERCISES B

Find the following differences.

1.   45
   − 13
   32

2.   43
   − 20
   23

3.   84
   − 43
   41

4.   581
   − 124
   457

5.   566
   − 157
   409

6.   511
   − 208
   303

7.   430
   − 319
   111

8.   270
   − 115
   155

9.   208
   − 40
   168

10.   600
   − 103
   497

11.   500
   − 128
   372

12.   900
   − 303
   597

13.   3728
   − 1154
   2574

14.   5432
   − 2734
   2698

15.   6005
   − 251
   5754

16.   9030
   − 675
   8355

17.   25,132
   − 14,627
   10,505

18.   42,135
   − 11,329
   30,806

Check the following subtraction problems.

19.   46
   − 38
   18
   incorrect (should be 8)

20.   407
   − 229
   178
   correct

21.   700
   − 296
   504
   incorrect (should be 404)

22.   3075
   − 346
   2729
   correct

23.   5040
   − 2504
   3546
   incorrect (should be 2536)

Find the difference and check your work.

24.   208
   − 129
   79

25.   601
   − 392
   209

26.   7302
   − 488
   6814

27.   4700
   − 1308
   3392

28.   8000
   − 6989
   1011

Solve.

29. A new car weighs 3105 pounds, while a new truck with a camper weighs 6970 pounds. First estimate the difference in weight, and then find the exact difference.
   estimated difference: 4000 pounds;
   exact difference: 3865 pounds

30. A new printing press can print 5890 brochures in one day, while an older model can print only 2995 brochures each day. First estimate the difference in the number of brochures per day, and then find the exact difference.
   estimated difference: 3000 brochures;
   exact difference: 2895 brochures

31. A new stereo system sells for $2650 plus sales tax in the amount of $106.

    (a) What is the cost of the system, including the sales tax?   **$2756**

    (b) If you make a down payment of $750, what amount remains to be paid?   **$2006**

32. A salesperson has a monthly expense account of $550. Suppose the amount already spent this month includes $125 for gasoline, $210 for food, and $86 for the telephone.

    (a) What is the total spent thus far this month? **$421**

    (b) What is the balance remaining in the expense account?   **$129**

1.6 EXERCISES C

In the following exercises do what is in the parentheses first.

1. $(3 + 2) - 4$ 1

2. $(8 + 6) - (9 + 1)$
[Answer: 4]

3. $(8 - 5) + 3$ 6

4. $(6 - 2) - (9 - 8)$
[Answer: 3]

1.7 MORE APPLICATIONS

This section contains a variety of word problems that involve finding sums and differences. Follow these steps when solving word problems.

To Solve a Word Problem

1. Read the problem several times to be certain you understand what is being asked. Draw a picture whenever possible.

2. Identify what is given and what must be found, then make a plan of attack.

3. Solve, using your plan and the given information.

4. Ask yourself if your answer is reasonable. If it is, check your work. If not, return to Step 1.

HINT

Deciding if an answer is reasonable is important and can help you avoid making errors. For example, suppose the problem asks for the price of a new car. If you obtain an answer of $200, your answer is clearly unreasonable, and you should begin again.

EXAMPLE 1   AN APPLICATION IN BANKING

Carlos has two passbook savings accounts, one with $2350 and the other with $1745. He also has three certificates of deposit with values of $5000, $2500, and $1250. What is the total of these cash assets?

    Total the amounts in the five different accounts at the right.

$$
\begin{array}{r}
2350 \\
1745 \\
5000 \\
2500 \\
+\ 1250 \\
\hline
12845
\end{array}
$$

This is a reasonable total, one which checks using the method of reverse-order addition. Thus, Carlos has $12,845 in cash assets.

PRACTICE EXERCISE 1

At a recent concert, ticket prices were $7 for adults, $4 for students between 12 and 20 years of age, and $2 for children 12 years and younger. Becky is 13 years old and her sister Cindy is 9. If their parents take them to the concert, how much will the four of them pay?

Answer: $20

EXAMPLE 2	AN APPLICATION IN SPORTS

The players' scores in a recent basketball game are shown in the table. What was the team score?

Total the numbers in the points column.

First time   *Second time*

			Player	Points
21	21		Hudson	21
17	17		Betton	17
12	12		Herman	12
10	10		Williams	10
8	8		Ingram	8
7	7		Hurd	7
7	7		Payne	7
3	3		Spencer	3
3	3		Rodriguez	3
438 Incorrect	88 Correct			

Since 438 is clearly unreasonable for the score of a team in a basketball game, reread the problem. Again it is clear that the nine numbers must be added. In adding a second time, the error is found. The actual sum is 88, which does check.

PRACTICE EXERCISE 2

When Rosa purchased a new car, the price she paid was posted on a window sticker like the one shown here. How much did Rosa pay for her car?

Base price	$6500
Power package	$1100
Radio	$ 125
Air conditioner	$ 700
Special paint	$ 95
Tax and license	$ 436
TOTAL	

Answer: $8956

EXAMPLE 3	READING A MAP

Figure 1.19 shows a sketch of a map of part of the western United States.

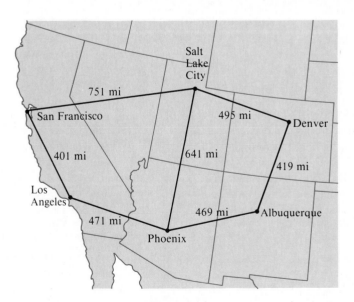

Figure 1.19

PRACTICE EXERCISE 3

The sketch below shows a map of the eastern United States.

(a) How far is it from Cleveland to Atlanta through Baltimore?

(a) How far is it from Denver to Phoenix through Albuquerque?

From Denver to Albuquerque is 419 mi, and from Albuquerque to Phoenix is 469 mi. Find the following sum.

$$\begin{array}{r} 419 \\ + \ 469 \\ \hline 888 \end{array}$$

Since this is reasonable, and does check, it is 888 mi from Denver to Phoenix through Albuquerque.

(b) Is it shorter to go from Los Angeles to Salt Lake City through San Francisco or through Phoenix?

The distance from Los Angeles to Salt Lake City through San Francisco is found by adding 401 and 751. The distance from Los Angeles to Salt Lake City through Phoenix is found by adding 471 and 641.

$$\begin{array}{r} 401 \\ + \ 751 \\ \hline 1152 \end{array} \quad \text{Through San Francisco} \qquad \begin{array}{r} 471 \\ + \ 641 \\ \hline 1112 \end{array} \quad \text{Through Phoenix}$$

Since these sums appear reasonable, and do check, it is shorter to go by way of Phoenix.

(c) The Jacksons, who live in Los Angeles, plan a trip to Denver for their summer vacation. Going, they will take the southern route through Phoenix and Albuquerque. They plan to return through Salt Lake City and San Francisco. How far will they travel?

Add the six numbers 471, 469, 419, 495, 751, and 401. The total, 3006, seems reasonable and does check. Thus, the family will travel 3006 mi on their vacation.

$$\begin{array}{r} 471 \\ 469 \\ 419 \\ 495 \\ 751 \\ + \ 401 \\ \hline 3006 \end{array}$$

(b) Is it shorter to go from Baltimore to Nashville through Atlanta or through Cleveland and Louisville?

(c) The Crums, who live in Louisville, plan a trip to Baltimore. Going, they plan to take the direct route, and they plan to return home through Atlanta and Nashville. How far will they travel on their trip?

Answers: (a) 988 mi   (b) It is shorter to go through Cleveland and Louisville (856 mi) than through Atlanta (887 mi). (c) 1653 mi

---

EXAMPLE 4   AN APPLICATION IN GEOGRAPHY

The height of Mount Everest is twenty-nine thousand, twenty-eight feet, and the height of Mount McKinley is twenty thousand, three hundred twenty feet. How much higher is Mount Everest than Mount McKinley?

First visualize the problem as shown in Figure 1.20.

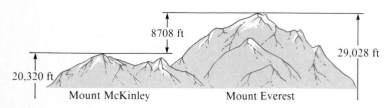

Figure 1.20

PRACTICE EXERCISE 4

The height of Pike's Peak is fourteen thousand, one hundred ten feet, and the height of Long's Peak is fourteen thousand, two hundred fifty-six feet. How much higher is Long's Peak than Pike's Peak?

Subtract 20,320 (the height of Mount McKinley) from 29,028 (the height of Mount Everest).

$$
\begin{array}{r}
{\scriptstyle 8\ 10} \\
2\,9\,\not{0}\,28 \\
-\ 20320 \\
\hline
8708
\end{array}
$$

This difference is certainly reasonable, and it checks. Thus, Mount Everest is 8708 ft higher than Mount McKinley.

Answer: 146 feet

EXAMPLE 5   An Application in Recreation

On a two-week camping trip, a group of students took along 42 six-packs of soda. The first week they drank 18 six-packs, and the second week they drank another 21 six-packs. How many six-packs were left at the end of the trip?

First find the total number of six-packs that were drunk during the two-week period.

$$
\begin{array}{r}
18 \\
+\ 21 \\
\hline
39
\end{array}
$$

Then, to find out how many six-packs were left, subtract this total from the number of six-packs taken along on the trip.

$$
\begin{array}{r}
42 \\
-\ 39 \\
\hline
3
\end{array}
$$

In view of what is given, 3 seems reasonable.

3	Number of six-packs left
18	Number drunk first week
+ 21	Number drunk second week
42	Number taken on trip

Thus, 3 checks and the group had 3 six-packs of soda left over.

PRACTICE EXERCISE 5

Mike weighs 171 pounds and Arn weighs 185 pounds. Larry weighs 5 pounds less than Arn. What is the total weight of the three men?

Answer: 536 pounds

1.7 EXERCISES A

Solve.

1. A theater is divided into three sections. There are 285 seats in the section on the right, 430 seats in the center section, and 265 seats in the section on the left. How many people will the theater seat?   **980**

2. Sam, a salesman, drove 135 mi on Monday, 207 mi on Tuesday, 175 on Wednesday, 87 on Thursday, and 325 on Friday. How far did Sam drive that week?   **929 mi**

3. The figure shown is called a **rectangle.** It has length 10 ft and width 4 ft. (Note: **ft** is the abbreviation for feet.) The **perimeter** of a rectangle is the distance around the figure, found by adding the measurements of all its sides. Find the perimeter of the rectangle shown.

10 ft

4 ft

28 ft

4. Tana has a garden in the shape of the triangle shown in the figure. (Note: **m** is the abbreviation for meters.) If she plans to put a fence around the garden, how many meters of fencing will she need? [*Hint:* Find the perimeter of the triangle.]

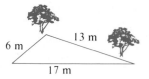

6 m   13 m   17 m

36 m

5. A bowler rolled games of 179, 203, and 188. What was her total for the three-game series?   **570**

6. Enrollments in the five colleges of a university are: Arts and Science, 3810; Business, 1608; Education, 1425; Engineering, 865; and Forestry, 239. What is the total university enrollment?   **7947 students**

Exercises 7–9 refer to the information given in the following table.

Game Statistics

Home Team			Visiting Team		
Player	Points	Fouls	Player	Points	Fouls
Hanaki	25	4	Kirk	18	3
Bonnett	19	3	Mutter	18	2
Condon	13	0	Westmoreland	15	4
Bell	7	5	Anderson	9	1
Walter	6	2	Ratliff	7	2
Meyer	3	1	Romero	1	3
Shaff	0	5	Thoreson	1	0

7. How many points were scored by the home team?
73 points

8. How many points were scored by the visiting team?
69 points

9. Which team had the most fouls?
The home team had more (20) than the visiting team (15).

10. Bob had $387 in his savings account. If he withdrew $175, how much was left in the account?   **$212**

11. The Henderson well in western Arizona is 6483 ft deep. The Lupine well in southern Colorado is 3976 ft deep. How much deeper is the Henderson well?   **2507 ft**

12. Before leaving on a trip, Luisa noticed that the odometer reading on her car was 38,427. When she returned home from the trip, the reading was 41,003. How many miles did she drive?   **2576 mi**

13 At the beginning of June, Mr. Hernandez had a balance of $625 in his checking account. During the month he made deposits of $300 and $235. He wrote checks for $23, $18, $113, $410, and $32. What was his balance at the end of the month?   **$564**

14. Janet buys a radio for $80 and a portable TV for $140. If the tax on these two purchases totals $11, how much change will she receive if she pays with a check for $300?   **$69**

Exercises 15–16 refer to the map below.

15 What is the shortest route from Salt Lake City to Seattle? How much shorter is it than the next shortest route?

16. The Hardy family, who live in Portland, plan a trip to Cheyenne on their summer vacation. How much shorter is it to go through Butte than through Reno?   **40 mi**

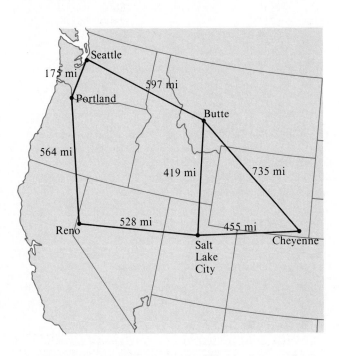

The route through Butte is 251 mi shorter than through Reno and Portland.

17. A bulk tanker contains 1478 gallons of gasoline. Three deliveries are made, one of 230 gallons, a second of 435 gallons, and the third of 170 gallons. How much gasoline remains in the tanker?   **643 gallons**

18 Sales at Darrell's Men's Wear were $1245 on Friday. If this is $305 more than on Thursday, and $785 less than on Saturday, what were the sales on Thursday? On Saturday?   **Thursday: $940; Saturday: $2030**

19. A truck weighs 4845 pounds when empty. When it is loaded with a camper, it weighs 7525 pounds. What is the weight of the camper?

20. The *margin of profit* when an item is sold is the list price of the item less the cost of the item. During a recent clearance sale, a store owner sold a bedroom set at cost, for $1150. If she had sold the set at its list price, $1595, what would the margin of profit have been?   **$445**

4845 pounds

7525 pounds

2680 pounds

21. On opening day of fall practice, 123 football players were present. If only 65 can eventually make the team, how many players must be cut from the squad?   **58 players**

22 Mike has $25 and Jim has $43. Larry has $12 more than Jim, and Henri has $9 less than Mike and Jim together. How much money do the four men have together?   **$182**

FOR REVIEW

Check the following subtraction problems.

23.
$$\begin{array}{r} 83 \\ -\ 47 \\ \hline 46 \end{array}$$
incorrect
(should be 36)

24.
$$\begin{array}{r} 649 \\ -\ 283 \\ \hline 526 \end{array}$$
incorrect
(should be 366)

25.
$$\begin{array}{r} 7003 \\ -\ 421 \\ \hline 6582 \end{array}$$
correct

26.
$$\begin{array}{r} 8050 \\ -\ 2569 \\ \hline 6591 \end{array}$$
incorrect
(should be 5481)

ANSWERS: 1. 980 2. 929 mi 3. 28 ft 4. 36 m 5. 570 6. 7947 7. 73 points 8. 69 points 9. The home team had more (20) than the visiting team (15). 10. $212 11. 2507 ft 12. 2576 mi 13. $564 14. $69 15. The route through Butte is 251 mi shorter than the route through Reno and Portland. 16. 40 mi 17. 643 gallons 18. Thursday: $940; Saturday: $2030 19. 2680 pounds 20. $445 21. 58 players 22. $182 23. incorrect (should be 36) 24. incorrect (should be 366) 25. correct 26. incorrect (should be 5481)

1.7 EXERCISES B

Solve.

1. Professor Mutter teaches three classes. There are 42 students in his Math 151 class, 32 in his Math 136 class, and 38 in his Math 130 class. How many students does Professor Mutter have?   **112**

2. In the first five games of the season, running back Allan Clark picked up 137 yd, 215 yd, 120 yd, 154 yd, and 183 yd. What was his total yardage through the first five games?   **809 yd**

3. The figure shown here is called a *square*. (Note: **cm** is the abbreviation for centimeters.) A **square** is a rectangle with four equal sides. Find the perimeter of the square shown.   **44 cm**

11 cm

4. Dan's scorecard for playing the front nine holes of golf is shown here. How many shots did Dan take on the front nine?

Hole	1	2	3	4	5	6	7	8	9
Score	3	5	6	4	4	5	4	3	5

39 shots

5. Peggy has a garden in the shape shown below. If she plans to put a fence around the garden, how many feet of fencing will she need?   **23 ft**

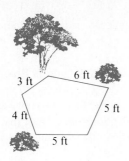

3 ft   6 ft
4 ft   5 ft
5 ft

6. On a five-day trip, the Watson family drove 135 mi, 380 mi, 427 mi, 265 mi, and 641 mi. What was the total mileage driven on the trip?   **1848 mi**

Exercises 7–8 refer to the map of Lake Powell shown below.

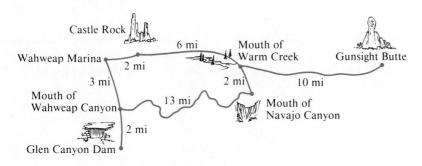

Castle Rock
Wahweap Marina   6 mi   Mouth of Warm Creek   Gunsight Butte
2 mi
3 mi   2 mi   10 mi
Mouth of Wahweap Canyon   13 mi
Mouth of Navajo Canyon
2 mi
Glen Canyon Dam

7. Which distance is greater, Castle Rock to Gunsight Butte, or Glen Canyon Dam to the mouth of the Navajo Canyon? How much greater?
Castle Rock to Gunsight Butte is greater by 1 mi.

8. What distance is traveled by a boat that leaves Wahweap Marina and makes a circular trip past the mouths of Wahweap Canyon, Navajo Canyon, and Warm Creek and returns to the marina by way of Castle Rock?   **26 mi**

9. Sarah purchases 9 square yards of one type of cloth, 14 square yards of another, and 7 square yards of a third. How many square yards of material does she buy?   **30 sq yds**

10. On a fishing trip, Mick caught 32 fish and Pete caught 17 fish. How many more fish did Mick catch than Pete?   **15 more fish**

11. Ms. Balushi has a subcompact car and a luxury sedan. She figures that over the period of one year, it will cost $735 to drive the luxury sedan to work while only $265 to drive the subcompact car. How much can she save by driving the subcompact car?
$470

12. At the beginning of the month, Mr. Lopez had a balance of $735 in his checking account. If he made deposits of $400 and $375, and wrote checks for $87, $180, $112, $415, and $7 during the month, what was his balance at the end of the month?
$709

13. Barb bought a coat for $74 and a purse for $17. If the tax on these two purchases totals $5, how much change will she receive if she pays with a check for $100?   **$4**

14. Bridal Veil Falls is 1957 feet high. Niagara Falls is 193 feet high. How much higher is Bridal Veil Falls than Niagara Falls?   **1764 ft**

Exercises 15–18 refer to the following backpacker's trail map.

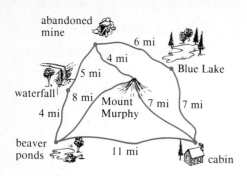

15. How far is it from the cabin to the abandoned mine by way of Mount Murphy? **11 mi**

16. If a hiker makes a complete loop around Mount Murphy, how far will she hike? **33 mi**

17. What is the shortest hike from Blue Lake to the beaver ponds?
15 mi (through the abandoned mine and the waterfall)

18. What is the shortest hike from the waterfall to the cabin? **15 mi (through the beaver ponds)**

19. Lori bought a car for $3800. If she spent $625 on repairs and then sold it for $5100, how much money did she make on the deal? **$675**

20. Rob earns $724 a week. If he has $98 withheld for federal taxes, $43 withheld for Social Security taxes, and $140 withheld and deposited in his credit union, what is his net take-home pay each week?
$443

21. Antwine bought a coat on sale. If the original price of the coat was $185 and it was discounted $40, how much change did he receive if he paid with two $100 bills? **$55**

22. At the start of a trip, the odometer on a car read 48,263 miles. At the end of the trip it read 51,191 miles. What distance was traveled on the trip?
2928 mi

FOR REVIEW

Check the following subtraction problems.

23.
$$\begin{array}{r} 62 \\ -\ 39 \\ \hline 23 \end{array}$$
correct

24.
$$\begin{array}{r} 828 \\ -\ 493 \\ \hline 435 \end{array}$$
incorrect (should be 335)

25.
$$\begin{array}{r} 4002 \\ -\ 938 \\ \hline 3064 \end{array}$$
correct

26.
$$\begin{array}{r} 2090 \\ -\ 1357 \\ \hline 1743 \end{array}$$
incorrect (should be 733)

1.7 EXERCISES C

Solve.

1. Let n represent the number of cars a dealer had available for sale one month. If she sold 60 during the month and had 120 left on the lot, find n.
[Answer: $n = 180$]

2. Let n represent the number of boys in an eighth-grade class. If there are 35 total students including 17 girls, find n. **$n = 18$**

CHAPTER 1 REVIEW

KEY WORDS

1.1 The **natural** or **counting numbers** are
1, 2, 3, 4, ...

The **whole numbers** are the natural numbers together with zero: 0, 1, 2, 3, ...

A **numeral** is a symbol used to represent a number.

A **number line** is used to "picture" numbers.

The first ten whole numbers, 0, 1, 2, 3, 4, 5, 6, 7, 8, and 9, are also called **digits.**

Our number system is a **place-value system** because the value of each digit depends on its place or position in a numeral.

1.2 Two numbers to be added are called **addends** and the result is called their **sum.**

1.3 Addition is a **binary operation,** which means that only *two* numbers are added at a time.

1.4 Numbers can be **rounded** or approximated by numbers to the nearest ten, hundred, thousand, and so on.

1.5 In a subtraction problem, the number being taken away is called the **subtrahend,** the num-ber it is taken from is the **minuend,** and the result is the **difference.** That is,

$$minuend - subtrahend = difference.$$

The **related addition sentence** to $7 - 3 = 4$, for example, is $4 + 3 = 7$.

Similarly, the **related subtraction sentence** to $4 + 3 = 7$, for example, is $7 - 3 = 4$ or $7 - 4 = 3$.

KEY CONCEPTS

1.1
1. The whole numbers occur in a natural order. If one whole number is to the left of a second whole number on a number line, it is *less than* the second. The symbol $<$ represents "less than," and $>$ represents "greater than."

2. The number 4379, written in standard notation, is a short form of expanded notation

$$4000 + 300 + 70 + 9.$$

3. Zero is used as a place-holder to show the difference between numbers such as 5032 and 532.

4. Write numbers such as 123 as "one hundred twenty-three" and *not* as "one hundred *and* twenty-three."

1.2
1. The *sum* of two numbers can be thought of as the number of objects in a combined collection.

2. Addition problems may be written either horizontally or vertically.

3. Changing the order of addition does not change the sum by the commutative law. For example, $2 + 9 = 9 + 2$.

1.3
1. The addition procedure can be shortened by using the process of carrying.

2. Regrouping sums does not change the answer by the associative law. For example, $(2 + 3) + 7 = 2 + (3 + 7)$.

3. Addition may be checked by the reverse-order method.

1.4 Rounded numbers can be used to estimate sums.

1.5
1. The *difference* of two numbers can be thought of as the number of objects which remain when objects are removed or taken away from a collection.

2. Subtraction can be defined by using the related addition sentence.

1.6
1. Borrowing in subtraction problems is the reverse of carrying in addition problems.

2. Subtraction can be checked by adding the difference to the subtrahend.

1.7 After you understand a word problem, make a plan of attack and identify what is given and what must be found. A picture or sketch may be helpful. Always make sure that any answer seems reasonable.

REVIEW EXERCISES

Part I

1.1
1. What is the symbol for a number called?
   numeral

2. Give the first five natural numbers.
   1, 2, 3, 4, 5

3. Give the first five whole numbers.
   0, 1, 2, 3, 4

4. What are the ten digits?
   0, 1, 2, 3, 4, 5, 6, 7, 8, 9

5. Why is the number system called a place-value system?
   because the value of each digit is determined by its place or position in the numeral

Answer Exercises 6–11 using the given number line.

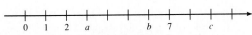

6. What number is paired with *a*?   **3**

7. What number is paired with *c*?   **9**

Place the correct symbol (< or >) between the given numbers, remembering that a, b, and c are from the number line above.

8. $2 \underline{\ \ <\ \ } 7$ **9.** $b \underline{\ \ >\ \ } a$ **10.** $b \underline{\ \ <\ \ } c$ **11.** $a \underline{\ \ <\ \ } c$

12. Write 13,247,165 in expanded notation.
10,000,000 + 3,000,000 + 200,000 + 40,000 + 7000 + 100 + 60 + 5

13. Write 400,000 + 3000 + 400 + 5 in standard notation.   403,405

14. Write a word name for 27,405,036.   twenty-seven million, four hundred five thousand, thirty-six

In the number 145,237,098 what digit represents each of the following places?

15. Millions   5   **16.** Hundreds   0   **17.** Ten thousands   3   **18.** Ones   8

1.2   **19.** Consider the addition problem 7 + 3 = 10.
(a) What are 7 and 3 called?   **(b)** What is 10 called?   **(c)** What is the symbol + called?
    addends   sum   plus sign

20. Complete the statement using the commutative law of addition. $5 + 4 = \underline{\ \ 4+5\ \ }$

21. When a given number is added to 0, what is the sum?   the given number

1.3   *Find the following sums and check your work.*

22.   68
     + 97
      165

23.   392
     + 79
      471

24.   457
    + 168
     625

25.   2507
   + 5926
    8433

26.   5678
   + 8765
   14,443

27.   28
     31
   + 54
    113

28.   425
    384
  + 293
   1102

29.   565
    137
    248
  + 309
   1259

30.   5654
    327
  + 4032
  10,013

31. 10,137
   4380
    296
   + 14
  14,827

32. Why is the operation of addition called a binary operation?   because *two* numbers are added at a time

33. What is the name of the law illustrated by (3 + 2) + 7 = 3 + (2 + 7)?   associative law of addition

1.4   *Round to the nearest ten.*

34. 73   70   **35.** 285   290   **36.** 4176   4180

Round to the nearest hundred.

37. 451   500   **38.** 83   100   **39.** 2550   2600

Round to the nearest thousand.

40. 7350   7000   **41.** 875   1000   **42.** 12,627   13,000

43. Estimate the sum by rounding to the nearest ten.

     1657
     2349
  + 836
   4850

44. Estimate the sum by rounding to the nearest hundred.

     1657
     2349
  + 836
   4800

Without finding the exact sum, determine if the given sum seems correct or incorrect.

45. 25
          37
          51
        + 14
         ———
         227   incorrect (should be 127)

46. 328
          19
        + 476
         ———
         823   correct

47. 3291
          407
        + 6930
         ————
         10,628   correct

48. During a two-day holiday, 9942 people visited the Grand Canyon National Park on the first day and 8032 visited it the second. First estimate the total number of visitors during the two days, and then find the exact number.   estimated total: 18,000 people; exact total: 17,974

1.5 **49.** Consider the subtraction problem $54 - 23 = 31$.   **(a)** What is the number 54 called?   **(b)** What is the number 23 called?   **(c)** What is the number 31 called?   **(d)** What is the symbol ''−'' called?
   (a) minuend   **(b)** subtrahend   **(c)** difference   **(d)** minus sign

1.6 *Find the following differences and check your work.*

50. 85
        − 32
         ——
         53

51. 63
        − 47
         ——
         16

52. 238
        − 106
         ———
         132

53. 654
        − 235
         ———
         419

54. 406
        − 139
         ———
         267

55. 352
        − 87
         ———
         265

56. 605
        − 70
         ———
         535

57. 6307
        − 2178
         ————
         4129

58. 37,003
        − 10,526
         ——————
         26,477

59. 40,000
        − 17,989
         ——————
         22,011

1.7 **60.** A nursery has a stock consisting of 42 willows, 70 crabapples, 38 black pines, and 18 weeping birches. How many trees are in the inventory?
   168

61. What is the perimeter of this figure?

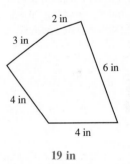

19 in

62. During one year, a magazine had 437,250 subscribers. The following year, an additional 45,593 families subscribed. How many subscribers did the magazine have that year?
483,843 subscribers

63. It costs Ms. Spinelli $735 to drive her car to work during the year. If she carpools with two others, it only costs $278 a year. How much can she save by forming a carpool? $457

64. At the beginning of the month, Sid had a balance of $693 in his checking account. During the month he made deposits of $450 and $580. He wrote checks for $137, $320, $18, $5, and $79. What was his balance at the end of the month? $1164

65. The Bonnetts bought a house for $55,280. After making several improvements which totaled $4375, they sold the house for $66,900. How much profit did they make on the sale?
$7245

Part II

Find the following sums or differences and check your work.

66. 307
        − 198
         ———
         109

67. 347
          206
        + 493
         ————
         1046

68. 2000
        − 743
         ————
         1257

69. 1478
          296
        + 3075
         ————
         4849

70. 35,001
        − 7296
         ——————
         27,705

71. Write 41,075 in expanded notation.
40,000 + 1000 + 70 + 5

72. Write 10,000,000 + 16,000 + 200 + 5 in standard notation. **10,016,205**

73. What is the name of the law illustrated by 5 + 11 = 11 + 5? **commutative law of addition**

74. We know that 8 + 0 = 8. What is 0 called in this sentence? **additive identity**

75. Round 4351 to the nearest **(a)** ten **(b)** hundred **(c)** thousand.
(a) **4350** (b) **4400** (c) **4000**

76. Estimate the following sum by rounding to the nearest hundred.

$$\begin{array}{r} 2641 \\ 375 \\ + 1529 \\ \hline 4500 \end{array}$$

77. Place the appropriate symbol (< or >) between the given numbers. 9 __>__ 0

78. Estimate the following difference by rounding to the nearest hundred.

$$\begin{array}{r} 6317 \\ - 4250 \\ \hline 2000 \end{array}$$

Solve.

79. Peter's monthly gross earnings are $2875. Suppose that $320 are taken out for federal income tax, $54 for state income tax, and $75 for an IRA.

(a) What is the total taken out of his gross pay for these three items? **$449**

(b) What is Peter's take-home pay? **$2426**

80. It takes 130 kilowatt-hours of electricity to operate a stereo for a year, 510 kilowatt-hours to operate a TV for a year, and 4225 kilowatt-hours to operate a spa for a year.

(a) How many kilowatt-hours are required to operate all three for a year?
4865 kilowatt-hours

(b) How many more kilowatt-hours are required to operate the spa than the other two together? **3585 kilowatt-hours**

81. The Shipps are retired and receive a Social Security payment of $491 each month. Next month they will receive an increase in benefits to $689 per month. First estimate the increase, and then find the actual increase.
estimated increase: $200; actual increase: $198

82. During the month of November, a basketball player scored 22, 18, 11, 27, and 13 points in five games. First estimate the total points scored in these five games, and then find the exact number of points.
estimated total: 90 points; actual total: 91 points

ANSWERS: 1. numeral 2. 1, 2, 3, 4, 5 3. 0, 1, 2, 3, 4 4. 0, 1, 2, 3, 4, 5, 6, 7, 8, 9 5. because the value of each digit is determined by its place or position in the numeral 6. 3 7. 9 8. $2 < 7$ 9. $b > a$ 10. $b < c$ 11. $a < c$ 12. 10,000,000 + 3,000,000 + 200,000 + 40,000 + 7000 + 100 + 60 + 5 13. 403,405 14. twenty-seven million, four hundred five thousand, thirty-six 15. 5 16. 0 17. 3 18. 8 19. (a) addends (b) sum (c) plus sign 20. 4 + 5 21. the given number 22. 165 23. 471 24. 625 25. 8433 26. 14,443 27. 113 28. 1102 29. 1259 30. 10,013 31. 14,827 32. because *two* numbers are added at a time 33. associative law of addition 34. 70 35. 290 36. 4180 37. 500 38. 100 39. 2600 40. 7000 41. 1000 42. 13,000 43. 4850 44. 4800 45. incorrect (should be 127) 46. correct 47. correct 48. estimated total: 18,000 people; exact total: 17,974 49. (a) minuend (b) subtrahend (c) difference (d) minus sign 50. 53 51. 16 52. 132 53. 419 54. 267 55. 265 56. 535 57. 4129 58. 26,477 59. 22,011 60. 168 61. 19 in 62. 483,843 subscribers 63. $457 64. $1164 65. $7245 66. 109 67. 1046 68. 1257 69. 4849 70. 27,705 71. 40,000 + 1000 + 70 + 5 72. 10,016,205 73. commutative law of addition 74. additive identity 75. (a) 4350 (b) 4400 (c) 4000 76. 4500 77. 9 > 0 78. 2000 79. (a) $449 (b) $2426 80. (a) 4865 kilowatt-hours (b) 3585 kilowatt-hours 81. estimated increase: $200; actual increase: $198 82. estimated total: 90 points; actual total: 91 points

CHAPTER 1 TEST

1. Give the first 4 counting numbers.

1. _____ 1, 2, 3, 4 _____

2. How do we read the symbol <?

2. _____ "is less than" _____

3. The statement $(2 + 3) + 7 = 2 + (3 + 7)$ illustrates what law of addition?

3. _____ associative law _____

4. Consider the number line.

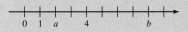

   0 1 a 4 b

4. (a) __2__ (b) __8__ (c) $a < b$

   (a) What number is paired with a?
   (b) What number is paired with b?
   (c) Place the appropriate symbol (< or >) between a and
      b. a __?__ b.

5. Write 35,408,438 in expanded notation.

5. $30,000,000 + 5,000,000 + 400,000$
   $+ 8000 + 400 + 30 + 8$

6. Write $300,000 + 80,000 + 400 + 7$ in standard notation.

6. _____ 380,407 _____

7. Write a word name for 1,659,288.

7. one million, six hundred fifty-nine
   thousand, two hundred eighty-eight

Find the following sums.

8.   51
   $+ 69$

9.   427
   $+ 88$

10.   5139
   $+ 4108$

11.   6213
     107
      40
   $+ 6857$

8. _____ 120 _____

9. _____ 515 _____

10. _____ 9247 _____

11. _____ 13,217 _____

12. Estimate   8142
           650
       $+ 99$

   (a) by rounding to the nearest ten
   (b) by rounding to the nearest hundred.

12. (a) __8890__ (b) __8900__

Find the following differences.

13.	93	14.	427	15.	502	16.	8205
	− 56		− 78		− 295		− 2136

13. _____ 37 _____

14. _____ 349 _____

15. _____ 207 _____

16. _____ 6069 _____

Solve.

17. Juan purchased three books for $18, $35, and $19. How much did he pay for the three books?

17. _____ $72 _____

18. Mr. Hagood has a pickup and a subcompact car. He figures that over the period of one year, it will cost $887 to drive the pickup to work while only $403 to drive the subcompact car. First estimate his savings by driving the car, and then find the exact savings.

18. estimated savings: $500; exact savings: $484

19. A bulk tanker contains 1985 gallons of fuel. Three deliveries are made, one of 265 gallons, a second of 340 gallons, and the third of 125 gallons. How much fuel remains in the tanker?

19. _____ 1255 gallons _____

20. Find the perimeter of a rectangle of length 16 feet and width 7 feet.

20. _____ 46 ft _____

21. Abby had a balance of $397 in her checking account. She deposited two checks worth $265 and $147. She then wrote three checks for $35, $105, and $211. What was the balance in her account after these transactions?

21. _____ $458 _____

Multiplying and Dividing Whole Numbers

BASIC MULTIPLICATION FACTS

STUDENT GUIDEPOSTS

1 Multiplication as Repeated Addition

2 Properties of Multiplication

3 The Multiplication Table

4 Multiples of a Number

---

1 MULTIPLICATION AS REPEATED ADDITION

In Chapter 1 we learned that subtraction was closely related to addition. Similarly, multiplication can be thought of as an extension of addition. For example, suppose we want to find the cost of four pieces of 5¢ candy. One way would be to find the sum

$$5 + 5 + 5 + 5 = 20.$$

Since repeated sums like this occur frequently, we should look for a shorter method for finding them. This shorter method is called **multiplication.** In our example, we are adding 5 to itself 4 times. In other words, we are finding

4 times 5.

The symbol $\times$, called the **multiplication sign** or **times sign,** means "times" in multiplication statements. The two numbers being multiplied are called **factors** (sometimes the **multiplier** and the **multiplicand**) and the result is called their **product.** Thus, in

$$4 \times 5 = 20,$$

4 and 5 are the factors (4 is the multiplier and the multiplicand is 5) whose product is 20. We can picture the product 4×5 as in Figure 2.1.

4 rows each with 5 objects results in 4×5 or 20 objects

Figure 2.1

Remember that in 4×5, 4 tells us the number of 5's that must be added together to obtain 20. That is,

$$4 \times 5 = \underbrace{5 + 5 + 5 + 5}_{\text{four 5's}} = 20.$$

Multiplication problems are written either horizontally or vertically.

Horizontal display   *Vertical display*

$$4 \times 5 = 20$$

factors   product

$$
\begin{array}{r}
5 \\
\times\ 4 \\
\hline
20
\end{array}
$$ ← factors

← product

Several other symbols or notations are used in multiplication. In place of the times symbol, $\times$, a raised dot is often used. For example,

$$4 \cdot 5 = 20 \quad \text{is the same as} \quad 4 \times 5 = 20.$$

Also, parentheses can be used around one or both of the factors with the multiplication signs $\times$ and $\cdot$ left out.

$$(4)(5) = 20 \quad \text{or} \quad 4(5) = 20 \quad \text{or} \quad (4)5 = 20$$

EXAMPLE 1 MULTIPLYING BY REPEATED ADDITION	PRACTICE EXERCISE 1

Multiply.

(a) $3 \times 2 = \underbrace{2 + 2 + 2}_{\text{three 2's}} = 6$

$$
\begin{array}{r}
2 \\
\times\ 3 \\
\hline
6
\end{array}
$$

(b) $4 \times 6 = \underbrace{6 + 6 + 6 + 6}_{\text{four 6's}} = 24$

$$
\begin{array}{r}
6 \\
\times\ 4 \\
\hline
24
\end{array}
$$

(c) $6 \cdot 4 = \underbrace{4 + 4 + 4 + 4 + 4 + 4}_{\text{six 4's}} = 24$

$$
\begin{array}{r}
4 \\
\times\ 6 \\
\hline
24
\end{array}
$$

(d) $(5)(0) = \underbrace{0 + 0 + 0 + 0 + 0}_{\text{five 0's}} = 0$

$$
\begin{array}{r}
0 \\
\times\ 5 \\
\hline
0
\end{array}
$$

Multiply.

(a) $6 \times 3 =$
$3 + 3 + 3 + 3 + 3 + 3 = $ _____

(b) $5 \times 7 =$
$7 + 7 + 7 + 7 + 7 = $ _____

(c) $7 \cdot 5 =$
$5 + 5 + 5 + 5 + 5 + 5 + 5 = $ _____

(d) $(3)(0) = $ _____

Answers: (a) 18   (b) 35   (c) 35
(d) 0

---

❷ PROPERTIES OF MULTIPLICATION

Parts **(b)** and **(c)** of Example 1 suggest that changing the order of multiplication does not change the product, since 4×6 and 6×4 are both 24. From a visual standpoint, consider a page with 4 rows, each containing 6 objects, as shown in Figure 2.2(a). When the page is turned as in Figure 2.2(b), the result has 6 rows, each containing 4 objects. Since the number of objects remains the same, $4 \times 6 = 6 \times 4$. This demonstrates an important property of multiplication.

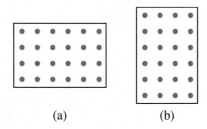

(a)   (b)

Figure 2.2

The Commutative Law of Multiplication

The **commutative law of multiplication** tells us that the order in which two (or more) numbers are multiplied does not change the product.

Also, part (**d**) of Example 1 shows that any whole number times zero is zero. This is sometimes called the **zero property of multiplication.**

EXAMPLE 2 MULTIPLYING BY ONE	**PRACTICE EXERCISE 2**

Multiply.

(**a**) $7 \times 1 = \underbrace{1 + 1 + 1 + 1 + 1 + 1 + 1}_{\text{seven 1's}} = 7$

(**b**) $1 \cdot 7 = 7$
$\downarrow$
one 7

PRACTICE EXERCISE 2

Multiply.

(**a**) $5 \times 1 =$
$1 + 1 + 1 + 1 + 1 = $ _____

(**b**) $1 \cdot 5 = $ _____

Answers: (a) 5 (b) 5

We see from Example 2 that $7 \times 1 = 1 \times 7 = 7$. This is a special case of another important property of multiplication. If 1 is multiplied by any whole number, the product is the *identical* number. As a result, 1 is often called the **multiplicative identity.**

❸ THE MULTIPLICATION TABLE

Finding products by repeated addition takes time. By memorizing certain basic facts, we can find the product of any two one-digit numbers quickly. These facts are summarized in a Multiplication Table below. To find 8×3 using the table, start at 8 in the left column and move straight across to the right until you are below the 3 in the top row. You have arrived at the number 24 which is 8×3.

×	0	1	2	3	4	5	6	7	8	9
0	0	0	0	0	0	0	0	0	0	0
1	0	1	2	3	4	5	6	7	8	9
2	0	2	4	6	8	10	12	14	16	18
3	0	3	6	9	12	15	18	21	24	27
4	0	4	8	12	16	20	24	28	32	36
5	0	5	10	15	20	25	30	35	40	45
6	0	6	12	18	24	30	36	42	48	54
7	0	7	14	21	28	35	42	49	56	63
8	0	8	16	24	32	40	48	56	64	72
9	0	9	18	27	36	45	54	63	72	81

Multiplication Table

❹ MULTIPLES OF A NUMBER

A **multiple** of a number is the product of the number and a whole number. To find the multiples of a number, simply multiply it by each of the whole numbers, beginning with zero.

EXAMPLE 3 **MULTIPLES OF A NUMBER**		**PRACTICE EXERCISE 3**

(a) Multiples of 2

$0 \times 2 = 0$
$1 \times 2 = 2$
$2 \times 2 = 4$
$3 \times 2 = 6$
$4 \times 2 = 8$
$\vdots$

(b) Multiples of 5

$0 \times 5 = 0$
$1 \times 5 = 5$
$2 \times 5 = 10$
$3 \times 5 = 15$
$4 \times 5 = 20$
$\vdots$

(a) List the multiples of 3.

(b) List the multiples of 10.

Answers: (a) 0, 3, 6, 9, 12, . . .
(b) 0, 10, 20, 30, 40, . . .

HINT

In the Multiplication Table, the numbers in any row are multiples of the number at the left of that row. Also, the numbers in any column are multiples of the number at the top of that column.

2.1 EXERCISES A

1. Consider the multiplication problem $5 \times 8 = 40$. **(a)** What is 5 called? **(b)** What is 8 called? **(c)** What is 40 called?
(a) multiplier or factor (b) multiplicand or factor (c) product

2. The fact that $2 \times 7 = 7 \times 2$ illustrates what law of multiplication? **commutative law**

3. The number 1 times any number always gives what number for the product? **that number**

4. The numbers 0, 3, 6, 9, 12, and 15 are all multiples of what number? **3**

You should be able to complete Exercises 5–54 in 90 seconds or less.

5. $\begin{array}{r} 0 \\ \times\,0 \\ \hline 0 \end{array}$	**6.** $\begin{array}{r} 7 \\ \times\,1 \\ \hline 7 \end{array}$	**7.** $\begin{array}{r} 2 \\ \times\,4 \\ \hline 8 \end{array}$	**8.** $\begin{array}{r} 6 \\ \times\,0 \\ \hline 0 \end{array}$	**9.** $\begin{array}{r} 3 \\ \times\,7 \\ \hline 21 \end{array}$	**10.** $\begin{array}{r} 0 \\ \times\,2 \\ \hline 0 \end{array}$	**11.** $\begin{array}{r} 8 \\ \times\,9 \\ \hline 72 \end{array}$	**12.** $\begin{array}{r} 5 \\ \times\,4 \\ \hline 20 \end{array}$	**13.** $\begin{array}{r} 3 \\ \times\,8 \\ \hline 24 \end{array}$	**14.** $\begin{array}{r} 3 \\ \times\,6 \\ \hline 18 \end{array}$
15. $\begin{array}{r} 2 \\ \times\,5 \\ \hline 10 \end{array}$	**16.** $\begin{array}{r} 2 \\ \times\,9 \\ \hline 18 \end{array}$	**17.** $\begin{array}{r} 3 \\ \times\,3 \\ \hline 9 \end{array}$	**18.** $\begin{array}{r} 1 \\ \times\,2 \\ \hline 2 \end{array}$	**19.** $\begin{array}{r} 0 \\ \times\,7 \\ \hline 0 \end{array}$	**20.** $\begin{array}{r} 1 \\ \times\,5 \\ \hline 5 \end{array}$	**21.** $\begin{array}{r} 6 \\ \times\,7 \\ \hline 42 \end{array}$	**22.** $\begin{array}{r} 8 \\ \times\,3 \\ \hline 24 \end{array}$	**23.** $\begin{array}{r} 8 \\ \times\,1 \\ \hline 8 \end{array}$	**24.** $\begin{array}{r} 5 \\ \times\,7 \\ \hline 35 \end{array}$
25. $\begin{array}{r} 6 \\ \times\,8 \\ \hline 48 \end{array}$	**26.** $\begin{array}{r} 7 \\ \times\,2 \\ \hline 14 \end{array}$	**27.** $\begin{array}{r} 7 \\ \times\,9 \\ \hline 63 \end{array}$	**28.** $\begin{array}{r} 3 \\ \times\,5 \\ \hline 15 \end{array}$	**29.** $\begin{array}{r} 8 \\ \times\,5 \\ \hline 40 \end{array}$	**30.** $\begin{array}{r} 9 \\ \times\,2 \\ \hline 18 \end{array}$	**31.** $\begin{array}{r} 6 \\ \times\,9 \\ \hline 54 \end{array}$	**32.** $\begin{array}{r} 4 \\ \times\,0 \\ \hline 0 \end{array}$	**33.** $\begin{array}{r} 1 \\ \times\,6 \\ \hline 6 \end{array}$	**34.** $\begin{array}{r} 1 \\ \times\,7 \\ \hline 7 \end{array}$
35. $\begin{array}{r} 4 \\ \times\,7 \\ \hline 28 \end{array}$	**36.** $\begin{array}{r} 7 \\ \times\,6 \\ \hline 42 \end{array}$	**37.** $\begin{array}{r} 0 \\ \times\,3 \\ \hline 0 \end{array}$	**38.** $\begin{array}{r} 8 \\ \times\,4 \\ \hline 32 \end{array}$	**39.** $\begin{array}{r} 8 \\ \times\,7 \\ \hline 56 \end{array}$	**40.** $\begin{array}{r} 5 \\ \times\,0 \\ \hline 0 \end{array}$	**41.** $\begin{array}{r} 1 \\ \times\,1 \\ \hline 1 \end{array}$	**42.** $\begin{array}{r} 4 \\ \times\,9 \\ \hline 36 \end{array}$	**43.** $\begin{array}{r} 0 \\ \times\,5 \\ \hline 0 \end{array}$	**44.** $\begin{array}{r} 6 \\ \times\,2 \\ \hline 12 \end{array}$

45. 9	**46.** 2	**47.** 0	**48.** 9	**49.** 9	**50.** 3	**51.** 5	**52.** 1	**53.** 4	**54.** 9
$\times 1$	$\times 0$	$\times 6$	$\times 5$	$\times 4$	$\times 2$	$\times 3$	$\times 8$	$\times 4$	$\times 8$
9	0	0	45	36	6	15	8	16	72

ANSWERS: 1. (a) multiplier or factor (b) multiplicand or factor (c) product   2. commutative law   3. that number
4. 3 5. 0 6. 7 7. 8 8. 0 9. 21 10. 0 11. 72 12. 20 13. 24 14. 18 15. 10 16. 18 17. 9 18. 2 19. 0
20. 5 21. 42 22. 24 23. 8 24. 35 25. 48 26. 14 27. 63 28. 15 29. 40 30. 18 31. 54 32. 0 33. 6
34. 7 35. 28 36. 42 37. 0 38. 32 39. 56 40. 0 41. 1 42. 36 43. 0 44. 12 45. 9 46. 0 47. 0 48. 45
49. 36 50. 6 51. 15 52. 8 53. 16 54. 72

2.1 EXERCISES B

1. Consider the multiplication problem $7 \times 9 = 63$. **(a)** What is 63 called? **(b)** What is 9 called? **(c)** What is 7 called?
  (a) product (b) multiplicand or factor (c) multiplier or factor

2. Use the commutative law of multiplication to complete the following: $9 \times 5 = \underline{5 \times 9}$.

3. The number 0 times any number always gives what number for the product?   0

4. The numbers 0, 7, 14, 21, 28, and 35 are all multiples of what number?   7

You should be able to complete Exercises 5–54 in 90 seconds or less.

5. $8 \cdot 1 = 8$	**6.** $9 \cdot 2 = 18$	**7.** $2 \cdot 0 = 0$	**8.** $8 \cdot 8 = 64$	**9.** $7 \cdot 9 = 63$
10. $6 \cdot 1 = 6$	**11.** $4 \cdot 2 = 8$	**12.** $3 \cdot 5 = 15$	**13.** $6 \cdot 7 = 42$	**14.** $9 \cdot 9 = 81$
15. $2 \cdot 6 = 12$	**16.** $7 \cdot 7 = 49$	**17.** $6 \cdot 9 = 54$	**18.** $7 \cdot 2 = 14$	**19.** $4 \cdot 1 = 4$
20. $0 \cdot 2 = 0$	**21.** $9 \cdot 8 = 72$	**22.** $7 \cdot 1 = 7$	**23.** $6 \cdot 8 = 48$	**24.** $9 \cdot 5 = 45$
25. $0 \cdot 0 = 0$	**26.** $3 \cdot 4 = 12$	**27.** $4 \cdot 0 = 0$	**28.** $1 \cdot 8 = 8$	**29.** $0 \cdot 6 = 0$
30. $5 \cdot 0 = 0$	**31.** $5 \cdot 5 = 25$	**32.** $6 \cdot 3 = 18$	**33.** $1 \cdot 2 = 2$	**34.** $0 \cdot 7 = 0$
35. $0 \cdot 8 = 0$	**36.** $8 \cdot 9 = 72$	**37.** $0 \cdot 5 = 0$	**38.** $5 \cdot 7 = 35$	**39.** $9 \cdot 3 = 27$
40. $0 \cdot 4 = 0$	**41.** $1 \cdot 7 = 7$	**42.** $7 \cdot 4 = 28$	**43.** $3 \cdot 2 = 6$	**44.** $3 \cdot 0 = 0$
45. $8 \cdot 6 = 48$	**46.** $6 \cdot 5 = 30$	**47.** $5 \cdot 8 = 40$	**48.** $0 \cdot 3 = 0$	**49.** $9 \cdot 7 = 63$
50. $5 \cdot 9 = 45$	**51.** $3 \cdot 7 = 21$	**52.** $6 \cdot 4 = 24$	**53.** $9 \cdot 0 = 0$	**54.** $4 \cdot 8 = 32$

2.1 EXERCISES C

1. If a and b represent whole numbers, use the commutative law to complete. $a \times b = \underline{b \times a}$

2. If a represents a whole number, then **(a)** $a \times 1 = \underline{a}$ **(b)** $a \times 0 = \underline{0}$.

3. If your teacher assigns the multiples of 4 in 2.1 Exercises A, which problems should you work?
  4, 8, 12, 16, 20, 24, 28, 32, 36, 40, 44, 48, 52

2.2 MULTIPLYING WHOLE NUMBERS

STUDENT GUIDEPOSTS

① Multiplying a Two-Digit Number by a One-Digit Number

② Multiplying a Many-Digit Number by a One-Digit Number

③ Multiplying a Number by a Many-Digit Number

④ Reverse-Order Check of Multiplication

⑤ Applications of Multiplying Whole Numbers

① MULTIPLYING A TWO-DIGIT NUMBER BY A ONE-DIGIT NUMBER

The Multiplication Table in Section 2.1 includes all products of two one-digit numbers. We now consider the product of a one-digit number and a two-digit number, for example, 2×43. We know that

$$2 \times 43 = \underbrace{43 + 43}_{\text{two 43's}} = 86.$$

In expanded form,

$$
\begin{aligned}
2 \times 43 = 2 \times (40 + 3) &= 2 \times (4 \text{ tens} + 3 \text{ ones}) \\
&= (4 \text{ tens} + 3 \text{ ones}) + (4 \text{ tens} + 3 \text{ ones}) \\
&= 4 \text{ tens} + 4 \text{ tens} + 3 \text{ ones} + 3 \text{ ones} \\
&= 8 \text{ tens} + 6 \text{ ones} = 86.
\end{aligned}
$$

There is a simpler way to find the product as shown in the following two steps.

$$
\begin{array}{r}
43 \\
\times\ 2 \\
\hline
86
\end{array}
$$

② Multiply 4, the tens digit in 43, by 2 and write the result in the tens column ⟶ 86 ⟵ ① Multiply 3, the ones digit in 43, by 2 and write the result in the ones column

---

EXAMPLE 1 MULTIPLYING A TWO-DIGIT BY A ONE-DIGIT NUMBER

Find the product.

$$
\begin{array}{r}
32 \\
\times\ 3 \\
\end{array}
$$

$$
\begin{array}{r}
32 \\
\times\ 3 \\
\hline
96
\end{array}
$$

② 9 is the product of 3 and the tens digit in 32 ⟶ 96 ⟵ ① 6 is the product of 3 and the ones digit in 32

PRACTICE EXERCISE 1

Find the product.

$$
\begin{array}{r}
34 \\
\times\ 2 \\
\end{array}
$$

Answer: 68

---

② MULTIPLYING A MANY-DIGIT NUMBER BY A ONE-DIGIT NUMBER

The procedure shown above can be extended to some products of one-digit numbers and many-digit numbers.

EXAMPLE 2 MULTIPLYING A THREE-DIGIT BY A ONE-DIGIT NUMBER

Find the product.
$$\begin{array}{r} 423 \\ \times\ 2 \\ \hline \end{array}$$

③ Multiply 4, the hundreds digit in 423, by 2 and write the result in the hundreds column

$$\begin{array}{r} 423 \\ \times\ 2 \\ \hline 846 \end{array}$$

① Multiply 3, the ones digit in 423, by 2 and write the result in the ones column

② Multiply 2, the tens digit in 423, by 2 and write the result in the tens column

PRACTICE EXERCISE 2

Find the product.
$$\begin{array}{r} 1233 \\ \times\ 2 \\ \hline \end{array}$$

Answer: 2466

In the examples above, each individual step resulted in a one-digit number. In some problems, such as 2×73, individual steps may give 2-digit numbers.

$$\begin{array}{r} 73 \\ \times 2 \\ \hline 146 \end{array}$$

② Multiply tens digit, 7, by 2 and write result here

① Multiply ones digit, 3, by 2 and write result here

EXAMPLE 3 A ONE-DIGIT TIMES A THREE-DIGIT NUMBER

Find the product.
$$\begin{array}{r} 723 \\ \times\ 3 \\ \hline \end{array}$$

$$\begin{array}{r} 723 \\ \times\ 3 \\ \hline 2169 \end{array}$$

③ $3 \times 7 = 21$

① $3 \times 3 = 9$

② $3 \times 2 = 6$

PRACTICE EXERCISE 3

Find the product.
$$\begin{array}{r} 831 \\ \times\ 2 \\ \hline \end{array}$$

Answer: 1662

In Example 3, only the final step had two digits. We must now look at problems in which a two-digit number occurs sooner, for example, 2×38. Using expanded notation,

$$2 \times 38 = 2(30 + 8) = 2(3 \text{ tens} + 8 \text{ ones})$$
$$= (3 \text{ tens} + 8 \text{ ones}) + (3 \text{ tens} + 8 \text{ ones})$$
$$= 6 \text{ tens} + \boxed{16 \text{ ones}}$$
$$= 6 \text{ tens} + \boxed{10 \text{ ones} + 6 \text{ ones}}$$
$$= 6 \text{ tens} + 1 \text{ ten} + 6 \text{ ones} \qquad \mathbf{10\ ones = 1\ ten}$$
$$= 7 \text{ tens} + 6 \text{ ones} = 76.$$

There is a simpler way to find this product. First, multiply 2 times the ones digit, 8, and obtain 16. Write 6 in the ones place and, to remember that 1 ten must still be recorded, write a small 1 above the tens digit in 38.

$$\begin{array}{r} {\scriptstyle 1} \\ 38 \\ \times\ 2 \\ \hline 6 \end{array}$$

Next, multiplying 2 times 3, the tens digit, gives 6 which stands for 6 tens. These 6 tens added to the 1 ten obtained earlier gives 7 tens. Put the 7 in the tens place.

$$
\begin{array}{r}
1 \\
38 \\
\times\ 2 \\
\hline
76
\end{array}
$$

EXAMPLE 4 MULTIPLYING USING REMINDER NUMBERS	PRACTICE EXERCISE 4

Find the products.

(a)
$$
\begin{array}{r}
96 \\
\times\ 7
\end{array}
$$

$$
\begin{array}{r}
4\ \ \ \\
96 \\
\times\ 7 \\
\hline
672
\end{array}
$$

① $7 \times 6 = 42$; record 2, write 4 as a reminder
② $7 \times 9 = 63$; $63 + 4 = 67$; 67 tens = 60 tens + 7 tens = 6 hundreds + 7 tens
③ Record 7 in tens column and 6 in hundreds column

(b)
$$
\begin{array}{r}
2905 \\
\times\ 5
\end{array}
$$

$$
\begin{array}{r}
4\ 2\ \ \ \\
2905 \\
\times\ 5 \\
\hline
14525
\end{array}
$$

① $5 \times 5 = 25$; record 5, write reminder 2
② $5 \times 0 = 0$; $0 + 2 = 2$; record 2
③ $5 \times 9 = 45$; record 5, write reminder 4
④ $5 \times 2 = 10$; $10 + 4 = 14$; record 14

Find the products.

(a)
$$
\begin{array}{r}
75 \\
\times\ 8
\end{array}
$$

(b)
$$
\begin{array}{r}
3056 \\
\times\ 4
\end{array}
$$

Answers: (a) 600   (b) 12,224

❸ MULTIPLYING A NUMBER BY A MANY-DIGIT NUMBER

At the beginning of this section, we found the product $2 \times 43 = 86$. Since changing the order of multiplication does not change the product, according to the commutative law, we also know that $43 \times 2 = 86$. Suppose we attempt to multiply in that order. We first multiply the ones digit.

$$
\begin{array}{r}
2 \\
\times\ 43 \\
\hline
6
\end{array}
$$

Next we multiply 4 times 2. Remember that in 43, 4 represents 4 tens. Thus, we are actually multiplying (4 tens) $\times 2 = 8$ tens $= 80$. We show this by writing 80 below 6.

$$
\begin{array}{r}
2 \\
\times\ 43 \\
\hline
6 \\
80
\end{array}
$$

We draw a line below 80 and add the 8 tens and 6 ones.

$$
\begin{array}{r}
2 \\
\times\ 43 \\
\hline
6 \\
80 \\
\hline
86
\end{array}
$$
  First product
  Second product
  Sum of first and second products

Let us summarize what we have learned.

To Multiply a Number by a Many-Digit Number

1. Find the first product in exactly the same way as when multiplying a many-digit number by a one-digit number.
2. To form the second product, place a 0 in the ones column, multiply the multiplicand by the tens digit in the multiplier. Place the result to the left of 0.
3. If the multiplier has three or more digits, a third product is formed by placing zeros in the ones and tens columns and multiplying the multiplicand by the appropriate digit. This process is continued for more digits.
4. Add all products to find the final product.

EXAMPLE 5   MULTIPLYING BY A TWO-DIGIT NUMBER

Find the products.

(a)
```
   78
 × 96
```

```
  7
  4
  78
× 96
 468    ① First product: 6 × 78 = 468
7020    ② Second product: Place 0, with 9 × 78 = 702
7488    ③ Add first and second products
```

Notice that the first reminder number, 4, written when finding 6×8, was crossed out when the second reminder number, 7, was written.

(b)
```
   438
 ×  52
```

```
   438
 ×  52
  ₁876    ① 2 × 438 = 876
 21900    ② Place 0; 5 × 438 = 2190
 22776    ③ Add products
```

Multiplying a number by a 3-digit number uses the same approach.

PRACTICE EXERCISE 5

Find the products.

(a)
```
   69
 × 87
```

(b)
```
   629
 ×  43
```

Answers: (a) 6003   (b) 27,047

EXAMPLE 6   MULTIPLYING BY A THREE-DIGIT NUMBER

Find the products.

(a)
```
   325
 × 618
```

The first two rows of this product are found the same way as in previous examples. The third row has two 0's to the right with 6×325 to the left.

PRACTICE EXERCISE 6

Find the products.

(a)
```
   437
 × 516
```

```
    13
    2̸4̸
   325
  × 618
  ─────
  2600          First product: 8 × 325
₁₁3250          Second product: 0 with 1 × 325
195000          Third product: 00 with 6 × 325
──────
200850          The desired product is the sum of the three products
```

(b)   647
         × 305

```
        12
        2̸3̸
       647
     × 305
     ─────
      3235      5 × 647
      0000      0 with 0 × 647
    194100      00 with 3 × 647
    ──────
    197335      Add products
```

Since the second row does not affect the sum, we can save time by omitting it from our work. Thus, we would show the product in the following way.

```
       647
     × 305
     ─────
      3235
    194100      Keep 00 with 3 × 647
    ──────
    197335
```

(b)   835
         × 209

Answers: (a) 225,492
(b) 174,515

HINT

With practice, and by paying close attention to the columns in the products, we may omit the extra zeros shown to the right in each row. For example, we might write the product in Example 6(**a**) in the following form.

```
       325
     × 618
     ─────
      2600
       325
      1950
     ──────
    200850
```

---

④ REVERSE-ORDER CHECK OF MULTIPLICATION

One method for checking a multiplication problem is based on the fact that changing the order of multiplication does not change the product. Suppose we found the following product.

```
       23
     ×  47
     ────
      161
       92
     ────
     1081
```

To check our work, we interchange the multiplier and the multiplicand and find the new product.

$$
\begin{array}{r}
47 \\
\times\ 23 \\
\hline
141 \\
94 \\
\hline
1081
\end{array}
$$

Since we know that 47×23 is the same as 23×47, the answers should be equal. This check is the **reverse-order check of multiplication.**

EXAMPLE 7   USING THE REVERSE-ORDER CHECK

Find the product of 329 and 54. Check your work using the reverse-order check.

Multiply:
$$
\begin{array}{r}
329 \\
\times\ 54 \\
\hline
1316 \\
1645 \\
\hline
17766
\end{array}
$$

Check:
$$
\begin{array}{r}
54 \\
\times\ 329 \\
\hline
486 \\
108 \\
162 \\
\hline
17766
\end{array}
$$

Factors are interchanged

The check on the right may look strange. We usually write the number with more digits above the one with fewer digits in a multiplication problem. However, this example shows that the multiplication method works either way.

PRACTICE EXERCISE 7

Find the product of 457 and 35. Check your work using the reverse-order check.

$$
\begin{array}{r}
457 \\
\times\ 35 \\
\hline
\end{array}
$$

Check:
$$
\begin{array}{r}
35 \\
\times\ 457 \\
\hline
\end{array}
$$

Answer: 15,995

---

5 APPLICATIONS OF MULTIPLYING WHOLE NUMBERS

Some real-life applied problems involve multiplying whole numbers.

EXAMPLE 8   AN APPLICATION IN EDUCATION

There are 32 mathematics classes with 25 students enrolled in each. What is the total number of students in these classes?

To solve this problem multiply 32 by 25.

$$
\begin{array}{r}
32 \\
\times\ 25 \\
\hline
160 \\
640 \\
\hline
800
\end{array}
$$

Thus, there are 800 students in the 32 classes.

PRACTICE EXERCISE 8

There are 52 weeks in a year. How many weeks are there in 8 years?

Answer: 416 weeks

2.2 EXERCISES A

Find the product.

1.	2.	3.	4.	5.
$\begin{array}{r} 25 \\ \times\ 2 \\ \hline 50 \end{array}$	$\begin{array}{r} 22 \\ \times\ 3 \\ \hline 66 \end{array}$	$\begin{array}{r} 33 \\ \times\ 3 \\ \hline 99 \end{array}$	$\begin{array}{r} 36 \\ \times\ 2 \\ \hline 72 \end{array}$	$\begin{array}{r} 53 \\ \times\ 3 \\ \hline 159 \end{array}$

6. 38
× 2
‾‾‾
76

7. 93
× 3
‾‾‾
279

8. 443
× 2
‾‾‾
886

9. 503
× 3
‾‾‾
1509

10. 523
× 3
‾‾‾
1569

11. 527
× 3
‾‾‾
1581

12. 496
× 7
‾‾‾
3472

13. 287
× 8
‾‾‾
2296

14. 700
× 3
‾‾‾
2100

15. 3213
× 2
‾‾‾
6426

16. 5003
× 3
‾‾‾
15,009

17. 9325
× 4
‾‾‾
37,300

18. 8495
× 7
‾‾‾
59,465

19. 43
× 12
‾‾‾
516

20. 31
× 23
‾‾‾
713

21. 33
× 22
‾‾‾
726

22. 73
× 45
‾‾‾
3285

23. 37
× 40
‾‾‾
1480

24. 80
× 80
‾‾‾
6400

25. 603
× 42
‾‾‾
25,326

26. 839
× 42
‾‾‾
35,238

27. 425
× 312
‾‾‾
132,600

28. 803
× 601
‾‾‾
482,603

29. 900
× 900
‾‾‾
810,000

30. 777
× 888
‾‾‾
689,976

Find the product and check using the reverse-order check.

31. 47
× 19
‾‾‾
893

32. 161
× 37
‾‾‾
5957

33. 545
× 139
‾‾‾
75,755

Solve.

34 Playing tennis burns up 230 calories every hour. How many calories would be burned up by a player in a match taking 4 hours? **920 calories**

35. One bottle of soda holds 12 fluid ounces. How many fluid ounces of soda are in a case of 24 bottles? **288 fluid ounces**

36. A water tank contains 50,000 gallons of water. When opened, an outlet valve will release 40 gallons of water every minute.

 (a) How many gallons of water will be released in 48 minutes? **1920 gallons**

 (b) How many gallons of water remain in the tank after 48 minutes? **48,080 gallons**

Exercises 37–42 will help you prepare for the next section. Perform operations inside parentheses first.

37. Find $5 + 3$. **8**

38. Find 2×5. **10**

39. Find 2×3. **6**

40. Find $2 \times (5 + 3)$. **16**

41. Find $(2 \times 5) + (2 \times 3)$. **16**

42. Compare the results of Exercises 40 and 41. What do you discover?
$2 \times (5 + 3) = 16$ and
$(2 \times 5) + (2 \times 3) = 16$

ANSWERS: 1. 50 2. 66 3. 99 4. 72 5. 159 6. 76 7. 279 8. 886 9. 1509 10. 1569 11. 1581 12. 3472 13. 2296 14. 2100 15. 6426 16. 15,009 17. 37,300 18. 59,465 19. 516 20. 713 21. 726 22. 3285 23. 1480 24. 6400 25. 25,326 26. 35,238 27. 132,600 28. 482,603 29. 810,000 30. 689,976 31. 893 32. 5957 33. 75,755 34. 920 calories 35. 288 fluid ounces 36. (a) 1920 gallons (b) 48,080 gallons 37. 8 38. 10 39. 6 40. 16 41. 16 42. $2 \times (5 + 3) = 16$ and $(2 \times 5) + (2 \times 3) = 16$

2.2 EXERCISES B

Find the products.

| 1. | 13
$\times\ 3$
39 | 2. | 41
$\times\ 2$
82 | 3. | 57
$\times\ 1$
57 | 4. | 18
$\times\ 3$
54 | 5. | 61
$\times\ 7$
427 |

| 6. | 72
$\times\ 3$
216 | 7. | 231
$\times\ 3$
693 | 8. | 321
$\times\ 4$
1284 | 9. | 223
$\times\ 4$
892 | 10. | 605
$\times\ 4$
2420 |

| 11. | 783
$\times\ 5$
3915 | 12. | 999
$\times\ 7$
6993 | 13. | 656
$\times\ 9$
5904 | 14. | 979
$\times\ 2$
1958 | 15. | 7013
$\times\ 2$
14,026 |

| 16. | 6257
$\times\ 2$
12,514 | 17. | 6987
$\times\ 1$
6987 | 18. | 6993
$\times\ 9$
62,937 | 19. | 52
$\times\ 11$
572 | 20. | 14
$\times\ 42$
588 |

| 21. | 64
$\times\ 15$
960 | 22. | 83
$\times\ 92$
7636 | 23. | 70
$\times\ 38$
2660 | 24. | 40
$\times\ 99$
3960 | 25. | 748
$\times\ 19$
14,212 |

| 26. | 554
$\times\ 78$
43,212 | 27. | 851
$\times\ 223$
189,773 | 28. | 387
$\times\ 430$
166,410 | 29. | 954
$\times\ 702$
669,708 | 30. | 803
$\times\ 150$
120,450 |

Find the product and check using the reverse-order check.

31.   83
   $\times\ 48$
   3984

32.   452
   $\times\ 12$
   5424

33.   893
   $\times\ 452$
   403,636

Solve.

34. There are 365 days in a year. How many days are there in 5 years?   **1825 days**

35. What is the cost of 15 stereos at $298 each?   **$4470**

36. The manufacturer of woodburning stoves has 247 stoves in stock. The company can make 12 stoves each day.

(a) How many stoves can be made in 25 days?   **300 stoves**

(b) Assuming no stoves are sold out of stock, how many stoves will be in stock 25 days from now?   **547 stoves**

Exercises 37–42 will help you prepare for the next section. Perform operations inside parentheses first.

37. Find $2 + 7$.   **9**

38. Find 4×2.   **8**

39. Find 4×7.   **28**

40. Find $4 \times (2 + 7)$.   **36**

41. Find $(4 \times 2) + (4 \times 7)$.   **36**

42. Compare the results of Exercises 40 and 41. What do you discover?
$4 \times (2 + 7) = 36$ **and**
$(4 \times 2) + (4 \times 7) = 36$

2.2 EXERCISES C

Find the products.

1.   4312
   × 648
[Answer: 2,794,176]

2.   4006
   × 6004
24,052,024

3.   7592
   × 4362
[Answer: 33,116,304]

4.   4932
   × 2715
13,390,380

5. A communications satellite travels 31,200 miles in one orbit of the earth. If the satellite makes 6 orbits per day, how many miles will it travel in 1 year? [Use 365 days for 1 year.]   **68,328,000 miles**

2.3 PROPERTIES OF MULTIPLICATION AND SPECIAL PRODUCTS

STUDENT GUIDEPOSTS

1 Associative Law of Multiplication

2 Distributive Laws

3 Multiplying by 10, 100, or 1000

4 Using Rounded Numbers to Estimate and Check Products

1 ASSOCIATIVE LAW OF MULTIPLICATION

Multiplication, like addition, is a binary operation (that is, we multiply two numbers at a time). To find the product of three or more numbers, we can multiply in steps. For example, consider the product $3 \times 2 \times 5$. We can first multiply 3×2, then multiply the result by 5.

$$(3 \times 2) \times 5 = 6 \times 5 = 30$$

Or we can multiply 2×5 first, then multiply the result by 3.

$$3 \times (2 \times 5) = 3 \times 10 = 30$$

The parentheses show which products are found first. The same result is obtained regardless of how the factors are grouped. That is,

$$(3 \times 2) \times 5 = 3 \times (2 \times 5),$$

which is an example of an important property of multiplication.

Associative Law of Multiplication

The **associative law of multiplication** tells us that the product of more than two numbers is not changed by grouping. No matter how many numbers are multiplied, they are always multiplied two at a time.

H I N T

Because of the commutative and associative laws of multiplication, when you multiply two or more numbers, the order of the numbers or the grouping of the numbers does not affect the product.

② DISTRIBUTIVE LAWS

An important relationship between multiplication and addition is shown in the following problem.

$$3 \times (4 + 5) = 3 \times 9 \qquad \text{Operate within parentheses first}$$
$$= 27 \qquad \text{Find the product of 3 and 9}$$

Next, find the following.

$$(3 \times 4) + (3 \times 5) = 12 + 15 \qquad \text{Operate within the parentheses first}$$
$$= 27 \qquad \text{Find the sum of 12 and 15}$$

As a result,

$$3 \times (4 + 5) = (3 \times 4) + (3 \times 5).$$

In a sense, the expression to the right of the equal sign is formed by "distributing" multiplication by 3 over the sum of 4 and 5. That is why this property is called the **distributive law** of multiplication over addition.

The multiplication procedure in Section 2.2 actually involves repeated use of the distributive law. Consider

$$3 \times 42 = 3 \times (40 + 2) \qquad \text{Expanded notation}$$
$$= (3 \times 40) + (3 \times 2) \qquad \text{Distributive law}$$
$$= 120 + 6 \qquad \text{Multiply}$$
$$= 126. \qquad \text{Add}$$

Notice that when the problem is written vertically, 3×2 is found in the first row and 3×40 in the second row.

$$
\begin{array}{rl}
42 & \\
\underline{\times 3} & \\
6 & (3 \times 2) \\
\underline{120} & \underline{+(3 \times 40)} \\
126 & (3 \times 40) + (3 \times 2)
\end{array}
$$

③ MULTIPLYING BY 10, 100, OR 1000

Multiplication problems involving the numbers 10, 100, and 1000 occur frequently. Recall that **multiples of 10** are numbers such as

$$10, \quad 20, \quad 30, \quad 40, \quad 50.$$

Multiples of 100 include

$$100, \quad 200, \quad 300, \quad 400, \quad 500, \quad \text{and so on,}$$

while **multiples of 1000** include

$$1000, \quad 2000, \quad 3000, \quad 4000, \quad 5000, \quad \text{and so on.}$$

Products that involve multiples of 10, 100, or 1000 can be found quickly by using the following rule.

To Multiply a Number:
1. by **10**, attach **one 0** at the end of the number,
2. by **100**, attach **two 0's** at the end of the number,
3. by **1000**, attach **three 0's** at the end of the number.

EXAMPLE 1   MULTIPLYING BY 10, 100, OR 1000

Find each product.

(a) $10 \cdot 5 = 50$ Attach 0 to 5

(b) $10 \cdot 435 = 4350$ Attach 0 to 435

(c) $100 \cdot 4 = 400$ Attach two 0's to 4

(d) $100 \cdot 497 = 49{,}700$ Attach two 0's to 497

(e) $1000 \cdot 9 = 9000$ Attach three 0's to 9

(f) $1000 \cdot 138 = 138{,}000$ Attach three 0's to 138

PRACTICE EXERCISE 1

Find each product.

(a) $8 \cdot 10$

(b) $648 \cdot 10$

(c) $3 \cdot 100$

(d) $476 \cdot 100$

(e) $7 \cdot 1000$

(f) $929 \cdot 1000$

Answers: (a) 80   (b) 6480
(c) 300   (d) 47,600   (e) 7000
(f) 929,000

When multiplying two multiples of 10, 100, or 1000, similar shortcuts are possible.

EXAMPLE 2   MULTIPLYING WITH MULTIPLES OF 10, 100, OR 1000

Find each product.

(a)   40   Multiply $7 \cdot 4$ and attach two (the total number of
     $\times\ 70$ zeros in the factors) zeros
     2800

(b)   700   Multiply $9 \cdot 7$ and attach three zeros ($2 + 1 = 3$)
     $\times\ 90$
     63000

In each case, the total number of zeros attached to the product is the sum of the zeros in the factors.

PRACTICE EXERCISE 2

Find each product.

(a)   400
      $\times\ 700$

(b)   9000
      $\times\ \ \ 400$

Answers: (a) 280,000
(b) 3,600,000

---

❹ USING ROUNDED NUMBERS TO ESTIMATE AND CHECK PRODUCTS

In Section 2.2 we used the reverse-order method to check a multiplication problem. The obvious drawback of this technique is that it takes just as long to check as to find the original product. For this reason, it is helpful to have another method. Rounded approximations for the factors can be used for **estimating products** and discovering any major errors.

EXAMPLE 3   CHECKING MULTIPLICATION BY ESTIMATING

Find the product of 476 and 31; then check your work by estimating the product.

     Multiply: 476   Check: 500   476 rounds to 500
          $\times\ 31$ $\times\ 30$ 31 rounds to 30
          476   15000
         1428
         14756

   Since 15,000 is "close" to 14,756, we would assume that no major error has been made.

PRACTICE EXERCISE 3

Find the product of 295 and 42; then check your work by estimating the product.

Answer: Rounding 295 to 300 and 42 to 40, the estimated product is $40 \cdot 300 = 12{,}000$ (the actual product is 12,390).

|HINT|

When rounding numbers to find estimated products, round all four-digit numbers to the nearest thousand, all three-digit numbers to the nearest hundred, and all two-digit numbers to the nearest ten.

Rounded numbers can also be used to estimate solutions to applied problems.

| EXAMPLE 4 ESTIMATING A PRODUCT IN AN APPLIED PROBLEM | PRACTICE EXERCISE 4 |

There are 4985 students in a school, and each spent $41 for athletic fees. Without finding the exact amount, estimate the total amount of money spent on athletic fees by these students.

Since 4985 is about 5000 and $41 is about $40, we can estimate the total by multiplying 5000 by 40.

$$\left.\begin{array}{r} 5000 \\ \underline{40} \\ 200000 \end{array}\right\} \quad \text{Mental work}$$

Thus, about $200,000 was spent on these fees.

If one piece of pie has 205 calories, without finding the exact number, estimate the number of calories in 89 pieces.

Answer: 18,000 calories (using 200 × 90)

2.3 EXERCISES A

1. Use the associative law of multiplication to complete the following: (3 × 5) × 7 = _____.   3 × (5 × 7)

2. Use the distributive law to complete the following: 4 × (5 + 7) = _____.   (4 × 5) + (4 × 7)

3. (a) Evaluate 7 × (2 + 9). (b) Evaluate (7 × 2) + (7 × 9). (c) Why are your answers to (a) and (b) the same?
   (a) 77 (b) 77 (c) because of the distributive law

4. (a) Evaluate 9 × (3 × 5). (b) Evaluate (9 × 3) × 5. (c) Why are your answers to (a) and (b) the same?
   (a) 135 (b) 135 (c) because of the associative law

5. (a) Evaluate 8 × 11. (b) Evaluate 11 × 8. (c) Why are your answers to (a) and (b) the same?
   (a) 88 (b) 88 (c) because of the commutative law

Find each product.

6. 10 · 9	7. 42 · 10	8. 387 · 10	9. 7 · 100	10. 100 · 77
90	420	3870	700	7700

11. 100 · 427	12. 1000 · 5	13. 1000 · 42	14. 78 · 1000	15. 1000 · 529
42,700	5000	42,000	78,000	529,000

16. 70	17. 800	18. 700	19. 6000	20. 7000
× 60	× 400	× 20	× 30	× 500
4200	320,000	14,000	180,000	3,500,000

Check by estimating the product.

21.
$$\begin{array}{r} 78 \\ \times\,21 \\ \hline 78 \\ 156 \\ \hline 1638 \end{array}$$
$20 \times 80 = 1600$;
appears to be correct

22.
$$\begin{array}{r} 409 \\ \times\,38 \\ \hline 3272 \\ 1227 \\ \hline 15542 \end{array}$$
$40 \times 400 = 16{,}000$;
appears to be correct

23.
$$\begin{array}{r} 813 \\ \times\,491 \\ \hline 813 \\ 7317 \\ 3252 \\ \hline 499183 \end{array}$$
$500 \times 800 = 400{,}000$;
appears to be incorrect (correct product is 399,183)

Solve.

24. There are 29 volumes in a set of reference books, and each volume has 995 pages. Without finding the exact number, estimate the total number of pages in the set. $30 \times 1000 = 30{,}000$ pages

25 A store had 103 customers on Saturday, and the average amount spent by each was $289. Without finding the exact amount, estimate the total receipts for Saturday. $100 \times 300 = \$30{,}000$

26. Jeff's car can get 28 miles to a gallon of gasoline. Without finding the exact number of miles, estimate the distance Jeff can drive on 42 gallons of gasoline.
$30 \times 40 = 1200$ miles

27. A block layer can lay 46 cinder blocks an hour. Estimate the number of blocks that he can lay in a 52-hour workweek. [Do not find the exact number.]
$50 \times 50 = 2500$ blocks

FOR REVIEW

Find the product and check your work using the reverse-order check.

28.
$$\begin{array}{r} 161 \\ \times\,37 \\ \hline 5957 \end{array}$$

29.
$$\begin{array}{r} 545 \\ \times\,139 \\ \hline 75{,}755 \end{array}$$

ANSWERS: 1. $3 \times (5 \times 7)$ 2. $(4 \times 5) + (4 \times 7)$ 3. (a) 77 (b) 77 (c) because of the distributive law   4. (a) 135 (b) 135 (c) because of the associative law   5. (a) 88 (b) 88 (c) because of the commutative law   6. 90   7. 420   8. 3870   9. 700   10. 7700   11. 42,700   12. 5000   13. 42,000   14. 78,000   15. 529,000   16. 4200   17. 320,000   18. 14,000   19. 180,000   20. 3,500,000   21. $20 \times 80 = 1600$; appears to be correct   22. $40 \times 400 = 16{,}000$; appears to be correct   23. $500 \times 800 = 400{,}000$; appears to be incorrect (correct product is 399,183)   24. 30,000 pages   25. $30,000   26. $30 \times 40 = 1200$ miles   27. $50 \times 50 = 2500$ blocks   28. 5957   29. 75,755

2.3 EXERCISES B

1. Use the associative law of multiplication to complete the following: $2 \times (4 \times 9) = $ _____. $(2 \times 4) \times 9$

2. Use the distributive law to complete the following: $(2 \times 5) + (2 \times 8) = $ _____. $2 \times (5 + 8)$

3. (a) Evaluate $10 \times (2 + 3)$. **(b)** Evaluate $(10 \times 2) + (10 \times 3)$. **(c)** Why are your answers to **(a)** and **(b)** the same?
(a) 50 (b) 50 (c) because of the distributive law

4. (a) Evaluate $3 \times (4 \times 10)$. (b) Evaluate $(3 \times 4) \times 10$. (c) Why are your answers to (a) and (b) the same?

(a) **120** (b) **120** (c) **because of the associative law**

5. (a) Evaluate 12×4. (b) Evaluate 4×12. (c) Why are your answers to (a) and (b) the same?

(a) **48** (b) **48** (c) **because of the commutative law**

Find each product.

6. $10 \cdot 5$
50

7. $71 \cdot 10$
710

8. $593 \cdot 10$
5930

9. $100 \cdot 8$
800

10. $43 \cdot 100$
4300

11. $100 \cdot 645$
64,500

12. $1000 \cdot 3$
3000

13. $1000 \cdot 57$
57,000

14. $84 \cdot 1000$
84,000

15. $259 \cdot 1000$
259,000

16. 70
  $\times\ 40$
  2800

17. 400
  $\times\ 700$
  280,000

18. 500
  $\times\ 30$
  15,000

19. 9000
  $\times\ 40$
  360,000

20. 8000
  $\times\ 400$
  3,200,000

Check by estimating the product.

21. 61
  $\times\ 88$
  488
  488
  4368

$90 \times 60 = 5400;$ appears to be incorrect (correct product is 5368)

22. 678
  $\times\ 51$
  678
  3390
  34578

$50 \times 700 = 35,000;$ appears to be correct

23. 290
  $\times\ 603$
  870
  1740
  18270

$600 \times 300 = 180,000;$ appears to be incorrect (correct product is 174,870)

Solve.

24. A man burns up 395 calories each day during his exercise period. Without finding the exact number, about how many calories will he burn up in 105 days? $400 \times 100 = 40,000$ **calories**

25. A store has 7958 bottles of pills in its inventory. If the average number of pills per bottle is 98, approximately how many pills are in the store? [Do not find the exact number.] $8000 \times 100 = 800,000$ **pills**

26. Diane can type 85 words a minute. Estimate the number of words that she can type in 7 hours. [Do not find the actual number of words. Use the fact that 7 hours is 420 minutes.]

$90 \times 400 = 36,000$ **words**

27. A merry-go-round in a carnival makes 11 revolutions a minute. Without finding the exact number, estimate the number of revolutions made in 103 minutes. $10 \times 100 = 1000$ **revolutions**

FOR REVIEW

Find the product and check your work using the reverse-order check.

28. 452
  $\times\ 12$
  5424

29. 893
  $\times\ 452$
  403,636

2.3 EXERCISES C

Let a, b, and c represent any three whole numbers. Complete each law.

1. Associative law: $a \times (b \times c) = $ _____ $(a \times b) \times c$ _____.

2. Commutative law: $c \times b = $ _____ $b \times c$ _____.

3. Distributive law: $a \times (b + c) = $ _____ $(a \times b) + (a \times c)$ _____.

2.4   BASIC DIVISION FACTS

══════════ STUDENT GUIDEPOSTS ══════════

1 Division as Repeated Subtraction   **3** Division Involving the Number 1

2 Related Multiplication Sentence   **4** Division Involving the Number 0

1 DIVISION AS REPEATED SUBTRACTION

The final arithmetic operation we study is division. Recall that multiplication can be thought of as repeated addition. Similarly, division can be thought of as repeated subtraction.

The word *division* means "separation into parts." Consider dividing 15 by 3. We use ÷ as the **division symbol** and write

$$15 \div 3,$$

(read "15 divided by 3"). In the example, 15, the number being divided, is called the **dividend,** and 3, the dividing number, is the **divisor.** Suppose we have 15 objects and separate or divide them into 3 equal parts. As shown in Figure 2.3, each equal part has 5 objects in it.

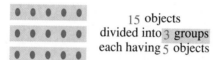

15 objects
divided into 3 groups
each having 5 objects

Figure 2.3

We write $15 \div 3 = 5$, and call 5 the **quotient** of 15 and 3.

We could also look at the problem as repeated subtraction. That is, $15 \div 3$ could be thought of as the number of times 3 can be subtracted from 15 until zero remains.

$$
\begin{array}{r}
15 \\
-\ 3 \\
\hline
12
\end{array}
$$ First subtraction of 3

$$
\begin{array}{r}
-\ 3 \\
\hline
9
\end{array}
$$ Second subtraction of 3

$$
\begin{array}{r}
-\ 3 \\
\hline
6
\end{array}
$$ Third subtraction of 3

$$
\begin{array}{r}
-\ 3 \\
\hline
3
\end{array}
$$ Fourth subtraction of 3

$$
\begin{array}{r}
-\ 3 \\
\hline
0
\end{array}
$$ Fifth subtraction of 3

We have subtracted 3 from 15 a total of 5 times. That is, $15 \div 3 = 5$. The three ways to write a division problem are illustrated by

$$15 \div 3 = 5, \quad \frac{15}{3} = 5, \quad \text{and } 3\overline{)15}.$$

═══════════ **HINT** ═══════════

In general, division problems are written in one of the following three ways.

$$\text{dividend} \div \text{divisor} = \text{quotient} \qquad \frac{\text{dividend}}{\text{divisor}} = \text{quotient} \qquad \text{divisor}\overline{)\text{dividend}}^{\text{quotient}}$$

| **EXAMPLE 1   DIVIDING BY REPEATED SUBTRACTION** | **PRACTICE EXERCISE 1** |

Find the quotient $20 \div 5$ by repeated subtraction.

Since 5 can be subtracted from 20 four times,

$$20 \div 5 = 4.$$

$$
\begin{array}{r}
20 \\
-\ 5 \\
\hline
15
\end{array}
$$
First subtraction

$$
\begin{array}{r}
-\ 5 \\
\hline
10
\end{array}
$$
Second subtraction

$$
\begin{array}{r}
-\ 5 \\
\hline
5
\end{array}
$$
Third subtraction

$$
\begin{array}{r}
-\ 5 \\
\hline
0
\end{array}
$$
Fourth subtraction

PRACTICE EXERCISE 1

Find the quotient by repeated subtraction.

$$\frac{30}{6}$$

Answer: 5; 6 can be subtracted 5 times

2 RELATED MULTIPLICATION SENTENCE

Notice that

$$20 \div 5 = 4 \qquad \text{Division sentence}$$

and

$$5 \times 4 = 20. \qquad \text{Multiplication sentence}$$

For a given division sentence, there is a **related multiplication sentence.**

The Basic Principle of Division

For every division sentence

$$(\text{dividend}) \div (\text{divisor}) = (\text{quotient}),$$

there is a related multiplication sentence

$$(\text{divisor}) \times (\text{quotient}) = (\text{dividend}).$$

HINT

To obtain a related multiplication sentence for a division problem, move the divisor to the other side of the equal sign and multiply.

| **EXAMPLE 2   FINDING THE RELATED MULTIPLICATION SENTENCE** | **PRACTICE EXERCISE 2** |

Give a related multiplication sentence for each division sentence.

(a) $14 \div 7 = 2$

Since 14 is the dividend, 7 the divisor, and 2 the quotient, a related multiplication sentence is $7 \times 2 = 14$.

Notice that by the commutative law, $2 \times 7 = 14$ is also a related multiplication sentence.

(b) $\dfrac{36}{4} = 9$

A related multiplication sentence is $4 \times 9 = 36$.

PRACTICE EXERCISE 2

Give a related multiplication sentence for each division sentence.

(a) $25 \div 5 = 5$

(b) $6\overline{)42}$ with quotient 7

Answers: (a) $5 \times 5 = 25$
(b) $6 \times 7 = 42$

Using the division principle we can learn the basic division facts. When we know them well enough, it will not be necessary to think of the related multiplication sentence.

| EXAMPLE 3 USING THE BASIC DIVISION (MULTIPLICATION) FACTS | PRACTICE EXERCISE 3 |

Divide.

(a) $18 \div 3 = 6$ This is true since $6 \times 3 = 18$

(b) $35 \div 7 = 5$ This is true since $5 \times 7 = 35$

(c) $8)\overline{64}^{\,8}$ This is true since $8 \times 8 = 64$

(d) $\dfrac{54}{9} = 6$ This is true since $6 \times 9 = 54$

Divide.

(a) $36 \div 9$

(b) $49 \div 7$

(c) $8)\overline{72}$

(d) $\dfrac{48}{6}$

Answers: (a) 4 (b) 7 (c) 9
(d) 8

❸ DIVISION INVOLVING THE NUMBER 1

The basic principle of division is helpful for special division problems. In particular,

$$8 \div 8 = 1 \quad \text{since} \quad 8 \cdot 1 = 8$$

$$\frac{12}{12} = 1 \quad \text{since} \quad 12 \cdot 1 = 12$$

$$35)\overline{35}^{\,1} \quad \text{since} \quad 35 \cdot 1 = 35.$$

In general,

(any nonzero whole number) ÷ (itself) = 1.

Also,
$$5 \div 1 = 5 \quad \text{since} \quad 1 \cdot 5 = 5$$

$$\frac{9}{1} = 9 \quad \text{since} \quad 1 \cdot 9 = 9$$

$$1)\overline{13}^{\,13} \quad \text{since} \quad 1 \cdot 13 = 13.$$

These are examples of the following general rule.

(any whole number) ÷ 1 = (that whole number)

❹ DIVISION INVOLVING THE NUMBER 0

When we divide 6 by 2, we see that

$$2)\overline{6}^{\,3} \quad \text{means} \quad 2 \times 3 = 6.$$

Suppose we try to divide 6 by 0 in the same manner. Let's use **?** for the quotient.

$$0)\overline{6}^{\,?} \quad \text{means} \quad 0 \times ? = 6.$$

Then **?** must represent what number? Suppose **?** is 6, then

$$? \times 0 = 6 \times 0 = 0, \quad \text{not } 6.$$

What if **?** is 0? Then,

$$\boxed{?} \times 0 = \boxed{0} \times 0 = 0, \quad \text{not } 6.$$

What if **?** is 2? Then,

$$\boxed{?} \times 0 = \boxed{2} \times 0 = 0, \quad \text{not } 6.$$

In fact, what if **?** is any number? Then,

$$\boxed{?} \times 0 = \boxed{\text{(any number)}} \times 0 = 0, \quad \text{not } 6.$$

Since there is no number we can use for **?** to make **?** $\times$ 0 equal to 6, we conclude that 0 cannot be used as a divisor of 6, or for that matter, as a divisor of any number.

Division by Zero

Division by 0 is undefined. That is, we agree not to divide any number by 0.

Although division *by* zero is impossible, zero can be *divided by* any whole number (except 0). For example,

$$0 \div 3 = 0 \qquad \text{since} \qquad 3 \cdot 0 = 0$$

$$\frac{0}{12} = 0 \qquad \text{since} \qquad 12 \cdot 0 = 0$$

$$9\overline{)0}^{\,0} \qquad \text{since} \qquad 9 \cdot 0 = 0.$$

In general,

$$0 \div (\text{any nonzero whole number}) = 0.$$

The order of division, like that of subtraction, cannot be changed. A single example shows this. Since

$$8 \div 4 = 2 \qquad \text{but} \qquad 4 \div 8 \text{ is not a whole number,}$$

we see that $8 \div 4 \neq 4 \div 8$. Similarly,

$$(8 \div 4) \div 2 = 2 \div 2 = 1 \qquad \text{but} \qquad 8 \div (4 \div 2) = 8 \div 2 = 4.$$

Thus, $(8 \div 4) \div 2 \neq 8 \div (4 \div 2)$.

HINT

Division is neither commutative nor associative. That is, changing the order of or grouping symbols in a division problem should never be done.

2.4 EXERCISES A

1. Consider the division problem $6 \div 2 = 3$. **(a)** What is 6 called? **(b)** What is 2 called? **(c)** What is 3 called?
(a) dividend (b) divisor (c) quotient

2. In any division problem, the product of the divisor and the quotient is equal to what number?   dividend

3. If $\square$ is any whole number, $\square \div 1 = \underline{\boxed{\square}}$.

4. If $\square$ is any whole number except 0, $0 \div \square = \underline{\quad 0 \quad}$.

You should be able to complete Exercises 5–49 in 90 seconds or less. Find each quotient.

5. $18 \div 2 = 9$ **6.** $28 \div 4 = 7$ **7.** $30 \div 6 = 5$ **8.** $16 \div 4 = 4$ **9.** $8 \div 4 = 2$

10. $5 \overline{)25}$ 5 **11.** $2 \overline{)2}$ 1 **12.** $6 \overline{)18}$ 3 **13.** $2 \overline{)6}$ 3 **14.** $8 \overline{)72}$ 9

15. $9 \div 1 = 9$ **16.** $27 \div 9 = 3$ **17.** $16 \div 8 = 2$ **18.** $7 \div 7 = 1$ **19.** $0 \div 8 = 0$

20. $\dfrac{12}{6} = 2$ **21.** $\dfrac{1}{1} = 1$ **22.** $\dfrac{36}{9} = 4$ **23.** $\dfrac{42}{6} = 7$ **24.** $\dfrac{24}{4} = 6$

25. $5 \div 5 = 1$ **26.** $18 \div 9 = 2$ **27.** $16 \div 2 = 8$ **28.** $45 \div 5 = 9$ **29.** $48 \div 6 = 8$

30. $3 \overline{)15}$ 5 **31.** $2 \overline{)12}$ 6 **32.** $5 \overline{)40}$ 8 **33.** $5 \overline{)0}$ 0 **34.** $4 \overline{)36}$ 9

35. $6 \div 1 = 6$ **36.** $24 \div 3 = 8$ **37.** $64 \div 8 = 8$ **38.** $63 \div 7 = 9$ **39.** $20 \div 5 = 4$

40. $\dfrac{0}{1} = 0$ **41.** $\dfrac{48}{8} = 6$ **42.** $\dfrac{54}{6} = 9$ **43.** $\dfrac{0}{2} = 0$ **44.** $\dfrac{8}{8} = 1$

45. $24 \div 8 = 3$ **46.** $6 \div 3 = 2$ **47.** $36 \div 6 = 6$ **48.** $21 \div 3 = 7$ **49.** $35 \div 5 = 7$

FOR REVIEW

Find the products.

50. $\begin{array}{r} 500 \\ \times\ 30 \\ \hline 15{,}000 \end{array}$

51. $\begin{array}{r} 600 \\ \times\ 900 \\ \hline 540{,}000 \end{array}$

52. $\begin{array}{r} 4000 \\ \times\ 20 \\ \hline 80{,}000 \end{array}$

Check by estimating the product.

53. $\begin{array}{r} 92 \\ \times\ 19 \\ \hline 828 \\ 92 \\ \hline 1748 \end{array}$ $90 \times 20 = 1800;$ appears to be correct

54. $\begin{array}{r} 743 \\ \times\ 102 \\ \hline 1486 \\ 743 \\ \hline 8916 \end{array}$ $700 \times 100 = 70{,}000;$ appears to be incorrect (correct product is 75,786)

ANSWERS: 1. (a) dividend (b) divisor (c) quotient   2. dividend   3. $\square$ 4. 0   5. 9   6. 7   7. 5   8. 4   9. 2   10. 5   11. 1   12. 3   13. 3   14. 9   15. 9   16. 3   17. 2   18. 1   19. 0   20. 2   21. 1   22. 4   23. 7   24. 6   25. 1   26. 2   27. 8   28. 9   29. 8   30. 5   31. 6   32. 8   33. 0   34. 9   35. 6   36. 8   37. 8   38. 9   39. 4   40. 0   41. 6   42. 9   43. 0   44. 1   45. 3   46. 2   47. 6   48. 7   49. 7   50. 15,000   51. 540,000   52. 80,000   53. $90 \times 20 = 1800$; appears to be correct   54. $700 \times 100 = 70,000$; appears to be incorrect (correct product is 75,786)

2.4 EXERCISES B

1. Consider the division problem $54 \div 9 = 6$. **(a)** What is 6 called? **(b)** What is 9 called? **(c)** What is 54 called?
   (a) quotient (b) divisor (c) dividend

2. In any division problem, the dividend is equal to the product of the quotient and what number?   divisor

3. Any nonzero whole number divided by itself is equal to what number?   1

4. What whole number can never be used as a divisor?   0

You should be able to complete Exercises 5–49 in 90 seconds or less. Find each quotient.

5. $\dfrac{20}{4} = 5$ **6.** $\dfrac{48}{6} = 8$ **7.** $\dfrac{12}{4} = 3$ **8.** $\dfrac{24}{4} = 6$ **9.** $\dfrac{28}{7} = 4$

10. $16 \div 4 = 4$ **11.** $56 \div 7 = 8$ **12.** $6 \div 2 = 3$ **13.** $4 \div 2 = 2$ **14.** $7 \div 7 = 1$

15. $\dfrac{0}{5} = 0$ **16.** $\dfrac{12}{3} = 4$ **17.** $\dfrac{63}{7} = 9$ **18.** $\dfrac{4}{4} = 1$ **19.** $\dfrac{0}{2} = 0$

20. $5\overline{)15}$ 3   **21.** $2\overline{)16}$ 8   **22.** $9\overline{)63}$ 7   **23.** $9\overline{)36}$ 4   **24.** $7\overline{)49}$ 7

25. $\dfrac{27}{3} = 9$ **26.** $\dfrac{35}{5} = 7$ **27.** $\dfrac{30}{5} = 6$ **28.** $\dfrac{9}{3} = 3$ **29.** $\dfrac{21}{3} = 7$

30. $6 \div 3 = 2$ **31.** $18 \div 3 = 6$ **32.** $42 \div 7 = 6$ **33.** $24 \div 8 = 3$ **34.** $0 \div 9 = 0$

35. $\dfrac{32}{4} = 8$ **36.** $\dfrac{1}{1} = 1$ **37.** $\dfrac{0}{7} = 0$ **38.** $\dfrac{18}{9} = 2$ **39.** $\dfrac{35}{7} = 5$

40. $1\overline{)0}$ 0   **41.** $9\overline{)72}$ 8   **42.** $1\overline{)6}$ 6   **43.** $3\overline{)15}$ 5   **44.** $1\overline{)3}$ 3

45. $\dfrac{9}{1} = 9$ **46.** $\dfrac{56}{8} = 7$ **47.** $\dfrac{25}{5} = 5$ **48.** $\dfrac{9}{9} = 1$ **49.** $\dfrac{18}{2} = 9$

FOR REVIEW

Find the products.

50. $\begin{array}{r} 700 \\ \times\ 70 \\ \hline 49{,}000 \end{array}$

51. $\begin{array}{r} 800 \\ \times\ 800 \\ \hline 640{,}000 \end{array}$

52. $\begin{array}{r} 1000 \\ \times\ 500 \\ \hline 500{,}000 \end{array}$

Check by estimating the product.

53. $\begin{array}{r} 28 \\ \times\ 69 \\ \hline 252 \\ 128 \\ \hline 1532 \end{array}$ $30 \times 70 = 2100$; appears to be incorrect (correct product is 1932)

54. $\begin{array}{r} 802 \\ \times\ 489 \\ \hline 7218 \\ 6416 \\ 3208 \\ \hline 392{,}178 \end{array}$ $800 \times 500 = 400{,}000$; appears to be correct

2.4 EXERCISES C

Let n represent a whole number. Find n by writing the related multiplication or division sentence.

1. $n \div 5 = 6$
[Answer: 30]

2. $n \times 8 = 24$ 3

3. $n \div 7 = 5$
[Answer: 35]

4. $n \times 6 = 42$ 7

Assume that a, b, and c represent whole numbers in Exercises 5–10. Complete each statement.

5. $a \div a = \underline{1}$

6. $a \div 1 = \underline{a}$

7. $0 \div a = \underline{0}$
(Assume a is not 0.)

8. Are $a \div b$ and $b \div a$ equal?   no

9. Are $(a \div b) \div c$ and $a \div (b \div c)$ equal?   no

10. What does $a \div 0$ equal?   undefined

2.5 DIVIDING WHOLE NUMBERS

=== STUDENT GUIDEPOSTS ===

1 Dividing by a One-Digit Number
2 Remainder in a Division Problem
3 Dividing by a Two-Digit Number
4 Dividing by a Three-Digit Number
5 Checking Division

1 DIVIDING BY A ONE-DIGIT NUMBER

The basic division facts allow us to divide special two-digit numbers by one-digit numbers. We now consider more general problems like the one below. If we try to find the quotient

$$86 \div 2$$

by repeated subtraction, we will have to subtract 2 a great number of times. We can shorten the process by subtracting many 2's at once, as follows.

```
      86
   −  20     Ten 2's subtracted
      66
   −  20     Ten more 2's subtracted
      46
   −  20     Ten more 2's subtracted
      26
   −  20     Ten more 2's subtracted
       6
   −   6     Subtract three 2's
       0
```

The total number of 2's subtracted is

$$10 + 10 + 10 + 10 + 3 = 43.$$

As a result, we see that

$$86 \div 2 = 43 \quad \text{or} \quad 2\overline{)86}^{\,43}.$$

In the previous display, we see that if the tens digit, 8, is divided by 2, we obtain the tens digit, 4, of the quotient. Similarly, dividing the ones digit, 6, by 2, we obtain the ones digit, 3, of the quotient.

$$8 \div 2 = 4 \longrightarrow \underset{2\overline{)86}}{43} \longleftarrow 3 = 6 \div 2$$

When dividing the tens digit and the ones digit, we use basic division facts.

EXAMPLE 1 DIVIDING BY A ONE-DIGIT NUMBER	**PRACTICE EXERCISE 1**

Find the quotient. $3\overline{)96}$

$$\begin{array}{c} \rule{1em}{0pt} 3 = 9 \div 3 \quad \text{(tens digit)} \\ \downarrow \\ \underset{3\overline{)96}}{32} \longleftarrow 2 = 6 \div 3 \quad \text{(ones digit)} \end{array}$$

Divide tens digit first, then divide ones digit.

Find the quotient.

$$3\overline{)690}$$

Answer: 230

In Example 1, the quotient in each step was a basic division fact. We next consider problems in which this is not the case. For example, suppose we find the quotient

$$3\overline{)48}.$$

First we use repeated subtraction.

$$\begin{array}{rl} 48 & \\ -\ 30 & \text{Subtract ten 3's} \\ \hline 18 & \\ -\ 18 & \text{Subtract six 3's} \\ \hline 0 & \end{array}$$

Adding 10 and 6 gives the quotient, 16. Had we written the problem as before, we would have

$$\underset{3\overline{)48}}{16}.$$

Since $4 \div 3$ is not a basic division fact, we must find this quotient in several steps.

① Find the largest number less than 4 which is divisible by 3. This number is 3, and $3 \div 3 = 1$. Place 1 above the 4.

$$\overset{1}{3\,\overline{)\,48}}$$

② Multiply 1 times 3 and place the product below 4.
③ Draw a line below 3 and subtract.

$$\begin{array}{r} 1 \\ 3\overline{)48} \\ \underline{3} \\ 1 \end{array}$$

④ Bring down the eight.
⑤ Divide 18 by 3 and place the quotient, 6, to the right of 1.

$$\begin{array}{r} 16 \\ 3\overline{)48} \\ \underline{3} \\ 18 \end{array}$$

⑥ Multiply 6 times 3 and place the result below 18.
⑦ Draw a line below 18 and subtract.

$$\begin{array}{r} 16 \\ 3\overline{)48} \\ \underline{3} \\ 18 \\ \underline{18} \\ 0 \end{array}$$

In the second step, the 1 in the quotient stands for 1 ten, so the product 1×3 represents 3 tens or 30. Compare this to the result of repeated subtraction.

② REMAINDER IN A DIVISION PROBLEM

When we divided 48 by 3 above, the final difference, 0, is called the **remainder.** In general, the remainder can be any number less than the divisor.

EXAMPLE 2 DIVISION WITH A NONZERO REMAINDER	PRACTICE EXERCISE 2

Find the quotient. Follow the numbered steps. $4\overline{)537}$

⑤ $12 \div 4 = 3$

① $4 \div 4 = 1 \longrightarrow$ 134 $\longleftarrow$ ⑨ $16 \div 4 = 4$

$4\overline{)537}$

② $1 \times 4 = 4 \longrightarrow$ 4

③ $5 - 4 = 1 \longrightarrow$ 13 $\longleftarrow$ ④ Bring down 3

⑥ $3 \times 4 = 12 \longrightarrow$ 12

⑦ $13 - 12 = 1 \longrightarrow$ 17 $\longleftarrow$ ⑧ Bring down 7

⑩ $4 \times 4 = 16 \longrightarrow$ 16

1 $\longleftarrow$ ⑪ $17 - 16 = 1$ (remainder)

Find the quotient.

$3\overline{)473}$

Answer: 157 R2

HINT

If you obtain a remainder that is greater than or equal to the divisor, you have made a mistake. In the final division step, the divisor went into the dividend more times. Remember that your remainder must always be less than the divisor.

In many cases the first digit of the dividend is less than the divisor.

EXAMPLE 3 DIVIDING INTO THE FIRST TWO DIGITS	PRACTICE EXERCISE 3

Find the quotient. $3\overline{)263}$

Since there is no number less than or equal to 2 which is divisible by 3, use the first *two* digits of the dividend in the first step. Find the largest number less than or equal to 26 which is divisible by 3. That number, 24, when divided by 3 gives a quotient of 8. Place 8 above 6 and continue.

① $24 \div 3 = 8 \longrightarrow$ 87 $\longleftarrow$ ⑤ $21 \div 3 = 7$

$3\overline{)26\,3}$

② $8 \times 3 = 24 \longrightarrow$ 24

③ $26 - 24 = 2 \longrightarrow$ 23 $\longleftarrow$ ④ Bring down 3

⑥ $7 \times 3 = 21 \longrightarrow$ 21

2 $\longleftarrow$ ⑦ $23 - 21 = 2$ (remainder)

Find the quotient.

$6\overline{)656}$

Answer: 109 R2

It is easy to extend the process to dividends with more than three digits.

| EXAMPLE 4   DIVIDING A FOUR-DIGIT NUMBER | PRACTICE EXERCISE 4 |

Find the quotient. $5\overline{)3825}$

⑤ $30 \div 5 = 6$

① $35 \div 5 = 7$ ──────┐ ┌────── ⑨ $25 \div 5 = 5$

$$765$$
$$5\overline{)3825}$$

② $7 \times 5 = 35$ ⟶ $\underline{35}$
③ $38 - 35 = 3$ ⟶ $3\mathbf{2}$ ⟵ ④ Bring down 2
⑥ $6 \times 5 = 30$ ⟶ $\underline{30}$
⑦ $32 - 30 = 2$ ⟶ $2\mathbf{5}$ ⟵ ⑧ Bring down 5
⑩ $5 \times 5 = 25$ ⟶ $\underline{25}$
                          0 ⟵ ⑪ $25 - 25 = 0$

Find the quotient.

$2\overline{)4056}$

Answer: 2028

With practice you will learn a shortcut illustrated in the next example.

| EXAMPLE 5   A SHORTCUT OCCURRING IN SOME PROBLEMS | PRACTICE EXERCISE 5 |

Find the quotient. $7\overline{)3528}$

$$5\overset{0}{}4$$
$$7\overline{)3528}$$
$$\underline{35}$$
$$0\,28 \qquad \text{Since 7 cannot divide 2, bring down 8}$$
$$\qquad\qquad \text{and write 0 above 2}$$
$$\underline{28}$$
$$0$$

Find the quotient.

$8\overline{)4824}$

Answer: 603

▰▰▰▰▰▰▰▱ **C A U T I O N** ▱▰▰▰▰▰▰▰▰

If you forget to write the digit 0 in a quotient, such as in Example 5, your answer will be wrong. Remember that 0 is an important placeholder in our number system.

▰▰▰▰▰▰▰

❸ DIVIDING BY A TWO-DIGIT NUMBER

Finding quotients when the divisor has more than one digit is a bit more challenging. Consider

$$25\overline{)825}.$$

The quotient can be found directly using what we have learned about dividing by one-digit numbers. We first decide which of the first few digits of the dividend (825) make a number greater than or equal to the divisor (25). In our example, that number is 82. The largest number less than 82 which is divisible by 25 is 75. The quotient, 3, is placed above the last digit in 82.

$$\overset{3}{25\overline{)8\,2\,5}} \qquad 82 \geq 25 \text{ and } 75 \div 25 = 3$$

Next multiply 3×25, place the result below 82 and subtract. The difference is 7. Bring down 5.

$$
\begin{array}{r}
3 \\
25\overline{)825} \\
75 \\
\hline
75
\end{array}
$$

Finally divide 75 by 25. The quotient is 3 which is placed above 5. Multiply 3×25, place the product below 75, and subtract to obtain 0.

$$
\begin{array}{r}
33 \\
25\,\overline{)825} \\
75 \\
\hline
75 \\
75 \\
\hline
0
\end{array}
$$

Dividing by a two-digit number sometimes involves a bit of guesswork and luck as we try to find each digit in the quotient. For example, suppose we want to find the quotient $960 \div 42$. We might start by looking at the first digits in 960 and 42 and asking

What number less than or equal to 9 can be divided by 4?

Since $8 \leq 9$, and $8 \div 4 = 2$, we try 2. This technique, while not foolproof, gives us a starting point, as shown in the next example.

EXAMPLE 6 DIVIDING BY A TWO-DIGIT NUMBER	PRACTICE EXERCISE 6

Find the quotient. $32\overline{)739}$

The number less than or equal to 7 which is divisible by 3 is 6, and since $6 \div 3 = 2$, begin by placing 2 above 3 in 739.

$$
\begin{array}{l}
①\ 6 \div 3 = 2 \longrightarrow 2 \\
\qquad\qquad\qquad 32\overline{)739} \\
②\ 2 \times 32 = 64 \longrightarrow \quad 64 \\
③\ 73 - 64 = 9 \longrightarrow \quad 99 \longleftarrow ④\ \text{Bring down 9}
\end{array}
$$

To divide 99 by 32, find the largest number less than or equal to 9 which is divisible by 3. That number is 9 and $9 \div 3 = 3$. Thus, place 3 above 9 in 739 and proceed as before.

$$
\begin{array}{r}
23 \\
3\,2\overline{)739} \\
64 \\
\hline
9\,9 \\
3 \times 32 = 96 \longrightarrow \quad 96 \\
\hline
3 \longleftarrow\ 99 - 96 = 3\ \text{(remainder)}
\end{array}
$$

PRACTICE EXERCISE 6

Find the quotient.

$$41\overline{)861}$$

Answer: 21

This method can be applied to dividends with more digits.

EXAMPLE 7 DIVIDING BY A TWO-DIGIT NUMBER	PRACTICE EXERCISE 7

Find the quotient. $67\overline{)5294}$

The number less than or equal to 52 which is divisible by 6 is 48, so we try the quotient, 8.

$$
\begin{array}{r}
8 \\
67\overline{)5294} \\
8 \times 67 = 536 \longrightarrow \quad 536
\end{array}
$$

PRACTICE EXERCISE 7

Find the quotient.

$$46\overline{)2181}$$

But $536 > 529$ so we cannot subtract. (As we said earlier, this method is not foolproof.) We try a smaller quotient, 7.

$$7 \times 67 = 469 \longrightarrow \begin{array}{r} 7 \\ 67{\overline{)5294}} \\ \underline{469} \\ 604 \end{array}$$

Now we try 60 divided by 6. But since this quotient is the two-digit number 10, we "back off" and use the largest one-digit number, 9.

$$9 \times 67 = 603 \longrightarrow \begin{array}{r} 79 \\ 67{\overline{)5294}} \\ \underline{469} \\ 604 \\ \underline{603} \\ 1 \end{array}$$

Answer: **47 R19**

④ DIVIDING BY A THREE-DIGIT NUMBER

An **algorithm** is a procedure or method used to perform a certain task. The algorithm for finding quotients that we have shown involves "trial-and-error." Sometimes estimating can make "trial-and-error" faster. The number of trials that you make before finding the desired quotient will decrease with practice.

This algorithm can be used to find quotients involving three-digit divisors.

EXAMPLE 8 DIVIDING BY A THREE-DIGIT NUMBER	PRACTICE EXERCISE 8

Find the quotient. $562{\overline{)116527}}$

$$10 \div 5 = 2 \longrightarrow \qquad \begin{array}{r} \text{412 cannot be divided by 562} \\ 207 \longleftarrow 35 \div 5 = 7 \\ 562{\overline{)116527}} \\ \underline{1124} \\ 412 \\ \underline{000} \\ 4127 \\ \underline{3934} \\ 193 \end{array}$$

The work can be shortened by bringing down the next digit, 7, when we realize that 412 cannot be divided by 562. But be sure to place 0 in the quotient.

$$\begin{array}{r} \text{Remember this 0} \\ 207 \\ 562{\overline{)116527}} \\ \underline{1124} \\ 4127 \\ \underline{3934} \\ 193 \end{array}$$

Find the quotient.

$$451{\overline{)93808}}$$

Answer: **208**

HINT

Some students may question the necessity of learning the division algorithms since division problems can be solved more quickly with a calculator. Although this is true, the techniques here are needed for understanding similar algorithms found in algebra.

⑤ CHECKING DIVISION

As with other arithmetic operations, there should be a way to check if a quotient is correct. Consider the following division problem.

$$\text{divisor} \longrightarrow 4\overline{)25} \quad \begin{array}{l} \underline{6} \longleftarrow \text{quotient} \\ \longleftarrow \text{dividend} \\ \underline{24} \\ 1 \longleftarrow \text{remainder} \end{array}$$

Observe that

$$(6 \times 4) + 1 = 25$$
$$24 + 1 = 25.$$

This relationship is true in general and provides a check for division.

Check for Division

In any division problem

$$\text{(quotient} \times \text{divisor)} + \text{remainder} = \text{dividend.}$$

Suppose we check the division problem given in Example 7. Since the quotient is 79, the divisor is 67, the remainder is 1, and the dividend is 5294, we have the following:

$$\overbrace{(79 \times 67)}^{\text{(quotient} \times \text{divisor)}} + \overbrace{1}^{\text{+ remainder}} = (5293) + 1$$
$$= 5294 \longleftarrow \text{dividend}$$

Thus, the work was correct.

The final example in this section illustrates how rounded numbers can be used to find approximate solutions to applied division problems.

EXAMPLE 9   ESTIMATING IN A DIVISION APPLICATION

Manuel wants to divide 117 pieces of candy among 9 children. First find the approximate number each will receive; and then find the exact number.

Since 117 is about 120 and 9 is about 10, we can mentally divide 120 by 10 to obtain approximately 12 pieces for each. To find the exact number, divide 117 by 9. The quotient is 13, so each child receives exactly 13 pieces.

PRACTICE EXERCISE 9

There are 1092 weeks in 21 years. First find the approximate number of weeks in one year; then find the exact number.

Answer: Approximate number is 50 (using 1000 ÷ 20); exact number is 52 (using 1092 ÷ 21)

2.5 EXERCISES A

Find each quotient.

1. $2\overline{)46}$ 23   **2.** $3\overline{)93}$ 31   **3.** $2\overline{)446}$ 223   **4.** $4\overline{)408}$ 102   **5.** $2\overline{)468}$ 234

6. $2\overline{)58}$ 29   **7.** $5\overline{)90}$ 18   **8.** $3\overline{)438}$ 146   **9.** $6\overline{)966}$ 161   **10.** $3\overline{)225}$ 75

11. $6\overline{)546}$ 91   **12.** $5\overline{)905}$ 181   **13.** $5\overline{)545}$ 109   **14.** $6\overline{)654}$ 109   **15.** $4\overline{)4076}$ 1019

16 $2\overline{)6054}$ 3027   **17.** $4\overline{)2836}$ 709   **18.** $5\overline{)4535}$ 907   **19.** $32\overline{)672}$ 21   **20.** $63\overline{)756}$ 12

21 $23\overline{)2231}$ 97   **22.** $28\overline{)1316}$ 47   **23** $473\overline{)13244}$ 28   **24.** $623\overline{)35511}$ 57   **25.** $755\overline{)31710}$ 42

Find each quotient and remainder. Check your answer.

26. $7\overline{)67}$ **27.** $6\overline{)714}$ **28.** $42\overline{)737}$ **29** $83\overline{)7553}$ **30.** $466\overline{)10736}$
    Q: 9; R: 4   Q: 119; R: 0   Q: 17; R: 23   Q: 91; R: 0   Q: 23; R: 18

Solve.

31. A man buys 9 suits for $1782. First find the approximate cost of each suit, and then find the exact cost.
approximate cost: $180; exact cost: $198

32 There are 4011 calories in 21 candy bars. First find the approximate number of calories in each bar, and then find the exact number.
approximate number: 200 calories;
exact number: 191 calories

ANSWERS: 1. 23   2. 31   3. 223   4. 102   5. 234   6. 29   7. 18   8. 146   9. 161   10. 75   11. 91   12. 181
13. 109   14. 109   15. 1019   16. 3027   17. 709   18. 907   19. 21   20. 12   21. 97   22. 47   23. 28   24. 57   25. 42
26. Q: 9; R: 4 [(9 × 7) + 4 = 63 + 4 = 67]   27. Q: 119; R: 0 [(119 × 6) + 0 = 714 + 0 = 714]   28. Q: 17; R: 23
[(17 × 42) + 23 = 714 + 23 = 737]   29. Q: 91; R: 0 [(91 × 83) + 0 = 7553 + 0 = 7553]   30. Q: 23; R: 18
[(23 × 466) + 18 = 10,718 + 18 = 10,736]   31. approximate cost: $180 (using 1800 ÷ 10); exact cost: $198 (using
1782 ÷ 9)   32. approximate number: 200 calories (using 4000 ÷ 20); exact number: 191 calories (using 4011 ÷ 21)

2.5 EXERCISES B

Find each quotient.

1. $2\overline{)64}$ 32   **2.** $6\overline{)60}$ 10   **3.** $3\overline{)639}$ 213   **4.** $3\overline{)609}$ 203   **5.** $2\overline{)96}$ 48

6. $3\overline{)84}$ 28   **7.** $6\overline{)84}$ 14   **8.** $4\overline{)712}$ 178   **9.** $2\overline{)986}$ 493   **10.** $4\overline{)384}$ 96

11. $2\overline{)134}$ 67   **12.** $3\overline{)327}$ 109   **13.** $4\overline{)208}$ 52   **14.** $3\overline{)3645}$ 1215   **15.** $6\overline{)6066}$ 1011

16. $3\overline{)6309}$ 2103   **17.** $8\overline{)4064}$ 508   **18.** $9\overline{)5418}$ 602   **19.** $45\overline{)720}$ 16   **20.** $43\overline{)3698}$ 86

21. $52\overline{)3796}$ 73   **22.** $67\overline{)3484}$ 52   **23.** $175\overline{)8575}$ 49   **24.** $89\overline{)31773}$ 357   **25.** $347\overline{)92996}$ 268

Find each quotient and remainder. Check your answer.

26. $9\overline{)93}$
Q: 10; R: 3

27. $23\overline{)1081}$
Q: 47; R: 0

28. $67\overline{)5728}$
Q: 85; R: 33

29. $253\overline{)14173}$
Q: 56; R: 5

30. $342\overline{)68742}$
Q: 201; R: 0

Solve.

31. A man buys 11 radios at a total cost of $396. First find the approximate cost of 1 radio, then find the exact cost.

approximate cost: $40; actual cost: $36

32. A man's heart beats 2016 times during an exercise period of 21 minutes. First find the approximate number of beats per minute, and then find the actual number (which in this case is also really an approximation).

approximate number: 100 beats per minute; actual number: 96 beats per minute

2.5 EXERCISES C

Find the quotient and remainder.

1. $7\overline{)18114}$ Q: 2587; R: 5

2. $27\overline{)177696}$
[Answer: Q: 6581; R: 9]

3. $208\overline{)106029}$ Q: 509; R: 157

2.6 MORE APPLICATIONS

STUDENT GUIDEPOSTS

1. Total Value Relationship
2. Area of a Rectangle
3. Business Applications
4. Calculating Miles Per Gallon
5. Unit Pricing

Many word problems involve multiplication and division of whole numbers. As with addition and subtraction word problems, you should make sure that your answer is reasonable.

1 TOTAL VALUE RELATIONSHIP

A basic principle used in many applied problems involves finding a *total value*. The total value of a quantity made up of several parts is the value per part multiplied by the number of parts.

$$\text{(total value)} = \text{(value per part)} \cdot \text{(number of parts)}.$$

For example, if you have 35 dimes, the total value of these coins is given by:

(total value of 35 dimes) = (value of 1 dime) · (number of dimes)

$$= \quad (10¢) \quad \cdot \quad (35)$$

$$= \quad 350¢.$$

We use this principle in the next example.

| **EXAMPLE 1** TOTAL VALUE IN A BUSINESS PROBLEM | **PRACTICE EXERCISE 1** |

During a sale on men's sweaters, a store owner sold 43 sweaters at $18 a sweater. How much money did he take in on the sweater sale?

For **1** sweater sold, he receives $18 \cdot 1 = \$18$.

For **2** sweaters sold, he receives $18 \cdot 2 = \$36$.

For **3** sweaters sold, he receives $18 \cdot 3 = \$54$.

For **43** sweaters sold,

$$\text{(total value of 43 sweaters sold)} = \text{(value per sweater)} \cdot \text{(number of sweaters)}$$
$$\$774 \qquad\qquad = \qquad\qquad \$18 \qquad \cdot \qquad 43.$$

That is, we multiply 18 by 43 to obtain $774. Since $774 seems reasonable (40 sweaters at $20 each would bring in $800; so 774 appears to check), the store owner took in $774 during the sale.

Lucy had a job babysitting which paid $3 per hour. How much was she paid for sitting a total of 14 hours?

Answer: $42

❷ AREA OF A RECTANGLE

The formula for finding the area of a rectangle has many applications involving multiplication.

> ### Area of a Rectangle
>
> The area of a rectangle is given by the formula
>
> **Area = (length) · (width).**

The rectangle in Figure 2.4 has area given by

$$\text{Area} = \text{(length)} \cdot \text{(width)}$$
$$= 3 \cdot 2$$
$$= 6.$$

1 square in	1 square in	1 square in
1 square in	1 square in	1 square in

2 in (width)

3 in (length)

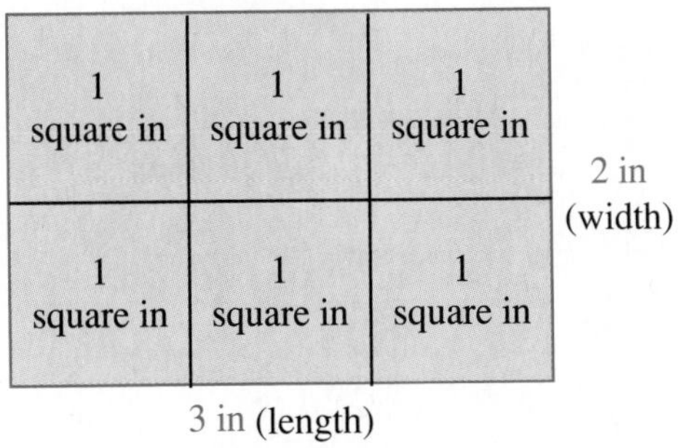

Figure 2.4

Since the rectangle contains 6 squares which have 1 inch sides, the measure of the area of the rectangle is given in square inches (sometimes abbreviated **sq in**). Thus, the rectangle has area 6 sq in. Had the sides of the rectangle been 3 ft and 2 ft, the area would have been 6 sq ft.

EXAMPLE 2 AREA AND PERIMETER OF A RECTANGLE	**PRACTICE EXERCISE 2**

A room is 4 yd wide and 5 yd long (**yd** is the abbreviation for yards).

(a) Find the floor area of the room.

In many word problems, it is helpful to make a sketch which describes the situation. A sketch of the floor of the room is shown in Figure 2.5. The area of the floor is

$$\text{Area} = (\text{length}) \cdot (\text{width})$$

$$= 5 \cdot 4 = 20 \text{ sq yd.}$$

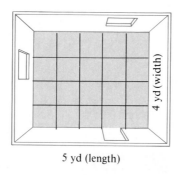

4 yd (width)

5 yd (length)

Figure 2.5

(b) Find the perimeter of the room.

The perimeter is the distance around the room found by adding the lengths of its sides. Thus, the perimeter is

$$5 + 5 + 4 + 4 = 18 \text{ yd.}$$

(c) If carpet costs $16 a square yard, how much would it cost to carpet the room?

One square yard of carpet costs $16. Then 20 sq yd at $16 a yard would cost

$$20 \cdot 16 = \$320.$$

The floor in a recreation hall is 20 yd long and 14 yd wide.

(a) Find the area of the floor.

(b) Find the perimeter of the hall.

(c) If carpeting costs $13 a square yard, how much would it cost to carpet the hall?

Answers: (a) 280 sq yd   (b) 68 yd
(c) $3640

❸ BUSINESS APPLICATIONS

Many applications of multiplication can be found in business. Forms such as invoices, shipping orders, packing slips, and purchase orders use multiplication. A sample purchase order is shown in the next example.

| **EXAMPLE 3** An Application in Business | | **Practice Exercise 3** |

Complete the purchase order in Figure 2.6.

NORTHERN PENNSYLVANIA UNIVERSITY
General Stores

Customer's Order No. ___11937-Q___   Date ___March 20___   19_89_

Name ___Mathematics Department___

Address ___Box 5717   MAT 115___

Sold By A.S.	Cash	C.O.D.	Charge	On Acct. ✔	Mdse. Retd.	Paid Out	

Quantity	Description	Unit Price	Amount
50 reams	Multi-purpose Paper	$2	
6 cans	Duplicating Fluid	$3	
3	Staplers	$4	
8 pkg	Writing Pads	$3	
1	Paper Trimmer	$42	
	TOTAL		

Rec'd. by _____

Figure 2.6

Complete the purchase order.

Quantity	Description	Unit price	Amount
12	Shirts	$14	**(a)**
36	Pair socks	$ 4	**(b)**
50	Pair slacks	$24	**(c)**
25	Belts	$13	**(d)**
	TOTAL		**(e)**

Quantity	Description	Unit Price	Amount
50 reams	Multi-purpose Paper	$2	$100
6 cans	Duplicating Fluid	$3	18
3	Staplers	$4	12
8 pkg	Writing Pads	$3	24
1	Paper Trimmer	$42	42
	TOTAL		$196

Figure 2.7

To complete this form, compute the numbers that go into the Amount column. The price of each ream of Multi-purpose Paper is $2 so the cost of 50 reams is $50 \times \$2$, which is $100. Enter this cost in the Amount column. Compute and enter the remaining costs the same way. See Figure 2.7.

Find the total cost of all items by adding the separate costs in the Amount column ($100 + 18 + 12 + 24 + 42$). Enter this sum, $196, next to TOTAL as shown.

Answers: (a) $168   (b) $144
(c) $1200   (d) $325   (e) $1837

HINT

Read an applied problem carefully and be aware that a given problem might involve more than one arithmetic operation. In Example 3, for instance, to complete the purchase order we first find five products and then find their sum.

④ CALCULATING MILES PER GALLON

In the next example we provide a formula for finding the number of miles that a vehicle can be driven using one gallon of gasoline.

EXAMPLE 4 CALCULATING MILES PER GALLON

In order to check the gas mileage on his car, Mike noted that the odometer reading was 18,432 when he filled the tank with gas. At 18,696 he refilled the tank with 11 gallons of gas. What was his mileage per gallon?

The miles per gallon (mpg) is found by:

$$\text{mpg} = \frac{\text{number of miles driven}}{\text{number of gallons used}}$$

To find the number of miles driven, subtract odometer readings.

$$
\begin{array}{r}
18696 \\
- \ 18432 \\
\hline
264 \quad \text{Miles driven}
\end{array}
$$

Thus,

$$\text{mpg} = \frac{264}{11} = 24.$$

Since 24 mpg seems reasonable and does check ($24 \times 11 = 264$ and $18{,}432 + 264 = 18{,}696$), Mike is getting 24 miles per gallon of gas.

PRACTICE EXERCISE 4

On a recent trip, Sharon traveled from El Paso, Texas, to Cheyenne, Wyoming, a distance of 750 miles, and returned to El Paso. If she used a total of 75 gallons of gas on the trip, how many miles per gallon did she get?

Answer: 20 mpg

The next two examples provide other types of applications involving arithmetic operations.

EXAMPLE 5 MAKING PAYMENTS ON A PURCHASE

Toni agreed to purchase a used car from her parents for $1100, by paying $80 down and the rest in equal monthly payments for one year. How much were her monthly payments if she is not being charged any interest?

Because she paid $80 down, the difference $1100 − $80 = $1020 is the amount she has to pay over 12 months (1 year = 12 months). To find the amount paid each month, divide 1020 by 12.

$$
\begin{array}{r}
85 \\
12\overline{)1020} \\
\underline{96} \\
60 \\
\underline{60} \\
0
\end{array}
\qquad
\text{Check:}
\qquad
\begin{array}{r}
85 \\
\times \ 12 \\
\hline
170 \\
85 \\
\hline
1020
\end{array}
\qquad \text{and} \qquad
\begin{array}{r}
1020 \\
+ \ 80 \\
\hline
1100
\end{array}
$$

Since this amount seems reasonable and does check, her monthly payment is $85.

PRACTICE EXERCISE 5

Lennie purchased four new tires for his car. If each tire cost $75 and the tax on the purchase of the four was $12, what was the total cost of the tires? If the dealer agreed to let him pay for the tires in 3 equal monthly payments, how much does he pay each month?

Answer: $312; $104

| EXAMPLE 6   DISTRIBUTING PRIZE MONEY | PRACTICE EXERCISE 6 |

A seven-man basketball squad won a local tournament that had a $500 prize for the winning team. The seven players and the coach agree to divide the money equally (in dollars) among themselves with any remainder going to the coach. How much does the coach receive?

$$
\begin{array}{r}
62 \\
8\overline{)500} \\
48 \\
\hline
20 \\
16 \\
\hline
4 \quad \text{Remainder}
\end{array}
$$

Check: $(62 \times 8) + 4 = 496 + 4$
$= 500$

Each of the seven players should receive $62 with the coach receiving $62 plus the $4 left over. Thus, the coach receives $66. (Does the answer seem reasonable?)

The twelve members of a fraternity bought a raffle ticket and won the $1000 prize. They agree to split the prize among themselves equally (in dollars) with any remainder going into their party fund. How much does each receive and how much is deposited in the fund?

Answer: $83 to each member and $4 to the fund

⑤ UNIT PRICING

The wise shopper has learned to compare prices by determining *unit prices*. For example, if a can of beans weighs 15 oz (**oz** stands for ounces) and costs 45¢, the unit price of the beans is

$$\frac{45}{15} = 3¢ \text{ per ounce.}$$

If a different brand of beans sells for 52¢ but weighs 13 oz, its unit price is

$$\frac{52}{13} = 4¢ \text{ per ounce.}$$

Assuming equal quality of the products, the first can is clearly the better buy. Unit pricing gives us excellent examples of division problems.

2.6 EXERCISES A

Solve.

1. A compact car has a 12-gallon gas tank and can travel 29 miles on one gallon of gas. How far can it travel on a full tank?   **348 mi**

2. If a can of cola contains 86 calories, how many calories are in a six-pack of cola?   **516 calories**

3. There are 365 days in a year.
   (a) How many days are there in 5 years?
   (b) How many hours are there in a year? [1 day = 24 hours]
   (c) How many hours are there in 5 years?
   (a) **1825 days** (b) **8760 hours** (c) **43,800 hours**

4 A hardware store sells 35 snow blowers at $329 each and 40 chain saws at $289 each.
   (a) How much money does the store take in on the sale of snow blowers?
   (b) How much money does the store take in on the sale of chain saws?
   (c) How much money does the store take in on the sale of both snow blowers and chain saws?
   (a) **$11,515** (b) **$11,560** (c) **$23,075**

5. Suppose that 225 lift tickets are sold per day at the Sugar Loaf Ski Area.

   (a) How many tickets are sold in a 14-day period?

   (b) If each ticket costs $7, how much money is taken in during a 14-day period?

   (a) **3150 tickets** (b) **$22,050**

6. A rectangular vegetable garden is 14 m wide and 26 m long (**m** stands for meter).

   (a) What is the area of the garden?

   (b) What is the perimeter of the garden?

   (a) **364 sq m** (b) **80 m**

7. A woman buys 3 suits at $175 each, 5 blouses at $21 each, and 6 scarves at $8 each. How much has she spent? **$678**

8 Barbara wants to put a rectangular mirror measuring 8 ft by 5 ft on the wall of her living room. If a mirror costs $3 a square foot, how much will the mirror cost? **$120**

9. A rectangular room measures 9 yd long and 5 yd wide. How much will it cost to carpet the room over a foam pad if carpet costs $20 a sq yd, padding costs $2 a sq yd, and installation costs are included in the price of carpet and padding? **$990**

10. A sketch of the floor plan of a family room is shown in the figure. Mr. Kirk plans to cover the floor with vinyl floor covering costing $9 a sq yd. How much will it cost to complete the job? [*Hint:* Divide the floor plan into two rectangular areas.] **$441**

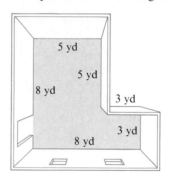

11. Complete the following purchase order.

Quantity	Description	Unit Price	Amount
3	Back Packs	$90	(a) $270
1	Tent	140	(b) $140
4	Sleeping Bags	110	(c) $440
4	Ground Pads	15	(d) $60
1	Pack Stove	20	(e) $20
8	Complete Meals	7	(f) $56
		TOTAL	(g) $986

12. Sam and Carolyn have a home mortgage which requires a monthly payment of $520 for 30 years.

    (a) How many monthly payments must they make over the 30-year period?

    (b) How much will they have paid in total at the end of 30 years?

    (a) 360 payments   (b) $187,200

13 A domed football stadium is divided into 20 sections of seats. Suppose that 5 sections each seat 400 spectators, 8 sections each seat 1200 spectators, and the remaining sections each seat 950 spectators.

    (a) What is the seating capacity of the dome?

    (b) If every seat is sold for the championship game, and the price of each ticket is $5, how much money is taken in for the game?

    (a) 18,250 seats   (b) $91,250

14. Karen can type 85 words per minute. How long will it take her to type 21,760 words?   256 minutes

15. At the start of a trip, Mr. Creighten's odometer read 24,273. When he returned from the trip, it read 25,561. If he used 56 gallons of gas, what was his gas mileage?   23 mpg

16. A certain brand of toothpaste sells for 80¢. How many tubes can be purchased for $6, and how much change is left? [*Hint:* Use 600¢ for $6.]
7 tubes; 40¢ in change

17 Which is the better buy, a 12-lb bag of dog food selling for $3.72 (372¢) or a 9-lb bag of dog food selling for $2.88 (288¢)?
The 12-lb bag at 31¢ per lb is a better buy than the 9-lb bag at 32¢ per lb.

18. Eleven people pool their money and buy a lottery ticket. The prize is $1500 and they agree, if they win, to divide the money equally (in dollars) and draw straws for any amount left over after the split. What is the most that a winner could receive?
$140

19. A Toyota get 26 mpg. How many gallons would it take to make a 2444-mile trip?   94 gallons

20 The enrollment at a university is 12,100 students. If the faculty numbers 550, what is the number of students per teacher?   22

21. Dr. Yard buys 225 shares of stock for a total cost of $31,050. What was the cost of each share? $138

22. The gate receipts at a concert were $87,850. If tickets sold for $7 each, how many were in attendance?
12,550

23. Assuming equal quality, which is the better buy, a 43-oz box of soap for $1.29 (129¢) or a 78-oz box of soap for $1.56 (156¢)?
At 2¢ per oz, the 78-oz box is a better buy than the 43-oz box at 3¢ per oz.

FOR REVIEW

Find each quotient and check your answer.

24. $8\overline{)1880}$ Q: 235; R: 0   **25.** $37\overline{)7750}$ Q: 209; R: 17   **26.** $259\overline{)107744}$ Q: 416; R: 0

Exercise 27 will help you prepare for the next section.

27. List the multiples of 2 between 1 and 15.   2, 4, 6, 8, 10, 12, 14

ANSWERS: 1. 348 mi 2. 516 calories 3. (a) 1825 days (b) 8760 hours (c) 43,800 hours 4. (a) $11,515 (b) $11,560 (c) $23,075 5. (a) 3150 tickets (b) $22,050 6. (a) 364 sq m (b) 80 m 7. $678 8. $120 9. $990 10. $441 11. (a) $270 (b) $140 (c) $440 (d) $60 (e) $20 (f) $56 (g) $986 12. (a) 360 payments (b) $187,200 13. (a) 18,250 seats (b) $91,250 14. 256 minutes 15. 23 mpg 16. 7 tubes; 40¢ in change 17. The 12-lb bag at 31¢ per lb is a better buy than the 9-lb bag at 32¢ per lb. 18. $140 19. 94 gallons 20. 22 21. $138 22. 12,550 23. At 2¢ per oz, the 78-oz box is a better buy than the 43-oz box at 3¢ per oz. 24. Q: 235; R: 0. [(235 × 8) + 0 = 1880 + 0 = 1880] 25. Q: 209; R: 17. [(209 × 37) + 17 = 7733 + 17 = 7750] 26. Q: 416; R: 0. [(416 × 259) + 0 = 107,744 + 0 = 107,744] 27. 2, 4, 6, 8, 10, 12, 14

2.6 EXERCISES B

Solve.

1. A boat has a 45-gallon gas tank. If the boat can travel 4 miles on one gallon of gas, how far can it go on a full tank of gas?　**180 mi**

2. During a sale on CB radios, Chuck sold 18 radios at $78 each. How much money did he take in? **$1404**

3. There are 365 days in a year.

(a) How many days are there in 7 years?

(b) How many hours are there in 7 years?

(c) How many minutes are there in 7 years?

(a) **2555 days** (b) **61,320 hours** (c) **3,679,200 minutes**

4. A room is 12 ft wide and 15 ft long.

(a) Find the floor area of the room.

(b) Find the perimeter of the room.

(c) If carpet costs $2 a sq ft, how much would it cost to carpet the room?

(a) **180 sq ft** (b) **54 ft** (c) **$360**

5. Suppose that 120 tickets are sold per day to see *The Mummy* at The Tourist Trap.

(a) How many tickets are sold in a 30-day period?

(b) If each ticket costs $4, how much money is taken in during a 30-day period?

(a) **3600 tickets** (b) **$14,400**

6. A wall is 20 ft long and 8 ft high.

(a) What is the area of the wall?

(b) What is the perimeter of the wall?

(a) **160 sq ft** (b) **56 ft**

7. Betsy buys 3 sweaters at $39 each, 2 skirts at $42 each, and 7 pairs of earrings at $4 a pair. How much has she spent?　**$229**

8. The flat roof of a mobile home measures 60 ft long and 15 ft wide. If it costs $2 a square foot to put sealer on the roof, how much will the seal job cost? **$1800**

9. A rectangular patio measures 30 ft long by 12 ft wide. How much will it cost to put outdoor carpeting on the patio if carpet costs $3 a square foot and installation costs are $1 a square foot?　**$1440**

10. A sketch of the floor plan of the first floor in the Andrews' home is shown in the figure. How much will it cost to lay carpeting, costing $20 a square yard, over padding, costing $3 a square yard, over the entire floor?　**$4094**

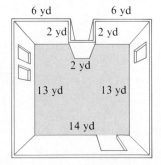

11. Complete the following purchase order.

Quantity	Description	Unit Price	Amount
120	Dress Shirts	$17	(a) $2040
85	Ties	8	(b) $680
36	Suits	170	(c) $6120
24	Belts	13	(d) $312
		TOTAL	(e) $9152

12. The Orlandos have a house mortgage which requires a monthly payment of $480 for 25 years.

    (a) How many monthly payments must they make over the 25-year period?

    (b) How much will they have paid in total at the end of 25 years?

    (a) **300 payments** (b) **$144,000**

13. A basketball arena is divided into 16 sections of seats. Suppose that 8 sections each seat 300 spectators, 4 sections will each seat 420 spectators, and the remaining sections will each seat 565 spectators.

    (a) What is the seating capacity of the arena?

    (b) If every seat is sold for $4 each, how much money will the athletic department take in?

    (a) **6340 seats** (b) **$25,360**

14. Max used 1470 bricks to cover a wall. If there are 35 rows of bricks, how many bricks are in each row?
42 bricks

15. When Anita filled the gas tank on her car before leaving on a trip, she observed that the odometer reading was 27,154. When she returned from the trip, she refilled the tank and noted that the odometer reading was then 27,518. If she purchased 14 gallons of gas, what was her mileage per gallon?
26 mpg

16. A certain brand of cereal sells for 95¢. How many boxes can be purchased for $12, and how much change is left?
12 boxes; 60¢ in change

17. Dyane purchased a used car from a friend for $1500 by paying $150 down and the rest in equal monthly payments over a period of 18 months (no interest was charged). How much were her monthly payments?   **$75**

18. Which is a better buy, a 9-lb bag of kitty litter selling for 99¢ or a 13-lb bag selling for $1.30 (130¢)?
The 13-lb bag at 10¢ per lb is better than than 9-lb bag at 11¢ per lb.

19. The receipts at a play were $15,300. If tickets sold for $6 each, how many people were in attendance?
2550 people

20. The enrollment at a community college is 18,200 students. If the faculty number 728, what is the number of students per teacher?   **25**

21. Mr. Wilkins divides $135,000 among his daughters. If each receives $45,000, how many daughters does he have?   **3 daughters**

22. Dr. Johnson, a chemist, pours 384 milliliters of acid into 4 flasks in such a way that each flask has the same amount. How much acid is in each flask?
96 milliliters

23. Assuming equal quality, which is a better buy, a 5-lb bag of dog food for $2.00 (200¢), or an 8-lb bag of dog food for $3.04 (304¢)?
The 8-lb bag at 38¢ per lb is a better buy than the 5-lb bag at 40¢ per lb.

FOR REVIEW

Find each quotient and check your answer.

24. $7\overline{)2282}$ Q: 326; R: 0

25. $23\overline{)1499}$ Q: 65; R: 4

26. $157\overline{)32105}$ Q: 204; R: 77

Exercise 27 will help you prepare for the next section.

27. List the multiples of 3 between 1 and 20.

3, 6, 9, 12, 15, 18

2.6 EXERCISES C

1. Let *n* represent the number of people at a concert. If tickets were sold for $11 each and $2816 was collected, find *n*.   *n* = 256

2. Let *n* represent the number of shirts that Sherman bought. The shirts cost $16 each. If he paid a total of $68 for the shirts plus a $20 pair of pants, find *n*. [Answer: *n* = 3]

2.7 DIVISIBILITY AND PRIMES

=== STUDENT GUIDEPOSTS ===

1 Divisors of a Whole Number

2 Prime and Composite Numbers

3 Finding the Prime Factorization of a Number

4 Divisibility Tests

1 DIVISORS OF A WHOLE NUMBER

In this section we consider several topics related to multiplication and division of whole numbers. Recall that a **multiple** of a whole number is the product of the number and some other whole number. For example, some multiples of 3 are

$$0 = 3 \times 0, \quad 3 = 3 \times 1, \quad 6 = 3 \times 2, \quad 9 = 3 \times 3, \quad 12 = 3 \times 4.$$

Another way of saying the same thing (if we exclude zero) is that one number is a **divisor** or **factor** of a second if the second is a multiple of the first. For example,

$$6 = 3 \times 2.$$

6 is a multiple of both 3 and 2   3 and 2 are both divisors or factors of 6

EXAMPLE 1 FINDING MULTIPLES OF A NUMBER	PRACTICE EXERCISE 1

Find the first ten multiples of 6.

Multiply 6 by each of the first ten whole numbers.

$0 \times 6 = 0$	$5 \times 6 = 30$
$1 \times 6 = 6$	$6 \times 6 = 36$
$2 \times 6 = 12$	$7 \times 6 = 42$
$3 \times 6 = 18$	$8 \times 6 = 48$
$4 \times 6 = 24$	$9 \times 6 = 54$

Find the first ten multiples of 5.

Answer: 0, 5, 10, 15, 20, 25, 30, 35, 40, 45

EXAMPLE 2   FINDING DIVISORS OF A NUMBER

Show that 1, 2, 3, 4, 6, 8, 12, and 24 are all divisors of 24.

Since 24 is a multiple of each of these numbers, as shown below,

$$1 \times 24 = 24 \qquad 2 \times 12 = 24 \qquad 3 \times 8 = 24 \qquad 4 \times 6 = 24$$

each is a divisor of 24.

PRACTICE EXERCISE 2

Show that 1, 2, 5, 10, 25, and 50 are all divisors of 50.

Answer: Since $1 \times 50 = 50$, $2 \times 25 = 50$, and $5 \times 10 = 50$, 50 is a multiple of each number, which means each of the numbers is a divisor of 50.

In Example 2, notice that 1, 2, 3, 4, 6, 8, 12, and 24 all divide 24 evenly; that is, the remainder is zero. This is true in general.

Divisor of a Number

One number is a divisor of another if division results in a remainder of zero.

HINT

To find the divisors of a number such as 15, we would need to divide 15 by each of the numbers less or equal to 15: 1, 2, 3, 4, 5, 6, 7, 8, 9, 10, 11, 12, 13, 14, and 15, and look at the remainders. This process would show that the only divisors of 15 are 1, 3, 5, and 15, since dividing 15 by these numbers, and only these numbers, results in a remainder of 0. Actually, there is no need to try 8, 9, 10, 11, 12, 13, and 14 since they are greater than half of 15.

❷ PRIME AND COMPOSITE NUMBERS

Certain natural numbers have only two divisors, 1 and the number itself. For example, the only divisors of 7 are 1 and 7.

Prime Numbers

A natural number that has *exactly two different* divisors, 1 and itself, is called a **prime number.**

EXAMPLE 3   RECOGNIZING PRIME NUMBERS

Are the following numbers prime?

(a) 1   is *not* prime since 1 can only be written as $1 = 1 \times 1$. Thus, 1 does not have two *different* divisors.

(b) 2   is prime. The only divisors of 2 are 1 and 2 ($2 = 1 \times 2$). Thus, 2 is the first prime number.

(c) 4   is *not* prime since $4 = 2 \times 2$.

PRACTICE EXERCISE 3

Are the following numbers prime?

(a) 3

(b) 5

(c) 9

Answers: (a) 3 is prime.   (b) 5 is prime.   (c) 9 is not prime since $9 = 3 \times 3$.

Composite Numbers

A natural number greater than 1 that is not a prime is a **composite number.**

H I N T

Every composite number can be written as a product of two natural numbers *other than* the number itself and 1.

EXAMPLE 4   RECOGNIZING COMPOSITE NUMBERS

Show that the following numbers are composite.

(a) 4 is composite since $4 = 2 \times 2$.

(b) 6 is composite since $6 = 2 \times 3$.

(c) 8 is composite since $8 = 2 \times 4$.

(d) 9 is composite since $9 = 3 \times 3$.

PRACTICE EXERCISE 4

Show that the following numbers are composite.

(a) 10

(b) 12

(c) 18

(d) 25

Answers: (a) $10 = 2 \times 5$
(b) $12 = 3 \times 4$ (c) $18 = 2 \times 9$
(d) $25 = 5 \times 5$

H I N T

The number 1 is neither prime nor composite. It is the only natural number with this property.

❸ FINDING THE PRIME FACTORIZATION OF A NUMBER

Every composite number can be written as a product of primes. In Example 4, 4 was written as 2×2, 6 as 2×3, and 9 as 3×3. Although 8 was written as 2×4, we could also write 8 as $2 \times 2 \times 2$. In all cases, the composite number has been written as a product of prime factors, called the **prime factorization** of the number.

Prime Factorization of a Composite Number

Each composite number can be written as a product of primes, and except for the order in which the primes are written, this factorization is the only one possible.

Although we know that a composite number can be expressed as a product of primes, we do not yet have a good way to find these prime factors. The best way is to try to divide the number by each of the primes

$$2, \quad 3, \quad 5, \quad 7, \quad 11, \quad 13, \quad 17, \quad 19, \quad 23, \ldots,$$

starting with 2 and continuing in order of size as long as necessary. When a remainder is zero, the prime is a factor of the number. The process is repeated using the new quotient as the dividend. For example, to find the prime factors of 60, start by dividing 60 by 2.

$$60 \div 2 = 30, \qquad \text{so} \qquad 60 = 2 \cdot 30$$

Now try possible prime divisors of 30, starting again with 2.

$$30 \div 2 = 15, \qquad \text{so} \qquad 60 = 2 \cdot 30 = 2 \cdot 2 \cdot 15$$

Next look for prime divisors of 15. Since the remainder when 15 is divided by 2 is 1, not zero, move on to the next prime, 3.

$$15 \div 3 = 5 \qquad \text{so} \qquad 60 = 2 \cdot 2 \cdot 15 = 2 \cdot 2 \cdot 3 \cdot 5$$

Since all factors listed (2, 2, 3, and 5) are prime, we say the prime factorization of 60 is $2 \cdot 2 \cdot 3 \cdot 5$.

One way to display the prime factorization of 60 shown above is to use a *factor tree* like the following.

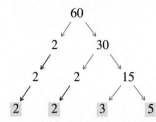

Divide 60 by 2, write down 2 and the quotient, 30

Divide 30 by 2; on the next row, write both 2's and the new quotient, 15

2 does not divide 15, so try the next larger prime, 3

These are the prime factors

This process can be continued for larger numbers. Multiplying the numbers across each row of the factor tree gives the original number, with the bottom row giving the prime factors.

Another technique that can be used to obtain the same prime factorization of 60 is shown below.

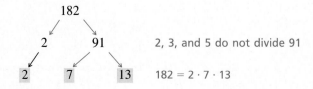

The prime factorization of 60 is $2 \cdot 2 \cdot 3 \cdot 5$.

The information given in color in the above display is used for explanation and would not be written if you use this technique.

EXAMPLE 5 FINDING PRIME FACTORIZATIONS	**PRACTICE EXERCISE 5**

Find the prime factors of each number.

(a) 28

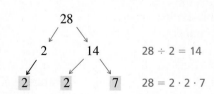

$28 \div 2 = 14$

$28 = 2 \cdot 2 \cdot 7$

(b) 182

182

2   91   2, 3, and 5 do not divide 91

2   7   13 $182 = 2 \cdot 7 \cdot 13$

Find the prime factors of each number.

(a) 20

(b) 120

(c) 825

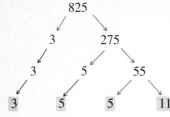

3 is the smallest prime that divides 825

$825 = 3 \cdot 5 \cdot 5 \cdot 11$

Alternatively, we find the same prime factorization using the following.

$$3 \underline{| 825}$$
$$\quad 5 \underline{| 275} \quad \leftarrow 275 = 825 \div 3$$
$$\qquad 5 \underline{| 55} \quad \leftarrow 55 = 275 \div 5$$
$$\qquad\quad 11 \quad \leftarrow 11 = 55 \div 5$$

Again we obtain $825 = 3 \cdot 5 \cdot 5 \cdot 11$.

(d) 3185

$$5 \underline{| 3185}$$
$$\quad 7 \underline{| 637}$$
$$\qquad 7 \underline{| 91}$$
$$\qquad\quad 13$$

Thus, $3185 = 5 \cdot 7 \cdot 7 \cdot 13$.

(c) 490

(d) 3003

Answers: (a) $20 = 2 \cdot 2 \cdot 5$
(b) $120 = 2 \cdot 2 \cdot 2 \cdot 3 \cdot 5$
(c) $490 = 2 \cdot 5 \cdot 7 \cdot 7$
(d) $3003 = 3 \cdot 7 \cdot 11 \cdot 13$

④ DIVISIBILITY TESTS

To find out whether a given number (prime or composite) is a divisor of another without actually dividing, we can use simple tests called *divisibility tests*. We shall give tests for the numbers

2,   3,   5,   and   10.

Divisibility Test for 2

A number is divisible by 2 if its ones digit is 0, 2, 4, 6, or 8.

Numbers divisible by 2 are called **even numbers.** The first ten even whole numbers are 0, 2, 4, 6, 8, 10, 12, 14, 16, and 18. Numbers that are not even, such as 1, 3, 5, 7, 9, 11, and 13, are called **odd numbers.**

EXAMPLE 6   TESTING DIVISIBILITY BY 2

Are the following numbers divisible by 2?

(a) 235 is not divisible by 2 since the ones digit is 5.

(b) 4284 is divisible by 2 since the ones digit is 4.

PRACTICE EXERCISE 6

Are the following numbers divisible by 2?

(a) 350

(b) 3273

Answers: (a) Yes; its ones digit is 0.   (b) No; its ones digit is 3 and not 0, 2, 4, 6, or 8.

Divisibility Test for 3

A number is divisible by 3 if the sum of its digits is divisible by 3.

Since the sum of the digits of a number is smaller than the original number, it is easier to find out if the sum is divisible by 3.

EXAMPLE 7   TESTING DIVISIBILITY BY 3

Are the following numbers divisible by 3?

(a) 24 is divisible by 3 since $2 + 4 = 6$ and 6 is divisible by 3.

(b) 3241 is not divisible by 3 since $3 + 2 + 4 + 1 = 10$ and 10 is not divisible by 3.

PRACTICE EXERCISE 7

Are the following numbers divisible by 3?

(a) 73

(b) 6594

Answers: (a) No; $7 + 3 = 10$, and 10 is not divisible by 3. (b) Yes; $6 + 5 + 9 + 4 = 24$, and 24 is divisible by 3.

Divisibility Test for 5

A number is divisible by 5 if its ones digit is 0 or 5.

EXAMPLE 8   TESTING DIVISIBILITY BY 5

Are the following numbers divisible by 5?

(a) 320 is divisible by 5 since the ones digit is 0.

(b) 4304 is not divisible by 5 since the ones digit is 4.

PRACTICE EXERCISE 8

Are the following numbers divisible by 5?

(a) 625

(b) 3948

Answers: (a) Yes; its ones digit is 5. (b) No; its ones digit, 8, is neither 0 nor 5.

HINT

For a number to be divisible by 6, it must be divisible by both 2 and 3, since $6 = 2 \times 3$. Thus, a divisibility test for 6 is a combination of the tests for 2 and 3.

Divisibility Test for 10

A number is divisible by 10 if its ones digit is 0.

EXAMPLE 9 Testing Divisibility by **10**

Are the following numbers divisible by 10?

(a) 240 is divisible by 10 since its ones digit is 0.

(b) 325,361 is not divisible by 10 since its ones digit is 1, not 0.

PRACTICE EXERCISE 9

Are the following numbers divisible by 10?

(a) 8395

(b) 23,486,280

Answers: (a) No; its ones digit is 5 not 0.   (b) Yes; its ones digit is 0.

2.7 EXERCISES A

Answer true or false. If the answer is false, explain why.

1. Since $15 = 3 \times 5$, 15 is a multiple of both 3 and 5.   **true**

2. A natural number with exactly two different divisors, 1 and itself, is called a composite number.
false (prime number)

3. The only even prime number is 2.   **true**

Find all the divisors of each number.

4. 18   **1, 2, 3, 6, 9, 18**

5. 28   **1, 2, 4, 7, 14, 28**

6. 56   **1, 2, 4, 7, 8, 14, 28, 56**

Find the first ten multiples of each number.

7. 4
0, 4, 8, 12, 16, 20, 24, 28, 32, 36

8. 7
0, 7, 14, 21, 28, 35, 42, 49, 56, 63

9. 100
0, 100, 200, 300, 400, 500, 600, 700, 800, 900

Find the prime factorization of each number.

10. 55 $55 = 5 \cdot 11$

11. 65 $65 = 5 \cdot 13$

12. 78 $78 = 2 \cdot 3 \cdot 13$

13 140 $140 = 2 \cdot 2 \cdot 5 \cdot 7$

14. 169 $169 = 13 \cdot 13$

15. 252 $252 = 2 \cdot 2 \cdot 3 \cdot 3 \cdot 7$

16. 294 $294 = 2 \cdot 3 \cdot 7 \cdot 7$

17. 429 $429 = 3 \cdot 11 \cdot 13$

18. 693 $693 = 3 \cdot 3 \cdot 7 \cdot 11$

19. 1300 $1300 = 2 \cdot 2 \cdot 5 \cdot 5 \cdot 13$

20. 3675 $3675 = 3 \cdot 5 \cdot 5 \cdot 7 \cdot 7$

21. 5100 $5100 = 2 \cdot 2 \cdot 3 \cdot 5 \cdot 5 \cdot 17$

Determine whether each number is prime or composite.

22. 13   **prime**

23. 27   **composite**

24. 31   **prime**

25. 49   **composite**

26. 51   **composite**

27. 97   **prime**

Use the following numbers to answer Exercises 28–31.

120, 126, 240, 147, 130, 104, 70, 135, 72, 110, 88, 4125

28. Which numbers are divisible by 2?
120, 126, 240, 130, 104, 70, 72, 110, 88

29. Which numbers are divisible by 3?
120, 126, 240, 147, 135, 72, 4125

30. Which numbers are divisible by 5?
120, 240, 130, 70, 135, 110, 4125

31. Which numbers are divisible by 10?
120, 240, 130, 70, 110

FOR REVIEW

Solve.

32. In a particular organization, there are 5 times as many women as men. If there are 465 women, how many men are there?   **93 men**

33. Burford purchased a TV from his roommate for $614. He paid $50 down and the rest in equal monthly payments over a period of one year. If no interest was charged, what were his monthly payments?   **$47 per month**

34. A hall measures 2 yd wide by 10 yd long, and a living room measures 12 yd long by 7 yd wide. If it costs $20 a square yard for carpeting, how much will it cost to carpet the hall and living room?
$2080

35. The Connors have a second mortgage which requires a monthly payment of $200 for 12 years.

   (a) How many monthly payments must they make over the 12-year period?

   (b) How much will they have paid in total at the end of 12 years?

   (a) 144 (b) $28,800

Exercises 36–38 will help you prepare for the next section. Find each product.

36. $2 \cdot 2 \cdot 2 \cdot 2 \cdot 2$ 32

37. $3 \cdot 3 \cdot 3 \cdot 3$ 81

38. $5 \cdot 5$ 25

ANSWERS: 1. true   2. false (prime number)   3. true   4. 1, 2, 3, 6, 9, 18   5. 1, 2, 4, 7, 14, 28   6. 1, 2, 4, 7, 8, 14, 28, 56   7. 0, 4, 8, 12, 16, 20, 24, 28, 32, 36   8. 0, 7, 14, 21, 28, 35, 42, 49, 56, 63   9. 0, 100, 200, 300, 400, 500, 600, 700, 800, 900   10. $55 = 5 \cdot 11$ 11. $65 = 5 \cdot 13$ 12. $78 = 2 \cdot 3 \cdot 13$ 13. $140 = 2 \cdot 2 \cdot 5 \cdot 7$ 14. $169 = 13 \cdot 13$ 15. $252 = 2 \cdot 2 \cdot 3 \cdot 3 \cdot 7$ 16. $294 = 2 \cdot 3 \cdot 7 \cdot 7$ 17. $429 = 3 \cdot 11 \cdot 13$ 18. $693 = 3 \cdot 3 \cdot 7 \cdot 11$ 19. $1300 = 2 \cdot 2 \cdot 5 \cdot 5 \cdot 13$ 20. $3675 = 3 \cdot 5 \cdot 5 \cdot 7 \cdot 7$ 21. $5100 = 2 \cdot 2 \cdot 3 \cdot 5 \cdot 5 \cdot 17$ 22. prime   23. composite   24. prime   25. composite   26. composite   27. prime   28. 120, 126, 240, 130, 104, 70, 72, 110, 88   29. 120, 126, 240, 147, 135, 72, 4125   30. 120, 240, 130, 70, 135, 110, 4125   31. 120, 240, 130, 70, 110   32. 93 men   33. $47 per month   34. $2080   35. (a) 144 (b) $28,800   36. 32   37. 81   38. 25

2.7 EXERCISES B

Answer true or false. If the answer is false, explain why.

1. Since $12 = 2 \times 6$, 2 and 6 are divisors of 12.   **true**

2. The multiples of 2 are odd numbers.   **false (even numbers)**

3. The only natural number which is neither prime nor composite is 1.   **true**

Find all divisors of each number.

4. 21
   1, 3, 7, 21

5. 36
   1, 2, 3, 4, 6, 9, 12, 18, 36

6. 54
   1, 2, 3, 6, 9, 18, 27, 54

Find the first ten multiples of each number.

7. 3
    0, 3, 6, 9, 12, 15, 18, 21, 24, 27

8. 6
    0, 6, 12, 18, 24, 30, 36, 42, 48, 54

9. 1000
    0, 1000, 2000, 3000, 4000, 5000, 6000, 7000, 8000, 9000

Find the prime factorization of each number.

10. 66 $66 = 2 \cdot 3 \cdot 11$

11. 82 $82 = 2 \cdot 41$

12. 96 $96 = 2 \cdot 2 \cdot 2 \cdot 2 \cdot 2 \cdot 3$

13. 135 $135 = 3 \cdot 3 \cdot 3 \cdot 5$

14. 273 $273 = 3 \cdot 7 \cdot 13$

15. 420 $420 = 2 \cdot 2 \cdot 3 \cdot 5 \cdot 7$

16. 630 $630 = 2 \cdot 3 \cdot 3 \cdot 5 \cdot 7$

17. 945 $945 = 3 \cdot 3 \cdot 3 \cdot 5 \cdot 7$

18. 1750 $1750 = 2 \cdot 5 \cdot 5 \cdot 5 \cdot 7$

19. 1050 $1050 = 2 \cdot 3 \cdot 5 \cdot 5 \cdot 7$

20. 5733 $5733 = 3 \cdot 3 \cdot 7 \cdot 7 \cdot 13$

21. 5610 $5610 = 2 \cdot 3 \cdot 5 \cdot 11 \cdot 17$

Determine whether each number is prime or composite.

22. 17   prime

23. 26   composite

24. 31   prime

25. 55   composite

26. 63   composite

27. 213   composite

Use the following numbers to answer Exercises 28–31.

815, 150, 156, 42, 784, 125, 343, 90, 210, 66, 56, 3280

28. Which numbers are divisible by 2?
    150, 156, 42, 784, 90, 210, 66, 56, 3280

29. Which numbers are divisible by 3?
    150, 156, 42, 90, 210, 66

30. Which numbers are divisible by 5?
    815, 150, 125, 90, 210, 3280

31. Which numbers are divisible by 10?
    150, 90, 210, 3280

FOR REVIEW

Solve.

32. A map has a scale of 35 miles to the inch. If two cities are actually 280 miles apart, how far apart are they on the map?   **8 inches**

33. A man has only $20 bills in his wallet. If he buys 3 pairs of shoes at $42 each, how many bills must he use? How much change will he receive?
7 bills with $14 in change

34. An auditorium is divided into three sections. There are 20 rows, each with 12 seats, in the first section. The second section consists of 30 rows, each with 18 seats. The third section has the same number of seats as the first section. How many seats are in the auditorium?   **1020**

35. A student pays $120 a month for rent and $75 a month for food during the 9-month school year.
(a) How much does she spend on rent?
(b) How much does she spend on food?
(c) How much does she spend on rent and food together?
(a) $1080 (b) $675 (c) $1755

Exercises 36–38 will help you prepare for the next section. Find each product.

36. $3 \cdot 3 \cdot 3$ 27

37. $2 \cdot 2 \cdot 2 \cdot 2 \cdot 2 \cdot 2 \cdot 2$ 128

38. $5 \cdot 5 \cdot 5 \cdot 5$ 625

2.7 EXERCISES C

Find the prime factorization of each number.

1. 13,650 $2 \cdot 3 \cdot 5 \cdot 5 \cdot 7 \cdot 13$

2. 17,199 $3 \cdot 3 \cdot 3 \cdot 7 \cdot 7 \cdot 13$

3. 39,270
[Answer: $2 \cdot 3 \cdot 5 \cdot 7 \cdot 11 \cdot 17$]

2.8 POWERS, ROOTS, AND ORDER OF OPERATIONS

=== STUDENT GUIDEPOSTS ===

1 Powers of a Number

3 Order of Operations

2 Perfect Squares and Square Roots

4 Evaluating Using Parentheses

1 POWERS OF A NUMBER

Often we are asked to multiply the same number by itself several times. For example, the number 64 expressed as a product of primes is

$$64 = 2 \cdot 2 \cdot 2 \cdot 2 \cdot 2 \cdot 2.$$

Such expanded products can be abbreviated by using *exponents*. We write

$$2^6 \quad \text{for} \quad \underbrace{2 \cdot 2 \cdot 2 \cdot 2 \cdot 2 \cdot 2}_{\text{six 2's}}.$$

base   exponent   six 2's

The expression 2^6, read ''2 to the sixth,'' is called the **sixth power** of 2; 6 is an **exponent,** and 2 is the **base.** An exponent tells how many times a base is used as a factor. Suppose we find the first five powers of 2.

The **first power** of 2 is 2^1 or 2.
The **second power** of 2, also read ''two **squared,**'' is $2^2 = 2 \cdot 2 = 4$.
The **third power** of 2, also read ''two **cubed,**'' is $2^3 = 2 \cdot 2 \cdot 2 = 8$.
The **fourth power** of 2 is $2^4 = 2 \cdot 2 \cdot 2 \cdot 2 = 16$.
The **fifth power** of 2 is $2^5 = 2 \cdot 2 \cdot 2 \cdot 2 \cdot 2 = 32$.

EXAMPLE 1 EVALUATING NUMBERS HAVING EXPONENTS

Find the value by writing without exponents.

(a) 5^3 We have $5^3 = 5 \cdot 5 \cdot 5 = 125$.

(b) 4^5 We have $4^5 = 4 \cdot 4 \cdot 4 \cdot 4 \cdot 4 = 1024$.

PRACTICE EXERCISE 1

Find the value by writing without exponents.

(a) 3^7

(b) 7^3

Answers: (a) 2187   (b) 343

EXAMPLE 2 WRITING NUMBERS USING EXPONENTS

Write each product using exponents.

(a) $6 \cdot 6 \cdot 6 \cdot 6$ Since there are four 6's, the base is 6 and the exponent is 4. Thus, $6 \cdot 6 \cdot 6 \cdot 6 = 6^4$.

(b) $10 \cdot 10 \cdot 10$ There are three 10's, so we have 10^3.

PRACTICE EXERCISE 2

Write each product using exponents.

(a) $5 \cdot 5 \cdot 5 \cdot 5 \cdot 5 \cdot 5$

(b) $100 \cdot 100 \cdot 100 \cdot 100$

Answers: (a) 5^6 (b) 100^4

EXAMPLE 3 PRIME FACTORIZATION USING EXPONENTS

Write the prime factorization for each number using exponents.

(a) 100   Using a factor tree, we have $100 = 2 \cdot 2 \cdot 5 \cdot 5$. Using exponents, $2 \cdot 2$ becomes 2^2 and $5 \cdot 5$ becomes 5^2. Thus, $100 = 2^2 \cdot 5^2$.

(b) 540   Using a factor tree, we have $540 = 2 \cdot 2 \cdot 3 \cdot 3 \cdot 3 \cdot 5$. Using exponents, $540 = 2^2 \cdot 3^3 \cdot 5$.

PRACTICE EXERCISE 3

Write the prime factorization for each number using exponents.

(a) 1323

(b) 2200

Answers: (a) $3^3 \cdot 7^2$
(b) $2^3 \cdot 5^2 \cdot 11$

EXAMPLE 4 EVALUATING PRODUCTS HAVING EXPONENTS

Find the value of each product.

(a) $2^5 \cdot 5^2 = \underbrace{2 \cdot 2 \cdot 2 \cdot 2 \cdot 2} \cdot \underbrace{5 \cdot 5}$

$\qquad = \quad\quad 32 \quad\quad \cdot \; 25$

$\qquad = 800$

(b) $2 \cdot 3^4 \cdot 7^2 = \underbrace{2} \cdot \underbrace{3 \cdot 3 \cdot 3 \cdot 3} \cdot \underbrace{7 \cdot 7}$

$\qquad\qquad = 2 \cdot \quad\quad 81 \quad\quad \cdot \; 49$

$\qquad\qquad = 7938$

PRACTICE EXERCISE 4

Find the value of each product.

(a) $3^3 \cdot 5^3$

(b) $2^3 \cdot 3 \cdot 7^2 \cdot 13$

Answers: (a) 3375   (b) 15,288

❷ PERFECT SQUARES AND SQUARE ROOTS

The opposite of finding powers is taking *roots*. For example, we know that the square of 3 is 9, $3^2 = 9$. We call 3 a *square root* of 9, and 9 is called a *perfect square*. That is,

$$3^2 = 9.$$

9 is the square of 3, making 9 a perfect square

3 is a square root of 9

Perfect Squares and Square Roots

When a whole number is squared, the result is a **perfect square**. The number which is being squared is a **square root** of the perfect square.

EXAMPLE 5 FINDING SQUARE ROOTS

Find the square root of each number.

(a) 25   Since $25 = 5^2$, 5 is a square root of 25.

(b) 64   Since $64 = 8^2$, 8 is a square root of 64.

(c) 0   Since $0 = 0^2$, 0 is a square root of 0.

PRACTICE EXERCISE 5

Find the square root of each number.

(a) 16

(b) 81

(c) 1

Answers: (a) 4   (b) 9   (c) 1

HINT

Remember that to find the square root of a number, you must ask:

"What number squared is equal to the given number?"

The symbol $\sqrt{}$, called a **radical,** means "square root." For example, $\sqrt{25}$ is read "the square root of 25." Similarly, $\sqrt{64}$ is read "the square root of 64." By Example 5 we see that

$$\sqrt{25} = 5 \quad \text{and} \quad \sqrt{64} = 8.$$

CAUTION

Do not confuse the terms *square* and *square root.* For example,

$$16 = 4^2.$$

4 is a *square root* of 16

16 is the *square* of 4

EXAMPLE 6   FINDING SQUARE ROOTS

Find each square root.

(a) $\sqrt{9}$ Since $9 = 3^2$, $\sqrt{9} = 3$.

(b) $\sqrt{100}$ Since $100 = 10^2$, $\sqrt{100} = 10$.

PRACTICE EXERCISE 6

Find each square root.

(a) $\sqrt{36}$

(b) $\sqrt{144}$

Answers: (a) 6 (not 6^2)   (b) 12

CAUTION

A common mistake is to say that $\sqrt{9} = 3^2$. Notice that $3^2 = 9$ so that $\sqrt{9}$ is just 3, *not* 3^2.

The first sixteen perfect squares and their whole number square roots are listed in the following table.

Perfect square	Whole number square root
0	$\sqrt{0} = 0$
1	$\sqrt{1} = 1$
4	$\sqrt{4} = 2$
9	$\sqrt{9} = 3$
16	$\sqrt{16} = 4$
25	$\sqrt{25} = 5$
36	$\sqrt{36} = 6$
49	$\sqrt{49} = 7$
64	$\sqrt{64} = 8$
81	$\sqrt{81} = 9$
100	$\sqrt{100} = 10$
121	$\sqrt{121} = 11$
144	$\sqrt{144} = 12$
169	$\sqrt{169} = 13$
196	$\sqrt{196} = 14$
225	$\sqrt{225} = 15$

③ ORDER OF OPERATIONS

Some problems involve several operations. For example, when we checked a division problem such as

$$\begin{array}{r} 4 \\ 6\overline{)25} \\ \underline{24} \\ 1 \end{array}$$

we wrote $(6 \times 4) + 1 = 24 + 1 = 25$. The parentheses show that 6 and 4 are multiplied first, and then the product is added to 1. If we had omitted the parentheses and written

$$6 \times 4 + 1,$$

there could be some confusion about the *order* of operations. Should we multiply 6×4 first and then add 1, or should we add $4 + 1$ first, and then multiply the sum by 6 to obtain $6 \cdot 5 = 30$? The parentheses were necessary to avoid this confusion. Another way to eliminate this problem is to agree on a definite **order of operations** when addition, subtraction, multiplication, division, finding powers, and taking roots occur in a problem.

Order of Operations
1. Find all powers and square roots, in any order, first.
2. Multiply and divide, in order, from left to right.
3. Add and subtract, in order, from left to right.

EXAMPLE 7 USING ORDER OF OPERATIONS

Perform the indicated operations.

(a) $2 + 3 \cdot 4$

$$
\begin{aligned}
2 + \boxed{3 \cdot 4} &= 2 + \boxed{12} \quad &\text{Multiply first} \\
&= 14 \quad &\text{Then add}
\end{aligned}
$$

(b) $5 \cdot 6 - 12 \div 4$

$$
\begin{aligned}
\boxed{5 \cdot 6} - \boxed{12 \div 4} &= 30 - 3 \quad &\text{Multiply and divide first} \\
&= 27 \quad &\text{Then subtract}
\end{aligned}
$$

(c) $\sqrt{64} - 2^3$

$$
\begin{aligned}
\boxed{\sqrt{64}} - \boxed{2^3} &= 8 - 8 \quad &\text{Take root and cube first} \\
&= 0 \quad &\text{Then subtract}
\end{aligned}
$$

(d) $4^2 \div 2 + \sqrt{9} - 1$

$$
\begin{aligned}
\boxed{4^2} \div 2 + \boxed{\sqrt{9}} - 1 &= \boxed{16} \div 2 + 3 - 1 \quad &\text{Find power and root first} \\
&= \boxed{8} + 3 - 1 \quad &\text{Divide next} \\
&= 11 - 1 \quad &\text{Add next} \\
&= 10 \quad &\text{Subtract last}
\end{aligned}
$$

PRACTICE EXERCISE 7

Perform the indicated operations.

(a) $5 \cdot 7 - 10$

(b) $18 \div 3 + 2 \cdot 4$

(c) $5^2 - \sqrt{36}$

(d) $\sqrt{81} + 6^2 \div 9 - 2$

Answers: (a) 25   (b) 14   (c) 19 (d) 11

④ EVALUATING USING PARENTHESES

When parentheses are used, they affect the order of operations.

> **Evaluating Using Parentheses**
>
> Always operate within parentheses first. Then follow the standard order of operations.

EXAMPLE 8 OPERATIONS INVOLVING PARENTHESES

Perform the indicated operations.

(a) $(5 - 1) \cdot 3 + 7$

$$(5 - 1) \cdot 3 + 7 = 4 \cdot 3 + 7 \qquad \text{Subtract inside parentheses first}$$
$$= 12 + 7 \qquad \text{Multiply before adding}$$
$$= 19$$

(b) $\sqrt{9}(6 \div 2) + 5^2$

$$\sqrt{9}(6 \div 2) + 5^2 = \sqrt{9}(3) + 5^2 \qquad \text{Divide inside the parentheses first}$$
$$= 3 \cdot 3 + 25 \qquad \text{Take root and square next}$$
$$= 9 + 25 \qquad \text{Multiply before adding}$$
$$= 34$$

(c) $(3 \cdot 5)^2$

$$(3 \cdot 5)^2 = (15)^2 \qquad \text{Multiply inside parentheses first}$$
$$= 225 \qquad \text{Then square}$$

(d) $3 \cdot 5^2$

$$3 \cdot 5^2 = 3 \cdot 25 \qquad \text{Square first}$$
$$= 75 \qquad \text{Then multiply}$$

(e) $(2 + 3)^3$

$$(2 + 3)^3 = (5)^3 \qquad \text{Add inside parentheses first}$$
$$= 125 \qquad \text{Then cube}$$

(f) $2^3 + 3^3$

$$2^3 + 3^3 = 8 + 27 \qquad \text{Cube first}$$
$$= 35 \qquad \text{Then add}$$

PRACTICE EXERCISE 8

Perform the indicated operations.

(a) $2(6 + 1) - 3$

(b) $3^2 - \sqrt{4}(10 \div 5)$

(c) $(2 \cdot 7)^2$

(d) $2 \cdot 7^2$

(e) $(1 + 4)^3$

(f) $1^3 + 4^3$

Answers: (a) 11 (b) 5 (c) 196
(d) 98 (e) 125 (f) 65

CAUTION

Parts **(c)** and **(d)** in Example 8 show that

$$3 \cdot 5^2 \neq (3 \cdot 5)^2,$$

and parts **(e)** and **(f)** show that

$$(2 + 3)^3 \neq 2^3 + 3^3.$$

A common mistake when working with exponents is to assume that expressions like those above are equal. Remember to evaluate within parentheses before finding a power.

2.8 EXERCISES A

1. The number 3^4 is given. **(a)** What is the number 4 called? **(b)** What is the number 3 called? **(c)** What power of 3 is this?   (a) exponent (b) base (c) fourth

2. What is the second power of a number called?   the square of the number

3. When parentheses are not used, which operation is performed first, addition or multiplication?   multiplication

Find the value by writing without exponents.

4. 2^8 256

5. 10^4 10,000

6. 8^3 512

Write each product using exponents.

7. $3 \cdot 3 \cdot 3 \cdot 3$ 3^4

8. $9 \cdot 9 \cdot 9 \cdot 9 \cdot 9$ 9^5

9. $4 \cdot 4 \cdot 4 \cdot 4 \cdot 4 \cdot 4$ 4^6

Find the prime factorization of each number using exponents.

10. 252 $2^2 \cdot 3^2 \cdot 7$

11. 2646 $2 \cdot 3^3 \cdot 7^2$

12 4400 $2^4 \cdot 5^2 \cdot 11$

Evaluate each product without using exponents.

13. $2 \cdot 3^3 \cdot 11$ 594

14. $3^2 \cdot 5^2 \cdot 7^2$ 11,025

15. $3 \cdot 5^3 \cdot 13^2$ 63,375

Find the square root of each number.

16. 49   7

17. 100   10

18. 121   11

Find the square of each number.

19. 49   2401

20. 100   10,000

21. 121   14,641

Find each square root.

22. $\sqrt{16}$ 4

23. $\sqrt{1}$ 1

24. $\sqrt{169}$ 13

Perform the indicated operations.

25. $3 + 9 - 5$ 7

26. $14 \div 7 - 1$ 1

27. $4 - 8 \div 2 + 1$ 1

28. $7 + 2 \cdot 3 - 4$ 9

29. $36 \div 6 - 3 \cdot 2$ 0

30. $5 + 2^2$ 9

31. $\sqrt{25} - 2^2$ 1

32 $5^2 - \sqrt{49}$ 18

33. $3^2 + 5 - \sqrt{4} + 1$ 13

34 $\sqrt{25} + 15 - 2^2 \cdot 5$ 0

35. $2^5 - \sqrt{100} + 7$ 29

36. $2^3 \div \sqrt{16} + 3$ 5

37. $(9 - 2) \cdot 3 + 4$ 25

38. $(2 + 3) \div 5 - 0$ 1

39. $6 + 2(5 - 3)$ 10

40. $(5 + 1) \cdot (4 - 2) - 3$ 9

41 $\sqrt{4}(9 \div 3) + 2^3$ 14

42. $2(\sqrt{4} \div 2) - 2$ 0

43 $(3 \cdot 7)^2$ 441   **44** $3 \cdot 7^2$ 147   **45.** $3^2 \cdot 7^2$ 441

46 $(3 + 7)^2$ 100   **47** $3^2 + 7^2$ 58   **48.** $2(3 + 7)^2$ 200

FOR REVIEW

49. Find all the divisors of 68.   1, 2, 4, 17, 34, 68   **50.** Is 85 prime or composite?   composite

51. Is 38 divisible by 2?   yes   **52.** Is 38 divisible by 3?   no

53. Is 425 divisible by 5?   yes   **54.** Is 425 divisible by 10?   no

55. Find the prime factorization of 1650. $2 \cdot 3 \cdot 5 \cdot 5 \cdot 11$

ANSWERS: 1. (a) exponent (b) base (c) fourth   2. the square of the number   3. multiplication   4. 256   5. 10,000
6. 512   7. 3^4 8. 9^5 9. 4^6 10. $2^2 \cdot 3^2 \cdot 7$ 11. $2 \cdot 3^3 \cdot 7^2$ 12. $2^4 \cdot 5^2 \cdot 11$ 13. 594   14. 11,025   15. 63,375   16. 7
17. 10   18. 11   19. 2401   20. 10,000   21. 14,641   22. 4   23. 1   24. 13   25. 7   26. 1   27. 1   28. 9   29. 0
30. 9   31. 1   32. 18   33. 13   34. 0   35. 29   36. 5   37. 25   38. 1   39. 10   40. 9   41. 14   42. 0   43. 441
44. 147   45. 441   46. 100   47. 58   48. 200   49. 1, 2, 4, 17, 34, 68   50. composite ($85 = 5 \cdot 17$)   51. yes   52. no
53. yes   54. no   55. $2 \cdot 3 \cdot 5 \cdot 5 \cdot 11$

2.8   EXERCISES B

1. The number 5^4 is given. **(a)** What is the number 5 called? **(b)** What is the number 4 called? **(c)** What power of 5 is this?   (a) base (b) exponent (c) fourth

2. What is the third power of a number called?   the cube of the number

3. When parentheses are not used, which operation is performed first, subtraction or division?   division

Find the value by writing without exponents.

4. 3^5 243   **5.** 5^3 125   **6.** 10^5 100,000

Write each product using exponents.

7. $2 \cdot 2 \cdot 2$ 2^3 **8.** $5 \cdot 5 \cdot 5 \cdot 5 \cdot 5 \cdot 5 \cdot 5$ 5^7 **9.** $8 \cdot 8 \cdot 8 \cdot 8$ 8^4

Find the prime factorization of each number using exponents.

10. 270 $2 \cdot 3^3 \cdot 5$ **11.** 1008 $2^4 \cdot 3^2 \cdot 7$ **12.** 11,880 $2^3 \cdot 3^3 \cdot 5 \cdot 11$

Evaluate each product without using exponents.

13. $2^2 \cdot 3 \cdot 13$ 156   **14.** $3^2 \cdot 5^3 \cdot 7$ 7875   **15.** $3 \cdot 7^2 \cdot 13^2$ 24,843

Find the square root of each number.

16. 36   6   **17.** 81   9   **18.** 144   12

Find the square of each number.

19. 36   1296

20. 81   6561

21. 144   20,736

Find each square root.

22. $\sqrt{64}$ 8

23. $\sqrt{169}$ 13

24. $\sqrt{225}$ 15

Perform the indicated operations.

25. $5 + 8 - 3$ 10

26. $18 \div 9 - 1$ 1

27. $6 - 12 \div 4 + 7$ 10

28. $9 + 3 \cdot 5 - 7$ 17

29. $49 \div 7 - 2 \cdot 2$ 3

30. $\sqrt{25} + 75$ 80

31. $\sqrt{81} - 2^3$ 1

32. $6^2 - \sqrt{121}$ 25

33. $4^2 + 3 - \sqrt{9} + 2$ 18

34. $\sqrt{49} + 5 - 2^2 \cdot 3$ 0

35. $2^6 - \sqrt{144} + 5$ 57

36. $3^2 \div \sqrt{9} + 1$ 4

37. $(8 - 4) \cdot 2 + 6$ 14

38. $(3 + 1) \div 2 - 0$ 2

39. $5 + 2(7 - 3)$ 13

40. $(2 + 3) \cdot (6 - 1) - 2$ 23

41. $\sqrt{25}(6 \div 2) + 3^2$ 24

42. $3(\sqrt{16} \div 2) - 6$ 0

43. $(2 \cdot 11)^2$ 484

44. $2 \cdot 11^2$ 242

45. $2^2 \cdot 11^2$ 484

46. $(2 + 6)^2$ 64

47. $2^2 + 6^2$ 40

48. $3(2 + 6)^2$ 192

FOR REVIEW

49. Find all divisors of 80.
    1, 2, 4, 5, 8, 10, 16, 20, 40, 80

50. Is 83 prime or composite?   prime

51. Is 147 divisible by 2?   no

52. Is 147 divisible by 3?   yes

53. Is 147 divisible by 5?   no

54. Is 2350 divisible by 10?   yes

55. Find the prime factorization of 4875. $3 \cdot 5 \cdot 5 \cdot 5 \cdot 13$

2.8 EXERCISES C

Perform the indicated operations.

1. $\sqrt{64}(6 \div 2 - 1) - 8 \div 4$ 14

2. $9 \div 3 \cdot 5 - 8 \div 2 + \sqrt{4}$
[Answer: 13]

3. $\sqrt{1 + 8} \cdot \sqrt{6 - 2} + (5 - 2)^2 \div 3$
[Answer: 9]

4. $(4^2 - 3^2) \div 7 + (\sqrt{25} - \sqrt{4})$ 4

CHAPTER 2 REVIEW

KEY WORDS

2.1 In a **multiplication** problem, such as $2 \times 4 = 8$, $\times$ is called the **multiplication sign** or **times sign**, 2 and 4 are **factors**, 2 is the **multiplier**, 4 is the **multiplicand**, and 8 is the **product**.

A **multiple** of a number is the product of the number and a whole number.

2.4 In a division problem, such as $12 \div 3 = 4$, $\div$ is called the **division symbol**, 12 is the **dividend**, 3 is the **divisor**, and 4 is the **quotient**.

The final difference in a division problem is called the **remainder**, a number that is less than the divisor.

An **algorithm** is a method or procedure used to perform a certain task.

2.7 One number is a **divisor** of a second if the second is a multiple of the first.

A **prime number** is a natural number that has exactly two different divisors, 1 and itself.

A **composite number** is a natural number greater than 1 that is not prime.

An **even number** is divisible by 2. Remember that 0 is an even number.

A number that is not even is called an **odd number.**

2.8 In the expression 3^5, 5 is called the **exponent**, 3 is called the **base**, and 3^5 is called the **fifth power** of 3.

The second power of a number is called the **square** of the number.

The third power of a number is called the **cube** of the number.

When a whole number is squared, the result is called a **perfect square** and the number being squared is a **square root** of the perfect square.

The symbol $\sqrt{}$ is called a **radical** and is used to indicate square roots.

KEY CONCEPTS

2.1 When two whole numbers are multiplied, the multiplier shows the number of times the multiplicand is to be added. For example,

$$4 \times 3 = 3 + 3 + 3 + 3.$$

four 3's in the sum

2. By the commutative law, changing the order of multiplication does not change the product. For example,

$$2 \times 5 = 5 \times 2.$$

3. The number 1 is the multiplicative identity since

$$1 \times (\text{any number}) = \text{that number.}$$

4. Any whole number times 0 is 0.

5. Any whole number times 1 is that whole number.

2.2 1. When multiplying a many-digit number by a one-digit number, write down the ones digit and add the tens digit to the next product.

2. When multiplying by a many-digit number, remember to use zeros as placeholders or else to move each new product over one more place to the left.

3. To check a multiplication problem, interchange the multiplier and the multiplicand and find the new product (the reverse-order check).

2.3 1. By the associative law, regrouping the factors in a multiplication problem does not change the product. For example,

$$(2 \times 3) \times 5 = 2 \times (3 \times 5).$$

2. The distributive law illustrates an important property relating multiplication and addition. For example,

$$2(3 + 5) = (2 \times 3) + (2 \times 5).$$

3. To multiply a number by 10, attach one 0 at the end of the number. To multiply by 100, attach two 0's, and to multiply by 1000, attach three 0's.

4. Products can be estimated by rounding the multiplier and multiplicand and finding the product of the results.

2.4 1. A related multiplication sentence for

$$10 \div 2 = 5 \quad \text{is} \quad 2 \times 5 = 10.$$

2. Any whole number, except 0, divided by itself is 1.

3. Any whole number divided by 1 is that whole number.

4. Division by zero is undefined.

5. Zero divided by any whole number, except 0, is 0.

2.5 A division problem can be checked by using (quotient × divisor) + remainder = dividend.

2.6 **1.** (total value) = (value per part) · (number of parts)

2. The area of a rectangle is given by the formula

$$\text{Area} = (\text{length}) \cdot (\text{width}).$$

3. Miles per gallon (mpg) is found by

$$\text{mpg} = \frac{\text{number of miles driven}}{\text{number of gallons used}}.$$

2.7 Each composite number can be written as a product of primes called the prime factorization of the number. A factor tree is useful in finding these primes.

2.8 **1.** Do not confuse the terms square and square root. For example,

4 is the **square root** of 16

16 is the **square** of 4.

2. Operations should be performed in the following order.
(a) Operate within parentheses.
(b) Evaluate powers and roots.
(c) Multiply and divide from left to right.
(d) Add and subtract from left to right.

REVIEW EXERCISES

Part I

2.1 **1.** Consider the multiplication problem 7 × 9 = 63. **(a)** What is 7 called? **(b)** What is 9 called? **(c)** What is 63 called? **(a) multiplier (or factor) (b) multiplicand (or factor) (c) product**

2. The fact that 4 × 3 = 3 × 4 illustrates what law of multiplication? **commutative law**

3. Any number multiplied by 1 always gives what number for the product? **that number**

4. The numbers 0, 4, 8, 12, 16, and 20 are the first six multiples of what number? **4**

2.2 *Find the following products and check using the reverse-order check.*

5. 347 × 5 = 1735 **6.** 6007 × 6 = 36,042 **7.** 4398 × 7 = 30,786 **8.** 41 × 23 = 943 **9.** 79 × 65 = 5135

10. 723 × 31 = 22,413 **11.** 479 × 628 = 300,812 **12.** 848 × 207 = 175,536 **13.** 4039 × 705 = 2,847,495 **14.** 6493 × 3000 = 19,479,000

2.3 **15.** The fact that (4 × 7) × 9 = 4 × (7 × 9) illustrates what law of multiplication? **associative law**

16. The fact that 3 × (4 + 1) = (3 × 4) + (3 × 1) illustrates what law of multiplication and addition? **distributive law**

Check by estimating the product.

17. 86 × 31: 86, 258, 2666; 90 · 30 = 2700; appears to be correct

18. 307 × 513: 921, 307, 1835, 187491; 300 · 500 = 150,000; appears to be incorrect (correct product is 157,491).

2.4 **19.** Consider the division problem $12 \div 4 = 3$.

    (a) What is 12 called?   dividend

    (b) What is 4 called?   divisor

    (c) What is 3 called?   quotient

20. What is the related multiplication sentence for $12 \div 4 = 3$? $4 \times 3 = 12$

Complete the following.

21. $5 \div 5 = \underline{1}$ **22.** $5 \div 1 = \underline{5}$ **23.** $0 \div 5 = \underline{0}$ **24.** $5 \div 0 = \underline{\text{undefined}}$

2.5 *Find each quotient and check your work.*

    25. $4\overline{)292}$ Q: 73; R: 0   **26.** $7\overline{)6531}$ Q: 933; R: 0   **27.** $6\overline{)2793}$ Q: 465; R: 3

    28. $62\overline{)806}$ Q: 13; R: 0   **29.** $47\overline{)1252}$ Q: 26; R: 30   **30.** $83\overline{)10375}$ Q: 125; R: 0

2.6 *Solve.*

31. Mr. Kwan buys a new car for $8420. If he pays $500 down and the balance in 36 equal monthly payments, how much will he pay each month? (Do not worry about interest at this time.)
$220

32. A nine-man softball team wins a tournament and receives a prize of $1000. They agree to split the prize money evenly (in dollars) with the remainder to be given to the captain. How much does the captain receive? $112

33. Tickets for a rock concert, each selling for $8, went on sale at 9:00 A.M. At 1:00 the tickets were all sold. There were 3200 tickets sold the first hour, 2950 the second, 2400 the third, and 1500 the fourth.

    (a) How many tickets were sold?   10,050 tickets

    (b) What were the total receipts? $80,400

34. Dick wants to carpet his dining and living rooms with carpet selling for $18 a square yard. A sketch of the floor plan of the rooms is shown in the figure. How much will the carpet cost? [*Hint:* Draw a line dividing the area into two rectangles.] $1116

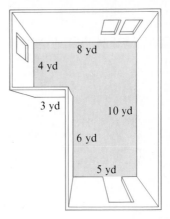

35. Assuming equal quality, which is the better buy, a 12-oz box of cat food for 84¢ or a 13-oz box of food for 78¢?

The 13-oz box for 78¢ is a better buy at 6¢ per ounce than the 12-oz box for 84¢ at 7¢ per ounce.

2.7 **36.** The numbers 2 and 5 are divisors or factors of 10 since $2 \times 5 = 10$. What is 10 relative to 2 and 5?   multiple

37. What do we call a natural number that has exactly two different divisors, 1 and itself?   prime

38. What is a number called if it is greater than 1 and not prime?   composite

39. What is the only natural number which is neither prime nor composite?   1

40. What do we call a natural number that is divisible by 2?   even

41. How many even prime numbers are there?   one (only the number 2)

42. Find all divisors of 105.
1, 3, 5, 7, 15, 21, 35, 105

43. Find the first five multiples of 13.
0, 13, 26, 39, 52

Find the prime factorization of each number.

44. 195 $195 = 3 \cdot 5 \cdot 13$

45. 112 $112 = 2 \cdot 2 \cdot 2 \cdot 2 \cdot 7$

46. 490 $490 = 2 \cdot 5 \cdot 7 \cdot 7$

47. 1089 $1089 = 3 \cdot 3 \cdot 11 \cdot 11$

48. 2730 $2730 = 2 \cdot 3 \cdot 5 \cdot 7 \cdot 13$

49. 3400 $3400 = 2 \cdot 2 \cdot 2 \cdot 5 \cdot 5 \cdot 17$

Use the numbers 140, 273, 105, *and* 150 *to answer Exercises 50–53.*

50. Which numbers are divisible by 2?
140, 150

51. Which numbers are divisible by 3?
273, 105, 150

52. Which numbers are divisible by 5?
140, 105, 150

53. Which numbers are divisible by 10?
140, 150

2.8 **54.** Is 50 a perfect square?   no

55. What is the square of 9?   81

56. What is the square root of 9?   3

57. Evaluate 6^3 without using an exponent.
$6 \cdot 6 \cdot 6 = 216$

58. Write $5 \cdot 5 \cdot 5 \cdot 5 \cdot 5 \cdot 5$ using an exponent.
5^6

59. Evaluate $2^3 \cdot 11^2$ without using exponents.
968

60. Find the prime factorization of 11,000 using exponents. $2^3 \cdot 5^3 \cdot 11$

Find each square root.

61. $\sqrt{81}$ 9

62. $\sqrt{144}$ 12

Perform the indicated operations.

63. $2 - 1 + 3$ 4

64. $3 - 8 \div 4 + 2$ 3

65. $\sqrt{36} - 2^2$ 2

66. $(7 - 3) \cdot 2 + 1$ 9

67. $(5 \cdot 4)^2$ 400

68. $5 \cdot 4^2$ 80

Part II

Solve.

69. A collection of 308 people are to be divided into subgroups of 11. First find the approximate number of people in each subgroup, and then find the exact number.
approximate number: 30 people; exact number: 28 people

70. An average of 10,990 people live in each of 11 counties. First estimate the number of people living in these counties; then find the actual number based on these figures.
estimated number: 110,000 people; exact number: 120,890 people

Perform the indicated operations.

71. 623
   $\times\ 8$
   ___
   4984

72.   52
   $\times\ 43$
   ___
   2236

73. $43\overline{)1204}$
   Q: 28; R: 0

74. $27\overline{)1515}$
   Q: 56; R: 3

75. $28 \div 1$ 28

76. $0 \div 28$ 0

77. $28 \div 28$ 1

78. $8 - 5 + 6 - 2$ 7

79. $\sqrt{81} - 2^2$ 5

80. $(4 - 2) \cdot 6 - 3$ 9

81. $3^2 \cdot 2^3$ 72

82. $(3 \cdot 2)^4$ 1296

83. What is the square of 16?   256

84. What is the square root of 16?   4

85. Is 4311 divisible by 3?   yes

86. Find the prime factorization of 3234.
   $2 \cdot 3 \cdot 7 \cdot 7 \cdot 11$

87. Check by estimating the product.
   91
   $\times\ 48$
   ___
   728
   364
   ___
   4368
   $90 \cdot 50 = 4500$; **appears to be correct**

88. Estimate the following product by rounding to the nearest hundred.
   708
   $\times\ 489$
   ___
   350,000

Solve.

89. The total enrollment in the four classes (freshmen, sophomores, juniors, and seniors) at State University is 8227. If there are 2436 freshmen, 2107 sophomores, and 1939 juniors, how many seniors are there in attendance?   **1745 seniors**

90. On Monday, the hens at the Chicken Farm laid 4942 eggs, and on Tuesday they laid 4802 eggs. How many 12-egg cartons will be needed to hold the eggs laid on these two days?
   812 cartons

ANSWERS: 1. (a) multiplier (or factor) (b) multiplicand (or factor) (c) product   2. commutative law   3. that number   4. 4   5. 1735   6. 36,042   7. 30,786   8. 943   9. 5135   10. 22,413   11. 300,812   12. 175,536   13. 2,847,495   14. 19,479,000   15. associative law   16. distributive law   17. $90 \cdot 30 = 2700$; appears to be correct   18. $300 \cdot 500 = 150,000$; appears to be incorrect (correct product is 157,491)   19. (a) dividend (b) divisor (c) quotient   20. $4 \times 3 = 12$ 21. 1   22. 5   23. 0   24. undefined   25. Q: 73; R: 0   26. Q: 933; R: 0   27. Q: 465; R: 3   28. Q: 13; R: 0   29. Q: 26; R: 30   30. Q: 125; R: 0   31. $220   32. $112   33. (a) 10,050 tickets (b) $80,400   34. $1116   35. The 13-oz box for 78¢ is a better buy at 6¢ per ounce than the 12-oz box for 84¢ at 7¢ per ounce.   36. multiple   37. prime   38. composite   39. 1   40. even   41. one (only the number 2)   42. 1, 3, 5, 7, 15, 21, 35, 105   43. 0, 13, 26, 39, 52   44. $195 = 3 \cdot 5 \cdot 13$ 45. $112 = 2 \cdot 2 \cdot 2 \cdot 2 \cdot 7$ 46. $490 = 2 \cdot 5 \cdot 7 \cdot 7$ 47. $1089 = 3 \cdot 3 \cdot 11 \cdot 11$ 48. $2730 = 2 \cdot 3 \cdot 5 \cdot 7 \cdot 13$ 49. $3400 = 2 \cdot 2 \cdot 2 \cdot 5 \cdot 5 \cdot 17$ 50. 140, 150   51. 273, 105, 150   52. 140, 105, 150   53. 140, 150   54. no   55. 81   56. 3   57. $6 \cdot 6 \cdot 6 = 216$ 58. 5^6 59. 968   60. $2^3 \cdot 5^3 \cdot 11$ 61. 9   62. 12   63. 4   64. 3   65. 2   66. 9   67. 400   68. 80   69. approximate number: 30 people; exact number: 28 people   70. estimated number: 110,000 people; exact number: 120,890 people   71. 4984   72. 2236   73. Q: 28; R: 0   74. Q: 56; R: 3   75. 28   76. 0   77. 1   78. 7   79. 5   80. 9   81. 72   82. 1296   83. 256   84. 4   85. yes   86. $2 \cdot 3 \cdot 7 \cdot 7 \cdot 11$ 87. $90 \cdot 50 = 4500$; appears to be correct   88. 350,000   89. 1745 seniors   90. 812 cartons

CHAPTER 2 TEST

1. What law of multiplication is illustrated by the following?

$$2 \times 9 = 9 \times 2$$

2. What do we call a natural number greater than 1 that is not a prime?

3. When is a number divisible by 2?

Find the products.

4. 329
    × 4

5. 4037
    × 209

6. 9695
    × 400

7. Estimate the product 6103×296 by using rounded numbers.

Complete the following.

8. $15 \div 15 = \underline{\quad ? \quad}$

9. $0 \div 15 = \underline{\quad ? \quad}$

10. $15 \div 1 = \underline{\quad ? \quad}$

11. $15 \div 0 = \underline{\quad ? \quad}$

Find the quotients.

12. $5\overline{)1180}$

13. $23\overline{)3556}$

14. $32\overline{)14668}$

1. _____commutative law_____

2. _____composite_____

3. _____when its ones digit is 0, 2, 4, 6, or 8_____

4. _____1316_____

5. _____843,733_____

6. _____3,878,000_____

7. _____1,800,000_____

8. _____1_____

9. _____0_____

10. _____15_____

11. _____undefined_____

12. _____Q: 236; R: 0_____

13. _____Q: 154; R: 14_____

14. _____Q: 458; R: 12_____

126

15. Find the prime factorization of 350.

15. _____ $2 \cdot 5 \cdot 5 \cdot 7$ or $2 \cdot 5^2 \cdot 7$ _____

16. Wanda buys 3 coats for $130 each, 6 blouses for $27 each, and 5 pairs of earrings for $14 each. How much has she spent?

16. _____ $622 _____

17. Dick wants to put carpeting costing $30 a square yard on the floor in his bedroom. If the room measures 5 yd long and 4 yd wide, how much will the carpeting cost?

17. _____ $600 _____

18. What is the square root of 25?

18. _____ 5 _____

19. What is the square of 25?

19. _____ 625 _____

Find each square root.

20. $\sqrt{64}$

20. _____ 8 _____

21. $\sqrt{25}$

21. _____ 5 _____

Perform the indicated operations.

22. $4 + 2 - 1$

22. _____ 5 _____

23. $(8 - 2) \div 3 + 7$

23. _____ 9 _____

24. $2 \cdot 3^3$

24. _____ 54 _____

25. $(2 \cdot 3)^3$

25. _____ 216 _____

26. $\sqrt{49} - 2^2$

26. _____ 3 _____

27. A car dealer has 51 cars on a lot, and each is valued at $11,150. Without finding the exact amount, estimate the total value of the cars on the lot.

27. _____ $550,000 _____

Multiplying and Dividing Fractions

3.1 FRACTIONS

1 FRACTIONS

In the first two chapters we looked at whole numbers. Whole numbers are used to name ''whole'' or entire quantities such as 3 (whole) miles, 15 (whole) dollars, or 1 (whole) day. At times, however, we may be interested in parts of a single quantity such as part of a mile, part of a dollar, or part of a day. For this, we use *fractions*.

Suppose we have a bar like the one in Figure 3.1.

Figure 3.1

This bar may be thought of as a piece of wood, a bar of butter, a bar of candy, or any similar quantity. If the bar is divided into two equal parts, each of the parts is called a **half** of the whole bar. See Figure 3.2. This quotient is a new number, called *one half,* and is written

$$\frac{1}{2}.$$

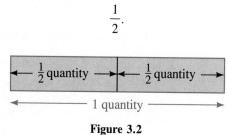

Figure 3.2

If we divide the bar into three equal parts as in Figure 3.3, each part is a **third** of the whole. The quotient

$$\frac{1}{3}$$

is a new number called *one third*. If the bar were a bar of candy to be split equally among 3 children, each child would receive $\frac{1}{3}$ bar. If there were 2 girls and 1 boy, the girls would receive 2 of the 3 equal parts, that is, *two thirds* of the bar. The shaded part in Figure 3.3 corresponds to $\frac{2}{3}$. The number $\frac{2}{3}$, read "2 over 3" or "2 divided by 3," is a quotient of two whole numbers, 2 and 3.

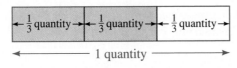

Figure 3.3

Any quotient of two whole numbers is called a **fraction.** The top number (the dividend) is called the **numerator** of the fraction, while the bottom number (the divisor) is called the **denominator.** In our example with the candy bar, we would have the following.

numerator $\longrightarrow$ 2 $\longleftarrow$ Number of parts the girls received
denominator $\longrightarrow$ 3 $\longleftarrow$ Bar divided into 3 equal parts

---

EXAMPLE 1   FRACTION OF A WHOLE

A pie is to be divided into equal parts and served to 5 boys and 1 girl. Suppose the pie is represented by a circle, as in Figure 3.4, which has been divided into 6 equal parts (5 + 1 = 6). Each part is a **sixth** of the pie, and each child receives $\frac{1}{6}$ of the pie. Since there are 5 boys, they will receive

numerator $\longrightarrow$ 5 $\longleftarrow$ Number of parts the boys receive
denominator $\longrightarrow$ 6 $\longleftarrow$ Number of equal parts of pie

of the pie. The shaded part of the pie corresponds to this fraction, while the unshaded part, $\frac{1}{6}$, corresponds to the part received by the girl.

Figure 3.4

EXAMPLE 2   FRACTION OF A WHOLE

The square in Figure 3.5 is divided into equal parts. What fraction of the square is shaded?

There are 4 equal parts with 3 of them shaded. Thus, the shaded fraction of the square is *three fourths,* written $\frac{3}{4}$.

PRACTICE EXERCISE 1

A man divided his land as shown in the figure. If his daughter got the shaded portion, what fraction of the land did she get?

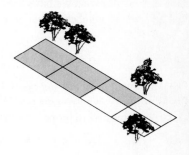

Answer: $\frac{5}{8}$

PRACTICE EXERCISE 2

The shaded part of the lot shown on the next page was used for a building. What fraction of the lot was used?

Figure 3.5

Answer: $\frac{1}{3}$

② PROPER AND IMPROPER FRACTIONS

All the fractions that we have considered thus far have been *proper fractions*. A **proper fraction** is one whose numerator is less than its denominator. For example,

$$\frac{1}{2}, \quad \frac{1}{3}, \quad \frac{2}{3}, \quad \frac{5}{6}, \quad \frac{3}{4}, \quad \frac{32}{67}, \quad \frac{101}{102}$$

are all proper fractions.

Some fractions are not proper fractions. For example, consider the bar in Figure 3.6 which is divided into five equal parts. All five parts are shaded. In this case, the fraction of the bar which is shaded is *five fifths,* or $\frac{5}{5}$. But since the whole bar is shaded,

$$1 = \frac{5}{5}. \qquad \text{One whole = five fifths}$$

$\frac{1}{5}$	$\frac{1}{5}$	$\frac{1}{5}$	$\frac{1}{5}$	$\frac{1}{5}$

Figure 3.6

This agrees with what we have learned about division of whole numbers. Recall that any whole number except zero divided by itself is 1.

$$\frac{2}{2} = 1, \qquad \frac{3}{3} = 1, \qquad \frac{4}{4} = 1, \qquad \frac{25}{25} = 1$$

The fractions above are special types of improper fractions. An **improper fraction** is one whose numerator is greater than or equal to its denominator. Other examples of improper fractions are:

$$\frac{3}{2}, \quad \frac{5}{4}, \quad \frac{7}{3}, \quad \frac{12}{12}, \quad \frac{9}{3}, \quad \frac{8}{7}.$$

Consider the fraction $\frac{4}{3}$, read *four thirds*. Let us use as a unit the bar divided into 3 equal parts in Figure 3.7.

1 bar

$\frac{1}{3}$	$\frac{1}{3}$	$\frac{1}{3}$

Figure 3.7

The question is, what fraction of the bar in Figure 3.7 is shaded in Figure 3.8? Four of the thirds are shaded, so that the shaded region represents $\frac{4}{3}$.

$\dfrac{4}{3}$ ⟵ Number of parts shaded
⟵ Number of parts in *one* unit

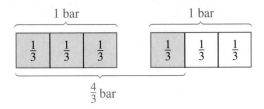

Figure 3.8

///////////////// **CAUTION** ///////////////

Notice that the unit in Figure 3.7 is *one* bar divided into *thirds*. Thus the fraction shaded is

$$\frac{4}{3} \quad \text{not} \quad \frac{4}{6}.$$

| EXAMPLE 3 IMPROPER FRACTION | PRACTICE EXERCISE 3 |

Consider the bar in Figure 3.9 as one whole unit.

| $\frac{1}{6}$ | $\frac{1}{6}$ | $\frac{1}{6}$ | $\frac{1}{6}$ | $\frac{1}{6}$ | $\frac{1}{6}$ |

Figure 3.9

What fraction of the bar shown in Figure 3.9 is shaded in Figure 3.10? What fraction is left unshaded?

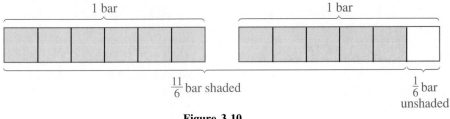

Figure 3.10

Since each bar has been divided into 6 equal parts and 11 of the parts are shaded, the shaded portion corresponds to $\frac{11}{6}$. The unshaded part represents $\frac{1}{6}$.

Shaded: $\dfrac{11}{6}$ ⟵ Number of parts shaded
⟵ Number of parts in *one* bar

Unshaded: $\dfrac{1}{6}$ ⟵ Number of parts not shaded
⟵ Number of parts in *one* bar

What fraction of the figure is shaded below?

Answer: $\frac{5}{4}$

We have seen that the whole number 1 can be written as a fraction such as $\frac{3}{3}$. Other whole numbers can also be written as fractions. For example,

$$\frac{4}{2} = 2, \qquad \frac{6}{2} = 3, \qquad \frac{8}{2} = 4.$$

In each case, if the numerator is divided by the denominator, the result is a whole number. The whole number 0 can also be represented by a fraction. For example,

$$\frac{0}{5} \quad \text{corresponds to 0.}$$

We can interpret $\frac{0}{5}$ as dividing an object into 5 equal parts and taking none of them.

❸ TYPES OF FRACTIONS

We now summarize what we have learned about fractions below.

Types of Fractions

Proper fractions

$\frac{4}{9}$ Numerator less than denominator

$\frac{0}{9} = 0$ Whole number zero

Improper fractions

$\frac{9}{4}$ Numerator greater than denominator

$\frac{4}{4} = 1$ Numerator equals denominator; whole number one

$\frac{12}{4} = 3$ Whole number 3

EXAMPLE 4 FRACTION OF A NUMBER OF OBJECTS

A dozen bottles of soda were taken to a picnic. If seven bottles were drunk, what fractional part was left?

Note that 7 of the 12 bottles are empty and 5 are full ($12 - 7 = 5$). Thus, $\frac{5}{12}$ of the original amount was left over.

PRACTICE EXERCISE 4

Wendy Reed made two dozen cookies. Her son Paul ate 13 of them before dinner. What fraction of the cookies was left?

Answer: $\frac{11}{24}$

EXAMPLE 5 FRACTION OF A NUMBER OF PEOPLE

During a basketball game, 3 of the 10 players on a team fouled out. What fractional part of the team fouled out?

Of the 10 players, 3 fouled out, or $\frac{3}{10}$ of the team.

PRACTICE EXERCISE 5

A basketball player made 9 of 11 shots during a game. What fraction of her shots did she make?

Answer: $\frac{9}{11}$

3.1   EXERCISES A

Give the numerator and the denominator of each fraction.

1. $\dfrac{5}{7}$

numerator: 5;
denominator: 7

2. $\dfrac{11}{4}$

numerator: 11;
denominator: 4

3. $\dfrac{0}{8}$

numerator: 0;
denominator: 8

4. $\dfrac{12}{12}$

numerator: 12;
denominator: 12

What fractional part of each figure is shaded?

5.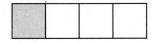

$\dfrac{1}{4}$

6.

$\dfrac{1}{3}$

7.

$\dfrac{2}{5}$

8.

$\dfrac{2}{4}$

9.

$\dfrac{3}{8}$

10.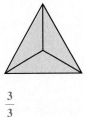

$\dfrac{3}{3}$

11.

$\dfrac{1}{2}$

12.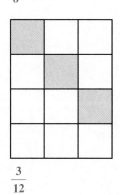

$\dfrac{3}{12}$

13.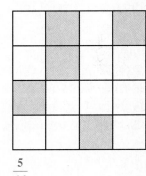

$\dfrac{5}{16}$

14 The bar

$\boxed{\dfrac{1}{4}}\boxed{\dfrac{1}{4}}\boxed{\dfrac{1}{4}}\boxed{\dfrac{1}{4}}$

represents one whole unit. What fraction of the unit is shaded below? Consider the whole figure. There is one answer, and it is an improper fraction.

$\dfrac{5}{4}$

15. The figure

$\boxed{\dfrac{1}{6}}\boxed{\dfrac{1}{6}}\boxed{\dfrac{1}{6}}\boxed{\dfrac{1}{6}}\boxed{\dfrac{1}{6}}\boxed{\dfrac{1}{6}}$

represents one whole unit. What fraction of the unit is shaded below? Consider the whole figure. There is one answer, and it is an improper fraction.

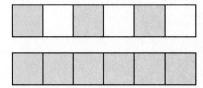

$\dfrac{9}{6}$

16. The circle

represents one whole unit. What fraction of the unit is shaded below?

$\dfrac{11}{8}$

17. The triangle

represents one whole unit. What fraction of the unit is shaded below?

$\dfrac{5}{3}$

18. The numerator of the fraction $\frac{4}{7}$ stands for how many of the equal parts of a whole?   4

19. A class is divided into 8 equal groups and 3 of the groups are chosen for a special project.

  (a) What fractional part of the class is chosen? $\frac{3}{8}$

  (b) What fractional part is not chosen? $\frac{5}{8}$

20. A baseball player gets 4 hits in 9 times at bat during a double header. The number of hits is what fraction of his total at bats? $\frac{4}{9}$

21. Two pies are each cut into 6 pieces, and each of 7 people eats one piece.

  (a) What fractional part of one pie was eaten? $\frac{7}{6}$

  (b) What fractional part of one pie was not eaten? $\frac{5}{6}$

22. A student received $300 from home. She spent $30 for books, $75 for a bicycle, $40 for utilities, and put the rest into savings. What fractional part of the $300 went for **(a)** books? **(b)** the bicycle? **(c)** utilities? **(d)** savings?

  (a) $\frac{30}{300}$ (b) $\frac{75}{300}$ (c) $\frac{40}{300}$ (d) $\frac{155}{300}$

Express each fraction as a whole number.

23. $\dfrac{12}{3}$ 4

24. $\dfrac{0}{5}$ 0

25. $\dfrac{43}{43}$ 1

26. $\dfrac{26}{13}$ 2

ANSWERS: 1. numerator: 5; denominator: 7 2. numerator: 11; denominator: 4 3. numerator: 0; denominator: 8 4. numerator: 12; denominator: 12 5. $\frac{1}{4}$ 6. $\frac{1}{3}$ 7. $\frac{2}{5}$ 8. $\frac{2}{4}$ 9. $\frac{3}{8}$ 10. $\frac{3}{3}$ 11. $\frac{1}{2}$ 12. $\frac{3}{12}$ 13. $\frac{5}{16}$ 14. $\frac{5}{4}$ 15. $\frac{9}{6}$ 16. $\frac{11}{8}$ 17. $\frac{5}{3}$ 18. 4 19. (a) $\frac{3}{8}$ (b) $\frac{5}{8}$ 20. $\frac{4}{9}$ 21. (a) $\frac{7}{6}$ (b) $\frac{5}{6}$ 22. (a) $\frac{30}{300}$ (b) $\frac{75}{300}$ (c) $\frac{40}{300}$ (d) $\frac{155}{300}$ 23. 4 24. 0 25. 1 26. 2

3.1 EXERCISES B

Give the numerator and the denominator of each fraction.

1. $\dfrac{9}{17}$ numerator: 9;
denominator: 17

2. $\dfrac{25}{3}$ numerator: 25;
denominator: 3

3. $\dfrac{7}{1}$ numerator: 7;
denominator: 1

4. $\dfrac{23}{23}$ numerator: 23;
denominator: 23

What fractional part of each figure is shaded?

5.

$\dfrac{4}{5}$

6.

$\dfrac{1}{4}$

7.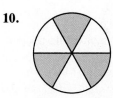

$\dfrac{3}{8}$

8.

$\dfrac{3}{3}$

9.

$\dfrac{1}{2}$

10.

$\dfrac{3}{6}$

11.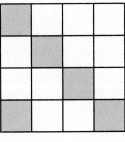

$\dfrac{5}{16}$

12.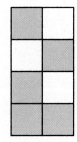

$\dfrac{5}{8}$

13.

$\dfrac{8}{12}$

14. The square

$\boxed{\begin{array}{c} \frac{1}{2} \\ \frac{1}{2} \end{array}}$

represents one whole unit. What fraction of the unit is shaded below? Consider the whole figure. There is one answer, and it is an improper fraction.

$\dfrac{3}{2}$

15. The bar

represents one whole unit. What fraction of the unit is shaded below? Consider the whole figure. There is one answer, and it is an improper fraction.

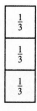

$\dfrac{5}{3}$

16. The circle

represents one whole unit. What fraction of the unit is shaded below?

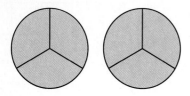

$\dfrac{6}{3}$

17. The bar

$\frac{1}{8}$	$\frac{1}{8}$	$\frac{1}{8}$	$\frac{1}{8}$	$\frac{1}{8}$	$\frac{1}{8}$	$\frac{1}{8}$	$\frac{1}{8}$

represents one whole unit. What fraction of the unit is shaded below?

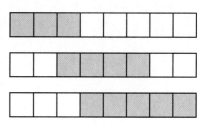

$\dfrac{12}{8}$

18. The numerator of the fraction $\dfrac{11}{7}$ stands for how many of the equal parts of a whole?   **11**

19. A committee of 13 women is divided into two sub-committees, one with 7 members which will study a particular project, and the other which will search for a guest speaker. What fraction of the committee has been chosen to find the speaker? $\frac{6}{13}$

20. A basketball player made 7 free throws in 11 attempts during the first half of a game. The number of free throws made is what fraction of his total attempts? $\frac{7}{11}$

21. Three pies are each cut into 6 pieces. Each of 11 people eat one piece.

   (a) What fractional part of one pie was eaten? $\frac{11}{6}$

   (b) What fractional part of one pie was not eaten? $\frac{7}{6}$

22. Maria Lopez had $500 in her checking account. She spent $100 for food, $125 for clothes, $50 for a night out, and put the rest in savings. What fractional part of the $500 went for **(a)** food? **(b)** clothes? **(c)** the night out? **(d)** savings?

   (a) $\frac{100}{500}$ **(b)** $\frac{125}{500}$ **(c)** $\frac{50}{500}$ **(d)** $\frac{225}{500}$

Express each fraction as a whole number.

23. $\dfrac{20}{4}$ **5**

24. $\dfrac{18}{1}$ **18**

25. $\dfrac{0}{101}$ **0**

26. $\dfrac{500}{500}$ **1**

3.1 EXERCISES C

Let a and b represent natural numbers with a > b. Tell which of the following fractions are proper and which are improper.

1. $\dfrac{a}{b}$ improper

2. $\dfrac{b}{a}$

   [Answer: proper]

3. $\dfrac{a+b}{a-b}$

   [Answer: improper]

4. $\dfrac{2a}{a}$ improper

3.2 RENAMING FRACTIONS

STUDENT GUIDEPOSTS

1. Equal Fractions
2. Building and Reducing Fractions
3. Reducing to Lowest Terms by Factoring into Primes
4. Dividing Out Common, Nonprime Factors

1 EQUAL FRACTIONS

Every fraction has many names. This can be shown by the three bars in Figure 3.11. In each, exactly the same part of the bar is shaded. In **(a),** the amount shaded is

$$\frac{1}{3} \text{ of the bar.}$$

In **(b),** $\frac{2}{6}$ of the bar is shaded,

and in **(c),** $\frac{4}{12}$ of the bar is shaded.

(a)
(b)
(c)

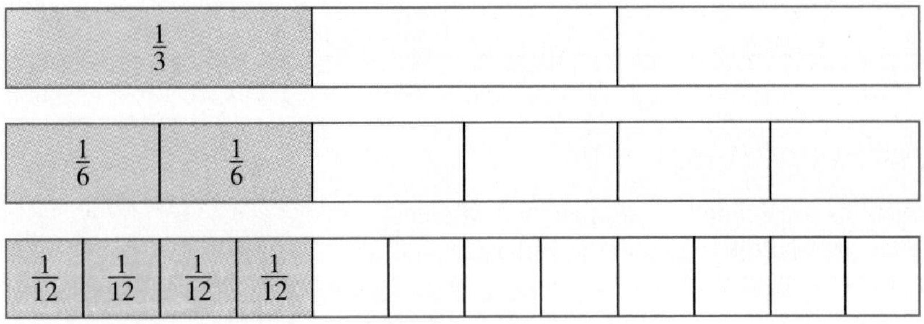

Figure 3.11

Since the same part of the bar is shaded in all three cases,

$$\frac{1}{3}, \quad \frac{2}{6}, \quad \text{and} \quad \frac{4}{12}$$

are all names for the same fraction. Fractions which are names for the same number are **equal fractions.** Thus,

$$\frac{1}{3} = \frac{2}{6} = \frac{4}{12}.$$

Knowing that $\frac{2}{6} = \frac{4}{12}$

leads us to a test for equality of fractions. Multiply as indicated.

$$2 \cdot 12 = 24 \qquad \frac{2}{6} \times \frac{4}{12} \qquad 6 \cdot 4 = 24$$

This gives the equality $2 \cdot 12 = 6 \cdot 4$.

These products are called **cross products.**

Equal Fractions
Two fractions are equal when the cross products are equal.

We can use this to test for equality of any two fractions.

EXAMPLE 1 TESTING FOR EQUALITY

Are the two fractions equal?

(a) $\dfrac{3}{4}$ and $\dfrac{6}{8}$

$3 \cdot 8 = 24$ $\quad \dfrac{3}{4} \diagdown \dfrac{6}{8} \quad$ $4 \cdot 6 = 24$ $\qquad$ Thus, $\dfrac{3}{4} = \dfrac{6}{8}$.

(b) $\dfrac{2}{9}$ and $\dfrac{6}{27}$ $\qquad$ Since $2 \cdot 27 = 54$ and $9 \cdot 6 = 54$,

$$\frac{2}{9} = \frac{6}{27}.$$

(c) $\dfrac{5}{9}$ and $\dfrac{4}{7}$ $\qquad$ Since $5 \cdot 7 = 35$ and $9 \cdot 4 = 36$,

$$\frac{5}{9} \neq \frac{4}{7}.$$

PRACTICE EXERCISE 1

Are the two fractions equal?

(a) $\dfrac{9}{12}$ and $\dfrac{15}{20}$

(b) $\dfrac{20}{50}$ and $\dfrac{16}{40}$

(c) $\dfrac{9}{11}$ and $\dfrac{5}{6}$

Answers: (a) yes (b) yes (c) no

❷ BUILDING AND REDUCING FRACTIONS

You probably noticed that equal fractions are related in another way besides having equal cross products. As an example, multiplying both the numerator and denominator of $\frac{1}{3}$ by 2 gives $\frac{2}{6}$.

$$\frac{1 \times 2}{3 \times 2} = \frac{2}{6}$$

Similarly, multiplying both the numerator and denominator of $\frac{1}{3}$ by 4 gives $\frac{4}{12}$.

$$\frac{1 \times 4}{3 \times 4} = \frac{4}{12}$$

These are examples of *building fractions*.

Building Fractions
If both the numerator and denominator of a fraction are *multiplied* by the same whole number (except 0), an **equal fraction** is formed.

EXAMPLE 2 BUILDING FRACTIONS

(a) Find a fraction with denominator 20 that is equal to $\frac{1}{4}$.

We are looking for a fraction of the form $\dfrac{?}{20}$ such that

$$\frac{1}{4} = \frac{?}{20}.$$

PRACTICE EXERCISE 2

(a) Find a fraction with denominator 16 that is equal to $\frac{5}{2}$.

To build an equal fraction, the numerator and denominator of $\frac{1}{4}$ must be multiplied by the *same* number. To get 20 for a denominator, that number must be 5, since $4 \times 5 = 20$.

$$\frac{1}{4} = \frac{1 \times 5}{4 \times 5} = \frac{5}{20}$$

(b) Find a fraction with numerator 18 that is equal to $\frac{3}{2}$.

Find $\dfrac{18}{?}$ such that

$$\frac{3}{2} = \frac{18}{?}.$$

What number multiplied by 3 gives 18? That number is 6.

$$\frac{3}{2} = \frac{3 \times 6}{2 \times 6} = \frac{18}{12}$$

(b) Find a fraction with numerator 30 that is equal to $\frac{5}{8}$.

Answers: (a) $\frac{40}{16}$ (b) $\frac{30}{48}$

The reverse of building fractions is *reducing fractions*. Starting with the fraction $\frac{5}{20}$ and dividing both the numerator and the denominator by 5 gives

$$\frac{5 \div 5}{20 \div 5} = \frac{1}{4},$$

which is equal to $\frac{5}{20}$. This is an example of the following.

Reducing Fractions

If both the numerator and denominator of a fraction are *divided* by the same whole number (except 0), an **equal fraction** is formed.

EXAMPLE 3   REDUCING FRACTIONS

(a) Find a fraction with denominator 3 that is equal to $\frac{14}{21}$.

We are looking for a fraction of the form $\dfrac{?}{3}$ such that

$$\frac{14}{21} = \frac{?}{3}.$$

To get an equal fraction, the numerator and denominator of $\frac{14}{21}$ must be divided by the *same* number. To get 3 for a denominator, that number must be 7, since $21 \div 7 = 3$.

$$\frac{14}{21} = \frac{14 \div 7}{21 \div 7} = \frac{2}{3}$$

(b) Find a fraction with numerator 2 that is equal to $\frac{22}{11}$.

Find $\dfrac{2}{?}$ such that

$$\frac{22}{11} = \frac{2}{?}.$$

PRACTICE EXERCISE 3

(a) Find a fraction with denominator 5 that is equal to $\frac{49}{35}$.

(b) Find a fraction with numerator 6 that is equal to $\frac{72}{12}$.

Divide numerator and denominator by 11, because $22 \div 11 = 2$.

$$\frac{22}{11} = \frac{22 \div 11}{11 \div 11} = \frac{2}{1}$$

Notice that this fraction is really the whole number 2.

Answers: (a) $\frac{7}{5}$ (b) $\frac{6}{1} = 6$

❸ REDUCING TO LOWEST TERMS BY FACTORING INTO PRIMES

We have seen that every fraction has many names. For example, the fraction $\frac{1}{4}$ can be expressed as

$$\frac{1}{4}, \quad \frac{2}{8}, \quad \frac{3}{12}, \quad \frac{4}{16}, \quad \frac{5}{20}, \quad \text{and so on.}$$

In this list, every fraction to the right of $\frac{1}{4}$ has been *built up* from $\frac{1}{4}$ by multiplying numerator and denominator by some number. On the other hand, $\frac{1}{4}$ can be obtained by *reducing* each fraction to the right of it, by dividing numerator and denominator by some number.

Of all the possible names for the fraction $\frac{1}{4}$, 1 and 4 are the *smallest* numbers we can use in the name. In a sense, $\frac{1}{4}$ is the best name for the number since it cannot be reduced.

Reduced to Lowest Terms
When 1 is the only nonzero whole number divisor of both the numerator and denominator of a fraction, the fraction is **reduced to lowest terms.**

Notice that

$\dfrac{2}{8}$ is not in lowest terms since 2 is a divisor of 2 and 8,

$\dfrac{3}{12}$ is not in lowest terms since 3 is a divisor of 3 and 12.

However,

$\dfrac{1}{4}$ is in lowest terms since 1 is the only divisor of 1 and 4.

Reducing a fraction to lowest terms is simplified by using prime numbers. Recall from Section 2.7 that a whole number greater than 1 is *prime* if its only divisors are itself and 1. Also, recall that every whole number greater than 1 is either prime or can be expressed as a product of primes.

To Reduce a Fraction to Lowest Terms
1. Factor the numerator and denominator into primes.
2. Divide out all prime factors common to both the numerator and denominator.
3. Multiply remaining primes, first in the numerator and then in the denominator.

EXAMPLE 4 REDUCING TO LOWEST TERMS	PRACTICE EXERCISE 4

Reduce $\frac{4}{6}$ to lowest terms.

    Factor the numerator and denominator into primes.

$$\frac{4}{6} = \frac{2 \cdot 2}{3 \cdot 2}$$

Since 2 is a prime factor common to both the numerator and denominator, divide it out of both, giving an *equal fraction*.

$$\frac{(2 \cdot 2) \div 2}{(3 \cdot 2) \div 2} = \frac{2 \cdot (2 \div 2)}{3 \cdot (2 \div 2)}$$

$$= \frac{2 \cdot 1}{3 \cdot 1} = \frac{2}{3}$$ A number divided by itself equals 1

A shorthand way of writing this division is by simply crossing out the common factors.

$$\frac{4}{6} = \frac{2 \cdot \overset{1}{\cancel{2}}}{3 \cdot \underset{1}{\cancel{2}}} = \frac{2 \cdot 1}{3 \cdot 1} = \frac{2}{3}$$

Reduce $\frac{10}{25}$ to lowest terms.

Answer: $\frac{2}{5}$

HINT

Notice that we inserted small "ones" in our work in Example 4. It is important to get in the habit of writing them so that you will not make mistakes when reducing fractions such as the following:

$$\frac{3}{15} = \frac{\overset{1}{\cancel{3}}}{5 \cdot \underset{1}{\cancel{3}}} = \frac{1}{5}.$$

Failure to write the 1s could lead you to the answer $\frac{0}{5}$ or 0, which is wrong.

    The process of dividing out common factors by crossing them out, as shown in Example 4 and in the following examples, is sometimes called *canceling factors*. However, we will use the words *divide out*.

EXAMPLE 5 DIVIDING OUT COMMON PRIME FACTORS	PRACTICE EXERCISE 5

Reduce to lowest terms by factoring into primes.

(a) $\dfrac{12}{42} = \dfrac{2 \cdot 2 \cdot 3}{2 \cdot 3 \cdot 7}$ Factor numerator and denominator into primes

$\quad\quad = \dfrac{\overset{1}{\cancel{2}} \cdot 2 \cdot \overset{1}{\cancel{3}}}{\underset{1}{\cancel{2}} \cdot \underset{1}{\cancel{3}} \cdot 7}$ Divide out common prime factors 2 and 3

$\quad\quad = \dfrac{2}{7}$

Reduce to lowest terms by factoring into primes.

(a) $\dfrac{98}{63}$

(b) $\dfrac{5}{20} = \dfrac{5}{2 \cdot 2 \cdot 5}$ Factor numerator and denominator into primes

$= \dfrac{\overset{1}{\cancel{5}}}{2 \cdot 2 \cdot \cancel{5}}$ Divide out common prime factor 5

$= \dfrac{1}{4}$ Multiply remaining factors in the denominator

Remember to write the small 1s to avoid getting the wrong answer $\frac{0}{4}$ or 0.

(c) $\dfrac{210}{273} = \dfrac{2 \cdot \overset{1}{\cancel{3}} \cdot 5 \cdot \overset{1}{\cancel{7}}}{\underset{1}{\cancel{3}} \cdot \underset{1}{\cancel{7}} \cdot 13} = \dfrac{2 \cdot 5}{13} = \dfrac{10}{13}$

(b) $\dfrac{7}{35}$

(c) $\dfrac{330}{770}$

Answers: (a) $\frac{14}{9}$ (b) $\frac{1}{5}$ (c) $\frac{3}{7}$

④ DIVIDING OUT COMMON, NONPRIME FACTORS

In Example 5 **(a)** we could have noticed that 6 was a factor of both the numerator and denominator and written

$$\dfrac{12}{42} = \dfrac{2 \cdot 6}{7 \cdot 6} = \dfrac{2 \cdot \overset{1}{\cancel{6}}}{7 \cdot \underset{1}{\cancel{6}}} = \dfrac{2}{7}.$$

This technique does not require factoring into primes and can sometimes save a few steps.

| **EXAMPLE 6   DIVIDING OUT COMMON, NONPRIME FACTORS** | **PRACTICE EXERCISE 6** |

Reduce to lowest terms.

(a) $\dfrac{4}{20} = \dfrac{\overset{1}{\cancel{4}}}{5 \cdot \underset{1}{\cancel{4}}} = \dfrac{1}{5}$

(b) $\dfrac{16}{40} = \dfrac{2 \cdot \overset{1}{\cancel{8}}}{5 \cdot \underset{1}{\cancel{8}}} = \dfrac{2}{5}$

(c) $\dfrac{36}{120} = \dfrac{3 \cdot \overset{1}{\cancel{12}}}{10 \cdot \underset{1}{\cancel{12}}} = \dfrac{3}{10}$

Reduce to lowest terms.

(a) $\dfrac{5}{40}$

(b) $\dfrac{18}{63}$

(c) $\dfrac{20}{150}$

Answers: (a) $\frac{1}{8}$ (b) $\frac{2}{7}$ (c) $\frac{2}{15}$

//////////// CAUTION ////////////

When reducing fractions by dividing out common, nonprime factors, be sure to reduce the fraction to lowest terms. For example, in Example 6 **(b),** had we only divided out the common factor 4,

$$\frac{16}{40} = \frac{4 \cdot \overset{1}{\cancel{4}}}{10 \cdot \underset{1}{\cancel{4}}} = \frac{4}{10},$$

we would not be finished since the fraction is not yet in lowest terms.

//////////

Before continuing, you should review the divisibility tests for 2, 3, 5, and 10 in Chapter 2. Reducing fractions can be simplified by using these tests.

| **EXAMPLE 7** APPLICATION IN EDUCATION | **PRACTICE EXERCISE 7** |

On a math test, Marvin got 60 correct out of 80 problems. On his history test, he got 75 correct out of 100 problems. What fractional part of each test did he get correct? On which test did he perform better?

On the math test, 60 correct out of 80 problems results in $\frac{60}{80}$ of the problems being correct. On the history test, he had $\frac{75}{100}$ correct. But

$$60 \cdot 100 = 6000 = 80 \cdot 75.$$

Thus,
$$\frac{60}{80} = \frac{75}{100}$$

since the cross-products are equal. Also, note that each fraction reduces to $\frac{3}{4}$. Thus, Marvin performed the same on the two tests.

Nina had 21 correct out of 30 on a math placement test and 70 out of 100 on an English test. What fractional part of each test did she get correct? On which test did she perform better?

Answer: Math, $\frac{7}{10}$; English, $\frac{7}{10}$; she scored the same on both tests.

3.2 EXERCISES A

The shaded portions of the bars illustrate that what fractions are equal?

1.

$$\frac{1}{4} = \frac{2}{8} = \frac{4}{16}$$

2.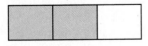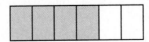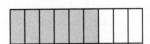

$$\frac{2}{3} = \frac{4}{6} = \frac{6}{9}$$

Decide which fractions are equal by using the cross-product rule.

3. $\frac{9}{12}$ and $\frac{3}{4}$

$9 \cdot 4 = 12 \cdot 3$; so $\frac{9}{12} = \frac{3}{4}$

4. $\frac{2}{5}$ and $\frac{3}{7}$

$2 \cdot 7 \neq 5 \cdot 3$; so $\frac{2}{5} \neq \frac{3}{7}$

5. $\frac{6}{16}$ and $\frac{3}{8}$

$6 \cdot 8 = 16 \cdot 3$; so $\frac{6}{16} = \frac{3}{8}$

6. $\frac{6}{8}$ and $\frac{3}{2}$

$6 \cdot 2 \neq 8 \cdot 3$; so $\frac{6}{8} \neq \frac{3}{2}$

7. $\frac{12}{32}$ and $\frac{4}{11}$

$12 \cdot 11 \neq 32 \cdot 4$; so $\frac{12}{32} \neq \frac{4}{11}$

8 $\frac{8}{9}$ and $\frac{96}{108}$

$8 \cdot 108 = 9 \cdot 96$; so $\frac{8}{9} = \frac{96}{108}$

In each of the following, find the numerator of the new fraction having the given denominator.

9. $\frac{3}{4} = \frac{?}{12}$ 9

10. $\frac{3}{2} = \frac{?}{6}$ 9

11. $\frac{1}{2} = \frac{?}{22}$ 11

12. $\frac{5}{8} = \frac{?}{16}$ 10

13 $\frac{9}{18} = \frac{?}{6}$ 3

14. $\frac{5}{20} = \frac{?}{4}$ 1

15. $\frac{20}{18} = \frac{?}{9}$ 10

16. $\frac{8}{12} = \frac{?}{3}$ 2

17. $\frac{6}{18} = \frac{?}{3}$ 1

In each of the following, find the denominator of the new fraction having the given numerator.

18. $\frac{2}{7} = \frac{4}{?}$ 14

19. $\frac{1}{9} = \frac{3}{?}$ 27

20. $\frac{3}{5} = \frac{9}{?}$ 15

21. $\frac{12}{20} = \frac{3}{?}$ 5

22. $\frac{9}{2} = \frac{27}{?}$ 6

23 $\frac{36}{14} = \frac{18}{?}$ 7

Reduce each fraction to lowest terms.

24. $\frac{4}{10}$ $\frac{2}{5}$

25. $\frac{15}{10}$ $\frac{3}{2}$

26. $\frac{6}{18}$ $\frac{1}{3}$

27. $\frac{16}{8}$ $\frac{2}{1}$ or 2

28. $\frac{39}{52}$ $\frac{3}{4}$

29. $\frac{12}{30}$ $\frac{2}{5}$

30. $\frac{121}{22}$ $\frac{11}{2}$

31 $\frac{210}{105}$ $\frac{2}{1}$ or 2

32. $\frac{42}{18}$ $\frac{7}{3}$

Solve.

33. During the season, a starting ball player made 144 free throws out of 162 attempts. A reserve player went to the line only 9 times but made 8 shots. Which player made the greater fractional part of his free-throw attempts?

   Each player made $\frac{8}{9}$ of his shots.

34. During May, one scientific equipment dealer sold 21 microscopes from his inventory of 77 microscopes. A second dealer sold 18 microscopes from his inventory of 66 microscopes. Which dealer sold the greater fractional part of his stock?

   Each dealer sold $\frac{3}{11}$ of his stock.

FOR REVIEW

What fractional part of each figure is shaded?

35.
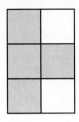

$$\frac{4}{6} = \frac{2}{3}$$

36.

$$\frac{4}{8} = \frac{1}{2}$$

Express each fraction as a whole number.

37. $\dfrac{17}{17}$ 1   **38.** $\dfrac{0}{99}$ 0   **39.** $\dfrac{100}{5}$ 20   **40.** $\dfrac{84}{12}$ 7

ANSWERS: 1. $\frac{1}{4} = \frac{2}{8} = \frac{4}{16}$ 2. $\frac{2}{3} = \frac{4}{6} = \frac{6}{9}$ 3. $9 \cdot 4 = 12 \cdot 3$, so $\frac{9}{12} = \frac{3}{4}$ 4. $2 \cdot 7 \neq 5 \cdot 3$; so $\frac{2}{5} \neq \frac{3}{7}$ 5. $6 \cdot 8 = 16 \cdot 3$; so $\frac{6}{16} = \frac{3}{8}$ 6. $6 \cdot 2 \neq 8 \cdot 3$; so $\frac{6}{8} \neq \frac{3}{2}$ 7. $12 \cdot 11 \neq 32 \cdot 4$; so $\frac{12}{32} \neq \frac{4}{11}$ 8. $8 \cdot 108 = 9 \cdot 96$; so $\frac{8}{9} = \frac{96}{108}$ 9. 9   10. 9   11. 11   12. 10   13. 3   14. 1   15. 10   16. 2   17. 1   18. 14   19. 27   20. 15   21. 5   22. 6   23. 7   24. $\frac{2}{5}$ 25. $\frac{3}{2}$ 26. $\frac{1}{3}$ 27. $\frac{2}{1}$ or 2   28. $\frac{3}{4}$ 29. $\frac{2}{5}$ 30. $\frac{11}{2}$ 31. $\frac{2}{1}$ or 2   32. $\frac{7}{3}$ 33. The starter made $\frac{144}{162} = \frac{8}{9}$ of his attempts, and the reserve made $\frac{8}{9}$ of his shots. Thus, each made the same fractional part of his attempts.   34. $\frac{21}{77}$ and $\frac{18}{66}$ are the same fractional parts (both equal $\frac{3}{11}$ reduced to lowest terms).   35. $\frac{4}{6}$ $\left(\text{or } \frac{2}{3} \text{ reduced}\right)$ 36. $\frac{4}{8}$ $\left(\text{or } \frac{1}{2} \text{ reduced}\right)$ 37. 1   38. 0   39. 20   40. 7

3.2 EXERCISES B

The shaded portions of the bars illustrate that what fractions are equal?

1.
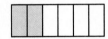

$$\frac{1}{3} = \frac{2}{6} = \frac{4}{12}$$

2.

$$\frac{3}{5} = \frac{6}{10} = \frac{9}{15}$$

Decide which fractions are equal by using the cross-product rule.

3. $\dfrac{7}{8}$ and $\dfrac{21}{24}$
$7 \cdot 24 = 8 \cdot 21$; so $\dfrac{7}{8} = \dfrac{21}{24}$

4. $\dfrac{5}{12}$ and $\dfrac{6}{13}$
$5 \cdot 13 \neq 12 \cdot 6$; so $\dfrac{5}{12} \neq \dfrac{6}{13}$

5. $\dfrac{9}{6}$ and $\dfrac{18}{12}$
$9 \cdot 12 = 6 \cdot 18$; so $\dfrac{9}{6} = \dfrac{18}{12}$

6. $\dfrac{13}{15}$ and $\dfrac{27}{30}$
$13 \cdot 30 \neq 15 \cdot 27$; so $\dfrac{13}{15} \neq \dfrac{27}{30}$

7. $\dfrac{11}{12}$ and $\dfrac{132}{144}$
$11 \cdot 144 = 12 \cdot 132$; so $\dfrac{11}{12} = \dfrac{132}{144}$

8. $\dfrac{4}{5}$ and $\dfrac{400}{500}$
$4 \cdot 500 = 5 \cdot 400$; so $\dfrac{4}{5} = \dfrac{400}{500}$

In each of the following, find the numerator of the new fraction having the given denominator.

9. $\dfrac{3}{8} = \dfrac{?}{24}$ 9

10. $\dfrac{9}{7} = \dfrac{?}{35}$ 45

11. $\dfrac{1}{11} = \dfrac{?}{55}$ 5

12. $\dfrac{?}{5} = \dfrac{8}{20}$ 2

13. $\dfrac{4}{12} = \dfrac{?}{3}$ 1

14. $\dfrac{21}{14} = \dfrac{?}{2}$ 3

15. $\dfrac{25}{75} = \dfrac{?}{3}$ 1

16. $\dfrac{15}{12} = \dfrac{?}{4}$ 5

17. $\dfrac{100}{500} = \dfrac{?}{5}$ 1

In each of the following, find the denominator of the new fraction having the given numerator.

18. $\dfrac{6}{?} = \dfrac{12}{10}$ 5

19. $\dfrac{4}{?} = \dfrac{24}{18}$ 3

20. $\dfrac{28}{4} = \dfrac{7}{?}$ 1

21. $\dfrac{25}{15} = \dfrac{5}{?}$ 3

22. $\dfrac{36}{24} = \dfrac{6}{?}$ 4

23. $\dfrac{100}{20} = \dfrac{5}{?}$ 1

Reduce each fraction to lowest terms.

24. $\dfrac{9}{18}$ $\dfrac{1}{2}$

25. $\dfrac{24}{16}$ $\dfrac{3}{2}$

26. $\dfrac{20}{4}$ $\dfrac{5}{1}$ or 5

27. $\dfrac{36}{48}$ $\dfrac{3}{4}$

28. $\dfrac{24}{60}$ $\dfrac{2}{5}$

29. $\dfrac{14}{126}$ $\dfrac{1}{9}$

30. $\dfrac{55}{33}$ $\dfrac{5}{3}$

31. $\dfrac{42}{6}$ $\dfrac{7}{1}$ or 7

32. $\dfrac{96}{528}$ $\dfrac{2}{11}$

Solve.

33. In one shipment of tires 4 of the 200 were defective. The next shipment of 700 tires contained 14 which were defective. Which shipment had the greater fraction of defective tires?

Both shipments had $\frac{1}{50}$ defective tires.

34. In June a poll showed that 25 of 30 people surveyed favored the new freeway. The next month 55 of 66 people favored the freeway. During which month did the greater fraction favor the new freeway?

In both months, $\frac{5}{6}$ favored the freeway.

FOR REVIEW

What fractional part of each figure is shaded?

35.

$\dfrac{2}{3}$

36.

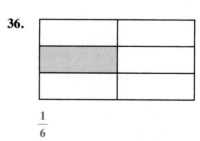

$\dfrac{1}{6}$

Express each fraction as a whole number.

37. $\dfrac{217}{217}$ 1

38. $\dfrac{0}{581}$ 0

39. $\dfrac{666}{18}$ 37

40. $\dfrac{2000}{200}$ 10

3.2 EXERCISES C

Let a and b represent natural numbers. Reduce to lowest terms. [Hint: $a^2 = a \cdot a$]

1. $\dfrac{3^3 4^2}{3^2 4}$

[Answer: 12]

2. $\dfrac{5^4 a^2}{5^2 a}$ $25a$

3. $\dfrac{27a^3}{81a^7}$

$\left[\text{Answer: } \frac{1}{3a^4}\right]$

4. $\dfrac{a^2 b^3}{ab}$ ab^2

5. $\dfrac{5a^3 b^4}{10a^2 b^6}$ $\dfrac{a}{2b^2}$

6. $\dfrac{42a^4 b^2}{14ab^3}$

$\left[\text{Answer: } \frac{3a^3}{b}\right]$

3.3 MULTIPLYING FRACTIONS

STUDENT GUIDEPOSTS

1 Meaning of Multiplication of Fractions

2 Rule for Multiplying Fractions

3 Properties of Multiplication

4 Powers and Roots of Fractions

1 MEANING OF MULTIPLICATION OF FRACTIONS

The simplest arithmetic operation on fractions is multiplication. Notice that the word *of* after a fraction means "times."

What is one half of 8?

Obviously, half of 8 is 4. Let us multiply $\frac{1}{2}$ times 8 as follows.

$$\frac{1}{2} \cdot 8 = \frac{1}{2} \cdot \frac{8}{1} \qquad \text{Write 8 as the fraction } \tfrac{8}{1}$$

$$= \frac{1 \cdot 8}{2 \cdot 1} \qquad \begin{array}{l}\text{Multiply numerators}\\ \text{Multiply denominators}\end{array}$$

$$= \frac{8}{2} \qquad \text{Divide result}$$

$$= 4$$

The final answer is 4, which agrees with our intuitive notion of $\frac{1}{2}$ of 8.

HINT

Two facts are useful:

1. The word *of* means "times" when it follows a fraction.

2. The product of two fractions $= \dfrac{\text{product of their numerators}}{\text{product of their denominators}}$.

Suppose we look at a second example and find the product of $\frac{1}{2}$ and $\frac{1}{3}$. The shaded part of the bar in Figure 3.12 corresponds to the fraction $\frac{1}{3}$. If we multiply $\frac{1}{2} \cdot \frac{1}{3}$, we recognize this as

$$\frac{1}{2} \text{ of } \frac{1}{3}.$$

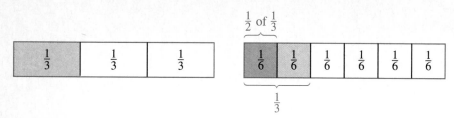

Figure 3.12   Figure 3.13

Just as we found $\frac{1}{2}$ of 8 (which we recognized as 4), suppose we find $\frac{1}{2}$ of $\frac{1}{3}$ by taking one half of the shaded part in Figure 3.12. Figure 3.13 shows that

$$\frac{1}{2} \text{ of } \frac{1}{3} \text{ is equal to } \frac{1}{6}.$$

Finding the product of $\frac{1}{2}$ and $\frac{1}{3}$ by multiplying numerators, multiplying denominators, and dividing the results gives the same answer.

$$\frac{1}{2} \cdot \frac{1}{3} = \frac{1 \cdot 1}{2 \cdot 3} = \frac{1}{6}$$

EXAMPLE 1 MULTIPLYING FRACTIONS	PRACTICE EXERCISE 1

Find each product.

(a) $\dfrac{1}{3} \cdot \dfrac{1}{7} = \dfrac{1 \cdot 1}{3 \cdot 7} = \dfrac{1}{21}$ Multiply numerators, multiply denominators and divide the products

(b) $\dfrac{2}{3} \cdot \dfrac{5}{9} = \dfrac{2 \cdot 5}{3 \cdot 9} = \dfrac{10}{27}$

Find each product.

(a) $\dfrac{1}{6} \cdot \dfrac{1}{8}$

(b) $\dfrac{3}{8} \cdot \dfrac{7}{5}$

Answers: (a) $\frac{1}{48}$ (b) $\frac{21}{40}$

② RULE FOR MULTIPLYING FRACTIONS

In Example 1, we multiplied fractions by first multiplying the numerators, then multiplying the denominators, and then dividing the numerators' product by the denominators' product. For example,

$$\frac{3}{8} \cdot \frac{4}{5} = \frac{3 \cdot 4}{8 \cdot 5} = \frac{12}{40}.$$

We know from Section 3.2 that to reduce $\frac{12}{40}$ to lowest terms, we need to factor both the numerator and the denominator into primes and divide out common prime factors.

$$\frac{12}{40} = \frac{\overset{1}{\cancel{2}} \cdot \overset{1}{\cancel{2}} \cdot 3}{\underset{1}{\cancel{2}} \cdot \underset{1}{\cancel{2}} \cdot 2 \cdot 5} = \frac{3}{10}$$

Multiplying and then reducing fractions can be time consuming. To reduce the number of steps, and thus have less chance of making an error, use the following rule.

To Multiply Two (or More) Fractions

1. Factor all numerators and denominators into primes.
2. Place all numerator factors over all denominator factors.
3. Divide out all prime factors common to both the numerator and denominator.
4. Multiply the remaining primes, first in the numerator, and then in the denominator.

If we return to the original example of $\frac{3}{8} \cdot \frac{4}{5}$ and use these rules, we get

$$\frac{3}{8} \cdot \frac{4}{5} = \frac{3}{2 \cdot 2 \cdot 2} \cdot \frac{2 \cdot 2}{5}$$

$$= \frac{3 \cdot \overset{1}{\cancel{2}} \cdot \overset{1}{\cancel{2}}}{\underset{1}{\cancel{2}} \cdot \underset{1}{\cancel{2}} \cdot 2 \cdot 5}$$

$$= \frac{3}{10}.$$

This gives us the same answer we obtained above, but has fewer steps. Notice that we could have factored out the common nonprime factor of 4 to shorten our work even further.

$$\frac{3}{8} \cdot \frac{4}{5} = \frac{3 \cdot \overset{1}{\cancel{4}}}{\underset{2}{\cancel{8}} \cdot 5}$$

$$= \frac{3 \cdot 1}{2 \cdot 5}$$

$$= \frac{3}{10}$$

EXAMPLE 2 MULTIPLYING FRACTIONS	PRACTICE EXERCISE 2

Find each product.

(a) $\frac{2}{9} \cdot \frac{3}{4} = \frac{2}{3 \cdot 3} \cdot \frac{3}{2 \cdot 2}$ Factor into primes

$= \dfrac{\overset{1}{\cancel{2}} \cdot \overset{1}{\cancel{3}}}{3 \cdot \underset{1}{\cancel{3}} \cdot \underset{1}{\cancel{2}} \cdot 2}$ Put all numerator factors over all denominator factors and divide out common prime factors

$= \dfrac{1}{3 \cdot 2}$ Remember that the numerator is 1, not 0

$= \dfrac{1}{6}$

(b) $\frac{2}{7} \cdot 14 = \frac{2}{7} \cdot \frac{14}{1}$ Write 14 as the fraction $\frac{14}{1}$

$= \frac{2}{7} \cdot \frac{2 \cdot 7}{1}$ Factor

Find each product.

(a) $\frac{3}{7} \cdot \frac{5}{6}$

(b) $\frac{8}{13} \cdot 39$

$$= \frac{2 \cdot 2 \cdot \overset{1}{7}}{\underset{1}{7 \cdot 1}} \qquad \text{Indicate product and divide out 7}$$

$$= \frac{4}{1} = 4$$

(c) $\dfrac{2}{15} \cdot \dfrac{3}{5} = \dfrac{2}{3 \cdot 5} \cdot \dfrac{3}{5} = \dfrac{2 \cdot \overset{1}{\cancel{3}}}{\underset{1}{\cancel{3} \cdot 5 \cdot 5}} = \dfrac{2}{25}$

(c) $\dfrac{7}{18} \cdot \dfrac{9}{2}$

By recognizing that 15 divided by 3 is 5, we can shorten our work as follows.

$$\frac{2}{\underset{5}{\cancel{15}}} \cdot \frac{\overset{1}{\cancel{3}}}{5} = \frac{2}{5 \cdot 5} = \frac{2}{25}$$

Answers: (a) $\frac{5}{14}$ (b) **24**   (c) $\frac{7}{4}$

---

③ PROPERTIES OF MULTIPLICATION

Multiplication of fractions has the same properties as multiplication of whole numbers. The commutative law of multiplication states that changing the order of a product does not change the result. For example,

$$\frac{2}{3} \cdot \frac{7}{8} = \frac{7}{8} \cdot \frac{2}{3}.$$

The associative law of multiplication states that regrouping products does not change the result. For example,

$$\left(\frac{2}{3} \cdot \frac{7}{8} \right) \cdot \frac{3}{4} = \frac{2}{3} \cdot \left(\frac{7}{8} \cdot \frac{3}{4} \right).$$

The number 1, written as $\frac{1}{1}$, is the multiplicative identity, since we can multiply any fraction by 1 and obtain that same fraction. For example,

$$1 \cdot \frac{2}{3} = \frac{2}{3} \cdot 1 = \frac{2}{3}.$$

The associative law allows us to multiply three fractions without using parentheses.

EXAMPLE 3   MULTIPLYING THREE FRACTIONS

Find the product of $\frac{4}{9}$, $\frac{3}{14}$, and $\frac{7}{6}$.

$$\frac{4}{9} \cdot \frac{3}{14} \cdot \frac{7}{6} = \frac{2 \cdot 2}{3 \cdot 3} \cdot \frac{3}{2 \cdot 7} \cdot \frac{7}{2 \cdot 3} \qquad \text{Factor all numerators and denominators into primes}$$

$$= \frac{\cancel{2} \cdot \cancel{2} \cdot \cancel{3} \cdot \cancel{7}}{3 \cdot 3 \cdot \cancel{2} \cdot \cancel{7} \cdot \cancel{2} \cdot 3} \qquad \text{Indicate products and divide out common prime factors}$$

$$= \frac{1}{3 \cdot 3} \qquad \text{Remember that the numerator is 1}$$

$$= \frac{1}{9}$$

PRACTICE EXERCISE 3

Find the product of $\frac{7}{5}$, $\frac{10}{21}$, and $\frac{3}{8}$.

Answer: $\frac{1}{4}$

④ POWERS AND ROOTS OF FRACTIONS

In Section 2.8 we studied powers and roots of whole numbers. We can find powers and roots of fractions in a similar manner.

EXAMPLE 4   POWERS OF FRACTIONS

Find the value of each power.

$$\overset{\longrightarrow \text{two } \frac{2}{3}\text{'s}}{}$$

(a) $\left(\dfrac{2}{3}\right)^2 = \overbrace{\dfrac{2}{3}\cdot\dfrac{2}{3}} = \dfrac{2\cdot 2}{3\cdot 3} = \dfrac{4}{9}$

(b) $\left(\dfrac{5}{4}\right)^2 = \dfrac{5}{4}\cdot\dfrac{5}{4} = \dfrac{5\cdot 5}{4\cdot 4} = \dfrac{25}{16}$

PRACTICE EXERCISE 4

Find the value of each power.

(a) $\left(\dfrac{5}{7}\right)^2$

(b) $\left(\dfrac{8}{3}\right)^2$

Answers: (a) $\frac{25}{49}$ (b) $\frac{64}{9}$

EXAMPLE 5   ROOTS OF FRACTIONS

Find the value of each radical expression.

(a) $\sqrt{\dfrac{4}{9}}$ Since $\dfrac{2}{3}\cdot\dfrac{2}{3} = \dfrac{4}{9}$, $\sqrt{\dfrac{4}{9}} = \dfrac{2}{3}$. $\frac{2}{3}$ is a square root of $\frac{4}{9}$

(b) $\sqrt{\dfrac{121}{25}}$ Since $\dfrac{11}{5}\cdot\dfrac{11}{5} = \dfrac{121}{25}$, $\sqrt{\dfrac{121}{25}} = \dfrac{11}{5}$.

PRACTICE EXERCISE 5

Find the value of each radical expression.

(a) $\sqrt{\dfrac{25}{16}}$

(b) $\sqrt{\dfrac{144}{49}}$

Answers: (a) $\frac{5}{4}$ (b) $\frac{12}{7}$

Many word problems involve finding products of fractions.

EXAMPLE 6   APPLICATION TO AREA

A farmer has a plot of land in the form of a rectangle $\frac{1}{3}$ mi wide and $\frac{7}{8}$ mi long. What is the area of the parcel of land?

First, make a sketch of the information as in Figure 3.14. Recall that the area of a rectangle is the length times the width.

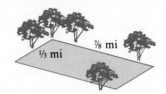

Figure 3.14

$$\text{Area} = \dfrac{7}{8}\cdot\dfrac{1}{3} = \dfrac{7}{2\cdot 2\cdot 2}\cdot\dfrac{1}{3}$$

$$= \dfrac{7\cdot 1}{2\cdot 2\cdot 2\cdot 3} = \dfrac{7}{24}$$

PRACTICE EXERCISE 6

A picture is $\frac{3}{4}$ yd wide and $\frac{10}{7}$ yd long. What is the area of the picture?

Thus, the area of the land is $\frac{7}{24}$ square miles. Figure 3.15 shows how this answer can be interpreted.

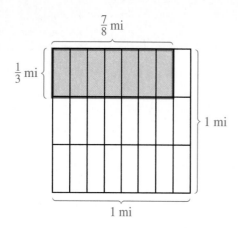

Figure 3.15

Notice that 1 square mile has been divided into 24 equal parts and that 7 of these parts, or $\frac{7}{24}$ of one square mile, corresponds to the area of the parcel of land.

Answer: $\frac{15}{14}$ sq yd

EXAMPLE 7 APPLICATION TO MAP READING	PRACTICE EXERCISE 7

On a map, 1 inch represents 32 miles. How many miles does $\frac{3}{4}$ of an inch represent?

Find $\frac{3}{4}$ of 32 miles. Used in this way *of* means *multiply*.

$$\frac{3}{4} \text{ of } 32$$

$$\frac{3}{4} \cdot 32 = \frac{3}{4} \cdot \frac{32}{1} = \frac{3}{2 \cdot 2} \cdot \frac{2 \cdot 2 \cdot 2 \cdot 2 \cdot 2}{1}$$

$$= \frac{3 \cdot \overset{1}{\cancel{2}} \cdot \overset{1}{\cancel{2}} \cdot 2 \cdot 2 \cdot 2}{\underset{1}{\cancel{2}} \cdot \underset{1}{\cancel{2}} \cdot 1}$$

$$= \frac{3 \cdot 2 \cdot 2 \cdot 2}{1}$$

$$= \frac{24}{1} = 24$$

Thus, $\frac{3}{4}$ of an inch represents 24 mi. If we notice that 4 divides into 32, we can shorten our work.

$$\frac{3}{4} \cdot \frac{32}{1} = \frac{3}{\underset{1}{\cancel{4}}} \cdot \frac{\overset{8}{\cancel{32}}}{1} = \frac{24}{1} = 24$$

If a large loaf of bread weighs 3 pounds, how much would $\frac{2}{3}$ of a loaf weigh?

Answer: 2 pounds

EXAMPLE 8 APPLICATION TO LABOR	PRACTICE EXERCISE 8

Maria makes \$35 for working one full day. How much should she receive for working $\frac{3}{5}$ of a day?

  Calculate $\frac{3}{5}$ of \$35.

$$\frac{3}{5} \text{ of } 35$$
$$\downarrow \downarrow \downarrow$$
$$\frac{3}{5} \cdot 35 = \frac{3}{5} \cdot \frac{35}{1} = \frac{3}{\cancel{5}} \cdot \frac{\overset{7}{\cancel{35}}}{1}$$
$$= \frac{3 \cdot 7}{1} = \frac{21}{1} = 21$$

Thus, she should receive \$21 for $\frac{3}{5}$ of a day's work.

Frank is paid \$12 per hour for construction work. How much should he be paid if he works $\frac{7}{3}$ hours?

Answer: \$28

3.3 EXERCISES A

Answer true or false. If the answer is false, explain why.

1. The phrase "$\frac{3}{4}$ of a number" translates to $\frac{3}{4} \cdot$ (that number).   true

2. The product of two fractions can be found by multiplying all numerators and dividing the result by the product of all denominators.   true

3. When multiplying fractions, the final product should always be reduced to lowest terms.   true

4. The fact that $\left(\frac{2}{3} \cdot \frac{3}{4}\right) \cdot \frac{2}{5} = \frac{2}{3} \cdot \left(\frac{3}{4} \cdot \frac{2}{5}\right)$ illustrates the commutative law.
    false; this illustrates the associative law

5. The fact that $1 \cdot \frac{7}{8} = \frac{7}{8}$ illustrates the associative law.
    false; this illustrates the multiplicative identity

6. The fact that $\frac{2}{3} \cdot \frac{8}{9} = \frac{8}{9} \cdot \frac{2}{3}$ illustrates the commutative law.   true

Find each product.

7. $\frac{2}{9} \cdot \frac{1}{3}$ $\frac{2}{27}$

8. $\frac{4}{3} \cdot \frac{3}{10}$ $\frac{2}{5}$

9. $\frac{1}{4} \cdot \frac{8}{9}$ $\frac{2}{9}$

10. $\frac{6}{7} \cdot \frac{14}{3}$ 4

11. $\frac{8}{9} \cdot \frac{3}{4}$ $\frac{2}{3}$

12. $\frac{3}{5} \cdot 40$ 24

13. $7 \cdot \frac{3}{21}$ 1

14. $\frac{6}{5} \cdot \frac{1}{3}$ $\frac{2}{5}$

15. $\frac{6}{35} \cdot \frac{20}{12}$ $\frac{2}{7}$

16. $\frac{21}{6} \cdot \frac{4}{7}$ 2

17. $\frac{37}{19} \cdot 0$ 0

18. $\frac{7}{5} \cdot 10$ 14

19. $\dfrac{7}{8} \cdot \dfrac{4}{35}$ $\dfrac{1}{10}$

20. $\dfrac{18}{84} \cdot \dfrac{36}{27}$ $\dfrac{2}{7}$

21. $\dfrac{35}{9} \cdot \dfrac{3}{28}$ $\dfrac{5}{12}$

22. $\dfrac{2}{3} \cdot \dfrac{10}{9} \cdot \dfrac{6}{5}$ $\dfrac{8}{9}$

23. $\dfrac{3}{8} \cdot \dfrac{24}{9} \cdot 7$ 7

24. $\dfrac{160}{169} \cdot \dfrac{13}{8} \cdot \dfrac{26}{5}$ 8

Find the value of each power.

25. $\left(\dfrac{3}{4}\right)^2$ $\dfrac{9}{16}$

26. $\left(\dfrac{1}{9}\right)^2$ $\dfrac{1}{81}$

27. $\left(\dfrac{7}{3}\right)^2$ $\dfrac{49}{9}$

Find the value of each radical.

28. $\sqrt{\dfrac{9}{16}}$ $\dfrac{3}{4}$

29. $\sqrt{\dfrac{1}{49}}$ $\dfrac{1}{7}$

30. $\sqrt{\dfrac{144}{25}}$ $\dfrac{12}{5}$

Solve.

31. Claude owns a plot of land in the shape of a rectangle $\frac{4}{7}$ km wide by $\frac{7}{8}$ km long (**km** is the abbreviation for *kilometer*). What is the area of his plot?

32. On a map, 1 inch represents 27 miles. How many miles does $\frac{4}{3}$ inch represent?

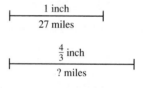

36 mi

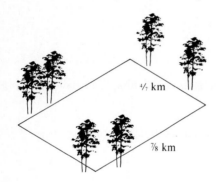

$\dfrac{1}{2}$ sq km

33. Max earned $225 for working one five-day week. How much would he earn working three days? [*Hint:* Three days is $\frac{3}{5}$ of one five-day week.] $135

34. Ramona took out a loan for two thirds the cost of a remodeling job. If the job cost $1230, how much did she borrow? $820

FOR REVIEW

Decide which fractions are equal by using cross products.

35. $\dfrac{4}{5}$ and $\dfrac{5}{6}$

$4 \cdot 6 \neq 5 \cdot 5$; so $\dfrac{4}{5} \neq \dfrac{5}{6}$

36. $\dfrac{4}{16}$ and $\dfrac{3}{12}$

$4 \cdot 12 = 16 \cdot 3$; so $\dfrac{4}{16} = \dfrac{3}{12}$

37. $\dfrac{9}{5}$ and $\dfrac{7}{4}$

$9 \cdot 4 \neq 5 \cdot 7$; so $\dfrac{9}{5} \neq \dfrac{7}{4}$

Reduce each fraction to lowest terms.

38. $\dfrac{12}{18}$ $\dfrac{2}{3}$ **39.** $\dfrac{65}{91}$ $\dfrac{5}{7}$ **40.** $\dfrac{273}{182}$ $\dfrac{3}{2}$

Find the missing term in each fraction.

41. $\dfrac{16}{6} = \dfrac{?}{3}$ 8   **42.** $\dfrac{30}{5} = \dfrac{12}{?}$ 2   **43.** $\dfrac{9}{42} = \dfrac{?}{14}$ 3

ANSWERS: 1. true 2. true 3. true 4. false (associative law) 5. false (multiplicative identity) 6. true 7. $\frac{2}{27}$
8. $\frac{2}{5}$ 9. $\frac{2}{9}$ 10. 4 11. $\frac{2}{3}$ 12. 24 13. 1 14. $\frac{2}{5}$ 15. $\frac{2}{7}$ 16. 2 17. 0 18. 14 19. $\frac{1}{10}$ 20. $\frac{2}{7}$ 21. $\frac{5}{12}$ 22. $\frac{8}{9}$ 23. 7
24. 8 25. $\frac{9}{16}$ 26. $\frac{1}{81}$ 27. $\frac{49}{9}$ 28. $\frac{3}{4}$ 29. $\frac{1}{7}$ 30. $\frac{12}{5}$ 31. $\frac{1}{2}$ sq km 32. 36 mi 33. \$135 34. \$820 35. $4 \cdot 6 \neq$
$5 \cdot 5$; so $\frac{4}{5} \neq \frac{5}{6}$ 36. $4 \cdot 12 = 16 \cdot 3$; so $\frac{4}{16} = \frac{3}{12}$ 37. $9 \cdot 4 \neq 5 \cdot 7$; so $\frac{9}{5} \neq \frac{7}{4}$ 38. $\frac{2}{3}$ 39. $\frac{5}{7}$ 40. $\frac{3}{2}$ 41. 8 42. 2 43. 3

3.3 EXERCISES B

Answer true or false. If the answer is false, explain why.

1. To reduce the number of steps in multiplying fractions, divide out common factors before multiplying numerators and multiplying denominators.   **true**

2. When common factors are divided out before multiplying numerators and denominators, the resulting fraction will be reduced to lowest terms.   **true**

3. To square a fraction, square the numerator and square the denominator.   **true**

4. The fact that $\frac{1}{2} \cdot \frac{3}{7} = \frac{3}{7} \cdot \frac{1}{2}$ illustrates the associative law.   **false; this illustrates the commutative law**

5. The fact that $\left(\frac{2}{9} \cdot \frac{5}{13}\right) \cdot \frac{4}{7} = \frac{2}{9} \cdot \left(\frac{5}{13} \cdot \frac{4}{7}\right)$ illustrates the commutative law.
   false; this illustrates the associative law

6. Any fraction multiplied by 1 gives that fraction as the product.   **true**

Find each product.

7. $\dfrac{7}{8} \cdot \dfrac{1}{2}$ $\dfrac{7}{16}$ **8.** $\dfrac{9}{4} \cdot \dfrac{4}{3}$ 3   **9.** $\dfrac{6}{5} \cdot \dfrac{10}{3}$ 4

10. $\dfrac{2}{7} \cdot \dfrac{21}{8}$ $\dfrac{3}{4}$ **11.** $\dfrac{3}{8} \cdot 40$ 15   **12.** $9 \cdot \dfrac{17}{18}$ $\dfrac{17}{2}$

13. $\dfrac{20}{35} \cdot \dfrac{7}{4}$ 1   **14.** $\dfrac{2}{3} \cdot \dfrac{5}{6}$ $\dfrac{5}{9}$ **15.** $\dfrac{62}{9} \cdot 0$ 0

16. $\dfrac{7}{8} \cdot \dfrac{24}{42}$ $\dfrac{1}{2}$ **17.** $\dfrac{27}{26} \cdot \dfrac{13}{9}$ $\dfrac{3}{2}$ **18.** $10 \cdot \dfrac{9}{2}$ 45

19. $\dfrac{35}{40} \cdot \dfrac{8}{7}$ 1   **20.** $\dfrac{99}{13} \cdot \dfrac{39}{11}$ 27   **21.** $\dfrac{32}{25} \cdot \dfrac{15}{16}$ $\dfrac{6}{5}$

22. $\dfrac{1}{3} \cdot \dfrac{20}{9} \cdot \dfrac{6}{5}$ $\dfrac{8}{9}$

23. $6 \cdot \dfrac{3}{7} \cdot \dfrac{28}{12}$ 6

24. $\dfrac{169}{5} \cdot \dfrac{25}{13} \cdot \dfrac{1}{13}$ 5

Find the value of each power.

25. $\left(\dfrac{1}{7}\right)^2$ $\dfrac{1}{49}$

26. $\left(\dfrac{5}{8}\right)^2$ $\dfrac{25}{64}$

27. $\left(\dfrac{10}{3}\right)^2$ $\dfrac{100}{9}$

Find the value of each radical.

28. $\sqrt{\dfrac{25}{9}}$ $\dfrac{5}{3}$

29. $\sqrt{\dfrac{1}{100}}$ $\dfrac{1}{10}$

30. $\sqrt{\dfrac{169}{121}}$ $\dfrac{13}{11}$

Solve.

31. A rectangular field is $\dfrac{3}{7}$ mi wide and $\dfrac{7}{9}$ mi long. What is the area of the field? $\dfrac{1}{3}$ sq mi

32. On a map, 1 inch represents 12 miles. How many miles are represented by $\dfrac{5}{6}$ inches?   **10 mi**

33. Tuition for one semester amounts to $2352. If Jack paid two thirds of this amount at early registration, how much is left to be paid?   **$784**

34. Susan's annual salary is $28,500. What is her monthly salary?   **$2375**

FOR REVIEW

Decide which fractions are equal by using cross products.

35. $\dfrac{7}{9}$ and $\dfrac{9}{11}$

$7 \cdot 11 \neq 9 \cdot 9$; so $\dfrac{7}{9} \neq \dfrac{9}{11}$

36. $\dfrac{20}{15}$ and $\dfrac{8}{6}$

$20 \cdot 6 = 15 \cdot 8$; so $\dfrac{20}{15} = \dfrac{8}{6}$

37. $\dfrac{3}{12}$ and $\dfrac{2}{6}$

$3 \cdot 6 \neq 12 \cdot 2$; so $\dfrac{3}{12} \neq \dfrac{2}{6}$

Reduce each fraction to lowest terms.

38. $\dfrac{24}{15}$ $\dfrac{8}{5}$

39. $\dfrac{81}{90}$ $\dfrac{9}{10}$

40. $\dfrac{175}{350}$ $\dfrac{1}{2}$

Find the missing term in each fraction.

41. $\dfrac{20}{8} = \dfrac{?}{2}$ 5

42. $\dfrac{4}{1} = \dfrac{12}{?}$ 3

43. $\dfrac{22}{?} = \dfrac{88}{44}$ 11

3.3 EXERCISES C

Let a and b represent natural numbers. Find the products.

1. $\dfrac{9}{a} \cdot \dfrac{a^2}{36}$ $\dfrac{a}{4}$

2. $\dfrac{7a^2}{6b^3} \cdot \dfrac{36ab}{35a^2}$

$\left[\text{Answer: } \dfrac{6a}{5b^2}\right]$

3. $\dfrac{2^2 a^4}{3^3 b^4} \cdot \dfrac{3a^2 b^3}{4ab}$ $\dfrac{a^5}{9b^2}$

3.4 DIVIDING FRACTIONS

════════ STUDENT GUIDEPOSTS ════════

❶ Reciprocal of a Fraction   ❸ Related Division Sentence
❷ Method of Dividing Fractions

❶ RECIPROCAL OF A FRACTION

Division of fractions is closely related to multiplication. In fact, a division problem can be changed to an equivalent multiplication problem using a *reciprocal*.

Reciprocal of a Fraction

Two fractions are **reciprocals** if their product is 1.

Consider the fractions $\frac{3}{4}$ and $\frac{4}{3}$.

$$\frac{3}{4} \cdot \frac{4}{3} = \frac{\overset{1}{\cancel{3}} \cdot \overset{1}{\cancel{4}}}{\underset{1}{\cancel{4}} \cdot \underset{1}{\cancel{3}}} = \frac{1}{1} = 1$$

Thus $\frac{3}{4}$ and $\frac{4}{3}$ are reciprocals of each other. Notice that $\frac{4}{3}$ is obtained when we interchange the numerator and denominator of $\frac{3}{4}$. This is true of all reciprocals and gives an easy way to find the reciprocal of a number.

Finding Reciprocals

To find the reciprocal of a fraction, interchange the numerator and denominator. Thus the reciprocal of $\frac{a}{b}$ is $\frac{b}{a}$.

EXAMPLE 1 FINDING RECIPROCALS	PRACTICE EXERCISE 1
Find the reciprocal of each number.	Find the reciprocal of each number.
(a) $\dfrac{3}{8}$ The reciprocal of $\frac{3}{8}$ is $\frac{8}{3}$.	**(a)** $\dfrac{13}{11}$
(b) $\dfrac{1}{3}$ The reciprocal of $\frac{1}{3}$ is $\frac{3}{1}$ or 3.	**(b)** $\dfrac{1}{20}$
(c) 5 Since 5 can be written as $\frac{5}{1}$, its reciprocal is $\frac{1}{5}$.	**(c)** 9
(d) 0 0 has no reciprocal since 0 is $\frac{0}{1}$ and $\frac{1}{0}$ is undefined.	**(d)** 1
	Answers: (a) $\frac{11}{13}$ (b) 20 (c) $\frac{1}{9}$ (d) 1

❷ METHOD OF DIVIDING FRACTIONS

The fact that the product of reciprocals is 1 gives us a way to convert a division problem into a multiplication problem. For example, consider

$$\frac{2}{3} \div \frac{4}{9} \quad \text{or} \quad \frac{\frac{2}{3}}{\frac{4}{9}}.$$

Recall from Section 3.2 that when the numerator and denominator of a fraction are multiplied by the same nonzero whole number, an equal fraction is formed. The same holds true when both numerator and denominator are multiplied by the same nonzero fraction. Suppose we multiply numerator and denominator of

$$\frac{\frac{2}{3}}{\frac{4}{9}}$$

by $\frac{9}{4}$, the reciprocal of the denominator, $\frac{4}{9}$.

$$\frac{\frac{2}{3} \cdot \frac{9}{4}}{\frac{4}{9} \cdot \frac{9}{4}} = \frac{\frac{2}{3} \cdot \frac{9}{4}}{\frac{\overset{1}{\cancel{4}} \cdot \overset{1}{\cancel{9}}}{\underset{1}{\cancel{9}} \cdot \underset{1}{\cancel{4}}}} = \frac{\frac{2}{3} \cdot \frac{9}{4}}{1}$$

Just as $\quad \dfrac{5}{1} = 5, \quad \dfrac{10}{1} = 10, \quad$ and $\quad \dfrac{127}{1} = 127,$

so, also, does

$$\frac{\frac{2}{3} \cdot \frac{9}{4}}{1} = \frac{2}{3} \cdot \frac{9}{4}.$$

We have just shown that the division problem

$$\frac{2}{3} \div \frac{4}{9}$$

can be converted to the multiplication problem

$$\frac{2}{3} \cdot \frac{9}{4}.$$

This is true in general.

To Divide Two Fractions

1. Replace the divisor by its reciprocal and change the division sign to multiplication.
2. Proceed as in multiplication.

EXAMPLE 2 **DIVIDING FRACTIONS**		**PRACTICE EXERCISE 2**

Find each quotient.

(a) $\dfrac{7}{12} \div \dfrac{14}{9} = \dfrac{7}{12} \cdot \dfrac{9}{14}$ Replace divisor by its reciprocal and multiply

$= \dfrac{7}{2 \cdot 2 \cdot 3} \cdot \dfrac{3 \cdot 3}{2 \cdot 7}$ Factor

$= \dfrac{\overset{1}{7} \cdot \overset{1}{\cancel{3}} \cdot 3}{2 \cdot 2 \cdot \underset{1}{\cancel{3}} \cdot 2 \cdot \underset{1}{\cancel{7}}}$ Divide out common factors

$= \dfrac{3}{2 \cdot 2 \cdot 2} = \dfrac{3}{8}$

(b) $\dfrac{20}{3} \div 5 = \dfrac{20}{3} \div \dfrac{5}{1}$ 5 can be written as $\frac{5}{1}$

$= \dfrac{20}{3} \cdot \dfrac{1}{5}$ The reciprocal of $\frac{5}{1}$ is $\frac{1}{5}$

$= \dfrac{2 \cdot 2 \cdot 5}{3} \cdot \dfrac{1}{5}$

$= \dfrac{2 \cdot 2 \cdot \overset{1}{\cancel{5}} \cdot 1}{3 \cdot \underset{1}{\cancel{5}}} = \dfrac{2 \cdot 2 \cdot 1}{3} = \dfrac{4}{3}$

Find each quotient.

(a) $\dfrac{8}{15} \div \dfrac{2}{5}$

(b) $\dfrac{14}{9} \div 7$

Answers: (a) $\frac{4}{3}$ (b) $\frac{2}{9}$

Before considering applied problems, remember that $10 \div 2$ can be interpreted as the number of 2's in 10. There are five 2's in 10. Similarly,

$$\frac{2}{5} \div \frac{1}{10}$$

means "How many $\frac{1}{10}$ths are in $\frac{2}{5}$?" Consider Figure 3.16.

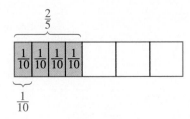

Figure 3.16

By counting we see that there are four $\frac{1}{10}$ths in $\frac{2}{5}$. But notice that

$$\frac{2}{5} \div \frac{1}{10} = \frac{2}{5} \cdot \frac{10}{1} = \frac{2 \cdot 10}{5 \cdot 1}$$

$$= \frac{2 \cdot 2 \cdot \overset{1}{\cancel{5}}}{\underset{1}{\cancel{5}} \cdot 1} = 4.$$

Thus, using the division rule we find the number of $\frac{1}{10}$ths in $\frac{2}{5}$ without relying on a figure.

EXAMPLE 3 APPLICATION OF DIVISION	PRACTICE EXERCISE 3

A piece of string 12 m long is to be cut into pieces each of length $\frac{3}{4}$ m. How many pieces can be cut?

Find the number of $\frac{3}{4}$'s in 12. That is, find $12 \div \frac{3}{4}$.

$$12 \div \frac{3}{4} = \frac{12}{1} \cdot \frac{4}{3} = \frac{3 \cdot 4}{1} \cdot \frac{4}{3} = \frac{\overset{1}{\cancel{3}} \cdot 4 \cdot 4}{1 \cdot \underset{1}{\cancel{3}}} = \frac{16}{1} = 16$$

Thus, 16 pieces of length $\frac{3}{4}$ m can be cut.

A butcher wants to cut a roast weighing 14 lb into steaks that each weigh $\frac{2}{3}$ lb. How many steaks will he get?

Answer: 21 steaks

HINT

Notice that 12 was not factored completely to primes in Example 3. Since the only factor (besides 1) in the denominator was 3, there was no need to factor 12 beyond $3 \cdot 4$. With practice, you can take shortcuts such as this.

❸ RELATED DIVISION SENTENCE

Many division problems are more easily solved by first considering a multiplication sentence and then changing it to the **related division sentence.** For example, suppose we know that $\frac{1}{2}$ of a trip is 220 miles, and want to know how long the whole trip is. If we know that

$$\frac{1}{2} \text{ of the trip is 220,}$$

we can write

$$\frac{1}{2} \text{ of } \square \text{ is } 220$$
$$\downarrow \downarrow \downarrow \ \downarrow \quad \downarrow$$
$$\frac{1}{2} \cdot \square = 220. \qquad \text{Multiplication sentence}$$

We can find out what $\square$ is by writing the related division sentence.

$$\square = \frac{220}{\frac{1}{2}} \qquad \text{Related division sentence}$$

$$\frac{220}{\frac{1}{2}} = 220 \cdot \frac{2}{1} = \frac{220}{1} \cdot \frac{2}{1} = \frac{440}{1} = 440$$

Thus $\qquad \frac{1}{2} \cdot 440 = 220 \qquad$ Multiplication sentence

and $\qquad \frac{220}{\frac{1}{2}} = 440 \qquad$ Related division sentence

are two different forms of the same sentence. We use this information in the next examples.

| EXAMPLE 4   Using the Related Division Sentence | PRACTICE EXERCISE 4 |

Find the missing number.

(a) $\frac{2}{3}$ of a number is 42.

$$\frac{2}{3} \text{ of } \square \text{ is } 42$$
$$\downarrow \downarrow \downarrow \ \downarrow \downarrow$$
$$\frac{2}{3} \cdot \square = 42 \qquad \text{Multiplication sentence}$$

$$\square = \frac{42}{\frac{2}{3}} \qquad \text{Related division sentence}$$

$$= \frac{42}{1} \cdot \frac{3}{2}$$

$$= \frac{21 \cdot \overset{1}{\cancel{2}} \cdot 3}{1 \cdot \underset{1}{\cancel{2}}} = 63$$

(b) $\frac{5}{9}$ of the weight is 20 lb.

$$\frac{5}{9} \text{ of } \square \text{ is } 20$$
$$\downarrow \downarrow \downarrow \ \downarrow \downarrow$$
$$\frac{5}{9} \cdot \square = 20 \qquad \text{Multiplication sentence}$$

$$\square = \frac{20}{\frac{5}{9}} \qquad \text{Related division sentence}$$

$$= \frac{20}{1} \cdot \frac{9}{5}$$

$$= \frac{4 \cdot \overset{1}{\cancel{5}} \cdot 9}{1 \cdot \underset{1}{\cancel{5}}} = 36 \text{ lb}$$

Find the missing number.

(a) $\frac{9}{2}$ of a number is 30.

(b) $\frac{7}{8}$ of the length is 350 cm.

Answers: (a) $\frac{20}{3}$ (b) 400 cm

| EXAMPLE 5   Application of Division | PRACTICE EXERCISE 5 |

After driving 160 miles, the Carlsons had completed $\frac{5}{7}$ of their trip. What was the total length of their trip?

    We know that $\frac{5}{7}$ *of* the trip *is* 160 miles. Remember *of* becomes *times* and *is* becomes *equals* in the multiplication sentence.

$$\frac{5}{7} \text{ of } \square \text{ is } 160$$
$$\downarrow \downarrow \downarrow \ \downarrow \ \downarrow$$
$$\frac{5}{7} \cdot \square = 160 \qquad \text{Multiplication sentence}$$

$$\square = \frac{160}{\frac{5}{7}} \qquad \text{Related division sentence}$$

After working 16 hours, Carla had completed $\frac{8}{11}$ of a job. How long will it take do the total job?

$$\frac{8}{11} \text{ of the job is } 16$$
$$\downarrow \downarrow \ \downarrow \ \downarrow \downarrow$$
$$\frac{8}{11} \cdot \square = 16$$

$$= \frac{160}{1} \cdot \frac{7}{5}$$

$$= \frac{\overset{1}{\cancel{5}} \cdot 32 \cdot 7}{1 \cdot \underset{1}{\cancel{5}}} = 224$$

Thus, the total length of the trip was 224 miles.   **Answer: 22 hr**

3.4 EXERCISES A

Find the reciprocal of each number.

1. $\frac{2}{3}$ $\frac{3}{2}$ **2.** $\frac{4}{5}$ $\frac{5}{4}$ **3.** $\frac{1}{7}$ 7   **4.** 3 $\frac{1}{3}$ **5.** 0   none

Find each quotient.

6. $\frac{1}{3} \div \frac{5}{6}$ $\frac{2}{5}$ **7.** $\frac{2}{3} \div \frac{1}{9}$ 6   **8.** $\frac{2}{3} \div \frac{4}{9}$ $\frac{3}{2}$ **9.** $\frac{3}{7} \div \frac{9}{28}$ $\frac{4}{3}$

10. $7 \div \frac{14}{3}$ $\frac{3}{2}$ **11.** $\frac{5}{4} \div \frac{1}{2}$ $\frac{5}{2}$ **12.** $\frac{20}{9} \div \frac{10}{15}$ $\frac{10}{3}$ **13.** $\frac{8}{11} \div 4$ $\frac{2}{11}$

14 $\frac{13}{15} \div \frac{39}{5}$ $\frac{1}{9}$ **15.** $\frac{2}{7} \div \frac{2}{7}$ 1   **16.** $\frac{4}{7} \div 44$ $\frac{1}{77}$ **17.** $\frac{20}{27} \div \frac{35}{36}$ $\frac{16}{21}$

Find the missing number.

18. $\frac{3}{8} \cdot \square = 18$ 48   **19.** $\frac{11}{2} \cdot \square = 33$ 6   **20.** $\square \cdot \frac{7}{3} = 35$ 15

21. $\frac{10}{7}$ of $\square$ is 20   14   **22** $\frac{10}{9}$ of $\square$ is 25 $\frac{45}{2}$ **23.** $\frac{1}{8}$ of $\square$ is $\frac{2}{9}$ $\frac{16}{9}$

24. $\frac{2}{11}$ of $\square$ is 10   55   **25.** $\frac{8}{1}$ of $\square$ is 56   7   **26.** $\frac{10}{9}$ of $\square$ is $\frac{8}{3}$ $\frac{12}{5}$

Solve.

27. A piece of rope 10 m long is to be cut into pieces each of length $\frac{2}{3}$ m. How many pieces can be cut?

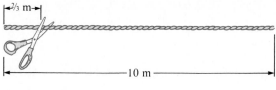

15 pieces

28 After driving 365 miles, the Johnsons had completed $\frac{5}{6}$ of their trip. What was the total length of the trip?

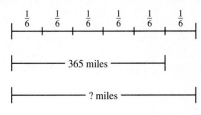

438 mi

29. A piece of wire $\frac{7}{8}$ m long is to be cut into 14 equal pieces. What is the length of each piece? $\frac{1}{16}$ **m**

30. How many pills, each containing $\frac{9}{2}$ grain of an antibiotic, must be taken to make an 18-grain dosage?
4 pills

FOR REVIEW

31. The value of a house is $75,000. For insurance purposes, the value of the contents of the house are estimated to be $\frac{1}{5}$ of the value of the house. What is the estimated value of the contents? [*Hint:* $\frac{1}{5}$ *of* value of house = value of contents. What does *of* translate to?]   **$15,000**

32. Pam receives an inheritance of $8000. If she invests $\frac{3}{16}$ of this amount in bonds and $\frac{5}{8}$ of it in a mutual fund, how much does she invest in each category?
$1500 in bonds; $5000 in the mutual fund

Find each product.

33. $\frac{4}{11} \cdot \frac{121}{32}$ $\frac{11}{8}$

34. $13 \cdot \frac{7}{169}$ $\frac{7}{13}$

35. $\frac{2}{7} \cdot \frac{28}{3} \cdot \frac{9}{4}$ 6

ANSWERS: 1. $\frac{3}{2}$ **2.** $\frac{5}{4}$ **3.** 7 **4.** $\frac{1}{3}$ **5.** 0 has no reciprocal **6.** $\frac{2}{5}$ **7.** 6 **8.** $\frac{3}{2}$ **9.** $\frac{4}{3}$ **10.** $\frac{3}{2}$ **11.** $\frac{5}{2}$ **12.** $\frac{10}{3}$ **13.** $\frac{2}{11}$
14. $\frac{1}{9}$ **15.** 1 **16.** $\frac{1}{77}$ **17.** $\frac{16}{21}$ **18.** 48 **19.** 6 **20.** 15 **21.** 14 **22.** $\frac{45}{2}$ **23.** $\frac{16}{9}$ **24.** 55 **25.** 7 **26.** $\frac{12}{5}$ **27.** 15 pieces
28. 438 mi **29.** $\frac{1}{16}$ m **30.** 4 pills **31.** $15,000 **32.** $1500 in bonds; $5000 in the mutual fund **33.** $\frac{11}{8}$ **34.** $\frac{7}{13}$
35. 6

3.4   EXERCISES B

Find the reciprocal of each number.

1. $\frac{4}{7}$ $\frac{7}{4}$

2. $\frac{11}{3}$ $\frac{3}{11}$

3. $\frac{1}{15}$ 15

4. 6 $\frac{1}{6}$

5. $\frac{100}{99}$ $\frac{99}{100}$

Find each quotient.

6. $\frac{1}{5} \div \frac{2}{15}$ $\frac{3}{2}$

7. $\frac{3}{4} \div \frac{1}{8}$ 6

8. $\frac{2}{3} \div \frac{1}{6}$ 4

9. $\frac{4}{9} \div \frac{1}{3}$ $\frac{4}{3}$

10. $9 \div \dfrac{18}{7}$ $\dfrac{7}{2}$ **11.** $\dfrac{7}{4} \div \dfrac{1}{2}$ $\dfrac{7}{2}$ **12.** $\dfrac{5}{7} \div \dfrac{15}{28}$ $\dfrac{4}{3}$ **13.** $\dfrac{8}{13} \div 4$ $\dfrac{2}{13}$

14. $\dfrac{13}{17} \div \dfrac{39}{34}$ $\dfrac{2}{3}$ **15.** $\dfrac{4}{3} \div \dfrac{4}{3}$ 1 **16.** $\dfrac{6}{11} \div 36$ $\dfrac{1}{66}$ **17.** $\dfrac{20}{16} \div \dfrac{10}{8}$ 1

Find the missing number.

18. $\dfrac{2}{9} \cdot \square = 12$ 54   **19.** $\dfrac{13}{7} \cdot \square = 39$ 21   **20.** $\square \cdot \dfrac{3}{20} = 27$ 180

21. $\dfrac{5}{16}$ of $\square$ is 15   48   **22.** $\dfrac{8}{7}$ of $\square$ is 20 $\dfrac{35}{2}$ **23.** $\dfrac{1}{6}$ of $\square$ is $\dfrac{7}{13}$ $\dfrac{42}{13}$

24. $\dfrac{4}{15}$ of $\square$ is 30 $\dfrac{225}{2}$ **25.** $\dfrac{6}{1}$ of $\square$ is 24   4   **26.** $\dfrac{15}{7}$ of $\square$ is $\dfrac{5}{14}$ $\dfrac{1}{6}$

Solve.

27. Beth needs $\frac{3}{8}$ of a yard of material to make a pillow. How many pillows can she make from 18 yards of material?   **48 pillows**

28. A tank holds 220 gallons of fuel when it is $\frac{4}{5}$ full. What is the capacity of the tank?   **275 gallons**

29. If five people divide a $\frac{3}{4}$-pound box of candy, how much should each receive? $\frac{3}{20}$ **lb**

30. When Joey hammers a nail into a board, he sinks it $\frac{2}{9}$ of an inch with each blow. How many times will he have to hit a 4-inch nail to drive it completely into the board?   **18**

FOR REVIEW

31. The day before spring break, only $\frac{3}{5}$ of the students come to Professor Condon's class. If she has 60 students enrolled, how many are in attendance?
36 students

32. A container contains $\frac{2}{3}$ of a gallon of milk when full. How much milk is in the container when it is $\frac{1}{4}$ full?
$\frac{1}{6}$ **gallon**

Find each product.

33. $\dfrac{5}{13} \cdot \dfrac{52}{55}$ $\dfrac{4}{11}$ **34.** $\dfrac{19}{144} \cdot 12$ $\dfrac{19}{12}$ **35.** $\dfrac{3}{4} \cdot \dfrac{42}{6} \cdot \dfrac{5}{14}$ $\dfrac{15}{8}$

3.4 EXERCISES C

Let a represent a nonzero fraction. Find a, by writing the related division sentence. The letter a is used in place of the symbol $\square$ in the examples.

1. $5 \cdot a = 7$

$\left[\text{Answer: } \frac{7}{5}\right]$

2. $\dfrac{2}{3} \cdot a = \dfrac{14}{9}$ $\dfrac{7}{3}$

3. $\dfrac{a}{\frac{3}{4}} = \dfrac{1}{3}$

$\left[\text{Answer: } \frac{1}{4}\right]$

3.5 MIXED NUMBERS

STUDENT GUIDEPOSTS

1 Mixed Numbers

2 Changing an Improper Fraction to a Mixed Number

3 Changing a Mixed Number to an Improper Fraction

1 MIXED NUMBERS

Many of the answers to word problems are improper fractions which may look a bit strange. For example, $\frac{9}{2}$ miles is not as familiar to us as $4\frac{1}{2}$ miles. For this reason, improper fractions are often written as *mixed numbers*.

A **mixed number** is the sum of a whole number and a proper fraction. The improper fraction $\frac{7}{6}$ can be represented as in Figure 3.17.

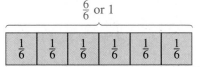

Figure 3.17

Since all 6 parts of the first bar and 1 additional part of the second are shaded,

$$\frac{7}{6} \quad \text{is} \quad 1 + \frac{1}{6}. \qquad \text{1 bar} + \tfrac{1}{6} \text{ of another bar}$$

To write $1 + \dfrac{1}{6}$ as a mixed number, omit the plus sign and write

$$1\frac{1}{6}, \text{ read ``one and one sixth.''}$$

2 CHANGING AN IMPROPER FRACTION TO A MIXED NUMBER

To change an improper fraction to a mixed number, think of the improper fraction as a division problem. For example, $\frac{7}{6}$ is

$$7 \div 6.$$

Perform the division.

$$\text{divisor} \longrightarrow 6\overline{)7} \begin{array}{l} 1 \longleftarrow \text{quotient} \\ \longleftarrow \text{dividend} \\ \underline{6} \\ 1 \longleftarrow \text{remainder} \end{array}$$

Recall that this division problem can be checked as follows:

$$7 = 6 \cdot 1 + 1 \qquad \text{(dividend)} = \text{(divisor)} \cdot \text{(quotient)} + \text{(remainder)}$$

Dividing through by 6, $\dfrac{7}{6} = \dfrac{\cancel{6} \cdot 1}{\cancel{6}} + \dfrac{1}{6} = 1 + \dfrac{1}{6} = 1\dfrac{1}{6}.$

This is a way to obtain a mixed number from an improper fraction.

To Change an Improper Fraction to a Mixed Number

1. Reduce the fraction to lowest terms (if necessary).
2. Divide the numerator by the denominator to obtain the quotient and remainder.
3. The mixed number is

$$\text{quotient} + \frac{\text{remainder}}{\text{divisor}}.$$

EXAMPLE 1 CHANGING IMPROPER FRACTIONS TO MIXED NUMBERS

Change each improper fraction to a mixed number.

(a) $\dfrac{15}{4}$

$$\text{divisor} \longrightarrow 4\overline{)15} \begin{array}{l} \longleftarrow \text{quotient} \\ \longleftarrow \text{dividend} \end{array}$$
$$\dfrac{12}{3} \longleftarrow \text{remainder}$$

Thus, the mixed number is $3\frac{3}{4}$, read "three and three fourths."

(b) $\dfrac{121}{35}$

$$\text{divisor} \longrightarrow 35\overline{)121} \begin{array}{l} \longleftarrow \text{quotient} \\ \longleftarrow \text{dividend} \end{array}$$
$$\dfrac{105}{16} \longleftarrow \text{remainder}$$

Thus, $\dfrac{121}{35} = 3\dfrac{16}{35}$.

(c) $\dfrac{27}{6}$

In this case, first reduce $\frac{27}{6}$ to lowest terms.

$$\frac{27}{6} = \frac{\overset{1}{\cancel{3}} \cdot 9}{\underset{1}{\cancel{3}} \cdot 2} = \frac{9}{2} \qquad 2\overline{)9} \\ \dfrac{8}{1}$$

Thus, $\dfrac{27}{6} = \dfrac{9}{2} = 4\dfrac{1}{2}$.

PRACTICE EXERCISE 1

Change each improper fraction to a mixed number.

(a) $\dfrac{23}{7}$

(b) $\dfrac{114}{23}$

(c) $\dfrac{35}{15}$

Answers: (a) $3\frac{2}{7}$ (b) $4\frac{22}{23}$
(c) $2\frac{1}{3}$

❸ CHANGING A MIXED NUMBER TO AN IMPROPER FRACTION

We now want to change a mixed number into an improper fraction. From Figure 3.18, we can see that the mixed number $3\frac{1}{2}$ is the same as the improper fraction $\frac{7}{2}$.

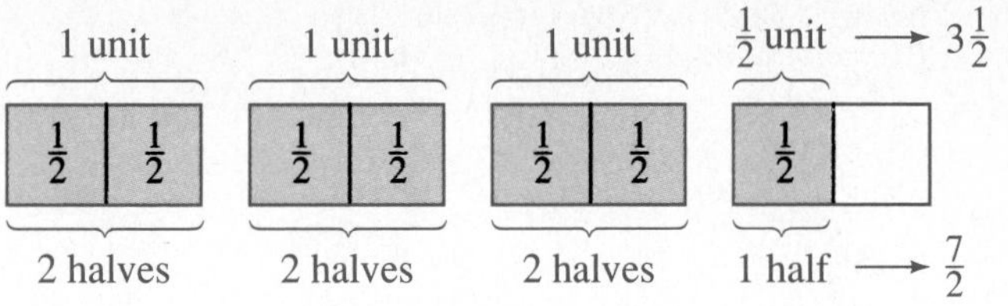

Figure 3.18

Each whole unit has 2 halves and there are 3 whole units plus 1 half unit.

$$2 \cdot 3 + 1 = 7.$$

Putting this result over the denominator, 2, gives the improper fraction $\frac{7}{2}$. The same result can be obtained by the following procedure.

$$3\frac{1}{2} \text{ over} = \frac{2 \cdot 3 + 1}{2} = \frac{7}{2}$$

To Change a Mixed Number to an Improper Fraction

1. Multiply the denominator by the whole number.
2. Add the result to the numerator of the fraction.
3. This sum, over the denominator, is the improper fraction.

EXAMPLE 2 CHANGING MIXED NUMBERS TO IMPROPER FRACTIONS

Change each mixed number to an improper fraction.

(a) $3\frac{1}{4}$

$$3\frac{1}{4} \text{ over} = \frac{4 \cdot 3 + 1}{4} = \frac{12 + 1}{4} = \frac{13}{4}$$

(b) $17\frac{2}{3}$

$$17\frac{2}{3} = \frac{3 \cdot 17 + 2}{3} = \frac{51 + 2}{3} = \frac{53}{3}$$

PRACTICE EXERCISE 2

Change each mixed number to an improper fraction.

(a) $7\frac{2}{5}$

(b) $22\frac{1}{4}$

Answers: (a) $\frac{37}{5}$ (b) $\frac{89}{4}$

3.5 EXERCISES A

Change each improper fraction to a mixed number.

1. $\frac{7}{5}$ $1\frac{2}{5}$

2. $\frac{17}{3}$ $5\frac{2}{3}$

3. $\frac{23}{8}$ $2\frac{7}{8}$

4. $\dfrac{45}{6}$ $\quad 7\dfrac{1}{2}$

5. $\dfrac{71}{10}$ $\quad 7\dfrac{1}{10}$

6. $\dfrac{19}{2}$ $\quad 9\dfrac{1}{2}$

7. $\dfrac{135}{4}$ $\quad 33\dfrac{3}{4}$

8 $\dfrac{257}{9}$ $\quad 28\dfrac{5}{9}$

9. $\dfrac{142}{11}$ $\quad 12\dfrac{10}{11}$

Change each mixed number to an improper fraction.

10. $3\dfrac{1}{3}$ $\quad \dfrac{10}{3}$

11. $4\dfrac{3}{5}$ $\quad \dfrac{23}{5}$

12. $7\dfrac{1}{8}$ $\quad \dfrac{57}{8}$

13. $10\dfrac{3}{4}$ $\quad \dfrac{43}{4}$

14. $9\dfrac{7}{11}$ $\quad \dfrac{106}{11}$

15. $21\dfrac{2}{3}$ $\quad \dfrac{65}{3}$

16. $5\dfrac{7}{8}$ $\quad \dfrac{47}{8}$

17 $32\dfrac{7}{10}$ $\quad \dfrac{327}{10}$

18. $66\dfrac{2}{3}$ $\quad \dfrac{200}{3}$

FOR REVIEW

Solve.

19. A tank holds 560 gallons of water when it is $\frac{5}{8}$ full. What is the capacity of the tank? **896 gallons**

20. How many steaks each weighing $\frac{3}{8}$ pound can be cut from a 15-pound roast? **40 steaks**

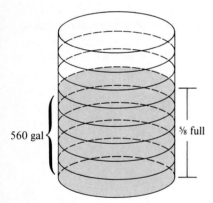

560 gal ⅝ full

Find each quotient.

21. $\dfrac{10}{21} \div \dfrac{20}{14}$ $\quad \dfrac{1}{3}$

22. $\dfrac{5}{13} \div 10$ $\quad \dfrac{1}{26}$

23. $38 \div \dfrac{19}{2}$ $\quad 4$

ANSWERS: 1. $1\frac{2}{5}$ 2. $5\frac{2}{3}$ 3. $2\frac{7}{8}$ 4. $7\frac{1}{2}$ 5. $7\frac{1}{10}$ 6. $9\frac{1}{2}$ 7. $33\frac{3}{4}$ 8. $28\frac{5}{9}$ 9. $12\frac{10}{11}$ 10. $\frac{10}{3}$ 11. $\frac{23}{5}$ 12. $\frac{57}{8}$ 13. $\frac{43}{4}$
14. $\frac{106}{11}$ 15. $\frac{65}{3}$ 16. $\frac{47}{8}$ 17. $\frac{327}{10}$ 18. $\frac{200}{3}$ 19. 896 gallons 20. 40 steaks 21. $\frac{1}{3}$ 22. $\frac{1}{26}$ 23. 4

3.5 EXERCISES B

Change each improper fraction to a mixed number.

1. $\dfrac{8}{5}$ $1\dfrac{3}{5}$

2. $\dfrac{19}{3}$ $6\dfrac{1}{3}$

3. $\dfrac{27}{8}$ $3\dfrac{3}{8}$

4. $\dfrac{51}{6}$ $8\dfrac{1}{2}$

5. $\dfrac{86}{10}$ $8\dfrac{3}{5}$

6. $\dfrac{17}{2}$ $8\dfrac{1}{2}$

7. $\dfrac{173}{4}$ $43\dfrac{1}{4}$

8. $\dfrac{269}{9}$ $29\dfrac{8}{9}$

9. $\dfrac{312}{11}$ $28\dfrac{4}{11}$

Change each mixed number to an improper fraction.

10. $3\dfrac{2}{3}$ $\dfrac{11}{3}$

11. $5\dfrac{2}{5}$ $\dfrac{27}{5}$

12. $6\dfrac{3}{8}$ $\dfrac{51}{8}$

13. $12\dfrac{1}{4}$ $\dfrac{49}{4}$

14. $11\dfrac{7}{11}$ $\dfrac{128}{11}$

15. $26\dfrac{1}{3}$ $\dfrac{79}{3}$

16. $3\dfrac{7}{8}$ $\dfrac{31}{8}$

17. $4\dfrac{7}{10}$ $\dfrac{47}{10}$

18. $76\dfrac{2}{3}$ $\dfrac{230}{3}$

FOR REVIEW

Solve.

19. A wire 20 inches long is to be cut into pieces each of length $\frac{5}{8}$ inch. How many pieces can be cut?
32 pieces

20. After hiking 26 miles, the campers had completed $\frac{2}{3}$ of a hike. What was the total length of the hike?
39 mi

Find each quotient.

21. $\dfrac{8}{15} \div \dfrac{22}{6}$ $\dfrac{8}{55}$

22. $\dfrac{9}{11} \div 18$ $\dfrac{1}{22}$

23. $108 \div \dfrac{12}{7}$ 63

3.5 EXERCISES C

Change each improper fraction to a mixed number.

1. $\dfrac{1055}{7}$ $150\dfrac{5}{7}$

2. $\dfrac{2615}{23}$
$\left[\text{Answer: } 113\dfrac{16}{23}\right]$

Change each mixed number to an improper fraction.

3. $201\dfrac{3}{17}$ $\dfrac{3420}{17}$

4. $463\dfrac{51}{103}$
$\left[\text{Answer: } \dfrac{47,740}{103}\right]$

3.6 MULTIPLYING AND DIVIDING MIXED NUMBERS

═══ STUDENT GUIDEPOSTS ═══
① Method for Multiplying and Dividing Mixed Numbers
② Applications of Mixed Numbers

① METHOD FOR MULTIPLYING AND DIVIDING MIXED NUMBERS

Mixed numbers can be multiplied and divided just like proper and improper fractions. Change any mixed numbers to improper fractions and proceed as in Sections 3.3 and 3.4.

To Multiply or Divide Mixed Numbers

1. Change any mixed numbers to improper fractions.
2. Proceed as when multiplying or dividing fractions.

EXAMPLE 1 MULTIPLYING MIXED NUMBERS

Multipy.

(a) $2\frac{1}{2} \cdot 3\frac{1}{4}$

First change to improper fractions.

$$2\frac{1}{2} = \frac{2 \cdot 2 + 1}{2} = \frac{5}{2} \quad \text{and} \quad 3\frac{1}{4} = \frac{4 \cdot 3 + 1}{4} = \frac{13}{4}$$

$$2\frac{1}{2} \cdot 3\frac{1}{4} = \frac{5}{2} \cdot \frac{13}{4} = \frac{65}{8} = 8\frac{1}{8} \quad \text{Change improper fraction to a mixed number}$$

(b) $7\frac{3}{8} \cdot 4 = \frac{8 \cdot 7 + 3}{8} \cdot \frac{4}{1} = \frac{59}{8} \cdot \frac{4}{1} = \frac{59 \cdot \overset{1}{\cancel{4}}}{2 \cdot \underset{1}{\cancel{4}} \cdot 1}$

$$= \frac{59}{2} = 29\frac{1}{2}$$

(c) $\left(5\frac{1}{4}\right)^2$

Find $\left(5\frac{1}{4}\right)\left(5\frac{1}{4}\right)$.

$$5\frac{1}{4} = \frac{4 \cdot 5 + 1}{4} = \frac{21}{4} \quad \text{Change to improper fraction}$$

$$\left(5\frac{1}{4}\right)^2 = \left(\frac{21}{4}\right)^2 = \left(\frac{21}{4}\right)\left(\frac{21}{4}\right)$$

$$= \frac{21 \cdot 21}{4 \cdot 4} = \frac{441}{16} = 27\frac{9}{16}$$

PRACTICE EXERCISE 1

Multiply.

(a) $1\frac{1}{5} \cdot 5\frac{2}{7}$

(b) $12 \cdot 10\frac{2}{9}$

(c) $\left(7\frac{2}{3}\right)^2$

Answers: (a) $6\frac{12}{35}$ (b) $122\frac{2}{3}$ (c) $58\frac{7}{9}$

///////////// **C A U T I O N** /////////////

It is tempting to try to multiply

$$2\frac{1}{2} \cdot 3\frac{1}{4}$$

by multiplying the whole numbers $(2 \cdot 3 = 6)$ and then multiplying the fractions $\left(\frac{1}{2} \cdot \frac{1}{4} = \frac{1}{8}\right)$. That would make the product $6\frac{1}{8}$. But from Example 1(a), this product is $8\frac{1}{8}$. Thus, the parts of mixed numbers *should not* be multiplied or divided separately.

/////////////

EXAMPLE 2   DIVIDING MIXED NUMBERS

Divide.

(a) $4\frac{2}{3} \div 3\frac{1}{2}$

First change to improper fractions.

$$4\frac{2}{3} = \frac{3 \cdot 4 + 2}{3} = \frac{14}{3} \quad \text{and} \quad 3\frac{1}{2} = \frac{2 \cdot 3 + 1}{2} = \frac{7}{2}$$

$$4\frac{2}{3} \div 3\frac{1}{2} = \frac{14}{3} \div \frac{7}{2} = \frac{14}{3} \cdot \frac{2}{7} \quad \text{Multiply by } \tfrac{2}{7}\text{, the reciprocal of } \tfrac{7}{2}$$

$$= \frac{2 \cdot 7}{3} \cdot \frac{2}{7}$$

$$= \frac{2 \cdot \overset{1}{7} \cdot 2}{3 \cdot \underset{1}{7}} = \frac{4}{3} = 1\frac{1}{3}$$

(b) $5\frac{3}{5} \div 2 = \frac{28}{5} \div 2$

$$= \frac{28}{5} \cdot \frac{1}{2} = \frac{\overset{1}{2} \cdot 14 \cdot 1}{5 \cdot \underset{1}{2}} = \frac{14}{5} = 2\frac{4}{5}$$

(c) $7 \div 8\frac{2}{5} = 7 \div \frac{42}{5} = 7 \cdot \frac{5}{42} = \frac{\overset{1}{7} \cdot 5}{\underset{1}{7} \cdot 6} = \frac{5}{6}$

PRACTICE EXERCISE 2

Divide.

(a) $2\frac{4}{5} \div 1\frac{3}{4}$

(b) $10\frac{5}{8} \div 5$

(c) $4 \div 7\frac{1}{3}$

Answers: (a) $1\frac{3}{5}$ (b) $2\frac{1}{8}$ (c) $\frac{6}{11}$

---

② APPLICATIONS OF MIXED NUMBERS

Many applied problems can be expressed better with mixed numbers than with improper fractions. For example, we would normally say that Michael can hike $4\frac{1}{2}$ miles and not that he can hike $\frac{9}{2}$ miles.

| EXAMPLE 3 APPLICATION TO RECREATION | PRACTICE EXERCISE 3 |

Michael can hike $4\frac{1}{2}$ miles in one hour. At the same rate, how far can he hike in 6 hours?

Since he can go $4\frac{1}{2}$ miles in one hour, he can go 6 times that distance in 6 hours.

$$6 \cdot 4\frac{1}{2} = 6 \cdot \frac{9}{2} \quad \text{Change to an improper fraction}$$

$$= \frac{6}{1} \cdot \frac{9}{2}$$

$$= \frac{\overset{1}{\cancel{2}} \cdot 3 \cdot 9}{\underset{1}{\cancel{2}}} = 27$$

Thus, Michael can go 27 miles in 6 hours.

One bag of fruit weighs $8\frac{3}{4}$ pounds. How much would 12 of these bags weigh?

Answer: 105 pounds

| EXAMPLE 4 APPLICATION TO SEWING | PRACTICE EXERCISE 4 |

A uniform requires $2\frac{5}{8}$ yards of material. How many uniforms can be made from $57\frac{3}{4}$ yards of material?

Since each uniform requires $2\frac{5}{8}$ yards, then $2\frac{5}{8}$ times the number of uniforms must be $57\frac{3}{4}$.

$$2\frac{5}{8} \cdot \square = 57\frac{3}{4}$$

$$\square = 57\frac{3}{4} \div 2\frac{5}{8} \quad \text{Related division sentence}$$

$$= \frac{231}{4} \div \frac{21}{8}$$

$$= \frac{231}{4} \cdot \frac{8}{21}$$

$$= \frac{11 \cdot \overset{1}{\cancel{21}} \cdot \overset{1}{\cancel{4}} \cdot 2}{\underset{1}{\cancel{4}} \cdot \underset{1}{\cancel{21}}} = 22$$

Thus, 22 uniforms can be made.

How many $2\frac{2}{5}$-lb packages of fish can be obtained from 132 lb of fish?

Answer: 55 packages

3.6 EXERCISES A

In the following exercises, give all answers as mixed numbers.

Multiply.

1. $3\frac{1}{5} \cdot 4\frac{2}{3}$ $14\frac{14}{15}$

2. $5\frac{4}{7} \cdot 2\frac{2}{13}$ 12

3. $12\frac{2}{9} \cdot 4\frac{1}{5}$ $51\frac{1}{3}$

4. $6\frac{4}{5} \cdot 11$ $74\frac{4}{5}$

5. $\left(3\frac{1}{2}\right)^2$ $12\frac{1}{4}$

6. $\left(2\frac{2}{5}\right)^2$ $5\frac{19}{25}$

7. $3 \cdot 1\frac{1}{2} \cdot 2\frac{1}{3}$ $10\frac{1}{2}$

8 $2\frac{3}{5} \cdot 6\frac{1}{4} \cdot 12$ 195

9. $11\frac{1}{2} \cdot 5\frac{1}{4} \cdot \frac{2}{7}$ $17\frac{1}{4}$

Divide.

10. $3\frac{1}{6} \div 2\frac{1}{2}$ $1\frac{4}{15}$

11. $8\frac{5}{6} \div 4\frac{1}{9}$ $2\frac{11}{74}$

12. $3 \div 7\frac{4}{5}$ $\frac{5}{13}$

13. $16\frac{1}{2} \div 8$ $2\frac{1}{16}$

14. $\frac{7}{8} \div 5\frac{1}{4}$ $\frac{1}{6}$

15. $10\frac{1}{5} \div \frac{3}{10}$ 34

16. $9\frac{2}{7} \div 1\frac{2}{3}$ $5\frac{4}{7}$

17. $3\frac{1}{8} \div 30$ $\frac{5}{48}$

18. $80 \div 4\frac{2}{7}$ $18\frac{2}{3}$

Solve.

19. A wheel on a child's car is turning at a rate of $105\frac{1}{2}$ revolutions per minute. If the car is driven for 14 minutes, how many revolutions does the wheel make?   **1477 revolutions**

20. Cindy is paid $1\frac{3}{4}$ dollars per hour for babysitting. If she sits for a period of $4\frac{1}{2}$ hours, how much is she paid? $7\frac{7}{8}$ **dollars**

21. A car traveled 420 miles on $12\frac{4}{5}$ gallons of gas. How many miles per gallon did it get?
$32\frac{13}{16}$ **mpg**

22 A pompon girl's uniform requires $3\frac{1}{8}$ yards of material. How many uniforms can be made from 25 yards of material?   **8 uniforms**

23. The weight of one cubic foot of water is $62\frac{1}{2}$ lb. How much does $3\frac{1}{3}$ cubic feet of water weigh?
$208\frac{1}{3}$ **lb**

24. How many $5\frac{1}{2}$-gram samples can be obtained from 176 grams of a chemical?   **32 samples**

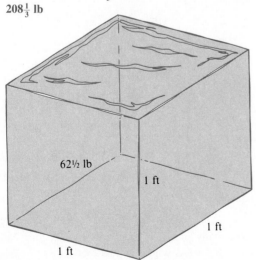

62½ lb

1 ft

1 ft

1 ft

1 cubic foot of water

FOR REVIEW

Change each improper fraction to a mixed number.

25. $\frac{9}{7}$ $1\frac{2}{7}$ **26.** $\frac{18}{5}$ $3\frac{3}{5}$ **27.** $\frac{123}{15}$ $8\frac{1}{5}$

Change each mixed number to an improper fraction.

28. $5\frac{2}{11}$ $\frac{57}{11}$ **29.** $10\frac{6}{13}$ $\frac{136}{13}$ **30.** $81\frac{3}{20}$ $\frac{1623}{20}$

ANSWERS: 1. $14\frac{14}{15}$ 2. 12 3. $51\frac{1}{3}$ 4. $74\frac{4}{5}$ 5. $12\frac{1}{4}$ 6. $5\frac{19}{25}$ 7. $10\frac{1}{2}$ 8. 195 9. $17\frac{1}{4}$ 10. $1\frac{4}{15}$ 11. $2\frac{11}{74}$ 12. $\frac{5}{13}$ 13. $2\frac{1}{16}$ 14. $\frac{1}{6}$ 15. 34 16. $5\frac{4}{7}$ 17. $\frac{5}{48}$ 18. $18\frac{2}{3}$ 19. 1477 revolutions 20. $7\frac{7}{8}$ dollars 21. $32\frac{13}{16}$ mpg 22. 8 uniforms 23. $208\frac{1}{3}$ lb 24. 32 samples 25. $1\frac{2}{7}$ 26. $3\frac{3}{5}$ 27. $8\frac{1}{5}$ 28. $\frac{57}{11}$ 29. $\frac{136}{13}$ 30. $\frac{1623}{20}$

3.6 EXERCISES B

In the following exercises, give all answers as mixed numbers.

Multiply.

1. $1\frac{3}{4} \cdot 2\frac{1}{7}$ $3\frac{3}{4}$ **2.** $6\frac{1}{8} \cdot 3\frac{3}{14}$ $19\frac{11}{16}$ **3.** $2\frac{2}{5} \cdot 8\frac{1}{3}$ 20

4. $5\frac{2}{9} \cdot 12$ $62\frac{2}{3}$ **5.** $\left(2\frac{1}{3}\right)^2$ $5\frac{4}{9}$ **6.** $\left(6\frac{1}{2}\right)^2$ $42\frac{1}{4}$

7. $4 \cdot 2\frac{1}{3} \cdot 5\frac{1}{4}$ 49   **8.** $1\frac{1}{10} \cdot 5\frac{5}{7} \cdot 14$ 88   **9.** $\frac{2}{11} \cdot 7\frac{1}{3} \cdot 10\frac{1}{4}$ $13\frac{2}{3}$

Divide.

10. $5\frac{5}{6} \div 1\frac{1}{2}$ $3\frac{8}{9}$ **11.** $7\frac{4}{5} \div 3\frac{1}{4}$ $2\frac{2}{5}$ **12.** $5 \div 7\frac{1}{4}$ $\frac{20}{29}$

13. $13\frac{1}{2} \div 5$ $2\frac{7}{10}$ **14.** $\frac{4}{7} \div 2\frac{2}{9}$ $\frac{9}{35}$ **15.** $9\frac{3}{5} \div \frac{4}{15}$ 36

16. $8\frac{1}{15} \div 7\frac{1}{5}$ $1\frac{13}{108}$ **17.** $6\frac{3}{4} \div 60$ $\frac{9}{80}$ **18.** $20 \div 5\frac{5}{6}$ $3\frac{3}{7}$

Solve.

19. A motor rotates at $221\frac{1}{2}$ revolutions per second. How many revolutions will it make in one minute?   **13,290 revolutions**

20. Bill worked for $5\frac{1}{5}$ hours as a babysitter. How much was he paid if he received $1\frac{1}{4}$ dollars per hour? $6\frac{1}{2}$ dollars

21. A truck traveled 378 miles on $15\frac{2}{5}$ gallons of fuel. How many miles per gallon did it get? $24\frac{6}{11}$ mpg

22. How many pieces of steel rod $3\frac{1}{3}$ feet long can be cut from a rod which is 110 feet long?   33

23. The area of a room is its length times its width. What is the area of a room which is $8\frac{3}{4}$ yards long and $6\frac{2}{5}$ yards wide?   **56 square yards**

24. There are 26 pieces of candy of the same size in a box. The total weight of the candy is $5\frac{1}{5}$ pounds. How much does each piece of candy weigh?

$\frac{1}{5}$ **lb**

FOR REVIEW

Change each improper fraction to a mixed number.

25. $\frac{11}{7}$ $1\frac{4}{7}$

26. $\frac{21}{4}$ $5\frac{1}{4}$

27. $\frac{218}{12}$ $18\frac{1}{6}$

Change each mixed number to an improper fraction.

28. $4\frac{1}{12}$ $\frac{49}{12}$

29. $11\frac{3}{5}$ $\frac{58}{5}$

30. $101\frac{1}{8}$ $\frac{809}{8}$

3.6 EXERCISES C

Perform the indicated operations.

1. $36\frac{2}{13} \cdot 27\frac{5}{8}$ $998\frac{3}{4}$

2. $105\frac{7}{25} \div 62\frac{8}{15}$

$\left[\text{Answer: } 1\frac{229}{335}\right]$

3.7 MORE APPLICATIONS

This section considers a variety of applications of fractions and mixed numbers. Having different types of problems in the same section will give practice at deciding which operation should be used.

| EXAMPLE 1 APPLICATION TO CONSTRUCTION | PRACTICE EXERCISE 1 |

A rectangular room is $12\frac{1}{2}$ feet long and $10\frac{1}{5}$ feet wide. How many square feet of tile are needed to cover the floor of the room? If tile costs $1\frac{1}{5}$ dollars for one square foot, how much will it cost to tile the floor?

Since the area of the floor is the length of the room times its width, this problem requires multiplication. See Figure 3.19.

A rectangular room is $6\frac{2}{3}$ yards long and $4\frac{1}{4}$ yards wide. What is the area of the room? If carpet costs $16\frac{1}{2}$ dollars per square yard installed, how much will it cost to carpet the room?

$$\text{Area} = \text{length} \cdot \text{width}$$

$$= 12\frac{1}{2} \cdot 10\frac{1}{5}$$

$$= \frac{25}{2} \cdot \frac{51}{5}$$

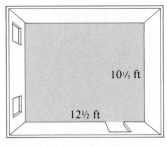

10⅕ ft

12½ ft

Figure 3.19

$$= \frac{\overset{5}{\cancel{25}}}{2} \cdot \frac{51}{\underset{1}{\cancel{3}}}$$

$$= \frac{255}{2} = 127\frac{1}{2}$$

Thus, $127\frac{1}{2}$ square feet of tile are required. The total cost is found by multiplying the number of square feet by the cost for each square foot.

$$127\frac{1}{2} \cdot 1\frac{1}{5} = \frac{255}{2} \cdot \frac{6}{5}$$

$$= \frac{51 \cdot \overset{1}{\cancel{5}} \cdot \overset{1}{\cancel{2}} \cdot 3}{\underset{1}{\cancel{2}} \cdot \underset{1}{\cancel{5}}}$$

$$= 153$$

The cost to tile the floor is 153 dollars.

Answer: $28\frac{1}{3}$ square yards, $467\frac{1}{2}$ dollars

---

EXAMPLE 2 HOUSEHOLD APPLICATION

Howard bought $27\frac{1}{3}$ pounds of nuts. If he put $18\frac{2}{3}$ pounds in the freezer, what fraction of the nuts did he freeze?

To find a fraction of the nuts, divide the number of pounds frozen by the total number of pounds of nuts.

$$18\frac{2}{3} \div 27\frac{1}{3} = \frac{56}{3} \div \frac{82}{3} \qquad \text{Change to improper fractions}$$

$$= \frac{56}{3} \cdot \frac{3}{82}$$

$$= \frac{28 \cdot \overset{1}{\cancel{2}} \cdot \overset{1}{\cancel{3}}}{\underset{1}{\cancel{3}} \cdot \underset{1}{\cancel{2}} \cdot 41} = \frac{28}{41}$$

The fraction of the nuts which were frozen is $\frac{28}{41}$.

PRACTICE EXERCISE 2

After working for several hours, Andrew had processed $42\frac{1}{2}$ pounds of an 80-pound shipment of cheese. What fraction of the cheese had he processed?

Answer: $\frac{17}{32}$

EXAMPLE 3 MARKETING APPLICATION

A fruit market sells grapes in bags. If $3\frac{1}{3}$ pounds of grapes are put into each bag, how many bags can be prepared from a shipment of 120 pounds of grapes?

If we knew the number of bags we could multiply $3\frac{1}{3}$ times that number to obtain 120 pounds.

$$3\frac{1}{3} \cdot \square = 120$$

$$\square = 120 \div 3\frac{1}{3} \qquad \text{Related division sentence}$$

$$= 120 \div \frac{10}{3}$$

$$= \frac{120}{1} \cdot \frac{3}{10}$$

PRACTICE EXERCISE 3

When a fuel tank is $\frac{2}{5}$ full, it contains 320 gallons. How many gallons of fuel will the tank hold?

$$\frac{2}{5} \text{ of } \square \text{ is } 320$$

$$\downarrow \ \downarrow \ \downarrow \ \downarrow \ \downarrow$$

$$\frac{2}{5} \cdot \square = 320$$

$$= \frac{12 \cdot \overset{1}{\cancel{10}} \cdot 3}{1 \cdot \cancel{10}} = 36$$

Thus, 36 bags can be prepared for sale.

Answer: **800 gallons**

EXAMPLE 4   REVOLUTIONS OF A RECORD

A record revolves at a rate of $33\frac{1}{3}$ revolutions per minute. If it takes $15\frac{1}{2}$ minutes to play one song, how many revolutions does the record make during the song?

Look at a simpler case first: If the song lasted 2 minutes, the record would make $2 \cdot 33\frac{1}{3}$ revolutions. Since the song takes $15\frac{1}{2}$ minutes, the record makes $15\frac{1}{2} \cdot 33\frac{1}{3}$ revolutions.

$$15\frac{1}{2} \cdot 33\frac{1}{3} = \frac{2 \cdot 15 + 1}{2} \cdot \frac{3 \cdot 33 + 1}{3}$$

$$= \frac{31}{2} \cdot \frac{100}{3}$$

$$= \frac{31 \cdot \overset{1}{\cancel{2}} \cdot 50}{\underset{1}{\cancel{2}} \cdot 3}$$

$$= \frac{1550}{3} = 516\frac{2}{3}$$

During the song, the record makes $516\frac{2}{3}$ revolutions.

PRACTICE EXERCISE 4

A flywheel turns at the rate of $102\frac{1}{2}$ revolutions per second. How many revolutions will it make in $5\frac{2}{5}$ seconds?

Answer: **$553\frac{1}{2}$ revolutions**

Mixed numbers give a better idea of the size of a number than improper fractions. For example, it is easier to understand

$$516\frac{2}{3} \text{ revolutions } \quad \text{than} \quad \frac{1550}{3} \text{ revolutions.}$$

CAUTION

It is easy to confuse the mixed number notation with multiplication. Often, when no symbol is used, multiplication is assumed. This is not true with mixed numbers.

$$516\frac{2}{3} \quad \text{is} \quad 516 + \frac{2}{3} \quad not \quad 516 \cdot \frac{2}{3}$$

To show this,

$$516 + \frac{2}{3} = 516\frac{2}{3} = \frac{1550}{3},$$

but

$$516 \cdot \frac{2}{3} = \frac{\overset{172}{\cancel{516}} \cdot 2}{\underset{1}{\cancel{3}}} = 172 \cdot 2 = 344.$$

3.7 EXERCISES A

Solve.

1. A tank holds 240 gallons when it is $\frac{2}{3}$ full. How many gallons does it hold when full?

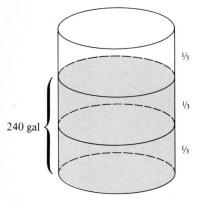

360 gallons

2. A tank holds 240 gallons when it is full. How many gallons does it contain when it is $\frac{2}{3}$ full?

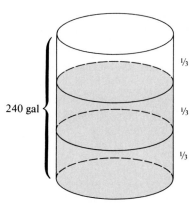

160 gallons

3. If $\frac{5}{6}$ of a class consists of boys and 42 students are in the class, how many students are boys? How many are girls?   35 boys, 7 girls

4. Walter ate $2\frac{1}{5}$ pounds of a $3\frac{1}{2}$-pound box of candy the first day. What fraction of the candy had he eaten? $\frac{22}{35}$

5. Twila can run $7\frac{1}{2}$ miles in one hour. At the same rate how far can she run in 2 hours?   15 mi

6. Sam added $2\frac{1}{2}$ cups of flour to a cake recipe. He later found out that this was $1\frac{1}{2}$ times the amount that he should have added. How much should he have used? $1\frac{2}{3}$ cups

7. A recipe calls for $1\frac{1}{4}$ cups of sugar. How much sugar should be used if the recipe is doubled? How much if $\frac{1}{3}$ the recipe is to be used?
$2\frac{1}{2}$ cups, $\frac{5}{12}$ cup

8. A man's shirt requires $2\frac{3}{5}$ yards of material. How many shirts can be made from 52 yards of material?
20 shirts

9. A domed stadium has rectangular floor dimensions of 120 yards by 70 yards. If $\frac{9}{10}$ of the floor area is to be covered with synthetic turf, how many square yards of turf must be laid? If the turf costs $15 a square yard installed, how much will the project cost?   7560 sq yd of turf at a cost of $113,400

10 A family has an income of $18,000 a year. If $\frac{1}{5}$ the income is spent for food, $\frac{1}{4}$ for housing, $\frac{1}{10}$ for clothing, and $\frac{3}{8}$ for taxes, how much is spent for each?
food: $3600; housing: $4500; clothing: $1800; taxes: $6750

11. Brenda worked $10\frac{2}{3}$ hours and received $136. What was her hourly rate of pay?   **$12\frac{3}{4}$ dollars per hour**

12. A tank holds 560 gallons of water when it is full. When a leak was discovered, $\frac{5}{8}$ of the water had drained out. How much water was left in the tank?
210 gallons

13 A record makes 45 revolutions per minute. How long will it take to make $321\frac{1}{2}$ revolutions?
$7\frac{13}{90}$ **minutes**

14. A record makes $33\frac{1}{3}$ revolutions per minute. How many revolutions are made in $6\frac{3}{5}$ minutes?
220 revolutions

FOR REVIEW

Perform the indicated operations.

15. $2\frac{1}{9} \cdot 3\frac{3}{5}$ $7\frac{3}{5}$

16. $5\frac{3}{7} \div 4\frac{2}{5}$ $1\frac{18}{77}$

17. $\left(6\frac{2}{3}\right)^2$ $44\frac{4}{9}$

18. $3 \cdot 5\frac{1}{7} \cdot 4\frac{2}{3}$ 72

19. $9\frac{3}{5} \div 10$ $\frac{24}{25}$

20. $16 \div 7\frac{1}{3}$ $2\frac{2}{11}$

ANSWERS: 1. 360 gallons   2. 160 gallons   3. 35 boys, 7 girls   4. $\frac{22}{35}$ 5. 15 mi   6. $1\frac{2}{3}$ cups   7. $2\frac{1}{2}$ cups, $\frac{5}{12}$ cup
8. 20 shirts   9. 7560 sq yd of turf at a cost of $113,400   10. food: $3600; housing: $4500; clothing: $1800;
taxes: $6750   11. $12\frac{3}{4}$ dollars per hour   12. 210 gallons   13. $7\frac{13}{90}$ minutes   14. 220 revolutions   15. $7\frac{3}{5}$ 16. $1\frac{18}{77}$
17. $44\frac{4}{9}$ 18. 72   19. $\frac{24}{25}$ 20. $2\frac{2}{11}$

3.7   EXERCISES B

Solve.

1. A container is $\frac{3}{10}$ full and has 15 quarts in it. How much will it hold when full?   **50 quarts**

2. A container holds 15 quarts when it is full. How many quarts does it contain when it is $\frac{3}{10}$ full?
$4\frac{1}{2}$ **quarts**

3. A bag contains 120 marbles which are either red or blue. If $\frac{3}{5}$ of the marbles are blue, how many of each color are in the bag?   **blue: 72; red: 48**

4. There were $7\frac{1}{3}$ gallons of punch prepared for a party. If $6\frac{2}{5}$ gallons were drunk, what fraction of the punch was consumed? $\frac{48}{55}$

5. Hank rides his bike at a rate of $12\frac{2}{3}$ miles per hour. How far can he go in $6\frac{1}{2}$ hours? $82\frac{1}{3}$ **mi**

6. A chemist added $5\frac{1}{4}$ liters of alcohol to a solution. She later discovered that this was $2\frac{1}{2}$ times as much as she should have added. How much should have been added? $2\frac{1}{10}$ **liters**

7. To make a party punch, $2\frac{3}{4}$ quarts of ice cream are required. How much ice cream should be added if three times the punch recipe is to be made? How much if $\frac{1}{5}$ the recipe is to be made?
$8\frac{1}{4}$ **quarts;** $\frac{11}{20}$ **quarts**

8. A connecting wire must be $8\frac{1}{4}$ inches long. How many of these wires can be cut from a roll of wire 198 inches long?   **24**

9. A large living room is 12 yards wide and 16 yards long. If $\frac{5}{8}$ of the floor area is to be carpeted, how many square yards are required? If carpet costs $24 per square yard installed, how much will the job cost?   **120 sq yd of carpet at a cost of $2880**

10. Gloria Moore has $2400 for college expenses. If she spends $\frac{1}{4}$ of the money for clothing, $\frac{1}{8}$ for books, $\frac{1}{6}$ for travel, and $\frac{2}{5}$ for tuition, how much is spent for each?
clothing: $600; books: $300; travel: $400; tuition: $960

11. Nick worked $72\frac{4}{5}$ hours and received $673\frac{2}{5}$. What was his hourly rate of pay?   **9\frac{1}{4}$ per hour**

12. A water truck holds 250 gallons when it is full. After $\frac{3}{5}$ of the tank is drained out, how much is left in the tank?   **100 gallons**

13. A record makes $33\frac{1}{3}$ revolutions each minute. How long will it take to make $137\frac{1}{2}$ revolutions?
$4\frac{1}{8}$ **minutes**

14. A pulley on a machine is turning at a rate of $214\frac{1}{2}$ revolutions per minute. How many revolutions will it make in $10\frac{2}{3}$ minutes?   **2288 revolutions**

FOR REVIEW

Perform the indicated operation.

15. $5\frac{2}{3} \cdot 9\frac{3}{5}$ $54\frac{2}{5}$

16. $1\frac{3}{8} \div 9\frac{3}{7}$ $\frac{7}{48}$

17. $\left(1\frac{1}{10}\right)^2$ $1\frac{21}{100}$

18. $\frac{5}{8} \cdot 9\frac{3}{5} \cdot 4$ 24

19. $12\frac{2}{3} \div 20$ $\frac{19}{30}$

20. $42 \div 11\frac{1}{3}$ $3\frac{12}{17}$

3.7 EXERCISES C

Solve. Let a represent a natural number.

1. If a is multiplied by $\frac{3}{4}$ the result is $\frac{1}{12}$. Find a. $\frac{1}{9}$

2. If a is divided by $\frac{5}{6}$ the result is $\frac{3}{25}$. Find a.
$\left[\text{Answer: } \frac{1}{10}\right]$

CHAPTER 3 REVIEW

KEY WORDS

3.1 A **fraction** is the quotient of two whole numbers.

The **numerator** is the top number in a fraction.

The **denominator** is the bottom number in a fraction.

A **proper fraction** is one whose numerator is less than its denominator.

An **improper fraction** is one whose numerator is greater than or equal to its denominator.

3.2 Fractions that are names for the same number are **equal fractions.**

A fraction is **reduced to lowest terms** when 1 is the only nonzero whole number which is a divisor of both the numerator and denominator.

3.3 The **commutative law of multiplication** states that changing the order of a product does not change the result.

The **associative law of multiplication** states that regrouping products does not change the result.

The **multiplicative identity** tells us we can multiply any number by 1 and always obtain that same number.

3.4 Two fractions are **reciprocals** if their product is 1.

3.5 A **mixed number** is the sum of a whole number and a proper fraction.

KEY CONCEPTS

3.1 **1.** Any fraction like $\frac{0}{9}$ is equal to 0.

2. Any fraction like $\frac{4}{4}$ is equal to 1.

3.2 **1.** Fractions are equal when their cross products are equal.

2. When the numerator and denominator of a fraction are multiplied or divided by the same whole number (except 0), an equal fraction is formed.

3. To reduce a fraction to lowest terms, factor the numerator and denominator into primes and divide out all common factors.

3.3 **1.** When multiplying fractions, divide out common factors of the numerator and denominator first before multiplying the remaining factors.

2. The word *of* translates to "times" when it follows a fraction in a word problem.

3. Fractions satisfy the commutative and associative laws of multiplication, and 1 is the multiplicative identity.

3.4 **1.** The reciprocal of a fraction is found by interchanging the numerator and the denominator of the fraction.

2. To divide fractions, multiply the dividend by the reciprocal of the divisor.

3. Zero is the only number that does not have a reciprocal.

4. To solve some applied problems, we first write a multiplication sentence then change to the related division sentence.

3.5 **1.** A mixed number like $3\frac{1}{2}$ is $3 + \frac{1}{2}$ $\left(\text{not } 3 \text{ times } \frac{1}{2}\right)$.

2. Improper fractions can be changed to mixed numbers by dividing.

3. A mixed number such as $5\frac{2}{7}$ can be changed to an improper fraction as follows:

$$5\frac{2}{7} = \frac{7 \cdot 5 + 2}{7} = \frac{37}{7}$$

3.6 **1.** Change mixed numbers to improper fractions in order to multiply or divide.

2. You should not multiply whole number parts and fraction parts separately when multiplying mixed numbers.

3.7 In solving applied problems determine which of the following is required.

(a) Finding a fractional part

(b) Multiplying fractions

(c) Dividing fractions

REVIEW EXERCIS

Part I

3.1 **1.** What is the denominator of the fraction $\frac{2}{9}$?   9

2. Is the fraction $\frac{21}{2}$ a proper or an improper fraction?   improper

3. Is $\frac{7}{7}$ a proper or an improper fraction?
improper

4. The fraction $\frac{0}{11}$ is the same as what number?
0

What fractional part of each figure is shaded?

5.

$$\frac{2}{4} \text{ or } \frac{1}{2}$$

6.

$$\frac{1}{3}$$

7.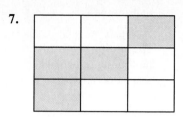

$$\frac{4}{9}$$

8. Let the figure

be one whole unit. What fraction of the unit is shaded below?

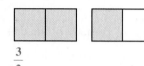

$$\frac{3}{2}$$

9. A ball player made 11 freethrows in 14 attempts during a game. The number of shots she made is what fraction of the shots that she attempted?

$$\frac{11}{14}$$

Express each fraction as a whole number.

10. $\dfrac{9}{9}$ 1

11. $\dfrac{31}{0}$ undefined

12. $\dfrac{0}{15}$ 0

3.2 *Determine which fractions are equal by using the cross-product rule.*

13. $\dfrac{8}{11}$ and $\dfrac{5}{7}$ $\dfrac{8}{11} \neq \dfrac{5}{7}$

14. $\dfrac{3}{12}$ and $\dfrac{5}{20}$ $\dfrac{3}{12} = \dfrac{5}{20}$

In each of the following, find the missing term of the fraction.

15. $\dfrac{5}{8} = \dfrac{?}{24}$ 15

16. $\dfrac{3}{9} = \dfrac{1}{?}$ 3

17. $\dfrac{4}{?} = \dfrac{32}{8}$ 1

Reduce each fraction to lowest terms.

18. $\dfrac{20}{30}$ $\dfrac{2}{3}$

19. $\dfrac{180}{144}$ $\dfrac{5}{4}$

20. $\dfrac{42}{28}$ $\dfrac{3}{2}$

3.3 **21.** The word *of* translates to what operation when it follows a fraction in a problem? multiplication

22. The fact that $\frac{1}{3} \cdot \frac{7}{8} = \frac{7}{8} \cdot \frac{1}{3}$ illustrates which of the properties of multiplication? commutative law

Find each product.

23. $\dfrac{2}{3} \cdot \dfrac{18}{7}$ $\dfrac{12}{7}$ or $1\dfrac{5}{7}$

24. $5 \cdot \dfrac{3}{25}$ $\dfrac{3}{5}$

25. $\dfrac{6}{5} \cdot \dfrac{4}{7} \cdot \dfrac{5}{12}$ $\dfrac{2}{7}$

Find the value of each power or root.

26. $\left(\dfrac{7}{6}\right)^2$ $\dfrac{49}{36}$

27. $\sqrt{\dfrac{121}{25}}$ $\dfrac{11}{5}$

Solve.

28. There are 65 members of a photography club, and four-fifths of the members own a Pantex camera. How many members do not own this kind of camera?   **13 members**

3.4 **29.** What is the reciprocal of $\frac{13}{7}$? $\frac{7}{13}$

30. What is the only number that does not have a reciprocal?   **0**

Find each quotient.

31. $\frac{2}{9} \div \frac{4}{27}$ $\frac{3}{2}$

32. $12 \div \frac{3}{4}$ **16**

33. $\frac{7}{3} \div 21$ $\frac{1}{9}$

Find the missing number.

34. $\frac{5}{7} \cdot \square = 25$ **35**

35. $\frac{2}{9}$ of $\square$ is $\frac{4}{25}$ $\frac{18}{25}$

Solve.

36. When a tank is $\frac{1}{6}$ full, it contains 130 gallons of water. What is the capacity of the tank?   **780 gallons**

3.5 **37.** What do we call the sum of a whole number and a proper fraction?   **mixed number**

38. What is the understood operation between 3 and $\frac{7}{8}$ in the mixed number $3\frac{7}{8}$?   **addition**

Change to a mixed number.

39. $\frac{127}{4}$ $31\frac{3}{4}$

40. $\frac{243}{11}$ $22\frac{1}{11}$

41. $\frac{97}{2}$ $48\frac{1}{2}$

Change to an improper fraction.

42. $3\frac{9}{13}$ $\frac{48}{13}$

43. $21\frac{1}{5}$ $\frac{106}{5}$

44. $16\frac{2}{3}$ $\frac{50}{3}$

3.6 *Perform the indicated operations.*

45. $2\frac{3}{4} \cdot 5\frac{1}{8}$ $14\frac{3}{32}$

46. $7\frac{2}{3} \div 4\frac{3}{5}$ $1\frac{2}{3}$

47. $4\frac{2}{11} \cdot 8\frac{1}{4}$ $34\frac{1}{2}$

48. $5 \cdot 7\frac{2}{3} \cdot 10\frac{1}{5}$ **391**

49. $18 \div 12\frac{2}{5}$ $1\frac{14}{31}$

50. $22\frac{1}{2} \div 15$ $1\frac{1}{2}$

Solve.

51. A wheel turns at $32\frac{1}{2}$ revolutions per second. How many revolutions will it make in $10\frac{1}{5}$ seconds? $331\frac{1}{2}$ **revolutions**

52. A wheel turns at $32\frac{1}{2}$ revolutions per second. How long will it take to make 120 revolutions? $3\frac{9}{13}$ **seconds**

3.3– *Solve.*
3.7

53. On a map 1 inch represents 35 miles. How many miles does $1\frac{3}{5}$ in represent?   **56 mi**

54. A rectangular room is $2\frac{1}{8}$ yards wide and $4\frac{1}{3}$ yards long. How much will it cost to carpet the room using carpet which sells for $13\frac{1}{2}$ dollars per sq yd? $124\frac{5}{16}$ **dollars**

55. A container holds 186 gallons of fuel when it is $\frac{2}{3}$ full. What is the capacity of the container?
279 gallons

56. A container holds 186 gallons of fuel when full. If $\frac{2}{3}$ of a full tank is drained off, how much fuel remains? **62 gallons**

57. A merry-go-round makes $8\frac{3}{4}$ revolutions per minute. How many revolutions will be made during a $3\frac{3}{5}$ minute ride? **$31\frac{1}{2}$ revolutions**

58. An Oldsmobile Omega traveled 480 miles on $13\frac{3}{4}$ gallons of gas. How many miles per gallon did it get? **$34\frac{10}{11}$ mpg**

Part II

Find the missing number.

59. $\frac{3}{4} \cdot \square = 12$ **16**

60. $\frac{1}{7}$ of $\square$ is $\frac{5}{21}$ **$\frac{5}{3}$**

Perform the indicated operation.

61. $\left(\frac{2}{7}\right)^2$ **$\frac{4}{49}$**

62. $\sqrt{\frac{144}{4}}$ **6**

63. $\frac{3}{5} \cdot \frac{2}{9} \cdot \frac{5}{4}$ **$\frac{1}{6}$**

64. $\frac{16}{5} \div \frac{4}{25}$ **20**

65. $6\frac{2}{3} \cdot 8\frac{1}{2}$ **$56\frac{2}{3}$**

66. $12\frac{3}{4} \div 4\frac{1}{2}$ **$2\frac{5}{6}$**

Find the missing term of each fraction.

67. $\frac{5}{12} = \frac{?}{36}$ **15**

68. $\frac{8}{?} = \frac{24}{15}$ **5**

Solve.

69. A Buick traveled 234 miles on $10\frac{2}{5}$ gallons of gas. How many miles per gallon did it get? **$22\frac{1}{2}$ mpg**

70. On a map 1 inch represents 50 miles. How many miles are represented by $5\frac{3}{4}$ inches? **$287\frac{1}{2}$ mi**

Express each fraction as a whole number.

71. $\frac{45}{45}$ **1**

72. $\frac{0}{21}$ **0**

Determine which fractions are equal by using the cross-product rule.

73. $\frac{5}{30}$ and $\frac{2}{12}$ **$\frac{5}{30} = \frac{2}{12}$**

74. $\frac{11}{9}$ and $\frac{21}{18}$ **$\frac{11}{9} \neq \frac{21}{18}$**

ANSWERS: 1. 9 2. improper 3. improper 4. 0 5. $\frac{2}{4}$ or $\frac{1}{2}$ 6. $\frac{1}{3}$ 7. $\frac{4}{9}$ 8. $\frac{3}{2}$ 9. $\frac{11}{14}$ 10. 1 11. undefined 12. 0 13. $\frac{8}{11} \neq \frac{5}{7}$ 14. $\frac{3}{12} = \frac{5}{20}$ 15. 15 16. 3 17. 1 18. $\frac{2}{3}$ 19. $\frac{5}{4}$ 20. $\frac{3}{2}$ 21. multiplication 22. commutative law 23. $\frac{12}{7}$ or $1\frac{5}{7}$ 24. $\frac{3}{5}$ 25. $\frac{2}{7}$ 26. $\frac{49}{36}$ 27. $\frac{11}{5}$ 28. 13 members 29. $\frac{7}{13}$ 30. 0 31. $\frac{3}{2}$ 32. 16 33. $\frac{1}{9}$ 34. 35 35. $\frac{18}{25}$ 36. 780 gallons 37. mixed number 38. addition 39. $31\frac{3}{4}$ 40. $22\frac{1}{11}$ 41. $48\frac{1}{2}$ 42. $\frac{48}{13}$ 43. $\frac{106}{5}$ 44. $\frac{50}{3}$ 45. $14\frac{3}{32}$ 46. $1\frac{2}{3}$ 47. $34\frac{1}{2}$ 48. 391 49. $1\frac{14}{31}$ 50. $1\frac{1}{2}$ 51. $331\frac{1}{2}$ revolutions 52. $3\frac{9}{13}$ seconds 53. 56 mi 54. $124\frac{5}{16}$ dollars 55. 279 gallons 56. 62 gallons 57. $31\frac{1}{2}$ revolutions 58. $34\frac{10}{11}$ mpg 59. 16 60. $\frac{5}{3}$ 61. $\frac{4}{49}$ 62. 6 63. $\frac{1}{6}$ 64. 20 65. $56\frac{2}{3}$ 66. $2\frac{5}{6}$ 67. 15 68. 5 69. $22\frac{1}{2}$ mpg 70. $287\frac{1}{2}$ mi 71. 1 72. 0 73. $\frac{5}{30} = \frac{2}{12}$ 74. $\frac{11}{9} \neq \frac{21}{18}$

1. What kind of fraction has the numerator less than the denominator?

   1. _____proper_____

2. Which law is illustrated by $\frac{2}{3} \cdot \left(\frac{3}{4} \cdot \frac{7}{8}\right) = \left(\frac{2}{3} \cdot \frac{3}{4}\right) \cdot \frac{7}{8}$?

   2. ____associative law____

3. What fractional part of the figure is shaded?

   3. _____$\frac{1}{6}$_____

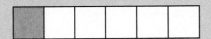

4. Determine if $\frac{9}{2}$ and $\frac{36}{8}$ are equal by using the cross-product rule.

   4. _____$\frac{9}{2} = \frac{36}{8}$_____

Find the missing term.

5. $\frac{3}{5} = \frac{?}{35}$

   5. _____21_____

6. $\frac{9}{11} = \frac{27}{?}$

   6. _____33_____

7. Reduce $\frac{105}{45}$ to lowest terms.

   7. _____$\frac{7}{3}$_____

Perform the indicated operations.

8. $\frac{9}{5} \cdot \frac{20}{39}$

   8. _____$\frac{12}{13}$_____

9. $\frac{3}{8} \div \frac{21}{2}$

   9. _____$\frac{1}{28}$_____

10. $\frac{5}{8} \cdot \frac{12}{5} \cdot \frac{2}{11}$

    10. _____$\frac{3}{11}$_____

11. $\frac{9}{7} \div \frac{3}{35}$

    11. _____15_____

12. $\left(\frac{3}{8}\right)^2$

    12. _____$\frac{9}{64}$_____

13. $\sqrt{\frac{49}{144}}$

    13. _____$\frac{7}{12}$_____

14. Find the missing number.

$\frac{4}{5}$ of □ is 22

14. _____ $\frac{55}{2}$ or $27\frac{1}{2}$ _____

15. Change $\frac{48}{20}$ to a mixed number.

15. _____ $2\frac{2}{5}$ _____

16. Change $7\frac{2}{11}$ to an improper fraction.

16. _____ $\frac{79}{11}$ _____

Perform the indicated operations and give the answer as a mixed number.

17. $5\frac{1}{5} \cdot 4\frac{1}{4}$

17. _____ $22\frac{1}{10}$ _____

18. $7\frac{1}{3} \div 9\frac{2}{9}$

18. _____ $\frac{66}{83}$ _____

19. $7 \cdot 4\frac{3}{8} \cdot 10\frac{2}{7}$

19. _____ 315 _____

20. $6 \div 2\frac{4}{5}$

20. _____ $2\frac{1}{7}$ _____

Solve.

21. A container holds 18 gallons when it is $\frac{8}{9}$ full. How many gallons does it contain when it is full?

21. _____ $20\frac{1}{4}$ gallons _____

22. A record makes $33\frac{1}{3}$ revolutions per minute. How many revolutions does it make in $16\frac{1}{2}$ minutes?

22. _____ 550 revolutions _____

Adding and Subtracting Fractions

4.1 ADDING AND SUBTRACTING LIKE FRACTIONS

STUDENT GUIDEPOSTS

1. Like and Unlike Fractions
2. Adding Like Fractions
3. Subtracting Like Fractions
4. Combination Problems and Applications

1 LIKE AND UNLIKE FRACTIONS

Fractions which have the same denominator are called **like fractions.** For example, the fractions

$$\frac{1}{6}, \quad \frac{5}{6}, \quad \frac{7}{6}, \quad \frac{9}{6}, \quad \frac{11}{6}$$

are like fractions. Fractions with different denominators are called **unlike fractions.** Thus,

$$\frac{2}{3}, \quad \frac{1}{5}, \quad \frac{3}{2}, \quad \frac{1}{4}, \quad \frac{7}{8}$$

are all unlike fractions.

2 ADDING LIKE FRACTIONS

We begin the study of addition of fractions with like fractions. If we want to add

$$\frac{1}{8} + \frac{5}{8},$$

the sum becomes obvious when the problem is rewritten as

1 eighth + 5 eighths.

Just as

1 book + 5 books = 6 books,

1 eighth + 5 eighths = 6 eighths.

That is,

$$\frac{1}{8} + \frac{5}{8} = \frac{6}{8}.$$

187

We simply add numerators $(1 + 5)$ and place the result over the common denominator, 8. The sum, $\frac{6}{8}$, should then be reduced to lowest terms, $\frac{3}{4}$. The same is true in general.

To Add Like Fractions

1. Add their numerators.
2. Place this sum over the common denominator.
3. Reduce the fraction to lowest terms, if possible.

EXAMPLE 1 ADDING LIKE FRACTIONS

Find each sum.

(a) $\dfrac{3}{4} + \dfrac{7}{4} = \dfrac{3 + 7}{4}$ Place sum of numerators over denominator, 4

$\phantom{\dfrac{3}{4} + \dfrac{7}{4}} = \dfrac{10}{4}$ Add

$\phantom{\dfrac{3}{4} + \dfrac{7}{4}} = \dfrac{\cancel{2} \cdot 5}{\cancel{2} \cdot 2} = \dfrac{5}{2}$ Reduce to lowest terms

(b) $\dfrac{7}{30} + \dfrac{1}{30} + \dfrac{12}{30} = \dfrac{7 + 1 + 12}{30}$ Add numerators

$\phantom{\dfrac{7}{30} + \dfrac{1}{30} + \dfrac{12}{30}} = \dfrac{20}{30}$

$\phantom{\dfrac{7}{30} + \dfrac{1}{30} + \dfrac{12}{30}} = \dfrac{\cancel{10} \cdot 2}{\cancel{10} \cdot 3} = \dfrac{2}{3}$ Reduce

In **(b)**, we took a shortcut and did not factor numerator and denominator to primes when reducing $\frac{20}{30}$, because it was obvious that 10 was a common factor which could be divided out.

PRACTICE EXERCISE 1

Find each sum.

(a) $\dfrac{2}{5} + \dfrac{1}{5}$

(b) $\dfrac{11}{42} + \dfrac{17}{42} + \dfrac{5}{42}$

Answers: (a) $\frac{3}{5}$ (b) $\frac{11}{14}$

---

❸ SUBTRACTING LIKE FRACTIONS

It is as easy to subtract like fractions as to add them. For example, to find

$$\frac{7}{8} - \frac{3}{8},$$

we think of

$$7 \text{ eighths} - 3 \text{ eighths} = 4 \text{ eighths}$$

and subtract as we would if we were subtracting

$$7 \text{ books} - 3 \text{ books} = 4 \text{ books}.$$

Thus,

$$\frac{7}{8} - \frac{3}{8} = \frac{4}{8}, \quad \text{which reduces to} \quad \frac{1}{2}.$$

To Subtract Like Fractions

1. Subtract their numerators.
2. Place this difference over the common denominator.
3. Reduce the fraction to lowest terms, if possible.

EXAMPLE 2 SUBTRACTING LIKE FRACTIONS	**PRACTICE EXERCISE 2**

Find each difference.

(a) $\dfrac{5}{6} - \dfrac{1}{6} = \dfrac{5-1}{6}$ Place difference of numerators over common denominator, 6

$\phantom{\dfrac{5}{6} - \dfrac{1}{6}} = \dfrac{4}{6}$ Subtract

$\phantom{\dfrac{5}{6} - \dfrac{1}{6}} = \dfrac{\cancel{2} \cdot 2}{\cancel{2} \cdot 3} = \dfrac{2}{3}$ Reduce to lowest terms

(b) $\dfrac{7}{11} - \dfrac{7}{11} = \dfrac{7-7}{11}$ Subtract numerators and place difference over common denominator, 11

$\phantom{\dfrac{7}{11} - \dfrac{7}{11}} = \dfrac{0}{11} = 0$

Find each difference.

(a) $\dfrac{7}{17} - \dfrac{2}{17}$

(b) $\dfrac{9}{15} - \dfrac{4}{15}$

Answers: (a) $\frac{5}{17}$ (b) $\frac{1}{3}$

CAUTION

Never add or subtract fractions by adding or subtracting denominators. For example,

$$\frac{2}{5} + \frac{1}{5} \quad \textit{is not the same as} \quad \frac{2+1}{5+5} = \frac{3}{10},$$

since

$$\frac{2}{5} + \frac{1}{5} = \frac{3}{5}.$$

Also,

$$\frac{2}{5} - \frac{1}{5} \quad \textit{is not the same as} \quad \frac{2-1}{5-5},$$

which is undefined.

---

❹ COMBINATION PROBLEMS AND APPLICATIONS

We can find combinations of sums and differences of like fractions by adding or subtracting in order from left to right. An example will help to make this clear.

| **EXAMPLE 3** COMBINATION PROBLEM | **PRACTICE EXERCISE 3** |

Perform the indicated operations.

$$\frac{4}{13} + \frac{11}{13} - \frac{3}{13} = \frac{4 + 11 - 3}{13}$$ Combine numerators as indicated

$$= \frac{15 - 3}{13}$$ Add first from left

$$= \frac{12}{13}$$ Subtract

Perform the indicated operations.

$$\frac{7}{23} + \frac{9}{23} - \frac{4}{23}$$

Answer: $\frac{12}{23}$

| **EXAMPLE 4** APPLICATION TO RECREATION | **PRACTICE EXERCISE 4** |

On Monday Murphy swam $\frac{9}{20}$ of a mile and on Tuesday he swam $\frac{7}{20}$ of a mile. What is the total distance that he swam on the two days?

 The total distance is the sum of the two distances $\frac{9}{20}$ and $\frac{7}{20}$.

$$\frac{9}{20} + \frac{7}{20} = \frac{9 + 7}{20} = \frac{16}{20} = \frac{4}{5}$$

Thus, he swam a total of $\frac{4}{5}$ of a mile.

Joe typed for $\frac{5}{4}$ hours one day and $\frac{15}{4}$ hours the next day. How many total hours did he type during the two days?

Answer: 5 hours

| **EXAMPLE 5** APPLICATION TO ENTERTAINMENT | **PRACTICE EXERCISE 5** |

A recipe for a party drink calls for $\frac{3}{10}$ of a gallon of 7-Up, $\frac{7}{10}$ of a gallon of lemonade, and $\frac{1}{10}$ of a gallon of orange sherbet.

(a) How many gallons of drink can be made from the recipe?
  Add:

$$\frac{3}{10} + \frac{7}{10} + \frac{1}{10} = \frac{3 + 7 + 1}{10} = \frac{11}{10}.$$

Thus, the recipe will make $\frac{11}{10}$ or $1\frac{1}{10}$ gallons of drink.

(b) If the recipe is doubled, how much of each ingredient is needed and how much drink will be made?

Multiply the amount of each ingredient by 2.

$$2 \cdot \frac{3}{10} = \frac{2}{1} \cdot \frac{3}{10} = \frac{6}{10} = \frac{3}{5}$$ Amount of 7-Up

$$2 \cdot \frac{7}{10} = \frac{2}{1} \cdot \frac{7}{10} = \frac{14}{10} = \frac{7}{5}$$ Amount of lemonade

$$2 \cdot \frac{1}{10} = \frac{2}{1} \cdot \frac{1}{10} = \frac{2}{10} = \frac{1}{5}$$ Amount of orange sherbet

A mixture of nuts calls for $\frac{1}{8}$ pound of almonds, $\frac{7}{8}$ pound of peanuts, and $\frac{5}{8}$ pound of walnuts.

(a) How many pounds are in the final mixture?

(b) If four times the recipe is to be made, how many pounds of each nut is required, and how many pounds are in the final mixture?

Adding,

$$\frac{3}{5} + \frac{7}{5} + \frac{1}{5} = \frac{3 + 7 + 1}{5} = \frac{11}{5}.$$

Thus, if the recipe is doubled, there will be $\frac{11}{5}$ or $2\frac{1}{5}$ gallons of drink which is reasonable since $2 \cdot \frac{11}{10} = \frac{2 \cdot 11}{2 \cdot 5} = \frac{11}{5}$.

Answers: (a) $\frac{13}{8}$ pounds (b) $\frac{1}{2}$ pound of almonds, $\frac{7}{2}$ pounds of peanuts, $\frac{5}{2}$ pounds of walnuts, $\frac{13}{2}$ pounds of nuts

---

EXAMPLE 6 APPLICATION TO WORK

If it takes $\frac{11}{24}$ of a day to do a job and Mike has been working for $\frac{7}{24}$ of a day, how much longer will it take him to complete the job?

Subtract the time worked from the total time required to complete the job.

$$\frac{11}{24} - \frac{7}{24} = \frac{4}{24} = \frac{1}{6}$$

Thus, it will take Mike $\frac{1}{6}$ of a day to finish the job.

PRACTICE EXERCISE 6

A contractor estimated that it would take about $\frac{24}{5}$ weeks to do a job. After work has gone on for $\frac{18}{5}$ weeks, how many weeks are left to complete the job?

Answer: $\frac{6}{5}$ weeks

4.1 EXERCISES A

Find the sum.

1. $\frac{2}{7} + \frac{3}{7}$ $\frac{5}{7}$

2. $\frac{1}{4} + \frac{5}{4}$ $\frac{3}{2}$

3. $\frac{3}{12} + \frac{11}{12}$ $\frac{7}{6}$

4. $\frac{1}{6} + \frac{5}{6}$ 1

5. $\frac{3}{2} + \frac{9}{2}$ 6

6. $\frac{7}{16} + \frac{13}{16}$ $\frac{5}{4}$

7. $\frac{3}{8} + \frac{5}{8} + \frac{7}{8}$ $\frac{15}{8}$

8 $\frac{2}{35} + \frac{18}{35} + \frac{25}{35}$ $\frac{9}{7}$

9. $\frac{17}{24} + \frac{23}{24} + \frac{8}{24}$ 2

Find the difference.

10. $\frac{7}{6} - \frac{5}{6}$ $\frac{1}{3}$

11. $\frac{9}{11} - \frac{3}{11}$ $\frac{6}{11}$

12. $\frac{5}{9} - \frac{2}{9}$ $\frac{1}{3}$

13. $\frac{23}{25} - \frac{8}{25}$ $\frac{3}{5}$

14 $\frac{28}{35} - \frac{18}{35}$ $\frac{2}{7}$

15. $\frac{73}{81} - \frac{10}{81}$ $\frac{7}{9}$

Perform the indicated operations.

16. $\frac{2}{13} + \frac{9}{13} - \frac{5}{13}$ $\frac{6}{13}$

17. $\frac{9}{25} - \frac{3}{25} + \frac{4}{25}$ $\frac{2}{5}$

18. $\frac{22}{12} - \frac{10}{12} - \frac{6}{12}$ $\frac{1}{2}$

19. $\frac{8}{3} + \frac{14}{3} - \frac{1}{3}$ 7

20 $\frac{7}{10} - \frac{1}{10} + \frac{3}{10}$ $\frac{9}{10}$

21. $\frac{9}{5} - \frac{6}{5} - \frac{3}{5}$ 0

Solve.

22. Jim hiked from Skunk Creek to Hidden Valley, a distance of $\frac{11}{15}$ mi, and from there to Indian Cave, a distance of $\frac{8}{15}$ mi. How far did he hike? $1\frac{4}{15}$ **mi**

23 Becky had 3 days to do a job. If she did $\frac{5}{12}$ of it the first day and $\frac{3}{12}$ of it the second, how much must she do the third day to complete the job? [*Hint:* 1 job can be written as $\frac{12}{12}$ job.] $\frac{1}{3}$ **of the job**

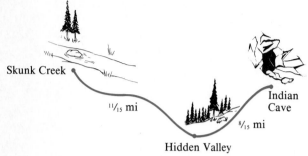

Skunk Creek

Indian Cave

$^{11}/_{15}$ mi

$^{8}/_{15}$ mi

Hidden Valley

24. To get the exact color he wanted, Bjorn mixed $\frac{1}{12}$ of a gallon of white paint with $\frac{7}{12}$ of a gallon of yellow and $\frac{8}{12}$ of a gallon of red paint. How much paint did he get? $\frac{4}{3}$ **gallons**

25. If Bjorn in Exercise 24 decides to triple the amount used, how much of each color will he need, and how much paint will result?

$\frac{1}{4}$ **gallon white;** $\frac{7}{4}$ **gallons yellow; 2 gallons red; 4 gallons total**

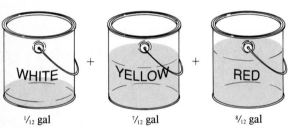

WHITE   +   YELLOW   +   RED

$^{1}/_{12}$ gal $^{7}/_{12}$ gal $^{8}/_{12}$ gal

ANSWERS: 1. $\frac{5}{7}$ 2. $\frac{3}{2}$ 3. $\frac{7}{4}$ 4. 1 5. 6 6. $\frac{5}{4}$ 7. $\frac{15}{8}$ 8. $\frac{9}{7}$ 9. 2 10. $\frac{1}{3}$ 11. $\frac{6}{11}$ 12. $\frac{1}{3}$ 13. $\frac{3}{5}$ 14. $\frac{2}{7}$ 15. $\frac{7}{9}$
16. $\frac{6}{13}$ 17. $\frac{2}{5}$ 18. $\frac{1}{2}$ 19. 7 20. $\frac{9}{10}$ 21. 0 22. $1\frac{4}{15}$ mi 23. $\frac{1}{3}$ of the job $\left(\frac{5}{12}+\frac{3}{12}=\frac{8}{12},\ \text{then } \frac{12}{12}-\frac{8}{12}=\frac{4}{12}=\frac{1}{3}\right)$
24. $\frac{4}{3}$ gallons 25. $\frac{3}{12}=\frac{1}{4}$ gallon of white; $\frac{21}{12}=\frac{7}{4}$ gallons of yellow; and $\frac{24}{12}=2$ gallons of red add up to $\frac{48}{12}=4$ gallons of paint.

4.1 EXERCISES B

Find the sum.

1. $\dfrac{3}{5}+\dfrac{4}{5}$ $\dfrac{7}{5}$

2. $\dfrac{5}{8}+\dfrac{3}{8}$ 1

3. $\dfrac{2}{15}+\dfrac{11}{15}$ $\dfrac{13}{15}$

4. $\dfrac{2}{9}+\dfrac{4}{9}$ $\dfrac{2}{3}$

5. $\dfrac{7}{3}+\dfrac{11}{3}$ 6

6. $\dfrac{9}{20}+\dfrac{3}{20}$ $\dfrac{3}{5}$

7. $\dfrac{1}{4}+\dfrac{3}{4}+\dfrac{5}{4}$ $\dfrac{9}{4}$

8. $\dfrac{5}{48}+\dfrac{9}{48}+\dfrac{1}{48}$ $\dfrac{5}{16}$

9. $\dfrac{7}{60}+\dfrac{51}{60}+\dfrac{2}{60}$ 1

Find the difference.

10. $\dfrac{4}{9}-\dfrac{1}{9}$ $\dfrac{1}{3}$

11. $\dfrac{8}{13}-\dfrac{6}{13}$ $\dfrac{2}{13}$

12. $\dfrac{11}{15}-\dfrac{2}{15}$ $\dfrac{3}{5}$

13. $\dfrac{17}{12} - \dfrac{11}{12}$ $\dfrac{1}{2}$ **14.** $\dfrac{42}{55} - \dfrac{17}{55}$ $\dfrac{5}{11}$ **15.** $\dfrac{72}{105} - \dfrac{37}{105}$ $\dfrac{1}{3}$

Perform the indicated operations.

16. $\dfrac{7}{17} + \dfrac{5}{17} - \dfrac{6}{17}$ $\dfrac{6}{17}$ **17.** $\dfrac{10}{30} - \dfrac{7}{30} + \dfrac{2}{30}$ $\dfrac{1}{6}$ **18.** $\dfrac{14}{6} - \dfrac{5}{6} - \dfrac{1}{6}$ $\dfrac{4}{3}$

19. $\dfrac{7}{15} + \dfrac{17}{15} - \dfrac{11}{15}$ $\dfrac{13}{15}$ **20.** $\dfrac{23}{28} - \dfrac{20}{28} + \dfrac{3}{28}$ $\dfrac{3}{14}$ **21.** $\dfrac{8}{12} - \dfrac{2}{12} - \dfrac{5}{12}$ $\dfrac{1}{12}$

Solve.

22. There was $\frac{5}{24}$ of a gallon of water in a container. How many gallons were in the container after $\frac{7}{24}$ gallon was added? $\frac{1}{2}$ **gallon**

23. It takes $\frac{25}{36}$ of an hour to do a job. Beth worked $\frac{7}{36}$ of an hour one day and $\frac{11}{36}$ of an hour the next day. How many hours would she need to work the third day to complete the job? $\frac{7}{36}$ **hour**

24. Nancy has $\frac{1}{9}$ gallon of white paint, $\frac{7}{9}$ gallon of blue paint, and $\frac{5}{9}$ gallon of green paint. How much paint would she have if she mixed them all together?

$\frac{13}{9}$ **gallons**

25. If Nancy in Exercise 24 decided that she only wants $\frac{1}{4}$ of the amount of each color, how much of each would she mix and how much paint would result?

$\frac{1}{36}$ **gallon white,** $\frac{7}{36}$ **gallon blue,**
$\frac{5}{36}$ **gallon green,** $\frac{13}{36}$ **gallons total**

4.1 EXERCISES C

Let a represent a natural number. Find the sum or difference.

1. $\dfrac{16}{a} + \dfrac{11}{a}$

$\left[\text{Answer: } \frac{27}{a}\right]$

2. $\dfrac{16}{a} - \dfrac{11}{a}$ $\dfrac{5}{a}$

3. Burford added $\frac{1}{4}$ and $\frac{3}{4}$ as follows.

$$\frac{1}{4} + \frac{3}{4} = \frac{1+3}{4+4} = \frac{4}{8} = \frac{1}{2}$$

What is wrong with his work?   **He should not add denominators. The actual sum is 1, not $\frac{1}{2}$.**

4.2 LEAST COMMON MULTIPLES

STUDENT GUIDEPOSTS

1 Listing Method for Finding the LCM

2 Prime Factorization Method for Finding the LCM

3 Special Algorithm for Finding the LCM

4 LCM of Denominators

❶ LISTING METHOD FOR FINDING THE LCM

Recall from Section 2.1 that the multiples of a number are obtained by multiplying that number by each whole number. Consider the numbers 6 and 8. The nonzero multiples of 6 are

6, 12, 18, **24**, 30, 36, 42, **48**, 54, 60, 66, **72**, . . . ,

and the nonzero multiples of 8 are

8, 16, **24**, 32, 40, **48**, 56, 64, **72**, ...

Any number which is common to both lists of nonzero multiples is called a common multiple. The common multiples of 6 and 8 are

24, 48, 72, ...

The smallest of the common multiples of two (or more) counting numbers is called the **least common multiple,** or **LCM,** of the numbers. Since 24 is the smallest number in the list of common multiples of 6 and 8, we say that 24 is the LCM of 6 and 8. This method of finding the LCM of two or more numbers is called the **listing method.** When using the listing method, it is a good idea to use a chart to organize your work as shown below.

Multiples of 6	6	12	18	**24**	30	36	42
Multiples of 8	8	16	**24**				

EXAMPLE 1   FINDING THE LCM USING THE LISTING METHOD

Find the LCM of the given numbers.

(a) 12 and 30

Multiples of 12:	12,	24,	36,	48,	**60**,	72,	84
Multiples of 30:	30,	**60**					

We stop writing multiples of 30 once we find a multiple that is also in the first list. Since the smallest number common to both lists is 60, we say the LCM of 12 and 30 is 60.

(b) 6, 10, and 15

Multiples of 6:	6,	12,	18,	24,	**30**,	36,	42
Multiples of 10:	10,	20,	**30**,	40,	50,	60	
Multiples of 15:	15,	**30**					

Since 30 is in the first two lists, we can stop after 30 in the third list. The LCM of 6, 10, and 15 is 30.

PRACTICE EXERCISE 1

Find the LCM of the given numbers.

(a) 14 and 18

(b) 5, 8, and 20

Answers: (a) 126   (b) 40

❷ PRIME FACTORIZATION METHOD FOR FINDING THE LCM

Recall that every counting number greater than 1 is either prime or can be factored into primes. Any multiple (including the LCM) of two or more natural numbers must have as factors (divisors) all primes which are factors of each of the natural numbers. This leads to another efficient method for finding LCMs.

Suppose we factor both 6 and 8 into primes.

$$6 = 2 \cdot 3$$
$$8 = 2 \cdot 2 \cdot 2 = 2^3$$

The LCM of 6 and 8 must have $2 \cdot 3$ as a factor since the LCM is a multiple of 6. It must also have $2 \cdot 2 \cdot 2$ or 2^3 as a factor since it must be a multiple of 8. The smallest number which fits these restrictions is

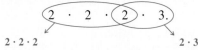

$$2 \cdot 2 \cdot 2 \qquad\qquad 2 \cdot 3$$

Thus, the LCM is $2 \cdot 2 \cdot 2 \cdot 3 = 2^3 \cdot 3 = 24$. This illustrates a general method for finding LCMs.

To Find the LCM Using Prime Factors

1. Factor each number into primes using exponents.
2. If there are no common prime factors, the LCM is the product of all the prime factors.
3. If there are common prime factors, the LCM is the product of the highest power of each prime factor.

To use the rule above to find the LCM of 6 and 8, first factor each number into primes using exponents.

$$6 = 2 \cdot 3$$
$$8 = 2 \cdot 2 \cdot 2 = 2^3$$

Since these two numbers have a common prime factor of 2, we use the highest power of each prime factor. We highlight where each prime occurs the greatest number of times.

$$6 = 2 \cdot \mathbf{3} \qquad\qquad \text{One 2, one 3}$$
$$8 = \mathbf{2 \cdot 2 \cdot 2} = 2^{\mathbf{3}} \qquad \text{Three 2s}$$

Now we can find the LCM by taking the product of 2^3 and 3.

$$\text{LCM} = \mathbf{2 \cdot 2 \cdot 2 \cdot 3} = 2^3 \cdot 3 = 24$$

EXAMPLE 2   FINDING THE LCM USING PRIME FACTORS

Find the LCM of the numbers in Example 1 by using prime factors.

(a) 12 and 30

$$12 = \mathbf{2 \cdot 2} \cdot \mathbf{3} = 2^2 \cdot 3 \qquad \text{Two 2's, one 3}$$
$$30 = 2 \cdot 3 \cdot \mathbf{5} \qquad\qquad \text{One 2, one 3, one 5}$$

The LCM must be the product of two 2's, one 3, and one 5. Thus, the LCM of 12 and 30 is $2 \cdot 2 \cdot 3 \cdot 5 = 60$.

PRACTICE EXERCISE 2

Find the LCM using prime factors.

(a) 14 and 18

(b) 6, 10, and 15

$$6 = \mathbf{2} \cdot \mathbf{3} \quad \boxed{\text{One 2, one 3}}$$
$$10 = 2 \cdot \mathbf{5} \quad \text{One 2, } \boxed{\text{one 5}}$$
$$15 = 3 \cdot 5 \quad \text{One 3, one 5}$$

Thus, the LCM of 6, 10, and 15 is $2 \cdot 3 \cdot 5 = 30$.

(b) 5, 8, and 20

Answers: (a) 126 (b) 40

Compare the results of Example 2 with those of Example 1.

| EXAMPLE 3 FINDING THE LCM USING PRIME FACTORS | PRACTICE EXERCISE 3 |

Find the LCM using prime factors.

(a) 7 and 28

$$7 = \mathbf{7} \qquad\qquad \boxed{\text{One 7}}$$
$$28 = \mathbf{2} \cdot \mathbf{2} \cdot 7 = 2^2 \cdot 7 \quad \boxed{\text{Two 2's}}, \text{ one 7}$$

The LCM $= 2 \cdot 2 \cdot 7 = 2^2 \cdot 7 = 28$. Note that the LCM of 7 and 28 is one of the two numbers itself. This will happen when one number is a multiple of the other (28 is a multiple of 7).

(b) 6, 10, and 25

$$6 = \mathbf{2} \cdot \mathbf{3} \qquad \boxed{\text{One 2}}, \text{ one 3}$$
$$10 = 2 \cdot 5 \qquad \text{One 2, one 5}$$
$$25 = \mathbf{5} \cdot \mathbf{5} = 5^2 \quad \boxed{\text{Two 5's}}$$

The LCM is $2 \cdot 3 \cdot 5 \cdot 5 = 2 \cdot 3 \cdot 5^2 = 150$.

Find the LCM using prime factors.

(a) 13 and 39

(b) 8, 12, and 30

Answers: (a) 39 (b) 120

| EXAMPLE 4 APPLICATION OF LCM | PRACTICE EXERCISE 4 |

A man wishes to apply ceramic tiles to a portion of a wall. He plans to use three types of tiles, one 5 inches long, another 6 inches long, and a third 14 inches long. What is the shortest length of wall space he can cover if the first row contains only 14-in tiles, the second row contains only 6-in tiles, and the third contains only 5-in tiles? [Assume that the tiles are not cut and are laid with no space between them.]

The length of the first row is a multiple of 14 inches, the length of the second row is a multiple of 6 inches, and the length of the third row is a multiple of 5 inches. Thus, the shortest distance that can be covered by these three lengths of tile is the LCM of 5, 6, and 14.

$$5 = \mathbf{5} \qquad \boxed{\text{One 5}}$$
$$6 = \mathbf{2} \cdot \mathbf{3} \qquad \boxed{\text{One 2}, \text{ one 3}}$$
$$14 = 2 \cdot \mathbf{7} \qquad \text{One 2, } \boxed{\text{one 7}}$$

The LCM is $2 \cdot 3 \cdot 5 \cdot 7 = 210$. The shortest distance than can be covered is 210 inches by using 15 of the 14-in tiles ($210 \div 14 = 15$) in the bottom row, 35 of the 6-in tiles ($210 \div 6 = 35$) in the next row, and 42 of the 5-in tiles ($210 \div 5 = 42$) in the third row.

Three types of pipe are to be laid in the same ditch. One pipe is 3 feet long, another 4 feet long, and the third is 10 feet long. What is the shortest distance from the start to where the three will have a seam at the same point?

Answer: 60 feet

3 SPECIAL ALGORITHM FOR FINDING THE LCM (OPTIONAL)

Although the prime factorization method is probably best for use in later mathematics courses, the following method works well for whole numbers. To explain the procedure we will find the LCM of 36 and 45 using this method.

The basic idea is to divide these numbers by the prime numbers to see which primes are factors of each. It is best to start with the lowest prime and work up. Remember the first few prime numbers are

$$2, \ 3, \ 5, \ 7, \ 11.$$

Try to divide 36 and 45 by 2.

2	36	45
	18	45

Two divides 36 but not 45. We put the quotient $18 = 36 \div 2$ below 36 and bring down 45 unchanged. Two will also divide 18.

2	36	45
2	18	45
	9	45

Since $18 \div 2 = 9$, we put 9 and 45 on the next line. Now 2 will not divide 9 or 45 so we try the next prime, 3.

2	36	45
2	18	45
3	9	45
	3	15

Since $9 \div 3 = 3$ and $45 \div 3 = 15$, the next line has a 3 and a 15. Divide by 3 again.

2	36	45
2	18	45
3	9	45
3	3	15
	1	5

We complete the process by dividing by 5.

2	36	45
2	18	45
3	9	45
3	3	15
5	1	5
	1	1

When the bottom row is all ones we stop the process and the LCM is the product of the prime factors in the left column since each divided one of the two numbers. Thus the LCM of 36 and 45 is

$$\text{LCM} = 2 \cdot 2 \cdot 3 \cdot 3 \cdot 5 = 180.$$

In the next example we find the LCM of three numbers using this procedure.

EXAMPLE 5 FINDING THE LCM BY SPECIAL ALGORITHM

Find the LCM of 7, 9, and 12.

2	7	9	12	Divide by 2
2	7	9	6	Divide by 2 again
3	7	9	3	Divide by 3
3	7	3	1	Divide by 3 again
7	7	1	1	Divide by 7
	1	1	1	

Since the last row is all ones, the process stops and the

$$\text{LCM} = 2 \cdot 2 \cdot 3 \cdot 3 \cdot 7 = 252.$$

PRACTICE EXERCISE 5

Find the LCM of 6, 11, and 14.

Answer: 462

---

④ LCM OF DENOMINATORS

We conclude this section by finding the LCM of the denominators of a pair of fractions and expressing each as an equal fraction having the LCM as denominator. This will prepare you for addition and subtraction of fractions in Section 4.3.

EXAMPLE 6 FINDING THE LCM OF DENOMINATORS

Find the LCM of the denominators of $\frac{2}{15}$ and $\frac{3}{20}$ and express each as an equal fraction with the LCM as denominator.

We must find the LCM of 15 and 20.

$$15 = 3 \cdot 5 \qquad \text{One 3 , one 5}$$
$$20 = 2 \cdot 2 \cdot 5 = 2^2 \cdot 5 \qquad \text{Two 2s , one 5}$$

Thus the LCM $= 2 \cdot 2 \cdot 3 \cdot 5 = 2^2 \cdot 3 \cdot 5 = 60$. To express $\frac{2}{15}$ with denominator 60, we must multiply numerator and denominator by 2^2, the missing factors which make the denominator equal $2^2 \cdot 3 \cdot 5$ or 60.

$$\frac{2}{15} = \frac{2 \cdot 2^2}{15 \cdot 2^2} = \frac{8}{60}$$

Now multiply numerator and denominator of $\frac{3}{20}$ by 3, the missing factor which makes the denominator equal $2^2 \cdot 3 \cdot 5$ or 60.

$$\frac{3}{20} = \frac{3 \cdot 3}{20 \cdot 3} = \frac{9}{60}$$

With the LCM $= 60$, then $\frac{2}{15} = \frac{8}{60}$ and $\frac{3}{20} = \frac{9}{60}$.

PRACTICE EXERCISE 6

Find the LCM of the denominators of $\frac{3}{14}$ and $\frac{7}{18}$ and express each as an equal fraction with the LCM as denominator.

Answer: LCM = 126; $\frac{3}{14} = \frac{27}{126}$; $\frac{7}{18} = \frac{49}{126}$

4.2 EXERCISES A

Use the listing method to find the LCM.

1. 24 and 18   72

2. 15 and 9   45

3. 20 and 30   60

4. 3 and 8   24

5. 10, 12, and 20   60

6. 8, 9, and 12   72

Use prime factors to find the LCM.

7. 36 and 15   180

8. 30 and 28   420

9. 13 and 9   117

10. 2, 5, and 7   70

11. 121 and 22   242

12. 4, 6, and 9   36

13. 52 and 66   1716

14 18, 24, and 30   360

15. 22, 55, and 121   1210

Use any method to find the LCM.

16. 20 and 45   180

17. 18 and 28   252

18. 25 and 30   150

19. 10, 15, and 27   270

20. 12, 14, and 28   84

21. 9, 20, and 25   900

Find the LCM of the denominators and express each fraction as an equal fraction with the LCM for a denominator.

22. $\dfrac{1}{8}$ and $\dfrac{7}{12}$

LCM = 24; $\dfrac{1}{8} = \dfrac{3}{24}$; $\dfrac{7}{12} = \dfrac{14}{24}$

23. $\dfrac{3}{4}$ and $\dfrac{5}{6}$

LCM = 12; $\dfrac{3}{4} = \dfrac{9}{12}$; $\dfrac{5}{6} = \dfrac{10}{12}$

24. $\dfrac{5}{28}$ and $\dfrac{11}{42}$

LCM = 84; $\dfrac{5}{28} = \dfrac{15}{84}$; $\dfrac{11}{42} = \dfrac{22}{84}$

25. $\dfrac{18}{25}$, $\dfrac{1}{15}$, and $\dfrac{1}{3}$

LCM = 75; $\dfrac{18}{25} = \dfrac{54}{75}$;

$\dfrac{1}{15} = \dfrac{5}{75}$; $\dfrac{1}{3} = \dfrac{25}{75}$

26. $\dfrac{3}{8}$, $\dfrac{2}{9}$, and $\dfrac{5}{12}$

LCM = 72; $\dfrac{3}{8} = \dfrac{27}{72}$

$\dfrac{2}{9} = \dfrac{16}{72}$; $\dfrac{5}{12} = \dfrac{30}{72}$

27. $\dfrac{3}{10}$, $\dfrac{7}{100}$, and $\dfrac{19}{1000}$

LCM = 1000; $\dfrac{3}{10} = \dfrac{300}{1000}$;

$\dfrac{7}{100} = \dfrac{70}{1000}$; $\dfrac{19}{1000} = \dfrac{19}{1000}$

Solve.

28. Mercury, Venus, and Earth revolve around the sun once every 3, 7, and 12 months, respectively. If the three planets are now in the same straight line, what is the smallest number of months that must pass before they line up again?   **84 months**

29 A blocklayer has three lengths of blocks: 8 inches, 9 inches, and 14 inches. He plans to lay a wall of each type of block so that the three walls are the same length. Neglecting the mortar seams, what is the shortest length of wall possible?   **504 in**

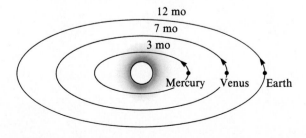

FOR REVIEW

Perform the indicated operations.

30. $\dfrac{7}{12} + \dfrac{2}{12}$ $\dfrac{3}{4}$

31. $\dfrac{13}{11} - \dfrac{6}{11}$ $\dfrac{7}{11}$

32. $\dfrac{4}{7} + \dfrac{1}{7} + \dfrac{2}{7}$ 1

33. $\dfrac{11}{6} - \dfrac{7}{6}$ $\dfrac{2}{3}$

34. $\dfrac{18}{15} - \dfrac{13}{15} + \dfrac{1}{15}$ $\dfrac{2}{5}$

35. $\dfrac{6}{3} - \dfrac{2}{3} - \dfrac{1}{3}$ 1

36. Dennis walked $\frac{13}{4}$ miles due north. He then walked $\frac{7}{4}$ miles due south. Finally he walked $\frac{10}{4}$ miles due north.
   (a) How far north is he from his starting point?   **(b)** How far did he walk? $\frac{15}{2}$ mi
      4 mi

ANSWERS: 1. 72   2. 45   3. 60   4. 24   5. 60   6. 72   7. 180   8. 420   9. 117   10. 70   11. 242   12. 36
13. 1716   14. 360   15. 1210   16. 180   17. 252   18. 150   19. 270   20. 84   21. 900   22. LCM = 24; $\frac{1}{8} = \frac{3}{24}$; $\frac{7}{12} = \frac{14}{24}$
23. LCM = 12; $\frac{3}{4} = \frac{9}{12}$; $\frac{5}{6} = \frac{10}{12}$ 24. LCM = 84; $\frac{5}{28} = \frac{15}{84}$; $\frac{11}{42} = \frac{22}{84}$ 25. LCM = 75; $\frac{18}{25} = \frac{54}{75}$; $\frac{1}{15} = \frac{5}{75}$; $\frac{1}{3} = \frac{25}{75}$
26. LCM = 72; $\frac{3}{8} = \frac{27}{72}$, $\frac{2}{9} = \frac{16}{72}$, $\frac{5}{12} = \frac{30}{72}$ 27. LCM = 1000; $\frac{3}{10} = \frac{300}{1000}$; $\frac{7}{100} = \frac{70}{1000}$; $\frac{19}{1000} = \frac{19}{1000}$ 28. 84 months   29. 504 in
30. $\frac{3}{4}$ 31. $\frac{7}{11}$ 32. 1   33. $\frac{2}{3}$ 34. $\frac{2}{5}$ 35. 1   36. (a) 4 mi (b) $\frac{15}{2}$ mi

4.2 EXERCISES B

Use the listing method to find the LCM.

1. 15 and 6   30

2. 30 and 40   120

3. 45 and 30   90

4. 5 and 16   80

5. 5, 8, and 12   120

6. 7, 21, and 18   126

Use prime factors to find the LCM.

7. 18 and 20   180

8. 30 and 21   210

9. 13 and 12   156

10. 3, 8, and 9   72

11. 169 and 26   338

12. 9, 15, and 22   990

13. 45 and 42   630

14. 20, 26, and 34   4420

15. 49, 35, and 65   3185

Use any method to find the LCM.

16. 10 and 55   110

17. 15 and 42   210

18. 40 and 75   600

19. 9, 12, and 20   180

20. 8, 10, and 36   360

21. 12, 20, and 42   420

Find the LCM of the denominators and express each fraction as an equal fraction with the LCM for a denominator.

22. $\dfrac{3}{10}$ and $\dfrac{4}{15}$

LCM = 30; $\dfrac{3}{10} = \dfrac{9}{30}$; $\dfrac{4}{15} = \dfrac{8}{30}$

23. $\dfrac{5}{14}$ and $\dfrac{1}{4}$

LCM = 28; $\dfrac{5}{14} = \dfrac{10}{28}$; $\dfrac{1}{4} = \dfrac{7}{28}$

24. $\dfrac{6}{55}$ and $\dfrac{5}{44}$

LCM = 220; $\dfrac{6}{55} = \dfrac{24}{220}$; $\dfrac{5}{44} = \dfrac{25}{220}$

25. $\dfrac{2}{3}, \dfrac{5}{18},$ and $\dfrac{7}{30}$

LCM = 90; $\dfrac{2}{3} = \dfrac{60}{90}$;

$\dfrac{5}{18} = \dfrac{25}{90}$; $\dfrac{7}{30} = \dfrac{21}{90}$

26. $\dfrac{7}{27}, \dfrac{3}{4},$ and $\dfrac{1}{24}$

LCM = 216; $\dfrac{7}{27} = \dfrac{56}{216}$

$\dfrac{3}{4} = \dfrac{162}{216}$; $\dfrac{1}{24} = \dfrac{9}{216}$

27. $\dfrac{7}{20}, \dfrac{9}{200}, \dfrac{11}{2000}$

LCM = 2000; $\dfrac{7}{20} = \dfrac{700}{2000}$;

$\dfrac{9}{200} = \dfrac{90}{2000}$; $\dfrac{11}{2000} = \dfrac{11}{2000}$

Solve.

28. Three satellites in the same plane revolve around the earth once every 6, 12, and 14 hours. If the three satellites are in the same straight line, what is the least number of hours that must pass before they line up again?   **84 hr**

29. A bricklayer has three lengths of bricks: 4 inches, 5 inches, and 6 inches. He plans to lay a wall of each type of brick so that the three walls are the same length. Neglecting mortar seams, what is the shortest length of wall possible?   **60 in**

FOR REVIEW

Perform the indicated operations.

30. $\dfrac{3}{11} + \dfrac{1}{11}$ **$\dfrac{4}{11}$**

31. $\dfrac{13}{9} - \dfrac{7}{9}$ **$\dfrac{2}{3}$**

32. $\dfrac{21}{48} + \dfrac{17}{48} + \dfrac{5}{48}$ **$\dfrac{43}{48}$**

33. $\dfrac{7}{20} - \dfrac{2}{20}$ **$\dfrac{1}{4}$**

34. $\dfrac{8}{5} - \dfrac{1}{5} + \dfrac{3}{5}$ **2**

35. $\dfrac{13}{11} - \dfrac{2}{11} - \dfrac{7}{11}$ **$\dfrac{4}{11}$**

36. Jane has 3 days to do a job. She did $\frac{7}{15}$ of the job on the first day and $\frac{2}{15}$ on the second.

    (a) How much had she done in two days?   **(b)** How much must she do the third day to complete the job?

    $\dfrac{3}{5}$

                                                                    $\dfrac{2}{5}$

4.2 EXERCISES C

Find the LCM.

1. 126 and 735   [Answer: 4410]

2. 60, 198, and 495   **1980**

4.3 ADDING AND SUBTRACTING UNLIKE FRACTIONS

> ━━━━━━━━ STUDENT GUIDEPOSTS ━━━━━━━━
>
> **❶** Least Common Denominator (LCD)   **❷** Method for Adding and Subtracting Unlike Fractions

❶ LEAST COMMON DENOMINATOR (LCD)

In Section 4.2 we found the least common multiple (LCM) of the denominators of fractions. The LCM of the denominators of two or more fractions is called the **least common denominator (LCD)** of the fractions. The first step in adding or subtracting unlike fractions is to find the LCD and change unlike fractions to like fractions with the LCD as denominator.

❷ METHOD FOR ADDING AND SUBTRACTING UNLIKE FRACTIONS

Suppose we use the problem

$$\frac{1}{3} + \frac{3}{4}$$

to illustrate the addition method. You can probably tell by inspection that the least common denominator (LCD) is 12. As we did at the end of Section 4.2, change each fraction to an equal fraction with the LCD = 12 as denominator.

$$\frac{1}{3} = \frac{1 \cdot 4}{3 \cdot 4} = \frac{4}{12}$$ Multiply numerator and denominator by 4, the missing factor which makes the denominator 12.

$$\frac{3}{4} = \frac{3 \cdot 3}{4 \cdot 3} = \frac{9}{12}$$ Multiply numerator and denominator by 3, the missing factor which makes the denominator 12.

Thus, $\frac{1}{3} + \frac{3}{4} = \frac{4}{12} + \frac{9}{12} = \frac{4 + 9}{12} = \frac{13}{12}$.

To Add or Subtract Unlike Fractions

1. Rewrite the sum or difference with each denominator written as a prime or factored into primes.

2. Find the LCD (the LCM of all denominators).

3. Multiply the numerator and denominator of each fraction by all factors present in the LCD but missing in the denominator of the particular fraction.

4. Add or subtract the resulting like fractions.

EXAMPLE 1   ADDING AND SUBTRACTING FRACTIONS

Perform the indicated operation.

(a) $\dfrac{2}{3} + \dfrac{1}{4} = \dfrac{2}{3} + \dfrac{1}{2 \cdot 2}$ Rewrite denominators as primes; LCD is $2 \cdot 2 \cdot 3$

$= \dfrac{2 \cdot 2 \cdot 2}{2 \cdot 2 \cdot 3} + \dfrac{1 \cdot 3}{2 \cdot 2 \cdot 3}$ Multiply by missing factors so denominators equal $2 \cdot 2 \cdot 3$

$= \dfrac{8}{12} + \dfrac{3}{12}$ Simplify and add

$= \dfrac{8 + 3}{12}$

$= \dfrac{11}{12}$

(b) $\dfrac{5}{9} + \dfrac{1}{6} = \dfrac{5}{3 \cdot 3} + \dfrac{1}{2 \cdot 3}$ Rewrite denominators as primes; LCD is $2 \cdot 3 \cdot 3$

$= \dfrac{2 \cdot 5}{2 \cdot 3 \cdot 3} + \dfrac{1 \cdot 3}{2 \cdot 2 \cdot 3}$ Multiply by missing factors so denominators equal $2 \cdot 3 \cdot 3$

$= \dfrac{10}{18} + \dfrac{3}{18}$ Simplify and add

$= \dfrac{10 + 3}{18}$

$= \dfrac{13}{18}$

PRACTICE EXERCISE 1

Perform the indicated operations.

(a) $\dfrac{2}{5} + \dfrac{1}{6}$

(b) $\dfrac{1}{4} + \dfrac{3}{10}$

With practice, we may be able to take some shortcuts. For example, if we know that 18 is the LCD of 9 and 6, we can skip the factoring step and add as follows.

(c) $7 - \dfrac{5}{11}$

$$\frac{5}{9} + \frac{1}{6} = \frac{5 \cdot 2}{9 \cdot 2} + \frac{1 \cdot 3}{6 \cdot 3} \qquad \text{Multiply by missing factors so}$$
$$\text{denominators equal 18}$$

$$= \frac{10}{18} + \frac{3}{18} = \frac{13}{18}$$

(c) $2 - \dfrac{3}{7} = \dfrac{2}{1} - \dfrac{3}{7} \qquad \text{LCD} = 7$

$$= \frac{2 \cdot 7}{1 \cdot 7} - \frac{3}{7} \qquad \text{Multiply by missing factor}$$

$$= \frac{14 - 3}{7} \qquad \text{Simplify and subtract}$$

$$= \frac{11}{7}$$

Answers: (a) $\frac{17}{30}$ (b) $\frac{11}{20}$ (c) $\frac{72}{11}$

Sometimes fractions are added vertically. For example, $\frac{2}{3}$ and $\frac{1}{4}$ could be added as follows.

$$\frac{2}{3} = \frac{2}{3} = \frac{2 \cdot 2 \cdot 2}{2 \cdot 2 \cdot 3} = \frac{8}{12} \qquad \text{The LCD} = 2 \cdot 2 \cdot 3 = 12$$

$$+\frac{1}{4} = \frac{1}{2 \cdot 2} = \frac{1 \cdot 3}{2 \cdot 2 \cdot 3} = \frac{3}{12}$$

$$\frac{11}{12} \qquad 8 + 3 = 11$$

Compare this with Example 1**(a)**.

EXAMPLE 2   **ADDING AND SUBTRACTING FRACTIONS**

PRACTICE EXERCISE 2

Perform the indicated operation.

(a) $\dfrac{7}{12} - \dfrac{2}{9} = \dfrac{7}{2 \cdot 2 \cdot 3} - \dfrac{2}{3 \cdot 3}$ Rewrite denominators
    LCD $= 2 \cdot 2 \cdot 3 \cdot 3 = 36$

$$= \frac{7 \cdot 3}{2 \cdot 2 \cdot 3 \cdot 3} - \frac{2 \cdot 2 \cdot 2}{2 \cdot 2 \cdot 3 \cdot 3} \qquad \begin{array}{l}\text{Multiply by missing}\\ \text{factors so denominators}\\ \text{equal } 2 \cdot 2 \cdot 3 \cdot 3\end{array}$$

$$= \frac{21 - 8}{36} \qquad \text{Simplify and subtract}$$

$$= \frac{13}{36}$$

Perform the indicated operations.

(a) $\dfrac{11}{15} - \dfrac{3}{20}$

(b) $\dfrac{2}{3} - \dfrac{1}{6} = \dfrac{2}{3} - \dfrac{1}{2 \cdot 3}$ LCD $= 2 \cdot 3 = 6$

$\qquad = \dfrac{2 \cdot 2}{2 \cdot 3} - \dfrac{1}{2 \cdot 3}$ Multiply by missing factor

$\qquad = \dfrac{4 - 1}{6}$ Simplify and subtract

$\qquad = \dfrac{3}{6}$

$\qquad = \dfrac{1 \cdot \cancel{3}}{2 \cdot \cancel{3}}$ Reduce to lowest terms

$\qquad = \dfrac{1}{2}$

(c) $\dfrac{7}{9} + 3 = \dfrac{7}{3 \cdot 3} + \dfrac{3}{1}$ LCD $= 3 \cdot 3 = 9$

$\qquad = \dfrac{7}{3 \cdot 3} + \dfrac{3 \cdot 3 \cdot 3}{1 \cdot 3 \cdot 3}$ Multiply by missing factors

$\qquad = \dfrac{7}{9} + \dfrac{27}{9}$ Simplify and add

$\qquad = \dfrac{7 + 27}{9}$

$\qquad = \dfrac{34}{9}$

(b) $\dfrac{7}{9} - \dfrac{5}{18}$

(c) $\dfrac{4}{11} + 1$

Answers: (a) $\frac{7}{12}$ (b) $\frac{1}{2}$ (c) $\frac{15}{11}$

As with addition, fractions can also be subtracted vertically. For example, $\frac{7}{12} - \frac{2}{9}$ can be written as follows.

$$
\begin{array}{ccccccc}
\dfrac{7}{12} & = & \dfrac{7}{2 \cdot 2 \cdot 3} & = & \dfrac{7 \cdot 3}{2 \cdot 2 \cdot 3 \cdot 3} & = & \dfrac{21}{36} \\[2mm]
-\dfrac{2}{9} & = & -\dfrac{2}{3 \cdot 3} & = & -\dfrac{2 \cdot 2 \cdot 2}{2 \cdot 2 \cdot 3 \cdot 3} & = & -\dfrac{8}{36} \\[2mm]
& & & & & & \dfrac{13}{36}
\end{array}
$$

The LCD $= 2 \cdot 2 \cdot 3 \cdot 3 = 36$

$21 - 8 = 13$

Compare this with Example 2**(a).**

Sums and differences of more than two fractions can also be found.

EXAMPLE 3 OPERATIONS ON THREE FRACTIONS	PRACTICE EXERCISE 3

Perform the indicated operations.

(a) $\dfrac{1}{3} + \dfrac{1}{4} + \dfrac{5}{6} = \dfrac{1}{3} + \dfrac{1}{2 \cdot 2} + \dfrac{5}{2 \cdot 3}$ Factor denominators: LCD $= 2 \cdot 2 \cdot 3 = 12$

$\qquad = \dfrac{2 \cdot 2 \cdot 1}{2 \cdot 2 \cdot 3} + \dfrac{1 \cdot 3}{2 \cdot 2 \cdot 3} + \dfrac{2 \cdot 5}{2 \cdot 2 \cdot 3}$ Supply missing factors

$\qquad = \dfrac{4 + 3 + 10}{2 \cdot 2 \cdot 3} = \dfrac{17}{12}$

Perform the indicated operations.

(a) $\dfrac{1}{5} + \dfrac{3}{10} + \dfrac{5}{14}$

(b) $\dfrac{13}{15} - \dfrac{1}{5} - \dfrac{1}{2} = \dfrac{13}{3 \cdot 5} - \dfrac{1}{5} - \dfrac{1}{2}$ LCD $= 2 \cdot 3 \cdot 5 = 30$

$\qquad = \dfrac{2 \cdot 13}{2 \cdot 3 \cdot 5} - \dfrac{2 \cdot 3 \cdot 1}{2 \cdot 3 \cdot 5} - \dfrac{1 \cdot 3 \cdot 5}{2 \cdot 3 \cdot 5}$

$\qquad = \dfrac{26 - 6 - 15}{2 \cdot 3 \cdot 5}$

$\qquad = \dfrac{5}{2 \cdot 3 \cdot 5} = \dfrac{\cancel{5}}{2 \cdot 3 \cdot \cancel{5}} = \dfrac{1}{6}$ Reduce to lowest terms

(b) $\dfrac{19}{20} - \dfrac{1}{4} - \dfrac{2}{3}$

Answers: (a) $\dfrac{6}{7}$ (b) $\dfrac{1}{30}$

EXAMPLE 4 APPLICATION OF ADDITION

In making three different kinds of pastries, Mr. Chandler used $\frac{3}{4}$ cup of flour for the first batch, $\frac{7}{8}$ cup for the second, and $\frac{5}{3}$ cups for the third. How much flour did he use?

Add the three fractions.

$\dfrac{3}{4} + \dfrac{7}{8} + \dfrac{5}{3} = \dfrac{3}{2 \cdot 2} + \dfrac{7}{2 \cdot 2 \cdot 2} + \dfrac{5}{3}$ LCD $= 2 \cdot 2 \cdot 2 \cdot 3 = 24$

$\qquad = \dfrac{3 \cdot 2 \cdot 3}{2 \cdot 2 \cdot 2 \cdot 3} + \dfrac{7 \cdot 3}{2 \cdot 2 \cdot 2 \cdot 3} + \dfrac{2 \cdot 2 \cdot 2 \cdot 5}{2 \cdot 2 \cdot 2 \cdot 3}$

$\qquad = \dfrac{18 + 21 + 40}{24} = \dfrac{79}{24} = 3\dfrac{7}{24}$

Thus, Mr. Chandler used $3\frac{7}{24}$ cups of flour.

PRACTICE EXERCISE 4

Wilma ran $\frac{7}{8}$ mi, $\frac{3}{4}$ mi, and $\frac{9}{10}$ mi during one training session. How many miles did she run?

Answer: $\frac{101}{40}$ or $2\frac{21}{40}$ mi

//////////////// **C A U T I O N** ////////////////

Never add or subtract fractions by adding or subtracting numerators and denominators. For example,

$\qquad \dfrac{2}{5} + \dfrac{1}{4}$ *is not the same as* $\dfrac{2 + 1}{5 + 4} = \dfrac{3}{9} = \dfrac{1}{3}$,

since

$\qquad \dfrac{2}{5} + \dfrac{1}{4} = \dfrac{2 \cdot 4}{5 \cdot 4} + \dfrac{1 \cdot 5}{4 \cdot 5} = \dfrac{8 + 5}{20} = \dfrac{13}{20}.$

//////////

4.3 EXERCISES A

Add.

1. $\dfrac{2}{3} + \dfrac{1}{2}$ $\dfrac{7}{6}$

2. $\dfrac{11}{12} + \dfrac{5}{6}$ $\dfrac{7}{4}$

3. $\dfrac{3}{10} + \dfrac{5}{12}$ $\dfrac{43}{60}$

4. $\dfrac{7}{10} + \dfrac{2}{3}$ $\dfrac{41}{30}$

5. $\dfrac{7}{12} + \dfrac{1}{18}$ $\quad \dfrac{23}{36}$

6. $\dfrac{11}{21} + \dfrac{2}{35}$ $\quad \dfrac{61}{105}$

7. $\dfrac{3}{28} + \dfrac{13}{70}$ $\quad \dfrac{41}{140}$

8 $4 + \dfrac{4}{5}$ $\quad \dfrac{24}{5}$

9. $\dfrac{7}{8} + 3$ $\quad \dfrac{31}{8}$

10. $\begin{array}{r}\dfrac{5}{12} \\ + \dfrac{3}{4} \\ \hline \dfrac{7}{6}\end{array}$

11 $\begin{array}{r}\dfrac{1}{6} \\ + \dfrac{3}{5} \\ \hline \dfrac{23}{30}\end{array}$

12. $\begin{array}{r}\dfrac{7}{15} \\ + \dfrac{4}{25} \\ \hline \dfrac{47}{75}\end{array}$

Subtract.

13. $\dfrac{3}{8} - \dfrac{1}{4}$ $\quad \dfrac{1}{3}$

14. $\dfrac{3}{4} - \dfrac{2}{3}$ $\quad \dfrac{1}{12}$

15. $\dfrac{7}{11} - \dfrac{2}{7}$ $\quad \dfrac{27}{77}$

16. $\dfrac{7}{15} - \dfrac{13}{35}$ $\quad \dfrac{2}{21}$

17 $4 - \dfrac{4}{5}$ $\quad \dfrac{16}{5}$

18. $1 - \dfrac{8}{9}$ $\quad \dfrac{1}{9}$

19. $\dfrac{5}{24} - \dfrac{1}{8}$ $\quad \dfrac{1}{12}$

20. $\dfrac{17}{12} - \dfrac{9}{16}$ $\quad \dfrac{41}{48}$

21. $\dfrac{19}{15} - 1$ $\quad \dfrac{4}{15}$

22. $\begin{array}{r}\dfrac{4}{7} \\ - \dfrac{1}{14} \\ \hline \dfrac{1}{2}\end{array}$

23 $\begin{array}{r}\dfrac{8}{15} \\ - \dfrac{3}{20} \\ \hline \dfrac{23}{60}\end{array}$

24. $\begin{array}{r}\dfrac{29}{12} \\ - \dfrac{2}{5} \\ \hline \dfrac{5}{12}\end{array}$

Perform the indicated operations.

25. $\dfrac{2}{3} + \dfrac{3}{4} + \dfrac{1}{6}$ $\quad \dfrac{19}{12}$

26. $\dfrac{3}{5} + \dfrac{1}{3} + \dfrac{7}{10}$ $\quad \dfrac{49}{30}$

27. $\dfrac{14}{15} - \dfrac{2}{5} - \dfrac{1}{3}$ $\quad \dfrac{1}{5}$

28. $\dfrac{7}{20} - \dfrac{1}{4} + \dfrac{3}{8}$ $\quad \dfrac{19}{40}$

29 $\dfrac{8}{15} + \dfrac{1}{12} - \dfrac{5}{20}$ $\quad \dfrac{11}{30}$

30. $\dfrac{7}{3} - 2 + \dfrac{1}{7}$ $\quad \dfrac{10}{21}$

Solve.

31. To obtain the right shade of paint for his living room, Alphonse mixed $\frac{7}{8}$ of a gallon of white paint with $\frac{2}{3}$ of a gallon of yellow and $\frac{3}{5}$ of a gallon of blue. How much paint did he have? $\frac{257}{120}$ **gallons**

32. Suppose Alphonse in Exercise 31 used 2 gallons of the paint that he mixed to paint his living room. How much paint did he have left? $\frac{17}{120}$ **of a gallon**

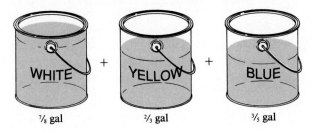

WHITE + YELLOW + BLUE

⅞ gal   ⅔ gal   ⅗ gal

FOR REVIEW

Find the LCM.

33. 27 and 35   945

34. 10 and 45   90

35. 12, 14, and 21   84

The following exercises review material from Section 3.6. We review multiplication and division of mixed numbers to give a contrast with addition and subtraction of mixed numbers in the next section.

36. $3\frac{2}{3} \cdot 6\frac{1}{11}$ $22\frac{1}{3}$

37. $5\frac{1}{2} \div 8\frac{3}{4}$ $\frac{22}{35}$

38. $10 \div 7\frac{6}{7}$ $1\frac{3}{11}$

ANSWERS: 1. $\frac{7}{6}$ 2. $\frac{7}{4}$ 3. $\frac{43}{60}$ 4. $\frac{41}{30}$ 5. $\frac{23}{36}$ 6. $\frac{61}{105}$ 7. $\frac{41}{140}$ 8. $\frac{24}{5}$ 9. $\frac{31}{8}$ 10. $\frac{7}{6}$ 11. $\frac{23}{30}$ 12. $\frac{47}{75}$ 13. $\frac{1}{8}$ 14. $\frac{1}{12}$
15. $\frac{27}{77}$ 16. $\frac{2}{21}$ 17. $\frac{16}{5}$ 18. $\frac{1}{9}$ 19. $\frac{1}{12}$ 20. $\frac{41}{48}$ 21. $\frac{4}{15}$ 22. $\frac{1}{2}$ 23. $\frac{23}{60}$ 24. $\frac{5}{12}$ 25. $\frac{19}{12}$ 26. $\frac{49}{30}$ 27. $\frac{1}{5}$ 28. $\frac{19}{40}$ 29. $\frac{11}{30}$
30. $\frac{10}{21}$ 31. $\frac{257}{120}$ gallons 32. $\frac{17}{120}$ of a gallon 33. 945 34. 90 35. 84 36. $22\frac{1}{3}$ 37. $\frac{22}{35}$ 38. $1\frac{3}{11}$

4.3 EXERCISES B

Add.

1. $\frac{1}{3} + \frac{1}{4}$ $\frac{7}{12}$

2. $\frac{5}{12} + \frac{1}{6}$ $\frac{7}{12}$

3. $\frac{1}{10} + \frac{7}{12}$ $\frac{41}{60}$

4. $\frac{2}{7} + \frac{3}{8}$ $\frac{37}{56}$

5. $\frac{9}{16} + \frac{5}{6}$ $\frac{67}{48}$

6. $\frac{9}{22} + \frac{4}{33}$ $\frac{35}{66}$

7. $\frac{21}{80} + \frac{11}{24}$ $\frac{173}{240}$

8. $\frac{3}{10} + 7$ $\frac{73}{10}$

9. $3 + \frac{9}{11}$ $\frac{42}{11}$

10.
$$\frac{7}{10}$$
$$+ \frac{2}{5}$$
$$\overline{\frac{11}{10}}$$

11.
$$\frac{5}{8}$$
$$+ \frac{5}{12}$$
$$\overline{\frac{25}{24}}$$

12.
$$\frac{8}{35}$$
$$+ \frac{11}{21}$$
$$\overline{\frac{79}{105}}$$

Subtract.

13. $\dfrac{5}{8} - \dfrac{1}{3}$ $\dfrac{7}{24}$

14. $\dfrac{8}{11} - \dfrac{2}{7}$ $\dfrac{34}{77}$

15. $\dfrac{11}{15} - \dfrac{6}{35}$ $\dfrac{59}{105}$

16. $\dfrac{13}{14} - \dfrac{2}{21}$ $\dfrac{5}{6}$

17. $5 - \dfrac{3}{7}$ $\dfrac{32}{7}$

18. $7 - \dfrac{3}{8}$ $\dfrac{53}{8}$

19. $\dfrac{1}{7} - \dfrac{1}{13}$ $\dfrac{6}{91}$

20. $\dfrac{7}{24} - \dfrac{9}{40}$ $\dfrac{1}{15}$

21. $5 - \dfrac{34}{7}$ $\dfrac{1}{7}$

22. $\begin{array}{r} \dfrac{7}{10} \\ -\ \dfrac{2}{5} \\ \hline \dfrac{3}{10} \end{array}$

23. $\begin{array}{r} \dfrac{13}{22} \\ -\ \dfrac{5}{33} \\ \hline \dfrac{29}{66} \end{array}$

24. $\begin{array}{r} 4 \\ -\ \dfrac{17}{5} \\ \hline \dfrac{3}{5} \end{array}$

Perform the indicated operations.

25. $\dfrac{1}{4} + \dfrac{1}{5} + \dfrac{3}{10}$ $\dfrac{3}{4}$

26. $\dfrac{3}{8} + \dfrac{1}{6} + \dfrac{5}{12}$ $\dfrac{23}{24}$

27. $\dfrac{20}{21} - \dfrac{2}{3} - \dfrac{2}{7}$ 0

28. $\dfrac{13}{30} - \dfrac{2}{5} + \dfrac{7}{10}$ $\dfrac{11}{15}$

29. $\dfrac{7}{16} + \dfrac{9}{24} - \dfrac{7}{10}$ $\dfrac{9}{80}$

30. $\dfrac{11}{2} - 5 - \dfrac{3}{11}$ $\dfrac{5}{22}$

31. There were three candidates from the Western Party in an election. One got $\frac{1}{5}$ of the votes, another $\frac{1}{12}$ of the votes, and the third $\frac{3}{8}$ of the votes. What total fraction of the votes did the Western Party receive? $\frac{79}{120}$

32. Two candidates from the Northern Party received the remaining votes in the election in Exercise 31. What fraction of the votes did the Northern Party get? $\frac{41}{120}$

FOR REVIEW

Find the LCM.

33. 22 and 77   154

34. 75 and 95   1425

35. 6, 10, and 15   30

The following exercises review material from Section 3.6. We review multiplication and division of mixed numbers to give a contrast with addition and subtraction of mixed numbers in the next section.

36. $4\dfrac{1}{2} \cdot 10\dfrac{3}{8}$ $46\dfrac{11}{16}$

37. $7\dfrac{4}{5} \div 3\dfrac{6}{7}$ $2\dfrac{1}{45}$

38. $8\dfrac{4}{5} \div 4$ $2\dfrac{1}{5}$

4.3 EXERCISES C

Let a and b represent natural numbers. Perform the indicated operations.

1. $\dfrac{3}{a} + \dfrac{2}{b}$ $\dfrac{3b + 2a}{ab}$

2. $\dfrac{5}{ab} - \dfrac{2}{b}$ $\left[\text{Answer: } \frac{5 - 2a}{ab}\right]$

4.4 ADDING AND SUBTRACTING MIXED NUMBERS

STUDENT GUIDEPOSTS

1 Adding Mixed Numbers **3** Applications of Mixed Numbers
2 Subtracting Mixed Numbers

1 ADDING MIXED NUMBERS

In Chapter 3 we defined a mixed number as the sum of a whole number and a proper fraction. For example,

$$6\frac{3}{4} = 6 + \frac{3}{4}.$$

In Section 3.6 we changed mixed numbers to improper fractions before multiplying or dividing. This same procedure works for addition, but most problems are more easily done by adding the whole number parts and then adding the fraction parts.

EXAMPLE 1 ADDING MIXED NUMBERS	PRACTICE EXERCISE 1

Add $3\frac{5}{8}$ and $2\frac{1}{6}$.

$$3\frac{5}{8} = 3\frac{5}{2 \cdot 2 \cdot 2} = 3\frac{5 \cdot 3}{2 \cdot 2 \cdot 2 \cdot 3} = 3\frac{15}{24}$$

$$+ 2\frac{1}{6} = 2\frac{1}{2 \cdot 3} \quad = 2\frac{1 \cdot 2 \cdot 2}{2 \cdot 2 \cdot 2 \cdot 3} = 2\frac{4}{24}$$

$$5\frac{19}{24}$$

add whole numbers   add fractions

Add $7\frac{1}{10}$ and $3\frac{4}{15}$.

Answer: $10\frac{11}{30}$

EXAMPLE 2 FRACTIONS ADDING TO AN IMPROPER FRACTION	PRACTICE EXERCISE 2

Add $425\frac{3}{8}$ and $211\frac{2}{3}$.

$$425\frac{3}{8} = \quad 425\frac{3}{2 \cdot 2 \cdot 2} = \quad 425\frac{3 \cdot 3}{2 \cdot 2 \cdot 2 \cdot 3} = \quad 425\frac{9}{24}$$

$$+ 211\frac{2}{3} = + 211\frac{2}{3} \quad = +211\frac{2 \cdot 2 \cdot 2 \cdot 2}{2 \cdot 2 \cdot 2 \cdot 3} = + 211\frac{16}{24}$$

$$636\frac{25}{24}$$

add whole numbers   add fractions

Add $324\frac{2}{5}$ and $447\frac{3}{4}$.

Since $\frac{25}{24}$ is an improper fraction, change it to the mixed number $1\frac{1}{24}$ and rewrite the answer as follows.

$$636\frac{25}{24} = 636 + 1\frac{1}{24} \qquad \frac{25}{24} \text{ is } 1\frac{1}{24}$$

$$= \boxed{636 + 1} + \frac{1}{24} \qquad \text{Rewrite mixed number as a sum}$$

$$= \boxed{637} + \frac{1}{24} \qquad \text{Add whole numbers}$$

$$= 637\frac{1}{24} \qquad \text{Final answer}$$

Answer: $772\frac{3}{20}$

Use the same method to add three or more mixed numbers.

| EXAMPLE 3 ADDING THREE MIXED NUMBERS | PRACTICE EXERCISE 3 |

Add $10\frac{1}{3}$, $5\frac{5}{6}$, and $3\frac{3}{4}$.

Add $8\frac{1}{4}$, $3\frac{5}{12}$, and $1\frac{5}{6}$.

$$10\frac{1}{3} = 10\frac{1 \cdot 4}{3 \cdot 4} = 10\frac{4}{12} \qquad \text{LCD is 12}$$

$$5\frac{5}{6} = 5\frac{5 \cdot 2}{6 \cdot 2} = 5\frac{10}{12}$$

$$+ \; 3\frac{3}{4} = 3\frac{3 \cdot 3}{4 \cdot 3} = 3\frac{9}{12}$$

$$\rule{3cm}{0.4pt}$$

$$18\frac{23}{12}$$

add whole numbers   add fractions

Change $\frac{23}{12}$ to $1\frac{11}{12}$ and add to 18.

$$18 + 1\frac{11}{12} = 18 + 1 + \frac{11}{12}$$

$$= 19\frac{11}{12}$$

Answer: $13\frac{1}{2}$

2 SUBTRACTING MIXED NUMBERS

As with addition, mixed numbers can be subtracted by subtracting the whole numbers and the fractions separately.

| **EXAMPLE 4   SUBTRACTING MIXED NUMBERS** | **PRACTICE EXERCISE 4** |

Subtract: $48\frac{3}{4} - 27\frac{1}{3}$.

$$48\frac{3}{4} = \quad 48\frac{3}{2 \cdot 2} = \quad 48\frac{3 \cdot 3}{2 \cdot 2 \cdot 3} = \quad 48\frac{9}{12}$$

$$-27\frac{1}{3} = \; -27\frac{1}{3} \; = \; -27\frac{2 \cdot 2 \cdot 1}{2 \cdot 2 \cdot 3} = \; -27\frac{4}{12}$$

$$21\frac{5}{12}$$

subtract whole numbers   subtract fractions

Subtract: $121\frac{5}{8} - 102\frac{1}{5}$.

Answer: $19\frac{17}{40}$

HINT

An extra step is sometimes needed for this kind of subtraction. Sometimes the fraction being subtracted from is smaller than the fraction being subtracted. When this occurs, borrow.

| **EXAMPLE 5   BORROWING WHEN SUBTRACTING MIXED NUMBERS** | **PRACTICE EXERCISE 5** |

Subtract.

(a) $57\frac{1}{4}$

$-33\frac{1}{2}$

When subtracting the fractions, we try

$$\frac{1}{4} - \frac{1}{2} = \frac{1}{4} - \frac{2}{4} = \frac{1 - 2}{4},$$

which we do not know how to do. Therefore, we must borrow 1 $\left(\text{in the form } \frac{4}{4}\right)$ from 57 and add it to $\frac{1}{4}$ $\left(\text{making } \frac{5}{4}\right)$ before we can subtract.

$$56\frac{\frac{1}{4} + \frac{4}{4}}{}$$

$$57\frac{1}{4} = \quad \cancel{57}\frac{1}{4} = \quad 56\frac{5}{4} = \quad 56\frac{5}{4}$$

$$-33\frac{1}{2} = \; -33\frac{1}{2} = \; -33\frac{1}{2} = \; -33\frac{2}{4}$$

$$23\frac{3}{4}$$

Subtract.

(a) $92\frac{1}{6}$

$-54\frac{2}{3}$

(b) $8\frac{9}{10}$

-6

(b) $6\dfrac{3}{4} = \quad 6\dfrac{3}{4}$

$\dfrac{-\ 2\quad = \ -2\dfrac{0}{4}}{\quad\quad\quad 4\dfrac{3}{4}}$ $2 = 2 + 0 = 2 + \dfrac{0}{4}$

$\dfrac{3}{4} - \dfrac{0}{4} = \dfrac{3 - 0}{4} = \dfrac{3}{4}$

(c) $6 \ = \ 5\dfrac{4}{4}$

$\dfrac{-\ 2\dfrac{3}{4} = \ -2\dfrac{3}{4}}{\quad\quad\quad 3\dfrac{1}{4}}$ Compare this with **(b)** above

(c) 15

$\dfrac{-\ 10\dfrac{7}{8}}{\qquad}$

Answers: (a) $37\dfrac{1}{2}$ (b) $2\dfrac{9}{10}$
(c) $4\dfrac{1}{8}$

As was true with whole numbers, subtraction of mixed numbers can be checked by addition.

These must be equal

$\quad\rightarrow 6$
$\quad - 2\dfrac{3}{4} \leftarrow$ Add these and place the sum below the dashed line
$\quad \ \ 3\dfrac{1}{4} \leftarrow$
$\ - - - -$
$\quad\rightarrow 5\dfrac{4}{4} = 6 \leftarrow$

❸ APPLICATIONS OF MIXED NUMBERS

Many applied problems involve the addition or subtraction of mixed numbers.

EXAMPLE 6   APPLICATION OF ADDITION

Mike weighs $158\dfrac{1}{2}$ lb, Den weighs $138\dfrac{3}{4}$ lb, and Murph weighs $172\dfrac{1}{8}$ lb. Find the combined weight of these men.
We must add the three weights.

$$158\dfrac{1}{2} = 158\dfrac{4}{8} \quad \text{LCD is 8}$$

$$138\dfrac{3}{4} = 138\dfrac{6}{8}$$

$$\underline{172\dfrac{1}{8} = 172\dfrac{1}{8}}$$

$$468\dfrac{11}{8} = 468 + 1 + \dfrac{3}{8} = 469\dfrac{3}{8}$$

Thus, the combined weight of the men is $469\dfrac{3}{8}$ lb.

PRACTICE EXERCISE 6

On a deep sea fishing trip Reva caught three fish weighing $122\dfrac{2}{5}$ lb, $97\dfrac{7}{10}$ lb, and $171\dfrac{4}{5}$ lb. What was the total weight of her fish?

Answer: $391\dfrac{9}{10}$ lb

| EXAMPLE 7   APPLICATION OF SUBTRACTION | PRACTICE EXERCISE 7 |

It took Barb $2\frac{3}{4}$ days to sew a set of draperies. Had she hired a professional seamstress, it would have taken $1\frac{1}{5}$ days. How much time could she have saved by using the professional?

The time she could have saved is the difference between the two times.

$$
\begin{array}{r}
2\dfrac{3}{4} = \quad 2\dfrac{3\cdot 5}{4\cdot 5} = \quad 2\dfrac{15}{20} \quad \text{LCD is 20}\\[2mm]
-1\dfrac{1}{5} = -1\dfrac{1\cdot 4}{5\cdot 4} = -1\dfrac{4}{20}\\[2mm]
\hline
1\dfrac{11}{20}
\end{array}
$$

Thus, she could have saved $1\frac{11}{20}$ days.

Randy had a job to do that required $6\frac{3}{8}$ days to finish. After he had worked for $4\frac{3}{4}$ days, how many days were left to complete the job?

Answer: $1\frac{5}{8}$

4.4 EXERCISES A

Add.

1. $5\dfrac{3}{8}$
$+\ 2\dfrac{1}{2}$
$\overline{\ 7\dfrac{7}{8}}$

2. $6\dfrac{1}{6}$
$+\ 5\dfrac{1}{5}$
$\overline{11\dfrac{11}{30}}$

3. $3\dfrac{2}{5}$
$+\ 4\dfrac{5}{6}$
$\overline{\ 8\dfrac{7}{30}}$

4. $8\dfrac{1}{12}$
$+\ 7\dfrac{3}{4}$
$\overline{15\dfrac{5}{6}}$

5. $2\dfrac{2}{7}$
$+\ 4\dfrac{5}{21}$
$\overline{\ 6\dfrac{11}{21}}$

6. $3\dfrac{3}{20}$
$+\ 2\dfrac{5}{24}$
$\overline{\ 5\dfrac{43}{120}}$

7. $6\dfrac{7}{33}$
$+\ 10\dfrac{3}{11}$
$\overline{16\dfrac{16}{33}}$

8 $15\dfrac{7}{8}$
$+\ 4$
$\overline{19\dfrac{7}{8}}$

9. 8
$+\ 7\dfrac{3}{10}$
$\overline{15\dfrac{3}{10}}$

10 $215\dfrac{7}{8}$
$+\ 147\dfrac{1}{2}$
$\overline{363\dfrac{3}{8}}$

11. $685\dfrac{2}{11}$
$+\ 296\dfrac{2}{3}$
$\overline{981\dfrac{28}{33}}$

12. $45\dfrac{5}{12}$
$+\ 88\dfrac{15}{16}$
$\overline{134\dfrac{17}{48}}$

Subtract.

13. $9\dfrac{2}{9}$
$-\ 3\dfrac{1}{3}$
$\overline{\ 5\dfrac{8}{9}}$

14. $11\dfrac{4}{5}$
$-\ 5\dfrac{2}{3}$
$\overline{\ 6\dfrac{2}{15}}$

15. $6\dfrac{3}{8}$
$-\ 5\dfrac{3}{4}$
$\overline{\ \dfrac{5}{8}}$

16. $10\dfrac{2}{5}$
$-\ 1\dfrac{7}{20}$
$\overline{\ 9\dfrac{1}{20}}$

17. $\begin{array}{r} 4\frac{8}{9} \\ -\ 3\frac{11}{15} \\ \hline 1\frac{7}{45} \end{array}$

18. $\begin{array}{r} 8\frac{1}{4} \\ -\ 3\frac{2}{3} \\ \hline 4\frac{7}{12} \end{array}$

19. $\begin{array}{r} 4\frac{11}{35} \\ -\ 2\frac{4}{21} \\ \hline 2\frac{13}{105} \end{array}$

20 $\begin{array}{r} 17\frac{6}{11} \\ -\ 8 \\ \hline 9\frac{6}{11} \end{array}$

21. $\begin{array}{r} 15 \\ -\ 4\frac{9}{16} \\ \hline 10\frac{7}{16} \end{array}$

22 $\begin{array}{r} 485\frac{9}{10} \\ -\ 316\frac{2}{5} \\ \hline 169\frac{1}{2} \end{array}$

23. $\begin{array}{r} 925\frac{1}{7} \\ -\ 495\frac{11}{14} \\ \hline 429\frac{5}{14} \end{array}$

24. $\begin{array}{r} 211\frac{3}{22} \\ -\ 201\frac{7}{33} \\ \hline 9\frac{61}{66} \end{array}$

Add.

25. $\begin{array}{r} 10\frac{2}{3} \\ 4\frac{1}{5} \\ +\ 7\frac{2}{15} \\ \hline 22 \end{array}$

26 $\begin{array}{r} 7\frac{1}{5} \\ 3\frac{2}{3} \\ +\ 1\frac{1}{15} \\ \hline 11\frac{14}{15} \end{array}$

27. $\begin{array}{r} 8\frac{3}{4} \\ 2\frac{1}{12} \\ +\ 4\frac{5}{6} \\ \hline 15\frac{2}{3} \end{array}$

28. $\begin{array}{r} 5\frac{8}{9} \\ 2\frac{1}{3} \\ +\ 1\frac{5}{12} \\ \hline 9\frac{23}{36} \end{array}$

Solve.

29. Juan took a hike over three different trails. If the three trails were $5\frac{1}{3}$ mi, $8\frac{3}{4}$ mi, and $7\frac{7}{8}$ mi long, what was the total distance that he hiked? $21\frac{23}{24}$ mi

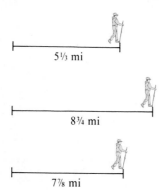

5⅓ mi

8¾ mi

7⅞ mi

30 It took Arn $7\frac{1}{8}$ days to type a manuscript. Had he not been broke, he could have hired a typist who could do the job in $1\frac{2}{3}$ days. How many days could have been saved? $5\frac{11}{24}$ days

31. Mr. Horn owns $25\frac{1}{4}$ acres of land in Colorado, $160\frac{2}{3}$ acres in Utah, and $185\frac{1}{6}$ acres in Florida. What is the total number of acres that he owns? $371\frac{1}{12}$ acres

32. It took $3\frac{5}{8}$ hours for Burford to do a job. With help he could have done the job in $2\frac{3}{4}$ hours. How much time could he have saved had he had help? $\frac{7}{8}$ hr

FOR REVIEW

Perform the indicated operations.

33. $\dfrac{4}{5} + \dfrac{1}{15} + \dfrac{7}{30}$

$1\dfrac{1}{10}$

34. $\dfrac{4}{7} + \dfrac{11}{14} - \dfrac{8}{21}$

$\dfrac{41}{42}$

35. $\dfrac{9}{5} - \dfrac{17}{15} - \dfrac{16}{25}$

$\dfrac{2}{75}$

36. $\dfrac{1}{10} + \dfrac{4}{35} - \dfrac{2}{21}$

$\dfrac{5}{42}$

The following exercises review material from Section 1.1 to help prepare you for the next section. Place the appropriate sign, $<$ or $>$, between the whole numbers.

37. $16 \underline{\ >\ } 4$
38. $2 \underline{\ <\ } 45$
39. $0 \underline{\ <\ } 6$
40. $8 \underline{\ >\ } 7$

ANSWERS: 1. $7\frac{7}{8}$ 2. $11\frac{11}{30}$ 3. $8\frac{7}{30}$ 4. $15\frac{5}{6}$ 5. $6\frac{11}{21}$ 6. $5\frac{43}{120}$ 7. $16\frac{16}{33}$ 8. $19\frac{7}{8}$ 9. $15\frac{3}{10}$ 10. $363\frac{3}{8}$ 11. $981\frac{28}{33}$
12. $134\frac{17}{48}$ 13. $5\frac{8}{9}$ 14. $6\frac{2}{15}$ 15. $\frac{5}{8}$ 16. $9\frac{1}{20}$ 17. $1\frac{7}{45}$ 18. $4\frac{7}{12}$ 19. $2\frac{13}{105}$ 20. $9\frac{6}{11}$ 21. $10\frac{7}{16}$ 22. $169\frac{1}{2}$ 23. $429\frac{5}{14}$
24. $9\frac{61}{66}$ 25. 22 26. $11\frac{14}{15}$ 27. $15\frac{2}{3}$ 28. $9\frac{23}{36}$ 29. $21\frac{23}{24}$ mi 30. $5\frac{11}{24}$ days 31. $371\frac{1}{12}$ acres 32. $\frac{7}{8}$ hr 33. $1\frac{1}{10}$
34. $\frac{41}{42}$ 35. $\frac{2}{75}$ 36. $\frac{5}{42}$ 37. $>$ 38. $<$ 39. $<$ 40. $>$

4.4 EXERCISES B

Add.

1. $2\dfrac{3}{10}$
$+ 3\dfrac{2}{5}$
$\overline{5\dfrac{7}{10}}$

2. $7\dfrac{1}{8}$
$+ 9\dfrac{1}{6}$
$\overline{16\dfrac{7}{24}}$

3. $4\dfrac{2}{11}$
$+ 3\dfrac{3}{4}$
$\overline{7\dfrac{41}{44}}$

4. $8\dfrac{4}{15}$
$+ 1\dfrac{4}{5}$
$\overline{10\dfrac{1}{15}}$

5. $4\dfrac{7}{8}$
$+ 2\dfrac{7}{10}$
$\overline{7\dfrac{23}{40}}$

6. $8\dfrac{5}{12}$
$+ 10\dfrac{7}{15}$
$\overline{18\dfrac{53}{60}}$

7. $6\dfrac{3}{13}$
$+ 9\dfrac{7}{26}$
$\overline{15\dfrac{1}{2}}$

8. $19\dfrac{4}{21}$
$+ 10$
$\overline{29\dfrac{4}{21}}$

9. 17
$+ 1\dfrac{5}{8}$
$\overline{18\dfrac{5}{8}}$

10. $106\dfrac{4}{5}$
$+ 317\dfrac{4}{15}$
$\overline{424\dfrac{1}{15}}$

11. $1060\dfrac{1}{7}$
$+ 1121\dfrac{2}{3}$
$\overline{2181\dfrac{17}{21}}$

12. $21\dfrac{7}{10}$
$+ 19\dfrac{19}{20}$
$\overline{41\dfrac{13}{20}}$

Subtract.

13. $8\dfrac{2}{5}$
$- 6\dfrac{8}{15}$
$\overline{1\dfrac{13}{15}}$

14. $15\dfrac{5}{7}$
$- 12\dfrac{2}{3}$
$\overline{3\dfrac{1}{21}}$

15. $7\dfrac{3}{10}$
$- 4\dfrac{4}{5}$
$\overline{2\dfrac{1}{2}}$

16. $3\dfrac{1}{6}$
$- 1\dfrac{7}{18}$
$\overline{1\dfrac{7}{9}}$

17.
$$
\begin{array}{r}
3\frac{3}{10} \\
-\,2\frac{4}{15} \\
\hline
1\frac{1}{30}
\end{array}
$$

18.
$$
\begin{array}{r}
8\frac{1}{8} \\
-\,4\frac{3}{5} \\
\hline
3\frac{21}{40}
\end{array}
$$

19.
$$
\begin{array}{r}
12\frac{11}{15} \\
-\,11\frac{5}{6} \\
\hline
\frac{9}{10}
\end{array}
$$

20.
$$
\begin{array}{r}
9\frac{7}{11} \\
-\,4 \\
\hline
5\frac{7}{11}
\end{array}
$$

21.
$$
\begin{array}{r}
11 \\
-\,7\frac{3}{14} \\
\hline
3\frac{11}{14}
\end{array}
$$

22.
$$
\begin{array}{r}
325\frac{4}{5} \\
-\,170\frac{1}{4} \\
\hline
155\frac{11}{20}
\end{array}
$$

23.
$$
\begin{array}{r}
428\frac{1}{5} \\
-\,147\frac{7}{8} \\
\hline
280\frac{13}{40}
\end{array}
$$

24.
$$
\begin{array}{r}
901\frac{1}{15} \\
-\,899\frac{20}{21} \\
\hline
1\frac{4}{35}
\end{array}
$$

Add.

25.
$$
\begin{array}{r}
1\frac{2}{11} \\
2\frac{3}{22} \\
+\,4\frac{1}{2} \\
\hline
7\frac{9}{11}
\end{array}
$$

26.
$$
\begin{array}{r}
4\frac{2}{3} \\
6\frac{5}{6} \\
+\,7\frac{1}{12} \\
\hline
18\frac{7}{12}
\end{array}
$$

27.
$$
\begin{array}{r}
4\frac{3}{4} \\
4\frac{3}{8} \\
+\,4\frac{3}{16} \\
\hline
13\frac{5}{16}
\end{array}
$$

28.
$$
\begin{array}{r}
9\frac{5}{6} \\
7\frac{1}{4} \\
+\,1\frac{2}{3} \\
\hline
18\frac{3}{4}
\end{array}
$$

Solve.

29. On Monday Debbie worked out for $2\frac{1}{4}$ hours, on Wednesday $3\frac{2}{5}$ hours, and on Friday $3\frac{1}{3}$ hours. What is the sum of her hours for these three days? $8\frac{59}{60}$ **hr**

30. Harry's paper for a history class has to be 16 pages long. How many more pages must he write if he has done $9\frac{2}{5}$ pages? $6\frac{3}{5}$ **pages**

31. On a trip the Williamses filled their gas tank three times. They required $16\frac{2}{5}$ gallons, $21\frac{7}{10}$ gallons, and $19\frac{1}{5}$ gallons. How many gallons of gasoline did they buy on the trip? $57\frac{3}{10}$ **gallons**

32. A contractor estimates that it will take $17\frac{1}{2}$ days to complete a job. How many days of work are left after $9\frac{3}{4}$ days have been completed? $7\frac{3}{4}$ **days**

FOR REVIEW

Perform the indicated operations.

33. $\dfrac{1}{6} + \dfrac{1}{8} + \dfrac{1}{10}$ $\dfrac{47}{120}$

34. $\dfrac{7}{12} + \dfrac{3}{4} - \dfrac{5}{8}$ $\dfrac{17}{24}$

35. $\dfrac{10}{7} - \dfrac{7}{10} - \dfrac{3}{14}$ $\dfrac{18}{35}$

36. $\dfrac{9}{11} + \dfrac{5}{33} - \dfrac{7}{22}$ $\dfrac{43}{66}$

The following exercises review material from Section 1.1 to help prepare you for the next section. Place the appropriate sign, < or >, between the whole numbers.

37. $1 \;\underline{\; < \;}\; 9$

38. $4 \;\underline{\; > \;}\; 0$

39. $16 \;\underline{\; < \;}\; 18$

40. $25 \;\underline{\; > \;}\; 15$

4.4 EXERCISES C

Perform the indicated operations.

1. $209\dfrac{4}{5} - 122\dfrac{4}{15} + 65\dfrac{1}{3}$ $152\dfrac{13}{15}$

2. $489\dfrac{3}{4} - 280\dfrac{5}{8} - 2\dfrac{1}{2}$ $\left[\text{Answer: } 206\tfrac{5}{8}\right]$

4.5 ORDER OF OPERATIONS AND COMPARING FRACTIONS

STUDENT GUIDEPOSTS

❶ Rule for the Order of Operations   **❷** Comparing Fractions

❶ RULE FOR THE ORDER OF OPERATIONS

The order of operations on whole numbers that we studied in Section 2.8 applies to fractions as well, and is reviewed below.

Order of Operations
Operations should be performed in the following order.

1. Operate within parentheses.
2. Find all powers and roots in any order.
3. Multiply and divide, in order, from left to right.
4. Add and subtract, in order, from left to right.

EXAMPLE 1 ORDER OF OPERATIONS

Evaluate each expression.

(a) $\left(\dfrac{5}{6} - \dfrac{1}{6}\right) \cdot \dfrac{3}{4} = \dfrac{4}{6} \cdot \dfrac{3}{4}$ Operate inside parentheses first

$\qquad = \dfrac{2}{3} \cdot \dfrac{3}{4}$ Reduce fraction

$\qquad = \dfrac{\overset{1}{\cancel{2}} \cdot \overset{1}{\cancel{3}}}{\cancel{3} \cdot \underset{1}{\cancel{2}} \cdot 2}$ Multiply

$\qquad = \dfrac{1}{2}$

PRACTICE EXERCISE 1

Evaluate each expression.

(a) $\dfrac{4}{15} \cdot \left(\dfrac{3}{8} + \dfrac{1}{8}\right)$

(b) $\dfrac{5}{12} + \dfrac{3}{8} \div \dfrac{3}{4} = \dfrac{5}{12} + \dfrac{3}{8} \cdot \dfrac{4}{3}$ Divide first

$= \dfrac{5}{12} + \dfrac{\overset{1}{\cancel{3}} \cdot \overset{1}{\cancel{4}}}{2 \cdot \underset{1}{\cancel{4}} \cdot \underset{1}{\cancel{3}}}$

$= \dfrac{5}{12} + \dfrac{1}{2} = \dfrac{5}{12} + \dfrac{6}{12}$ Then add, the LCD is 12

$= \dfrac{11}{12}$

(c) $\dfrac{5}{7} \cdot \dfrac{7}{2} - \dfrac{1}{11} \div \dfrac{3}{22} = \dfrac{5 \cdot 7}{7 \cdot 2} - \dfrac{1}{11} \cdot \dfrac{22}{3}$ Multiply and divide first

$= \dfrac{5 \cdot \overset{1}{\cancel{7}}}{\underset{1}{\cancel{7}} \cdot 2} - \dfrac{1 \cdot 2 \cdot \overset{1}{\cancel{11}}}{\underset{1}{\cancel{11}} \cdot 3}$

$= \dfrac{5}{2} - \dfrac{2}{3}$

$= \dfrac{15}{6} - \dfrac{4}{6}$ LCD is 6

$= \dfrac{11}{6} = 1\dfrac{5}{6}$ Subtract last

PRACTICE EXERCISE 1

(b) $\dfrac{5}{9} \div \dfrac{5}{3} - \dfrac{1}{10}$

(c) $\dfrac{9}{20} \div \dfrac{3}{10} - \dfrac{4}{11} \cdot \dfrac{11}{20}$

Answers: (a) $\dfrac{2}{15}$ (b) $\dfrac{7}{30}$
(c) $1\dfrac{3}{10}$

EXAMPLE 2 ORDER OF OPERATIONS

Evaluate each expression.

(a) $\left(\dfrac{9}{11} - \dfrac{3}{11}\right) \cdot \left(\dfrac{1}{2}\right)^2 = \dfrac{6}{11} \cdot \left(\dfrac{1}{2}\right)^2$ Evaluate inside parentheses first

$= \dfrac{6}{11} \cdot \dfrac{1}{4}$ Find square next

$= \dfrac{\overset{1}{\cancel{2}} \cdot 3 \cdot 1}{11 \cdot \underset{1}{\cancel{2}} \cdot 2}$ Multiply

$= \dfrac{3}{22}$

(b) $\sqrt{\dfrac{4}{9}} - \dfrac{1}{3} \cdot \dfrac{1}{2} = \dfrac{2}{3} - \dfrac{1}{3} \cdot \dfrac{1}{2}$ Find square root first

$= \dfrac{2}{3} - \dfrac{1}{6}$ Multiply

$= \dfrac{4}{6} - \dfrac{1}{6}$ Now subtract

$= \dfrac{3}{6} = \dfrac{1}{2}$ Reduce

PRACTICE EXERCISE 2

Evaluate each expression.

(a) $\left(\dfrac{2}{3}\right)^2 \div \left(\dfrac{4}{7} + \dfrac{5}{14}\right)$

(b) $\dfrac{3}{5} - \sqrt{\dfrac{16}{25}} \cdot \dfrac{1}{2}$

(c) $\left(\dfrac{3}{4}\right)^2 + \dfrac{5}{3} \div \dfrac{4}{3} - \dfrac{2}{5} \cdot \dfrac{5}{4} = \dfrac{9}{16} + \dfrac{5}{3} \div \dfrac{4}{3} - \dfrac{2}{5} \cdot \dfrac{5}{4}$ Square first

$= \dfrac{9}{16} + \dfrac{5}{3} \cdot \dfrac{3}{4} - \dfrac{2 \cdot 5}{5 \cdot 4}$ Divide and multiply

$= \dfrac{9}{16} + \dfrac{5 \cdot \overset{1}{\cancel{3}}}{\cancel{3} \cdot 4} - \dfrac{\overset{1}{\cancel{2}} \cdot \overset{1}{\cancel{5}}}{\cancel{5} \cdot \cancel{2} \cdot 2}$

$= \dfrac{9}{16} + \dfrac{5}{4} - \dfrac{1}{2}$

$= \dfrac{9}{16} + \dfrac{20}{16} - \dfrac{8}{16}$ Now add and subtract

$= \dfrac{21}{16} = 1\dfrac{5}{16}$

(c) $\dfrac{4}{15} - \dfrac{3}{5} \cdot \dfrac{2}{15} + \left(\dfrac{1}{5}\right)^2$

Answers: (a) $\frac{56}{117}$ (b) $\frac{1}{5}$ (c) $\frac{17}{75}$

❷ COMPARING FRACTIONS

Recall that one whole number is greater than another if the first is to the right of the second on a number line. Deciding which of two fractions is larger is just as simple, when the denominators are the same. Consider the fractions

$$\dfrac{2}{3} \quad \text{and} \quad \dfrac{3}{5}.$$

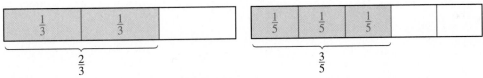

Figure 4.1

Compare the bars in Figure 4.1. It is not obvious which fraction is larger. Now find the LCD of these fractions.

$$\dfrac{2}{3} = \dfrac{2 \cdot 5}{3 \cdot 5} = \dfrac{10}{15} \quad \text{and} \quad \dfrac{3}{5} = \dfrac{3 \cdot 3}{3 \cdot 5} = \dfrac{9}{15}$$

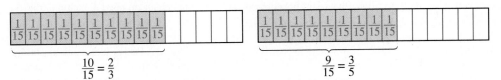

Figure 4.2

Comparing the bars in Figure 4.2 makes it clear that the shaded portion $\frac{10}{15}$ is larger than the shaded portion $\frac{9}{15}$. Thus

$$\dfrac{10}{15} > \dfrac{9}{15} \quad \text{and} \quad \dfrac{2}{3} > \dfrac{3}{5}.$$

But it is not necessary to compare fractional parts of bars since certainly

10 fifteenths is more than **9** fifteenths.

When two fractions have the same denominator, the larger fraction has the larger numerator.

To Tell Which of Two Fractions Is Larger
1. Find a common denominator and convert each fraction to an equal fraction having that denominator. 2. The larger fraction has the larger numerator.

This rule allows us to compare fractions using what we know about the order of whole numbers.

EXAMPLE 3   COMPARING FRACTIONS

Which of the two fractions is larger?

(a) $\dfrac{5}{6}$ or $\dfrac{3}{4}$

$$\frac{5}{6} = \frac{5}{2 \cdot 3} = \frac{2 \cdot 5}{2 \cdot 2 \cdot 3} = \frac{10}{12}$$

$$\frac{3}{4} = \frac{3}{2 \cdot 2} = \frac{3 \cdot 3}{2 \cdot 2 \cdot 3} = \frac{9}{12}$$

The LCD of 4 and 6 is $2 \cdot 2 \cdot 3 = 12$

Since $10 > 9$, $\dfrac{10}{12} > \dfrac{9}{12}$. Thus, $\dfrac{5}{6} > \dfrac{3}{4}$.

(b) $\dfrac{3}{20}$ or $\dfrac{1}{5}$

$$\frac{3}{20} = \frac{3}{2 \cdot 2 \cdot 5} = \frac{3}{30}$$

$$\frac{1}{5} = \frac{2 \cdot 2 \cdot 1}{2 \cdot 2 \cdot 5} = \frac{4}{20}$$

The LCD of 5 and 20 is $2 \cdot 2 \cdot 5 = 20$

Since $4 > 3$, $\dfrac{4}{20} > \dfrac{3}{20}$. Thus $\dfrac{1}{5} > \dfrac{3}{20}$.

PRACTICE EXERCISE 3

Which of the two fractions is larger?

(a) $\dfrac{7}{12}$ or $\dfrac{6}{11}$

(b) $\dfrac{1}{7}$ or $\dfrac{2}{17}$

Answers: (a) $\frac{7}{12} > \frac{6}{11}$ (b) $\frac{1}{7} > \frac{2}{17}$

EXAMPLE 4   APPLICATION TO RECREATION

A map of trails joining Blue Ridge Hill and the blackberry patch is shown in Figure 4.3. Is it farther from Blue Ridge Hill to the blackberry patch by way of Luna Pond or by way of White Rock Creek? How much farther?

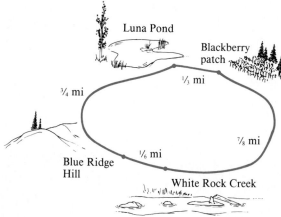

Figure 4.3

PRACTICE EXERCISE 4

In June, Chicago had two rainstorms, one which dropped $\frac{3}{8}$ inch and the other $1\frac{1}{4}$ inches of precipitation. During July, there were three storms with $\frac{1}{3}$, $\frac{4}{5}$, and $\frac{1}{8}$ inch of rain. During which of these months did Chicago have more precipitation? How much more?

The distance from Blue Ridge Hill to the blackberry patch by way of Luna Pond can be found by adding $\frac{3}{4}$ and $\frac{1}{3}$.

$$\frac{3}{4} + \frac{1}{3} = \frac{3 \cdot 3}{4 \cdot 3} + \frac{1 \cdot 4}{3 \cdot 4} = \frac{9}{12} + \frac{4}{12} = \frac{13}{12}$$

The distance by way of White Rock Creek is found by adding $\frac{1}{6}$ and $\frac{7}{8}$.

$$\frac{1}{6} + \frac{7}{8} = \frac{1}{2 \cdot 3} + \frac{7}{2 \cdot 2 \cdot 2} = \frac{2 \cdot 2 \cdot 1}{2 \cdot 2 \cdot 2 \cdot 3} + \frac{7 \cdot 3}{2 \cdot 2 \cdot 2 \cdot 3}$$

$$= \frac{4 + 21}{24} = \frac{25}{24}$$

Write $\frac{13}{12}$ with the same denominator as $\frac{25}{24}$.

$$\frac{13}{12} = \frac{13 \cdot 2}{12 \cdot 2} = \frac{26}{24}$$

Since $26 > 25$, $\frac{26}{24} > \frac{25}{24}$. Thus, $\frac{13}{12} > \frac{25}{24}$ so it is farther by way of Luna Pond.

To find out how much farther it is, subtract $\frac{25}{24}$ from $\frac{13}{12}$. Using the fractions above,

$$\frac{13}{12} - \frac{25}{24} = \frac{26}{24} - \frac{25}{24} = \frac{1}{24}.$$

Thus, it is $\frac{1}{24}$ of a mile farther.

Answer: $\frac{11}{30}$ inch more in June

4.5 EXERCISES A

Evaluate each expression.

1. $\left(\frac{5}{9} + \frac{4}{9}\right) \cdot \frac{11}{15}$

$\frac{11}{15}$

2. $\frac{4}{5} \div \frac{2}{3} - \frac{2}{5}$

$\frac{4}{5}$

3. $\frac{4}{3} \cdot \left(\frac{7}{8} - \frac{1}{4}\right)$

$\frac{5}{6}$

4. $\frac{2}{3} \div \frac{1}{9} \cdot \frac{1}{4}$

$1\frac{1}{2}$

5. $\frac{15}{16} - \frac{3}{4} \cdot \frac{1}{2}$

$\frac{9}{16}$

6. $\left(\frac{4}{7} - \frac{3}{14}\right) \div \frac{4}{21}$

$1\frac{7}{8}$

7. $\sqrt{\frac{1}{4}} - \left(\frac{1}{4}\right)^2$

$\frac{7}{16}$

8 $\left(\frac{3}{5}\right)^2 - \frac{1}{5} + \frac{6}{25}$

$\frac{2}{5}$

9. $\frac{2}{3} \cdot \frac{6}{11} - 2 \div \frac{11}{2}$ 0

10 $\sqrt{\frac{1}{9}} + \frac{6}{7} \div \frac{3}{14} - 3 \cdot \frac{1}{9}$ 4

11. $\left(\dfrac{4}{5} - \dfrac{1}{10}\right)^2 \div \dfrac{7}{10}$ $\dfrac{7}{10}$

12. $\dfrac{5}{6} \div \dfrac{2}{3} - \dfrac{1}{4} \div \dfrac{1}{2} - \left(\dfrac{1}{2}\right)^2$ $\dfrac{1}{2}$

Which of the two fractions is larger?

13. $\dfrac{4}{15}$ or $\dfrac{1}{4}$ $\dfrac{4}{15} > \dfrac{1}{4}$

14. $\dfrac{29}{21}$ or $\dfrac{25}{18}$ $\dfrac{29}{21} < \dfrac{25}{18}$

15. $\dfrac{4}{6}$ or $\dfrac{27}{42}$ $\dfrac{4}{6} > \dfrac{27}{42}$

16. $\dfrac{35}{11}$ or $\dfrac{41}{13}$ $\dfrac{35}{11} > \dfrac{41}{13}$

17 $\dfrac{17}{30}$ or $\dfrac{21}{40}$ $\dfrac{17}{30} > \dfrac{21}{40}$

18. $\dfrac{19}{8}$ or $\dfrac{81}{36}$ $\dfrac{19}{8} > \dfrac{81}{36}$

Answer true or false.

19. $\dfrac{1}{4} < \dfrac{1}{3}$ true

20. $\dfrac{3}{8} > \dfrac{4}{7}$ false

21. $\dfrac{25}{3} < \dfrac{35}{4}$ true

22. $3\dfrac{2}{5} > 3\dfrac{5}{12}$ false

23. $\dfrac{1}{2} \div \dfrac{1}{4} > \dfrac{1}{3} \div \dfrac{1}{9}$ false

24. $\dfrac{1}{2} < \dfrac{1}{3} < \dfrac{1}{4}$ false

Solve.

25. Fahrenheit temperature can be found by multiplying Celsius temperature by $\dfrac{9}{5}$ and adding 32°. If Celsius temperature is 25°, what is the Fahrenheit temperature?   77°

26 Carl did $\dfrac{1}{4}$ of a job on Monday morning and $\dfrac{7}{40}$ in the afternoon. He did $\dfrac{1}{4}$ of the job on Tuesday morning and $\dfrac{1}{5}$ on Tuesday afternoon. On which day did he do more of the job? How much more?
$\dfrac{1}{40}$ more on Tuesday

FOR REVIEW

Perform the indicated operations.

27. $3\dfrac{2}{3}$
$+ 2\dfrac{1}{6}$
$\overline{5\dfrac{5}{6}}$

28. $4\dfrac{2}{7}$
$- 3\dfrac{5}{14}$
$\overline{\dfrac{13}{14}}$

29. $416\dfrac{1}{9}$
$+ 201\dfrac{2}{3}$
$\overline{617\dfrac{7}{9}}$

30. $8\dfrac{3}{10}$
$2\dfrac{4}{5}$
$+ 5\dfrac{2}{5}$
$\overline{16\dfrac{1}{2}}$

ANSWERS: 1. $\dfrac{11}{15}$ 2. $\dfrac{4}{5}$ 3. $\dfrac{5}{6}$ 4. $1\dfrac{1}{2}$ 5. $\dfrac{9}{16}$ 6. $1\dfrac{7}{8}$ 7. $\dfrac{7}{16}$ 8. $\dfrac{2}{5}$ 9. 0 10. 4 11. $\dfrac{7}{10}$ 12. $\dfrac{1}{2}$ 13. $\dfrac{4}{15} > \dfrac{1}{4}$
14. $\dfrac{29}{21} < \dfrac{25}{18}$ 15. $\dfrac{4}{6} > \dfrac{27}{42}$ 16. $\dfrac{35}{11} > \dfrac{41}{13}$ 17. $\dfrac{17}{30} > \dfrac{21}{40}$ 18. $\dfrac{19}{8} > \dfrac{81}{36}$ 19. true 20. false 21. true 22. false 23. false
24. false 25. 77° 26. Carl did $\dfrac{1}{40}$ more of the job on Tuesday 27. $5\dfrac{5}{6}$ 28. $\dfrac{13}{14}$ 29. $617\dfrac{7}{9}$ 30. $16\dfrac{1}{2}$

4.5 EXERCISES B

Evaluate each expression.

1. $\left(\dfrac{9}{7} - \dfrac{3}{7}\right) \cdot \dfrac{7}{12}$ $\dfrac{1}{2}$

2. $\dfrac{2}{9} \cdot \dfrac{3}{4} - \dfrac{1}{6}$ 0

3. $\left(\dfrac{8}{7} - \dfrac{9}{14}\right) \div \dfrac{3}{4}$ $\dfrac{2}{3}$

4. $\dfrac{3}{5} \cdot \dfrac{5}{6} \div \dfrac{1}{4}$ 2

5. $\dfrac{9}{11} - \dfrac{2}{7} \div \dfrac{11}{7}$ $\dfrac{7}{11}$

6. $\dfrac{4}{21} \cdot \left(\dfrac{9}{14} - \dfrac{1}{7}\right)$ $\dfrac{2}{21}$

7. $\left(\dfrac{2}{3}\right)^2 - \sqrt{\dfrac{4}{81}}$ $\dfrac{2}{9}$

8. $\dfrac{5}{8} - \dfrac{1}{4} - \left(\dfrac{1}{2}\right)^2$ $\dfrac{1}{8}$

9. $\dfrac{15}{16} \div \dfrac{3}{8} - 4 \cdot \dfrac{1}{12}$ $2\dfrac{1}{6}$

10. $\dfrac{14}{20} \cdot \dfrac{8}{7} - \sqrt{\dfrac{1}{25}} + \dfrac{1}{3} \div \dfrac{5}{3}$ $\dfrac{4}{5}$

11. $\dfrac{12}{21} \div \left(\dfrac{7}{3} - \dfrac{5}{3}\right)^2$ $1\dfrac{2}{7}$

12. $\dfrac{6}{11} \div \dfrac{3}{22} - \left(\dfrac{1}{4}\right)^2 - \dfrac{5}{2} \div \dfrac{10}{3}$ $3\dfrac{3}{16}$

Which of the two fractions is larger?

13. $\dfrac{2}{5}$ or $\dfrac{3}{7}$ $\dfrac{2}{5} < \dfrac{3}{7}$

14. $\dfrac{19}{18}$ or $\dfrac{37}{36}$ $\dfrac{19}{18} > \dfrac{37}{36}$

15. $\dfrac{22}{33}$ or $\dfrac{14}{22}$ $\dfrac{22}{33} > \dfrac{14}{22}$

16. $\dfrac{49}{10}$ or $\dfrac{64}{15}$ $\dfrac{49}{10} > \dfrac{64}{15}$

17. $\dfrac{31}{50}$ or $\dfrac{13}{20}$ $\dfrac{31}{50} < \dfrac{13}{20}$

18. $\dfrac{25}{6}$ or $\dfrac{109}{28}$ $\dfrac{25}{6} > \dfrac{109}{28}$

Answer true or false.

19. $\dfrac{2}{5} < \dfrac{4}{9}$ true

20. $\dfrac{7}{11} > \dfrac{13}{20}$ false

21. $\dfrac{71}{8} > \dfrac{53}{6}$ true

22. $5\dfrac{9}{13} < 5\dfrac{17}{26}$ false

23. $\dfrac{3}{8} \cdot \dfrac{4}{9} < \dfrac{2}{3} \div 5$ false

24. $\dfrac{1}{2} > \dfrac{1}{3} > \dfrac{1}{4}$ true

Solve.

25. Celsius temperature can be found by subtracting 32° from Fahrenheit temperature and then multiplying by $\dfrac{5}{9}$. If Fahrenheit temperature is 95°, what is the Celsius temperature?   **35°**

26. Dorci hiked $5\dfrac{3}{16}$ miles on Thursday morning and $4\dfrac{1}{8}$ miles on Thursday afternoon. She went $4\dfrac{1}{6}$ miles on Friday morning and $5\dfrac{1}{8}$ miles on Friday afternoon. On which day did she cover more miles? How much more? $\dfrac{1}{48}$ **mi more on Thursday**

FOR REVIEW

Perform the indicated operations.

27.
$$\begin{array}{r} 5\dfrac{3}{8} \\ + 7\dfrac{1}{4} \\ \hline 12\dfrac{5}{8} \end{array}$$

28.
$$\begin{array}{r} 9\dfrac{2}{5} \\ - 8\dfrac{9}{10} \\ \hline \dfrac{1}{2} \end{array}$$

29.
$$\begin{array}{r} 925\dfrac{3}{10} \\ - 416\dfrac{3}{8} \\ \hline 508\dfrac{37}{40} \end{array}$$

30.
$$\begin{array}{r} 4\dfrac{1}{2} \\ 2\dfrac{3}{8} \\ + 9\dfrac{5}{24} \\ \hline 16\dfrac{1}{12} \end{array}$$

4.5 EXERCISES C

Evaluate each expression.

1. $\left(\dfrac{1}{3} - \dfrac{1}{4}\right)^2 \div \dfrac{5}{12} + \sqrt{\dfrac{6}{25} - \dfrac{2}{25}} - \dfrac{3}{5} \cdot \dfrac{1}{6}$ $\dfrac{19}{60}$

2. $\left[\left(\dfrac{4}{7} - \dfrac{3}{14}\right) \div \dfrac{1}{21}\right] - \sqrt{\left(\dfrac{2}{7} + \dfrac{1}{7}\right)^2}$

$\left[\text{Answer: } 7\dfrac{1}{14}\right]$

4.6 MORE APPLICATIONS

This section is designed to give practice with a variety of applications. Read each problem carefully to determine which operations are required and in what order.

EXAMPLE 1   CONSTRUCTION APPLICATION

Vince is building a cabin in the mountains. To minimize transportation cost he wants to buy boards from which he can cut pieces that are either all 10 inches in length, all 12 inches, or all 16 inches. What is the shortest board that he can buy and have no waste?

To prevent waste the board must be a length which is a multiple of each of the numbers 10, 12, and 16. The shortest such board will be of length the least common multiple (LCM).

$$10 = 2 \cdot 5 \qquad \text{Factor 10}$$
$$12 = 2 \cdot 2 \cdot 3 = 2^2 \cdot 3 \qquad \text{Factor 12}$$
$$16 = 2 \cdot 2 \cdot 2 \cdot 2 = 2^4 \qquad \text{Factor 16}$$

The LCM is $2 \cdot 2 \cdot 2 \cdot 2 \cdot 3 \cdot 5 = 2^4 \cdot 3 \cdot 5 = 240$. Thus the shortest board that will work is 240 inches or 20 ft long.

PRACTICE EXERCISE 1

An artist makes three kinds of geometric decorations out of wire. He uses 8 inches of wire for one kind, 14 inches of wire for another, and 21 inches for the third. What is the shortest length of wire from which he can cut pieces all of which are either 8 inches, 14 inches, or 21 inches, with no waste?

Answer: 168 inches, the LCM of 8, 14, and 21

EXAMPLE 2   APPLICATION TO AGRICULTURE

A grain dealer needed to fill a silo with wheat in three days. The first day deliveries filled $\frac{3}{8}$ of the silo. The second day an additional $\frac{1}{3}$ of the silo was filled. See Figure 4.4.

PRACTICE EXERCISE 2

A family business is owned by a father and his two children. The son owns $\frac{5}{14}$ of the business and the daughter owns $\frac{2}{7}$ of it.

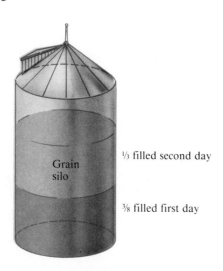

Grain silo

⅓ filled second day

⅜ filled first day

Figure 4.4

(a) How much of the silo was filled after two days?

The fraction in the silo is the sum of the two days' deliveries

$$\frac{3}{8} + \frac{1}{3} = \frac{3 \cdot 3}{8 \cdot 3} + \frac{1 \cdot 8}{3 \cdot 8}$$ LCD $= 8 \cdot 3$ since 8 and 3 have no common factors

$$= \frac{9}{24} + \frac{8}{24}$$

$$= \frac{9 + 8}{24} = \frac{17}{24}$$

Thus, $\frac{17}{24}$ of the silo was filled.

(b) What fraction of the silo must be filled the third day to complete the job?

First note that the fraction for a full silo is 1 and then subtract $\frac{17}{24}$ from 1.

$$1 - \frac{17}{24} = \frac{24}{24} - \frac{17}{24}$$ LCD $= 24$

$$= \frac{24 - 17}{24} = \frac{7}{24}$$

Thus, $\frac{7}{24}$ of the silo must be filled the third day.

(a) How much do the children own?

(b) How much does the father own?

Answers: **(a)** $\frac{9}{14}$ **(b)** $\frac{5}{14}$

EXAMPLE 3 APPLICATION TO WORK

Carrie worked $6\frac{1}{2}$ hours on Monday, $5\frac{2}{3}$ hours on Tuesday, and $2\frac{1}{4}$ hours on Wednesday. How many hours did she work during the three days?

To find her total hours, add the three mixed numbers.

$$6\frac{1}{2} = 6\frac{6}{12} \qquad \text{LCD} = 12$$

$$5\frac{2}{3} = 5\frac{8}{12}$$

$$+\ 2\frac{1}{4} = 2\frac{3}{12}$$

$$13\frac{17}{12} = 13 + 1 + \frac{5}{12} = 14\frac{5}{12}$$

Carrie worked $14\frac{5}{12}$ hours.

PRACTICE EXERCISE 3

On Thursday, Dennis ran $3\frac{2}{3}$ miles, on Friday he ran $4\frac{1}{2}$ miles, and on Saturday, $\frac{3}{4}$ mile. What is the total distance that he ran on the three days?

Answer: $8\frac{11}{12}$ mi

4.6 EXERCISES A

Solve.

1. Pam did $\frac{5}{8}$ of a job and Marsha did $\frac{1}{4}$ of the job. How much of the job was left to do? $\frac{1}{8}$

2. It takes $6\frac{3}{4}$ hours to roast a turkey. If it has been in the oven for $2\frac{1}{4}$ hours, how long will it be before it is done? $4\frac{1}{2}$ **hours**

3. A grain dealer received shipments of $7\frac{3}{8}$ tons, $9\frac{1}{6}$ tons, and $5\frac{1}{4}$ tons of wheat. How much wheat did she receive in the three shipments? $21\frac{19}{24}$ **tons**

4. If the dealer in Exercise 3 needed a total of 30 tons of wheat, how much more must be received? $8\frac{5}{24}$ **tons**

5. For a wiring job, an electronic technician needs lengths of 9 cm, 12 cm, and 15 cm. What is the shortest piece of wire from which he can cut pieces of the same length with no waste?   **180 cm**

6. In June, Pittsburgh had two rainstorms, one which dropped $\frac{5}{16}$ inch and the other $2\frac{1}{2}$ inches of precipitation. During August, there were three storms with $\frac{1}{4}$, $\frac{7}{8}$, and $1\frac{5}{12}$ inches of rain. During which month did Pittsburgh have more precipitation? How much more? $\frac{13}{48}$ **inches more in June**

7. Celsius temperature can be found by subtracting 32° from Fahrenheit temperature and then multiplying by $\frac{5}{9}$. If Fahrenheit temperature is 50°, what is the Celsius temperature?   **10°**

8. Three sides of a four-sided pen are $120\frac{1}{4}$ ft, $135\frac{2}{3}$ ft, and $160\frac{1}{5}$ ft. If the perimeter of the pen is $515\frac{4}{15}$ ft, what is the length of the fourth side? $99\frac{3}{20}$ **ft**

9 Lon worked $7\frac{3}{8}$ days, $6\frac{2}{5}$ days, $5\frac{4}{5}$ days, $8\frac{1}{4}$ days, and $7\frac{1}{2}$ days to complete a project. What was his total number of days worked? $35\frac{13}{40}$ **days**

10 What was Lon paid for working the days in Exercise 9 if he received $40 per day?   **$1413**

11. On Friday Carlos walked $1\frac{1}{5}$ mi and ran $\frac{5}{8}$ mi. On Saturday, he walked $\frac{3}{4}$ mi and ran $1\frac{1}{6}$ mi. On which day did he travel a greater distance? How much greater?
$\frac{11}{120}$ **mi farther on Saturday**

12 The stock of the Mina Corporation opened at $121\frac{3}{8}$. It increased $1\frac{3}{4}$ during the first hour of trading, but fell $3\frac{1}{8}$ before noon. It then increased $2\frac{7}{8}$ before closing time. What was the closing value of the stock? $122\frac{7}{8}$

FOR REVIEW

Evaluate each expression.

13. $\left(\frac{7}{9} - \frac{1}{3}\right) \div \frac{1}{3} \cdot \frac{1}{4}$ $\frac{1}{3}$

14. $\sqrt{\frac{25}{16} - \left(\frac{3}{4}\right)^2}$ $\frac{11}{16}$

15. $\frac{2}{5} \div \frac{3}{10} - \frac{4}{9} \cdot \frac{3}{2}$ $\frac{2}{3}$

Answer true or false.

16. $\frac{3}{8} < \frac{8}{24}$ **false**

17. $\frac{22}{7} < \frac{10}{3}$ **true**

18. $\frac{4}{5} \cdot \frac{1}{2} > \frac{5}{6} \div \frac{10}{3}$ **true**

ANSWERS: 1. $\frac{1}{8}$ 2. $4\frac{1}{2}$ hours   3. $21\frac{19}{24}$ tons   4. $8\frac{5}{24}$ tons   5. 180 cm   6. $\frac{13}{48}$ inches more in June   7. 10°   8. $99\frac{3}{20}$ ft   9. $35\frac{13}{40}$ days   10. $1413   11. $\frac{11}{120}$ mi farther on Saturday   12. $122\frac{7}{8}$ 13. $\frac{1}{3}$ 14. $\frac{11}{16}$ 15. $\frac{2}{3}$ 16. false   17. true   18. true

4.6 EXERCISES ·B

Solve.

1. In the morning Paul completed $\frac{5}{12}$ of his project for the day. By the middle of the afternoon he had done only another $\frac{1}{6}$ of the project. How much of the job was left to do? $\frac{5}{12}$

2. Victoria must wait $5\frac{5}{6}$ hours for the results of her test. If she has been waiting for $4\frac{1}{3}$ hours, how much longer must she wait?
$1\frac{1}{2}$ hours

3. A contractor for a construction job ordered three loads of gravel. One weighed $2\frac{1}{4}$ tons, one $5\frac{5}{6}$ tons, and the other $4\frac{5}{8}$ tons. What total weight of gravel did he receive? $12\frac{17}{24}$ tons

4. If the contractor in Exercise 3 really needed 20 tons of gravel, how much more must he order?
$7\frac{7}{24}$ tons

5. To construct a brick wall, three lengths of bricks are used, 6 inches, 14 inches, and 16 inches. What is the shortest length of wall that can be constructed if only one length of brick is used in any one row?
336 inches

6. During the first week of the spring thaw, two measurements of lake level were made. On the first, the lake level had increased $2\frac{3}{4}$ ft from the previous measurement. On the second, the level had increased another $3\frac{5}{12}$ ft. Three measurements the second week showed increases of $1\frac{5}{6}$ ft, $2\frac{2}{3}$ ft, and $2\frac{1}{4}$ ft. During which week did the level increase more? How much more?
$\frac{7}{12}$ ft more the second week

7. Fahrenheit temperature can be found by multiplying Celsius temperature by $1\frac{4}{5}$ and adding 32°. If Celsius temperature is 15°, what is the Fahrenheit temperature?   59°

8. Jo must have $216\frac{1}{2}$ points to win a contest. If for three projects she has received $25\frac{1}{4}$ points, $17\frac{1}{6}$ points, and $42\frac{3}{8}$ points, how many more points does she need to win? $131\frac{17}{24}$ points

9. Fran worked for $32\frac{3}{4}$ hours, $28\frac{5}{6}$ hours, $42\frac{1}{6}$ hours, and $19\frac{3}{4}$ hours over a four-week period. What was her total number of hours worked? $123\frac{1}{2}$ hours

10. Fran in Exercise 9 is paid $10 per hour. How much should she be paid for the four weeks' work?
$1235

11. On Wednesday a grocery distributor received two shipments of grapes, $395\frac{3}{8}$ lb and $215\frac{3}{4}$ lb. On Thursday two more shipments of $198\frac{1}{2}$ lb and $416\frac{1}{8}$ lb were received. On which day did he receive more grapes? How many pounds more?

$3\frac{1}{2}$ lb more on Thursday

12. The opening quote for stock of the Nina Corporation was $62\frac{1}{8}$. During the morning it dropped $4\frac{3}{4}$ but increased again $6\frac{5}{8}$ before noon. It then dropped $1\frac{1}{8}$ before closing time. What was the closing value of the stock? $62\frac{7}{8}$

FOR REVIEW

Evaluate each expression.

13. $\left(\frac{5}{8} - \frac{1}{4}\right) \cdot \frac{1}{9} \div \frac{3}{4}$ $\frac{1}{18}$

14. $\left(\frac{3}{5}\right)^2 - \sqrt{\frac{1}{25}}$ $\frac{4}{25}$

15. $\frac{3}{8} \cdot \frac{2}{9} - \frac{3}{10} \div 4\frac{1}{2}$ $\frac{1}{60}$

Answer true or false.

16. $\frac{7}{8} > \frac{17}{20}$ true

17. $\frac{35}{6} < \frac{43}{7}$ true

18. $\frac{9}{2} \div \frac{3}{8} > \frac{10}{3} \cdot \frac{15}{4}$ false

4.6   EXERCISES C

Let a represent a natural number.

1. If one shipment was $2\frac{1}{2}a$ units and a second shipment was $4\frac{2}{3}a$ units, what was the total of the two shipments?

$\left[\text{Answer: } 7\frac{1}{6}a \text{ units}\right]$

2. Lonna has $10\frac{3}{4}a$ units and Maria has $7\frac{2}{5}a$ units. How many more units does Lonna have? $3\frac{7}{20}a$ **units**

CHAPTER 4 REVIEW

KEY WORDS

4.1   **Like fractions** have the same denominator.
Unlike fractions have different denominators.

4.2   A whole number that is in the lists of nonzero multiples of two or more numbers is called a **common multiple.**

The **least common multiple (LCM)** of two or more counting numbers is the smallest of all common multiples of the numbers.

4.3   The **least common denominator (LCD)** is the least common multiple of the denominators of fractions in an addition or subtraction problem.

4.2   Three methods of finding LCM's are the listing method, finding prime factors, and the special optional algorithm.

4.3   Fractions to be added or subtracted must be changed to equal fractions with the same LCD. You should *not* find the LCD when multiplying or dividing fractions.

4.4   The method used for adding or subtracting mixed numbers is to add or subtract the whole number parts and the proper fraction parts separately. Do *not* try to multiply or divide mixed fractions using this method.

4.5   **1.** Operations should be performed in the following order.
  a. Operate within parentheses.
  b. Evaluate powers and roots.
  c. Multiply and divide from left to right.
  d. Add and subtract from left to right.

2. To tell which of two fractions is larger, change them to equal fractions with the same LCD and compare numerators.

KEY CONCEPTS

4.1   **1.** Like fractions are added or subtracted by adding or subtracting the numerators and placing the result over the common denominator.

2. *Never* add or subtract fractions by adding or subtracting denominators.

REVIEW EXERCISES

Part I

4.1   *Perform the indicated operations.*

1. $\dfrac{3}{22} + \dfrac{9}{22}$ $\dfrac{6}{11}$

2. $\dfrac{17}{18} - \dfrac{1}{18}$ $\dfrac{8}{9}$

3. $\dfrac{4}{25} + \dfrac{1}{25}$ $\dfrac{1}{5}$

4. $\dfrac{8}{35} - \dfrac{3}{35}$ $\dfrac{1}{7}$

5. $\dfrac{7}{10} + \dfrac{11}{10} - \dfrac{3}{10}$ $\dfrac{3}{2}$ or $1\frac{1}{2}$

6. $\dfrac{9}{17} - \dfrac{3}{17} - \dfrac{6}{17}$ 0

4.2 *Find the LCM of the given numbers.*

7. 40 and 8   40

8. 12 and 30   60

9. 15 and 35   105

10. 9 and 30   90

11. 120 and 14   840

12. 12, 14, and 18   252

4.3 *Perform the indicated operations.*

13. $\dfrac{5}{12} + \dfrac{3}{18}$ $\dfrac{7}{12}$

14. $3 + \dfrac{7}{9}$ $\dfrac{34}{9}$ or $3\dfrac{7}{9}$

15. $\dfrac{4}{7} - \dfrac{1}{6}$ $\dfrac{17}{42}$

16. $3 - \dfrac{4}{3}$ $\dfrac{5}{3}$ or $1\dfrac{2}{3}$

17. $\dfrac{12}{15} + \dfrac{2}{3} - \dfrac{1}{5}$ $\dfrac{19}{15}$ or $1\dfrac{4}{15}$

18. $\dfrac{5}{3} - 1 + \dfrac{2}{7}$ $\dfrac{20}{21}$

4.4 *Perform the indicated operations.*

19. $2\dfrac{1}{8}$
$+\ 7\dfrac{3}{4}$ $9\dfrac{7}{8}$

20. $5\dfrac{2}{7}$
$-\ 3\dfrac{2}{3}$ $1\dfrac{13}{21}$

21. $6\dfrac{2}{5}$
$+\ 8\dfrac{7}{8}$ $15\dfrac{11}{40}$

22. $9\dfrac{3}{7}$
$-\ 4\dfrac{5}{6}$ $4\dfrac{25}{42}$

23. $215\dfrac{1}{4}$
$+\ 375\dfrac{2}{11}$ $590\dfrac{19}{44}$

24. $627\dfrac{1}{2}$
$-\ 421\dfrac{5}{8}$ $205\dfrac{7}{8}$

4.5 *Evaluate each expression.*

25. $\left(\dfrac{2}{5} - \dfrac{1}{15}\right) \div \dfrac{5}{3}$ $\dfrac{1}{5}$

26. $\sqrt{\dfrac{4}{9}} + \left(\dfrac{1}{3}\right)^2$ $\dfrac{7}{9}$

27. $\dfrac{1}{8} \cdot \dfrac{2}{3} - \dfrac{1}{12}$ 0

28. $\dfrac{1}{2} \cdot \left(\dfrac{7}{9} + \dfrac{1}{3}\right) \div \dfrac{5}{6}$ $\dfrac{2}{3}$

29. $\left(\dfrac{3}{8} - \dfrac{1}{4}\right)^2 \cdot \dfrac{8}{9}$ $\dfrac{1}{72}$

30. $\dfrac{1}{6} \div \dfrac{5}{9} - \left(\dfrac{1}{5}\right)^2$ $\dfrac{13}{50}$

Which of the two fractions is larger?

31. $\dfrac{4}{17}$ or $\dfrac{1}{4}$ $\dfrac{1}{4}$ is larger

32. $\dfrac{8}{11}$ or $\dfrac{5}{7}$ $\dfrac{8}{11}$ is larger

33. $\dfrac{38}{9}$ or $\dfrac{47}{11}$ $\dfrac{47}{11}$ is larger

Answer true or false.

34. $\dfrac{7}{12} < \dfrac{10}{18}$ false

35. $\dfrac{25}{3} > \dfrac{41}{5}$ true

36. $7\dfrac{19}{21} < 7\dfrac{49}{51}$ true

4.6 *Solve.*

37. A contractor needs steel rods which measure 2 feet, 6 feet, and 9 feet. He will cut only one length from each rod. What is the shortest rod that he can order and be able to cut each length with no waste? **18 ft**

38. Carla worked for $6\frac{1}{2}$ hours, $8\frac{3}{4}$ hours, $9\frac{1}{6}$ hours, and $7\frac{1}{4}$ hours the first four days of the week. How many total hours did she work in the four days? **$31\frac{2}{3}$ hr**

Part II

39. It takes $3\frac{11}{15}$ weeks to complete a building project, and the contractor has been working for $2\frac{2}{5}$ weeks. How much longer will it be before the job is finished? **$1\frac{1}{3}$ weeks**

40. Robert ran $7\frac{3}{4}$ miles and $6\frac{2}{3}$ miles during one week. The next week he ran $5\frac{1}{2}$ miles, $3\frac{1}{3}$ miles, and $5\frac{1}{4}$ miles. During which week did he run farther? How much farther? **$\frac{1}{3}$ mi more the first week**

Find the LCM of the given numbers.

41. 16 and 22   **176**

42. 35 and 45   **315**

43. 6, 14, and 24   **168**

Place the appropriate sign, $<$ or $>$, between the two fractions.

44. $\frac{5}{6} > \frac{9}{11}$

45. $\frac{16}{7} < \frac{23}{9}$

46. $\frac{41}{37} < \frac{62}{55}$

Perform the indicated operations.

47. $3\frac{1}{2}$ $+ 7\frac{2}{3}$ = $11\frac{1}{6}$

48. $16\frac{1}{9}$ $- 12\frac{17}{18}$ = $3\frac{1}{6}$

49. $6\frac{2}{5}$ $9\frac{1}{10}$ $+ 12\frac{14}{15}$ = $28\frac{13}{30}$

50. $\left(\frac{2}{3} - \frac{1}{9}\right) \div \frac{11}{18}$ $\frac{10}{11}$

51. $\frac{6}{11} + \frac{13}{22} - \frac{5}{33}$ $\frac{65}{66}$

52. $\sqrt{\frac{25}{64}} - \left(\frac{1}{4}\right)^2$ $\frac{9}{16}$

ANSWERS: 1. $\frac{6}{11}$ 2. $\frac{8}{9}$ 3. $\frac{1}{5}$ 4. $\frac{1}{7}$ 5. $\frac{3}{2}$ or $1\frac{1}{2}$ 6. 0 7. 40 8. 60 9. 105 10. 90 11. 840 12. 252 13. $\frac{7}{12}$ 14. $\frac{34}{9}$ or $3\frac{7}{9}$ 15. $\frac{17}{42}$ 16. $\frac{5}{3}$ or $1\frac{2}{3}$ 17. $\frac{19}{15}$ or $1\frac{4}{15}$ 18. $\frac{20}{21}$ 19. $9\frac{7}{8}$ 20. $1\frac{13}{21}$ 21. $15\frac{11}{40}$ 22. $4\frac{25}{42}$ 23. $590\frac{19}{44}$ 24. $205\frac{7}{8}$ 25. $\frac{1}{5}$ 26. $\frac{7}{9}$ 27. 0 28. $\frac{2}{3}$ 29. $\frac{1}{72}$ 30. $\frac{13}{50}$ 31. $\frac{1}{4}$ is larger 32. $\frac{8}{11}$ is larger 33. $\frac{47}{11}$ is larger 34. false 35. true 36. true 37. 18 ft 38. $31\frac{2}{3}$ hours 39. $1\frac{1}{3}$ weeks 40. $\frac{1}{3}$ mi more the first week 41. 176 42. 315 43. 168 44. $>$ 45. $<$ 46. $<$ 47. $11\frac{1}{6}$ 48. $3\frac{1}{6}$ 49. $28\frac{13}{30}$ 50. $\frac{10}{11}$ 51. $\frac{65}{66}$ 52. $\frac{9}{16}$

Find the LCM of the given numbers.

1. 14 and 15

1. _____ 210 _____

2. 8, 12, and 20

2. _____ 120 _____

Perform the indicated operations.

3. $\dfrac{5}{21} + \dfrac{2}{21}$

3. _____ $\dfrac{1}{3}$ _____

4. $\dfrac{14}{15} - \dfrac{4}{15}$

4. _____ $\dfrac{2}{3}$ _____

5. $\dfrac{7}{12} + \dfrac{3}{8}$

5. _____ $\dfrac{23}{24}$ _____

6. $\dfrac{9}{20} - \dfrac{1}{6}$

6. _____ $\dfrac{17}{60}$ _____

7. $\dfrac{12}{7} - \dfrac{13}{14} + \dfrac{1}{2}$

7. _____ $\dfrac{9}{7}$ or $1\dfrac{2}{7}$ _____

8. $\dfrac{7}{8} - \dfrac{5}{12} - \dfrac{1}{6}$

8. _____ $\dfrac{7}{24}$ _____

9. $2\dfrac{5}{9}$

$+ 7\dfrac{1}{3}$

9. _____ $9\dfrac{8}{9}$ _____

10. $42\dfrac{7}{15}$

$- 33\dfrac{3}{5}$

10. _____ $8\dfrac{13}{15}$ _____

11. $3\dfrac{3}{4}$

$8\dfrac{1}{3}$

$+ 5\dfrac{7}{12}$

11. _____ $17\dfrac{2}{3}$ _____

12. $4\frac{3}{8}$

   $2\frac{3}{4}$

   $+\ 1\frac{1}{2}$

12. _____ $8\frac{5}{8}$ _____

Evaluate each expression.

13. $\sqrt{\frac{4}{9}} - \frac{2}{3} \cdot \frac{15}{20} + 4 \div \frac{1}{2}$

13. _____ $8\frac{1}{6}$ _____

14. $\left(\frac{8}{9} - \frac{2}{3}\right)^2 \div \frac{4}{9} - \frac{1}{2} \cdot \frac{1}{5}$

14. _____ $\frac{1}{90}$ _____

Answer true or false.

15. $\frac{17}{20} < \frac{27}{30}$

15. _____ true _____

16. $5\frac{12}{13} > 5\frac{43}{40}$

16. _____ false _____

Solve.

17. During one hour, Maria made three sales of material: $7\frac{1}{2}$ yd, $3\frac{1}{4}$ yd, and $2\frac{7}{8}$ yd. What was the total yardage sold?

17. _____ $13\frac{5}{8}$ yd _____

18. A repair job is estimated to require $6\frac{5}{6}$ hours to complete. If Hank has been working for $2\frac{11}{12}$ hours, how much longer will he have to work to finish the job?

18. _____ $3\frac{11}{12}$ hours _____

Decimals

5.1 DECIMAL NOTATION

1 Decimal Fractions and Decimals
2 Writing Decimals in Expanded Notation
3 Using Decimals for Amounts of Money

1 DECIMAL FRACTIONS AND DECIMALS

As we saw in Chapter 3, a fraction has many names. For example,

$$\frac{1}{2}, \quad \frac{2}{4}, \quad \frac{3}{6}, \quad \frac{4}{8}, \quad \frac{5}{10}, \quad \frac{25}{50}, \quad \frac{50}{100}$$

are all names for the fraction *one half*. A fraction which has a denominator equal to a power of 10 (10, 100, 1000, 10,000, and so forth) is called a **decimal fraction.** The following are decimal fractions.

$\frac{5}{10}$ is read **"five tenths."**

$\frac{3}{100}$ is read **"three hundredths."**

$\frac{47}{1000}$ is read **"forty-seven thousandths."**

$\frac{121}{10,000}$ is read **"one hundred twenty-one ten thousandths."**

Decimal fractions are usually referred to simply as **decimals.** Each decimal has two forms, a *fractional form,*

$$\frac{5}{10}, \quad \frac{3}{100}, \quad \frac{47}{1000}, \quad \frac{121}{10,000},$$

and a *decimal form* or *decimal notation,*

$$0.5, \quad 0.03, \quad 0.047, \quad \text{and} \quad 0.0121.$$

Recall that the number system is a place-value system. The decimal form of a number is part of this system, as shown in the following revised place-value chart first seen in Chapter 1. Notice that the value of any position is one tenth the value of the position to its immediate left.

Thousands	Hundreds	Tens	Ones	Tenths	Hundredths	Thousandths	Ten thousandths	Hundred thousandths	Millionths
1000	100	10	1	$\frac{1}{10}$	$\frac{1}{100}$	$\frac{1}{1000}$	$\frac{1}{10,000}$	$\frac{1}{100,000}$	$\frac{1}{1,000,000}$

When a number is written in decimal form, a period, called a **decimal point,** separates the ones digit from the tenths digit. It serves as a reference point: positions to the left of the decimal point end in *s* while those to the right end in *ths*. Consider the number 352.1783 shown below.

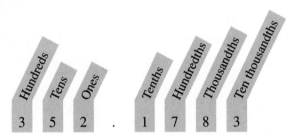

The decimal point separates the ones digit from the tenths digit.

This number is read   **three hundred fifty-two** *and* **one thousand seven hundred eighty-three ten thousandths.**

The three points to remember about word names for decimals are given below.

To Read a Decimal

1. The part of the number to the left of the decimal point is read just like a whole number.
2. The decimal point is read ''and.''
3. The part of the number to the right of the decimal point is read like a whole number but is followed by the name of the place filled by the farthest right digit.

EXAMPLE 1   WORD NAMES FOR DECIMALS

Give the word name for each decimal.

(a) .5 or 0.5 is read ''five tenths.''

     ↑

    tenths place

Both .5 and 0.5 are correct ways to write five tenths. The second form is preferred since it calls attention to the decimal point.

PRACTICE EXERCISE 1

Give the word name for each decimal.

(a) 0.3

(b) 3.5 is read "three and five tenths."

(c) 0.03 is read "three hundredths."
↑
hundredths place

(d) 4.03 is read "four and three hundredths."

(e) 0.008 is read "eight thousandths."
↑
thousandths place

(f) 21.008 is read "twenty-one and eight thousandths."

(g) 4.107 is read "four and one hundred seven thousandths."

(b) 5.9

(c) 0.08

(d) 2.01

(e) 0.005

(f) 57.005

(g) 6.242

Answers: (a) three tenths
(b) five and nine tenths   (c) eight
hundredths   (d) two and one
hundredth   (e) five thousandths
(f) fifty-seven and five thousandths
(g) six and two hundred forty-two
thousandths

If there are no digits (or possibly only zeros) to the right of the decimal point, the decimal is actually a whole number. In such cases, the decimal point is usually omitted. For example,

$$47, \quad 47., \quad \text{and} \quad 47.0$$

are all names for the whole number forty-seven.

EXAMPLE 2   WRITING NUMBERS IN DECIMAL NOTATION

Write each number in decimal notation.

(a) Three and nine tenths is written 3.9.
        ↓   ↓   ↓
        3   ·   9

(b) Thirteen and three hundredths is written 13.03.
         ↓   ↓   ↓
         13   ·   03

(c) Five thousand four and twenty-seven thousandths is written 5004.027.

(d) Forty-two thousandths is written 0.042.

PRACTICE EXERCISE 2

Write each number in decimal notation.

(a) Six and four tenths

(b) Thirty-five and two hundredths

(c) Eight thousand forty-five and six hundred eight thousandths

(d) Ninety-three ten thousandths

Answers: (a) 6.4   (b) 35.02
(c) 8045.608   (d) 0.0093

H I N T

A less formal method for reading decimals is commonly used. For example,

3.42	is often read	"three *point* four two,"
10.001	is often read	"one zero *point* zero zero one,"
6254.398	is often read	"six two five four *point* three nine eight."

We simply read each digit in order from left to right and say "point" when we reach the decimal point.

② WRITING DECIMALS IN EXPANDED NOTATION

When whole numbers were first studied, we discussed expanded notation. For example,

$$2375 \quad \text{is} \quad 2000 + 300 + 70 + 5$$

in expanded notation. This notation can be extended to all decimals, as shown below for the decimal 23.456.

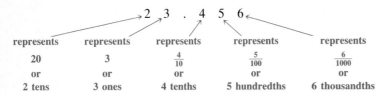

represents	represents	represents	represents	represents
20	3	$\frac{4}{10}$	$\frac{5}{100}$	$\frac{6}{1000}$
or	or	or	or	or
2 tens	3 ones	4 tenths	5 hundredths	6 thousandths

Thus, we can write 23.456 in expanded notation as

$$20 + 3 + \frac{4}{10} + \frac{5}{100} + \frac{6}{1000}.$$

EXAMPLE 3 WRITING DECIMALS IN EXPANDED NOTATION

Write each decimal in expanded notation.

(a) 432.78

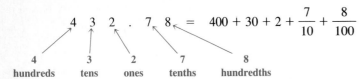

$$= \quad 400 + 30 + 2 + \frac{7}{10} + \frac{8}{100}$$

4	3	2	7	8
hundreds	tens	ones	tenths	hundredths

(b) $25.6794 = 20 + 5 + \frac{6}{10} + \frac{7}{100} + \frac{9}{1000} + \frac{4}{10,000}$

(c) $403.002 = 400 + 3 + \frac{2}{1000}$

(d) $3050.7020 = 3000 + 50 + \frac{7}{10} + \frac{2}{1000}$

The final zero to the right does not affect the value of the decimal and could be omitted.

PRACTICE EXERCISE 3

Write each decimal in expanded notation.

(a) 258.65

(b) 43.2963

(c) 605.007

(d) 1060.05030

Answers: (a) $200 + 50 + 8 + \frac{6}{10} + \frac{5}{100}$
(b) $40 + 3 + \frac{2}{10} + \frac{9}{100} + \frac{6}{1000} + \frac{3}{10,000}$
(c) $600 + 5 + \frac{7}{1000}$ (d) $1000 + 60 + \frac{5}{100} + \frac{3}{10,000}$

---

③ USING DECIMALS FOR AMOUNTS OF MONEY

Perhaps the most common use of decimal notation is for amounts of money. For example, we could read

$25.76 as twenty-five *and* seventy-six hundredths dollars.

However, since one hundredth of a dollar is one *cent* (1¢) we usually say

twenty-five dollars *and* seventy-six cents.

Notice that the word *and* is still identified with the decimal point. We must put the decimal point in the right place. For instance,

$$\$25 \quad \text{and} \quad \$0.25$$

do not represent the same amount: $25 or $25.00 is twenty-five dollars while $0.25 is twenty-five hundredths of a dollar or twenty-five cents (25¢).

EXAMPLE 4   WRITING WORD NAMES FOR MONEY

Write a word name for each amount of money.

(a) $4.76   Four and seventy-six hundredths dollars *or* four dollars and seventy-six cents

(b) $1327.03   One thousand three hundred twenty-seven and three hundredths dollars *or* one thousand three hundred twenty-seven dollars and three cents

(c) $41.00   Forty-one and zero hundredths dollars *or* forty-one dollars and no cents *or simply* forty-one dollars

---

PRACTICE EXERCISE 4

Write a word name for each amount of money.

(a) $3.98

(b) $2650.07

(c) $75.00

Answers: (a) Three and ninety-eight hundredths dollars *or* three dollars and ninety-eight cents
(b) Two thousand six hundred fifty and seven hundredths dollars *or* two thousand six hundred fifty dollars and seven cents
(c) Seventy-five and zero hundredths dollars *or* seventy-five dollars and no cents *or simply* seventy-five dollars

---

EXAMPLE 5   WRITING DECIMALS FOR AMOUNTS OF MONEY

Write a decimal for each amount of money.

(a) Three hundred twenty-seven and twelve hundredths dollars is written $327.12.

(b) Two dollars and thirty-seven cents is written $2.37.

(c) Seventy-six and no hundredths dollars is written $76.00.

PRACTICE EXERCISE 5

Write a decimal for each amount of money.

(a) Six hundred fifty-four and twenty-three hundredths dollars

(b) Ten dollars and ninety-five cents

(c) Eighty-four and no hundredths dollars

Answers: (a) $654.23   (b) $10.95
(c) $84.00

Writing checks requires writing word names for decimals. It has become customary to write fractions for the cents portion of an amount of money. For example, $3.48 is written

$$\text{three and } \frac{48}{100} \text{ dollars}$$

instead of

three and forty-eight hundredths dollars.

This is done on the sample check in Figure 5.1.

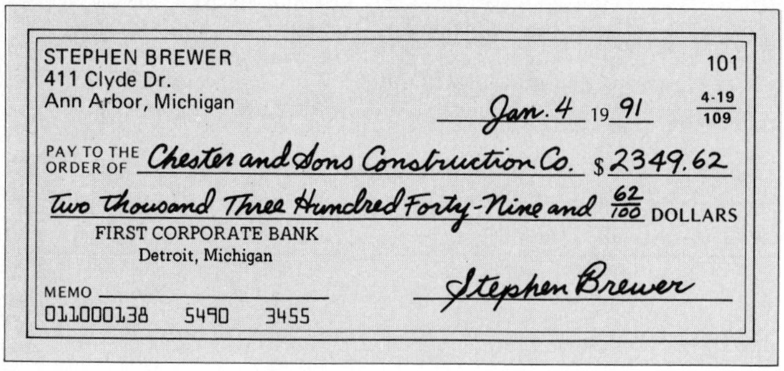

STEPHEN BREWER
411 Clyde Dr.
Ann Arbor, Michigan   101

                        Jan. 4 19 *91*   4-19
                                                 109

PAY TO THE
ORDER OF *Chester and Sons Construction Co.* $ *2349.62*

Two thousand Three Hundred Forty-Nine and 62/100 DOLLARS

FIRST CORPORATE BANK
    Detroit, Michigan

MEMO _____   *Stephen Brewer*
011000138   5490   3455

Figure 5.1

5.1 EXERCISES A

1. What are fractions called that have a denominator equal to a power of 10?   **decimal fractions or decimals**

2. What do we call the period used in decimal notation?   **decimal point**

Give the formal *word name for each decimal.*

3. 0.7
  seven tenths

4. 2.05
  two and five hundredths

5. 43.29
  **forty-three and
  twenty-nine hundredths**

6. 127.562
  **one hundred twenty-seven and five
  hundred sixty-two thousandths**

7. 302.008
  **three hundred two and
  eight thousandths**

8. 0.0009
  nine ten thousandths

Write each number in decimal notation.

9. four hundredths   **0.04**

10. two hundred twenty-eight and nine tenths   **228.9**

11. twenty-three and seven hundred forty-eight thousandths   **23.748**

12. eight hundred one and one hundred eight thousandths   **801.108**

13. three and one ten thousandth   **3.0001**

14. four hundred ninety-seven and six thousand forty-three ten thousandths   **497.6043**

Write each decimal in expanded notation.

15. 41.8 $40 + 1 + \dfrac{8}{10}$

16. 30.07 $30 + \dfrac{7}{100}$

17. 2050.703 $2000 + 50 + \dfrac{7}{10} + \dfrac{3}{1000}$

18. 40.1005 $40 + \dfrac{1}{10} + \dfrac{5}{10,000}$

Write a word name for the dollar amount using dollars and cents.

19. $1.47   one dollar and forty-seven cents

20. $23.04   twenty-three dollars and four cents

21. $162.00
one hundred sixty-two dollars [and zero (or no) cents]

22. $10.40   ten dollars and forty cents

Write a word name as it would appear on a check.

23. $427.68   Four hundred twenty-seven and $\frac{68}{100}$ dollars

24. $20.00   Twenty and $\frac{00}{100}$ dollars

25. $3.02   Three and $\frac{02}{100}$ dollars

FOR REVIEW

The following exercises will help you prepare for the next section. Write each improper fraction as a mixed number.

26. $\frac{9}{4}$ $2\frac{1}{4}$

27. $\frac{42}{11}$ $3\frac{9}{11}$

28. $\frac{103}{100}$ $1\frac{3}{100}$

Write each mixed number as an improper fraction.

29. $2\frac{1}{5}$ $\frac{11}{5}$

30. $5\frac{7}{10}$ $\frac{57}{10}$

31. $7\frac{23}{100}$ $\frac{723}{100}$

ANSWERS: 1. decimal fractions or decimals 2. decimal point 3. seven tenths 4. two and five hundredths
5. forty-three and twenty-nine hundredths 6. one hundred twenty-seven and five hundred sixty-two thousandths
7. three hundred two and eight thousandths 8. nine ten thousandths 9. 0.04 10. 228.9 11. 23.748 12. 801.108
13. 3.0001 14. 497.6043 15. $40 + 1 + \frac{8}{10}$ 16. $30 + \frac{7}{10}$ 17. $2000 + 50 + \frac{7}{10} + \frac{3}{1000}$ 18. $40 + \frac{1}{10} + \frac{5}{10,000}$ 19. one
dollar and forty-seven cents 20. twenty-three dollars and four cents 21. one hundred sixty-two dollars [and zero (or
no) cents] 22. ten dollars and forty cents 23. Four hundred twenty-seven and $\frac{68}{100}$ dollars 24. Twenty and $\frac{00}{100}$ dollars 25. Three and $\frac{02}{100}$ dollars 26. $2\frac{1}{4}$ 27. $3\frac{9}{11}$ 28. $1\frac{3}{100}$ 29. $\frac{11}{5}$ 30. $\frac{57}{10}$ 31. $\frac{723}{100}$

5.1 EXERCISES B

1. Decimal fractions have denominators that are powers of what number?   10

2. The decimal point in decimal notation separates which two digits?   ones and tenths

Give the formal word name for each decimal.

3. 0.9   nine tenths

4. 3.06   three and six hundredths

5. 52.39
fifty-two and thirty-nine hundredths

6. 147.053
one hundred forty-seven and
fifty-three thousandths

7. 602.007
six hundred two and
seven thousandths

8. 0.0005
five ten thousandths

Write each number in decimal notation.

9. seven hundredths   0.07

10. three hundred thirty-five and eight tenths   335.8

11. forty-two and eight hundred sixty-five thousandths   42.865

12. seven and two ten thousandths   7.0002

13. five hundred one and one hundred five thousandths   501.105

14. three hundred six and one thousand forty-two ten thousandths   306.1042

Write each decimal in expanded notation.

15. 32.5 $30 + 2 + \dfrac{5}{10}$ **16.** 60.03 $60 + \dfrac{3}{100}$ **17.** 51.006

$50 + 1 + \dfrac{6}{1000}$

18. 80.2006

$80 + \dfrac{2}{10} + \dfrac{6}{10,000}$

Write a word name for the dollar amount using dollars and cents.

19. $1.58
one dollar and
fifty-eight cents

20. $47.09
forty-seven dollars
and nine cents

21. $20.06
twenty dollars and
six cents

22. $535.00
five hundred thirty-five dollars
[and zero (or no) cents]

Write a word name as it would appear on a check.

23. $6.05
Six and $\frac{5}{100}$ dollars

24. $70.00
Seventy and $\frac{00}{100}$ dollars

25. $395.49
Three hundred ninety-five and
$\frac{49}{100}$ dollars

FOR REVIEW

The following exercises will help you prepare for the next section. Write each improper fraction as a mixed number.

26. $\dfrac{13}{4}$ $3\dfrac{1}{4}$

27. $\dfrac{28}{13}$ $2\dfrac{2}{13}$

28. $\dfrac{201}{100}$ $2\dfrac{1}{100}$

Write each mixed number as an improper fraction.

29. $3\dfrac{1}{4}$ $\dfrac{13}{4}$

30. $4\dfrac{7}{10}$ $\dfrac{47}{10}$

31. $8\dfrac{31}{100}$ $\dfrac{831}{100}$

5.1   EXERCISES C

1. The speed of light is approximately 273,213.6 feet per second. Give the formal word name for this number.
two hundred seventy-three thousand two hundred thirteen and six tenths

2. Write a word name for $10,428.64 as it would appear on a check.
Ten thousand four hundred twenty eight and $\frac{64}{100}$ dollars

5.2   CONVERSIONS BETWEEN FRACTIONS AND DECIMALS

STUDENT GUIDEPOSTS

1 Changing Decimals to Fractions
2 Changing Fractions to Decimals
3 Summary List of Equivalent Fractions and Decimals

❶ CHANGING DECIMALS TO FRACTIONS

Recall from Section 5.1 that any decimal fraction has two forms, a decimal form and a fractional form. We now consider how to change from one form to the other.

Converting decimals to fractions is simply a matter of thinking of the decimal in words and writing the fraction, as in the following table.

Decimal	Word Name	Fraction
0.7	seven tenths	$\dfrac{7}{10}$
0.23	twenty-three hundredths	$\dfrac{23}{100}$
0.423	four hundred twenty-three thousandths	$\dfrac{423}{1000}$
3.1	three and one tenth	$3\dfrac{1}{10}$

The resulting fraction (which may be part of a mixed number) should be reduced to lowest terms.

To Change a Decimal to a Fraction   (First Method)

1. Say the word name for the decimal.
2. Write the word name as a fraction or a mixed number.
3. Reduce the fraction to lowest terms.

EXAMPLE 1   DECIMALS TO FRACTIONS (FIRST METHOD)

Change each decimal to a fraction or mixed number.

(a) 0.6 is read "six tenths." Thus,

$$0.6 = \frac{6}{10} \qquad \text{Six tenths is } \tfrac{6}{10}$$

$$= \frac{\cancel{2} \cdot 3}{\cancel{2} \cdot 5} = \frac{3}{5}. \qquad \text{Reduce } \tfrac{6}{10} \text{ to lowest terms}$$

(b) 0.35 is read "thirty-five hundredths."

$$0.35 = \frac{35}{100}$$

$$= \frac{\cancel{5} \cdot 7}{\cancel{5} \cdot 20} = \frac{7}{20} \qquad \text{Reduce to lowest terms}$$

(c) 0.348 is read "three hundred forty-eight thousandths."

$$0.348 = \frac{348}{1000}$$

$$= \frac{\cancel{4} \cdot 87}{\cancel{4} \cdot 250} = \frac{87}{250}$$

PRACTICE EXERCISE 1

Change each decimal to a fraction or mixed number.

(a) 0.2

(b) 0.65

(c) 0.804

(d) 4.8 is read "four and eight tenths."

$$4.8 = 4 + \frac{8}{10}$$

$$= 4\frac{8}{10}$$

$$= 4\frac{\cancel{2} \cdot 4}{\cancel{2} \cdot 5}$$

$$= 4\frac{4}{5}$$

(d) 53.625

Answers: (a) $\frac{1}{5}$ (b) $\frac{13}{20}$ (c) $\frac{201}{250}$
(d) $53\frac{5}{8}$

Another method of conversion allows us to write any decimal as a proper or improper fraction. Consider 0.38, for example. Since *two* digits follow the decimal point, put *two* zeros after 1 to form the denominator of the fraction. When the decimal point is removed, the resulting number is the numerator of the desired fraction. Sometimes the fraction must be reduced.

$$0.38 = \frac{38}{100} = \frac{\cancel{2} \cdot 19}{\cancel{2} \cdot 50} = \frac{19}{50} \quad \text{Reduced}$$

two digits after the decimal point   two zeros after 1

To Change a Decimal to a Fraction (Second Method)

1. Count the digits following the decimal point; put that number of zeros after 1 in the denominator of the fraction.

2. Remove the decimal point; put the resulting whole number in the numerator of the fraction.

3. Reduce the fraction to lowest terms.

EXAMPLE 2   DECIMALS TO FRACTIONS (SECOND METHOD)

Change each decimal to a fraction or mixed number using the second method of conversion.

(a) 0.6

$$\frac{6}{10} \longleftarrow \text{Remove the decimal point and put 6 in the numerator}$$

$$\longleftarrow \left\{ \begin{array}{l} \text{Since one digit follows the decimal point in 0.6, 1} \\ \text{ followed by } \textit{one} \text{ zero, or 10, is the denominator} \end{array} \right.$$

$$= \frac{\cancel{2} \cdot 3}{\cancel{2} \cdot 5} = \frac{3}{5} \quad \text{Reduce to lowest terms}$$

(b) 0.35

$$\frac{35}{100} \begin{array}{l} \longleftarrow \text{Remove decimal point} \\ \longleftarrow \text{1 followed by two zeros (0.35 two digits)} \end{array}$$

$$= \frac{\cancel{5} \cdot 7}{\cancel{5} \cdot 20} = \frac{7}{20} \quad \text{Reduce to lowest terms}$$

PRACTICE EXERCISE 2

Change each decimal to a fraction or mixed number using the second method of conversion.

(a) 0.2

(b) 0.65

(c) 0.348

$$\frac{348}{1000} \longleftarrow \text{Remove decimal point}$$
$$\phantom{\frac{348}{1000}} \longleftarrow \text{1 followed by three zeros (0.348 three digits)}$$

$$= \frac{\cancel{4} \cdot 87}{\cancel{4} \cdot 250} = \frac{87}{250} \quad \text{Reduce}$$

(d) 4.8

$$\frac{48}{10} \longleftarrow \text{Remove decimal point}$$
$$\phantom{\frac{48}{10}} \longleftarrow \text{1 followed by one zero}$$

$$= \frac{\cancel{2} \cdot 24}{\cancel{2} \cdot 5} \quad \text{Reduce}$$

$$= \frac{24}{5} \text{ or } 4\frac{4}{5}$$

(c) 0.804

(d) 53.625

Answers: (a) $\frac{1}{5}$ (b) $\frac{13}{20}$ (c) $\frac{201}{250}$
(d) $53\frac{5}{8}$

❷ CHANGING FRACTIONS TO DECIMALS

Next we change fractions to decimals. Remember that one meaning for a fraction is division. For example,

$$\frac{2}{3} \quad \text{means} \quad 2 \div 3 \quad \text{or} \quad 3\overline{)2}.$$

In Chapter 2 we learned how to divide one whole number (dividend) by a smaller whole number (divisor). The same method can be used when the divisor is greater than the dividend, as in the fractions $\frac{2}{3}$ or $\frac{1}{4}$. The result is a decimal. Place a decimal point to the right of the dividend and attach zeros as necessary. Place a decimal point in the quotient directly above the decimal point in the dividend. Divide as if there were no decimals present. The following example illustrates.

To change $\frac{1}{4}$ to a decimal, divide 1 by 4.

$$
\begin{array}{c}
. \longleftarrow \text{③ Place decimal point in quotient above decimal point in dividend} \\
4\overline{)1.0} \longleftarrow \text{② Attach one zero} \\
\uparrow \\
\text{① Place decimal point after 1}
\end{array}
$$

Ignore the decimal points and divide 10 by 4. Then, attach a second 0 to the dividend, bring it down, and divide 4 into 20.

$$
\begin{array}{r}
.25 \\
4\overline{)1.00} \\
\underline{8}\downarrow \\
20 \\
\underline{20} \\
0
\end{array}
\qquad 20 \div 4 = 5
$$

Stop when the remainder is 0. The quotient is the desired decimal.

$$\frac{1}{4} = 0.25$$

We can check that the method is correct by reversing the steps and changing 0.25 to a fraction.

$$0.25 \quad \text{is} \quad \frac{25}{100} \quad \text{which reduces to} \quad \frac{1 \cdot 25}{4 \cdot 25} = \frac{1}{4}.$$

Suppose we change $\frac{2}{3}$ to a decimal.

$$
\begin{array}{r}
. \longleftarrow \text{②Place decimal point in quotient} \\
3\overline{)2.0} \longleftarrow \text{①Attach one zero after decimal point}
\end{array}
$$

We divide 20 by 3, then attach another zero, bring it down, and again divide 20 by 3.

$$
\begin{array}{r}
.66 \\
3\overline{)2.00} \\
\underline{1\ 8}\downarrow \\
20 \\
\underline{18} \\
2
\end{array}
$$

If we attach another zero and bring it down, we would again divide 20 by 3. By now, it is clear that this pattern will continue. Therefore, $\frac{2}{3}$ can be represented by

$$0.666 \cdots$$

The three dots show that the pattern of 6's continues. We often omit the dots and place a bar over the repeating digit (or digits) to shorten the notation.

$$0.\overline{6} \quad \text{means} \quad 0.666 \cdots$$

The two examples above illustrate the following important property of fractions.

Terminating and Repeating Decimals

Every fraction can be written as a decimal.

1. If the division process ends with a remainder of zero, the decimal is a **terminating decimal.**

2. If the division process does not end, there is a digit or block of digits which repeats, and the decimal is a **repeating decimal.**

Thus, 0.25 is a terminating decimal and $0.\overline{6}$ or $0.666 \cdots$ is a repeating decimal.

To Change a Fraction to a Decimal

1. Use the numerator as the dividend and the denominator as the divisor in a division problem.

2. Place decimal points after the numerator (dividend) and in the same position in the quotient.

3. Attach one zero at a time to the numerator (now the dividend) and divide as whole numbers.

4. Stop when a remainder of 0 is obtained or when a repeating pattern is found.

EXAMPLE 3 CHANGING FRACTIONS TO DECIMALS	**PRACTICE EXERCISE 3**

Change each fraction to a decimal.

Change each fraction to a decimal.

(a) $\dfrac{1}{2}$ Put in decimal points and attach a zero, then divide 10 by 2.

(a) $\dfrac{1}{5}$

$$\begin{array}{r} .5 \\ 2\overline{)1.0} \\ \underline{1\ 0} \\ 0 \end{array}$$

Since the remainder is 0, division stops. Thus, $\frac{1}{2}$ is the terminating decimal 0.5 $\left(\text{remember that } 0.5 = \frac{5}{10} = \frac{1 \cdot \cancel{5}}{2 \cdot \cancel{5}} = \frac{1}{2}\right)$.

(b) $\dfrac{3}{8}$

(b) $\dfrac{7}{8}$

$$\begin{array}{r} .375 \\ 8\overline{)3.000} \\ \underline{2\ 4} \\ 60 \\ \underline{56} \\ 40 \\ \underline{40} \\ 0 \end{array}$$

Thus, $\frac{3}{8}$ is the terminating decimal 0.375 $\left(\text{remember that } 0.375 = \frac{375}{1000} = \frac{3 \cdot \cancel{125}}{8 \cdot \cancel{125}} = \frac{3}{8}\right)$.

(c) $\dfrac{1}{3}$

(c) $\dfrac{2}{9}$

$$\begin{array}{r} .33 \\ 3\overline{)1.000} \\ \underline{9} \\ 10 \\ \underline{9} \\ 10 \end{array}$$

Obviously, the pattern of 3's will continue. Thus, $\frac{1}{3}$ is the repeating decimal

$$0.\overline{3} \quad \text{or} \quad 0.333\cdots$$

(d) $\dfrac{7}{33}$

(d) $\dfrac{12}{33}$

$$\begin{array}{r} .2121 \\ 33\overline{)7.0000} \\ \underline{6\ 6} \\ 40 \\ \underline{33} \\ 70 \\ \underline{66} \\ 40 \\ \underline{33} \\ 7 \end{array}$$

Again, we see a pattern: the block of two digits, 21, will continue to repeat. Thus,

$$\frac{7}{33} = 0.2121 \cdots = 0.\overline{21}.$$

Notice that the bar is placed over the *two* repeating digits, 2 and 1.

Answers: (a) 0.2 (b) 0.875
(c) 0.$\overline{2}$ (d) 0.$\overline{36}$

Improper fractions can be changed to decimals using the same procedure.

| **EXAMPLE 4** CHANGING IMPROPER FRACTIONS TO DECIMALS | **PRACTICE EXERCISE 4** |

Change each fraction to a decimal.

(a) $\dfrac{7}{4}$

$$4\overline{)7.0} \quad \begin{array}{l}\text{Place decimal points} \\ \text{Attach a zero}\end{array}$$

This time we divide 7 by 4 and place 1 in the quotient above 7 before the decimal point.

$$
\begin{array}{r}
1.75 \\
4\overline{)7.00} \\
\underline{4} \\
3\,0 \\
\underline{2\,8} \\
20 \\
\underline{20} \\
0
\end{array}
\quad \text{Place decimal points and attach a zero}
$$

Attach and bring down another zero

Thus, $\frac{7}{4}$ is equal to 1.75.

(b) $\dfrac{14}{3}$

$$
\begin{array}{r}
4.66 \\
3\overline{)14.00} \\
\underline{12} \\
2\,0 \\
\underline{1\,8} \\
20 \\
\underline{18} \\
2
\end{array}
$$

The pattern of 6's will repeat. Thus, $\frac{14}{3} = 4.666 \cdots = 4.\overline{6}$.

Change each fraction to a decimal.

(a) $\dfrac{11}{4}$

(b) $\dfrac{11}{3}$

Answers: (a) 2.75 (b) 3.$\overline{6}$

❸ SUMMARY LIST OF EQUIVALENT FRACTIONS AND DECIMALS

You have probably noticed that we changed terminating decimals to fractions but did not change repeating decimals to fractions. The method for making such conversions is beyond this level of work. Certain basic conversions occur often

enough that they should be memorized. These are included in the following list.
Memorize these now, before doing the exercises.

$$\frac{1}{2} = 0.5 \qquad \frac{1}{8} = 0.125 \qquad \frac{3}{4} = 0.75$$

$$\frac{1}{3} = 0.\overline{3} \qquad \frac{1}{9} = 0.\overline{1} \qquad \frac{3}{8} = 0.375$$

$$\frac{1}{4} = 0.25 \qquad \frac{1}{10} = 0.1 \qquad \frac{5}{8} = 0.625$$

$$\frac{1}{5} = 0.2 \qquad \frac{2}{3} = 0.\overline{6} \qquad \frac{7}{8} = 0.875$$

5.2 EXERCISES A

Change each decimal to a fraction (or mixed number).

1. 1.4 $1\frac{2}{5}$ or $\frac{7}{5}$ **2.** 3.25 $3\frac{1}{4}$ or $\frac{13}{4}$ **3.** 21.9 $21\frac{9}{10}$ or $\frac{219}{10}$ **4.** 4.05 $4\frac{1}{20}$ or $\frac{81}{20}$

5. 3.002 $3\frac{1}{500}$ or $\frac{1501}{500}$ **6.** 0.015 $\frac{3}{200}$ **7** 0.1302 $\frac{651}{5000}$ **8.** 493.72 $493\frac{18}{25}$ or $\frac{12{,}343}{25}$

Change each decimal to a fraction or mixed number using the second method as shown in Example 2.

9. 0.3 $\frac{3}{10}$ **10.** 0.65 $\frac{13}{20}$ **11.** 2.7 $2\frac{7}{10}$ or $\frac{27}{10}$

12. 0.305 $\frac{61}{200}$ **13.** 1.29 $1\frac{29}{100}$ or $\frac{129}{100}$ **14.** 2.005 $2\frac{1}{200}$ or $\frac{401}{200}$

Change each fraction to a decimal.

15. $\frac{3}{4}$ 0.75 **16.** $\frac{5}{6}$ $0.8\overline{3}$ **17.** $\frac{4}{9}$ $0.\overline{4}$ **18.** $\frac{7}{11}$ $0.\overline{63}$

19. $\frac{1}{16}$ 0.0625 **20.** $\frac{7}{3}$ $2.\overline{3}$ **21.** $\frac{11}{4}$ 2.75 **22** $\frac{19}{16}$ 1.1875

Write each of the following as a fraction, using the table in the text that you memorized.

23. $0.\overline{3}$ $\frac{1}{3}$ **24.** 0.375 $\frac{3}{8}$ **25.** 0.125 $\frac{1}{8}$ **26.** 1.5 $\frac{3}{2}$ **27.** $0.\overline{6}$ $\frac{2}{3}$

28. 0.875 $\frac{7}{8}$ **29.** 0.2 $\frac{1}{5}$ **30.** 0.625 $\frac{5}{8}$ **31.** $0.\overline{1}$ $\frac{1}{9}$ **32.** 0.25 $\frac{1}{4}$

33. 0.1 $\frac{1}{10}$ **34.** 1.25 $\frac{5}{4}$ **35.** 0.5 $\frac{1}{2}$ **36.** 0.75 $\frac{3}{4}$ **37.** $1.\overline{3}$ $\frac{4}{3}$

FOR REVIEW

38. Give the formal word name for 401.003.
four hundred one and three thousandths

39. Write twenty-three and one hundred seven thousandths in decimal notation.   **23.107**

40. Write a word name for $34.85 **(a)** as it would appear on a check; **(b)** using dollars and cents.
(a) **Thirty-four and $\frac{85}{100}$ dollars** (b) **thirty-four dollars and eighty-five cents**

The following exercises will help you prepare for the next section. Round each number to the nearest (a) ten, (b) hundred, and (c) thousand.

41. 1365
   (a) **1370** (b) **1400** (c) **1000**

42. 12,501
   (a) **12,500** (b) **12,500** (c) **13,000**

43. 7494
   (a) **7490** (b) **7500** (c) **7000**

ANSWERS: 1. $1\frac{2}{5}$ or $\frac{7}{5}$ 2. $3\frac{1}{4}$ or $\frac{13}{4}$ 3. $21\frac{9}{10}$ or $\frac{219}{10}$ 4. $4\frac{1}{20}$ or $\frac{81}{20}$ 5. $3\frac{1}{500}$ or $\frac{1501}{500}$ 6. $\frac{3}{200}$ 7. $\frac{651}{5000}$ 8. $493\frac{18}{25}$ or $\frac{12,343}{25}$ 9. $\frac{3}{10}$ 10. $\frac{13}{20}$ 11. $2\frac{7}{10}$ or $\frac{27}{10}$ 12. $\frac{61}{200}$ 13. $1\frac{29}{100}$ or $\frac{129}{100}$ 14. $2\frac{1}{200}$ or $\frac{401}{200}$ 15. 0.75 16. 0.8$\overline{3}$ (Place the bar over 3 only; 8 does not repeat.) 17. 0.$\overline{4}$ 18. 0.$\overline{63}$ 19. 0.0625 20. 2.$\overline{3}$ 21. 2.75 22. 1.1875 23. $\frac{1}{3}$ 24. $\frac{3}{8}$ 25. $\frac{1}{8}$ 26. $\frac{3}{2}$ 27. $\frac{2}{3}$ 28. $\frac{7}{8}$ 29. $\frac{1}{5}$ 30. $\frac{5}{8}$ 31. $\frac{1}{9}$ 32. $\frac{1}{4}$ 33. $\frac{1}{10}$ 34. $\frac{5}{4}$ 35. $\frac{1}{2}$ 36. $\frac{3}{4}$ 37. $\frac{4}{3}$ 38. four hundred one and three thousandths 39. 23.107 40. (a) Thirty-four and $\frac{85}{100}$ dollars (b) thirty four dollars and eighty-five cents 41. (a) 1370 (b) 1400 (c) 1000 42. (a) 12,500 (b) 12,500 (c) 13,000 43. (a) 7490 (b) 7500 (c) 7000

5.2 EXERCISES B

Change each decimal to a fraction (or mixed number).

1. 2.4 $2\frac{2}{5}$ or $\frac{12}{5}$

2. 5.75 $5\frac{3}{4}$ or $\frac{23}{4}$

3. 22.8 $22\frac{4}{5}$ or $\frac{114}{5}$

4. 7.03 $7\frac{3}{100}$ or $\frac{703}{100}$

5. 4.002 $4\frac{1}{500}$ or $\frac{2001}{500}$

6. 0.035 $\frac{7}{200}$

7. 0.3212 $\frac{803}{2500}$

8. 525.25 $525\frac{1}{4}$ or $\frac{2101}{4}$

Change each decimal to a fraction or mixed number using the second method as shown in Example 2.

9. 0.4 $\frac{2}{5}$

10. 0.85 $\frac{17}{20}$

11. 2.8 $2\frac{4}{5}$ or $\frac{14}{5}$

12. 0.205 $\frac{41}{200}$

13. 1.39 $1\frac{39}{100}$ or $\frac{139}{100}$

14. 3.007 $3\frac{7}{1000}$ or $\frac{3007}{1000}$

Change each fraction to a decimal.

15. $\frac{1}{4}$ 0.25

16. $\frac{3}{20}$ 0.15

17. $\frac{7}{9}$ 0.$\overline{7}$

18. $\frac{3}{11}$ 0.$\overline{27}$

19. $\frac{3}{16}$ 0.1875

20. $\frac{19}{5}$ 3.8

21. $\frac{13}{4}$ 3.25

22. $\frac{10}{3}$ 3.$\overline{3}$

Write each of the following as a decimal, using the table in the text that you memorized.

23. $\frac{2}{3}$ 0.$\overline{6}$

24. $\frac{1}{10}$ 0.1

25. $\frac{3}{2}$ 1.5

26. $\frac{1}{5}$ 0.2

27. $\frac{7}{8}$ 0.875

28. $\frac{5}{8}$ 0.625   **29.** $\frac{1}{2}$ 0.5   **30.** $\frac{4}{3}$ $1.\overline{3}$ **31.** $\frac{3}{4}$ 0.75   **32.** $\frac{3}{8}$ 0.375

33. $\frac{1}{8}$ 0.125   **34.** $\frac{1}{3}$ $0.\overline{3}$ **35.** $\frac{1}{4}$ 0.25   **36.** $\frac{1}{9}$ $0.\overline{1}$ **37.** $\frac{5}{4}$ 1.25

FOR REVIEW

38. Give a formal word name for 639.0007.
six hundred thirty-nine and seven ten thousandths

39. Write one hundred fifteen and two hundred sixty-nine ten thousandths in decimal notation.   115.0269

40. Write a word name for $82.89   **(a)** as it would appear on a check;   **(b)** using dollars and cents.
(a) eighty-two and $\frac{89}{100}$ dollars (b) eighty-two dollars and eighty-nine cents

The following exercises will help you prepare for the next section. Round each number to the nearest (a) ten, (b) hundred, and (c) thousand.

41. 2526
    (a) 2530 (b) 2500 (c) 3000

42. 17,495
    (a) 17,500 (b) 17,500 (c) 17,000

43. 6351
    (a) 6350 (b) 6400 (c) 6000

5.2   EXERCISES C

1. Change 236.525 to a mixed number. $236\frac{21}{40}$

2. Change 1000.0001 to a mixed number.
$1000\frac{1}{10,000}$

3. Change $\frac{1}{7}$ to a decimal.
[Answer: $0.\overline{142857}$]

4. Change $\frac{7}{13}$ to a decimal. $0.\overline{538461}$

5.3   ROUNDING AND ESTIMATING DECIMALS

STUDENT GUIDEPOSTS

1 Finding the Number of Decimal Places

2 Rounding Decimals

3 Estimating Decimal Computations

1 FINDING THE NUMBER OF DECIMAL PLACES

In Section 1.4 we learned how to round whole numbers to the nearest ten, hundred, thousand, and so forth. We now see how to round decimals to the nearest one, tenth, hundredth, and so forth, using the notion of *decimal places*.

Number of Decimal Places

If a number is written in decimal notation, the number of digits to the right of the decimal point is called the number of **decimal places** in the number.

EXAMPLE 1 FINDING THE NUMBER OF DECIMAL PLACES

Give the number of decimal places in each number.

(a) 3. **15** has *two* decimal places.

↑

two digits to the right of the decimal point

(b) 40. **397** has *three* decimal places.

↑

three digits

(c) 12 or 12. has zero or no decimal places.

(d) 0.600 has three decimal places.

(e) 2.6060 has four decimal places.

PRACTICE EXERCISE 1

Give the number of decimal places in each number.

(a) 5.2

(b) 37.029

(c) 75

(d) 0.350

(e) 6.0003

Answers: (a) one   (b) three
(c) zero   (d) three   (e) four

///////////// **CAUTION** ///////////

Compare the three decimals 0.6, 0.60, and 0.600, having one, two, and three decimal places, respectively. Although these numbers have the same value, by using zeros to the right of the digit 6, we are emphasizing a greater degree of precision or accuracy. If the numbers were used for measurements, 0.600 would indicate accuracy to the nearest thousandth, 0.60 to the nearest hundredth, and 0.6 to the nearest tenth.

///////////

❷ ROUNDING DECIMALS

The following table summarizes the relationship between the number of decimal places and the position corresponding to this number, and is helpful when rounding decimals.

Decimal Places	Last Position on the Right	Example
0	ones	5 or 5.
1	tenths	5.7
2	hundredths	5.76
3	thousandths	5.764
4	ten thousandths	5.7648

To Round a Decimal to a Particular Number of Places

1. Locate the position (ones, tenths, hundredths, etc.) corresponding to this number of places.

2. If the first digit to the right of this position is less than 5, drop it and all digits to the right of it. Do not change the digit in the place to which you are rounding.

3. If the first digit to the right of this position is greater than or equal to 5, increase the digit in the desired place by one and drop all digits to the right of it.

Other methods of rounding can be used. However, the method given here is perhaps the most useful and certainly the simplest. Notice that rounding decimals is much like rounding whole numbers.

EXAMPLE 2   ROUNDING TO ONE DECIMAL PLACE

Round 35.24 to one decimal place (to the nearest tenth).
   We are rounding to the tenths position.

$$35.\boxed{2}4$$

↑
round to this position

Since 4, the first digit to the right of 2, is less than 5, we drop 4 and do not change 2. Thus, rounded to one decimal place, 35.24 becomes 35.2.

Round 68.32 to one decimal place (to the nearest tenth).

Answer: 68.3

We often round to obtain approximate values for numbers. For example, 35.24 is approximately equal to 35.2. The symbols ≈ and ≐ stand for "is approximately equal to." Thus, from Example 2,

$$35.24 \approx 35.2 \quad \text{or} \quad 35.24 \doteq 35.2.$$

EXAMPLE 3   ROUNDING TO TWO DECIMAL PLACES

Round 2.3482 to two decimal places (to the nearest hundredth).
   Round to the hundredths position.

first digit to the right of rounding position

$$2.3\boxed{4}82$$

↑
round to this position

Since the first digit to the right of the hundredths position is 8, which is greater than 5, increase 4 to 5 and drop 8 and 2. Thus,

$$2.3482 \approx 2.35, \quad \text{rounded to two decimal places.}$$

Round 6.5473 to two decimal places (to the nearest hundredth).

Answer: 6.55

EXAMPLE 4   ROUNDING TO THE NEAREST TENTH

Round 53.85 to the nearest tenth (to one decimal place).

first digit is 5, so increase 8 to 9 and drop 5

$$53.\boxed{8}5$$

↑
round to this position

Thus, 53.85 ≈ 53.9, rounded to the nearest tenth.

Round 3.125 to the nearest hundredth (to two decimal places).

Answer: 3.13

Increases in a particular digit may need to be carried over to the next digit. This is shown in the next example.

EXAMPLE 5 ROUNDING ACROSS A DIGIT

Round 1.3895 to three decimal places (to the nearest thousandth).

drop 5 and increase 9 by 1

1.38 9 5

round to here

When 9 is increased by 1, the result is 10, so replace 9 by 0 and increase 8 to 9. Thus,

$$1.3895 \approx 1.390.$$

Do not give 1.39 for the answer; 1.39 has only two decimal places. The final zero in 1.390 must be present.

PRACTICE EXERCISE 5

Round 41.4975 to two decimal places (to the nearest hundredth).

Answer: 41.50

CAUTION

When rounding, do not round off one digit at a time from the right. For example, to round 2.6149 to the nearest hundredth, do not round to the nearest thousandth,

2.615

and then to the nearest hundredth,

2.62.

This is incorrect, since using our method,

$2.6149 \approx 2.61.$ Rounded to the nearest hundredth

EXAMPLE 6 ROUNDING TO THE NEAREST ONE

Round 21.654 to the nearest one (0 decimal places).

since 6 is greater than 5, increase 1 to 2
and drop 6, 5, and 4

2 1 .654

round to here

Thus, $21.654 \approx 22$, rounded to the nearest one (or whole number).

PRACTICE EXERCISE 6

Round 16.539 to the nearest one (0 decimal places).

Answer: 17

Decimals that do not terminate, but have repeating blocks of digits such as

$$0.333 \cdots \quad \text{or} \quad 0.\overline{3} \quad \text{and} \quad 0.212121 \cdots \quad \text{or} \quad 0.\overline{21},$$

can also be rounded. The procedure is identical to that for terminating decimals.

EXAMPLE 7 ROUNDING REPEATING DECIMALS	**PRACTICE EXERCISE 7**

(a) Round $0.\overline{3}$ to the nearest tenth.

$$0.\overline{3} = 0.\underline{3}33\cdots$$

— 3 < 5, so drop remaining digits

↑ round to here

Thus, $0.\overline{3} \approx 0.3$.

(b) Round $0.\overline{6}$ to the nearest thousandth.

— 6 > 5, so increase thousandths digit from 6 to 7 and drop remaining digits

$$0.\overline{6} = 0.66\underline{6}6\cdots$$

↑ round to here

Thus, $0.\overline{6} \approx 0.667$.

(c) Round $0.\overline{28}$ to the nearest tenth.

— 8 > 5 so increase tenths digit from 2 to 3 and drop remaining digits

$$0.\overline{28} = 0.\underline{2}82828\cdots$$

↑ round to here

Thus, $0.\overline{28} \approx 0.3$.

(d) Round $0.\overline{28}$ to the nearest hundredth.

— 2 < 5 so drop remaining digits

$$0.\overline{28} = 0.28\underline{2}828\cdots$$

↑ round to here

Thus, $0.\overline{28} \approx 0.28$.

(a) Round $0.\overline{2}$ to the nearest tenth.

(b) Round $0.\overline{8}$ to the nearest hundredth.

(c) Round $0.\overline{51}$ to the nearest hundredth.

(d) Round $0.\overline{51}$ to the nearest thousandth.

Answers: (a) 0.2   (b) 0.89   (c) 0.52   (d) 0.515

Calculations involving amounts of money must often be rounded to the nearest dollar (one) or to the nearest cent (hundredths). For example, suppose we obtain

$$\$57.432$$

as an answer. Since our smallest unit of money is the cent, or one hundredth of a dollar, we usually round such figures to the nearest cent and write

$$\$57.43 \quad \text{instead of} \quad \$57.432.$$

At times (when completing an income tax form, for example), we may round an amount of money to the nearest dollar. For example, to the nearest dollar,

$$\$57.43 \approx \$57.00 \text{ or } \$57,$$

and

$$\$112.50 \approx \$113.00 \quad \text{or} \quad \$113.$$

❸ ESTIMATING DECIMAL COMPUTATIONS

In the sections that follow we will consider sums, differences, products, and quotients of decimals. Often in real-life practical situations, an estimated result might be all that is needed. For example, if we were deciding whether to buy 50 pounds of ground beef at $1.97 per pound, we might want to know the approximate cost. Since $1.97 is about $2.00, the cost is approximately

$$50 \times \$2 = \$100.$$

Similarly, if we have borrowed $8897.50 and plan to pay it back in 30 equal monthly payments, the approximate monthly payment is

$$\$9000 \div 30 = \$300.$$

In the first situation, we rounded $1.97 to the nearest dollar, and in the second, we rounded to the nearest thousand dollars. Such estimates are useful in daily activities and can often be made mentally.

EXAMPLE 8 ESTIMATING IN A CONSUMER PROBLEM	PRACTICE EXERCISE 8

The Thoresons plan to carpet their living room using carpeting that costs $14.95 a sq yd. If the room is rectangular with length 8.1 yd and width 5.8 yd, what would be the approximate cost of the job?

The approximate area of the room is

$8 \times 6 = 48$ sq yd   Rounding 8.1 and 5.8 to the nearest one

Since each square yard of carpeting costs approximately $15 (rounding to the nearest one), the total cost is about

$$15 \times 48 = \$720.$$

The Randolphs borrowed $8095.50 and plan to pay it back in 30 equal monthly payments. Estimate their monthly payment.

Answer: $270 (using 8100 ÷ 30)

HINT

In any applied problem involving decimals, try to estimate the solution by rounding to the nearest whole numbers and use this estimation to help you decide whether the answer you obtain seems reasonable and accurate.

5.3 EXERCISES A

Give the number of decimal places in each number.

1. 32.1   1   **2.** 4.23   2   **3.** 6.453   3   **4.** 42.3965   4

5. 13   0 or none   **6.** 42.80   2   **7.** 4.280   3   **8.** 0.4280   4

Round each number to one decimal place (to the nearest tenth).

9. 3.21   3.2   **10.** 4.67   4.7   **11.** 3.256   3.3

12. 5.4249   5.4   **13.** 32.75   32.8   **14.** 151.951   152.0

Round each number to two decimal places (to the nearest hundredth).

15. 7.321   7.32

16. 3.487   3.49

17. 4.3209   4.32

18. 5.4249   5.42

19. 34.0085   34.01

20. 34.0995   34.10

Round each number to three decimal places (to the nearest thousandth).

21. 3.4579   3.458

22. 7.2571   7.257

23. 3.4906   3.491

24. 5.4249   5.425

25. 45.0085   45.009

26. 9.0307   9.031

Round each number to the nearest one.

27. 24.1   24

28. 63.5129   64

29. 5.4249   5

30. 9.003   9

31. 19.84   20

32. 39.099   39

*Round each repeating decimal to the nearest (**a**) tenth, (**b**) hundredth, and (**c**) thousandth.*

33. $0.\overline{4}$
  (a) 0.4
  (b) 0.44
  (c) 0.444

34 $0.\overline{35}$
  (a) 0.4
  (b) 0.35
  (c) 0.354

35. $0.\overline{123}$
  (a) 0.1
  (b) 0.12
  (c) 0.123

36. $0.\overline{7}$
  (a) 0.8
  (b) 0.78
  (c) 0.778

37. $0.\overline{93}$
  (a) 0.9
  (b) 0.94
  (c) 0.939

38. $0.\overline{455}$
  (a) 0.5
  (b) 0.46
  (c) 0.455

Round each amount of money to the nearest cent.

39. $4.237 $4.24

40. $6.235 $6.24

41. $7.231 $7.23

42. $8.105 $8.11

Round each amount of money to the nearest dollar.

43. $37.29 $37

44. $13.49 $13

45. $28.50 $29

46. $16.99 $17

Solve.

47 A bricklayer will use 290 bricks at a cost of 9.5¢ per brick. What is the approximate total cost of the bricks? $30 [3000¢]

48. What is the approximate cost of 95 pounds of beans selling for 41¢ per pound?
$40 [using 100 × 40 = 4000¢]

FOR REVIEW

49. Change 7.08 to a mixed number.
$7\frac{2}{25}$

50. Change $\frac{8}{9}$ to a decimal. $0.\overline{8}$

ANSWERS: 1. 1 2. 2 3. 3 4. 4 5. 0 or none 6. 2 7. 3 8. 4 9. 3.2 10. 4.7 11. 3.3 12. 5.4 13. 32.8 14. 152.0 15. 7.32 16. 3.49 17. 4.32 18. 5.42 19. 34.01 20. 34.10 21. 3.458 22. 7.257 23. 3.491 24. 5.425 25. 45.009 26. 9.031 27. 24 28. 64 29. 5 30. 9 31. 20 32. 39 33. (a) 0.4 (b) 0.44 (c) 0.444 34. (a) 0.4 (b) 0.35 (c) 0.354 35. (a) 0.1 (b) 0.12 (c) 0.123 36. (a) 0.8 (b) 0.78 (c) 0.778 37. (a) 0.9 (b) 0.94 (c) 0.939 38. (a) 0.5 (b) 0.46 (c) 0.455 39. $4.24 40. $6.24 41. $7.23 42. $8.11 43. $37 44. $13 45. $29 46. $17 47. $30 [3000¢] 48. $40 [using 100 × 40 = 4000¢] 49. $7\frac{2}{25}$ 50. $0.\overline{8}$

5.3 EXERCISES B

Give the number of decimal places in each number.

1. 63.1   1   **2.** 7.45   2   **3.** 8.425   3   **4.** 51.405   3

5. 15   0 or none   **6.** 71.90   2   **7.** 7.190   3   **8.** 0.7190   4

Round each number to one decimal place (to the nearest tenth).

9. 4.31   4.3   **10.** 5.87   5.9   **11.** 6.357   6.4

12. 2.6249   2.6   **13.** 41.06   41.1   **14.** 0.964   1.0

Round each number to two decimal places (to the nearest hundredth).

15. 9.432   9.43   **16.** 6.278   6.28   **17.** 12.4208   12.42

18. 13.5091   13.51   **19.** 3.0995   3.10   **20.** 0.0355   0.04

Round each number to three decimal places (to the nearest thousandth).

21. 4.3569   4.357   **22.** 2.3581   2.358   **23.** 6.3585   6.359

24. 76.0095   76.010   **25.** 8.0606   8.061   **26.** 7.4008   7.401

Round each number to the nearest one.

27. 26.1   26   **28.** 32.507   33   **29.** 7.499   7

30. 6.005   6   **31.** 29.78   30   **32.** 8.0099   8

Round each repeating decimal to the nearest (a) tenth, (b) hundredth, and (c) thousandth.

33. $0.\overline{9}$
  (a) 1.0 (b) 1.00 (c) 1.000

34. $0.\overline{17}$
  (a) 0.2 (b) 0.17 (c) 0.172

35. $0.\overline{234}$
  (a) 0.2 (b) 0.23 (c) 0.234

36. $0.\overline{1}$
  (a) 0.1 (b) 0.11 (c) 0.111

37. $0.8\overline{4}$
  (a) 0.8 (b) 0.85 (c) 0.848

38. $0.6\overline{55}$
  (a) 0.7 (b) 0.66 (c) 0.656

Round each amount of money to the nearest cent.

39. $7.382 $7.38   **40.** $4.535 $4.54   **41.** $9.448 $9.45   **42.** $10.205 $10.21

Round each amount of money to the nearest dollar.

43. $62.38 $62   **44.** $41.48 $41   **45.** $74.50 $75   **46.** $32.98 $33

Solve.

47. What is the approximate cost of 32 pounds of coffee selling for $2.95 per pound? $90

48. A clock radio sells for $39.95. If Bill has $243.65 in his checking account, approximately how many radios could he buy?   6 [using 240 ÷ 40]

FOR REVIEW

49. Change 2.016 to a mixed number. $2\frac{2}{125}$

50. Change $\frac{5}{33}$ to a decimal. $0.\overline{15}$

5.3 EXERCISES C

1. Change $\frac{4}{7}$ to a decimal and round the result to the nearest **(a)** tenth, **(b)** hundredth, and **(c)** thousandth.   (a) **0.6** (b) **0.57** (c) **0.571**

2. What is the approximate cost of 305 books each selling for $9.95?   **$3000 (using $10 × 300)**

3. John and Sue borrowed $6123.50 and plan to pay it back in 30 equal monthly payments. What will be their approximate monthly payment? [Answer: $200 (using $6000 ÷ 30)]

4. A paneling job requires 592 sheets of wood paneling at a cost of $20.75 per sheet. What is the approximate cost of the paneling?   **$12,600 (using $21 × 600)**

5. The Wards plan to carpet their living room using carpet costing $19.50 a square yard. If the room is rectangular in shape as shown to the right, what would be the approximate cost of the job? [Answer: $1080 (using 9 × 6 × $20)]

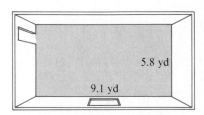

5.8 yd

9.1 yd

5.4 ADDING AND SUBTRACTING DECIMALS

STUDENT GUIDEPOSTS

1 Adding Two or More Decimals
2 Applications Adding Decimals
3 Subtracting Two Decimals
4 Applications Subtracting Decimals

1 ADDING TWO OR MORE DECIMALS

When we studied the basic operations on fractions, we began with multiplication and division, and then went on to the more difficult operations of addition and subtraction. For numbers in decimal notation, addition and subtraction are easier. For example, suppose we want to find the following sum.

$$\begin{array}{r} 2.35 \\ + 1.21 \\ \end{array}$$

We might convert each decimal to fractional form and then add the fractions. But if we add the columns in the original problem and move the decimal point straight down, we obtain the same sum much more easily.

$$\begin{array}{r} 2.35 \\ \underline{1.21} \\ 3.56 \\ \end{array}$$

This illustrates a general method.

To Add Two or More Decimals
1. Arrange the numbers in columns so that the decimal points line up.
2. Add in columns from right to left as if adding whole numbers. (Additional zeros may be placed to the right of the decimal points if needed.)
3. Place the decimal point in the sum in line with the other decimal points.

EXAMPLE 1   ADDING DECIMALS

Find each sum.

(a) 21.3 + 4.8

$$\begin{array}{r} \overset{1}{21.3} \\ +\ 4.8 \\ \hline 26.1 \end{array}$$

① Arrange the numbers in columns so that the decimal points line up
② Carry 1 after adding 8 and 3 just as in addition of whole numbers
③ Place the decimal point in line with others

(b) 37 + 2.5 + 10.03 + 425.008

$$\begin{array}{r} 37. \\ 2.5 \\ 10.03 \\ +\ 425.008 \\ \hline 474.538 \end{array} \quad \text{or} \quad \begin{array}{r} 37.000 \\ 2.500 \\ 10.030 \\ +\ 425.008 \\ \hline 474.538 \end{array}$$

Writing 0's to the right of the decimal points may help you keep the decimal points lined up vertically

PRACTICE EXERCISE 1

Find each sum.

(a) 154.5 + 60.7

(b) 3.8 + 68 + 40.07 + 653.009

Answers: (a) **215.2**   (b) **764.879**

❷ APPLICATIONS ADDING DECIMALS

You can see that adding decimals is much like adding whole numbers. Following are several examples of word problems that involve adding decimals. Make certain that your answers are reasonable.

EXAMPLE 2   ADDING DECIMALS IN A MILEAGE PROBLEM

Before leaving on a trip, Mike's odometer read 43,271.9. If he drove 827.6 miles on the trip, what did it read when he returned?

The return reading must be the beginning reading plus the number of miles traveled.

$$\begin{array}{r} 43271.9 \\ +\ 827.6 \\ \hline 44099.5 \end{array}$$

Thus, when he returned, his odometer read 44,099.5.

PRACTICE EXERCISE 2

On a two-day trip, Steve drove 489.3 miles the first day and 501.8 miles the second. How far did he drive on the trip?

Answer: **991.1 miles**

EXAMPLE 3 ADDING DECIMALS ON A TELEPHONE BILL

A good example of adding decimals can be found in a long-distance telephone bill. A typical bill is shown in Figure 5.2.

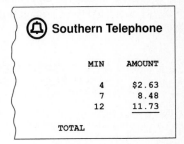

```
602 555-0801    FEB.25, 1991                    🔔 Mountain Bell

ITEMIZED CALLS

NO.   DATE   TIME    TO PLACE         TO AREA-NO.   MIN   AMOUNT
1     115    747P    MONTGOMERY,AL    205 999-4801   2     $1.46
2     123    1035A   NEWARK,NJ        503 888-3213   3      2.56
3     2 1    345P    LOS ANGELES,CA   301 777-4243   1      0.87
4     219    832A    BROOKLYN,NY      212 423-2131   5      3.27

                     TOTAL OF ITEMIZED CALLS EXCLUDING TAX $8.16
```

Figure 5.2

The total due on long-distance calls is $8.16, the sum of the decimals in the last column.

EXAMPLE 4 ADDING DECIMALS IN A DISTANCE PROBLEM

Consider the map in Figure 5.3. Starting in West Yorkville, Margot hiked the complete circular trail past Devon Falls, Mount Blanc, Sanditon, Brunswick, East Yorkville, and back to West Yorkville. How long was her hike?

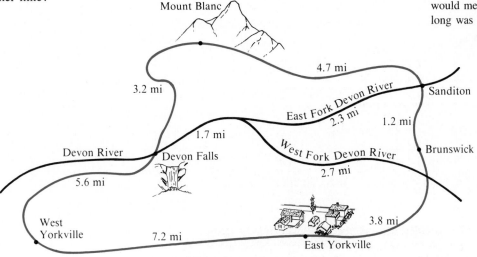

Figure 5.3

Add the distances on the map.

$$
\begin{array}{r}
5.6 \\
3.2 \\
4.7 \\
1.2 \\
3.8 \\
+\ \ 7.2 \\
\hline
25.7
\end{array}
$$

Thus, Margot hiked a distance of 25.7 miles.

PRACTICE EXERCISE 3

Part of a monthly telephone bill is given below. Find the total bill for long-distance calls.

```
🔔 Southern Telephone

     MIN      AMOUNT

      4       $2.63
      7        8.48
     12       11.73

   TOTAL
```

Answer: $22.84

PRACTICE EXERCISE 4

Use the map in Figure 5.3. Henry decided to hike along the river from Sanditon to Devon Falls and then on to West Yorkville where his wife would meet him with their car. How long was his hike?

Answer: 9.6 mi

| EXAMPLE 5   ADDING DECIMALS ON A DEPOSIT SLIP | PRACTICE EXERCISE 5 |

A typical bank checking account deposit slip is shown in Figure 5.4. The total deposit is found by adding all amounts in the right-hand column. Notice that a vertical line separates the dollars column from the cents column. It is used instead of a decimal point because it has been found that people are more likely to arrange the numbers correctly using this method.

Part of a bank checking account deposit slip is shown below. Find the total deposit.

DEPOSIT

	Dollars	Cents
CASH	3025	51
CHECKS	431	68
	15	06
	283	93
	4	58
TOTAL		

First Federal Bank
New Orleans, LA

Date August 2 19 91   DEPOSIT

Account No. **894-667-0523**

Sarah Williams
120 Charles St.
New Orleans, LA

	Dollars	Cents
CASH	13	27
CHECKS (List Seperately)	127	40
	312	80
	4	95
TOTAL DEPOSIT	458	42

Figure 5.4

Answer: $3760.76

❸ SUBTRACTING TWO DECIMALS

Subtraction of decimals is similar to addition. For example, suppose we want to find the following difference.

$$\begin{array}{r} 2.35 \\ -\ 1.21 \end{array}$$

By subtracting in columns, we obtain the difference.

$$\begin{array}{r} 2.35 \\ -\ 1.21 \\ \hline 1.14 \end{array}$$

The decimal point is moved straight down just as in an addition problem.

To Subtract Two Decimals

1. Arrange the numbers in columns so that the decimal points line up.

2. Subtract in columns from right to left as if subtracting whole numbers. (Additional zeros may be placed to the right of the decimal points if needed.)

3. Place the decimal point in the difference in line with the other decimal points.

| **EXAMPLE 6** SUBTRACTING DECIMALS | **PRACTICE EXERCISE 6** |

Find each difference.

(a) 43.7 − 12.5

Arrange the numbers in columns so that the decimal points line up, then subtract.

$$\begin{array}{r} 43.7 \\ -\ 12.5 \\ \hline 31.2 \end{array}$$

↑
place decimal point in line with others

(b) 121.37 − 48.2

$$\begin{array}{r} {\scriptstyle 1\ 11} \\ 1\ \cancel{2}\ \cancel{1}\ .37 \\ -\quad 4\ 8\ .\mathbf{2}0 \leftarrow \text{supply extra zero} \\ \hline 7\ 3\ .17 \end{array}$$

7 = 7 − 0
1 = 3 − 2
3 = 11 − 8
7 = 11 − 4

Notice that we had to borrow 1 ten from the 2 tens (or 20) in order to subtract 8.

(c) 48.52 − 19.67

$$\begin{array}{r} {\scriptstyle 17\quad 14} \\ {\scriptstyle 3\ 7\quad 4\ 12} \\ \cancel{4}\ \cancel{8}\ .\ \cancel{5}\ \cancel{2} \\ -\ 1\ 9\ .\ 6\ 7 \\ \hline 2\ 8\ .\ 8\ 5 \end{array}$$

5 = 12 − 7 (borrow 1 tenth from .5)
8 = 14 − 6 (borrow 1 from 8)
8 = 17 − 9 (borrow 1 ten from 40)
2 = 3 − 1

Borrowing is done as if we were subtracting whole numbers. Of course, with practice, the small numbers and the "crossing out" can be eliminated. The problem would then look like

$$\begin{array}{r} 48.52 \\ -\ 19.67 \\ \hline 28.85. \end{array}$$

Find each difference.

(a) 75.8 − 23.6

(b) 433.41 − 84.3

(c) 94.31 − 36.58

Answers: (a) 52.2   (b) 349.11
(c) 57.73

| **H I N T** |

Computationally, the only real difference between adding and subtracting decimals and adding and subtracting whole numbers is placing the decimal point. This is easy to do once the numbers have been written with the decimal points lined up.

❹ APPLICATIONS SUBTRACTING DECIMALS

Many word problems involve subtracting decimals. As before, use common sense and estimation to be sure that your answers are reasonable.

EXAMPLE 7   SUBTRACTING DECIMALS IN A DISTANCE PROBLEM

It is 13.6 miles from Wetumpka to Slapout and 32.1 miles from Wetumpka to Pine Level. How much farther is it from Wetumpka to Pine Level than to Slapout? See Figure 5.5.

Subtract 13.6 from 32.1.

$$
\begin{array}{r}
32.1 \\
- \ 13.6 \\
\hline
18.5
\end{array}
$$

It is 18.5 miles farther to Pine Level.

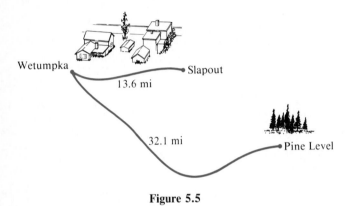

Wetumpka

Slapout

13.6 mi

32.1 mi

Pine Level

Figure 5.5

PRACTICE EXERCISE 7

Donna noticed that the odometer on her car read 35,126.8 miles when she filled her gas tank in Oklahoma City. She then drove to Houston where her odometer reading was 35,576.1. How far is it from Oklahoma City to Houston?

Answer: 449.3 miles

EXAMPLE 8   SUBTRACTING DECIMALS IN A TAX PROBLEM

Mr. Wong had a total income of $27,085.50 last year. He had deductions of $427.56, $2037.15, $402.03, and $1327.48. If taxable income is total income minus deductions, what was Mr. Wong's taxable income?

First find the total of all deductions.

$$
\begin{array}{r}
\$ \ 427.56 \\
2037.15 \\
402.03 \\
+ \ 1327.48 \\
\hline
\$4194.22
\end{array}
$$

Next subtract this total from his total income.

$$
\begin{array}{r}
\$27085.50 \\
- \ \ \ 4194.22 \\
\hline
\$22891.28
\end{array}
$$

Thus, Mr. Wong had a taxable income of $22,891.28.

PRACTICE EXERCISE 8

Dr. Weston bought a shirt for $22.95 and a pair of pants for $34.98. Tax on the two purchases was $1.15 and $1.75, respectively. He paid for the items with a $100 bill and received a $20 bill, a $10 bill, a $5 bill, four $1 bills, one dime, one nickel, and two pennies in change. Was the change correct?

Answer: Yes; the change should be $39.17.

Example 8 illustrates that more than one operation is often used when solving a word problem. This is also shown in the next example.

EXAMPLE 9   **OPERATIONS IN A CHECKING ACCOUNT**

To keep track of the balance in a checking account, most people use a check register similar to the one in Figure 5.6.

The "Balance Forward" of $608.27 is copied from the preceding page in the register. The first entry is a check written for $85.23. This amount is entered in the "Amount of Check" column and also in the working column under $608.27. The balance in the account, $523.04, is found by subtracting. Next, a deposit of $728.30 is entered in both the "Amount of Deposit" column and in the working column. Adding gives the new balance of $1251.34. The next check for $435.75 is entered as before, and the amount subtracted in the working column yields a new balance of $815.59. Thus, keeping a check register involves repeated additions and subtractions of decimals.

Check No.	Date	Checks Issued to or Deposit Description	Amount of Check	Amount of Deposit		Balance Forward
						608 | 27
301	6/22	Marvin's Market	85 | 23		Check or Deposit	85 | 23
					Balance	523 | 04
	6/25	Paycheck Deposited		728 | 30	Check or Deposit	728 | 30
					Balance	1251 | 34
302	7/1	Home Mortgage for July	435 | 75		Check or Deposit	435 | 75
					Balance	815 | 59

Figure 5.6

PRACTICE EXERCISE 9

On April 1, Cindi Franks had $958.12 in her checking account. During the month, she made two deposits, one of $327.50 and a second of $693.73. She wrote three checks in the amounts $483.16, $103.09, and $723.49. What was her balance at the end of the month?

Answer: **$669.61**

5.4 EXERCISES A

Find each sum.

1. 41.2 + 33.9   **75.1**

2. 4.37 + 2.09   **6.46**

3. 22.3 + 2.23 + 223   **247.53**

4. 0.003 + 2.107 + 135.1
137.21

5. 6.035 + 10.09 + 100.9
117.025

6. 4.3 + 5.01 + 0.005 + 41.376
50.691

7.   6.05
    21.3
+ 4.712
32.062

8.   32.079
    1.428
  403.6
+ 10.05
447.157

9.   105.30
   6.4193
  10.03
+ 4217.508
4339.2573

Arrange the following numbers in columns and find their sum.

10. $132.57, $41.03, $1.79, $10.00   **$185.39**

11. $4237.88, $1.23, $14.40, $103.05, $15.00
$4371.56

Parts of monthly telephone bills are shown in Exercises 12–13. Find the total bills for the long-distance calls.

12.

🔔 **Pacific Telephone**

MIN	AMOUNT
3	$1.57
5	6.23
1	0.37

TOTAL

$8.17

13.

🔔 **New England Telephone**

MIN	AMOUNT
10	$13.25
2	4.40
5	7.23
1	0.87

TOTAL

$25.75

Exercises 14–15 refer to the following map.

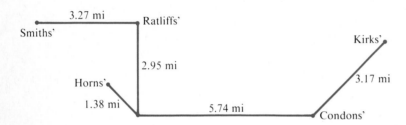

14. How far is it by road from the Smiths' to the Horns'?   **7.6 mi**

15. How far is it by road from the Kirks' to the Smiths'?
15.13 mi

Parts of bank checking account deposit slips are shown in Exercises 16–17. Find the total deposits.

16.

DEPOSIT

	Dollars	Cents
CASH	47	23
CHECKS	185	60
	13	07
	478	39
	6	50
TOTAL		

$730.79

17.

DEPOSIT

	Dollars	Cents
CASH	1030	47
CHECKS	427	33
	48	05
	11	27
	4	38
	617	07
TOTAL		

$2138.57

Solve.

18. Find the perimeter of the figure. (Perimeter is the distance around.)

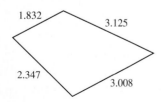

10.312

19. Find the sum of the following numbers: two thousand one hundred thirty-six and forty-one hundredths; five hundred three and two hundred seven thousandths; nine thousand one and one thousand nine ten thousandths. Give the answer in decimal notation and in words.
11,640.7179; eleven thousand six hundred forty and seven thousand one hundred seventy-nine ten thousandths

Find each difference.

20. $41.3 - 22.5$ **18.8**

21. $6.03 - 4.2$ **1.83**

22. $42.0 - 7.5$ **34.5**

23. $101.3 - 1.013$ **100.287**

24. $4.00 - 3.75$ **0.25**

25. $12.056 - 3.009$ **9.047**

26. 67.284
      $- 13.498$
      53.786

27. 28.4
       $- 3.45$
      24.95

28. 7
       $- 1.009$
      5.991

Solve.

29 The Marshall family had a total income of $31,255.75 last year. If they had deductions of $1237.40, $725.36, $2347.01, and $444.83, how much taxable income did they have?   **$26,501.15**

30. Becky received a check for $125.00. If she spent $27.45 for a pair of shoes, $45.30 for a dress, and $18.78 for books, how much did she have left?   **$33.47**

31. Complete the missing items in the following check register.

Check No.	Date	Checks Issued to or Deposit Description	Amount of Check	Amount of Deposit		Balance Forward 325 40		
420	3/2	Harry's Hair Hut	25 50		Check or Deposit		**(a)**	$25 50
					Balance		**(b)**	$299 90
421	3/3	Credit Union March Car Payment	165 35		Check or Deposit		**(c)**	$165 35
					Balance		**(d)**	$134 55
	3/4	Gift from Uncle Claude		225 00	Check or Deposit		**(e)**	$225 00
					Balance		**(f)**	$359 55

32. When Nicky was sick, he ran a temperature of 102.3° Fahrenheit. If normal body temperature is 98.6° Fahrenheit, how many degrees above normal was his temperature?   **3.7°**

33. When Den filled the gas tank on his car, the odometer read 32,478.8. The next time he filled it, the odometer read 32,920.3. How far had he driven?   **441.5 mi**

34. Lucy bought a record for $8.65 and paid with a $10 bill. She received one dime, one quarter, and one $1 bill in change. Was this the correct change?   **yes**

35. Subtract three hundred twenty-seven and eight hundredths from two thousand seven and seven hundred three thousandths. Give the difference in decimal notation and in words.
   1680.623; one thousand six hundred eighty and six hundred twenty-three thousandths

36 When Jerri was sick in the hospital, she received four antibiotic injections of 2.65 milligrams, 2.75 milligrams, 3.5 milligrams, and 4.0 milligrams.

(a) What was the total of the four injections?

12.9 milligrams

(b) If all injections were drawn from the same bottle that contained 30.5 milligrams of the antibiotic, how much remained in the bottle?

17.6 milligrams

37 In the figure below, find the length of x and y.

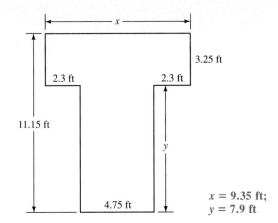

$x = 9.35$ ft;
$y = 7.9$ ft

38. The table to the right shows the amount of rainfall for three months during the summer. If the normal rainfall for this three-month period is 9.6 inches, was the rainfall above or below average and by how much?

Month	Rainfall (in inches)
June	2.1
July	4.3
August	3.7

above average by 0.5 inches

FOR REVIEW

Give the number of decimal places in each number.

39. 42.37   2

40. 103.301   3

41. 21.0506   4

42. 1139   0

43. Round 3.1594 to the nearest **(a)** tenth, **(b)** hundredth, **(c)** thousandth.

3.2   3.16   3.159

Use the following newspaper advertisements to answer Exercises 44–47.

AM/FM Stereo 3-In-One Compact Unit

• AM/FM stereo receiver
• Front loading cassette deck
• 6 ½ full range speaker system

$139⁸⁸

19″ Color TV
Quick Start Picture Tube

• 105 channel cable ready
• 100% solid state
• Infrared remote control

$283³³

44. Estimate the cost of one TV and one stereo. Then find the exact cost.
$420 [using 280 + 140]; $423.21

45. About how much more is the price of the TV than the price of the stereo? What is the exact difference in price?
$140 [using 280 − 140]; $143.45

46. About how many stereos could be purchased for $600?   4 [using 600 ÷ 150]

47. Estimate the total cost of purchasing 5 television sets. $1500 [using 5 × 300]

ANSWERS: 1. 75.1 2. 6.46 3. 247.53 4. 137.21 5. 117.025 6. 50.691 7. 32.062 8. 447.157 9. 4339.2573 10. $185.39 11. $4371.56 12. $8.17 13. $25.75 14. 7.6 mi 15. 15.13 mi 16. $730.79 17. $2138.57 18. 10.312 19. 11,640.7179; eleven thousand six hundred forty and seven thousand one hundred seventy-nine ten thousandths 20. 18.8 21. 1.83 22. 34.5 23. 100.287 24. 0.25 25. 9.047 26. 53.786 27. 24.95 28. 5.991 29. $26,501.15 30. $33.47 31. (a) $25.50 (b) $299.90 (c) $165.35 (d) $134.55 (e) $225.00 (f) $359.55 32. 3.7° 33. 441.5 mi 34. yes 35. 1680.623; one thousand six hundred eighty and six hundred twenty-three thousandths 36. (a) 12.9 milligrams (b) 17.6 milligrams 37. $x = 9.35$ ft; $y = 7.9$ ft 38. above the average by 0.5 inches 39. 2 40. 3 41. 4 42. 0 43. (a) 3.2 (b) 3.16 (c) 3.159 44. $420 [using 280 + 140]; $423.21 45. $140 [using 280 − 140]; $143.45 46. 4 [using 600 ÷ 150] 47. $1500 [using 5 × 300]

5.4 EXERCISES B

Find each sum.

1. 6.21 + 70.3   **76.51**

2. 2.59 + 1.07   **3.66**

3. 46.3 + 4.63 + 0.463
51.393

4. 4.032 + 40.32 + 403.2   **447.552**

5. 1.017 + 40.08 + 200.6   **241.697**

6. 2.5 + 6.02 + 0.009 + 23.765
32.294

7.
```
   9.01
  37.005
+  2.3
  48.315
```

8.
```
  65.028
   1.351
 609.5
+ 20.07
 695.949
```

9.
```
  121.73
    2.407
   52.1
+1035.113
 1211.35
```

Arrange the following numbers in columns and find their sum.

10. $20.00, $55.95, $1.32, $163.55   **$240.82**

11. $135.62, $827.03, $2.06, $11,427.21, $25.33
$12,417.25

Parts of monthly telephone bills are shown in Exercises 12–13. Find the total bills for the long-distance calls.

12.

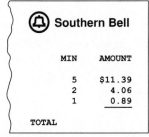

Southern Bell	
MIN	AMOUNT
5	$11.39
2	4.06
1	0.89
TOTAL	

$16.34

13.

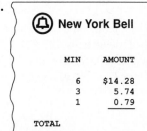

New York Bell	
MIN	AMOUNT
6	$14.28
3	5.74
1	0.79
TOTAL	

$20.81

Exercises 14–15 refer to the following map.

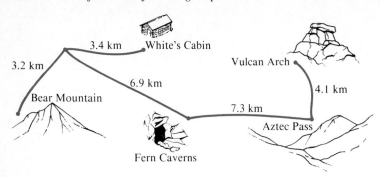

3.4 km White's Cabin

Vulcan Arch

3.2 km

6.9 km

4.1 km

Bear Mountain

7.3 km

Aztec Pass

Fern Caverns

14. Following the trails, how far is it from Vulcan Arch to White's Cabin? **21.7 km**

15. Following the trails, how far is it from Aztec Pass to Bear Mountain? **17.4 km**

Parts of bank checking account deposit slips are shown in Exercises 16–17. Find the total deposits.

16.

DEPOSIT		
	Dollars	Cents
CASH	62	09
CHECKS	477	51
	1035	20
	6	05
TOTAL		

$1580.85

17.

DEPOSIT		
	Dollars	Cents
CASH	148	75
CHECKS	10	96
	1069	47
	285	32
TOTAL		

$1514.50

18. Find the perimeter of the figure.

6.415

4.119

6.607

3.987

4.251

25.379

19. Find the sum of the following numbers: three thousand and twenty-one hundredths; four hundred two and six thousandths; fifty-six and two hundred forty-nine ten thousandths. Give the answer in decimal notation and in words.

3458.2409; three thousand four hundred fifty-eight and two thousand four hundred nine ten thousandths

Find each difference.

20. $42.1 - 31.5$ **10.6**

21. $7.05 - 6.2$ **0.85**

22. $51.0 - 6.3$ **44.7**

23. $202.5 - 2.025$ **200.475**

24. $3.00 - 1.25$ **1.75**

25. $15.055 - 2.009$ **13.046**

26.   97.317
       $- 18.529$
       78.788

27.   38.6
       $- 9.75$
       28.85

28.   8
       $- 1.003$
       6.997

Solve.

29. The Lopez family had a total income of $42,621.35 last year. If they had deductions of $2427.51, $425.30, $612.12, and $1230.07, how much taxable income did they have? **$37,926.35**

30. Michelle plans a trip of 1285.5 miles. If she drove 195.6 mi on Monday, 283.2 mi on Tuesday, and 641.7 mi on Wednesday, how far must she drive on Thursday to complete the trip? **165 mi**

31. Complete the missing items in the check register shown below.

Check No.	Date	Checks Issued to or Deposit Description	Amount of Check	Amount of Deposit		Balance Forward 614 30	
315	10/2	Graydon's Grocery Store	32 47		Check or Deposit		
					Balance		
	10/2	Deposit Week's Wages		176 19	Check or Deposit		
					Balance		
316	10/5	County Farm Insurance	110 59		Check or Deposit		
					Balance		

(a) $32 47
(b) $581 83
(c) $176 19
(d) $758 02
(e) $110 59
(f) $647 43

32. When Ralph had the flu he ran a temperature of 103.1° Fahrenheit. If normal body temperature is 98.6° Fahrenheit, how many degrees above normal was his temperature? **4.5°**

33. When Laurie filled the gas tank on her truck, the odometer read 19,627.2. The next time she filled it, the odometer read 19,810.1. How far had she driven? **182.9 mi**

34. Jack bought a stereo for $289.75, and paid with three $100 bills and three quarters. He received one $10 bill, one $1 bill, and one quarter in change. Was this the correct change?

No; he should have received one $10 bill and one $1 bill

35. Subtract two hundred fifty-seven and three thousandths from four thousand six and one tenth. Give the difference in decimal notation and in words.

3749.097; three thousand seven hundred forty-nine and ninety-seven thousandths

36. The times recorded for the members of the ASU 400-meter relay team were 10.5 sec, 10.8 sec, 11.2 sec, and 10.3 sec.

(a) What was the combined time for the event?
42.8 sec
(b) The team's best time prior to this race was 42.9 sec. Did the team better its time; and if so, by how much?
yes; by 0.1 sec

37. In the figure below find the length of x and y.

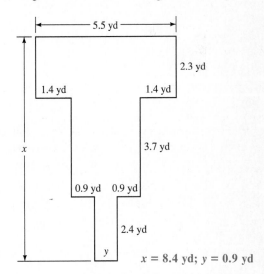

$x = 8.4$ yd; $y = 0.9$ yd

38. The table to the right shows the number of gallons of gasoline used by a delivery truck during a four-month period. During the same period last year a total of 642.3 gallons were used. How many gallons more or less were used this year than last?

Month	Number of gallons
January	132.5
February	147.8
March	201.3
April	187.6

26.9 gallons more

FOR REVIEW

Give the number of decimal places in each number.

39. 26.3   1   **40.** 4.005   3   **41.** 6.0109   4   **42.** 247   0

43. Round 6.2584 to the nearest **(a)** tenth, **(b)** hundredth, **(c)** thousandth.

          6.3   6.26   6.258

Use the following newspaper advertisements to answer Exercises 44–47.

14 Day Programmable Remote VHS Videocassette Recorder with Stereo Sound

• Automatic scan tuner
• Shuttle Search for easy location of desired program segment

$597²⁷

Microwave Oven with 600 watts of cooking power

• Microwave cookbook included

New Low Price

$178³²

44. Estimate the cost of one video recorder and one microwave oven. Then find the exact cost.
$800 [using 600 + 200]; $775.59

45. About how much more is the recorder than the oven? What is the exact difference in price?
$400 [using 600 − 200]; $418.95

46. About how many video recorders could be purchased for $3000?   **5 [using 3000 ÷ 600]**

47. Estimate the total cost of purchasing 6 microwave ovens.   **$1200 [using 6 × 200]**

5.4   EXERCISES C

Solve.

1. The average of three numbers is their sum divided by 3. Marvin owns three cats weighing 9.1 lb, 3.3 lb, and 11.6 lb. What is the average weight of his cats?
8 lb

2. Find the length of x in the figure at right if both rectangles in the figure are squares (all four sides are equal).
[Answer: 5.05 cm]

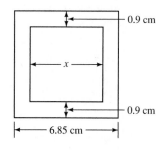

5.5 MULTIPLYING DECIMALS

1 BASIC MULTIPLICATION

Multiplying decimals is very much like multiplying whole numbers once we decide what to do with the decimal points. Suppose we multiply

$$0.7 \times 0.21$$

as fractions. We know

$$0.7 = \frac{7}{10} \quad \text{and} \quad 0.21 = \frac{21}{100}.$$

Then

$$0.7 \times 0.21 = \frac{7}{10} \times \frac{21}{100}$$

$$= \frac{7 \times 21}{10 \times 100} = \frac{147}{1000}$$

$$= 0.147.$$

Now write the problem vertically. Ignoring the decimal points, multiply as if multiplying whole numbers.

$$\begin{array}{r} 0.21 \\ \times\ 0.7 \\ \hline 147 \end{array}$$ Decimal point not yet placed

Placing the decimal point in the product depends on the number of decimal places in the numbers multiplied. The factor 0.21 has two decimal places and 0.7 has one, so the product has three decimal places, the sum of the decimal places in the factors.

$$\begin{array}{r} 0.\boxed{21} \\ \times\ 0.\boxed{7} \\ \hline 0.\boxed{147} \end{array}$$ 2 decimal places
1 decimal place
3 decimal places (2 + 1 = 3)

To Multiply Two Decimals

1. Ignore the decimal points and multiply as if the numbers were whole numbers.

2. Find the sum of the decimal places in the two factors.

3. Place the decimal point so that the product has the same number of decimal places as the sum in Step **2**. If the whole number product obtained in Step **1** has zeros on the end, they must be counted when placing the decimal point.

| EXAMPLE 1 MULTIPLYING DECIMALS | PRACTICE EXERCISE 1 |

Find each product.

Find each product.

(a) 0.6×1.7

$$
\begin{array}{r}
1.\boxed{7} \quad \text{1 decimal place} \\
\times\ 0.\boxed{6} \quad \text{1 decimal place} \\
\hline
1.\boxed{02} \quad \text{2 decimal places in product } (1 + 1 = 2)
\end{array}
$$

↑
place decimal point here

(a) 0.4×2.8

(b) 0.2×0.3

$$
\begin{array}{r}
0.3 \quad \text{1 decimal place} \\
\times\ 0.2 \quad \text{1 decimal place} \\
\hline
0.06 \quad \text{2 decimal places } (1 + 1 = 2)
\end{array}
$$

↑
**place decimal point here after inserting
a 0 to the left of the 6.**

Notice that the product is 0.06 and *not* 0.6 or 0.60.

(b) 0.3×0.3

(c) 0.6×2.45

(c) 0.4×1.35

$$
\begin{array}{r}
1.\boxed{35} \quad \text{2 decimal places} \\
\times\ 0.\boxed{4} \quad \text{1 decimal place} \\
\hline
0.\boxed{540} \quad \text{3 decimal places in product } (2 + 1 = 3)
\end{array}
$$

↑   ↑
count the zero for placing decimal point

place decimal point here

(d) 1.2×4.007

(d) 1.3×6.002

$$
\begin{array}{r}
6.\boxed{002} \quad \text{3 decimal places} \\
\times\ 1.\boxed{3} \quad \text{1 decimal place} \\
\hline
18006 \\
6002 \\
\hline
7.8026 \quad \text{4 decimal places in product } (3 + 1 = 4)
\end{array}
$$

↑
place decimal point here

(e) 10.05×23

(e) 42×11.03

$$
\begin{array}{r}
11.\boxed{03} \quad \text{2 decimal places} \\
\times\ 42 \quad \text{0 decimal places} \\
\hline
2206 \\
4412 \\
\hline
463.26 \quad \text{2 decimal places in product } (2 + 0 = 2)
\end{array}
$$

Answers: (a) **1.12**   (b) **0.09**
(c) **1.470 or 1.47**   (d) **4.8084**
(e) **231.15**

HINT

Incorrect placing of the decimal point in the answer to a problem is a common error. Estimating can give a quick check. For example, suppose we had to multiply

$$3.2 \times 21.037.$$

We could round 3.2 to 3 and 21.037 to 21. Since $3 \times 21 = 63$, we would expect the product to be somewhere near 63.

$$
\begin{array}{r}
21.037 \\
\times\ 3.2 \\
\hline
42074 \\
63101 \\
\hline
67.3084
\end{array}
$$

— decimal point *must* be placed here according to the estimated product

---

| **EXAMPLE 2** PLACING THE DECIMAL POINT BY ESTIMATION | **PRACTICE EXERCISE 2** |

Find 2.07×495.3 and place the decimal point by estimating the product.

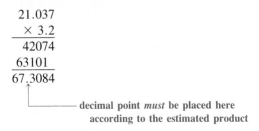

estimate

$$
\begin{array}{rcl}
495.3 & \longrightarrow & 500 \\
\times\ 2.07 & \longrightarrow & \times\ 2 \\
\hline
34671 & & 1000 \leftarrow \text{estimated product} \\
99060 & & \\
\hline
1025.271 & &
\end{array}
$$

— since estimated product is 1000, decimal point must be placed here

Notice that counting decimal places would give the same placement. The work shown in color should be done mentally.

Find 29.1×195.07 and place the decimal point by estimating the product.

Answer: 5676.537 [using $30 \times 200 = 6000$, the decimal point must be placed as shown]

---

② MULTIPLYING DECIMALS BY A POWER OF 10

Multiplying decimals by a power of 10 (10, 100, 1000, 10,000, and so forth), can be simplified by merely moving the decimal point. For example, consider the two products below.

$$
\begin{array}{rr}
12.428 & 12.428 \\
\times\ 10 & \times\ 1000 \\
\hline
124.280 & 12,428.000
\end{array}
$$

place decimal point here   place decimal point here

When we multiplied 12.428 by 10, the product 124.28 could be obtained by moving the decimal point one place to the right in 12.428. Similarly, the product of 12.428 and 1000 could be obtained by moving the decimal point three places to the right. In general, we have the following rule.

To Multiply a Decimal by a Power of 10
Move the decimal point to the *right* the same number of decimal places as the number of zeros in the power of 10.

EXAMPLE 3 MULTIPLYING BY POWERS OF 10

Find each product.

(a) 2.135×10

Since there is one zero in 10, move the decimal point in 2.135 one place to the right.

$$2.135 \times 10 = 21{,}35$$
$$\text{1 zero} \quad \text{right 1 place}$$

(b) 2.135×1000

Move the decimal point 3 places to the right.

$$2.135 \times 1000 = 2135{,}$$
$$\text{3 zeros} \quad \text{right 3 places}$$

(c) $2.135 \times 100{,}000$

Move the decimal point 5 places to the right.

$$2.135 \times 100{,}000 = 213500{,} = 213{,}500$$
$$\text{5 zeros} \quad \text{right 5 places}$$

Notice that we needed to attach two zeros when moving the decimal point.

PRACTICE EXERCISE 3

Find each product.

(a) 10×48.215

(b) 100×48.215

(c) $10{,}000 \times 48.215$

Answers: (a) **482.15** (b) **4821.5**
(c) **482,150**

❸ CONVERTING DOLLARS TO CENTS

A familiar application of multiplying by 100 involves converting dollars to cents. For example,

$1.98 converted to cents is 198¢.

Multiplying 1.98 by 100 gives 198.

To Convert Dollars to Cents
Discard the $ sign, move the decimal point two places to the right (multiply by 100), and attach a ¢ sign to the right of the result.

EXAMPLE 4 CONVERTING DOLLARS TO CENTS

Convert from dollars to cents.

(a) $45.35 Discard the $ sign, move the decimal point two places to the right, and attach the ¢ symbol. The result is 4535¢.

(b) $0.25 = 25¢

PRACTICE EXERCISE 4

Convert from dollars to cents.

(a) $295.39

(b) $0.30

Answers: (a) **29,539¢** (b) **30¢**

❹ MULTIPLYING DECIMALS BY 0.1, 0.01, AND 0.001

When a decimal is multiplied by 0.1, 0.01, or 0.001, the product can also be found by moving the decimal point, this time to the left.

To Multiply a Decimal by 0.1, 0.01, or 0.001

Move the decimal point to the *left* the same number of decimal places as decimal places in 0.1, 0.01, or 0.001.

EXAMPLE 5 MULTIPLYING BY 0.1, 0.01, AND 0.001

Find the product.

(a) 0.1 × 2.135

Since 0.1 has 1 decimal place, the decimal point in 2.135 is moved 1 place to the left.

0.1 × 2.135 = 0.2135

1 decimal place   left 1 place

2.135	3 decimal places
0.1	1 decimal place
0.2135	3 + 1 = 4 decimal places

The product to the right shows the accuracy of this method.

(b) 0.01 × 2.135

Move the decimal point 2 places to the left.

0.01 × 2.135 = 0.02135

2 decimal places   left 2 places

Notice that an extra zero was attached.

(c) 0.001 × 2.135

Move the decimal point 3 places to the left.

0.001 × 2.135 = 0.002135

3 decimal places   left 3 places

This time two zeros were supplied.

PRACTICE EXERCISE 5

Find each product.

(a) 0.1 × 48.215

(b) 0.01 × 48.215

(c) 0.001 × 48.215

Answers: (a) **4.8215** (b) **0.48215** (c) **0.048215**

❺ CONVERTING CENTS TO DOLLARS

Multiplying by 0.01 converts from cents to dollars. For example,

198¢ converted to dollars is $1.98.

Multiplying 198 by 0.01 gives 1.98.

To Convert Cents to Dollars

Discard the ¢ sign, move the decimal point two places to the left (multiply by 0.01), and attach a $ sign at the left of the result.

EXAMPLE 6 CONVERTING CENTS TO DOLLARS

Convert from cents to dollars.

(a) 7955¢   Discard the ¢ sign, move the decimal point two places to the left (remember that 7955 is the same as 7955.), and attach a $ sign. The result is $79.55.

(b) 8¢ = $0.08

Notice that the extra zeros are supplied in the result.

PRACTICE EXERCISE 6

Convert from cents to dollars.

(a) 12,995¢

(b) 15¢

Answers: (a) **$129.95**   (b) **$0.15**

---

⑥ APPLICATIONS MULTIPLYING DECIMALS

Many applied problems are solved by multiplying decimals.

EXAMPLE 7 MULTIPLYING DECIMALS IN A CONSUMER PROBLEM

Bob bought 8 books at a cost of $2.95 each. How much did he spend?
   Multiply $2.95 by 8.

$$\begin{array}{r} \$2.95 \\ \times\ 8 \\ \hline \$23.60 \end{array}$$

2 decimal places
0 decimal places
2 decimal places (2 + 0 = 2)

Thus, Bob spent $23.60 for the books. Since 8 books at $3 each would cost $24 [24 = 8 × 3], our work appears correct.

PRACTICE EXERCISE 7

Shawn Herman makes a car payment of $324.30 per month for 36 months. How much will he pay altogether?

Answer: **$11,674.80**

EXAMPLE 8 CALCULATING WEEKLY EARNINGS

Alberta is paid $7.23 per hour. Her hourly time card for one week is shown in Figure 5.7. How much was she paid that week?
   To find out how much she was paid, we need to know the total number of hours that she worked.

Day	Hours worked
Mon	5.7
Tues	3.25
Wed	4.9
Thur	6.5
Fri	7.8

$$\begin{array}{r} 5.7 \\ 3.25 \\ 4.9 \\ 6.5 \\ +\ 7.8 \\ \hline 28.15 \end{array}$$

Figure 5.7

Since she worked 28.15 hours at $7.23 per hour, multiply.

$$\begin{array}{r} 28.15 \\ \times\ 7.23 \\ \hline 8445 \\ 5630\ \ \\ 19705\ \ \ \\ \hline 203.5245 \end{array}$$

Rounded to the nearest cent, Alberta earned $203.52. Since 30 hours at $7 an hour would result in $210, our work appears to be correct.

PRACTICE EXERCISE 8

A used car salesman earns a salary of 0.04 times his total sales. During one week, he sold three cars for $2450.75, $3675.40, and $4425.35, respectively. How much did he earn that week?

Answer: **$422.06**

| **EXAMPLE 9 MULTIPLYING DECIMALS IN A PHYSICS PROBLEM** | **PRACTICE EXERCISE 9** |

The air pressure at sea level is 14.7 pounds per square inch of surface area. The surface area of the body of a man is approximately 2225 square inches. To the nearest pound, what is the total air pressure on his body at sea level?

Multiply the pressure per square inch by the total number of square inches of surface area.

$$
\begin{array}{r}
2225 \\
\times\ 14.7 \\
\hline
15575 \\
8900 \\
2225 \\
\hline
32707.5
\end{array}
$$

$0 + 1 = 1$ decimal place

Thus, the total pressure on the man's body (to the nearest pound) is 32,708 pounds, a result that may surprise you.

Rounded to the nearest cent, what is the cost in dollars and cents of 22.5 gallons of gasoline at 131.9¢ per gallon?

Answer: $29.68

5.5 EXERCISES A

Find the products. Check placement of the decimal point by estimating the product.

1. 14.03
$\times\ 0.5$
7.015

2. 6.58
$\times\ 1.2$
7.896

3. 12.05
$\times\ 2.3$
27.715

4. 3.004
$\times\ 1.7$
5.1068

5. 5.107
$\times\ 8.4$
42.8988

6. 4.0108
$\times\ 2.6$
10.42808

7. 12.411
$\times\ 5.4$
67.0194

8. 12.28
$\times\ 13$
159.64

Without making actual computations, use an estimate to decide which of the given answers is correct.

9. 4.1×39.047 **(a)** 1.60093 **(b)** 16.0093 **(c)** 160.093 **(d)** 1600.93
(c)

10. 2.15×403.7 **(a)** 8.67955 **(b)** 86.7955 **(c)** 867.955 **(d)** 8679.55
(c)

11. 32.3×48.1 **(a)** 1553.63 **(b)** 15,536.3 **(c)** 155,363.0 **(d)** 1,553,630.0
(a)

12. 203.475×9.407 **(a)** 191.409 **(b)** 1914.09 **(c)** 19,140.9 **(d)** 191,409
(b)

Find the products.

13. 15.237×10 152.37

14. 15.237×100 1523.7

15. $15.237 \times 10,000$ 152,370

16. 15.237×0.1 1.5237

17. 15.237×0.01 0.15237

18. 15.237×0.001 0.015237

Convert from dollars to cents.

19. $5.45   545¢

20. $21.98   2198¢

21. $0.79   79¢

Convert from cents to dollars.

22. 250¢ $2.50

23. 7936¢ $79.36

24. 4¢ $0.04

Solve.

25. Joe bought 5 paperback books for $2.75 each. How much did he spend?   **$13.75**

26. Carlos is paid $8.19 per hour. His timecard for one week is shown in the figure. How much was he paid that week?

Day	Hours worked
Mon	6.3
Tues	4.7
Wed	8.2
Thur	6.9
Fri	7.4

$274.37

27. The EPA-estimated miles per gallon for a midsized car is 21.8 mpg. The capacity of its gas tank is 19.4 gallons. To the nearest mile, what is the driving range of the car?   **423 mi**

28. The air pressure at sea level is 14.7 pounds per square inch of surface area. The surface area of a student's body is approximately 2085 square inches. To the nearest pound, what is the total air pressure on his body at sea level?   **30,650 pounds**

29. Pedro buys 12.5 pounds of peaches at $0.48 per pound and 9.6 pounds of hamburger at $1.37 per pound. How much change will he receive if he pays with a check for $25.00?   **$5.85**

30. Find the area of a rectangle, to the nearest hundredth of a square centimeter, if its length is 14.38 cm and its width is 9.41 cm.   **135.32 sq cm**

31 Barbara plans to cover a rectangular wall with vinyl covering costing $0.67 a sq ft. If the wall is 13.5 ft long and 7.6 ft high, how much will the project cost?   **$68.74**

32. Steve Hellmann's contract specifies that he be paid $14.38 per hour for a 40-hour work week. For all hours over 40, he must be paid time and a half (1.5 times the normal hourly rate). Last week Mr. Hellmann worked 51 hours. How much was he paid?   **$812.47**

33. It costs $17.50 a day plus $0.18 per mile to rent a car at Cheapo Car Rentals. How much would it cost to rent the car for two days if it is driven a total of 240 miles?   **$78.20**

34. Dave worked 41 hours one week and was paid $7.87 an hour. What was his approximate pay for the week? His exact pay?
$320 [using 40 × 8]; $322.67

35 Herman wishes to buy four new tires for his car. He is interested in the tires shown in the advertisement to the right. What would be the approximate cost of the tires? The exact cost?
$160 [using 4 × 40]; $155.00

All Season Radials
with White Walls

EVERYDAY LOW PRICES
$38^{75}

FOR REVIEW

36. Mr. and Mrs. Bonnett stayed at the Frontier Hotel in Las Vegas on a 3-day/2-night package. Their room cost $139.95, and they charged $37.28 in food and beverage to their room account. If Mr. Bonnett paid the bill with two $100 bills, how much change did he receive?   **$22.77**

37. Find the perimeter of the garden sketched in the figure below.

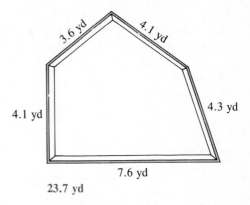

3.6 yd
4.1 yd
4.1 yd
4.3 yd
7.6 yd
23.7 yd

38. On a five-day trip, a family traveled 2450.3 miles. On the first day they traveled 233.2 mi, on the second 537.3 mi, on the third 641.4 mi, and on the fourth 596.3 mi.

(a) How far did they travel on the fifth day?
442.1 mi

(b) What is the answer to **(a)** rounded to the nearest mile?
442 mi

(c) What is the answer to **(a)** if all distances are rounded to the nearest mile before any calculation?   443 mi

(d) Compare the answers to **(b)** and **(c).**
Since answers differ, it is better not to round until the last step (unless you are estimating an answer).

ANSWERS: 1. 7.015 2. 7.896 3. 27.715 4. 5.1068 5. 42.8988 6. 10.42808 7. 67.0194 8. 159.64 9. (c) 10. (c) 11. (a) 12. (b) 13. 152.37 14. 1523.7 15. 152,370 16. 1.5237 17. 0.15237 18. 0.015237 19. 545¢ 20. 2198¢ 21. 79¢ 22. $2.50 23. $79.36 24. $0.04 25. $13.75 26. $274.37 27. 423 mi 28. 30,650 pounds 29. $5.85 30. 135.32 sq cm 31. $68.74 32. $812.47 33. $78.20 34. $320 [using 40 × 8]; $322.67 35. $160 [using 4 × 40]; $155.00 36. $22.77 37. 23.7 yd 38. (a) 442.1 mi (b) 442 mi (c) 443 mi (d) Since answers differ, it is better not to round until the last step (unless you are estimating an answer).

5.5 EXERCISES B

Find the products. Check placement of the decimal point by estimating the product.

1. 21.02
   × 0.7
   14.714

2. 2.59
   × 1.5
   3.885

3. 14.06
   × 5.4
   75.924

4. 2.001
   × 1.2
   2.4012

5. 3.109
   × 5.6
   17.4104

6. 3.0109
   × 4.3
   12.94687

7. 13.611
   × 3.2
   43.5552

8. 14.56
   × 18
   262.08

Without making actual computations, use an estimate to decide which of the given answers is correct.

9. 6.2 × 51.075   **(a)** 3.16665   **(b)** 31.6665   **(c)** 316.665   **(d)** 3166.65
   (c)

10. 4.25 × 291.8   **(a)** 1.24015   **(b)** 12.4015   **(c)** 124.015   **(d)** 1240.15
   (d)

11. 41.5 × 62.3   **(a)** 2585.45   **(b)** 258.545   **(c)** 25.8545   **(d)** 2.58545
   (a)

12. 195.625 × 3.508   **(a)** 68.62525   **(b)** 686.2525   **(c)** 6862.525   **(d)** 68625.25
   (b)

Find the products.

13. 341.59 × 10   **3415.9**

14. 341.59 × 1000   **341,590**

15. 341.59 × 100   **34,159**

16. 341.59 × 0.1   **34.159**

17. 341.59 × 0.001   **0.34159**

18. 341.59 × 0.01   **3.4159**

Convert from dollars to cents.

19. $8.23   **823¢**

20. $45.15   **4515¢**

21. $0.67   **67¢**

Convert from cents to dollars.

22. 705¢   **$7.05**

23. 2009¢   **$20.09**

24. 12¢   **$0.12**

Solve.

25. Dyane bought 6 greeting cards at $1.25 each. How much did she spend?   **$7.50**

26. Peter is paid $11.53 per hour. His hourly time card for one week is shown below. How much was he paid that week (to the nearest cent)?

Day	Hours worked
Mon	7.2
Tues	6.5
Wed	8.1
Thur	6.9
Fri	5.4

$393.17

27. What is the driving range (to the nearest mile) of a car with EPA-estimated miles per gallon of 26.2 and a gas tank with capacity of 14.5 gallons?
   380 miles

28. The air pressure at sea level is 14.7 pounds per square inch of surface area. If the surface area of a young child is 1125 square inches, to the nearest pound, what is the total air pressure on his body at sea level?   **16,538 pounds**

29. Rosa buys 11.3 pounds of bananas at $0.35 per pound and 8.2 pounds of ground meat for $1.09 per pound. How much change will she receive if she pays with a $20 bill?   **$7.11**

30. Find the area of a rectangle, to the nearest tenth of a square foot, if its length is 6.34 ft and its width is 2.59 ft.   **16.4 sq ft**

31. What is the price of a mirror measuring 5.2 ft long and 3.7 ft wide if the cost per square foot is $0.85? **$16.35**

32. Ms. Morgan is paid $12.35 per hour for a 40-hour work week. For all hours over 40 she is paid time-and-a-half as overtime. If she worked 53 hours one week, how much was she paid?   **$734.83**

33. It costs $18.25 a day plus $0.22 per mile to rent a car at Local Car Rentals. How much would it cost to rent the car for three days if it is driven 395 miles? **$141.65**

34. It costs $12.05 per hour to operate a particular machine. What is the approximate cost of running the machine for 48 hours? The exact cost?
$600 [using 12 × 50]; $578.40

35. Mr. and Mrs. Washington decide to purchase a vacuum cleaner, advertised at right, for each of their five children. What would be the approximate cost of the cleaners? The exact cost?
$300 [using 5 × 60]; $299.40

$59⁸⁸

Upright Vacuum Cleaner with
12″ Beater Bar Brush

FOR REVIEW

36. The Bennings spent three nights at the Desert Oasis Hotel in Palm Springs. The price of the room was $105.50 per night, and they charged $62.39 in food and beverages to their room. If Mr. Benning paid the bill with four $100 bills, how much change did he receive?   **$21.11**

37. Find the perimeter of the lot sketched in the figure.

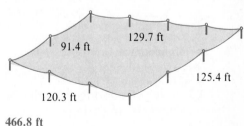

129.7 ft

91.4 ft

125.4 ft

120.3 ft

466.8 ft

38. On a four-day vacation, a family spent a total of $1123.06. They spent $418.19 on the first day, $176.43 on the second day, and $212.05 on the third day.

    (a) How much did they spend on the fourth day?
        $316.39

    (b) What is the answer to **(a)** rounded to the nearest dollar?   **$316**

    (c) What is the answer to **(a)** if all amounts are rounded to the nearest dollar before making any calculations?   **$317**

    (d) Compare the answers to **(b)** and **(c)**.
        Since answers differ, we should never round until the last step (unless you are estimating an answer).

5.5 EXERCISES C

Solve.

1. The cost of heating fuel is 98.9¢ per gallon. To the nearest cent, how much will 38.5 gallons cost? [Answer: $38.08]

2. Terry McGinnis plans to carpet her family room with carpeting which costs $18.95 a square yard. If the room is rectangular in shape, 7.2 yards long and 4.9 yards wide, what would be the approximate cost of the job? The exact cost?
$700 [using 7 × 5 × 20]; $668.56

3. The Hagoods must pay $8.35 in property tax for every $1000 of assessed value on their home. How much tax do they pay if the assessed value of their home is $102,000?   **$851.70**

4. Find the area of the region below by dividing the region into two rectangles.

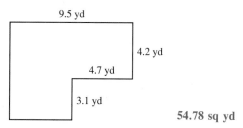

54.78 sq yd

5.6 DIVIDING DECIMALS

STUDENT GUIDEPOSTS

1 Dividing a Decimal by a Whole Number

2 Dividing a Decimal by a Decimal

3 Finding Quotients Rounded to a Given Decimal Place

4 Dividing Decimals by a Power of 10

5 Dividing Decimals by 0.1, 0.01, and 0.001

6 Quotients With a Repeating Block of Digits

7 Finding Quotients of Two Whole Numbers

8 Applications Dividing Decimals

❶ DIVIDING A DECIMAL BY A WHOLE NUMBER

Dividing decimals is much like dividing whole numbers, except for placing the decimal point. Converting fractions to decimals as was done in Section 5.2 is a special case of this division process.

Suppose we divide

$$2.7 \div 3$$

using fractions. We know

$$2.7 = 2\frac{7}{10} = \frac{27}{10}.$$

Then

$$2.7 \div 3 = \frac{27}{10} \div \frac{3}{1}$$

$$= \frac{27}{10} \times \frac{1}{3}$$

$$= \frac{9 \cdot \cancel{3} \cdot 1}{10 \cdot \cancel{3}}$$

$$= \frac{9}{10}$$

$$= 0.9.$$

Thus, 2.7 divided by 3 must be 0.9. Now, suppose we write the quotient as follows.

$$3\overline{)2.7}$$

Ignoring the decimal point and dividing as we divided whole numbers gives the following result.

$$
\begin{array}{r}
9 \\
3\overline{)2.7} \\
\underline{2\ 7} \\
0
\end{array}
\quad \text{Decimal point not yet placed}
$$

Moving the decimal point in the dividend (2.7) straight up results in the desired quotient, 0.9.

$$
\begin{array}{r}
.9 \\
3\overline{)2{\uparrow}7} \\
\underline{2\ 7} \\
0
\end{array}
$$

This illustrates a general procedure.

To Divide a Decimal by a Whole Number
1. Place a decimal point directly above the decimal point in the dividend.
2. Divide as if both numbers were whole numbers.

EXAMPLE 1 DIVIDING DECIMALS BY A WHOLE NUMBER	PRACTICE EXERCISE 1

Find the quotients.

(a) $1.12 \div 7$

Place a decimal point in the quotient directly above the decimal point in the dividend, then divide as if the dividend were a whole number.

$$
\begin{array}{r}
.16 \\
7\overline{)1.12} \\
\underline{7} \\
42 \\
\underline{42} \\
0
\end{array}
$$

The quotient is 0.16.

(b) $11.34 \div 21$

$$
\begin{array}{r}
.54 \\
21\overline{)11.34} \\
\underline{10\ 5} \\
84 \\
\underline{84} \\
0
\end{array}
$$
Place decimal point
Divide as if whole numbers

The quotient is 0.54.

(c) $113.95 \div 43$

$$
\begin{array}{r}
2.65 \\
43\overline{)113.95} \\
\underline{86} \\
27\ 9 \\
\underline{25\ 8} \\
2\ 15 \\
\underline{2\ 15} \\
0
\end{array}
$$
Place decimal point and divide

The quotient is 2.65.

Find the quotients.

(a) $1.08 \div 4$

(b) $15.04 \div 32$

(c) $149.15 \div 19$

Answers: **(a)** 0.27 **(b)** 0.47 **(c)** 7.85

❷ DIVIDING A DECIMAL BY A DECIMAL

Division problems with decimals for both the dividend and the divisor can be changed into problems with whole number divisors. For example, suppose we write the division problem

$$0.35\overline{)1.498}$$

as a fraction.

$$\frac{1.498}{0.35}$$

Multiplying numerator and denominator by 100

$$\frac{1.498 \times 100}{0.35 \times 100} = \frac{149.8}{35}$$

Since the first digit to the right of the hundredths position is 3, the quotient is approximately 15.94, rounded to the nearest hundredth. That is,

$$2.551 \div 0.16 \approx 15.94.$$

Answers: (a) 0.7   (b) 14.57

HINT

In Section 5.5 we placed decimals and checked decimal products by estimating the answer. The same technique can be used when dividing decimals.

EXAMPLE 4   ESTIMATING TO LOCATE DECIMAL PLACE

Find $156.3 \div 12$ and place the decimal point by estimating the answer.
Since the quotient is about $150 \div 10$, the result should be about 15.

$$
\begin{array}{r}
13\overset{\displaystyle\downarrow}{.}025 \\
12\overline{)156.300} \\
\underline{12} \\
36 \\
\underline{36} \\
30 \\
\underline{24} \\
60 \\
\underline{60} \\
0
\end{array}
$$

decimal point *must* be placed here
according to estimated quotient

PRACTICE EXERCISE 4

Find $17.064 \div 8$ and place the decimal point by estimating the quotient.

Answer: 2.133 [Since 17.064 is about 16 and $16 \div 8$ is 2, the quotient must be about 2.]

---

④ DIVIDING DECIMALS BY A POWER OF 10

The quotient in Example 4 was approximated by dividing 150 by 10. Dividing a decimal by 10 is the same as multiplying the decimal by 0.1 since

$$150 \div 10 = 150 \times \frac{1}{10} = 150 \times 0.1 \,.$$

As a result, from Section 5.5, dividing by 10 (multiplying by 0.1) is simply a matter of moving the decimal point one place to the left. Similar rules apply when dividing by any power of 10 such as 10, 100, 1000, or 10,000.

To Divide a Decimal by a Power of 10

Move the decimal point to the *left* the same number of decimal places as the number of zeros in the power of 10.

EXAMPLE 5 DIVIDING DECIMALS BY A POWER OF 10

Find each quotient.

(a) $43.21 \div 100$

Since there are 2 zeros in 100, move the decimal point in 43.21 two places to the left.

$$43.21 \div 100 = 0.4321$$
2 zeros   left 2 places

(b) $43.21 \div 1000$ Move the decimal point 3 places to the left.

$$43.21 \div 1000 = 0.04321$$
3 zeros   left 3 places

Notice that an extra zero was needed when moving the decimal point.

(c) $43.21 \div 100,000$

Move the decimal point 5 places to the left.
$$43.21 \div 100,000 = 0.0004321$$

PRACTICE EXERCISE 5

Find each quotient.

(a) $652.5 \div 100$

(b) $652.5 \div 1000$

(c) $652.5 \div 100,000$

Answers: (a) **6.525** (b) **0.6525** (c) **0.006525**

---

⑤ DIVIDING DECIMALS BY 0.1, 0.01, AND 0.001

Suppose we divide 1.235 by 0.1. Dividing by 0.1 is the same as multiplying by 10.

$$1.235 \div 0.1 = 1.235 \div \frac{1}{10} = 1.235 \times \frac{10}{1} = 1.235 \times 10$$

As a result, from Section 5.5, dividing by 0.1 (multiplying by 10) is the same as moving the decimal point one place to the right. Similar rules apply when dividing by 0.01 or 0.001. This is summarized in the following rule.

To Divide a Decimal by 0.1, 0.01, or 0.001

Move the decimal point to the *right* the same number of decimal places as decimal places in 0.1, 0.01, or 0.001.

EXAMPLE 6 DIVIDING DECIMALS BY 0.1, 0.01, OR 0.001

Find each quotient.

(a) $54.32 \div 0.1$

Since 0.1 has one decimal place, move the decimal point in 54.32 one place to the right.

$$54.32 \div 0.1 = 543.2$$
right 1 place

PRACTICE EXERCISE 6

Find each quotient.

(a) $439.7 \div 0.1$

(b) $54.32 \div 0.001$

Move the decimal point three places to the right since 0.001 has three decimal places.

$$54.32 \div 0.001 = 54320.$$

right 3 places

Notice that a zero must be attached to obtain the quotient 54,320.

(b) $439.7 \div 0.01$

Answers: (a) **4397** (b) **43.970**

HINT

The difference between multiplying a decimal by a power of 10 and dividing by a power of 10 is simply reversing the direction we move the decimal point. This is easy to remember since multiplication and division are *opposite* processes. The same remarks apply to multiplying and dividing by 0.1, 0.01, or 0.001.

---

⑥ QUOTIENTS WITH A REPEATING BLOCK OF DIGITS

Some division problems result in quotients which have a repeating block of digits, as mentioned in Section 5.2. When this repetition is discovered, stop the process and place a bar over the repeating digits.

EXAMPLE 7   A QUOTIENT WITH A REPEATING BLOCK OF DIGITS

Find $2.19 \div 0.99$.

$$
\begin{array}{r}
2.2121 \\
0.99\overline{)2.19.0000} \\
\underline{1\,98} \\
21\,0 \\
\underline{19\,8} \\
1\,20 \\
\underline{99} \\
210 \\
\underline{198} \\
120 \\
\underline{99} \\
21
\end{array}
$$

We can see that the digits 21 will continue to repeat after the decimal point. Thus,

$$2.19 \div 0.99 = 2.\overline{21}.$$

PRACTICE EXERCISE 7

Find $1.74 \div 0.33$.

Answer: $5.\overline{27}$

---

⑦ FINDING QUOTIENTS OF TWO WHOLE NUMBERS

When one whole number is divided by another, the quotient is sometimes given as a decimal rounded to a particular degree of accuracy.

EXAMPLE 8 DIVIDING WHOLE NUMBERS

Find $431 \div 29$, rounded to the nearest hundredth.

```
           14.862
      29)431.000    Add three zeros
         29
        ──
        141
        116
        ───
         25 0
         23 2
         ────
          1 80
          1 74
          ────
            60
            58
            ──
             2
```

Since the first digit to the right of the hundredths position is 2, the quotient is approximately 14.86, rounded to the nearest hundredth.

PRACTICE EXERCISE 8

Find $243 \div 21$, rounded to the nearest tenth.

Answer: 11.6

8 APPLICATIONS DIVIDING DECIMALS

Many applied problems are solved by dividing one decimal by another. As always, estimate the answer and ask yourself if the actual answer seems reasonable.

EXAMPLE 9 CALCULATING MONTHLY PAYMENTS

How many months will it take to pay off a loan of $894.18 if the monthly payments are $42.58?

Divide the total to be paid ($894.18) by the amount paid each month ($42.58).

```
               21.
      42.58)894.18.
           851 6
           ─────
            42 58
            42 58
            ─────
                0
```

Thus, it will take 21 months to pay off the loan. Since $900 \div 40 = 22.5$, this answer seems reasonable.

PRACTICE EXERCISE 9

Fern took out a loan for $7752.96 which is to be paid off in 36 equal monthly payments. How much is each payment?

Answer: $215.36

EXAMPLE 10 FINDING A GRADE POINT AVERAGE

Finding the grade point average (GPA) of a student is an important application of decimals. Calculate a GPA using a 4-point scale, that is, a perfect GPA (every grade an A) is 4.00. Each letter grade is assigned a value: A = 4, B = 3, C = 2, D = 1, and F = 0. The first step is to find the *course value,* that is, the product of the credits and the grade value, as shown in Figure 5.8.

PRACTICE EXERCISE 10

Burford earned a C in a 4-credit English course, a D in a 3-credit biology course, a B in a 2-credit physical education course, and an F in a 4-credit history course. Calculate Burford's GPA (rounded to the nearest hundredth) on a 4-point scale.

Course	Credits	Grade	Value	Course Value
Math	4	B	3	$4 \cdot 3 = 12$
English	3	A	4	$3 \cdot 4 = 12$
Music	2	A	4	$2 \cdot 4 = 8$
History	3	C	2	$3 \cdot 2 = 6$
Physics	3	D	1	$3 \cdot 1 = 3$
Totals	15			41

Figure 5.8

$$\text{GPA} = \frac{\text{sum of course values}}{\text{total credits}} = \frac{41}{15} = 2.733$$

The student's GPA, rounded to the nearest hundredth, is 2.73.

Answer: **1.31**

5.6 EXERCISES A

Find the quotients.

1. $6\overline{)5.4}$ 0.9

2. $8\overline{)18.4}$ 2.3

3. $12\overline{)43.2}$ 3.6

4. $25\overline{)171.5}$ 6.86

5. $1.4\overline{)1.12}$ 0.8

6. $1.4\overline{)0.112}$ 0.08

7. $0.27\overline{)0.378}$ 1.4

8 $0.31\overline{)2.232}$ 7.2

9. $4.2\overline{)8.988}$ 2.14

10. $0.015\overline{)0.0945}$ 6.3

11. $1.6\overline{)56}$ 35

12. $25\overline{)69}$ 2.76

Find the quotient, rounded to the nearest tenth.

13. $1.9\overline{)2.36}$ 1.2

14. $48\overline{)315}$ 6.6

15. $1.02\overline{)3.185}$ 3.1

Find the quotient, rounded to the nearest hundredth.

16. $1.9\overline{)2.36}$ 1.24

17 $48\overline{)315}$ 6.56

18. $1.02\overline{)3.185}$ 3.12

Without making actual calculations, use an estimate to decide which of the given answers is correct.

19. $243.41 \div 19$ **(a)** 1.28111   **(b)** 12.8111   **(c)** 128.111   **(d)** 1281.11
(b)

20. $625.03 \div 25.1$ **(a)** 24.9016   **(b)** 2.49016   **(c)** 2490.16   **(d)** 24,901.6
(a)

21. $357.103 \div 62.8$ **(a)** 0.56864   **(b)** 5.6864   **(c)** 56.864   **(d)** 568.64
(b)

22. $403.229 \div 6.417$ **(a)** 0.062838   **(b)** 0.62838   **(c)** 6.2838   **(d)** 62.838
(d)

Find the quotients.

23. $32.71 \div 10$ **3.271**

24. $41.38 \div 100$ **0.4138**

25. $5.329 \div 10,000$ **0.0005329**

26. $621.43 \div 100,000$ **0.0062143**

27. $8.319 \div 1000$ **0.008319**

28. $0.0327 \div 0.1$ **0.327**

29. $426.5 \div 0.001$ **426,500**

30. $0.627 \div 0.1$ **6.27**

31. $4.398 \div 0.01$ **439.8**

Find the quotient. (Each quotient is a repeating decimal.)

32 $1.8\overline{)21.2}$ **11.$\overline{7}$**

33. $1.05\overline{)2.135}$ **2.0$\overline{3}$**

34. $0.36\overline{)2.58}$ **7.1$\overline{6}$**

Solve.

35. How many months will it take to pay off a loan of $1176.30 if the monthly payments are $65.35?
18 months

36. The Rappaports used 260 stepping stones to build a patio. If the total cost of the stones was $107.15, how much did each stone cost (to the nearest tenth of a cent)?   **41.2¢ or $0.412**

37 The assessed value of a parcel of property can be found by dividing the amount of tax by the tax rate. Margaret's property was taxed at a rate of 0.147, and she paid a property tax of $382.40. Find the assessed value of her property.   **$2601.36**

38. The balance owed on three rooms of furniture amounts to $3157.42. If Tana wishes to pay off the account in 24 equal monthly payments, how much should each payment be?   **$131.56**

39. The following table shows Sam's grades during one semester. Complete the table and calculate his GPA, correct to the nearest hundredth.

Course	Credits	Grade	Value	Course Value
History	3	D	1	
English	3	A	4	
Algebra	5	B	3	
Chemistry	3	F	0	
Art	2	C	2	

GPA = 2.13

40. Which is the better buy, 9 apples for $1.62 or 13 apples for $2.21?
13 apples at 17¢ each is better than 9 apples at 18¢ each

FOR REVIEW

Find the products.

41.
$$\begin{array}{r} 3.02 \\ \times\ 0.5 \\ \hline 1.51 \end{array}$$

42.
$$\begin{array}{r} 7.38 \\ \times\ 1.2 \\ \hline 8.856 \end{array}$$

43.
$$\begin{array}{r} 7.011 \\ \times\ 2.01 \\ \hline 14.09211 \end{array}$$

44. 691.35×100 69,135   **45.** 691.35×0.1 69.135   **46.** 691.35×0.01 6.9135

47. Convert \$12.45 to cents.   1245¢

48. Convert 17¢ to dollars.   \$0.17

49. Marvin bought 8 complete backpacking meals at a cost of \$6.78 each. If tax on the purchases amounted to \$2.17, and he paid with three \$20 bills, how much change did he receive?   \$3.59

50. It costs \$98.00 a week plus 16.5¢ (\$0.165) per mile to rent a car at Horn's Rentals. How much would it cost to rent the car for one week if it is driven a total of 683 miles?   \$210.70

51. A blocklayer earns \$13.78 per hour for a 40-hour work week and time and a half for each hour of overtime. If Pete worked 52 hours during the week laying blocks, how much was he paid?   \$799.24

52. Suppose all we knew was that Pete earned the total amount in the answer to Exercise 51 by working 52 hours. What would we say was his hourly rate?   \$15.37

The following exercises will help you prepare for the next section. Place the correct symbol ($<$ or $>$) between each pair of numbers.

53. $0 < 7$

54. $\frac{3}{2} > \frac{5}{4}$

55. $6 < \frac{25}{4}$

Perform the indicated operations.

56. $2 + 8 \div 4 - 1$ 3

57. $\left(\frac{1}{2} + \frac{1}{4}\right) \cdot 2 - 1$ $\frac{1}{2}$

58. $\sqrt{9} \div 4 + \frac{1}{4}$ 1

ANSWERS: 1. 0.9 2. 2.3 3. 3.6 4. 6.86 5. 0.8 6. 0.08 7. 1.4 8. 7.2 9. 2.14 10. 6.3 11. 35 12. 2.76 13. 1.2 14. 6.6 15. 3.1 16. 1.24 17. 6.56 18. 3.12 19. (b) 20. (a) 21. (b) 22. (d) 23. 3.271 24. 0.4138 25. 0.0005329 26. 0.0062143 27. 0.008319 28. 0.327 29. 426,500 30. 6.27 31. 439.8 32. 11.$\overline{7}$ 33. 2.0$\overline{3}$ 34. 7.1$\overline{6}$ 35. 18 months 36. 41.2¢ or \$0.412 37. \$2601.36 38. \$131.56 39. GPA = 2.13 40. 13 apples at 17¢ each is better than 9 apples at 18¢ each 41. 1.51 42. 8.856 43. 14.09211 44. 69,135 45. 69.135 46. 6.9135 47. 1245¢ 48. \$0.17 49. \$3.59 50. \$210.70 51. \$799.24 52. \$15.37 (He earned \$799.24 for 52 hours of work, so divide \$799.24 by 52.) 53. < 54. > 55. < 56. 3 57. $\frac{1}{2}$ 58. 1

5.6 EXERCISES B

Find the quotients.

1. $9\overline{)23.4}$ 2.6   **2.** $5\overline{)0.45}$ 0.09   **3.** $14\overline{)37.8}$ 2.7   **4.** $35\overline{)89.6}$ 2.56

5. $1.7\overline{)7.31}$ 4.3   **6.** $1.7\overline{)0.731}$ 0.43   **7.** $0.41\overline{)0.8487}$ 2.07   **8.** $0.52\overline{)3.536}$ 6.8

9. $5.1\overline{)116.79}$ 22.9   **10.** $0.022\overline{)0.1056}$ 4.8   **11.** $1.3\overline{)74.1}$ 57   **12.** $32\overline{)86}$ 2.6875

Find the quotient, rounded to the nearest tenth.

13. $1.8\overline{)2.47}$ 1.4   **14.** $41\overline{)365}$ 8.9   **15.** $1.05\overline{)4.172}$ 4.0

Find the quotient, rounded to the nearest hundredth.

16. $1.8\overline{)2.47}$ 1.37   **17.** $41\overline{)365}$ 8.90   **18.** $1.05\overline{)4.172}$ 3.97

Without making actual calculations, use an estimate to decide which of the given answers is correct.

19. $321.7 \div 21$　　　　　(a) 1.53190　　　　(b) 15.3190　　　　(c) 153.190　　　　(d) 1531.90
　(b)

20. $489.02 \div 92.2$　　　(a) 530.390　　　　(b) 53.0390　　　　(c) 5.30390　　　　(d) 0.0530390
　(c)

21. $449.207 \div 49.3$　　(a) 0.091117　　　(b) 0.91117　　　　(c) 9.1117　　　　(d) 91.117
　(c)

22. $702.116 \div 9.328$　(a) 0.0752697　　(b) 0.752697　　　(c) 7.52697　　　(d) 75.2697
　(d)

Find the quotients.

23. $64.85 \div 100$　　**0.6485**　　　**24.** $31.65 \div 10$　　**3.165**　　　**25.** $1.357 \div 1000$　　**0.001357**

26. $435.07 \div 10,000$　　**0.043507**　　**27.** $921.05 \div 100,000$　　**0.0092105**　　**28.** $0.0569 \div 0.1$　　**0.569**

29. $327.5 \div 0.001$　　**327,500**　　**30.** $0.0659 \div 0.01$　　**6.59**　　**31.** $2.678 \div 0.001$　　**2678**

Find the quotient. (Each quotient is a repeating decimal.)

32. $4.5\overline{)1.45}$　　**$0.3\overline{2}$**　　　　**33.** $33\overline{)189}$　　**$5.\overline{72}$**　　　　**34.** $2.16\overline{)11.16}$　　**$5.1\overline{6}$**

Solve.

35. How many months will it take to pay off a loan of $4054.68 if the monthly payments are $112.63?
36 months

36. The Giffords used 320 tiles to build a patio. If the total cost of the tiles was $45.75, how much did each tile cost (to the nearest tenth of a cent)?
14.3¢

37. The assessed value of a lot can be found by dividing the amount of tax by the tax rate. If Denise's lot was taxed at a rate of 0.151, and she paid a property tax of $468.10, find the assessed value of her property.
$3100.00

38. If Johnny used 112.6 gallons of gas on a trip of 2780 miles, to the nearest tenth, what was his mpg?
24.7 mpg

39. The following table shows Wes's grades during the fall term. Complete the table and calculate his GPA, correct to the nearest tenth.

Course	Credits	Grade	Value	Course Value
English	3	B	3	
Chemistry	4	C	2	
French	3	A	4	
Geometry	3	B	3	
History	2	F	0	

GPA = 2.53

40. Which is the better buy, a 32-ounce bottle of ketchup for $1.60 or a 48-ounce bottle for $2.88?
The 32-ounce bottle at 5¢ an ounce is a better buy than the 48-ounce bottle at 6¢ an ounce.

FOR REVIEW

Find the products

41.
$$\begin{array}{r} 6.08 \\ \times\ 0.4 \\ \hline 2.432 \end{array}$$

42.
$$\begin{array}{r} 4.41 \\ \times\ 2.3 \\ \hline 10.143 \end{array}$$

43.
$$\begin{array}{r} 2.022 \\ \times\ 3.01 \\ \hline 6.08622 \end{array}$$

44. 21.683×10 **216.83**   **45.** 21.683×0.01 **0.21683**   **46.** 21.683×1000 **21,683**

47. Convert \$3.86 to cents.   **386¢**

48. Convert 6439¢ to dollars.   **\$64.39**

49. Susan bought 7 TV dinners at a cost of \$2.79 each. If tax on the purchases amounted to \$0.98, and she paid with three \$10 bills, how much change did she receive?   **\$9.49**

50. It costs \$158.00 a week plus 18.5¢ (\$0.185) per mile to rent a motor home at Wannabegger Rentals. How much would it cost to rent a motor home for three weeks if it is driven a total of 1800 miles? **\$807.00**

51. Jess earns \$12.35 per hour for a 40-hour work week, and time and a half for each hour of overtime. How much should Jess be paid this week if he worked a total of 54 hours?   **\$753.35**

52. Suppose all we knew was that Jess earned the total amount in the answer to Exercise 51 by working 54 hours. What would we say was his hourly rate? **\$13.95**

The following exercises will help you prepare for the next section. Place the correct symbol ($<$ or $>$) between each pair of numbers.

53. $\dfrac{1}{2} > 0$

54. $\dfrac{9}{8} < \dfrac{5}{4}$

55. $3 > \dfrac{44}{15}$

Perform the indicated operations.

56. $3 + 9 \times 2 - 5$ **16**

57. $\left(\dfrac{3}{8} - \dfrac{1}{4}\right) \cdot 4 - \dfrac{1}{2}$ **0**

58. $5^2 - \sqrt{16} + 2$ **23**

5.6 EXERCISES C

Solve.

1. On his two-week summer vacation, Sam Passamonte drove 715.6 miles the first week and 527.3 miles the second week. He purchased gas three times during the vacation, the last time when he arrived home. If he left home with a full tank and bought 17.2 gallons, 21.3 gallons, and 19.7 gallons, to the nearest tenth, how many miles did he get to the gallon?
[Answer: 21.4 mpg]

2. A water tank is filled with water. The tank, when empty, weighs 187.6 pounds and, when full, weighs 839.4 pounds. If one cubic foot of water weighs 62.5 pounds, how many cubic feet of water does the tank hold? Give answer rounded to the nearest tenth of a cubic foot.   **10.4 cubic feet**

5.7 ORDER OF OPERATIONS AND COMPARING DECIMALS

STUDENT GUIDEPOSTS

1 Operations Involving Both Fractions and Decimals

2 Order of Operations for Decimals

3 Arranging Decimals in Order of Size

4 Comparing Unit Prices

❶ OPERATIONS INVOLVING BOTH FRACTIONS AND DECIMALS

In Chapters 3 and 4 we considered operations on fractions, and in this chapter we considered the same operations on decimals. At times, fractions and decimals may occur in the same problem.

| EXAMPLE 1   MULTIPLYING A FRACTION AND A DECIMAL | PRACTICE EXERCISE 1 |

Find the product. $\dfrac{3}{4} \times 4.28$

We could change $\frac{3}{4}$ to decimal form, 0.75, and then multiply.

$$
\begin{array}{r}
4.28 \\
\times\ 0.75 \\
\hline
2140 \\
2996 \\
\hline
3.2100
\end{array}
$$

Or, we could change 4.28 to fractional form,

$$4.28 = 4\frac{28}{100} = \frac{100 \cdot 4 + 28}{100} = \frac{428}{100},$$

and then multiply.

$$\frac{3}{4} \times \frac{428}{100} = \frac{3}{4} \times \frac{4 \cdot 107}{100}$$

$$= \frac{3 \cdot \cancel{4} \cdot 107}{\cancel{4} \cdot 100} = \frac{321}{100} = 3\frac{21}{100}$$

Practice Exercise 1

Find the quotient.

$$4.9 \div \frac{7}{8}$$

Answer: $5\dfrac{3}{5}$ or 5.6

In general, problems which involve both decimals and fractions may be solved either by changing all numbers to decimals or by changing all numbers to fractions. When fractions are changed to decimals we must often make approximations.

| EXAMPLE 2   ADDING A FRACTION AND A DECIMAL | PRACTICE EXERCISE 2 |

Find the sum. $\dfrac{1}{3} + 5.75$

Since $\frac{1}{3} = 0.\overline{3} = 0.333\ldots$, we can only *approximate* the sum using decimals. As a result, it might be better to change 5.75 to fractional form.

$$5.75 = 5\frac{75}{100} = 5\frac{3}{4} = \frac{4 \cdot 5 + 3}{4} = \frac{23}{4}$$

Practice Exercise 2

Find the difference.

$$2.25 - \frac{2}{3}$$

Thus,

$$\frac{1}{3} + 5.75 = \frac{1}{3} + \frac{23}{4} = \frac{1}{3} + \frac{23}{2 \cdot 2} \qquad \text{LCD} = 2 \cdot 2 \cdot 3 = 12$$

$$= \frac{2 \cdot 2 \cdot 1}{2 \cdot 2 \cdot 3} + \frac{23 \cdot 3}{2 \cdot 2 \cdot 3} \qquad \text{Supply missing numbers}$$

$$= \frac{2 \cdot 2 + 23 \cdot 3}{2 \cdot 2 \cdot 3} \qquad \text{Add numerators over LCD}$$

$$= \frac{4 + 69}{12}$$

$$= \frac{73}{12}$$

$$= 6\frac{1}{12}.$$

If we had used the decimal 0.33 to approximate $\frac{1}{3}$, we would have found the sum as follows.

$$\frac{1}{3} + 5.75 \approx 0.33 + 5.75$$

$$= 6.08$$

Check: Does $6\frac{1}{12} = 6.08$?

Since $\frac{1}{12} = 0.08\overline{3}$ (dividing 1 by 12),

$$6\frac{1}{12} = 6.08\overline{3}.$$

Thus, using 0.33 to approximate $\frac{1}{3}$ makes $\frac{1}{3} + 5.75 \approx 6.08$ correct to the nearest hundredth. In some problems, this approximation may be accurate enough, and we can avoid adding fractions.

Answer: $1\frac{7}{12}$ [It is best to change 2.25 to a fraction.]

EXAMPLE 3   APPLICATION TO SEWING

A certain fabric costs \$4.29 per yard. How much will $3\frac{1}{8}$ yards cost?

Multiply $3\frac{1}{8} \times 4.29$. Since $\frac{1}{8} = 0.125$ has a terminating decimal form, find the product using decimals.

$$\begin{array}{r} 3.125 \\ \times\ 4.29 \\ \hline 28125 \\ 6250 \\ 12500 \\ \hline 13.40625 \end{array} \qquad 3\frac{1}{8} = 3.125$$

Rounding to the nearest cent, the fabric will cost \$13.41.

PRACTICE EXERCISE 3

Brad works $8\frac{5}{6}$ hours repairing a television set. If he receives \$13.75 per hour, how much will he be paid?

Answer: \$121.46

❷ ORDER OF OPERATIONS FOR DECIMALS

The order of operations was first considered in Section 2.8 relative to whole numbers and again in Section 4.5 when fractions were involved. The same order applies to decimals and to decimals mixed with fractions. For easy reference and review, the basic rules are repeated here.

Order of Operations
Operations should be performed in the following order.

1. Operate within parentheses.
2. Evaluate all powers and roots in any order.
3. Multiply and divide, in order, from left to right.
4. Add and subtract, in order, from left to right.

EXAMPLE 4 USING THE ORDER OF OPERATIONS ON DECIMALS

Perform the indicated operations.

(a) $2.5 + 3.7 - 1.9 = 6.2 - 1.9$ Add first

$= 4.3$ Then subtract

(b) $(1.8)(3.4) - 2.24 = 6.12 - 2.24$ Multiply first

$= 3.88$ Then subtract

(c) $6.75 \div (1.5 + 3.5) = 6.75 \div 5$ Add inside parentheses first

$= 1.35$ Then divide

(d) $(1.2)^2 + 2.8 \div 1.4 = 1.44 + 2.8 \div 1.4$ Square first

$= 1.44 + 2$ Divide next

$= 3.44$ Then add

PRACTICE EXERCISE 4

Perform the indicated operations.

(a) $4.32 \div 1.6 + 6.7$

(b) $15.45 - (2.3)(4.1)$

(c) $4.3 \cdot (6.7 - 5.1)$

(d) $8.76 - (2.4)^2 \div 0.8$

Answers: (a) **9.4** (b) **6.02**
(c) **6.88** (d) **1.56**

EXAMPLE 5 ORDER OF OPERATIONS ON FRACTIONS AND DECIMALS

Perform the indicated operations.

(a) $\dfrac{1}{2} + (3.5)(2.4) = \dfrac{1}{2} + 8.4$ Multiply first

$= 0.5 + 8.4$ Change $\frac{1}{2}$ to a decimal

$= 8.9$ Then add

(b) $\left(\dfrac{1}{4}\right)^2 + (1.5)^2 - 2.3 = \dfrac{1}{16} + 2.25 - 2.3$ Square first

$= 0.0625 + 2.25 - 2.3$ Change $\frac{1}{16}$ to a decimal

$= 2.3125 - 2.3$ Add first

$= 0.0125$ Then subtract

(c) $3.2 \div \dfrac{1}{4} - 6\dfrac{4}{5} = 3.2 \times 4 - 6.8$ Multiply by the reciprocal of $\frac{1}{4}$, 4

$= 12.8 - 6.8$ Multiply first

$= 6$ Then subtract

PRACTICE EXERCISE 5

Perform the indicated operations.

(a) $\left(\dfrac{2}{3}\right)\left(\dfrac{3}{5}\right) \div 0.8$

(b) $4.8 \div \left(\dfrac{1}{4}\right)^2 + 6.9$

(c) $\left(\dfrac{7}{8}\right)(0.12) - 0.005$

Answers: (a) **0.5** (b) **83.7**
(c) **0.1**

❸ ARRANGING DECIMALS IN ORDER OF SIZE

The next topic involves comparing decimals. For example, we might want to arrange

$$0.3, \quad 0.31, \quad 0.301$$

in order of size from left to right, the smallest first. To make such comparisons, first write each number with the same number of decimal places attaching zeros as needed.

$$0.300, \quad 0.310, \quad 0.301$$

Since the largest number of decimal places in the given numbers is three (in 0.301), we have written each number with three decimal places. We ignore the decimal points and compare sizes as if comparing whole numbers. Left to right, smallest first, the arrangement is

$$300, \quad 301, \quad 310.$$

Thus, the original decimals arranged by size are

$$0.3, \quad 0.301, \quad 0.31.$$

EXAMPLE 6 ARRANGING DECIMALS IN ORDER OF SIZE	**PRACTICE EXERCISE 6**

Arrange the following decimals in order of size, with the smallest on the left.

$$0.05, \quad 3.5, \quad 3.05, \quad 3.4, \quad 0.5$$

First write all numbers with two decimal places.

$$0.05, \quad 3.50, \quad 3.05, \quad 3.40, \quad 0.50$$

Ignoring decimals, arrange

$$5, \quad 350, \quad 305, \quad 340, \quad 50.$$

The two-place decimals, arranged by size, are

$$0.05, \quad 0.50, \quad 3.05, \quad 3.40, \quad 3.50,$$

so the original decimals are arranged by size as follows.

$$0.05, \quad 0.5, \quad 3.05, \quad 3.4, \quad 3.5$$

Practice Exercise 6

Arrange the following decimals in order of size with the smallest on the left.

$$1.06, \quad 1.6, \quad 1.16, \quad 1.106, \quad 0.16$$

Answer: **0.16, 1.06, 1.106, 1.16, 1.6**

Decimals can be compared with fractions by changing the fractions to decimals.

EXAMPLE 7 ARRANGING FRACTIONS AND DECIMALS	**PRACTICE EXERCISE 7**

Arrange the following numbers in order of size with the smallest on the left.

$$0.26, \quad 2.6, \quad \frac{1}{4}, \quad \frac{2}{5}, \quad 0.39$$

Practice Exercise 7

Arrange the following numbers in order of size with the smallest on the left.

$$0.76, \quad 7\frac{1}{2}, \quad \frac{3}{4}, \quad 0.7, \quad \frac{4}{5}$$

First change the fractions to decimals. Since $\frac{1}{4} = 0.25$ and $\frac{2}{5} = 0.4$, the list becomes:

$$0.26, \quad 2.6, \quad 0.25, \quad 0.4, \quad 0.39$$

Next write all the numbers with two decimal places.

$$0.26, \quad 2.60, \quad 0.25, \quad 0.40, \quad 0.39$$

Ignoring the decimals, we must arrange

$$26, \quad 260, \quad 25, \quad 40, \quad 39.$$

The two-place decimals arranged by size are

$$0.25, \quad 0.26, \quad 0.39, \quad 0.40, \quad 2.60,$$

so the original numbers are arranged by size as follows:

$$\frac{1}{4}, \quad 0.26, \quad 0.39, \quad \frac{2}{5}, \quad 2.6.$$

Answer: $0.7, \frac{3}{4}, 0.76, \frac{4}{5}, 7\frac{1}{2}$

④ COMPARING UNIT PRICES

Comparing sizes of decimals is important in unit-pricing problems. The **unit price** of an item is

$$\text{unit price} = \frac{\text{total price}}{\text{number of units}}.$$

For example, if a 16-ounce (abbreviated **oz**) can of juice sells for 89¢,

$$\text{unit price} = \frac{\text{total price}}{\text{number of units}} = \frac{89¢}{16 \text{ oz}} \approx 5.56¢ \text{ per oz.}$$

Also, if a box of 300 facial tissues sells for 58¢, the unit price is

$$\frac{\text{total price}}{\text{number of units}} = \frac{58¢}{300} \approx 0.19¢ \text{ per tissue.}$$

The wise shopper compares unit prices of various brands of an item.

EXAMPLE 8 COMPARING UNIT PRICES

The following table contains information about four brands of detergent. Comparing unit prices, we see that Brand Z is the best buy (assuming that the quality of all products is the same).

Brand	Ounces	Total Price	Unit Price
W	18	$0.87	$0.87 \div 18 \approx \$0.048$
X	20	$0.93	$0.93 \div 20 \approx \$0.047$
Y	16	$0.78	$0.78 \div 16 \approx \$0.049$
Z	25	$1.16	$1.16 \div 25 \approx \$0.046$

PRACTICE EXERCISE 8

A 360-sheet roll of paper towels sells for 69¢ while a 440-sheet roll sells for 79¢. Assuming equal quality, which is the better buy?

Answer: The 440-sheet roll at approximately 0.18¢ per sheet is a better buy than the 360-sheet roll at 0.19¢ per sheet.

5.7 EXERCISES A

Perform the indicated operations by changing all decimals to fractions.

1. $\frac{1}{4} + 2.6$ $2\frac{17}{20}$

2 $3.25 - \frac{3}{8}$ $2\frac{7}{8}$

3. $5\frac{1}{7} + 7.4$ $12\frac{19}{35}$

4. $\frac{2}{3} \times 1.75$ $1\frac{1}{6}$

5. $4.1 \div \frac{5}{6}$ $4\frac{23}{25}$

6. $2\frac{2}{3} \times 3.25$ $8\frac{2}{3}$

Perform the indicated operations by first changing all fractions to decimals, correct to two decimal places. Give answers to the nearest hundredth.

7. $\frac{1}{4} + 2.6$ 2.85

8 $3.25 - \frac{3}{8}$ 2.87

9. $5\frac{1}{7} + 7.4$ 12.54

10. $\frac{2}{3} \times 1.75$ 1.17

11. $4.1 \div \frac{5}{6}$ 4.94

12. $2\frac{2}{3} \times 3.25$ 8.68

Solve.

13. Benny walked 3.2 hours at a rate of $2\frac{1}{2}$ miles per hour. How far did he walk? **8 miles**

14. Kerry earns \$5.85 per hour working as a dance instructor. She substituted for Janine one week and received $2\frac{1}{2}$ times her usual hourly rate. What was her pay per hour that week? **\$14.63**

15. A tailor has 44 yards of material. If $3\frac{2}{3}$ yards are required to make a coat, how many coats can be made? **12 coats**

16. Two hikers gave the number of miles that they hiked on Saturday as 7.25 miles and $6\frac{7}{8}$ miles. How much farther did the one hike than the other?
0.375 or $\frac{3}{8}$ mile

Perform the indicated operations. Answers may vary somewhat due to rounding differences.

17. $3.7 - 2.1 + 8.8$
10.4

18. $(4.3)(6.5) + 7.29$
35.24

19. $9.45 \div (1.4)(2.5)$
2.7

20 $(6.2)^2 - 6.6 \div 1.1$
32.44

21. $\frac{1}{4} + (3.5)(6.4)$
22.65

22. $(2.5)^2 + \left(\frac{1}{3}\right)^2 - 1.9$
4.46$\overline{1}$

23. $7.2 \div \frac{1}{2} - 2\frac{1}{3}$
12.0$\overline{6}$

24. $(3.7)^2 \div \left(\frac{1}{4}\right)^2 + \frac{1}{8}$
219.165

Arrange the given numbers in order of size with the smallest on the left.

25. 0.085, 0.85, 8.5, 8.05
0.085, 0.85, 8.05, 8.5

26. 0.047, 0.048, 0.08, 0.04
0.04, 0.047, 0.048, 0.08

27. 0.51, $\dfrac{1}{2}$, $\dfrac{7}{16}$, 0.15, 5.1

   0.15, $\dfrac{7}{16}$, $\dfrac{1}{2}$, 0.51, 5.1

㉘ $\dfrac{2}{3}$, 0.65, 6.05, $\dfrac{7}{10}$, 6.5

   0.65, $\dfrac{2}{3}$, $\dfrac{7}{10}$, 6.05, 6.5

Solve.

29. A $5\frac{1}{2}$-pound ham sells for $17.38. What is the price per pound (the unit price)?
$3.16 per pound

30. If $16\frac{1}{4}$ ounces of drink sell for 87¢, what is the price per ounce (the unit price) to the nearest tenth of a cent?   **5.4¢ per ounce**

In Exercises 31–32, complete the unit price column in each table and decide which item is the best buy.

31.

Brand	Ounces	Total Price	Unit Price
A	12	$1.27	$0.106
B	8	$0.83	$0.104
C	16	$1.61	$0.101

Brand C is the best buy.

32.

Brand	Number of Tissues	Total Price	Unit Price
Western	200	47¢	0.235¢
Eastern	250	56¢	0.224¢
Southern	400	93¢	0.233¢

Eastern is the best buy.

FOR REVIEW

33. The table below shows Wilma's grades during one quarter. Complete the table and calculate her GPA.

Course	Credits	Grade	Value	Course Value
Geology	4	A	4	
Calculus	4	B	3	
Western Art	3	D	1	
Sociology	3	A	4	
Psychology	3	C	2	

GPA = 2.88

34. At the start of a trip, Gordy noticed that his odometer reading was 32,043.6. At the end of the trip it was 35,581.0. If he used 165.3 gallons of gas, how many miles per gallon did he get, to the nearest tenth?   **21.4 mpg**

35. It costs $17.85 a day plus 15¢ per mile for every mile driven over 100 to rent a car. How much will it cost to rent the car for three days if it is driven a total of 437 miles?   **$104.10**

ANSWERS: 1. $2\frac{17}{20}$ 2. $2\frac{7}{8}$ 3. $12\frac{19}{35}$ 4. $1\frac{1}{6}$ 5. $4\frac{23}{25}$ 6. $8\frac{2}{3}$ 7. 2.85   8. 2.87   9. 12.54   10. 1.17   11. 4.94   12. 8.68   13. 8 miles   14. $14.63   15. 12   16. 0.375 or $\frac{3}{8}$ mile   17. 10.4   18. 35.24   19. 2.7   20. 32.44   21. 22.65   22. $4.46\overline{1}$ 23. $12.0\overline{6}$ 24. 219.165   25. 0.085, 0.85, 8.05, 8.5   26. 0.04, 0.047, 0.048, 0.08   27. 0.15, $\frac{7}{16}$, $\frac{1}{2}$, 0.51, 5.1   28. 0.65, $\frac{2}{3}$, $\frac{7}{10}$, 6.05, 6.5   29. $3.16 per pound   30. 5.4¢ per ounce   31. Brand A is $0.106 per ounce; Brand B is $0.104 per ounce; Brand C is $0.101 per ounce. Thus, Brand C is the best buy.   32. Western is 0.235¢ per tissue; Eastern is 0.224¢ per tissue; Southern is 0.233¢ per tissue. Thus, Eastern is the best buy.   33. GPA = 2.88   34. 21.4 mpg   35. $104.10

5.7 EXERCISES B

Perform the indicated operations by changing all decimals to fractions.

1. $\dfrac{3}{4} + 4.2$ $4\dfrac{19}{20}$

2. $2.25 - \dfrac{7}{8}$ $1\dfrac{3}{8}$

3. $6.2 + 3\dfrac{4}{5}$ 10

4. $\dfrac{1}{3} \times 6.25$ $2\dfrac{1}{12}$ **5.** $8.4 \div \dfrac{1}{6}$ $50\dfrac{2}{5}$ **6.** $2.75 \times 4\dfrac{1}{3}$ $11\dfrac{11}{12}$

Perform the indicated operations by first changing all fractions to decimals, correct to two decimal places. Give answers to the nearest hundredth.

7. $\dfrac{3}{4} + 4.2$ **4.95**   **8.** $2.25 - \dfrac{7}{8}$ **1.38**   **9.** $6.2 + 3\dfrac{4}{5}$ **10**

10. $\dfrac{1}{3} \times 6.25$ **2.06**   **11.** $8.4 \div \dfrac{1}{6}$ **49.41**   **12.** $2.75 \times 4\dfrac{1}{3}$ **11.91**

Solve.

13. A certain fabric costs $3.79 per yard. How much will $5\dfrac{3}{8}$ yards cost?   **$20.37**

14. Kathy Shipp has $295.50 in a passbook savings account and $3\dfrac{1}{4}$ times that much in a money market fund. How much does she have in the money market fund?   **$960.38**

15. Claude paid $27,423.52 for $51\dfrac{3}{4}$ acres of land. Rounded to the nearest cent, what was the cost per acre?   **$529.92**

16. One tank holds 21.6 gallons of fuel and another holds $12\dfrac{1}{8}$ gallons. What is the combined fuel capacity of the two tanks? $33\dfrac{29}{40}$ **gallons**

Perform the indicated operations. Answers may vary somewhat due to rounding differences.

17. $5.9 + 6.2 - 8.7$
3.4

18. $12.8 + (3.9)(4.2)$
29.18

19. $3.85 \div (1.4)(2.5)$
1.1

20. $(4.9)^2 - 8.4 \div 1.2$
17.01

21. $\dfrac{1}{8} + (2.1)^2$
4.535

22. $(4.1)^2 + \left(\dfrac{1}{4}\right)^2 - 2.7$
14.1725

23. $6.8 \div \dfrac{1}{4} + 3\dfrac{1}{5}1$
30.4

24. $(1.6)^2 \div \left(\dfrac{1}{2}\right)^2 - \dfrac{3}{8}$
9.865

Arrange the given numbers in order of size with the smallest on the left.

25. 0.074, 7.04, 0.74, 70.4
0.074, 0.74, 7.04, 70.4

26. 0.95, 0.095, 0.59, 5.9
0.095, 0.59, 0.95, 5.9

27. $\dfrac{1}{4}$, 0.23, $\dfrac{1}{5}$, 2.25, 0.02
0.02, $\dfrac{1}{5}$, 0.23, $\dfrac{1}{4}$, 2.25

28. 0.32, 3.2, $\dfrac{3}{10}$, $\dfrac{1}{3}$, 3.02
$\dfrac{3}{10}$**, 0.32, $\dfrac{1}{3}$, 3.02, 3.2**

Solve.

29. A $12\dfrac{1}{2}$-ounce bottle of shampoo costs $1.00. What is the price per ounce (unit price)?
$0.08 per ounce

30. A roll of aluminum foil costing 39¢ contains 25.5 square feet of foil. To the nearest tenth of a cent, what is the price per square foot?
1.5¢ per square foot

In Exercises 31–32 complete the unit price column in each table and decide which item is the best buy.

31.

Brand	Grams	Total Price	Unit Price
X	40	84¢	2.1¢
Y	35	77¢	2.2¢
Z	30	57¢	1.9¢

Best buy is Brand Z.

32.

Brand of Corn	Ounces	Total Price	Unit Price
High Quality	19	$0.85	4.5¢
Top Crop	$14\dfrac{1}{2}$	$0.76	5.2¢
Generic	12.9	$0.69	5.3¢

Best buy is High Quality.

FOR REVIEW

33. The table below shows Norman's grades during the spring quarter. Complete the table and calculate his GPA.

Course	Credits	Grade	Value	Course Value
GLG 101	4	B	3	
MAT 105	3	A	4	
SOC 240	3	B	3	
PSY 150	3	C	2	
PE 100	1	A	4	

GPA = 3.07

34. Karen bought 2 sweaters at $12.95 each and a pair of shoes for $31.90. The sales tax on her purchases amounted to $2.76. She put the items on layaway by paying $\frac{1}{3}$ down. How much must she pay when she picks up the items?   **$52.37**

35. It costs $18.50 a day to rent a paint sprayer. If paint costs $12.95 a gallon and four gallons will be required, how much will it cost Juan to paint his fence if he uses the sprayer for $1\frac{1}{2}$ days?   **$79.55**

5.7 EXERCISES C

1. On a recent trip George used $28\frac{1}{2}$ gallons of gas which was purchased at 122.9¢ per gallon. How much did George spend for gas on the trip?
[Answer: $35.03]

2. Complete the unit price column and decide which brand of soap is the best buy.

Brand	Ounces	Total Price	Unit Price
Klean	5	36¢	7.2¢
Skrub	6	39¢	6.5¢
Shine	4.5	35¢	7.8¢
Zust	$3\frac{7}{8}$	29¢	7.5¢

Best buy is Skrub.

5.8 THE CALCULATOR

STUDENT GUIDEPOSTS

❶ Basic Operations on a Calculator   **❷** Special Functions on a Calculator

❶ BASIC OPERATIONS ON A CALCULATOR

You may have wondered why calculators have not been discussed until now. The reason is simple. The basic skills of arithmetic must be well understood. While the calculator can help perform complex arithmetic calculations, it does *not* tell what calculations to perform.

Although simple computations can easily be done by hand, lengthy computations, especially those involving decimals, are greatly simplified by using a calculator.

Since there are many calculators on the market, and since they all have different features, it is difficult to give a thorough discussion of how to use any particular one of them. The best advice is to study the instruction manual which comes with your model.

Although differences exist, most calculators consist of three major parts:

1. A **display** on which numbers are shown.
2. A **keyboard** with buttons called **keys,** which include **number keys** (the ten digits and a decimal point), and **operation keys** (+, −, ×, ÷, and =).
3. A **register** in which numbers are stored (not displayed) and used in computations.

Number keys place numbers in the display, and operation keys transfer numbers to the register or calculate results using numbers already in the register. Also, there is a **clear key** which "clears" the calculator of number or operation entries. The clear key removes all numbers used in previous calculations and places the number 0 in the display so that a new calculation can begin.

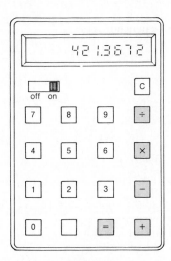

Figure 5.9

A typical calculator is shown in Figure 5.9. The operation keys appear as ▨, the decimal point key is ⸫·⸬, the number keys appear as ☐, and [C] is the clear key. The number

$$421.3672$$

is shown in the display and is entered into the calculator by pressing the following keys in order from left to right.

$$\boxed{4}\ \boxed{2}\ \boxed{1}\ \boxed{·}\ \boxed{3}\ \boxed{6}\ \boxed{7}\ \boxed{2}$$

Making calculations is merely a matter of pressing the right keys in the right order. For example, to find the sum

$$3 + 8$$

the keys to press, and the order in which they are pressed, are shown below.

$\boxed{3}$	$\boxed{+}$	$\boxed{8}$	$\boxed{=}$
First	*Second*	*Third*	*Fourth*
This places the number 3 in the display.	This transfers 3 to the register where it is ready to be added to the next number.	This places 8 in the display.	This adds the number in the display (8) to the number in the register (3) and places the result (11) in the display.

To find the difference

$$15 - 6,$$

press the following keys in order.

$$\boxed{1}\ \boxed{5}\ \boxed{-}\ \boxed{6}\ \boxed{=}$$

The display will show the difference, which is 9.

To find the product

$$12.3 \times 1.05,$$

press the following keys in the order shown.

$$\boxed{1}\ \boxed{2}\ \boxed{\cdot}\ \boxed{3}\ \boxed{\times}\ \boxed{1}\ \boxed{\cdot}\ \boxed{0}\ \boxed{5}\ \boxed{=}$$

The display will show the product, which is 12.915.

Lengthy division problems are greatly simplified by using a calculator. For example, to find

$$4.23738 \div 4.13,$$

press, in order,

$$\boxed{4}\ \boxed{\cdot}\ \boxed{2}\ \boxed{3}\ \boxed{7}\ \boxed{3}\ \boxed{8}\ \boxed{\div}\ \boxed{4}\ \boxed{\cdot}\ \boxed{1}\ \boxed{3}\ \boxed{=}\ .$$

The quotient, 1.026, is shown in the display.

A calculator can find sums of more than two addends. For example, to find

$$1.3 + 2.06 + 4.32 + 11.5$$

press, in order,

$$\boxed{1}\ \boxed{\cdot}\ \boxed{3}\ \boxed{+}\ \boxed{2}\ \boxed{\cdot}\ \boxed{0}\ \boxed{6}\ \boxed{+}\ \boxed{4}\ \boxed{\cdot}\ \boxed{3}\ \boxed{2}\ \boxed{+}\ \boxed{1}\ \boxed{1}\ \boxed{\cdot}\ \boxed{5}\ \boxed{=}\ .$$

The sum, 19.18, will appear in the display.

Most calculators are limited by the number of digits that can be displayed, usually eight.

EXAMPLE 1 BASIC OPERATIONS ON A CALCULATOR

Perform the following calculations using your calculator.

(a) $5.3 + 7.6$

The sequence of keys to press is shown below.

$$\boxed{5}\ \boxed{\cdot}\ \boxed{3}\ \boxed{+}\ \boxed{7}\ \boxed{\cdot}\ \boxed{6}\ \boxed{=}$$

The sum, 12.9, will appear in the display.

(b) $8.5 - 1.9$

$$\boxed{8}\ \boxed{\cdot}\ \boxed{5}\ \boxed{-}\ \boxed{1}\ \boxed{\cdot}\ \boxed{9}\ \boxed{=}$$

The difference, 6.6, will appear in the display when the above keys are pressed.

(c) 12.9×18.1

$$\boxed{1}\ \boxed{2}\ \boxed{\cdot}\ \boxed{9}\ \boxed{\times}\ \boxed{1}\ \boxed{8}\ \boxed{\cdot}\ \boxed{1}\ \boxed{=}$$

The product is 233.49.

(d) $3.948 \div 3.76$

$$\boxed{3}\ \boxed{\cdot}\ \boxed{9}\ \boxed{4}\ \boxed{8}\ \boxed{\div}\ \boxed{3}\ \boxed{\cdot}\ \boxed{7}\ \boxed{6}\ \boxed{=}$$

The quotient is 1.05.

PRACTICE EXERCISE 1

Perform the following calculations using your calculator.

(a) $2.1 + 3.5$

(b) $6.7 - 5.9$

(c) 14.5×8.6

(d) $69.1932 \div 5.29$

(e) $10.7 - 2.4 + 8.5$

$$\boxed{1}\boxed{0}\boxed{\cdot}\boxed{7}\boxed{-}\boxed{2}\boxed{\cdot}\boxed{4}\boxed{+}\boxed{8}\boxed{\cdot}\boxed{5}\boxed{=}$$

The answer is 16.8.

(f) $\dfrac{3}{4} + 2.6$

$$\boxed{3}\boxed{\div}\boxed{4}\boxed{+}\boxed{2}\boxed{\cdot}\boxed{6}\boxed{=}$$

When the first four keys have been pressed, the display will show 0.75, which is $\frac{3}{4}$ as a decimal. Continuing gives the sum 3.35.

(e) $2.5 + 3.6 - 1.8$

(f) $\dfrac{1}{4} + 3.7$

Answers: (a) **5.6** (b) **0.8**
(c) **124.7** (d) **13.08** (e) **4.3**
(f) **3.95**

❷ SPECIAL FUNCTIONS ON A CALCULATOR

Some calculators are more sophisticated and have other special keys. We will discuss four of these keys now. The first is $\boxed{x^2}$, the key used for squaring a number. Simply enter the number to be squared and press $\boxed{x^2}$.

| **EXAMPLE 2** SQUARING A NUMBER ON A CALCULATOR | **PRACTICE EXERCISE 2** |

Square each number.

(a) 12

The steps used to find $(12)^2$ are:

$$\boxed{1}\boxed{2}\boxed{x^2}$$

The display will show 144. Of course you could find $(12)^2$ as follows,

$$\boxed{1}\boxed{2}\boxed{\times}\boxed{1}\boxed{2}\boxed{=}$$

but the squaring key cuts down the number of entries considerably.

(b) 31.75

Use these steps.

$$\boxed{3}\boxed{1}\boxed{\cdot}\boxed{7}\boxed{5}\boxed{x^2}$$

The answer in the display is 1008.0625.

Square each number.

(a) 21

(b) 18.09

Answers: (a) **441** (b) **327.2481**

HINT

Since a calculator display is (usually), limited to eight digits, when a number with many decimal places is squared, the calculator will give only a rounded approximation of the result. For example, if you square 10.62358, the result should have 10 decimal places, but your calculator will round this off and give 112.86045, correct to five decimal places.

The next special key we consider is the square root key, $\boxed{\sqrt{}}$, used for finding the square root of a number. For many numbers, the calculator will give only an approximation of the square root.

EXAMPLE 3 FINDING SQUARE ROOTS ON A CALCULATOR

Find the square root of each number.

(a) 16

We know already that $\sqrt{16} = 4$, but use your calculator to verify this by following these steps.

$$\boxed{1}\ \boxed{6}\ \boxed{\sqrt{}}$$

The display shows 4.

(b) 13.25

Use these steps.

$$\boxed{1}\ \boxed{3}\ \boxed{.}\ \boxed{2}\ \boxed{5}\ \boxed{\sqrt{}}$$

The display will show 3.6400549, which is $\sqrt{13.25}$ correct to seven decimal places. In a practical situation, we might use 3.6 or 3.64 to approximate $\sqrt{13.25}$.

PRACTICE EXERCISE 3

Find the square root of each number.

(a) 81

(b) 28.76

Answers: (a) 9   (b) 5.3628351

CAUTION

In Example 3(b) we found $\sqrt{13.25}$ to be 3.6400549. This means that $(3.6400549)^2$ should be 13.25. Suppose we use 3.64 as an approximation of $\sqrt{13.25}$. If we square 3.64 by pressing

$$\boxed{3}\ \boxed{.}\ \boxed{6}\ \boxed{4}\ \boxed{x^2},$$

the display shows 13.2496, which is a close approximation of 13.25. Remember that when using a calculator, approximate values are often obtained so don't be concerned if your answers differ slightly from those obtained by someone else.

The third special key $\boxed{y^x}$ is the exponential key and is used to find powers of numbers other than squares (for which you will use $\boxed{x^2}$). When using this key, the base y is always entered first followed by the exponent x. The following example will make this clear.

EXAMPLE 4 FINDING POWERS ON A CALCULATOR

Find each power.

(a) 2^4

We know that $2^4 = 2 \cdot 2 \cdot 2 \cdot 2 = 16$ but use your calculator to verify this by following these steps.

$$\boxed{2}\ \underset{\substack{\text{This puts 2}\\\text{in for } y}}{\boxed{y^x}}\ \underset{\substack{\text{This puts 4}\\\text{in for } x}}{\boxed{4}}\ \boxed{=}$$

The display shows 16, which is 2^4.

PRACTICE EXERCISE 4

Find each power.

(a) 5^3

(b) $(1.09)^{10}$

Follow these steps.

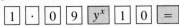

The display shows 2.3673637, which is $(1.09)^{10}$.

(b) $(4.36)^7$

Answers: (a) 125   (b) 29,950.593

Finally, some calculators have parentheses keys $($ and $)$. These keys are used to group operations and maintain the order of operations we have learned. For example, consider

$$10 \div 2 + 3 \quad \text{and} \quad 10 \div (2 + 3).$$

We know that we divide before adding so

$$10 \div 2 + 3 = 5 + 3 = 8.$$

However, with the parentheses in the second problem, we must add first and then divide 10 by the result.

$$10 \div (2 + 3) = 10 \div 5 = 2$$

Since your calculator uses the same order of operations, if you follow these steps

$$\boxed{1}\,\boxed{0}\,\boxed{\div}\,\boxed{2}\,\boxed{+}\,\boxed{3}\,\boxed{=},$$

the display gives 8. To find $10 \div (2 + 3)$, use the parentheses keys

$$\boxed{1}\,\boxed{0}\,\boxed{\div}\,\boxed{(}\,\boxed{2}\,\boxed{+}\,\boxed{3}\,\boxed{)}\,\boxed{=}$$

and the display shows 2, which we know is the correct result based on our order of operations.

EXAMPLE 5   USING THE PARENTHESES KEYS

Calculate each number.

(a) $6 \times (8 + 3)$

Use these steps.

$$\boxed{6}\,\boxed{\times}\,\boxed{(}\,\boxed{8}\,\boxed{+}\,\boxed{3}\,\boxed{)}\,\boxed{=}$$

The display shows 66. Verify this without using a calculator.

(b) $12.8 \div (6.1 - 2.3) + 5.5$

Follow these steps.

The answer in the display is 8.8684211, which is approximately 8.9, rounded to the nearest tenth.

PRACTICE EXERCISE 5

Calculate each number.

(a) $20 \div (7 + 3)$

(b) $4.3 \times (7.1 + 2.8) - 3.6$

Answers: (a) 2   (b) 38.97

HINT

This brief introduction to the calculator, together with your operator's manual, should help you discover the power and usefulness of calculating devices. However, do not become so dependent on your calculator that you reach for it to make simple calculations such as $2 + 5$, $11 - 4$, $12 \div 3$, $\sqrt{9}$, 4^2, and so forth. Such calculations can be performed mentally in far less time than it would take to make the appropriate entries. However, more complicated calculations, especially those involving decimals such as $14.362 \div 1.089$, $(6.4508)^2$, $\sqrt{4.78}$, and so forth, are greatly simplified with a calculator. In other words, you should be "calculator wise" and learn when to use and when *not* to use your calculator. Approximating calculations can also help you discover possible errors made by pressing a wrong key.

5.8 EXERCISES A

What calculation is made by pressing the keys as shown from left to right, and what is the result?

1. $\boxed{1}\,\boxed{0}\,\boxed{+}\,\boxed{3}\,\boxed{5}\,\boxed{=}$ $10 + 35 = 45$

2. $\boxed{6}\,\boxed{2}\,\boxed{-}\,\boxed{3}\,\boxed{1}\,\boxed{=}$ $62 - 31 = 31$

3. $\boxed{3}\,\boxed{8}\,\boxed{\times}\,\boxed{2}\,\boxed{4}\,\boxed{=}$ $38 \times 24 = 912$

4. $\boxed{5}\,\boxed{5}\,\boxed{\div}\,\boxed{1}\,\boxed{1}\,\boxed{=}$ $55 \div 11 = 5$

5. $\boxed{2}\,\boxed{.}\,\boxed{3}\,\boxed{+}\,\boxed{1}\,\boxed{.}\,\boxed{4}\,\boxed{=}$
$2.3 + 1.4 = 3.7$

6. $\boxed{4}\,\boxed{.}\,\boxed{7}\,\boxed{+}\,\boxed{.}\,\boxed{9}\,\boxed{=}$
$4.7 + 0.9 = 5.6$

7. $\boxed{6}\,\boxed{.}\,\boxed{1}\,\boxed{2}\,\boxed{\times}\,\boxed{5}\,\boxed{.}\,\boxed{3}\,\boxed{=}$
$6.12 \times 5.3 = 32.436$

8. $\boxed{1}\,\boxed{.}\,\boxed{3}\,\boxed{\div}\,\boxed{4}\,\boxed{.}\,\boxed{5}\,\boxed{=}$
$1.3 \div 4.5 = 0.2\overline{8}$

9. $\boxed{6}\,\boxed{2}\,\boxed{.}\,\boxed{3}\,\boxed{4}\,\boxed{\div}\,\boxed{4}\,\boxed{2}\,\boxed{.}\,\boxed{3}\,\boxed{5}\,\boxed{7}\,\boxed{=}$ $62.34 \div 42.357 = 1.4717756$

10. $\boxed{1}\,\boxed{.}\,\boxed{5}\,\boxed{+}\,\boxed{3}\,\boxed{.}\,\boxed{6}\,\boxed{2}\,\boxed{+}\,\boxed{5}\,\boxed{4}\,\boxed{.}\,\boxed{7}\,\boxed{+}\,\boxed{8}\,\boxed{0}\,\boxed{=}$ $1.5 + 3.62 + 54.7 + 80 = 139.82$

11. $\boxed{1}\,\boxed{.}\,\boxed{6}\,\boxed{x^2}$ $(1.6)^2 = 2.56$

12. $\boxed{2}\,\boxed{5}\,\boxed{.}\,\boxed{3}\,\boxed{7}\,\boxed{\sqrt{}}$
$\sqrt{25.37} = 5.0368641$

13. $\boxed{1}\,\boxed{1}\,\boxed{y^x}\,\boxed{4}\,\boxed{=}$ $(11)^4 = 14{,}641$

14. $\boxed{6}\,\boxed{\div}\,\boxed{(}\,\boxed{2}\,\boxed{+}\,\boxed{1}\,\boxed{)}\,\boxed{=}$
$6 \div (2 + 1) = 2$

15. $\boxed{6}\,\boxed{\div}\,\boxed{2}\,\boxed{+}\,\boxed{1}\,\boxed{=}$ $6 \div 2 + 1 = 4$

16. $\boxed{2}\,\boxed{5}\,\boxed{\sqrt{}}\,\boxed{+}\,\boxed{3}\,\boxed{x^2}\,\boxed{=}$
$\sqrt{25} + 3^2 = 14$

Perform the following calculations using your calculator.

17. 15.2
  $\underline{+\ 1.03}$
  16.23

18. 402.11
  $\underline{+\ 6.85}$
  408.96

19. 542.309
  $\underline{+\ 62.155}$
  604.464

20.   18.6
       − 9.2
          9.4

21.   703.12
       −   41.35
           661.77

22.   625.157
       − 142.668
          482.489

23.   21.7
       × 4.2
          91.14

24.   62.37
       × 4.71
          293.7627

25.   603.97
       ×   78.56
           47,447.883

26. $3.2\overline{)8.2}$ 2.5625

27. $6.21\overline{)0.467613}$ 0.0753

28. $40.3\overline{)209.765}$ 5.2050868

29. $\dfrac{1}{4} - 0.05$ 0.2

30. $7 + \dfrac{5}{16}$ 7.3125

31. $\dfrac{2}{5} \div 1.2$ $0.\overline{3}$

32. $0.2 + 5.8 - 3.1$ 2.9

33. $2.6 \times 5.3 \div 1.2$ $11.48\overline{3}$

34 $5 \div 1.2 + 6.5$ $10.\overline{6}$

35. $(6.98)^2$ 48.7204

36. $\sqrt{4.29}$ 2.0712315

37. $(10.3)^5$ 115,927.41

38. $4.8 \div (3.2 - 1.7)$ 3.2

39. $(2.6 + 4.3) \times (1.3 - 0.8)$ 3.45

40. $(7.04)^2 - \sqrt{8.15}$ 46.70678

41. $(3.8)^8 \div (6.3 + 10.4)$ 2603.4683

Solve.

42. The sales tax on an item is found by multiplying the cost by 0.0775. How much tax would be charged on a new car which costs $6745.83?   **$522.80**

43. To find the property tax on an apartment building, the assessed value of the building is multiplied by the tax rate, 0.163. How much tax would have to be paid on an apartment building which has an assessed value of $823,475.80?   **$134,226.56**

44. Find the area of a rectangle which is 4.237 m long and 3.075 m wide.   **13.028775 sq m**

45. To find the monthly interest which must be paid on his home mortgage, Carlos must multiply the mortgage balance by 0.007285. If his balance in June is $38,276.55, how much interest must he pay?   **$278.84**

FOR REVIEW

46. Arrange the given numbers in order of size with the smallest on the left.

$$3.1, \quad \dfrac{3}{10}, \quad 0.31, \quad 0.013, \quad \dfrac{1}{3}$$

0.013, $\frac{3}{10}$, 0.31, $\frac{1}{3}$, 3.1

47. A 12-ounce bottle of soda sells for 45¢. What is the price per ounce (unit price)?   **3.75¢ per ounce**

ANSWERS: 1. $10 + 35 = 45$ 2. $62 - 31 = 31$ 3. $38 \times 24 = 912$ 4. $55 \div 11 = 5$ 5. $2.3 + 1.4 = 3.7$ 6. $4.7 + 0.9 = 5.6$ 7. $6.12 \times 5.3 = 32.436$ 8. $1.3 \div 4.5 = 0.2\overline{8}$ 9. $62.34 \div 42.357 = 1.4717756$ 10. $1.5 + 3.62 + 54.7 + 80 = 139.82$ 11. $(1.6)^2 = 2.56$ 12. $\sqrt{25.37} = 5.0368641$ 13. $(11)^4 = 14{,}641$ 14. $6 \div (2 + 1) = 2$ 15. $6 \div 2 + 1 = 4$ 16. $\sqrt{25} + 3^2 = 14$ 17. 16.23 18. 408.96 19. 604.464 20. 9.4 21. 661.77 22. 482.489 23. 91.14 24. 293.7627 25. $47{,}447.883$ 26. 2.5625 27. 0.0753 28. 5.2050868 29. 0.2 30. 7.3125 31. $0.\overline{3}$ 32. 2.9 33. $11.48\overline{3}$ 34. $10.\overline{6}$ 35. 48.7204 36. 2.0712315 37. $115{,}927.41$ 38. 3.2 39. 3.45 40. 46.70678 41. 2603.4683 42. $\$522.80$ 43. $\$134{,}226.56$ 44. 13.028775 sq m 45. $\$278.84$ 46. $0.013, \frac{3}{10}, 0.31, \frac{1}{3}, 3.1$ 47. $3.75¢$ per ounce

5.8 EXERCISES B

What calculation is being made by pressing the keys as shown left to right, and what is the result?

1. $\boxed{2}\,\boxed{0}\,\boxed{+}\,\boxed{4}\,\boxed{6}\,\boxed{=}$ $20 + 46 = 66$ 2. $\boxed{8}\,\boxed{2}\,\boxed{-}\,\boxed{7}\,\boxed{3}\,\boxed{=}$ $82 - 73 = 9$

3. $\boxed{4}\,\boxed{1}\,\boxed{\times}\,\boxed{3}\,\boxed{5}\,\boxed{=}$ $41 \times 35 = 1435$ 4. $\boxed{5}\,\boxed{6}\,\boxed{\div}\,\boxed{8}\,\boxed{=}$ $56 \div 8 = 7$

5. $\boxed{6}\,\boxed{\cdot}\,\boxed{5}\,\boxed{+}\,\boxed{2}\,\boxed{\cdot}\,\boxed{9}\,\boxed{=}$ $6.5 + 2.9 = 9.4$ 6. $\boxed{9}\,\boxed{\cdot}\,\boxed{8}\,\boxed{+}\,\boxed{\cdot}\,\boxed{3}\,\boxed{=}$ $9.8 + 0.3 = 10.1$

7. $\boxed{4}\,\boxed{\cdot}\,\boxed{3}\,\boxed{5}\,\boxed{\times}\,\boxed{1}\,\boxed{\cdot}\,\boxed{2}\,\boxed{=}$ $4.35 \times 1.2 = 5.22$ 8. $\boxed{5}\,\boxed{\cdot}\,\boxed{2}\,\boxed{\div}\,\boxed{1}\,\boxed{\cdot}\,\boxed{3}\,\boxed{=}$ $5.2 \div 1.3 = 4$

9. $\boxed{9}\,\boxed{0}\,\boxed{\cdot}\,\boxed{0}\,\boxed{4}\,\boxed{3}\,\boxed{8}\,\boxed{5}\,\boxed{\div}\,\boxed{2}\,\boxed{8}\,\boxed{\cdot}\,\boxed{4}\,\boxed{0}\,\boxed{5}\,\boxed{=}$ $90.04385 \div 28.405 = 3.17$

10. $\boxed{5}\,\boxed{\cdot}\,\boxed{6}\,\boxed{-}\,\boxed{3}\,\boxed{\cdot}\,\boxed{0}\,\boxed{5}\,\boxed{+}\,\boxed{2}\,\boxed{\cdot}\,\boxed{9}\,\boxed{9}\,\boxed{=}$ $5.6 - 3.05 + 2.99 = 5.54$

11. $\boxed{4}\,\boxed{\cdot}\,\boxed{2}\,\boxed{8}\,\boxed{x^2}$ $(4.28)^2 = 18.3184$ 12. $\boxed{3}\,\boxed{\cdot}\,\boxed{4}\,\boxed{0}\,\boxed{9}\,\boxed{\sqrt{}}$
$\sqrt{3.409} = 1.8463477$

13. $\boxed{8}\,\boxed{\cdot}\,\boxed{5}\,\boxed{y^x}\,\boxed{4}\,\boxed{=}$ $(8.5)^4 = 5220.0625$ 14. $\boxed{1}\,\boxed{8}\,\boxed{\div}\,\boxed{(}\,\boxed{3}\,\boxed{+}\,\boxed{6}\,\boxed{)}\,\boxed{=}$
$18 \div (3 + 6) = 2$

15. $\boxed{1}\,\boxed{8}\,\boxed{\div}\,\boxed{3}\,\boxed{+}\,\boxed{6}\,\boxed{=}$ $18 \div 3 + 6 = 12$ 16. $\boxed{7}\,\boxed{x^2}\,\boxed{-}\,\boxed{4}\,\boxed{\cdot}\,\boxed{2}\,\boxed{\sqrt{}}\,\boxed{=}$
$7^2 - \sqrt{4.2} = 46.95061$

Perform the following calculations using your calculator.

17. $\begin{array}{r} 19.3 \\ + 1.08 \\ \hline 20.38 \end{array}$

18. $\begin{array}{r} 709.44 \\ + 5.87 \\ \hline 715.31 \end{array}$

19. $\begin{array}{r} 652.993 \\ + 43.875 \\ \hline 696.868 \end{array}$

20. $\begin{array}{r} 19.3 \\ - 5.7 \\ \hline 13.6 \end{array}$

21. $\begin{array}{r} 204.53 \\ - 54.88 \\ \hline 149.65 \end{array}$

22. $\begin{array}{r} 476.259 \\ - 341.563 \\ \hline 134.696 \end{array}$

23. $\begin{array}{r} 42.8 \\ \times 6.5 \\ \hline 278.2 \end{array}$

24. $\begin{array}{r} 59.47 \\ \times 2.38 \\ \hline 141.5386 \end{array}$

25. $\begin{array}{r} 705.84 \\ \times 23.75 \\ \hline 16763.7 \end{array}$

26. $6.5)\overline{30.55}$ 4.7 27. $4.58)\overline{0.273884}$ 0.0598 28. $20.7)\overline{109.36224}$ 5.2832

29. $\frac{1}{5} - 0.19$ 0.01 30. $12 + \frac{15}{16}$ 12.9375 31. $\frac{3}{8} \div 2.5$ 0.15

32. $0.9 + 4.07 - 2.58$ 2.39 33. $4.7 \times 2.9 \div 1.02$ 13.362745 34. $6 \div 2.4 + 7.3$ 9.8

35. $(4.09)^2$ **16.7281**   **36.** $\sqrt{11.37}$ 3.3719431   **37.** $(21.4)^6$ **96,046,742**

38. $6.3 \times (1.7 + 2.5)$ **26.46**   **39.** $(4.3 - 1.9) \times (8.5 + 2.7)$ **26.88**

40. $(21.08)^2 - \sqrt{14.3}$ **440.58487**   **41.** $(2.18)^6 \div (9.2 - 4.7)$ **23.852091**

Solve.

42. The sales tax on a purchase can be found by multiplying the selling price by 0.0425. How much tax would be charged on a new boat which sells for $24,325.65?   **$1033.84**

43. The property tax on a large apartment building can be found by multiplying the tax rate, 0.197, by the assessed value of the building. How much tax would have to be paid if the assessed value is $179,840.50?   **$35,428.58**

44. Find the area of a rectangular garden which measures 7.235 yards long and 5.108 yards wide.
36.95638 sq yd

45. To find the amount of interest paid in a given month, multiply the monthly balance by 0.008275. How much interest will Daphne pay in August if her loan balance is $42,179.85?   **$349.04**

FOR REVIEW

46. Arrange the given numbers in order of size with the smallest on the left.

$$0.88, \quad \frac{6}{7}, \quad \frac{13}{14}, \quad 0.92, \quad \frac{9}{10}$$

$\frac{6}{7}$, **0.88**, $\frac{9}{10}$, **0.92**, $\frac{13}{14}$

47. Janet earns $12.30 an hour. How much will she be paid for working $8\frac{3}{4}$ hours?   **$107.63**

5.8 EXERCISES C

Solve.

1. The area of a rectangle is 106.005 square yards, and its width is 9.25 yards. Divide the area by the width to find the length of the rectangle.
[Answer: 11.46 yd]

2. During five consecutive weeks, Mr. Whorton discovered that the overhead expenses on his business were $497.36, $643.21, $825.13, $503.07, and $601.58. What were the total expenses and the average expense per week?   **$3070.35; $614.07**

3. In a certain triangle, two sides are 4.28 cm and 7.15 cm. The third side is given by
$$\sqrt{(4.28)^2 + (7.15)^2}.$$
Find the third side, correct to the nearest hundredth of a centimeter.
[Answer: 8.33 cm]

4. The amount of money in a certain savings account at the end of 10 years when $1500 is invested initially is given by $(1.02)^{40}(1500)$. Find the amount.
$3312.06

CHAPTER 5 REVIEW

KEY WORDS

5.1 A **decimal fraction** is a fraction with denominator a power of 10. Decimal fractions are usually called simply **decimals.** The period used in a decimal is called a **decimal point.**

5.2 If the division process ends with a remainder of zero, the decimal is a **terminating decimal.**

If the division process does not end, there is a repeating block of digits, and the decimal is a **repeating decimal.**

5.3 The number of digits to the right of the decimal point is called the number of **decimal places** in the number.

5.8 On a calculator, the **display** shows numbers, the **register** stores numbers used in computations, the **number keys** are the buttons for the ten digits and the decimal point, and the **operation keys** are used to perform operations, such as adding, subtracting, multiplying, and dividing.

5.4 1. To add decimals, line up the decimal points and add like whole numbers.

2. To subtract decimals, line up the decimal points and subtract like whole numbers.

5.5 1. The product of two decimals has the same number of decimal places as the sum of the decimal places in the two factors.

2. To multiply by a power of 10, move the decimal point to the *right* the same number of decimal places as the number of zeros in the power of 10.

3. To multiply by 0.1, 0.01, or 0.001, move the decimal point to the *left* the same number of decimal places as decimal places in 0.1, 0.01, or 0.001.

5.6 1. When dividing a decimal by a decimal, move the decimal point in the divisor and dividend the same number of places to obtain a whole number divisor. The decimal point in the quotient is directly above the new decimal point in the dividend.

2. To divide by a power of 10, move the decimal point to the *left* the same number of decimal places as the number of zeros in the power of 10.

3. To divide by 0.1, 0.01, or 0.001, move the decimal point to the *right* the same number of decimal places as decimal places in 0.1, 0.01, 0.001.

5.7 The unit price of an item is

$$\text{unit price} = \frac{\text{total price}}{\text{number of units}}.$$

5.8 Only use a calculator for complex problems; do not become so dependent on it that you use it for simple calculations that can be found mentally.

KEY CONCEPTS

5.1 1. Each decimal has a fractional form and a decimal form.

2. A decimal point separates the ones digit from the tenths digit.

5.2 Every fraction can be written as a decimal that either terminates or repeats.

5.3 When rounding a decimal, if the first digit to the right of the desired position is less than 5, drop it and all digits to the right of it. If the first digit to the right is greater than or equal to 5, increase the digit in the desired position by one and drop all digits to the right of it. *Never* round off one digit at a time.

REVIEW EXERCISES

Part I

5.1 1. What de we call fractions that have a power of 10 in the denominator?
decimal fractions or decimals
Give the formal *word name for each decimal.*

2. 23.4
twenty-three and four tenths

3. 7.0005
seven and five ten thousandths

Write each number in decimal notation.

4. four and seven hundredths   4.07

5. two hundred thirty-three and three hundred two ten thousandths   233.0302

6. Write $27.16 using a word name as it would appear on a check.
Twenty-seven and $\frac{16}{100}$ dollars

5.2 *Change each decimal to a fraction (or mixed number).*

7. 0.63 $\frac{63}{100}$

8. 2.7 $2\frac{7}{10}$

9. 5.125 $5\frac{1}{8}$

Change each fraction to a decimal.

10. $\frac{1}{9}$ $0.\overline{1}$

11. $\frac{5}{3}$ $1.\overline{6}$

12. $\frac{3}{16}$ 0.1875

5.3 *Give the number of decimal places in each number.*

13. 2.103   3

14. 4.003   3

15. 0.01010   5

Round each decimal to the given number of decimal places or position.

16. 3.255; two
3.26
17. 6.4908; thousandth
6.491
18. 26.05; tenth
26.1
19. $37.51; dollar
$38
20. $0.\overline{84}$; tenth
0.8
21. $0.\overline{84}$; hundredth
0.85
22. $0.\overline{84}$; three
0.848
23. $0.426; cent
$0.43

Solve.

24. What is the approximate cost of 19 pounds of grapes at 99¢ per pound?
$20.00 [using 20 × 100¢]

25. Alvin has a new car loan in the amount of $7895.60. If he plans to pay off the loan in 40 equal monthly payments, what is the approximate amount he must pay each month?
$200 [using 8000 ÷ 40]

5.4 **26.** Find the sum.

6.1 + 6.01 + 0.601 + 0.0601 + 60.1
72.8711

27. The daily receipts for one week at Tony's Tastee Tacos were $421.36, $347.20, $289.17, $655.27, and $504.08. What was the total for the week? $2217.08

28. Find the total deposit on the following portion of a deposit slip. $2908.28

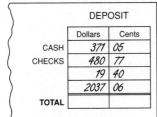

29. Find the perimeter of the given figure.

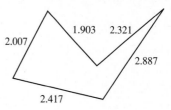

11.535

Find each difference.

30. 42.6 − 3.08   39.52

31. 100 − 74.36   25.64

32. 2 − 1.003   0.997

33. Ev received a check for $235.96. He spent $45.30 for a pair of shoes and $38.07 for a power saw. How much did he have left?
$152.59

34. Becky bought a sweater for $18.35 and a pair of earrings for $8.29. The sales tax on her purchases amounts to $1.07. How much change will she receive if she pays with two $20 bills?
$12.29

5.5   *Find each product.*

35.   13.06
        $\times$ 2.7
        35.262

36.   3.1108
        $\times$ 2.06
        6.408248

37.   0.4001
        $\times$ 1.3
        0.52013

38. The EPA-estimated miles per gallon for a car is 23.7 mpg. Its gas tank capacity is 11.6 gallons. To the nearest mile, what is the driving range of the car?   **275 mi**

39. Mr. Swokowski is paid $11.37 per hour. His timecard for one week is shown in the figure. How much was he paid that week?

Day	Hours worked
M	8.1
T	5.3
W	4.7
T	10.3
F	7.8

$411.59

Without making the actual computation, use an estimate to decide which of the given answers is correct.

40. 2.9 $\times$ 24.007   **(a)** 0.696   **(b)** 6.96   **(c)** 69.6   **(d)** 696
        (c)

Solve.

41. A car gets 28.9 mpg and has a gas tank with capacity 12.2 gallons. What is the approximate driving range of the car? The exact driving range? Isn't the exact driving range also an *approximate* driving range?

approximate range: 360 mi [using 30 $\times$ 12];
exact range: 352.6 mi; yes

42. What is the approximate cost of 9 pounds of candy selling for $1.87 per pound? The exact cost?   **$18.00 [using 9 $\times$ 2]; $16.83**

Find the products.

43. 10.785 $\times$ 1000   **10,785**

44. 10.785 $\times$ 0.1   **1.0785**

45. 10.785 $\times$ 0.01   **0.10785**

5.6   *Find each quotient.*

46. 2.9$\overline{)6.177}$ **2.13**

47. 42$\overline{)222.6}$ **5.3**

48. 0.032$\overline{)1.3312}$ **41.6**

49. 4.13$\overline{)15.6}$ (to the nearest tenth)   **3.8**

50. 0.089$\overline{)4.62}$ (to the nearest hundredth)   **51.91**

51. Terry drove 746.2 miles and used 34.9 gallons of gas. To the nearest tenth, how many miles per gallon did he get?   **21.4 mpg**

52. The table shows Betty's grades during one quarter. Complete the table and calculate her GPA correct to the nearest hundredth.

Course	Credits	Grade	Value	Course Value
Math	4	B	3	
French	1	F	0	
Biology	4	C	2	
English	3	A	4	
Russian	3	D	1	

GPA = 2.33

53. What is the approximate cost per pound of 9 pounds of candy selling for $18.45? The exact cost?   **$2 [20 $\div$ 10]; $2.05**

54. The Morins plan to carpet their master bedroom. If the room is rectangular in shape and measures 4.8 yards long by 3.9 yards wide, and the total cost of the job is $383.76, what is the approximate cost of one square yard of carpet? The exact cost?   **$20; $20.50**

Without making actual computations, use an estimate to decide which of the given answers is correct.

55. 401.87 $\div$ 39.6   **(a)** 1.015   **(b)** 10.15   **(c)** 101.5   **(d)** 1015
        (b)

Find the quotients.

56. $43.89 \div 100$ 0.4389   **57.** $1.497 \div 1000$ 0.001497   **58.** $286.1 \div 10$ 28.61

5.7   *Find the sums (to the nearest hundredth).*

59. $\dfrac{2}{3} + 7.25$ 7.92   **60.** $6\dfrac{1}{4} + 8.39$ 14.64   **61.** $\dfrac{1}{6} + 3.78$ 3.95

62. Clyde earns $12.43 an hour. How much will he be paid for working $7\dfrac{1}{4}$ hours?   **$90.12**

63. If $18\dfrac{1}{2}$ ounces of cereal sell for $1.09, what is the price per ounce (the unit price) to the nearest tenth of a cent?   **5.9¢**

64. The Meyers paid $113,750 for their house. If the house has 2000 sq ft, how much did they pay per sq ft?   **$56.88**

65. Arrange 0.74, 7.04, 7.4, 0.074, and 70.4 in order of size with the smallest on the left.   **0.074, 0.74, 7.04, 7.4, 70.4**

66. Complete the unit price column in the table and determine which item is the best buy.

Brand	Pounds	Total price	Unit price
X	7	$2.99	$0.427
Y	11	$4.71	$0.428
Z	8	$3.32	$0.415

Brand Z is the best buy.

Perform the indicated operations.

67. $6.3 \div \dfrac{1}{3} - 1\dfrac{1}{2}$ 17.4

68. $(2.6)^2 - \dfrac{7}{8} \times \dfrac{3}{14}$ 6.5725

5.8   *What calculation is made by pressing the keys as shown from left to right, and what is the result?*

69. $\boxed{8}\ \boxed{\cdot}\ \boxed{7}\ \boxed{\div}\ \boxed{1}\ \boxed{2}\ \boxed{=}$
$8.7 \div 12 = 0.725$

70. $\boxed{2}\ \boxed{1}\ \boxed{\cdot}\ \boxed{6}\ \boxed{-}\ \boxed{4}\ \boxed{\cdot}\ \boxed{3}\ \boxed{7}\ \boxed{=}$
$21.6 - 4.37 = 17.23$

71. $\boxed{1}\ \boxed{\cdot}\ \boxed{0}\ \boxed{0}\ \boxed{9}\ \boxed{\times}\ \boxed{3}\ \boxed{\cdot}\ \boxed{5}\ \boxed{=}$
$1.009 \times 3.5 = 3.5315$

72. $\boxed{6}\ \boxed{+}\ \boxed{2}\ \boxed{\cdot}\ \boxed{5}\ \boxed{-}\ \boxed{4}\ \boxed{=}$
$6 + 2.5 - 4 = 4.5$

73. $\boxed{2}\ \boxed{\cdot}\ \boxed{3}\ \boxed{\times}\ \boxed{6}\ \boxed{+}\ \boxed{8}\ \boxed{\cdot}\ \boxed{9}\ \boxed{=}$
$2.3 \times 6 + 8.9 = 22.7$

74. $\boxed{9}\ \boxed{\times}\ \boxed{9}\ \boxed{\times}\ \boxed{9}\ \boxed{=}$
$9 \times 9 \times 9 = 729$

75. $\boxed{4}\ \boxed{\cdot}\ \boxed{2}\ \boxed{x^2}$
$(4.2)^2 = 17.64$

76. $\boxed{3}\ \boxed{\cdot}\ \boxed{2}\ \boxed{7}\ \boxed{\sqrt{\ }}$
$\sqrt{3.27} = 1.8083141$

77. $\boxed{1}\ \boxed{\cdot}\ \boxed{5}\ \boxed{y^x}\ \boxed{6}\ \boxed{=}$
$(1.5)^6 = 11.390625$

78. $\boxed{6}\ \boxed{\div}\ \boxed{(}\ \boxed{2}\ \boxed{+}\ \boxed{3}\ \boxed{)}\ \boxed{-}\ \boxed{1}\ \boxed{=}$
$6 \div (2 + 3) - 1 = 0.2$

Solve.

79. The sales tax on an item is found by multiplying the cost by 0.0425. How much tax would be charged on a new motor home which costs $14,137.58?   **$600.85**

80. Find the area of a rectangle 5.621 m long by 3.749 m wide.   **21.073129 sq m**

Part II

Perform the indicated operations.

81. $\begin{array}{r} 4.1035 \\ \times\ 2.3 \\ \hline 9.43805 \end{array}$

82. $\begin{array}{r} 6.15 \\ 12.034 \\ +\ 9.2 \\ \hline 27.384 \end{array}$

83. $\begin{array}{r} 142.75 \\ -\ 38.016 \\ \hline 104.734 \end{array}$

84. 1.1596×100 115.96

85. $1.1596 \div 100$ 0.011596

86. 1.1596×0.1 0.11596

87. $1.1596 \div 0.001$ 1159.6

88. $53\overline{)113.42}$ 2.14

89. $0.35\overline{)4.68}$ (to the nearest tenth)
13.4

90. $3\frac{3}{4} + 2.48$ **6.23**

91. $4.67 - 1\frac{1}{5}$ **3.47**

92. $(2.6)^2 \times 4\frac{1}{2}$ **30.42**

93. Round 6.485 to the nearest **(a)** hundredth, **(b)** tenth, and **(c)** unit.   **(a) 6.49 (b) 6.5 (c) 6**

94. Change $\frac{11}{3}$ to a decimal.   **3.$\overline{6}$**

95. Change 0.85 to a fraction.   **$\frac{17}{20}$**

96. Write $102.38 using a word name as it would appear on a check.
One hundred two and $\frac{38}{100}$ dollars

97. Estimate the product 6098.5×41.89.
240,000 [using 6000 × 40]

Solve.

98. For every $1000 of assessed value on a store, the owner must pay $4.89 in real estate taxes. How much tax must be paid on a store with an assessed value of $125,600?   **$614.18**

99. Beth received checks of $20.50, $16.75, and $42.30. Out of this total she bought a sweater for $29.95. How much money did she have left?   **$49.60**

100. Jackie bought 3 books for $8.95 each and paid with two $20 bills. How much change did she get back?   **$13.15**

101. Which is a better buy, a 24-ounce bottle of soap for $1.20 or a 38-ounce bottle for $2.28?
The 24-ounce bottle at 5¢ an ounce is better than the 38-ounce bottle at 6¢ an ounce.

102. Jeff traveled 55 mph for $5\frac{1}{2}$ hours. If he used 12.8 gallons of gas during this time, what was his mileage per gallon, to the nearest tenth?
23.6 mpg

103. What is the approximate cost of 40 pounds of steak at $4.95 per pound? What is the exact cost?
approximate cost: $200; actual cost: $198

Without making the actual computation, use an estimate to decide which of the given answers is correct.

104. $398.097 \div 49.7$ **(a)** 0.801   **(b)** 8.01   **(c)** 80.1   **(d)** 801
(b)

Use a calculator to perform the following operations.

105. 62.308×401.375 **25,008.874**

106. $0.0435\overline{)2.6513}$ **60.949425**

107. $(3.4)^2 \div 1.7$ **6.8**

108. $(1.05)^5 \times 1050$ **1340.0957**

109. $6.35 \times (4.1 - 3.8)$ **1.905**

110. $\sqrt{6.9 \times 11.4}$ **8.8690473**

ANSWERS: 1. decimal fractions or decimals 2. twenty-three and four tenths 3. seven and five ten thousandths 4. 4.07 5. 233.0302 6. Twenty-seven and $\frac{16}{100}$ dollars 7. $\frac{63}{100}$ 8. $2\frac{7}{10}$ 9. $5\frac{1}{8}$ 10. $0.\overline{1}$ 11. $1.\overline{6}$ 12. 0.1875 13. 3 14. 3 15. 5 16. 3.26 17. 6.491 18. 26.1 19. $38 20. 0.8 21. 0.85 22. 0.848 23. $0.43 24. $20.00 [using 20 × 100¢] 25. $200 [using 8000 ÷ 40] 26. 72.8711 27. $2217.08 28. $2908.28 29. 11.535 30. 39.52 31. 25.64 32. 0.997 33. $152.59 34. $12.29 35. 35.262 36. 6.408248 37. 0.52013 38. 275 mi 39. $411.59 40. (c) 41. approximate range: 360 mi [using 30 × 12]; exact range: 352.6 mi; yes 42. $18.00 [using 9 × 2]; $16.83 43. 10,785 44. 1.0785 45. 0.10785 46. 2.13 47. 5.3 48. 41.6 49. 3.8 50. 51.91 51. 21.4 mpg 52. GPA = 2.33 53. $2 [20 ÷ 10]; $2.05 54. $20; $20.50 55. (b) 10.15 56. 0.4389 57. 0.001497 58. 28.61 59. 7.92 60. 14.64 61. 3.95 62. $90.12 63. 5.9¢ 64. $56.88 65. 0.074, 0.74, 7.04, 7.4, 70.4 66. Brand X is $0.427; Brand Y is $0.428; Brand Z is $0.415. Thus, Brand Z is the best buy. 67. 17.4 68. 6.5725 69. 8.7 ÷ 12 = 0.725 70. 21.6 − 4.37 = 17.23 71. 1.009 × 3.5 = 3.5315 72. 6 + 2.5 − 4 = 4.5 73. 2.3 × 6 + 8.9 = 22.7 74. 9 × 9 × 9 = 729 75. $(4.2)^2 = 17.64$ 76. $\sqrt{3.27} = 1.8083141$ 77. $(1.5)^6 = 11.390625$ 78. 6 ÷ (2 + 3) − 1 = 0.2 79. $600.85 80. 21.073129 sq m 81. 9.43805 82. 27.384 83. 104.734 84. 115.96 85. 0.011596 86. 0.11596 87. 1159.6 88. 2.14 89. 13.4 90. 6.23 91. 3.47 92. 30.42 93. (a) 6.49 (b) 6.5 (c) 6 94. 3.$\overline{6}$ 95. $\frac{17}{20}$ 96. One hundred two and $\frac{38}{100}$ dollars 97. 240,000 [using 6000 × 40] 98. $614.18 99. $49.60 100. $13.15 101. The 24-ounce bottle at 5¢ an ounce is better than the 38-ounce bottle at 6¢ an ounce. 102. 23.6 mpg 103. approximate cost: $200; actual cost: $198 104. (b) 105. 25,008.874 106. 60.949425 107. 6.8 108. 1340.0957 109. 1.905 110. 8.8690473

1. Give the formal word name for 73.005.

1. seventy-three and five thousandths

2. Write "ninety-two and six hundredths" in decimal notation.

2. 92.06

3. Change 0.16 to a fraction reduced to lowest terms.

3. $\frac{4}{25}$

4. Change $\frac{7}{16}$ to a decimal.

4. 0.4375

5. Round 6.355 to two decimal places.

5. 6.36

6. Round 45.05 to the nearest tenth.

6. 45.1

7. Round $0.\overline{74}$ to the nearest hundredth.

7. 0.75

Perform the indicated operations.

8. $9.1 + 0.91 + 9.01 + 91.0$

8. 110.02

9. $11 - 5.003$

9. 5.997

Solve.

10. Mr. and Mrs. Forbes stayed at the Sands Hotel in Reno. Their room cost $126.87 and they charged $48.53 in food and beverages to their room account. If Mr. Forbes paid the bill with two $100 bills, how much change did he receive?

10. $24.60

11. What is the *approximate* cost per pound of dog food if a 32-pound sack sells for $15.36? What is the *exact* cost per pound?

11. 50¢; 48¢

Perform the indicated operations.

12. 71.05
    × 1.8

13. 4.7)47.376

14. 6.357 × 100

15. 6.357 × 0.1

16. 6.357 ÷ 1000

17. 6.357 ÷ 0.1

12. _____127.89_____

13. _____10.08_____

14. _____635.7_____

15. _____0.6357_____

16. _____0.006357_____

17. _____63.57_____

Solve.

18. Hector buys 14.5 pounds of apples at $0.43 per pound, and 11.2 pounds of sausage at $1.59 per pound. To the nearest cent, how much do these purchases total?

18. _____$24.04_____

19. Dinah drove 956.3 miles and used 38.6 gallons of gas. To the nearest tenth, how many miles per gallon did she get?

19. _____24.8 mpg_____

20. A bricklayer uses 415 bricks at a cost of 9.3¢ per brick. What is the approximate cost of the bricks? The exact cost (to the nearest cent)?

20. _____$36.00; $38.60_____

21. Perform the indicated operations.

$$(3.5)(4.1) - \left(\frac{1}{2}\right)^2$$

21. _____14.1_____

22. What calculation is being made by pressing the keys shown from left to right? Determine the answer.

$$\boxed{8}\ \boxed{\cdot}\ \boxed{0}\ \boxed{3}\ \boxed{\div}\ \boxed{2}\ \boxed{\cdot}\ \boxed{5}\ \boxed{=}$$

22. _____8.03 ÷ 2.5 = 3.212_____

Applications of Ratio, Proportion, and Percent

6.1 RATIO AND PROPORTION

─── STUDENT GUIDEPOSTS ───

1 Ratio   **3** Solving Proportions

2 Proportion

1 RATIO

The word *ratio* is commonly used in many practical situations. For example, an excellent recipe for honey-butter is two cups of butter and three cups of honey. In this recipe, the *ratio* of butter to honey is 2 to 3. A **ratio** is a way to compare two numbers, like 2 parts of butter for every 3 parts of honey. A ratio can also be expressed as a quotient of two numbers. For example, the ratio of the number 5 to the number 8 is the quotient or fraction $\frac{5}{8}$. We call $\frac{5}{8}$ the **fraction form** for the ratio of 5 to 8. Another way of writing this ratio is

$$5:8, \text{ read ``the ratio of 5 to 8.''}$$

EXAMPLE 1   WRITING RATIOS IN FRACTION FORM

Write each ratio in fraction form.

(a) The ratio of 7 to 10 $\dfrac{7}{10}$

(b) The ratio 10:5 $\dfrac{10}{5}$ which reduces to $\frac{2}{1}$ (see Chapter 3.)

PRACTICE EXERCISE 1

Write each ratio in fraction form.

(a) The ratio of 5 to 11

(b) The ratio of 4 to 12

Answers: (a) $\frac{5}{11}$ (b) $\frac{4}{12}$ or $\frac{1}{3}$ (reduced)

Ratios occur in many applications such as tax rates, rate of speed, gas mileage, and unit cost. These are shown in the following table.

Application	*Ratio Involved*
\$20 tax on a \$400 purchase	$\dfrac{\$20}{\$400} = \dfrac{\$1}{\$20}$ means \$1 tax on each \$20 spent
300 miles in 6 hours	$\dfrac{300 \text{ mi}}{6 \text{ hr}} = \dfrac{50 \text{ mi}}{1 \text{ hr}} = 50\dfrac{\text{mi}}{\text{hr}} = 50 \text{ mph}$

100 miles on 5 gallons of gas

$$\frac{100 \text{ mi}}{5 \text{ gal}} = \frac{20 \text{ mi}}{1 \text{ gal}} = 20\frac{\text{mi}}{\text{gal}} = 20 \text{ mpg}$$

$8 for 4 pounds of meat

$$\frac{\$8}{4 \text{ lb}} = \$2 \text{ per lb}$$

40 children and 5 adults

$$\frac{40 \text{ children}}{5 \text{ adults}} = 8 \text{ children per adult}$$

HINT

The word *per* is often used to describe ratios; it means "for each." For example, $2 per lb means $2 for each pound. Also, *mph* is the abbreviation for "miles per hour," and *mpg* stands for "miles per gallon."

EXAMPLE 2   WRITING RATIOS

A company employs 18 women and 12 men.

(a) What is the ratio of men to women in the company?

$$\frac{12 \text{ men}}{18 \text{ women}} = \frac{12}{18} = \frac{2}{3}$$

The ratio of men to women is $\frac{2}{3}$ or 2 to 3 or 2:3.

(b) What is the ratio of women employed to the total number of employees?

There are $18 + 12 = 30$ total employees.

$$\frac{18 \text{ women}}{30 \text{ employees}} = \frac{18}{30} = \frac{3}{5}$$

The ratio of women to total employees is $\frac{3}{5}$ or 3 to 5 or 3:5.

PRACTICE EXERCISE 2

There are 200 floor-level seats and 120 balcony seats in an auditorium.

(a) What is the ratio of balcony seats to floor seats?

(b) What is the ratio of floor seats to the total number of seats?

Answers: (a) $\frac{3}{5}$ or 3 to 5 or 3:5
(b) $\frac{5}{8}$ or 5 to 8 or 5:8

EXAMPLE 3   RATIOS IN A GEOMETRY PROBLEM

Consider the triangle in Figure 6.1. Find the ratio of the length of the shortest side to the length of the longest side.

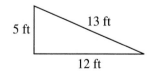

Figure 6.1

Since the shortest side is 5 ft and the longest side is 13 ft, the desired ratio is $\frac{5}{13}$ or 5 to 13 or 5:13.

PRACTICE EXERCISE 3

Find the ratio of the length of the longest side to the length of the next longest side in the triangle in Figure 6.1.

Answer: $\frac{13}{12}$ or 13 to 12 or 13:12

❷ PROPORTION

Any statement that says two quantities are equal is called an **equation.** For example, $2 + 1 = 3$ and $\frac{1}{2} = 0.5$ are equations. An equation which states that two ratios are equal is called a **proportion.** For example,

$$\frac{1}{2} = \frac{4}{8}$$

is a proportion. When a pair of numbers such as 1 and 2 has the same ratio as a pair such as 4 and 8, the numbers are called **proportional.** Recall from Section 3.2 that two fractions (ratios) are equal whenever their cross products are equal. The cross products in the above proportion are:

$$2 \cdot 4 \quad \longleftarrow \quad \frac{1}{2} \quad \times \quad \frac{4}{8} \quad \longrightarrow \quad 1 \cdot 8.$$

Since $2 \cdot 4 = 1 \cdot 8$, we know that the fractions are equal.

Proportions can be used to solve many problems. For example, if 3 hours are required to travel 165 miles, we might ask: "How many hours will be required to travel 275 miles?" The proportion

first time $\xrightarrow{\text{in hours}}$ $\quad \dfrac{3}{165} = \dfrac{\square}{275} \quad$ $\xleftarrow{\text{in hours}}$ second time

first distance $\xrightarrow{\text{in miles}}$ $\qquad\qquad\quad$ $\xleftarrow{\text{in miles}}$ second distance

completely describes this problem. That is, the ratio of the first time to the first distance must equal the ratio of the second time to the second distance. We need to find the missing number (represented by the $\square$). In section 3.2 we were able to find such numbers in similar problems by inspection. There is a more direct method. Since

$$\frac{3}{165} = \frac{\square}{275},$$

we know that the cross products are equal. Thus,

$$3 \cdot 275 = 165 \cdot \square$$
$$825 = 165 \cdot \square.$$

To determine $\square$, we use the related division sentence first introduced in Section 3.4. Recall that every multiplication sentence such as $2 \cdot 3 = 6$ has two related division sentences

$$3 = 6 \div 2 = \frac{6}{2} \quad \text{and} \quad 2 = 6 \div 3 = \frac{6}{3}.$$

In the work above, the multiplication sentence or equation

$$825 = 165 \cdot \square$$

can be replaced with the related division sentence or equation

$$\square = \frac{825}{165},$$

which reduces to $\square = 5$. Thus,

$$\frac{\square}{275} \text{ becomes } \frac{5}{275}, \qquad \text{Replace } \square \text{ with 5}$$

and we see that 5 is indeed the desired number since

$$\frac{3}{165} = \frac{5}{275}. \qquad \text{Both cross products are 825}$$

Often a letter (instead of □) is used to represent the missing number in a proportion. Suppose we use a. Then

$$\frac{3}{165} = \frac{\square}{275}$$

can be written

$$\frac{3}{165} = \frac{a}{275},$$

and we need to find the value of a. The method is exactly the same as above, using a instead of □.

$$3 \cdot 275 = 165 \cdot a \qquad \text{Cross products are equal}$$

$$875 = 165 \cdot a \qquad \text{Multiply 3 and 275}$$

$$a = \frac{825}{165} \qquad \text{Related division sentence}$$

$$a = 5 \qquad \text{Divide 825 by 165}$$

❸ SOLVING PROPORTIONS

The process of finding a missing number in a given proportion is called **solving the proportion.** The preceding discussion suggests a method for doing this.

To Solve a Proportion Involving a Missing Number a

1. Set the cross products equal to each other (that is, **equate** the cross products).
2. Then a is equal to the numerical cross product divided by the multiplier of a. That is, the related division sentence gives the desired value of a.

EXAMPLE 4 SOLVING PROPORTIONS

Solve the proportions.

(a) $$\frac{a}{4} = \frac{12}{16}$$

$$16 \cdot a = 4 \cdot 12 \qquad \text{Cross products are equal}$$

$$16 \cdot a = 48 \qquad \text{Simplify}$$

$$a = \frac{48}{16} \qquad \text{Divide by 16, the multiplier of a; this is the related}$$
$$\text{division sentence}$$

$$a = 3$$

Thus, the missing value is 3, and

$$\frac{3}{4} = \frac{12}{16}. \qquad \begin{array}{l} \text{To check, find the cross products:} \\ 3 \cdot 16 = 48 \text{ and } 4 \cdot 12 = 48 \end{array}$$

(b) $$\frac{2}{a} = \frac{4}{10}$$

$$2 \cdot 10 = 4 \cdot a \qquad \text{Equate cross products}$$

$$20 = 4 \cdot a \qquad \text{Simplify}$$

$$\frac{20}{4} = a \qquad \text{Divide 20 by 4}$$

$$5 = a$$

PRACTICE EXERCISE 4

Solve the proportions.

(a) $$\frac{a}{2} = \frac{15}{10}$$

(b) $$\frac{6}{21} = \frac{2}{a}$$

Thus, the missing value is 5, and

$$\frac{2}{5} = \frac{4}{10}.$$ To check, find the cross products:
   $2 \cdot 10 = 20$ and $5 \cdot 4 = 20$

Answers: (a) 3   (b) 7

HINT

Here is another way to remember how to find the missing number a in a proportion. Consider the proportion $\frac{a}{4} = \frac{12}{16}$ in Example 4(a). We found the cross-product equation $16 \cdot a = 4 \cdot 12$, or $16 \cdot a = 48$. Since $16 \cdot a$ and 48 are equal numbers, if they are both divided by the same number 16, the multiplier of a, we could assume that the quotients are also equal. That is,

$$\frac{16 \cdot a}{16} = \frac{48}{16}.$$

Since $\frac{16 \cdot a}{16} = \frac{16}{16} \cdot \frac{a}{1} = 1 \cdot a = a$, by dividing both sides of the cross-product equation by 16 we obtain the related division sentence $a = \frac{48}{16} = 3$. We will study this process in more detail in Chapter 9.

EXAMPLE 5   A PROPORTION IN A PROBLEM IN POLITICS

In an election the winning candidate won by a ratio of 5 to 4. If she received 600 votes, how many votes did the losing candidate receive?

If a represents the number of votes received by the losing candidate, the proportion

winning ratio $\rightarrow \dfrac{5}{4} = \dfrac{600}{a}$ $\leftarrow$ votes for winner
$\leftarrow$ votes for loser

describes the problem.

$5 \cdot a = 4 \cdot 600$ Equate cross products
$5 \cdot a = 2400$ Simplify
$a = \dfrac{2400}{5}$ Divide by 5
$a = 480$ Simplify

Thus, the loser received 480 votes.

PRACTICE EXERCISE 5

In Professor Walter's algebra class, the ratio of girls to boys is 4 to 3. If there are 18 boys in the class, how many girls are there?

Let a represent the number of girls in the class.

$$\frac{4}{3} = \frac{a}{18}$$

Answer: 24 girls

6.1 EXERCISES A

Write the fraction form for the given ratios.

1. 1 to 8 $\dfrac{1}{8}$

2. 2 to 7 $\dfrac{2}{7}$

3. 6 to 4 $\dfrac{6}{4}$ or $\dfrac{3}{2}$

4. 12 to 3 $\dfrac{12}{3}$ or $\dfrac{4}{1}$

5. 6 to 1 $\dfrac{6}{1}$

6. 4 to 4 $\dfrac{4}{4}$ or $\dfrac{1}{1}$

Write the ratio that is indicated by the given fraction.

7. $\dfrac{1}{4}$ 1 to 4

8. $\dfrac{3}{5}$ 3 to 5

9. $\dfrac{5}{2}$ 5 to 2

10. $\dfrac{13}{6}$ 13 to 6

11. $\dfrac{6}{13}$ 6 to 13

12. $\dfrac{9}{9}$ 9 to 9 or 1 to 1

Write as a ratio and reduce.

13. 200 miles in 4 hours
50 mph

14. $10 for 5 pounds of candy
$2 per pound

15. 24 children in 8 families
3 children per family

16. 125 miles on 5 gallons of gas
25 mpg

17. 75¢ for 3 candy bars
25¢ per candy bar

18. 88 ounces in 11 glasses
8 ounces per glass

19. 2700 words on 12 pages
225 words per page

20. $1350 earned in 3 weeks
$450 per week

21. 750 television sets in 500 households (reduce to a one-household comparison)
1.5 televisions per household

22. 10 cups of flour in 4 cakes (reduce to a one-cake comparison)
2.5 cups of flour per cake

Solve the proportions.

23. $\dfrac{a}{3}=\dfrac{14}{21}$ 2

24. $\dfrac{9}{a}=\dfrac{3}{1}$ 3

25. $\dfrac{4}{7}=\dfrac{a}{28}$ 16

26. $\dfrac{40}{35}=\dfrac{2}{a}$ $\dfrac{7}{4}$

27. $\dfrac{a}{13}=\dfrac{3}{39}$ 1

28. $\dfrac{12}{a}=\dfrac{4}{1}$ 3

29. $\dfrac{60}{72}=\dfrac{a}{6}$ 5

30. $\dfrac{2}{1}=\dfrac{1}{a}$ $\dfrac{1}{2}$

Solve.

31. If it takes 4 hours to travel 208 miles, how many hours would it take to travel 364 miles?   **7 hours**

32. Representative Wettaw won an election by a ratio of 8 to 5. If he received 10,400 votes, how many votes did his opponent receive?   **6500 votes**

33 If a wire 70 ft long weighs 84 pounds, how much will 110 ft of the same wire weigh?   **132 pounds**

34. If $\frac{1}{2}$ inch on a map represents 10 miles, how many miles are represented by $6\frac{1}{2}$ inches?   **130 miles**

35. If 3 coats cost a total of $156, how many coats can be bought for $364?   **7 coats**

36 A ranger wishes to estimate the number of antelope in a preserve. He catches 58 antelope, tags their ears, and returns them to the preserve. Some time later, he catches 29 antelope and discovers that 7 of them are tagged. Estimate the number of antelope in the preserve.   **about 240 antelope**

37. A baseball pitcher gave up 60 earned runs in 180 innings. Estimate the number of earned runs he will give up every 9 innings. This number is called his *earned run average*.   3

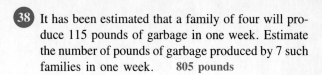 It has been estimated that a family of four will produce 115 pounds of garbage in one week. Estimate the number of pounds of garbage produced by 7 such families in one week.   **805 pounds**

FOR REVIEW

The following exercises will help you prepare for the next section. Change each fraction or mixed number to a decimal.

39. $\dfrac{13}{4}$ 3.25

40. $2\dfrac{5}{11}$ $2.\overline{45}$

41. $\dfrac{4}{5}$ 0.8

Change each decimal to a fraction or mixed number.

42. 0.3 $\dfrac{3}{10}$

43. 0.14 $\dfrac{7}{50}$

44. 5.6 $5\dfrac{3}{5}$

45. Change $\dfrac{4}{7}$ to a decimal, rounded to the nearest hundredth.   0.57

ANSWERS: 1. $\frac{1}{8}$ 2. $\frac{2}{7}$ 3. $\frac{6}{4}$ or $\frac{3}{2}$ 4. $\frac{12}{3}$ or $\frac{4}{1}$ 5. $\frac{6}{1}$ 6. $\frac{4}{4}$ or $\frac{1}{1}$ 7. 1 to 4 8. 3 to 5 9. 5 to 2 10. 13 to 6 11. 6 to 13 12. 9 to 9 or 1 to 1 13. 50 mph 14. \$2 per pound 15. 3 children per family 16. 25 mpg 17. 25¢ per candy bar 18. 8 ounces per glass 19. 225 words per page 20. \$450 per week 21. 1.5 televisions per household 22. 2.5 cups of flour per cake 23. 2 24. 3 25. 16 26. $\frac{7}{4}$ 27. 1 28. 3 29. 5 30. $\frac{1}{2}$ 31. 7 hours 32. 6500 votes 33. 132 pounds 34. 130 miles 35. 7 coats 36. about 240 antelope 37. 3 38. 805 pounds 39. 3.25 40. $2.\overline{45}$ 41. 0.8 42. $\frac{3}{10}$ 43. $\frac{7}{50}$ 44. $5\frac{3}{5}$ 45. 0.57

6.1 EXERCISES B

Write the fraction form for the given ratios.

1. 1 to 12 $\dfrac{1}{12}$

2. 3 to 7 $\dfrac{3}{7}$

3. 9 to 3 $\dfrac{9}{3}$ or $\dfrac{3}{1}$

4. 15 to 6 $\dfrac{15}{6}$ or $\dfrac{5}{2}$

5. 13 to 1 $\dfrac{13}{1}$

6. 6 to 6 $\dfrac{6}{6}$ or $\dfrac{1}{1}$

Write the ratio that is indicated by the given fractions.

7. $\dfrac{1}{5}$ 1 to 5

8. $\dfrac{4}{7}$ 4 to 7

9. $\dfrac{7}{2}$ 7 to 2

10. $\dfrac{15}{4}$ 15 to 4

11. $\dfrac{4}{15}$ 4 to 15

12. $\dfrac{8}{8}$ 8 to 8 or 1 to 1

Write as a ratio and reduce.

13. 120 feet in 4 seconds
   30 feet per second

14. \$12 for 4 pounds of nuts
   \$3 per pound

15. 320 fish in 16 aquariums
   20 fish per aquarium

16. 720 kilometers on 60 liters of gasoline
   12 kilometers per liter

17. \$7 for 4 tubes of toothpaste
   \$1.75 per tube

18. 360 gallons in 3 tanks
   120 gallons per tank

19. 250 chairs to 150 people
5 chairs for every 3 people

20. $2700 in 4 savings accounts
$675 per account

21. 450 cars in 200 families (reduce to a one-family comparison)   **2.25 cars per family**

22. 18 hours to split 4 cords of wood (reduce to a one-cord comparison)   **4.5 hours per cord**

Solve the proportions.

23. $\dfrac{a}{2} = \dfrac{11}{22}$ **1**

24. $\dfrac{7}{a} = \dfrac{14}{10}$ **5**

25. $\dfrac{3}{11} = \dfrac{a}{22}$ **6**

26. $\dfrac{50}{45} = \dfrac{10}{a}$ **9**

27. $\dfrac{a}{14} = \dfrac{28}{56}$ **7**

28. $\dfrac{15}{a} = \dfrac{3}{5}$ **25**

29. $\dfrac{56}{64} = \dfrac{a}{8}$ **7**

30. $\dfrac{3}{1} = \dfrac{1}{a}$ $\dfrac{1}{3}$

Solve.

31. If it takes 6 hours to travel 210 miles, how many hours will it take to travel 385 miles?   **11 hours**

32. Peter Horn won a city council election by a ratio of 7 to 4. If he received 4522 votes, how many votes did his opponent receive?   **2584 votes**

33. If a tree 15 ft tall casts a shadow 4 ft long, how long will the shadow of a 40 ft building be at the same instant? $10\frac{2}{3}$ **ft**

34. If $\frac{1}{4}$ inch on a map represents 20 miles, how many inches are required to represent 150 miles?
$1\frac{7}{8}$ **inches**

35. If 4 cassette tapes cost a total of $31.80, how many tapes could be bought for $71.55?   **9 tapes**

36. A game and fish officer wishes to estimate the number of trout in a lake. He catches 94 trout, tags their fins, and returns them to the lake. Some time later, he catches 50 trout and discovers that 12 of them are tagged. Estimate the number of trout in the lake. **about 392 trout**

37. A softball pitcher gave up 24 earned runs in 210 innings. The number of earned runs given up every 9 innings is called a pitcher's earned run average. Calculate her earned run average (to the nearest hundredth).   **1.03**

38. It has been estimated that a family of three will produce 85 pounds of garbage in one week. Estimate the number of pounds of garbage produced by 10 such families in one week.   **850 pounds**

FOR REVIEW

The following exercises will help you prepare for the next section. Change each fraction or mixed number to a decimal.

39. $\dfrac{13}{8}$ **1.625**

40. $1\dfrac{8}{9}$ $1.\overline{8}$

41. $\dfrac{3}{20}$ **0.15**

Change each decimal to a fraction or mixed number.

42. 0.7 $\dfrac{7}{10}$

43. 0.32 $\dfrac{8}{25}$

44. 2.05 $2\dfrac{1}{20}$

45. Change $\frac{3}{7}$ to a decimal, rounded to the nearest hundredth.   **0.43**

6.1 EXERCISES C

1. In a recent survey of 6274 adults, it was found that 3726 were female and 2548 were male. Find each ratio and use a calculator to express the fraction as a decimal rounded to the nearest hundredth.

   (a) What is the ratio of males to total adults in the survey? **0.41**

   (b) What is the ratio of females to total adults in the survey? **0.59**

   (c) What is the male-to-female-ratio? **0.68**

   (d) What is the female-to-male-ratio? **1.46**

2. A car travels 468.9 miles on 21.7 gallons of gas. Traveling at the same rate, how many gallons of gas would the car need to go 750 miles (to the nearest tenth)?
   [Answer: 34.7 gallons]

6.2 PERCENT AND PERCENT CONVERSIONS

STUDENT GUIDEPOSTS

1 The Meaning of Percent

2 Changing a Percent to a Fraction

3 Changing a Percent to a Decimal

4 Changing a Decimal to a Percent

5 Changing a Fraction to a Percent

1 THE MEANING OF PERCENT

The idea of percent and the percent symbol (%) are used widely in our society. We may pay a five-percent (5%) sales tax, read that Mr. Gomez won the election with 52% of the vote, or hear that a basketball player hit 80% of his free throws.

The word **percent** means ''per hundred.'' That is, it refers to the number of parts in each one hundred parts. For example, when the tax rate is 5%, we pay 5¢ tax on each 100¢ purchase.

$$\text{Percent: } 5\% \qquad \text{Fraction or ratio: } \frac{5}{100}$$

In Figure 6.2, the five shaded coins (5¢) correspond to 5% tax on the one hundred coins (100¢).

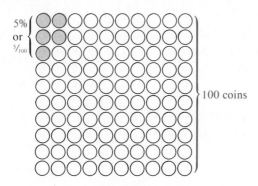

Figure 6.2

Thus, 5% is $\dfrac{5}{100}$ or $5 \times \dfrac{1}{100}$ or $5 \times 0.01.$

❷ CHANGING A PERCENT TO A FRACTION

If Mr. Gomez received 52 out of every 100 votes, then he got 52% of the vote.

$$\text{Percent: } 52\% \qquad \text{Fraction or ratio: } \frac{52}{100}$$

HINT

The words "out of" can be replaced with a division bar. Notice above how "52 out of every 100 votes" became the fraction $\frac{52}{100}$.

A basketball player could have hit 8 out of 10 (or 80 out of 100) free throws which we shall see is 80%.

$$\text{Percent: } 80\% \qquad \text{Fraction or ratio: } \frac{80}{100} = \frac{8}{10}$$

We also hear on the news that the price of gasoline is 250% of its price several years ago.

$$\text{Percent: } 250\% \qquad \text{Fraction or ratio: } \frac{250}{100} = 2.5$$

Thus, 250% means that the price of gasoline is 2.5 times what it was several years ago.

There are also percents less than 1%. For example, the cost of living might increase 0.8% during the month of December.

$$\text{Percent: } 0.8\% \qquad \text{Fraction or ratio: } \frac{0.8}{100} = \frac{(0.8)(10)}{(100)(10)} = \frac{8}{1000}$$

In each of these examples, the percent was changed to a fraction by removing the percent symbol (%) and dividing by 100. Since dividing by 100 is the same as multiplying by $\frac{1}{100}$, we can form the following rule.

To Change a Percent to a Fraction

1. Replace the % symbol with either $\left(\times \frac{1}{100}\right)$ or $(\div 100)$.
2. Evaluate and then reduce the fraction.

EXAMPLE 1 CHANGING PERCENTS TO FRACTIONS

Change each percent to a fraction.

(a) $38\% = 38 \times \dfrac{1}{100} = \dfrac{38}{100} = \dfrac{19 \cdot \cancel{2}}{50 \cdot \cancel{2}} = \dfrac{19}{50}$ Replace % with $\left(\times \frac{1}{100}\right)$ and reduce the fraction

(b) $500\% = 500 \times \dfrac{1}{100} = \dfrac{500}{100} = \dfrac{5 \cdot \cancel{100}}{1 \cdot \cancel{100}} = \dfrac{5}{1} = 5$

(c) $0.3\% = 0.3 \div 100 = \dfrac{0.3}{100} = \dfrac{(0.3)(10)}{(100)(10)}$ Multiply by $\frac{10}{10}$ to get a whole number in the numerator; this is the same as moving the decimal point one place to the right in both the numerator and denominator

$\qquad\qquad = \dfrac{3}{1000}$

PRACTICE EXERCISE 1

Change each percent to a fraction.

(a) 25%

(b) 300%

(c) 0.5%

(d) $2\frac{3}{4}\% = \left(2\frac{3}{4}\right)\left(\frac{1}{100}\right)$

$= \frac{11}{4} \cdot \frac{1}{100} = \frac{11}{400}$ Note: $2\frac{3}{4} = \frac{4 \cdot 2 + 3}{4} = \frac{11}{4}$

(e) $66\frac{2}{3}\% = \left(66\frac{2}{3}\right)\left(\frac{1}{100}\right)$

$= \left(\frac{200}{3}\right)\left(\frac{1}{100}\right)$ Note: $66\frac{2}{3} = \frac{3 \cdot 66 + 2}{3} = \frac{200}{3}$

$= \frac{2 \cdot \cancel{100}}{3 \cdot \cancel{100}} = \frac{2}{3}$

(f) $100\% = 100 \div 100 = \frac{100}{100} = 1$

(d) $3\frac{1}{5}\%$

(e) $33\frac{1}{3}\%$

(f) 1000%

Answers: (a) $\frac{1}{4}$ (b) 3 (c) $\frac{1}{200}$
(d) $\frac{4}{125}$ (e) $\frac{1}{3}$ (f) 10

❸ CHANGING A PERCENT TO A DECIMAL

Percents are frequently given as decimals. Changing to a decimal is exactly the same as changing to a fraction; simply remove the % symbol and multiply by $\frac{1}{100}$, which is 0.01 in decimal form.

To Change a Percent to a Decimal
Replace the % symbol with ($\times$ 0.01) and multiply.

HINT

Multiplying a number by 0.01 is the same as moving the decimal point two places to the left. Thus, to change 5% to a decimal write 0.05.

EXAMPLE 2 CHANGING PERCENTS TO DECIMALS	PRACTICE EXERCISE 2

Change each percent to a decimal.

(a) $29\% = 29 \times 0.01$ Remember that 29. is the same as 29, then move
$= 0.29. = 0.29$ the decimal point two places to the left

(b) $39.5\% = 39.5 \times 0.01$ Move the decimal point two places to the left
$= 0.39.5 = 0.395$

(c) $0.7\% = (0.7)(0.01)$
$= 0.00.7 = 0.007$ Put two zeros before 7

(d) $823\% = (823)(0.01) = 8.23. = 8.23$

Change each percent to a decimal.

(a) 65%

(b) 26.5%

(c) 0.3%

(d) 550%

(e) $2\frac{3}{4}\% = 2.75\%$ $\frac{3}{4} = 0.75$

$\qquad = (2.75)(0.01)$

$\qquad = 0.\underset{\smile}{02}.75 = 0.0275$

(f) $0.01\% = (0.01)(0.01) = 0.\underset{\smile}{00}.01 = 0.0001$

(e) $7\frac{3}{5}\%$

(f) 1%

Answers: (a) **0.65** (b) **0.265**
(c) **0.003** (d) **5.5** (e) **0.076**
(f) **0.01**

HINT

Remember that the % symbol means ''divide by 100 or multiply by $\frac{1}{100} = 0.01$.'' Thus, after the % symbol is removed and the change to decimal notation is made, you will have a smaller number. It may help if you keep in mind a simple example, such as

$$50\% = 0.5 \quad \text{or} \quad 50\% \text{ means } \frac{1}{2}.$$

④ CHANGING A DECIMAL TO A PERCENT

Some practical problems require changing from percent to fractions and decimals. Others require converting from a decimal or a fraction to a percent. To change a decimal to a percent we multiply by 100, the reverse of dividing by 100, which was used to change a percent to a decimal.

To Change a Decimal to a Percent
Multiply the decimal by 100 and attach the % symbol.

HINT

Multiplying a number by 100 is the same as moving the decimal point two places to the right. Thus, to change a decimal to a percent, move the decimal point two places to the right and attach the % symbol.

EXAMPLE 3   CHANGING DECIMALS TO PERCENTS

Change each decimal to a percent.

(a) $0.31 = (0.31)(100)\% = 31.\% = 31\%$ Move the decimal point two places to the right

(b) $3.25 = (3.25)(100)\% = 325.\% = 325\%$

(c) $1 = (1)(100)\% = 100.\% = 100\%$ Attach two zeros when moving the decimal point

(d) $0.01 = (0.01)(100)\% = 01.\% = 1\%$

(e) $0.007 = (0.007)(100)\% = 00.7\% = 0.7\%$

(f) $56.2 = (56.2)(100)\% = 5620.\% = 5620\%$

PRACTICE EXERCISE 3

Change each decimal to a percent.

(a) 0.62

(b) 2.75

(c) 2

(d) 0.02

(e) 0.002

(f) 35.3

Answers: (a) **62%** (b) **275%**
(c) **200%** (d) **2%** (e) **0.2%**
(f) **3530%**

HINT

To keep from moving the decimal point the wrong way, keep in mind a simple example, such as $0.5 = 50\%$.

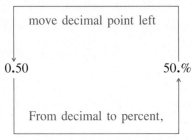

From percent to decimal,

move decimal point left

0.50   50.%

From decimal to percent,

move decimal point right

Consider the first letter of the words **D**ecimal and **P**ercent. Since **D** comes before **P** in the alphabet, the letters **DP** can remind us that to change from **D**ecimal to **P**ercent, the decimal point is moved from left to right while to change from **P**ercent to **D**ecimal, it is moved from right to left.

5 CHANGING A FRACTION TO A PERCENT

We change fractions to percents by multiplying the fraction by 100 and attaching the % symbol, just as was done with decimals.

To Change a Fraction to a Percent

Multiply the fraction by 100 and attach the % symbol.

EXAMPLE 4 CHANGING FRACTIONS TO PERCENTS

Change each fraction to a percent.

(a) $\dfrac{4}{5} = \dfrac{4}{5} \times 100\%$

$= \dfrac{4 \times \overset{20}{\cancel{100}}}{\cancel{5}}\%$

$= (4 \times 20)\% = 80\%$

(b) $\dfrac{1}{8} = \dfrac{1}{8} \times 100\% = \dfrac{100}{8}\%$

$= 12.5\%$ Divide 100 by 8

This could also be written as $12\frac{1}{2}\%$.

(c) $\dfrac{2}{3} = \dfrac{2}{3} \times 100\% = \dfrac{200}{3}\%$

$= 66.\overline{6}\%$ Divide 200 by 3

Since $0.\overline{6} = \frac{2}{3}$, we could also write this as $66\frac{2}{3}\%$.

PRACTICE EXERCISE 4

Change each fraction to a percent.

(a) $\dfrac{3}{10}$

(b) $\dfrac{5}{8}$

(c) $\dfrac{1}{3}$

Answers: (a) 30% (b) 62.5% or $62\frac{1}{2}\%$ (c) $33.\overline{3}\%$ or $33\frac{1}{3}\%$

In Example 4(**c**) we could have rounded the percent to 66.7% (correct to the nearest tenth of a percent) or to 66.67% (correct to the nearest hundredth of a percent) in a practical situation.

| EXAMPLE 5   FINDING APPROXIMATE PERCENTS | PRACTICE EXERCISE 5 |

Change $\frac{2}{7}$ to an approximate percent by rounding to the nearest percent, nearest tenth of a percent, and nearest hundredth of a percent.

```
      .28571
  7)2.00000       Divide 2 by 7
    1 4
    60
    56
    40
    35
    50
    49
    10
     7
     3
```

$\frac{2}{7} \approx 0.29 = 29\%$ To the nearest percent

$\frac{2}{7} \approx 0.286 = 28.6\%$ To the nearest tenth of a percent

$\frac{2}{7} \approx 0.2857 = 28.57\%$ To the nearest hundredth of a percent

Change $\frac{5}{7}$ to an approximate percent by rounding to the nearest percent, nearest tenth of a percent, and nearest hundredth of a percent.

Answer: 71%, 71.4%; 71.43%

Some fractions, decimals, and percents which occur often are listed in the table. The numbers on each line are equal. You should memorize this table for use in the exercises.

Fraction	Decimal	Percent	Fraction	Decimal	Percent
$\frac{1}{20}$	0.05	5%	$\frac{1}{2}$	0.5	50%
$\frac{1}{10}$	0.1	10%	$\frac{3}{5}$	0.6	60%
$\frac{1}{8}$	0.125	12.5%	$\frac{5}{8}$	0.625	62.5%
$\frac{1}{6}$	$0.1\overline{6}$	$16.\overline{6}\% = 16\frac{2}{3}\%$	$\frac{2}{3}$	$0.\overline{6}$	$66.\overline{6}\% = 66\frac{2}{3}\%$
$\frac{1}{5}$	0.2	20%	$\frac{7}{10}$	0.7	70%
$\frac{1}{4}$	0.25	25%	$\frac{3}{4}$	0.75	75%
$\frac{3}{10}$	0.3	30%	$\frac{4}{5}$	0.8	80%
$\frac{1}{3}$	$0.\overline{3}$	$33.\overline{3}\% = 33\frac{1}{3}\%$	$\frac{5}{6}$	$0.8\overline{3}$	$83.\overline{3}\% = 83\frac{1}{3}\%$
$\frac{3}{8}$	0.375	37.5%	$\frac{7}{8}$	0.875	87.5%
$\frac{2}{5}$	0.4	40%	$\frac{9}{10}$	0.9	90%
			1	1.0	100%

Table of Equal Fractions, Decimals, and Percents

6.2 EXERCISES A

1. The word *percent* means "per ___hundred___."

2. To change a percent to a fraction drop the % symbol and divide by 100 or multiply by ___$\frac{1}{100}$___.

3. To change a percent to a decimal replace the % symbol with ($\times$ 0.01) and multiply. This is the same as moving the decimal point two places to the ___left___.

Change each percent to a fraction.

4. 87% $\frac{87}{100}$ 5. 16% $\frac{4}{25}$ 6. 50% $\frac{1}{2}$ 7. 1% $\frac{1}{100}$

8. 117% $\frac{117}{100}$ 9. 125% $\frac{5}{4}$ 10. 1000% 10   11. $\frac{1}{4}$% $\frac{1}{400}$

12. 0.7% $\frac{7}{1000}$ 13. 12.5% $\frac{1}{8}$ 14. $16\frac{2}{3}$% $\frac{1}{6}$ 15. 0.05% $\frac{1}{2000}$

Change each percent to a decimal.

16. 92% 0.92   17. 60% 0.6   18. 1% 0.01   19. 145% 1.45

20. 1000% 10   21. 0.1% 0.001   22. 37.35% 0.3735   23. 200% 2

24. $\frac{1}{4}$% 0.0025   25. 800% 8   26. 4392.5% 43.925   27. $6\frac{3}{4}$% 0.0675

28. In the final two games of the season, a quarterback completed 58 passes in 100 attempts.
    (a) What fractional part of his attempts did he complete? $\frac{58}{100}$ or $\frac{29}{50}$ (reduced)
    (b) Write the fraction in (a) as a percent. **58%**
    (c) Write the percent in (b) as a decimal. **0.58**

29. On a recent placement exam, 45 students out of 100 who took the exam placed in algebra.
    (a) What fractional part of the students placed in algebra? $\frac{45}{100}$ or $\frac{9}{20}$ (reduced)
    (b) Write the fraction in (a) as a percent. **45%**
    (c) Write the percent in (b) as a decimal. **0.45**

30. Of all college basketball players, fewer than 0.8% will play in the NBA. Write 0.8% as a decimal. **0.008**

Change each decimal to a percent.

31. 0.07 7%   32. 0.375 37.5%   33. 3.75 375%

34. 37.5 3750%   35. 375 37,500%   36. 0.009 0.9%

37. 0.0009 0.09%   38. 0.32 32%   39. 0.032 3.2%

Change each fraction to a percent.

40. $\dfrac{13}{10}$ 130%

41. $\dfrac{7}{4}$ 175%

42. $\dfrac{5}{3}$ $166\dfrac{2}{3}\%$

43. $\dfrac{7}{50}$ 14%

44. $\dfrac{3}{25}$ 12%

45. $\dfrac{1}{1000}$ 0.1%

46 $\dfrac{100}{3}$ $3333\dfrac{1}{3}\%$

47. $\dfrac{7}{2}$ 350%

48. $\dfrac{106}{200}$ 53%

Find the approximate percent to the nearest tenth of a percent.

49 $\dfrac{4}{7}$ 57.1%

50. $\dfrac{2}{9}$ 22.2%

51. $\dfrac{15}{13}$ 115.4%

Complete the table given one value in each row.

	Fraction	Decimal	Percent			Fraction	Decimal	Percent
52.	$\dfrac{3}{10}$	0.3	30%		**53.**	$\dfrac{3}{5}$	0.6	60%
54.	$\dfrac{5}{8}$	0.625	62.5%		**55.**	1	1.0	100%
56.	$\dfrac{7}{8}$	0.875	87.5%		**57.**	$\dfrac{2}{3}$	$0.\overline{6}$	$66\dfrac{2}{3}\%$
58.	$\dfrac{1}{20}$	0.05	5%		**59.**	$\dfrac{1}{3}$	$0.\overline{3}$	$33\dfrac{1}{3}\%$
60.	$\dfrac{5}{6}$	$0.8\overline{3}$	$83\dfrac{1}{3}\%$		**61.**	$\dfrac{1}{2}$	0.5	50%
62.	$\dfrac{1}{6}$	$0.1\overline{6}$	$16\dfrac{2}{3}\%$		**63.**	$\dfrac{1}{8}$	0.125	12.5%

Give the answers to the following exercises as a fraction, as a decimal, and as a percent.

64. Of 30 students in an algebra class, 18 are girls. What portion of the students is girls?
$\dfrac{18}{30}$ or $\dfrac{3}{5}$ (reduced); 0.6; 60%

65. What portion of the students in Exercise 64 is boys?
$\dfrac{12}{30}$ or $\dfrac{2}{5}$ (reduced); 0.4% 40%

66. A chemist mixed 12 parts of acid with 13 parts of distilled water to obtain a particular solution. What part of the solution is acid? $\dfrac{12}{25}$; 0.48; 48%

67. What part of the solution in Exercise 66 is distilled water? $\dfrac{13}{25}$; 0.52; 52%

FOR REVIEW

68. Write the fraction form for the ratio 7 to 8.

$\frac{7}{8}$

69. Write the ratio that is indicated by $\frac{12}{11}$.

12 to 11

Solve the proportions.

70. $\frac{12}{a} = \frac{6}{2}$ 4

71. $\frac{7}{1} = \frac{1}{a}$ $\frac{1}{7}$

Solve.

72. If it takes 8 pounds of fertilizer to cover 3000 square feet of lawn, how many pounds will be necessary to cover 7500 square feet?   **20 pounds**

73. If it takes 8 pounds of fertilizer to cover 300 square feet of lawn, how many square feet can be covered by 35 pounds of fertilizer?   **1312.5 square feet**

The following exercises review topics covered in Section 3.4 to help you prepare for the next section. Solve.

74. If one-half of a number is 225, find the number.
450

75. After driving 720 miles, the Blows had completed $\frac{3}{5}$ of their trip. How long was their trip?

1200 miles

ANSWERS: 1. hundred 2. $\frac{1}{100}$ 3. left 4. $\frac{87}{100}$ 5. $\frac{4}{25}$ 6. $\frac{1}{2}$ 7. $\frac{1}{100}$ 8. $\frac{117}{100}$ 9. $\frac{5}{4}$ 10. 10 11. $\frac{1}{400}$ 12. $\frac{7}{1000}$ 13. $\frac{1}{8}$ 14. $\frac{1}{6}$ 15. $\frac{1}{2000}$ 16. 0.92 17. 0.6 18. 0.01 19. 1.45 20. 10 21. 0.001 22. 0.3735 23. 2 24. 0.0025 25. 8 26. 43.925 27. 0.0675 28. (a) $\frac{58}{100}$ $\left(\text{or } \frac{29}{50}\text{, reduced to lowest terms}\right)$ (b) 58% (c) 0.58 29. (a) $\frac{45}{100}$ $\left(\text{or } \frac{9}{20}\text{, reduced to lowest terms}\right)$ (b) 45% (c) 0.45 30. 0.008 31. 7% 32. 37.5% 33. 375% 34. 3750% 35. 37,500% 36. 0.9% 37. 0.09% 38. 32% 39. 3.2% 40. 130% 41. 175% 42. $166\frac{2}{3}$% 43. 14% 44. 12% 45. 0.1% 46. $3333\frac{1}{3}$% 47. 350% 48. 53% 49. 57.1% 50. 22.2% 51. 115.4% 52. 0.3; 30% 53. $\frac{3}{5}$; 60% 54. $\frac{5}{8}$; 0.625 55. 1.0; 100% 56. $\frac{7}{8}$; 87.5% 57. $\frac{2}{3}$; $0.\overline{6}$ 58. 0.05; 5% 59. $\frac{1}{3}$; $33\frac{1}{3}$% 60. $\frac{5}{6}$; $0.8\overline{3}$ 61. 0.5; 50% 62. $\frac{1}{6}$; $16\frac{2}{3}$% 63. $\frac{1}{8}$; 0.125 64. $\frac{18}{30}$ or $\frac{3}{5}$ (reduced); 0.6; 60% 65. $\frac{12}{30}$ or $\frac{2}{5}$ (reduced); 0.4; 40% 66. $\frac{12}{25}$; 0.48; 48% 67. $\frac{13}{25}$; 0.52; 52% 68. $\frac{7}{8}$ 69. 12 to 11 70. 4 71. $\frac{1}{7}$ 72. 20 pounds 73. 1312.5 square feet 74. 450 75. 1200 miles

---

6.2   EXERCISES B

1. The word which means "per hundred" is _____percent_____.

2. To change a percent to a decimal, remove the % symbol and move the decimal point to the left _____two_____ places.

3. To change a percent to a fraction multiply by 0.01 or divide by _____100_____.

Change each percent to a fraction.

4. 98% $\frac{49}{50}$

5. 22% $\frac{11}{50}$

6. 20% $\frac{1}{5}$

7. 6% $\frac{3}{50}$

8. 110% $\frac{11}{10}$

9. 375% $\frac{15}{4}$

10. 2000%   20

11. $\frac{3}{4}$% $\frac{3}{400}$

12. 0.2% $\frac{1}{500}$

13. 23.6% $\frac{59}{250}$

14. $83\frac{1}{3}$% $\frac{5}{6}$

15. 0.15% $\frac{3}{2000}$

Change each percent to a decimal.

16. 73%   0.73   **17.** 80%   0.8   **18.** 8%   0.08   **19.** 235%   2.35

20. 5000%   50   **21.** 0.8%   0.008   **22.** 62.45%   0.6245   **23.** 700%   7

24. $\frac{3}{4}$%   0.0075   **25.** 100%   1   **26.** 7325.4%   73.254   **27.** $10\frac{2}{5}$%   0.104

28. In a sample of 100 radios, 2 were found to be defective.
  (a) What fractional part of the radios was defective? $\frac{2}{100}$ or $\frac{1}{50}$ (reduced)
  (b) Write the fraction in (a) as a percent.   2%
  (c) Write the percent in (b) as a decimal.   0.02

29. Of 100 students in Weitzel Elementary School, 25 were found to have at least one cavity in a tooth.
  (a) What fractional part of the students had a cavity? $\frac{25}{100}$ or $\frac{1}{4}$ (reduced)
  (b) Write the fraction in (a) as a percent.   25%
  (c) Write the percent in (b) as a decimal.   0.25

30. During the month of June, of the cars sold at Tyrrell Chevrolet-Buick, 42.5% were Buicks. Write 42.5% as a decimal.   0.425

Change each decimal to a percent.

31. 0.09   9%   **32.** 0.435   43.5%   **33.** 4.35   435%

34. 43.5   4350%   **35.** 435   43,500%   **36.** 0.008   0.8%

37. 0.0008   0.08%   **38.** 0.62   62%   **39.** 0.062   6.2%

Change each fraction to a percent.

40. $\frac{25}{10}$ 250%   **41.** $\frac{11}{4}$ 275%   **42.** $\frac{4}{3}$ $133\frac{1}{3}$%

43. $\frac{11}{50}$ 22%   **44.** $\frac{4}{25}$ 16%   **45.** $\frac{1}{2000}$ 0.05%

46. $\frac{400}{3}$ $13,333.\overline{3}$%   **47.** $\frac{13}{2}$ 650%   **48.** $\frac{48}{200}$ 24%

Find the approximate percent to the nearest tenth of a percent.

49. $\frac{6}{7}$ 85.7%   **50.** $\frac{5}{9}$ 55.6%   **51.** $\frac{17}{13}$ 130.8%

Complete the table given one value in each row.

	Fraction	Decimal	Percent		Fraction	Decimal	Percent
52.	$\frac{3}{4}$	0.75	75%	**53.**	$\frac{1}{5}$	0.2	20%
54.	$\frac{9}{10}$	0.9	90%	**55.**	$\frac{1}{10}$	0.1	10%

	Fraction	Decimal	Percent			Fraction	Decimal	Percent
56.	$\frac{4}{5}$	0.8	80%	**57.**		$\frac{1}{4}$	0.25	25%
58.	$\frac{1}{8}$	0.125	12.5%	**59.**		$\frac{7}{10}$	0.7	70%
60.	$\frac{3}{8}$	0.375	37.5%	**61.**		$\frac{5}{6}$	$0.8\bar{3}$	$83.\bar{3}\%$
62.	$\frac{2}{5}$	0.4	40%	**63.**		$\frac{1}{3}$	$0.\bar{3}$	$33\frac{1}{3}\%$

Give the answers to the following exercises as a fraction, as a decimal, and as a percent.

64. On a 200-mile trip, Velma drove 140 miles and her friend Ingrid drove 60 miles. What portion of the trip did Velma drive?
$\frac{140}{200}$ or $\frac{7}{10}$ (reduced); 0.7; 70%

65. What portion of the trip in Exercise 64 did Ingrid drive?
$\frac{60}{200}$ or $\frac{3}{10}$ (reduced); 0.3; 30%

66. Of the 80 classes taught by the mathematics department at a community college, 55 were taught during the day and the rest were taught in the evening. What part of the classes was taught during the day?
$\frac{55}{80}$ or $\frac{11}{16}$ (reduced); 0.6875; 68.75%

67. What part of the classes in Exercise 66 was taught in the evening?
$\frac{25}{80}$ or $\frac{5}{16}$ (reduced); 0.3125; 31.25%

FOR REVIEW

68. Write the fraction form for the ratio 4 to 1.
$\frac{4}{1}$

69. Write the ratio which is indicated by $\frac{10}{10}$.
10 to 10 or 1 to 1

Solve the proportions.

70. $\frac{35}{a} = \frac{7}{1}$ 5

71. $\frac{1}{12} = \frac{a}{1}$ $\frac{1}{12}$

Solve.

72. If it takes 9 ribbons to type 3500 pages of a manuscript, how many ribbons are needed to type 700 pages?   **1.8 ribbons**

73. If it takes 9 ribbons to type 3500 pages of manuscript, how many pages can be typed using 3 ribbons?   **$1166.\bar{6}$ pages**

The following exercises review topics covered in Section 3.4 to help you prepare for the next section. Solve.

74. If one-fourth of a number is 110, find the number.
440

75. After working for 7 hours, Sarah had completed $\frac{2}{5}$ of a job. How long will it take to do the whole job?
17.5 hours

6.2 EXERCISES C

1. The price of a coat is reduced 35%.

(a) What fractional part of the original price amounts to the reduction?

(b) What fractional part of the original price is the new sale price?

(c) What percent of the original price must a buyer pay?

$\left[\text{Answer: (a) } \frac{7}{20} \text{ (b) } \frac{13}{20} \text{ (c) } 65\%\right]$

2. The sales-tax rate in Washburn, Indiana, is 3.5%. Write 3.5% as a decimal and as a fraction.
0.035; $\frac{7}{200}$

3. Each employee in the State Retirement System must contribute 6.35% of his salary to the retirement fund. Write 6.35% as a decimal and as a fraction.
0.0635; $\frac{127}{2000}$

Use a calculator to write each fraction as a percent rounded to the nearest hundredth of a percent.

4. $\frac{23}{41}$ **56.10%**

5. $\frac{7}{101}$ **6.93%**

6. $\frac{123}{477}$

[Answer: 25.79%]

7. $\frac{853}{329}$ **259.27%**

6.3 PERCENT PROBLEMS

═══════════════ **STUDENT GUIDEPOSTS** ═══════════════

❶ The Basic Percent Problem

❷ Solving Percent Problems (Method 1)

❸ Using Proportions to Solve Percent Problems (Method 2)

❹ Applications of Percent

❶ THE BASIC PERCENT PROBLEM

In Section 3.4 we solved problems of the type:

> On a two-day trip of 440 miles, Rob drove $\frac{1}{2}$ the distance the first day. How far did he drive the first day?

Letting □ represent the distance driven the first day, and recalling that the word "of" translates to multiplication, and the word "is" translates to equals,

$$\boxed{\text{distance 1st day}} \quad \text{is} \quad \frac{1}{2} \quad \text{of} \quad 440$$

$$\downarrow \qquad\qquad \downarrow \quad \downarrow \;\; \downarrow \quad \downarrow$$

$$\Box \qquad\quad = \left(\frac{1}{2}\right) \cdot (440).$$

Thus,

$$\Box = \frac{1}{2} \cdot 440 = 220 \text{ miles.}$$

Using the notion of percent, the same problem might be stated as follows:

> On a two-day trip of 440 miles, Rob drove 50% of the distance the first day. How far did he drive the first day?

By changing 50% to the fraction $\frac{1}{2}$, this problem can be solved as above. Similarly, 50% could be changed to the decimal 0.5.

$$\boxed{\text{distance 1st day}} \quad \text{is} \quad 50\% \;\; \text{of} \;\; 440$$

$$\downarrow \qquad\qquad \downarrow \;\; \downarrow \quad \downarrow \quad \downarrow$$

$$\Box \qquad\quad = (0.5) \cdot (440)$$

Thus,

$$\Box = (0.5)(440) = 220 \text{ miles.}$$

This problem is an example of the basic percent problem that takes the form

$$A \text{ is } P\% \text{ of } B \qquad \text{or} \qquad P\% \text{ of } B \text{ is } A.$$

When we solve a percent problem we must find one of A, P, or B, given the other two. In a percent problem, we can identify A, P, and B using the following:

A is the *amount* and it is next to the word *is*,
P is the *percent* and is followed by the % symbol,
B is the *base* and it follows the word *of*.

For example, consider the following statement.

$$48 \text{ is } 25\% \text{ of } 192$$
$$A = P\% \cdot B$$

This equation is called the **basic percent equation** and will be used in solving percent problems.

❷ SOLVING PERCENT PROBLEMS (METHOD 1)

We now illustrate the first method for solving a basic percent problem, ''A is $P\%$ of B,'' by finding the amount A when the percent P and base B are given.

EXAMPLE 1   SOLVING FOR THE AMOUNT A

Solve by using the basic percent equation $A = P\% \cdot B$.

(a) What is 5% of $22.40?

In this case, A is the unknown, $P = 5$, and $B = 22.40$. Thus we have

$$\boxed{\text{What number}} \text{ is } 5\% \text{ of } 22.40 ?$$
$$A = P\% \cdot B$$
$$A = (0.05) \cdot 22.40$$

Notice that *is* translates to $=$, *of* translates to $\cdot$ (times or multiplication), and 5% has been changed to decimal form, 0.05. To find A we simply multiply.

$$A = (0.05) \cdot (22.40) = 1.12$$

Thus, A is $1.12, that is, $1.12 is 5% of $22.40.

(b) 120% of 450 is what number?

This translates as follows:

$$120\% \text{ of } 450 \text{ is } \boxed{\text{what number}} ?$$
$$P\% \cdot B = A$$
$$1.2 \cdot 450 = A$$
$$540 = A \qquad \text{Multiply 1.2 by 450}$$

Thus, A is 540, that is, 120% of 450 is 540.

PRACTICE EXERCISE 1

Solve.

(a) What is 10% of $120.20?

(b) 210% of 70 is what number?

Answers: (a) **$12.02** (b) **147**

CAUTION

In any percent problem, $P\%$ must always be changed to a decimal or fraction form in order to carry out the necessary calculations. When you remove the % symbol, be sure to multiply by 0.01. For instance, in Example 1(a) it would have been wrong to write

$$A = (5) \cdot (22.40).$$

HINT

Part (a) of Example 1 shows that the percent of a number is smaller than the number when the percent is less than 100%, while part (b) shows that the percent of a number is larger than the number when the percent is greater than 100%. These facts can help us determine whether an answer to a problem is reasonable or not.

Solving for P or B in a percent problem is a bit more challenging than solving for A since in both cases the related division sentence must be used. Suppose now that B is the unknown. For example, consider

$$8 \text{ is } 40\% \text{ of what number?}$$

This sentence can be translated to the basic percent equation $A = P\% \cdot B$ just like we did before.

$$
\begin{array}{ccccc}
8 & \text{is} & 40\% & \text{of} & \boxed{\text{what number}}\,? \\
\downarrow & \downarrow & \downarrow & \downarrow & \downarrow \\
A & = & P\% & \cdot & B \\
\downarrow & & \downarrow & & \downarrow \\
8 & = & 0.40 & \cdot & B
\end{array}
$$

$A = 8$ and $P\% = 40\%$, which becomes 0.4

We can find B by writing the related division sentence

$$\frac{8}{0.4} = B. \qquad \text{Divide 8 by the multiplier of } B, 0.4$$

Dividing 8 by 0.4 gives

$$B = 20. \qquad \frac{8}{0.4} = 20$$

| EXAMPLE 2 SOLVING FOR THE BASE B | PRACTICE EXERCISE 2 |

Solve by using the basic percent equation $A = P\% \cdot B$.

Solve.

(a) 60 is 20% of what number?

(a) 392 is 70% of what number?

In this case, B is the unknown, $P = 20$, and $A = 60$. Thus, we have

$$
\begin{array}{ccccc}
60 & \text{is} & 20\% & \text{of} & \boxed{\text{what number}}\,? \\
\downarrow & \downarrow & \downarrow & \downarrow & \downarrow \\
A & = & P\% & \cdot & B \\
\downarrow & & \downarrow & & \downarrow \\
60 & = & (0.2) & \cdot & B
\end{array}
$$

Translate to the related division sentence and divide.

$$B = \frac{60}{0.2} = 300$$

(b) 10.5% of what number is 89.25?

Thus, 60 is 20% of 300. Does this seem reasonable?

(b) 2.5% of what number is 12.5?

Although the order of wording has been reversed, this is the same type of problem.

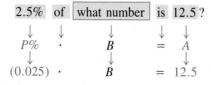

$$
\begin{array}{ccccc}
2.5\% & \text{of} & \boxed{\text{what number}} & \text{is} & 12.5\,? \\
\downarrow & \downarrow & \downarrow & \downarrow & \downarrow \\
P\% & \cdot & B & = & A \\
\downarrow & & \downarrow & & \downarrow \\
(0.025) & \cdot & B & = & 12.5
\end{array}
$$

The related division sentence gives

$$B = \frac{12.5}{0.025} = 500.$$

Thus, 2.5% of 500 is 12.5. Is this reasonable?

Answers: (a) 560 (b) 850

Finally, we consider finding the percent $P\%$ when the amount A and base B are given. For example, suppose we are asked:

What percent of 30 is 6?

First we translate to the basic percent equation $A = P\% \cdot B$, or in this case $P\% \cdot B = A$.

$$
\begin{array}{ccccc}
\boxed{\text{What percent}} & \text{of} & 30 & \text{is} & 6\,? \\
\downarrow & & \downarrow & \downarrow & \downarrow \\
P\% & \cdot & B & = & A \\
\downarrow & & \downarrow & & \downarrow \\
P\% & \cdot & 30 & = & 6
\end{array}
$$

Since $P\%$ is $P(0.01)$ changed to a decimal, substituting we have:

$$P(0.01) \cdot 30 = 6 \qquad P\% \text{ is } P(0.01)$$
$$P(0.3) = 6 \qquad (0.01) \cdot 30 = 0.3$$
$$P = \frac{6}{0.3} \qquad \text{Translate to the related division sentence}$$
$$P = 20 \qquad \frac{6}{0.3} = 20$$

Thus, $P\%$ is 20%, and 20% of 30 is 6.

EXAMPLE 3 SOLVING FOR THE PERCENT $P\%$

Solve by using the basic percent equation $P\% \cdot B = A$.

(a) What percent of 300 is 4.5?
Find the following:

$$
\begin{array}{ccccc}
\boxed{\text{What percent}} & \text{of} & 300 & \text{is} & 4.5\,? \\
\downarrow & & \downarrow & \downarrow & \downarrow \\
P\% & \cdot & B & = & A \\
\downarrow & & \downarrow & & \downarrow \\
P(0.01) & \cdot & (300) & = & 4.5 \qquad P\% = P(0.01) \\
& & P(3) & = & 4.5 \qquad (0.01)(300) = 3
\end{array}
$$

Translate to the related division sentence.

$$P = \frac{4.5}{3} \qquad \text{Related division sentence}$$
$$P = 1.5 \qquad \frac{4.5}{3} = 1.5$$

Thus, $P\%$ is 1.5%, and 1.5% of 300 is 4.5. Does this seem reasonable?

PRACTICE EXERCISE 3

Solve.

(a) What percent of 150 is 7.5?

(b) 320 is what percent of 240?
  Translate the following:

$$\underset{\downarrow}{\underset{\downarrow}{320}}\ \underset{\downarrow}{\text{is}}\ \boxed{\text{what percent}}\ \text{of}\ \boxed{240}\ ?$$

$$\begin{array}{ccccc}
\downarrow & & \downarrow & & \downarrow \\
A & = & P\% & \cdot & B \\
\downarrow & & \downarrow & & \downarrow \\
320 & = & P(0.01) & \cdot & 240
\end{array}$$

$$320 = P(2.4) \qquad (0.01)(240) = 2.4$$

$$\frac{320}{2.4} = P \qquad \text{Related division sentence}$$

$$133.\overline{3} = P \qquad \text{Divide 320 by 2.4}$$

Thus, $P\%$ is $133.\overline{3}\%$ or $133\frac{1}{3}\%$. Since 320 was larger than 240, it is reasonable to assume that the desired percent should be greater than 100%.

(b) 280 is what percent of 112?

Answers: (a) 5% (b) 250%

---

❸ USING PROPORTIONS TO SOLVE PERCENT PROBLEMS (METHOD 2)

We now consider a second method that can be used to solve percent problems. Since percent is the number of parts per hundred, a percent such as 60% can be written as a ratio in the following way:

$$\frac{60}{100} \quad \begin{array}{l} \leftarrow \text{Sixty parts} \\ \leftarrow \text{Per hundred} \end{array}$$

Also, the fraction $\frac{3}{5}$, changed to percent, is 60% (dividing 3 by 5 and converting to percent). We may express this fact by writing the proportion

$$\frac{3}{5} = \frac{60}{100}.$$

Suppose we are asked to solve the following percent problem:

What is 60% of 5?

Since, as shown above,

$$\frac{3}{5} = \frac{60}{100},$$

we know that 3 is 60% of 5. If we did not know the answer, we could solve the proportion

$$\frac{A}{5} = \frac{60}{100}$$

for A, the amount, to obtain 3.
  Now we ask another question:

3 is 60% of what number?

This can be written as the proportion

$$\frac{3}{B} = \frac{60}{100},$$

where B, the base, is now the number to be found. (Of course, in this case we know it is 5.)

Finally, consider the question:

What percent of 5 is 3?

In this case we have the proportion

$$\frac{3}{5} = \frac{P}{100}$$

where P, the percent, is what we are looking for (the 60 in 60%).

In general, combining what we have learned from these three percent questions, we have

$$\text{Amount} \rightarrow \frac{A}{B} = \frac{P}{100}. \leftarrow \text{Percent}$$
$$\text{Base} \quad \rightarrow$$

We call this the **basic percent proportion.** Every percent problem can be stated using this proportion where either A, B, or P is missing.

HINT

When reading a percent problem, notice that the amount, A, is next to the word *is* and the base, B, is next to the word *of*. Thus, the percent proportion

$$\frac{A}{B} = \frac{P}{100}$$

can be remembered by writing

$$\frac{\text{is}}{\text{of}} = \frac{\%}{100}.$$

We now solve the percent problems in Examples 1, 2, and 3 using percent proportions.

EXAMPLE 4   SOLVING FOR THE AMOUNT A

Solve by using the basic percent proportion.

(a) What is 5% of $22.40?

Start with the basic percent proportion

$$\frac{A}{B} = \frac{P}{100}$$

with A the unknown, $B = 22.40$, and $P = 5$.

$$\frac{A}{22.40} = \frac{5}{100}$$

$100A = (5)(22.40)$ Equate cross products

$A = \frac{(5)(22.40)}{100}$ Divide by the multiplier of A, 100

$A = 1.12$ Simplify

Thus, A is $1.12.

PRACTICE EXERCISE 4

Solve.

(a) What is 10% of $120.20?

(b) 120% of 450 is what number?

Substitute 120 for P and 450 for B in the percent proportion

$$\frac{A}{B} = \frac{P}{100}.$$

$$\frac{A}{450} = \frac{120}{100}$$

$100A = (120)(450)$ Equate cross products

$A = \dfrac{(120)(450)}{100}$ Divide by 100

$A = 540$ Simplify

Thus, 120% of 450 is 540.

(b) 210% of 70 is what number?

Answers: (a) $12.02   (b) 147

//////////// **CAUTION** ////////////

When using the percent proportion *do not* write the percent as a decimal or a fraction. This is a difference between the percent proportion and the percent equation.

//////////

EXAMPLE 5   SOLVING FOR THE BASE B

Solve by using the basic percent proportion.

(a) 60 is 20% of what number?

Substitute 60 for A and 20 for P in the basic percent proportion

$$\frac{A}{B} = \frac{P}{100}.$$

$$\frac{60}{B} = \frac{20}{100}$$

$(60)(100) = (20)B$ Equate cross products

$\dfrac{(60)(100)}{20} = B$ Divide by 20

$300 = B$

Thus, 60 is 20% of 300.

(b) 2.5% of what number is 12.5?

Substitute 12.5 for A and 2.5 for P in the basic percent proportion.

$$\frac{12.5}{B} = \frac{2.5}{100}$$

$(12.5)(100) = (2.5)B$ Equate cross products

$\dfrac{(12.5)(100)}{2.5} = B$ Divide by 2.5

$500 = B$

Thus, 2.5% of 500 is 12.5.

PRACTICE EXERCISE 5

Solve.

(a) 392 is 70% of what number?

(b) 10.5% of what number is 89.25?

Answers: (a) 560   (b) 850

| EXAMPLE 6   SOLVING FOR THE PERCENT $P\%$ | PRACTICE EXERCISE 6 |

Solve by using the basic percent proportion.

(a) What percent of 300 is 4.5?

Substitute 4.5 for A and 300 for B in the percent proportion

$$\frac{A}{B} = \frac{P}{100}.$$

$$\frac{4.5}{300} = \frac{P}{100}$$

$(4.5)(100) = (300)P$ Equate cross products

$$\frac{(4.5)(100)}{300} = P$$ Divide by 300

$$1.5 = P$$

Thus, 1.5% of 300 is 4.5.

(b) 320 is what percent of 240?

Substitute 320 for A and 240 for B in the percent proportion.

$$\frac{320}{240} = \frac{P}{100}$$

$(320)(100) = (240)P$ Equate cross products

$$\frac{(320)(100)}{240} = P$$ Divide by 240

$$133.\overline{3} = P$$

Thus, $P\%$ is $133.\overline{3}\%$ or $133\frac{1}{3}\%$.

Solve.

(a) What percent of 150 is 7.5?

(b) 280 is what percent of 112?

Answers: (a) **5%**   (b) **250%**

HINT

Consider both the equation method (Method 1) and the proportion method (Method 2) for solving percent problems and use the one that you prefer or that your instructor recommends. In the examples that follow, we present both methods in a side-by-side approach to help you decide which to use.

④ APPLICATIONS OF PERCENT

We conclude this section with three examples of typical percent applications.

| EXAMPLE 7   A PERCENT APPLICATION IN CHEMISTRY | PRACTICE EXERCISE 7 |

If 14 g (grams) of pure sulfuric acid is mixed with water and the resulting solution is 28% acid, what is the total weight of the solution?

We know that 14 g is 28% of the solution. Thus, we must answer the question:

14 is 28% of what number?

Minnie spends $120 a week to feed her family. If this is 24% of the total weekly family income, what is the weekly income of her family?

Method 1 (Equation Method)   Method 2 (Proportion Method)

$$A = P\% \cdot B$$

$$\frac{A}{B} = \frac{P}{100}$$

$$14 = (0.28)B$$

$$\frac{14}{0.28} = B$$

$$\frac{14}{B} = \frac{28}{100}$$

$$50 = B$$

$$(14)(100) = (28)B$$

$$\frac{(14)(100)}{28} = B$$

$$50 = B$$

Thus, 14 g is 28% of 50 g of the solution.

Answer: $500

EXAMPLE 8 A PERCENT APPLICATION IN FAMILY BUDGETING

The income of the Ross family is $1250 per month. If they spend 22% of their income for food, how much do they pay for food each month?

We must find 22% of the income, $1250. Thus the question is:

What is 22% of $1250?

Method 1   Method 2

$$A = P\% \cdot B$$

$$\frac{A}{B} = \frac{P}{100}$$

$$A = (0.22)(1250)$$

$$A = 275$$

$$\frac{A}{1250} = \frac{22}{100}$$

$$100A = (22)(1250)$$

$$A = \frac{(22)(1250)}{100} = 275$$

Thus, $275 is 22% of $1250, and the family spends $275 for food each month. If we come up with $2750 for an answer, since $2750 is more than $1250 but 22% is less than 100%, we would have known we had made an error.

PRACTICE EXERCISE 8

At a recent concert attended by 840 people, 45% of those in attendance were children. How many children were at the concert?

Answer: 378 children

EXAMPLE 9 AN APPLICATION OF PERCENT IN SPORTS

A basketball player hit 16 out of 20 free throws in a game. What was her shooting percent?

The question to answer is

16 is what percent of 20?

Method 1   Method 2

$$A = P\% \cdot B$$

$$\frac{A}{B} = \frac{P}{100}$$

$$16 = (P\%)(20)$$

$$16 = P(0.01)(20)$$

$$16 = P(0.2)$$

$$\frac{16}{20} = \frac{P}{100}$$

$$\frac{16}{0.2} = P$$

$$(16)(100) = (20)P$$

$$80 = P$$

$$\frac{(16)(100)}{20} = P = 80$$

Thus, her shooting percent was 80%. That is, 16 is 80% of 20.

PRACTICE EXERCISE 9

During the World Series, Reggie Jackson had 10 hits in 30 times at bat. What was his batting percentage in the series?

Answer: $33.\overline{3}\%$ (Batting percentages or averages are often expressed as three-place decimals so we might give 0.333 for the answer.)

6.3 EXERCISES A

Solve the percent problems. Before checking, ask yourself if your answer seems reasonable.

1. What is 20% of 150?   30

2. What is 140% of 20?   28

3. 70% of 600 is what number?   420

4. 10 is 20% of what number?   50

5. 200 is 40% of what number?   500

6. 4.5 is 150% of what number?   3

7. What percent of 50 is 10?   20%

8. What percent of 10 is 50?   500%

9. 75 is what percent of 225? $33\frac{1}{3}\%$

10. What is 35% of 70?   24.5

11. 6.2 is what percent of 24.8?   25%

12. 0.15 is 60% of what number?   0.25

13. There were 1600 votes cast in an election. Mr. Lawrence received 62% of the votes. How many votes did he receive? [*Hint:* What is 62% of 1600?]
992 votes

14. Flagstaff, Arizona, received 120 inches of snow one year. This was 150% of the normal snowfall. What is the normal annual snowfall in Flagstaff? [*Hint:* 120 is 150% of what number?]   80 in

15 A baseball team won 62% of its games and played a total of 50 games. How many games did the team win?   31 games

16. In an acid solution weighing 120 g there are 35 g of acid. To the nearest percent, what is the percent of acid in the solution?   29%

17. Ms. McShane received 2860 of the votes in an election. This was about 58% of the votes cast. Approximately how many votes were cast?
4931 votes

18 A student answered 22 questions correctly on a test having 30 questions. To the nearest percent, what percent of her answers were correct?   73%

19. What is the commission on a sale of $3600 if the commission rate is 8%? $288

20. If an account worth $1500 increased to $1800 in one year, what was the interest rate for the year?
20%

21. Wanda received an increase in salary of $2640. If this was a 12% raise, what was her original salary?
$22,000

22 Henry's weight decreased from 220 pounds to 187 pounds on a diet. What was the percent decrease in his weight?   15%

FOR REVIEW

23. Change 35% to a fraction. $\frac{7}{20}$

24. Change 13.5% to a decimal.   0.135

25. Tonya has deposited $2000 in a savings account which earns 8.75% interest. Write 8.75% as a decimal.
0.0875

Change to percents.

26. 0.17   17%

27. 0.005   0.5%

28. 0.05   5%

29. 0.5   50%

30. $\frac{3}{8}$ 37.5%

31. $\frac{7}{9}$ $77.\overline{7}\%$

32. In a two-person school board election, Dr. Yard received 2130 votes out of a total of 3360 votes cast.

   (a) What fractional part of the votes did Dr. Yard receive? $\frac{71}{112}$

   (b) Express the fraction in **(a)** as a decimal rounded to the nearest thousandth.   0.634

   (c) Express the decimal in **(b)** as a percent rounded to the nearest tenth of a percent.   63.4%

   (d) What percent of the votes (to the nearest tenth of a percent) did Dr. Yard's opponent receive?   36.6%

ANSWERS: 1. 30   2. 28   3. 420   4. 50   5. 500   6. 3   7. 20%   8. 500%   9. $33\frac{1}{3}$%   10. 24.5   11. 25%
12. 0.25   13. 992 votes   14. 80 in   15. 31 games   16. 29%   17. 4931 votes   18. 73%   19. $288   20. 20%
21. $22,000   22. 15%   23. $\frac{7}{20}$ 24. 0.135   25. 0.0875   26. 17%   27. 0.5%   28. 5%   29. 50%   30. 37.5%
31. 77.$\overline{7}$%   32. (a) $\frac{71}{112}$ (b) 0.634 (c) 63.4% (d) 36.6%

6.3   EXERCISES B

Solve the percent problems. Before checking, ask yourself if your answer seems reasonable.

1. What is 16% of 400?   64

2. What is 420% of 5?   21

3. 35% of 900 is what number?   315

4. 30 is 5% of what number?   600

5. 120 is 0.2% of what number?   60,000

6. 12.5 is 625% of what number?   2

7. What percent of 80 is 24?   30%

8. What percent of 520 is 130?   25%

9. 15 is what percent of 8?   187.5%

10. 55% of 4000 is what number?   2200

11. 0.6 is 12% of what number?   5

12. 9.8 is 280% of what number?   3.5

13. If there is a 0.05% impurity rate in a water sample of 820 grams, how many grams of impurities are in the sample?   **0.41 grams**

14. During one year Phoenix had a total of 168 days when the temperature exceeded 100°. This was 120% of normal. Normally how many days per year does the temperature exceed 100° in Phoenix?
140 days

15. There were 10,850 votes cast for the two people in an election, and the winner received 62% of the total. How many votes did she receive?
6727 votes

16. If 6 liters of acid are mixed with 9 liters of water, what is the percent of acid in the solution?   **40%**

17. Burford answered 66 questions correctly on a test and received a score of 44%. How many questions were on the test?   **150**

18. At a recent campus sold-out showing of the movie *Rambo First Blood Part II*, 485 members of the audience were under 30 years of age. If the theater seats 615, to the nearest tenth of a percent, what percent of those present was under 30 years old?
78.9%

19. What is the commission rate if Rollie earns a commission of $102.24 on sales of $568?   **18%**

20. From one year to the next, the price of ground beef went from 89¢ per pound to $1.09 per pound. What was the percent increase (to the nearest tenth of a percent)?   **22.5%**

21. Patrick received a raise of 14%. What is his new salary if his former salary was $36,000?   **$41,040**

22. A retailer bought a lamp for $35.00 and marked up the price 60%. What was the markup and what was the selling price?   **$21.00; $56.00**

FOR REVIEW

23. Change 82% to a fraction. $\frac{41}{50}$

24. Change $6\frac{1}{2}$% to a decimal.   0.065

25. Rod purchased a bike on sale at 30% off the regular price. Change 30% to a decimal.   0.3

Change to percents.

26. 0.31   31%

27. 0.004   0.4%

28. 0.04   4%

29. 0.4   40%   **30.** $\dfrac{7}{8}$ 87.5%   **31.** $\dfrac{1}{6}$ 16.$\overline{6}$%

32. Of the 1092 hours of prime-time television programming in a recent year, it was estimated that 648 hours were viewed by children under the age of 16.

   (a) What fractional part of the total prime-time hours was viewed by children? $\frac{54}{91}$

   (b) Express the fraction in **(a)** as a decimal rounded to the nearest thousandth.   **0.593**

   (c) Express the decimal in **(b)** as a percent rounded to the nearest tenth.   **59.3%**

   (d) What percent of the total hours (to the nearest tenth of a percent) was not viewed by children?   **40.7%**

6.3 EXERCISES C

Use a calculator and give answers to the nearest hundredth or nearest hundredth of a percent.

1. What is 12.7% of 257?
   [Answer: 32.64]

2. 27 is 6.3% of what number?
   428.57

3. What percent of 83 is 122?
   146.99%

4. During one year the Browns spent 10.3% of their income on entertainment and vacations. If they spent $3485.20 on these two items, to the nearest cent, what was their income?
   [Answer: $33,836.89]

Write each of the following as a percent proportion and solve. Give answers rounded to the nearest tenth or nearest tenth of a percent.

5. 6.9 is what percent of 72.7?
   [Answer: 9.5%]

6. 27.3 is 14.2% of what number?
   192.3

7. What is 102.6% of 38.5?
   39.5

6.4 TAX PROBLEMS

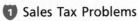

STUDENT GUIDEPOSTS

1 Sales Tax Problems   **3** Income Tax Problems

2 Social Security Tax Problems

In the next five sections we will study a variety of applications of percent all of which can be translated into the basic percent equation

$$A = P\% \cdot B$$

or the basic percent proportion

$$\frac{A}{B} = \frac{P}{100}.$$

We will give a dual presentation in many of the examples. Use the method you prefer.

1 SALES TAX PROBLEMS

In most states a tax is charged on purchases made in retail stores. The rate of this tax, called a **sales tax,** varies from location to location. In a state with a **sales-tax rate** of 5%, the tax on a purchase of $10.00 is given by sales tax = 5% of $10.00 .   Use $A = P\% \cdot B$

= (0.05) · ($10.00)

= $0.50

When you buy an item for $10.00, you must pay a

total price = price before tax + sales tax .

= $10.00   + $0.50

= $10.50

We now present a summary of the necessary concepts related to a sales-tax problem.

To Solve a Sales-Tax Problem

1. Use the basic percent equation

$$A \quad = \quad P\% \quad \cdot \quad B$$

sales tax = tax rate (0.01) · price before tax

or use the basic percent proportion

$$\frac{A}{B} = \frac{P}{100}$$

$$\frac{\text{sales tax}}{\text{price before tax}} = \frac{\text{tax rate}}{100}.$$

2. The total price is given by

total price = price before tax + sales tax.

HINT

In many sales-tax problems, a calculator can be useful since you are performing decimal calculations.

EXAMPLE 1   FINDING SALES TAX AND TOTAL PRICE

If the sales-tax rate is 5% in Tucson, Arizona, what is the sales tax on a TV selling for $625.40? What is the total price?

Find the sales tax.

Method 1 (Equation Method)

$$A \quad = \quad P \quad \% \cdot \quad B$$

sales tax = tax rate(0.01) · price before tax

= (0.05)   · ($625.40)

= $31.27

PRACTICE EXERCISE 1

If the sales-tax rate is 4% in Blue Key, Florida, what is the sales tax on a pair of water skis selling for $315.50? What is the total price including tax?

Method 2 (Proportion Method)

$$\frac{A}{B} = \frac{P}{100}$$

$$\frac{\text{sales tax}}{\text{price before tax}} = \frac{\text{tax rate}}{100}$$

$$\frac{\text{sales tax}}{\$625.40} = \frac{5}{100}$$

$$(\text{sales tax})(100) = (5)(\$625.40)$$

$$\text{sales tax} = \frac{(5)(\$625.40)}{100}$$

$$= \$31.27$$

Now add the price before tax and the sales tax to obtain the total price.

$$\text{total price} = \text{price before tax} + \text{sales tax}$$
$$\downarrow \qquad\qquad \downarrow \qquad \downarrow$$
$$= \qquad \$625.40 \quad + \quad \$31.27$$
$$= \$656.67$$

Thus, one must pay a total of $656.67 for the TV set.

Answer: $12.62; $328.12

///////////////// **CAUTION** ///////////////////

Always check to see if your answer seems reasonable. A 5% tax is $5 for each $100. If we had found the tax to be $312.70, we would have realized that this was much too big. (The decimal point is too far to the right.) Similarly, we know that an answer of $3.13 is wrong because the decimal point is too far to the left.

///////////////

The sales-tax rate in a particular city can be found if we know the price before tax and the amount of sales tax paid.

| **EXAMPLE 2** FINDING THE SALES-TAX RATE | **PRACTICE EXERCISE 2** |

What is the sales-tax rate if the tax is $2.50 on the purchase of a $62.50 sport coat?

The question asked is: $2.50 is what percent of $62.50?

What is the sales-tax rate in Denver if the tax is $24.77 on the purchase of a desk with a selling price of $495.40?

Method 1

$$A = P\% \cdot B$$

$$\$2.50 = P\% \cdot \$62.50$$
$$\downarrow \qquad \downarrow \qquad \downarrow$$
$$2.50 = P(0.01)(62.50)$$

$$2.50 = P(0.625)$$

$$\frac{2.50}{0.625} = P$$

$$4 = P$$

Method 2

$$\frac{A}{B} = \frac{P}{100}$$

$$\frac{\$2.50}{\$62.50} = \frac{P}{100}$$

$$(2.50)(100) = (62.50)P$$

$$\frac{(2.50)(100)}{62.50} = P$$

$$4 = P$$

Thus, the sales-tax rate is 4%. Since

$$(0.04)(62.50) = 2.50, \text{ 4% does check.}$$

Answer: 5%

Finally, when the sales-tax rate and the sales tax are given, we can find the price of an item before tax.

| EXAMPLE 3 FINDING THE PRICE OF AN ITEM | PRACTICE EXERCISE 3 |

A tax of $1.26 was charged on a purchase of cosmetics in a city where the tax rate is 3.5%.

(a) What was the price of the cosmetics before tax?

The question asked is: $1.26 is 3.5% of what number?

| Method 1 | Method 2 |

Method 1

$$A = P\% \cdot B$$
$$\$1.26 = (3.5)(0.01)B$$
$$1.26 = (0.035)B$$
$$\frac{1.26}{0.035} = B$$
$$36 = B$$

Method 2

$$\frac{A}{B} = \frac{P}{100}$$
$$\frac{\$1.26}{B} = \frac{3.5}{100}$$
$$(1.26)(100) = (3.5)B$$
$$\frac{(1.26)(100)}{3.5} = B = 36$$

The cosmetics sold for $36.00 before tax.

(b) What was the total price paid?

total price = price before tax + sales tax

$$= \quad \$36.00 \quad + \quad \$1.26 \ = \$37.26$$

A tax of $5.76 was charged on a purchase of a pair of boots in Dallas where the tax rate is 4.5%.

(a) What was the selling price (price before tax) of the boots?

(b) What was the total price paid?

Answers: (a) $128.00
(b) $133.76

② SOCIAL SECURITY TAX PROBLEMS

The Social Security tax is used by the federal government to provide income for retired people. In 1989 the tax rate was 7.51% on the first $48,000 of earnings. A wage earner who made $48,000 or more in 1989 paid 7.51% of $48,000, which was

$$(0.0751)(\$48,000) = \$3604.80,$$

in Social Security taxes. A person who made less than $48,000 paid 7.51% of whatever was earned. We summarize this in the following rule.

To Find the Amount of Social Security Tax (1989)

1. If total earnings are less than $48,000:
Use the percent equation

$$A \quad = \quad P \quad \% \quad \cdot \quad B$$
$$\text{Social Security tax} = (7.51)(0.01) \cdot \text{total earnings}$$

or use the percent proportion

$$\frac{A}{B} = \frac{P}{100}.$$
$$\frac{\text{Social Security tax}}{\text{total earnings}} = \frac{7.51}{100}$$

2. If total earnings are $48,000 or more, the Social Security tax is fixed at $3604.80.

| EXAMPLE 4 FINDING SOCIAL SECURITY TAX | PRACTICE EXERCISE 4 |

Clara earned \$39,600 in 1989. How much Social Security tax did she pay?

Benito earned \$22,480 in 1989. How much Social Security tax did he pay?

Since Clara made less than \$48,000, use Method 1 or Method 2.

Method 1

$$A = P\% \cdot B$$

$$\text{S. S. tax} = (7.51)(0.01) \cdot \text{total earnings}$$

$$= (7.51)(0.01) \cdot (\$39,600)$$

$$= (0.0751)(\$39,600)$$

$$= \$2973.96$$

Method 2

$$\frac{A}{B} = \frac{P}{100}$$

$$\frac{\text{S. S. tax}}{\text{total earnings}} = \frac{7.51}{100}$$

$$\frac{\text{S. S. tax}}{\$39,600} = \frac{7.51}{100}$$

$$(\text{S. S. tax})(100) = (7.51)(\$39,600)$$

$$\text{S. S. tax} = \frac{(7.51)(\$39,600)}{100}$$

$$= \$2973.96$$

Thus, Clara had to pay \$2973.96 in Social Security tax in 1989. Clearly a calculator is helpful in solving problems like this.

Answer: **\$1688.25 (to the nearest cent)**

❸ INCOME TAX PROBLEMS

Income tax is usually the biggest tax that a person pays each year. Many taxpayers find the tax by looking up their taxable income (total earnings less deductions) in a table and reading the tax. However, for some, determining federal or state income taxes requires use of percent.

The tax rate schedule for single taxpayers from the 1989 federal income tax forms is shown on the next page. (You can obtain other tables at your post office.) Suppose your taxable income was \$19,580. Look down the two columns on the left until you find the two numbers between which your income falls. In this case, \$19,580 is between \$18,550 and \$44,900. The tax is calculated by adding 28% of the difference between \$19,580 and \$18,550 to \$2,782.50. First subtract \$18,550 from \$19,580.

$$\$19,580 - \$18,550 = \$1030$$

Take 28% of \$1030,

$$(0.28)(\$1030) = \$288.40,$$

and add this amount to \$2,782.50 to obtain the total income tax to be paid.

$$\$2,782.50 + \$288.40 = \$3070.90$$

Thus, \$3070.90 is your income tax in 1989.

1989 Tax Rate Schedule

Schedule X — Single

If line 5 is:		The tax is:	of the amount over—
Over—	But not over—		
$0	$18,550	------- 15%	$0
18,550	44,900	$2,782.50 + 28%	18,550
44,900	93,130	10,160.50 + 33%	44,900
93,130	-------	Use Worksheet below to figure your tax.	

EXAMPLE 5   FINDING INCOME TAX

Paul is a single taxpayer who uses the tax rate schedule. In 1989 his taxable income was $9800. How much income tax did he pay?

Paul's taxable income, $9800, was between $0 and $18,550. He paid 15% of his taxable income which was

$$(0.15)($9800) = $1470.$$

PRACTICE EXERCISE 5

Rose is a single taxpayer who used the tax rate schedule in 1989. If her taxable income was $46,000, how much income tax did she pay?

Answer: $10,523.50

6.4   EXERCISES A

Solve the following problems.

1. What is the sales tax on a price before tax of $200 if the tax rate is 5%? What is the total cost?
   $10; $210

2. The price before tax of a dress is $62.80 and the tax rate is 5%. What is the tax and the total price?
   $3.14; $65.94

3. Find the tax and total price of a shirt if the price before tax is $15.20, and the tax rate is 2.5%.
   $0.38; $15.58

4. If $0.43 tax is charged on a price of $21.50, what is the tax rate?   **2%**

5. The tax rate is 2.5% and the tax is $3.05. What is the price before tax?   **$122**

6. If the tax on a coffee pot selling for $34.50 before tax is $0.52, what is the tax rate (to the nearest tenth of a percent)?   **1.5%**

7. The tax rate in Hudson, New York, is 5% and the tax charged on an item is $12.50. What is its price before tax?   **$250**

8. A used car lists for $6827.50. If the sales-tax rate is 4.5%, find the tax and the total price.
   $307.24, $7134.74

9. Susan had an income from wages of $15,800 in 1989. How much Social Security tax did she pay? **$1186.58**

10 Toni made $52,350 in 1989. What Social Security tax did she pay? **$3604.80 (she only paid tax on the first $48,000)**

11. Mark was a single taxpayer with a taxable income of $24,200 in 1989. What income tax did he pay? **$4364.50**

12. On total wages of $20,800 in 1989, Jim had a taxable income of $16,600. How much income tax did he pay on his taxable income? How much Social Security tax did he pay on his total wages? **$2490; $1562.08**

FOR REVIEW

Solve the percent problems.

13. What is 18% of 245? **44.1**

14. 5.2 is 8% of what number? **65**

15. 676 is what percent of 520? **130%**

16. In a recent poll, 52% of those contacted were female. If 182 females were in the poll, how many were contacted? **350**

17. Maria answered 76 of 88 questions correctly on a test. What was her percent score to the nearest percent? **86%**

ANSWERS: 1. $10; $210 2. $3.14; $65.94 3. $0.38; $15.58 4. 2% 5. $122 6. 1.5% 7. $250 8. $307.24; $7134.74 9. $1186.58 10. $3604.80 (she only paid tax on the first $48,000) 11. $4364.50 12. $2490; $1562.08 13. 44.1 14. 65 15. 130% 16. 350 17. 86%

6.4 EXERCISES B

Solve the following problems.

1. What is the sales tax on a price before tax of $300 if the tax rate is 4%? What is the total cost? **$12.00; $312.00**

2. If a TV sells for $620 (before tax) and the sales-tax rate is 5%, what is the tax? What is the total cost? **$31.00; $651.00**

3. Find the tax and total price of a toaster if the price before tax is $22.40, and the tax rate is 3.5%? **$0.78; $23.18**

4. If a tax of $18.90 is charged on a purchase of $525.00, what is the tax rate? **3.6%**

5. The sales-tax rate in Mountain, Colorado, is 4%. If the tax on a pair of ski boots was $3.58, what was the price before tax? **$89.50**

6. The tax on a sofa costing $685.50 before tax is $41.13. What is the tax rate? **6%**

7. The tax rate in Pacifica, California, is 6%, and the tax charged on an item is $33.63. What is the price before tax of the item? **$560.50**

8. A new boat lists for $17,435.75. If the sales-tax rate is 5.5%, find the tax and the total cost of the boat. **$958.97; $18,394.72**

9. The Social Security tax rate was 7.51% in 1989. How much Social Security tax did Claude pay on a salary of $18,200? **$1366.82**

10. Margie had an income of $54,650 in 1989. How much Social Security tax did she pay? **$3604.80 (she only paid tax on the first $48,000)**

11. Ruben was a single taxpayer with a taxable income of $33,200 in 1989. How much income tax did he pay? **$6884.50**

12. On total wages of $31,200 in 1989, Kent had a taxable income of $23,400. How much income tax did he pay on his taxable income? How much Social Security tax did he pay? **$4140.50; $2343.12**

FOR REVIEW

Solve the percent problems.

13. What is 120% of 65?
78

14. 40.8 is 12% of what number?
340

15. 48 is what percent of 32?
150%

16. It has been estimated that 18% of the residents of a county cannot read or write. If there are 26,580 people living in the county, about how many of them cannot read or write?   4784

17. Because of illness, Troy could only work 27 hours of his normal 40-hour work week. What percent of the normal week did he work?   67.5%

6.4 EXERCISES C

Solve.

1. A washer normally selling for $389.95 is put on sale at 20% off the regular price. If the sales-tax rate is 4.5%, to the nearest cent, how much will the washer cost including the sales tax? $326.00

2. Paul buys a typewriter for $459.95 and a desk for $189.98. If the sales tax on the purchase of these items is $26.65, to the nearest tenth of a percent, what is the sales-tax rate?
[Answer: 4.1%]

6.5 COMMISSION AND DISCOUNT PROBLEMS

STUDENT GUIDEPOSTS

1 Commission Problems   **2** Discount Problems

We next look at two quantities which, like sales tax, are calculated by taking a percent of the selling price.

1 COMMISSION PROBLEMS

Some salespersons receive all or part of their income as a percent of their sales. This is called a **commission.** The **commission rate** is the percent of the sales that the person receives.

To Solve a Commission Problem

Use the percent equation

$$A = P \% \cdot B$$

commission = commission rate(0.01) · total sales

or use the percent proportion

$$\frac{A}{B} = \frac{P}{100}$$

$$\frac{\text{commission}}{\text{total sales}} = \frac{\text{commission rate}}{100}.$$

EXAMPLE 1 FINDING A COMMISSION

Elaine sells furniture and receives a 20% commission on the selling price. In April, her sales totaled $8360. What was her commission?

Method 1

$$A \quad = \quad P \quad \% \quad \cdot \quad B$$

$$\text{commission} = \boxed{\text{commission rate}(0.01)} \cdot \boxed{\text{total sales}}$$

$$\qquad = \quad 20(0.01) \quad \cdot \quad \$8360$$

$$= (0.20)(\$8360)$$

$$= \$1672$$

Method 2

$$\frac{A}{B} = \frac{P}{100}$$

$$\frac{\text{commission}}{\text{total sales}} = \frac{\text{commission rate}}{100}$$

$$\frac{\text{commission}}{\$8360} = \frac{20}{100}$$

$$(100)\text{commission} = (20)(\$8360)$$

$$\text{commission} = \frac{(20)(\$8360)}{100}$$

$$= \$1672$$

Thus, Elaine received a commission of $1672.00 for April.

PRACTICE EXERCISE 1

Rich receives a 3% commission on all life insurance policy sales. In September his sales totaled $195,400. What was his commission?

Answer: $5862

As with sales-tax problems, we can find the commission rate if we know the total sales and the commission.

EXAMPLE 2 FINDING A COMMISSION RATE

Shane received a $123 commission on sales of $820 in men's clothing. What was his commission rate?

The question to answer is: $123 is what percent of $820?

Method 1

$$A \quad = \quad P \quad \% \quad \cdot \quad B$$

$$\boxed{\text{commission}} = \text{commission rate}(0.01) \cdot \boxed{\text{total sales}}$$

$$123 \quad = \text{commission rate}(0.01) \cdot \quad (820)$$

$$123 = \text{commission rate } (8.2)$$

$$\frac{123}{8.2} = \text{commission rate}$$

$$15 = \text{commission rate}$$

PRACTICE EXERCISE 2

Adrienne received a commission of $98.75 on sales of $790 in an appliance store. What was her commission rate?

Method 2

$$\frac{A}{B} = \frac{P}{100}$$

$$\frac{\text{commission}}{\text{total sales}} = \frac{\text{commission rate}}{100}$$

$$\frac{123}{820} = \frac{\text{commission rate}}{100}$$

$$(123)(100) = (820)\text{commission rate}$$

$$\frac{(123)(100)}{820} = \text{commission rate}$$

$$15 = \text{commission rate}$$

Shane's commission rate is 15%.

Answer: 12.5%

| **EXAMPLE 3**   FINDING TOTAL SALES | **PRACTICE EXERCISE 3** |

What are the total sales on which Mario received a commission of $336 if his commission rate is 12%?

    Since the commission is known but not the amount of the sales, the question is: $336 is 12% of what number?

If Louis received a commission of $46.04 working in a hardware store where he is paid at a commission rate of 8%, what were his total sales?

Method 1

$$A = P \% \cdot B$$

$$\text{commission} = \text{commission rate}(0.01) \cdot \text{total sales}$$

$$\downarrow \qquad \qquad \downarrow \qquad \qquad \downarrow$$

$$\$336 = 12(0.01) \cdot \text{total sales}$$

$$\$336 = (0.12)\text{total sales}$$

$$\frac{\$336}{0.12} = \text{total sales}$$

$$\$2800 = \text{total sales}$$

Method 2

$$\frac{A}{B} = \frac{P}{100}$$

$$\frac{\text{commission}}{\text{total sales}} = \frac{\text{commission rate}}{100}$$

$$\frac{\$336}{\text{total sales}} = \frac{12}{100}$$

$$(\$336)(100) = (12)\text{total sales}$$

$$\frac{(\$336)(100)}{12} = \text{total sales}$$

$$\$2800 = \text{total sales}$$

The sales total is $2800.

Answer: $575.50

EXAMPLE 4   Monthly Income Involving a Commission

An automobile salesperson receives a monthly salary of $500 plus 2% of the first $20,000 sales and 4% of sales above $20,000. What is the monthly income on sales of $38,000?

First, calculate 2% of $20,000.

$$(0.02)(\$20,000) = \$400$$

Next, calculate 4% of the difference $38,000 − $20,000 = $18,000.

$$(0.04)(\$18,000) = \$720$$

The total income is $500 plus the two commissions.

$$\$500 + \$400 + \$720 = \$1620$$

The total income for the month is $1620.

PRACTICE EXERCISE 4

A real estate salesman receives a monthly salary of $850 plus 0.75% of the first $100,000 in sales and 1.5% of sales above $100,000. How much did he earn in a month when he sold $225,000 worth of property?

Answer: $3925

❷ DISCOUNT PROBLEMS

A **discount** is a reduction in the regular price of an item when it is put on sale. The **discount rate** is the percent of the regular price that the item is reduced. The **sale price** is the price after the discount. For example, if an appliance store reduces a $600 refrigerator to a sale price of $450, the discount is

$$\$600 − \$450 = \$150.$$

The discount rate relative to the original price is

$$\frac{150}{600} = \frac{1 \cdot \cancel{150}}{4 \cdot \cancel{150}} = \frac{1}{4} = 0.25$$

which is 25%. We are usually given the regular price and the discount rate and are asked to find this discount. In this problem, given the $600 and the 25%, we could have calculated the discount.

$$(0.25)(\$600) = \$150$$

To Solve a Discount Problem

1. Use the percent equation

$$A \quad = \quad P \quad \% \quad \cdot \quad B$$
$$\text{discount} = \text{discount rate}(0.01) \cdot \text{regular price}$$

or use the percent proportion

$$\frac{A}{B} = \frac{P}{100}$$

$$\frac{\text{discount}}{\text{regular price}} = \frac{\text{discount rate}}{100}.$$

2. The sale price is given by

$$\text{sale price} = \text{regular price} − \text{discount}.$$

EXAMPLE 5 FINDING A DISCOUNT	**PRACTICE EXERCISE 5**

What is the discount and the sale price of a coat which is regularly priced $69.50 if the discount rate is 20%?

First find the discount.

Method 1

$$A = P\% \cdot B$$

$$\text{discount} = \text{discount rate}(0.01) \cdot \text{regular price}$$

$$= 20(0.01) \cdot \$69.50$$

$$= (0.20)(\$69.50)$$

$$= \$13.90$$

Method 2

$$\frac{A}{B} = \frac{P}{100}$$

$$\frac{\text{discount}}{\text{regular price}} = \frac{\text{discount rate}}{100}$$

$$\frac{\text{discount}}{\$69.50} = \frac{20}{100}$$

$$(100)\text{discount} = (20)(\$69.50)$$

$$\text{discount} = \frac{(20)(\$69.50)}{100}$$

$$= \$13.90$$

Then

$$\text{sale price} = \text{regular price} - \text{discount}$$

$$= 69.50 - 13.90$$

$$= 55.60.$$

During the sale, the coat can be bought for $55.60.

What is the discount and the sale price of a camera which is regularly priced $139.95 if the discount rate is 40%?

Answer: $55.98; $83.97

EXAMPLE 6 FINDING A SALE PRICE	**PRACTICE EXERCISE 6**

A retailer first discounted a $450 washing machine 10%. When the machine was not purchased, he discounted it 30% of the sale price. What was the second sale price and the overall rate of discount?

The first discount is

$$(0.10)(\$450) = \$45,$$

and the first sale price is

$$\$450 - \$45 = \$405.$$

Since the second discount is taken on the sale price of $405, find 30% of $405.

$$(0.30)(\$405) = \$121.50$$

The second discount is $121.50, and the second sale price is

$$\$405.00 - \$121.50 = \$283.50.$$

A jeweler put a diamond ring on sale at 15% off the regular price of $850.00. A week later he discounted it again 20% of the sale price. What was the final sale price and the overall rate of discount?

Since the second discount was taken on $405 and not on $450, the discount rates cannot be added. Add the discounts.

$$\$45.00 + \$121.50 = \$166.50$$

The total discount is $166.50 on a regular price of $450. The discount rate is given by:

$$\$166.50 = \text{discount rate}(0.01) \cdot \$450$$

$$\frac{\$166.50}{\$4.50} = \text{discount rate}$$

$$37 = \text{discount rate}$$

The overall discount rate is 37%, which is less than the 40% obtained by adding 10% and 30%. The reason for the difference is that the 30% discount was taken on $405 instead of on $450.

Answer: $578; 32% [Note that the overall discount is not the sum of the two discounts which is 35%.]

6.5 EXERCISES A

Solve the following problems.

1. An automobile salesperson receives a 2% commission on the price of each car sold. What is the commission on an $8500 sale?   **$170**

2. Mary receives a 15% commission on all clothing sales. If her sales were $3872.16 during the month of April, what was her commission?   **$580.82**

3. Alphonse was paid a commission of $142.65 on sales of $475.50. What was his commission rate? **30%**

4 Angela received a commission of $639.12 on her furniture sales for the month of June. If the commission rate is 12%, what were her total sales? **$5326**

5. The Sun Realty Company receives a 6% commission on all real estate sales. On July 16 the company sold a house for $82,560 and a lot for $21,320. What was the total commission on these sales? **$6232.80**

6. Lynne receives a 1.5% commission each month on all farm equipment sales of $20,000 or less. For all sales over $20,000, she is paid a 3% commission on the amount over $20,000. What is her commission on sales of $45,280 during the month of May? **$1058.40**

7. Considering the total commission that Lynne received in Exercise 6, what is her overall commission rate?

2.3% $\left(\text{to the nearest } \frac{1}{10}\%\right)$

8. A retailer discounted all the coats in her shop by 25%. What is the discount and the sale price on a coat regularly priced $128.60?
$32.15; $96.45

9. A refrigerator normally sells for $720, but an appliance dealer put it on sale at an 18% discount. What is the discount and the sale price?
$129.60; $590.40

10 If a $40 pair of shoes is on sale for $34, what is the discount rate?   15%

11. The price of a shirt is decreased by $3.50. If this is a 20% discount, what is the normal price of the shirt? $17.50

12. In a closeout sale, a carpet dealer has reduced all carpet prices by 40%. If the normal price of a pattern is $16.90 per square yard, how much would it cost to buy the carpet for a room which is 4 yards wide and 5 yards long? $202.80

13. Find the discount rate if a car normally priced at $6250 is on sale for $5687.50.   9%

14 In the spring a department store reduced the price of coats at a discount rate of 25%. When some coats were left, the store reduced their price 20% of the sale price. What was the second sale price of a coat that was originally priced $142.80? $85.68

15. What was the overall rate of discount on the coat in Exercise 14?   40%

16. A toaster normally selling for $42.50 is on sale for 12% off the normal price. What total price must Sean pay if, after the discount is made, a 5% sales tax is added? $39.27

FOR REVIEW

17. The sales-tax rate in Twin Falls, Idaho, is 4%. How much tax will Paula pay on the purchase of three posters each selling for $2.98 before tax? What is the total price of the posters? $0.36; $9.30

18. Jeff was charged $3.29 in sales tax on the purchase of a pair of basketball shoes. What is the sales-tax rate if the before tax selling price of the shoes was $65.80?   5%

19. Maria had an income from wages of $19,600 in 1989. How much Social Security tax did she pay if the rate was 7.51% that year? **$1471.96**

20. In 1989 Juanita had a taxable income of $15,743. If she used the tax table given in Section 6.4, how much income tax did she pay? **$2361.45**

ANSWERS: (Answers may vary slightly due to round-off differences.) 1. $170 2. $580.82 3. 30% 4. $5326 5. $6232.80 6. $1058.40 7. 2.3% $\left(\text{to the nearest } \frac{1}{10}\%\right)$ 8. $32.15; $96.45 9. $129.60; $590.40 10. 15% 11. $17.50 12. $202.80 13. 9% 14. $85.68 15. 40% 16. $39.27 17. $0.36; $9.30 18. 5% 19. $1471.96 20. $2361.45

6.5 EXERCISES B

Solve the following problems.

1. The Land Realty Company charges a commission of 6% on all sales. How much commission will the company receive for selling a ranch for $260,000? **$15,600**

2. Beth receives a 12% commission on all sales. If her sales totaled $2475.50 one week, what was her commission? **$297.06**

3. Jill received a commission of $402.60 on sales of $7320.00. What was her commission rate? **5.5%**

4. Art received a commission of $9817.50 on sales of computer hardware for the month of December. If his commission rate is 15%, what were his total sales? **$65,450.00**

5. The Vacation Center pays its employees an 8% commission on all sales. During June Shannon sold a motorhome for $36,580 and a travel trailer for $14,230. What was her total commission on these sales? **$4064.80**

6. Wayne receives a 2% commission on sales of $24,000 or less. For sales over $24,000, he receives 4% of the amount over $24,000. What is his total commission on sales of $36,200? **$968**

7. Considering the total commission that Wayne receives in Exercise 6, what is his overall commission rate (to the nearest tenth of a percent)? **2.7%**

8. Dana put all swim suits in her store on sale at 40% off the regular price. What is the discount and sale price of a suit normally priced at $35.00? **$14; $21**

9. A video recorder normally sells for $650, but it is put on sale at a 20% discount. What is the discount and the sale price? **$130; $520**

10. What is the discount rate if an appliance normally selling for $42.50 is on sale for $33.15? **22%**

11. The price of a steam iron is decreased by $7.35. If this is a 30% discount, what is the normal selling price of the iron? **$24.50**

12. Carpet normally selling for $19.80 per square yard is being reduced by 30%. How much would it cost to carpet a room which is 3 yards wide and 4 yards long? **$166.32**

13. Find the rate of discount if a truck, originally priced at $13,450, is sold for $11,567. **14%**

14. A set of dishes was first discounted 20%. When the dishes did not sell, they were reduced by 30% of the sale price. What was the second sale price if the dishes originally sold for $160.50? **$89.88**

15. What is the overall rate of discount on the dishes in Exercise 14? **44%**

16. A jacket normally selling for $35 is on sale for 15% off the normal price. What total price must Amber pay if, after the discount is made, a 4% sales tax is added? **$30.94**

FOR REVIEW

17. The sales-tax rate in Orchard, Washington, is 3%. How much tax will Bunni be charged on the purchase of four books each selling for $3.98 before tax? What is the total price of the books?
$0.48; $16.40

18. Stephanie was charged $1.42 in sales tax on the purchase of a gift for her father. What is the sales-tax rate if the gift was priced at $35.50 before tax?
4%

19. The Schulzes had an income from wages of $29,500 in 1989. How much Social Security tax did they pay if the rate was 7.51% that year? **$2215.45**

20. In 1989 Lupe had a taxable income of $38,200. If she used the tax table given in Section 6.4, how much tax did she pay? **$8284.50**

6.5 EXERCISES C

1. Jenny sold a sweater for $39.95, a pair of pants for $34.50, and a jacket for $79.98. If she received a commission of $19.30 on these sales, to the nearest tenth of a percent, what is the commission rate?
12.5%

2. A sofa was first discounted 25%. When it did not sell, it was reduced again by 30% of the sale price. When again it did not sell, it was reduced a third time by 40% of the second sale price. What was the third sale price if the sofa originally was priced at $785.50? To the nearest tenth of a percent, what was the overall rate of discount?
[Answer: $247.43, 68.5%]

6.6 INTEREST PROBLEMS

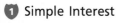

STUDENT GUIDEPOSTS

1 Simple Interest

2 Equal Payments on a Loan

3 Compound Interest

1 SIMPLE INTEREST

The money paid for the use of money is called **interest.** When we borrow money, we pay interest to the lender. When we invest money in a bank account we receive interest from the bank. The money borrowed or invested is called the **principal** and the money paid for the use is a percent of the principal. This percent is called the **rate of interest.** For example, suppose you borrow $500 for one year and the interest rate is 12% per year. To find the interest charged, find 12% of $500.

$$(0.12)(\$500) = \$60$$

At the end of the year you must return the $500 plus the interest, $60, for a total of

$$\$500 + \$60 = \$560.$$

If you had needed the $500 for two years, the interest rate would have been

$$2(0.12) = 0.24 = 24\%.$$

The total interest for two years is

$$(0.24)(\$500) = \$120.$$

At the end of two years you would pay

$$\$500 + \$120 = \$620.$$

Interest calculated this way is called **simple interest.** Interest is charged only on the principal, and not on the interest earned. In general, the following formula is used to calculate simple interest. Remember that the interest rate must always be changed from a percent to a decimal.

Simple Interest

If P represents a principal, R is a yearly rate of interest converted to a decimal, T is the time in years, and I is simple interest, then

$$I = P \cdot R \cdot T.$$

In addition, if A is the amount to be repaid or amount in the account, then

$$A = P + I.$$

EXAMPLE 1   THE AMOUNT TO REPAY ON A LOAN

Find the interest on $1500 for 2 years if the interest rate is 9% per year. How much must be repaid at the end of the 2-year period?

First find the simple interest. We are given that $P = \$1500$, $R = 9\% = 0.09$, and $T = 2$. Substitute these values.

$$
\begin{aligned}
I &= \quad P \quad \cdot \quad R \quad \cdot \; T \\
&= (\$1500) \cdot (0.09) \cdot (2) \\
&= \$270
\end{aligned}
$$

The interest is $270. Next we find A, the amount to be repaid.

$$
\begin{aligned}
A &= \quad P \quad + \quad I \\
&= \$1500 + \$270 \\
&= \$1770
\end{aligned}
$$

At the end of 2 years, $1770 must be repaid.

PRACTICE EXERCISE 1

Find the interest on $3500 for 3 years if the interest rate is 12% per year. How much must be repaid at the end of the 3-year period?

Answer: $1260; $4760

EXAMPLE 2   INTEREST ON A TIME LESS THAN ONE YEAR

What interest must be paid on a 6-month loan of $2800 if the interest rate is 16% per year? What amount must be repaid in 6 months?

Since 6 months is half of a year, we use 0.5 for T, along with $2800 for P, and $16\% = 0.16$ for R.

$$
\begin{aligned}
I &= \quad P \quad \cdot \quad R \quad \cdot \quad T \\
&= (\$2800) \cdot (0.16) \cdot (0.5) \\
&= \$224
\end{aligned}
$$

The interest for 6 months is $224. Next we find A, the amount to be repaid.

$$
\begin{aligned}
A &= \quad P \quad + \quad I \\
&= \$2800 + \$224 \\
&= \$3024
\end{aligned}
$$

Thus, in 6 months, $3024 must be repaid.

PRACTICE EXERCISE 2

What interest must be paid on an 8-month loan of $1400 if the interest rate is 9% per year? What amount must be repaid in 8 months?

Since 8 months is $\frac{8}{12}$ or $\frac{2}{3}$ of a year, use $\frac{2}{3}$ for the number of years when finding the simple interest.

Answer: $84; $1484

EXAMPLE 3 MONTHLY INTEREST ON A SAVINGS ACCOUNT

Heather put $3622.16 in a savings account for one month. If the interest paid is 9% per year, how much interest will Heather be paid at the end of the month?

Since 1 month is $\frac{1}{12}$ of a year, use $\frac{1}{12}$ for T, along with $3622.16 for P, and 9% = 0.09 for R.

$$I = P \cdot R \cdot T$$
$$= (\$3622.16) \cdot (0.09) \cdot \left(\frac{1}{12}\right)$$
$$= \$27.17 \quad \text{Rounded to the nearest cent}$$

Heather will be paid $27.17 in interest at the end of the month.

PRACTICE EXERCISE 3

Wally put $4150.15 in a savings account for two months. If the interest paid is 18% per year, how much interest will Wally earn at the end of two months (to the nearest cent)?

Answer: $124.50

② EQUAL PAYMENTS ON A LOAN

On some purchases the customer pays the amount of the purchase, together with simple interest charged, in monthly payments. For example, suppose you buy a refrigerator costing $750, pay $50 down, and wish to pay off the rest, with interest, over 18 months. The amount after the down payment is

$$\$750 - \$50 = \$700.$$

If the interest rate R is 12% = 0.12 per year and you have 18 months to pay, then $T = 1.5$ (18 months is 12 months + 6 months = 1 year + 0.5 year = 1.5 years), and the principal is $P = \$700$.

$$I = P \cdot R \cdot T$$
$$= (\$700) \cdot (0.12) \cdot (1.5)$$
$$= \$126$$

Add the principal and interest to obtain the amount A to be repaid.

$$A = P + I$$
$$= \$700 + \$126$$
$$= \$826$$

Thus, the amount to be paid in 18 equal payments is $826. To find the amount of each payment, divide $826 by 18.

$$\frac{\$826}{18} = \$45.89 \quad \text{Monthly payment, rounded to nearest cent}$$

To Find the Amount of Each Payment on a Loan

1. Find the total interest over the period.
2. Add the amount borrowed to the interest to find the total due.
3. Divide the total due by the number of payments.

| **EXAMPLE 4** **FINDING A MONTHLY PAYMENT** | **PRACTICE EXERCISE 4** |

Sam buys a television set for $587.50. He pays 10% down and the rest in 24 equal monthly payments. If the simple interest is 16%, what are his monthly payments?

First, find his down payment by taking 10% of $587.50.

$$(0.10)(\$587.50) = \$58.75$$

The balance due is found by subtracting the down payment from the price of the set.

$$\$587.50 - \$58.75 = \$528.75 \quad \text{Balance due}$$

Use the balance due, $528.75 as the principal P, $16\% = 0.16$ as R, and 2 for T (T is always in years and 24 months is 2 years).

$$I = \quad P \quad \cdot \quad R \quad \cdot T$$
$$= (\$528.75) \cdot (0.16) \cdot (2)$$
$$= \$169.20$$

Then the total due to be paid in the monthly payments is

$$\$528.75 + \$169.20 = \$697.95.$$

Divide the total due by 24 to find the amount of each monthly payment.

$$\text{monthly payment} = \frac{\$697.95}{24} = \$29.08 \quad \text{Rounded to nearest cent}$$

Sam must pay $29.08 each month for 24 months to pay off the television set.

The Wilsons purchase two rooms of furniture for $2850.80. They pay 20% down and the remainder in 36 equal monthly payments. If the simple interest charged is 12%, what are their monthly payments?

Answer: $86.16

❸ COMPOUND INTEREST

For many loans or savings accounts, interest is paid on interest as well as the principal. This type of interest is called **compound interest.** For example, if you borrow $1000 for a year at 10% per year simple interest, the interest paid is

$$(0.10)(\$1000) = \$100.$$

If the 10% per year is compounded every 6 months (semiannually), the interest is calculated every 6 months and added to the principal. Since there are 2 compounding periods every year, the yearly interest rate 10% is divided by 2 to obtain an interest rate of

$$\frac{10\%}{2} = 5\%$$

for every 6-month period. The interest on $1000 for the first 6 months is

$$(0.05)(\$1000) = \$50.$$

With compound interest, the principal on which interest is paid for the second 6 months is the original principal plus the first 6-month interest,

$$\$1000 + \$50 = \$1050.$$

The interest on $1050 for the second 6 months is

$$(0.05)(\$1050) = \$52.50.$$

The total interest for the year is the sum of the interests on the first and second 6-month periods.

$$\$50.00 + \$52.50 = \$102.50.$$

Notice that the compound interest amounted to $2.50 more than the simple interest.

This process could be continued to find the interest compounded semiannually over more than one year.

To Solve a Compound Interest Problem

1. Determine the interest rate for a compounding period by dividing the yearly rate by the number of periods per year.
2. Find the interest for the first period.
3. Use the sum of the interest found in Step 2 and the principal as the principal for the next period.
4. Continue until all periods have been accounted for.

Interest can be compounded over different periods. The most common types are shown in the table below where R is the annual or yearly interest rate.

Type of compounding	Number of compounding periods each year	Interest rate per period
compounded annually	1	R
compounded semiannually	2	$\dfrac{R}{2}$
compounded quarterly	4	$\dfrac{R}{4}$
compounded monthly	12	$\dfrac{R}{12}$

HINT

Remember that when interest is compounded, the interest used for each compounding period is

$$\frac{\text{Annual Interest Rate}}{\text{Number of Periods Per Year}}.$$

For example, if the annual interest rate is 12%, and compounding is semi-annually, quarterly, or monthly, use 6%, 3%, or 1%, respectively, as the interest rate each period.

EXAMPLE 5 INTEREST COMPOUNDED ANNUALLY

Richard invests $5000 in an account which pays 8% compounded annually. How much will he have in his account at the end of 2 years?

Calculate the interest for the first year. Since there is only one period per year, the interest rate is 8% = 0.08.

$$(0.08)(\$5000) = \$400$$

PRACTICE EXERCISE 5

Lori invests $3000 in an account which pays 11% interest compounded annually. How much will she have in the account at the end of 3 years?

The principal on which the second year interest is calculated is thus

$$\$5000 + \$400 = \$5400.$$

The interest for the second year is

$$(0.08)(\$5400) = \$432.$$

The total in the account at the end of 2 years is

$$\$5400 + \$432 = \$5832.$$

Answer: **$4102.89**

| **EXAMPLE 6** INTEREST COMPOUNDED SEMIANNUALLY | **PRACTICE EXERCISE 6** |

Joe borrowed \$600 for $1\frac{1}{2}$ years. Interest is 12% per year compounded semiannually. What amount must he pay back in $1\frac{1}{2}$ years?

Since semiannually means twice a year, the interest rate for a compounding period is

$$\frac{12\%}{2} = 6\%,$$

and the number of periods is 3. Interest for the first 6 months (the first period) is

$$(0.06)(\$600) = \$36.$$

The principal to be used for the second period is

$$\$600 + \$36 = \$636.$$

Interest for the second period is

$$(0.06)(\$636) = \$38.16.$$

The principal to be used for the third period is

$$\$636.00 + \$38.16 = \$674.16.$$

Interest for the third period is

$$(0.06)(\$674.16) = \$40.45. \quad \text{To the nearest cent}$$

The final amount to be paid is

$$\$674.16 + \$40.45 = \$714.61.$$

PRACTICE EXERCISE 6

Verna borrowed \$1500 for 1 year. Interest on the loan is 12% per year, compounded quarterly (four times a year). What amount must she pay back at the end of the year?

Remember that 12% per year compounded quarterly is 3% (one fourth of 12%) per period.

Answer: **$1688.26**

6.6 EXERCISES A

Solve the following problems.

1. Hans borrowed \$600 for one year at a bank where the interest rate is 12%. How much interest did he have to pay at the end of the year?   **$72**

2. Hannah put \$750 in a savings account which pays 6% interest per year. What did she have in her account at the end of the year?   **$795**

3. Peter bought a snow blower for \$450 and paid \$50 down. The remaining amount he paid off at the end of 6 months. If interest is 14% per year (7% for 6 months), how much did he have to pay when he paid off the loan?   **$428**

4 Lisa borrowed \$800 for expenses at the beginning of the school year. If simple interest was 12% per year, how much did she have to pay if she paid off the loan at the end of 9 months?   **$872**

5. Harold bought a $2526.50 stereo system and paid 10% down. The remaining money plus interest was to be paid at the end of 2 years. If interest is 14% per year simple interest, how much interest (to the nearest cent) did Harold have to pay?   **$636.68**

6. Lynn borrowed $5482.50 for 2 years at 12% simple interest. What did she have to pay at the end of the period to pay off the loan?   **$6798.30**

7. If Lynn in Exercise 6 had paid off the loan in 24 equal monthly payments, what would her monthly payment have been (to the nearest cent)? **$283.26**

8. Robert will pay off a $1213.40 debt in 18 monthly payments. Simple interest is 16% per year. What are his payments (to the nearest cent)?   **$83.59**

9. Susan borrowed $500 from a loan shark who charges 15% per month simple interest. What interest did she have to pay for one year? What was her monthly payment if she paid off the loan in 12 months?   **$900; $116.67**

10. Sylvester invests $2500 in an account which pays 7% per year compounded annually. What does he have in the account at the end of 2 years? **$2862.25**

11. Maria borrows $625.50 at 12% compounded semi-annually. What must she pay (to the nearest cent) to pay off the loan at the end of 18 months?   **$744.98**

12. Patricia has $7000 to invest in an account which pays 10% compound interest. How much is in her account at the end of one year if the interest is compounded semiannually? How much is in her account at the end of one year if the compounding is quarterly (4 times per year)? **$7717.50; $7726.69**

FOR REVIEW

13. Joseph receives a 20% commission on all appliances he sells. His total sales for June were $7218.50. What was his commission?   **$1443.70**

14. Michael's, a shop for men, put suits on sale at a 33% discount. What is the sale price of a $285 suit? **$190.95**

ANSWERS: 1. $72   2. $795   3. $428   4. $872   5. $636.68   6. $6798.30   7. $283.26   8. $83.59   9. $900; $116.67
10. $2862.25   11. $744.98   12. $7717.50; $7726.69   13. $1443.70   14. $190.95

6.6 EXERCISES B

Solve the following problems.

1. Brad borrowed $1800 for one year at his credit union where the interest rate is 10%. How much interest did he pay at the end of the year?   **$180**

2. Boyd put $900 in a savings account which pays 9% interest per year. What was the value of his account at the end of one year?   **$981**

3. Katherine bought a microwave oven for $480 and paid $80 down. She paid off the remaining amount plus interest at the end of 6 months. If the interest rate she was charged was 8%, how much did she pay?   **$416**

4. Dave Markee borrowed $700 to fix his car. If he was charged 12% interest per year, how much did he have to pay back at the end of 8 months?   **$756**

5. Colleen bought a coat for $252.20 and paid 20% down. The remaining amount is to be paid off at the end of 6 months. If the simple interest rate is 10% per year, what will she have to pay to pay off the loan?   **$211.85**

6. Calvin borrowed $628 for 2 years. If the simple interest rate is 12% per year, what amount will he have to pay at the end of the 2-year period?   **$778.72**

7. If Calvin in Exercise 6 pays off the loan in 24 equal monthly payments, how much will he pay each month?   **$32.45**

8. Bruce will pay off a $782.50 debt in 18 monthly payments. If simple interest is 14% per year, what is his monthly payment?   **$52.60**

9. Vinnie, the loan shark, charges 20% *per month* simple interest. How much interest will he collect on a loan of $1000 which is paid off in one year? If Vinnie collects the loan with interest in 12 equal monthly payments, how much does he receive each month?   **$2400; $283.33**

10. Maria invests $5800 in an account that pays 8% per year compounded semiannually. How much is in her account at the end of a year?   **$6273.28**

11. Virgil borrows $840.50 at 10% interest compounded semiannually. What must he pay (to the nearest cent) to pay off the loan at the end of 30 months?   **$1072.71**

12. William borrows $540.00 at 12% compound interest per year. What total amount must he pay at the end of the year if the compounding is semiannually? How much if the compounding is quarterly?   **$606.74; $607.78**

FOR REVIEW

13. Bridgette earned $123.65 in commissions on sales in a dress shop. If she works on a 10% commission rate, what were her total sales?   **$1236.50**

14. Mr. Miller, a senior citizen, receives a discount of 15% on all bills for food at the Village Waffle Shoppe. If the bill comes to $123.40 when he takes his family out to eat, how much must he pay?   **$104.89**

6.6 EXERCISES C

1. Find the simple interest earned on $13,455 at 11.25% per year for 20 months.   **$2522.81**

2. Carl invests $8575 in an account which pays 11.5% interest compounded semiannually. How much is in his account at the end of 18 months? [Answer: $10,140.87]

It can be shown that the amount A in an account if a principal P is deposited at 12% interest compounded monthly for 5 years is given by

$$A = P(1.01)^{60}.$$

Use a calculator with a $\boxed{y^x}$ *key and this formula to find the amount in an account with the value of P given in Exercises 3–5.*

3. $P = \$1000$ **$1816.70**   4. $P = \$5000$ **$9083.48**   5. $P = \$10,000$ **$18,166.97**

6.7 PERCENT INCREASE AND DECREASE PROBLEMS

The percent by which a quantity increases is called the **percent increase.** Similarly, the percent by which a quantity decreases is called the **percent decrease.** For example, if a salary increased 7%, we find the increase by taking 7% of the *original* salary. Thus, if you earn $20,000 a year and receive a 7% raise, the increase in salary or your raise is

$$(0.07)(\$20,000) = \$1400$$

and your new salary will be

$$
\begin{aligned}
\text{new salary} &= \text{former salary} + \text{raise} \\
&= \quad\quad \$20,000 \quad\ + \$1400 \\
&= \$21,400.
\end{aligned}
$$

If school enrollment decreases by 6%, we find the decrease by taking 6% of the *original* number of students. For example, if BSU had 10,500 students last year and enrollment went down 6% this year, the decrease in enrollment was

$$(0.06)(10,500) = 630 \text{ students.}$$

The enrollment this year is

$$
\begin{aligned}
\text{this year's enrollment} &= \text{last year's enrollment} - \text{decrease} \\
&= \quad\quad\quad 10,500 \quad\quad - \quad 630 \\
&= 9870 \text{ students.}
\end{aligned}
$$

These are examples of percent increase and percent decrease problems.

To Solve a Percent Increase (Decrease) Problem

1. Use the percent equation

$$
\begin{array}{ccccc}
A & = & P\% & \cdot & B
\end{array}
$$

increase (decrease) = percent increase (decrease) · original amount

or use the percent proportion

$$\frac{A}{B} = \frac{P}{100}$$

$$\frac{\text{increase (decrease)}}{\text{original amount}} = \frac{\text{percent increase (decrease)}}{100}.$$

2. The new amount is given by

new amount = original amount + increase

or

new amount = original amount − decrease.

EXAMPLE 1 PERCENT DECREASE APPLICATION

A labor union votes to take a 5% decrease in wages in order to keep the company from going bankrupt. If the average salary before the decrease was $17,500 per year, what is the new salary?

First find the decrease in salary.

PRACTICE EXERCISE 1

Due to declining enrollments, teachers in the Kachina District are faced with a 3% decrease in wages. If the average salary before the decrease is $14,700, what will be the new average salary?

Method 1

$$A \quad = \quad P\% \quad \cdot \quad B$$

decrease = percent decrease · original salary

$$= \quad (0.05) \quad \cdot \quad (\$17,500)$$

$$= \$875$$

Method 2

$$\frac{A}{B} = \frac{P}{100}$$

$$\frac{\text{decrease}}{\text{original salary}} = \frac{\text{percent decrease}}{100}$$

$$\frac{\text{decrease}}{\$17,500} = \frac{5}{100}$$

$$\text{decrease}(100) = (5)(\$17,500)$$

$$\text{decrease} = \frac{(5)(\$17,500)}{100}$$

$$= \$875$$

Thus, the decrease in salary is $875. Next find the new salary.

new salary = original salary − decrease in salary

$$= \quad \$17,500 \quad - \quad \$875$$

$$= \$16,625$$

The new average salary is $16,625.

Answer: $14,259

EXAMPLE 2 PERCENT INCREASE APPLICATION

State University expects an increase in enrollment of 2.5% next year. If the present enrollment is 15,276, what will be the approximate number of students next year?

First find the increase in enrollment.

Method 1

$$A \quad = \quad P\% \quad \cdot \quad B$$

increase = percent increase · original enrollment

$$= \quad (0.025) \quad \cdot \quad (15,276)$$

$$= 381.9$$

Method 2

$$\frac{A}{B} = \frac{P}{100}$$

$$\frac{\text{increase}}{\text{original enrollment}} = \frac{\text{percent increase}}{100}$$

$$\frac{\text{increase}}{15,276} = \frac{2.5}{100}$$

$$\text{increase}(100) = (2.5)(15,276)$$

$$\text{increase} = \frac{(2.5)(15,276)}{100}$$

$$= 381.9$$

PRACTICE EXERCISE 2

The placement director expects an increase of 3.5% in the number of students taking the placement exam in mathematics during the fall semester. If 2480 students took the exam last fall, what will be the approximate number of students tested this fall?

We must round to the nearest whole number so the increase will be approximately 382 students. The new enrollment is found by adding this year's enrollment to the increase.

$$
\begin{array}{rl}
15276 & \text{Original enrollment} \\
+\quad 382 & \text{Enrollment increase} \\
\hline
15658 & \text{New enrollment}
\end{array}
$$

There will be approximately 15,658 students next year.

Answer: 2567 students

HINT

Remember that percent increase or decrease is always taken on the original quantity. To find the new quantity *add* the increase to the original amount or *subtract* the decrease from the original amount.

EXAMPLE 3   FINDING A PERCENT DECREASE

The price of a dozen eggs dropped from 75¢ to 67¢. What was the percent decrease?

First find the amount of the decrease.

$$75\cent - 67\cent = 8\cent$$

Since 8¢ is the amount of the decrease, and 75¢ is the original price, we must find the percent decrease, P.

Method 1

$$
\begin{array}{ccccc}
A & = & P\% & \cdot & B \\
\text{decrease} & = & \text{percent decrease} & \cdot & \text{original price} \\
\downarrow & & \downarrow & & \downarrow \\
8 & = & P(0.01) & \cdot & (75)
\end{array}
$$

$$8 = P(0.75)$$

$$\frac{8}{0.75} = P$$

$$10.\overline{6} = P$$

Method 2

$$\frac{A}{B} = \frac{P}{100}$$

$$\frac{\text{decrease}}{\text{original price}} = \frac{\text{percent decrease}}{100}$$

$$\frac{8}{75} = \frac{P}{100}$$

$$(8)(100) = (75)P$$

$$\frac{(8)(100)}{75} = P$$

$$10.\overline{6} = P$$

Thus, the percent decrease was $10.\overline{6}\%$ or $10\frac{2}{3}\%$.

PRACTICE EXERCISE 3

The price of a six-pack of soda dropped from \$2.25 to \$1.85. What was the percent decrease, to the nearest tenth of a percent?

Answer: 17.8%

| **EXAMPLE 4** FINDING AN ORIGINAL PRICE | PRACTICE EXERCISE 4 |

The price of gasoline increased 10%. The amount of this increase was 12¢ per gallon. What was the original price of a gallon of gasoline?

   We know the percent increase and the amount of the increase, so we must find the original price.

Method 1

$$A \quad = \quad P\% \quad \cdot \quad B$$

$$\text{increase} = \boxed{\text{percent increase}} \cdot \text{original price}$$

$$12 \quad = \quad (0.10) \quad \cdot \text{original price}$$

$$\frac{12}{0.10} = \text{original price}$$

$$120 = \text{original price}$$

Method 2

$$\frac{A}{B} = \frac{P}{100}$$

$$\frac{\text{increase}}{\text{original price}} = \frac{\text{percent increase}}{100}$$

$$\frac{12}{\text{original price}} = \frac{10}{100}$$

$$(12)(100) = (10)\text{original price}$$

$$\frac{(12)(100)}{10} = \text{original price}$$

$$120 = \text{original price}$$

Thus, the original price of a gallon of gasoline was 120¢ or $1.20.

Answer: 90¢ per gallon

| **EXAMPLE 5** FINDING A PERCENT INCREASE | PRACTICE EXERCISE 5 |

The Smith family bought a house two years ago for $85,000. It is now valued at $102,680. To the nearest percent, what is the percent increase in value?

   To find the amount of the increase, subtract the original value from the new value.

$102680	New value
− 85000	Original value
$ 17680	Value increase

We must find the percent increase, P.

Method 1

$$A \quad = \quad P\% \quad \cdot \quad B$$

$$\boxed{\text{increase}} = \text{percent increase} \cdot \boxed{\text{original value}}$$

$$17{,}680 = \quad P(0.01) \quad \cdot \quad 85{,}000$$

$$17{,}680 = P(850)$$

$$\frac{17{,}680}{850} = P$$

$$20.8 = P$$

Method 2

$$\frac{A}{B} = \frac{P}{100}$$

$$\frac{\text{increase}}{\text{original value}} = \frac{\text{percent increase}}{100}$$

$$\frac{17,680}{85,000} = \frac{P}{100}$$

$$(17,680)(100) = (85,000)P$$

$$\frac{(17,680)(100)}{85,000} = P$$

$$20.8 = P$$

Thus, to the nearest percent, the percent increase is about 21%.

Answer: 54.2%

CAUTION

In problems such as the one in Example 5, be sure to compare the increase in value to the *original* value, not the new value. In such cases always ask:

increase is what percent of *original value*?

An interesting application of percent increase and decrease has to do with changes in automobile fuel economy. Such things as temperature, humidity, wind conditions, and road conditions have an effect on the amount of fuel used.

Suppose we assume the percent increases and decreases in the following tables for the given conditions. An increase in fuel economy means more miles per gallon (mpg) and a decrease means fewer mpg.

Condition	% Increase
tailwind	15%
smooth road	12%
high humidity	6%

Condition	% Decrease
headwind	17%
rough road	25%
low temperature	10%

EXAMPLE 6 APPLICATIONS OF FUEL ECONOMY

(a) If Michael's car normally gets 30 mpg, what should it get with a tailwind?

First use the table to find that a tailwind increases economy by 15%. Find 15% of 30 mpg.

$$(0.15)(30) = 4.5$$

A tailwind gives Michael an additional 4.5 mpg for a total of

$$30 + 4.5 = 34.5 \text{ mpg}.$$

Thus, Michael's car gets about 34.5 mpg with a tailwind.

(b) Suppose Mary drives on a rough road under high humidity conditions. What is the percent change in economy? If her normal mileage is 20 mpg, what is her mileage under the given conditions?

PRACTICE EXERCISE 6

(a) If Michael's car normally gets 30 mpg, what should it get when he drives on a smooth road?

Since the rough road means a 25% decrease and high humidity means a 6% increase, find the difference:

$$25\% - 6\% = 19\%.$$

Under these conditions there is a 19% decrease in economy. Taking 19% of 20,

$$(0.19)(20) = 3.8 \text{ mpg},$$

the amount of decrease is 3.8 mpg. The mileage under these conditions is

$$20 - 3.8 = 16.2 \text{ mpg}.$$

Thus, Mary's mileage is approximately 16.2 mpg.

(b) If Mary drives against a headwind but on a smooth road, what is the percent change in economy? What is her mileage under these conditions if normally she gets 20 mpg?

Answers: (a) 33.6 mpg   (b) 5% decrease; 19 mpg

6.7 EXERCISES A

Solve the following problems.

1. Laura was given a 9% increase in salary. What is her new salary if her old salary was $28,250?
$30,792.50

2. The average salary of the workers for Farms, Incorporated, was $12,320. After the union negotiated a 12% increase, what is the average new salary?
$13,798.40

3. The executives of the Paper Corporation agree to take an 8% salary cut in order to help the financial situation of the company. If the president was making $125,300 per year, what is his new salary?
$115,276

4. City College expects a decrease in enrollment of 3.5% next year. If the current enrollment is 23,254, what will be the approximate enrollment next year?
22,440

5 The population of East, Texas, increased from 329 to 415. To the nearest percent, what was the percent increase?   26%

6. Over a period of years, the price of a gallon of gasoline increased from $0.32 to $1.27. To the nearest percent, what was the percent increase?   297%

7. Because of oversupply, the price of lettuce dropped from 92¢ per head to 39¢ per head. To the nearest tenth of a percent, what was the percent decrease?
57.6%

8 Forrest received a 15% increase in salary. The amount of the increase was $1824. What was his original salary? $12,160

9. By attending a weight-control class, Mary lost 24 pounds. If she now weighs 115 pounds, what was her original weight and her percent decrease to the nearest tenth of one percent?   **139; 17.3%**

10. The value of a new car decreases 27% the first year. If the decrease amounted to $2300, what was the original value?   **$8518.52**

Refer to the mileage tables in this section for Exercises 11–14.

11. The normal gasoline mileage of a car is 30 mpg. What is its mileage on a smooth road?   **33.6 mpg**

12. Henry normally gets 8 mpg with his truck. If he drives on a rough road, what mileage will he get?   **6 mpg**

13. Jane is driving with a tailwind but the temperature is low. What will be her mileage if she normally gets 42 mpg?   **44.1 mpg**

14. Sue is driving into a headwind under conditions of high humidity. She has a decrease of 4.4 mpg in her mileage. What is her normal mileage?   **40 mpg**

15. Driving at high speeds reduces your mileage. Michael normally gets about 33 mpg, but by driving fast, he only got 27 mpg. To the nearest tenth of a percent, what was his percent decrease?   **18.2%**

16 By keeping his car in good mechanical condition, Arnie can get an increase in mileage of 13%. Good driving habits will give a 9% increase in mileage. By keeping his car in good condition and driving correctly, what mileage can Arnie get if his normal rate is 30 mpg?   **36.6 mpg**

FOR REVIEW

17. Sara buys a stereo for $728.50 and puts 10% down, paying the rest in 18 monthly payments. If the simple interest is 16%, what are her payments (to the nearest cent)?   **$45.17**

18. Henry puts $1500 in a savings account which pays 8% compounded semiannually. What will be the value of his account at the end of 2 years?   **$1754.79**

ANSWERS: 1. $30,792.50   2. $13,798.40   3. $115,276   4. 22,440   5. 26%   6. 297%   7. 57.6%   8. $12,160
9. 139; 17.3%   10. $8518.52   11. 33.6 mpg   12. 6 mpg   13. 44.1 mpg   14. 40 mpg   15. 18.2%   16. 36.6 mpg
17. $45.17   18. $1754.79

6.7 EXERCISES B

Solve the following problems.

1. Maxine received a salary of $18,560 one year. The next year she received a raise of 8%. What was her new salary?   **$20,044.80**

2. The employees of the Soft Shoe Company negotiated an 11% salary increase. If their average salary was $14,200, what is it now?   **$15,762**

3. Several years ago, the price of gasoline was $0.40 per gallon. What was the price after a 190% increase?   **$1.16**

4. A major corporation is forced to decrease its number of employees by 4.5% next year. If it now employs 3540 workers, how many will be employed next year?   **approximately 3381**

5. Northeast University's enrollment increased from 24,300 to 27,800. To the nearest tenth, what was the percent increase?   **14.4%**

6. Due to overstocking, a retailer is forced to decrease the price of an item from $1.85 to $1.25. To the nearest tenth of a percent, what is the percent decrease?   **32.4%**

7. A car dealer lowered the price of a new car from $14,800 to $13,320. To the nearest percent, what was the percent decrease?   **10%**

8. A new car decreased in value a total of $1920 the first year. If this was a 24% decrease, what was the original value?   **$8000**

9. After six weeks of basic training, a new recruit in the army lost 14 pounds. If he now weighs 176 pounds, what was his original weight and the percent decrease in weight (to the nearest tenth of a percent)?   **190 pounds; 7.4%**

10. The value of a Navajo Indian rug increased 15% during the first year. If the increase amounted to $114.87, what was the original price?   **$765.80**

Refer to the mileage tables in this section for Exercises 11–14.

11. The normal gasoline mileage for a motorhome is 12 mpg. What is the mileage with a tailwind? **13.8 mpg**

12. Danielle normally gets 26 mpg with her car. If she drives during a period of low temperature, what mileage will she get?   **23.4 mpg**

13. Rufus is driving with a headwind on a smooth road. What will his mileage be if he usually gets 18 mpg? **17.1 mpg**

14. Randy is driving into a headwind, on a rough road, and at a low temperature. If he has a decrease of 13 mpg in his mileage, what is his normal mileage? **25 mpg**

15. By driving at the speed limit, you can improve your mileage from what would be obtained by driving over the limit. If Marla got 14 mpg while exceeding the limit and 19 mpg while driving at the speed limit, what was the percent increase (to the nearest tenth of a percent)?   **35.7%**

16. When a car is tuned properly, its gas mileage can be improved by 5%. If Ronnie was getting 24 mpg before a tune-up, what mileage can he expect to get after the car has been serviced?   **25.2 mpg**

FOR REVIEW

17. Den bought a set of tools for $565.50 by putting 20% down and paying the rest in 12 monthly payments. If he is charged 14% simple interest, what are his payments?   **$42.98**

18. Montana plans to put $3000 into a savings account for two years. She can choose one account which pays 11% simple interest or another which pays 10% interest compounded semiannually. Which account will return the most interest?
The account with 11% simple interest will return more interest ($660) than the other account ($646.52).

6.7 EXERCISES C

Solve.

1. A new car decreases in value by approximately 27.5% each year. If a car is purchased for $12,685.35, what is its approximate value two years later?   **$6667.74**

2. By decreasing the thermostat from 72° to 68°, a company can reduce its heating bill by 35.5%. If the bill last month, prior to turning down the thermostat, was $478.63, what can the company expect to pay for heat this month?
[Answer: $308.72]

6.8 MARKUP AND MARKDOWN PROBLEMS

A special type of percent increase problem involves marking up prices in a business. Any wholesale or retail business must sell an item for more than its cost. This extra amount goes to pay operating expenses and give the owners a return on their investment. In many businesses, the difference between the cost and selling price is a fixed percent of the cost.

The amount by which the selling price is increased over the cost is called the **markup** on the item. The markup is divided by the cost to get the **percent markup.** For example, if the cost of an item is $4 and the selling price is $5, then the markup on the item is

$$\$5 - \$4 = \$1.$$

The percent markup is found by dividing $1 (markup) by $4 (cost).

$$\frac{\text{markup}}{\text{cost}} = \frac{1}{4} = 0.25 = 25\%$$

Thus, the percent markup is 25%.

In most businesses, the percent markup is given and the amount of the markup must be found. Suppose a 30% markup is needed to make a business prosper. When a pair of shoes costing $20 is to be sold, the sales staff can calculate the markup as follows:

$$(0.30)(\$20) = \$6.$$

The markup is $6 and the selling price is

$$\$20 + \$6 = \$26.$$

Thus, the store will sell the pair of shoes for $26.

On the other hand, suppose a wholesaler bought too many of a particular item. He might have to reduce his selling price to sell them. This is a percent decrease or a discount problem, and the **percent markdown** is based on the selling price of the item. For example, a 30% reduction would be found by taking 30% of the selling price. The amount of the price reduction is called the **markdown.**

To Solve a Markup (Markdown) Problem

1. Use the percent equation

A	$=$	$P\%$	$\cdot$	B

markup (markdown) = percent markup (markdown) · cost (price)

   or use the percent proportion

$$\frac{A}{B} = \frac{P}{100}$$

$$\frac{\text{markup (markdown)}}{\text{cost (price)}} = \frac{\text{percent markup (markdown)}}{100}.$$

2. To find the new selling price use

selling price = cost + markup

or

new selling price = old selling price − markdown.

HINT

In a markup problem, the markup is always a percent of the original cost, *not* a percent of the new selling price.

EXAMPLE 1 USING MARKUP TO FIND THE SELLING PRICE

An appliance wholesaler has found that his percent markup must be 25%. If he buys a washing machine for $285, what is the selling price?

Method 1

$$A = P\% \cdot B$$

markup = percent markup · cost

markup = (0.25) · $285

= $71.25

Method 2

$$\frac{A}{B} = \frac{P}{100}$$

$$\frac{\text{markup}}{\text{cost}} = \frac{\text{percent markup}}{100}$$

$$\frac{\text{markup}}{\$285} = \frac{25}{100}$$

$$\text{markup}(100) = (25)(\$285)$$

$$\text{markup} = \frac{(25)(\$285)}{100}$$

$$= \$71.25$$

The markup is $71.25, and the selling price is the sum of the markup and the cost.

$285.00	Cost
+ 71.25	Markup
$356.25	Selling price

The dealer must sell the washing machine for $356.25.

PRACTICE EXERCISE 1

In order to stay in business, the owner of a clothing store must have a percent markup of 85%. If he buys a suit for $125.00, what is the selling price of the suit?

Answer: $231.25

EXAMPLE 2 FINDING A MARKUP AND THE PERCENT MARKUP

The Shoe Store buys one type of shoe for $26.50 per pair and sells them for $34.45. What is the markup and the percent markup?

Find the markup first by finding the difference between the selling price and the cost.

$34.45	
− 26.50	
$ 7.95	Selling price − cost = markup

The markup is $7.95. The percent markup is found by answering the question: $7.95 is what percent of $26.50?

PRACTICE EXERCISE 2

A retailer buys a radio for $34.50 and sells it for $48.30. What is the markup and the percent markup?

Let P be the percent markup.

Method 1	Method 2

Method 1

$$A = P\% \cdot B$$

markup = percent markup · cost

$$7.95 = P(0.01) \cdot 26.50$$
$$7.95 = P(0.265)$$
$$\frac{7.95}{0.265} = P$$
$$30 = P$$

Method 2

$$\frac{A}{B} = \frac{P}{100}$$

$$\frac{\text{markup}}{\text{cost}} = \frac{\text{percent markup}}{100}$$

$$\frac{7.95}{26.50} = \frac{P}{100}$$

$$(7.95)(100) = (26.50)P$$

$$\frac{(7.95)(100)}{26.50} = P$$

$$30 = P$$

The percent markup is 30%.

Answer: $13.80; 40%

EXAMPLE 3 FINDING THE COST AND SELLING PRICE

PRACTICE EXERCISE 3

A clothing wholesaler has a 40% markup. If the markup on a suit is $86.50, what is the cost? What is the selling price?

A store owner operates on a 45% markup. If the markup on a stereo is $108, what is the cost? What is the selling price?

Method 1

$$A = P\% \cdot B$$

markup = percent markup · cost

$$86.50 = (0.40) \cdot \text{cost}$$
$$\frac{86.50}{0.40} = \text{cost}$$
$$216.25 = \text{cost}$$

Method 2

$$\frac{A}{B} = \frac{P}{100}$$

$$\frac{\text{markup}}{\text{cost}} = \frac{\text{percent markup}}{100}$$

$$\frac{86.50}{\text{cost}} = \frac{40}{100}$$

$$(86.50)(100) = \text{cost}(40)$$

$$\frac{(86.50)(100)}{40} = \text{cost}$$

$$216.25 = \text{cost}$$

The cost to the wholesaler is $216.25.

$$\text{selling price} = \text{cost} + \text{markup}$$
$$= \$216.25 + \$86.50$$
$$= \$302.75$$

Thus, the selling price is $302.75.

Answer: $240; $348

EXAMPLE 4 USING MARKDOWN TO FIND A NEW SELLING PRICE

PRACTICE EXERCISE 4

The wholesale dealer in Example 3 must reduce the selling price of his suits by 30% in order to make room for new suits. What is the markdown and the new selling price of the suit that sold for $302.75?

When the stereo did not sell, the owner in Practice Exercise 3 was forced to reduce the selling price by 30%. What was the markdown and the new selling price of the stereo which sold for $348?

Method 1

$$A = P\% \cdot B$$

markdown = percent markdown · price

$$\text{markdown} = (0.30) \cdot \$302.75$$
$$= \$90.83 \quad \text{Nearest cent}$$

Method 2

$$\frac{A}{B} = \frac{P}{100}$$

$$\frac{\text{markdown}}{\text{price}} = \frac{\text{percent markdown}}{100}$$

$$\frac{\text{markdown}}{\$302.75} = \frac{30}{100}$$

$$\text{markdown}(100) = (\$302.75)(30)$$

$$\text{markdown} = \frac{(\$302.75)(30)}{100}$$

$$= \$90.83$$

Then we have

new selling price = old selling price − markdown.

$$= \$302.75 - \$90.83$$

$$= \$211.92$$

The suit now sells for $211.92.

Answer: $104.40; $243.60

Notice that the new selling price in Example 4 is actually less than the cost given in Example 3.

$$\begin{array}{r} \$216.25 \\ -\ \ 211.92 \\ \hline \$\ \ \ 4.33 \end{array}$$

The suit was sold for $4.33 less than the cost. (The dealer lost on the sale.) This means that the 30% markdown ($90.83) is more than the 40% markup ($86.50). That is,

30% of $302.75   ($90.83)

is more than

40% of $216.25   ($86.50).

This shows that we need to keep in mind the amount on which the percent is taken as well as the percent rate when working with markup and markdown problems.

EXAMPLE 5 MARKUP AND MARKDOWN TOGETHER

The Green Grocery Company bought lettuce for 30¢ per head. The markup on the cost was 80%. After a few days, the price of the remaining lettuce was reduced by 60% of the selling price.

(a) What was the markup and the first selling price?

To find the markup, take 80% of 30¢.

$$(0.80)(30¢) = 24¢$$

The first selling price was

$$30¢ + 24¢ = 54¢.$$

PRACTICE EXERCISE 5

Booker's Books bought a book for $3.00. The markup on cost was 75%. When the book did not sell, it was reduced 40% of the selling price.

(a) What was the markup and the first selling price?

(b) What was the markdown and the second selling price?

The markdown is 60% of 54¢ (selling price).

$$(0.60)(54¢) = 32¢ \text{ (to the nearest cent)}$$

To find the second selling price, subtract 32¢ from 54¢.

$$54¢ - 32¢ = 22¢$$

(c) What was the loss on the lettuce sold after the markdown?

In order to sell the lettuce before it spoiled, the grocer sold it for 22¢. This was a loss of

$$30¢ - 22¢ = 8¢$$

per head. The grocer sold at an 8¢ loss to prevent a loss of 30¢ on each head spoiled.

(d) What percent of the cost was the loss?

The question we must answer is:

8¢ is **what percent** of 30¢?

$$
\begin{array}{ccccc}
A & = & P\% & \cdot & B \\
8 & = & P(0.01) & \cdot & 30 \\
8 & = & P(0.3) & & \\
\end{array}
$$

$$\frac{8}{0.3} = P \quad \text{Related division sentence}$$

$$26.\overline{6} = P$$

Thus, the percent loss based on the cost of the lettuce was $26.\overline{6}\%$.

(b) What was the markdown and the second selling price?

(c) What was the profit on the book sold after the markdown?

(d) What percent of the cost was the profit?

Answers: (a) $2.25; $5.25
(b) $2.10; $3.15   (c) $0.15
(d) 5%

HINT

You have probably noticed that in many of the percent applications we have considered, a basic relationship is used which can be summarized by using $A = P\% \cdot B$ or $\dfrac{A}{B} = \dfrac{P}{100}$ in the following forms.

$$\text{Change} = \text{percent} \cdot \text{original amount}$$

or

$$\frac{\text{change}}{\text{original amount}} = \frac{\text{percent}}{100}$$

Now that you have experience with the basic terminology used in the various applications, you may want to remember this one formula and use it to cover a multitude of situations. For example, in a markup problem, *change* is *markup*, *percent* is *percent markup*, and *original amount* is *cost*.

6.8 EXERCISES A

Solve the following problems.

1. If the markup is 20%, what is the selling price of a set of glasses that costs $26?   **$31.20**

2. Ace Appliances bought a refrigerator for $216.00 and sold it for $280.80. What were the markup and the percent markup?   **$64.80; 30%**

3. Roberta's Interior Design made $3.50 per yard on carpet sold to a customer. If the markup was 40%, what was the cost and what was the selling price?   **$8.75; $12.25**

4. To raise money for a project, the Student Club bought flowers for $2.00 per dozen and sold them on campus for $5.00 per dozen. What was the percent markup?   **150%**

5. Because of a high spoilage rate, the markup on fruit must be 120%. What is the selling price (to the nearest cent) of peaches that cost 22¢ per pound?   **48¢**

6. Fred's TV Shop bought a color television set for $582 and sold it for $727.50. What was the percent markup?   **25%**

7. The selling price of a toaster is $36.20. It is put on sale with a markdown (discount) rate of 30%. What is the reduced selling price?   **$25.34**

8. A coat which sells for $125 is put on sale for $95. What is the markdown rate?   **24%**

9. The markdown on a loaf of bread is 36¢ and the markdown rate is 40%. What was the old selling price? What is the new selling price?   **90¢; 54¢**

10. The Tough Tool company bought a set of tools for $120. The markup was 20%. When the set did not sell, the price was reduced at a markdown rate of 15%. What was the first selling price and what was the second selling price?   **$144; $122.40**

11. If the tool set in Exercise 10 sold at the second price, what was the profit? What percent of the cost was the profit?   **$2.40; 2%**

12. Mario's Men's Shop bought a suit for $180. The markup was 30%. Because of changing styles, the suit was sold at a discount of 25%. What was the loss? What percent of the cost was the loss?   **$4.50; 2.5%**

FOR REVIEW

13. Under normal conditions, Martha's car gets 28.5 mpg. While driving against a headwind, the mileage was 25.3 mpg. To the nearest tenth, what was the percent decrease?   **11.2%**

14. Jenny got a 12% raise. If her old salary was $16,250, what is her new salary?   **$18,200**

ANSWERS: 1. $31.20 2. $64.80; 30% 3. $8.75; $12.25 4. 150% 5. 48¢ 6. 25% 7. $25.34 8. 24% 9. 90¢; 54¢ 10. $144; $122.40 11. $2.40; 2% 12. $4.50; 2.5% 13. 11.2% 14. $18,200

6.8 EXERCISES B

Solve the following problems.

1. Hard Hardware bought a shipment of nails for $820.40. If the markup is 15%, what is the selling price? **$943.46**

2. Oranges cost a grocer 30¢ per pound and the markup is 70%. What is the selling price? **51¢**

3. The Fresh Grocery Store buys tomatoes for 32¢ per pound and sells them for 68¢ per pound. What is the percent markup? **112.5%**

4. The Video Store made $12.90 on the sale of a video tape. If the markup was 30%, what were the cost and the selling price of the tape? **$43.00; $55.90**

5. Fast Eddie bought a watch for $8.70 and sold it at a markup rate of 250%. What was the selling price of the watch? **$30.45**

6. The Fan Shop buys a ceiling fan for $125 and sells it for $195. What is the percent markup? **56%**

7. The selling price of a washer is $345.80. It is put on sale at a markdown rate of 35%. What is the reduced selling price? **$224.77**

8. A camera which sells for $289.75 is put on sale for $195.50. To the nearest tenth of a percent, what is the markdown rate? **32.5%**

9. The markup on books is 25%. There was a $2.05 markup on a novel. What was the original cost? **$8.20**

10. Lettuce was bought for 30¢ per head and sold at an 80% markup. The lettuce that did not sell the first day was reduced at a markdown rate of 50%. What was the first selling price and what was the second selling price? **54¢; 27¢**

11. If the lettuce in Exercise 10 sold at the second price, what was the loss? What percent of the cost was the loss? **3¢; 10% loss**

12. A ball costing $10 has a markup rate of 40% and is then put on sale at a markdown of 20%. What is the profit? What is the percent profit based on the cost? **$1.20; 12% profit**

FOR REVIEW

13. By driving at high speeds, Mark reduced his mileage by 18%. However, a tailwind gave him a 5% increase in mileage. Under these conditions, what was his mileage if normally it is 30 mpg? **26.1 mpg**

14. Last year Bill earned $42,560.00, and this year his salary will be $47,241.60. What was the percent raise that Bill received? **11%**

6.8 EXERCISES C

Solve.

1. Some store owners operate on a 50%–10% markup rate. The cost of an item is first marked up 50%, and the resulting price is marked up an additional 10% to obtain the selling price. Under these conditions, what would be the selling price of a men's suit which costs the owner $105? **$173.25**

2. A lamp which was purchased by a dealer for $28.70 was to be sold at a 70.5% markup. When it did not sell, the dealer reduced the price at a markdown rate of 50%. Did the dealer make a profit or take a loss when the lamp was then sold? What was the percent profit or loss (to the nearest tenth of a percent)? [Answer: loss of $4.23, 14.7%]

CHAPTER 6 REVIEW

KEY WORDS

6.1   A **ratio** is a comparison of two numbers using division.

An equation which states that two ratios are equal is a **proportion.**

When two pairs of numbers have the same ratio, the numbers are **proportional.**

Solving a proportion is the process of finding a missing number in a proportion.

6.2   The word **percent** means ''per hundred'' and refers to the number of parts in one hundred parts.

6.3   In the basic percent problem ''*A* is *P*% of *B*,'' *A* is the **amount,** *P* is the **percent,** and *B* is the **base.**

6.4   A tax charged on purchases is a **sales tax** and the percent of tax is the **sales-tax rate.**

6.5   Income received as a percent of total sales is called a **commission** and the percent used is the **commission rate.**

A **discount** is a reduction in the regular price of an item based on a **discount rate** which is a percent of the *regular* price. The **sale price** is the price after the discount.

6.6   Money paid for the use of money is called **interest.** The money borrowed or invested is the **principal,** and the **rate of interest** is the percent applied to the principal.

Interest found using only the principal and not using previous interest along with the principal is called **simple interest.**

Interest earned or paid on previous interest along with the principal is called **compound interest.**

6.7   The percent by which a quantity increases (decreases) is called the **percent increase (decrease).**

6.8   The amount by which the selling price of an item is increased over the cost is called **markup,** and the percent used to calculate this increase is the **percent markup.**

The **markdown** of an item is the price reduction based on the **markdown rate** applied to the selling price of the item.

KEY CONCEPTS

6.1   **1.** Ratios can be expressed in three ways, 2:3 or 2 to 3 or $\frac{2}{3}$. The latter form is the fractional form of the ratio.

2. To solve a proportion involving a missing number *a*, equate the cross products and divide the numerical cross product by the multiplier of *a*.

6.2   **1.** To change a percent to a fraction, replace the % symbol with either $\left(\times \frac{1}{100} \right)$ or ($\div$ 100). Then evaluate and reduce the fraction.

2. To change a percent to a decimal, replace the % symbol with ($\times$ 0.01) and multiply. This is the same as moving the decimal point two places to the left.

3. To change a decimal to a percent, multiply the decimal by 100 and attach the % symbol. This is the same as moving the decimal point two places to the right.

4. To change a fraction to a percent, multiply the fraction by 100 and attach the % symbol.

6.3   **1.** The basic percent equation is

$$A = P\% \cdot B.$$

To solve for *P* or *B*, use the related division sentences

$$P\% = \frac{A}{B} \quad \text{or} \quad B = \frac{A}{P\%}.$$

2. The basic percent proportion is

$$\frac{A}{B} = \frac{P}{100}.$$

Solve for any of *A*, *B*, or *P* by solving the proportion.

6.4 **1.** sales tax = tax rate (0.01) · price before tax

or

$$\frac{\text{sales tax}}{\text{price before tax}} = \frac{\text{tax rate}}{100}$$

and

total price = price before tax + sales tax

2. In 1989, the Social Security-tax rate was 7.51% on the first $48,000 of earnings.

Social Security tax = (7.51)(0.01) · total earnings

or

$$\frac{\text{Social Security tax}}{\text{total earnings}} = \frac{7.51}{100}$$

3. Use the tax-rate schedule given in Section 6.4 to calculate income taxes paid in 1989 by a single taxpayer.

6.6 **1.** If P is a principal, R is a yearly rate of interest converted to a decimal, T is the time in years, and I is simple interest, then $I = P \cdot R \cdot T$. If A is the amount to be repaid or the amount in the account, then $A = P + I$.

2. To find the amount of each payment on a loan, divide the sum of the principal and interest by the number of payments.

3. In a compound interest problem, the interest rate used for each period is

$$\frac{\text{Annual Interest Rate}}{\text{Number of Periods Per Year}}.$$

For example, if the annual interest rate is 10%, if we compound annually (1 time a year), semiannually (2 times a year), or quarterly (4 times a year), the interest rate per period is 10%, 5%, and 2.5%, respectively.

4. In a compound interest problem, the principal for the second period is the original principal plus the interest charged for the first period. This process continues over each period.

6.7 **1.** increase (decrease) = percent increase (decrease) · original amount

or

$$\frac{\text{increase (decrease)}}{\text{original amount}} = \frac{\text{percent increase (decrease)}}{100}$$

and

new amount = original amount + increase

or

new amount = original amount − decrease

2. Remember that in a percent increase (decrease) problem, the percent is taken on the original amount *not* the new amount.

6.8 **1.** markup (markdown) = percent markup (markdown) · cost (price)

or

$$\frac{\text{markup (markdown)}}{\text{cost (price)}} = \frac{\text{percent markup (markdown)}}{100}$$

and

selling price = cost + markup

or

new selling price = price − markdown

2. Percent markup is based on cost.

3. Percent markdown is based on selling price.

REVIEW EXERCISES

Part I

6.1 **1.** Write the fraction form for the ratio 8 to 5.
$\frac{8}{5}$

2. Write the ratio which is indicated by the fraction $\frac{1}{8}$.   **1:8 or 1 to 8**

Solve the proportions.

3. $\dfrac{a}{8} = \dfrac{1}{2}$ **4**

4. $\dfrac{7}{21} = \dfrac{6}{a}$ **18**

5. A boat can travel 140 miles on 35 gallons of gas. How far can it travel on 56 gallons of gas?
224 miles

6. If $\frac{1}{2}$ inch on a map represents 20 miles, how many miles will be represented by $5\frac{1}{2}$ inches?
220 miles

6.2   **7.** *Percent* means "per _____ hundred _____."

Change from percents to fractions.

8. 23% $\dfrac{23}{100}$ **9.** 324% $\dfrac{81}{25}$ **10.** 0.6% $\dfrac{3}{500}$ **11.** $6\dfrac{1}{4}\%$ $\dfrac{1}{16}$

Change from percents to decimals.

12. 84%   0.84   **13.** 267%   2.67   **14.** $\dfrac{3}{4}\%$ 0.0075   **15.** 0.3%   0.003

Change to percents.

16. 0.63   63%   **17.** 0.032   3.2%   **18.** 3.2   320%   **19.** $\dfrac{3}{5}$ 60%

Solve the following percent problems.

6.3   **20.** What is 30% of 620?   186

21. 6 is 0.5% of what number?   1200

22. What percent of 72 is 9?   12.5%

23. On a test with 300 questions, Susan got 86% correct. How many did she answer correctly?   258

24. A basketball player hit 19 of 26 shots. To the nearest percent, what percent of his shots did he make?   73%

25. There are 16 grams of alcohol in a solution which is 25% alcohol. How much does the solution weigh?   64 grams

6.4   **26.** What is the sales tax and total cost of a $17.50 shirt if the tax rate is 6%? $1.05; $18.55

27. If a sales tax of $13.97 is charged on the purchase of a $254 dryer, what is the sales-tax rate?   5.5%

28. Raymond had an income from wages of $17,350 in 1989. The Social Security tax rate was 7.51% on all wages up to $48,000 that year. What Social Security tax did Raymond pay? $1302.99

29. Diane had a taxable income of $47,300 in 1989. Use the tax table in Section 6.4 to determine how much income tax she paid. $10,952.50

6.5   **30.** Peter got a commission of $1530 for selling an antique car. If the selling price was $10,200, what was his commission rate?   15%

31. Wanda receives 2% commission on all car sales of $25,000 or less. She receives 3% commission on all sales above $25,000. What is her commission on sales of $42,500? $1025

32. A lawn mower normally sells for $216. At the end of the summer, it was put on sale at a 30% discount. What was the sale price? $151.20

33. Find the discount rate if a $622.50 ring is on sale for $510.45.   18%

6.6   **34.** Phillip bought a watch for $285. He paid 10% down and paid the remaining amount, plus interest, at the end of 6 months. If simple interest is 14% per year, how much did he pay at the end of 6 months? $274.46

35. William invests $3500 in an account which pays 8% compounded semiannually. How much will he have in the account at the end of 2 years? $4094.50

6.7   **36.** State University expects a 4.5% decrease in students next year. If the current enrollment is 26,820, what will be the enrollment next year?   25,613

37. The price of tomatoes increased from 42¢ per pound to 67¢ per pound. To the nearest tenth of a percent, what was the percent increase?   59.5%

6.8   **38.** A dealer bought a microwave oven for $682.50 and sold it for $819.00. What was the percent markup?   20%

39. An electric blanket which normally sells for $42.00 is put on sale for $29.40. What is the markdown rate?   30%

Part II

Solve.

40. Sam's car usually gets 28 mpg. He is driving on a rough road which will decrease his mileage by 14%, but a tailwind is increasing his mileage by 9%. What mileage should he get?
26.6 mpg

41. If the markup on a motorcycle is 15%, what is the selling price if the cost is $1020?
$1173

42. The average salary of the workers at Steel, Incorporated, was $19,280. After the union negotiated an 11% increase, what was the average salary?   **$21,400.80**

43. Shelley borrowed $722.50 to buy a stereo. If simple interest is 15% per year, what are her monthly payments if she pays for 18 months?
$49.17

44. Faye receives a commission of 20% on clothing sales. She sold $2868.50 worth of clothing during the month of May. What was her commission?   **$573.70**

45. The sales-tax rate in Newton, Vermont, is 4%. If a tax of $2.58 is charged on a dress, what is the price of the dress?   **$64.50**

46. The Shirt Shop bought one type of shirt for $14.50. The markup was 28%. Because of oversupply, the shirts were put on sale at a markdown of 25%. What was the percent of profit or loss based on the original cost?
4% (loss)

47. Bernadine makes $428 a week as a secretary. Her boss told her that starting next week she will receive a raise of $8\frac{1}{2}$% per week. What will be her new weekly salary?   **$464.38**

48. If $1000 increased to $1120 in a savings account one year, what was the simple interest rate for the year?
12%

49. If the sales-tax rate in Stephentown, New York, is 6%, what tax would be charged on a purchase of $12.50?   **$0.75**

Complete the table given one value in each row.

	Fraction	Decimal	Percent		Fraction	Decimal	Percent
50.	$\frac{1}{4}$	0.25	25%	**51.**	$\frac{21}{50}$	0.42	42%
52.	$\frac{7}{20}$	0.35	35%	**53.**	$\frac{13}{10}$	1.3	130%
54.	$\frac{4}{3}$	$1.\overline{3}$	$133.\overline{3}$% or $133\frac{1}{3}$%	**55.**	$\frac{1}{200}$	0.005	0.5%

56. Solve the proportion.

$$\frac{a}{3} = \frac{5}{15} \qquad 1$$

57. A car can go 450 miles on 18 gallons of gas. How far can the car go on 32 gallons of gas?
800 miles

58. 6 is what percent of 20?
30%

59. What is 12% of 37?
4.44

60. 80.75 is 85% of what number?   **95**

ANSWERS: 1. $\frac{8}{5}$ 2. 1:8 or 1 to 8   3. 4   4. 18   5. 224 miles   6. 220 miles   7. hundred   8. $\frac{23}{100}$ 9. $\frac{81}{25}$ 10. $\frac{3}{500}$
11. $\frac{1}{16}$ 12. 0.84   13. 2.67   14. 0.0075   15. 0.003   16. 63%   17. 3.2%   18. 320%   19. 60%   20. 186   21. 1200
22. 12.5%   23. 258   24. 73%   25. 64 grams   26. $1.05; $18.55   27. 5.5%   28. $1302.99   29. $10,952.50
30. 15%   31. $1025   32. $151.20   33. 18%   34. $274.46   35. $4094.50   36. 25,613   37. 59.5%   38. 20%
39. 30%   40. 26.6 mpg   41. $1173   42. $21,400.80   43. $49.17   44. $573.70   45. $64.50   46. 4% (loss)
47. $464.38   48. 12%   49. $0.75   50. 0.25; 25%   51. $\frac{21}{50}$; 42%   52. $\frac{7}{20}$; 0.35   53. $\frac{13}{10}$; 130%   54. $1.\overline{3}$; $133.\overline{3}$% or
$133\frac{1}{3}$%   55. $\frac{1}{200}$; 0.005   56. 1   57. 800 miles   58. 30%   59. 4.44   60. 95

1. Write the fraction form for the ratio 13 to 7.

1. $\dfrac{13}{7}$

2. Solve the proportion. $\dfrac{12}{a} = \dfrac{72}{18}$

2. 3

3. A steel cable 35 feet long weighs 105 pounds. How much will 93 feet of the same cable weigh?

3. 279 pounds

4. Change 95% to a fraction in lowest terms.

4. $\dfrac{19}{20}$

5. Change 0.8% to a decimal.

5. 0.008

6. Change 5.06 to a percent.

6. 506%

7. Change $\dfrac{9}{5}$ to a percent.

7. 180%

Solve.

8. 12 is what percent of 60?

8. 20%

9. What is 18% of 120?

9. 21.6

10. 160 is 400% of what number?

10. 40

11. The Social Security tax rate was 7.51% in 1989. How much Social Security tax did Winston pay on a salary of $15,400?

11. $1156.54

12. A toaster with a normal price of $29.50 is to be discounted 30%. What is the sale price?

12. _____ $20.65 _____

13. Herbert bought a TV for $720 and paid 25% down. If simple interest is 13%, how much will he have to pay at the end of a year to pay off the debt?

13. _____ $610.20 _____

14. While driving with a headwind, Tina's car got 22.8 mpg instead of the normal 27.9 mpg. To the nearest tenth, what was the percent decrease?

14. _____ 18.3% _____

15. A refrigerator was bought for $820 and put on sale at a markup rate of 20%. It was then reduced at a markdown rate of 15%. What was the percent profit or loss based on cost?

15. _____ 2% profit _____

16. Wylie Smith, sports information director, makes a profit for NAU of 40¢ on the sale of each basketball program. If the markup rate is 25%, what is the cost of each program, and what is the selling price?

16. _____ $1.60; $2.00 _____

17. It costs $29.90 for a tune-up at Andy's Garage. If Andy offers a Saturday special at 30% off the regular price, how much will a tune-up cost on Saturday?

17. _____ $20.93 _____

18. If an account worth $12,500 increased to $14,000 in one year, what was the simple interest rate for the year?

18. _____ 12% _____

Introduction to Statistics

7.1 CIRCLE GRAPHS AND BAR GRAPHS

❶ STATISTICS AND NUMERICAL DATA

Statistics is the branch of mathematics that deals with collecting, organizing, and summarizing numerical data in order to describe or interpret it. The world of sports relies on statistics such as batting averages, pass completions, free-throw percentages, and points scored. The number of people under age 30 in the United States, the average study time of junior college students, and the Michigan election results are all statistics.

Tables provide one way to present data in a manner that is easy to interpret. Suppose Tanya, a college student, has decided to keep a record of her monthly expenses. She might summarize the data in the table.

Monthly expenses	
Item	*Amount spent*
Rent	$200
Food	$150
Entertainment	$100
Clothes	$ 35
Other	$ 15
TOTAL	$500

❷ CIRCLE GRAPHS

Sometimes it is hard to interpret a large set of data. One way to simplify the process is to "picture" the data in some way. A **circle graph** is often used to represent parts of a whole unit. For example, the data collected by Tanya and presented in the table above could be organized and displayed in a circle graph. The circle graph in Figure 7.1 represents her entire budget, and each pie-shaped part, or sector, represents the number of dollars she spent on each item.

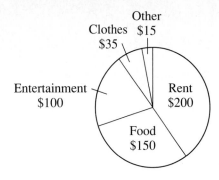

Figure 7.1 Circle Graph

By reading information from a circle graph, various comparisons can be made in the form of ratios. Consider the circle graph in Figure 7.1 again. Since the total amount Tanya spent during the month was $500, the ratio of rent to total expenses is

$$\frac{\$200}{\$500} = \frac{2}{5},$$

and the ratio of clothing expenses to total expenses is

$$\frac{\$35}{\$500} = \frac{7}{100}.$$

Converting these ratios to percents, Tanya sees that she spent 40% of her budget on rent and 7% on clothes.

Tanya can also find the ratio of one expense to another. For example, the ratio of food expenses to entertainment expenses is

$$\frac{\$150}{\$100} = \frac{3}{2},$$

and the ratio of rent to clothing expenses is

$$\frac{\$200}{\$35} = \frac{40}{7}.$$

Thus, for every $3 spent on food, Tanya spends $2 on entertainment, and for every $40 spent on rent, she spends $7 on clothes.

| **EXAMPLE 1** **READING A CIRCLE GRAPH** | **PRACTICE EXERCISE 1** |

The circle graph in Figure 7.2 shows the distribution of Mr. Green's monthly income.

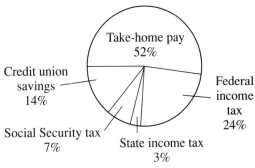

Figure 7.2 Circle Graph

Use the circle graph in Figure 7.2 to answer the following.

(a) If Mr. Green's income is $2800 each month, what amount does he pay in federal income tax?

(a) If Mr. Green's income is $2800 each month, determine his take-home pay.

　　　Since the percent of take-home pay is 52%, find

$$52\% \text{ of } \$2800.$$
$$\downarrow \qquad \downarrow \qquad \downarrow$$
$$0.52 \quad \cdot \quad 2800 = 1456$$

Thus, Mr. Green takes home $1456 each month.

(b) What is the ratio of federal income tax to state income tax?
　　The ratio is

$$\frac{24\%}{3\%} = \frac{8}{1}.$$

Thus, for every $8 paid in federal income tax Mr. Green pays $1 in state income tax.

(b) What is the ratio of credit union savings to Social Security tax?

Answers: (a) $672　(b) $\frac{2}{1}$

❸ BAR GRAPHS

Another way to present data is with a **bar graph.** Suppose 100 people have been polled regarding their preference among four brands of toothpaste, Brand A, Brand B, Brand C, and Brand D. The results of the poll are given in the bar graph in Figure 7.3.

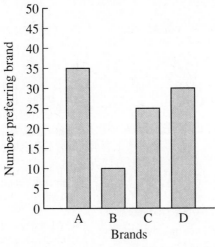

7.3　Bar Graph

　　The graph shows that 35 people chose Brand A, 10 chose Brand B, 25 chose Brand C, and 30 chose Brand D.

　　Bar graphs are quite useful when comparing data. It is easy to see at a glance which brand was most preferred (Brand A) and which was least preferred (Brand B). Also, Brand A was preferred to Brand B by a ratio of

$$\frac{35}{10} = \frac{7}{2}.$$

That is, for every 7 people who chose Brand A, 2 people chose Brand B.

EXAMPLE 2 READING A BAR GRAPH

Use the bar graph in Figure 7.3 to determine the ratio of people preferring Brand C to those preferring Brand B.

Since 25 people preferred Brand C and 10 people preferred Brand B, the ratio is $\frac{25}{10} = \frac{5}{2}$.

PRACTICE EXERCISE 2

Use the bar graph in Figure 7.3 to determine the ratio of those preferring Brand A over Brand C.

Answer: $\frac{7}{5}$

A double-bar graph is often used to compare two sets of data. The double-bar graph in Figure 7.4 shows the total enrollment in beginning algebra classes compared to intermediate algebra classes at a community college during a four-year period.

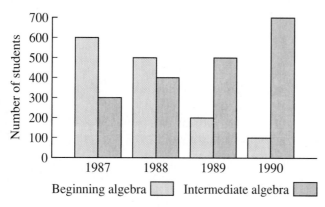

Beginning algebra ☐   Intermediate algebra ☐

Figure 7.4 Double-Bar Graph

EXAMPLE 3 READING A DOUBLE-BAR GRAPH

Use the double-bar graph in Figure 7.4 to answer the following.

(a) What was the number of beginning algebra students in 1988?

Since the bar on the left in 1988 rises to the 500 level, there were 500 students in beginning algebra in 1988.

(b) Which year had the greatest number of intermediate algebra students?

The right bar rises highest in 1990. Thus, the greatest number of students in intermediate algebra occurred in 1990. In fact, that number was 700 students.

(c) What could you say about the number of beginning algebra students over the four-year period?

Since the number went from 600 to 500 to 200 to 100, we can see that the number of beginning algebra students declined over the four-year period.

PRACTICE EXERCISE 3

Use the double-bar graph in Figure 7.4 to answer the following.

(a) What was the number of intermediate algebra students in 1989?

(b) Which year had more beginning algebra students, 1988 or 1989?

(c) What could you say about the number of intermediate algebra students over the four-year period?

Answers: (a) **500**   (b) **1988**
(c) **The number increased.**

7.1   EXERCISES A

Exercises 1–8 refer to the circle graph below which shows a breakdown of the amounts spent by the Warrens on an addition to their house.

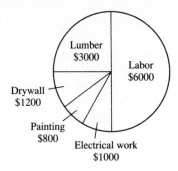

1. What was the total cost of the addition?   **$12,000**

2. What was the largest expense item on the project?   **labor**

3. What was the ratio of the cost of labor to the cost of lumber? $\frac{2}{1}$ **or 2 to 1**

4. What was the ratio of the cost of painting to the cost of drywall? $\frac{2}{3}$ **or 2 to 3**

5. What was the ratio of the cost of lumber to the cost of electrical work? $\frac{3}{1}$ **or 3 to 1**

6. Lumber accounted for what percent of the total cost?   **25%**

7. Labor accounted for what percent of the total cost?   **50%**

8. Electrical work accounted for what percent of the total cost? $8.\overline{3}\%$ **or** $8\frac{1}{3}\%$

Exercises 9–20 refer to the circle graph below which shows the distribution of the 42,500 students enrolled in various schools in Coconino County.

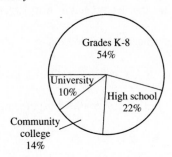

9. Which sector represents the largest number of students?   **grades K–8**

10. Which sector represents the smallest number of students?   **university students**

11. How many students are in grades K–8?   **22,950**

12. How many students are in high school?   **9350**

13. How many students are enrolled in the community college?   **5950**

14. How many students are enrolled in the university?   **4250**

15 How many students are in grades K–8 and in high school?   **32,300**

16. How many students are in the community college and in the university?   **10,200**

17. What is the ratio of high school students to university students? $\frac{11}{5}$ **or 11 to 5**

18. What is the ratio of students in grades K–8 to students in high school? $\frac{27}{11}$ **or 27 to 11**

19. What is the ratio of high school students to students in the university and the community college?
$\frac{11}{12}$ **or 11 to 12**

20. How many students are in school beyond the high school level?   **10,200**

Exercises 21–32 refer to the bar graph below which shows the number of students by major course of study in the College of Arts and Science.

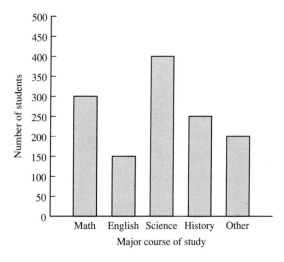

21. How many students are in the College of Arts and Science?   **1300**

22 What is the ratio of math majors to the total number of students in the college? $\frac{3}{13}$ **or 3 to 13**

23. What percent of students in the college are math majors (to the nearest tenth of a percent)?   **23.1%**

24. What is the ratio of science majors to the total number of students in the college? $\frac{4}{13}$ **or 4 to 13**

25 What percent of students in the college are science majors (to the nearest tenth of a percent)?   **30.8%**

26. What percent of students are math and science majors (to the nearest tenth of a percent)?   **53.8%**

27. What is the ratio of science majors to history majors? $\frac{8}{5}$ **or 8 to 5**

28. What is the ratio of English majors to math majors? $\frac{1}{2}$ **or 1 to 2**

29. What percent of students have a major other than math, science, English, or history (to the nearest tenth of a percent)?   **15.4%**

30. What is the total of math and English majors?   **450**

31. What percent of the students are math and English majors (to the nearest tenth of a percent)?   **34.6%**

32. What is the ratio of math and science majors to English and history majors? $\frac{7}{4}$ **or 7 to 4**

Exercises 33–40 refer to the double-bar graph below which shows the number of male and female athletes at a university over a four-year period.

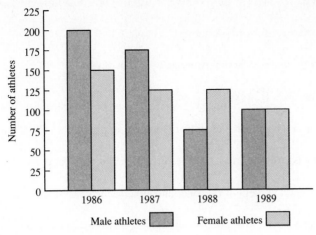

Male athletes   Female athletes

33. Which year had the greatest number of male athletes?   **1986**

34. Which year had the smallest number of male athletes?   **1988**

35. Which year had the greatest number of female athletes?   **1986**

36. Which year had the smallest number of female athletes?   **1989**

37. What was the ratio of male athletes to female athletes in 1986? $\frac{4}{3}$ **or 4 to 3**

38. What was the ratio of female athletes to male athletes in 1988? $\frac{5}{3}$ **or 5 to 3**

39. Which year had the same number of male and female athletes?   **1989**

40. Would it be fair to conclude from the graph that some athletics programs were forced to be dropped? Explain.
Yes; the number of males and females seemed to decline over the period.

ANSWERS: 1. $12,000   2. labor   3. $\frac{2}{1}$ or 2 to 1   4. $\frac{2}{3}$ or 2 to 3   5. $\frac{3}{1}$ or 3 to 1   6. 25%   7. 50%   8. $8.\overline{3}$% or $8\frac{1}{3}$%   9. grades K–8   10. university students   11. 22,950   12. 9350   13. 5950   14. 4250   15. 32,300   16. 10,200   17. $\frac{11}{5}$ or 11 to 5   18. $\frac{27}{11}$ or 27 to 11   19. $\frac{11}{12}$ or 11 to 12   20. 10,200   21. 1300   22. $\frac{3}{13}$ or 3 to 13   23. 23.1%   24. $\frac{4}{13}$ or 4 to 13   25. 30.8%   26. 53.8%   27. $\frac{8}{5}$ or 8 to 5   28. $\frac{1}{2}$ or 1 to 2   29. 15.4%   30. 450   31. 34.6%   32. $\frac{7}{4}$ or 7 to 4   33. 1986   34. 1988   35. 1986   36. 1989   37. $\frac{4}{3}$ or 4 to 3   38. $\frac{5}{3}$ or 5 to 3   39. 1989   40. Yes; the number of males and females seemed to decline over the period.

7.1   EXERCISES B

Exercises 1–8 refer to the circle graph below which gives the approximate land area of the seven continents in millions of square miles.

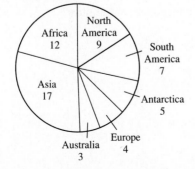

1. What is the approximate total land area of the seven continents?   57 million square miles

2. What is the largest continent by area?   Asia

3. What is the ratio of the area of Africa to the area of North America? $\frac{4}{3}$ or 4 to 3

4. What is the ratio of the area of Australia to the area of South America? $\frac{3}{7}$ or 3 to 7

5. What is the ratio of the area of Europe to the area of Asia? $\frac{4}{17}$ or 4 to 17

6. To the nearest tenth of a percent, the area of Asia is what percent of the total land area?   29.8%

7. To the nearest tenth of a percent, the area of North America is what percent of the total land area?   15.8%

8. North and South America together account for what percent of the total land area (to the nearest tenth of a percent)?
   28.1%

Exercises 9–20 refer to the circle graph below which shows how a university spends its funds totalling $25,000,000 in a given year.

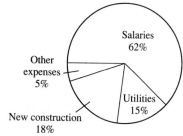

9. Which sector represents the greatest expenditure?   salaries

10. Which sector represents the smallest expenditure?   other expenses

11. How much was spent on salaries? $15,500,000

12. How much was spent on utilities? $3,750,000

13. How much was spent on new construction? $4,500,000

14. How much was spent on other expenses? $1,250,000

15. How much was spent on salaries and utilities? $19,250,000

16. How much was spent on utilities and new construction? $8,250,000

17. What is the ratio of utility expenses to new-construction expenses? $\frac{5}{6}$ or 5 to 6

18. What is the ratio of salaries to utilities? $\frac{62}{15}$ or 62 to 15

19. What is the ratio of salaries to the expenses for new construction and utilities? $\frac{62}{33}$ or 62 to 33

20. How much was spent in all categories besides salaries? $9,500,000

Exercises 21–32 refer to the bar graph at the right which shows the number of new cars sold by a dealer during a five-month period.

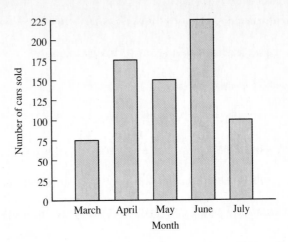

21. How many cars were sold in this five-month period?   **725**

22. What is the ratio of cars sold in March to the total number sold in the five-month period? $\frac{3}{29}$ **or 3 to 29**

23. What percent of the cars sold in the five-month period were sold in March (to the nearest tenth of a percent)?
10.3%

24. What is the ratio of cars sold in June to the total number sold in the five-month period? $\frac{9}{29}$ **or 9 to 29**

25. What percent of the cars sold in the five-month period were sold in June (to the nearest tenth of a percent)?
31.0%

26. What percent of the cars sold in the five-month period were sold in June and July (to the nearest tenth of a percent)?   **44.8%**

27. What is the ratio of cars sold in June to cars sold in May? $\frac{3}{2}$ **or 3 to 2**

28. What is the ratio of cars sold in March to cars sold in July? $\frac{3}{4}$ **or 3 to 4**

29. What percent of the cars sold in the five-month period were sold in a month other than July (to the nearest tenth of a percent)?   **86.2%**

30. What is the total of the cars sold in March and April?   **250**

31. What percent of the cars sold in the five-month period were sold in March and April (to the nearest tenth of a percent)?   **34.5%**

32. What is the ratio of the cars sold in March, April, and May to the cars sold in June and July? $\frac{16}{13}$ **or 16 to 13**

Exercises 33–40 refer to the double-bar graph at the right which shows the quarterly profits for a large corporation in millions of dollars for the years 1989 and 1990.

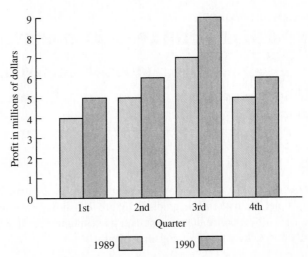

33. Which quarter showed the greatest profit in 1989?   **3rd**

34. Which quarter showed the smallest profit in 1989?   **1st**

35. Which quarter showed the greatest profit in 1990?   **3rd**

36. Which quarter showed the smallest profit in 1990?   **1st**

37. What was the profit in the 2nd quarter of 1989?   **5 million dollars**

38. What was the profit in the 3rd quarter of 1990?   **9 million dollars**

39. What was the total profit in all four quarters in 1990?   **26 million dollars**

40. Would it be fair to say that during both years there was a steady increase in profits through the first three quarters and a decline during the fourth quarter? Explain.   **Yes, this is clearly reflected in the graph.**

7.1 EXERCISES C

A die was rolled 600 times and the number of times that each number (1, 2, 3, 4, 5, or 6) came up was recorded. The results of this experiment are shown in the bar graph at the right. The statistical probability *of obtaining a particular number is defined as the ratio of the number of times that number came up on the die to the total number of times the die is rolled. Use this information in Exercises 1–6 to give the statistical probability of rolling the particular number.*

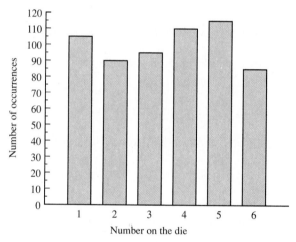

1. 1 $\left[\text{Answer: } \frac{7}{40}\right]$

2. 2 $\frac{3}{20}$

3. 3 $\frac{19}{120}$

4. 4 $\frac{11}{60}$

5. 5 $\frac{23}{120}$

6. 6 $\frac{17}{120}$

7.2 BROKEN-LINE GRAPHS AND HISTOGRAMS

STUDENT GUIDEPOSTS

1 Broken-Line Graphs

2 Histograms

3 Frequency Polygons

1 BROKEN-LINE GRAPHS

A **broken-line graph** is often used to show a trend in data collected over a period of time. For example, the broken-line graph in Figure 7.5 shows the number of traffic citations issued by a police department over a six-month period. Each dot shows the number of tickets given during that month.

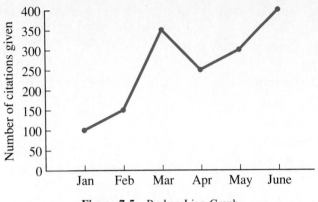

Figure 7.5 Broken-Line Graph

EXAMPLE 1 READING A BROKEN-LINE GRAPH

Refer to the broken-line graph in Figure 7.5.

(a) In which month were 350 tickets given out?

Since the dot above March is at the 350 level, 350 tickets were issued in March.

(b) How would you briefly describe the trend in giving out tickets over the six-month period?

Basically the number of tickets given increased from month-to-month with the exception of March to April.

PRACTICE EXERCISE 1

Refer to the broken-line graph in Figure 7.5.

(a) How many citations were issued in April?

(b) How many tickets were given out during the six-month period?

Answers: **(a) 250**   **(b) 1550**

Two broken-line graphs in the same figure are often used to show a comparison between two sets of data. For example, the broken-line graphs in Figure 7.6 show the quarterly profits realized by a company during the years 1988 and 1989. By showing both graphs together, it is easy to see at a glance how one year compared with the other.

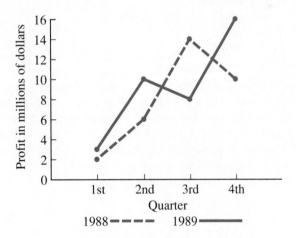

Figure 7.6 Double-Line Graph

EXAMPLE 2 READING A DOUBLE-LINE GRAPH

Refer to the graph in Figure 7.6.

(a) What was the profit made during the first quarter of 1989?

Since the dot on the solid line is halfway between the 2 and 4 million marks, the profit this quarter was 3 million dollars.

PRACTICE EXERCISE 2

Refer to the graph in Figure 7.6.

(a) What was the profit made during the 3rd quarter of 1988?

(b) Between which two consecutive quarters in 1988 was there a decrease in profit?

    The dashed line graph goes down from 14 million dollars to 10 million dollars between the 3rd and 4th quarters of 1988.

(c) In which year did the company show the greater profit?

    Adding up the quarterly profits in 1988 we obtain

$$2 + 6 + 14 + 10 = 32 \text{ million dollars.}$$

    Adding up the quarterly profits in 1989 we obtain

$$3 + 10 + 8 + 16 = 37 \text{ million dollars.}$$

    Thus, the greater profit was made in 1989.

(b) Between which two consecutive quarters in 1989 did the profit increase the most?

(c) Which year had the better profit change between the 3rd and 4th quarters?

Answers: (a) 14 million dollars (b) between the 3rd and 4th quarters (c) 1989

② HISTOGRAMS

Suppose sixty men in an organization list their heights in inches. The resulting data might be somewhat difficult to understand due to the large number of heights. To make the data easier to interpret, the heights could be divided into groups called **class intervals.** We might use 66 inches–68 inches as one interval, 68 inches–70 inches as another, and so forth. The number of heights in each class interval is called the class frequency. The table below summarizes the data collected.

Height in inches (class intervals)	Number of men (class frequency)
66–68	4
68–70	12
70–72	24
72–74	18
74–76	2

Suppose we assume that a person whose height is exactly 68 inches is in the class interval 68–70 and not in 66–68, and similarly for persons whose height is 70 inches, 72 inches, and 74 inches. Then each of the sixty men has his height in exactly one of the class intervals. A special type of bar graph, a **histogram** shown in Figure 7.7, is used to picture the information in the table. The width of each bar represents the range of values in that class interval, and the length or height of the bar corresponds to the **class frequency.**

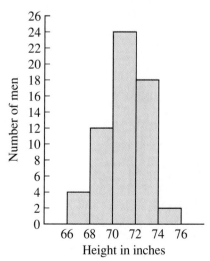

Figure 7.7 Histogram

| EXAMPLE 3 READING A HISTOGRAM | PRACTICE EXERCISE 3 |

Use the histogram in Figure 7.7 to answer the following.

(a) How many men are 70 inches or taller?

We must add the class frequencies for the class intervals 70–72, 72–74, and 74–76.

$$24 + 18 + 2 = 44$$

There are 44 men 70 inches or taller.

(b) Find the ratio of men 68 to 70 inches tall to those 72 to 74 inches tall.

Since there are 12 men with heights between 68 and 70 inches and 18 men with heights between 72 and 74 inches, the ratio is

$$\frac{12}{18} = \frac{2}{3} \text{ or } 2 \text{ to } 3.$$

(c) What percent of the men in the organization are between 68 and 70 inches in height?

Since there are 12 men 68 to 70 inches tall out of a total of 60 men in the organization, the question is: 12 is what percent of 60?

Method 1 (Equation Method)

$$\begin{array}{ccc} A = & P\% & \cdot B \\ \downarrow & \downarrow & \downarrow \\ 12 = & P(0.01) & \cdot 60 \end{array}$$

$$12 = P(0.60)$$

The related division sentence is:

$$P = \frac{12}{0.60}$$

$$P = 20$$

Method 2 (Proportion Method)

$$\frac{A}{B} = \frac{P}{100}$$

$$\frac{12}{60} = \frac{P}{100}$$

$$12(100) = 60(P)$$

The related division sentence is:

$$P = \frac{12(100)}{60}$$

$$P = 20$$

Thus, 20% of the men are between 68 and 70 inches tall.

Use the histogram in Figure 7.7 to answer the following.

(a) How many men are shorter than 72 inches?

(b) Find the ratio of men with heights between 66 and 68 inches to those between 74 and 76 inches.

(c) What percent of the men in the organization are between 68 and 74 inches in height?

Answers: (a) 40   (b) $\frac{2}{1}$ or 2 to 1
(c) 90%

❸ FREQUENCY POLYGONS

Sometimes data that is divided into class intervals is displayed using a special kind of broken-line graph. For instance, suppose we find the midpoint of each class interval, called the **class midpoint,** in the example above. The class midpoint of the class interval 66–68 inches is 67 inches, and the remaining class midpoints are 69 inches, 71 inches, 73 inches, and 75 inches. Locate the class midpoints in the histogram in Figure 7.7, and place a dot at the corresponding point on the top of each bar. When these dots are connected, as shown in Figure 7.8, the resulting broken-line graph is called a **frequency polygon.**

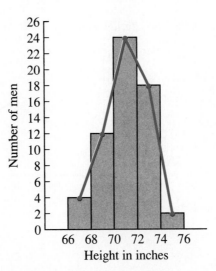

Figure 7.8 Frequency Polygon

HINT

Histograms and frequency polygons are used when data collected consists of a wide range of values such as numbers obtained by measurements.

7.2 EXERCISES A

The number of personal computers sold by a dealer in a six-week period is shown in the broken-line graph at the right. Use this graph in Exercises 1–10.

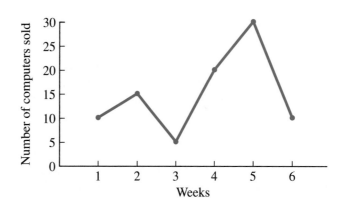

1. How many computers were sold in week 2?   **15**

2. During which week(s) were 10 computers sold?   **weeks 1 and 6**

3. How many computers were sold during the six-week period?   **90**

4. What was the ratio of computers sold in week 5 to week 3? $\frac{6}{1}$ **or 6 to 1**

5. During which week did the dealer sell the most computers?   **week 5**

6. During which week did the dealer sell the least computers?   **week 3**

7. Between which two consecutive weeks did the sales increase the most?   **Between weeks 3 and 4**

8. Between which two consecutive weeks did the sales decrease the most?   **Between weeks 5 and 6**

9. What percent of the total sales were made during week 5? $33.\overline{3}\%$ **or** $33\frac{1}{3}\%$

10. What percent of the total sales were made during week 1? $11.\overline{1}\%$ **or** $11\frac{1}{9}\%$

The double-line graph at the right shows the comparison between new subscribers signed up for cable television during a four-month period in 1988 and 1989. Use this graph in Exercises 11–18.

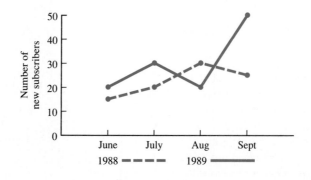

11. How many new subscribers were there in June of 1988?   15

12. How many new subscribers were there in August of 1989?   20

13. How many new subscribers were there in the four months of 1988?   90

14. How many new subscribers were there in the four months of 1989?   120

15. To the nearest percent, what was the percent increase or decrease in new subscribers from 1988 to 1989 during the four-month period?   **33% increase**

16. What was the ratio of new subscribers in June 1988 to June 1989? $\frac{3}{4}$ **or 3 to 4**

17. What was the ratio of new subscribers in July 1989 to August 1989? $\frac{3}{2}$ **or 3 to 2**

18. Between which two consecutive months of 1989 did the company have the greatest increase in new subscribers?   **August and September**

The hourly wages of all the employees of Mutter Manufacturing Company are presented in the histogram at the right. Use this information in Exercises 19–28. Assume that $5 is in the class interval $5–$7 not in $3–$5, similarly for $7, $9, and $11.

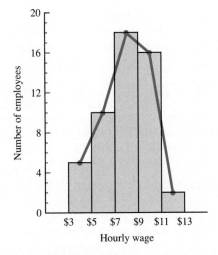

19. How many employees earn at least $3 and less than $5 an hour?   5

20. How many employees earn at least $7 and less than $9 an hour?   18

21. How many employees earn $7 an hour or more?   36

22 How many employees earn less than $9 an hour?   33

23. How many employees are there in Mutter Manufacturing Company?   51

24. What is the ratio of employees earning $9 or more an hour to the total number of employees? $\frac{6}{17}$ **or 6 to 17**

25 What percent of the employees earn $9 or more an hour (to the nearest tenth of a percent)?   35.3%

26. What is the ratio of employees earning less than $7 an hour to those who earn $9 or more an hour? $\frac{5}{6}$ **or 5 to 6**

27. What values would be used as the class midpoints of the class intervals to construct the frequency polygon for this data?   **$4, $6, $8, $10, $12**

28. Draw the frequency polygon in the figure above.   **shown above**

FOR REVIEW

Exercises 29–32 refer to the circle graph below which shows the distribution of Chuck's gross monthly income of $2500.

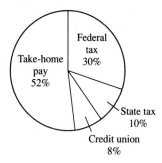

29. What is Chuck's monthly take-home pay? $1300

30. How much federal tax does Chuck pay each month? $750

31. How much state tax does Chuck pay each month? $250

32. What is the ratio of federal tax to take-home pay? $\frac{15}{26}$ or 15 to 26

ANSWERS: 1. 15   2. weeks 1 and 6   3. 90   4. $\frac{6}{1}$ or 6 to 1   5. week 5   6. week 3   7. Between weeks 3 and 4
8. Between weeks 5 and 6   9. 33.$\overline{3}$% or 33$\frac{1}{3}$%   10. 11.$\overline{1}$% or 11$\frac{1}{9}$%   11. 15   12. 20   13. 90   14. 120   15. 33%
increase   16. $\frac{3}{4}$ or 3 to 4   17. $\frac{3}{2}$ or 3 to 2   18. August and September   19. 5   20. 18   21. 36   22. 33   23. 51
24. $\frac{6}{17}$ or 6 to 17   25. 35.3%   26. $\frac{5}{6}$ or 5 to 6   27. $4, $6, $8, $10, $12   28.
29. $1300   30. $750   31. $250   32. $\frac{15}{26}$ or 15 to 26

7.2   EXERCISES B

The number of yards gained passing by quarterback Greg Wyatt in five home games of the 1989 season is shown in the broken-line graph below. Use this graph in Exercises 1–10.

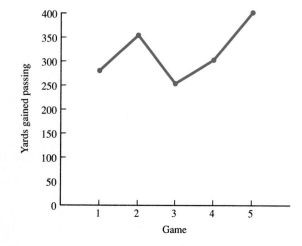

1. How many yards passing did Greg have in game 1?   **275 yd**

2. During which game did Greg have 400 yards passing?   **game 5**

3. How many total yards passing did Greg have in the five home games?   **1575 yd**

4. What was the ratio of yards passing in game 2 to game 3? $\frac{7}{5}$ **or 7 to 5**

5. In which game did Greg have the most yards passing?   **game 5**

6. In which game did Greg have the fewest yards passing?   **game 3**

7. Between which two consecutive home games did his yards passing increase the most?
   Between games 4 and 5

8. Between which two consecutive home games did his yards passing decrease?   **Between games 2 and 3**

9. To the nearest tenth of a percent, what percent of Greg's total passing yards in the five home games was gained in week 5?   **25.4%**

10. To the nearest tenth of a percent, what percent of Greg's total passing yards in the five home games was gained in week 1?   **17.5%**

The double-line graph below shows the number of entrants (in thousands) to Sunset Crater National Monument during four months in the summers of 1988 and 1989. Use this graph in Exercises 11–18.

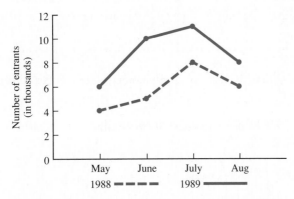

11. How many monument entrants were there in June of 1989?   **10,000**

12. How many monument entrants were there in June of 1988?   **5000**

13. How many monument entrants were there in the four-month period in 1988?   **23,000**

14. How many monument entrants were there in the four-month period in 1989?   **35,000**

15. To the nearest percent, during the four-month period what was the percent increase in monument entrants from 1988 to 1989?   **52%**

16. What was the ratio of monument entrants in June 1989 to June 1988? $\frac{2}{1}$ **or 2 to 1**

17. What was the ratio of monument entrants in July 1988 to June 1988? $\frac{8}{5}$ **or 8 to 5**

18. Between which two consecutive months in 1989 did the number of monument entrants increase the most?
    Between May and June

During a past period of years, the weather service has recorded the annual precipitation received by Morgan, Colorado, and the results are given in the histogram at the right. Use this information in Exercises 19–28. Assume that 5 inches is in the class interval 5 inches–10 inches, not in 0 inches–5 inches, similarly for 10 inches, 15 inches, and 20 inches.

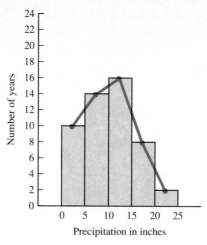

19. In how many years was the precipitation total less than 5 inches?   **10 yr**

20. In how many years was the precipitation total at least 10 inches and less than 15 inches?   **16 yr**

21. In how many years did Morgan receive 10 inches or more of precipitation?   **26 yr**

22. In how many years did Morgan receive less than 15 inches of precipitation?   **40 yr**

23. What was the total number of years in the recording period?   **50 yr**

24. What is the ratio of the number of years of 15 to 20 inches of precipitation to the number of years of 20 to 25 inches? $\frac{4}{1}$ **or 4 to 1**

25. In what percent of the total number of years did Morgan receive 20 inches or more of precipitation?   **4%**

26. What is the ratio of the number of years of less than 5 inches of precipitation to the number of years of 10 inches or more of precipitation? $\frac{5}{13}$ **or 5 to 13**

27. What values would be used as the class midpoints of the class intervals to construct the frequency polygon for this data?   **2.5 in, 7.5 in, 12.5 in, 17.5 in, 22.5 in**

28. Draw the frequency polygon in the figure above.   **shown above**

FOR REVIEW

Exercises 29–32 refer to the bar graph at the right which shows the number of students who preferred various cable television networks in a recent survey.

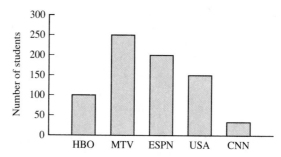

29. How many students preferred HBO?   **100**

30. How many students were in the survey?   **725**

31. To the nearest tenth of a percent, what percent of the students preferred ESPN?   **27.6%**

32. What was the ratio of students preferring MTV to CNN? $\frac{10}{1}$ **or 10 to 1**

7.2 EXERCISES C

The letters in our alphabet occur with different frequencies in the words that we use. The letter "e" occurs most frequently and the letter "z" occurs least frequently. In fact, it has been shown statistically that in a large sample of words, the words on a page in a text for example, approximately 13% of the letters will be "e", 7% will be "a", 6% will be "s", and 8% will be "n". Select a page in one of your textbooks (a page with many words), and count the number of times e, a, s, and n occur. Also, count the total number of letters on the page.

1. Make a table showing the number of times each of the four letters occurs and transfer the information to a bar graph.

2. Which of these letters occurred most frequently?

3. Which of these letters occurred least frequently?

4. How do the percents of occurrence for the letters e, a, s, and n in your data compare to the percents given above?

7.3 MEAN, MEDIAN, AND MODE

STUDENT GUIDEPOSTS

1 Mean (Average)
2 Median
3 Mode

1 MEAN (AVERAGE)

In Sections 7.1 and 7.2 numerical data were presented using circle graphs, bar graphs, broken-line graphs, and histograms. Often it is necessary to look for one number or statistic to characterize a whole collection of numbers. For example, suppose Jenny has scores of 69, 79, and 98 on three chemistry tests. To find out how she is doing in the course, Jenny might calculate the *mean* or *average* of her scores.

Mean (Average)

The **mean** or **average** of several values is the ratio of the sum of the values to the number of values.

$$\text{mean (average)} = \frac{\text{sum of values}}{\text{number of values}}$$

To find the mean of 69, 79, and 98, add these three scores and divide by 3 (the number of scores).

$$\text{mean} = \frac{69 + 79 + 98}{3} \quad \begin{matrix} \leftarrow \text{Sum of the scores} \\ \leftarrow \text{Number of scores} \end{matrix}$$

$$= \frac{246}{3} = 82$$

Making 69, 79, and 98 on three tests is the same as making

82,   82,   and   82

on three tests. The mean or average gives Jenny a better idea of how well she is doing in chemistry. Her professor may use the average to compute her course grade.

EXAMPLE 1 FINDING A MEAN IN A BUSINESS

The daily sales at George's Pie Place were $285.85, $192.50, $222.82, $315.16, $427.45, and $396.80 for a six-day week. To get a better idea of how well his business is doing, George would like to know the average daily sales for the week. What is the average?

Add the sales for the 6 days and divide by 6.

$285.85
192.50
222.82
315.16
427.45
396.80
$1840.58

$$\frac{\$1840.58}{6} \approx \$306.76$$ To the nearest cent

George's shop is averaging about $306.76 per day in sales.

PRACTICE EXERCISE 1

City College is playing State College in a basketball game. The heights in inches of the starting five players for City are 80, 72, 82, 79, and 85. The starting five players for State have heights in inches of 81, 79, 76, 76, and 82. On the average, which team is taller?

Answer: City, with an average height of 79.6 inches, is taller than State, with an average of 78.8 inches.

In Example 1, if George had known that the total sales for the 6-day week were $1840.58, he would only have had to divide by 6. The next example illustrates this type of problem.

EXAMPLE 2 FINDING A MEAN IN INDUSTRY

Suppose tests by General Motors show that one of its cars will go 420 miles on a tank of gasoline. What is the average miles per gallon (mpg) of the car if the tank holds 12.5 gal?

In this problem the total miles traveled on the 12.5 gal is given. Thus, all that is needed to find the average is to divide 420 mi by 12.5 gal.

$$\frac{420 \text{ mi}}{12.5 \text{ gal}} = 33.6 \frac{\text{mi}}{\text{gal}} = 33.6 \text{ mpg}$$

PRACTICE EXERCISE 2

The manufacturer of wood-burning stoves produced 78 stoves during a period of 12 days. What was the average daily production?

Answer: 6.5 stoves per day

❷ MEDIAN

The mean of a collection of numbers can often be influenced by one very large or very small number (compared to the others). As a result, a mean is often a poor measure of the characteristics of a set of data. For example, the mean of the numbers

1,   2,   3,   4,   90

is

$$\frac{1 + 2 + 3 + 4 + 90}{5} = \frac{100}{5} = 20.$$

Since four of the five numbers are very small compared to the fifth number, to use 20 to describe this set of data might be somewhat misleading. Actually, a number closer to 1, 2, 3, or 4 than 20 would provide a better description. Another way to describe a set of data is to look at the value in the middle after the numbers are arranged by size. For example, consider the following:

$$1, \quad 2, \quad 3, \quad 4, \quad 90.$$
$$\uparrow$$
$$\text{middle}$$

The value in the middle is 3, and certainly 3 gives a better description of the data set than the mean 20. The middle value of a collection of numbers that is arranged in order of size is the *median* of the data.

Suppose Russell made scores of 72, 65, and 88 on his first three tests. To find the middle score, or median score, arrange them in order of increasing value.

$$65, \quad 72, \quad 88$$
$$\uparrow$$
$$\text{middle}$$

The median score is 72.

But what about the middle score if Russell had four test scores, 72, 65, 88, and 95? Once again arrange them in order.

$$65, \quad 72, \quad 88, \quad 95$$
$$\uparrow$$
$$\text{middle}$$

In this case, there is no middle score. To approximate a middle score, average the *two* scores in the middle.

$$\text{median} = \frac{72 + 88}{2} = 80$$

Median

The middle value of several values is called the **median.** To find the median:

1. Arrange the values in increasing order from left to right.

2. If there is an odd number of values, the median is the middle one.

3. If there is an even number of values, the median is the average of the *two* values in the middle.

EXAMPLE 3   FINDING A MEDIAN

Find the median of the given numbers.

(a) 5, 12, 2, 14, 8, 7, 9

First arrange the numbers in order.

$$2, \quad 5, \quad 7, \quad 8, \quad 9, \quad 12, \quad 14$$

There are seven numbers, and the middle one is the fourth from the left. The median is 8.

(b) 125, 105, 216, 205, 110, 175

Arrange the numbers in order.

$$105, \quad 110, \quad 125, \quad 175, \quad 205, \quad 216$$

The two numbers in the middle are 125 and 175. Find their average.

$$\frac{125 + 175}{2} = 150$$

Thus, 150 is the median of the numbers.

PRACTICE EXERCISE 3

Find the median of the given numbers.

(a) 9, 20, 2, 18, 27

(b) 206, 140, 180, 375

Answers: (a) **18**   (b) **193**

EXAMPLE 4 AN APPLICATION OF MEAN AND MEDIAN

In Greenborough, the number of traffic accidents per day is recorded. The following are the numbers for one week.

$$12, \quad 18, \quad 22, \quad 10, \quad 8, \quad 15, \quad 27$$

Find the average number of accidents per day and the median number per day.

First find the average.

$$\frac{12 + 18 + 22 + 10 + 8 + 15 + 27}{7} = \frac{112}{7} = 16$$

To find the median, arrange the numbers in order.

$$8, \quad 10, \quad 12, \quad 15, \quad 18, \quad 22, \quad 27$$

The median is 15 which is close to the mean of 16.

PRACTICE EXERCISE 4

Jeanine made scores of 22, 24, 15, 27, 19, and 30 on six quizzes in geology lab. Find the average score and the median score on these quizzes.

Answer: average score: 22.8, median score: 23

H I N T

Remember to arrange the numbers by size before selecting the middle one as the median. For example, the median of

$$5, \quad 1, \quad 2, \quad 3, \quad 8$$

is 3, not 2. This is easy to see after rewriting the numbers as 1, 2, 3, 5, 8.

EXAMPLE 5 FINDING THE MEAN AND MEDIAN USING A TABLE

The Lopez Engineering Company has 30 employees. Six of them make $18,500 per year, 5 make $19,200 per year, 2 make $22,800 per year, 10 make $24,500 per year, 4 make $32,800 per year, 2 make $38,000 per year, and 1 makes $52,400 per year. What is the mean or average salary of the 30 employees? What is the median salary?

A good way to find the sum of the salaries is to set up a table.

PRACTICE EXERCISE 5

A new car dealer has 40 cars in stock. Five are valued at $18,000, 12 at $15,000, 10 at $12,000, 7 at $9000, and 6 at $8000. What is the mean or average value of a car on his lot? What is the median value?

Salary	Number of Employees	Salary Times Number of Employees
18,500	6	111,000
19,200	5	96,000
22,800	2	45,600
24,500	10	245,000
32,800	4	131,200
38,000	2	76,000
52,400	1	52,400
Totals	30	$757,200

When each salary is multiplied by the number of employees making that salary and the results are added, the total is $757,200. This is divided by 30, the total number of employees, to get the mean or average salary.

$$\frac{\$757,200}{30} = \$25,240$$

Since there are 30 employees, the median salary is the average of the middle two salaries once the salaries are arranged by size. This will be the average of the 15th and 16th salaries in the arrangement. Without actually listing all the salaries, we can find the 15th and 16th ones by looking at the table. Use the column for the number of employees and count down 15. Since the 15th and 16th positions are included in the 10 employees in the fourth row from the top, the 15th and 16th salary values are both $24,500, making the average of the two also $24,500. Thus, the median salary is $24,500.

Answer: mean = $12,525; median = $12,000

❸ MODE

The final measure used to describe a set of data is called the *mode*. Consider the numbers

$$1, \quad 5, \quad 101, \quad 101.$$

The mean is

$$\frac{1 + 5 + 101 + 101}{4} = \frac{208}{4} = 52,$$

and the median is

$$\frac{5 + 101}{2} = \frac{106}{2} = 53.$$

But does 52 or 53 give an accurate reflection of the data set? Since two of the four numbers are 101, much greater than 52 or 53, it might be better to describe this data set using 101, the number that occurs most frequently in the set.

Mode
The value that occurs most frequently in a set of values is the **mode** of the data. If there are two values that occur more frequently than all the others, the data has two modes and is called **bimodal.** If no value occurs more than once in the set, the data has **no mode.**

EXAMPLE 6 FINDING A MODE	PRACTICE EXERCISE 6

Find the mode of the given numbers.

(a) 5, 8, 1, 8, 7, 9, 8, 5

Since 8 occurs three times, 5 occurs twice, and all the rest of the numbers occur only once, the mode is 8.

(b) 2, 7, 3, 7, 8, 8, 1, 5

Since 7 and 8 both occur twice and all other numbers occur only once, the set of numbers has two modes, 7 and 8. It is a bimodal set.

(c) 3, 9, 7, 5, 2, 1, 11

Since every number occurs exactly one time, there is no mode.

Find the mode of the given numbers.

(a) 1, 1, 1, 5, 2, 2, 9, 8

(b) 6, 6, 2, 1, 5, 2, 9, 10

(c) 2, 8, 5, 12, 9, 7

Answers: (a) 1   (b) 6 and 2 (bimodal)   (c) no mode

7.3 · EXERCISES A

Find the mean of the given numbers. Give answers to the nearest tenth when appropriate.

1. 7, 19, 13   13

2. 7, 9, 8, 12   9

3. 12, 22, 18, 25, 9   17.2

4. 102, 95, 116, 118, 125, 106   110.3

5 7.2, 6.1, 8.6, 9.2, 5.5, 7.2, 8.8   7.5

6. 42, 43, 58, 55, 61, 41, 48, 48   49.5

Find the median.

7. 7, 19, 13   13

8. 7, 9, 8, 12   8.5

9. 12, 22, 25, 18, 9   18

10. 102, 95, 116, 118, 125, 106   111

11 7.2, 6.1, 8.6, 9.2, 5.5, 7.2, 8.8   7.2

12. 42, 43, 58, 55, 61, 41, 48, 48   48

Find the mode.

13. 5, 6, 1, 6, 3, 2   6

14. 2, 3, 9, 2, 9, 7, 9, 11   9

15. 8, 3, 8, 5, 7, 3, 2   8 and 3 (bimodal)

16. 2, 11, 17, 9, 6, 41   no mode

17. 4.3, 2.7, 4.3, 12.8, 15.1   4.3

18. 149, 150, 211, 150, 435, 147   150

Solve.

19. Suppose the Ford Motor Company's tests show that one of their cars will go 480 miles on a tank of gasoline. If the tank holds 13.6 gal, what is the average miles per gallon (mpg)?   35.3 mpg

20. A basketball player scored 252 points in 18 games. What is his average per game?   14

21. During a six-day week, the sales at Joseph's Shop for Men were $482.16, $316.58, $299.81, $686.47, $819.72, and $992.16. What were the average daily sales? $599.48

22. The heights of the starting five players on the Upstate University basketball team are 78 inches, 82 inches, 72 inches, 84 inches, and 73 inches. What is the average height? What is the median height?
77.8; 78

23 The table at the right gives information on the yearly salaries of the employees at one plant of Mutter Manufacturing. Find the average salary. Find the median salary. What is the mode of the salaries?
$23,135.71; $20,800; $20,800

Salary	Number of Employees	Salary Times Number of Employees
16,200	8	
20,800	10	
28,600	7	
32,000	2	
46,000	1	
Totals		

FOR REVIEW

The profit (in millions of dollars) made by the ABC Corporation over a six-year interval is shown in the broken-line graph below. Use this graph in Exercises 24–30.

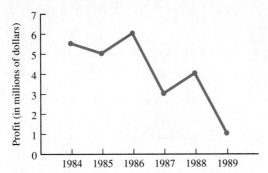

24. What profit was made in 1985?   **5 million dollars**

25. What profit was made in 1984?   **5.5 million dollars**

26. What was the total profit made in 1986 and 1987?   **9 million dollars**

27. What was the total profit made during the six-year interval?   **24.5 million dollars**

28. What was the ratio of the profit in 1986 to that in 1987? $\frac{2}{1}$ **or 2 to 1**

29. What percent (to the nearest tenth) of the total profit in the six-year interval was made in 1985?   **20.4%**

30. What percent (to the nearest tenth) of the total profit in the six-year interval was made in the years 1984 and 1985?   **42.9%**

ANSWERS: 1. 13   2. 9   3. 17.2   4. 110.3   5. 7.5   6. 49.5   7. 13   8. 8.5   9. 18   10. 111   11. 7.2   12. 48
13. 6   14. 9   15. 8 and 3 (bimodal)   16. no mode   17. 4.3   18. 150   19. 35.3 mpg   20. 14   21. $599.48
22. 77.8; 78   23. $23,135.71; $20,800; $20,800   24. 5 million dollars   25. 5.5 million dollars   26. 9 million dollars
27. 24.5 million dollars   28. $\frac{2}{1}$ or 2 to 1   29. 20.4%   30. 42.9%

7.3 EXERCISES B

Find the mean of the given numbers. Give answers to the nearest tenth when appropriate.

1. 11, 21, 19   17

2. 8, 12, 14, 22   14

3. 14, 35, 16, 19, 43   25.4

4. 130, 87, 102, 96, 145, 110   111.7

5. 6.3, 4.7, 8.2, 9.6, 7.1, 8.3, 2.5   6.7

6. 51, 57, 62, 65, 71, 83, 90, 97   72

Find the median.

7. 5, 12, 37   12

8. 6, 9, 10, 14   9.5

9. 5, 21, 27, 17, 19   19

10. 107, 93, 110, 84, 87, 130   100

11. 6.3, 4.7, 8.2, 9.6, 7.1, 8.3, 2.5   7.1

12. 51, 57, 62, 65, 90, 83, 71, 97   68

Find the mode.

13. 11, 17, 5, 17, 6, 21, 17   17

14. 6, 3, 6, 8, 11, 3, 21, 3   3

15. 2, 12, 5, 2, 8, 5, 7, 1   **2 and 5 (bimodal)**

16. 3, 9, 21, 8, 7, 26, 5   **no mode**

17. 4.8, 7.6, 8.1, 9.3, 7.6, 2.4   **7.6**

18. 159, 276, 423, 159, 342, 177   **159**

Solve.

19. Tests show that a motorhome will travel 535.5 miles on a tank of gasoline. If the tank holds 42.5 gallons, what is the average miles per gallon (mpg)?
12.6 mpg

20. Quarterback Craig Austin completed 438 passes in 11 games. What is the average number of completions per game (to the nearest tenth)?   **39.8**

21. The number of TV sets sold per day by a dealer over a 12-day period is 2, 4, 6, 2, 10, 3, 6, 8, 12, 8, 22, and 12. What is the median number of sets sold per day? What is the average number of sets sold per day (to the nearest tenth)?   **7; 7.9**

22. Maria scored 73, 85, 65, 92, 78, and 87 on six tests in math. What is her average score? What is her median score?   **80; 81.5**

23. The Mutter Manufacturing record of sick-leave days available to employees is recorded below. What is the average number of days available? What is the median number? What is the mode?   **24; 21; 21**

Number of Days Sick Leave Available	Number of Employees	Number of Days Times Number of Employees
14	4	
21	12	
28	8	
35	4	
Totals		

FOR REVIEW

The number of years of teaching experience by the faculty at CSU is shown in the histogram at the right. Use this graph in Exercises 24–30.

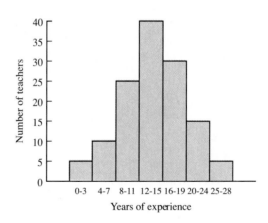

24. How many teachers have 0–3 years of experience?   **5**

25. How many teachers have 8–15 years of experience?   **65**

26. How many teachers have 12 or more years of experience?   **90**

27. How many teachers have 7 or less years of experience?   **15**

28. How many total teachers are represented by this data?   **130**

29. What percent (to the nearest tenth) of the total number of teachers have 16 or more years of experience?   **38.5%**

30. What is the ratio of the number of teachers with 16–19 years of experience to those with 7 or less years of experience? $\frac{2}{1}$ **or 2 to 1**

7.3 EXERCISES C

Solve.

1. Suppose Marni made scores of 65, 72, and 88 on the first three tests in math. What score did she make on the fourth test if the average score on the four was 80?
[Answer: 95]

2. Suppose Burford made scores of 32, 38, 51, and 82 on four of five exams. If the median score of the five exams is 51, could the score on Burford's fifth test have been 35?

   No. If it had been 35, the median would have been 38 not 51.

Which of the mean, median, or mode do you think is the best measure to describe each set of numbers in Exercises 3–6?

3. 1, 1, 1, 1, 20, 100   **mode**

4. 1, 1, 99, 99
[Answer: mode (bimodal)]

5. 1, 2, 3, 1000   **median**

6. 1, 2, 3, 4, 5   **mean or median**

CHAPTER 7 REVIEW

KEY WORDS

7.1 **Statistics** is a branch of mathematics that deals with collecting, organizing, and summarizing numerical data in order to describe or interpret it.

A **circle graph** uses pie-shaped sectors of a circle to represent different parts of a quantity.

A **bar graph** is used to compare various parts of a data set.

7.2 A **broken-line graph** can be used to show trends in data collected over a period of time.

A **histogram** is a special type of bar graph used to display data that is divided into groups called **class intervals**. The **class frequency** of an interval is represented by the length or height of each bar.

7.3 The **mean** or **average** of several values is the sum of the values divided by the number of values.

The **median** of a data set is the middle value or the average of the two middle values when the values are arranged in order of size.

The **mode** of a set of values is the value that occurs most frequently in the set.

KEY CONCEPTS

7.1 Numerical data can be displayed or pictured using a circle graph or a bar graph.

7.2 A histogram is often used when a large number of data items can best be described by dividing the data into groups.

7.3 1. Before finding the median of a collection of numbers, be sure to arrange the numbers in order of size.

2. If there is an even number of values, the median is the mean (average) of the two middle values.

3. A bimodal set of data has two values with the greatest frequency of occurrence. If all numbers in a collection occur exactly once, the data set has no mode.

REVIEW EXERCISES

Part I

7.1 *Exercises 1–4 refer to the circle graph at the right which shows preference in vacation locations of 300 adults surveyed in a random poll. Give answers rounded to the nearest tenth.*

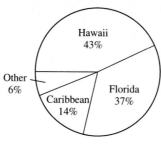

1. How many of those surveyed preferred Hawaii?   129

2. How many of those surveyed preferred the Caribbean?   42

3. What is the ratio of those preferring Hawaii to those preferring Florida? $\frac{43}{37}$ or 43 to 37

4. How many of those surveyed preferred a vacation location other than Hawaii?   171

7.2 *Exercises 5–10 refer to the broken-line graph at the right which shows the number of new homes sold in a subdivision over a six-month period.*

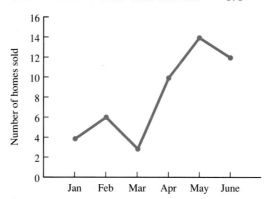

5. How many homes were sold in April?   10

6. How many homes were sold in May and June together?   26

7. Between which two consecutive months did sales of new homes show the greatest increase?
Between March and April

8. How many homes were sold in the six-month period?   49

9. What was the ratio of homes sold in April to those sold in March? $\frac{10}{3}$ or 10 to 3

10. What percent (to the nearest tenth) of the homes sold in the six-month period were sold in January?
8.2%

Exercises 11–16 refer to the histogram at the right which relates the number of cars to miles per gallon in an EPA survey. Assume that a car making 5 mpg, for example, is included in the range 5–10 mpg and not 0–5 mpg, similarly for 10 mpg, 15 mpg, and so forth.

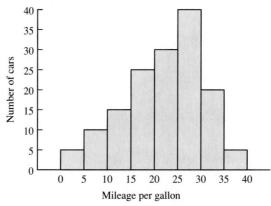

11. How many cars got at least 5 mpg and less than 10 mpg?   10

12. How many cars got 30 mpg or better?   25

13. How many cars got less than 20 mpg?   55

14. How many cars were represented in the survey?   150

15. What percent of the total number of cars got 35 mpg or better? $3.\overline{3}\%$ or $3\frac{1}{3}\%$

16. What is the ratio of cars that got 30 mpg or better to those that got less than 5 mpg? $\frac{5}{1}$ or 5 to 1

7.3 *Find the mean and median. Give answers to the nearest tenth.*

17. 16, 12, 19, 22, 8   15.4; 16

18. 375, 206, 311, 222, 320, 312   291; 311.5

19. Maxine's car will go 589 miles on 15.5 gallons of gasoline. What is Maxine's average miles per gallon?
38 mpg

Find the mode.

20. 6, 5, 1, 3, 5, 7, 12, 5   5

21. 2, 9, 3, 2, 7, 3, 4, 12   2 and 3 (bimodal)

22. The Thai Villa Restaurant had sales of $1020, $926, $820, $772, $1428, and $1362 over a six-day period. What were the average daily sales to the nearest dollar? What was the median? $1055; $973

23. The following table gives information of the monthly salaries of the employees of Lloyd's Shoppe. Find the average salary. Find the median salary. What is the mode of the salaries?
$917; $860; $860

Salary	Number of Employees	Salary Times Number of Employees
$ 820	4	
$ 860	8	
$ 920	5	
$ 980	2	
$1620	1	
Totals		

Part II

Exercises 24–27 refer to the bar graph at the right, which gives the number of adults choosing a particular sport as their favorite to view on television.

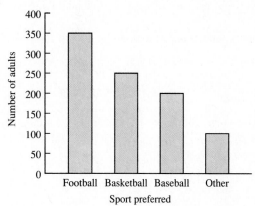

24. How many adults were in the survey?   900

25. What is the ratio of those preferring basketball to the total number surveyed? $\frac{5}{18}$ or 5 to 18

26. What percent of those surveyed preferred basketball (to the nearest tenth of a percent)?   27.8%

27. What is the ratio of those who prefer football to those who prefer baseball? $\frac{7}{4}$ or 7 to 4

Find the mean, median, and mode.

28. 12, 3, 9, 5, 5   6.8; 5; 5

29. 6.1, 2.5, 3.8, 9.2, 100.7   24.46; 6.1; no mode

30. 18, 12, 17, 29, 40, 12 $21.\overline{3}$; 17.5; 12

31. 4, 7, 4, 9, 7, 6, 8, 10   6.875; 7; 4 and 7

Exercises 32–40 refer to the double-bar graph below which compares the number of students enrolled in four math courses at State University in 1988 and 1989.

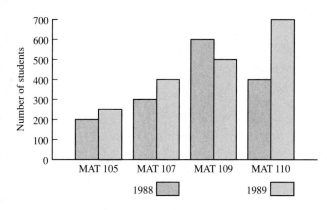

1988 ☐   1989 ☐

32. How many students were in MAT 107 in 1989?   400

33. How many students were in MAT 109 in 1988?   600

34. How many students were in these four math courses in 1988?   1500

35. How many students were in these four math courses in 1989?   1850

36. What was the ratio of students in MAT 110 in 1989 to those in MAT 110 in 1988? $\frac{7}{4}$ or 7 to 4

37. What percent of students in all four classes in 1988 were in MAT 109?   40%

38. During which year were there more students in MAT 109?   1988

39. What was the percent increase in students taking MAT 110 from 1988 to 1989?   75%

40. How many total students took MAT 105 in these two years?   450

ANSWERS: 1. 129   2. 42   3. $\frac{43}{37}$ or 43 to 37   4. 171   5. 10   6. 26   7. Between March and April   8. 49   9. $\frac{10}{3}$ or 10 to 3   10. 8.2%   11. 10   12. $\overline{25}$ 13. 55   14. 150   15. $3.\overline{3}$% or $3\frac{1}{3}$%   16. $\frac{5}{1}$ or 5 to 1   17. 15.4; 16   18. 291; 311.5   19. 38 mpg   20. 5   21. 2 and 3 (bimodal)   22. $1055; $973   23. $917; $860; $860   24. 900   25. $\frac{5}{18}$ or 5 to 18   26. 27.8%   27. $\frac{7}{4}$ or 7 to 4   28. 6.8; 5; 5   29. 24.46; 6.1; no mode   30. $21.\overline{3}$; 17.5; 12   31. 6.875; 7; 4 and 7   32. 400   33. 600   34. 1500   35. 1850   36. $\frac{7}{4}$ or 7 to 4   37. 40%   38. 1988   39. 75%   40. 450

Problems 1–3 refer to the circle graph below which shows the distribution of responses from 1000 adults about television viewing preference.

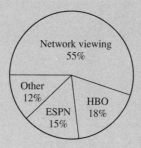

1. How many of those polled preferred HBO?

    1. _____180_____

2. How many of those polled preferred ESPN?

    2. _____150_____

3. What is the ratio of those who preferred network viewing to ESPN?

    3. $\dfrac{11}{3}$ or 11 to 3

Problems 4–8 refer to the bar graph below which shows the preferences of 150 adults in type of restaurant food.

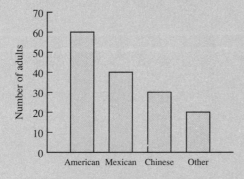

4. How many of those questioned preferred Mexican food?

    4. _____40_____

5. How many of those questioned preferred food other than American food?

    5. _____90_____

6. What is the ratio of those preferring American food to those preferring Chinese food?

    6. $\dfrac{2}{1}$ or 2 to 1

7. What percent of those interviewed preferred Mexican food?

    7. $26.\overline{6}\%$

8. What percent of those interviewed preferred some kind of food besides American?

    8. _____60%_____

The weights of all the members of a football team are presented in the histogram below. Use this information in problems 9–12. Assume that 175 pounds is in the class 175–200 pounds, not in 150–175 pounds, similarly for 200 pounds, 225 pounds, and 250 pounds.

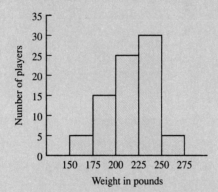

9. How many players weigh at least 200 and less than 225 pounds?

9. _____ 25

10. How many players are on the team?

10. _____ 80

11. What is the ratio of players weighing 225 pounds or more to the total number of players on the team?

11. _____ $\frac{7}{16}$ or 7 to 16

12. What percent of the players weigh less than 200 pounds?

12. _____ 25%

13. Find the mean of the given numbers.

         11,   22,   32,   17,   18

13. _____ 20

14. Find the median of the given numbers.

         11,   22,   32,   17,   18

14. _____ 18

15. Find the mode of the given numbers.

         14, 8, 9, 17, 8, 17, 23, 8, 5

15. _____ 8

16. The weights of the starting four defensive linemen for PSU are 246 pounds, 253 pounds, 257 pounds, and 262 pounds.
    (a) What is the average weight?   **(b)** What is the median weight?
    (c) What is the mode of the weights?

16. (a) _____ 254.5 pounds

       (b) _____ 255 pounds

       (c) _____ no mode

Measurement and Geometry

8.1 THE ENGLISH SYSTEM

STUDENT GUIDEPOSTS
---

1 English Units of Measure   3 Household Units

2 Unit Fraction   4 Compound Units

1 ENGLISH UNITS OF MEASURE

A standard system of units of measure is a useful tool for people living and working together. This chapter discusses two systems of measurement, the English system and the metric system. While many countries throughout the world use the metric system, the United States uses the English system.

Units of Measure: English System	
Length	1 foot (ft) = 12 inches (in)
	1 yard (yd) = 3 feet (ft)
	1 mile (mi) = 5280 feet (ft)
Volume	1 pint (pt) = 16 fluid ounces (fl oz)
	1 quart (qt) = 2 pints (pt)
	1 gallon (gal) = 4 quarts (qt)
Weight	1 pound (lb) = 16 ounces (oz)
	1 ton = 2000 pounds (lb)
Time	1 minute (min) = 60 seconds (sec)
	1 hour (hr) = 60 minutes (min)
	1 day (da) = 24 hours (hr)
	1 week (wk) = 7 days (da)
	1 year (yr) = 365 days (da)
	= 52 weeks (wk)
	= 12 months (mo)

As the table shows, each type of measurement (length, volume, and so on) has several units in the English system. The weight of this book might be given in ounces, while the weight of a student would usually be given in pounds.

❷ UNIT FRACTION

To work with units of measure, we need to know how to convert from one unit to another. A good way is to use **unit fractions,** fractions which have value 1. Some examples of unit fractions are

$$\frac{12 \text{ in}}{1 \text{ ft}}, \quad \frac{1 \text{ ft}}{12 \text{ in}}, \quad \frac{4 \text{ qt}}{1 \text{ gal}}, \quad \frac{1 \text{ gal}}{4 \text{ qt}}, \quad \frac{16 \text{ oz}}{1 \text{ lb}}, \quad \frac{1 \text{ lb}}{16 \text{ oz}}.$$

Each of the fractions has value 1 because the numerator and denominator are equal. It is important to show the unit of measure as well as the number. Obviously, $\frac{12}{1}$ *is not* 1, but $\frac{12 \text{ in}}{1 \text{ ft}}$ *is* 1 since

$$1 \text{ ft} = 12 \text{ in.}$$

Thus, to change 4 yd to feet, write

$$
\begin{aligned}
4 \text{ yd} &= 4 \text{ yd} \times 1 && \text{Multiply by 1} \\
&= \frac{4 \text{ yd}}{1} \times \frac{3 \text{ ft}}{1 \text{ yd}} && \text{Divide out yards} \\
&= 4 \times 3 \text{ ft} \\
&= 12 \text{ ft.}
\end{aligned}
$$

With this method, the yards are divided out as if they were numbers.

HINT

The fact that the yards do divide out tells us that we were right to multiply by 3 ft instead of dividing by 3 ft. If we had tried to multiply

$$\frac{4 \text{ yd}}{1} \times \frac{1 \text{ yd}}{3 \text{ ft}},$$

the yards would not have divided out and the change of units could not be made.

EXAMPLE 1 CONVERTING FEET TO INCHES	PRACTICE EXERCISE 1

Convert 16 feet to inches (16 ft = _____ in).

Since

$$1 \text{ ft} = 12 \text{ in,}$$

we need to use one of the unit fractions

$$\frac{12 \text{ in}}{1 \text{ ft}} \quad \text{or} \quad \frac{1 \text{ ft}}{12 \text{ in}}.$$

To divide out the ft in 16 ft, use $\frac{12 \text{ in}}{1 \text{ ft}}$.

$$
\begin{aligned}
16 \text{ ft} &= 16 \text{ ft} \times 1 && \text{Multiply by 1} \\
&= \frac{16 \text{ ft}}{1} \times \frac{12 \text{ in}}{1 \text{ ft}} && \text{Divide out ft} \\
&= 16 \times 12 \text{ in} \\
&= 192 \text{ in}
\end{aligned}
$$

Convert 23 feet to inches.

Answer: 276 in

```
┌─────────────────────────────────────────────────┐   ┌──────────────────────────────┐
│ EXAMPLE 2  CONVERTING INCHES TO FEET              │   │ PRACTICE EXERCISE 2          │
└─────────────────────────────────────────────────┘   └──────────────────────────────┘
```

Convert 72 inches to feet (72 in = _____ ft).
For this problem use the unit fraction $\frac{1\,ft}{12\,in}$.

Convert 108 inches to feet.

$$72\text{ in} = 72\text{ in} \boxed{\times 1} \qquad \text{Multiply by 1}$$

$$= \frac{72\text{ in}}{1} \times \boxed{\frac{1\text{ ft}}{12\text{ in}}} \qquad \text{Divide out in}$$

$$= \frac{72}{12}\text{ ft}$$

$$= 6\text{ ft}$$

Answer: 9 ft

```
┌─────────────────────────────────────────────────┐   ┌──────────────────────────────┐
│ EXAMPLE 3  CONVERSION IN WEIGHT AND VOLUME        │   │ PRACTICE EXERCISE 3          │
└─────────────────────────────────────────────────┘   └──────────────────────────────┘
```

Complete the conversion in units.

Complete the conversion in units.

(a) 22 lb = _____ oz.
The unit fraction is $\frac{16\,oz}{1\,lb}$.

(a) 15 lb = _____ oz.

$$22\text{ lb} = 22\text{ lb} \boxed{\times 1} \qquad \text{Multiply by 1}$$

$$= \frac{22\text{ lb}}{1} \times \boxed{\frac{16\text{ oz}}{1\text{ lb}}} \qquad \text{Divide out lb}$$

$$= 22 \times 16\text{ oz}$$

$$= 352\text{ oz}$$

(b) 144 fl oz = _____ pt.

(b) 96 fl oz = _____ pt.
With the unit fraction $\frac{1\,pt}{16\,fl\,oz}$, the fl oz divide out.

$$96\text{ fl oz} = 96\text{ fl oz} \boxed{\times 1}$$

$$= \frac{96\text{ fl oz}}{1} \times \boxed{\frac{1\text{ pt}}{16\text{ fl oz}}}$$

$$= \frac{96}{16}\text{ pt}$$

$$= 6\text{ pt}$$

(c) 9 pt = _____ qt.

(c) 6 pt = _____ qt.

$$6\text{ pt} = 6\text{ pt} \boxed{\times 1}$$

$$= \frac{6\text{ pt}}{1} \times \boxed{\frac{1\text{ qt}}{2\text{ pt}}}$$

$$= \frac{6}{2}\text{ qt}$$

$$= 3\text{ qt}$$

Answers: (a) 240   (b) 9
(c) 4.5

Parts (b) and (c) of Example 3 together change 96 fl oz to 3 qt. Some unit conversions may need two or more unit fractions.

| EXAMPLE 4 Using Several Unit Fractions | PRACTICE EXERCISE 4 |

Complete the change in units.

5 wk = _____ min.

$$5 \text{ wk} = 5 \text{ wk} \times 1$$

$$= \frac{5 \text{ wk}}{1} \times \frac{7 \text{ da}}{1 \text{ wk}} \qquad \text{Convert 5 wk to 35 da}$$

$$= \frac{35 \text{ da}}{1} \times \frac{24 \text{ hr}}{1 \text{ da}} \qquad \text{Convert 35 da to 840 hr}$$

$$= \frac{840 \text{ hr}}{1} \times \frac{60 \text{ min}}{1 \text{ hr}} \qquad \text{Convert 840 hr to 50,400 min}$$

$$= 840 \times 60 \text{ min}$$

$$= 50,400 \text{ min}$$

The steps can be combined.

$$5 \text{ wk} = \frac{5 \text{ wk}}{1} \times \frac{7 \text{ da}}{1 \text{ wk}} \times \frac{24 \text{ hr}}{1 \text{ da}} \times \frac{60 \text{ min}}{1 \text{ hr}}$$

$$= 5 \times 7 \times 24 \times 60 \text{ min}$$

$$= 50,400 \text{ min}.$$

Complete the change in units.

8 wk = _____ min

Answer: 80,640

③ HOUSEHOLD UNITS

Some units of measure used in the kitchen are given in this table.

Household Units
1 tablespoon (tbsp) = 3 teaspoons (tsp)
1 cup = 16 tablespoons (tbsp)
= 8 fluid ounces (fl oz)
1 pint (pt) = 2 cups

| EXAMPLE 5 Converting Household Units | PRACTICE EXERCISE 5 |

Complete the change in units.

(a) 7 tbsp = _____ tsp.

$$7 \text{ tbsp} = 7 \text{ tbsp} \times 1$$

$$= \frac{7 \text{ tbsp}}{1} \times \frac{3 \text{ tsp}}{1 \text{ tbsp}} \qquad \frac{3 \text{ tsp}}{1 \text{ tbsp}} \text{ is a unit fraction}$$

$$= 7 \times 3 \text{ tsp} = 21 \text{ tsp}$$

(b) 40 tbsp = _____ pt.

$$40 \text{ tbsp} = \frac{40 \text{ tbsp}}{1} \times \frac{1 \text{ cup}}{16 \text{ tbsp}} \times \frac{1 \text{ pt}}{2 \text{ cups}}$$

$$= 40 \times \frac{1}{16} \times \frac{1}{2} \text{ pt} = \frac{40}{(16)(2)} \text{ pt}$$

$$= \frac{2 \cdot 2 \cdot 2 \cdot 5}{2 \cdot 2 \cdot 2 \cdot 2 \cdot 2} \text{ pt} = \frac{5}{4} \text{ pt} = 1\frac{1}{4} \text{ pt}$$

Complete the change in units.

(a) 15 tbsp = _____ tsp.

(b) 120 tbsp = _____ pt.

Answers: (a) 45 (b) 3.75

4 COMPOUND UNITS

Remember from Chapter 6 that *percent* means "per hundred." We can also say that

$$\frac{12 \text{ in}}{1 \text{ ft}} \quad \text{means} \quad 12 \text{ inches per foot.}$$

Thus, 50 miles per hour can be written as

$$\frac{50 \text{ mi}}{1 \text{ hr}}.$$

Most of the time the 1 in the denominator is not written. 50 miles per hour may be written in any of the following ways.

$$\frac{50 \text{ mi}}{\text{hr}} \qquad 50\frac{\text{mi}}{\text{hr}} \qquad 50 \text{ mi/hr} \qquad 50 \text{ mph}$$

Other units can be written these ways.

$$12 \text{ feet per second} \quad \text{means} \quad \frac{12 \text{ ft}}{\text{sec}} = 12\frac{\text{ft}}{\text{sec}} = 12 \text{ ft/sec}$$

$$8 \text{ gallons per minute} \quad \text{means} \quad \frac{8 \text{ gal}}{\text{min}} = 8\frac{\text{gal}}{\text{min}} = 8 \text{ gal/min}$$

EXAMPLE 6   CONVERTING UNITS OF RATE

Complete the change in units.

$$30\frac{\text{mi}}{\text{hr}} = \underline{\quad\quad} \frac{\text{ft}}{\text{sec}}.$$

Combine the steps.

$$30\frac{\text{mi}}{\text{hr}} = \frac{30 \text{ mi}}{\text{hr}} = \frac{30 \text{ mi}}{\text{hr}} \times \frac{5280 \text{ ft}}{1 \text{ mi}} \times \frac{1 \text{ hr}}{60 \text{ min}} \times \frac{1 \text{ min}}{60 \text{ sec}} \qquad \begin{array}{l}\text{Divide out}\\ \text{mi, hr,}\\ \text{and min}\end{array}$$

$$= \frac{30 \times 5280 \text{ ft}}{60 \times 60 \text{ sec}}$$

$$= \frac{44 \text{ ft}}{\text{sec}} = 44\frac{\text{ft}}{\text{sec}}$$

PRACTICE EXERCISE 6

Complete the change in units.

$$90\frac{\text{mi}}{\text{hr}} = \underline{\quad\quad} \frac{\text{ft}}{\text{sec}}$$

Answer: 132

EXAMPLE 7   APPLICATION OF CONVERSION

Claudia wants to buy 15 inches of fringe which costs $1.08 per yard. How much must she pay?

First convert dollars per yard to dollars per foot, then convert dollars per foot to dollars per inch.

$$\frac{1.08 \text{ dollars}}{\text{yd}} = \frac{1.08 \text{ dollars}}{1 \text{ yd}} \times \frac{1 \text{ yd}}{3 \text{ ft}} \times \frac{1 \text{ ft}}{12 \text{ in}} \qquad \text{Unit fractions}$$

$$= \frac{1.08 \text{ dollars}}{3 \times 12 \text{ in}}$$

$$= \frac{0.03 \text{ dollars}}{\text{in}}$$

PRACTICE EXERCISE 7

Metal tubing costs $3.60 per yard. How much must Larry pay for 28 inches?

Since Claudia wants 15 in of the fringe, multiply 15 in by $\frac{0.03 \text{ dollars}}{\text{in}}$.

$$\frac{15 \text{ in}}{1} \times \frac{0.03 \text{ dollars}}{\text{in}} = \$0.45$$

She must pay \$0.45 for the fringe.

Answer: **\$2.80**

| EXAMPLE 8   RATE APPLICATION | PRACTICE EXERCISE 8 |

Herman is filling a tank with water from a larger tank. If 2 pints per minute flow through the filling tube, how many gallons per minute flow through?

Change pints per minute to gallons per minute.

$$2\frac{\text{pt}}{\text{min}} = \frac{2 \text{ pt}}{\text{min}} = \frac{2 \text{ pt}}{1 \text{ min}} \times \frac{1 \text{ qt}}{2 \text{ pt}} \times \frac{1 \text{ gal}}{4 \text{ qt}}$$

$$= \frac{2 \text{ gal}}{2 \times 4 \text{ min}} = \frac{1 \text{ gal}}{4 \text{ min}} = \frac{1}{4}\frac{\text{gal}}{\text{min}}$$

Thus, $\frac{1}{4}$ of a gallon flows through the tube per minute.

Bernice is told that a machine can fill 120 pints in one minute. How many gallons per minute is this?

Answer: **15 gal per min**

8.1 EXERCISES A

Study the lists of equal units before completing the following.

1. 5 ft = ____60____ in

2. 72 in = ____6____ ft

3. 72 in = ____2____ yd

4. 21 da = ____3____ wk

5. 64 oz = ____4____ lb

6. 36 hr = ____1.5____ da

7. 72 fl oz = ____4.5____ pt

8. 4500 lb = ____2.25____ ton

9. 6 wk = ____42____ da

10. 3 mi = ____15,840____ ft

11. 18 qt = ____4.5____ gal

12. 16 mo = ____$1\frac{1}{3}$____ yr

13. 1 hr = ____3600____ sec

(14) 5400 sec = ____1.5____ hr

15. 96 in = ____$2\frac{2}{3}$____ yd

16. 7 yd = ____252____ in

17. 3 ton = ____96,000____ oz

18. 96 fl oz = ____3____ qt

19. 2 mi = ____126,720____ in

20. 3 wk = ____30,240____ min

21. 7200 sec = ____$\frac{1}{12}$____ da

22. $15\frac{\text{mi}}{\text{hr}}$ = ____22____ $\frac{\text{ft}}{\text{sec}}$

(23) $88\frac{\text{ft}}{\text{sec}}$ = ____60____ $\frac{\text{mi}}{\text{hr}}$

24. $5\frac{\text{cents}}{\text{in}}$ = ____1.80____ $\frac{\text{dollars}}{\text{yd}}$

25. $2\frac{\text{gal}}{\text{min}}$ = ____$\frac{2}{15}$____ $\frac{\text{qt}}{\text{sec}}$

26. $200\frac{\text{lb}}{\text{da}}$ = ____0.7____ $\frac{\text{ton}}{\text{wk}}$

27. $32\frac{\text{oz}}{\text{hr}}$ = ____$\frac{1}{30}$____ $\frac{\text{lb}}{\text{min}}$

Solve.

28. George buys 30 in of material which costs $2.16 per yard. How much does he have to pay?   **$1.80**

29. The speed limit is 55 miles per hour and Maria is driving at 5280 feet per minute. Is she driving over the speed limit? If so, by how much?   **yes; $5\frac{mi}{hr}$**

30. Richard is filling a tank at the rate of 4 pints per minute. How many gallons per hour is this?
$30\frac{gal}{hr}$

31 A farmer needs to buy 3 miles of wire. What is the total cost if the wire costs 20¢ per foot?   **$3168**

FOR REVIEW

The following exercises review material from Chapter 4 to help prepare you for the next section.

32.
$$6\frac{1}{2}$$
$$+9\frac{1}{3} \quad 15\frac{5}{6}$$

33.
$$42\frac{2}{5}$$
$$-16\frac{3}{4} \quad 25\frac{13}{20}$$

34.
$$162\frac{5}{12}$$
$$-117\frac{17}{18} \quad 44\frac{17}{36}$$

ANSWERS: 1. 60 2. 6 3. 2 4. 3 5. 4 6. 1.5 7. 4.5 8. 2.25 9. 42 10. 15,840 11. 4.5 12. $1\frac{1}{3}$ 13. 3600
14. 1.5 15. $2\frac{2}{3}$ 16. 252 17. 96,000 18. 3 19. 126,720 20. 30,240 21. $\frac{1}{12}$ 22. 22 23. 60 24. 1.80 25. $\frac{2}{15}$
26. 0.7 27. $\frac{1}{30}$ 28. $1.80 29. yes; $5\frac{mi}{hr}$ 30. $30\frac{gal}{hr}$ 31. $3168 32. $15\frac{5}{6}$ 33. $25\frac{13}{20}$ 34. $44\frac{17}{36}$

8.1 EXERCISES B

Complete the change in units.

1. 9 ft = _____ 108 _____ in

2. 132 in = _____ 11 _____ ft

3. 132 in = _____ $3\frac{2}{3}$ _____ yd

4. 63 da = _____ 9 _____ wk

5. 88 oz = _____ 5.5 _____ lb

6. 96 hr = _____ 4 _____ da

7. 108 fl oz = _____ 6.75 _____ pt

8. 15,000 lb = _____ 7.5 _____ ton

9. 13 wk = _____ 91 _____ da

10. 10 mi = _____ 52,800 _____ ft

11. 27 qt = _____ 6.75 _____ gal

12. 37 mo = _____ $3\frac{1}{12}$ _____ yr

13. $\frac{1}{2}$ hr = _____ 1800 _____ sec

14. 400 sec = _____ $\frac{1}{9}$ _____ hr

15. 144 in = _____ 4 _____ yd

16. 12 yd = _____ 432 _____ in

17. 0.1 ton = _____ 3200 _____ oz

18. 160 fl oz = _____ 5 _____ qt

19. 0.4 mi = _____ 25,344 _____ in

20. $\frac{1}{2}$ wk = _____ 5040 _____ min

21. 86,400 sec = _____ 1 _____ da

22. $75\dfrac{\text{mi}}{\text{hr}} = \underline{}\ \dfrac{\text{ft}}{\text{sec}}$ 110

23. $11\dfrac{\text{ft}}{\text{sec}} = \underline{}\ \dfrac{\text{mi}}{\text{hr}}$ 7.5

24. $20\dfrac{\text{cents}}{\text{in}} = \underline{}\ \dfrac{\text{dollars}}{\text{yd}}$ 7.20

25. $30\dfrac{\text{gal}}{\text{min}} = \underline{}\ \dfrac{\text{qt}}{\text{sec}}$ 2

26. $700\dfrac{\text{lb}}{\text{da}} = \underline{}\ \dfrac{\text{ton}}{\text{wk}}$ 2.45

27. $640\dfrac{\text{oz}}{\text{hr}} = \underline{\phantom{\frac{2}{3}}}\ \dfrac{\text{lb}}{\text{min}}$ $\dfrac{2}{3}$

Solve.

28. Sirloin steak is priced at \$2.79 per pound. How much would the Moores have to pay for 35 lb?
\$97.65

29. Burt is driving at the rate of 55 feet per second. If he is driving where the speed limit is 40 miles per hour, how does his speed compare to the limit?
2.5 mi/hr below the limit

30. A water pump will deliver 210 gallons per minute. How many quarts per second is this? $14\frac{\text{qt}}{\text{sec}}$

31. Electrical wiring costs 5¢ per inch. How much would 25 yards of this wire cost?   **\$45**

FOR REVIEW

The following exercises review material from Chapter 4 to help prepare you for the next section.

32. $\begin{array}{r} 14\frac{3}{7} \\ +\ 8\frac{5}{14} \\ \hline \end{array}$ $22\frac{11}{14}$

33. $\begin{array}{r} 22\frac{1}{3} \\ -\ 11\frac{4}{5} \\ \hline \end{array}$ $10\frac{8}{15}$

34. $\begin{array}{r} 216\frac{29}{60} \\ -\ 129\frac{17}{20} \\ \hline \end{array}$ $86\frac{19}{30}$

8.1   EXERCISES C

Use a calculator to find the following change in units. Give answers correct to the nearest tenth.

1. $627\ \text{in} = \underline{}\ \text{yd}$ 17.4

2. $55\dfrac{\text{mi}}{\text{hr}} = \underline{}\ \dfrac{\text{ft}}{\text{sec}}$
[Answer: 80.7]

3. $143\ \text{da} = \underline{}\ \text{yr}$ 0.4

4. Mount McKinley in Alaska is the highest mountain in the United States at 20,320 feet above sea level. Give the height in miles.
[Answer: 3.8]

8.2   ARITHMETIC OF MEASUREMENT NUMBERS

================== STUDENT GUIDEPOSTS ==================

1 Simplifying Measurement Units   **4** Multiplying Measurement Units

2 Adding Measurement Units   **5** Dividing Measurement Units

3 Subtracting Measurement Units

1 SIMPLIFYING MEASUREMENT UNITS

Suppose the answer to a time-conversion problem is 64 minutes. When a time answer is more than 1 hour (60 min), the result is usually written in hours and minutes.

$$64 \text{ min} = \textbf{60 min} + 4 \text{ min}$$
$$= \textbf{1 hr } 4 \text{ min} \qquad \text{Plus sign left out}$$

This is read "one hour four minutes." This form of the answer gives a better idea of the size of some measurements, especially when the numbers are large. For example, we have a better idea of the time if we are told a job will take 8 hr rather than 480 min.

Other measurement units can also be simplified. For example, to change 32 inches to feet and inches, write 32 as the sum of a multiple of 12 and a number less than 12.

$$32 \text{ in} = \textbf{24 in} + 8 \text{ in}$$
$$= \textbf{2 ft } 8 \text{ in} \qquad 24 \text{ in} = 2 \text{ ft}$$

This is read "two feet eight inches."

EXAMPLE 1 SIMPLIFYING UNITS	PRACTICE EXERCISE 1

Simplify the following measurements.

(a) 3 hr 92 min

To change 92 minutes to hours and minutes, write 92 as the sum of a multiple of 60 and a number less than 60.

$$3 \text{ hr } \textbf{92 min} = 3 \text{ hr} + \textbf{60 min} + 32 \text{ min} \qquad 92 = 60 + 32$$
$$= 3 \text{ hr} + 1 \text{ hr} + 32 \text{ min} \qquad 60 \text{ min} = 1 \text{ hr}$$
$$= 4 \text{ hr } 32 \text{ min} \qquad \text{Add 3 hr and 1 hr}$$

(b) 3 yd 8 ft

$$3 \text{ yd } \textbf{8 ft} = 3 \text{ yd} + \textbf{6 ft} + 2 \text{ ft} \qquad 8 = 6 + 2$$
$$= 3 \text{ yd} + 2 \text{ yd} + 2 \text{ ft} \qquad 6 \text{ ft} = 2 \text{ yd}$$
$$= 5 \text{ yd } 2 \text{ ft}$$

Simplify the following measurements.

(a) 7 hr 116 min

(b) 5 yd 14 ft

Answers: **(a)** 8 hr 56 min
(b) 9 yd 2 ft

EXAMPLE 2 EXPANDING UNITS	PRACTICE EXERCISE 2

Write 125 inches as yards, feet, and inches.

First, change 125 inches to feet and inches by writing it as the sum of a multiple of 12 and a number less than 12.

$$125 \text{ in} = \textbf{120 in} + 5 \text{ in} \qquad 120 \text{ is a multiple of 12}$$
$$= \textbf{10}(12) \text{ in} + 5 \text{ in}$$
$$= 10 \text{ ft } 5 \text{ in} \qquad 12 \text{ in} = 1 \text{ ft}$$

Now change 10 ft to yards and feet.

$$10 \text{ ft } 5 \text{ in} = \textbf{9 ft} + 1 \text{ ft} + 5 \text{ in} \qquad 10 = 9 + 1$$
$$= 3 \text{ yd } 1 \text{ ft } 5 \text{ in} \qquad 9 \text{ ft} = 3 \text{ yd}$$

Write 400 inches as yards, feet, and inches.

Answer: **11 yd 0 ft 4 in =**
11 yd 4 in

EXAMPLE 3 SIMPLIFYING VOLUME UNITS	PRACTICE EXERCISE 3

Simplify 9 gal 18 qt 9 pt.

First change 9 pints to quarts and pints.

$$9 \text{ pt} = \textbf{8 pt} + 1 \text{ pt}$$
$$= 4 \text{ qt } 1 \text{ pt} \qquad 2 \text{ pt} = 1 \text{ qt}$$

Simplify 4 gal 34 qt 23 pt.

Adding 4 qt to 18 qt gives 22 qt. To change 22 qt to gallons and quarts, write 22 as the sum of a multiple of 4 and a number less than 4.

$$22 \text{ qt } 1 \text{ pt} = 20 \text{ qt} + 2 \text{ qt} + 1 \text{ pt}$$
$$= 5(4 \text{ qt}) + 2 \text{ qt} + 1 \text{ pt}$$
$$= 5 \text{ gal } 2 \text{ qt } 1 \text{ pt} \qquad 4 \text{ qt} = 1 \text{ gal}$$

Since

$$9 \text{ gal} + 5 \text{ gal} = 14 \text{ gal},$$

the final answer is 14 gal 2 qt 1 pt.

Answer: 15 gal 1 qt 1 pt

❷ ADDING MEASUREMENT UNITS

In these simplifications, we have been adding measurements. In Example 3 we wrote

$$9 \text{ gal} + 5 \text{ gal} = 14 \text{ gal}.$$

Only measurements in the same unit can be added. Thus, 9 gal can be added to 5 gal, but 14 gal cannot be added to 2 qt without first converting one of the units to the other.

In some cases, we need to add measurements such as 8 ft 5 in and 6 ft 4 in. This problem is like adding mixed numbers. We can change both to inches and add or we can add them as mixed measurement units.

EXAMPLE 4   ADDING UNITS

Add.

(a) 8 ft 5 in and 6 ft 4 in
  Add each unit separately.

$$\begin{array}{r} 8 \text{ ft} \quad 5 \text{ in} \\ +\ 6 \text{ ft} \quad 4 \text{ in} \\ \hline \text{sum of feet} \longrightarrow \quad 14 \text{ ft} \quad 9 \text{ in} \longleftarrow \text{sum of inches} \end{array}$$

The sum is 14 ft 9 in.

(b) 3 hr 28 min and 7 hr 54 min

$$\begin{array}{r} 3 \text{ hr} \quad 28 \text{ min} \\ +\ 7 \text{ hr} \quad 54 \text{ min} \\ \hline \text{sum of hours} \longrightarrow \quad 10 \text{ hr} \quad 82 \text{ min} \longleftarrow \text{sum of minutes} \end{array}$$

Since 82 min is more than 1 hr, simplify the sum.

$$10 \text{ hr } \mathbf{82 \text{ min}} = 10 \text{ hr} + \mathbf{60 \text{ min} + 22 \text{ min}}$$
$$= 10 \text{ hr} + 1 \text{ hr} + 22 \text{ min} \qquad 60 \text{ min} = 1 \text{ hr}$$
$$= 11 \text{ hr } 22 \text{ min}$$

The sum is 11 hr 22 min.

PRACTICE EXERCISE 4

Add.

(a) 17 ft 7 in and 5 ft 2 in

(b) 14 hr 42 min and 5 hr 37 min

**Answers: (a) 22 ft 9 in
(b) 20 hr 19 min**

❸ SUBTRACTING MEASUREMENT UNITS

To subtract measurement units, subtract only like units.

EXAMPLE 5   SUBTRACTING UNITS

Subtract 6 ft 4 in from 8 ft 5 in.

$$
\begin{array}{r}
8\text{ ft}\quad 5\text{ in}\\
-\ 6\text{ ft}\quad 4\text{ in}\\
\hline
2\text{ ft}\quad 1\text{ in}
\end{array}
$$

difference of feet ⟶   ⟵ difference of inches

The difference is 2 ft 1 in.

PRACTICE EXERCISE 5

Subtract 10 ft 4 in from 14 ft 11 in.

Answer: 4 ft 7 in

As with subtraction of mixed numbers, sometimes we need to borrow. With measurement units, this means changing one unit of the larger measure to units of the smaller.

EXAMPLE 6   SUBTRACTING WITH BORROWING

Subtract.
$$
\begin{array}{r}
17\text{ ft }2\text{ in}\\
-\ \ 9\text{ ft }8\text{ in}
\end{array}
$$

Since 2 inches is less than the 8 inches to be subtracted, borrow 1 ft (12 in) from 17 ft.

$$
\begin{aligned}
17\text{ ft }2\text{ in} &= 16\text{ ft} + 1\text{ ft} + 2\text{ in}\\
&= 16\text{ ft} + 12\text{ in} + 2\text{ in}\\
&= 16\text{ ft }14\text{ in}
\end{aligned}
$$

Now subtract.

$$
\begin{array}{r}
17\text{ ft }2\text{ in} = \quad 16\text{ ft}\quad 14\text{ in}\\
-\ \ 9\text{ ft }8\text{ in} = -\ \ 9\text{ ft}\quad 8\text{ in}\\
\hline
7\text{ ft}\quad 6\text{ in}
\end{array}
$$

The difference is 7 ft 6 in.

PRACTICE EXERCISE 6

Subtract.
$$
\begin{array}{r}
25\text{ ft }1\text{ in}\\
-\ 11\text{ ft }9\text{ in}
\end{array}
$$

Answer: 13 ft 4 in

❹ MULTIPLYING MEASUREMENT UNITS

Some multiplication problems involve measurement units. For example, suppose it requires 2 yd 1 ft 7 in of material to make one suit. To learn how much material is required for 4 suits, multiply 2 yd 1 ft 7 in by 4. As with addition and subtraction, all units could be converted to inches and multiplied by 4, but the same result can be reached by multiplying 2 yd by 4, 1 ft by 4, and 7 in by 4, and then simplifying the result.

$$
\begin{aligned}
4 \times (2\text{ yd }1\text{ ft }7\text{ in}) &= 4(2\text{ yd} + 1\text{ ft} + 7\text{ in})\\
&= 8\text{ yd} + 4\text{ ft} + 28\text{ in}\qquad \text{Distributive property}
\end{aligned}
$$

Now simplify. First change 28 in to feet: 28 in = 24 in + 4 in = 2 ft + 4 in. Then, rewrite the product.

$$
8\text{ yd} + 4\text{ ft} + 28\text{ in} = 8\text{ yd} + 4\text{ ft} + \mathbf{2}\textbf{ ft} + \mathbf{4}\textbf{ in}\qquad
$$
Replace 28 in by 2 ft + 4 in

$$= 8 \text{ yd} + 6 \text{ ft} + 4 \text{ in} \qquad \text{Add 4 ft and 2 ft}$$
$$= 8 \text{ yd} + 2 \text{ yd} + 4 \text{ in} \qquad \text{3 ft = 1 yd}$$
$$= 10 \text{ yd } 4 \text{ in}$$

Thus, 10 yd 4 in is needed for the 4 suits. This could be written as 10 yd 0 ft 4 in.

EXAMPLE 7   MULTIPLYING BY A NON-MEASUREMENT NUMBER

If a tank holds 7 gal 3 qt 1 pt of liquid, how much would be used to fill 6 tanks of the same size?

Multiply 7 gal 3 qt 1 pt by 6.

$$6 \times (7 \text{ gal } 3 \text{ qt } 1 \text{ pt}) = 6(7 \text{ gal} + 3 \text{ qt} + 1 \text{ pt})$$
$$= 42 \text{ gal} + 18 \text{ qt} + 6 \text{ pt}$$
$$= 42 \text{ gal} + 18 \text{ qt} + 3 \text{ qt} \qquad \text{2 pt = 1 qt}$$
$$= 42 \text{ gal} + 21 \text{ qt} \qquad \text{18 qt + 3 qt = 21 qt}$$
$$= 42 \text{ gal} + 20 \text{ qt} + 1 \text{ qt} \qquad \text{20 qt = 5(4 qt)}$$
$$= 42 \text{ gal} + 5 \text{ gal} + 1 \text{ qt} \qquad \text{4 qt = 1 gal}$$
$$= 47 \text{ gal } 1 \text{ qt} \qquad \begin{array}{l} \text{42 gal + 5 gal} \\ = \text{47 gal} \end{array}$$

PRACTICE EXERCISE 7

For a survival test each person was given 1 gal 2 qt 1 pt of water. How much water was needed to supply the 15 people taking the test?

Answer: 24 gal 1 qt 1 pt

It is possible to multiply measurement units by measurement units.

EXAMPLE 8   MULTIPLYING MEASUREMENT UNITS

Find the area of a rectangle which is 4 ft wide and 6 ft long.

Recall from Chapter 2 that

$$\text{area} = \text{length} \times \text{width}$$
$$= 6 \text{ ft} \times 4 \text{ ft}$$

To work the problem find not only 6×4 but also ft $\times$ ft.

$$\text{area} = 6 \text{ ft} \times 4 \text{ ft}$$
$$= (6 \times 4)(\text{ft} \times \text{ft})$$
$$= 24 \text{ ft}^2$$
$$= 24 \text{ sq ft}$$

The area is 24 sq ft. (The unit sq ft or ft^2 is used for area. We will say more about it later in this chapter.)

PRACTICE EXERCISE 8

Find the area of a rectangular garden which is 10 yd wide and 17 yd long.

Answer: 170 yd^2

---

5 DIVIDING MEASUREMENT UNITS

To divide a measurement by an ordinary number, first divide the largest unit by the number. If there is a remainder, add it to the next unit and divide again.

| **EXAMPLE 9   DIVIDING BY A NON-MEASUREMENT NUMBER** | **PRACTICE EXERCISE 9** |

Divide 17 lb 6 oz by 3.

Divide 42 lb 13 oz by 5.

First divide 17 lb by 3.

$$\begin{array}{r} 5 \text{ lb} \\ 3\overline{)17 \text{ lb}} \\ \underline{15} \\ 2 \text{ lb} \end{array}$$

Thus, 17 lb ÷ 3 = 5 lb with a remainder of 2 lb. Change 2 lb to ounces.

$$2 \text{ lb} = 2(16) \text{ oz} = 32 \text{ oz}$$

Add the 6 oz from the original measurement to 32 oz.

$$32 \text{ oz} + 6 \text{ oz} = 38 \text{ oz}$$

Now divide 38 oz by 3.

$$\begin{array}{r} 12 \text{ oz} \\ 3\overline{)38 \text{ oz}} \\ \underline{3} \\ 8 \\ \underline{6} \\ 2 \text{ oz} \end{array}$$

The answer to this division is 12 oz with remainder 2 or $12\frac{2}{3}$ oz. The final answer to the problem (17 lb 6 oz) ÷ 3 is

$$5 \text{ lb } 12\frac{2}{3} \text{ oz.}$$

Answer: 8 lb 9 oz

Some problems involve division of a measurement by a measurement. Suppose you want to know how many shirts costing $16 each can be bought for $48.

$$\frac{48 \text{ dollars}}{16 \text{ dollars}} = \frac{3 \times \cancel{16 \text{ dollars}}}{\cancel{16 \text{ dollars}}} = 3$$

The dollars cancel and the answer is the whole number (with no measurement unit) 3. You can buy 3 shirts for $48.

| **EXAMPLE 10   DIVIDING MEASUREMENT UNITS** | **PRACTICE EXERCISE 10** |

How many 4-ft shelves can be cut from a board 16 ft long?

One uniform can be made from 3 yd of material. How many uniforms can be made from 51 yd of the material?

Divide 16 ft by 4 ft.

$$\frac{16 \text{ ft}}{4 \text{ ft}} = \frac{4 \times \cancel{4 \text{ ft}}}{\cancel{4 \text{ ft}}} = 4$$

Thus, 4 shelves can be cut from the board.

Answer: 17 uniforms

8.2   EXERCISES A

Simplify.

1. 83 min
   1 hr 23 min

2. 3 hr 92 min
   4 hr 32 min

3. 5 yd 8 ft
   7 yd 2 ft

4. 8 gal 17 qt
12 gal 1 qt

5. 14 lb 73 oz
18 lb 9 oz

6. 4 yd 7 ft 23 in
6 yd 2 ft 11 in

Perform the indicated operations and simplify.

7.   8 lb   3 oz
    + 4 lb 10 oz
    12 lb 13 oz

8.   5 hr 52 min
    + 13 hr 41 min
    19 hr 33 min

9.   3 gal 3 qt
    + 4 gal 2 qt
    8 gal 1 qt

10.   6 yd 2 ft 11 in
    + 7 yd 1 ft   9 in
    14 yd 1 ft 8 in

11.   7 yd 2 ft
    − 3 yd 1 ft
    4 yd 1 ft

12.   8 hr 30 min
    − 2 hr 26 min
    6 hr 4 min

13.   8 gal 1 qt
    − 4 gal 3 qt
    3 gal 2 qt

14.   5 da 3 hr   8 min
    − 4 da 6 hr 40 min
    20 hr 28 min

15.   9 yd 2 ft   3 in
    − 2 yd 2 ft 10 in
    6 yd 2 ft 5 in

16. 3 × (5 hr 16 min)
    15 hr 48 min

17. 5 × (4 gal 3 qt)
    23 gal 3 qt

18. 4 × (2 yd 2 ft 11 in)
    11 yd 2 ft 8 in

19. 6 × (2 hr 18 min 14 sec)
    13 hr 49 min 24 sec

20. 3 in × 12 in
    36 sq in

21. 7 yd × 12 yd
    84 sq yd

22. (5 lb 12 oz) ÷ 2
    2 lb 14 oz

23. (7 hr 33 min) ÷ 3
    2 hr 31 min

24. 18 ft ÷ 3 ft
    6

Solve.

25. How many $6 ski caps can be bought for $24?
4

26. A chemist has 5 gal 3 qt of one liquid to be mixed with 4 gal 2 qt of another. How much will he have after he mixes them?   10 gal 1 qt

27. Henry took 6 gal 2 qt of water on a camping trip. After using 2 gal 4 qt, how much water did he have left?   3 gal 2 qt

28. If it takes 1 lb 12 oz of flour to make one loaf of bread, how much will it take to make 7 loaves?
12 lb 4 oz

29. Find the area of a rectangle which is 16 inches long and 5 inches wide.   80 sq in

30. If it takes 1 lb 12 oz of flour to make one loaf of bread, how much will it take to make a loaf one half the size?   14 oz

FOR REVIEW

Complete the unit conversion.

31. 16 ton = _____32,000_____ lb

32. $44 \dfrac{\text{ft}}{\text{sec}} = $ _____30_____ $\dfrac{\text{mi}}{\text{hr}}$

33. $50 \dfrac{\text{cents}}{\text{in}} = $ _____6_____ $\dfrac{\text{dollars}}{\text{ft}}$

ANSWERS: 1. 1 hr 23 min   2. 4 hr 32 min   3. 7 yd 2 ft   4. 12 gal 1 qt   5. 18 lb 9 oz   6. 6 yd 2 ft 11 in
7. 12 lb 13 oz   8. 19 hr 33 min   9. 8 gal 1 qt   10. 14 yd 1 ft 8 in   11. 4 yd 1 ft   12. 6 hr 4 min   13. 3 gal 2 qt
14. 20 hr 28 min   15. 6 yd 2 ft 5 in   16. 15 hr 48 min   17. 23 gal 3 qt   18. 11 yd 2 ft 8 in   19. 13 hr 49 min 24 sec
20. 36 sq in   21. 84 sq yd   22. 2 lb 14 oz   23. 2 hr 31 min   24. 6   25. 4   26. 10 gal 1 qt   27. 3 gal 2 qt
28. 12 lb 4 oz   29. 80 sq in   30. 14 oz   31. 32,000   32. 30   33. 6

8.2 EXERCISES B

Simplify.

1. 137 min
2 hr 17 min

2. 7 hr 72 min
8 hr 12 min

3. 4 yd 19 ft
10 yd 1 ft

4. 2 gal 25 qt
8 gal 1 qt

5. 6 lb 105 oz
12 lb 9 oz

6. 5 yd 14 ft 42 in
10 yd 2 ft 6 in

Perform the indicated operations and simplify.

7. 9 lb 11 oz
+ 5 lb 1 oz
14 lb 12 oz

8. 12 hr 39 min
+ 10 hr 32 min
23 hr 11 min

9. 9 gal 3 qt
+ 7 gal 3 qt
17 gal 2 qt

10. 12 yd 1 ft 5 in
+ 18 yd 2 ft 10 in
31 yd 1 ft 3 in

11. 4 yd 2 ft
− 1 yd 1 ft
3 yd 1 ft

12. 19 hr 49 min
− 15 hr 32 min
4 hr 17 min

13. 17 gal 1 qt
− 13 gal 2 qt
3 gal 3 qt

14. 22 da 2 hr 36 min
− 21 da 22 hr 48 min
3 hr 48 min

15. 6 yd 1 ft 1 in
− 3 yd 2 ft 2 in
2 yd 1 ft 11 in

16. 4 × (3 hr 10 min)
12 hr 40 min

17. 10 × (3 gal 2 qt)
35 gal

18. 6 × (7 yd 1 ft 8 in)
45 yd 1 ft

19. 3 × (5 hr 40 min 25 sec)
17 hr 1 min 15 sec

20. 14 in × 22 in
308 sq in

21. 5 yd × 9 yd
45 sq yd

22. (7 lb 8 oz) ÷ 3
2 lb 8 oz

23. (10 hr 20 min) ÷ 4
2 hr 35 min

24. 33 ft ÷ 11 ft
3

Solve.

25. Steak costs $3 per pound. How many pounds can Ralph buy with $18?   **6 lb**

26. For a company picnic, Ward brought 7 gal 2 qt of iced tea. If Sarah brought 10 gal 3 qt, how much tea did they have?   **18 gal 1 qt**

27. If the people at the picnic in Exercise 26 drank 15 gal 3 qt of the tea, how much was left?
2 gal 2 qt

28. Billy Joe has a chili recipe that requires 2 lb 6 oz of ground beef. How much meat should he buy to make 5 times his recipe?   **11 lb 14 oz**

29. Vicki needs to paint a wall which is 8 feet high and 12 feet wide. How many square feet of wall is this to paint?   **96 sq ft**

30. A cookie recipe requires 1 lb 2 oz of flour. How much flour would be used if only one third of the recipe is to be made?   **6 oz**

FOR REVIEW

Complete the unit conversion.

31. 0.5 ton = ___1000___ lb

32. $33 \frac{\text{ft}}{\text{sec}} = \underline{22.5} \frac{\text{mi}}{\text{hr}}$

33. $200 \frac{\text{cents}}{\text{in}} = \underline{24} \frac{\text{dollars}}{\text{ft}}$

8.2 EXERCISES C

Solve.

1. If material costs $2.70 per yard, how much does 8 feet of the material cost?
[Answer: $7.20]

2. Wire costs $13.85 per yard. How much does the wire cost per foot (to the nearest cent)?   **$4.62**

8.3 THE METRIC SYSTEM

STUDENT GUIDEPOSTS

1 Length in Meters
2 Volume in Liters
3 Weight in Grams
4 Units of Measure

Most of the world uses the metric system of measurement. It is a decimal system, and changes in units can be made by moving the decimal point. Thus, it is much easier to convert from one unit to another in the metric system than it is in the English system (just as easy as changing $1.27 to 127¢).

1 LENGTH IN METERS

The standard of length in the metric system is the **meter.** The opening in a normal doorway is about 2 meters high. Thus, a meter is about half the height of an opening in a doorway. A twin bed is about a meter wide and most newborn babies are about $\frac{1}{2}$ meter tall.

The actual length of the scale in Figure 8.1 is 0.1 meter. The distance between each mark is 0.01 meter.

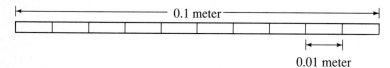

0.1 meter

0.01 meter

Figure 8.1

2 VOLUME IN LITERS

If we make a box as shown in Figure 8.2 (not actual size), the box holds one **liter.** This is the standard unit for capacity in the metric system. If you are thirsty on a summer day, you can probably drink a liter of liquid. It is a little less than three cans of soft drink.

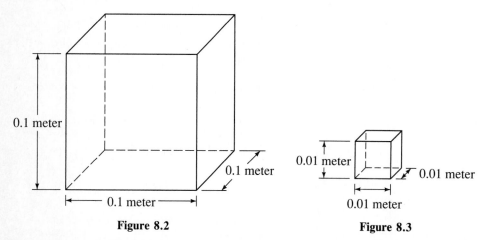

0.1 meter
0.1 meter
0.1 meter

Figure 8.2

0.01 meter
0.01 meter
0.01 meter

Figure 8.3

❸ WEIGHT IN GRAMS

The standard for mass or weight at sea level is the **gram.** Figure 8.3 shows a cube of side 0.01 meter. If this cube were filled with water, the weight of the water would be 1 gram. One large paper clip (or two small ones) weighs about one gram and a large steak might weigh 1000 grams.

❹ UNITS OF MEASURE

As with the English system, the various size units such as 1000 grams have names, and the naming is standard. For example, the names for 1000 meters, 1000 liters, and 1000 grams all start with the same prefix, *kilo*. Thus,

$$1000 \text{ meters} = 1 \text{ } kilo\text{meter,}$$
$$1000 \text{ liters} \;\; = 1 \text{ } kilo\text{liter,}$$
$$1000 \text{ grams} = 1 \text{ } kilo\text{gram.}$$

The following table lists the prefixes and symbols used in the metric system.

Number of units	1000	100	10	1	$\frac{1}{10}$ or 0.1	$\frac{1}{100}$ or 0.01	$\frac{1}{1000}$ or 0.001
Prefix	kilo	hecto	deka	unit	deci	centi	milli
Length	km	hm	dam	meter(m)	dm	cm	mm
Volume	kl	hl	dal	liter(L)	dl	cl	ml
Weight	kg	hg	dag	gram(g)	dg	cg	mg

The different size units are used in different applications. For example, the size of a room is given in meters, the distance between Miami and Washington, D.C., in kilometers, and bolt sizes in millimeters. In Figure 8.4, the paper clip and bolt are shown actual size.

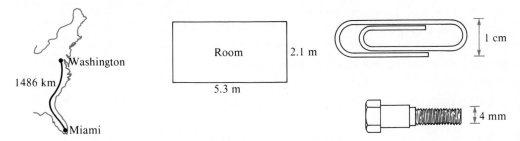

Figure 8.4

In Figure 8.5 (drawn to actual size), Figure 8.1 has been repeated using the new names we have learned and a few more details. This should give you a good idea of the sizes of some units of length and their relationship. A meter is 10 times the total length of the scale in Figure 8.5. A kilometer is about the length of five city blocks.

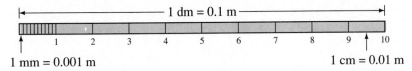

Figure 8.5

The most commonly used measures of length are the kilometer, meter, centimeter, and millimeter. These same prefixes—kilo, centi, and milli—are used for volume and weight also.

In volume units, the liter is used to measure gasoline and beverages. The kiloliter measures the capacity of large tanks, and the milliliter measures liquids in a laboratory.

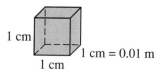

1 cm = 0.01 m

Figure 8.6

The cube in Figure 8.6, which is 1 cm on a side, has volume 1 ml and this much water weighs 1 g. Thus, the gram is a small unit of measurement. It is used in the laboratory where an even smaller unit, the milligram, is also needed. The unit that would suit many common weight applications is the kilogram. As has been mentioned, a large steak weighs about 1 kg (1000 g).

Now that we have some idea of the size of metric units, we turn to conversion problems. Instead of using decimal conversion units such as 1 mm = 0.001 m, it will be easier to write all unit fractions in terms of 10, 100, or 1000. The following table gives the equal units.

Length	Volume	Weight
1 km = 1000 m	1 kl = 1000 L	1 kg = 1000 g
1 hm = 100 m	1 hl = 100 L	1 hg = 100 g
1 dam = 10 m	1 dal = 10 L	1 dag = 10 g
1 m = 10 dm	1 L = 10 dl	1 g = 10 dg
1 m = 100 cm	1 L = 100 cl	1 g = 100 cg
1 m = 1000 mm	1 L = 1000 ml	1 g = 1000 mg

HINT

Remember that multiplying and dividing by powers of ten can be done by moving the decimal point. Consider these examples.

$$16.2 \times 100 = 1620 \qquad \text{Move two places to the right}$$

$$16.2 \div 100 = 0.162 \qquad \text{Move two places to the left}$$

EXAMPLE 1 CONVERTING METERS TO CENTIMETERS

5.2 m = _____ cm.

Since 1 m = 100 cm, the unit fraction is $\frac{100 \text{ cm}}{1 \text{ m}}$.

$$5.2 \text{ m} = 5.2 \text{ m} \times 1$$
$$= \frac{5.2 \text{ m}}{1} \times \frac{100 \text{ cm}}{1 \text{ m}}$$
$$= 5.2 \times 100 \text{ cm}$$
$$= 520 \text{ cm}$$

PRACTICE EXERCISE 1

13.25 m = _____ cm.

5.2 is multiplied by 100 which moves the decimal point 2 places to the right. That is,

$$5.2 \text{ m} = 520. \text{ cm.}$$
$$\overset{\smile}{}2$$

Answer: 1325

| **EXAMPLE 2** CONVERTING MILLILITERS TO DECILITERS | **PRACTICE EXERCISE 2** |

362 ml = _____ dl.

 The unit fractions are $\frac{1\text{ L}}{1000\text{ ml}}$ and $\frac{10\text{ dl}}{1\text{ L}}$.

$$362 \text{ ml} = \frac{362 \text{ ml}}{1} \times \frac{1 \text{ L}}{1000 \text{ ml}} \times \frac{10 \text{ dl}}{1 \text{ L}}$$

$$= 362 \times \frac{1}{1000} \times \frac{10}{1} \text{ dl}$$

$$= 362 \times \frac{10}{1000} \text{ dl}$$

$$= 362 \times \frac{1}{100} \text{ dl}$$

$$= 3.62 \text{ dl}$$

362 is divided by 100 which moves the decimal point 2 places to the left.

64,800 ml = _____ dl.

Answer: 648

 In Example 2, multiplying by one unit fraction moved the decimal point 3 places to the left and multiplying by the other moved it 1 place to the right. The same change of units could have been made by moving the decimal point 2 places to the left. This method of conversion is shown in the next example.

| **EXAMPLE 3** CONVERTING BY MOVING DECIMAL PLACES | **PRACTICE EXERCISE 3** |

362 ml = _____ dl.

 First show all the units in a diagram.

1000 L	100 L	10 L		0.1 L	0.01 L	0.001 L
(kl)	**(hl)**	**(dal)**	**1 L**	**(dl)**	**(cl)**	**(ml)**

Start at **ml** and make 2 moves to the *left* to get to **dl.** Thus, to change ml to dl move the decimal point 2 places to the *left*.

$$362 \text{ ml} = 3.62 \text{ dl} = 3.62 \text{ dl}$$

0.057 ml = _____ dl.

Answer: 0.00057

| **EXAMPLE 4** CONVERTING HECTOGRAMS TO CENTIGRAMS | **PRACTICE EXERCISE 4** |

0.0256 hg = _____ cg.

1000 g	100 g	10 g		0.1 g	0.01 g	0.001 g
(kg)	**(hg)**	**(dag)**	**1 g**	**(dg)**	**(cg)**	**(mg)**

0.000882 hg = _____ cg.

Start at hg and move 4 times to the *right* to get to cg. Thus, to change hg to cg, move the decimal point 4 places to the *right*.

$$0.0246 \text{ hg} = \underset{4}{00256.} \text{ cg} = 256 \text{ cg}$$

Answer: 8.82

EXAMPLE 5 CONVERTING MILLIMETERS TO HECTOMETERS	**PRACTICE EXERCISE 5**

3280 mm = _____ hm.

426,000 mm = _____ hm.

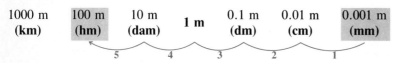

Start at mm and make 5 moves to the *left* to get to hm. Thus, to change mm to hm, move the decimal point 5 places to the *left*.

$$3280 \text{ mm} = \underset{5}{0.03280} \text{ hm}$$

Answer: 4.26

The time units for the metric system are the same as for the English system. However, the temperature scale is different. Both metric and English temperature will be discussed in the next section when we convert from one system to the other.

8.3 EXERCISES A

Answer true or false. If the answer is false, tell why.

1. The system of measurement used in most of the world is the metric system.   true

2. The opening in a normal doorway is about 10 meters high.   false (about 2 m)

3. A liter is the space in a box which is 0.1 meter or one decimeter on a side.   true

4. The weight of water that can be put in a box which is 0.01 meter (1 centimeter) on a side is one gram.   true

5. A large steak might weigh about one gram.   false (about one kilogram)

6. To change units of length in the metric system, just move the decimal point.   true

Complete the following by moving the decimal point as shown in the examples.

7. 1 km = _____1000_____ m

8. 1 hl = _____100_____ L

9. 1 dag = _____10_____ g

10. 1 m = _____10_____ dm

11. 1 mg = _____0.001_____ g

12. 1 g = _____1000_____ mg

13. 62 g = _____0.062_____ kg

14. 3.2 cm = _____0.032_____ m

15. 16 dal = _____160_____ L

16. 10.3 cm = _____0.00103_____ hm

17 0.00721 kg = _____0.721_____ dag

18. 63,000 ml = _____0.063_____ kl

19. 1700 dm = _____1.7_____ hm

20. 16.25 cg = _____162.5_____ mg

21. 0.000212 hl = _____21.2_____ ml

22. 92.6 mm = _____9.26_____ cm

㉓ 4728 dg = _____4.728_____ hg

24. 5,000,000 cl = _____5000_____ dal

25. 0.0052 hm = _____520_____ mm

26. 0.07 dag = _____700_____ mg

27. 3200 dl = _____0.32_____ kl

28. 0.0953 km = _____953_____ dm

29. 42 kg = _____420_____ hg

30. 7.2 hl = _____72,000_____ cl

31. $30 \frac{cm}{sec}$ = _____1080_____ $\frac{m}{hr}$

㉜ $20 \frac{cg}{da}$ = _____1.4_____ $\frac{g}{wk}$

33. $360 \frac{kl}{hr}$ = _____6,000,000_____ $\frac{ml}{min}$

Perform the indicated operations.

34.   2.05 m
     + 7.97 m   10.02 m

35.   382.6 ml
     + 23.8 ml   406.4 ml

36.   983 kg
     − 724 kg   259 kg

37.   75.36 cm
     − 28.71 cm   46.65 cm

38. 16.3 m
     × 5   81.5 m

39. 16.3 m
     × 5 m   81.5 m²

Solve.

40. How many 22-ml samples can be obtained from 330 ml of a chemical solution?   15

㊶ If steak costs $6 per kilogram, how much would 600 grams of steak cost? $3.60

42. Gabi is traveling at $60 \frac{km}{hr}$. How many meters per minute is this? $1000 \frac{m}{min}$

43. A rancher needs 10 km of wire. How much will it cost him if the wire is 40¢ per meter? $4000

FOR REVIEW

Perform the indicated operations and simplify.

44.   8 ft 11 in
     + 1 ft 3 in
     ‾‾‾‾‾‾‾‾‾‾‾‾
     10 ft 2 in

45.   8 lb 12 oz
     − 3 lb 15 oz
     ‾‾‾‾‾‾‾‾‾‾‾‾
     4 lb 13 oz

46.   12 gal 2 qt 1 pt
     − 6 gal 3 qt 1 pt
     ‾‾‾‾‾‾‾‾‾‾‾‾‾‾‾‾‾
     5 gal 3 qt

47. 5 × (7 yd 2 ft 10 in)
     39 yd 2 ft 2 in

48. (11 hr 40 min) ÷ 5
     2 hr 20 min

49. 42 ft ÷ 7 ft   6

ANSWERS: 1. true 2. false (about 2 m) 3. true 4. true 5. false (about one kilogram) 6. true 7. 1000 8. 100 9. 10 10. 10 11. 0.001 12. 1000 13. 0.062 14. 0.032 15. 160 16. 0.00103 17. 0.721 18. 0.063 19. 1.7 20. 162.5 21. 21.2 22. 9.26 23. 4.728 24. 5000 25. 520 26. 700 27. 0.32 28. 953 29. 420 30. 72,000 31. 1080 32. 1.4 33. 6,000,000 34. 10.02 m 35. 406.4 ml 36. 259 kg 37. 46.65 cm 38. 81.5 m 39. 81.5 m² (m × m = m²) 40. 15 41. $3.60 42. $1000 \frac{m}{min}$ 43. $4000 44. 10 ft 2 in 45. 4 lb 13 oz 46. 5 gal 3 qt 47. 39 yd 2 ft 2 in 48. 2 hr 20 min 49. 6

8.3 EXERCISES B

Answer true or false. If the answer is false, tell why.

1. The United States is one of the few countries in the world that uses the English system.   **true**

2. The opening in a normal doorway is about 2 meters high.   **true**

3. The liter is used to measure gasoline in the metric system.   **true**

4. One gram of meat is more than one pound of meat.   **false**

5. A large steak might weigh about one kilogram.   **true**

6. Units of weight in the metric system can be changed by moving the decimal point.   **true**

Complete the following by moving the decimal point as shown in the examples.

7. 40 km = _____40,000_____ m

8. 6 hl = _____600_____ L

9. 0.05 dag = _____0.5_____ g

10. 50 m = _____500_____ dm

11. 0.061 mg = _____0.000061_____ g

12. 0.061 g = _____61_____ mg

13. 420 g = _____0.42_____ kg

14. 2300 cm = _____23_____ m

15. 2.5 dal = _____25_____ L

16. 6320 cm = _____0.632_____ hm

17. 8.7 kg = _____870_____ dag

18. 493,000 ml = _____0.493_____ kl

19. 0.75 dm = _____0.00075_____ hm

20. 0.88 cg = _____8.8_____ mg

21. 5.5 hl = _____550,000_____ ml

22. 250 mm = _____25_____ cm

23. 98,750 dg = _____98.75_____ hg

24. 0.44 cl = _____0.00044_____ dal

25. 927 hm = _____92,700,000_____ mm

26. 0.000082 dag = _____0.82_____ mg

27. 200 dl = _____0.02_____ kl

28. 0.000024 km = _____0.24_____ dm

29. 0.081 kg = _____0.81_____ hg

30. 4400 hl = _____44,000,000_____ cl

31. $2\,\dfrac{cm}{sec}$ = _____72_____ $\dfrac{m}{hr}$

32. $500\,\dfrac{cg}{da}$ = _____35_____ $\dfrac{g}{wk}$

33. $0.06\,\dfrac{kl}{hr}$ = _____1000_____ $\dfrac{ml}{min}$

Perform the indicated operations.

34.   8.09 km
   + 3.66 km   **11.75 km**

35.   24.89 dl
   + 13.77 dl   **38.66 dl**

36.   402 g
   − 389 g   **13 g**

37.   4.735 m
   − 1.908 m   **2.827 m**

38. 213.2 cm
   × 8   **1705.6 cm**

39. 213.2 cm
   × 8 cm   **1705.6 cm²**

Solve.

40. Dr. Warren has 42 g of salt. How many 7-dg samples can she get from her supply?   **60**

41. Peaches are marked at $1.20 per kilogram. How much will 4000 g of peaches cost?   **$4.80**

42. A tank is being filled at the rate of 25 liters per minute. What is this in kiloliters per hour? $1.5\,\frac{kl}{hr}$

43. A perfume is worth $2.60 per milliliter. How much would 5 liters cost?   **$13,000**

FOR REVIEW

Perform the indicated operations and simplify.

44. 29 ft 7 in
   + 18 ft 6 in
   ──────────
   48 ft 1 in

45. 2 lb 3 oz
   − 1 lb 7 oz
   ──────────
   12 oz

46. 8 gal 1 qt
   − 2 gal 3 qt 1 pt
   ──────────
   5 gal 1 qt 1 pt

47. $8 \times (4 \text{ yd } 1 \text{ ft } 6 \text{ in})$
   36 yd

48. $(15 \text{ hr } 36 \text{ min}) \div 6$
   2 hr 36 min

49. $132 \text{ ft} \div 11 \text{ ft}$
   12

8.3 EXERCISES C

Complete the following using an appropriate diagram.

1. $50 \dfrac{\text{dl}}{\text{min}} = \underline{\quad} \dfrac{\text{hl}}{\text{hr}}$
   [Answer: 3]

2. $140 \dfrac{\text{kg}}{\text{wk}} = \underline{\quad} \dfrac{\text{dag}}{\text{hr}}$
   $83.\overline{3}$

3. $120 \dfrac{\text{cm}}{\text{hr}} = \underline{\quad} \dfrac{\text{km}}{\text{sec}}$
   0.0000003

4. If 1 mi = 1.61 km and a river is flowing at a rate of $4.2\frac{\text{mi}}{\text{hr}}$, how fast is it flowing in kilometers per hour?
   $\left[\text{Answer: } 6.762\frac{\text{km}}{\text{hr}}\right]$

8.4 CONVERSIONS BETWEEN MEASUREMENT SYSTEMS

STUDENT GUIDEPOSTS

1 Converting Units of Length, Volume, and Weight

2 Speed in Both Systems

3 Temperature in Both Systems

1 CONVERTING UNITS OF LENGTH, VOLUME, AND WEIGHT

If the United States ever changes to the metric system, it will not be necessary to change units from one system to another. But while both systems are in use it is important to know the relationship between the two.

In the table below are some useful conversion units of measure. Notice that these units are all approximations.

Length, Volume, and Weight		
	Metric to English	*English to Metric*
Length units	1 km ≈ 0.621 mi	1 mi ≈ 1.61 km
	1 m ≈ 39.37 in	1 in ≈ 2.54 cm
Volume units	1 L ≈ 1.06 qt	1 qt ≈ 0.946 L
Weight units	1 kg ≈ 2.20 lb	1 lb ≈ 454 g

HINT

Since the conversion units are approximate, we can get slightly different answers by converting different ways. For example,

$$111 \text{ km} \approx \frac{111 \text{ km}}{1} \times \frac{0.621 \text{ mi}}{1 \text{ km}}$$

$$\approx 68.93 \text{ mi.}$$

However,

$$111 \text{ km} \approx 111 \text{ km} \times \frac{1 \text{ mi}}{1.61 \text{ km}}$$

$$\approx 68.94 \text{ mi.}$$

Do not be concerned with small differences like this in answers.

Since we are familiar with the English units, these conversions will help us get a better idea of the size of the metric units. For example, on a quarter-mile track, a kilometer is about 2.5 times around. If two towns are 10 miles apart, they are about 16 km apart. A meter (39.37 in) is a little more than a yard (36 in). Figure 8.7 shows the relationship between centimeters and inches.

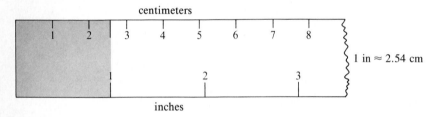

Figure 8.7

There is a little more milk in a liter container than there is in a quart, as shown in Figure 8.8.

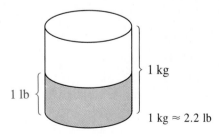

Figure 8.8

We now know that the 1-kg steak mentioned in Section 8.3 weighs 2.2 lb. Figure 8.9 shows the relative sizes of 1 lb and 1 kg of the same material.

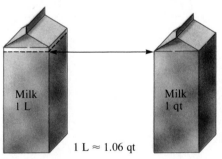

Figure 8.9

The following examples show some conversions between the two systems. In most of these problems a calculator is helpful.

| EXAMPLE 1 CONVERTING MEASURES OF LENGTH | PRACTICE EXERCISE 1 |

(a) 20 km = _____ mi.

Since 0.621 mi ≈ 1 km, the unit fraction is $\frac{0.621 \text{ mi}}{1 \text{ km}}$.

$$20 \text{ km} \approx \frac{20 \text{ km}}{1} \times \boxed{\frac{0.621 \text{ mi}}{1 \text{ km}}} \qquad \text{Multiply by the unit fraction}$$

$$= 20 \times (0.621) \text{ mi}$$

$$\approx 12.4 \text{ mi} \qquad \text{Rounded to the nearest tenth}$$

(b) 35 mi = _____ km.

Multiply by the unit fraction $\frac{1.61 \text{ km}}{1 \text{ mi}}$.

$$35 \text{ mi} \approx \frac{35 \text{ mi}}{1} \times \boxed{\frac{1.61 \text{ km}}{1 \text{ mi}}} \qquad \text{Multiply by the unit fraction}$$

$$= 35 \times 1.61 \text{ km}$$

$$\approx 56.4 \text{ km} \qquad \text{Rounded to the nearest tenth}$$

(c) $18\frac{3}{4}$ in = _____ cm.

The unit fraction is $\frac{2.54 \text{ cm}}{1 \text{ in}}$.

$$18\frac{3}{4} = 18.75 \text{ in}$$

$$\approx \frac{18.75 \text{ in}}{1} \times \frac{2.54 \text{ cm}}{1 \text{ in}}$$

$$= (18.75)(2.54) \text{ cm}$$

$$\approx 47.63 \text{ cm} \qquad \text{Rounded to the nearest hundredth}$$

PRACTICE EXERCISE 1

(a) 44 km = _____ mi.

(b) 120 mi = _____ km.

(c) $216\frac{1}{2}$ in = _____ cm.

Answers: **(a)** 27.3 **(b)** 193.2 **(c)** 549.9

| EXAMPLE 2 CONVERSION OF SPEED | PRACTICE EXERCISE 2 |

Joe is driving at a rate of $95\frac{\text{km}}{\text{hr}}$. What is his speed in miles per hour?

$$95\frac{\text{km}}{\text{hr}} \approx \frac{95 \text{ km}}{\text{hr}} \times \frac{0.621 \text{ mi}}{1 \text{ km}}$$

$$\approx 59\frac{\text{mi}}{\text{hr}} \qquad \text{Rounded to the nearest mile per hour}$$

PRACTICE EXERCISE 2

Vera is driving $50\frac{\text{km}}{\text{hr}}$ in a $30\frac{\text{mi}}{\text{hr}}$ zone. Approximately how many miles per hour is she traveling over the speed limit?

Answer: She is about $1\frac{\text{mi}}{\text{hr}}$ over the speed limit.

2 SPEED IN BOTH SYSTEMS

Figure 8.10 shows a comparison of some common speed limits. Approximations are to the nearest kilometer per hour. A speed of $55\frac{\text{mi}}{\text{hr}}$ is about $89\frac{\text{km}}{\text{hr}}$.

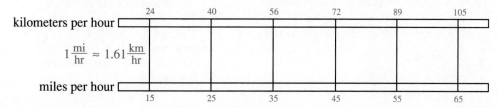

Figure 8.10

EXAMPLE 3 CONVERSION OF MILLILITERS TO QUARTS

520 ml = _____ qt.

Since the unit fraction that we know is $\frac{1.06 \text{ qt}}{1 \text{ L}}$, first change 520 ml to liters.

1 L	0.1 L (dl)	0.01 L (cl)	0.001 L (ml)

$$520 \text{ ml} = 0.520 \text{ L}$$

$$0.520 \text{ L} \approx \frac{0.52 \text{ L}}{1} \times \frac{1.06 \text{ qt}}{1 \text{ L}}$$

$$= (0.52)(1.06) \text{ qt}$$

$$\approx 0.55 \text{ qt} \qquad \text{Rounded to the nearest hundredth}$$

PRACTICE EXERCISE 3

1020 ml = _____ qt.

Answer: 1.08

EXAMPLE 4 APPLICATION OF VOLUME CONVERSION

The gasoline tank on Roberto's car holds 12.6 gal. How many liters will it hold?

$$12.6 \text{ gal} = \frac{12.6 \text{ gal}}{1} \times \frac{4 \text{ qt}}{1 \text{ gal}} \qquad \text{Change 12.6 gal to quarts}$$

$$= (12.6)(4) \text{ qt}$$

$$= 50.4 \text{ qt}$$

$$50.4 \text{ qt} \approx \frac{50.4 \text{ qt}}{1} \times \frac{0.946 \text{ L}}{1 \text{ qt}} \qquad \text{Change quarts to liters}$$

$$= (50.4)(0.946) \text{ L}$$

$$\approx 47.7 \text{ L} \qquad \text{To the nearest tenth}$$

PRACTICE EXERCISE 4

Deb has a foreign car with a 40-L gas tank. How many gallons does the tank hold?

Answer: 10.6 gal

EXAMPLE 5 WEIGHT CONVERSION

68.2 hg = _____ lb.

The unit fraction is $\frac{2.20 \text{ lb}}{1 \text{ kg}}$. First, change 62.8 hg to kilograms.

1000 g (1 kg)	100 g (1 hg)

68.2 hg = 6.82 kg

$$6.82 \text{ kg} \approx \frac{6.82 \text{ kg}}{1} \times \frac{2.20 \text{ lb}}{1 \text{ kg}} \qquad \text{Change kilograms to pounds}$$

$$= (6.82)(2.20) \text{ lb}$$

$$= 15.0 \text{ lb} \qquad \text{To the nearest tenth}$$

PRACTICE EXERCISE 5

2240 dg = _____ lb.

Answer: 0.49

EXAMPLE 6 APPLICATION OF WEIGHT CONVERSION

A football team's starting fullback weighs 200 lb. How many kilograms does he weigh?

$$200 \text{ lb} \approx \frac{200 \text{ lb}}{1} \times \frac{454 \text{ g}}{1 \text{ lb}}$$

$$= (200)(454) \text{ g}$$

$$= 90,800 \text{ g}$$

PRACTICE EXERCISE 6

A foreign basketball player trying to break into the NBA reported his weight as 97 kg. How many pounds does he weigh?

Change 90,800 g to kilograms.

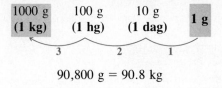

$$90{,}800 \text{ g} = 90.8 \text{ kg}$$

Answer: 213 lb

The following table gives comparisons of common weights of people in pounds and kilograms (to the nearest kilogram).

Pounds	Kilograms
100	45
125	57
150	68
175	79
200	91
225	102
250	114

3 TEMPERATURE IN BOTH SYSTEMS

The temperature scale used in the metric system is the **Celsius** (°C) scale. Water freezes at 0°C (32°F) and boils at 100°C (212°F). Figure 8.11 shows a comparison between degrees **Fahrenheit** (°F) and degrees Celsius. (In the past Celsius was called *centigrade*.)

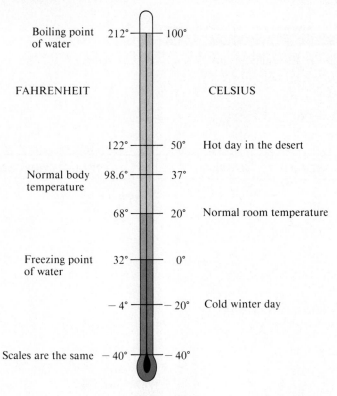

Figure 8.11

The equal temperatures on the scale in Figure 8.11 can be found once we know the relationship between the two scales. For example, using the following rules we can find that 20°C = 68°F.

To Convert from °C to °F

Multiply C, the temperature in °C, by $\frac{9}{5} = 1.8$ and add 32.

$$F = \frac{9}{5}C + 32 \qquad \text{or} \qquad F = 1.8C + 32$$

EXAMPLE 7 CONVERTING CELSIUS TO FAHRENHEIT	PRACTICE EXERCISE 7

20°C = _____ °F.

$$F = \frac{9}{5}C + 32$$

$$F = \frac{9}{5}20 + 32 \qquad C = 20$$

$$= \frac{9 \cdot 20}{5} + 32 \qquad \text{Multiply before adding}$$

$$= \frac{9 \cdot 4 \cdot 5}{5} + 32$$

$$= 36 + 32 = 68$$

Thus, 20°C is 68°F.

35°C = _____ °F.

Answer: 95

To Convert from °F to °C

First subtract 32 from F, the temperature in °F, then multiply by $\frac{5}{9}$.

$$C = \frac{5}{9}(F - 32)$$

EXAMPLE 8 CONVERTING FAHRENHEIT TO CELSIUS	PRACTICE EXERCISE 8

122°F = _____ °C.

$$C = \frac{5}{9}(F - 32)$$

$$= \frac{5}{9}(122 - 32) \qquad F = 122$$

$$= \frac{5}{9}(90) \qquad \text{Subtract inside parentheses first}$$

$$= \frac{5 \cdot \cancel{9} \cdot 10}{\cancel{9}} = 50 \qquad \text{Then multiply}$$

Thus, 122°F is 50°C.

86°F = _____ °C.

Answer: 30

| **EXAMPLE 9**   APPLICATION OF TEMPERATURE CONVERSION | **PRACTICE EXERCISE 9** |

You are going to a ballgame and the weatherman says the temperature will be 30°C. What kind of clothes should you wear?

If you have a better understanding of °F, change °C to °F using

$$F = \frac{9}{5} \cdot C + 32.$$

$$F = \frac{9}{5} \cdot 30 + 32 \qquad C = 30$$

$$= \frac{9 \cdot 30}{5} + 32$$

$$= \frac{9 \cdot \cancel{5} \cdot 6}{\cancel{5}} + 32$$

$$= 54 + 32 = 86$$

Thus, 30°C = 86°F, and you should wear light clothing.

Would you go swimming in water which has a temperature of 25°C?

Answer: 77°F water would be fine for swimming.

8.4 EXERCISES A

1. Which is longer, 1 mi or 1 km?   **1 mile**

2. Which is longer, 3 cm or 1 in?   **3 cm**

3. Which is more, 1 qt or 1 L?   **1 L**

4. Which is more, 1 kg or 3 lb?   **3 lb**

Complete.

5. 16 km = _____**9.9**_____ mi

6. 128 mi = _____**206**_____ km

7. 2.6 m = _____**102.4**_____ in

8. 72 in = _____**183**_____ cm

9. 3 km = _____**9837**_____ ft

10. 3.2 ft = _____**97.5**_____ cm

11. $80 \frac{km}{hr} = $ _____**50**_____ $\frac{mi}{hr}$

12. $20 \frac{mi}{hr} = $ _____**32**_____ $\frac{km}{hr}$

13. 20 qt = _____**18.9**_____ L

14 40 L = _____**10.6**_____ gal

15. 50 kg = _____**110**_____ lb

16. 5 lb = _____**2270**_____ g

17. 0.25 kg = _____**8.8**_____ oz

18. 625 lb = _____**284**_____ kg

19 40°C = _____**104**_____ °F

20. 50°F = _____**10**_____ °C

21. 200°C = _____**392**_____ °F

22 200°F = _____**$93\frac{1}{3}$**_____ °C

Solve.

23. Patricia bought a French car with the capacity of the gas tank given as 48 L. How many gallons of gasoline will the tank hold?   **12.7 gal**

24. Winston was driving in Mexico where the speed limit was $100 \frac{km}{hr}$. What is this in miles per hour?   **$62\frac{mi}{hr}$**

25. The average weight of the starting line of a football team is 258 lb. What is this in kilograms?
117 kg

26. Would you put your hand in water at 95°C?
no (95°C = 203°F)

FOR REVIEW

Complete the following using an appropriate diagram.

27. 720 mm = _____72_____ cm

28. 0.45 g = _____450_____ mg

29. 0.0021 kl = _____2100_____ ml

ANSWERS: (Some answers are rounded.) 1. 1 mile 2. 3 cm 3. 1 L 4. 3 lb 5. 9.9 6. 206 7. 102.4 8. 183 9. 9837 10. 97.5 11. 50 12. 32 13. 18.9 14. 10.6 15. 110 16. 2270 17. 8.8 18. 284 19. 104 20. 10 21. 392 22. 93 23. 12.7 gal 24. 62 $\frac{mi}{hr}$ 25. 117 kg 26. No (95°C = 203°F) 27. 72 28. 450 29. 2100

8.4 EXERCISES B

1. Which is longer, 1 dm or 1 in? **1 dm**

2. Which is longer, 10 mi or 17 km? **17 km**

3. Which is more, 1 gal or 4 L? **4 L**

4. Which is more, 500 g or 1 lb? **500 g**

Complete.

5. 125 km = _____77.6_____ mi

6. 48 mi = _____77.3_____ km

7. 0.056 m = _____2.2_____ in

8. 9.5 in = _____24.1_____ cm

9. 0.05 km = _____164_____ ft

10. 0.15 ft = _____4.57_____ cm

11. 25 $\frac{km}{hr}$ = _____15.5_____ $\frac{mi}{hr}$

12. 100 $\frac{mi}{hr}$ = _____161_____ $\frac{km}{hr}$

13. 0.5 qt = _____0.47_____ L

14. 8.5 L = _____2.25_____ gal

15. 0.2 kg = _____0.44_____ lb

16. $\frac{1}{2}$ lb = _____227_____ g

17. 0.0046 kg = _____0.16_____ oz

18. 85 lb = _____38.6_____ kg

19. 15°C = _____59_____ °F

20. 77°F = _____25_____ °C

21. 1000°C = _____1832_____ °F

22. 1000°F = _____$537\frac{7}{9}$_____ °C

Solve.

23. How many liters will a 19-gal gas tank hold?
72 L

24. Patricia's speedometer reads in kilometers per hour. How fast can she legally drive where the speed limit is 45 $\frac{mi}{hr}$? **72 $\frac{km}{hr}$**

25. Willy must have his weight down to 82 kg to compete in Europe. What is this in pounds?
180 lb

26. Instructions call for a temperature control to be at 65°C. What is this in °F?
149°F

FOR REVIEW

Complete the following using an appropriate diagram.

27. 0.5 mm = _____0.05_____ cm

28. 122 g = _____122,000_____ mg

29. 0.111 kl = _____111,000_____ ml

8.4 EXERCISES C

Use a calculator to complete the following, and give answers rounded to the nearest hundredth.

1. 17.65 pt = _____ L
[Answer: 8.35]

2. $3\frac{1}{9}$ m = _____ **122.48** in

3. 4.65 kg = _____ **163.68** oz

4. 75 $\frac{km}{hr}$ = _____ **68.31** $\frac{ft}{sec}$

5. 77.2°C = _____ °F
[Answer: 170.96]

6. 15.5°F = _____ **−9.17** °C

CHAPTER 8 REVIEW

KEY WORDS

8.1 The **English system** is the measurement system used in the United States.

A **unit fraction** is a fraction that has value 1 and is used to make measurement conversions.

8.3 The **metric system** is the system of measurement used in most of the countries of the world. The **meter** is the basic unit for length,

the **liter** is the basic unit for volume, and the **gram** is the basic unit for weight.

8.4 The **Celsius** scale is the temperature scale used in the metric system.

The **Fahrenheit** scale is the temperature scale used in the English system.

KEY CONCEPTS

8.3 **1.** In the metric system, conversions in units can be made by moving the decimal point.

2. A meter is about half the height of a doorway opening. A liter is the volume of a cube 0.1 m (1 dm) on a side. A gram is the weight of water that can be put in a cube 0.01 m (1 cm) on a side.

3.

Prefix	Symbol	Multiply standard unit by
kilo	k	1000
hecto	h	100
deka	da	10
deci	d	$\frac{1}{10} = 0.1$
centi	c	$\frac{1}{100} = 0.01$
milli	m	$\frac{1}{1000} = 0.001$

8.4 **1.** To change °C to °F, use $F = \frac{9}{5}C + 32$.

2. To change °F to °C, use $C = \frac{5}{9}(F - 32)$.

REVIEW EXERCISES

Part I

8.1 *Complete.*

1. 7 ft = _____ **84** in

2. 30 in = _____ **2.5** ft

3. 48 fl oz = _____ **3** pt

4. 90 min = _____ **1.5** hr

5. 3 ton = _____ **6000** lb

6. 18 pt = _____ **2.25** gal

8.2 *Perform the indicated operations and simplify.*

7. 10 ft 3 in
  + 7 ft 2 in
  _____
  17 ft 5 in

8. 10 ft 3 in
  − 7 ft 2 in
  _____
  3 ft 1 in

9. 18 hr 22 min
  + 5 hr 56 min
  _____
  24 hr 18 min

10. 18 hr 22 min
  − 5 hr 56 min
  _____
  12 hr 26 min

11. 12 yd 2 ft 2 in
  + 6 yd 2 ft 8 in
  _____
  19 yd 1 ft 10 in

12. 12 yd 2 ft 2 in
  − 6 yd 2 ft 8 in
  _____
  5 yd 2 ft 6 in

13. $3 \times$ (5 gal 3 qt 1 pt)
  17 gal 2 qt 1 pt

14. 8 ft $\times$ 3 ft
  24 ft^2

15. (8 lb 13 oz) $\div$ 3
  2 lb 15 oz

16. 18 dollars $\div$ 3 dollars
  6

8.3 *Complete.*

17. 860 m = _____0.86_____ km

18. 0.582 m = _____58.2_____ cm

19. 483 ml = _____4.83_____ dl

20. 48,300 mg = _____0.0483_____ kg

21. 0.00035 km = _____3.5_____ dm

22. 5.6 L = _____560_____ cl

Perform the indicated operations.

23. 15.2 m
  + 75.3 m
  _____
  90.5 m

24. 831.6 L
  − 416.9 L
  _____
  414.7 L

25. 42.8 g
  $\times$ 7
  _____
  299.6 g

8.4 *Complete.*

26. 12 km = _____7.452_____ mi

27. 420 in = _____10.668_____ m

28. $50 \dfrac{\text{mi}}{\text{hr}}$ = _____80.5_____ $\dfrac{\text{km}}{\text{hr}}$

29. 22 qt = _____20.8_____ L

30. 18 L = _____4.77_____ gal

31. 16 lb = _____7264_____ g

32. Would water at 65°C feel cool to your hand?
  no (It would feel hot: 65°C = 149°F.)

33. If your Italian car has a 54-liter gas tank, how many gallons will it hold?
  14.31 gal

Part II

Complete.

34. 0.032 kg = _____1.13_____ oz

35. 7.8 ft = _____237.7_____ cm

36. 114°F = _____46_____ °C

37. 6.35 dg = _____0.0635_____ dag

38. 9900 ml = _____0.099_____ hl

39. 0.0008 g = _____0.08_____ cg

40. 1 hr = _____3600_____ sec

41. 3200 oz = _____0.1_____ ton

42. $22 \dfrac{\text{ft}}{\text{sec}}$ = _____15_____ $\dfrac{\text{mi}}{\text{hr}}$

43. If it takes 2 lb 4 oz of flour to make a loaf of bread, how much will it take to make 5 loaves?
  11 lb 4 oz

44. An electrician needs 60 yd of wire. What will be the total cost if the wire is 50¢ per foot?
  $90

ANSWERS: (Some answers are rounded.) 1. 84 2. 2.5 3. 3 4. 1.5 5. 6000 6. 2.25 7. 17 ft 5 in 8. 3 ft 1 in 9. 24 hr 18 min 10. 12 hr 26 min 11. 19 yd 1 ft 10 in 12. 5 yd 2 ft 6 in 13. 17 gal 2 qt 1 pt 14. 24 ft^2 15. 2 lb 15 oz 16. 6 17. 0.86 18. 58.2 19. 4.83 20. 0.0483 21. 3.5 22. 560 23. 90.5 m 24. 414.7 L 25. 299.6 g 26. 7.452 27. 10.668 28. 80.5 29. 20.8 30. 4.77 31. 7264 32. no (It would feel hot: 65°C = 149°F.) 33. 14.31 gal 34. 1.13 35. 237.7 36. 46 37. 0.0635 38. 0.099 39. 0.08 40. 3600 41. 0.1 42. 15 43. 11 lb 4 oz 44. $90

Complete.

1. 40 oz = _____ lb

2. 2800 ml = _____ hl

3. 6.5 kg = _____ lb

4. 65°C = _____ °F

5. 2.3 hr = _____ min

6. $270 \dfrac{mi}{hr} = $ _____ $\dfrac{ft}{sec}$

7. 2.8 m = _____ in

8. 6.1 km = _____ cm

Perform the indicated operations and simplify.

9.   12 yd 2 ft 3 in
    − 3 yd 2 ft 8 in

10.   14 hr 22 min 6 sec
    − 9 hr 38 min 16 sec

11.   10 yd 2 ft 3 in
    + 12 yd 2 ft 11 in

12. 18 m × 9 m

13. 28 lb ÷ 7 lb

14. A small storage tank has a capacity of 4.3 gal. How many liters of a chemical will it hold?

15. For an experiment to work, the temperature must be above 20°C. If the temperature is 59°F, will the experiment work?

1. _____ 2.5 _____

2. _____ 0.028 _____

3. _____ 14.3 _____

4. _____ 149 _____

5. _____ 138 _____

6. _____ 396 _____

7. _____ 110.2 _____

8. _____ 610,000 _____

9. _____ 8 yd 2 ft 7 in _____

10. _____ 4 hr 43 min 50 sec _____

11. _____ 23 yd 2 ft 2 in _____

12. _____ 162 m² _____

13. _____ 4 _____

14. _____ 16.3 L _____

15. _____ no; 59°F = 15°C _____

Geometry

9.1 LOGICAL SYSTEMS

1 Inductive Reasoning   **2** Deductive Reasoning

1 INDUCTIVE REASONING

To help us understand geometry, we begin by considering the two basic ways that we reason, or think. The first can be illustrated by studying the following list of numbers:

$$4, \ 11, \ 18, \ 25.$$

What is the next number in this list? Most of us will try to discover what 4, 11, 18, and 25 have in common, and shortly realize that $4 + 7 = 11$, $11 + 7 = 18$, $18 + 7 = 25$, so it would seem that each number after 4 is obtained from the one in front of it by adding 7. As a result, we would probably conclude that the next number should be 32 because $25 + 7 = 32$. Our reasoning process involved considering several *specific* observations, and based on these, formulating the *general* conclusion that the list would continue in the same pattern if we always added 7 to one number to obtain the next.

Inductive Reasoning

We use **inductive reasoning** when we reach a general conclusion based on a limited collection of specific observations.

Natural and social scientists frequently use inductive reasoning. When a laboratory experiment is performed several times with the same result, the physicist might form a general conclusion based on this experimentation. Or a sociologist might collect information from a limited number of people and attempt to draw a general conclusion about the total population. Also, when a new medicine is tested on a sample of several hundred people, the test results might lead a scientist to draw conclusions about the drug's effectiveness. In all these cases, a general conclusion is drawn from specific observations.

Does inductive reasoning always lead to the same conclusion? The answer is no, not always. If you return to our earlier example of four numbers, in which we concluded that the next number must be 32, you will see that another conclusion is also possible. The next number in the series can be 1, followed by 8, 15, 22, etc. This sequence gives the dates of the Fridays in the year 1991, starting with Friday, January 4, 1991!

As you can see, although inductive reasoning is widely used, there are no guarantees that the conclusion drawn is always correct or that it is the only possible conclusion. The primary flaw with inductive reasoning is that we cannot be sure of what will generally be true based on a limited number of cases.

EXAMPLE 1   USING INDUCTIVE REASONING

Use inductive reasoning to determine the next element in each list. Remember that there might be more than one answer.

(a) 2, 4, 8, 16, 32

We might conclude that the next number is 64 because each number after the first is twice the one before it.

(b) o, t, t, f, f, s, s, e

This one is a bit more difficult. This is a list of the first letters in the words *one, two, three, four,* etc. Thus, the next letter would be *n*, the first letter in the word *nine*.

EXAMPLE 2   APPLICATION OF INDUCTIVE REASONING

To determine the preferred candidate in the upcoming student body presidential election, a student decides to interview the first 50 students who pass by the entrance to the student union building. He discovers that 42 prefer Abby Bayona and 8 prefer Shawn Herman. Based on these 50 specific observations, he concludes that Abby Bayona will win the election. Is this conclusion necessarily correct? Can you think of situations in which the results of his poll might not accurately reflect the feelings of the total student body?

Obviously the interviewer's conclusion is not necessarily correct. For example, the limited sample of 50 students interviewed might have included sorority sisters of Abby Bayona who happened to be attending a meeting in the union building.

PRACTICE EXERCISE 1

Use inductive reasoning to determine the next element in each list.

(a) 1, 4, 9, 16, 25

(b)

M   ♡   8   M   ᠥ
1   2   3   4   5

Answers: (a) One answer is 36 since each number is the square of the numbers 1, 2, 3, 4, and 5. That is, $1 = 1^2$, $4 = 2^2$, $9 = 3^2$, $16 = 4^2$, $25 = 5^2$, and $36 = 6^2$. (b) The next element is ∞ since each element is one of the numbers 1, 2, 3, 4, and 5 placed next to its mirror image.

PRACTICE EXERCISE 2

Can you suggest possible changes in the data that might occur if only male students were included in the survey?

Answer: Male students might tend to give Shawn Herman more votes.

② DEDUCTIVE REASONING

Although inductive reasoning might not always lead to the same conclusion, it is still an important process, one that is used even by mathematicians. For example, consider the following observations:

$$1 + 1 = 2, \quad 1 + 3 = 4, \quad 3 + 3 = 6, \quad 3 + 5 = 8,$$
$$5 + 1 = 6, \quad 7 + 3 = 10.$$

After examining these equations, a mathematician might conclude that the sum of two odd integers is always an even integer. Certainly these six observations do not prove that this statement is true. What if we listed several hundred such observations with the same results? Would that have proved the statement? The mathematician would say you cannot prove a general statement by giving any number of specific cases, unless, of course, the number of specific cases is limited. The mathematician usually requires a different kind of proof, one that is based on *deductive reasoning*.

Deductive Reasoning

We use **deductive reasoning** when we reach a specific conclusion based on a collection of generally accepted assumptions.

When we reason deductively, we start with one or more **premises,** which are previously known facts, and attempt to arrive at a conclusion that necessarily follows if the premises are accepted. An example of deductive reasoning is given below.

Premise: Terri is a happy person.
Premise: Every happy person is pleasant.
Conclusion: Terri is pleasant.

Because Terri is a happy person, and every happy person is pleasant, it follows logically that Terri, as one of the happy persons, must also be pleasant.

The next example gives practice in recognizing deductive and inductive reasoning.

EXAMPLE 3   COMPARING INDUCTIVE AND DEDUCTIVE REASONING

Determine if each conclusion follows logically from the premises, and state whether the reasoning is inductive or deductive.

(a) *Premise:* My coat is tan.
Premise: Bob's coat is tan.
Premise: Di's coat is tan.
Conclusion: All coats are tan.

Because we are reasoning from three specific examples and drawing a general conclusion, the process involves inductive reasoning. It is obvious that the conclusion does not logically follow from the premises.

(b) *Premise:* All athletes are in good physical condition.
Premise: Shelly is an athlete.
Conclusion: Shelly is in good physical condition.

PRACTICE EXERCISE 3

Determine if each conclusion follows logically from the premises, and state whether the reasoning is inductive or deductive.

(a) *Premise:* This year is leap year.
Conclusion: Next year will not be leap year.

(b) *Premise:* The Boston Celtics won the NBA Championship in the years 1960–66.
Premise: The Boston Celtics won the NBA Championship in the years 1968–69
Conclusion: The Celtics won the NBA Championship every year during the 1960s.

In this case, if we accept the premises, then the conclusion logically follows. This is an example of deductive reasoning. Notice that the conclusion may be true or false; we do not actually know whether Shelly is in good condition. However, this is beside the point because the conclusion logically follows from the premises. Certainly, if the premises are true and a conclusion follows, then the conclusion is also true. In mathematics we assume premises are true so that deduced conclusions are also true statements.

Answers: (a) Using deductive reasoning, the conclusion follows logically from the given premise and the accepted but unstated premise that leap year occurs every four years. (b) The conclusion does not necessarily follow from the premises. It was reached by inductive reasoning. It is a false statement because in 1967 the Philadelphia 76ers won the NBA Championship.

We now give an illustration of a practical application of deductive reasoning. Assume the following warranty is given with a new radio.

> This radio is warranted for one year from the date of purchase against defects in materials or workmanship. During this period any such defects will be repaired or the radio will be replaced at the company's option without charge. This warranty is void in the case of misuse or negligence.

Assume the premises stated in this warranty and answer the following questions.

(a) Bill purchased a radio and when he got home and opened the box, he discovered that the case was cracked. What could Bill conclude?

Because Bill had just purchased the radio, one year had not gone by. He had not been negligent, nor had he misused the radio. He concluded that the company would either repair or replace the radio.

(b) Beth purchased a radio and fourteen months later the digital display burned out. What could Beth conclude?

Because fourteen months exceeds the time period of the warranty, Beth concluded that the company would not be required by the warranty to repair or replace the radio.

(c) Jamie purchased a radio and two months later left it outside after listening to it while sunbathing. That night it rained, and the next day the radio would not play. What could Jamie conclude?

Although the radio became defective during the warranty period, the defect was due to Jamie's negligence. The company would not be required by the warranty to repair or replace it.

9.1 EXERCISES A

Use inductive reasoning in Exercises 1–6 to determine the next element in each list.

1. 3, 8, 13, 18, 23   28

2. 14, 12, 10, 8, 6   4

3. 1, 3, 9, 27, 81   243

4 1, 2, 4, 8, 16   32

5. 1, 1, 2, 3, 5, 8, 13, 21   34

6. S, M, T, W, T, F   S

In Exercises 7–11, determine if each conclusion follows logically from the premises, and state whether the reasoning is inductive or deductive.

7. *Premise:* If you are a mathematics major, then you can compute discounts on sale items.
   Premise: Becky is a mathematics major.
   Conclusion: Becky can compute discounts on sale items.
   follows logically, deductive reasoning

8. *Premise:* If you are a mathematics major, then you can compute discounts on sale items.
   Premise: Becky can compute discounts on sale items.
   Conclusion: Becky is a mathematics major.   **does not follow logically; many people can compute discounts on sale items but they are not necessarily mathematics majors**

9. *Premise:* It rained on Tuesday.
   Premise: It rained on Wednesday.
   Conclusion: It will rain on Thursday.
   does not follow logically; inductive reasoning

10. *Premise:* If I buy a car, then it will be a Buick.
    Premise: If I receive a check from my father, then I will buy a car.
    Premise: I received a check from my father.
    Conclusion: I will buy a Buick.
    follows logically; deductive reasoning

11. *Premise:* If it is a frog, then it is green.
    Premise: If it hops, then it is a frog.
    Conclusion: If it hops, then it is green.
    follows logically; deductive reasoning

12. Suppose we add the following premise to those given in Exercise 11. What conclusion can you reach?
    Premise: It hops.
    Conclusion: **It is green.**

The puzzles in Exercises 13–17 are classics, and solving them requires a certain amount of deductive reasoning. Some of these puzzles are quite challenging. Do not be discouraged if you have trouble finding solutions immediately. We hope the puzzles will get you thinking, and also provide a bit of entertainment.

13. Mary has two current coins in her purse. Together they total 55¢. One is not a nickel. What are the two coins? (*Hint:* We do *not* say that neither coin is a nickel.)
    a nickel and a 50¢ piece

14. If you take 5 apples from 8 apples what have you got?
    You have 5 apples.

15. We know there are 12 one-cent stamps in a dozen, but how many two-cent stamps are in a dozen?
    12

16. What arithmetic symbol can be placed between 2 and 3 to form a number greater than 2 and less than 3?
    a decimal point

17. A boat will carry 200 pounds maximum. How can a man weighing 200 pounds and his two daughters, each of whom weighs 100 pounds, use the boat to reach an island?   **The two daughters cross, and one returns with the boat and stays while the man crosses. The other daughter returns with the boat and picks up the remaining daughter.**

ANSWERS: 1. 28 2. 4 3. 243 4. 32 5. 34 6. S 7. follows logically; deductive reasoning 8. does not follow logically; many people can compute discounts on sale items but they are not necessarily mathematics majors 9. does not follow logically; this is inductive reasoning 10. follows logically; deductive reasoning 11. follows logically; deductive reasoning 12. Conclusion: It is green. 13. a nickel and a 50¢ piece (We are told that *one* is not a nickel, not *both* are not nickels.) 14. You have 5 apples. (We are not asked what was left.) 15. 12 16. A decimal point, since 2.3 is greater than 2 and less than 3. 17. The two daughters cross, and one returns with the boat and stays while the man crosses. The other daughter returns with the boat and picks up the remaining daughter.

9.1 EXERCISES B

Use inductive reasoning in Exercises 1–6 to determine the next element in each list.

1. 2, 8, 14, 20, 26   32

2. 24, 19, 14, 9   4

3. 1, 5, 25, 125   625

4. 40, 20, 10, 5 $\frac{5}{2}$

5. 1, 3, 4, 7, 11, 18, 29   47

6. J, F, M, A, M   J

In Exercises 7–11, determine if each conclusion follows logically from the premises, and state whether the reasoning is inductive or deductive.

7. *Premise:* If you are a home buyer, then you make payments.
Premise: Doug is a home buyer.
Conclusion: Doug makes payments.
follows logically; deductive reasoning

8. *Premise:* If you are a home buyer, then you make payments.
Premise: Doug makes payments.
Conclusion: Doug is a home buyer.
does not follow logically; many people make payments who are not necessarily buying a home

9. *Premise:* Last year I won money in Las Vegas.
Premise: The year before I won money in Las Vegas.
Conclusion: I will win in Las Vegas this year.
does not follow logically; inductive reasoning

10. *Premise:* If you are going to be an engineer, then you will study mathematics.
Premise: If you study mathematics, then you will get a good job.
Premise: Roy is going to be an engineer.
Conclusion: Roy will get a good job.
follows logically; deductive reasoning

11. *Premise:* If you are an ogg, then you are an arg.
Premise: If you are a pon, then you are an ogg.
Conclusion: If you are a pon, then you are an arg.
follows logically; deductive reasoning

12. Suppose we add the following premise to those given in Exercise 11. What conclusion can you reach?
Premise: You are a pon.
Conclusion: **You are an arg.**

The puzzles in Exercises 13–17 are classics, and solving them requires a certain amount of deductive reasoning. Some of these puzzles are quite challenging. Do not be discouraged if you have trouble finding solutions immediately. We hope the puzzles will get you thinking, and also provide a bit of entertainment.

13. A younger dog and an older dog are in the back yard. The younger dog is the older dog's daughter, but the older dog is not the younger dog's mother. Explain.
The older dog is the younger dog's father.

14. A rancher had 20 cattle. All but 12 died. How many did he have left?
He had 12 left.

15. A museum fired an archaeologist who claimed she found a coin dated 300 B.C. Why?
It would have been impossible in 300 B.C. to state the date as we know it.

16. The number of marbles in a jar doubles every minute and is full in 10 minutes. When was the jar half full?
9 minutes

17. How many times can you subtract 5 from 25?
once; the next time you are subtracting 5 from 20, *not* 25

9.1 EXERCISES C

1. A judge wishing to convict a defendant puts two pieces of paper in a hat. He tells the jury that if the defendant draws the piece marked ''guilty'' he will be convicted, but if he draws the piece marked ''innocent'' he will be set free. The hitch is that the judge wrote ''guilty'' on both pieces of paper. But when the crafty defendant showed the jury one piece of paper, the judge was forced to let him go free. How did the defendant outwit the judge?

   The defendant told the jury to look at the piece of paper remaining in the hat and let them draw the appropriate conclusion.

2. You have 3 sacks, each containing 3 coins. Two of the sacks contain real coins and each coin weighs 1 pound. The third contains counterfeit coins and each weighs 1 pound 1 ounce. A scale is available, but it can be used one time, and one time only, to obtain a particular measure of weight. How might you use the scale to determine which sack contains the counterfeit coins? (*Note:* You cannot add or subtract coins to a total since any change of reading on the scale up or down will cause it to zero out.)

   Identify the sacks as sack 1, sack 2, and sack 3. Take 1 coin from sack 1, 2 coins from sack 2, and 3 coins from sack 3. Place the coins on the scale. The reading will be 6 pounds 1 ounce, 6 pounds 2 ounces, or 6 pounds 3 ounces. The number of ounces identifies the sack with the counterfeit coins.

9.2 LINES AND ANGLES

STUDENT GUIDEPOSTS

1. Segments, Lines, and Rays
2. Coinciding, Intersecting, and Parallel Lines
3. Angles
4. Measuring Angles
5. Classifying Angles
6. Perpendicular Lines
7. Parallel Lines Cut by a Transversal

1 SEGMENTS, LINES, AND RAYS

A **point** is a precise location in space often symbolized by a dot and labeled with a capital letter, such as the point A shown at the right. •A

Figure 9.1 displays four geometric figures. An arrowhead indicates that the curve continues in that direction without end.

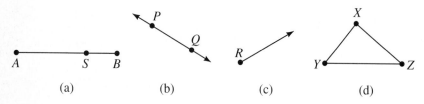

(a) (b) (c) (d)

Figure 9.1 Geometric Figures

A **segment** is a geometric figure made up of two **endpoints** and all points between them. The segment with endpoints A and B in Figure 9.1(a) can be denoted by either $\overline{AB}$ or $\overline{BA}$. The point S is said to be **on the segment** $\overline{AB}$. Figure 9.1(b) represents a (straight) **line,** which consists of the points on the segment $\overline{PQ}$ together with all points beyond P and Q. We denote this line by $\overleftrightarrow{PQ}$ or $\overleftrightarrow{QP}$.

Any two distinct points determine exactly one line. The two distinct points E and F determine the unique line $\overleftrightarrow{EF}$ shown in Figure 9.2. For convenience, we often use a small letter such as l to represent a line.

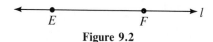

Figure 9.2

A **ray** is that part of a line lying on one side of a point P on the line, and includes the point P. A ray is shown in color in Figure 9.3. By selecting another point Q on the ray, we have a way to denote the ray by $\overrightarrow{PQ}$. Unlike the notations used for segments and lines, the rays $\overrightarrow{PQ}$ and $\overrightarrow{QP}$ are different. A ray has only one endpoint, which is always written first. Taken together, the rays $\overrightarrow{PQ}$ and $\overrightarrow{QP}$ make up the line $\overleftrightarrow{PQ}$. Another example of a ray is shown in Figure 9.1(c).

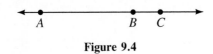

Figure 9.3   Ray

| **EXAMPLE 1   SEGMENTS, LINES, AND RAYS** | **PRACTICE EXERCISE 1** |

Use Figure 9.4 to answer each question. Deductive reasoning is used with premises consisting of the definitions given above.

Figure 9.4

(a) Is point C on $\overline{AB}$?   No

(b) Is point C on $\overleftrightarrow{AB}$?   Yes

(c) Is point C on $\overrightarrow{BA}$?   No

(d) What are the endpoints of $\overline{CA}$? A and C

(e) What are the endpoints of $\overrightarrow{CA}$?   Only C

(f) What are the endpoints of $\overleftrightarrow{CA}$?   There are no endpoints.

(g) Are $\overline{AB}$ and $\overline{BA}$ the same?   Yes

(h) Are $\overrightarrow{AB}$ and $\overrightarrow{BA}$ the same?   No

Draw two points X and Y and place point Z on $\overleftrightarrow{XY}$ but not on $\overline{XY}$. Use deductive reasoning and this figure to answer the following questions.

(a) Is Z on $\overrightarrow{YX}$?

(b) Is Z on $\overline{YX}$?

(c) Is Y on $\overrightarrow{ZX}$?

(d) What are the endpoints of $\overline{XY}$?

(e) What are the endpoints of $\overrightarrow{XY}$?

(f) What are the endpoints of $\overleftrightarrow{ZX}$?

(g) Are $\overline{YZ}$ and $\overline{ZY}$ the same?

(h) Are $\overrightarrow{ZX}$ and $\overrightarrow{XZ}$ the same?

Answers: **(a) yes** **(b) no**
(c) yes **(d)** X and Y
(e) only X
(f) **There are no endpoints.**
(g) yes **(h) no**

② COINCIDING, INTERSECTING, AND PARALLEL LINES

A **plane** is a surface with the property that any two points in it can be joined by a line, all points of which are also contained in the surface. Intuitively, a plane can be thought of as any flat surface such as a blackboard or a desktop. If two lines in a plane represent the same line, they are **coinciding lines.** When two lines share exactly one common point, they are **intersecting lines.** When two distinct lines in a plane do not intersect, the lines are called **parallel lines.** These three situations are illustrated in Figure 9.5

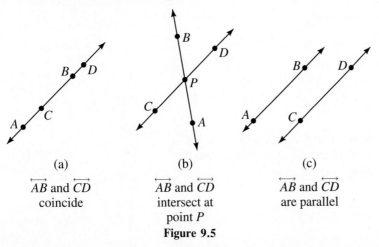

$\overleftrightarrow{AB}$ and $\overleftrightarrow{CD}$
coincide

$\overleftrightarrow{AB}$ and $\overleftrightarrow{CD}$
intersect at
point P

$\overleftrightarrow{AB}$ and $\overleftrightarrow{CD}$
are parallel

Figure 9.5

Two segments, two rays, or a segment and a ray can also be called **parallel** if they are on parallel lines. Thus, in Figure 9.5(c) $\overline{AB}$ is parallel to $\overline{CD}$, $\overline{AB}$ is parallel to $\overrightarrow{DC}$, and $\overline{AB}$ is parallel to $\overleftrightarrow{CD}$.

❸ ANGLES

An **angle** is a geometric figure consisting of two rays that share a common endpoint, called the **vertex** of the angle. In Figure 9.6, the two rays $\overrightarrow{AC}$ and $\overrightarrow{AB}$ form the angle denoted by $\angle BAC$, $\angle CAB$, or simply $\angle A$. The rays $\overrightarrow{AC}$ and $\overrightarrow{AB}$ are the **sides** of $\angle A$.

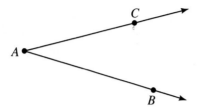

Figure 9.6 $\angle CAB$

We are familiar with measuring the length of segments by some suitable unit of measure such as inch, centimeter, foot, or meter. In order to measure an angle, we need a measuring unit. The most common unit is the degree (°). An angle with measure 0° is formed by two coinciding rays such as $\overrightarrow{AB}$ and $\overrightarrow{AC}$ in Figure 9.7. As the ray $\overrightarrow{AB}$ rotates in a counterclockwise direction from ray $\overrightarrow{AC}$, the two rays form larger and larger angles, as shown in Figure 9.8.

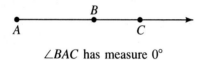

$\angle BAC$ has measure 0°

Figure 9.7

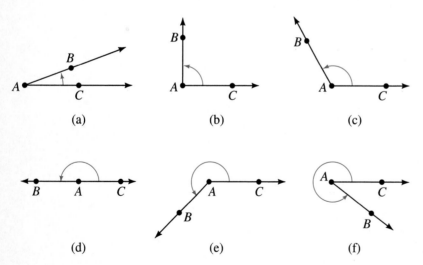

Figure 9.8 Angles

If $\overrightarrow{AB}$ is allowed to rotate completely around until it coincides with ray $\overrightarrow{AC}$ again, the resulting angle is said to measure 360°. Thus, an angle of measure 1° is formed by making $\frac{1}{360}$ of a complete rotation, as in Figure 9.9

Figure 9.9 Angle Measuring 1°

❹ MEASURING ANGLES

The approximate measure of an angle in degrees can be found by using a protractor (see Figure 9.10).

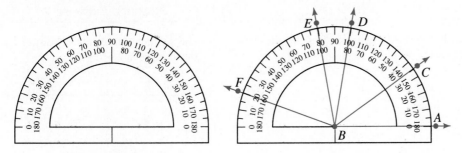

Figure 9.10 Protractor   **Figure 9.11** Measuring Angles

Figure 9.11 shows how a protractor is used to measure various angles. $\angle ABC$ has measure 35°, $\angle ABD$ has measure 80°, $\angle ABE$ has measure 100°, and $\angle ABF$ has measure 160°. Rather than say "$\angle ABC$ has measure 35°," we simply write $\angle ABC = 35°$. Also, when two angles have the same measure, we simply say that the angles are **equal.**

❺ CLASSIFYING ANGLES

When two rays forming an angle are in opposite directions, such as in Figure 9.8(d), the resulting angle is a **straight angle** and has measure 180°. When a ray $\overrightarrow{AB}$ is rotated through one-fourth of a complete revolution from $\overrightarrow{AC}$, as in Figure 9.8(b), the resulting angle is a **right angle** and has measure 90°. Right angles are often marked with a square corner at the vertex, such as ⌐ . Angles measuring between 0° and 90°, such as the angle in Figure 9.8(a), are **acute angles.** Angles measuring between 90° and 180°, such as the angle in Figure 9.8(c), are **obtuse angles.**

Two angles with the same measure are **equal** or **congruent.** Two angles whose measures total 90° are **complementary,** and two angles whose measures total 180° are **supplementary.**

When two lines intersect, four angles are formed. The intersecting lines $\overleftrightarrow{AB}$ and $\overleftrightarrow{CD}$ in Figure 9.12 form the four angles $\angle APD$, $\angle DPB$, $\angle BPC$, and $\angle CPA$. $\angle APC$ and $\angle DPB$ are **vertical angles** to each other. The same is true for $\angle APD$ and $\angle BPC$.

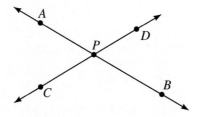

Figure 9.12 Vertical Angles

⑥ PERPENDICULAR LINES

Vertical angles have the same measure. Thus, in Figure 9.12, $\angle APC$ and $\angle DPB$ are equal, as are $\angle APD$ and $\angle BPC$. When the four angles formed by two intersecting lines all have measure 90°, that is, when they are all right angles, the lines are **perpendicular.**

| EXAMPLE 2   ANGLES AND LINES | PRACTICE EXERCISE 2 |

The following statements refer to Figure 9.13. The premises are the definitions, and we use deductive reasoning to make each statement.

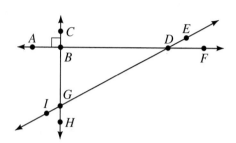

Figure 9.13

(a) $\overleftrightarrow{AB}$ and $\overleftrightarrow{DF}$ are coinciding lines.

(b) $\overleftrightarrow{AF}$ and $\overleftrightarrow{IE}$ are intersecting lines.

(c) $\overrightarrow{GI}$ and $\overrightarrow{GH}$ are the sides of $\angle IGH$.

(d) $\angle IGD$ is a straight angle.

(e) $\angle ABC$ is a right angle.

(f) $\angle IGB$ and $\angle BGD$ are supplementary.

(g) $\angle EDF$ is an acute angle.

(h) $\angle EDB$ is an obtuse angle.

(i) $\angle IGH$ and $\angle BGD$ are vertical angles.

(j) $\angle EDB$ and $\angle FDG$ are equal (they are vertical angles).

(k) $\overleftrightarrow{AF}$ and $\overleftrightarrow{CH}$ are perpendicular.

Use the following figure and answer true or false in (a)–(k).

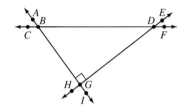

(a) $\overrightarrow{HD}$ and $\overleftrightarrow{GE}$ are coinciding lines.

(b) $\overleftrightarrow{AI}$ and $\overleftrightarrow{HE}$ are parallel lines.

(c) $\overrightarrow{AB}$ and $\overrightarrow{DB}$ are the sides of $\angle ABD$.

(d) $\angle BGI$ is a straight angle.

(e) $\angle BGD$ is a right angle.

(f) $\angle BDE$ and $\angle EDF$ are complementary.

(g) $\angle ABD$ is an acute angle.

(h) $\angle BDE$ is an obtuse angle.

(i) $\angle ABC$ and $\angle CBG$ are vertical angles.

(j) $\angle DGI$ and $\angle IGH$ are equal.

(k) $\overleftrightarrow{CB}$ and $\overleftrightarrow{HG}$ are intersecting lines.

Answers: (a) true   (b) false
(c) false   (d) true   (e) true
(f) false   (g) false   (h) true
(i) false   (j) true   (k) true

---

⑦ PARALLEL LINES CUT BY A TRANSVERSAL

A line that intersects each of two parallel lines is called a **transversal.** The two lines are said to be "cut" by the transversal and several angles are formed. In Figure 9.14, transversal $\overleftrightarrow{AD}$ cuts parallel lines $\overleftrightarrow{GF}$ and $\overleftrightarrow{HE}$ to form four **interior angles,** $\angle GBC$, $\angle FBC$, $\angle BCH$, and $\angle BCE$, and four **exterior angles,** $\angle ABG$, $\angle ABF$, $\angle HCD$, and $\angle ECD$. Since $\angle GBC$ and $\angle BCE$ are on alternate sides of the transversal, they are called **alternate interior angles.** Similarly, $\angle FBC$ and $\angle BCH$ are alternate interior angles. In Figure 9.14, $\angle ABF$ and $\angle BCE$ are **corresponding angles.** Three other pairs of corresponding angles are $\angle ABG$ and $\angle BCH$, $\angle GBC$ and $\angle HCD$, and $\angle FBC$ and $\angle ECD$.

Alternate interior angles always have the same measure. Thus, referring to Figure 9.14 again, ∠*FBC* and ∠*BCH* have the same measure, as do ∠*GBC* and ∠*BCE*. Also, corresponding angles always have the same measure so that, for example, ∠*ABF* has the same measure as ∠*BCF* in Figure 9.14. (You can draw several figures and use inductive reasoning to conclude that these angles are probably equal, but deductive reasoning is needed to actually prove the statements.)

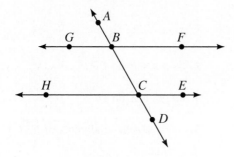

Figure 9.14

EXAMPLE 3 TRANSVERSAL CUTTING PARALLEL LINES

The following statements refer to Figure 9.15, in which $\overleftrightarrow{AB}$ is parallel to $\overleftrightarrow{DE}$. We use deductive reasoning to make the following statements.

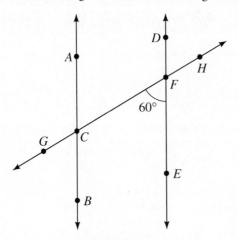

Figure 9.15

(a) $\overleftrightarrow{GH}$ is a transversal.

(b) ∠*DFC* is an interior angle.

(c) ∠*ACG* is an exterior angle.

(d) ∠*EFC* and ∠*ACF* are alternate interior angles.

(e) ∠*DFC* has the same measure as ∠*FCB*.

(f) ∠*HFE* has the same measure as ∠*FCB*.

(g) ∠*DFC* has measure 120°.

(h) ∠*ACG* has measure 120°.

(i) ∠*GCB* has measure 60°.

PRACTICE EXERCISE 3

Use the following figure, in which $\overleftrightarrow{PS}$ is parallel to $\overleftrightarrow{WU}$, and answer true or false in (a)–(i).

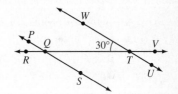

(a) $\overleftrightarrow{QT}$ is a transversal.

(b) ∠*VTU* is an interior angle.

(c) ∠*RQS* is an exterior angle.

(d) ∠*WTV* and ∠*QTU* are alternate interior angles.

(e) ∠*UTQ* and ∠*PQT* have the same measure.

(f) ∠*VTU* and ∠*PQR* have the same measure.

(g) ∠*WTV* has measure 150°.

(h) ∠*RQS* has measure 30°.

(i) ∠*PQT* has measure 150°.

Answers: (a) true   (b) false
(c) true   (d) false   (e) true
(f) true   (g) true   (h) false
(i) true

9.2 EXERCISES A

Use the figure below and deductive reasoning to answer each question in Exercises 1–16.

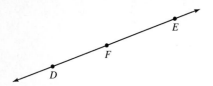

1. Is point *E* on $\overline{DF}$?   no

2. Is point *E* on $\overrightarrow{FD}$?   no

3. Is point *E* on $\overleftrightarrow{DF}$?   yes

4. Is point *E* on $\overrightarrow{DF}$?   yes

5. What are the endpoints of $\overrightarrow{FE}$?   only *F*

6. What are the endpoints of $\overleftrightarrow{FE}$?   none

7. What are the endpoints of $\overline{FE}$?   *F* and *E*

8. Are $\overline{DE}$ and $\overline{ED}$ the same?   yes

9. Are $\overrightarrow{DE}$ and $\overrightarrow{ED}$ the same?   no

10. Are $\overleftrightarrow{DE}$ and $\overleftrightarrow{ED}$ the same?   yes

11. Do $\overleftrightarrow{DE}$ and $\overleftrightarrow{DF}$ coincide?   yes

12. What is the measure of ∠*DFE*?   180°

13. What is the measure of ∠*EDF*?   0°

14. What is the vertex of ∠*EFD*?   *F*

15. Is ∠*EFD* a right angle?   no

16 Are ∠*EDF* and ∠*DFE* supplementary?   yes

Use the figure with the protractor to determine the measures of the given angles.

17. ∠*ABC*   20°

18. ∠*ABE*   95°

19. ∠*ABD*   55°

20. ∠*ABF*   170°

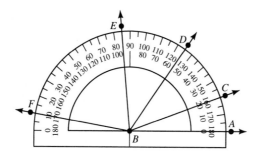

Use the figure below and deductive reasoning to answer true *or* false *in Exercises 21–34.*

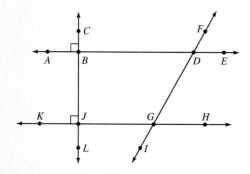

21. ∠*DBJ* is a right angle.   true

22. ∠*BDG* is an acute angle.   true

23. ∠*DGH* and ∠*JGI* are obtuse angles.   false

24. ∠*FDB* and ∠*EDG* have the same measure.   true

25. $\overleftrightarrow{AE}$ is parallel to $\overleftrightarrow{KH}$.   true

26. $\overleftrightarrow{CB}$ is perpendicular to $\overleftrightarrow{GH}$.   true

27. The measure of ∠GJL is 90°.   true

28. $\overrightarrow{GD}$ and $\overrightarrow{JG}$ are the sides of ∠JGD.   false

29. ∠IGJ and ∠JGF are supplementary.   true

30. $\overline{BE}$ is parallel to $\overrightarrow{JK}$.   true

31. $\overleftrightarrow{IF}$ is a transversal that cuts parallel lines $\overleftrightarrow{AE}$ and $\overleftrightarrow{KH}$.
true

32. ∠FDE and ∠BDG are corresponding angles.
false

33. ∠EDG and ∠DGJ are alternate interior angles.
true

34. ∠BDG and ∠DGH have the same measure.
true

FOR REVIEW

In Exercises 35–36, determine if each conclusion follows logically from the premises, and state whether the reasoning is inductive or deductive.

35. *Premise:* If the measure of ∠A is 70°, then the measure of ∠B is 20°.
Premise: The measure of ∠A is 70°.
Conclusion: The measure of ∠B is 20°.
follows logically; deductive reasoning

36. *Premise:* Point C is on $\overrightarrow{AB}$.
Premise: Point D is on $\overrightarrow{AB}$.
Premise: Point E is on $\overrightarrow{AB}$.
Conclusion: Point F is on $\overrightarrow{AB}$.
does not follow logically; inductive reasoning

ANSWERS: 1. no 2. no 3. yes 4. yes 5. only the point F 6. There are no endpoints. 7. F and E 8. yes
9. no 10. yes 11. yes 12. 180° 13. 0° 14. F 15. no 16. yes 17. 20° 18. 95° 19. 55° 20. 170° 21. true
22. true 23. false 24. true 25. true 26. true 27. true 28. false 29. true 30. true 31. true 32. false
33. true 34. true 35. follows logically; deductive reasoning 36. does not follow logically; inductive reasoning

9.2   EXERCISES B

Use the figure below and deductive reasoning to answer each question in Exercises 1–16.

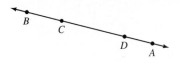

1. Is point C on $\overline{AB}$?   yes

2. Is point B on $\overrightarrow{DC}$?   yes

3. Is point A on $\overleftrightarrow{BD}$?   yes

4. Is A on $\overrightarrow{DC}$?   no

5. What are the endpoints of $\overrightarrow{DC}$?
only D

6. What are the endpoints of $\overleftrightarrow{DC}$?
none

7. What are the endpoints of $\overline{DC}$?
D and C

8. Are $\overline{BD}$ and $\overline{DB}$ the same?
yes

9. Are $\overrightarrow{BD}$ and $\overrightarrow{DB}$ the same?
no

10. Are $\overleftrightarrow{BD}$ and $\overleftrightarrow{DB}$ the same?
yes

11. Do $\overleftrightarrow{BA}$ and $\overleftrightarrow{DC}$ coincide?
yes

12. What is the measure of ∠DBC?
0°

13. What is the measure of ∠DCB?
180°

14. What is the vertex of ∠DCB?
C

15. Is ∠ADC a straight angle?
yes

16. Are ∠BCD and ∠CBD complementary?
no

Use the figure with the protractor to determine the measures of the given angles.

17. ∠*PQR*   35°

18. ∠*PQT*   110°

19. ∠*PQS*   75°

20. ∠*PQU*   155°

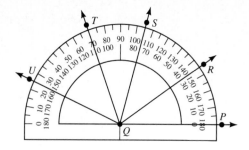

Use the figure below and deductive reasoning to answer true *or* false *in Exercises 21–34.*

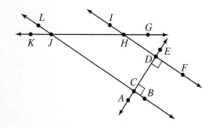

21. ∠*HDC* is a straight angle.   **false**

22. ∠*JHD* is an obtuse angle.   **true**

23. ∠*IHJ* and ∠*GHD* are vertical angles.   **true**

24. ∠*LJH* and ∠*KJC* have the same measure.   **true**

25. $\overleftrightarrow{IF}$ is parallel to $\overleftrightarrow{LJ}$.   **true**

26. $\overleftrightarrow{AC}$ is perpendicular to $\overleftrightarrow{IH}$.   **true**

27. The measure of ∠*JCA* is 180°.   **false**

28. $\overrightarrow{HG}$ and $\overrightarrow{HF}$ are the sides of ∠*GHF*.   **true**

29. ∠*JHD* and ∠*DHG* are complementary.   **false**

30. $\overleftrightarrow{CL}$ is perpendicular to $\overleftrightarrow{IF}$.   **false**

31. $\overleftrightarrow{KG}$ is a transversal that cuts parallel lines $\overleftrightarrow{IF}$ and $\overleftrightarrow{LB}$.
true

32. ∠*GHD* and ∠*HJC* are corresponding angles.
true

33. ∠*DHJ* and ∠*HJC* are alternate interior angles.
false

34. ∠*IHJ* and ∠*LJK* have the same measure.
true

FOR REVIEW

In Exercises 35–36, determine if each conclusion follows logically from the premises, and state whether the reasoning is inductive or deductive.

35. *Premise:* ∠*A* = ∠*B*
Premise: ∠*A* = ∠*C*
Premise: ∠*A* = ∠*D*
Conclusion: ∠*A* = ∠*E*
does not follow logically; inductive reasoning

36. *Premise: If* $\overleftrightarrow{AB}$ *is perpendicular to* $\overleftrightarrow{CD}$, *then* $\overleftrightarrow{AB}$ *is parallel to* $\overleftrightarrow{EF}$.
Premise: $\overleftrightarrow{AB}$ *is perpendicular to* $\overleftrightarrow{CD}$.
Conclusion: $\overleftrightarrow{AB}$ *is parallel to* $\overleftrightarrow{EF}$.
follows logically; deductive reasoning

9.2 EXERCISES C

Use the figure below and deductive reasoning to answer the questions in Exercises 1–8. Lines $\overleftrightarrow{BG}$ and $\overleftrightarrow{MI}$ are parallel.

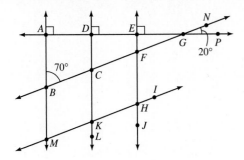

1. What is the measure of ∠*BCK*?   70°

2. What is the measure of ∠*EFG*?   70°

3. What is the measure of ∠*FHI*?   70°

4. What is the measure of ∠*HKL*?   110°

5. What is the measure of ∠*EGF*?   20°

6. What is the sum of the measures of ∠*GEF*, ∠*EFG*, and ∠*FGE*?   180°

7. What is the sum of the measures of ∠*GDC*, ∠*DCG*, and ∠*CGD*?   180°

8. What is the sum of the measures of ∠*CBM*, ∠*BMK*, ∠*MKC*, and ∠*KCB*?   360°

9.3 TRIANGLES

❶ TRIANGLES

In this section we present an introduction to the study of triangles and their properties. Consider three noncollinear points *A*, *B*, and *C* and the corresponding segments $\overline{AB}$, $\overline{BC}$, and $\overline{CA}$, as shown in Figure 9.16. The resulting geometric figure is a **triangle** that has sides $\overline{AB}$, $\overline{BC}$, and $\overline{CA}$, has **vertices** (plural of **vertex**) *A*, *B*, and *C*, and is denoted by △*ABC* (△ represents the word "triangle").

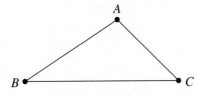

Figure 9.16 Triangle

❷ CLASSIFYING TRIANGLES

Triangles are often classified by their angles, as shown in Figure 9.17. They can also be classified by their sides, in Figure 9.18. We can also describe triangles with a combination of these terms, as in Figure 9.19.

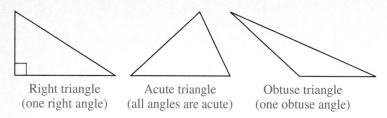

Right triangle
(one right angle)

Acute triangle
(all angles are acute)

Obtuse triangle
(one obtuse angle)

Figure 9.17

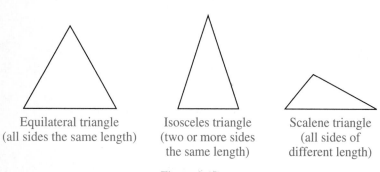

Equilateral triangle
(all sides the same length)

Isosceles triangle
(two or more sides
the same length)

Scalene triangle
(all sides of
different length)

Figure 9.18

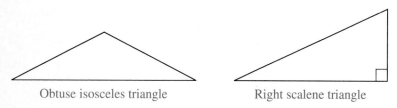

Obtuse isosceles triangle

Right scalene triangle

Figure 9.19

In Chapter 17, after we study radicals, we will discuss some of the properties of isosceles right triangles and right triangles with acute angles of 30° and 60°. For now, we restrict ourselves to some general properties of triangles.

Sum of Angles

The sum of the measures of the angles in any triangle is 180°.

EXAMPLE 1   ANGLES OF A TRIANGLE

Consider △*ABC* in Figure 9.20. If ∠*A* has measure 35° and ∠*B* has measure 70°, find the measure of ∠*C*.

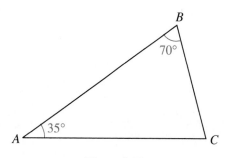

Figure 9.20

PRACTICE EXERCISE 1

In right △*ABC*, one acute angle has measure 42°. What is the measure of the other acute angle?

The sum of the measures of $\angle A$ and $\angle B$ is $70° + 35° = 105°$. For the sum of the three measures to equal $180°$, $\angle C$ must measure $180° - 105° = 75°$.

Answer: 48°

③ CONGRUENT TRIANGLES

Two triangles are **congruent** if they can be made to coincide by placing one on top of the other, either directly or by flipping one of them over. In Figure 9.21, $\triangle ABC$ and $\triangle DEF$ are congruent, since $\triangle ABC$ could be placed on top of $\triangle DEF$. $\triangle ABC$ and $\triangle GHI$ are also congruent, since if $\triangle GHI$ were flipped over it could be made to coincide with $\triangle ABC$. On the other hand, $\triangle ABC$ and $\triangle JKL$ are not congruent, since they cannot be made to coincide either directly or by flipping one of them over.

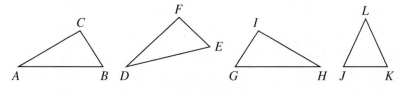

Figure 9.21

More precisely, two triangles will be congruent if the measures of the three sides of one triangle are equal to the measures of the three sides of the other, and the measures of the three angles of one are equal to the measures of the three angles of the other. Fortunately, in practical situations we do not need to establish equality relative to all six parts to show that two triangles are congruent. Three ways to show congruence are summarized as follows.

Congruent Triangles

Two triangles are congruent when one of the following can be shown.

1. Each of the three sides of one triangle is equal in length to a side of the other triangle (congruence by side-side-side, or SSS).

2. Two sides of one triangle are equal in length to two sides of the other triangle, and the angle formed by these sides in one triangle is equal in measure to the angle formed by the corresponding sides in the other triangle (congruence by side-angle-side, or SAS).

3. Two angles of one triangle are equal in measure to two angles of the other triangle, and the side between these two angles in one triangle is equal in length to the corresponding side in the other triangle (congruence by angle-side-angle, or ASA).

EXAMPLE 2   CONGRUENT TRIANGLES

Use deductive reasoning to state why the triangles in Figure 9.22 on the following page are congruent.

PRACTICE EXERCISE 2

State why the given triangles on the following page are congruent.

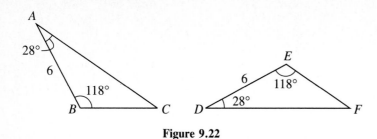

Figure 9.22

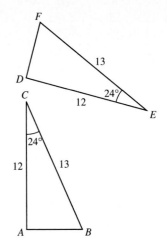

Since the measures of $\overline{AB}$ and $\overline{DE}$ are both 6, and $\overline{AB}$ and $\overline{DE}$ are located between the corresponding equal angles, $\angle A$ and $\angle D$, and $\angle B$ and $\angle E$, $\triangle ABC$ is congruent to $\triangle DEF$ by ASA.

Answer: The triangles are congruent by SAS.

🛡 SIMILAR TRIANGLES

Congruent triangles have the same size and shape. Triangles that have the same shape but may differ in size are called **similar.** Congruent triangles are always similar, but similar triangles need not be congruent. For example, the triangles in Figure 9.23 are similar but clearly not congruent.

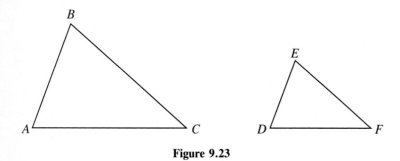

Figure 9.23

To determine whether two triangles are similar, we need only compare the angles.

Similar Triangles
Two triangles are similar when the angles of one are equal in measure to the angles of the other.

Since the sum of the angles of any triangle equals 180°, we know that whenever two angles of one triangle are equal to two angles of a second, the remaining angles are also equal, making the triangles similar.

| EXAMPLE 3 SIMILAR TRIANGLES | PRACTICE EXERCISE 3 |

Use deductive reasoning to state why the triangles in Figure 9.24 are similar.

State why the triangles are similar.

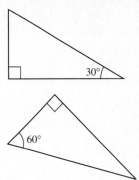

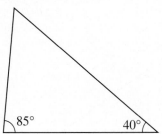

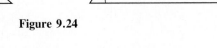

Figure 9.24

Since $85° + 40° = 125°$ and $180° - 125° = 55°$, the angles of the triangles are equal. Thus, the triangles are similar.

Answer: Both triangles are right triangles. Since $90° + 30° = 120°$ and $180° - 120° = 60°$, the angles of the triangles are equal.

In Section 9.5 we will discuss the perimeter and area of a triangle.

9.3 EXERCISES A

1. $\triangle ABC$ is given.

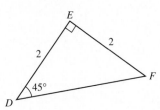

(a) What is the measure of $\angle C$?   **25°**
(b) Classify the triangle by its angles.   **obtuse triangle**
(c) Classify the triangle by its sides.   **scalene triangle**

2. $\triangle DEF$ is given.

(a) What is the measure of $\angle F$?   **45°**
(b) Classify the triangle by its angles   **right triangle**
(c) Classify the triangle by its sides.   **isosceles triangle**

Use deductive reasoning to decide whether the given triangles are congruent. If so, state why.

3.

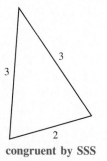

congruent by SSS

4.

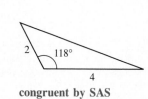

congruent by SAS

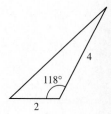

5.

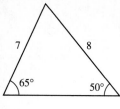
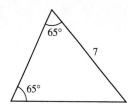

not congruent (similar)

6.

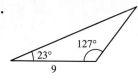

congruent by ASA

Use deductive reasoning to decide whether the given triangles are similar.

7.

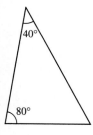

similar

8.

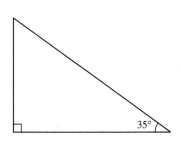

similar

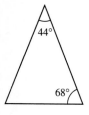 **9**

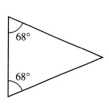

not similar

10.

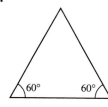
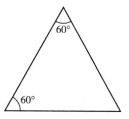

similar

11. Is it possible for an isosceles triangle to be a right triangle?   yes

12. Are all isosceles triangles equilateral?   no

FOR REVIEW

Answer true *or* false *in Exercises 13–16.*

13. A straight angle has measure 180°.   true

14. Two angles whose measures total 180° are complementary.   false

15. Vertical angles have the same measure.   true

16. Alternate interior angles are always complementary.   false

ANSWERS: 1. (a) 25° (b) obtuse triangle (c) scalene triangle 2. (a) 45° (b) right triangle (c) isosceles triangle
3. congruent by SSS 4. congruent by SAS 5. The triangles are similar but not congruent. 6. congruent by ASA
7. They are similar. 8. They are similar. 9. They are not similar. 10. They are similar. 11. yes 12. no
13. true 14. false (supplementary) 15. true 16. false (equal)

9.3 EXERCISES B

1. △*ABC* is given.

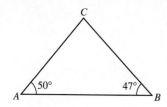

(a) What is the measure of ∠*C*?   83°
(b) Classify the triangle by its angles.   **acute triangle**
(c) Classify the triangle by its sides.   **scalene triangle**

2. △*DEF* is given.

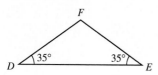

(a) What is the measure of ∠*F*?   110°
(b) Classify the triangle by its angles.   **obtuse triangle**
(c) Classify the triangle by its sides.   **isosceles triangle**

Use deductive reasoning to decide whether the given triangles are congruent. If so, state why.

3.

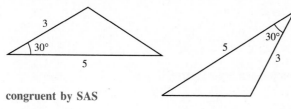

congruent by SAS

4.

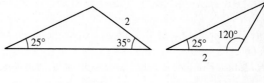

not congruent (similar)

5.

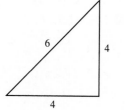

congruent by SSS

6.

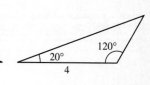

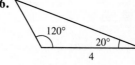

congruent by ASA

Use deductive reasoning to decide whether the given triangles are similar.

7.

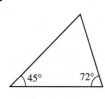

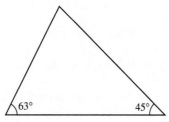

similar

8.

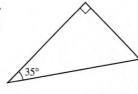

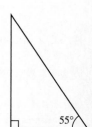

similar

9.

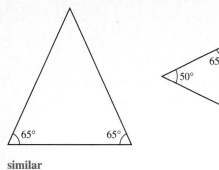

similar

10.

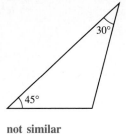

not similar

11. Is it possible for an equilateral triangle to be a right triangle?   no

12. Are all equilateral triangles isosceles?   yes

FOR REVIEW

Answer true *or* false *in Exercises 13–16. If the answer is false, tell why.*

13. An acute angle has measure 90°.   false

14. Two angles whose measures total 360° are complementary.   false

15. Corresponding angles always have the same measure.   true

16. Interior angles on the same side of a transversal are supplementary.   true

9.3 EXERCISES C

Use the figure below and deductive reasoning to answer the questions in Exercises 1–6.

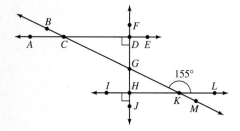

1. What is the measure of ∠GKH?   25°

2. What is the measure of ∠DGC?   65°

3. What is the measure of ∠BCA?   25°

4. Is △DGC similar to △GHK?   yes

5. Is △DGC congruent to △GHK?
   not necessarily

6. What is the measure of ∠DGK?   115°

9.4 CIRCLES

1 RADIUS, DIAMETER, AND CIRCUMFERENCE

A **circle** is a figure consisting of all points located the same distance r from a fixed point O called its **center.** In Figure 9.25 the segment $\overline{OA}$ or r is called a **radius** of the circle. The segment $\overline{DE}$, with length d, is called a **diameter** of the circle. The distance around the circle is called the **circumference** C of the circle. Since the diameter is twice the radius, we can write

$$d = 2r \quad \text{and} \quad r = \frac{d}{2}.$$

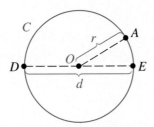

Figure 9.25   Circle

EXAMPLE 1 DIAMETER OF A CIRCLE	PRACTICE EXERCISE 1

Find the diameter of the circle with radius 2.3 cm, shown in Figure 9.26.

$$d = 2r = 2(2.3 \text{ cm})$$
$$= 4.6 \text{ cm}$$

Figure 9.26

Find the radius of a circle with diameter 22.4 meters.

Answer: 11.2 m

Suppose that we measure the diameter d and the circumference C of the circles formed by the tops of several cans. For example, assume that for one can C is 6.28 inches and d is 2.00 inches. Then

$$\frac{C}{d} = \frac{6.28 \text{ in}}{2.00 \text{ in}} = 3.14.$$

Measurements from other cans will give us approximately the same result. We can then use inductive reasoning to assume that the ratio of the circumference of any circle to its diameter is a constant approximately equal to 3.14. The number $\frac{22}{7}$ is also used as an approximation of this number, which is called π (the Greek letter *pi*). That is, $\frac{C}{d} = \pi$ or $C = \pi d$ for any circle.

Circumference of a Circle

The circumference of a circle with diameter d and radius r is

$$C = \pi d = 2\pi r.$$

EXAMPLE 2 FINDING CIRCUMFERENCE	PRACTICE EXERCISE 2

Find the circumference of a circle with diameter 32 mm. Use 3.14 for π to give an approximation of the actual value.

$$C = \pi d = \pi(32 \text{ mm}) = 32\pi \text{ mm}$$

The approximate value of C is

$$C \approx 3.14(32 \text{ mm})$$
$$\approx 100.5 \text{ mm.} \quad \text{Rounded to the nearest tenth}$$

Find the circumference of a circle with radius 63 inches. Use $\frac{22}{7}$ for π.

Answer: 396 in

---

❷ CENTRAL ANGLES AND ARCS

A **central angle** of a circle is an angle with its vertex at the center of the circle. In Figure 9.27, $\angle AOB$ is a central angle that intercepts **minor arc** $\overset{\frown}{ACB}$, the shorter part of the circle between A and B, and **major arc** $\overset{\frown}{ADB}$, shown in color.

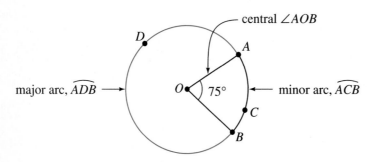

Figure 9.27

The **measure of an arc** is given by the measure of the central angle that determines it. Since the measure of central $\angle AOB$ in Figure 9.27 is 75°, the measure of minor arc $\overset{\frown}{ACB}$ is also 75°. Then the measure of major arc $\overset{\frown}{ADB}$ is 285°, since $360° - 75° = 285°$.

❸ CHORDS AND INSCRIBED ANGLES

A segment whose endpoints are on a circle is called a **chord** of the circle. Any chord that contains the center is also a diameter of the circle. Three distinct points A, B, and C on a circle determine **inscribed angle** $\angle ABC$, with sides determined by the chords $\overline{AB}$ and $\overline{BC}$ as shown in Figure 9.28. The inscribed angle $\angle ABC$ determines **intercepted arc** $\overset{\frown}{ADC}$ (shown in color).

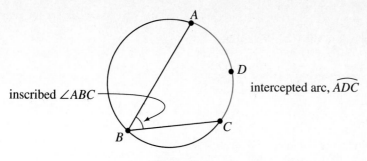

Figure 9.28

inscribed ∠*ABC*

intercepted arc, $\overarc{ADC}$

Consider the central ∠*AOC* measuring 60° in Figure 9.29, with inscribed ∠*ABC*. The measure of ∠*ABC* is 30°, one-half the measure of its intercepted arc, $\overarc{ADC}$, which has measure 60° (the measure of ∠*AOC*). This is true in general.

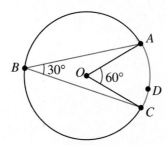

Figure 9.29

Measure of Inscribed Angle

The measure of any inscribed angle is one-half the measure of its intercepted arc.

EXAMPLE 3 MEASURING ARCS

Use Figure 9.30 and deductive reasoning to answer the following questions.

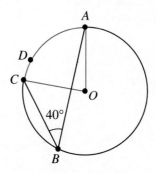

Figure 9.30

(a) What is the measure of $\overarc{ADC}$?

The measure of $\overarc{ADC}$ is the same as the measure of central ∠*AOC*. Since the measure of inscribed ∠*ABC* is one-half the measure of ∠*AOC*, the measure of ∠*AOC* is 80°. Thus, the measure of $\overarc{ADC}$ is 80°.

PRACTICE EXERCISE 3

Use the figure below to answer the following questions.

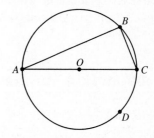

(a) What is the measure of $\overarc{ADC}$?

(b) What is the measure of $\overset{\frown}{ABC}$?

The measure of $\overset{\frown}{ABC}$ is 360° less the measure of $\overset{\frown}{ADC}$. Thus, the measure of $\overset{\frown}{ABC}$ is $360° - 80° = 280°$.

(b) What is the measure of inscribed $\angle ABC$?

Answers: **(a)** 180° **(b)** 90°

❹ SEMICIRCLES

A chord that passes through the center of a circle is a diameter, and an arc whose endpoints are the endpoints of a diameter is called a **semicircle.**

Suppose we inscribe an angle in a semicircle as shown in Figure 9.31.

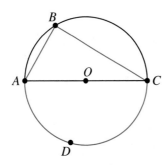

Figure 9.31

We can use deductive reasoning to prove that any angle inscribed in a semicircle is a right angle. First we note that since $\overset{\frown}{ADC}$ is one-half of the circle, it has measure 180°. Therefore,

$$\angle ABC = \frac{1}{2}\overset{\frown}{ADC}$$

$$= \frac{1}{2}(180°)$$

$$= 90°$$

Thus, $\angle ABC$ is a right angle.

❺ SECANTS AND TANGENTS

A line can intersect a circle in two points or exactly one point. A **secant** is a line that intersects a circle in two points. A **tangent** is a line that intersects a circle in exactly one point.

EXAMPLE 4 DETERMINING SECANTS AND TANGENTS

Refer to Figure 9.32 to answer each question.

(a) Which lines are secants? Since $\overleftrightarrow{PB}$, $\overleftrightarrow{PD}$, and $\overleftrightarrow{QC}$ intersect the circle in two points, they are secants.

(b) Which lines are tangents? Since $\overleftrightarrow{PA}$, $\overleftrightarrow{PF}$, and $\overleftrightarrow{QE}$ intersect the circle in exactly one point, they are tangents.

(c) Are there lines in Figure 9.32 that are neither secants nor tangents? Yes. $\overleftrightarrow{PQ}$ does not intersect the circle at all.

PRACTICE EXERCISE 4

Refer to the figure to answer each question.

(a) Which lines are secants?

(b) Which lines are tangents?

(c) Which lines are neither tangents nor secants?

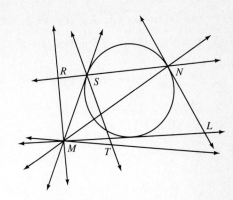

Figure 9.32

Answers: (a) $\overleftrightarrow{MN}$, $\overleftrightarrow{SN}$, and $\overleftrightarrow{ST}$
(b) $\overleftrightarrow{LN}$, $\overleftrightarrow{MS}$, and $\overleftrightarrow{LM}$
(c) $\overleftrightarrow{MT}$ and $\overleftrightarrow{MR}$

9.4 EXERCISES A

Find the diameter of the circle with the given radius.

1. $r = 11$ in   **22 in**

2. $r = 5.85$ cm   **11.7 cm**

3. $r = \dfrac{3}{4}$ ft $\dfrac{3}{2}$ ft

Find the radius of the circle with the given diameter.

4. $d = 24$ m   **12 m**

5. $d = \dfrac{5}{2}$ in $\dfrac{5}{4}$ in

6. $d = 7.24$ mm   **3.62 mm**

Find the circumference of the circle. Leave the answer using π.

7. $r = 7$ yd   **14π yd**

8. $r = \dfrac{7}{2}$ ft   **7π ft**

9 $r = 7.7$ mi   **15.4π mi**

Find the circumference of the circle. Use 3.14 for π.

10. $r = 6.8$ km
    42.7 km

11. $d = 92$ in
    288.9 in

12 $r = 0.05$ cm
    0.314 cm

Use the figure and deductive reasoning to answer the questions in Exercises 13–16.

13. What is the measure of $\overset{\frown}{ADC}$?
    120°

14. What is the measure of $\overset{\frown}{ABC}$?
    240°

15. What is the measure of inscribed $\angle ABC$?
    60°

16. What is the measure of the arc intercepted by $\angle ABC$?   **120°**

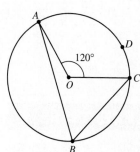

Use the figure and deductive reasoning to answer the questions in Exercises 17–20.

17. What is the measure of central ∠*AOC*?
   50°

18. What is the measure of $\overset{\frown}{ADC}$?
   50°

19. What is the measure of $\overset{\frown}{ABC}$?
   310°

20. What is the measure of the arc intercepted by ∠*ABC*?   **50°**

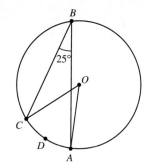

21. If a circle passes through all three vertices of right △*ABC* with ∠*C* = 90°, what is the measure of $\overset{\frown}{ACB}$?
   180°

Use the figure and deductive reasoning in Exercises 22–24.

22. Find the lines that are secants.
   $\overleftrightarrow{AB}$ **and** $\overleftrightarrow{AC}$

23. Find the lines that are tangents.
   $\overleftrightarrow{AE}$ **and** $\overleftrightarrow{BC}$

24. Find the lines that are neither secants nor tangents. $\overleftrightarrow{BD}$

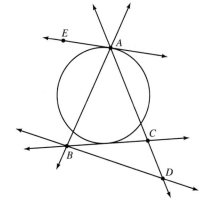

FOR REVIEW

25. In a triangle two angles have measures 82° and 54°. What is the measure of the third angle?   **44°**

26. In a right triangle one acute angle has measure 16°. What is the measure of the other acute angle?   **74°**

27. In an isosceles triangle one angle has measure 108°. What are the measures of the other angles?   **36° and 36°**

28. What are the measures of the angles in an equilateral triangle?   **60°, 60°, and 60°**

ANSWERS: 1. 22 in   2. 11.7 cm   3. $\frac{3}{2}$ ft   4. 12 m   5. $\frac{5}{4}$ in   6. 3.62 mm   7. 14π yd   8. 7π ft   9. 15.4π mi
10. 42.7 km   11. 288.9 in   12. 0.314 cm   13. 120°   14. 240°   15. 60°   16. 120°   17. 50°   18. 50°   19. 310°
20. 50°   21. 180°   22. $\overleftrightarrow{AB}$ and $\overleftrightarrow{AC}$ 23. $\overleftrightarrow{AE}$ and $\overleftrightarrow{BC}$ 24. $\overleftrightarrow{BD}$ 25. 44°   26. 74°   27. 36° and 36°   28. 60°, 60°, and
60°

9.4   EXERCISES B

Find the diameter of the circle with the given radius.

1. *r* = 28 cm   **56 cm**

2. *r* = 2.75 ft   **5.5 ft**

3. $r = \dfrac{7}{6}$ in $\dfrac{7}{3}$ in

Find the radius of the circle with the given diameter.

4. $d = 120$ mm   60 mm

5. $d = \dfrac{2}{3}$ in $\dfrac{1}{3}$ in

6. $d = 28.6$ m   14.3 m

Find the circumference of the circle. Leave the answer using π.

7. $r = 14$ m 28π m

8. $r = \dfrac{14}{11}$ cm $\dfrac{28}{11}\pi$ cm

9. $r = 5.67$ yd 11.34π yd

Find the circumference of the circle. Use 3.14 for π.

10. $d = 32.8$ ft
103.0 ft

11. $r = 420$ mm
2637.6 mm

12. $d = 1.64$ in
5.15 in

Use the figure and deductive reasoning to answer the questions in Exercises 13–16.

13. What is the measure of $\overset{\frown}{ADC}$?
64°

14. What is the measure of $\overset{\frown}{ABC}$?
296°

15. What is the measure of $\angle AOC$?
64°

16. What is the measure of the arc intercepted by $\angle ABC$?   64°

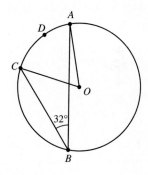

Use the figure and deductive reasoning to answer the questions in Exercises 17–20.

17. What is the measure of central $\angle AOC$?
180°

18. What is the measure of $\overset{\frown}{ABC}$?
180°

19. What is the measure of inscribed $\angle ABC$?
90°

20. What is the measure of the arc intercepted by $\angle ABC$?   180°

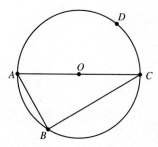

21. If A, B, and C are points on a circle and $\angle ACB = 90°$, what can be said about $\overline{AB}$ in relation to the circle?
It is a diameter of the circle.

Use the figure and deductive reasoning in Exercises 22–24.

22. Find the lines that are secants.
$\overleftrightarrow{PQ}$, $\overleftrightarrow{PR}$, and $\overleftrightarrow{QT}$

23. Find the lines that are tangents.
$\overleftrightarrow{RS}$

24. Find the lines that are neither secants nor tangents. $\overleftrightarrow{RQ}$

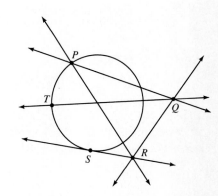

FOR REVIEW

25. Two of the angles of a triangle have measures 105° and 55°. What is the measure of the third angle?   20°

26. In a right triangle one acute angle has measure 47°. What is the measure of the other acute angle?   43°

27. If one of the angles of an isosceles triangle has measure 60°, what can be said about the triangle?
It is equilateral.

28. What is the measure of the equal angles in an isosceles right triangle?   45°

9.4 EXERCISES C

In Exercises 1–3, assume that the earth is 93,000,000 *miles from the sun and that the orbit of the earth is a circle. Use* 3.14 *for* π.

1. What distance does the earth travel in one year?
584,000,000 mi

2. What distance does the earth travel in one day?
1,600,000 mi

3. What distance does the earth travel in one minute?
1111 mi

4. In the figure, $\overline{AB} \parallel \overline{CD}$. Use deductive reasoning to prove that $\overset{\frown}{AC} = \overset{\frown}{BD}$.
Since $\overline{AB}$ and $\overline{CD}$ are parallel lines cut by the transversal $\overline{AD}$, the alternate interior angles $\angle ADC$ and $\angle BAD$ are equal. Also, equal inscribed angles intercept equal arcs. Thus, $\overset{\frown}{AC} = \overset{\frown}{BD}$.

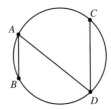

9.5 PERIMETER AND AREA

================= **STUDENT GUIDEPOSTS** =================

① Rectangles   **⑤** Triangles

② Squares   **⑥** Trapezoids

③ Square Units of Measure   **⑦** Circles

④ Parallelograms

① RECTANGLES

In this section we review formulas for finding the perimeters and areas of six geometric figures. These formulas provide problem-solving tools for numerous applications in algebra. We begin with the **rectangle,** a four-sided figure having four right angles and parallel opposite sides of equal length. Figure 9.33 shows a rectangle with **length** l and **width** w.

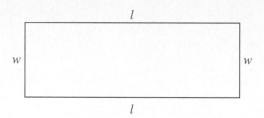

Figure 9.33   Rectangle

The **perimeter** of a geometric figure is the distance around the figure. Thus, the perimeter of a rectangle is

$$l + w + l + w = l + l + w + w$$
$$= 2l + 2w.$$

Perimeter of a Rectangle

The perimeter P of a rectangle is equal to twice the length plus twice the width.

$$P = 2l + 2w$$

The perimeter of the rectangle in Figure 9.34 is given by

$$P = 2l + 2w$$
$$= 2(5 \text{ cm}) + 2(3 \text{ cm})$$
$$= 10 \text{ cm} + 6 \text{ cm} = 16 \text{ cm}.$$

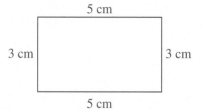

Figure 9.34

EXAMPLE 1   PERIMETER OF A RECTANGLE

Find the length of a fence needed to enclose a rectangular pasture that is 72 m wide and 90 m long.

The perimeter of the pasture is the length of the fence needed.

$$P = 2l + 2w$$
$$= 2(90 \text{ m}) + 2(72 \text{ m})$$
$$= 180 \text{ m} + 144 \text{ m}$$
$$= 324 \text{ m}$$

Thus, 324 m of fence are needed.

PRACTICE EXERCISE 1

A farmer needs to fence a rectangular pasture that is 520 ft long and 470 ft wide. How long must the fence be?

Answer: 1980 ft

❷ SQUARES

A special type of rectangle, called a **square,** is shown in Figure 9.35. All its sides have the same length. If s is the length of a side, then the perimeter of a square is

$$P = 2l + 2w$$
$$= 2s + 2s$$
$$= 4s.$$

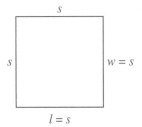

Figure 9.35   Square

Perimeter of a Square
The perimeter of a square is equal to four times the length of a side. $$P = 4s$$

EXAMPLE 2   PERIMETER OF A SQUARE

Find the perimeter of a square whose sides are 15.3 m. See Figure 9.36.

$$P = 4s$$
$$= 4(15.3 \text{ m})$$
$$= 61.2 \text{ m}$$

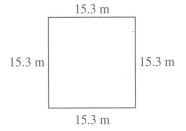

Figure 9.36

PRACTICE EXERCISE 2

A square has side $3\frac{3}{4}$ ft. Find the perimeter.

Answer: 15 ft

---

❸ SQUARE UNITS OF MEASURE

Two special squares are shown in Figure 9.37. The area of a square that has side 1 cm is one **square centimeter** and is written 1 cm^2. Likewise, if the side is 1 in, the area is one **square inch,** 1 in^2. Similarly, if some other unit of length is used, the area is 1 square unit, or 1 unit^2.

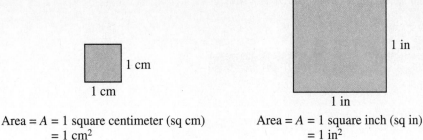

Area = A = 1 square centimeter (sq cm)   Area = A = 1 square inch (sq in)
          = 1 cm^2 = 1 in^2

Figure 9.37

Square units of measure are used for the areas of rectangles. The rectangle in Figure 9.38 has area 10 cm^2 since there are 10 of the 1-cm^2 squares contained in it. In general, this is what is meant by *area*. Notice that 10 cm^2 can be found by multiplying 5 cm by 2 cm.

$$10 \text{ cm}^2 = 5 \text{ cm} \cdot 2 \text{ cm}$$

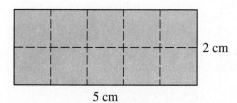

Figure 9.38

Area of a Rectangle

The area A of a rectangle is equal to the length times the width.

$$A = lw$$

EXAMPLE 3 AREA OF A RECTANGLE

Find the area of the rectangle with width 6 m and length 12 m, shown in Figure 9.39. (Sometimes we say ''the rectangle that is 6 m by 12 m.'')

$$A = lw$$
$$= (12 \text{ m})(6 \text{ m})$$
$$= (12)(6) \text{ m}^2 = 72 \text{ m}^2$$

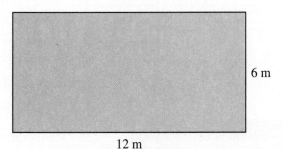

Figure 9.39

The area is 72 square meters.

PRACTICE EXERCISE 3

Find the area of a rectangle that is 17 cm by 20 cm.

Answer: **340 cm^2**

Since all sides of a square have the same measure s, the area of a square is

$$lw = ss = s^2.$$

Area of a Square

The area of a square is equal to the square of a side.

$$A = s^2$$

EXAMPLE 4   AREA OF A COMPOSITE REGION

Find the area of the region in Figure 9.40.

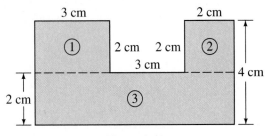

Figure 9.40

The region can be divided into two rectangles and a square.

① is 2 cm by 3 cm   Area ① = 3 cm × 2 cm = 6 cm²
② is 2 cm by 2 cm   Area ② = 2 cm × 2 cm = 4 cm²
③ is 2 cm by 8 cm   Area ③ = 8 cm × 2 cm = 16 cm²

The total area is the sum of the three areas.

$$A = 6 \text{ cm}^2 + 4 \text{ cm}^2 + 16 \text{ cm}^2 = 26 \text{ cm}^2$$

PRACTICE EXERCISE 4

Find the area of the region shown.

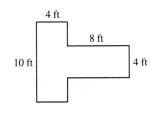

Answer: 72 ft²

---

④ PARALLELOGRAMS

A four-sided figure whose opposite sides are parallel is a **parallelogram.** A rectangle is a special case of this type of figure. Several parallelograms are shown in Figure 9.41

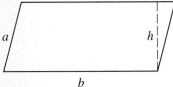

b = base a = side h = height

Figure 9.41 · Parallelograms

Perimeter of a Parallelogram

The perimeter of a parallelogram with side a and base b is given by

$$P = 2a + 2b.$$

EXAMPLE 5 PERIMETER OF A PARALLELOGRAM

Find the perimeter of the parallelogram shown in Figure 9.42.

$$P = 2(\textbf{10.5 m}) + 2(\textbf{6.5 m})$$
$$= 34 \text{ m}$$

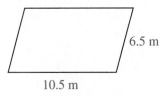

10.5 m

Figure 9.42

PRACTICE EXERCISE 5

Find the perimeter of a parallelogram with base $7\frac{1}{2}$ ft and side $4\frac{1}{2}$ ft.

Answer: 24 ft

To discover how to find the area of a parallelogram, cut off one end and put it on the other end. This forms a rectangle (Figure 9.43) with length $l = b$ and width $w = h$. Thus, the area of the parallelogram (which is the area of the rectangle) is equal to the base times the height.

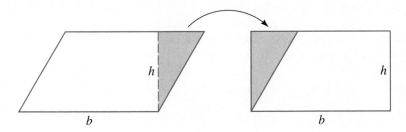

Figure 9.43

Area of a Parallelogram

The area of a parallelogram with base b and height h is given by

$$A = bh.$$

EXAMPLE 6 AREA OF A PARALLELOGRAM

Find the area of the parallelogram shown in Figure 9.44 on the following page, with height 15.2 cm and base 20.5 cm.

$$A = bh$$
$$= (\textbf{20.5 cm})(\textbf{15.2 cm})$$
$$= (20.5)(15.2) \text{ cm}^2$$
$$= 311.6 \text{ cm}^2$$

PRACTICE EXERCISE 6

Find the area of a parallelogram with base 10.8 mm and height 12.5 mm.

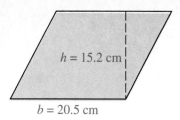

Figure 9.44

Answer: 135 mm^2

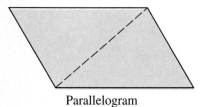

Parallelogram

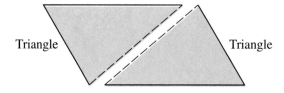

Triangle   Triangle

Figure 9.45

5 TRIANGLES

If the parallelogram in Figure 9.45 is cut along the dotted line, the area of each of the two triangles formed is one-half the area of the parallelogram. Thus, a triangle can be made into a parallelogram by adding to it a triangle of the same size. This is shown in Figure 9.46, where triangle *B* is the same size and shape as the original triangle *A*. Since the heights of the triangle and the parallelogram are the same, the area of the triangle must be one-half the area of the parallelogram.

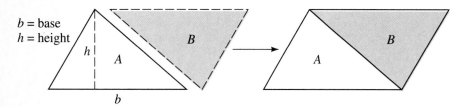

Figure 9.46   Triangles

Area of a Triangle

The area of a triangle with base *b* and height *h* is

$$A = \frac{1}{2}bh.$$

EXAMPLE 7   APPLICATION OF AREA

A wall to be painted is in the shape shown in Figure 9.47.

(a) How many square feet must be painted?

PRACTICE EXERCISE 7

The wall shown in the figure is to be painted.

First find the area of the rectangle.

$$A = lw = 10 \text{ ft} \cdot 8 \text{ ft} = 80 \text{ ft}^2$$

The base of the triangle is 10 ft and the height is 6 ft.

$$A = \frac{1}{2}bh = \frac{1}{2}(10 \text{ ft})(6 \text{ ft}) = 30 \text{ ft}^2$$

The total area to be covered is

$$80 \text{ ft}^2 + 30 \text{ ft}^2 = 110 \text{ ft}^2.$$

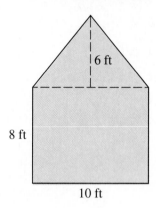

8 ft

6 ft

10 ft

Figure 9.47

(b) How much paint will it take if one gallon covers 220 ft^2?

Multiply 110 ft^2 by $\frac{1 \text{ gal}}{220 \text{ ft}^2}$.

$$\frac{110 \text{ ft}^2}{1} \cdot \frac{1 \text{ gal}}{220 \text{ ft}^2} = \frac{110 \text{ gal}}{2(110)} = \frac{1}{2} \text{ gal}$$

It will take $\frac{1}{2}$ gallon of paint to cover the area.

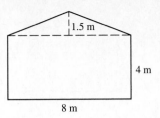

1.5 m

4 m

8 m

(a) What is the area to be painted?

(b) If one liter of paint covers 19 m^2, how many liters of paint are required?

Answers: (a) 38 m^2 (b) 2 L

🛑 TRAPEZOIDS

A **trapezoid** is a four-sided figure with at least two parallel sides. See Figure 9.48. The parallel lines are called the *bases* and *h* is the *height*.

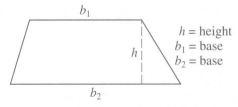

b_1

h

b_2

h = height
b_1 = base
b_2 = base

Figure 9.48   Trapezoid

If we take a trapezoid the same shape and size as the given one, turn it over and put the two trapezoids together, as in Figure 9.49, it forms a parallelogram. The area of the parallelogram is equal to the base times the height. The base is $b_1 + b_2$ and the height is h. The area of the trapezoid is one-half their product.

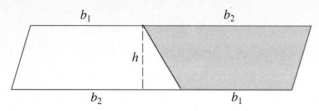

Figure 9.49

Area of a Trapezoid

The area A of a trapezoid with height h and bases b_1 and b_2 is

$$A = \frac{1}{2}(b_1 + b_2)h.$$

EXAMPLE 8 AREA OF A TRAPEZOID	PRACTICE EXERCISE 8

Find the area of the trapezoid shown in Figure 9.50.

$$A = \frac{1}{2}(b_1 + b_2)h$$

$$= \frac{1}{2}(24 \text{ mm} + 16 \text{ mm})(12 \text{ mm})$$

$$= \frac{1}{2}(40 \text{ mm})(12 \text{ mm})$$

$$= \frac{1}{2}(40)(12) \text{ mm}^2 = 240 \text{ mm}^2$$

Find the area of the trapezoid with bases 6 inches and 14 inches and height 9 inches.

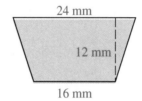

24 mm

12 mm

16 mm

Figure 9.50

Answer: 90 in^2

➐ CIRCLES

In Section 9.4 we discussed circles. We calculated circumferences but did not determine the area of a circle.

Area of a Circle

The area A of a circle is given by

$$A = \pi r^2.$$

We can get some idea of why this is true by looking at Figure 9.51. Suppose we cut the circle in Figure 9.51(a) along all the radii (plural of radius). When we put the top half of the circle above the bottom half as in Figure 9.51(b) and then slide them together as in Figure 9.51(c), the figure formed is approximately a parallelogram with height r. Its base is πr since $2\pi r$ is the circumference and half the circle is used on top and half on the bottom. Thus,

$$A = bh$$
$$= (\pi r)(r) = \pi r^2.$$

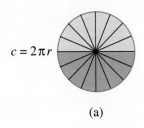

(a)

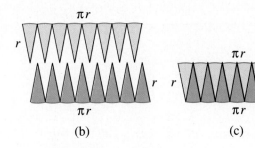

(b)   (c)

Figure 9.51

EXAMPLE 9 **APPLICATION OF AREA OF A CIRCLE**	**PRACTICE EXERCISE 9**

In the machine part shown in Figure 9.52, each circular hole has radius 3 cm. Find the area of metal that is left. Use 3.14 for π.

A triangle with base 10 inches and height 5 inches has three holes of radius 1 inch drilled through it. What is the area left in the triangle after the holes are drilled? Use 3.14 for π.

First calculate the area of the trapezoid.

$$A = \frac{1}{2}(b_1 + b_2)h$$

$$= \frac{1}{2}(16 \text{ cm} + 24 \text{ cm})(14 \text{ cm})$$

$$= \frac{1}{2}(40 \text{ cm})(14 \text{ cm})$$

$$= \frac{1}{2}(40)(14) \text{ cm}^2$$

$$= (20)(14) \text{ cm}^2 = 280 \text{ cm}^2$$

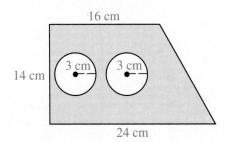

Figure 9.52

Now find the area of each circle.

$$A = \pi r^2 \approx 3.14(3 \text{ cm})^2$$
$$= (3.14)(9) \text{ cm}^2 = 28.26 \text{ cm}^2$$

Thus, the two circles have a combined area of

$$2(28.26 \text{ cm}^2) = 56.52 \text{ cm}^2.$$

The area of metal is the area of the trapezoid minus the area of the circles.

$$
\begin{array}{r}
280.00 \\
-\ \ 56.52 \\
\hline
223.48
\end{array}
$$

The area left is 223.48 cm^2.

Answer: 15.58 in^2

9.5 EXERCISES A

Find the perimeter of the rectangle (or square) with the given width and length.

1. 6 ft by 11 ft
34 ft

2. 2.5 m by 3.6 m
12.2 m

3. 102 mm by 30.5 mm
265 mm

4. 6.8 mi by $9\frac{1}{2}$ mi
32.6 mi

5. 7.2 cm by 7.2 cm
28.8 cm

6. $22\frac{1}{2}$ mi by $22\frac{1}{2}$ mi
90 mi

Find the area of the rectangle (or square) with the given width and length.

7. 6 ft by 11 ft
66 ft^2

8. 2.5 m by 3.6 m
9.0 m^2

9. 102 mm by 30.5 mm
3111 mm^2

10. 6.8 mi by $9\frac{1}{2}$ mi
64.6 mi^2

11. 7.2 cm by 7.2 cm
51.84 cm^2

12. $22\frac{1}{2}$ mi by $22\frac{1}{2}$ mi
506.25 mi^2

Solve.

13. Find the perimeter and area of the given figure.
16 m; 9 m^2

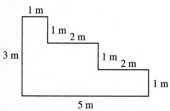

14. Fencing costs $3.20 per yard. If a pasture 62 yd by 205 yd is to be fenced, how much will it cost?
$1708.80

15. A rose garden is 7.2 m by 10.8 m. If a square with side 2.5 m is used in the middle of the garden for a fountain, what area remains for roses?
71.51 m^2

16 The area shown is to be carpeted. How much will it cost if carpet costs $16.50 per square yard?
$445.50

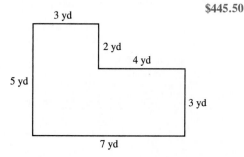

Find the area of each figure.

17.
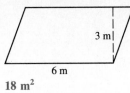
6 m
3 m|
18 m²

18.
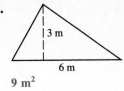
3 m
6 m
9 m²

19.
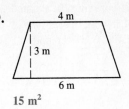
4 m
3 m
6 m
15 m²

20.
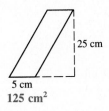
25 cm
5 cm
125 cm²

21.
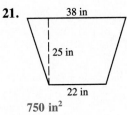
38 in
25 in
22 in
750 in²

22.
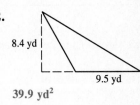
8.4 yd
9.5 yd
39.9 yd²

23.
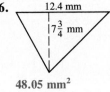
5½ mi
7.2 mi
39.6 mi²

24.
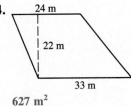
24 m
22 m
33 m
627 m²

25.
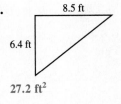
8.5 ft
6.4 ft
27.2 ft²

26.
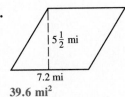
12.4 mm
7¾ mm
48.05 mm²

27.
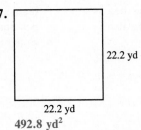
22.2 yd
22.2 yd
492.8 yd²

28.
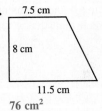
7.5 cm
8 cm
11.5 cm
76 cm²

Find the perimeter of each figure.

29.
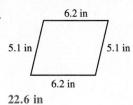
6.2 in
5.1 in   5.1 in
6.2 in
22.6 in

30.
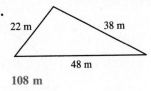
22 m   38 m
48 m
108 m

31.
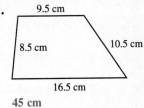
9.5 cm
8.5 cm   10.5 cm
16.5 cm
45 cm

32 Find the area of the figure.   **322 cm²**
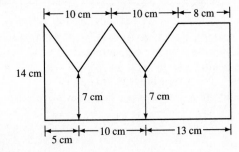
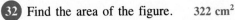
10 cm   10 cm   8 cm
14 cm
7 cm   7 cm
5 cm   10 cm   13 cm

33. The recreation area sketched in the figure is to be carpeted. How much will it cost if carpet is $14.50 per square yard?   **$4988**
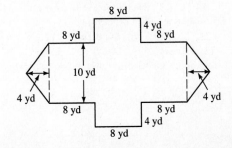
8 yd
4 yd
8 yd   8 yd
10 yd
4 yd   4 yd
8 yd
8 yd   8 yd
4 yd
8 yd

34. For a play, 10 walls shaped like the one shown in the figure must be painted on both sides. How much will it cost if a gallon of paint covers 160 ft^2 and costs $12.50 per gallon? $50

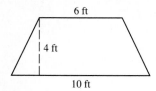

Find the area of the circle. Leave the answer using π.

35. $r = 7$ yd   **49π yd^2**

36. $r = \dfrac{7}{2}$ ft $\dfrac{49}{2}\pi$ ft^2

37 $r = 7.7$ mi
59.29π mi^2

Find the area of the circle. Use 3.14 for π.

38. $r = 6.8$ km
145.2 km^2

39. $d = 92$ in
6644.2 in^2

40 $r = 0.05$ cm
0.00785 cm^2

Solve.

41 How much more area is there for water to pass through in a $\frac{3}{4}$-inch diameter water hose than there is in a $\frac{1}{2}$-inch diameter water hose? Use 3.14 for π.
0.246 in^2

42. A 12-inch diameter pizza costs $4.50. A 16-inch diameter pizza costs $7.50. Which pizza costs less per square inch?   **12-in: $0.0398 per in^2; 16-in: $0.0373 per in^2; 16-in pizza costs less.**

FOR REVIEW

Use the figure and deductive reasoning in Exercises 43–45.

43. What is the measure of $\angle AOB$?
180°

44. What is the measure of $\overset{\frown}{ADB}$?
180°

45. What is the measure of $\angle ACB$?
90°

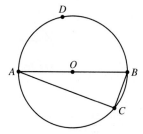

ANSWERS: (Some answers are rounded.) 1. 34 ft 2. 12.2 m 3. 265 mm 4. 32.6 mi 5. 28.8 cm 6. 90 mi
7. 66 ft^2 8. 9.0 m^2 9. 3111 mm^2 10. 64.6 mi^2 11. 51.84 cm^2 12. 506.25 mi^2 13. 16 m; 9 m^2 14. $1708.80
15. 71.51 m^2 16. $445.50 17. 18 m^2 18. 9 m^2 19. 15 m^2 20. 125 cm^2 21. 750 in^2 22. 39.9 yd^2 23. 39.6 mi^2
24. 627 m^2 25. 27.2 ft^2 26. 48.05 mm^2 27. 492.8 yd^2 (to the nearest tenth) 28. 76 cm^2 29. 22.6 in 30. 108 m
31. 45 cm 32. 322 cm^2 33. $4988 34. $50 35. 49$\pi$ yd^2 36. $\frac{49}{2}\pi$ ft^2 37. 59.29π mi^2 38. 145.2 km^2
39. 6644.2 in^2 40. 0.00785 cm^2 41. 0.246 in^2 42. 12-in: $0.0398 per in^2; 16-in: $0.0373 per in^2; 16-in pizza costs
less. 43. 180° 44. 180° 45. 90°

9.5 EXERCISES B

Find the perimeter of the rectangle (or square) with the given width and length.

1. 10 ft by 17 ft   **54 ft**

2. 8.2 m by 9.5 m   **35.4 m**

3. 2500 mm by 45.5 mm
5091 mm

4. 2.2 mi by $6\frac{1}{2}$ mi   **17.4 mi**

5. 40.2 cm by 40.2 cm   **160.8 cm**

6. $10\frac{1}{2}$ mi by $10\frac{1}{2}$ mi   **42 mi**

Find the area of the rectangle (or square) with the given width and length.

7. 10 ft by 17 ft   **170 ft²**

8. 8.2 m by 9.5 m   **77.9 m²**

9. 2500 mm by 45.5 mm
113,750 mm²

10. 2.2 mi by $6\frac{1}{2}$ mi   **14.3 mi²**

11. 40.2 cm by 40.2 cm
1616.04 cm²

12. $10\frac{1}{2}$ mi by $10\frac{1}{2}$ mi
110.25 mi²

Solve.

13. The play area shown needs to have a new fence around it. How long will the fence need to be?

28 m

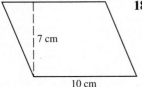

14. A rancher needs to fence an area which is 2500 yd by 4200 yd. If fencing costs $2.60 per yard, how much will it cost?
$34,840

15. A flower garden is in the shape of a square which is 22.5 m on a side. If a walk 1.5 m wide is put across the garden parallel to one of the sides, how much area is left for flowers?
472.5 m²

16. The area shown is carpeted with carpeting costing $24.50 per square yard. How much did it cost?
$2156

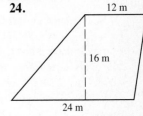

Find the area of each figure.

17.

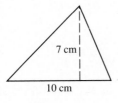

70 cm²

18.

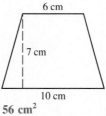

35 cm²

19.

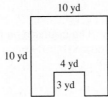

56 cm²

20.

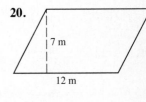

84 m²

21.

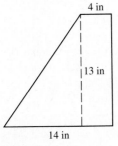

117 in²

22.

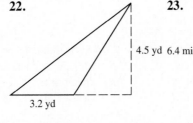

7.2 yd²

23.

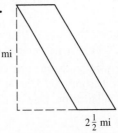

16 mi²

24.

288 m²

25.

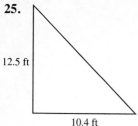

12.5 ft

10.4 ft

65 ft²

26.

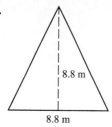

8.8 m

8.8 m

38.72 m²

27.

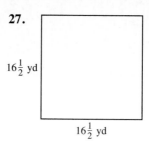

16½ yd

16½ yd

272¼ yd²

28.

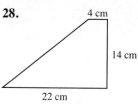

4 cm

14 cm

22 cm

182 cm²

Find the perimeter of each figure.

29.

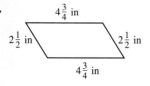

4¾ in

2½ in   2½ in

4¾ in

14½ in

30.

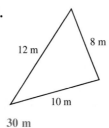

12 m   8 m

10 m

30 m

31.

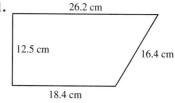

26.2 cm

12.5 cm   16.4 cm

18.4 cm

73.5 cm

32. Find the area of the figure.   **135 cm²**

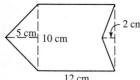

5 cm 10 cm   2 cm

12 cm

33. The patio area shown is to be covered with outdoor carpeting. What will be the total cost if the carpeting can be installed for $22.50 per square meter?
$5580

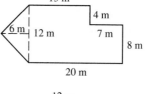

13 m

6 m 12 m   4 m
                7 m
                    8 m

20 m

34. The sheet of metal shown is to be insulated on both sides. If 6 m² of insulation cost $10.20, how much will it cost to do the job?
$163.20

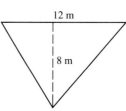

12 m

8 m

Find the area of the circle. Leave the answer using π.

35. $r = 14$ m   **196π m²**

36. $r = \dfrac{14}{11}$ cm $\dfrac{196}{121}\pi$ cm²

37. $r = 5.67$ yd   **32.15π yd²**

Find the area of the circle. Use 3.14 for π.

38. $d = 32.8$ ft
   844.5 ft²

39. $r = 420$ mm
   553,896 mm²

40. $d = 1.64$ in
   2.11 in²

Solve.

41. A circular garden has radius 9 m. If a 1-m-wide circular walk is put around it, what is the area of the walk? Use 3.14 for π. (*Hint:* Find the area of a circle with 10-m radius and subtract the area of the garden.)
59.66 m²

42. A 14-inch-diameter pizza costs $6.00 and a 16-inch-diameter pizza costs $8.00. Which pizza costs less per square inch? Use 3.14 for π.
14-in: $0.039 per in²; 16-in: $0.040 per in²; 14-in costs less.

FOR REVIEW

Use the figure and deductive reasoning in Exercises 43–45.

43. What is the measure of $\overset{\frown}{ADB}$?
80°

44. What is the measure of $\overset{\frown}{ACB}$?
280°

45. What is the measure of $\angle ACB$?
40°

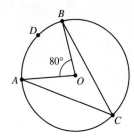

9.5 EXERCISES C

Solve.

1. A patio is in the shape of a trapezoid with bases 8.1 yd and 6.7 yd and height 5.8 yd. Assuming no waste, to the nearest cent how much will it cost to cover the patio with outdoor carpeting that costs $2.15 a square foot?
[Answer: $830.50]

2. A sheet of metal is in the shape of a triangle with base 1500 cm and height 800 cm, and has a piece removed from the center of it in the shape of a parallelogram with base 125 cm and height 85 cm. How much will it cost to cover the metal on both sides with rustproofing paint that costs $0.45 per square meter? $53.04

3. Find the area (to the nearest hundredth) of the metal in the machine part shown, which has three circular holes drilled in it. Use 3.14 for π.

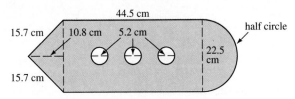

[Answer: 1257.77 cm^2]

4. Find the exterior perimeter (to the nearest tenth) of the machine part in Exercise 3.   **155.7 cm**

5. The shaded portion of the circle at right is called a **sector** of the circle, and determined by central $\angle AOB$. The area of a sector is given by the formula $A = \frac{\pi}{360} r^2 a$ where a is the measure of central $\angle AOB$. Use this formula to find the area (to the nearest hundredth) of the sector determined by the given central $\angle AOB$ in a circle of radius 5 cm. Use 3.14 for π.

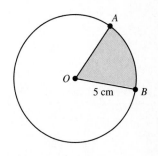

   (a) $\angle AOB = 180°$
      39.25 cm^2

   (b) $\angle AOB = 90°$
      19.63 cm^2

   (c) $\angle AOB = 35°$
      7.63 cm^2

9.6 VOLUME AND SURFACE AREA

1 VOLUME

To measure length we use units such as 1 cm and 1 inch. When we talk about area we use units such as 1 cm² and 1 in². To find volume we will use cubic units such as 1 cm³ and 1 in³.

The **volume** of a cube that is 1 cm by 1 cm by 1 cm is one **cubic centimeter** and is written 1 cm³. Similarly, a cube that is 1 in by 1 in by 1 in is one **cubic inch,** 1 in³. See Figure 9.53. If some other unit were used for measuring, the volume of the cube would be 1 cubic unit, or 1 unit³.

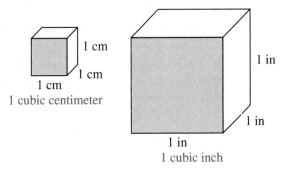

1 cubic centimeter

1 cubic inch

Figure 9.53

2 RECTANGULAR SOLIDS

The cubic unit of measure is used for the volume of solids. The **rectangular solid** in Figure 9.54(a) has volume 20 cm³ since there are 10 cm³ on the bottom layer and 10 cm³ on the top layer. In Figure 9.54(b) there is one more layer (10 cm³) than in Figure 9.54(a), so the volume is 30 cm³. These volumes can be found without a diagram by multiplying the area of the base by the height.

$$20 \text{ cm}^3 = \underbrace{5 \text{ cm} \cdot 2 \text{ cm}}_{\text{area of base}} \cdot \underbrace{2 \text{ cm}}_{\text{height}}$$

$$30 \text{ cm}^3 = \underbrace{5 \text{ cm} \cdot 2 \text{ cm}}_{\text{area of base}} \cdot \underbrace{3 \text{ cm}}_{\text{height}}$$

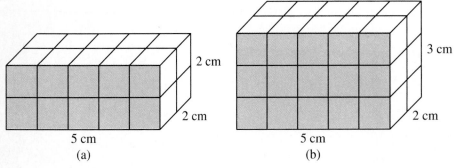

(a)   (b)

Figure 9.54

Volume of a Rectangular Solid

The volume V of a rectangular solid is equal to the area of the base lw times the height h.

$$V = lwh$$

EXAMPLE 1   VOLUME OF A RECTANGULAR SOLID

A highway construction truck has a bed that is 5.5 yd long, 2.5 yd wide, and 2 yd deep. How many cubic yards of gravel will it hold?

Find the volume of the truck bed.

$$
\begin{aligned}
V &= lwh \\
&= (5.5 \text{ yd})(2.5 \text{ yd})(2 \text{ yd}) \\
&= (5.5)(2.5)(2) \text{ yd}^3 \\
&= 27.5 \text{ yd}^3
\end{aligned}
$$

The truck will hold 27.5 yd^3 of gravel.

PRACTICE EXERCISE 1

A trailer for hauling cotton is 18 ft by 8 ft by 10 ft. What volume of cotton will it hold?

Answer: 1440 ft^3

---

❸ CUBES

A **cube** is a rectangular solid for which $l = w = h$. We talk about *edges* of the cube, and call the length of each edge e.

Volume of a Cube

The volume V of a cube is equal to the cube of the edge.

$$V = e^3 \qquad (e^3 \text{ is } e \text{ times } e \text{ times } e)$$

EXAMPLE 2   VOLUME OF A CUBE

Find the volume of the cube in Figure 9.55 which has an edge of 1.5 ft.

$$
\begin{aligned}
V &= (1.5 \text{ ft})^3 \\
&= (1.5 \text{ ft})(1.5 \text{ ft})(1.5 \text{ ft}) \\
&= 3.375 \text{ ft}^3
\end{aligned}
$$

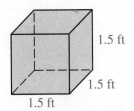

1.5 ft

1.5 ft

1.5 ft

Figure 9.55

PRACTICE EXERCISE 2

Find the volume of a cube which has an edge of $3\frac{1}{2}$ inches.

Answer: $42\frac{7}{8}$ in^3

④ SURFACE AREA

Another type of problem involving a rectangular solid is finding the area of the surface. If we can do this, we can find the amount of material needed to make a rectangular tank or the amount of paint needed to paint a building. The **surface area** of an object is the area of all its surfaces. Consider the solid in Figure 9.56, which was used to count cubic centimeters earlier. Now, we count square centimeters on its surface. There are (5 cm)(3 cm) = 15 cm² on the front face. But the back is just like the front, so there are

$$2(15 \text{ cm}^2) = 30 \text{ cm}^2 \qquad 2lh$$

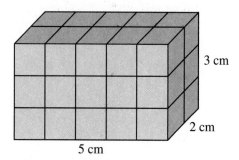

3 cm

2 cm

5 cm

Figure 9.56

on the front and back together. On the two ends there are

$$2(2 \text{ cm})(3 \text{ cm}) = 12 \text{ cm}^2. \qquad 2wh$$

On the bottom there are

$$(5 \text{ cm})(2 \text{ cm}) = 10 \text{ cm}^2,$$

with the same number on the top. The total area of the top and bottom is

$$2(10 \text{ cm}^2) = 20 \text{ cm}^2. \qquad 2lw$$

The total surface area is the sum of all surfaces.

Area of front and back:	30 cm²	2lh
Area of two ends:	12 cm²	2wh
Area of bottom and top:	20 cm²	2lw
Total surface area:	62 cm²	

Surface Area of a Rectangular Solid

The total surface area A of a rectangular solid is

$$A = 2lh + 2wh + 2lw.$$

EXAMPLE 3   SURFACE AREA OF A RECTANGULAR SOLID

Find the total surface area of a rectangular solid 6 m by 3 m by 7 m.

$$A = 2lh + 2wh + 2lw$$
$$= 2(6 \text{ m})(7 \text{ m}) + 2(3 \text{ m})(7 \text{ m}) + 2(6 \text{ m})(3 \text{ m})$$
$$= 2(6)(7)\text{m}^2 + 2(3)(7)\text{m}^2 + 2(6)(3)\text{m}^2$$
$$= 84 \text{ m}^2 + 42 \text{ m}^2 + 36 \text{ m}^2 = 162 \text{ m}^2$$

PRACTICE EXERCISE 3

Find the surface area of a rectangular solid 12 ft by 16 ft by 20 ft.

Answer: 1504 ft²

| EXAMPLE 4   APPLICATION OF SURFACE AREA | PRACTICE EXERCISE 4 |

The four vertical sides of a storage building are to be painted. One gallon of paint will cover 150 ft^2 of area. How much paint will be used if the building is 16 ft by 12 ft by 8 ft?

The outside walls of a workshop are to be covered with siding that costs $1.50 per square foot. How much will it cost if the building is 16 ft wide, 18 ft long, and 8 ft high?

The area of the front and back is twice the length times the height.

$$2(16 \text{ ft})(8 \text{ ft}) = 256 \text{ ft}^2$$

The area of the two ends is twice the width times the height.

$$2(12 \text{ ft})(8 \text{ ft}) = 192 \text{ ft}^2$$

The total area is the sum of these two areas.

Area of front and back:	256 ft^2
Area of two ends:	192 ft^2
Total area:	448 ft^2

Since the paint covers 150 ft^2 per gal, the unit fraction is $\dfrac{1 \text{ gal}}{150 \text{ ft}^2}$.

$$\frac{448 \text{ ft}^2}{1} \times \frac{1 \text{ gal}}{150 \text{ ft}^2} = \frac{448}{150} \text{ gal}$$

$$\approx 2.99 \text{ gal} \qquad \text{To the nearest hundredth}$$

Thus, it will take about 3 gal to paint the building.

Answer: $816

Surface Area of a Cube

The surface area of a cube with edge e is given by

$$A = 6e^2.$$

| EXAMPLE 5   SURFACE AREA OF A CUBE | PRACTICE EXERCISE 5 |

Find the surface area of a cube with edge 14 cm.

Find the surface area of a cube with edge 200 mm.

$$A = 6e^2$$
$$= 6(14 \text{ cm})^2$$
$$= 6(196) \text{ cm}^2$$
$$= 1176 \text{ cm}^2$$

Answer: 240,000 mm^2

5 CYLINDERS

A **cylinder** (see Figure 9.57) is a solid with circular ends of the same radius. For example, a can of beans is in the shape of a cylinder.

Remember that the volume of a rectangular solid is the area of the base times the height. The same is true for a cylinder. Since the base is a circle with area πr^2, we have the following formula.

Volume of a Cylinder

The volume V of a cylinder is equal to the area of the base, πr^2, times the height, h.

$$V = \pi r^2 h$$

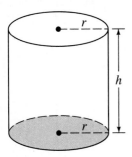

Figure 9.57 Cylinder

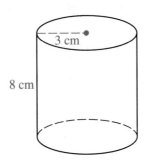

Figure 9.58

EXAMPLE 6 VOLUME OF A CYLINDER

Find the volume of the cylinder in Figure 9.58, which is 8 cm high and has a base of radius 3 cm. Use 3.14 for π.

Since $r = 3$ cm and $h = 8$ cm,

$$V = \pi r^2 h$$
$$\approx 3.14(3 \text{ cm})^2(8 \text{ cm})$$
$$= (3.14)(9)(8) \text{ cm}^3 = 226.08 \text{ cm}^3.$$

PRACTICE EXERCISE 6

Find the volume of a cylinder with radius of the base 6 cm and height 20 cm. Use 3.14 for π.

Answer: 2260.8 cm^3

EXAMPLE 7 APPLICATION OF VOLUME

A can of blueberry pie filling has diameter 3 in and height 4.5 in. How many cans of filling are needed to fill a 9-in diameter pie pan 1 in deep?

First find the volume of pie filling in one can.

$$V = \pi r^2 h$$
$$\approx 3.14(1.5 \text{ in})^2(4.5 \text{ in})$$
$$= (3.14)(2.25)(4.5) \text{ in}^3$$
$$\approx 31.8 \text{ in}^3 \qquad \text{Volume in one can to nearest tenth}$$

Now find the volume of pie filling in a 9-in diameter (4.5-in radius) pie pan.

$$V = \pi r^2 h$$
$$\approx 3.14(4.5 \text{ in})^2(1 \text{ in}) \qquad \text{Filling is 1 in deep}$$
$$= (3.14)(20.25)(1) \text{ in}^3$$
$$\approx 63.6 \text{ in}^3 \qquad \text{Volume in pie pan to nearest tenth}$$

PRACTICE EXERCISE 7

Water is stored in a cylindrical tank with diameter of the base 12 m and height 10 m. How many water trucks having cylindrical tanks with diameter 1.6 m and length 5 m can be filled from the storage tank?

Since one can has 31.8 in^3 of filling, multiply by the unit fraction
$\dfrac{1 \text{ can}}{31.8 \text{ in}^3}$.

$$\frac{63.6 \text{ in}^3}{1} \times \frac{1 \text{ can}}{31.8 \text{ in}^3} = \frac{63.6}{31.8} \text{ can} = 2 \text{ cans}$$

It will take 2 cans of filling to make the pie.

Answer: 112.5

To see how to find the surface area of a cylinder, consider a can without top or bottom. See Figure 9.59(a). If it is cut along the seam and pressed flat, as in Figure 9.59(b), a rectangle is formed. The length of the rectangle is the circumference of the circle and the width is the height. Thus, the surface area of the side of the can is $2\pi rh$.

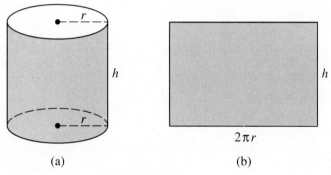

(a)   (b)

Figure 9.59

Since the area of the top of the can is πr^2, and the area of the bottom is also πr^2, we have the following:

Surface Area of a Cylinder

The surface area A of a cylinder is given by

$$A = 2\pi rh + 2\pi r^2.$$

EXAMPLE 8   SURFACE AREA OF A CYLINDER

Find the surface area of a cylinder with radius 7 ft and height 9 ft.

$$A = 2\pi rh + 2\pi r^2$$
$$\approx 2(3.14)(7 \text{ ft})(9 \text{ ft}) + 2(3.14)(7 \text{ ft})^2$$
$$= 2(3.14)(7)(9) \text{ ft}^2 + 2(3.14)(49) \text{ ft}^2$$
$$= 703.36 \text{ ft}^2$$

PRACTICE EXERCISE 8

Find the surface area of a cylinder with radius 16 ft and height 4 ft.

Answer: 2009.6 ft^2

⑥ SPHERES

A **sphere** (Figure 9.60) is a solid in the shape of a ball. The radius of a sphere is the distance from its center to the surface. The diameter is twice the radius. If a sphere is cut through the center, the cut surface is a circle.

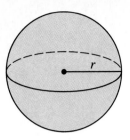

Figure 9.60 Sphere

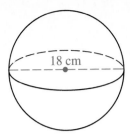

18 cm

Figure 9.61

Volume of a Sphere

The volume V of a sphere is given by

$$V = \frac{4}{3}\pi r^3.$$

EXAMPLE 9   VOLUME OF A SPHERE

Find the volume of the sphere in Figure 9.61 with diameter 18 cm.

If the diameter is 18 cm, then the radius r is $\frac{18}{2}$ cm or 9 cm.

$$V = \frac{4}{3}\pi r^3 \approx \frac{4}{3}(3.14)(9 \text{ cm})^3$$

$$= \frac{4}{3}(3.14)729 \text{ cm}^3$$

$$= 3052.08 \text{ cm}^3$$

PRACTICE EXERCISE 9

Find the volume of a sphere with diameter 10 ft.

Answer: 523.3 ft^3

Surface Area of a Sphere

The surface area A of a sphere is given by

$$A = 4\pi r^2.$$

EXAMPLE 10   APPLICATION OF SURFACE AREA

A spherical tank with radius 3.2 m is to be covered with insulation. If the insulation costs $2.50 per square meter, how much will the material cost?

$$A = 4\pi r^2$$

$$\approx 4(3.14)(3.2 \text{ m})^2$$

$$= 4(3.14)(10.24) \text{ m}^2$$

$$\approx 128.61 \text{ m}^2 \qquad \text{To the nearest hundredth}$$

Since the insulation costs $2.50 per square meter, the cost is

$$\frac{128.61 \text{ m}^2}{1} \times \frac{\$2.50}{1 \text{ m}^2} \approx \$321.53. \qquad \text{To the nearest cent}$$

PRACTICE EXERCISE 10

A spherical tank with diameter 6.4 m is to be prepared for painting. If the labor costs are $4.50 per square meter, how much will the job cost?

Answer: $578.76

9.6 EXERCISES A

Find the volume of the rectangular solid with the given dimensions.

1. 10 m by 5 m by 6 m
300 m³

2. 2.3 cm by 1.2 cm by 4.5 cm
12.42 cm³

③ $9\frac{1}{2}$ in by 3.4 in by $8\frac{3}{4}$ in
282.625 in³

Find the volume of the cube with the given edge length.

4. 8 m   512 m³

5. 10.2 cm   1061.208 cm³

6. $1\frac{3}{5}$ in   4.096 in³

Find the total surface area of each rectangular solid.

7. 6 ft by 2 ft by 5 ft
104 ft²

8. $1\frac{1}{2}$ in by $\frac{1}{2}$ in by 4 in

$17\frac{1}{2}$ in²

9. 10.2 yd by $8\frac{1}{2}$ yd by 4.5 yd

341.7 yd²

⑩ 15.5 cm by 6.4 cm by 20.2 cm

1083.16 cm²

Find the surface area of the cube with the given edge length.

11. 5 ft   150 ft²

12. 21 mm   2646 mm²

13. $1\frac{1}{2}$ yd $13\frac{1}{2}$ yd²

Solve.

14. Frank's truck is to be used to carry topsoil. If the truck bed is 2 yd by 1.5 yd by 0.8 yd, how many cubic yards of topsoil will it hold?
2.4 yd³

⑮ A tank is in the shape of a cube. How many grams of water will it hold if the edge of the cube is 22 cm? Note that 1 cm³ of water weighs 1 g.
10,648 g

⑯ How many square centimeters of glass did it take to make the bottom and sides of the tank in Exercise 15?
2420 cm²

17. Elizabeth owns a rectangular building that is 30 ft by 12 ft by 8 ft. How much will it cost to put siding on the building if the siding costs $0.75 per square foot?
$504

Find the volume of each cylinder.

18. $r = 6$ cm, $h = 5$ cm
565.2 cm³

⑲ $d = 24.4$ in, $h = 30$ in
14,020.7 in³

20. $d = 11.4$ m, $h = 4.4$ m
448.9 m³

Find the surface area of each cylinder.

21. $r = 6$ cm, $h = 5$ cm
414.48 cm²

㉒ $d = 24.4$ in, $h = 30$ in
3233.2 in²

23. $d = 11.4$ m, $h = 4.4$ m
361.5 m²

Find the volume of each sphere.

24. $r = 12$ mm
7234.56 mm³

㉕ $r = \frac{3}{4}$ in
1.766 in³

26. $d = 5.6$ yd
91.91 yd³

Find the surface area of each sphere.

27. $r = 12$ mm
    1808.64 mm²

28. $r = \dfrac{3}{4}$ in
    7.065 in²

29 $d = 5.6$ yd
    98.47 yd²

Solve.

30 A can of cherry pie filling has diameter 8 cm and height 12 cm. How many cans (to the nearest tenth) are needed to fill a 20-cm diameter pie pan 3 cm deep?
1.6 cans

31. A cylindrical storage tank has radius 3.8 ft and height 9.8 ft. How many gallons (to the nearest tenth) of paint are needed to paint the tank (including top and bottom) if one gallon covers 150 ft²?
2.2 gal

32. A spherical tank has radius 8.6 m. If a rust-preventing material costs $1.50 per square meter, what is the cost to rustproof the tank?
$1393.41

33. A spherical tank with radius 32 cm is filled with gasoline. How many liters are in the tank? Note that 1 cm³ = 0.001 L.
137.19 L

FOR REVIEW

Find the area of metal left on each of the machine parts shown, which have circular holes drilled in them. Use 3.14 for π.

34.

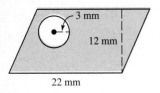

235.74 mm²

35

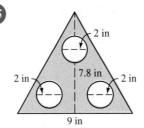

25.68 in²

36.

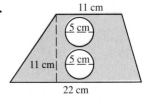

142.25 cm²

ANSWERS: 1. 300 m³   2. 12.42 cm³   3. 282.625 in³   4. 512 m³   5. 1061.208 cm³   6. 4.096 in³   7. 104 ft²
8. $17\frac{1}{2}$ in²   9. 341.7 yd²   10. 1083.16 cm²   11. 150 ft²   12. 2646 mm²   13. $13\frac{1}{2}$ yd²   14. 2.4 yd³   15. 10,648 g
16. 2420 cm²   17. $504   18. 565.2 cm³   19. 14,020.7 in³ (nearest tenth)   20. 448.9 m³ (nearest tenth)
21. 414.48 cm²   22. 3233.2 in² (nearest tenth)   23. 361.5 m² (nearest tenth)   24. 7234.56 mm³   25. 1.766 in³ (nearest
thousandth)   26. 91.91 yd³ (nearest hundredth)   27. 1808.64 mm²   28. 7.065 in²   29. 98.47 yd² (nearest hundredth)
30. 1.6 cans   31. 2.2 gal   32. $1393.41   33. 137.19 L (nearest hundredth)   34. 235.74 mm²   35. 25.68 in²
36. 142.25 cm²

9.6 EXERCISES B

Find the volume of the rectangular solid with the given dimensions.

1. 6 m by 12 m by 15 m
    1080 m³

2. 9.2 cm by 1.5 cm by 7.5 cm
    103.5 cm³

3. $6\dfrac{1}{2}$ in by 5.5 in by $3\dfrac{1}{5}$ in
    114.4 in³

Find the volume of the cube with the given edge length.

4. 12 m   1728 m³

5. 0.1 cm   0.001 cm³

6. $2\frac{1}{6}$ in $10\frac{37}{216}$ in³

Find the total surface area of each rectangular solid.

7. 3 ft by 5 ft by 12 ft   222 ft²

8. $4\frac{1}{2}$ in by $\frac{1}{4}$ in by 6 in $59\frac{1}{4}$ in²

9. 8.6 yd by $4\frac{1}{5}$ yd by 7.5 yd   264.24 yd²

10. 12.2 cm by 100 cm by 6.5 cm   3898.6 cm²

Find the surface area of the cube with the given edge length.

11. 17 ft   1734 ft²

12. 0.01 mm   0.0006 mm²

13. $4\frac{1}{6}$ yd $104\frac{1}{6}$ yd²

Solve.

14. A large earth mover has a bed which is 5 yd by 4 yd by 10 yd. How many cubic yards of dirt will it hold?
200 yd³

15. A water tank is 22.5 cm by 10 cm by 5.2 cm. How many grams of water will it hold? Note that 1 cm³ of water weighs 1 g.
1170 g

16. A box which is a cube with edge 8.5 dm is to be covered with foil. How much foil is required?
433.5 dm²

17. A rectangular tank 4 m by 6 m by 5 m (5 m is the height) needs to be insulated on the sides and top. If insulation costs $2.15 per square meter, how much will it cost?
$266.60

Find the volume of each cylinder.

18. $r = 9$ cm, $h = 15$ cm
3815.1 cm³

19. $d = 6.8$ m, $h = 10.5$ m
381.1 m³

20. $d = 20.4$ in, $h = 8.5$ in
2776.8 in³

Find the surface area of each cylinder.

21. $r = 9$ cm, $h = 15$ cm
1356.5 cm²

22. $d = 6.8$ m, $h = 10.5$ m
296.8 m²

23. $d = 20.4$ in, $h = 8.5$ in
1197.8 in²

Find the volume of each sphere.

24. $r = 30$ mm   113,040 mm³

25. $r = \frac{2}{5}$ in   0.2679 in³

26. $d = 9.8$ yd   492.6 yd³

Find the surface area of each sphere.

27. $r = 30$ mm   11,304 mm²

28. $r = \frac{2}{5}$ in   2.010 in²

29. $d = 9.8$ yd   301.6 yd²

Solve.

30. Mercury is stored in a cylinder of radius 16 cm and height 20 cm. How many cylindrical tubes with radius 0.5 cm and height 100 cm can be filled from a full supply?
204.8

31. A cylindrical tank is to be made from sheet metal. If the tank is to have radius 2.2 ft and height 12 ft, what will the metal cost if it is $16.50 per square foot?
$3237.09

32. A spherical tank with diameter 36 ft is to be insulated. If insulation costs $0.50 per square foot, how much will the job cost?
$2034.72

33. How many liters of chemical can be stored in a sphere with radius 650 mm? Note that 1 mm³ = 0.000001 L.
1149.8 L

FOR REVIEW

Find the area of metal left on each of the machine parts shown, which have circular holes drilled in them. Use 3.14 *for* π.

34.

2 cm   3 cm

6 cm

14.86 cm²

35.

14 mm   4 mm

4 mm

12 mm

58.88 mm²

36.

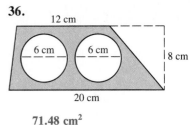

12 cm

6 cm   6 cm   8 cm

20 cm

71.48 cm²

9.6 EXERCISES C

1. Find the volume of the wood remaining when the rectangular block shown at right has a hole bored through it. Use 3.14 for π and give the answer rounded to the nearest hundredth.
[Answer: 0.46 ft³]

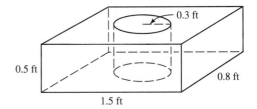

0.3 ft

0.5 ft

0.8 ft

1.5 ft

2. Find the total surface area of the wood remaining in the block in Exercise 1.   **5.08 ft²**

CHAPTER 9 REVIEW

KEY WORDS

9.1 We use **inductive reasoning** when we reach a general conclusion based on a limited collection of specific observations.

We use **deductive reasoning** when we reach a specific conclusion based on a collection of generally accepted assumptions.

9.2 A **segment** is a geometric figure made up of two endpoints and all points between them.

A **line** consists of all the points on a segment together with all the points beyond the endpoints of the segment.

A **ray** is that part of a line lying on one side of point P on the line, together with the point P.

A **plane** is a surface with the property that any two points in it can be joined by a line, all points of which are also contained in the surface.

Parallel lines are distinct lines in a plane that do not intersect.

An **angle** is a geometric figure consisting of two rays that share a common endpoint called the **vertex** of the angle.

A **straight** angle has measure 180°.

A **right** angle has measure 90°.

An **acute** angle has measure between 0° and 90°.

An **obtuse** angle has measure between 90° and 180°.

Two angles are **complementary** if their measures total 90°.

Two angles are **supplementary** if their measures total 180°.

Vertical angles are non-adjacent angles formed by intersecting lines.

When the four angles formed by two intersecting lines all have measure 90°, the lines are **perpendicular.**

A **transversal** is a line that intersects each of two parallel lines.

9.3 A **triangle** is a three-sided geometric figure.

Two triangles are **congruent** if they can be made to coincide by placing one on top of the other.

Two triangles are **similar** if the angles of one are equal to the angles of the other.

9.4 A **circle** is a figure with all points the same distance from a point called the **center** of the circle.

The **radius** of a circle is the distance from the center to any point on the circle.

The **diameter** of a circle is twice the radius.

The **circumference** of a circle is the distance around the circle.

A **secant** is a line that intersects a circle in two points.

A **tangent** is a line that intersects a circle in exactly one point.

9.5 A **rectangle** is a four-sided figure with opposite sides parallel and adjacent sides perpendicular.

The **perimeter** of a figure is the distance around the figure.

A **square** is a rectangle with all sides equal.

A **square unit** is the area of a square that measures one unit on each side.

A **parallelogram** is a four-sided figure whose opposite sides are parallel and equal.

A **trapezoid** is a four-sided figure with two parallel sides.

9.6 A **cube** is a rectangular solid with length, width, and height the same.

A **cubic unit** is the volume of a cube that has measure one unit on each edge.

The **surface area** of an object is the area of all of its surfaces.

A **cylinder** is a solid with circular ends of the same radius.

A **sphere** is a solid in the shape of a ball.

KEY CONCEPTS

9.2
1. Angles are measured in degrees (°).
2. Vertical angles are equal in measure.
3. Alternate interior angles are equal in measure.
4. Corresponding angles are equal in measure.

9.3
1. Similar triangles have all corresponding angles equal.
2. Triangles can be shown to be congruent by SSS, SAS, or ASA.

9.4
1. For a circle: $d = 2r$ and $C = \pi d = 2\pi r$.
2. The measure of an arc is given by the measure of the central angle that determines it.
3. The measure of any inscribed angle is one-half the measure of its intercepted arc.

4. An angle inscribed in a semicircle is a right angle.

9.5 Rectangle: $P = 2l + 2w$, $A = lw$
Square: $P = 4s$, $A = s^2$
Parallelogram: $A = bh$
Triangle: $A = \frac{1}{2}bh$
Trapezoid: $A = \frac{1}{2}(b_1 + b_2)h$
Circle: $d = 2r$, $C = \pi d = 2\pi r$, $A = \pi r^2$

9.6 Rectangular solid: $V = lwh$,
$A = 2lh + 2wh + 2lw$
Cube: $V = e^3$, $A = 6e^2$
Cylinder: $V = \pi r^2 h$, $A = 2\pi rh + 2\pi r^2$
Sphere: $V = \frac{4}{3}\pi r^3$, $A = 4\pi r^2$

REVIEW EXERCISES

Part I

9.1 *Use inductive reasoning in Exercises 1–3 to determine the next element in each list.*

1. 2, 7, 12, 17, 22   27

2. A, C, E, G, I   K

3. $1, \frac{1}{2}, \frac{1}{4}, \frac{1}{8}, \frac{1}{16}$ $\frac{1}{32}$

In Exercises 4–7, determine if each conclusion follows logically from the premises, and state whether the reasoning is inductive or deductive.

4. *Premise:* If I wash my car, then it will rain.
Premise: I washed my car.
Conclusion: It will rain.
    follows logically; deductive reasoning

5. *Premise:* If I wash my car, then it will rain.
Premise: It is raining.
Conclusion: I washed my car.
    does not follow logically; fallacy

6. *Premise:* Bob Begay owns a VCR.
Premise: Bob Jones owns a VCR.
Premise: Bob Santos owns a VCR.
Conclusion: Everyone named Bob owns a VCR.
    does not follow logically; inductive reasoning

7. *Premise:* If it is Sunday, then tomorrow is Monday.
Premise: If tomorrow is Monday, then it is a holiday.
Conclusion: If it is Sunday, tomorrow is a holiday.
    follows logically; deductive reasoning

8. Explain why
Conclusion: Tomorrow is a holiday.
does not follow logically from the premises in Exercise 7.
    We cannot conclude that tomorrow is a holiday, only that if it is Sunday, then tomorrow is a holiday.

9.2,
9.3 *Use the figure below and deductive reasoning to answer* true *or* false *in Exercises 9–18. If the answer is false, tell why.*

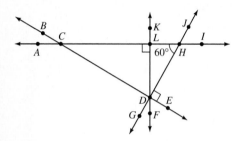

9. ∠CDH is a straight angle.   false

10. ∠LDB has measure 60°.   true

11. $\overleftrightarrow{FK}$ is perpendicular to $\overleftrightarrow{AC}$.   true

12. The measure of ∠CDE is 180°.   true

13. ∠LCD and ∠LHD are supplementary.   false

14. ∠BCA has measure 30°.   true

15. $\overrightarrow{CD}$ is parallel to $\overline{HI}$.   false

16. △LHD is similar to △CLD.   true

17. △CDH is congruent to △CLD.   false

18. △CDH is a right triangle.   true

9.4 **19.** Find the diameter of the circle with radius 2.6 cm.   5.2 cm

20. Find the radius of the circle with diameter $\frac{4}{5}$ yd. $\frac{2}{5}$ yd

21. Find the approximate circumference of the circle with diameter 9.2 ft. Use 3.14 for π.   28.9 ft

Refer to the figure to the right and use deductive reasoning to answer each question in Exercises 22–26.

22. What is ∠AOB called with respect to the circle?   central angle

23. What is ∠ACB called with respect to the circle?   inscribed angle

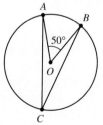

24. What is the measure of $\overarc{AB}$?   50°

25. What is the measure of $\overarc{ACB}$?   310°

26. What is the measure of ∠ACB?   25°

9.5 *Find the perimeter and area of the rectangle or square.*

27. 9 cm by 20 cm
58 cm; 180 cm²
Find the area of each figure.

28. $6\frac{1}{2}$ ft by 4.2 ft
21.4 ft; 27.3 ft²

29. 3.2 mi by 3.2 mi
12.8 mi; 10.24 mi²

30.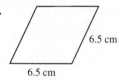
58.9 m²

31.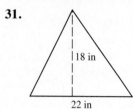
198 in²

32.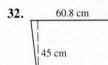
2497.5 cm²

Find the perimeter of each figure.

33.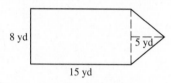
26 m

34.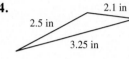
7.85 in

35.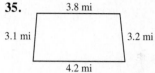
14.3·mi

36. What will it cost to carpet the area shown if carpet costs $22.50 per square yard?   **$3150**

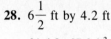

9.6 *Find the volume of each rectangular solid.*

37. 2.2 ft by 1.5 ft by 6.4 ft
21.12 ft³
Find the volume of each cube with the given edge length.

38. 52 cm by 22 cm by 8 cm
9152 cm³

39. 16 m   **4096 m³**

40. 8.2 in   **551.368 in³**

Find the surface area of each rectangular solid.

41. 2.2 ft by 1.5 ft by 6.4 ft
53.96 ft²
Find the surface area of each cube with the given edge length.

42. 52 cm by 22 cm by 8 cm
3472 cm²

43. 16 m   **1536 m²**

44. 8.2 m   **403.44 in²**

Find the volume of each cylinder. Use 3.14 for π.

45. $r = 8$ cm, $h = 12$ cm
2411.52 cm³
Find the surface area of each cylinder.

46. $d = 2.2$ in, $h = 1.5$ in
5.7 in³

47. $r = 8$ cm, $h = 12$ cm
1004.8 cm²
Find the volume of each sphere.

48. $d = 2.2$ in, $h = 1.5$ in
17.96 in²

49. $r = 10$ ft
4186.7 ft³
Find the surface area of each sphere.

50. $d = 8.8$ m
356.6 m³

51. $r = 10$ ft
1256 ft²

52. $d = 8.8$ m
243.2 m²

Part II

Find the circumference and area of each circle. Use 3.14 for π, and round answers to the nearest tenth.

53. $r = 6.8$ km
42.7 km; 145.2 km²

54. $d = 22$ in
69.1 in; 379.9 in²

55. $r = 8\dfrac{1}{2}$ ft   **53.4 ft, 226.9 ft²**

56. A 12-in-diameter pizza costs $5.00 and a 16-in-diameter pizza costs $8.00. Which pizza costs less per square inch?
16-in costs less (12-in: $0.044 per in²; 16-in: $0.040 per in²)

57. A spherical tank with radius 12 ft is to be insulated. If insulation costs $0.50 per square foot, how much will it cost to insulate the tank?
$904.32

58. How many grams of water will a tank 10 cm by 8 cm by 2.5 cm hold? (*Hint:* Water weighs 1 gm per cubic centimeter.]
200 g

59. How many square centimeters of glass did it take to make the sides and bottom of the tank in Exercise 58?
170 cm²

60. The area shown is to be carpeted. How much will it cost if carpet is $19.50 per square yard?   **$741**

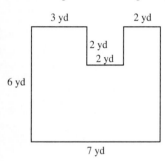

61. Are the given triangles congruent? Explain.
yes; ASA

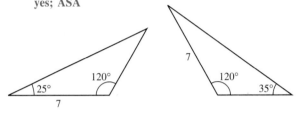

Use the figure to work Exercises 62–64.

62. Find the lines that are secants. $\overleftrightarrow{AC}$ **and** $\overleftrightarrow{PB}$

63. Find the lines that are tangents. $\overleftrightarrow{PQ}$ **and** $\overleftrightarrow{PA}$

64. Find the lines that are neither secants nor tangents. $\overleftrightarrow{BQ}$

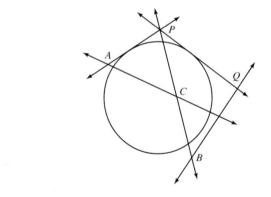

65. If 3 cats kill 3 mice in 3 minutes, how long will it take 100 cats to kill 100 mice?   **3 minutes**

66. There are 5 white socks and 5 black socks in a drawer. How many socks must you remove without looking, to be assured of having a pair that match?
3

ANSWERS: 1. 27 2. K 3. $\frac{1}{32}$ 4. follows logically; deductive reasoning 5. does not follow logically; the premise did not say that if it rains, then I washed my car. 6. does not follow logically; inductive reasoning 7. follows logically; deductive reasoning 8. We cannot conclude that tomorrow is a holiday, only that if it is Sunday, then tomorrow is a holiday. 9. false (right angle) 10. true 11. true 12. true 13. false (complementary) 14. true 15. false ($\overleftrightarrow{CD}$ and $\overleftrightarrow{HI}$ intersect at C) 16. true 17. false (only one side equal) 18. true 19. 5.2 cm 20. $\frac{2}{5}$ yd 21. 28.9 ft 22. central angle 23. inscribed angle 24. 50° 25. 310° 26. 25° 27. 58 cm; 180 cm² 28. 21.4 ft; 27.3 ft² 29. 12.8 mi; 10.24 mi² 30. 58.9 m² 31. 198 in² 32. 2497.5 cm² 33. 26 m 34. 7.85 in 35. 14.3 mi 36. $3150 37. 21.12 ft³ 38. 9152 cm³ 39. 4096 m³ 40. 551.368 in³ 41. 53.96 ft² 42. 3472 cm² 43. 1536 m² 44. 403.44 in² 45. 2411.52 cm³ 46. 5.7 in³ 47. 1004.8 cm² 48. 17.96 in² 49. 4186.7 ft³ 50. 356.6 m³ 51. 1256 ft² 52. 243.2 m² 53. 42.7 km; 145.2 km² 54. 69.1 in; 379.9 in² 55. 53.4 ft; 226.9 ft² 56. 16-in costs less (12-in: $0.044 per square inch; 16-in: $0.040 per square inch) 57. $904.32 58. 200 g 59. 170 cm² 60. $741 61. yes; ASA 62. $\overleftrightarrow{AC}$ and $\overleftrightarrow{PB}$ 63. $\overleftrightarrow{PQ}$ and $\overleftrightarrow{PA}$ 64. $\overleftrightarrow{BQ}$ 65. 3 minutes 66. 3

1. Use inductive reasoning to give the next element in the following list: 125, 25, 5, 1.

1. _____ $\dfrac{1}{5}$ _____

2. Does the conclusion below follow logically from the premises? What type of reasoning is being used?

   Premise: Michael Jordan eats oatmeal.
   Premise: Larry Bird eats oatmeal.
   Premise: Magic Johnson eats oatmeal.
   Conclusion: All great basketball players eat oatmeal.

2. __no; inductive reasoning__

Use the figure below and deductive reasoning to answer true *or* false *in problems 3–6.*

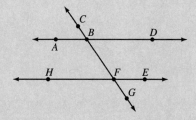

3. $\angle ABD$ is a straight angle.

3. _____ true _____

4. $\angle BFE$ and $\angle BFH$ are complementary.

4. _____ false _____

5. $\overleftrightarrow{AB}$ is parallel to $\overleftrightarrow{FG}$.

5. _____ false _____

6. $\angle EFG$ and $\angle ABC$ are equal.

6. _____ true _____

7. Are the given triangles congruent?

7. _____ yes _____

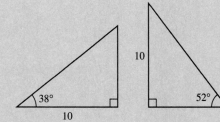

Use the figure below and deductive reasoning to answer problems 8–10.

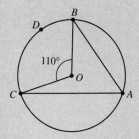

8. What is the measure of $\overset{\frown}{BDC}$?

8. _____ 110°

9. What is the measure of $\overset{\frown}{BAC}$?

9. _____ 250°

10. What is the measure of $\angle BAC$?

10. _____ 55°

Problems 11–20 refer to the figure below. Answer true *or* false.

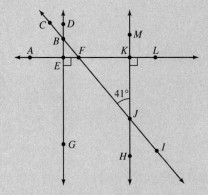

11. $\angle FBE$ is an obtuse angle.

11. _____ false

12. $\angle MKF$ is a right angle.

12. _____ true

13. $\overleftrightarrow{EG}$ is parallel to $\overrightarrow{HM}$.

13. _____ true

14. $\overleftrightarrow{AE}$ is perpendicular to $\overline{JH}$.

14. _____ true

15. $\angle IJH$ has measure 41°.

15. _____ true

16. $\angle FBE$ has measure 41°.

16. _____ true

17. $\angle BFE$ and $\angle BFK$ are complementary.

17. _____ false

18. $\triangle KJF$ is a right triangle.

18. _____ true

19. ∠*CBD* has measure 41°.

19. _____true_____

20. △*BFE* is similar to △*JKF*.

20. _____true_____

21. The area shown is to be carpeted. How much will it cost if carpet is $22.50 per square yard?

21. _____$945_____

5 yd

3 yd

4 yd

3 yd

22. Find the circumference of a circle with radius 2.6 ft. Use 3.14 for π.

22. _____16.3 ft_____

23. Find the area of a circle with diameter 9.2 m. Use 3.14 for π.

23. _____66.4 m²_____

24. Find the **(a)** volume and **(b)** surface area of a cube with edge 3.6 ft.

24. (a) _____46.7 ft³_____

(b) _____77.8 ft²_____

25. A 14-in-diameter pizza costs $7.50 and a 16-in pizza costs $8.50. Which costs less per square inch?

25. _____16-in_____

26. Find the area of the given figure.

26. _____12 ft²_____

6 ft

4 ft

27. Find the perimeter of the given figure.

27. _____31 m_____

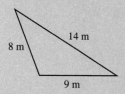

14 m

8 m

9 m

28. Find the volume of a cylinder with radius 8 cm and height 10 cm.

28. _____2009.6 cm³_____

29. Find the surface area of a sphere with radius 20 inches.

29. _____5024 in²_____

Solve.

30. Gravel costs $1.50 per cubic meter. How much would a truckload cost if the bed of the truck is 4.2 m by 5.0 m by 2.5 m?

30. _____$78.75_____

Introduction to Algebra

10.1 VARIABLES, EXPONENTS, AND ORDER OF OPERATIONS

STUDENT GUIDEPOSTS

1. Variables
2. Exponential Notation
3. Order of Operations
4. Symbols of Grouping
5. Evaluating Algebraic Expressions
6. Formulas

① VARIABLES

In algebra, we often use letters such as a, b, x, y, A and B to represent numbers. A letter that can be replaced by various numbers is called a **variable.** Lengthy verbal expressions can often be symbolized by brief algebraic expressions using variables, as in the following examples.

| Three | times | a natural number | becomes | $3 \cdot n$ |
| 3 | $\cdot$ | n | | |

| Seven | plus | twice | a natural number | becomes | $7 + 2 \cdot n$ |
| 7 | $+$ | $2 \cdot$ | n | | |

In each example, the variable n represents a natural number. Other notations for $3 \cdot n$ are $3(n)$, $(3)(n)$, and $3n$, with the last form preferred. The product of two natural numbers m and n would most likely be represented by mn, but could also be expressed by $m \cdot n$, $m(n)$, and $(m)(n)$. We usually avoid using the symbol $\times$ for multiplication since it can be confused with the letter x used as a variable.

② EXPONENTIAL NOTATION

When a variable or number is multiplied by itself several times, as in

$$2 \cdot 2 \cdot 2 \cdot 2 \quad \text{or} \quad x \cdot x \cdot x,$$

we can use **exponential notation** to avoid writing long strings of **factors** (the individual numbers in the expressed product). For example, we write $2 \cdot 2 \cdot 2 \cdot 2$ as 2^4,

$$\underbrace{2 \cdot 2 \cdot 2 \cdot 2}_{\text{4 factors}} = 2^{4} \leftarrow \text{exponent}$$
$$\underset{\text{base}}{\uparrow}$$

where 2 is called the **base,** 4 the **exponent,** and 2^4 the **exponential expression** (read 2 to the **fourth power**). Similarly, $x \cdot x \cdot x = x^3$ is called the **third power** or **cube** of x. The **square** or **second power** of a is a^2. The **first power** of a is a^1, which we write simply as a.

Exponential Notation

If a is any number and n is a natural number,

$$a^n = \underbrace{a \cdot a \cdot a \cdots a.}_{n \text{ factors}}$$

EXAMPLE 1   USING EXPONENTS

Write in exponential notation.

(a) $\underbrace{7 \cdot 7 \cdot 7}_{3 \text{ factors}} = 7^3$

(b) $\underbrace{a \cdot a \cdot a \cdot a \cdot a \cdot a}_{6 \text{ factors}} = a^6$

(c) $\underbrace{3 \cdot 3}_{\substack{2 \\ \text{factors}}} \cdot \underbrace{x \cdot x \cdot x \cdot x \cdot x}_{5 \text{ factors}} = 3^2 x^5$

PRACTICE EXERCISE 1

Write in exponential notation.

(a) $4 \cdot 4 \cdot 4 \cdot 4 \cdot 4 \cdot 4 \cdot 4$

(b) $z \cdot z \cdot z \cdot z$

(c) $6 \cdot 6 \cdot 6 \cdot y \cdot y \cdot y \cdot y$

Answers: (a) 4^7 (b) z^4
(c) $6^3 y^4$

---

③ ORDER OF OPERATIONS

When exponents or powers are used together with the operations of addition, subtraction, multiplication, and division, the resulting numerical expressions can be confusing unless we agree to an order of operations. For example,

$2 \cdot 5^2$ *could* equal 10^2 or 100,	If we first multiply then square
or $2 \cdot 5^2$ *could* equal $2 \cdot 25$ or 50.	If we first square then multiply

Similarly,

$2 \cdot 3 + 4$ *could* equal $2 \cdot 7$ or 14,	If we first add then multiply
or $2 \cdot 3 + 4$ *could* equal $6 + 4$ or 10.	If we first multiply then add

According to the following rule, we see that the second procedure in each of these examples is the correct one to use.

To Evaluate a Numerical Expression

1. Evaluate all powers, in any order, first.
2. Do all multiplications and divisions in order from left to right.
3. Do all additions and subtractions in order from left to right.

EXAMPLE 2   EVALUATING NUMERICAL EXPRESSIONS

Evaluate each numerical expression.

(a) $2 + 3 \cdot 4 = 2 + 12 = 14$ Multiply first, then add

(b) $20 - 3^2 = 20 - 9 = 11$ Square first, then subtract

PRACTICE EXERCISE 2

Evaluate each numerical expression.

(a) $5 \cdot 2 + 1$

(b) $6 + 2^3$

(c) $5 \cdot 6 - 12 \div 3 = 30 - 4 = 26$ Multiply and divide first, then subtract

(d) $25 \div 5 + 3 \cdot 2^3 = 25 \div 5 + 3 \cdot 8$ Cube first

$\qquad\qquad\qquad = 5 + 24$ Divide and multiply second

$\qquad\qquad\qquad = 29$ Add last

(c) $18 \div 9 + 3 \cdot 2$

(d) $16 \cdot 2 - 3 \cdot 2^2$

Answers: (a) **11**   (b) **14**   (c) **8**
(d) **20**

❹ SYMBOLS OF GROUPING

Suppose that we want to evaluate three times the sum of 2 and 5. If we write $3 \cdot 2 + 5$ and use the above rule, we will obtain $6 + 5$ or 11. However, it is clear from the first sentence that we want 3 times 7 or 21 for the result. **Symbols of grouping,** such as parentheses ( ), square brackets [ ], or braces { }, can help us symbolize the problem correctly as $3 \cdot (2 + 5)$. The grouping symbols contain the expression that must be evaluated first. In this case, we must add before multiplying:

$$3 \cdot (2 + 5) = 3 \cdot (7) = 21.$$

Generally, we omit the dot and write $3(2 + 5)$ for $3 \cdot (2 + 5)$. Also, instead of writing $3 \cdot (7)$, we will remove the parentheses and write $3 \cdot 7$.

One other method of grouping is to use a fraction bar. For example, in

$$\frac{4 \cdot 5}{7 + 3},$$

the fraction bar acts like parentheses since the expression is the same as

$$(4 \cdot 5) \div (7 + 3).$$

We first multiply 4 and 5, then add 7 and 3, and finally we divide the results. Thus,

$$\frac{4 \cdot 5}{7 + 3} = \frac{20}{10} = 2.$$

To Evaluate a Numerical Expression Involving Grouping Symbols

Evaluate all expressions within the grouping symbols first, and begin with the innermost symbols of grouping if more than one set of symbols is present.

EXAMPLE 3   **EVALUATING NUMERICAL EXPRESSIONS**

Evaluate each numerical expression.

(a) $3 + (2 \cdot 5) = 3 + 10 = 13$ Evaluate inside the parentheses first

(b) $(4 + 5)2 + 3 = (9)2 + 3$ Work inside the parentheses first

$\qquad\qquad\qquad = 18 + 3$ Multiply before adding

$\qquad\qquad\qquad = 21$

(c) $[3(8 + 2) + 1]4 = [3(10) + 1]4$ Innermost grouping symbol first

$\qquad\qquad\qquad = [30 + 1]4$ Multiply before adding inside brackets

$\qquad\qquad\qquad = [31]4$ Combine numbers inside brackets before multiplying

$\qquad\qquad\qquad = 124$

PRACTICE EXERCISE 3

Evaluate each numerical expression.

(a) $3 \cdot (2 + 5)$

(b) $(6 - 1)3 - 2$

(c) $2[2(7 + 2) - 1]$

Answers: (a) **21**   (b) **13**   (c) **34**

We now summarize the order to be followed when evaluating a numerical expression.

Order of Operations

1. Evaluate within grouping symbols first, beginning with the innermost set if more than one set is used.
2. Evaluate all powers.
3. Perform all multiplications and divisions in order from left to right.
4. Perform all additions and subtractions in order from left to right.

| EXAMPLE 4 EVALUATING USING ALL RULES | PRACTICE EXERCISE 4 |

Evaluate each numerical expression.

(a) $2^3 + 4^3 = 8 + 64 = 72$ Cube first, then add

(b) $(2 + 4)^3 = 6^3 = 216$ Add first inside parentheses, then cube

(c) $(3 \cdot 5)^2 = 15^2 = 225$ Multiply first, then square

(d) $3 \cdot 5^2 = 3 \cdot 25 = 75$ Square first, then multiply

(e) $5^2 - 4^2 = 25 - 16 = 9$ Square first, then subtract

(f) $(5 - 4)^2 = 1^2 = 1$ Subtract first, then square

(g) $3^2 - 15 \div 5 + 5 \cdot 2 = 9 - 15 \div 5 + 5 \cdot 2$ Square first, then divide
$= 9 - 3 + 10$ and multiply in order
$= 16$ Subtract and add in order

(h) $(7 - 3)^2 = 4^2 = 16$ Subtract first, then square

(i) $7^2 - 3^2 = 49 - 9 = 40$ Square first, then subtract

Evaluate each numerical expression.

(a) $1^2 + 6^2$

(b) $(1 + 6)^2$

(c) $(2 \cdot 7)^2$

(d) $2 \cdot 7^2$

(e) $8^2 - 6^2$

(f) $(8 - 6)^2$

(g) $2^3 - 20 \div 4 + 4 \cdot 3$

(h) $(9 + 2)^2$

(i) $9^2 + 2^2$

Answers: **(a)** 37   **(b)** 49   **(c)** 196
(d) 98   **(e)** 28   **(f)** 4   **(g)** 15
(h) 121   **(i)** 85

C A U T I O N

Example 4 indicates common errors to avoid when working with exponents. For example,

$$2^3 + 4^3 \neq (2 + 4)^3, \quad (3 \cdot 5)^2 \neq 3 \cdot 5^2, \quad 5^2 - 4^2 \neq (5 - 4)^2.$$

The symbol $\neq$ means "is *not* equal to." In general, if a, b, and n are numbers,

$$a^n + b^n \neq (a + b)^n, \quad (ab)^n \neq ab^n, \quad a^n - b^n \neq (a - b)^n.$$

5 EVALUATING ALGEBRAIC EXPRESSIONS

An **algebraic expression** contains variables as well as numbers. We can use the order of operations for evaluating numerical expressions to evaluate algebraic expressions when specific values for the variables are given.

To Evaluate an Algebraic Expression

1. Replace each variable (letter) with the specified value.
2. Proceed as in evaluating numerical expressions.

EXAMPLE 5 EVALUATING AN ALGEBRAIC EXPRESSION

Evaluate $2(a + b) - c$ when $a = 3$, $b = 4$, and $c = 5$.

$$
\begin{aligned}
2(a + b) - c &= 2(3 + 4) - 5 &&\text{Replace each letter with given value} \\
&= 2(7) - 5 &&\text{Evaluate inside parentheses first} \\
&= 14 - 5 = 9
\end{aligned}
$$

PRACTICE EXERCISE 5

Evaluate $3(x - y) + z$, when $x = 4$, $y = 2$, and $z = 1$.

Answer: 7

EXAMPLE 6 EVALUATING AN ALGEBRAIC EXPRESSION

Evaluate $5[12 - 3(a + 1) + b] - c$ when $a = 2$, $b = 7$, and $c = 4$.

$$
\begin{aligned}
5[12 - 3(a + 1) + b] - c &= 5[12 - 3(2 + 1) + 7] - 4 &&\text{Replace variables with numbers} \\
&= 5[12 - 3(3) + 7] - 4 &&\text{Multiply before} \\
&= 5[12 - 9 + 7] - 4 &&\text{adding or} \\
&= 5[10] - 4 = 50 - 4 = 46 &&\text{subtracting}
\end{aligned}
$$

PRACTICE EXERCISE 6

Evaluate $w - 2[3(u - 1) + v]$, when $u = 3$, $v = 1$, and $w = 14$.

Answer: 0

EXAMPLE 7 EVALUATING AN ALGEBRAIC EXPRESSION

Evaluate $\dfrac{ab - 1}{c}$ for the given values.

(a) $a = 2$, $b = 3$, and $c = 0$

$$
\frac{ab - 1}{c} = \frac{(2)(3) - 1}{0} \qquad \text{Division by zero undefined}
$$

That is, this expression is undefined when $c = 0$.

(b) $a = 3$, $b = \dfrac{1}{3}$, and $c = 5$

$$
\frac{ab - 1}{c} = \frac{(3)\left(\frac{1}{3}\right) - 1}{5} = \frac{1 - 1}{5} = \frac{0}{5} = 0 \qquad \tfrac{0}{5} = 0
$$

PRACTICE EXERCISE 7

Evaluate $\dfrac{3x}{yz - x}$ for the given values.

(a) $x = 0$, $y = 5$, $z = 8$

(b) $x = 6$, $y = 2$, $z = 3$

Answers: (a) 0   (b) undefined

EXAMPLE 8 EVALUATING AN ALGEBRAIC EXPRESSION

Evaluate the following when $a = 2$, $b = 1$, $c = 3$.

(a) $3a^2 = 3(2)^2 \qquad$ Not $(3 \cdot 2)^2 = 6^2 = 36$

$\qquad = 3 \cdot 4 = 12$

(b) $2ab^2c^3 = 2(2)(1)^2(3)^3$

$\qquad = 2(2)(1)(27) = 4 \cdot 27 = 108$

(c) $(2c)^2 - 2c^2 = (2 \cdot 3)^2 - 2(3)^2 \qquad$ Watch the substitution

$\qquad = 6^2 - 2 \cdot 9$

$\qquad = 36 - 18 = 18$

(d) $a^a + c^c = 2^2 + 3^3$

$\qquad = 4 + 27 = 31$

PRACTICE EXERCISE 8

Evaluate the following when $u = 5$, $v = 3$, and $w = 1$.

(a) $2u^2$

(b) $3uv^2w$

(c) $(3u)^2 - 3u^2$

(d) $v^v - w^w$

Answers: (a) 50   (b) 135
(c) 150   (d) 26

6 FORMULAS

A **formula** is an algebraic statement that relates two or more quantities. For example,

$$d = rt$$

is a formula that relates the distance, d, traveled by an object moving at an average rate, r, for a period of time, t. Other familiar formulas, related to geometric figures, such as

$A = lw$ Area of a rectangle in terms of its length and width

and $P = 2l + 2w$ Perimeter of a rectangle in terms of its length and width

are summarized on the inside cover of the text. In order to use a particular formula in an applied problem, we must first understand the meaning of the variables, and next, evaluate the algebraic expression (formula) for the given values.

EXAMPLE 9 EVALUATE FORMULAS

If we invest P dollars (called the principal) at an interest rate, r, compounded annually for a period of t years, it will grow to an amount, A, given by

$$A = P(1 + r)^t.$$

If a principal of \$1000 is invested at 12% interest, compounded annually, how much will be in the account at the end of two years?

$A = P(1 + r)^t$ Start with the formula

$= 1000(1 + 0.12)^2$ Substitute 1000 for P, 0.12 for r, and 2 for t

$= 1000(1.12)^2$

$= 1000(1.2544)$

$= 1254.4$

Thus, there will be \$1254.40 in the account. Notice that 12% was converted to a decimal, 0.12, in order to solve this problem.

PRACTICE EXERCISE 9

The perimeter of a rectangle, P, is given by $P = 2l + 2w$, where l is its length and w is its width. What is the perimeter of a rectangle having a length of 15 inches and a width of 8 inches?

Answer: 46 inches

10.1 EXERCISES A

1. What is a letter used to represent a number called?   **variable**

2. Given the exponential expression x^7, **(a)** what is the base? x **(b)** what is the exponent? 7

3. In the expression $3x$, what is the exponent on x?   1

Write in exponential notation.

4. $8 \cdot 8 \cdot 8 \cdot 8 \cdot 8$ 8^5

5. $(2x)(2x)$ $(2x)^2$

6. $2 \cdot x \cdot x$ $2x^2$

7. $(b + c)(b + c)(b + c)$ $(b + c)^3$

8. $7aaaaaaa$ $7a^7$

9. $6 \cdot 6 \cdot 6 \cdot y \cdot y \cdot z \cdot z \cdot z \cdot z$ $6^3y^2z^4$

Write without using exponents.

10. $a^2b^3c^4$ *aabbbcccc*

11. $3y^3$ *3yyy*

12. $(3y)^3$ *(3y)(3y)(3y)*

13. 1^{51} 1

14. $a^3 - c^3$ *aaa − ccc*

15. $\left(\dfrac{2}{3}\right)^3 x^2$

$\left(\dfrac{2}{3}\right)\left(\dfrac{2}{3}\right)\left(\dfrac{2}{3}\right)xx$ *or* $\dfrac{8}{27}xx$

Evaluate.

16. $3 \cdot 2^2$ 12

17. $(3 \cdot 2)^2$ 36

18. $(3 + 2)^2$ 25

19. $3^2 + 2^2$ 13

20. $(5 - 2)^3$ 27

21. $5^3 - 2^3$ 117

22. $8 - 2^3$ 0

23. $(8 - 2)^3$ 216

24. $(7 - 7)^3$ 0

25. $9 - 2 \cdot 3 + 1$ 4

26. $(9 - 2) \cdot 3 + 4 \cdot 0$ 21

27. $\dfrac{2(3 - 1) - 4}{5}$ 0

28. $\dfrac{4(3 + 5) - 10}{0}$ **undefined**

29. $12 \div 4 + 2 \cdot 3 - 1$ 8

30. $2[8 - 2(4 - 1) + 5]$ 14

31. $15 - 3\{2(5 - 4) + 3\}$ 0

32. $10 - 2[8 - 2(7 - 3)]$ 10

33. $16 \div 4 \cdot 2 \div 4 - 2$ 0

34. Evaluate the cube of the difference five minus two.   27

35. Evaluate five cubed minus two cubed.   117

36. Evaluate ten minus three squared.   1

37. Evaluate the square of the difference ten minus three.   49

Evaluate when a = 2, b = 3, c = 5, d = 0, and x = 12.

38. $6a + b$ 15

39. $6(a + b)$ 30

40. $3ab \div d$ **undefined**

41. $2b^2$ 18

42. $(2b)^2$ 36

43. $(c - b)^2$ 4

44. $c^2 - b^2$ 16

45. $(a + b)^3$ 125

46. $a^3 + b^3$ 35

47. $3a^3 + 2$ 26

48. $(3a)^3 + 2$ 218

49. $a^a + b^b$ 31

50. $\dfrac{2abcd - 3d}{x}$ 0

51 $x - [a(b + 1) - c]$ 9

52. $5a + 2[c + x \div b]$ 28

Solve.

53. The perimeter of a rectangle, P, is given by $P = 2l + 2w$, where l is its length and w is its width. What is the perimeter of a rectangle of length 20 feet and width 13 feet?   **66 ft**

54. The distance an automobile travels, d, at an average rate of speed, r, for a period of time, t, is given by $d = rt$. How far does a car travel in 11 hours at an average speed of 55 mph?   **605 mi**

55. Use $A = P(1 + r)^t$ to find the amount of money in an account at the end of two years if a principal of $500 is invested at 14% interest, compounded annually.   **$649.80**

56. The area of a square, A, is given by $A = s^2$, where s is the length of a side. What is the area of a square with side 3.5 cm?   **12.25 cm^2**

57. The area of a triangle, A, is given by $A = \frac{1}{2}bh$, where b is the length of its base and h is its height. Find the area of a triangle with height 15 m and base 8 m   **60 m^2**

58. The temperature measured in degrees Celsius, °C, can be obtained from degrees Fahrenheit, °F, by using $C = \frac{5}{9}(F - 32)$. Find C when F is 200°.   **93.$\overline{3}$°**

59. The surface area of a cube, A, is given by $A = 6e^2$ where e is the length of an edge. Find the surface area of a cube with an edge of $2\frac{1}{2}$ inches.   **37.5 in^2**

60 The surface area of a cylinder, A, with height, h, and base radius, r, is given by $A = 2\pi rh + 2\pi r^2$. Use 3.14 for π and find the surface area of a cylinder with a radius of 2 cm and a height of 10 cm.   **150.72 cm^2**

ANSWERS: 1. variable 2. (a) x (b) 7 3. 1 4. 8^5 5. $(2x)^2$ 6. $2x^2$ 7. $(b + c)^3$, *not* $b^3 + c^3$ 8. $7a^7$ 9. $6^3y^2z^4$
10. *aabbbcccc* 11. $3yyy$ 12. $(3y)(3y)(3y)$ 13. 1 14. *aaa − ccc* 15. $\left(\frac{2}{3}\right)\left(\frac{2}{3}\right)\left(\frac{2}{3}\right)xx$ or $\frac{8}{27}xx$ 16. 12 17. 36 18. 25
19. 13 20. 27 21. 117 22. 0 23. 216 24. 0 25. 4 26. 21 27. 0 28. undefined 29. 8 30. 14 31. 0
32. 10 33. 0 34. 27 35. 117 36. 1 37. 49 38. 15 39. 30 40. undefined 41. 18 42. 36 43. 4 44. 16
45. 125 46. 35 47. 26 48. 218 49. 31 50. 0 51. 9 52. 28 53. 66 ft 54. 605 miles 55. $649.80
56. 12.25 cm^2 57. 60 m^2 58. 93.$\overline{3}$° 59. 37.5 in^2 60. 150.72 cm^2

10.1 EXERCISES B

1. When evaluating a numerical expression involving more than one set of grouping symbols, always evaluate within which set of symbols first?   **innermost**

2. Given the exponential expression a^6, **(a)** What is the base? *a*   **(b)** what is the exponent? **6**

3. In the expression $8y$, what is the exponent on y?   **1**

Write in exponential notation.

4. $c \cdot c \cdot c \cdot c \cdot c$ c^5

5. $3 \cdot 3 \cdot y \cdot y \cdot y \cdot y \cdot y \cdot y$ $3^2 y^6$

6. $(3a)(3a)$ $(3a)^2$

7. $3 \cdot a \cdot a$
   $3a^2$

8. $(x - y)(x - y)(x - y)$
   $(x - y)^3$

9. $5 \cdot 5 \cdot 5 \cdot a \cdot a \cdot c \cdot c \cdot c$
   $5^3 a^2 c^3$

Write without using exponents.

10. $a^3 b^4 c^2$ *aaabbbbcc*

11. $2z^3$ *2zzz*

12. $(2z)^3$ $(2z)(2z)(2z)$

13. $x^2 - y^2$ *xx − yy*

14. $a^3 + b^3$ *aaa + bbb*

15. $\left(\dfrac{1}{2}\right)^3 a^2$

$\left(\dfrac{1}{2}\right)\left(\dfrac{1}{2}\right)\left(\dfrac{1}{2}\right)$ *aa* or $\dfrac{1}{8}$ *aa*

Evaluate.

16. $2 \cdot 7^2$ 98

17. $(2 \cdot 7)^2$ 196

18. $(2 + 7)^2$ 81

19. $2^2 + 7^2$ 53

20. $(4 - 3)^2$ 1

21. $4^2 - 3^2$ 7

22. $4 - 2^2$ 0

23. $(4 - 2)^2$ 4

24. $(3 - 3)^3$ 0

25. $14 - 5 + 2 \cdot 6$ 21

26. $0 \cdot (5 - 2) + 0$ 0

27. $\dfrac{2(5 + 1) - 3}{3}$ 3

28. $\dfrac{5(2 + 8) - 19}{0}$ undefined

29. $25 \div 5 + 2 \cdot 4 - 1$ 12

30. $3[10 - 2(3 - 1) + 6]$ 36

31. $20 - 2[3(6 - 5) + 4]$ 6

32. $30 - 5[7 - 2(6 - 3)]$ 25

33. $20 \div 5 \cdot 2 \div 4 - 2$ 0

34. Evaluate the square of the sum of eleven and three.   196

35. Evaluate eleven squared plus three squared.   130

36. Evaluate twelve minus two cubed.   4

37. Evaluate the cube of the difference twelve minus two.   1000

Evaluate when $x = 3$, $y = 2$, $z = 4$, $w = 0$, and $a = 24$.

38. $5x + y$ 17

39. $5(x + y)$ 25

40. $4xy \div w$ undefined

41. $3y^2$ 12

42. $(3y)^2$ 36

43. $(z - y)^2$ 4

44. $z^2 - y^2$ 12

45. $(x + z)^3$ 343

46. $x^3 + z^3$ 91

47. $2z^3 + 1$ 129

48. $(2z)^3 + 1$ 513

49. $x^y + y^x$ 17

50. $\dfrac{3wxyz - 5w}{a}$ 0

51. $a - [x(y + 3) - z]$ 13

52. $3x + 4[x + z \div y]$ 29

Solve.

53. The perimeter of a square P, with side of length s is given by $P = 4s$. What is the perimeter of a square with side of length 15 feet?   60 ft

54. The area of a rectangle, A, is given by $A = lw$, where l is its length and w its width. If a rectangle is 2 cm wide and 7 cm long, what is its area?
   14 cm^2

55. Use the formula $A = P(l + r)^t$ to find the amount of money in an account at the end of two years if a principal of $2000 is invested at 8% interest, compounded annually.   **$2332.80**

56. The area of a circle, A, with radius r, is given by $A = \pi r^2$. Use 3.14 for π and find the area of a circle with radius 20 m.   **1256 m^2**

57. The volume, V, of a box with length l, width w, and height h, is given by $V = lwh$. If the box is 9 inches long, 6 inches wide, and 11 inches high, what is its volume?   **594 in³**

58. The temperature measured in degrees Fahrenheit, °F, can be obtained from degrees Celsius, °C, by using $F = \frac{9}{5}C + 32$. Find F when C is 20°.
68°

59. The surface area of a sphere, A, with radius r, is given by $A = 4\pi r^2$. Use 3.14 for π and find the surface area of a sphere with radius 50 cm.
31,400 cm²

60. The surface area of a cylinder, A, is given by $A = 2\pi rh + 2\pi r^2$, where r is the radius of its base and h is its height. Find A if $r = 10$ inches, $h = 8$ inches, and $\pi = 3.14$.   **1130.4 in²**

10.1   EXERCISES C

Evaluate.

1. $2(5 - 3)^2 - [7 - (8 - 6)^2]$ 5   **2.** $13 - [11 - (7 - 4)^2]^3$ 5   **3.** $[(6 \div 3)^2 + 9 \cdot (3 - 1)^2]^2$
[Answer: 1600]

Evaluate when $a = 1.35$, $b = 2.75$, and $c = 6.2$. A calculator would be helpful in these exercises.

4. $abc + b^2$ 30.58   **5.** $c^2 - a^2 + b^2$ 44.18   **6.** $2c^2 - (a + b)$ 72.78

10.2   RATIONAL NUMBERS AND THE NUMBER LINE

STUDENT GUIDEPOSTS

1 Number Line
2 Less Than and Greater Than
3 Integers
4 Equal (Unequal) Numbers
5 Rational Numbers
6 Absolute Values

1 NUMBER LINE

Many ideas in algebra can be better understood if we "picture" them in some way. Numbers are often pictured using a *number line*. Consider the set of whole numbers. We draw a line like the one in Figure 10.1, and select some unit of length. Then, starting at an arbitrary point that we label 0, and call the **origin,** we mark off unit lengths to the right, labeling the points 1, 2, 3, ... The result is a **number line** displaying the whole numbers, shown in Figure 10.1. Every whole number is paired with a point on this line. The arrowhead points to the direction in which the whole numbers continue to be identified.

Figure 10.1   Number Line

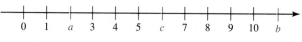

What whole numbers are paired with a, b, and c on the number line in Figure 10.2?

Point a is paired with 2, point b with 11, and point c with 6.

PRACTICE EXERCISE 1

What whole numbers are paired with x, y, and z on the following number line?

Answer: x paired with 4, y paired with 3, z paired with 1

② LESS THAN AND GREATER THAN

The whole numbers occur in a natural order. For example, we know that 3 is less than (has a smaller value than) 8, and that 8 is greater than (has a larger value than) 3. Note on the number line in Figure 10.2 that 3 is to the left of 8, while 8 is to the right of 3. The symbol $<$ means **"is less than."** We write "3 is less than 8" as

$$3 < 8.$$

Similarly, the symbol $>$ means **"is greater than."** We write "8 is greater than 3" as

$$8 > 3.$$

Notice that $3 < 8$ and $8 > 3$ have the same meaning even though they are read differently.

Order of Whole Numbers

Suppose a and b are any two whole numbers.

1. If a is to the left of b on a number line, then $a < b$.
2. If a is to the right of b on a number line, then $a > b$.

EXAMPLE 2   ORDERING WHOLE NUMBERS

Place the correct symbol ($<$ or $>$) between the numbers in the given pairs. Use part of a number line if necessary.

(a) 2   7
$2 < 7$. Notice that 2 is to the left of 7 on the number line in Figure 10.3.

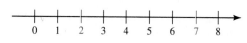

Figure 10.3

(b) 12   9
$12 > 9$. Notice that 12 is to the right of 9 on the number line in Figure 10.4.

PRACTICE EXERCISE 2

Place the correct symbol ($<$ or $>$) between the given pairs of numbers.

(a) 3   11

(b) 15   7

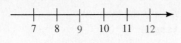

Figure 10.4

Notice in Example 2 that when $<$ and $>$ are used, *the symbol always points to the smaller of the two numbers*.

The number lines we have considered thus far display no numbers to the left of 0. What meaning could we give to a number to the left of 0? One example of such a number is a temperature of 5° below zero on a cold day, which is sometimes represented as $-5°$ (read "negative 5°"). Others are shown in the following table.

Measurement	Number
5° above zero	5° (or $+5°$)
5° below zero	$-5°$
100 ft above sea level	100 ft (or $+100$ ft)
100 ft below sea level	-100 ft
\$16 deposit into an account	\$16 (or $+$\$16)
\$16 check written on an account	$-$\$16

❸ INTEGERS

When we put a negative sign in front of each of the counting numbers, we obtain the **negative integers.** The collection of negative integers together with the counting numbers (sometimes called **positive integers**) and zero is called the set of **integers.**

$$\{\ldots, -3, -2, -1, 0, 1, 2, 3, \ldots\} \qquad \text{Integers}$$

The whole numbers are often called **nonnegative integers.** The number line in Figure 10.5 shows negative integers, positive integers, and zero. Note that we can write 1 as $+1$, 2 as $+2$, and so on. Also, we read -1 as "negative 1," for example.

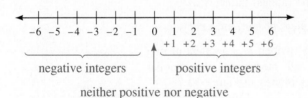

Figure 10.5 Integer Number Line

❹ EQUAL (UNEQUAL) NUMBERS

When using the number line, we say that two numbers are **equal** (for example, $3 + 2 = 5$) if they correspond to the same point on a number line. If two numbers correspond to different points on a number line, they are **unequal.** We use $\neq$ to represent "is unequal to." We can expand our definitions of *less than* and *greater than* for whole numbers to include the integers. That is, one integer is **less than** a second if it is to the left of the second on a number line, and one integer is **greater than** a second if it is to the right of the second on a number line. If one number is **less than or equal to** another number, we use the symbol $\le$ (similarly $\ge$ represents **greater than or equal to**). For example,

$$-4 \le 4, \quad -2 \le 0, \quad 0 \le 3, \quad 2 \le 2, \quad \text{and} \quad -1 \le 1.$$

Also, all negative integers are less than zero (left of zero), all positive integers are greater than zero (right of zero), and any positive integer is greater than any negative integer.

EXAMPLE 3 ORDER RELATIONS BETWEEN NUMBERS	PRACTICE EXERCISE 3

Some relationships between numbers on the number line in Figure 10.6 follow.

Figure 10.6

In Words	*In Symbols*
(a) 1 is less than a	$1 < a$
(b) c is greater than d	$c > d$
(c) e is unequal to -5	$e \ne -5$
(d) d is equal to -4	$d = -4$
(e) b is less than or equal to b	$b \le b$
(f) a is greater than or equal to 2	$a \ge 2$

Use Figure 10.6 and write the following in symbols.

(a) d is less than 0

(b) a is greater than or equal to -1

(c) b is equal to 7

(d) e is less than or equal to e

(e) a is greater than d

(f) c is unequal to -8

Answers: (a) $d < 0$ (b) $a \ge -1$
(c) $b = 7$ (d) $e \le e$ (e) $a > d$
(f) $c \ne -8$

⑤ RATIONAL NUMBERS

In addition to the integers, the fractions we studied in Chapter 1, together with their negatives, can be shown or **plotted** on a number line. All of the fractions to the right of zero are called **positive rational numbers.** Those to the left of zero are called **negative rational numbers,** and together, along with zero, they form the set of **rational numbers.** We can think of the rational numbers as

{all numbers that can be written as a quotient of two integers}.

Remember from Chapter 3 that any integer can be written as a quotient of two integers, for example $-2 = -\frac{2}{1}$ and $1 = \frac{1}{1}$, so all integers are also rational numbers. Some examples of rational numbers are plotted in Figure 10.7

Figure 10.7 Comparing Rational Numbers

Just as with integers, we can define the order relationships of *less than* and *greater than* for rational numbers. For example, we see that

$$-\frac{1}{2} < \frac{1}{2}, \qquad \frac{3}{2} < \frac{9}{4}, \qquad 2\frac{7}{8} > -\frac{3}{2}, \qquad \text{and} \qquad -\frac{9}{4} > -3\frac{1}{3}$$

by considering the number line in Figure 10.7.

A very important property of our number system states that if two numbers are unequal, then one of them must be less than the other. Given two integers, it is easy to determine if they are equal and, if not, which is less than (or which is greater than) the other. However, this is not quite so easy when considering two rational numbers.

Ordering Rational Numbers

Given two positive rational numbers $\dfrac{a}{b}$ and $\dfrac{c}{d}$,

1. $\dfrac{a}{b} = \dfrac{c}{d}$ whenever $ad = bc$.

2. $\dfrac{a}{b} > \dfrac{c}{d}$ whenever $ad > bc$.

3. $\dfrac{a}{b} < \dfrac{c}{d}$ whenever $ad < bc$.

The product ad is called the **first cross product,** and the product bc is called the **second cross product.** Thus, two positive fractions are equal if the cross products are equal, and the order of two unequal fractions is the same as the order of the first and second cross products. Similar results can be stated for negative rational numbers.

EXAMPLE 4   ORDERING RATIONAL NUMBERS

Place the appropriate symbol =, >, or < between the given pairs of fractions.

(a) $\dfrac{4}{6}$ $\dfrac{28}{42}$

Since $4 \cdot 42 = 168$ and $6 \cdot 28 = 168$, the fractions are equal.

$$\frac{4}{6} = \frac{28}{42}.$$

(b) $\dfrac{24}{7}$ $\dfrac{39}{11}$

Since $24 \cdot 11 = 264$ and $7 \cdot 39 = 273$, and $264 < 273$,

$$\frac{24}{7} < \frac{39}{11}.$$

(c) $\dfrac{15}{9}$ $\dfrac{31}{20}$

Since $15 \cdot 20 = 300$ and $9 \cdot 31 = 279$, and $300 > 279$,

$$\frac{15}{9} > \frac{31}{20}.$$

PRACTICE EXERCISE 4

Place the appropriate symbol, =, >, or <, between the given pairs of fractions.

(a) $\dfrac{14}{3}$ $\dfrac{23}{5}$

(b) $\dfrac{21}{13}$ $\dfrac{18}{11}$

(c) $\dfrac{9}{28}$ $\dfrac{36}{112}$

Answers: (a) >   (b) <   (c) =

⑥ ABSOLUTE VALUES

In Chapters 3 and 4 we worked the four basic operations of addition, subtraction, multiplication, and division on the nonnegative rational numbers. When operating on *all* rational numbers we use the notion of absolute value.

Absolute Value of a Number
The **absolute value** of a number x is the distance from zero to x on a number line. We denote the absolute value of x by $|x|$, and read the symbol as "the absolute value of x."

EXAMPLE 5   ABSOLUTE VALUE USING A NUMBER LINE

(a) As shown in Figure 10.8, 3 is 3 units from 0. Thus, $|3| = 3$.

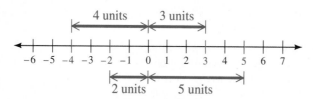

Figure 10.8   Absolute Value Using a Number Line

(b) 5 is 5 units from 0. Thus, $|5| = 5$.

(c) 0 is 0 units from 0. Thus, $|0| = 0$.

(d) -2 is 2 units from 0. Thus, $|-2| = 2$.

(e) -4 is 4 units from 0. Thus, $|-4| = 4$.

PRACTICE EXERCISE 5

Find the following absolute values.

(a) $|-8|$

(b) $|8|$

(c) $|0.05|$

(d) $|-0.05|$

(e) $\left|-\dfrac{3}{8}\right|$

Answers: (a) 8   (b) 8   (c) 0.05
(d) 0.05   (e) $\frac{3}{8}$

When finding the absolute value of a number, remember that the absolute value of a positive number or zero is the number itself. The absolute value of a negative number is the positive number formed by removing the minus sign. Thus, *the absolute value of a number is always greater than or equal to zero, never negative.*

10.2   EXERCISES A

Answer true *or* false.

1. {0, 1, 2, 3, . . .} is called the set of nonnegative integers.   **true**

2. If a number x is located to the left of a number y on a number line, then $x > y$.   **false**

3. The numeral -7 is read "negative seven."   **true**

4. A temperature of 32° below zero could be written as $+32°$.   **false**

5. The distance from zero to a number y on a number line is called the absolute value of y.   **true**

6. An elevation of 4500 feet above sea level would be denoted by -4500.   **false**

7. A withdrawal of $200 from a savings account could be denoted by $-\$200$.   **true**

Place the correct symbol $=$, $>$, or $<$ between the given pairs of numbers.

8. $7 > 1$

9. $0 < 3$

10. $-2 < 0$

11. $2 < 9$

12. $-3 < 2$

13. $\dfrac{1}{2} > -\dfrac{1}{2}$

14. $-\dfrac{3}{4} < \dfrac{1}{4}$

15. $-5 = -5$

Is the given statement true or false?

16. $-12 \le 0$ **true**

17. $6 \le 6$ **true**

18. $0 \le -3$ **false**

19. $-5 \le -5$ **true**

20. $-2 \le -7$ **false**

21. $\dfrac{2}{3} \le -\dfrac{2}{3}$ **false**

22. $4\dfrac{1}{2} \ge 4.5$ **true**

23. $-0.5 \le -\dfrac{1}{2}$ **true**

24. Plot the numbers $\dfrac{2}{3}$, $-\dfrac{7}{8}$, $\dfrac{5}{4}$, $-\dfrac{7}{4}$, $\dfrac{5}{2}$, and $-\dfrac{10}{3}$ on the given number line.

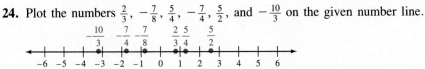

25. Insert the appropriate symbol $=$, $>$, or $<$ between the two positive fractions.

    (a) If $ad = bc$, then $\dfrac{a}{b} = \dfrac{c}{d}$.

    (b) If $ad < bc$, then $\dfrac{a}{b} < \dfrac{c}{d}$.

    (c) If $ad > bc$, then $\dfrac{a}{b} > \dfrac{c}{d}$.

    (d) The products ad and bc are called _____**cross products**_____.

Place the appropriate symbol $=$, $>$, or $<$ between the given pairs of fractions.

㉖ $\dfrac{17}{43} > \dfrac{28}{79}$

27. $\dfrac{35}{11} > \dfrac{41}{13}$

28. $\dfrac{4}{11} < \dfrac{7}{19}$

29. $\dfrac{13}{50} = \dfrac{39}{150}$

30. $\dfrac{5}{7} < \dfrac{60}{83}$

31. $\dfrac{2}{9} < \dfrac{21}{91}$

Evaluate the absolute values.

32. $|17|$ **17**

33. $|-1|$ **1**

34. $|0|$ **0**

35. $\left|-\dfrac{3}{4}\right|$ $\dfrac{3}{4}$

36. $|3.1|$ **3.1**

37. $\left|-\dfrac{5}{2}\right|$ $\dfrac{5}{2}$

38. $|45|$ **45**

39. $|-0.8|$ **0.8**

FOR REVIEW

Evaluate when x = 5, y = 2, and z = 10.

40. $(xy)^2$ 100

41. xy^2 20

42. $\dfrac{z}{xy}$ 1

43. $\dfrac{2x + y^2}{z}$ 1.4

44. $(x - y)^2$ 9

45. $x^2 - y^2$ 21

ANSWERS: 1. true   2. false   3. true   4. false   5. true   6. false   7. true   8. >   9. <   10. <   11. <   12. <
13. >   14. <   15. =   16. true   17. true   18. false   19. true   20. false   21. false   22. true   23. true
24.   25. (a) =   (b) <   (c) >   (d) cross products   26. >   27. >   28. <
29. =   30. <   31. <   32. 17   33. 1   34. 0   35. $\frac{3}{4}$ 36. 3.1   37. $\frac{5}{2}$ 38. 45   39. 0.8   40. 100   41. 20   42. 1
43. 1.4   44. 9   45. 21

10.2 EXERCISES B

Answer true or false.

1. {. . . , −3, −2, −1} is called the set of negative integers.   **true**

2. If a number w is located to the right of a number v on a number line, then $w < v$.   **false**

3. The numeral −11 is read ''negative eleven.''   **true**

4. The low temperature last winter in Twin Falls, Idaho of 38° below zero could be written −38°.   **true**

5. The absolute value of z, denoted by $|z|$, is the distance from zero to z on a number line.   **true**

6. An airplane flying at an altitude of 25,000 feet could be described as flying at an altitude of −25,000 feet.
false

7. A check in the amount of $35.00 written on an account could be denoted by −$35.00.   **true**

Place the correct symbol =, >, or < between the given pairs of numbers.

8. 4 < 9

9. 5 > 0

10. 0 > −7

11. 3 < 12

12. −5 < 1

13. $\dfrac{2}{3}$ > $-\dfrac{2}{3}$

14. −0.2 < 0.2

15. −6 = −6

Is the given statement true or false?

16. −3 ≥ −4   **true**

17. −1 ≥ −1   **true**

18. 0 ≤ −1   **false**

19. −8 ≤ −8   **true**

20. −5 ≤ −40   **false**

21. $\dfrac{1}{4}$ ≥ $-\dfrac{1}{4}$ **true**

22. $3\dfrac{1}{5}$ ≤ 3.2   **true**

23. −0.1 ≤ $-\dfrac{1}{10}$ **true**

24. Plot the numbers $\frac{1}{3}$, $-\frac{3}{4}$, $\frac{7}{4}$, $-\frac{9}{4}$, $\frac{7}{2}$, and $-\frac{11}{3}$ on a number line.

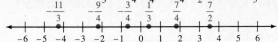

25. Insert the appropriate symbol $=$, $>$, or $<$ between the two positive fractions.

(a) If $xw < yv$ then $\dfrac{x}{y}$ $<$ $\dfrac{v}{w}$.

(b) If $xw = yv$ then $\dfrac{x}{y}$ $=$ $\dfrac{v}{w}$.

(c) If $xw > yv$ then $\dfrac{x}{y}$ $>$ $\dfrac{v}{w}$.

(d) The cross products used to determine order on the two rational numbers $\dfrac{x}{y}$ and $\dfrac{v}{w}$ are $\underline{\hspace{0.3em}xw\hspace{0.3em}}$ and $\underline{\hspace{0.3em}yv\hspace{0.3em}}$.

Place the appropriate symbol $=$, $>$, or $<$ between the given pairs of fractions.

26. $\dfrac{3}{5}$ $<$ $\dfrac{7}{10}$

27. $\dfrac{36}{11}$ $<$ $\dfrac{77}{20}$

28. $\dfrac{19}{43}$ $=$ $\dfrac{57}{129}$

29. $\dfrac{6}{7}$ $>$ $\dfrac{47}{55}$

30. $\dfrac{121}{9}$ $<$ $\dfrac{55}{3}$

31. $\dfrac{27}{201}$ $>$ $\dfrac{13}{121}$

Evaluate the absolute values.

32. $|25|$ 25

33. $|-2|$ 2

34. $\left|-\dfrac{1}{2}\right|$ $\dfrac{1}{2}$

35. $|-0|$ 0

36. $|8.7|$ 8.7

37. $\left|-\dfrac{9}{4}\right|$ $\dfrac{9}{4}$

38. $|112|$ 112

39. $|-0.03|$ 0.03

FOR REVIEW

Evaluate when $a = 7$, $b = 4$, and $c = 3$.

40. $(bc)^3$ 1728

41. bc^3 108

42. $\dfrac{b+c}{a}$ 1

43. $\dfrac{2a+b}{c^2}$ 2

44. $(a-b)^3$ 27

45. $a^3 - b^3$ 279

10.2 EXERCISES C

Place the appropriate symbol $=$, $>$, or $<$ between the pairs of fractions.

1. $-\dfrac{14}{19}$ $<$ $-\dfrac{2}{3}$

2. $-\dfrac{21}{8}$ $>$ $-\dfrac{31}{11}$

3. $-\dfrac{77}{14}$ $=$ $-\dfrac{11}{2}$

10.3 ADDITION AND SUBTRACTION OF RATIONAL NUMBERS

STUDENT GUIDEPOSTS

1 Adding Numbers with Like Signs
2 Adding Numbers with Unlike Signs
3 Additive Identity
4 Additive Inverse (Negative)
5 Commutative Law of Addition
6 Associative Law of Addition
7 Adding More Than Two Numbers
8 Subtracting Numbers

In earlier chapters we reviewed the four basic operations on the nonnegative rational numbers. Now we extend our treatment to addition and subtraction of all rational numbers. Consider the addition problem $3 + 2$ on the number line in Figure 10.9.

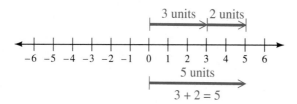

Figure 10.9

To find $3 + 2$, we start at 0, move 3 units to the *right,* and then move 2 more units to the *right.* We are then 5 units to the *right* of zero. Thus, $3 + 2 = 5$.

We can also find $3 + (-2)$ on a number line. We first move 3 units to the *right* and then 2 units to the *left.* This puts us 1 unit to the *right* of 0, as shown in Figure 10.10. Thus, $3 + (-2) = 1$.

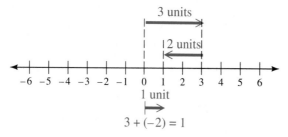

Figure 10.10

Intuitively, if we think of numbers as temperatures, an increase of 3° $(+3)$ followed by a second increase of 2° $(+2)$ results in a total increase of 5° $(+5)$. Similarly, an increase of 3° $(+3)$ followed by a decrease of 2° (-2) results in a net increase of 1° $(+1)$.

Notice that when adding numbers using a number line, we move to the *right when a number is positive* and to the *left when it is negative.*

EXAMPLE 1   ADDING NUMBERS WITH LIKE SIGNS

Find $(-3) + (-2)$.

Draw a number line. Move 3 units to the left for -3 and then 2 more units to the left for -2. As seen in Figure 10.11, we end up 5 units to the left of 0, at -5. Thus, $(-3) + (-2) = -5$.

PRACTICE EXERCISE 1

Find $(-11) + (-8)$.

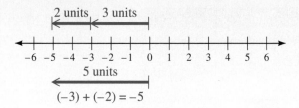

Figure 10.11

Answer: -19

| **EXAMPLE 2** **Adding Numbers with Unlike Signs** | **Practice Exercise 2** |

Find $(-3) + 2$.

Draw a number line. Move 3 units to the left and then 2 units to the right. As Figure 10.12 shows, we are then at -1. Thus, $(-3) + 2 = -1$.

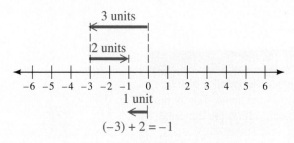

Figure 10.12

Find $(-11) + 8$.

Answer: -3

| **EXAMPLE 3** **Adding Numbers with Unlike Signs** | **Practice Exercise 3** |

Find $2 + (-3)$.

Draw a number line. Move 2 units to the right and then 3 units to the left. (See Figure 10.13.) This puts us at -1. Thus, $2 + (-3) = -1$.

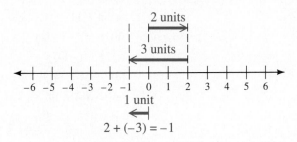

Figure 10.13

Find $8 + (-11)$.

Answer: -3

---

❶ ADDING NUMBERS WITH LIKE SIGNS

The number-line method of adding numbers shows us what addition means, but it takes too much time. For example, we found $(-3) + (-2)$ to be -5 using a number line (Figure 10.11). However, we could have added $3 + 2$ and attached a minus sign:

$$(-3) + (-2) = -(3 + 2) = -5.$$

This is an example of the following rule.

Adding Numbers Having Like Signs
To add two numbers having like signs add their absolute values. The sum has the same sign as the numbers being added.

② ADDING NUMBERS WITH UNLIKE SIGNS

When the signs of the numbers being added are not the same, we need another rule. For $3 + (-2) = 1$ (Figure 10.10), we could have found the difference between 3 and 2 and attached a plus sign:

$$3 + (-2) = +(3 - 2) = +1 = 1.$$

Also, for $(-3) + 2 = 2 + (-3) = -1$ (Figures 10.12 and 10.13), we could have subtracted 2 from 3 and attached a minus sign:

$$(-3) + 2 = 2 + (-3) = -(3 - 2) = -1.$$

These are examples of the following rule.

Adding Numbers Having Unlike Signs
To add two numbers having unlike signs, subtract the smaller absolute value from the larger absolute value. The result has the same sign as the number with the larger absolute value. If the absolute values are the same, the sum is zero.

Remember that if a number (other than zero) has no sign, this is the same as having a plus sign attached. For example, 5 and +5 are the same.

EXAMPLE 4   ADDING RATIONAL NUMBERS

Add using the preceding rules.

(a) $6 + 7 = 13$

(b) $(-6) + (-7) = -(6 + 7) = -13$

(c) $6 + (-7) = -(7 - 6) = -1$ $|-7| = 7 > 6 = |6|$

(d) $(-6) + 7 = +(7 - 6) = +1 = 1$ $|-6| = 6 < 7 = |7|$

(e) $12 + (-12) = 0$ $|12| = 12 = |-12|$

PRACTICE EXERCISE 4

Add.

(a) $5 + 14$

(b) $(-5) + (-14)$

(c) $5 + (-14)$

(d) $(-5) + 14$

(e) $(-14) + 14$

Answers: (a) **19**   (b) **−19**   (c) **−9**   (d) **9**   (e) **0**

③ ADDITIVE IDENTITY

As was the case with whole numbers, the number 0 plays a special role in this system since for any number a,

$$a + 0 = 0 + a = a.$$

Intuitively, any number a remains identically the same when added to 0. For this reason, 0 is called the **additive identity.**

④ ADDITIVE INVERSE (NEGATIVE)

Example 4(e) illustrates an important property of our number system. If a is any number, we call $-a$ the **additive inverse,** or **negative** of a, which means

$$a + (-a) = (-a) + a = 0.$$

EXAMPLE 5 ADDITIVE IDENTITY AND INVERSES	PRACTICE EXERCISE 5

Add.

(a) $0 + 7 = 7 + 0 = 7$ ⠀⠀0 is the additive identity

(b) $7 + (-7) = 0$ ⠀⠀7 and -7 are additive inverses

(c) $(-2.3) + (-4.1) = -(2.3 + 4.1) = -6.4$

(d) $(-3.5) + 1.7 = -(3.5 - 1.7) = -1.8$

(e) $\dfrac{1}{2} + \left(-\dfrac{3}{4}\right) = -\left(\dfrac{3}{4} - \dfrac{1}{2}\right) = -\left(\dfrac{3}{4} - \dfrac{2}{4}\right) = -\dfrac{1}{4}$

Add.

(a) $13 + 0$

(b) $(-13) + 0$

(c) $(-4.02) + (-1.35)$

(d) $(-6.35) + 2.11$

(e) $\left(-\dfrac{2}{9}\right) + \dfrac{4}{11}$

Answers: (a) 13 ⠀(b) -13
(c) -5.37 ⠀(d) -4.24 ⠀(e) $\frac{14}{99}$

⑤ COMMUTATIVE LAW OF ADDITION

Two additional properties of addition involve changing the order of addition, and rearranging grouping symbols in sums of three (or more) numbers. If a and b are two numbers, the **commutative law of addition** states that

$$a + b = b + a.$$

The order of addition can be changed by the commutative law. For example, $16 + 5 = 5 + 16$.

⑥ ASSOCIATIVE LAW OF ADDITION

Given another number c, the **associative law of addition** states that

$$(a + b) + c = a + (b + c).$$

The grouping symbols can be rearranged by the associative law. For example, $(5 + 2) + 3 = 5 + (2 + 3)$.

EXAMPLE 6 COMMUTATIVE AND ASSOCIATIVE LAWS	PRACTICE EXERCISE 6

(a) Use the commutative law to complete the equation $5 + \frac{1}{4} = $ _____ .
The commutative law says the order of addition can be changed, so $5 + \frac{1}{4} = \frac{1}{4} + 5$.

(b) Use the associative law to complete the equation $\left(7 + \frac{1}{2}\right) + (-2) = $

_____ .
The associative law says the grouping symbols can be rearranged, so $\left(7 + \frac{1}{2}\right) + (-2) = 7 + \left[\frac{1}{2} + (-2)\right]$.

(a) Complete the following using the commutative law.
$\left(-\frac{1}{2}\right) + 9 = $ _____

(b) Complete the following using the associative law.
$(7 + 0.5) + 2 = $ _____

Answers: (a) $9 + \left(-\frac{1}{2}\right)$
(b) $7 + (0.5 + 2)$

⑦ ADDING MORE THAN TWO NUMBERS

Taken together, the commutative and associative laws of addition allow us to add several signed numbers by reordering and regrouping the numbers to obtain the sum in the most convenient manner.

Adding More Than Two Numbers

1. Add all positive numbers.
2. Add all negative numbers.
3. Now use the rule for adding numbers with unlike signs.

EXAMPLE 7 ADDING SEVERAL NUMBERS	PRACTICE EXERCISE 7

Add.

$(-2) + (-3) + 2 + 7 + (-5) + 6 + 9 + (-1)$
$= [2 + 7 + 6 + 9] + [(-2) + (-3) + (-5) + (-1)]$
$= [2 + 7 + 6 + 9] + [-(2 + 3 + 5 + 1)]$
$= [24] + [-11] = +[24 - 11] = 13$

Add.

$5 + (-2) + (-1) + 7 + (-9) + 6$

Answer: 6

With practice we should be able to skip many of the intermediate steps in the preceding examples and make computations mentally.

⑧ SUBTRACTING NUMBERS

Subtraction of numbers can be defined as addition of an additive inverse or a negative. For example, we can think of $5 - 3 = 2$ as $5 + (-3) = 2$.

Subtracting Numbers

To subtract one number from another, change the sign of the number being subtracted and use the rule for adding numbers. Thus, for a and b rational numbers,

$$a - b = a + (-b).$$

EXAMPLE 8 SUBTRACTING NUMBERS	PRACTICE EXERCISE 8

Subtract.

(a) $5 - (+3) = 5 + (-3)$ Change sign and add
$= 5 - 3 = 2$
(b) $5 - (-3) = 5 + 3 = 8$ Change sign and add
(c) $(-5) - 3 = (-5) + (-3)$ Change sign and add
$= -(5 + 3) = -8$
(d) $(-5) - (-3) = (-5) + 3$ Change sign and add
$= -(5 - 3) = -2$

Subtract.

(a) $17 - (+2)$

(b) $17 - (-2)$

(c) $(-17) - 2$

(d) $(-17) - (-2)$

(e) $(-5) - 0 = (-5) + (-0)$ Change sign and add

$= (-5) + 0$ The negative of zero is zero:
$-0 = 0$

$= -5$

(e) $0 - (-17)$

Answers: (a) 15   (b) 19
(c) -19 (d) -15 (e) 17

EXAMPLE 9   SUBTRACTING NUMBERS

Subtract.

(a) $5 - (+5) = 5 + (-5) = 0$

(b) $5 - (-5) = 5 + 5 = 10$

(c) $(-5) - 5 = (-5) + (-5) = -(5 + 5) = -10$

(d) $(-5) - (-5) = (-5) + 5 = 0$

(e) $(-5.8) - (-3.2) = (-5.8) + 3.2 = -(5.8 - 3.2) = -2.6$

(f) $\left(-\dfrac{2}{3}\right) - \dfrac{1}{6} = \left(-\dfrac{2}{3}\right) + \left(-\dfrac{1}{6}\right) = \left(-\dfrac{4}{6}\right) + \left(-\dfrac{1}{6}\right) = -\dfrac{5}{6}$

PRACTICE EXERCISE 9

Subtract.

(a) $12 - (+12)$

(b) $12 - (-12)$

(c) $(-12) - 12$

(d) $(-12) - (-12)$

(e) $(-1.2) - (-5.7)$

(f) $\left(-\dfrac{1}{8}\right) - \dfrac{3}{5}$

Answers: (a) 0   (b) 24   (c) -24
(d) 0   (e) 4.5   (f) $-\frac{29}{40}$

CAUTION

Although addition satisfies both the commutative and associative laws, subtraction does not. For example,

$3 = 5 - 2 \neq 2 - 5 = -3.$ Subtraction is *not* commutative

Likewise,

$(8 - 4) - 3 = 4 - 3 = 1,$

but   Subtraction is *not* associative

$8 - (4 - 3) = 8 - 1 = 7.$

EXAMPLE 10   APPLICATION TO PERSONAL BANKING

John had \$932 in his bank account on May 1. He deposited \$326 and \$791 during the month and wrote checks for \$816, \$315, and \$940 during the month. What was his balance at the end of May?

We can consider the deposits as positive numbers and the checks written as negative numbers. Thus, we must add 932, 326, 791, -816, -315, and -940.

```
 932              -816
 326              -315              -2071
+791              -940               2049
2049  Sum of deposits  -2071  Sum of checks   -22  Final sum
```

Thus, John is overdrawn (in the red) \$22 on his account.

PRACTICE EXERCISE 10

Marlo had a temperature of 100.1° at noon. It rose 2.3° before dropping 3.5° by 3:00 P.M. What was her temperature at 3:00 P.M.?

Answer: 98.9°

10.3 EXERCISES A

Perform the indicated operations.

1. $4 + 3$ 7

2. $(-4) + (-3)$ -7

3. $4 + (-3)$ 1

4. $(-4) + 3$ -1

5. $4 + 0$ 4

6. $0 + (-4)$ -4

7. $7 + (-7)$ 0

8. $(-7) + 7$ 0

9. $(-7) + (-7)$ -14

10. $7 + 7$ 14

11. $(-8) + (-5)$ -13

12. $9 + (-1)$ 8

13. $19 + (-25)$ -6

14. $(-16) + 14$ -2

15. $4 - 3$ 1

16. $(-4) - (-3)$ -1

17. $4 - (-3)$ 7

18. $(-4) - 3$ -7

19. $0 - 4$ -4

20. $(-4) - 0$ -4

21. $7 - (-7)$ 14

22. $(-7) - 7$ -14

23. $(-7) - (-7)$ 0

24. $7 - 7$ 0

25. $(-8) - (-5)$ -3

26. $9 - (-1)$ 10

27. $(-1) - 9$ -10

28. $14 - 21$ -7

29. $19 - (-25)$ 44

30. $(-16) - 14$ -30

31. $1.3 + (-2.5)$ -1.2

32. $(-1.3) + 2.5$ 1.2

33. $(-1.3) + (-2.5)$ -3.8

34. $1.3 - (-2.5)$ 3.8

35. $(-1.3) - 2.5$ -3.8

36. $(-1.3) - (-2.5)$ 1.2

37. $\dfrac{1}{2} + \left(-\dfrac{2}{5}\right)$ $\dfrac{1}{10}$

38. $\left(-\dfrac{1}{2}\right) + \dfrac{2}{5}$ $-\dfrac{1}{10}$

39. $\left(-\dfrac{1}{2}\right) + \left(-\dfrac{2}{5}\right)$ $-\dfrac{9}{10}$

40. $\dfrac{1}{2} - \left(-\dfrac{2}{5}\right)$ $\dfrac{9}{10}$

41. $\left(-\dfrac{1}{2}\right) - \dfrac{2}{5}$ $-\dfrac{9}{10}$

42. $\left(-\dfrac{1}{2}\right) - \left(-\dfrac{2}{5}\right)$ $-\dfrac{1}{10}$

43. $(-3) + 2 + (-7) + (-8) + 9$ -7

44. $5 + 8 + 9 + (-3) + (-7)$ 12

45. $(-8) + (-6) + 9 + (-3) + 5 + (-6)$ -9

46. $(-4) + (-2) + 6 + 7 + (-1)$ 6

47. Show by example that subtraction does not satisfy the commutative law. Can you find numbers a and b for which $a - b$ does equal $b - a$? $2 - 1 = 1 \neq -1 = 1 - 2$; if $a = b$ then $a - b = 0 = b - a$

Add.

48.
$$
\begin{array}{r}
23 \\
-16 \\
-36 \\
5 \\
\underline{81}
\end{array}
$$
   57

49.
$$
\begin{array}{r}
-16 \\
-81 \\
-14 \\
-70 \\
\underline{-93}
\end{array}
$$
   -274

50.
$$
\begin{array}{r}
-64 \\
25 \\
38 \\
-17 \\
\underline{-83}
\end{array}
$$
   -101

51 When a cold front came through Cut Bank, Montana, the temperature dropped from 35° above zero to 13° below zero ($-13°$). What was the total change in temperature? $-48°$

52. On March 1, Allen had $64 in his bank account. During the month, he wrote checks for $81, $108, and $192 and made deposits of $123 and $140. What was the value of his account at the end of the month? $-$54$

53. A helicopter is 600 ft above sea level and a submarine directly below it is 250 ft below sea level (-250 ft). How far apart are they?   **850 ft**

54. At 5:30 A.M., it was $-42°$F. By 2:00 P.M. the same day, the temperature had risen $57°$ to the recorded high temperature. Shortly thereafter, a cold front passed through dropping the temperature $41°$ by 5:00 P.M. What was the temperature at 5:00 P.M.? $-26°$

FOR REVIEW

Place the correct symbol $=$, $<$, or $>$ between the given pairs of numbers.

55. $-5 < -4$

56. $\dfrac{2}{3} < \dfrac{77}{110}$

57. $\dfrac{5}{12} = \dfrac{30}{72}$

58. $-4.7 < -1.2$

Evaluate the absolute values.

59. $|-81|$ 81

60. $\left|-\dfrac{5}{4}\right|$ $\dfrac{5}{4}$

61. $|17|$ 17

62. $|-7.25|$ 7.25

ANSWERS: 1. 7 2. -7 3. 1 4. -1 5. 4 6. -4 7. 0 8. 0 9. -14 10. 14 11. -13 12. 8 13. -6 14. -2 15. 1 16. -1 17. 7 18. -7 19. -4 20. -4 21. 14 22. -14 23. 0 24. 0 25. -3 26. 10 27. -10 28. -7 29. 44 30. -30 31. -1.2 32. 1.2 33. -3.8 34. 3.8 35. -3.8 36. 1.2 37. $\frac{1}{10}$ 38. $-\frac{1}{10}$ 39. $-\frac{9}{10}$ 40. $\frac{9}{10}$ 41. $-\frac{9}{10}$ 42. $-\frac{1}{10}$ 43. -7 44. 12 45. -9 46. 6 47. $2 - 1 = 1 \neq -1 = 1 - 2$; if $a = b$ then $a - b = 0 = b - a$ 48. 57 49. -274 50. -101 51. $-48°$ 52. $-\$54$ 53. 850 ft 54. $-26°$ 55. $<$ 56. $<$ 57. $=$ 58. $<$ 59. 81 60. $\frac{5}{4}$ 61. 17 62. 7.25

10.3   EXERCISES B

Perform the indicated operations.

1. $5 + 2$ 7

2. $(-5) + (-2)$ -7

3. $5 + (-2)$ 3

4. $(-5) + 2$ -3

5. $0 + 5$ 5

6. $(-5) + 0$ -5

7. $5 + (-5)$ 0

8. $(-5) + 5$ 0

9. $(-5) + (-5)$ -10

10. $5 + 5$ 10

11. $(-9) + (-3)$ -12

12. $(-2) + 11$ 9

13. $18 + (-23)$ -5

14. $(-15) + 19$ 4

15. $8 - 7$ 1

16. $(-8) - (-7)$ -1

17. $8 - (-7)$ 15

18. $(-8) - 7$ -15

19. $10 - 0$ 10

20. $0 - (-10)$ 10

21. $6 - (-6)$ 12

22. $(-6) - 6$ -12

23. $(-6) - (-6)$ 0

24. $6 - 6$ 0

25. $(-11) - (-4)$ -7

26. $(-13) - (-1)$ -12

27. $(-1) - 7$ -8

28. $(-12) - 24$ -36

29. $18 - (-27)$ 45

30. $(-17) - 11$ -28

31. $1.5 + (-3.8)$ -2.3

32. $(-1.5) + 3.8$ 2.3

33. $(-1.5) + (-3.8)$ -5.3

34. $1.5 - (-3.8)$ 5.3

35. $(-1.5) - 3.8$ -5.3

36. $(-1.5) - (-3.8)$ 2.3

37. $\dfrac{1}{4} + \left(-\dfrac{3}{8}\right)$ $-\dfrac{1}{8}$ **38.** $\left(-\dfrac{1}{4}\right) + \dfrac{3}{8}$ $\dfrac{1}{8}$ **39.** $\left(-\dfrac{1}{4}\right) + \left(-\dfrac{3}{8}\right)$ $-\dfrac{5}{8}$

40. $\dfrac{1}{4} - \left(-\dfrac{3}{8}\right)$ $\dfrac{5}{8}$ **41.** $\left(-\dfrac{1}{4}\right) - \dfrac{3}{8}$ $-\dfrac{5}{8}$ **42.** $\left(-\dfrac{1}{4}\right) - \left(-\dfrac{3}{8}\right)$ $\dfrac{1}{8}$

43. $(-5) + 4 + (-9) + (-10) + 6$ -14 **44.** $5 + 9 + (-3) + 1 + (-10)$ 2

45. $(-5) + (-7) + 10 + (-3) + 4 + (-7)$ -8 **46.** $(-6) + (-3) + 6 + 10 + (-2)$ 5

47. Show by example that subtraction does not satisfy the associative law. Can you find numbers, a, b, and c for which $(a - b) - c$ does equal $a - (b - c)$? $(2 - 1) - 3 = 1 - 3 = -2 \neq 4 = 2 - (-2) = 2 - (1 - 3)$; if $a = b = c = 0$, then $(0 - 0) - 0 = 0 - 0 = 0 = 0 - 0 = 0 - (0 - 0)$

Add.

48.   26   56
       -14
       -51
         3
      $\underline{92}$

49. -17 -195
       -20
       -43
        -5
     $\underline{-110}$

50. -78 -147
        32
        51
       -60
      $\underline{-92}$

51. In a five-hour period of time, the temperature in Empire, N.Y. dropped from 41° to 12° below zero. What was the total temperature change? $-53°$

52. On July 4, Uncle Sam had $97 in his checking account. During the next week he wrote checks for $15, $102, and $312, and made deposits of $145 and $170. What was the value of his account at the end of the week? $-$17$

53. A balloon is 1200 ft above sea level, and a diving bell directly below it is 340 ft beneath the surface of the water. How far apart are the two?   **1540 ft**

54. At 8:00 A.M., Leah Johnson had a temperature of 99.8°. It rose another 3.1° before falling 4.3° by noon. What was her temperature at noon?   **98.6°**

FOR REVIEW

Place the correct symbol =, <, or > between the given pairs of numbers.

55. $-6 > -7$ **56.** $\dfrac{3}{4} < \dfrac{96}{120}$ **57.** $\dfrac{7}{8} = \dfrac{77}{88}$ **58.** $0 > -11.5$

Evaluate the absolute values.

59. $\left|-\dfrac{9}{10}\right|$ $\dfrac{9}{10}$ **60.** $|-210|$ 210 **61.** $|42|$ 42 **62.** $|-0.01|$ 0.01

10.3 EXERCISES C

Perform the indicated operations.

1. $(-962) - (-508)$ -454 **2.** $-48.62 - 40.002$ -88.622 **3.** $64\dfrac{7}{24} - 98\dfrac{13}{16}$

$$\left[\text{Answer: } -34\dfrac{25}{48}\right]$$

10.4 MULTIPLICATION AND DIVISION OF RATIONAL NUMBERS

1 MULTIPLYING NUMBERS

In multiplying and dividing numbers, we need to decide what sign to give the product or quotient. To help us, we look for patterns in several products.

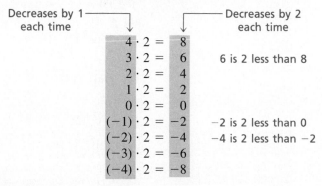

Decreases by 1 each time ——→   ——→ Decreases by 2 each time

$$4 \cdot 2 = 8$$
$$3 \cdot 2 = 6 \qquad \text{6 is 2 less than 8}$$
$$2 \cdot 2 = 4$$
$$1 \cdot 2 = 2$$
$$0 \cdot 2 = 0$$
$$(-1) \cdot 2 = -2 \qquad \text{-2 is 2 less than 0}$$
$$(-2) \cdot 2 = -4 \qquad \text{-4 is 2 less than -2}$$
$$(-3) \cdot 2 = -6$$
$$(-4) \cdot 2 = -8$$

We know the products $4 \cdot 2$, $3 \cdot 2$, $2 \cdot 2$, $1 \cdot 2$, and $0 \cdot 2$, and we can see that these products decrease by 2 each time. For this pattern to continue, a negative number times a positive number must be negative. For example,

$$(-1) \cdot 2 = -2, \quad (-2) \cdot 2 = -4, \quad (-3) \cdot 2 = -6, \quad (-4) \cdot 2 = -8.$$

Similarly, any positive number times any negative number must be negative. We use this to find a pattern in the following products.

Decreases by 1 each time ——→   ——→ Increases by 2 each time

$$4 \cdot (-2) = -8$$
$$3 \cdot (-2) = -6 \qquad \text{-6 is 2 more than -8}$$
$$2 \cdot (-2) = -4$$
$$1 \cdot (-2) = -2$$
$$0 \cdot (-2) = 0 \qquad \text{0 is 2 more than -2}$$
$$(-1) \cdot (-2) = 2$$
$$(-2) \cdot (-2) = 4$$
$$(-3) \cdot (-2) = 6$$
$$(-4) \cdot (-2) = 8$$

For this pattern to continue, a negative number times a negative number must be a positive number. For example,

$$(-1) \cdot (-2) = 2, \quad (-2) \cdot (-2) = 4, \quad (-3) \cdot (-2) = 6, \quad (-4) \cdot (-2) = 8.$$

Our results are summarized in the following rule.

> ### Multiplying Numbers
> 1. Multiply the absolute values of the numbers.
> 2. If both signs are positive or both are negative, the product is positive.
> 3. If one sign is positive and one is negative, the product is negative.
> 4. If one number (or both numbers) is zero, the product is zero.

When multiplying numbers, remember that *like signs have a positive product and unlike signs have a negative product*.

EXAMPLE 1 MULTIPLYING NUMBERS	**PRACTICE EXERCISE 1**

Multiply.

(a) $5 \cdot 2 = 10$ Like signs

(b) $(-5)(-2) = (5 \cdot 2) = 10$ Like signs

(c) $5 \cdot (-2) = -(5 \cdot 2) = -10$ Unlike signs

(d) $(-5) \cdot 2 = -(5 \cdot 2) = -10$ Unlike signs

(e) $0 \cdot (-8) = 0$

(f) $0 \cdot 0 = 0$

(g) $(3)\left(\dfrac{1}{3}\right) = 1$

Multiply.

(a) $4 \cdot 11$

(b) $(-4)(-11)$

(c) $4 \cdot (-11)$

(d) $(-4) \cdot 11$

(e) $(12) \cdot 0$

(f) $0 \cdot (-12)$

(g) $\left(\dfrac{3}{7}\right)\left(\dfrac{7}{3}\right)$

Answers: (a) 44   (b) 44
(c) −44   (d) −44   (e) 0   (f) 0
(g) 1

---

❷ MULTIPLICATIVE IDENTITY

The number 1 plays a role in multiplication similar to the one 0 plays in addition. Since for any number a,

$$a \cdot 1 = 1 \cdot a = a,$$

we see that a number remains identically the same when multiplied by 1. For this reason, 1 is sometimes called the **multiplicative identity.**

❸ MULTIPLICATIVE INVERSE (RECIPROCAL)

Example 1(g) illustrates a property of reciprocals that we have seen before. If a is any number except 0, we call $\frac{1}{a}$ the **multiplicative inverse** (or **reciprocal**) of a, which means that

$$a \cdot \frac{1}{a} = \frac{1}{a} \cdot a = 1.$$

EXAMPLE 2 MULTIPLICATIVE IDENTITY AND INVERSES	**PRACTICE EXERCISE 2**

Multiply.

(a) $(-5)\left(-\dfrac{1}{5}\right) = 1$ −5 and −$\frac{1}{5}$ are reciprocals or multiplicative inverses

Multiply.

(a) $\left(-\dfrac{1}{9}\right)(-9)$

(b) $1 \cdot (-7) = (-7) \cdot 1 = -7$ 1 is the multiplicative identity

(c) $(-2.1) \cdot 3.5 = -(2.1)(3.5) = -7.35$

(d) $(-2.1) \cdot (-3.5) = (2.1)(3.5) = 7.35$

(e) $\left(\frac{1}{4}\right)\left(-\frac{3}{5}\right) = -\left(\frac{1}{4}\right)\left(\frac{3}{5}\right) = -\frac{3}{20}$

(f) $\left(-\frac{1}{4}\right)\left(-\frac{3}{5}\right) = \left(\frac{1}{4}\right)\left(\frac{3}{5}\right) = \frac{3}{20}$

(b) $1 \cdot (-13)$

(c) $(-4.1) \cdot (-2.9)$

(d) $(-4.1) \cdot (2.9)$

(e) $\left(\frac{1}{7}\right)\left(-\frac{3}{8}\right)$

(f) $\left(-\frac{1}{7}\right)\left(-\frac{3}{8}\right)$

Answers: (a) 1   (b) -13
(c) 11.89   (d) -11.89 (e) $-\frac{3}{56}$
(f) $\frac{3}{56}$

❹ COMMUTATIVE LAW OF MULTIPLICATION

Two more properties of multiplication, similar to the ones for addition, involve changing the order of multiplication and rearranging grouping symbols in products of three (or more) numbers. If a and b are two numbers, the **commutative law of multiplication** states that

$$a \cdot b = b \cdot a.$$

The order of multiplication can be changed by the commutative law. For example, $4 \cdot 7 = 7 \cdot 4$.

❺ ASSOCIATIVE LAW OF MULTIPLICATION

Given another number c, the **associative law of multiplication** states that

$$(a \cdot b) \cdot c = a \cdot (b \cdot c).$$

The grouping symbols can be changed by the associative law. For example, $(2 \cdot 5) \cdot 3 = 2 \cdot (5 \cdot 3)$.

EXAMPLE 3   COMMUTATIVE AND ASSOCIATIVE LAWS

(a) Use the commutative law to complete the equation $7 \cdot \frac{3}{4} =$ _____ .
The commutative law says the order of multiplication can be changed, so $7 \cdot \frac{3}{4} = \frac{3}{4} \cdot 7$.

(b) Use the associative law to complete the equation $\left(3 \cdot \frac{1}{5}\right) \cdot (-5) =$
_____ . The associative law says the grouping symbols can be rearranged, so $\left(3 \cdot \frac{1}{5}\right) \cdot (-5) = 3 \cdot \left(\frac{1}{5} \cdot (-5)\right)$.

PRACTICE EXERCISE 3

(a) Use the commutative law to complete the following.
$(0.09) \cdot (1000) =$ _____

(b) Use the associative law to complete the following.
$\left(\frac{1}{2} \cdot \frac{3}{10}\right) \cdot (-6) =$ _____
Answers: (a) $(1000) \cdot (0.09)$
(b) $\left(\frac{1}{2}\right) \cdot \left(\frac{3}{10} \cdot (-6)\right)$

Often we will need to multiply more than two numbers. To do this, multiply in pairs and keep track of the sign. Notice in Example 4 that *products involving an odd number of minus signs are negative while those with an even number are positive*.

EXAMPLE 4   MULTIPLYING MORE THAN TWO NUMBERS

Multiply.

(a) $(3)(-2)(-4) = (-6)(-4) = 24$ Multiply 3 times
                                                 -2 first

PRACTICE EXERCISE 4

Multiply.

(a) $(2)(-1)(-4)(3)$

(b) $(-1)(-1)(-2)(-3)(-4) = (1)(-2)(-3)(-4)$ $(-1)(-1) = 1$
$\qquad\qquad\qquad\qquad\quad\; = (-2)(-3)(-4)$ $(1)(-2) = -2$
$\qquad\qquad\qquad\qquad\quad\; = (6)(-4)$ $(-2)(-3) = 6$
$\qquad\qquad\qquad\qquad\quad\; = -24$

(c) $(-2)^3 = (-2)(-2)(-2) = (4)(-2)$ $(-2)(-2) = 4$
$\qquad\qquad\qquad\qquad\quad\;\; = -8$

(b) $(-6)(-1)(-5)(-2)(-1)$

(c) $(-3)^4$

Answers: (a) 24   (b) -60
(c) 81

⑥ DIVIDING NUMBERS

We have seen that the only difference between multiplying positive numbers and multiplying rational numbers is finding the sign of the product. Recall that division can be thought of as multiplication by a reciprocal. For example,

$$15 \div 3 \text{ or } \frac{15}{3} \quad \text{ is the same as } \quad 15 \cdot \frac{1}{3}.$$

Thus, we can use the same rules for signs in division that we used for multiplication.

Dividing Numbers
1. Divide the absolute values of the numbers.
2. If both signs are positive or both are negative, the quotient is positive.
3. If one sign is positive and one sign is negative, the quotient is negative.
4. Zero divided by any number (except zero) is zero, and division of any number by 0 is undefined.

Remember that the rule of signs for division is the same as for multiplication: *like signs have a positive quotient and unlike signs have a negative quotient.*

EXAMPLE 5 DIVIDING NUMBERS	**PRACTICE EXERCISE 5**

Divide.

(a) $6 \div 3 = 2$ Like signs

(b) $(-6) \div (-3) = 6 \div 3 = 2$ Like signs

(c) $6 \div (-3) = -(6 \div 3) = -2$ Unlike signs

(d) $(-6) \div 3 = -(6 \div 3) = -2$ Unlike signs

(e) $0 \div (-3) = 0$

(f) $(-3) \div 0$ is undefined

(g) $\dfrac{1}{2} \div \left(-\dfrac{3}{5}\right) = \dfrac{1}{2} \cdot \left(-\dfrac{5}{3}\right) = -\left(\dfrac{1}{2} \cdot \dfrac{5}{3}\right) = -\dfrac{5}{6}$

Divide.

(a) $12 \div 4$

(b) $(-12) \div (-4)$

(c) $12 \div (-4)$

(d) $(-12) \div 4$

(e) $(-0.3) \div 0$

(f) $0 \div (-0.3)$

(g) $\left(-\dfrac{3}{5}\right) \div \dfrac{9}{10}$

Answers: (a) 3   (b) 3   (c) -3
(d) -3 (e) undefined   (f) 0
(g) $-\frac{2}{3}$

CAUTION

Although multiplication satisfies both the commutative and associative laws, division does not. For example,

$$3 = 6 \div 2 \neq 2 \div 6 = \frac{2}{6} = \frac{1}{3}. \qquad \text{Division is not commutative}$$

Likewise,

$$(8 \div 4) \div 2 = 2 \div 2 = 1$$

but   Division is not associative

$$8 \div (4 \div 2) = 8 \div 2 = 4.$$

EXAMPLE 6   APPLICATION TO PERSONAL BANKING

Willie wrote 3 checks for $75 each and 4 checks for $105 each. By how much did this change his bank balance?

The 3 checks for $75 could be calculated as follows:

$$(3)(-75) = -225.$$

There are 4 checks for $105, so

$$(4)(-105) = -420.$$

Thus, his account is changed by

$$-225 + (-420) = -225 - 420 = -645 \text{ dollars.}$$

His balance is $645 less.

PRACTICE EXERCISE 6

A diver descended below the surface of the ocean by diving 25 ft each minute for 11 minutes. How far below sea level was he at this time?

Answer: 275 ft (−275)

EXAMPLE 7   APPLICATION TO PERSONAL FINANCE

Mary, Sue, and Maria share an apartment with total rent of $720. By how much is each woman's account changed if they each write a check for an equal share of the rent?

Since the rent is a decrease in their accounts, we use −720. To find the shares, we divide −720 by 3:

$$-720 \div 3 = -240.$$

Thus, each woman's account is decreased by $240.

PRACTICE EXERCISE 7

During a 5-hour period, the temperature decreased 42°. What was the average decrease per hour?

Answer: 8.4° (−8.4°)

When evaluating numerical expressions involving rational numbers, follow the order of operations and rules of grouping given in Section 10.1. That is, evaluate within grouping symbols first, beginning with innermost, evaluate powers next, and then perform multiplication and division left to right followed by addition and subtraction.

EXAMPLE 8 ORDER OF OPERATIONS	PRACTICE EXERCISE 8

Evaluate each numerical expression.

(a) $(-3) + 5 \cdot (-1) - (-4)$

$\quad = (-3) + (-5) - (-4)$ Multiply first

$\quad = (-8) - (-4)$ Add first from left

$\quad = (-8) + 4$ Then subtract

$\quad = -4$

(b) $(-2)^3 + 1 = (-2)(-2)(-2) + 1$ $a^3 = aaa$

$\qquad\qquad\quad = (4)(-2) + 1$ $(-2)(-2) = 4$

$\qquad\qquad\quad = (-8) + 1$ Multiply first

$\qquad\qquad\quad = -7$ Then add

(c) $\dfrac{1}{2} - \left\{ \dfrac{1}{4} - \left[\left(-\dfrac{3}{4} \right) - \left(-\dfrac{1}{4} \right) \right] \right\}$

$\quad = \dfrac{1}{2} - \left\{ \dfrac{1}{4} - \left[\left(-\dfrac{3}{4} \right) + \dfrac{1}{4} \right] \right\}$

$\quad = \dfrac{1}{2} - \left\{ \dfrac{1}{4} - \left[-\dfrac{2}{4} \right] \right\}$ Innermost parentheses first

$\quad = \dfrac{1}{2} - \left\{ \dfrac{1}{4} + \dfrac{2}{4} \right\}$

$\quad = \dfrac{1}{2} - \left(\dfrac{3}{4} \right)$

$\quad = \dfrac{2}{4} - \dfrac{3}{4} = -\dfrac{1}{4}$

Evaluate each numerical expression.

(a) $5 + (-3) \cdot (-2) - 6$

(b) $(-4)^2 + (-1)^3$

(c) $-0.15 + [3.2 - (5.65 - 1.6)]$

Answers: (a) 5   (b) 15   (c) −1

❼ DOUBLE SIGN PROPERTIES

Notice in Example 8 that we use parentheses to avoid writing plus and minus signs right next to each other. The parentheses may be removed and the two signs replaced by one according to the following rule, part 4 of which is called the **double negative property.**

Double Sign Properties

If a is any number,

1. $+(+a) = +a$ 2. $+(-a) = -a$

3. $-(+a) = -a$ 4. $-(-a) = +a.$

Notice that when a plus sign precedes parentheses, we *do not* change the sign inside when the parentheses and plus sign are removed. However, when a minus sign precedes parentheses, we *do* change the sign inside when the parentheses and minus sign are removed.

| **EXAMPLE 9 DOUBLE SIGN PROPERTIES** | **PRACTICE EXERCISE 9** |

Write without parentheses.

(a) $+(+6) = +6 = 6$

(b) $+(-8) = -8$

(c) $-(+7) = -7$

(d) $-(-5) = +5 = 5$

(e) $-[-(-2)] = -[+2] = -2$ Remove innermost parentheses first

Write without parentheses.

(a) $+(+11)$

(b) $+(-19)$

(c) $-(+17)$

(d) $-(-15)$

(e) $-[+(-12)]$

Answers: (a) 11 (b) -19
(c) -17 (d) 15 (e) 12

| **EXAMPLE 10 EVALUATING EXPRESSIONS** | **PRACTICE EXERCISE 10** |

Evaluate when $a = -3$ and $b = -2$.

(a) $a - b = (-3) - (-2) = (-3) + 2 = -1$ Use parentheses when substituting

(b) $b - [-a] = (-2) - [-(-3)] = (-2) - [3]$
$= -2 - 3 = -5$

(c) $2a - 3b = 2(-3) - 3(-2) = -6 - (-6) = -6 + 6 = 0$

(d) $|a - 2b| = |(-3) - 2(-2)| = |(-3) - (-4)|$
$= |(-3) + 4| = |1| = 1$

Evaluate when $x = -5$ and $y = -1$.

(a) $x - y$

(b) $y - (-x)$

(c) $2x - 6y$

(d) $|y - 3x|$

Answers: (a) -4 (b) -6
(c) -4 (d) 14

| **EXAMPLE 11 EVALUATING EXPRESSIONS** | **PRACTICE EXERCISE 11** |

Evaluate $3[2(a + 2) - 3(b + c)]$ when $a = -2$, $b = 4$, and $c = -8$.

$3[2(a + 2) - 3(b + c)] = 3[2(-2 + 2) - 3(4 + (-8))]$ Replace variables with numbers

$= 3[2(0) - 3(-4)]$ Add inside parentheses first

$= 3[0 + 12]$

$= 3[12]$

$= 36$

Evaluate $-4[(u - v) - (w - u)]$ when $u = -3$, $v = 5$, and $w = -1$.

Answer: 40

10.4 EXERCISES A

Perform the indicated operations.

1. $(2)(6)$ 12

2. $(-2)(-6)$ 12

3. $(2)(-6)$ -12

4. $(-2)(6)$ -12

5. $0 \cdot (-6)$ 0

6. $(-5) \cdot 7$ -35

7. $(-3)(-9)$ 27

8. $(-10)(8)$ -80

9. $(9)(-12)$ -108

10. $(12)(-12)$ -144

11. $(-12)(-12)$ 144

12. $(-15) \cdot 20$ -300

13. $6 \div 2$ 3

14. $(-6) \div (-2)$ 3

15. $6 \div (-2)$ -3

16. $(-6) \div 2$ -3

17. $0 \div (-6)$ 0

18. $(-6) \div 0$ undefined

19. $\dfrac{99}{-11}$ -9

20. $\dfrac{-36}{-4}$ 9

21. $\dfrac{-12}{12}$ -1

22. $\dfrac{0}{-38}$ 0

23. $\dfrac{-48}{-16}$ 3

24. $\dfrac{52}{-13}$ -4

25. $(1.2)(-2.3)$ -2.76

26. $(-1.2)(-2.3)$ 2.76

27. $(-1.2)(2.3)$ -2.76

28. $\left(\dfrac{2}{5}\right)\left(-\dfrac{15}{4}\right)$ $-\dfrac{3}{2}$

29. $\left(-\dfrac{2}{5}\right)\left(-\dfrac{15}{4}\right)$ $\dfrac{3}{2}$

30. $\left(-\dfrac{2}{5}\right)\left(\dfrac{15}{4}\right)$ $-\dfrac{3}{2}$

31. $6.4 \div (-0.8)$ -8

32. $(-6.4) \div (-0.8)$ 8

33. $\dfrac{-6.4}{0.8}$ -8

34. $\dfrac{3}{5} \div \left(-\dfrac{9}{15}\right)$ -1

35. $\left(-\dfrac{3}{5}\right) \div \left(-\dfrac{9}{15}\right)$ 1

36. $\dfrac{-\dfrac{3}{5}}{\dfrac{9}{15}}$ -1

37. $(-1)(-1)(-1)$ -1

38. $(-1)(2)(-3)$ 6

39. $(-1)(-2)(-3)(-4)$ 24

40. $(-1)(-2)(-3)(4)(-5)(6)(-1)$ -720

41. $(-1)(-1)(-1)(-1)(-1)(-1)(-1)$ -1

42. Show by example that division does not satisfy the associative law. Can you find numbers a, b, and c for which $(a \div b) \div c$ does equal $a \div (b \div c)$? $(4 \div 2) \div 2 = 2 \div 2 = 1 \neq 4 = 4 \div 1 = 4 \div (2 \div 2)$; if $a = b = c = 1$, then $(1 \div 1) \div 1 = 1 \div 1 = 1 = 1 \div 1 = 1 \div (1 \div 1)$

43. Tony had to write four \$12 checks during the month of June. By how much did this change his bank balance? $-\$48$

44 Martha had 423 points in a contest, but then she received 20 penalty points 3 times. What was her total after the 3 penalties?   363 points

Evaluate each numerical expression.

45. $(-3)^3 + 27$ 0

46. $(-1)^5 + (-1)^3$ -2

47. $2 - [3 - (4 - 1)]$ 2

48. $4 + 2[(-1) - (-5)]$ 12

49. $(3 - 8) \cdot 4 + 1$ -19

50 $3 - \dfrac{2 + (-4)}{1 - 5}$ $\dfrac{5}{2}$

Write without parentheses using only one sign (or no sign).

51. $-(-3)$ 3

52. $-(+4)$ -4

53. $+(-x)$ $-x$

54. $-(-x)$ x

55. $-[+(-2)]$ 2

56. $-[-(+3)]$ 3

57. $-[-(-4)]$ -4

58. $+[-(-10)]$ 10

Evaluate the following expressions when $a = -1$, $b = 2$, and $c = -4$.

59. $a^2 - 1$ 0

60. $abc + b^2$ 12

61. $3(a + b) + c$ -1

62. $2(b - a) - c$ 10

63 $2[3(a + 1) + 2(4 + c)]$ 0

64. $a^2 + b^2 + c$ 1

FOR REVIEW

Perform the following operations.

65. $(-12) + 9$ -3

66. $(-12) + (-9)$ -21

67. $15 + (-8)$ 7

68. $(-12) - 9$ -21

69. $(-12) - (-9)$ -3

70. $15 - (-8)$ 23

Add.

71. -13 7 **72.** -28 -135 **73.** -92 -45 **74.** -102 -262
    12   -29 81   321
    26   -42 27 -421
    -18 -36 -61 -60

75 The Brisebois family owed the credit union $3000. They repaid $1200 and later borrowed $650. What number represents the status of their account? $-$2450

ANSWERS: 1. 12 2. 12 3. −12 4. −12 5. 0 6. −35 7. 27 8. −80 9. −108 10. −144 11. 144 12. −300 13. 3 14. 3 15. −3 16. −3 17. 0 18. undefined 19. −9 20. 9 21. −1 22. 0 23. 3 24. −4 25. −2.76 26. 2.76 27. −2.76 28. $-\frac{3}{2}$ 29. $\frac{3}{2}$ 30. $-\frac{3}{2}$ 31. −8 32. 8 33. −8 34. −1 35. 1 36. −1 37. −1 38. 6 39. 24 40. −720 41. −1 42. $(4 \div 2) \div 2 = 2 \div 2 = 1 \neq 4 = 4 \div 1 = 4 \div (2 \div 2)$, if $a = b = c = 1$, then $(1 \div 1) \div 1 = 1 \div 1 = 1 = 1 \div 1 = 1 \div (1 \div 1)$ 43. −$48 44. 363 points 45. 0 46. −2 47. 2 48. 12 49. −19 50. $\frac{5}{2}$ 51. 3 52. −4 53. $-x$ 54. x 55. 2 56. 3 57. −4 58. 10 59. 0 60. 12 61. −1 62. 10 63. 0 64. 1 65. −3 66. −21 67. 7 68. −21 69. −3 70. 23 71. 7 72. −135 73. −45 74. −262 75. −$2450

10.4   EXERCISES B

Perform the indicated operations.

1. $(5)(9)$ 45

2. $(-5)(-9)$ 45

3. $(5)(-9)$ −45

4. $(-5)(9)$ −45

5. $(-7)(0)$ 0

6. $(-2)(-7)$ 14

7. $(12)(-2)$ −24

8. $(-12)(2)$ −24

9. $(-6)(-13)$ 78

10. $(-6)(13)$ −78

11. $(-11)(-11)$ 121

12. $(-15)(30)$ −450

13. $9 \div 3$ 3

14. $(-9) \div (-3)$ 3

15. $9 \div (-3)$ −3

16. $(-9) \div 3$ −3

17. $0 \div (-5)$ 0

18. $(-5) \div 0$ undefined

19. $\dfrac{88}{-11}$ −8

20. $\dfrac{-24}{-8}$ 3

21. $\dfrac{-15}{-15}$ 1

22. $\dfrac{0}{-25}$ 0

23. $\dfrac{-100}{10}$ −10

24. $\dfrac{52}{-26}$ −2

25. $(1.5)(-2.2)$ −3.3

26. $(-1.5)(-2.2)$ 3.3

27. $(-1.5)(2.2)$ −3.3

28. $\left(\dfrac{3}{5}\right)\left(-\dfrac{10}{9}\right)$ $-\dfrac{2}{3}$

29. $\left(-\dfrac{3}{5}\right)\left(-\dfrac{10}{9}\right)$ $\dfrac{2}{3}$

30. $\left(-\dfrac{3}{5}\right)\left(\dfrac{10}{9}\right)$ $-\dfrac{2}{3}$

31. $4.8 \div (-0.6)$ −8

32. $(-4.8) \div (-0.6)$ 8

33. $\dfrac{-4.8}{-0.6}$ 8

34. $\dfrac{2}{5} \div \left(-\dfrac{6}{15}\right)$ −1

35. $\left(-\dfrac{2}{5}\right) \div \left(-\dfrac{6}{15}\right)$ 1

36. $\dfrac{-\dfrac{2}{5}}{\dfrac{6}{15}}$ −1

37. $(-1)(-1)(-1)(-1)$ 1

38. $(-1)(-2)(-3)$ −6

39. $(-1)(2)(-3)(4)$ 24

40. $(-1)(-2)(3)(-1)(5)(-4)$ 120

41. $(-1)(-1)(-1)(-1)(-1)(-1)(-1)(-1)(-1)$ −1

42. Show by example that division does not satisfy the commutative law. Can you find numbers a and b for which $a \div b$ does equal $b \div a$? $2 \div 1 \neq 1 \div 2; \ 1 \div 1 = 1 \div 1$

43. During each of 5 consecutive hours, the temperature dropped 7°. By how much did this change the temperature?   −35°

44. Hank Anderson received 330 points for a performance, but then was penalized 25 points each for four rule infractions. What was his point total after the penalties were imposed?   230 points

Evaluate each numerical expression.

45. $(-3) + 2 \cdot (-4)$ −11

46. $(-2)^2 + 3$ 7

47. $4 - [2 - (5 - 3)]$ 4

48. $6 + 3[(-2) - (-4)]$ 12   **49.** $(4 - 7) \cdot 5 + 2$ -13 **50.** $2 - \dfrac{4 + (-5)}{3 - 5}$ $\dfrac{3}{2}$

Write without parentheses using only one sign (or no sign).

51. $-(-5)$ 5   **52.** $+(-2)$ -2 **53.** $-(+a)$ $-a$ **54.** $-(-a)$ a

55. $-[+(-3)]$ 3   **56.** $-[-(+2)]$ 2   **57.** $-[-(-7)]$ -7 **58.** $+[-(-8)]$ 8

Evaluate the following expressions when $x = -2$, $y = 1$, and $z = -3$.

59. $3xy + z$ -9 **60.** $xyz + z^2$ 15   **61.** $3(y - x) - z$ 12

62. $x + \dfrac{y}{z}$ $-\dfrac{7}{3}$ **63.** $2[4(x + 2) + 7(z + 3)]$ 0   **64.** $x^2 + y^2 + z$ 2

FOR REVIEW

Perform the following operations.

65. $(-17) + 6$ -11 **66.** $(-17) + (-6)$ -23 **67.** $17 + (-6)$ 11

68. $(-17) - 6$ -23 **69.** $(-17) - (-6)$ -11 **70.** $17 - (-6)$ 23

Add.

71. $\begin{array}{r} -16 \\ 15 \\ 3 \\ -27 \\ \hline \end{array}$ -25 **72.** $\begin{array}{r} -40 \\ -27 \\ -19 \\ -52 \\ \hline \end{array}$ -138 **73.** $\begin{array}{r} -97 \\ 27 \\ 13 \\ -47 \\ \hline \end{array}$ -104 **74.** $\begin{array}{r} -205 \\ 420 \\ -318 \\ -\ 65 \\ \hline \end{array}$ -168

75. A football team lost 6 yards on first down, gained 11 yards on second down, and lost 4 yards on third down. If the series started on their 25 yard line, where was the ball placed on fourth down?   **26 yard line**

10.4 EXERCISES C

Evaluate each numerical expression.

1. $-2 - [3 - (4 - 2)^2]^2$ -3 **2.** $[17 - (-5 + 8)^2 - 6 \div 2]^2$ 25

3. $-[2^3 - (2 - 4)]^2 - \dfrac{16 - (-6)}{15 - 4}$ [Answer: -102]

A calculator would be helpful in Exercises 4–6.

4. $(3.65)^2 - (2.57 - 1.22)^2$ 11.5   **5.** $0.0123[7.2 - (5.5 - 4.6)]$ 0.07749

6. $\dfrac{1.63 - 0.7}{4.8} + \dfrac{(0.76)^2}{1 - (2.71 - 2.11)}$ 1.63775

10.5 THE DISTRIBUTIVE LAWS AND SIMPLIFYING EXPRESSIONS

═══════════════ STUDENT GUIDEPOSTS ═══════════════

1 Distributive Laws   **3** Collecting Like Terms

2 Terms, Factors, and Coefficients   **4** Simplified Expressions

1 DISTRIBUTIVE LAWS

The operations of addition (or subtraction) and multiplication are related by two very important properties called the **distributive laws.** Consider the numerical expression $2(3 + 5)$. Since the order of operations tells us to work within the parentheses first, we evaluate it as follows:

$$2 (3 + 5) = 2 (8) = 16.$$

However, in this case if we were to "distribute" the product of 2 over the sum of $3 + 5$,

$$2 (3 + 5) = 2 \cdot 3 + 2 \cdot 5 = 6 + 10 = 16,$$

we obtain the same result. Similarly,

$$4 (8 - 3) = 4 (5) = 20,$$

and

$$4 (8 - 3) = 4 \cdot 8 - 4 \cdot 3 = 32 - 12 = 20.$$

Thus multiplication can be "distributed over" addition and subtraction. These examples illustrate the following laws.

Distributive Laws of Multiplication Over Addition and Subtraction
If a, b and c are numbers,
1. $a(b + c) = ab + ac$ 2. $a(b - c) = ab - ac.$

Since multiplication is commutative, products are not affected by changing the order of multiplication. Thus we also have

$$(b + c)a = ba + ca \quad \text{and} \quad (b - c)a = ba - ca.$$

Also, multiplication distributes over sums with more than two terms. For example,

$$a(b + c + d) = ab + ac + ad.$$

EXAMPLE 1 EVALUATING USING THE DISTRIBUTIVE LAW

Evaluate each numerical expression in two ways.

(a) $7 (3 + 2) = 7 \cdot 3 + 7 \cdot 2 = 21 + 14 = 35$. Also,
$7 (3 + 2) = 7 (5) = 35$.

(b) $5 (8 - 3) = 5 \cdot 8 - 5 \cdot 3 = 40 - 15 = 25$. Also,
$5 (8 - 3) = 5 (5) = 25$.

PRACTICE EXERCISE 1

Evaluate each numerical expression in two ways.

(a) $4(1 + 9)$

(b) $3(2 - 7)$

(c) $6(7 + 3 - 5) = 6 \cdot 7 + 6 \cdot 3 - 6 \cdot 5 = 42 + 18 - 30 = 30$. Also,

$6(7 + 3 - 5) = 6(5) = 30$.

(d) $-7(3 + 2) = (-7)(3) + (-7)(2) = (-21) + (-14) = -35$. Also,

$-7(3 + 2) = (-7)(5) = -35$.

(e) $-2(-3 - 6) = (-2)(-3) - (-2)(6) = 6 - (-12) = 6 + 12 = 18$.

Also,

$-2(-3 - 6) = -2(-9) = (2)(9) = 18$.

(c) $5(4 - 6 + 3)$

(d) $-3(5 - 1)$

(e) $-9(-2 - 8)$

Answers: (a) **40**   (b) **−15**   (c) **5**
(d) **−12**   (e) **90**

The distributive laws are useful in computation, but they are more useful when we work with algebraic expressions. In Section 10.1, we introduced algebraic expressions involving sums, differences, products, or quotients of numbers and variables.

② TERMS, FACTORS, AND COEFFICIENTS

A part of an expression that is a product of numbers and variables and that is separated from the rest of the expression by plus (or minus) signs is called a **term.** The numbers and letters that are multiplied in a term are called **factors** of the term. The numerical factor is called the **(numerical) coefficient** of the term. For example,

$$3x, \qquad 4a + 7, \qquad 3x + 7y - z, \qquad 4x + 5a + 3 - 8x$$

are algebraic expressions with one, two, three, and four terms, respectively. In $4a + 7$, the term $4a$ has factors 4 and a, and 4 is the coefficient of the term.

Two terms are **similar** or **like terms** if they contain the same variables to the same powers. In $4x + 5a + 3 - 8x$, the terms $4x$ and $-8x$ are like terms. Note that the minus sign goes with the term and thus the coefficient of $-8x$ is -8.

If the terms of an expression have a common factor, the distributive laws can be used in reverse to **remove the common factor** by a process called **factoring.**

We use the distributive law in the following way

$$ab + ac = a(b + c).$$

EXAMPLE 2 FACTORING USING THE DISTRIBUTIVE LAW	PRACTICE EXERCISE 2

Use the distributive laws to factor.

(a) $3x + 3y = 3(x + y)$ — The distributive law in reverse order

(b) $4a - 4b = 4(a - b)$ — 4 and $(a - b)$ are factors of $4a - 4b$

(c) $5u + 5v - 5w = 5(u + v - w)$

(d) $3x + 6 = 3 \cdot x + 3 \cdot 2$ — 6 is $3 \cdot 2$

$\qquad = 3(x + 2)$ — Factor out 3

(e) $8a - 8 = 8 \cdot a - 8 \cdot 1$ — Express 8 as $8 \cdot 1$

$\qquad = 8(a - 1)$ — Factor out 8

(f) $-3x - 3y = (-3)x + (-3)y$ — Factor out -3

$\qquad = (-3)(x + y)$

Note that $-3x - 3y$ is not $-3(x - y)$ since $-3(x - y) = -3x + 3y$.

Use the distributive laws to factor.

(a) $7a + 7b$

(b) $3u - 3w$

(c) $11c - 11d + 11v$

(d) $6u + 6$

(e) $3w - 15$

(f) $-12a - 12b$

Answers: (a) $7(a + b)$ (b) $3(u - w)$
(c) $11(c - d + v)$ (d) $6(u + 1)$
(e) $3(w - 5)$ (f) $-12(a + b)$

In any factoring problem, we check by multiplying. For example, since

$$3(x + y) = 3x + 3y \quad \text{and} \quad 4(a - b) = 4a - 4b,$$

our factoring in the first two parts of Example 2 is correct.

❸ COLLECTING LIKE TERMS

When an expression contains like terms, it can be simplified by **collecting like terms.** This process is illustrated in the next example.

| **EXAMPLE 3** COLLECTING LIKE TERMS | **PRACTICE EXERCISE 3** |

Use the distributive laws to collect like terms.

(a) $5\,x + 7\,x = (5 + 7)\,x = 12x$ Factor out x

(b) $-8x + 2x = (-8 + 2)x = -6x$ Factor out x

(c) $-2a + 7a - 9a = (-2 + 7 - 9)a = -4a$ Factor out a

(d) $6y + 2y - y + 5 = 6 \cdot y + 2 \cdot y - 1 \cdot y + 5$ $-y = -1 \cdot y$

$ = (6 + 2 - 1)\,y + 5$ Factor y out of first three terms

$ = 7y + 5$

The terms $7y$ and 5 *cannot* be collected since they are not like terms. Thus, $7y + 5$ *is not* $12y$. You can see this more easily if you replace y by some number. For example, when $y = 2$.

$$7\,y + 5 = 7\,(2) + 5 = 14 + 5 = 19, \quad \text{but} \quad 12y = 12(2) = 24.$$

(e) $4a + 7b - a + 6b = 4a - a + 7b + 6b$ Commutative law

$ = 4 \cdot a - 1 \cdot a + 7 \cdot b + 6 \cdot b$ $-a = -1 \cdot a$

$ = (4 - 1)a + (7 + 6)b$ Distributive law

$ = 3a + 13b$

With practice some steps can be left out. We should be able to see, for example, that $4a - a = 3a$ and $7b + 6b = 13b$.

(f) $0.07x + x = (0.07)\,x + 1 \cdot x$ $x = 1 \cdot x$

$ = (0.07 + 1)\,x$ Distributive law

$ = 1.07x$

Use the distributive laws to collect like terms.

(a) $3u + 10u$

(b) $-9v + 5v$

(c) $-w + 2w - 5w$

(d) $3z - z + 2z + 13$

(e) $6x + 6y - x - 7y$

(f) $p + (0.14)p$

Answers: (a) $13u$ (b) $-4v$
(c) $-4w$ (d) $4z + 13$
(e) $5x - y$ (f) $(1.14)p$

❹ SIMPLIFIED EXPRESSIONS

When all like terms of an expression have been collected, we say that the expression has been **simplified.**

Using the distributive law and the fact that $-x = (-1) \cdot x$, we get

$$-(a + b) = (-1)(a + b) = (-1)(a) + (-1)(b) = -a - b,$$
$$-(a - b) = (-1)(a - b) = (-1)(a) - (-1)(b) = -a + b,$$
$$-(-a - b) = (-1)(-a - b) = (-1)(-a) - (-1)(b) = a + b.$$

These observations lead us to the next rule.

To Simplify an Expression by Removing Parentheses

1. When a negative sign precedes parentheses, remove the parentheses (and the negative sign in front of the parentheses) by changing the sign of every term within the parentheses.

2. When a plus sign precedes parentheses, remove the parentheses without changing any of the signs of the terms.

EXAMPLE 4 REMOVING PARENTHESES	**PRACTICE EXERCISE 4**

Simplify by removing parentheses.

(a) $-(x + 1) = -x - 1$ Change all signs

(b) $-(x - 1) = -x + 1$ Change all signs

(c) $-(-x + y + 5) = +x - y - 5$ Change all signs

$= x - y - 5$

(d) $+(-x + y + 5) = -x + y + 5$ Change *no* signs

(e) $x - (y - 3) = x - y + 3$ Change all signs within parentheses

Simplify by removing parentheses.

(a) $-(2 + y)$

(b) $-(-2 - y)$

(c) $-(5 - u - w)$

(d) $+(5 - u - w)$

(e) $a - (1 - b)$

Answers: (a) $-2 - y$ (b) $2 + y$
(c) $-5 + u + w$ (d) $5 - u - w$
(e) $a - 1 + b$

The process of removing parentheses is sometimes called **clearing parentheses.**

EXAMPLE 5 CLEARING PARENTHESES AND COLLECTING LIKE TERMS	**PRACTICE EXERCISE 5**

Clear parentheses and collect like terms.

(a) $3x + (2x - 7) = 3x + 2x - 7$ Do not change signs

$= (3 + 2)x - 7$ Collect like terms

$= 5x - 7$

(b) $y - (4y - 4) = y - 4y + 4$ Change all signs within parentheses

$= (1 - 4)y + 4$ Collect like terms

$= -3y + 4$

(c) $3 - (5a + 2) + 7a = 3 - 5a - 2 + 7a$ Change all signs within parentheses

$= 7a - 5a + 3 - 2$ Commutative law

$= 2a + 1$ Collect like terms

(d) $3x - (-2x - 7) + 5 = 3x + 2x + 7 + 5$ Change all signs within parentheses

$= (3 + 2)x + 7 + 5$

$= 5x + 12$

Clear parentheses and collect like terms.

(a) $a + (2 - 3a)$

(b) $b - (1 - b)$

(c) $2x - (3x + 4) - x$

(d) $5 - (-8 - 2w) + 8$

Answers: (a) $-2a + 2$ (b) $2b - 1$
(c) $-2x - 4$ (d) $2w + 21$

////////////// **CAUTION** //////////////

Change *all* signs within the parentheses.

$$-(-2x - 7) = 2x + 7, \quad not \quad 2x - 7.$$

//////////

EXAMPLE 6 **EVALUATING EXPRESSIONS**	**PRACTICE EXERCISE 6**

Evaluate the expressions when $a = -2$ and $b = -1$.

(a) $3\,a + 7 = 3\,(-2) + 7 = -6 + 7 = 1.$

Note how using parentheses at the substitution step helps us to avoid ambiguous statements. For example, without parentheses above we would have $3 \cdot -2 + 7$, which is confusing.

(b) $4\,a^2 = 4\,(-2)^2 = 4 \cdot 4 = 16$ Only -2 is squared. Compare this example with the next one

(c) $(4\,a)^2 = (4 \cdot (-2))^2 = (-8)^2 = 64$ From (b) and (c) we see that $4a^2 \neq (4a)^2$

(d) $-\,a^2 = -(-2)^2 = -(4) = -4$ Compare this with the next example

(e) $(-a)^2 = (-(-2))^2 = (2)^2 = 4$ From (d) and (e) we see that $-a^2 \neq (-a)^2$

(f) $-a - b = -(-2) - (-1) = 2 + 1 = 3$

(g) $-a^3 = -(-2)^3 = -(-8) = +8 = 8$

Evaluate the expressions when $u = -3$ and $w = -1$.

(a) $2u - 3$

(b) $3w^2$

(c) $(3w)^2$

(d) $-u^3$

(e) $(-u)^3$

(f) $-u - w$

(g) $-(-u + w^2)$

Answers: (a) -9 (b) 3 (c) 9 (d) 27 (e) 27 (f) 4 (g) -4

////////////// **CAUTION** //////////////

Two common mistakes to avoid are: (1) forgetting to change *all* signs when a minus sign appears in front of a set of parentheses (for example, $-(x - 2)$ is not $-x - 2$), and (2) making sign errors when evaluating expressions such as $-a^2$ (for example, $-a^2$ is not the same as $(-a)^2$).

//////////

10.5 EXERCISES A

1. (a) Compute $-2(3 - 8)$
    10

(b) Compute $(-2)(3) - (-2)(8)$
    10

(c) Why are these two equal?
    distributive law

2. (a) Compute $-3(4 + 2)$
    −18

(b) Compute $(-3)(4) + (-3)(2)$
    −18

(c) Why are these two equal?
    distributive law

3. In $2x - 3y + 7 - 9x$ the terms $2x$ and $-9x$ are called similar or ___**like**___ terms.

How many terms does each expression have?

4. $2a + b - 3$ **3**

5. $-2x - 3y + 4 - z$ **4**

6. $3w$ **1**

Multiply.

7. $4(x + y)$ $4x + 4y$

8. $5(x - y)$ $5x - 5y$

9. $10(x + 2y)$ $10x + 20y$

10. $10(x + 2y + 3z)$
$10x + 20y + 30z$

11. $2(2a + 1 + 6b)$
$4a + 2 + 12b$

12. $-5(x + y)$
$-5x - 5y$

Use the distributive laws to factor.

13. $4x + 4y$ $4(x + y)$

14. $5x - 5y$ $5(x - y)$

15. $10x + 20y$ $10(x + 2y)$

16. $10x + 20y + 30z$
$10(x + 2y + 3z)$

17. $4a + 2 + 12b$
$2(2a + 1 + 6b)$

18. $-5x - 5y$
$-5(x + y)$

Simplify by removing parentheses.

19. $-(y + 2)$ $-y - 2$

20. $-(-a - b)$ $a + b$

21. $-(-x + 2)$ $x - 2$

22. $-(2y - 2)$ $-2y + 4$

23. $x - (3 - b)$ $x - 3 + b$

24. $-(x - y - 4)$ $-x + y + 4$

25. $+(-x - y - z)$
$-x - y - z$

26. $-(1 + x) + (a - b)$
$-1 - x + a - b$

27. $+(u - v) - (w - 3)$
$u - v - w + 3$

Use the distributive laws and collect like terms.

28. $3x - 8x$ $-5x$

29. $-4z - 9z$ $-13z$

30. $1 - 2x + x$ $1 - x$

31. $y - 3 - 5y$
$-4y - 3$

32. $2x - y - 5x + y$
$-3x$

33. $a - b + a - b$
$2a - 2b$

34. $1 - x + 2y - 3x - y$
$1 - 4x + y$

35. $-a + 3 + b - 2a + b$
$-3a + 3 + 2b$

36. $u - 3 + 3v - 2u + 1$
$-u - 2 + 3v$

(37) $\dfrac{1}{2}x - \dfrac{2}{3}y - \dfrac{5}{2}x - \dfrac{1}{3}y$

$-2x - y$

38. $-\dfrac{3}{4}a + \dfrac{1}{4} + \dfrac{3}{4}a - \dfrac{1}{4}b$

$\dfrac{1}{4} - \dfrac{1}{4}b$

39. $2.1u - 3 - 5.8u$

$-3 - 3.7u$

40. $x - 3(x + 1)$
$-2x - 3$

41. $2a - (1 - 3a)$
$5a - 1$

42. $2(1 - 2u) + u$
$2 - 3a$

(43) $2x - (-x + 1) + 3$
$3x + 2$

44. $a + (2 - 5a) - 3$
$-4a - 1$

45. $7u - (-3u - 1) + 5$
$10u + 6$

46. $-2[x - 3(x + 1)]$
$4x + 6$

(47) $a - [3a - (1 - 2a)]$
$-4a + 1$

48. $2 - [3u - (-2 + 3u)]$
0

49. $3(x - 2) - 2[5y - (2x - y)]$
$7x - 12y - 6$

50. $2(3a - 4b) - (-4a + b) - (-4b - 5a)$
$15a - 5b$

51. $-[2u - (u - v)] - 3[(u - 2v) - 3v]$ $-4u + 14v$

Evaluate the following expressions when $x = -3$, $y = -1$ *and* $z = 2$.

52. $-2x - y$ 7

53. $2x^2$ 18

54. $(2x)^2$ 36

55 $-2x^2$ -18

56 $(-2x)^2$ 36

57. $-x - y + z$ 6

58. $x^2 - y^2 - z^2$ 4

59. $-3y^2 - (x + z)$ -2

60. $z^3 - x^3$ 35

61. $x^2 - 4yz$ 17

62. $(x - y)^3$ -8

63. $x^3 - y^3$ -26

64. $x^2y^2z^2$ 36

65. $2x + 3z + y + 1$ 0

66. $-(-x)$ -3

67. $|x + y|$ 4

68 $|x - y|$ 2

69. $|x^2 + y^2|$ 10

FOR REVIEW

Perform the indicated operations.

70. $(3)(-11)$ -33

71. $(-3)(-11)$ 33

72. $(-12) \div 6$ -2

73. $(-12) \div (-6)$ 2

74. $\left(-\dfrac{1}{3}\right)\left(-\dfrac{3}{4}\right)$ $\dfrac{1}{4}$

75. $(-6.3) \div (-0.9)$ 7

ANSWERS: 1. (a) 10 (b) 10 (c) distributive law 2. (a) -18 (b) -18 (c) distributive law 3. like 4. 3 5. 4 6. 1 7–12. Answers given in exercises 13–18 13–18. Answers given in exercises 7–12 19. $-y - 2$ 20. $a + b$ 21. $x - 2$ 22. $-2y + 2$ 23. $x - 3 + b$ 24. $-x + y + 4$ 25. $-x - y - z$ 26. $-1 - x + a - b$ 27. $u - v - w + 3$ 28. $-5x$ 29. $-13z$ 30. $1 - x$ 31. $-4y - 3$ 32. $-3x$ 33. $2a - 2b$ 34. $1 - 4x + y$ 35. $-3a + 3 + 2b$ 36. $-u - 2 + 3v$ 37. $-2x - y$ 38. $\frac{1}{4} - \frac{1}{4}b$ 39. $-3 - 3.7u$ 40. $-2x - 3$ 41. $5a - 1$ 42. $2 - 3u$ 43. $3x + 2$ 44. $-4a - 1$ 45. $10u + 6$ 46. $4x + 6$ 47. $-4a + 1$ 48. 0 49. $7x - 12y - 6$ 50. $15a - 5b$ 51. $-4u + 14v$ 52. 7 53. 18 54. 36 55. -18 56. 36 57. 6 58. 4 59. -2 60. 35 61. 17 62. -8 63. -26 64. 36 65. 0 66. -3 67. 4 68. 2 69. 10 70. -33 71. 33 72. -2 73. 2 74. $\frac{1}{4}$ 75. 7

10.5 EXERCISES B

1. (a) Compute $-3(5 - 2)$.
   -9

(b) Compute $(-3)(5) - (-3)(2)$.
   -9

(c) Why are these two equal?
   distributive law

2. (a) Compute $-4(6 + 2)$.
   -32

(b) Compute $(-4)(6) + (-4)(2)$.
   -32

(c) Why are these two equal?
   distributive law

3. In $2a + b + 8 - 10a$ the term $2a$ and $-10a$ are called like or __similar__ terms.

How many terms does each expression have?

4. $2a + 7 + 6w - 3$ 4

5. $4a + 8b$ 2

6. $2w - 3 + 7b + a + c$ 5

Multiply.

7. $4(x - y)$ $4x - 4y$

8. $a(x + y)$ $ax + ay$

9. $10(a + 4b)$ $10a + 40b$

10. $11(x + 2y + 3z)$
   $11x + 22y + 33z$

11. $2(3x + 1 + 4y)$
   $6x + 2 + 8y$

12. $-6(a + w)$
   $-6a - 6w$

Use the distributive laws to factor.

13. $4x - 4y$ $4(x - y)$

14. $ax + ay$ $a(x + y)$

15. $10a + 40b$ $10(a + 4b)$

16. $11x + 22y + 33z$
$11(x + 2y + 3z)$

17. $6x + 2 + 8y$
$2(3x + 1 + 4y)$

18. $-6a - 6w$
$-6(a + w)$

Simplify by removing parentheses.

19. $-(x + 3)$ $-x - 3$

20. $-(-x - y)$ $x + y$

21. $-(-w + 5)$ $w - 5$

22. $-(4a - 4)$ $-4a + 4$

23. $a - (2 - y)$ $a - 2 + y$

24. $-(a - b - 9)$ $-a + b + 9$

25. $+(-a - b - w)$
$-a - b - w$

26. $-(1 + a) + (w - c)$
$-1 - a + w - c$

27. $+(x - y) - (a - 8)$
$x - y - a + 8$

Use the distributive laws and collect like terms.

28. $-3x + 8x$ $5x$

29. $-3w - 10w$ $-13w$

30. $1 - 3x + 2x$ $1 - x$

31. $5y + 2 - 3y$ $2y + 2$

32. $x + y - x - y$ 0

33. $2y - u + y - 3u$ $3y - 4u$

34. $1 - 2a + b - a - 3b$
$1 - 3a - 2b$

35. $w + 3 - 2w + b - 4$
$-w + b - 1$

36. $z - 4 + 3z - a + 2$
$4z - a - 2$

37. $\dfrac{1}{4}a - \dfrac{1}{5}w + \dfrac{3}{4}a + \dfrac{2}{5}w$

$a + \dfrac{1}{5}w$

38. $-\dfrac{1}{3}x + \dfrac{1}{2} - \dfrac{2}{3}x + \dfrac{1}{3}y$

$-x + \dfrac{1}{3}y + \dfrac{1}{2}$

39. $2.5w - 3 + 4.1w$
$6.6w - 3$

40. $a - 2(a + 5)$
$-a - 10$

41. $2w - (4 - 5w)$
$7w - 4$

42. $3(1 - 5x) + x$
$3 - 14x$

43. $2b - (-b + 2) + 5$
$3b + 3$

44. $y + (3 - 2y) - 5$
$-y - 2$

45. $2y - (-5y - 2) + 7$
$7y + 9$

46. $-3[a - 2(a + 2)]$
$3a + 12$

47. $x - [4x - (3 - 4x)]$
$-7x + 3$

48. $4 - [2w - (-3 - 2w)]$
$1 - 4w$

49. $-6(x + 2y) - 5[8x - (-2x + y)]$
$-56x - 7y$

50. $-3(-a + 3b) + 5(-2a - 2b) - 2(3a - b)$
$-13a - 17b$

51. $-[(6u - v) - 5v] - 4[-u - (2u - v)]$ $6u + 2v$

Evaluate the following expressions when $a = -2$, $b = -1$, and $c = 3$.

52. $-3a - b$ 7

53. $3a^2$ 12

54. $(3a)^2$ 36

55. $-3a^2$ -12

56. $(-3a)^2$ 36

57. $-a - b + c$ 6

58. $a^2 - b^2 - c^2$ -6

59. $-2b^2 - (a + c)$ -3

60. $c^3 - a^3$ 35

61. $a^2 - 3bc$ 13

62. $(a - b)^3$ -1

63. $a^3 - b^3$ -7

64. $a^2b^2c^2$ 36

65. $2a + b - 3c + 5$ -9

66. $-(-a)$ -2

67. $|a + c|$ 1

68. $|a - c|$ 5

69. $|b^2 + c^2|$ 10

FOR REVIEW

Perform the indicated operations.

70. $(4)(-10)$ -40

71. $(-4)(-10)$ 40

72. $(-18) \div 9$ -2

73. $(-18) \div (-9)$ 2

74. $\left(-\dfrac{1}{5}\right)\left(-\dfrac{5}{6}\right)$ $\dfrac{1}{6}$

75. $(-5.4) \div (-0.9)$ 6

10.5 EXERCISES C

Use the distributive laws and collect like terms.

1. $3\{x - 2[x - (y - x)] - (2x - y)\}$
$-15x + 9y$

2. $4\{[5(a - 2b) - 3b] - 2[6a - (3b - a)]\}$
[Answer: $-36a - 28b$]

3. $7x - \{[6x - 2y - 3(2x - y)] - 3[-x - (3y - x)]\}$
$7x - 10y$

4. $-5\{[-4(a + 5b) - 2b] - 7[a - (-a - b)]\}$
[Answer: $90a + 145b$]

10.6 IRRATIONAL AND REAL NUMBERS

STUDENT GUIDEPOSTS

1 Irrational and Real Numbers

2 Perfect Squares and Square Roots

3 Principal Square Roots, Radicals, and Radicands

Thus far we have concentrated on working with the rational numbers. Remember that a **ratio**nal number is the quotient **(ratio)** of two integers. Equivalently a rational number is a number with a decimal form that either terminates or repeats a block of digits.

1 IRRATIONAL AND REAL NUMBERS

There are many numbers that cannot be expressed in this way. One of the most familiar of these is π, the number equal to the ratio of the circumference of any circle to its diameter. Such numbers are called *irrational numbers*. In decimal notation **irrational numbers** do not terminate nor do they have a repeating block of digits. The **real numbers** are found by taking the rational numbers together with the irrational numbers.

2 PERFECT SQUARES AND SQUARE ROOTS

Before identifying some of the more familiar irrational numbers, we look at the notion of perfect squares and their *square roots*.

Perfect Squares and Square Roots

When an integer is squared, the result is called a **perfect square.** Either of the identical factors of a perfect square number is called a **square root** of the number.

EXAMPLE 1 SQUARE ROOTS

(a) 4 is a perfect square since $4 = 2^2$. We call 2 a square root of 4. Since $4 = (-2)^2$, -2 is also a square root of 4.

(b) 25 is a perfect square since $25 = 5^2$. The two square roots of 25 are 5 and -5.

(c) 0 is a perfect square since $0 = 0^2$. Unlike other perfect squares, 0 has only one square root, namely itself, 0.

PRACTICE EXERCISE 1

(a) Is 1 a perfect square?

(b) Is 81 a perfect square?

(c) Is 40 a perfect square?

Answers: (a) Yes; $1 = 1^2$
(b) Yes; $81 = 9^2$ (c) No; no integer times itself is equal to 40.

Do not confuse the terms *square* and *square root!*

25 is the **square** of 5

$$5^2 = 5 \cdot 5 = 25$$

5 is a **square** root of 25

③ PRINCIPAL SQUARE ROOTS, RADICALS, AND RADICANDS

All perfect squares other than 0 have two square roots, one positive and one negative. When we refer to the square root of a number, do we mean the positive root or the negative root? To avoid confusion, we refer to the positive root as the **principal square root** and we use the symbol $\sqrt{}$, called a **radical,** to represent it. For example, the principal or positive square root of 4 is $\sqrt{4}$, or 2. The number under the radical (4 in this case) is called the **radicand.** Since the radical symbol only designates the principal or positive (possibly zero) square root of a number, we designate the negative square root by $-\sqrt{}$. For example, $-\sqrt{4} = -2$.

The first sixteen perfect squares and their square roots are listed in the following table.

Perfect square	Positive square root of N	Negative square root of N
N	$\sqrt{N}$	$-\sqrt{N}$
0	0	$-0 = 0$
1	1	-1
4	2	-2
9	3	-3
16	4	-4
25	5	-5
36	6	-6
49	7	-7
64	8	-8
81	9	-9
100	10	-10
121	11	-11
144	12	-12
169	13	-13
196	14	-14
225	15	-15

In addition to whole number perfect squares, there are also fractional perfect squares. A fraction that can be factored into the product of two identical fractional factors is called a **perfect square,** and each factor is called a **square root** of the fraction.

EXAMPLE 2 FRACTIONAL PERFECT SQUARES

(a) $\frac{4}{9}$ is a perfect square, with $\sqrt{\frac{4}{9}} = \frac{2}{3}$ and $-\sqrt{\frac{4}{9}} = -\frac{2}{3}$.

(b) $\frac{25}{81}$ is a perfect square, with $\sqrt{\frac{25}{81}} = \frac{5}{9}$ and $-\sqrt{\frac{25}{81}} = -\frac{5}{9}$.

PRACTICE EXERCISE 2

(a) Is $\frac{36}{121}$ a perfect square?

(b) Is $\frac{50}{200}$ a perfect square?

Answers: (a) Yes; $\frac{36}{121} = \left(\frac{6}{11}\right)^2$
(b) Yes; first reduce the fraction to $\frac{1}{4}$, and $\frac{1}{4} = \left(\frac{1}{2}\right)^2$.

Example 2 shows that a fraction is a perfect square if its numerator and denominator are whole number perfect squares. However, a fraction can be a perfect square without this feature. For example, if we reduce $\frac{8}{18}$ to lowest terms, we change it to a ratio of whole number perfect squares:

$$\sqrt{\frac{8}{18}} = \sqrt{\frac{2 \cdot 4}{2 \cdot 9}} = \sqrt{\frac{4}{9}} = \frac{2}{3}.$$

Thus far we have only discussed square roots of perfect squares. What about square roots of other numbers? This problem can be divided into two parts, square roots of positive numbers and square roots of negative numbers. Square roots of negative numbers are not real numbers and will not be discussed in this text. However, square roots of positive numbers that are not perfect squares supply us with many examples of irrational numbers. These include $\sqrt{2}$, $\sqrt{3}$, $\sqrt{5}$, $\sqrt{6}$, and $\sqrt{7}$ among many, many more.

Suppose we consider $\sqrt{2}$, the number which when squared is 2. We know that

$$\sqrt{1} = 1 \text{ and } \sqrt{4} = 2.$$

Since $1 < 2 < 4$, we would expect $\sqrt{1} < \sqrt{2} < \sqrt{4}$ so that

$$1 < \sqrt{2} < 2.$$

Since there is no integer between 1 and 2, $\sqrt{2}$ cannot be an integer. In more advanced work it can be shown that $\sqrt{2}$ is not a rational number either.

Most of our work in this text will be with rational numbers until Chapter 17 when we consider operations with radicals.

EXAMPLE 3 NUMBER SYSTEMS

Determine the numbers in the set $\{-5, 0, 4, \sqrt{7}, 5, -\pi, \frac{3}{2}, -1.8\}$ that belong to the specified set.

(a) Natural numbers: 4 and 5

(b) Whole numbers: 0, 4, and 5

(c) Integers: 0, 4, 5, and -5

(d) Rational numbers: 0, 4, 5, -5, $\frac{3}{2}$, and -1.8

(e) Irrational numbers: $\sqrt{7}$, $-\pi$

(f) Real numbers: all the numbers in the set are real numbers

---

PRACTICE EXERCISE 3

Determine the numbers in the set $\left\{-7, -\frac{3}{5}, -\sqrt{17}, 0, 11, \pi, \frac{11}{2}, 7.22\right\}$ that belong to the specified set.

(a) Natural numbers

(b) Whole numbers

(c) Integers

(d) Rational numbers

(e) Irrational numbers

(f) Real numbers

Answers: (a) 11   (b) 0, 11
(c) $-7, 0, 11$ (d) $-7, -\frac{3}{5}, 0, 11$, $\frac{11}{2}, 7.22$ (e) $-\sqrt{17}, \pi$ (f) all the numbers

We conclude with the diagram in Figure 10.14 that shows the relationships among the number systems we have discussed.

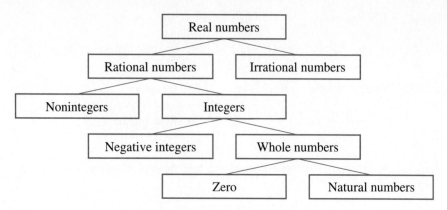

Figure 10.14 Number Systems

10.6 EXERCISES A

Evaluate the square roots.

1. $\sqrt{36}$ 6

2. $-\sqrt{49}$ -7

3. $\sqrt{225}$ 15

4. $-\sqrt{196}$ -14

5. $\sqrt{0}$ 0

6. $-\sqrt{1}$ -1

7. $-\sqrt{169}$ -13

8. $\sqrt{81}$ 9

9. $\sqrt{\dfrac{9}{25}}$ $\dfrac{3}{5}$

10. $\sqrt{\dfrac{64}{196}}$ $\dfrac{4}{7}$

11. $-\sqrt{\dfrac{49}{81}}$ $-\dfrac{7}{9}$

12. $-\sqrt{\dfrac{36}{225}}$ $-\dfrac{2}{5}$

13. $\sqrt{\dfrac{1}{144}}$ $\dfrac{1}{12}$

14. $\sqrt{\dfrac{4}{169}}$ $\dfrac{2}{13}$

15. $\sqrt{\dfrac{49}{196}}$ $\dfrac{1}{2}$

16. $\sqrt{\dfrac{9}{81}}$ $\dfrac{1}{3}$

17 $\sqrt{\dfrac{8}{18}}$ $\dfrac{2}{3}$

18. $-\sqrt{\dfrac{2}{50}}$ $-\dfrac{1}{5}$

19. $\sqrt{\dfrac{12}{48}}$ $\dfrac{1}{2}$

20. $\sqrt{\dfrac{75}{363}}$ $\dfrac{5}{11}$

21. (a) Compute $(16)^2$ **(b)** $\sqrt{256} =$
    256   16

22. (a) Compute $(17)^2$ **(b)** $\sqrt{289} =$
    289   17

23. (a) Compute $(18)^2$ **(b)** $\sqrt{324} =$
    324   18

24. (a) Compute $(19)^2$ **(b)** $\sqrt{361} =$
    361   19

25. If x is any whole number, what is $\sqrt{x^2}$? x

26. If x is any natural number, what is $-\sqrt{x^2}$? $-x$

27. (a) What is the square of 9?   **(b)** What is the principal square root of 9?
    81   3

28. (a) What is the square of $\frac{1}{4}$?   **(b)** What is the principal square root of $\frac{1}{4}$?
    $\frac{1}{16}$ $\frac{1}{2}$

29 Between what two positive integers is $\sqrt{15}$ located?　　**3 and 4**

30. Between what two positive integers is $\sqrt{150}$ located?　　**12 and 13**

In Exercises 31–36, list the numbers in the set $\left\{-\frac{7}{2}, -1, 0, -\sqrt{11}, -3, 2.66, 2\pi, \frac{\sqrt{2}}{2}\right\}$ *that belong to the specified set.*

31. Natural numbers　　**none**

32. Whole numbers　　**0**

33. Integers　　**−1, 0, −3**

34. Rational numbers
$-\frac{7}{2}, -1, 0, -3, 2.66$

35. Irrational numbers
$-\sqrt{11}, 2\pi, \frac{\sqrt{2}}{2}$

36. Real numbers
all the numbers

Of the natural numbers, the whole numbers, the integers, the rational numbers, and the real numbers, which is the smallest set that contains the number given in Exercises 37–44?

37. −3
integers

38. 0
whole numbers

39. $\sqrt{13}$
real numbers

40. −10
integers

41. 22
natural numbers

42. −1.326
rational numbers

43. $-\dfrac{11}{5}$
rational numbers

44. $3\sqrt{3}$
real numbers

FOR REVIEW

Remove parentheses and collect like terms.

45. $2a - (4a - 1)$　　$-2a + 1$

46. $-(1 - a) + (a + 1)$　　$2a$

47. $-[-(1 - a)]$　　$1 - a$

48. $2[y - (3y - 2)]$　　$-4y + 4$

49. $-3[-y + (1 - y)]$　　$6y - 3$

50. $-[-(-y)] + y$　　0

ANSWERS:　1. 6　2. −7　3. 15　4. −14　5. 0　6. −1　7. −13　8. 9　9. $\frac{3}{5}$　10. $\frac{4}{7}$　11. $-\frac{7}{9}$　12. $-\frac{2}{5}$　13. $\frac{1}{12}$　14. $\frac{2}{13}$　15. $\frac{1}{2}$　16. $\frac{1}{3}$　17. $\frac{2}{3}$　18. $-\frac{1}{5}$　19. $\frac{1}{2}$　20. $\frac{5}{11}$　21. (a) 256　(b) 16　22. (a) 289　(b) 17　23. (a) 324　(b) 18　24. (a) 361　(b) 19　25. x　26. $-x$　27. (a) 81　(b) 3　28. (a) $\frac{1}{16}$　(b) $\frac{1}{2}$　29. 3 and 4　30. 12 and 13　31. none　32. 0　33. −1, 0, −3　34. $-\frac{7}{2}$, −1, 0, −3, 2.66　35. $-\sqrt{11}$, 2π, $\frac{\sqrt{2}}{2}$　36. all the numbers　37. integers　38. whole numbers　39. real numbers　40. integers　41. natural numbers　42. rational numbers　43. rational numbers　44. real numbers　45. $-2a + 1$　46. $2a$　47. $1 - a$　48. $-4y + 4$　49. $6y - 3$　50. 0

10.6 EXERCISES B

Evaluate the square roots.

1. $\sqrt{196}$　　14

2. $-\sqrt{36}$　　−6

3. $\sqrt{49}$　　7

4. $-\sqrt{121}$　　−11

5. $\sqrt{144}$　　12

6. $-\sqrt{64}$　　−8

7. $-\sqrt{100}$　　−10

8. $\sqrt{169}$　　13

9. $\sqrt{\dfrac{4}{9}}$　　$\dfrac{2}{3}$

10. $\sqrt{\dfrac{25}{196}}$　　$\dfrac{5}{14}$

11. $-\sqrt{\dfrac{81}{121}}$　　$-\dfrac{9}{11}$

12. $-\sqrt{\dfrac{36}{49}}$　　$-\dfrac{6}{7}$

13. $\sqrt{\dfrac{1}{225}}$　　$\dfrac{1}{15}$

14. $\sqrt{\dfrac{4}{36}}$　　$\dfrac{1}{3}$

15. $\sqrt{\dfrac{49}{121}}$　　$\dfrac{7}{11}$

16. $\sqrt{\dfrac{4}{16}}$　　$\dfrac{1}{2}$

17. $\sqrt{\dfrac{8}{32}}$ $\dfrac{1}{2}$ **18.** $-\sqrt{\dfrac{50}{98}}$ $-\dfrac{5}{7}$ **19.** $\sqrt{\dfrac{12}{75}}$ $\dfrac{2}{5}$ **20.** $\sqrt{\dfrac{147}{363}}$ $\dfrac{7}{11}$

21. (a) Compute $(20)^2$ **(b)** $\sqrt{400} =$ **22. (a)** Compute $(21)^2$ **(b)** $\sqrt{441} =$
       400   20   441   21
23. (a) Compute $(24)^2$ **(b)** $\sqrt{576} =$ **24. (a)** Compute $(25)^2$ **(b)** $\sqrt{625} =$
       576   24   625   25

25. If a is any whole number, what is $\sqrt{a^2}$? a **26.** If a is any natural number what is $-\sqrt{a^2}$? $-a$

27. (a) What is the square of 16? **(b)** What is the principal square root of 16?
       256   4

28. (a) What is the square of $\dfrac{1}{9}$? **(b)** What is the principal square root of $\dfrac{1}{9}$?
       $\dfrac{1}{81}$ $\dfrac{1}{3}$

29. Between what two positive integers is $\sqrt{30}$ located?   **5 and 6**

30. Between what two positive integers is $\sqrt{200}$ located?   **14 and 15**

In Exercises 31–36, list the numbers in the set $\left\{6, \dfrac{5}{4}, -0.235, 0, \sqrt{22}, -4, \dfrac{\pi}{2}\right\}$ *that belong to the specified set.*

31. Natural numbers   6   **32.** Whole numbers   6, 0   **33.** Integers   6, 0, −4

34. Rational numbers   **35.** Irrational numbers   **36.** Real numbers
    $6, \dfrac{5}{4}, -0.235, 0, -4$ $\sqrt{22}, \dfrac{\pi}{2}$ all the numbers

Of the natural numbers, the whole numbers, the integers, the rational numbers, and the real numbers, which is the smallest set that contains the number given in Exercises 37–44?

37. 0   **38.** −16   **39.** 45   **40.** $\sqrt{23}$
    whole numbers   integers   natural numbers   real numbers

41. 63.21   **42.** 4π **43.** $6\sqrt{5}$ **44.** $-\dfrac{21}{5}$
    rational numbers   real numbers   real numbers
                                                                                 rational numbers

FOR REVIEW

Remove parentheses and collect like terms.

45. $4y - (6y - 3)$ $-2y + 3$ **46.** $-(2 - y) + (2 + y)$ $2y$ **47.** $-[-(3 - y)]$ $3 - y$

48. $3[w - (2w - 4)]$ $-3w + 12$ **49.** $-4[-w + (2 - w)]$ $8w - 8$ **50.** $-[-(-w)] - w$ $-2w$

10.6 EXERCISES C

Evaluate the square roots where a *is a natural number.*

1. $\sqrt{25a^2}$ $5a$ **2.** $\sqrt{\dfrac{16}{a^4}}$ $\dfrac{4}{a^2}$ **3.** $-\sqrt{\dfrac{625}{576}}$ $-\dfrac{25}{24}$ **4.** $-\sqrt{\dfrac{847}{1008}}$

$\left[\text{Answer: } -\dfrac{11}{12}\right]$

Use a calculator to find the square root of each number to two decimal places.

5. $\sqrt{2}$ 1.41   **6.** $\sqrt{7}$ 2.65   **7.** $\sqrt{26}$ 5.10   **8.** $\sqrt{191}$
                                                                                 [Answer: 13.82]

CHAPTER 10 REVIEW

KEY WORDS

10.1 A **variable** is a letter that can be replaced by various numbers.

An **exponent** is a power on a number or variable indicating how many times the number or variable is used as a factor.

The **base** of an exponential expression is the number or variable that is raised to a power.

10.2 The **integers** are . . . , -3, -2, -1, 0, 1, 2, 3, ...

The **rational numbers** are the numbers that can be written as the ratio of two integers.

10.3 The **additive identity** is 0.

The **additive inverse** or **negative** of a is $-a$.

10.4 The **multiplicative identity** is 1.

The **multiplicative inverse** or **reciprocal** of a ($a \neq 0$) is $\frac{1}{a}$.

10.5 A **term** is part of an expression that is a product of numbers and variables and that is separated from the rest of the expression by plus (or minus) signs.

The **factors** of a term are numbers or letters that are multiplied in the term.

The **coefficient** of a term is the numerical factor.

10.6 An **irrational number** is a number that is not rational and has a decimal representation that does not terminate or repeat.

The **real numbers** are the rational numbers together with the irrational numbers.

A **perfect square** results when an integer or a rational number is squared.

Either of the identical factors of a perfect square number is a **square root** of the number.

KEY CONCEPTS

10.1 **1.** In an expression such as $2y^3$, only y is cubed—not $2y$. That is, $2y^3$ is not the same as $(2y)^3$.

2. When simplifying a numerical expression,

First: Evaluate within grouping symbols.

Second: Evaluate all powers.

Third: Perform all multiplications and divisions in order from left to right.

Fourth: Perform all additions and subtractions in order from left to right.

10.2 **1.** If a and b are any numbers such that a is to the left of b on a number line, then $a < b$ or $b > a$.

2. The absolute value of a number x, denoted by $|x|$, is the distance from x to zero on a number line and as a result, is never negative.

10.3 If a, b, and c are any numbers,

(a) $a + 0 = 0 + a = a$ Additive identity

(b) $a + (-a) = (-a) + a = 0$ Additive inverses (negatives)

(c) $a + b = b + a$ Commutative law of addition

(d) $(a + b) + c = a + (b + c)$ Associative law of addition

10.4 If a, b, and c are any numbers,

(a) $a \cdot 1 = 1 \cdot a = a$ Multiplicative identity

(b) $a \cdot \frac{1}{a} = \frac{1}{a} \cdot a = 1$ ($a \neq 0$) Multiplicative inverses (reciprocals)

(c) $ab = ba$ Commutative law of multiplication

(d) $(ab)c = a(bc)$ Associative law of multiplication

10.5 **1.** The distributive laws,

$$a(b + c) = ab + ac$$

and

$$a(b - c) = ab - ac,$$

are used to factor and to collect like terms in algebraic expressions.

2. Only like terms can be combined. For example, $4y + 5 \neq 9y$ since $4y$ and 5 are *not* like terms.

3. A minus sign before a set of parentheses changes the sign of *every* term inside when the parentheses are removed. For example, $-(2 - a) = -2 + a$, *not* $-2 - a$.

10.6 The radical symbol by itself only designates the principal (positive or zero) square root of a number. For example, $\sqrt{9} = 3$ not -3. We must write $-\sqrt{9}$ to represent -3.

REVIEW EXERCISES

Part I

10.1 *Write in exponential notation.*

1. *aaaaa* a^5
2. $(3z)(3z)(3z)$ $(3z)^3$
3. $(a + w)(a + w)$ $(a + w)^2$

Write without using exponents.

4. b^7 *bbbbbbb*
5. $2x^3$ $2xxx$
6. $(2x)^3$ $(2x)(2x)(2x)$

Evaluate.

7. $2 \cdot 5 - 9 \div 3 + 1$ 8
8. $2 \cdot 3^3 - 3 \cdot 2^3$ 30

Evaluate when $a = 2$, $b = 5$, and $x = 4$.

9. $3a^2$ 12
10. $(3a)^2$ 36
11. 3^2a 18

12. $(b + x)^2$ 81
13. $b^2 + x^2$ 41
14. $abx - 10x$ 0

15. $2 + 5[x + (b - a)]$ 37
16. $2a + 3[x \div a \cdot b - 4]$ 22

10.2 *Find the absolute values.*

17. $|12|$ 12
18. $|-7|$ 7
19. $|-2.51|$ 2.51

Are the following true or false?

20. $-11 < -1$ true
21. $0 \leq -4$ false
22. $5 \leq 5$ true

Place the appropriate symbol, =, >, or <, between the pairs of fractions.

23. $\dfrac{2}{17} = \dfrac{10}{85}$
24. $\dfrac{13}{5} > \dfrac{28}{11}$
25. $\dfrac{6}{7} < \dfrac{11}{12}$

26. The lowest temperature ever recorded in Hawley Lake, Arizona was 41° below zero. How could this be written? $-41°$

10.3 *Perform the indicated operations.*

27. $(-2) + (-5)$ -7
28. $2 + (-5)$ -3
29. $(-2) - (-5)$ 3

30. $(-2) - 5$ -7
31. $2 - (-5)$ 7
32. $0 - (-2)$ 2

33. $(-2.3) + (-1.5)$ -3.8
34. $\left(\dfrac{1}{5}\right) - \left(-\dfrac{3}{10}\right)$ $\dfrac{1}{2}$
35. $\left(-\dfrac{1}{5}\right) - \left(-\dfrac{3}{10}\right)$ $\dfrac{1}{10}$

36. $(-5) + (-3) + 5 + (-8) + (-4) + (-6)$ -21

37. On first down the LSU football team made 7 yards. It lost 9 yards on second down, then picked up 11 yards on third down. If the series started at the LSU 30 yard line, where was the ball spotted on fourth down?   **39 yard line**

10.4 *Perform the indicated operations.*

38. $(-4)(2)$ **−8**

39. $(-14) \div 7$ **−2**

40. $(-14) \div (-7)$ **2**

41. $\dfrac{0}{-12}$ **0**

42. $\left(-\dfrac{1}{3}\right)\left(\dfrac{15}{8}\right)$ $-\dfrac{5}{8}$

43. $(-2.5)(-1.6)$ **4**

44. $(-2)(-2)(-1)(2)(-3)(-1)(-1)(-1)$
−24

45. When Bertha was on a crash diet, she lost 7 pounds in each of 6 consecutive weeks. Use a number to express her total weight loss.
−42 pounds

Evaluate the following:

46. $(-2) + (-8) \div (-4)$ **0**

47. $(-1)^3 + 5$ **4**

48. $3 - 2[1 - (4 - 3)]$ **3**

49. $(6 - 8) \cdot 2 - (-4)$ **0**

Write without parentheses using only one sign (or no sign).

50. $-(-9)$ **9**

51. $-(-y)$ **y**

52. $-[-(-8)]$ **−8**

53. $-[+(-x)]$ **x**

10.5 *Factor.*

54. $-4x + 12$
$-4(x - 3)$

55. $-2 - 6x - 10y$
$-2(1 + 3x + 5y)$

56. $-5x - 25y$
$-5(x + 5y)$

Collect like terms.

57. $3a - 2 - 7a + 4 - a$ $-5a + 2$

58. $-x + 2 - 3y + 4x + y$ $3x - 2y + 2$

Clear parentheses and collect like terms.

59. $y - (2y - 3) + y - (-y + 2)$
$y + 1$

60. $-3[2a - (4 - a)] - (-a - 3)$
$-8a + 15$

Evaluate when $b = -2$ and $c = 4$.

61. $7b^2$ **28**

62. $-7b^2$ **−28**

63. $(-7b)^2$ **196**

64. $(c - b)^2$ **36**

65. $-b^3$ **8**

66. $|b^2 - c|$ **0**

10.6 *Evaluate the square roots.*

67. $-\sqrt{169}$ **−13**

68. $\sqrt{\dfrac{48}{75}}$ $\dfrac{4}{5}$

69. $-\sqrt{\dfrac{27}{3}}$ **−3**

70. (a) What is the square of 25?   **(b)** What is the principal square root of 25?
     625   **5**

Of the natural numbers, the whole numbers, the integers, the rational numbers, and the real numbers, which is the smallest set that contains the number given.

71. 11
     natural numbers

72. -13
     integers

73. $\dfrac{4}{5}$
     rational numbers

74. $-\sqrt{17}$
     real numbers

75. -3.2
     rational numbers

76. 0
     whole numbers

Part II

Perform the indicated operations.

77. $\left(\dfrac{3}{4}\right) \div \left(-\dfrac{15}{8}\right)$ $-\dfrac{2}{5}$

78. $(0)(24)$ 0

79. $(-1)(-2)(-4)(-6)$ 48

80. $\left(\dfrac{1}{8}\right) + \left(-\dfrac{3}{4}\right)$ $-\dfrac{5}{8}$

81. $(6.8) - (9.2)$ -2.4

82. $\left(-\dfrac{2}{7}\right) - \left(-\dfrac{3}{14}\right)$ $-\dfrac{1}{14}$

Evaluate.

83. $(-3) - 14 \div 7$ -5

84. $(2 - 8) \cdot 5 + 4 \div 2$ -28

85. $|-1.7|$ 1.7

86. $|6 - 8|$ 2

87. $2^3 - (6 - 3) \cdot 5$ -7

88. $-[-(-8)]$ -8

89. $-\sqrt{64}$ -8

90. $\sqrt{\dfrac{32}{50}}$ $\dfrac{4}{5}$

91. $-\sqrt{\dfrac{169}{144}}$ $-\dfrac{13}{12}$

Evaluate when $x = -2$ and $y = -5$.

92. $9x^2$ 36

93. $-y^3$ 125

94. $|x - y^2|$ 27

Factor.

95. $-8x + 2$
 $-2(4x - 1)$

96. $-6x - 12$
 $-6(x + 2)$

97. $3a - 6b + 9c$
 $3(a - 2b + 3c)$

Place the appropriate symbol, $=$, $>$, or $<$, between the pairs of fractions.

98. $\dfrac{6}{7} > \dfrac{4}{5}$

99. $\dfrac{21}{11} > \dfrac{19}{10}$

100. $\dfrac{36}{26} = \dfrac{18}{13}$

Clear parentheses and collect like terms.

101. $x + (x - 3) - (2x + 1)$ -4

102. $-2[5 - (y - 2)] - 3(y + 1)$ $-y - 17$

103. Use $A = P(1 + r)^t$ to find the amount of money in an account at the end of two years if a principal of $5000 is invested at 15% interest, compounded annually.   **$6612.50**

ANSWERS: 1. a^5 2. $(3z)^3$ 3. $(a + w)^2$ 4. *bbbbbbb* 5. $2xxx$ 6. $(2x)(2x)(2x)$ 7. 8 8. 30 9. 12 10. 36 11. 18 12. 81 13. 41 14. 0 15. 37 16. 22 17. 12 18. 7 19. 2.51 20. true 21. false 22. true 23. $=$ 24. $>$ 25. $<$ 26. $-41°$ 27. -7 28. -3 29. 3 30. -7 31. 7 32. 2 33. -3.8 34. $\frac{1}{2}$ 35. $\frac{1}{10}$ 36. -21 37. 39 yard line 38. -8 39. -2 40. 2 41. 0 42. $-\frac{5}{8}$ 43. 4 44. -24 45. -42 pounds 46. 0 47. 4 48. 3 49. 0 50. 9 51. y 52. -8 53. x 54. $-4(x - 3)$ 55. $-2(1 + 3x + 5y)$ 56. $-5(x + 5y)$ 57. $-5a + 2$ 58. $3x - 2y + 2$ 59. $y + 1$ 60. $-8a + 15$ 61. 28 62. -28 63. 196 64. 36 65. 8 66. 0 67. -13 68. $\frac{4}{5}$ 69. -3 70. (a) 625 (b) 5 71. natural numbers 72. integers 73. rational numbers 74. real numbers 75. rational numbers 76. whole numbers 77. $-\frac{2}{5}$ 78. 0 79. 48 80. $-\frac{5}{8}$ 81. -2.4 82. $-\frac{1}{14}$ 83. -5 84. -28 85. 1.7 86. 2 87. -7 88. -8 89. -8 90. $\frac{4}{5}$ 91. $-\frac{13}{12}$ 92. 36 93. 125 94. 27 95. $-2(4x - 1)$ 96. $-6(x + 2)$ 97. $3(a - 2b + 3c)$ 98. $>$ 99. $>$ 100. $=$ 101. -4 102. $-y - 17$ 103. $6612.50

1. Write $a \cdot a \cdot a$ in exponential notation.

    1. _____ a^3

2. Evaluate. $42 \div 6 - (4 - 2) \cdot 3$

    2. _____ 1

Evaluate when $x = 3$, $y = 2$, and $z = 5$.

3. $3z^2$

    3. _____ 75

4. $(3z)^2$

    4. _____ 225

5. $(x + y)^3$

    5. _____ 125

6. $x^3 + y^3$

    6. _____ 35

7. $z + 2[(x + y) - z]$

    7. _____ 5

8. Use $A = P(1 + r)^t$ to find the amount of money in an account at the end of two years if a principal of $500 is invested at 9% interest, compounded annually.

    8. _____ $594.05

9. True or false: the number $\frac{2}{7}$ is called the negative of $\frac{7}{2}$.

    9. _____ false

10. Find the absolute value. $|-13|$

    10. _____ 13

11. Is the following true or false? $-15 < -5$

    11. _____ true

12. Place the appropriate symbol, $=$, $<$, or $>$, between the pair of fractions. $\frac{6}{13}$ $\frac{3}{5}$

    12. _____ $<$

13. The lowest temperature recorded in Polar, Idaho was 49° below zero. If the highest temperature was 92°, what is the difference between these extremes?

    13. _____ 141°

Perform the indicated operations.

14. $(-8) + (-14)$

    14. _____ -22

15. $5 - (-3)$

    15. _____ 8

16. $\left(\dfrac{1}{3}\right)\left(-\dfrac{3}{8}\right)$

16. _____ $-\dfrac{1}{8}$

17. $(-4.5) \div (-1.5)$

17. _____ 3

18. $(-3) + 8 - (-7) - 9 - (-2)$

18. _____ 5

19. $(-2)(-1)(-1)(-1)(3)(-3)$

19. _____ -18

20. Evaluate.
    $(2 - 7) \cdot 5 - (-6)$

20. _____ -19

21. Write $-[-(-6)]$ without parentheses using only one sign (or no sign).

21. _____ -6

22. Factor.
    $-3y + 15$

22. _____ $-3(y - 5)$

23. Collect like terms.
    $3y - 2b + 5 + b - 5y$

23. _____ $5 - b - 2y$

24. Clear parentheses and collect like terms.
    $y - (2y - 3) + y$

24. _____ 3

25. Evaluate $x^2 - y^2$ when $x = 4$ and $y = 1$.

25. _____ 15

26. Multiply. $-3(4y - 2)$

26. _____ $-12y + 6$

27. Evaluate.
    $\sqrt{121}$

27. _____ 11

28. Evaluate.
    $-\sqrt{\dfrac{50}{32}}$

28. _____ $-\dfrac{5}{4}$

Are the following true or false?

29. $\sqrt{81}$ is an integer.

29. _____ true

30. 4π is a rational number.

30. _____ false

581

Linear Equations and Inequalities

11.1 LINEAR EQUATIONS AND THE ADDITION-SUBTRACTION RULE

STUDENT GUIDEPOSTS
---

① Equations

② Linear Equations

③ Equivalent Equations

④ Addition-Subtraction Rule

① EQUATIONS

An **equation** is a statement that two quantities are equal. The two quantities are written with an equal sign ($=$) between them. Some equations are true, some are false, and for some the truth value cannot be determined. For example,

$1 + 2 = 3$ is true,
$2 + 5 = 3 - 7$ is false, and
$x + 2 = 9$ is neither true nor false since the value of x is not known.

Equations such as $x + 2 = 9$, which involve variables, are our primary concern in this chapter. If the variable can be replaced by a number that makes the resulting equation true, that number is called a **solution** of the equation. The process of finding all solutions is called **solving the equation.**

② LINEAR EQUATIONS

In this chapter we concentrate on **linear equations** in one variable in which the exponent on the variable is 1. For this reason, a linear equation is often called a **first-degree equation.** Every linear equation in one variable x can be written in the form

$$ax + b = 0$$

where a and b are known real numbers and $a \neq 0$. Some simple linear equations can be solved directly by inspection or observation. Others require more sophisticated techniques which are discussed in the material that follows. For example, it is easy to see that 5 is a solution to the equation $x = 5$, whereas finding the solution of $2x + 3 = 8 - x$ is not so obvious and will require something beyond simple inspection. The equation

$$x + 2 = 9$$

has 7 as a solution because when x is replaced by 7,

$$7 + 2 = 9 \quad \text{True}$$

is a true equation. Note that 3 is not a solution since

$$3 + 2 = 9 \quad \text{False}$$

is false. An equation such as this, which is true for some replacements of the variable and false for others, is a **conditional equation.** The equation

$$x + 1 = 1 + x$$

has many solutions. In fact every number is a solution of this equation. An equation like this, called an **identity,** has the entire set of real numbers for its solution set. The equation

$$x + 1 = x$$

has no solutions. An equation like this is called a **contradiction.**

③ EQUIVALENT EQUATIONS

Consider the two equations

$$x + 1 = 3 \quad \text{and} \quad x = 2.$$

Both equations have 2 as a solution. Although it is easier to see that 2 is a solution to the second, both can be solved by inspection. When two equations have exactly the same solutions, they are called **equivalent equations.** To solve an equation, we try to change it into an equivalent equation that can be solved by direct inspection. This usually means that the variable is isolated on one side, as in $x = 2$.

④ ADDITION-SUBTRACTION RULE

Following is the first rule for solving equations.

> ### Addition-Subtraction Rule
>
> An equivalent equation is obtained if the same quantity is added to or subtracted from both sides of an equation.
>   Suppose a, b, and c are real numbers.
>
>   If $a = b$, then $a + c = b + c$ and $a - c = b - c$.

For example, if we start with the true equation $5 = 5$ and then add 2 to both sides or subtract 3 from both sides,

$$5 + 2 = 5 + 2 \quad \text{and} \quad 5 - 3 = 5 - 3,$$

the resulting equations are also true since $7 = 7$ and $2 = 2$.

EXAMPLE 1 USING THE ADDITION-SUBTRACTION RULE	**PRACTICE EXERCISE 1**

Solve. $x - 2 = 7$

$x - 2 + 2 = 7 + 2$ Add 2 to both sides to isolate x on the left side

$\quad x + 0 = 9$

$\quad\quad x = 9$

The solution is 9.

Solve. $y - 8 = -1$

Answer: 7

| **EXAMPLE 2** USING THE ADDITION-SUBTRACTION RULE | **PRACTICE EXERCISE 2** |

Solve. $y + 3 = 17$

$y + 3 - 3 = 17 - 3$ Subtract 3 from both sides to isolate y on the
$y + 0 = 14$ left side
$y = 14$

The solution is 14.

Solve. $z + 9 = 4$

Answer: -5

Get in the habit of checking all indicated solutions to an equation. To do this, replace the variable throughout the original equation with the indicated solution to see whether the resulting equation is true.

| **EXAMPLE 3** USING THE ADDITION-SUBTRACTION RULE | **PRACTICE EXERCISE 3** |

Solve. $3 = x + 5$

$3 - 5 = x + 5 - 5$ Subtract 5 from both sides to isolate x on the
$-2 = x + 0$ right side
$-2 = x$

The solution is -2.
Check: $3 \overset{?}{=} -2 + 5$ Replace x with -2 in the original equation
$3 = 3$

The solution -2 does check.

Solve. $21 = z + 15$

Answer: 6

In Example 3, we isolated the variable on the right side of the equation instead of the left side as in the first two examples. Since $-2 = x$ and $x = -2$ are equivalent, we see that it makes no difference whether the isolated variable is on the left or on the right.

Equations involving fractions or decimals are solved in the same manner.

| **EXAMPLE 4** EQUATION WITH FRACTIONS | **PRACTICE EXERCISE 4** |

Solve. $\dfrac{1}{2} + y = -3$

$\dfrac{1}{2} - \dfrac{1}{2} + y = -3 - \dfrac{1}{2}$ Subtract $\frac{1}{2}$ from both sides

$0 + y = -\dfrac{6}{2} - \dfrac{1}{2}$ The LCD of -3 and $-\frac{1}{2}$ is 2

$y = \dfrac{-6 - 1}{2}$

$y = -\dfrac{7}{2}$ Subtract fractions

Solve. $\dfrac{3}{4} + x = \dfrac{1}{8}$

The solution is $-\frac{7}{2}$. To check this, substitute $-\frac{7}{2}$ for y in $\frac{1}{2} + y = -3$.

$$\frac{1}{2} + \left(-\frac{7}{2}\right) \stackrel{?}{=} -3$$

$$\frac{1}{2} - \frac{7}{2} \stackrel{?}{=} -3$$

$$\frac{1-7}{2} \stackrel{?}{=} -3$$

$$-\frac{6}{2} \stackrel{?}{=} -3$$

$$-3 = -3$$

Answer: $-\frac{5}{8}$

11.1   EXERCISES A

Give an example of each of the following.   **Answers will vary.**

1. A true equation
   3 = 2 + 1

2. A contradiction
   0 = 5

3. A linear equation
   x + 2 = 0

Solve the following equations by direct observation or inspection.

4. $x + 3 = 5$ **2**

5. $2y = 8$ **4**

6. $2 + z = 2 - z$ **0**

7. $2 + x = 2 + x$
   any number is a solution

8. $y + 5 = y + 1$
   no solution

9. $2z = 5z$
   0

Use the equation $y - 1 = 5$ to answer Exercises 10–12.

10. What is the variable?   *y*

11. What is the left side of the equation?   **y − 1**

12. Is it an identity?   **no**

Use the addition-subtraction rule to solve. Check all solutions.

13. $x + 3 = 7$ **4**

14. $y - 4 = 9$ **13**

15. $z + 12 = -4$ **−16**

16. $x - 9 = -6$ **3**

17. $y + 16 = 15$ **−1**

18. $z + 11 = 13$ **2**

19. $4 = x + 1$ **3**

20. $-7 = y - 3$ **−4**

21. $-3 = z + 2$ **−5**

22. $\frac{1}{3} + x = 5$ $\frac{14}{3}$

23. $\frac{3}{2} + y = -2$ $-\frac{7}{2}$

24. $-\frac{1}{4} + z = \frac{3}{4}$ **1**

25. $x - 2.1 = 3.4$ **5.5**

26. $-4.2 = y + 1.7$ **−5.9**

27. $-2.6 = z - 3.9$ **1.3**

28 $x + 2\frac{1}{2} = 5$ $2\frac{1}{2}$ **29.** $y - 3\frac{2}{3} = 4$ $7\frac{2}{3}$ **30.** $1\frac{3}{8} + z = \frac{1}{4}$ $-1\frac{1}{8}$

In Exercises 31–34, is the statement true *or* false? *If the statement is false, tell why.*

31. -2 is a solution to $x + 2 = 0$.   **true**   **32.** $y + 3 = 5$ is a linear equation.   **true**

33. Zero is a solution of $x = 2x$.   **true**   **34.** $x + 2 = x$ is an identity.   **false**

FOR REVIEW

Exercises 35–38 review material from Chapter 10 to help prepare for the next section. Simplify each expression.

35. $2(5x)$ $10x$ **36.** $(-1)(-z)$ z **37.** $(-4)\left(-\frac{1}{4}y\right)$ y **38.** $\left(-\frac{2}{5}\right)\left(-\frac{5}{2}x\right)$ x

---

ANSWERS 1–3. Answers will vary.   4. 2   5. 4   6. 0   7. any number is a solution.   8. no solution   9. 0   10. y
11. $y - 1$ 12. no   13. 4   14. 13   15. -16 16. 3   17. -1 18. 2   19. 3   20. -4 21. -5 22. $\frac{14}{3}$ 23. $-\frac{7}{2}$
24. 1   25. 5.5   26. -5.9 27. 1.3   28. $2\frac{1}{2}$ 29. $7\frac{2}{3}$ 30. $-1\frac{1}{8}$ 31. true   32. true   33. true
34. false (contradiction)   35. $10x$ 36. z 37. y 38. x

11.1 EXERCISES B

Give an example of each of the following.   **Answers will vary.**

1. A false equation $7 = 2 - 5$ **2.** An identity $x - 6 = x - 6$ **3.** An equation that is neither true nor false $x - 6 = 0$

Solve the following equations by direct observation or inspection.

4. $x + 5 = 12$ 7   **5.** $3y = 15$ 5   **6.** $4 + z = 4 - z$ 0

7. $1 + x = 1 + x$
   any number is a solution   **8.** $y + 2 = y + 8$
   no solution   **9.** $8z = 3z$ 0

Use the equation $x + 3 = 19$ to answer Exercises 10–12.

10. What is the right side of the equation?   **19**   **11.** What is the solution?   **16**   **12.** Is this a contradiction?   **no**

Solve and check.

13. $x + 2 = 12$ 10   **14.** $y - 3 = 10$ 13   **15.** $z + 13 = -2$ -15

16. $x - 11 = -9$ 2   **17.** $y + 17 = 18$ 1   **18.** $z + 9 = 18$ 9

19. $5 = x + 7$ -2 **20.** $-3 = y - 3$ 0   **21.** $-4 = z + 3$ -7

22. $\frac{1}{3} + x = 8$ $\frac{23}{3}$ **23.** $\frac{1}{3} + y = \frac{7}{3}$ 2   **24.** $-\frac{1}{4} + z = 1$ $\frac{5}{4}$

25. $x - 2.3 = 5.7$ 8.0   **26.** $-5.7 = y - 2.3$ -3.4 **27.** $-3.8 = z - 1.2$ -2.6

28. $x + 1\frac{3}{4} = 3$ $1\frac{1}{4}$ **29.** $y - 2\frac{1}{5} = 7$ $9\frac{1}{5}$ **30.** $3\frac{2}{3} + z = \frac{1}{9}$ $-3\frac{5}{9}$

In Exercises 31–34, is the statement true *or* false? *If the statement is false, tell why.*

31. -4 is a solution to $4 + x = 0$.   true

32. $z + 2 = 8$ is a linear equation.   true

33. A solution of $3y = y$ is 0.   true

34. $y + 1 = 1 - y$ is a contradiction.   false

FOR REVIEW

Exercises 35–38 review material from Chapter 10 to help prepare for the next section. Simplify each expression.

35. $8(4x)$ $32x$

36. $\frac{1}{3}(3y)$ y

37. $\left(-\frac{1}{2}\right)(-2x)$ x

38. $\left(-\frac{2}{3}\right)\left(-\frac{3}{2}z\right)$ z

11.1 EXERCISES C

Solve. A calculator would be helpful for these exercises.

1. $3.75 - x = 7\frac{2}{5}$

-3.65

2. $0.00125 + y = \dfrac{1}{1000}$

-0.00025

3. $125\frac{7}{8} = 95.2 - z$

[Answer: -30.675]

11.2 THE MULTIPLICATION-DIVISION RULE

In Section 11.1 we were careful to choose equations for which the coefficient of the variable was 1. When the coefficient of the variable is a number other than 1, we need to carry out an additional step using the following rule.

Multiplication-Division Rule

An equivalent equation is obtained if both sides of an equation are multiplied or divided by the same nonzero quantity.

Suppose a, b, and c are real numbers with $c \neq 0$.

$$\text{If } a = b, \text{ then } ac = bc \text{ and } \frac{a}{c} = \frac{b}{c}.$$

For example, if we start with the true equation $6 = 6$ and then multiply both sides by 2 or divide both sides by 2,

$$6 \cdot 2 = 6 \cdot 2 \quad \text{and} \quad \frac{6}{2} = \frac{6}{2},$$

the resulting equations are also true since $12 = 12$ and $3 = 3$.

EXAMPLE 1 USING THE MULTIPLICATION-DIVISION RULE	PRACTICE EXERCISE 1

Solve. $2x = 10$

$\dfrac{1}{2} \cdot 2x = \dfrac{1}{2} \cdot 10$ Multiply both sides by the reciprocal of 2, which is $\frac{1}{2}$

$1 \cdot x = 5$

$x = 5$

Solve. $9y = -27$

The solution is 5.

Check: $\quad 2 \cdot 5 \stackrel{?}{=} 10 \qquad$ Replace x with 5 in the original equation

$\qquad\qquad 10 = 10$

Answer: -3

Equivalently, in Example 1 we could divide both sides of the equation by 2.

$$2x = 10$$

$$\frac{2x}{2} = \frac{10}{2} \qquad \text{Divide both sides by 2}$$

$$\frac{2}{2} \cdot x = 5$$

$$1 \cdot x = 5$$

$$x = 5$$

| **EXAMPLE 2** USING THE MULTIPLICATION-DIVISION RULE | **PRACTICE EXERCISE 2** |

Solve. $\quad \dfrac{1}{3}x = 18$

Solve. $\quad \dfrac{1}{8}y = -1$

$3 \cdot \dfrac{1}{3}x = 3 \cdot 18 \qquad$ Multiply both sides by the reciprocal of $\frac{1}{3}$, which is 3

$\quad 1 \cdot x = 54$

$\qquad x = 54$

The solution is 54. Check.

Answer: -8

Notice in Examples 1 and 2 that we used the multiplication-division rule to make the coefficient of the variable equal to 1. This is the necessary final step in solving an equation. With practice, many of the steps may be performed mentally.

| **EXAMPLE 3** USING THE MULTIPLICATION-DIVISION RULE | **PRACTICE EXERCISE 3** |

Solve. $\quad \dfrac{x}{\frac{4}{5}} = 20$

Solve. $\quad \dfrac{z}{\frac{1}{7}} = -35$

In an equation of this type, simplify the left side first.

$$\frac{x}{\frac{4}{5}} = \frac{\frac{x}{1}}{\frac{4}{5}} = \frac{x}{1} \div \frac{4}{5} = \frac{x}{1} \cdot \frac{5}{4} = \frac{5}{4} \cdot x$$

Thus we are actually solving

$$\frac{5}{4} \cdot x = 20$$

$$\frac{4}{5} \cdot \frac{5}{4} \cdot x = \frac{4}{5} \cdot 20 \qquad \text{Multiply both sides by } \frac{4}{5}$$

$$1 \cdot x = 16 \qquad \frac{4}{5} \cdot 20 = \frac{4 \cdot 20}{5} = \frac{4 \cdot 4}{1} = 16$$

$$x = 16$$

The solution is 16.

Check: $\dfrac{16}{\frac{4}{5}} \stackrel{?}{=} 20$

$\dfrac{5}{4} \cdot 16 \stackrel{?}{=} 20$

$20 = 20$

Answer: -5

| **EXAMPLE 4**   **EQUATION WITH DECIMALS** | **PRACTICE EXERCISE 4** |

Solve. $-2.7x = 54$

$$\dfrac{-2.7x}{-2.7} = \dfrac{54}{-2.7} \qquad \text{Divide both sides by } -2.7$$

$$\dfrac{-2.7}{-2.7} \cdot x = \dfrac{54}{-2.7}$$

$$1 \cdot x = -20$$

$$x = -20$$

The solution is -20.
Check: $-2.7\,(-20) \stackrel{?}{=} 54$

$54 = 54$

Solve. $0.6y = -7.2$

Answer: -12

We can remember the multiplication-division rule with the statement, "multiply or divide both sides of the equation by the same quantity."

11.2   EXERCISES A

Use the multiplication-division rule to solve. Check all solutions.

1. $5x = 25$ 5

2. $56 = 8y$ 7

3. $-11z = 77$ -7

4. $12x = -84$ -7

5. $-72 = -y$ 72

6. $-2z = 2$ -1

7. $-5x = -50$ 10

8. $-60y = 360$ -6

9. $\dfrac{z}{1} = 10$ 10

10. $\dfrac{2}{3}x = 10$ 15

11. $\dfrac{y}{4} = -3$ -12

12. $-\dfrac{2}{5}z = 8$ -20

13. $\dfrac{x}{\frac{1}{3}} = 27$ 9

14. $\dfrac{y}{\frac{7}{2}} = -6$ -21

15 $\dfrac{z}{-\frac{1}{8}} = 16$ -2

16 $\dfrac{2x}{3} = 4$ 6

17. $\dfrac{3y}{2} = -6$ -4

18. $\dfrac{4z}{-3} = -8$ 6

19. $4x = -24.8$ -6.2

20. $1.2y = 7.2$ 6

21. $-0.8z = -6.4$ 8

22 $2\dfrac{1}{2}x = 10$ 4

23. $-3\dfrac{1}{3}y = 20$ -6

24. $-4\dfrac{1}{5}z = -63$ 15

FOR REVIEW

Solve.

25. $-8 + x = -12$ -4

26. $-\dfrac{3}{4} + y = -\dfrac{7}{4}$ -1

27. $7.4 + z = -3.2$ -10.6

28. Is $2 + x = x + 2$ an identity?   yes

29. Is 0 a solution of $x - 3 = -3 - x$?   yes

ANSWERS: 1. 5 2. 7 3. -7 4. -7 5. 72 6. -1 7. 10 8. -6 9. 10 10. 15 11. -12 12. -20 13. 9
14. -21 15. -2 16. 6 17. -4 18. 6 19. -6.2 20. 6 21. 8 22. 4 23. -6 24. 15 25. -4 26. -1
27. -10.6 28. yes 29. yes

11.2 EXERCISES B

Use the multiplication-division rule to solve. Check.

1. $7x = 49$ 7

2. $72 = 6y$ 12

3. $-11z = 44$ -4

4. $12x = -96$ -8

5. $-63 = -y$ 63

6. $-3z = 48$ -16

7. $-20x = -300$ 15

8. $30y = -240$ -8

9. $\dfrac{z}{1} = -30$ -30

10. $\dfrac{2}{3}x = 14$ 21

11. $\dfrac{y}{5} = -3$ -15

12. $-\dfrac{2}{7}z = 8$ -28

13. $\dfrac{x}{\frac{1}{2}} = 10$ 5

14. $\dfrac{y}{\frac{1}{9}} = 27$ 3

15. $\dfrac{z}{-\frac{2}{3}} = 12$ -8

16. $\dfrac{3x}{5} = 6$ 10

17. $\dfrac{4y}{3} = -2$ $-\dfrac{3}{2}$

18. $\dfrac{8z}{-5} = -16$ 10

19. $3x = -25.5$ -8.5 **20.** $1.2y = 8.4$ 7 **21.** $-1.2z = -4.8$ 4

22. $3\frac{2}{3}x = 22$ 6 **23.** $-4\frac{1}{4}y = 17$ -4 **24.** $-1\frac{3}{5}z = -32$ 20

FOR REVIEW

Solve.

25. $-21 + x = -25$ -4 **26.** $-\frac{5}{6} + y = \frac{7}{12}$ $\frac{17}{12}$ **27.** $1.9 + z = -6.7$ -8.6

28. Is $x + 1 = x - 1$ a contradiction?   yes   **29.** Is -6 a solution of $x - 6 = 0$?   no

11.2 EXERCISES C

Solve. A calculator might be helpful in these exercises.

1. $0.0568x = 0.006816$ 0.12 **2.** $9\frac{7}{12}y = -7\frac{3}{16}$ $-\frac{3}{4}$ **3.** $-15\frac{3}{4}z = 0.525$

[Answer: $-0.0\overline{3}$]

11.3 SOLVING EQUATIONS BY COMBINING RULES

STUDENT GUIDEPOSTS

1 Equations with Similar Terms

2 Equations with the Variable on Both Sides

3 Equations with Parentheses

4 Equations with Fractional or Decimal Coefficients

5 Summary of Methods

When we solve an equation, we isolate the variable on one side so we can find the solution by inspection. If we need to use both rules to do this, we generally use the addition-subtraction rule before the multiplication-division rule, as in the following example.

EXAMPLE 1 USING BOTH RULES

Solve. $3x + 5 = 11$

$3x + 5 - 5 = 11 - 5$ Use addition-subtraction rule first to subtract 5 from both sides

$3x + 0 = 6$

$3x = 6$

$\frac{1}{3} \cdot 3x = \frac{1}{3} \cdot 6$ Now use multiplication-division rule to multiply both sides by $\frac{1}{3}$

$1 \cdot x = 2$

$x = 2$

The solution is 2.

PRACTICE EXERCISE 1

Solve. $4y + 10 = 2$

Check: $3 \cdot 2 + 5 \overset{?}{=} 11$

$6 + 5 \overset{?}{=} 11$

$11 = 11$

Answer: -2

① EQUATIONS WITH SIMILAR TERMS

Sometimes the sides of an equation have similar terms. These terms should be combined before attempting to apply the rules.

EXAMPLE 2 EQUATION WITH SIMILAR TERMS	**PRACTICE EXERCISE 2**

Solve. $15 = 4y + 3y$

$15 = 7y$ Combine similar terms

$\dfrac{1}{7} \cdot 15 = \dfrac{1}{7} \cdot 7y$ Multiply both sides by $\frac{1}{7}$

$\dfrac{15}{7} = 1 \cdot y$

$\dfrac{15}{7} = y$

The solution is $\frac{15}{7}$. Check by substituting in the original equation.

Solve. $9z - 3z = 2$

Answer: $\frac{1}{3}$

② EQUATIONS WITH THE VARIABLE ON BOTH SIDES

When an equation has terms involving the variable on both sides, we use the addition-subtraction rule to get them together on the same side so they can then be combined.

EXAMPLE 3 VARIABLE ON BOTH SIDES	**PRACTICE EXERCISE 3**

Solve. $7x - 2 = 2x + 3$

$7x - 2 + 2 = 2x + 3 + 2$ Add 2 to both sides

$7x + 0 = 2x + 5$

$7x = 2x + 5$

$7x - 2x = 2x - 2x + 5$ Subtract $2x$ from both sides

$5x = 0 + 5$

$5x = 5$

$\dfrac{1}{5} \cdot 5x = \dfrac{1}{5} \cdot 5$ Multiply both sides by $\frac{1}{5}$

$1 \cdot x = 1$

$x = 1$

The solution is 1. Check.

Solve. $4 + 3y = -8 - y$

Answer: -3

To solve an equation, use the addition-subtraction rule to isolate all terms involving the variable on one side of the equation and then combine them. Next apply, if necessary, the multiplication-division rule. Always remember to combine similar terms before applying the rules.

| **EXAMPLE 4** Variable in Several Terms | **Practice Exercise 4** |

Solve. $2 - x + 10 = 3x + 2x + 24$

$$-x + 12 = 5x + 24 \qquad \text{Combine similar terms}$$
$$-x + 12 - 12 = 5x + 24 - 12 \qquad \text{Subtract 12 from both sides}$$
$$-x = 5x + 12$$
$$-x - 5x = 5x - 5x + 12 \qquad \text{Subtract } 5x \text{ from both sides}$$
$$-6x = 12$$
$$\left(-\frac{1}{6}\right)(-6x) = \left(-\frac{1}{6}\right)12 \qquad \text{Multiply both sides by } -\frac{1}{6}$$
$$1 \cdot x = -2$$
$$x = -2$$

The solution is -2.
Check: $2 - (-2) + 10 \overset{?}{=} 3(-2) + 2(-2) + 24$
$$2 + 2 + 10 \overset{?}{=} -6 - 4 + 24$$
$$14 = 14$$

Solve. $4y + 2 - y = 3 + 5y - 9$

Answer: **4**

---

❸ EQUATIONS WITH PARENTHESES

When solving an equation containing parentheses, remove the parentheses using the distributive property. The resulting equation can then be solved using previous methods.

| **EXAMPLE 5** Equation with Parentheses | **Practice Exercise 5** |

Solve. $3(x + 6) = 21$

$$3 \cdot x + 3 \cdot 6 = 21 \qquad \text{To remove parentheses, multiply both 6 and } x \text{ by 3}$$
$$3x + 18 = 21$$
$$3x + 18 - 18 = 21 - 18 \qquad \text{Subtract 18}$$
$$3x = 3$$
$$\frac{1}{3} \cdot 3x = \frac{1}{3} \cdot 3 \qquad \text{Multiply by } \frac{1}{3}$$
$$1 \cdot x = 1$$
$$x = 1$$

The solution is 1. Check.

Solve. $5(z + 2) = 10$

Answer: **0**

| **EXAMPLE 6** Equation with Parentheses | **Practice Exercise 6** |

Solve. $6z - (3z - 4) = 14$

$$6z - 3z + 4 = 14 \qquad \text{Remove parentheses by changing signs}$$

Solve. $4y - (y - 8) = -1$

$$3z + 4 = 14$$

$$3z + 4 - 4 = 14 - 4 \qquad \text{Subtract 4}$$

$$3z = 10$$

$$z = \frac{10}{3}$$

The solution is $\frac{10}{3}$.

Check: $6\left(\dfrac{10}{3}\right) - \left(3\left(\dfrac{10}{3}\right) - 4\right) \overset{?}{=} 14$

$$20 - (10 - 4) \overset{?}{=} 14$$

$$20 - 6 \overset{?}{=} 14$$

$$14 = 14$$

Answer: -3

EXAMPLE 7 EQUATION WITH PARENTHESES	PRACTICE EXERCISE 7

Solve. $2x - 4(2x - 4) = 22 - 3(x - 5)$

$$2x - 8x + 16 = 22 - 3x + 15 \qquad \text{Remove parentheses}$$

$$-6x + 16 = 37 - 3x \qquad \text{Combine like terms}$$

$$-6x + 16 - 16 = 37 - 3x - 16 \qquad \text{Subtract 16}$$

$$-6x = 21 - 3x$$

$$-6x + 3x = 21 - 3x + 3x \qquad \text{Add } 3x$$

$$-3x = 21$$

$$\left(-\frac{1}{3}\right)(-3x) = \left(-\frac{1}{3}\right)21 \qquad \text{Multiply by } -\frac{1}{3}$$

$$1 \cdot x = -7$$

$$x = -7$$

The solution is -7. Check.

Solve.

$$3z - 2(1 - z) = 7 - 5(z + 3)$$

Answer: $-\frac{3}{5}$

④ EQUATIONS WITH FRACTIONAL OR DECIMAL COEFFICIENTS

If an equation has fractional coefficients, it helps to simplify first by multiplying both sides by the least common denominator (LCD) of all fractions. For example, to solve

$$\frac{1}{2}x + \frac{3}{4} = \frac{5}{6},$$

first multiply both sides by the least common denominator 12.

$$12\left(\frac{1}{2}x + \frac{3}{4}\right) = 12 \cdot \frac{5}{6}$$

$$12 \cdot \frac{1}{2}x + 12 \cdot \frac{3}{4} = 12 \cdot \frac{5}{6} \qquad \text{Use distributive law}$$

$$6x + 9 = 10$$

We have less chance of making an error solving this equation than the original. The same remarks apply to equations with decimal coefficients. First eliminate the decimals by multiplying both sides by an appropriate power of 10. For example, to solve

$$1.5y + 3.25 = 6$$

multiply both sides by 100 to clear all decimals.

$$100(1.5y + 3.25) = 100 \cdot 6$$
$$150y + 325 = 600$$
$$150y = 275$$
$$y = \frac{275}{150} = \frac{11}{6}$$

We have combined several steps and performed some of the calculations mentally in the above example. With practice, you should be able to do the same.

⑤ SUMMARY OF METHODS

We conclude this section by summarizing how to solve linear equations.

To Solve a Linear Equation

1. Simplify both sides by clearing parentheses, fractions, or decimals. Collect like terms.

2. Use the addition-subtraction rule to isolate all variable terms on one side and all constant terms on the other side. Collect like terms when possible.

3. Use the multiplication-division rule to obtain a variable with coefficient of 1.

4. Check the solution by substituting in the original equation.

5. If an identity results, the original equation has every real number as a solution. If a contradiction results, there is no solution.

11.3 EXERCISES A

Solve each equation. Check all solutions.

1. $2x + 1 = 5$ 2

2. $3y - 1 = 8$ 3

3. $-4z + 3 = 5$ $-\dfrac{1}{2}$

4. $\dfrac{3}{4}x + \dfrac{1}{4} = 1$ 1

5. $\dfrac{2}{5} - y = \dfrac{3}{10}$ $\dfrac{1}{10}$

6. $12 = -1.2z + 24$ 10

7. $2x + 3x = 10$ 2

8. $3y - y = 7$ $\dfrac{7}{2}$

9. $\dfrac{1}{4}z + \dfrac{1}{2}z = -9$ -12

10. $9.2x - 3.1x = 12.2$ 2

⑪ $-\dfrac{1}{7}y + y = 18$ 21

12. $3z = 5 - 7z$ $\dfrac{1}{2}$

13. $32 - 8x = 3 - 9x$ -29 **14** $-7 + 21y + 23 = 16$ 0 **15.** $z + 3z = 8 - 2z + 10$ 3

16. $x + 9 + 6x = 2 + x + 1$ -1 **17** $6y - 4y + 1 = 12 + 2y - 11$ identity

18. $2.1x - 3.2 = -8.4x - 45.2$ -4 **19** $3y + \dfrac{5}{2}y + \dfrac{3}{2} = \dfrac{1}{2}y + \dfrac{5}{2}y$ $-\dfrac{3}{5}$

Remove parentheses and solve.

20. $3(2x + 1) = 21$ 3 **21.** $10 = 5(y - 20)$ 22 **22.** $2(5z + 1) = 42$ 4

23. $3x - 4(x + 2) = 5$ -13 **24.** $2y - (13 - 2y) = 59$ 18 **25.** $3z - (z + 4) = 0$ 2

26. $5(x + 4) - 4(x + 3) = 0$ -8 **27.** $4(y - 3) - 6(y + 1) = 0$ -9 **28.** $9z - (3z - 18) = 36$ 3

29 $5(2 - 3x) = 15 - (x + 7)$ $\dfrac{1}{7}$ **30.** $8y - (2y + 5) = 0$ $\dfrac{5}{6}$ **31.** $9z - (5z - 2) = 8$ $\dfrac{3}{2}$

32. $7(x - 5) = 10 - (x + 1)$ $\dfrac{11}{2}$ **33.** $(y - 9) - (y + 6) = 4y$ $-\dfrac{15}{4}$ **34.** $8(2z + 1) = 4(7z + 7)$ $-\dfrac{5}{3}$

35. $5 + 3(x + 2) = 3 - (x + 2)$ **36** $\dfrac{1}{3}(6x - 9) = \dfrac{1}{2}(8x - 4)$ **37** $5(x + 1) - 4x = x - 5$

   $-\dfrac{5}{2}$ $-\dfrac{1}{2}$ no solution

38 $3(2 - 4x) = 4(2x - 1) - 2(1 + x)$ $\dfrac{2}{3}$ **39.** $2 + y = 3 - 2[1 - 2(y + 1)]$ -1

40. $-2(z - 1) - 3(2z + 1) = -9$ 1 **41** $-4(-x + 1) + 3(2x + 3) - 7x = -10$ -5

FOR REVIEW

Solve.

42. $\dfrac{x}{-\frac{1}{5}} = 20$ -4

43. $\dfrac{y}{-5} = -2$ 10

44. $\dfrac{3z}{7} = -6$ -14

45. $-2\dfrac{3}{8}x = -19$ 8

46. $y - \dfrac{2}{3} = \dfrac{1}{9}$ $\dfrac{7}{9}$

47. $3.6 - z = 7.2$ -3.6

ANSWERS: 1. 2 2. 3 3. $-\frac{1}{2}$ 4. 1 5. $\frac{1}{10}$ 6. 10 7. 2 8. $\frac{7}{2}$ 9. -12 10. 2 11. 21 12. $\frac{1}{2}$ 13. -29 14. 0
15. 3 16. -1 17. every real number (identity) 18. -4 19. $-\frac{3}{5}$ 20. 3 21. 22 22. 4 23. -13 24. 18 25. 2
26. -8 27. -9 28. 3 29. $\frac{1}{7}$ 30. $\frac{5}{6}$ 31. $\frac{3}{2}$ 32. $\frac{11}{2}$ 33. $-\frac{15}{4}$ 34. $-\frac{5}{3}$ 35. $-\frac{5}{2}$ 36. $-\frac{1}{2}$ 37. no solution 38. $\frac{2}{3}$
39. -1 40. 1 41. -5 42. -4 43. 10 44. -14 45. 8 46. $\frac{7}{9}$ 47. -3.6

11.3 EXERCISES B

Solve and check.

1. $3x + 1 = 7$ 2

2. $4y - 2 = 10$ 3

3. $-2z + 5 = 9$ -2

4. $\dfrac{2}{3}x + \dfrac{1}{3} = 1$ 1

5. $\dfrac{3}{5} - y = \dfrac{7}{10}$ $-\dfrac{1}{10}$

6. $10 = -0.5z + 5$ -10

7. $3x + 7x = 40$ 4

8. $4y - y = 27$ 9

9. $\dfrac{1}{8}z + \dfrac{5}{8}z = -6$ -8

10. $1.3x + 2.7x = 4.4$ 1.1

11. $-\dfrac{1}{5}y + y = 8$ 10

12. $5z = 2 - 9z$ $\dfrac{1}{7}$

13. $25 - 7x = 4 - 10x$ -7

14. $-2 + 8y + 27 = 1$ -3

15. $z + 2z = 3 - 2z + 17$ 4

16. $x + 3 + 4x = 2 + x + 1$ 0

17. $3y - y + 10 = 14 + 2y - 4$ identity

18. $6.4 - 4.2x = 16.8x + 90.4$ -4

19. $2x - \dfrac{1}{3} + \dfrac{2}{3}x = \dfrac{4}{9} + \dfrac{1}{3}x$ $\dfrac{1}{3}$

Clear parentheses and solve.

20. $2(3x + 1) = 14$ 2

21. $9 = 3(4y - 1)$ 1

22. $7(2z - 1) = -35$ -2

23. $4x - 2(x + 1) = -1$ $\dfrac{1}{2}$

24. $7y - (4 - 3y) = 11$ $\dfrac{3}{2}$

25. $2z - (z + 1) = 0$ 1

26. $6(x + 2) - 2(2x + 1) = 0$ -5

27. $4(2y - 1) - 7(y + 2) = 0$ 18

28. $6z - (2z - 5) = 0$ $-\dfrac{5}{4}$

29. $3(2 - 4x) = 13 - (x + 1)$ $-\dfrac{6}{11}$

30. $7y - (2y + 15) = 0$ 3

31. $8z - (3z - 1) = 0$ $-\dfrac{1}{5}$

32. $4(x - 2) = 12 - (x + 3)$ $\dfrac{17}{5}$

33. $(y - 2) - (y + 2) = 4y$ -1

34. $9(2z - 1) = 3(z + 2)$ 1

35. $7 + 3(x + 1) = 5 - (x + 1)$ $-\dfrac{3}{2}$

36. $\dfrac{1}{2}(2y - 4) = \dfrac{2}{3}(9y + 3)$ $-\dfrac{4}{5}$

37. $6(z - 1) - 3z = 3z + 8$ **no solution**

38. $2(1 - 3x) = 5(2x + 1) - 3(1 + x)$ 0

39. $4 + y = 8 - 3[2 - (y + 2)]$ -2

40. $-3(z - 4) - 2(3z + 1) = -8$ 2

41. $-7(-x + 2) + 5(2x + 1) - 11x = 3$ 2

FOR REVIEW

Solve.

42. $-\dfrac{3}{5}x = 6$ -10

43. $\dfrac{y}{-\frac{1}{7}} = 14$ -2

44. $\dfrac{2z}{9} = -12$ -54

45. $-3\dfrac{1}{7}x = -88$

   28

46. $\dfrac{3}{4} - y = \dfrac{1}{2}$

   $\dfrac{1}{4}$

47. $8.4 + z = -9.2$

   -17.6

11.3 EXERCISES C

Solve and check.

1. $0.02x - (0.56x - 7.33) = 2.15 - (0.7x - 0.11)$
   -31.6875

2. $3\dfrac{2}{3}x + 7\dfrac{1}{9} - \left(\dfrac{x}{6} - 4\dfrac{1}{9}\right) = -\left(2\dfrac{1}{6} - 6\dfrac{2}{9}x\right)$
   $\left[\text{Answer: } \dfrac{241}{49}\right]$

3. $x(x - 3) - 3x[1 - (x - 1] = 4x(x - 5) + 22$
   2

4. $-3[1 - (x^2 - 2)] - 2x(1 - x) = -5[4 - x(x - 2)]$
   $\left[\text{Answer: } -\dfrac{11}{8}\right]$

11.4 THE LANGUAGE OF PROBLEM SOLVING

STUDENT GUIDEPOSTS

1 Problem Solving   **2** Translating Words into Equations

1 PROBLEM SOLVING

To solve a word problem using algebra, we need to do two things. First we need to translate the *words* of the problem into an algebraic equation, and second, we need to solve the equation. We have learned how to solve many simple equations; now we will concentrate on translating word problems into equations. Some common terms and their symbolic translations are given below.

Symbol	+	−	·	÷	=
Terms	sum	minus	times	divided by	equals
	sum of	less	of	quotient of	is
	plus	less than	product	ratio	is equal to
	and	diminished by	product of		is as much as
	added to	difference	multiplied by		is the same as
	increased by	difference between			the result is
	more than	subtracted from			
		decreased by			

❷ TRANSLATING WORDS INTO EQUATIONS

Any letter (we often use x) can be used to stand for the unknown or desired quantity. Some examples of translation follow.

A number increased by 3 is 12.
$$x + 3 = 12$$

Three times a number diminished by 6 is the same as 12.
$$3 \cdot x - 6 = 12$$

My age seven years ago was eleven.
$$x - 7 = 11$$

5% of a number is 10.
$$0.05 \cdot x = 10$$

Although we often use x as the variable in a translation, it is sometimes helpful to use another letter more indicative of the quantity it represents. For example, we might use t for time, w for wages, A for area, or M for Mary's age.

EXAMPLE 1 TRANSLATING PHRASES INTO SYMBOLS

Select a variable to represent each quantity and translate the phrase into symbols.

Word phrase	*Symbolic translation*
(a) The time plus 4 hours	$t + 4$
(b) My wages less $100 for taxes	$w - 100$
(c) One half of the area	$\frac{1}{2}A$
(d) Twice Mary's age in 3 years	$2(M + 3)$
(e) $2000 less the cost	$2000 - c$

PRACTICE EXERCISE 1

Select a variable and translate each phrase into symbols.

(a) My age increased by 6 years

(b) The price plus $10 in tax

(c) Twice the area

(d) Five miles less than the distance

(e) Five miles less the distance

Answer: (a) $a + 6$ (b) $p + 10$
(c) $2A$ (d) $d - 5$ (e) $5 - d$

| EXAMPLE 2 TRANSLATING SENTENCES INTO SYMBOLS | PRACTICE EXERCISE 2 |

Use x for the variable and translate each word sentence into symbols.

Word expression	Symbolic translation
(a) Five times a number is 20.	$5x = 20$
(b) The product of a number and 7 is 35.	$7x = 35$
(c) Twice a number, increased by 3, is 11.	$2x + 3 = 11$
(d) Six is 4 less than twice a number.	$6 = 2x - 4$
(e) Six is 4 less twice a number.	$6 = 4 - 2x$
(f) One tenth of a number is 13.	$\frac{1}{10}x = 13$
(g) Five times a number, minus twice the number, equals 10.	$5x - 2x = 10$
(h) A number subtracted from 5 is 12.	$5 - x = 12$
(i) 5 subtracted from a number is 12.	$x - 5 = 12$
(j) A number divided by 9 is equivalent to $\frac{2}{3}$.	$\frac{x}{9} = \frac{2}{3}$

Use x for the variable and translate into symbols.

(a) Twice a number is 50.

(b) The product of a number and 8 is 64.

(c) Three times a number, increased by 2, is 25.

(d) Ten is 2 less a number.

(e) Ten is 2 less than a number.

(f) Three fifths of a number is 9.

(g) Twice a number, decreased by four times the number, is the same as the number.

(h) Her age diminished by 3 is 1.

(i) Three diminished by her age is 1.

(j) His weight divided by 3 is the same as 24.

Answers: (a) $2x = 50$ (b) $8x = 64$
(c) $3x + 2 = 25$ (d) $10 = 2 - x$
(e) $10 = x - 2$ (f) $\frac{3}{5}x = 9$
(g) $2x - 4x = x$ (h) $x - 3 = 1$
(i) $3 - x = 1$ (j) $\frac{x}{3} = 24$

11.4 EXERCISES A

Select a variable to represent each quantity and translate the phrase into symbols.

1. The sum of a number and 7 $n + 7$

2. A number subtracted from 10 $10 - n$

3. The product of a number and 3 $3n$

4. A number divided by 13 $\frac{n}{13}$

5. The time plus 8 hours $t + 8$

6. The time 6 hours ago $t - 6$

7. His salary plus $200 $s + 200$

8. My wages less $50 $w - 50$

9. Twice the volume $2V$

10. $3000 less the cost $3000 - c$

11. A number added to its reciprocal $n + \frac{1}{n}$

12. Her score less 20 points $s - 20$

13. 300 more than twice the number of votes
$2v + 300$

14. 300 more than the number of votes, doubled
$2(v + 300)$

15. 4% of the selling price $0.04p$

16. Cost plus 8% of the cost $c + 0.08c$

Using x for the variable, translate the following into symbols.

17. 7 times a number is 25. $7x = 25$

18. A number increased by 8 equals 12.
$x + 8 = 12$

19. A number decreased by 15 is equal to 35.
$x - 15 = 35$

20. The product of a number and 6 is 42.
$6x = 42$

21. Twice a number, increased by 3, is 27.
$2x + 3 = 27$

22. Three times a number, diminished by 8, equals 24.
$3x - 8 = 24$

23. A number less 5 is 12.
$x - 5 = 12$

24. 8 is 4 less than three times a number.
$8 = 3x - 4$

25. A number diminished by 3 is 9.
$x - 3 = 9$

26. Seven times a number, less 2, is 31.
$7x - 2 = 31$

27. The product of a number and 3, decreased by 5, equals 34. $3x - 5 = 34$

28. When 10 is added to three times a number the result is the same as twice the number, plus 12.
$10 + 3x = 2x + 12$

29. 3% of a number is 12. $0.03x = 12$

30. The sum of two numbers is 24 and one of them is three times the other. $x + 3x = 24$

31. Seven times 2 less than a number is 31. (Compare with Exercise 26.) $7(x - 2) = 31$

32. Twice the sum of a number and 5 is three times the number. $2(x + 5) = 3x$

33. Seven more than a number is 13.
$7 + x = 13$

34. Four times my age in 3 years is 48.
$4(x + 3) = 48$

35. When 4 is divided by some number the quotient is the same as $\frac{1}{2}$. $\frac{4}{x} = \frac{1}{2}$

36. One half a number, less twice the reciprocal of the number is $\frac{3}{2}$. $\frac{1}{2}n - 2\frac{1}{n} = \frac{3}{2}$

FOR REVIEW

Solve and check.

37. $-\frac{1}{4}y + y = 27$ 36

38. $15 - 3z = 5 + 7z$ 1

39. $2(y - 3) = 2(4y + 1) - 3(y + 1)$ $-\frac{5}{3}$

40. $2 + x = 4 - 2[1 - (x - 3)]$ 6

ANSWERS: 1. $n + 7$ 2. $10 - n$ 3. $3n$ 4. $\frac{n}{13}$ 5. $t + 8$ 6. $t - 6$ 7. $s + 200$ 8. $w - 50$ 9. $2V$ 10. $3000 - c$
11. $n + \frac{1}{n}$ 12. $s - 20$ 13. $2v + 300$ 14. $2(v + 300)$ 15. $0.04p$ 16. $c + 0.08c$ 17. $7x = 25$ 18. $x + 8 = 12$
19. $x - 15 = 35$ 20. $6x = 42$ 21. $2x + 3 = 27$ 22. $3x - 8 = 24$ 23. $x - 5 = 12$ 24. $8 = 3x - 4$ 25. $x - 3 = 9$
26. $7x - 2 = 31$ 27. $3x - 5 = 34$ 28. $10 + 3x = 2x + 12$ 29. $0.03x = 12$ 30. $x + 3x = 24$ 31. $7(x - 2) = 31$
32. $2(x + 5) = 3x$ 33. $7 + x = 13$ 34. $4(x + 3) = 48$ 35. $\frac{4}{x} = \frac{1}{2}$ 36. $\frac{1}{2}n - 2\frac{1}{n} = \frac{3}{2}$ or $\frac{n}{2} - \frac{2}{n} = \frac{3}{2}$ 37. 36 38. 1
39. $-\frac{5}{3}$ 40. 6

11.4 EXERCISES B

Select a variable to represent each quantity and translate the phrase into symbols.

1. The sum of a number and 12 $n + 12$

2. A number subtracted from 33 $33 - n$

3. The product of a number and 5 $5n$

4. A number divided by 7 $\frac{n}{7}$

5. The time plus 3 hours $t + 3$

6. The time 14 hours ago $t - 14$

7. Her salary plus $500 $s + 500$

8. My wages less $75 $w - 75$

9. Three times the volume $3V$

10. $5000 less the cost $5000 - c$

11. A number less its reciprocal $n - \dfrac{1}{n}$

12. His score less 28 points $s - 28$

13. 100 more than triple the number of votes
$3v + 100$

14. 100 more than the number of votes, tripled
$3(v + 100)$

15. 24% of a number $0.24n$

16. Cost plus 10% of the cost $c + 0.10c$

Using x for the variable, translate the following into symbols.

17. 4 times a number is 32.
$4x = 32$

18. A number increased by 3 equals 14.
$x + 3 = 14$

19. A number decreased by 13 is equal to 49.
$x - 13 = 49$

20. The product of a number and 8 is 96.
$8x = 96$

21. Twice a number, increased by 4, is 86.
$2x + 4 = 86$

22. Three times a number, diminished by 17, equals 42.
$3x - 17 = 42$

23. A number less 4 is 21.
$x - 4 = 21$

24. 9 is 3 less than five times a number.
$9 = 5x - 3$

25. A number diminished by 2 is 18.
$x - 2 = 18$

26. Twelve times a number, less 17, is 39.
$12x - 17 = 39$

27. The product of a number and 7, decreased by 8, equals 29. $7x - 8 = 29$

28. When 15 is added to three times a number the result is the same as twice the number, plus 40.
$3x + 15 = 2x + 40$

29. 9% of a number is 15.
$0.09x = 15$

30. The sum of two numbers is 86, and one of them is three times the other. $x + 3x = 86$

31. Eleven times 4 less than a number is 25.
$11(x - 4) = 25$

32. Twice the sum of a number and 8 is four times the number. $2(x + 8) = 4x$

33. Six more than a number is 29.
$6 + x = 29$

34. Five times my age in 2 years is 200.
$5(x + 2) = 200$

35. When 7 is divided by some number the quotient is the same as $\frac{4}{9}$. $\dfrac{7}{x} = \dfrac{4}{9}$

36. One third a number, less three times the reciprocal of the number is $\frac{4}{3}$. $\dfrac{x}{3} - \dfrac{3}{x} = \dfrac{4}{3}$

FOR REVIEW

Solve and check.

37. $-\dfrac{1}{6}y + y = 75$ 90

38. $12 - 4z = 2z + 12$ 0

39. $3(y - 4) = 2(3y + 1) - (y - 3)$ $-\dfrac{17}{2}$

40. $1 + x = 3 - 2[2 - (x - 5)]$ 12

11.4 EXERCISES C

Using x for the variable, translate the following into symbols.

1. If two-thirds of eight more than a number is increased by five times the reciprocal of the number, the result is twice the number, decreased by 8.
$\frac{2}{3}(x + 8) + 5\left(\frac{1}{x}\right) = 2x - 8$

2. A number divided by six less than the number is the same as the product of the number and three more than the number. $\frac{x}{x-6} = x(x + 3)$

11.5 BASIC APPLICATIONS AND PERCENT

═══ STUDENT GUIDEPOSTS ═══

1 Method for Solving Applied Problems

2 Consecutive Integer Applications

3 Percent Translations

4 Commission Rate and Percent Increase

5 Interest Applications

1 METHOD FOR SOLVING APPLIED PROBLEMS

Success with applied problems (or word problems) comes with practice. Problem solving will be easier if you follow these steps.

To Solve an Applied Problem

1. Read the problem (perhaps several times) and determine what quantity (or quantities) you are asked to find.

2. Represent the unknown quantity (or quantities) using a single letter.

3. Determine which expressions are equal and write an equation.

4. Solve the equation and state the answer to the problem.

5. Check to see if the answer (or answers) satisfy the conditions of the problem.

EXAMPLE 1   NUMBER PROBLEM

Twice a number, increased by 13, is 71. What is the number?

Let x = the number. The problem may be symbolized as follows.

$$
\begin{array}{ccccc}
\text{twice} & \text{a number} & \text{increased by} & 13 & \text{is } 71 \\
\downarrow & \downarrow & \downarrow & \downarrow & \downarrow\ \downarrow \\
2\ \cdot & x & + & 13 & =\ 71
\end{array}
$$

$$2 \cdot x + 13 = 71$$
$$2x + 13 - 13 = 71 - 13$$
$$2x = 58$$
$$x = 29$$

The number is 29.

Check: Twice 29 plus 13 is indeed 71.

PRACTICE EXERCISE 1

Three times a number, decreased by 24, is 60. What is the number?

Answer: 28

▨▨▨▨▨▨▨ CAUTION ▨▨▨▨▨▨▨

Always be neat and complete when working word problems. Time-consuming errors happen when you try to take shortcuts. Writing out what the variable means helps in solving the problem.

▨▨▨▨▨▨▨

EXAMPLE 2 APPLICATION TO AGE	**PRACTICE EXERCISE 2**

Bob is three times as old as Dick, and the sum of their ages is 48 years. How old is each?

Let x = Dick's age,

$3x$ = Bob's age.

$$x + 3x = 48$$
$$4x = 48$$
$$x = 12$$
$$\text{and} \quad 3x = 36$$

Dick is 12 years old and Bob is 36 years old.
Check: $36 = 3 \cdot 12$ and $12 + 36 = 48$.

Practice Exercise 2: The sum of two numbers is 77. If one number is three-fourths of the other, find the two numbers

Answer: 33 and 44

---

② CONSECUTIVE INTEGER APPLICATIONS

Consecutive integers are integers which immediately follow each other in regular counting order. For example, 13, 14, and 15 are consecutive integers. In general, if x is an integer, the next consecutive integer is $x + 1$, and the next after that is $x + 2$. An example of consecutive *even* integers is 2, 4, 6. In this case, if x is the integer, the next consecutive even integer is $x + 2$, and the next after that is $x + 4$. The same is true for consecutive odd integers such as 3, 5, 7. If x is the first, $x + 2$ is the next and $x + 4$ follows.

EXAMPLE 3 CONSECUTIVE INTEGERS	**PRACTICE EXERCISE 3**

The sum of three consecutive integers is 96. Find the numbers.

Let x = first integer,

$x + 1$ = next consecutive integer (Why?),

$x + 2$ = third consecutive integer (Why?).

$$x + (x + 1) + (x + 2) = 96$$
$$3x + 3 = 96 \quad \text{Combine like terms}$$
$$3x = 93 \quad \text{Subtract 3}$$
$$x = 31$$
$$\text{and} \quad x + 1 = 32$$
$$\text{and} \quad x + 2 = 33$$

The numbers are 31, 32, and 33.
Check: $31 + 32 + 33 = 96$ and the three integers are consecutive.

Practice Exercise 3: Wendy and Diane were born in consecutive years. If the sum of their ages is 43 and Diane is the older, how old is each?

Answer: Wendy is 21 and Diane is 22.

EXAMPLE 4 CONSECUTIVE EVEN INTEGERS	**PRACTICE EXERCISE 4**

The sum of two consecutive even integers is 2 less than twice the larger. Find the numbers.

Let x = the first even integer,

$x + 2$ = the second (larger) even integer.

Practice Exercise 4: The sum of three consecutive odd integers is 135. Find the numbers.

$$x + (x + 2) = 2(x + 2) - 2$$

$2x + 2 = 2x + 4 - 2$	Clear parentheses
$2x + 2 = 2x + 2$	Combine like terms
$2 = 2$	Subtract $2x$

Since we obtained an identity, any value of x will make the equation true. Therefore, *any* two consecutive even integers solve the problem.

Answer: 43, 45, 47

❸ PERCENT TRANSLATIONS

Many applied problems involve percent and can be restated into one of three basic forms. The following examples illustrate these three types.

What is 25% of 40?
What percent of 60 is 12?
8 is 40% of what number?

Letting x represent the unknown number in each statement and recalling that the word *of* translates to *times,* these three statements can be translated into equations. The first becomes

What is 25% of 40?
x = 0.25 · 40.

Recall from Chapter 1 that 25% = 0.25

Solving, we have $x = (0.25)(40) = 10$. The second statement translates to

What percent of 60 is 12?
x · 60 = 12.

Solving, $x = \dfrac{12}{60} = 0.2.$

In percent notation, x becomes 20%.
   Finally, the third statement becomes

8 is 40% of what number?
8 = 0.40 · x.

Thus, $x = \dfrac{8}{0.40} = 20.$

EXAMPLE 5 SOLVING PERCENT PROBLEMS

(a) 75% of 128 is what?
The equation to solve is $(0.75)(128) = x$.

$$96 = x$$

(b) 18 is what percent of 48?
The equation to solve is $18 = x \cdot 48$.

$$\frac{18}{48} = x \quad \text{Divide by 48}$$

$$0.375 = x$$

In percent notation, $x = 37.5\%$.

PRACTICE EXERCISE 5

(a) 36% of 225 is what?

(b) 102 is what percent of 68?

Answers: (a) 81   (b) 150%

| **EXAMPLE 6** APPLICATION TO SPORTS | **PRACTICE EXERCISE 6** |

Manny Black had a batting percentage (percent of hits in times at bat) of 40% in a series. If he had 12 hits, how many times was he at bat?

In effect, we are asked: 12 is 40% of what?

Let x be the number of times at bat.

$$12 = (0.40)x$$

$$\frac{12}{0.40} = x \qquad \text{Divide by } 0.40$$

$$30 = x$$

Thus, he went to bat 30 times in the series.

For the season, basketball player Andy Hurd shot 90.5% from the free throw line. If he made 181 free throws, how many times did he go to the free throw line?

Answer: 200

4 COMMISSION RATE AND PERCENT INCREASE

Some salespersons receive all or part of their income as a percent of their sales. This is called a **commission.** The **commission rate** is the percent of the sales that the person receives. That is,

$$\text{(commission)} = \text{(commission rate)} \cdot \text{(total sales)}.$$

| **EXAMPLE 7** COMMISSION APPLICATION | **PRACTICE EXERCISE 7** |

Paul received a commission of $720 on the sale of a machine. If his commission rate is 8% of the selling price, what was the selling price?

The problem can be written: 8% of what is $720?

Let x be the selling price.

$$(0.08)x = 720$$

$$x = \frac{720}{0.08}$$

$$x = 9000$$

Thus, the machine sold for $9000.

Kerry received a commission of $462 on the sale of a new car. If her commission rate is 3% of the selling price, what was the selling price?

Answer: $15,400

| **EXAMPLE 8** PERCENT INCREASE APPLICATION | **PRACTICE EXERCISE 8** |

Due to inflation, the price of an item rose 25% over its price the year before. What was the price last year if the price this year is 60¢?

Problems such as this can be solved by using the equation

$$\text{(previous price)} + \text{(increase in price)} = \text{(new price)}.$$

Let x be last year's price.

$$x + (0.25) \cdot x = 60$$

$$1.25x = 60 \qquad\qquad \begin{aligned} x + (0.25) \cdot x &= 1 \cdot x + (0.25) \cdot x \\ &= (1 + 0.25) \cdot x = 1.25x \end{aligned}$$

$$x = \frac{60}{1.25} = 48$$

Last year's price was 48¢.

Due to overstocking, a dealer is forced to sell a typewriter at a 20% discount. What was the original price of the typewriter if he sells it for $288?

Answer: $360

⑤ INTEREST APPLICATIONS

Money that we pay to borrow money or the money earned on savings is called **interest.** The money borrowed or saved is called the **principal,** and the interest is calculated as a percent of the principal. This percent is called the **interest rate.** For example, suppose you borrow $500 for one year and the interest rate is 12% per year. The interest charged would be

$$(0.12)(\$500) = \$60.$$

If you needed the money for two years, the interest rate would be

$$2(0.12) = 0.24 = 24\%,$$

with the total interest for two years amounting to

$$(0.24)(\$500) = \$120.$$

This type of interest is called **simple interest** since interest is charged only on the principal and not on the interest itself.

EXAMPLE 9 APPLICATION TO SIMPLE INTEREST

What sum of money invested at 4% simple interest will amount to $988 in one year?

Interest problems like this can be solved by using the equation

$$(\text{amount invested}) + (\text{interest}) = (\text{total amount}).$$

Let x be the amount invested.

$$x + (0.04) \cdot x = 988 \qquad \text{Interest} = (0.04)x$$
$$(1.04) \cdot x = 988$$
$$x = \frac{988}{1.04} = 950$$

The amount invested is $950.

PRACTICE EXERCISE 9

Missie must have $1100 available at the end of one year. What amount must she invest now at 10% simple interest to have the desired amount at the end of the year?

Answer: $1000

11.5 EXERCISES A

Solve. (Some problems have been started.)

1. If 14 is added to four times a number, the result is 38. Find the number.

Let x = the desired number,
   $4x$ = four times the number.
6

2. The sum of the two numbers is 180. The larger number is five times the smaller number. Find the two numbers.

Let x = the smaller number.
150, 30

3. The sum of two numbers is 98. If one is six times the other, find the two numbers.   **14, 84**

4. Two-thirds of a number is 124; what is the number?

Let x = the desired number,
   $\frac{2}{3}x$ = two thirds the desired number.
186

5. Two-thirds of the human body is water. If your body contains 124 pounds of water, how much do you weigh? [*Hint:* Compare with Exercise 4.]   **186 lb**

6. The first of two numbers is four times the second. If the first is 30,168, what is the second?   **7542**

7. The area of Lake Superior is four times the area of Lake Ontario. If the area of Lake Superior is 30,168 mi², what is the area of Lake Ontario? [*Hint:* Compare with Exercise 6.]   **7542 mi²**

8. The sum of three consecutive integers is 78. Find the three integers.

Let x = the first integer,
$x + 1$ = the second integer,
_____ = the third integer.
25, 26, 27

9. The sum of two consecutive odd integers is 96. Find the two integers.

Let x = the first odd integer,
$x + 2$ = the next odd integer.
47, 49

10. The sum of two consecutive even integers is 162. Find the two integers.   **80, 82**

11. Maria is 4 years older than Juan. The sum of their ages is 32 years. How old is each?

Let x = Juan's age,
$x + 4$ = Maria's age.
Juan is 14, Maria is 18.

12. George is 5 years older than Sue. If the sum of their ages is 47, how old is each?
George is 26, Sue is 21.

13. How old is Marvin if you get 100 years when you multiply his age by 6 and add 16 years?   **14**

14. If three-fourths of a number is 27, what is the number?   **36**

15. The sum of two numbers is 9, and one is twice the other. What are the numbers?

Let x = the first number,
$2x$ = the second number.
3, 6

16. A board is 9 ft long. It is to be cut into two pieces in such a way that one piece is twice as long as the other. How long is each piece? [*Hint:* Compare with Exercise 15.]   **3 ft and 6 ft**

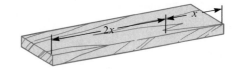

17 A board is 17 ft long. It is to be cut into two pieces in such a way that one piece is seven ft longer than the other. How long is each piece?   **5 ft and 12 ft**

18 Tony is four times as old as Angela and half as old as Theresa. The sum of the three ages is 39 years. How old is each?
Angela is 3, Tony is 12, Theresa is 24.

19. 45 is 15% of what?

$45 = 0.15 \cdot x$
300

20. 105 is 20% of what?   **525**

21. 25% of 164 is what?

$0.25 \cdot 164 = x$
41

22. 18% of 90 is what?   **16.2**

23. What is 30% of 420?   **126**

24. What percent of 90 is 15?

$x \cdot 90 = 15$
$16\frac{2}{3}\%$

25. What percent of 320 is 20? $6\frac{1}{4}\%$

26. 40 is what percent of 200?   **20%**

27. 25 is what percent of 500?   **5%**

28. 25 is 4% of what?   **625**

29. In 150-ml acid solution there are 45 ml of acid. What is the percent of acid in the solution?   **30%**

30. A basketball player hit 120 free throws in 150 attempts. What percent of her shots did she make? **80%**

31. Juan received a commission of $104.50 on the sale of a typewriter. If his commission rate is 11% of the selling price, what was the selling price?

Let x = selling price,
$0.11x$ = his commission on sales.
$950

32 Percy received $31.50 interest on his savings last year. If he was paid 6% simple interest, what amount did he have invested?   **$525**

33. A baseball player got 42 hits one season. If his batting average was 28%, how many times was he up to bat?   **150**

34 A dress is discounted 30% and the sale price is $51.45. What was the original price?   **$73.50**

35. If the sales-tax rate is 5% and the marked price plus tax of a mixer is $37.59, what is the marked price?

> Let x = marked price,
> $0.05x$ = tax on the marked price,
> $x + 0.05x$ = price plus tax.
> **$35.80**

36. The price of a package of gum rose 20% last year. If the present price is 42¢, what was the price last year?   **35¢**

37. If the present retail price of an item is 65¢ and due to increased overhead the price will have to be raised 20%, what will be the new price?   **78¢**

38 What sum invested at $4\frac{1}{2}$% simple interest will amount to $1254 in one year?   **$1200**

39. A baseball player got 21 hits in 70 times at bat. What was his batting average?   **30%**

40. A family spent $200 a month for food. This was 16% of their monthly income. What was their monthly income?   **$1250**

41. The sales-tax rate in Murphyville is 4%. How much tax would be charged on a purchase of $12.50?   **50¢**

42. What sum at 4% simple interest will amount to $1508 in 1 year?   **$1450**

43. The area of Greenland is 25% of the area of the United States. What is the area of Greenland if the area of the U.S. is 3,615,000 mi²?   **903,750 mi²**

44. The human brain is $2\frac{1}{2}$% of the total body weight. If Nick's brain weighs 4 lb, how much does Nick weigh?   **160 lb**

FOR REVIEW

Select a variable to represent each quantity and translate the phrase into symbols.

45. Price plus 10% of the price $p + 0.10p$

46. The reciprocal of a number less twice the number $\frac{1}{n} - 2n$

The following exercises review formulas from geometry that will help you prepare for the next section. These formulas with figures are on the inside cover of this text and were discussed in Chapter 9. A calculator will be helpful in some exercises.

47. A rectangle with length l and width w has area A given by $A = lw$. Find the area of a rectangle with length 4.5 ft and width 3.2 ft.   **14.4 ft²**

48. A triangle with base b and altitude h has area A given by $A = \frac{1}{2}bh$. Find the area of a triangle with base $\frac{5}{4}$ in and altitude $\frac{8}{3}$ in.   **$\frac{5}{3}$ in²**

49. A circle with radius r has circumference C given by $C = 2\pi r$. Find the circumference of a circle with diameter 9.2 yd, correct to the nearest tenth of a yard. **28.9 yd**

50. A trapezoid with bases b_1 and b_2 and altitude h has area A given by $A = \frac{1}{2}(b_1 + b_2)h$. Find the area, to the nearest tenth of a square centimeter, of a trapezoid with bases 8.6 cm and 12.3 cm and altitude 4.9 cm. **51.2 cm²**

ANSWERS: 1. 6 2. 150, 30 3. 14, 84 4. 186 5. 186 lb 6. 7542 7. 7542 mi² 8. 25, 26, 27 9. 47, 49 10. 80, 82 11. Juan is 14, Maria is 18. 12. George is 26 and Sue is 21. 13. 14 14. 36 15. 3, 6 16. 3 ft and 6 ft 17. 5 ft and 12 ft 18. Angela is 3, Tony is 12, Theresa is 24. 19. 300 20. 525 21. 41 22. 16.2 23. 126 24. $16\frac{2}{3}\%$ 25. $6\frac{1}{4}\%$ 26. 20% 27. 5% 28. 625 29. 30% 30. 80% 31. $950 32. $525 33. 150 34. $73.50 35. $35.80 36. 35¢ 37. 78¢ 38. $1200 39. 30% 40. $1250 41. 50¢ 42. $1450 43. 903,750 mi² 44. 160 lb 45. $p + 0.10p$ 46. $\frac{1}{n} - 2n$ 47. 14.4 ft² 48. $\frac{5}{3}$ in² 49. 28.9 yd 50. 51.2 cm²

11.5 EXERCISES B

1. If 9 is added to six times a number, the result is 39. Find the number. **5**

2. The sum of two numbers is 72. The larger number is five times the smaller number. Find the two numbers. **12, 60**

3. The sum of two numbers is 48. If one is five times the other, find the two numbers. **8, 40**

4. Three-fourths of a number is 288; what is the number? **384**

5. Three-fourths of the contents of a filled tank is water. If the tank is holding 288 gallons of water, how much fluid is in it? **384 gallons**

6. The first of two numbers is five times the second. If the first is 1250, what is the second? **250**

7. The number of acres in the Henderson farm is five times the number of acres in the Carlson farm. If the Henderson farm has 1250 acres, how many acres are there in the Carlson farm? **250 acres**

8. The sum of three consecutive odd integers is 195. Find the integers. **63, 65, 67**

9. The sum of two consecutive even integers is 90. What are the integers? **44, 46**

10. The sum of two consecutive integers is 203. Find the integers. **101, 102**

11. Barb is 9 years older than Larry. Twice the sum of their ages is 126. How old is each? **Barb is 36, Larry is 27.**

12. Hector is 7 years younger than Pedro. If the sum of their ages is 39, how old is each? **Hector is 16, Pedro is 23.**

13. How old is Burford if you get 99 years when you multiply his age by 10 and subtract 31 years? **13**

14. Two thirds of a number is 62; what is the number? **93**

15. The sum of two numbers is 168, and one is three times the other. What are the numbers? **42, 126**

16. A rope is 168 feet long. It is to be cut into two pieces in such a way that one piece is three times as long as the other. How long is each piece? **42 ft, 126 ft**

17. A steel rod is 22 m long. It is cut into two pieces in such a way that one piece is 6 m longer than the other. How long is each piece? **8 m, 14 m**

18. Hortense is five times as old as Wally and half as old as Lew. The sum of their ages is 80 years. How old is each? **Hortense is 25, Lew is 50, Wally is 5.**

19. 48 is 12% of what? **400**

20. 18 is 4% of what? **450**

21. 35% of 180 is what? **63**

22. 50% of 280 is what? **140**

23. What is 3% of 1250? **37.5**

24. What percent of 85 is 17? **20%**

25. What percent of 60 is 210? **350%**

26. 15 is what percent of 300? **5%**

27. 2.5 is what percent of 7.5? **$33\frac{1}{3}\%$**

28. 1230 is 60% of what? **2050**

29. In a 250-ml salt solution there are 15 ml of salt. What is the percent of salt in the solution? **6%**

30. A basketball player hit 12 shots in 20 attempts during one game. What was his shooting percent? **60%**

31. Blue Carpet Realty received a commission of $4950 on the sale of a house. If the commission rate is 6%, what was the selling price of the house? **$82,500**

32. Terry McGinnis received $845 interest on her savings last year. If she is paid 13% simple interest, what amount did she have invested? **$6500**

33. A quarterback completed 21 passes in one game for a completion percentage of 60%. How many passes did he attempt? **35**

34. A sport coat was discounted 20% and the sale price was $68. What was the original price? **$85**

35. Fred now makes $18,920 per year. What was his salary before he received a 10% raise? **$17,200**

36. The price of a pet mouse rose 30% last year. If the present price is $1.56, what was the price one year ago? **$1.20**

37. Due to inflation, the cost of an item rose $7.20. This was a 12% increase. What was the former price? The new price? **$60.00, $67.20**

38. What sum invested at $8\frac{1}{2}\%$ simple interest will amount to $1410.50 in one year? **$1300**

39. A prize fighter won 38 of his 40 professional fights. What was his percent of wins? **95%**

40. The Perez family spent $125 last week for food. This was 20% of their weekly income. What was their weekly income? **$625**

41. The sales-tax rate in Canyon City is 4%. How much tax would be charged on a purchase of $327? **$13.08**

42. What sum at 11% simple interest will amount to $7215 in 1 year? **$6500**

43. The area of the Wyler Ranch is 35% of the area of the Rodriguez Ranch. What is the area of the Wyler Ranch if the area of the Rodriguez Ranch is 1200 mi²? **420 mi²**

44. One year the population of the United States was 5% of the total world population. If the U.S. population was 200,000,000, what was the world population? **4,000,000,000**

FOR REVIEW

Select a variable to represent each quantity and translate the phrase into symbols.

45. Cost increased by 15% of the cost
$c + 0.15\,c$

46. A number less twice its reciprocal $n - \dfrac{2}{n}$

The following exercises review formulas from geometry that will help you prepare for the next section. These formulas with figures are on the inside cover of this text and are discussed in Chapter 9. A calculator will be helpful in some exercises.

47. The perimeter P of a rectangle with length l and width w is given by $P = 2l + 2w$. Find the perimeter of a rectangle with length 8.6 m and width 5.3 m. **27.8 m**

48. The area A of a parallelogram with base b and altitude h is given by $A = bh$. Find the area of a parallelogram with base 246 cm and altitude 188 cm. **46,248 cm²**

49. The volume V of a cylinder with radius r and height h is given by $V = \pi r^2 h$. Find the volume of a cylinder, correct to the nearest tenth of a cubic foot, if the radius is 5.4 ft and height 11.2 ft.　**1026.0 ft³**

50. The volume V of a sphere with radius r is given by $V = \frac{4}{3}\pi r^3$. Find the volume, to the nearest tenth of a cubic inch, of a sphere with radius 3.8 inches.　**229.8 in³**

11.5 EXERCISES C

Solve.

1. A rope 27 m long is to be cut into three pieces in such a way that the second piece is one-fourth the first and the third is 3 m longer than the second. Find the length of each piece.　**16 m, 4 m, 7 m**

2. A retailer uses a 30% markup to make a profit on suits. If Chris Katsaropoulos bought a suit for $124.02 and that included a 6% sales tax, what was the original cost of the suit to the dealer? [Answer: $90.00]

11.6 GEOMETRY AND MOTION PROBLEMS

STUDENT GUIDEPOSTS

1 Geometry Problems

2 Motion Problems

1 GEOMETRY PROBLEMS

Many geometry problems require knowledge of basic formulas that are summarized inside the back cover and discussed in Chapter 9. As you read through a problem, try to determine which formula is appropriate. It is always helpful to make a sketch, and label the given parts of the figure.

EXAMPLE 1 APPLICATION OF GEOMETRY

Find the width of a rectangle if its area is 192 ft² and its length is 16 ft.

To work this problem, we need to know that the area of a rectangle is given by $A = lw$. Make a sketch as in Figure 11.1.

Let w = width of rectangle.

$$A = lw$$
$$192 = 16\,w$$
$$12 = w$$

$A = lw$

$A = 192\text{ ft}^2$

w

16 ft

Figure 11.1

The width is 12 ft. Note that it is necessary to include the units in the answer.

PRACTICE EXERCISE 1

The area of a rectangular garden is 19.8 m² and its width is 3.6 m. Find the length of the garden.

Answer: 5.5 m

EXAMPLE 2 APPLICATION OF GEOMETRY

If the perimeter of a rectangular room is 32 ft and its length is three times its width, find its dimensions. Make a sketch as in Figure 11.2.

The perimeter of a rectangle is given by $P = 2l + 2w$.

Let w = width of room,
$3w$ = length of room.

$$P = 2l + 2w$$
$$32 = 2(3w) + 2(w)$$
$$32 = 6w + 2w$$
$$32 = 8w$$
$$4 = w$$
and $12 = 3w$

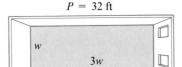

P = 32 ft

w

$3w$

Figure 11.2

The length of the room is 12 ft and the width is 4 ft.

EXAMPLE 3 GEOMETRY PROBLEM

The first angle of a triangle is three times as large as the second. The third angle is 40° larger than the first angle. What is the measure of each angle?

The sum of the measures of the angles of any triangle is 180°. Make a sketch as in Figure 11.3.

Let $3x$ = measure of the first angle,
x = measure of the second angle,
$3x + 40$ = measure of the third angle.

$$x + 3x + 3x + 40 = 180$$
$$7x + 40 = 180$$
$$7x = 140$$
$$x = 20$$
and $3x = 60$
and $3x + 40 = 100$

$3x + 40$

$3x$

x

Figure 11.3

The measures are 20°, 60°, and 100°.

EXAMPLE 4 APPLICATION OF GEOMETRY

The circumference of a circular flower bed is 182 ft. How many feet of pipe do we need to reach from the edge of the bed to a fountain in the center of the bed? The circumference of a circle is given by $C = 2\pi r$. [Use 3.14 for π]. Make a sketch as in Figure 11.4.

$$C = 2\pi r$$
$$182 = 2(3.14)r$$
$$\frac{182}{2(3.14)} = r$$

r

C = 182 ft

Figure 11.4

$28.98 \approx r$ To the nearest hundredth

We need about 29 ft of pipe.

PRACTICE EXERCISE 2

A farmer encloses a pasture with 20 miles of fence. If the width of the pasture is one-fourth the length, find the dimensions of the pasture.

Answer: 2 miles by 8 miles

PRACTICE EXERCISE 3

The second angle of a triangle is one-half as large as the first, and the third is 45° less than three times the first. Find the measure of each.

Answer: 50°, 25°, 105°

PRACTICE EXERCISE 4

The circumference of a circle is 46 in. Use 3.14 for π and find the approximate length (to the nearest tenth) of the diameter.

Answer: 14.6 in

In Example 4 we used the symbol ≈ to represent the phrase "is *approximately equal* to." It is important to use ≈ whenever an approximate or rounded number is indicated.

EXAMPLE 5 APPLICATION OF GEOMETRY	**PRACTICE EXERCISE 5**

In Figure 11.5 a cylindrical storage bin with radius 10 ft is to be built. How high will it need to be to store 10,200 ft³ of grain?

The volume of a cylinder is given by $V = \pi r^2 h$. [Use 3.14 for π.]

A cylindrical tank with radius 8 m has surface area 965 m². Use 3.14 for π and find the approximate height of the tank. The formula for surface area is $A = 2\pi rh + 2\pi r^2$. Let h = height of tank.

$V = \pi r^2 h$

Figure 11.5

$$V = \pi r^2 h$$
$$10{,}200 = (3.14)(10)^2 h$$
$$10{,}200 = 314h$$
$$\frac{10{,}200}{314} = h$$
$$32.48 \approx h \qquad \text{To the nearest hundredth}$$

The bin must be about 32.5 ft high.

Answer: 11.2 m

❷ MOTION PROBLEMS

Problems that involve distances and rates of travel often result in simple linear equations. The distance d that an object travels in a given time t at a fixed rate r is given by the formula

$$\textbf{(distance)} = \textbf{(rate)} \cdot \textbf{(time)} \quad \text{or} \quad d = rt.$$

For example, a boy walking at a rate of 4 mph for 3 hr will travel a distance of

$$d = rt = 4 \cdot 3 = 12 \text{ miles.}$$

Most types of motion problems depend in some way on the formula $d = rt$.

EXAMPLE 6   MOTION PROBLEM

An automobile is driven 336 mi in 7 hr. How fast (at what rate) was the car driven?

Let r = average rate of travel,
336 = distance d driven,
7 = time t of travel.
We need to solve the following equation:

$$336 = r(7)$$

$$\frac{336}{7} = r$$

$$48 = r.$$

Thus, the average rate of travel was 48 mph.

PRACTICE EXERCISE 6

Kevin rides his bike 162 km at an average rate of 36 km/hr. How long did he ride?

Answer: 4.5 hr

EXAMPLE 7   MOTION PROBLEM

Two hikers leave the same camp, one traveling east and the other traveling west. The one going east is hiking at a rate 1 mph faster than the other, and after 5 hr they are 35 mi apart. How fast is each hiking?

Make a sketch like the one in Figure 11.6.

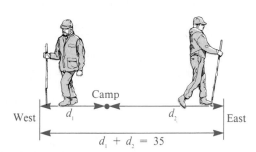

Figure 11.6

Let r = rate of hiker going west,
$r + 1$ = rate of hiker going east,
$d_1 = 5r$ = distance the westbound hiker travels,
$d_2 = 5(r + 1)$ = distance the eastbound hiker travels.
The equation we need to solve is

$$d_1 + d_2 = 5r + 5(r + 1) = 35$$

$$5r + 5r + 5 = 35$$

$$10r + 5 = 35$$

$$10r = 30$$

$$r = 3.$$

Thus, the westbound hiker is traveling 3 mph and the eastbound hiker is traveling 4 mph.

PRACTICE EXERCISE 7

Two boats leave an island at the same time, one sailing north and the other south. If one travels 5 mph faster than the other, and they are 140 mi apart in 4 hr, at what rate is each sailing?

Answer: 15 mph, 20 mph

11.6 EXERCISES A

Solve. If necessary, check the inside front cover for a list of formulas. (Some problems have been started.)

1. The perimeter of a rectangle is 60 in. If the length is 4 in more than the width, find the dimensions.

   Let w = width of rectangle,
   $w + 4$ = length of rectangle.
   17 in, 13 in

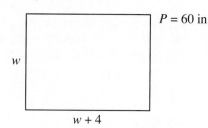

2. If one angle of a triangle is 60° more than the smallest angle and the third angle is six times the smallest angle, find the measure of each angle.

   Let x = smallest angle,
   $x + 60$ = another angle,
   $6x$ = third angle.
   15°, 75°, 90°

3. Find the length of a rectangle if its area is 104 in² and its width is 8 in.   **13 in**

4. Find the height of a triangle whose base is 12 cm and whose area is 84 cm².   **14 cm**

5. A circular patio has a radius of 14 ft. How many feet of edging material will be needed to enclose the patio? [Use $\pi \approx 3.14$.]   **87.9 ft**

6. In a triangle, the longest side is twice the shortest side, and the third side is 5 ft shorter than the longest side. Find the three sides if the perimeter is 45 ft.   **10 ft, 15 ft, 20 ft**

7. Find the height of a parallelogram whose base is 19 in and whose area is 133 in².   **7 in**

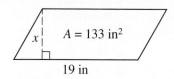

8. If two angles of a triangle are equal and the third angle is equal to the sum of the first two, find the measure of each.   **45°, 45°, 90°**

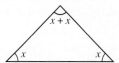

9. An isosceles triangle has two equal sides called legs and a third side called the base. If each leg of an isosceles triangle is four times the base and the perimeter is 108 m, find the length of the legs and base.   **48 m, 48 m, 12 m**

10. The perimeter of a rectangle is 630 in. Find the dimensions if the length is 25 in more than the width.   **145 in, 170 in**

11. Find the height of a trapezoid if the area is 247 m² and the bases are 16 m and 22 m. [*Hint: A = $\frac{1}{2}(b_1 + b_2)h$*]   **13 m**

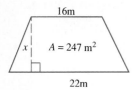

16m

x $A = 247\ m^2$

22m

12. Find the base of a parallelogram if the perimeter is 88 cm and a side is 20 cm.   **24 cm**

13. Find the area of a circular garden having a diameter of 96 yd.   **7234.6 yd²**

14 The volume of a grain storage silo in the shape of a rectangular solid is 4725 m³. If both the width and the height are 15 m, what is the length?   **21 m**

15. Find the height of a cylindrical tank with volume 3200 m³ and radius 8 m.   **15.9 m**

h

8 m

16. The perimeter of a square is 48 ft. What is the length of a side?   **12 ft**

17 A cube with edge 10 ft is submerged in a rectangular tank containing water. If the tank is 40 ft by 50 ft, how much does the level of water in the tank rise?   **0.5 ft**

18. Find the volume, rounded to the nearest tenth, of a sphere with radius 2 m.   **33.5 m³**

19 The sphere in Exercise 18 is dropped into a rectangular tank that is 10 m long and 8 m wide. To the nearest tenth, how much does this raise the water level?   **0.4 m**

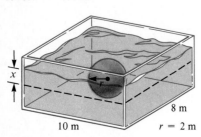

x

8 m

10 m $r = 2\ m$

20 A rancher wishes to enclose a pasture that is 3 mi long and 2 mi wide with a fence selling for 35¢ per linear foot. How much will the project cost?   **$18,480.00**

21. Mary Lou Mercer walks for 7 hours at a rate of 5 km/hr. How far does she walk?   **35 km**

22. Bob Packard runs a distance of 20 miles at a rate of 8 mph. How many minutes does he run? [*Hint: Note the units used.*]   **150 minutes**

23. A hiker crossed a valley by walking 5 mph the first 2 hr and 3 mph the next 4 hr. What was the total distance that she hiked?   **22 mi**

24. Two trains leave Omaha, one traveling east and the other traveling west. If one is moving 20 mph faster than the other, and if after 4 hr they are 520 mi apart, how fast is each going?   **75 mph, 55 mph**

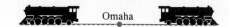

25 Two ships sail north from Bermuda. One is traveling at 22 knots and the other at 17 knots. How many nautical miles will they be from each other after 8 hours?   **40 nautical miles**

26. Two cars that are 550 mi apart and whose speeds differ by 10 mph are moving towards each other. What is the speed of each if they meet in 5 hr?   **50 mph, 60 mph**

FOR REVIEW

27. The sum of three consecutive odd integers is 135. What are the integers?   **43, 45, 47**

28. Twice a number, less 7, is the same as three times the number, less 30. What is the number?   **23**

29. Gus is 7 years older than Jan. If twice the sum of their ages is 66, how old is each?
Gus is 20, Jan is 13.

30. A wire is 42 in long. It is to be cut into three pieces in such a way that the second is twice the first and the third is equal to the sum of the lengths of the first two. How long is each piece?   **7 in, 14 in, 21 in**

31. 54 is 120% of what?   **45**

32. 30 is what percent of 150?   **20%**

33. If the cost of an item rose 8¢, a 16% increase, what was the old price? What is the new price?   **50¢, 58¢**

34. If the sales-tax rate is 3%, how much tax would be charged on a purchase of $18.00?   **54¢**

35. If a student answered 68 questions correctly on an 80-question exam, what was her percent score?   **85%**

36. Betsy earned $11,130 one year after she received a 6% raise. What was her former salary?   **$10,500**

ANSWERS: Some answers are rounded. 1. 17 in, 13 in 2. 15°, 75°, 90° 3. 13 in 4. 14 cm 5. 87.9 ft 6. 10 ft, 15 ft, 20 ft 7. 7 in 8. 45°, 45°, 90° 9. 48 m, 48 m, 12 m 10. 145 in, 170 in 11. 13 m 12. 24 cm 13. 7234.6 yd² 14. 21 m 15. 15.9 m 16. 12 ft 17. 0.5 ft 18. 33.5 m³ 19. 0.4 m 20. $18,480.00 21. 35 km 22. 150 minutes 23. 22 mi 24. 75 mph, 55 mph 25. 40 nautical miles 26. 50 mph, 60 mph 27. 43, 45, 47 28. 23 29. Gus is 20, Jan is 13. 30. 7 in, 14 in, 21 in 31. 45 32. 20% 33. 50¢, 58¢ 34. 54¢ 35. 85% 36. $10,500

11.6 EXERCISES B

Solve.

1. The perimeter of a rectangle is 52 ft. If the length is 8 ft more than the width, find the dimensions.
17 ft, 9 ft

2. If one angle of a triangle is 72° more than the smallest angle, and the third angle is seven times the smallest angle, find the measure of each angle.
12°, 84°, 84°

3. The area of a rectangle is 32.5 ft² and the width is 5 ft. What is the length? **6.5 ft**

4. If the area of a triangle is 78 cm² and its base is 12 cm, find its height. **13 cm**

5. A circular garden is to be enclosed with edging material. How many meters of edging will be needed if the radius of the garden is 5 m? [Use $\pi \approx 3.14$.]
31.4 m

6. In a triangle, the longest side is three times as long as the shortest side, and the third side is 2 ft shorter than the longest side. Find the three sides if the perimeter is 26 ft. **6 ft, 12 ft, 10 ft**

7. Find the height of a parallelogram with a base of 17 cm and an area of 93.5 cm². **5.5 cm**

8. If two angles of a triangle are equal and the third angle is equal to twice their sum, find the measure of each. **30°, 30°, 120°**

9. An isosceles triangle with a base of 13 in has a perimeter of 53 in. Find the length of its legs. **20 in**

10. The perimeter of a rectangular table is 90 in. Find its dimensions if the length is 15 in more than the width. **30 in, 15 in**

11. What is the height of a trapezoid with area 24 yd² and bases of length 4.5 yd and 3.5 yd? **6 yd**

12. Find the base of a parallelogram if the perimeter is 100 in and a side is 15 in. **35 in**

13. What is the area of a circular rug having diameter 82 inches? **5278.3 in²**

14. The volume of a rectangular solid is 26,250 cm³. If both the width and the height are 25 cm, what is the length? **42 cm**

15. Find the height of a cylindrical storage tank with volume 1105.3 ft³ and radius 4 ft. **22 ft**

16. The perimeter of a square is 72 yd. What is the length of a side? **18 yd**

17. A cube with edge 8 in is submerged in a rectangular tank containing water. If the tank is 10 in by 20 in, how much does the level of water in the tank rise?
2.56 in

18. Find the volume of a sphere with radius 8 cm.
2143.6 cm³

19. If the sphere in Exercise 18 is submerged in a rectangular tank 20 cm wide by 30 cm long, how much will the level of the water increase? **3.6 cm**

20. How much will it cost to enclose a garden with a picket fence if the garden is 12 yd long, 8 yd wide, and fencing costs 85¢ per linear foot? **$102**

21. A ship travels for 15 hr at 20 knots, where a knot is 1 nautical mile per hour. How many nautical miles has it traveled?   **300 nautical miles**

22. Jessie Meyer drove her boat a distance of 45 miles at a rate of 30 mph. How many minutes did she drive? **90 minutes**

23. Barry Silverstein traveled by car for 6 hr at a rate of 55 mph and then by boat for 2 hr at 24 mph. What was the total distance traveled?   **378 mi**

24. Two campers leave Rocky Mountain National Park, one traveling north and the other traveling south. After 3 hr they are 255 mi apart, and one has been traveling 5 mph faster than the other. How fast is each traveling?   **45 mph, 40 mph**

25. Two truckers leave Atlanta at the same time heading west. How far apart will they be in 13 hr if one is traveling at a rate of 60 mph and the other at 56 mph?   **52 mi**

26. Two families leave home at 8:00 A.M., planning to meet for a picnic at a point between them. One travels at a rate of 45 mph and the other travels at 55 mph. If they live 500 miles apart, at what time do they meet?   **1:00 P.M.**

FOR REVIEW

27. The sum of three consecutive even integers is 222. What are the integers?   **72, 74, 76**

28. Five times a number, decreased by 8, is the same as four times 1 more than the number. Find the number.   **12**

29. Lucy is seven years older than her sister. In 3 years she will be twice as old as her sister is now. How old is each now?   **Lucy is 17, her sister is 10.**

30. A 36-m steel rod is cut into three pieces. The second piece is twice as long as the first and the third is three times as long as the second. How long is each piece?   **4 m, 8 m, 24 m**

31. 540 is 150% of what?   **360**

32. 98 is what percent of 140?   **70%**

33. If the cost of a calculator rose $1.92 and this was a 12% increase, what was the old price? What is the new price?   **$16.00, $17.92**

34. If the sales-tax rate is 6%, how much tax will be charged on the purchase of a wallet costing $14.50? **87¢**

35. During July a car dealer sold 45 cars out of his total inventory of 54 cars. What percent of his inventory was sold?   **$83\frac{1}{3}\%$**

36. After an 8% salary increase, Henry makes $24,300. What was his salary before the raise?   **$22,500**

11.6 EXERCISES C

Solve.

1. The area shown required 124 m of fencing to enclose. Find the dimensions by finding the value of *x*.
12 m

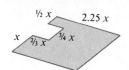

2. Pam drove her car at an average speed of 55 mph for one-half the distance of a trip and then traveled by boat at 20 mph the second half. How many miles did she travel if the total trip took 9 hours?
[Answer: 264 mi]

SUPPLEMENTARY APPLIED PROBLEMS

The following exercises use the techniques presented in the previous three sections.

1. Seven more than twice Herman's age is equal to 73. How old is Herman?   **33 yr**

2. The perimeter of Ivan Danielson's yard is 528 yd. If the yard is rectangular in shape with length 24 yd more than the width, what are the dimensions? **120 yd, 144 yd**

3. On a recent field trip, Cindy noticed that the bus was completely full and that there were 14 more girls than boys on the bus. If the bus has a capacity of 54 children, how many boys were on the bus?   **20**

4. Bob, Mary, and Sharon were born in consecutive years. If the sum of their ages is 36, how old is each?   **Bob is 11, Mary is 12, Sharon is 13.**

5 A pole is standing in a pond. If one-fourth of the height of the pole is in the sand at the bottom of the pond, 9 ft are in the water, and three-eighths of the total height is in the air above the water, what is the length of the pole?   **24 ft**

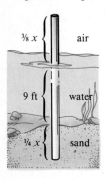

6. In an election with only two candidates, the winner received 50 votes more than twice the total for the loser. If a total of 1310 votes were cast, how many did each candidate receive?   **420, 890**

7. Four-fifths of the operating budget of a community college goes for faculty salaries. If the total of all faculty salaries is $4,000,000, what is the total operating budget?   **$5,000,000**

8. If $16.80 is charged as tax on the sale of a washing machine priced at $420, what is the tax rate? **4%**

9. A fuel tank contained 840 gallons of fuel. An additional quantity of fuel was added making a total of 1092 gallons. What was the percent increase?   **30%**

10 The measure of the second angle of a triangle is one-fourth the measure of the third and 30° more than the first. Find the measure of each.   **5°, 35°, 140°**

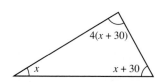

11. International Car Rental will rent a compact car for $12.50 per day plus 10¢ per mile driven. If Charlene rented a compact for two days and was billed a total of $73.00, how many miles did she drive?   **480 mi**

12. Tim Chandler bought a pair of running shoes on sale for $33.60. If the sale price was discounted 30% from the original price, what was the original price?   **$48.00**

13 A rectangular patio is twice as long as it is wide. If its length were decreased 8 ft and the width were increased 8 ft, it would be square. What are its dimensions?   **32 ft, 16 ft**

14. The relationship between °F and °C is given by the formula $F = \frac{9}{5}C + 32$. Determine °C for a temperature of 50°F.   **10°C**

15. How much simple interest will be earned on a deposit of $800 for one year at an interest rate of 12% per year?   **$96**

16. Barbara purchased a Ford Bronco for $16,500. If the options totaled $5800, what percent of the purchase price was the base price of the vehicle?   **64.8%**

17. Leah was able to do 40 more situps than Lynn on a recent fitness test. If the total for both girls was 290, how many situps did Lynn do?   **125**

18. The perimeter of a square is eleven cm less than five times the length of a side. Find the length of a side.   **11 cm**

19 The population of Deserted, Utah was 740 in 1990. This was a 20% decrease in the population of 1980. What was the population in 1980?   **925**

20 Gene won $10,000 in the Illinois Lottery. He invested part in a savings account that earned 14% simple interest, and the rest in a fund that paid 11% simple interest. If at the end of one year he earned an income of $1310 from the two, how much was invested at each rate?   **$7000 in savings, $3000 in the fund**

21. The Arnotes plan to sell their home. They must receive $82,000 after deducting the sales commission of 6% on the selling price. Rounded to the nearest dollar, at what selling price should the house be listed?   **$87,234**

22. Dr. Cotera wishes to enclose a rectangular plot of land with 140 yards of fencing in such a way that the length is four times the width. What will be the dimensions of the plot?   **14 yd, 56 yd**

23. On a recent vacation, Leona traveled one-third the total distance by car, 400 miles by boat, and one-half the total distance by air. What was the total distance traveled on the vacation?   **2400 mi**

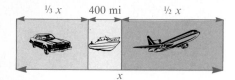

24 Becky must have an average of 90 on four tests in geology to get an A in the course. What is the lowest score she can make on the fourth test if her first three grades are 96, 78, and 91?   **95**

25. When ice floats in water, approximately $\frac{8}{9}$ of the height of the ice is below the surface of the water. If the tip of an iceberg is 35 feet above the surface of the water, what is the approximate depth of the iceberg?   **315 ft**

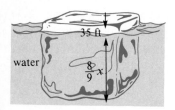

26. A boat traveled from Puerto Vallarta to an island and back again in a total time of 10.5 hours. How far is the island from Puerto Vallarta if the boat averaged 15 mph going to the island and 20 mph returning?   **90 mi**

27. In a school election, Morton beats his two opponents by 230 and 175 votes, respectively. If there were a total of 1635 votes cast, how many did Morton receive?   **680 votes**

28. A meteorite weighing 4.5 tons was discovered near Meteor Crater, Arizona, in 1918. If the meteorite contains about 87% iron, how many pounds of iron does it contain?   **7830 lb**

29. After draining their swimming pool to make several minor repairs, the Bonnetts plan to refill it using their garden hose. Suppose the hose lets water in at a rate of 20 gallons per minute and the pool holds 36,000 gallons. Will the pool be filled by noon on Sunday if the water is turned on at 9:00 A.M. Saturday?   **No. It will take 30 hours to fill the pool.**

30. To estimate the trout population in Lake Louise, 400 banded trout were released on June 1, and the lake was closed to all fishing. On June 10, a sample of 100 trout were caught, 6 of which were banded. As a result, the ranger estimated that 6% of all the trout in Lake Louise were banded. Approximately how many trout were in the lake?   **6667 trout**

ANSWERS: 1. 33 yr   2. 120 yd, 144 yd   3. 20   4. Bob is 11, Mary is 12, Sharon is 13.   5. 24 ft   6. 420, 890   7. $5,000,000   8. 4%   9. 30%   10. 5°, 35°, 140°   11. 480 mi   12. $48.00   13. 32 ft, 16 ft   14. 10°C   15. $96   16. 64.8%   17. 125   18. 11 cm   19. 925   20. $7000 in savings, $3000 in the fund   21. $87,234   22. 14 yd, 56 yd   23. 2400 mi   24. 95   25. 315 ft   26. 90 mi   27. 680 votes   28. 7830 lb   29. No. It will take 30 hours to fill the pool.   30. 6667 trout

11.7 SOLVING AND GRAPHING LINEAR INEQUALITIES

STUDENT GUIDEPOSTS

1. Graphing Equations
2. Graphing Inequalities
3. Solving Linear Inequalities
4. Addition-Subtraction Rule
5. Multiplication-Division Rule
6. Solving by a Combination of Rules
7. Method for Solving Linear Inequalities
8. Applications Involving Inequalities

Often in algebra we try to picture abstract ideas. This was done in Chapter 10, for instance, using a number line. Recall that a number line is marked off in unit lengths with each point on the line associated with a real number. The origin is identified with the number zero, positive numbers correspond to points to the right of zero, and negative numbers correspond to points to the left of zero. We can identify any real number with exactly one point on a number line, and, conversely, every point on a number line corresponds to exactly one real number.

1 GRAPHING EQUATIONS

Linear equations in one variable like those we studied earlier can be graphed on a number line by plotting the points that correspond to solutions of the equation.

EXAMPLE 1   GRAPHING SOLUTIONS TO EQUATIONS

Graph $2x + 1 = 3$.

First solve the equation.

$$2x + 1 - 1 = 3 - 1 \qquad \text{Subtract 1}$$
$$2x = 2$$
$$x = 1$$

The solution is $x = 1$. Plot the point corresponding to 1 on a number line, as in Figure 11.7.

Figure 11.7

PRACTICE EXERCISE 1

Graph $3x - 2 = 4$.

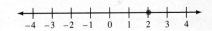

Answer: Graph the solution $x = 2$.

To graph $2x + 1 = 1 + 2x$, first solve the equation.

$$2x + 1 - 1 = 1 - 1 + 2x \qquad \text{Subtract 1}$$
$$2x = 2x$$
$$2x - 2x = 2x - 2x \qquad \text{Subtract } 2x$$
$$0 = 0$$

Since $0 = 0$ is always true, this equation is an identity and every number is a solution. Then every number must be plotted, and the graph of the solution is the entire number line shown in Figure 11.8.

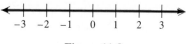

Figure 11.8

Suppose we graph $x + 1 = x$. If we solve the equation, we obtain the following result.

$$x - x + 1 = x - x \quad \text{Subtract } x$$
$$1 = 0$$

This equation is a contradiction because $1 \neq 0$. There are no solutions to the equation so we have no points to plot.

② GRAPHING INEQUALITIES

Graphing equations is a relatively simple procedure. The procedure for graphing **inequalities,** statements containing the symbols $<$, $\leq$, $>$, or $\geq$, is shown with the following figures.

In Figure 11.9, the graph of $x > 3$ has an open circle at 3. This means that the number 3 is not included among the solutions, while the colored line shows that all points to the right of 3 are included.

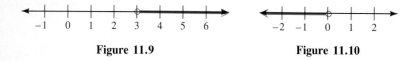

Figure 11.9   **Figure 11.10**

The graph of $x < 0$ is shown in Figure 11.10. In Figure 11.11, the graph of $x \geq -1$ has a solid circle at the point -1, indicating that -1 is included in the graph of $x \geq -1$. Finally, the graph of $x \leq 1$ is given in Figure 11.12.

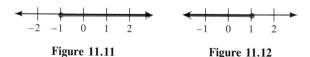

Figure 11.11   **Figure 11.12**

③ SOLVING LINEAR INEQUALITIES

Now that we know how to graph simple inequalities such as $x > 3$, $x < 0$, $x \geq -1$, and $x \leq 1$, we can start to solve and graph more complex ones. Statements containing the inequality symbols $<$, $\leq$, $>$, and $\geq$, such as

$$x + 1 < 7, \quad 2x - 1 \geq 3, \quad 3x + 1 \leq 4x - 5, \quad \text{and} \quad x + 5 > 2(x - 3)$$

are called **linear inequalities.** A **solution** to a linear inequality is a number that, when substituted for the variable, makes the inequality true. For example, -6, 0, and 5 are some of the solutions to $x + 1 < 7$ since

$$-6 + 1 < 7, \quad 0 + 1 < 7, \quad \text{and} \quad 5 + 1 < 7$$

are all true.

Solving linear inequalities is much like solving linear equations since most of the same rules apply. There is one exception that will be discussed shortly. Like equivalent equations, **equivalent inequalities** have exactly the same solutions. Solving an inequality is a matter of transforming it to an equivalent inequality for which the solution is obvious.

④ ADDITION-SUBTRACTION RULE

If we start with the true inequality

$$7 < 15$$

and add 4 to both sides, we obtain another true inequality.

$$7 + 4 < 15 + 4$$
$$11 < 19$$

Likewise, if we subtract 3 from both sides, we again obtain a true inequality.

$$7 - 3 < 15 - 3$$
$$4 < 12$$

These observations lead to the addition-subtraction rule for inequalities.

> ### Addition-Subtraction Rule
>
> An equivalent inequality is obtained if the same quantity is added to or subtracted from both sides of an inequality.
> Suppose a, b, and c are real numbers.
>
> If $a < b$ then $a + c < b + c$ and $a - c < b - c.$

We use this rule to solve inequalities in very much the same way that we used the similar addition-subtraction rule to solve equations. The key is to isolate the variable on one side of the inequality.

| **EXAMPLE 2**   ADDITION-SUBTRACTION RULE | PRACTICE EXERCISE 2 |

Solve and graph.

$$x + 1 < 7$$
$$x + 1 - 1 < 7 - 1 \quad \text{Subtract 1 from both sides}$$
$$x + 0 < 6$$
$$x < 6$$

All numbers less than 6 are solutions, and the graph is given in Figure 11.13. The fact that 6 is not part of the solution is indicated by the open circle.

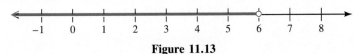

Figure 11.13

Solve and graph.

$$y + 7 > 3$$

Answer: Graph $y > -4.$

Most of the equations we have solved have had only one or two solutions. For inequalities, there are many solutions.

EXAMPLE 3 ADDITION-SUBTRACTION RULE	PRACTICE EXERCISE 3

Solve and graph.

$$x - 3 \geq 5$$
$$x - 3 + 3 \geq 5 + 3 \quad \text{Add 3 to both sides}$$
$$x + 0 \geq 8$$
$$x \geq 8$$

All numbers greater than or equal to 8 are solutions. Generally, we indicate the solution simply by writing $x \geq 8$. The graph is given in Figure 11.14. Notice that 8 is part of the solution, so the circle at the point corresponding to 8 is solid.

Figure 11.14

Solve and graph.

$$y - 15 \leq -12$$

Answer: Graph $y \leq 3$.

⑤ MULTIPLICATION-DIVISION RULE

If we start with the true inequality

$$6 < 14$$

and multiply both sides by 3, we obtain the true inequality

$$18 < 42.$$

However, if we multiply both sides by -2, we obtain the false inequality

$$-12 < -28. \qquad \text{This is false}$$

To obtain a true inequality here, we need to reverse the inequality symbol, that is, change $<$ to $>$. These observations lead to the multiplication-division rule for inequalities.

Multiplication-Division Rule

An equivalent inequality is obtained in the following situations.

1. Each side of the inequality is multiplied or divided by the same *positive* quantity.

$$\text{If } c > 0 \text{ and } a < b \text{ then } ac < bc \text{ and } \frac{a}{c} < \frac{b}{c}.$$

2. Each side of the inequality is multiplied or divided by the same *negative* quantity and the inequality symbol is reversed.

$$\text{If } c < 0 \text{ and } a < b \text{ then } ac > bc \text{ and } \frac{a}{c} > \frac{b}{c}.$$

EXAMPLE 4   MULTIPLICATION-DIVISION RULE

Solve and graph.

$$4x > 16$$

$$\frac{1}{4} \cdot 4x > \frac{1}{4} \cdot 16 \qquad \text{$\frac{1}{4}$ is positive so inequality symbol remains the same}$$

$$1 \cdot x > 4$$

$$x > 4$$

The graph is given in Figure 11.15.

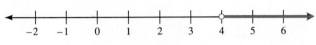

Figure 11.15

PRACTICE EXERCISE 4

Solve and graph.

$$\frac{1}{2}y < 3$$

Answer: Graph $y < 6$.

EXAMPLE 5   MULTIPLICATION-DIVISION RULE

Solve and graph.

$$-\frac{1}{2}x \le 7$$

$$\downarrow$$

$$(-2)\left(-\frac{1}{2}x\right) \ge (-2)(7) \qquad \text{Multiply both sides by -2 and } reverse \text{ the inequality symbol}$$

$$1 \cdot x \ge -14$$

$$x \ge -14$$

The graph is given in Figure 11.16.

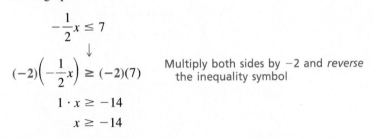

Figure 11.16

PRACTICE EXERCISE 5

Solve and graph.

$$-7y \ge -35$$

Answer: Graph $y \le 5$.

///////////// **CAUTION** ///////////

To multiply or divide by a negative number, always reverse the inequality symbol. Thus, if $-x < -5$, then $x > 5$ (not $x < 5$). This is the only substantial difference between solving an equation and solving an inequality.

///////////

6 SOLVING BY A COMBINATION OF RULES

We now solve inequalities that require both the addition-subtraction rule and the multiplication-division rule. As with solving equations, the basic idea is to isolate the variable on one side of the inequality.

EXAMPLE 6 COMBINATION OF RULES

Solve and graph.

$$2x + 5 < 7$$
$$2x + 5 - 5 < 7 - 5 \qquad \text{Subtract 5}$$
$$2x < 2$$
$$\frac{1}{2} \cdot 2x < \frac{1}{2} \cdot 2 \qquad \text{Multiply by } \frac{1}{2} \text{ or divide by 2; the inequality remains the same}$$
$$1 \cdot x < 1$$
$$x < 1$$

The graph is given in Figure 11.17.

Figure 11.17

PRACTICE EXERCISE 6

Solve and graph.

$$3y - 7 \geq 5$$

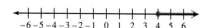

Answer: Graph $y \geq 4$.

EXAMPLE 7 COMBINATION OF RULES

Solve and graph.

$$6 - 4x \geq 2 - 3x$$
$$6 - 6 - 4x \geq 2 - 6 - 3x \qquad \text{Subtract 6}$$
$$-4x \geq -4 - 3x$$
$$-4x + 3x \geq -4 - 3x + 3x \qquad \text{Add } 3x$$
$$-x \geq -4$$
$$(-1)(-x) \leq (-1)(-4) \qquad \text{Multiply by } -1 \text{ and } \textit{reverse} \text{ inequality symbol}$$
$$x \leq 4$$

The graph is given in Figure 11.18.

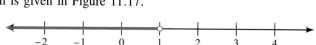

Figure 11.18

PRACTICE EXERCISE 7

Solve and graph.

$$9 - 8y < 6 - 5y$$

Answer: Graph $y > 1$.

Often an inequality involves parentheses. When this occurs, first remove all parentheses and then proceed as in previous cases.

EXAMPLE 8 INEQUALITY WITH PARENTHESES

Solve. $-6(2 + x) \geq 2(4x - 2)$

$$-12 - 6x \geq 8x - 4 \qquad \text{Remove parentheses}$$
$$-12 + 12 - 6x \geq 8x - 4 + 12 \qquad \text{Add 12}$$
$$-6x \geq 8x + 8$$
$$-6x - 8x \geq 8x - 8x + 8 \qquad \text{Subtract } 8x$$

PRACTICE EXERCISE 8

Solve and graph.

$$5(z - 3) > -3(7 - z)$$

$$-14x \geq 8$$

$$\downarrow$$

$$\left(-\frac{1}{14}\right)(-14x) \leq \left(-\frac{1}{14}\right) \cdot 8 \qquad \text{Multiply by } -\frac{1}{14} \text{ and } reverse$$
$$\text{inequality symbol}$$

$$x \leq -\frac{8}{14}$$

$$x \leq -\frac{4}{7}$$

The graph is given in Figure 11.19.

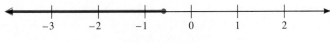

Figure 11.19

Answer: Graph $z > -3$.

⑦ METHOD FOR SOLVING LINEAR INEQUALITIES

The method for solving a linear inequality is summarized below.

To Solve a Linear Inequality
1. Simplify both sides of the inequality by clearing parentheses, fractions, and decimals.
2. Use the addition-subtraction rule to isolate all variable terms on one side and all constant terms on the other side. Collect like terms when possible.
3. Use the multiplication-division rule to obtain a variable with coefficient of 1. Be sure to reverse the inequality symbol whenever multiplying or dividing both sides by a negative number.

⑧ APPLICATIONS INVOLVING INEQUALITIES

Some applied problems result in inequalities.

EXAMPLE 9   EDUCATION APPLICATION

For Beth to get an A in her Spanish course, she must earn a total of 360 points on four tests each worth 100 points. If she got scores of 87, 96, and 91 on the first three tests, find (using an inequality) the range of scores she could make on the fourth test to get an A.

    Let s = score Beth must get on the fourth test. Since the total of her four test scores must be greater than or equal to 360, we must solve

$$87 + 96 + 91 + s \geq 360.$$
$$274 + s \geq 360$$
$$s \geq 86 \qquad \text{Subtract 274 from both sides}$$

Beth must make a score of 86 or better on the fourth test to get an A in the course.

PRACTICE EXERCISE 9

If Angie's age is tripled then increased by 5, the result is less than or equal to 25 plus her age. Angie is at most how old?

Answer: 10 yr ($a \leq 10$)

11.7 EXERCISES A

Graph the equation on a number line.

1. $2x - 3 = 1$

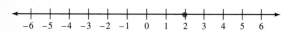

2. $x + 5 = x$ no solution

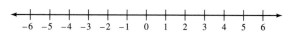

3. $x - 1 = x - 1$

4. $\dfrac{y}{\frac{1}{4}} = 20$

5. $-\dfrac{1}{10} - \dfrac{3}{5}y = \dfrac{1}{5}$

6. $a - 3 + 4a = 2 + a - 5$

7. $3(z - 1) - 2(z + 3) = -7$

8. $12 - x = 3 + 2[4 - (x - 1)]$

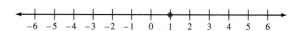

Graph the inequality on a number line.

9. $x > 2$

10. $x \leq 2$

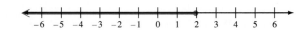

11. $y \leq -1$

12. $y > -1$

13. $2a > 10$

14. $-2a \leq -8$

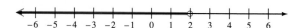

15. $1 - 3x < 8$

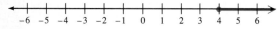

16. $2x + 3 > 5x - 3$

17. $-8y - 21 \geq 7 - 15y$

18. $25 \geq 5(3 - z) - 10$

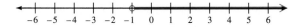

19. $4(a - 1) < 2(3a + 2)$

20. $-2a + 3 + a < 2(a + 3)$

Solve.

21. $x + 9 > 12$
$x > 3$

22. $x - 3 < 14$
$x < 17$

23. $x + 1 \geq -5$
$x \geq -6$

24. $z + 2.3 > 4.7$
$z > 2.4$

25. $3.9 > z - 5.2$
$9.1 > z$ $(z < 9.1)$

26. $z - \dfrac{3}{4} \leq \dfrac{2}{3}$
$z \leq \dfrac{17}{12}$

27 $2a + 1 > a - 3$
$a > -4$

28. $a - 4 \leq 2a + 5$
$-9 \leq a \quad (a \geq -9)$

29. $3a - 11 \geq 2a + 9$
$a \geq 20$

30. $3(b + 1) > 2b - 5$
$b > -8$

31. $3x > 12$
$x > 4$

32. $\frac{1}{3}x < -4$
$x < -12$

33 $-\frac{1}{4}y \geq 2 \quad y \leq -8$

34. $-3y < 7 \quad y > -\frac{7}{3}$

35. $2.1y > 4.2 \quad y > 2$

36. $-2 < -5z \quad \frac{2}{5} > z \quad \left(z < \frac{2}{5}\right)$

37. $-a \leq -7 \quad a \geq 7$

38. $-3.1z \leq 9.3 \quad z \geq -3$

39. $-\frac{2}{3}z > \frac{4}{3} \quad z < -2$

40. $3b \geq -\frac{1}{2} \quad b \geq -\frac{1}{6}$

41. $-3b \geq \frac{1}{2} \quad b \leq -\frac{1}{6}$

42. $1 - 3x \geq -8 \quad x \leq 3$

43. $-8y - 7 \geq 21 - 15y$
$y \geq 4$

44. $4(z - 3) \leq 3(2z + 4)$
$z \geq -12$

45 $3(2x + 3) - (3x + 2) < 12$
$x < \frac{5}{3}$

46. $5(x + 3) + 4 \geq x - 1$
$x \geq -5$

47. $y + 3 - 4y > 2(y + 12)$
$y < -\frac{21}{5}$

48. $3 - 2z < 5(z - 7)$

$z > \frac{38}{7}$

49. $0.05 + 3(z - 1.2) > 2z$
$z > 3.55$

50 $\frac{3}{4}z - \frac{3}{8} < \frac{3}{2} + \frac{1}{8}z$

$z < 3$

Are the following statements true *or* false? *If the statement is false, tell why.*

51. If $x < 9$, then $-2x < -18$.
false $(-2x > -18)$

52. If $x > 4$, then $3x > 12$.
true

53. If $x \leq 7$, then $x + 1 \leq 8$.
true

54. If $x < -1$, then $-x > 1$.
true

55. If $x < 3$, then $x - 7 > -4$.
false $(x - 7 < -4)$

56. If $-x < -10$, then $1 - x < -9$.
true

Solve.

57 The product of a number and 3 is greater than or equal to the number less 8. Find the numbers that satisfy this. $\quad x \geq -4$

58. If twice my age, increased by 7, is greater than or equal to 31 less my age, I am at least how old?
8 yr $(a > 8)$

59. Professor Packard will give an F to any student with a point total less than 180 in a course having three 100 point exams. Burford made 38 points on the first exam and 54 on the second. Determine the minimum score he could make on the third exam to avoid failing the course. **88 ($s \geq 88$)**

60 For Darrell Fosberg to win a trip to Bermuda, his new car sales must average 50 over the three-month period June, July, and August. If he sold 47 cars in June and 62 cars in July, how many cars must he sell in August to qualify for the trip?
at least 41 ($n \geq 41$)

FOR REVIEW

61. The length of a rectangular room is 5 yd more than its width. If its perimeter is 54 yd, find its dimensions. **11 yd, 16 yd**

62. Find the volume, rounded to the nearest tenth, of a sphere with radius 3 cm. [Use 3.14 for π.]
113.0 cm³

63. Roy drove 378 miles in 7 hr. What was his average speed? **54 mph**

64. Two families leave their homes at the same time planning to meet at a point between them. If one travels 55 mph, the other travels 50 mph, and they live 273 miles apart, how long will it take for them to meet? **2.6 hr**

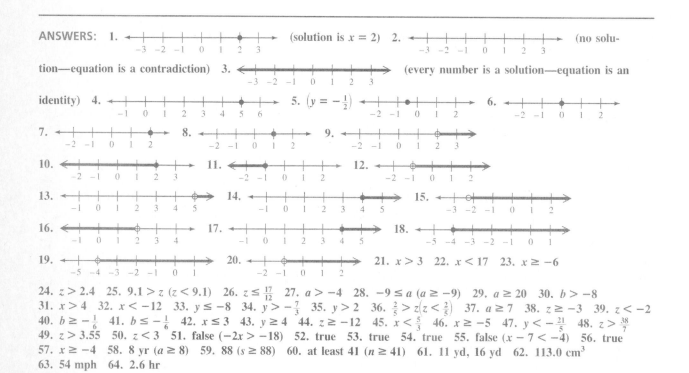

ANSWERS: **1.** (solution is $x = 2$) **2.** (no solution—equation is a contradiction) **3.** (every number is a solution—equation is an identity) **4.** **5.** ($y = -\frac{1}{2}$) **6.** **7.** **8.** **9.** **10.** **11.** **12.** **13.** **14.** **15.** **16.** **17.** **18.** **19.** **20.** **21.** $x > 3$ **22.** $x < 17$ **23.** $x \geq -6$ **24.** $z > 2.4$ **25.** $9.1 > z$ ($z < 9.1$) **26.** $z \leq \frac{17}{12}$ **27.** $a > -4$ **28.** $-9 \leq a$ ($a \geq -9$) **29.** $a \geq 20$ **30.** $b > -8$ **31.** $x > 4$ **32.** $x < -12$ **33.** $y \leq -8$ **34.** $y > -\frac{7}{3}$ **35.** $y > 2$ **36.** $\frac{2}{5} > z$ ($z < \frac{2}{5}$) **37.** $a \geq 7$ **38.** $z \geq -3$ **39.** $z < -2$ **40.** $b \geq -\frac{1}{6}$ **41.** $b \leq -\frac{1}{6}$ **42.** $x \leq 3$ **43.** $y \geq 4$ **44.** $z \geq -12$ **45.** $x < \frac{5}{3}$ **46.** $x \geq -5$ **47.** $y < -\frac{21}{5}$ **48.** $z > \frac{38}{7}$ **49.** $z > 3.55$ **50.** $z < 3$ **51.** false ($-2x > -18$) **52.** true **53.** true **54.** true **55.** false ($x - 7 < -4$) **56.** true **57.** $x \geq -4$ **58.** 8 yr ($a \geq 8$) **59.** 88 ($s \geq 88$) **60.** at least 41 ($n \geq 41$) **61.** 11 yd, 16 yd **62.** 113.0 cm³ **63.** 54 mph **64.** 2.6 hr

11.7   EXERCISES B

Graph the equation on a number line.

1. $2x - 5 = 1$

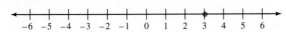

2. $4x + 3 = 4x - 3$ no solution

3. $x - 5 = x - 5$

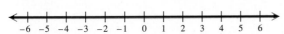

4. $\dfrac{y}{\frac{1}{3}} = 9$

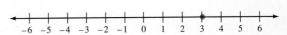

5. $-\dfrac{1}{4} - \dfrac{3}{8}y = \dfrac{1}{2}$

6. $a - 2 + 3a = 4 + a - 5$

7. $2(z - 2) - 3(z + 1) = -4$

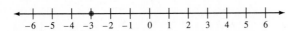

8. $7 - 2x = 3 + [2 - (x - 3)]$

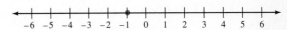

Graph the inequality on a number line.

9. $x > 1$

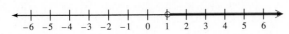

10. $x \le 1$

11. $y \le -3$

12. $y > -3$

13. $3a < 12$

14. $-3a > -9$

15. $1 - 2x < 4$

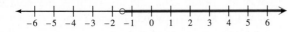

16. $3x + 1 < x - 5$

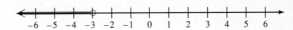

17. $-4y - 4 \ge 3 - 11y$

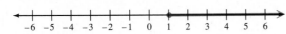

18. $3(z + 7) - 18 \le 2z$

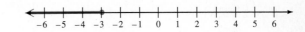

19. $5(a - 2) \ge 2(3a - 4)$

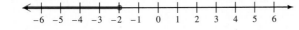

20. $-3a - 4 + 2a \ge 3(a + 2)$

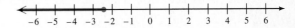

Solve.

21. $x + 3 > 2$ $x > -1$

22. $x - 4 \le 7$ $x \le 11$

23. $x + 3 \ge -4$ $x \ge -7$

24. $z + 1.9 < 3.9$ $z < 2$

25. $4.2 < z - 8.1$ $z > 12.3$

26. $z - \dfrac{2}{3} \le \dfrac{3}{4}$ $z \le \dfrac{17}{12}$

27. $a - 5 \le 2a + 3$
   $a \ge -8$

28. $2a + 11 > a - 3$
   $a > -14$

29. $4a - 12 \ge 3a - 1$
   $a \ge 11$

30. $2(b + 5) \geq b + 7$
$b \geq -3$

31. $2x > 14$
$x > 7$

32. $\frac{1}{3}x < -8$ $x < -24$

33. $-\frac{1}{3}y < 2$ $y > -6$

34. $-5y < 11$ $y > -\frac{11}{5}$

35. $2.1y \leq 6.3$ $y \leq 3$

36. $-5a > -30$ $a < 6$

37. $-a > -10$ $a < 10$

38. $-4.1z > 8.2$ $z < -2$

39. $-\frac{1}{3}z \leq \frac{2}{3}$ $z \geq -2$

40. $4b \geq -\frac{1}{5}$ $b \geq -\frac{1}{20}$

41. $-4b \geq \frac{1}{5}$ $b \leq -\frac{1}{20}$

42. $1 - 4x \geq -7$ $x \leq 2$

43. $-5y - 2 < 32 + 12y$
$y > -2$

44. $5(y - 2) > 4(2y + 1)$
$y < -\frac{14}{3}$

45. $2(3x + 1) - (5x + 4) \geq 2$
$z \geq 4$

46. $3x - 1 \leq x - (3x - 14)$
$x \leq 3$

47. $x - 2 + 2x \leq 2(x + 5)$
$x \leq 12$

48. $3 - 2(z + 1) \geq 5 - z$ $z \leq -4$

49. $0.01 + 2(z + 1.5) > 3z$
$z < 3.01$

50. $\frac{1}{4}z + \frac{3}{8} \geq \frac{1}{2} - \frac{3}{8}z$ $z \geq \frac{1}{5}$

Are the following statements true *or* false? *If the statement is false, tell why.*

51. If $x > 7$, then $-2x < -14$.   true

52. If $x < 5$, then $3x < 15$.   true

53. If $x \leq -1$, then $x + 1 \leq 0$.   true

54. If $-x \geq 1$, then $x \geq 1$.   false $(x \leq -1)$

55. If $x < 5$, then $x - 4 < 1$.   true

56. If $-x \leq -5$, then $2 - x \leq -3$.   true

Solve.

57. The product of a number and 5 is less than or equal to that number increased by 8. Find the numbers that satisfy this. $x \leq 2$

58. If Jeff's age is doubled, then diminished by 4, the result is greater than or equal to 5 plus his age. Jeff is at least how old?   9 yr $(a \geq 9)$

59. To get an A in chemistry, Paula must have at least 270 points out of a total of 300 in the course. She has made 87 points on the first test and 98 on the second. At least how many points must she get on the third and final exam to receive an A?
85 $(p \geq 85)$

60. Each team of four members in a tug-of-war can have no more than 600 lb total weight to qualify for the finals. Ron's team has three members weighing 120 lb, 140 lb, and 128 lb. What is the greatest amount Ron can weigh in order for his team to qualify?   212 lb $(w \leq 212)$

FOR REVIEW

61. The width of a rectangular table is 22 in less than its length. If its perimeter is 216 in, find its dimensions.   43 in, 65 in

62. A circular cylinder has volume 117.75 cm³ and radius 5 cm. What is the height of the cylinder? [Use 3.14 for π.]   1.5 cm

63. Martha hiked 25 miles at an average rate of 3 mph. What length of time did she hike? $8\frac{1}{3}$ hr

64. Two planes leave Atlanta at the same time, one heading east at 400 mph and the other heading west at 460 mph. How long will it take for them to be 3010 miles apart?   3.5 hr

11.7 EXERCISES C

Solve.

1. $\dfrac{x-5}{3} - \dfrac{2x+1}{2} \le x - (2x-1)$ $x \le \dfrac{19}{2}$

2. The temperature of a hot-spring bath in Mexico is advertised to stay between 35°C and 40°C. What is this range of temperature in degrees Fahrenheit?   [Answer: $95° \le F \le 104°$]

CHAPTER 11 REVIEW

KEY WORDS

11.1   An **equation** is a statement that two quantities are equal.

A **solution** to an equation is a number which makes the equation true.

A **linear equation** can be written in the form $ax + b = 0$.

A **conditional equation** is true for some re-placements of the variable and false for others.

An **identity** has the entire set of real numbers for its solution set.

A **contradiction** has no solutions.

Equivalent equations have exactly the same solutions.

11.5   **Consecutive integers** are integers which im-mediately follow each other in regular count-ing order.

A **commission** is income as a percent of total sales.

Interest is money we pay to borrow money or money earned on savings.

11.7   A **linear inequality** is a statement involving the inequality symbols $<$, $\le$, $>$, or $\ge$, such as $x + 1 < 7$, $2x - 1 \ge 3$, $3x + 1 \le 4x - 5$, and $x + 5 > 2(x - 3)$.

A **solution** to a linear inequality is a number that, when substituted for the variable, makes the inequality true.

Equivalent inequalities have exactly the same solutions.

KEY CONCEPTS

11.1   When using the addition-subtraction rule to solve an equation, add or subtract the same expression on both sides.

11.2   When using the multiplication-division rule to solve an equation, multiply by a number that makes the coefficient of x equal to 1. For ex-ample, to solve

$$\frac{x}{\frac{4}{5}} = 20$$

multiply by $\frac{4}{5}$, *not* by $\frac{5}{4}$.

11.3 1. Use the addition-subtraction rule before the multiplication-division rule when a combina-tion is necessary to solve an equation.

2. When solving an equation involving paren-theses, first use the distributive law to clear all parentheses.

11.5 1. Write out complete descriptions of the vari-ables and terms when solving applied prob-lems.

2. Two consecutive even (or odd) integers can be named x and $x + 2$. Two consecutive in-tegers can be named x and $x + 1$.

3. When solving percent-increase problems add the increase to the original amount. For ex-ample, an amount of money x plus 5% in-terest can be represented as $x + 0.05x = (1 + 0.05)x = (1.05)x$.

11.6 1. An accurate sketch is often helpful for solving a geometry problem.

    2. The formula $d = rt$, or (distance) = (rate) · (time), is used in many motion problems.

11.7 1. The same expression can be added or subtracted on both sides of an inequality.

    2. When you multiply or divide both sides of an inequality by a negative number, *reverse* the symbol of inequality.

For example, $-x < 5$
$$(-1)(-x) > (-1)(5)$$
$$x > -5.$$

3. When graphing $x \geq c$ or $x \leq c$ on a number line, the point corresponding to c is part of the graph, shown with a solid dot at c. When graphing $x > c$ or $x < c$, the point corresponding to c is *not* part of the graph, shown with an "open" dot at c.

REVIEW EXERCISES

Part I

11.1 1. Use the equation $2(x + 1) = 2x + 2$ to answer Exercises (a)–(f).

    (a) What is the variable? ____x____

    (b) What is the left side? ___$2(x + 1)$___

    (c) What is the right side? ___$2x + 2$___

    (d) What is the solution? ___every real number___

    (e) Is this an identity? ___yes___

    (f) Is this a contradiction? ___no___

Solve for x.

2. $x + 9 = 13$ 4

3. $\dfrac{2}{3} - x = \dfrac{4}{9}$ $\dfrac{2}{9}$

11.2 4. $-2.3x = 4.6$ -2

5. $\dfrac{x}{\frac{1}{3}} = 15$ 5

11.3 6. $2x + 3 = x + 3 - 4x$ 0

7. $x - \dfrac{1}{3} = -8x$ $\dfrac{1}{27}$

8. $3(x - 4) + 5 = 2(x + 1) - 3$ 6

9. $5(2x - 2) - 3(x - 4) = 0$ $-\dfrac{2}{7}$

10. $x - (x + 3) = x - 6$ 3

11. $20 - 3(x + 5) = 0$ $\dfrac{5}{3}$

11.4 *Letting x represent the unknown number, translate the following into symbols.*

12. A number increased by 5 is 7.
$x + 5 = 7$

13. Four times a number is 23.
$4x = 23$

14. Twice a number, increased by 7, is −12.
$2x + 7 = -12$

15. When 5 is added to six times a number, the result is the same as 3 less than the number. $6x + 5 = x - 3$

11.5 16. Fred is 12 years older than Bertha. Twice the sum of their ages is 100. How old is each?
Fred is 31, Bertha is 19.

17. The sum of three consecutive even integers is 138. Find the integers.   44, 46, 48

18. If the sales-tax rate is 4%, how much tax would be charged on a purchase of $420?   **$16.80**

19. 30% of what is 198?   660

20. 14% of 50 is what?   7

21. What percent of 900 is 585?   65%

22. The price of an item rose 15% last year. If the present price is $41.40, what was the price last year? $36

23. What sum of money invested at 12% simple interest will amount to $716.80 in 1 year?
$640

11.6 24. The diameter of a circle is 7.56 cm. What is the radius?   3.78 cm

25. The area of a rectangle is 25.2 ft^2 and the length is 6 ft. What is the width?   4.2 ft

26. The perimeter of a rectangle is 80 ft. If the length is 10 ft more than the width, what are the dimensions?   15 ft, 25 ft

27. Two trains leave the same city, one heading north and the other south. If one train is moving 5 mph faster than the other, and if after 4 hr they are 556 mi apart, how fast is each traveling?   72 mph, 67 mph

Graph the equation or inequality on a number line.

11.7 28. $3x + 1 = 10$

29. $\dfrac{y}{\frac{1}{8}} = 16$

30. $4 - z = 8 - [1 - (z - 3)]$

31. $x > 3$

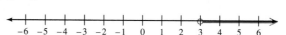

32. $-2a \le -2$

33. $3(y - 1) \le 2(y - 2)$

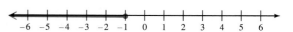

Solve.

34. $x - 3 > 4$ $x > 7$

35. $y + 2 \le -5$ $y \le -7$

36. In solving an inequality, when is the symbol of inequality reversed?
when both sides are multiplied or divided by a negative number

Solve.

37. $\dfrac{1}{5}x \le 35$ $x \le 175$

38. $-4y > 24$ $y < -6$

39. $6 - 2x > -4 + 3x$ $x < 2$

40. $5 + x > 3 - (x + 10)$
$x > -6$

41. $3(x + 2) \le 5x - 4$
$x \ge 5$

42. $2 - (x - 2) < 7(x - 3)$
$x > \dfrac{25}{8}$

Part II

Solve.

43. After receiving a 20% discount on the selling price, Holly paid $6.60 for a record. What was the price of the record before the discount?
$8.25

44. A sphere with radius 7 cm is submerged in a rectangular tank 20 cm wide and 25 cm long. How much will the level of the water rise in the tank?   **approximately 2.9 cm**

45. $-y = -4.7$ 4.7

46. $1\frac{1}{2} = z + 2$ $-\frac{1}{2}$

47. $3(x - 2) = 5 - (x + 1)$ $\frac{5}{2}$

48. $6x + 1 < 2(x - 3)$ $x < -\frac{7}{4}$

49. $-3x \geq \frac{6}{5}$ $x \leq -\frac{2}{5}$

50. $4 - 2(x - 1) \leq 6 - (x - 1)$
$x \geq -1$

Answer true *or* false. *If the statement is false, tell why.*

51. A statement that two quantities are equal is called an inequality.   false

52. The equation $x + 5 = 5 + x$ is an example of an identity.   true

53. The equation $x + 5 = x - 5$ is an example of a conditional equation.   false

54. Two equations which have exactly the same solutions are called equivalent.   true

Graph each inequality.

55. $x - 2 \leq -5$

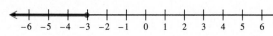

56. $-\frac{1}{4}y > 1$

57. $2 - 3(x + 1) < x + 3$

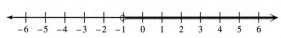

58. $3x - 6 \geq 4 - (x - 6)$

ANSWERS: 1. (a) x (b) $2(x + 1)$ (c) $2x + 2$ (d) every real number (e) yes (f) no 2. 4 3. $\frac{2}{9}$ 4. -2 5. 5
6. 0 7. $\frac{1}{27}$ 8. 6 9. $-\frac{2}{7}$ 10. 3 11. $\frac{5}{3}$ 12. $x + 5 = 7$ 13. $4x = 23$ 14. $2x + 7 = -12$ 15. $6x + 5 = x - 3$
16. Fred is 31, Bertha is 19. 17. 44, 46, 48 18. \$16.80 19. 660 20. 7 21. 65% 22. \$36 23. \$640
24. 3.78 cm 25. 4.2 ft 26. 15 ft, 25 ft 27. 72 mph, 67 mph 28.
29. 30. 31.
32. 33. 34. $x > 7$ 35. $y \leq -7$ 36. when both sides are multi-
plied or divided by a negative number 37. $x \leq 175$ 38. $y < -6$ 39. $x < 2$ 40. $x > -6$ 41. $x \geq 5$ 42. $x > \frac{25}{8}$
43. \$8.25 44. approximately 2.9 cm 45. 4.7 46. $-\frac{1}{2}$ 47. $\frac{5}{2}$ 48. $x < -\frac{7}{4}$ 49. $x \leq -\frac{2}{5}$ 50. $x \geq -1$ 51. false
(equality) 52. true 53. false (contradiction) 54. true 55.
56. 57. 58.

Solve.

1. $x - 5 = 10$

1. _____ 15 _____

2. $4x = 32$

2. _____ 8 _____

3. $x + \dfrac{3}{4} = \dfrac{5}{4}$

3. _____ $\dfrac{1}{2}$ _____

4. $5.1x = -10.2$

4. _____ -2 _____

5. $\dfrac{1}{4}x = 9$

5. _____ 36 _____

6. $\dfrac{x}{\frac{1}{5}} = 20$

6. _____ 4 _____

7. $3x - 5 = 8x + 10$

7. _____ -3 _____

8. $3(x + 2) - 5(x - 4) = 0$

8. _____ 13 _____

9. $4x - (x + 6) = 3$

9. _____ 3 _____

10. The sum of two consecutive odd integers is 168. What are the integers?

10. _____ 83, 85 _____

11. The price of a dress was $54 but the price was increased by 15%. What is the new price?

11. _____ $62.10 _____

12. The area of a triangle is 240 cm^2 and its height is 12 cm. What is the length of the base?

12. _____ 40 cm _____

13. Two trains that are 840 miles apart and whose speeds differ by 9 mph are traveling towards each other. If they will meet in 8 hours, at what speed is each traveling?

13. _____ 48 mph, 57 mph _____

Solve the inequalities.

14. $3x - 6 \leq 9x + 12$

14. _____ $x \geq -3$ _____

15. $3 - (2x - 5) > 6x + 4$

15. _____ $x < \dfrac{1}{2}$ _____

16. Graph the equation $2x + 5 = 9$ on the number line.

16.

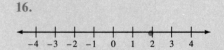

17. Graph the inequality $2(y + 1) \geq 1 - (y + 5)$ on the number line.

17.

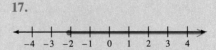

Graphing

12.1 THE RECTANGULAR COORDINATE SYSTEM

1 Rectangular Coordinate System   3 Quadrants

2 Ordered Pairs   4 Linear Equations in Two Variables

In Section 11.7 we graphed linear equations and inequalities in one variable by plotting points on a number line. Now we develop a system in which equations and inequalities in two variables can be graphed. Consider a horizontal number line and a vertical number line as shown in Figure 12.1.

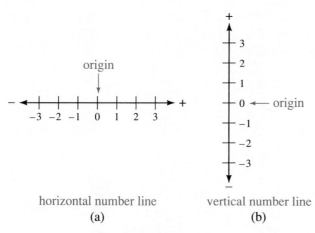

horizontal number line
(a)

vertical number line
(b)

Figure 12.1

1 RECTANGULAR COORDINATE SYSTEM

When the horizontal number line and the vertical number line are placed together so that the two origins coincide and the lines are perpendicular, as in Figure 12.2, the result is called a **rectangular** or **Cartesian coordinate system** (named after French mathematician René Descartes), or a **coordinate plane** (a plane is a flat surface that extends infinitely far in all directions).

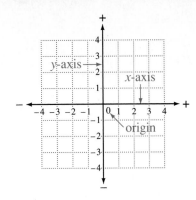

Figure 12.2   Rectangular or Cartesian Coordinate System

The horizontal number line is called the **x-axis,** and the vertical number line is called the **y-axis.** Collectively the number lines are referred to as **axes,** and the point of intersection of the lines is the **origin.**

❷ ORDERED PAIRS

Just as there is one and only one point on a number line associated with each number, there is one and only one point in a plane associated with each **ordered pair** of numbers. For example, the ordered pair (3, 2) is identified with a point in a coordinate plane as follows:

The first number, 3, the **x-coordinate** of the point, is associated with a value on the x-axis.

The second number, 2, the **y-coordinate,** is associated with a value on the y-axis.

The ordered pair (3, 2), is identified with the point where the vertical line through 3 on the x-axis intersects the horizontal line through 2 on the y-axis. See Figure 12.3. Note that the point (2, 3) is different from (3, 2).

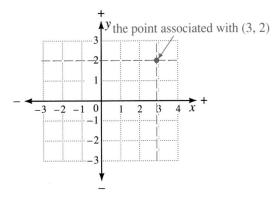

Figure 12.3   Plotting the Point (3, 2)

EXAMPLE 1 PLOTTING POINTS	PRACTICE EXERCISE 1

The points associated with (1, 3), (−2, 1), (−3, −2), and (2, −2) are given in the coordinate plane in Figure 12.4. For (1, 3) we go 1 unit right and 3 units up. We find (−2, 1) by going 2 units left and 1 unit up.

Indicate the points associated with (−1, 2), (3, −2), (−1, −1), and (1, 2) in the given coordinate plane.

(−3, −2) is 3 units left and 2 units down. (2, −2) is 2 units right and 2 units down.

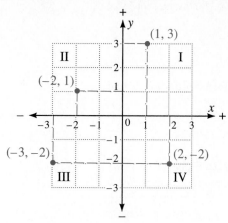

Figure 12.4

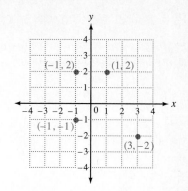

Answer: For (−1, 2) go 1 unit left and 2 units up; for (3, −2) go 3 units right and 2 down; for (−1, −1) go 1 unit left and 1 down; for (1, 2) go 1 unit right and 2 up.

③ QUADRANTS

The axes in a rectangular coordinate system divide the plane into four sections called **quadrants.** The first, second, third, and fourth quadrants are identified by the Roman numerals I, II, III, and IV in Figure 12.4. The *x*-coordinate (first) and the *y*-coordinate (second) have the following signs in each quadrant:

$$\text{I: } (+, +), \quad \text{II: } (−, +), \quad \text{III: } (−, −), \quad \text{IV: } (+, −)$$

We often use (x, y) to refer to a general ordered pair of numbers. The point P in the plane associated with the pair (x, y) has x-coordinate x and y-coordinate y. We plot a point P when we identify it with a given pair of numbers in a plane, and we often refer to "the point (x, y)" or write $P(x, y)$.

EXAMPLE 2   DETERMINING ORDERED PAIRS

The points A, B, C, D, E, F, G, and H in Figure 12.5 have coordinates $A(4, 1)$, $B(0, 2)$, $C(0, 0)$, $D(−2, 1)$, $E(−4, −2)$, $F(0, −3)$, $G(1, −2)$, and $H(3, 0)$.

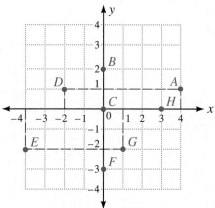

Figure 12.5

PRACTICE EXERCISE 2

Give the coordinates of each point.

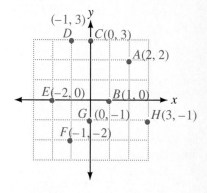

Answer: $A(2, 2)$, $B(1, 0)$, $C(0, 3)$, $D(−1, 3)$, $E(−2, 0)$, $F(−1, −2)$, $G(0, −1)$, $H(3, −1)$

❹ LINEAR EQUATIONS IN TWO VARIABLES

In Chapter 11 we solved linear equations in one variable. Such equations can always be written in the form

$$ax + b = 0. \qquad a, b \text{ constants and } a \neq 0$$

We now consider linear equations in two variables. A **linear equation in two variables** x and y, is an equation of the form

$$ax + by = c. \qquad a, b, c \text{ constants, } a \text{ and } b \text{ not both zero.}$$

A **solution** to a linear equation in two variables is an ordered pair of numbers which when substituted for the variables results in a true equation. In an *ordered pair* the x-value is always written first and the y-value is always written second.

We can show that the ordered pairs $(1, 3)$ and $(10, 0)$ are two solutions to the linear equation $x + 3y = 10$.

For the ordered pair $(1, 3)$, we must substitute 1 for x and 3 for y in the equation.

$$x + 3y = 10$$
$$1 + 3(3) = 10 \qquad x = 1 \text{ and } y = 3$$
$$1 + 9 = 10$$
$$10 = 10 \qquad \text{True}$$

Thus, $(1, 3)$ is a solution.

For the ordered pair $(10, 0)$, we substitute 10 for x and 0 for y.

$$x + 3y = 10$$
$$10 + 3(0) = 10 \qquad x = 10 \text{ and } y = 0$$
$$10 + 0 = 10$$
$$10 = 10 \qquad \text{True}$$

Thus, $(10, 0)$ is a solution. You might verify that $(4, 2)$ and $\left(0, \frac{10}{3}\right)$ are also solutions to this equation. However, $(9, 0)$ is *not* a solution.

$$x + 3y = 10$$
$$9 + 3(0) \overset{?}{=} 10 \qquad x = 9 \text{ and } y = 0$$
$$9 + 0 \overset{?}{=} 10$$
$$9 \neq 10 \qquad \text{False}$$

EXAMPLE 3 COMPLETING ORDERED PAIRS

Given the equation $3x + 2y = 6$, complete the ordered pairs so that they are solutions to the equation.

$$(0, \quad), \quad (\quad, 0), \quad (1, \quad), \quad (\quad, 1), \quad (-2, \quad)$$

PRACTICE EXERCISE 3

Given the equation $4x - 3y = 12$, complete the ordered pairs so that they are solutions to the equation.

$$(0, \quad), (\quad, 0), (2, \quad), (\quad, -2),$$
$$(-3, \quad)$$

To complete the ordered pair (0,   ), substitute 0 for x in $3x + 2y = 6$ and solve for y.

$$3(0) + 2y = 6$$

$$2y = 6$$

$$y = 3$$

Thus, the completed ordered pair is (0, 3).

To complete the ordered pair (, 0), substitute 0 for y in $3x + 2y = 6$ and solve for x.

$$3x + 2(0) = 6$$

$$3x = 6$$

$$x = 2$$

The ordered pair is (2, 0).

To complete the ordered pair (1,   ), substitute 1 for x and solve for y.

$$3(1) + 2y = 6$$

$$3 + 2y = 6$$

$$2y = 3$$

$$y = \frac{3}{2}$$

The completed ordered pair is $\left(1, \frac{3}{2}\right)$.

Similarly, substitute 1 for y and solve for x to complete (, 1), obtaining $\left(\frac{4}{3}, 1\right)$, and to complete (−2,   ), substitute −2 for x and solve to obtain (−2, 6).

Answer: $(0, -4)$, $(3, 0)$, $\left(2, -\frac{4}{3}\right)$, $\left(\frac{3}{2}, -2\right)$, $(-3, -8)$

EXAMPLE 4   APPLICATION TO BUSINESS

Mr. Paducci has a small business that manufactures wood-burning stoves. He has found that the cost, y, in dollars of producing a certain number, x, of stoves is given by the equation

$$y = 150x + 80.$$

Find the cost to produce 1, 2, and 5 stoves.

Complete the ordered pairs (1,   ), (2,   ), and (5,   ). Let $x = 1$ and solve for y.

$$y = 150(1) + 80$$

$$y = 230$$

Thus, (1,   ) becomes (1, 230). Similarly, (2,   ) becomes (2, 380) and (5,   ) becomes (5, 830). This means that it costs Mr. Paducci $230 to produce 1 stove, $380 to produce 2 stoves, and $830 to produce 5 stoves.

PRACTICE EXERCISE 4

Laura Hayes manufactures small appliances. She uses the equation $y = 45x + 60$ to find the cost, y, of making x appliances. Find the cost to produce 2 appliances, 7 appliances, and 20 appliances.

Answer: $150, $375, $960

12.1 EXERCISES A

1. Plot the points associated with the given pairs of numbers: $A(2, 4)$, $B(4, -1)$, $C(-3, 4)$, $D(-3, 0)$, $E(2, 0)$, $F(-2, -2)$, $G(1, -4)$, and $H(4, -4)$.

2. Give the coordinates of the points A, B, C, D, E, F, G, and H.
   $A(0, 0)$, $B(3, 1)$, $C(4, 0)$, $D(0, 3)$, $E(-3, 2)$, $F(-2, 0)$, $G(-4, -3)$, $H(1, -2)$

3. Plot the points associated with the given pairs of numbers: $M\left(\frac{1}{2}, 2\right)$, $N\left(-\frac{3}{2}, 3\right)$, $P\left(-2, -\frac{3}{4}\right)$, and $Q\left(3, -\frac{5}{2}\right)$.

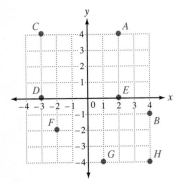
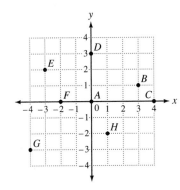
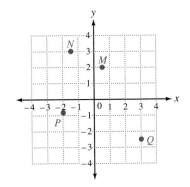

4. In which quadrant are the points M, N, P, and Q of Exercise 3 located?
   M:I; N:II; P:III; Q:IV

Give the quadrant in which each of the following is located.

5. $(-8, -8)$ III

6. $(6, -3)$ IV

7. $(4, 12)$ I

8. $(-2, 7)$ II

9. What are the four regions called into which a plane is separated by a Cartesian coordinate system?
   quadrants

10. What is the x-coordinate of the point named by the ordered pair $(-2, 5)$? -2

11. What is the y-coordinate of the point named by the ordered pair $(3, -8)$? -8

12. What are the coordinates of the origin? $(0, 0)$

13. What is another name for the horizontal axis in a Cartesian coordinate system?   *x*-axis

14. If the first coordinate of a point is positive and the second is negative, in which quadrant is the point located?   IV

15. If the x-coordinate of a point is negative and the y-coordinate is positive, in which quadrant is the point located?   II

Using the given equation, complete each ordered pair so that it will be a solution to the equation.

16. $x - y = 2$
    (a) $(0, -2)$ (b) $(2, 0)$
    (c) $(4, 2)$ (d) $(-1, -3)$

17 $x - 5 = 0$
   x cannot be 0
   (a) $(0, \quad)$ (b) $(5, 0)$
   (c) $(5, \quad)$ (d) $(5, -10)$
   any number

18. $x + y = 5$
    (a) $(0, 5)$ **(b)** $(5, 0)$
    (c) $(3, 2)$ **(d)** $(9, -4)$

19. $2x - y = 4$
    (a) $(0, -4)$ **(b)** $(2, 0)$
    (c) $(-2, -8)$ **(d)** $(5, 6)$

20. $x + 5y = 10$
    (a) $(0, 2)$ **(b)** $(10, 0)$
    (c) $(-10, 4)$ **(d)** $(-5, 3)$

21. $5x - 2y = 10$
    (a) $(0, -5)$ **(b)** $(2, 0)$
    (c) $(7, \dfrac{25}{2})$ **(d)** $(\dfrac{2}{5}, -4)$

22 The distance, y, traveled by a car averaging 55 mph over x hours is given by the equation

$$y = 55x.$$

Complete the ordered pairs $(1, \quad)$, $(5, \quad)$, and $(10, \quad)$. How far does the car travel **(a)** in 1 hour? **(b)** in 5 hours? **(c)** in 10 hours?   **(1, 55), (5, 275), (10, 550)**
(a) 55 miles (b) 275 miles (c) 550 miles

23. Draw the triangle with vertices $(3, 4)$, $(-3, 1)$, and $(1, -2)$.

24. Draw the rectangle with corners $(3, 1)$, $(1, 3)$, $(-3, -1)$, and $(-1, -3)$.

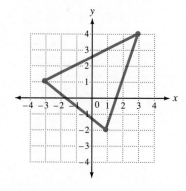

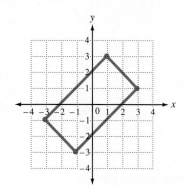

FOR REVIEW

Graph the equation on a number line.

25. $a + 3 = 3(a - 1)$

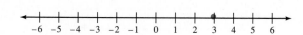

26. $3x + 8 = 8 + 3x$

Graph the inequality on a number line.

27. $2x \geq 7$

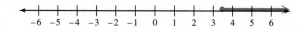

28. $3(y - 1) < 4 - (y - 1)$

ANSWERS:

1.

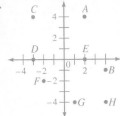

2. A(0, 0),
   B(3, 1),
   C(4, 0),
   D(0, 3),
   E(−3, 2),
   F(−2, 0),
   G(−4, −3),
   H(1, −2).

3.

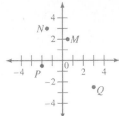

4. M:I, N:II, P:III, Q:IV   5. III   6. IV   7. I
8. II   9. quadrants   10. −2   11. −8
12. (0, 0)   13. x-axis   14. IV   15. II
16. (a) (0, −2) (b) (2, 0) (c) (4, 2)
(d) (−1, −3)   17. (a) x cannot be 0 (b) (5, 0)
(c) (5, any number) (d) (5, −10)
18. (a) (0, 5) (b) (5, 0) (c) (3, 2) (d) (9, −4)
19. (a) (0, −4) (b) (2, 0) (c) (−2, −8) (d) (5, 6)   20. (a) (0, 2) (b) (10, 0) (c) (−10, 4) (d) (−5, 3)   21. (a) (0, −5)
(b) (2, 0) (c) $\left(7, \frac{25}{2}\right)$ (d) $\left(\frac{2}{5}, -4\right)$ 22. (1, 55), (5, 275), (10, 550) (a) 55 miles (b) 275 miles (c) 550 miles

23.

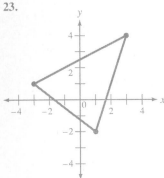

24.

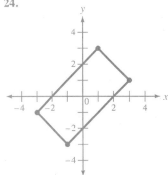

25.

26.

27.

28.

12.1   EXERCISES B

1. Plot the points associated with the given pairs of numbers:
   A(1, 3), B(2, −1), C(−3, 4), D(−1, 1), E(−3, −3), F(2, −3).

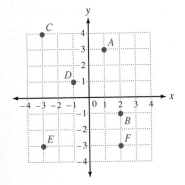

2. Give the coordinates of the points A, B, C, D, E, and F in the figure below.
   A(3, 0),   B(0, −4),   C(1, 4)
   D(−2, 3), E(−1, 0), F(−4, −2)

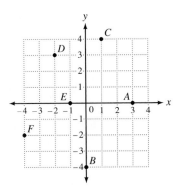

3. Plot the points associated with the given pairs of numbers:
   $M\left(-\frac{1}{2}, 1\right), N\left(\frac{3}{2}, 2\right), P\left(2, -\frac{3}{4}\right),$ and $Q\left(-3, -\frac{5}{2}\right).$

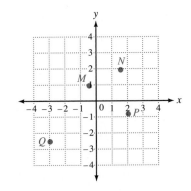

4. In which quadrants are the points M, N, P, and Q of Exercise 3 located?   **M:II; N:I; P:IV; Q:III**

Give the quadrant in which each of the following is located.

5. $(4, -1)$ **IV**   **6.** $(-4, 3)$ **II**   **7.** $(6, 6)$ **I**   **8.** $(-12, -4)$ **III**

9. What is another name for a Cartesian coordinate system?
   x, y-coordinate system, rectangular coordinate system, coordinate plane

10. What is the y-coordinate of the point named by the ordered pair $(-2, 5)$?   **5**

11. What is the x-coordinate of the point named by the ordered pair $(3, -8)$?   **3**

12. What is the name of the point with coordinates $(0, 0)$?   **origin**

13. What is another name for the vertical axis in a Cartesian coordinate system?   **y-axis**

14. If the first coordinate of a point is negative and the second is positive, in which quadrant is the point located?   **II**

15. If the x-coordinate of a point is negative and the y-coordinate is also negative, in which quadrant is the point located?   **III**

Using the given equation, complete each ordered pair so that it will be a solution to the equation.

16. $x + y = 2$
   (a) $(0, 2)$ **(b)** $(2, 0)$
   (c) $(3, -1)$ **(d)** $(4, -2)$

17. $y + 1 = 0$ **y cannot be 0**
   (a) $(0, -1)$ **(b)** $(\ , 0)$
   (c) $(3, -1)$ **(d)** $(\ , -1)$
   any number

18. $x - y = 3$
   (a) $(0, -3)$ **(b)** $(3, 0)$
   (c) $(2, -1)$ **(d)** $(-1, -4)$

19. $2x + y = 4$
   (a) $(0, 4)$ **(b)** $(2, 0)$
   (c) $(-2, 8)$ **(d)** $(-1, 6)$

20. $x - 5y = 10$
   (a) $(0, -2)$ **(b)** $(10, 0)$
   (c) $(-10, -4)$ **(d)** $(25, 3)$

21. $5x + 2y = 10$
   (a) $(0, 5)$ **(b)** $(2, 0)$
   (c) $(7, -\dfrac{25}{2})$ **(d)** $(\dfrac{18}{5}, -4)$

22. The cost, y, in dollars of producing a number, x, of items has been estimated by the equation

$$y = 300x + 15.$$

Complete the ordered pairs $(1, \ )$, $(3, \ )$, and $(10, \ )$. Give the cost of producing   **(a)** 1 item,   **(b)** 3 items,   **(c)** 10 items.   **$(1, 315)$, $(3, 915)$, $(10, 3015)$, (a) \$315 (b) \$915 (c) \$3015**

23. Draw the triangle with vertices $(-2, 7)$, $(1, 1)$, and $(-4, -5)$.

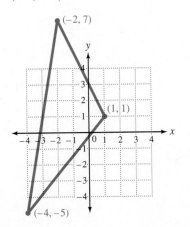

24. Draw the rectangle with corners $(-4, 2)$, $(-1, 5)$, $(5, -1)$, and $(2, -4)$.

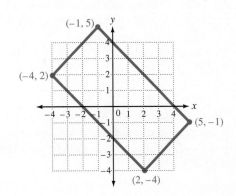

FOR REVIEW

Graph the equation on a number line.

25. $2a + 1 = 3(a - 2)$

26. $3x - 8 = 3x + 8$ no solution

Graph the inequality on a number line.

27. $3x < -5$

28. $2(y - 3) \geq 3 - (y - 6)$

12.1 EXERCISES C

Using the given equation, complete each ordered pair so that it will be a solution to the equation. A calculator might be helpful.

1. $0.012x - 1.565y = 1.878$
   (a) $(0, \quad)$ **(b)** $(156.5, 0)$
   (c) $(234.75, \quad)$ **(d)** $(469.5, 2.4)$
   [Answers: **(a)** $(0, -1.2)$; **(c)** $(234.75, 0.6)$]

2. $-\dfrac{3}{16}x + \dfrac{7}{12}y = 3\dfrac{5}{8}$

   (a) $(0, \dfrac{87}{14})$ **(b)** $(-\dfrac{58}{3}, 0)$

   (c) $(-\dfrac{5}{9}, \dfrac{169}{28})$ **(d)** $(-\dfrac{134}{9}, 1\dfrac{3}{7})$

12.2 GRAPHING LINEAR EQUATIONS

STUDENT GUIDEPOSTS

1 Graphing Equations

2 General Form of a Linear Equation

3 Graphing Using Intercepts

1 GRAPHING EQUATIONS

The **graph** of an equation in two variables x and y is the set of points in a Cartesian coordinate system that corresponds to solutions of the equation. Since there are usually infinitely many solutions, we cannot find and plot each pair. Generally, we plot enough points to see a pattern, and then connect these points with a line or curve to graph the equation.

An excellent way to display a collection of ordered-pair solutions to an equation such as

$$y = 2x + 5$$

is to make a table of values. Choose several values for x and substitute these values into the equation to compute the corresponding value for y. Place each y-value beside the x-value used to calculate it. (Calculations are usually done mentally or as scratch work.)

Substitution	Results in $y = 2x + 5$	x	y
$x = 0$	$y = 2(0) + 5 = 5$	0	5
$x = 1$	$y = 2(1) + 5 = 7$	1	7
$x = -1$	$y = 2(-1) + 5 = 3$	-1	3
$x = 2$	$y = 2(2) + 5 = 9$	2	9
$x = -2$	$y = 2(-2) + 5 = 1$	-2	1
$x = 3$	$y = 2(3) + 5 = 11$	3	11
$x = -3$	$y = 2(-3) + 5 = -1$	-3	-1

This table lists seven (of the infinitely many) solutions to the equation $y = 2x + 5$.

$$(0, 5), \quad (1, 7), \quad (-1, 3), \quad (2, 9), \quad (-2, 1), \quad (3, 11), \quad (-3, -1)$$

Now plot the points that correspond to these ordered-pair solutions in a rectangular coordinate system, as in Figure 12.6.

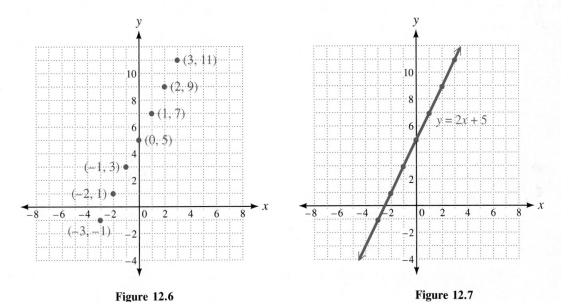

Figure 12.6   **Figure 12.7**

It appears that all seven points lie on a straight line. It is reasonable to assume that the graph of this equation is the straight line passing through these seven points, as in Figure 12.7.

To Graph an Equation in the Two Variables x and y

1. Make a table of values. These values represent the ordered-pair solutions of the equation.
2. Plot the points that correspond to the ordered-pair solutions in a Cartesian coordinate system.
3. Connect the points with a line or curve.

| EXAMPLE 1 GRAPHING BY PLOTTING POINTS | PRACTICE EXERCISE 1 |

Graph $y + x = 3$.

Before making a table, it is often helpful to solve the equation for one of the variables (usually for y):

$$y = -x + 3.$$

To make a table of values, choose several values for x and substitute each into the equation to compute the corresponding value of y. See the table at the side. Plot the points that correspond to the ordered pairs from the table:

$(0, 3), \quad (1, 2), \quad (-1, 4), \quad (2, 1), \quad (-2, 5), \quad (3, 0), \quad (-3, 6).$

Connecting these points gives us a straight line, as in Figure 12.8.

Graph $2x + y = -2$.

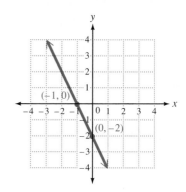

x	y
0	3
1	2
-1	4
2	1
-2	5
3	0
-3	6

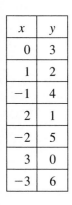

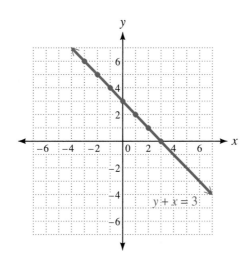

Figure 12.8

Answer: Straight line passing through $(0, -2)$ and $(-1, 0)$.

In the first part of this chapter, we graphed an equation in one variable on a number line. Usually, for graphing purposes, an equation in one variable such as

$$x = 5 \quad \text{or} \quad y + 2 = 0$$

is thought of as an equation in two variables with the coefficient of the missing variable equal to zero. That is,

$$x = 5 \text{ is the same as } x + 0 \cdot y = 5$$

and $\quad y + 2 = 0$ is the same as $0 \cdot x + y + 2 = 0$.

With this in mind, such equations can be graphed in a Cartesian coordinate system.

EXAMPLE 2   LINES PARALLEL TO THE AXES

PRACTICE EXERCISE 2

(a) Graph $x = 5$ in a rectangular coordinate system.

Solutions to this equation always have an x-coordinate of 5 and can have any number as y-coordinate. For example,

$$(5, 0), \quad (5, -1), \quad (5, 1), \quad (5, 2)$$

are all solutions, since $x = 5$ is the same as $x + 0 \cdot y = 5$, and we know

$$5 + 0 \cdot (\text{any number}) = 5 + 0 = 5.$$

Plot the points from the table and draw a line through them. The graph of $x = 5$ is a straight line parallel to the y-axis, 5 units to the right of the y-axis. See Figure 12.9.

(a) Graph $x + 3 = 0$.

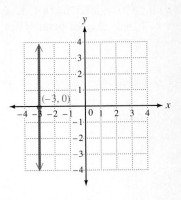

x	y
5	0
5	1
5	−1
5	2
5	−2
5	3
5	−3

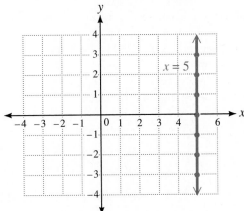

Figure 12.9

(b) Graph $y + 2 = 0$ in a Cartesian coordinate system.

We can write this equation as $y = -2$. Thus, solutions to this equation always have a y-coordinate of -2 and can have any number as x-coordinate. For example,

$$(0, -2), \quad (1, -2), \quad (-1, -2), \quad (2, -2)$$

are all solutions since $y + 2 = 0$ is the same as $0 \cdot x + y + 2 = 0$, and we know that

$$0 \cdot (\text{any number}) + (-2) + 2 = -2 + 2 = 0.$$

(b) Graph $y = \dfrac{5}{2}$.

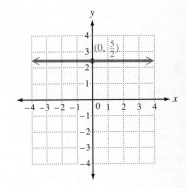

x	y
0	−2
1	−2
−1	−2
2	−2
−2	−2
3	−2
−3	−2

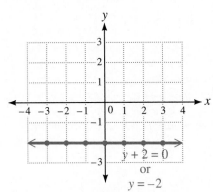

Figure 12.10

Plot the points given in the table and draw a line through them. The graph of $y + 2 = 0$ ($y = -2$) is a straight line parallel to the x-axis, 2 units below the x-axis. (See Figure 12.10).

Answers: (a) a straight line parallel to the y-axis, 3 units to the left of the y-axis   (b) a straight line parallel to the x-axis, $\frac{5}{2}$ units above the x-axis

If we are asked to graph an equation such as $x = 5$ or $y + 2 = 0$, we must first know whether it is considered an equation in one variable or an equation in two variables having a zero coefficient on the missing variable. In the first case, the graph would be a point on a number line, while in the second, the graph would be a straight line in a rectangular coordinate system. From this point on, we will look only at equations in two variables, and all graphs will be in the coordinate plane.

❷ GENERAL FORM OF A LINEAR EQUATION

In graphing we can sometimes minimize the number of points that we plot by knowing the nature of the graph. An equation of the form

$$ax + by + c = 0$$

is called a **linear equation in two variables.** The numbers a, b, and c are constant real numbers, a and b not both zero, and this form of the equation is called the **general form.** The graph of a linear equation is always a straight **line** and can be determined by plotting just two points. Linear equations are also called **first-degree equations** since the variables are raised to the first power only. The following table provides practice at identifying linear equations.

Equation	Nature	General form
$2x + y = 7$	linear	$2x + y - 7 = 0$
$x = -y - 8$	linear	$x + y + 8 = 0$
$2y = 3x - 5$	linear	$3x - 2y - 5 = 0$
$x + 5 = 0$	linear	$x + 0 \cdot y + 5 = 0$
$2y = 3$	linear	$0 \cdot x + 2y - 3 = 0$
$y = x^2 + 3$	not linear (x to second power)	
$x - y^3 + 7 = 0$	not linear (y to third power)	
$y = \frac{5}{x}$	not linear (cannot be put in the form $ax + by + c = 0$)	

❸ GRAPHING USING INTERCEPTS

Suppose we are given a linear equation such as

$$2x - 3y = 6.$$

Knowing that the graph of a linear equation is a straight line and that only two points are needed to determine a straight line, our work graphing linear equations can be shortened considerably. Rather than making a table of values that includes many solutions, we need only two solutions. In most instances, the two pairs that are easiest to find are the **intercepts.** The point at which a line crosses the x-axis, where y is zero, is the **x-intercept.** The point at which a line crosses the y-axis, where x is zero, is the **y-intercept.** To find the intercepts, fill in the following table.

x	y
0	
	0

The x-intercept, which is a point on the x-axis, has y-coordinate 0 while the y-intercept, which is a point on the y-axis, has x-coordinate 0. We substitute 0 for x in $2x - 3y = 6$ and solve for y to find the y-intercept.

$$2(0) - 3y = 6$$
$$0 - 3y = 6$$
$$-3y = 6$$
$$y = -2$$

Find the x-intercept by substituting 0 for y in $2x - 3y = 6$ and solving.

$$2x - 3(0) = 6$$
$$2x = 6$$
$$x = 3$$

The completed table

x	y
0	-2
3	0

displays the y-intercept $(0, -2)$ and the x-intercept $(3, 0)$. Plot these two inter-cepts and draw the straight line through them for the graph of $2x - 3y = 6$ in Figure 12.11. We can check our work by showing that $(-3, -4)$ satisfies the equation and is on the graph in Figure 12.11.

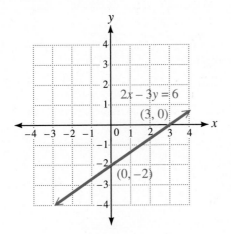

Figure 12.11

To Graph a Linear Equation $ax + by + c = 0$

1. If $a \neq 0$ and $b \neq 0$, find the x- and y-intercepts, plot them, and draw the line through them. If both intercepts are $(0, 0)$, find and plot another point.
2. If $a = 0$, $y = $ a constant and the graph is a line parallel to the x-axis.
3. If $b = 0$, $x = $ a constant and the graph is a line parallel to the y-axis.

EXAMPLE 3 GRAPHING USING INTERCEPTS

PRACTICE EXERCISE 3

(a) Graph $3x + 4y = 12$.

First find the x- and y-intercepts by completing the following table.

x	y
0	
	0

When $x = 0$, $4y = 12$, so that $y = 3$. When $y = 0$, $3x = 12$, so that $x = 4$. The completed table,

x	y
0	3
4	0

shows that the x-intercept is (4, 0) and the y-intercept is (0, 3). Plot (0, 3) and (4, 0) and connect the points to obtain the graph. See Figure 12.12.

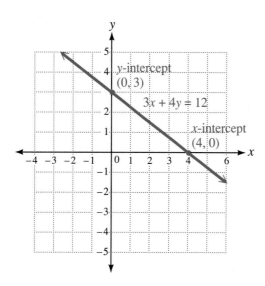

Figure 12.12

(b) Graph $x + 3y = 0$.

First find the x- and y-intercepts by completing the table.

x	y
0	
	0

(a) Graph $-6x + y = 6$.

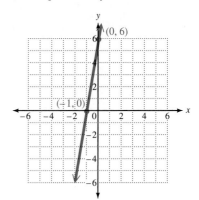

(b) Graph $4x - 3y = 0$.

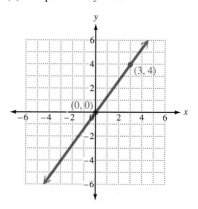

This time both intercepts are (0, 0), so we need to find another point on the line. When $x = 3$, $3 + 3y = 0$, so that $y = -1$. This leads to the table below.

x	y
0	0
3	−1

Plot (0, 0) and (3, −1) and connect the points to obtain the graph in Figure 12.13.

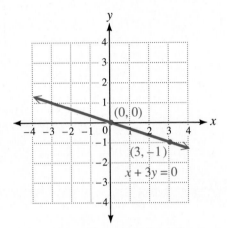

Figure 12.13

Answers: (a) straight line through x-intercept (−1, 0) and y-intercept (0, 6)   (b) straight line through x- and y-intercept (0, 0) and the point (3, 4)

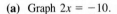

EXAMPLE 4   VERTICAL AND HORIZONTAL LINES

PRACTICE EXERCISE 4

(a) Graph $x = 2$.

In this case, with no y term, the solution will be (2, y) for any number y. Thus, the graph is the line through the x-intercept (2, 0) parallel to the y-axis, as in Figure 12.14.

(a) Graph $2x = -10$.

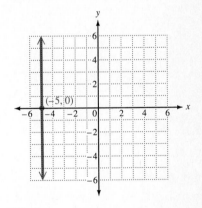

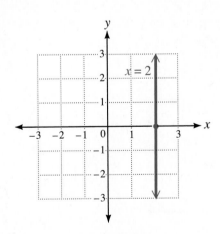

Figure 12.14

(b) Graph $2y = -2$.

The equation can be simplified to $y = -1$. In this case, with no x term, the solutions will be $(x, -1)$ for any number x. Thus, the graph is the line through the y-intercept $(0, -1)$ parallel to the x-axis, as shown in Figure 12.15.

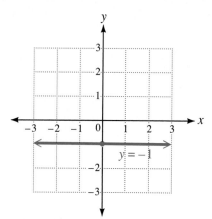

Figure 12.15

(b) Graph $y = -\dfrac{7}{2}$.

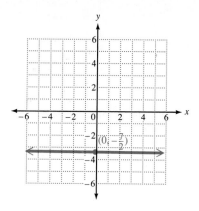

Answers: **(a)** **line through x-intercept $(-5, 0)$ parallel to the y-axis** **(b)** **line through y-intercept $\left(0, -\frac{7}{2}\right)$ parallel to the x-axis**

12.2 EXERCISES A

Each of the following is an equation in the two variables x and y. Make a table of values and graph each equation in a rectangular coordinate system.

1. $y = x + 2$

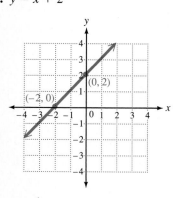

2. $x - y = 1$

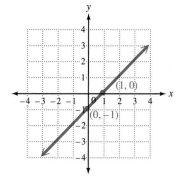

3. $y = 3x + 1$

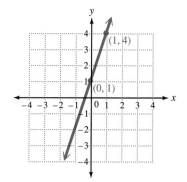

4. $y = \dfrac{1}{2}x - 1$

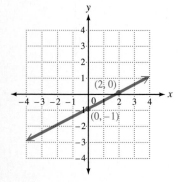

5. $y = 2$

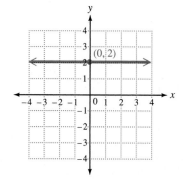

6. $x = -1$

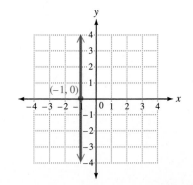

7. Which of the following are linear equations? Explain.

(a) $2x + y = -4$
linear

(b) $x - y^3 = 5$
not linear

(c) $2xy = 7$
not linear

(d) $\dfrac{3}{x} + y = 7$
not linear

(e) $x^2 + y^2 = 5$
not linear

(f) $x = y - 8$
linear

8. Why is a linear equation also called a first-degree equation?
the variables appear only to the first power

9. What is an equation of the form $ax + by + c = 0$ ($a \neq 0$ or $b \neq 0$) called?
the general form of a linear equation

10. An equation of the type $x = c$ (c a constant) has as its graph a line parallel to which axis?
y-axis

11. An equation of the type $y = c$ (c a constant) has as its graph a line parallel to which axis?
x-axis

12. What is a point where a graph crosses the x-axis called?
an x-intercept

13. What is a point where a graph crosses the y-axis called?
a y-intercept

14. To graph a general linear equation, only two points are necessary. What are the best points to use?
intercepts

Find the intercepts and graph the following.

15. $y + 2x - 4 = 0$

16. $3y - 2x = 12$

17. $y + x = 0$

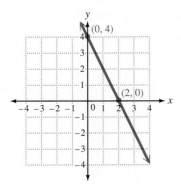

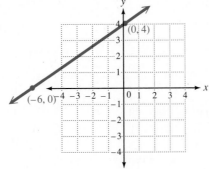

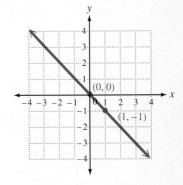

18. $y = 3$ **19.** $2x - 1 = 0$ **20.** $y + 2x = 0$

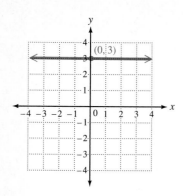 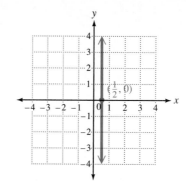 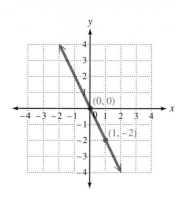

21. What are the intercepts of the line $x = 0$? What is the graph?
   x-intercept: $(0, 0)$; y-intercept: every point on the y-axis; the graph is the y-axis

FOR REVIEW

Using the given equation, complete each ordered pair so that it will be a solution to the equation.

22. $5x - 3y = 15$
   (a) $(0, -5)$ **(b)** $(3, 0)$
   (c) $(\frac{12}{5}, -1)$ **(d)** $(-1, \frac{-20}{3})$

23. $y = 3$
   (a) $(1, 3)$ **(b)** $(-1, 3)$
   (c) $(5, 3)$ **(d)** $(-3, 3)$

24. The annual salary, y, of a salesman is given in terms of his total sales, x, by the equation

$$y = \$10,000 + (0.1)x.$$

Complete the ordered pairs $(10,000, \ \ )$, $(20,000, \ \ )$, and $(50,000, \ \ )$. How much does the salesman earn if he sells **(a)** 10,000 items, **(b)** 20,000 items, **(c)** 50,000 items?
   (a) \$11,000 (b) \$12,000 (c) \$15,000

25. Find the area of the rectangle with corners $(4, -3)$, $(4, 5)$, $(-3, -3)$, and $(-3, 5)$.   **56 square units**

The following exercises will help you prepare for the next section. Starting with the point $A(1, 2)$, give the coordinates of point B.

26. B is 2 units up from A and then 1 unit to the right. $(2, 4)$

27. B is 2 units down from A and then 3 units to the right. $(4, 0)$

28. B is 3 units up from A and then 2 units to the left. $(-1, 5)$

29. B is 3 units down from A and then 4 units to the left. $(-3, -1)$

ANSWERS:

1.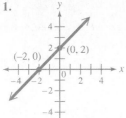

2.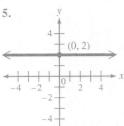

3.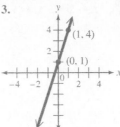

4.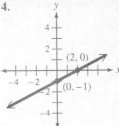

5.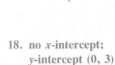

6.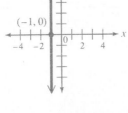

7. (a) and (f) are the only linear equations
8. the variables appear only to the first power
9. the general form of a linear equation
10. y-axis
11. x-axis
12. an x-intercept
13. a y-intercept
14. intercepts

15. x-intercept (2, 0);
    y-intercept (0, 4)

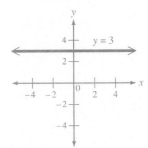

16. x-intercept (−6, 0);
    y-intercept (0, 4)

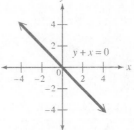

17. x-intercept (0, 0);
    y-intercept (0, 0);
    another point is (1, −1)

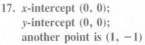

18. no x-intercept;
    y-intercept (0, 3)

19. x-intercept $\left(\frac{1}{2}, 0\right)$;
    no y-intercept

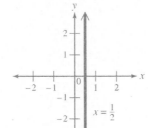

20. x-intercept (0, 0);
    y-intercept (0, 0);
    another point is (1, −2)

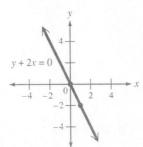

21. x-intercept (0, 0); y-intercept; every point on the y-axis; the graph is the y-axis   22. (a) (0, −5) (b) (3, 0)
(c) $\left(\frac{12}{5}, -1\right)$ (d) $\left(-1, \frac{-20}{3}\right)$ 23. (a) (1, 3) (b) (−1, 3) (c) (5, 3) (d) (−3, 3)   24. (10,000, 11,000),
(20,000, 12,000), (50,000, 15,000) (a) $11,000 (b) $12,000 (c) $15,000   25. 56 square units   26. (2, 4)   27. (4, 0)
28. (−1, 5)   29. (−3, −1)

12.2 EXERCISES B

Each of the following is an equation in the two variables x and y. Make a table of values and graph each equation in a rectangular coordinate system.

1. $y = x + 1$

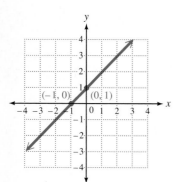

2. $y = 3x - 1$

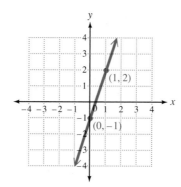

3. $y = \frac{1}{2}x + 1$

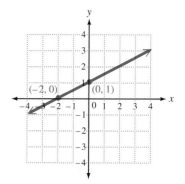

4. $x + y = -1$

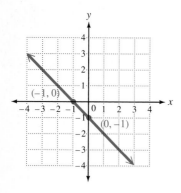

5. $x = 4$

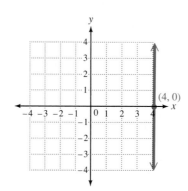

6. $y = -2$

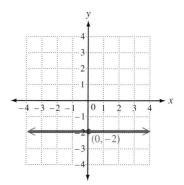

7. Which of the following are linear equations? Explain.

   (a) $3x - y = 8$
   linear

   (b) $x + y^2 = 7$
   not linear

   (c) $3xy = -1$
   not linear

   (d) $\frac{5}{y} + x = 3$
   not linear

   (e) $x^2 + y^2 = 16$
   not linear

   (f) $x = y - 3$
   linear

8. Why is a first-degree equation also called a linear equation?
   graph is a straight line

9. Give the general form of a linear equation.
$ax + by + c = 0$, a **and** b **not both zero**

10. Give the general form of the equation of a line parallel to the x-axis. $y = c$

11. Give the general form of the equation of a line parallel to the y-axis. $x = c$

12. The x-intercept of the graph of a line is a point on which axis? x**-axis**

13. The y-intercept of the graph of a line is a point on which axis? y**-axis**

14. In general, what are the two best points to use when graphing a linear equation?   **the intercepts**

Find the intercepts and graph the following.

15. $y - 2x + 4 = 0$

16. $3y + 2x = -12$

17. $2y + x = 0$

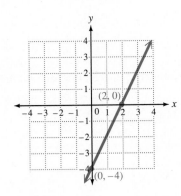
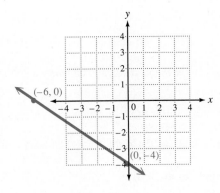

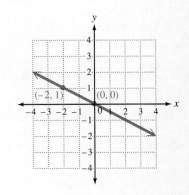

18. $y = -4$

19. $2x + 3 = 0$

20. $x - 3y = 0$

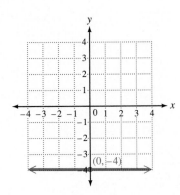
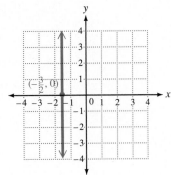
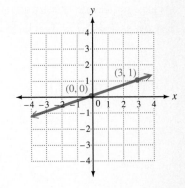

21. What are the intercepts of the line $y = 0$? What is the graph?
   x**-intercept: every point on the** x**-axis;** y**-intercept (0, 0); the graph is the** x**-axis**

FOR REVIEW

Using the given equation, complete each ordered pair so that it will be a solution to the equation.

22. $x - 3y = 6$

    (a) $(0, -2)$ **(b)** $(6, 0)$

    (c) $(3, -1)$ **(d)** $(-1, -\frac{7}{3})$

23. $x = -1$

    (a) $(-1, 1)$ **(b)** $(-1, -1)$

    (c) $(-1, 5)$ **(d)** $(-1, -3)$

24. The number, y, of deer that can live in a forest preserve is related to the number of acres, x, in the preserve. If this relationship is given by the equation

$$y = 0.5x + 3$$

complete the ordered pairs $(20, \ )$, $(100, \ )$, and $(200, \ )$. How many deer can the preserve sustain if it contains **(a)** 20 acres? **(b)** 100 acres? **(c)** 200 acres?

(a) 13 deer **(b) 53 deer** **(c) 103 deer**

25. Find the area of the triangle with vertices $(1, 2)$, $(-3, -2)$, and $(5, -2)$.   **16 square units**

The following exercises will help you prepare for the next section. Starting with the point A $(-3, 5)$ give the coordinates of point B.

26. B is 4 units up from A and then 3 units to the right.

    (0, 9)

27. B is 4 units down from A and then 2 units to the right.

    (−1, 1)

28. B is 1 unit up from A and then 4 units to the left.

    (−7, 6)

29. B is 6 units down from A and then 1 unit to the left.

    (−4, −1)

12.2 EXERCISES C

Graph.

1. $0.05x - 0.02y = 0.10$

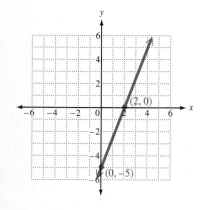

2. $\dfrac{2}{5}x + \dfrac{3}{10}y = \dfrac{6}{5}$

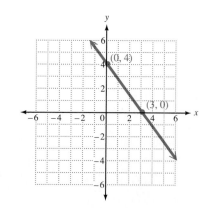

12.3 SLOPE OF A LINE

━━━━━━━ STUDENT GUIDEPOSTS ━━━━━━━

1 Definition of Slope   **3** Parallel Lines

2 Nature of the Slope of a Line   **4** Perpendicular Lines

1 DEFINITION OF SLOPE

The graph of a linear equation may be horizontal (parallel to the x-axis), may be vertical (parallel to the y-axis), may ''slope'' upward from lower to upper right, or may ''slope'' downward from upper left to lower right. The way that a line slopes or does not slope can be precisely defined.

The formal definition of slope uses any two points on the line. The coordinates of the points are written (x_1, y_1) and (x_2, y_2) where the subscript distinguishes between the two while identifying the x- and y-coordinates. We read x_1 as ''x-sub-one'' and y_1 as ''y-sub-one,'' for example. The *slope* of the line passing through (x_1, y_1) and (x_2, y_2) is defined as the ratio of the vertical change, *rise*, to the horizontal change, *run*, as we move from (x_1, y_1) to (x_2, y_2) along the line. See Figure 12.16. This ratio will be the same for any two points on the line.

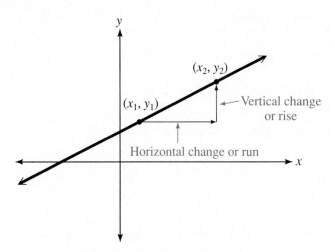

Figure 12.16   Rise and Run of a Line

We can define the rise and run as numbers by using the coordinates of the points. The ratio of these numerical values gives us a precise definition of slope. Refer to Figure 12.17.

The Slope of a Line

Let $P(x_1, y_1)$ and $Q(x_2, y_2)$ be two points on a nonvertical line. The **slope** of the line, denoted m, is given by the equation

$$m = \frac{y_2 - y_1}{x_2 - x_1} = \frac{\text{change in } y\text{-coordinates}}{\text{change in } x\text{-coordinates}} = \frac{\text{rise}}{\text{run}}.$$

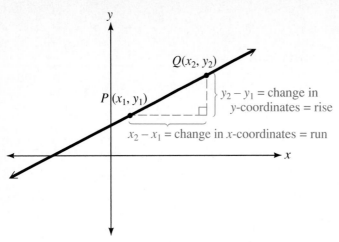

Figure 12.17 Slope $= \dfrac{y_2 - y_1}{x_2 - x_1} = \dfrac{\text{rise}}{\text{run}}$

| **EXAMPLE 1 SLOPE OF A LINE** | **PRACTICE EXERCISE 1** |

Find the slope of the line passing through the given points.

(a) (4, 3) and (1, 2).

See the graph in Figure 12.18. Suppose we identify point $P(x_1, y_1)$ with (1, 2) and point $Q(x_2, y_2)$ with (4, 3). The slope will then be given by

$$m = \frac{y_2 - y_1}{x_2 - x_1} = \frac{3 - 2}{4 - 1} = \frac{1}{3}.$$

What happens if we identify $P(x_1, y_1)$ with (4, 3) and, $Q(x_2, y_2)$ with (1, 2)? In this case we have

$$m = \frac{y_2 - y_1}{x_2 - x_1} = \frac{2 - 3}{1 - 4} = \frac{-1}{-3} = \frac{1}{3}.$$

Thus, we see that *the slope is the same regardless of how the two points are identified.*

Find the slope of the line passing through the given points.

(a) (5, 2) and (3, 7)

(b) (6, −3) and (−1, 8)

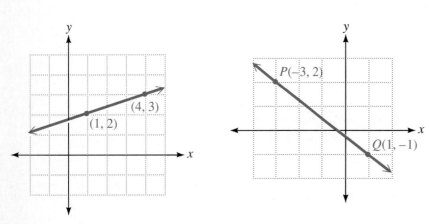

Figure 12.18   **Figure 12.19**

(b) $(-3, 2)$ and $(1, -1)$.

Let us identify $P(x_1, y_1)$ with $(-3, 2)$ and $Q(x_2, y_2)$ with $(1, -1)$ as in Figure 12.19. The slope is given by

$$m = \frac{y_2 - y_1}{x_2 - x_1} = \frac{(-1) - (2)}{(1) - (-3)} \qquad \text{Watch signs}$$

$$= \frac{-3}{1 + 3} = -\frac{3}{4}.$$

Answers: (a) $-\frac{5}{2}$ (b) $-\frac{11}{7}$

⚠⚠⚠⚠⚠⚠⚠ CAUTION ⚠⚠⚠⚠⚠⚠⚠⚠

Make sure that the coordinates are subtracted in the same order. *Do not compute*

$$\frac{y_2 - y_1}{x_1 - x_2}. \qquad \text{This is wrong}$$

EXAMPLE 2   ZERO AND UNDEFINED SLOPE

Find the slope of the line passing through the given points

(a) $(3, 2)$ and $(-1, 2)$.

Identifying $P(x_1, y_1)$ with $(3, 2)$ and $Q(x_2, y_2)$ with $(-1, 2)$ in Figure 12.20, we obtain

$$m = \frac{y_2 - y_1}{x_2 - x_1} = \frac{2 - 2}{-1 - 3} = \frac{0}{-4} = 0.$$

PRACTICE EXERCISE 2

Find the slope of the line passing through the given points.

(a) $(-3, -4)$ and $(5, -4)$

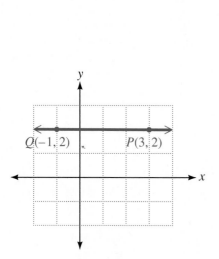

Figure 12.20

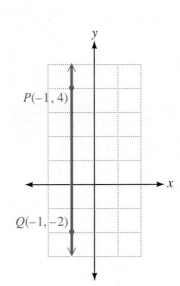

Figure 12.21

(b) $(-1, 4)$ and $(-1, -2)$.

Identifying $P(x_1, y_1)$ with $(-1, 4)$ and $Q(x_2, y_2)$ with $(-1, -2)$ in Figure 12.21, we obtain

$$m = \frac{y_2 - y_1}{x_2 - x_1} = \frac{-2 - 4}{-1 - (-1)} = \frac{-6}{-1 + 1} = \frac{-6}{0}, \text{ which is undefined.}$$

In this case we say that the slope of the line is undefined.

(b) $(3, -3)$ and $(3, 0)$

Answers: (a) 0   (b) undefined slope

② NATURE OF THE SLOPE OF A LINE

The nature of the slope of a line is summarized below.

Summary of the Slope of a Line

1. A line which slopes from lower left to upper right has a **positive slope**.
2. A line which slopes from upper left to lower right has a **negative slope**.
3. A horizontal line (parallel to the x-axis) has **zero slope**.
4. A vertical line (parallel to the y-axis) has **undefined slope**.

The graphs in Figure 12.22 show these four cases.

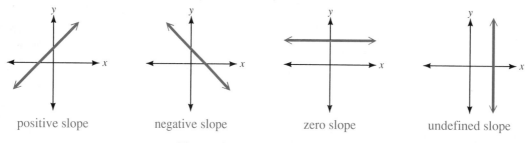

positive slope   negative slope   zero slope   undefined slope

Figure 12.22   Summary of Slopes

③ PARALLEL LINES

Slope can be used to determine when two lines are **parallel** (never intersect).

Parallel Lines

If two nonvertical lines have slopes m_1 and m_2 and $m_1 = m_2$, then the lines are parallel. (Equal slopes determine parallel lines).

EXAMPLE 3   PARALLEL LINES

Verify that the line l_1 through $(1, 8)$ and $(-2, -1)$ and the line l_2 through $(2, 4)$ and $(-1, -5)$ in Figure 12.23 are parallel.

The slope of l_1 is $m_1 = \dfrac{8 - (-1)}{1 - (-2)} = \dfrac{8 + 1}{1 + 2} = \dfrac{9}{3} = 3.$

PRACTICE EXERCISE 3

Is the line l_1 through $(4, -6)$ and $(-3, -1)$ parallel to the line l_2 through $(-2, 5)$ and $(7, -1)$?

The slope of l_2 is $m_2 = \dfrac{4 - (-5)}{2 - (-1)} = \dfrac{4 + 5}{2 + 1} = \dfrac{9}{3} = 3.$

Since $m_1 = m_2$, the lines are parallel.

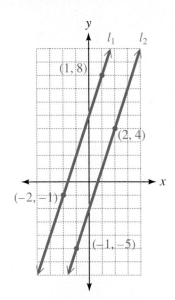

Figure 12.23

Answer: $m_1 = -\frac{5}{7}$, $m_2 = -\frac{2}{3}$, no

④ PERPENDICULAR LINES

Two lines are **perpendicular** if they intersect at right (90°) angles. The slopes of perpendicular lines have the following property.

Perpendicular Lines

If two nonvertical lines have slopes m_1 and m_2 and $m_1 m_2 = -1$, then the lines are perpendicular.

EXAMPLE 4   PERPENDICULAR LINES

Verify that the line l_1 through $(-1, 3)$ and $(2, -1)$ and the line l_2 through $(2, 1)$ and $(-2, -2)$ in Figure 12.24 are perpendicular.

The slope of l_1 is $m_1 = \dfrac{3 - (-1)}{-1 - 2} = \dfrac{3 + 1}{-3} = -\dfrac{4}{3}$

The slope of l_2 is $m_2 = \dfrac{1 - (-2)}{2 - (-2)} = \dfrac{1 + 2}{2 + 2} = \dfrac{3}{4}.$

Since $m_1 m_2 = \left(-\frac{4}{3}\right)\left(\frac{3}{4}\right) = -1$, l_1 and l_2 are perpendicular.

PRACTICE EXERCISE 4

Verify that the line l_1 through $(5, -2)$ and $(6, 7)$ is perpendicular to the line l_2 through $(-1, -3)$ and $(8, -4)$.

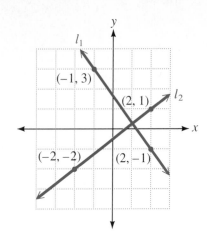

Figure 12.24 Perpendicular Lines

Answer: $m_1 = 9$, $m_2 = -\frac{1}{9}$, $m_1 m_2 = 9\left(-\frac{1}{9}\right) = -1$

12.3 EXERCISES A

Find the slope of the line passing through the given pair of points.

1. (3, 5) and (1, 3)   **1**

2. (−2, 3) and (−4, 1)   **1**

3. (−7, 1) and (3, 9) $\dfrac{4}{5}$

4. (2, 7) and (2, −3)   **undefined**

5. (3, 4) and (−1, 4)   **0**

6. (0, 2) and (5, 0) $-\dfrac{2}{5}$

Answer Exercises 7–10 with one of the following phrases: (**a**) *positive slope* (**b**) *negative slope* (**c**) *zero slope* (**d**) *undefined slope.*

7. A line parallel to the *y*-axis has ___**undefined slope**___.

8. A line that slopes from lower left to upper right has ___**positive slope**___.

9. A line perpendicular to the *y*-axis has ___**zero slope**___.

10. The *x*-axis has ___**zero slope**___.

In Exercises 11–13, m_1 and m_2 represent the slopes of two lines. State whether the lines are parallel, perpendicular, or neither.

11. $m_1 = -3$ and $m_2 = -3$
parallel

12. $m_1 = -2$ and $m_2 = \frac{1}{2}$
perpendicular

13. $m_1 = \frac{1}{5}$ and $m_2 = 5$
neither

14 Do the three points (2, 3), (0, 2), and (−2, 1) all lie on the same straight line? Explain using slopes.
Yes. The slope of the line through (2, 3) and (0, 2) is $\frac{1}{2}$, the same as the slope of the line through (0, 2) and (−2, 1).

Find the slope of the given line by first finding two points on the line then using the definition of slope.

15. $4x - y + 7 = 0$ **4**

16 $5x + 1 = 0$
undefined slope

17. $2 - y = 0$ **0**

18. $3x + 5y = 0$ $-\dfrac{3}{5}$

19. $x + y = 5$ -1

20. $2x - 3y + 1 = 0$ $\dfrac{2}{3}$

21. Verify that the line l_1 through $(-2, -1)$ and $(1, 5)$ and the line l_2 through $(4, 3)$ and $(-1, -7)$ are parallel.
They are parallel since both have slope 2.

22. Verify that the line l_1 through $(-1, 3)$ and $(2, 4)$ and the line l_2 through $(6, -1)$ and $(5, 2)$ are perpendicular.
They are perpendicular since $m_1 = \frac{1}{3}$, $m_2 = -3$, and $m_1 m_2 = -1$.

FOR REVIEW

Give the intercepts and graph the following.

23. $x - 6y = 2$

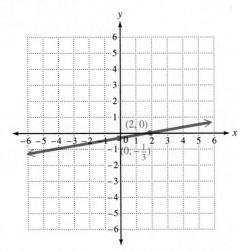

24. $5x + y = 0$

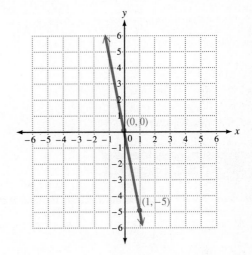

25. $x = -5$

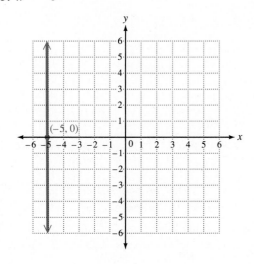

26. $y = 6$

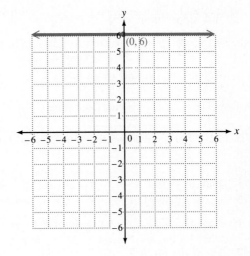

27. Is $\dfrac{3}{x} + \dfrac{2}{y} = 5$ a linear equation? Explain.
No. It cannot be written in the form $ax + by + c = 0$.

ANSWERS: 1. 1 2. 1 3. $\frac{4}{5}$ 4. undefined slope 5. 0 6. $-\frac{2}{5}$ 7. undefined slope 8. positive slope 9. zero slope 10. zero slope 11. parallel 12. perpendicular 13. neither 14. Yes. The slope of the line through (2, 3) and (0, 2) is $\frac{1}{2}$, the same as the slope of the line through (0, 2) and (−2, 1). 15. 4 16. undefined slope 17. 0 18. $-\frac{3}{5}$ 19. −1 20. $\frac{2}{3}$ 21. They are parallel since both have slope 2. 22. They are perpendicular since $m_1 = \frac{1}{3}$, $m_2 = -3$, and $m_1 m_2 = -1$.

23. *x*-intercept (2, 0);   24. *x*-intercept (0, 0);   25. *x*-intercept (−5, 0);   26. no *x*-intercept;
     y-intercept $\left(0, -\frac{1}{3}\right)$ *y*-intercept (0, 0);   no *y*-intercept   *y*-intercept (0, 6)
                          another point is (1, −5);

 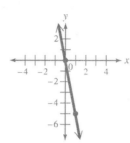 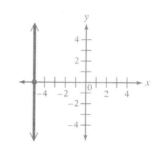 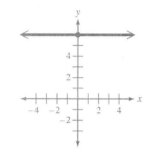

27. No. It cannot be written in the form $ax + by + c = 0$.

12.3 EXERCISES B

Find the slope of the line passing through the given pairs of points.

1. (4, 8) and (10, 2)   −1   **2.** (−2, 1) and (−6, 5)   −1   **3.** (6, 1) and (6, −5)   undefined

4. (4, −1) and (−2, −1)   0   **5.** (−3, −1) and (−2, −3)   −2   **6.** (0, 1) and (−2, 0) $\frac{1}{2}$

Answer Exercises 7–10 with one of the following phrases: (a) positive slope (b) negative slope (c) zero slope (d) undefined slope.

7. A line parallel to the *x*-axis has ___?___.   zero slope

8. A line that slopes from upper left to lower right has ___?___.   negative slope

9. A line perpendicular to the *x*-axis has ___?___.   undefined slope

10. The *y*-axis has ___?___.   undefined slope

In Exercises 11–13, m_1 and m_2 represent the slopes of two lines. State whether the lines are parallel, perpendicular, or neither.

11. $m_1 = \frac{1}{4}$ and $m_2 = \frac{1}{4}$ **12.** $m_1 = \frac{1}{5}$ and $m_2 = -5$ **13.** $m_1 = \frac{3}{2}$ and $m_2 = \frac{2}{3}$
    parallel   perpendicular   neither

14. Do the three points (1, 2), (−1, −1), and (3, 6) all lie on the same straight line? Explain using slopes.
    No. The slope of the line through (−1, −1) and (1, 2) is $\frac{3}{2}$, but the slope of the line through (1, 2) and (3, 6) is 2.

Find the slope of the given line by first finding two points on the line then using the definition of slope.

15. $x + 4y - 9 = 0$ $-\dfrac{1}{4}$ **16.** $3y - 7 = 0$ 0 **17.** $4 + x = 0$ undefined

18. $2x - 7y = 0$ $\dfrac{2}{7}$ **19.** $y - x = 3$ 1 **20.** $3x - 5y + 2 = 0$ $\dfrac{3}{5}$

21. Verify that the line l_1 through $(4, -2)$ and $(1, 1)$ and the line l_2 through $(3, 3)$ and $(7, -1)$ are parallel.
They are parallel since both have slope -1.

22. Verify that the line l_1 through $(-2, 4)$ and $(2, 5)$ and the line l_2 through $(3, 2)$ and $(4, -2)$ are perpendicular.   They are perpendicular since $m_1 = \frac{1}{4}$, $m_2 = -4$, and $m_1 m_2 = -1$.

FOR REVIEW

Give the intercepts and graph the following.

23. $6x + y = 3$

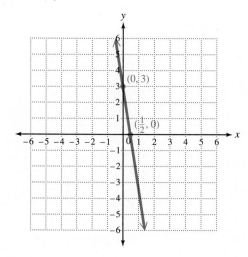

24. $3x + y = 0$

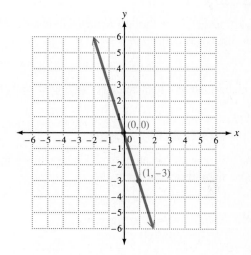

25. $y = 5$

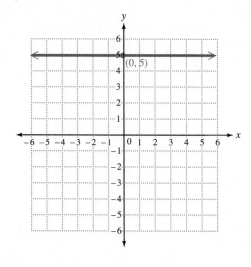

26. $x = -3$

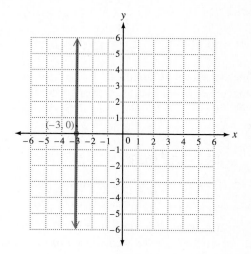

27. Is $3x^2 + 2y^2 = 5$ a linear equation? Explain.
No. Both x and y are squared.

12.3 EXERCISES C

1. Tell whether the lines $ax + c = 0$ and $by + c = 0$ are parallel, perpendicular, or neither.

perpendicular since $ax + c = 0$, is a vertical line and $by + c = 0$ is a horizontal line

2. Find the slope of the line with equation $ax + by + c = 0$. [*Hint: a, b,* and *c* are constants.] $-\dfrac{a}{b}$

12.4 FORMS OF THE EQUATION OF A LINE

STUDENT GUIDEPOSTS

1 General Form
2 Slope-Intercept Form
3 Graphing Using Slope
4 Point-Slope Form

1 GENERAL FORM

The equation of a line can be written in several forms. In Section 12.2 the **general form** was defined to be

$$ax + by + c = 0 \qquad a, b, c \text{ constants}$$

where *a* and *b* are not both zero. For example, consider the equation $x - 3y + 6 = 0$.

2 SLOPE-INTERCEPT FORM

Suppose we solve this equation for *y*. The result is a special form of the equation called the **slope-intercept form.**

$$x - 3y + 6 = 0$$
$$-3y = -x - 6$$
$$y = \frac{1}{3}x + 2 \qquad \text{Divide through by } -3$$

To explain why this is called the slope-intercept form first calculate *y* when $x = 0$.

$$y = \frac{1}{3}(0) + 2 = 2$$

Thus, the *y*-intercept of the line is (0, 2). Also, if *x* increases by 3, say from 0 to 3, *y* increases by 1. That is, *y* was 2 when $x = 0$, and when $x = 3$,

$$y = \frac{1}{3}(3) + 2 = 3.$$

Hence the slope of the line is

$$m = \frac{\text{change in } y}{\text{change in } x} = \frac{\text{rise}}{\text{run}} = \frac{1}{3}.$$

When an equation is in slope-intercept form the coefficient of *x* is the slope, and the constant term is the *y*-coordinate of the *y*-intercept of the line.

> ### Slope-Intercept Form of the Equation of a Line
>
> If the equation of a line is solved for y, the resulting equation is in **slope-intercept form**
>
> $$y = mx + b,$$
>
> where m is the slope of the line and $(0, b)$ is the y-intercept.

EXAMPLE 1 FINDING THE SLOPE AND y-INTERCEPT	**PRACTICE EXERCISE 1**

What are the slope and y-intercept of the line with equation $4x + 2y + 1 = 0$?

First solve for y to obtain the slope-intercept form.

$$4x + 2y + 1 = 0$$
$$2y = -4x - 1$$
$$y = -2x - \frac{1}{2} \qquad \text{Divide by 2}$$

$m = -2 = $ slope $\left(0, -\frac{1}{2}\right) = y\text{-intercept}$ $\left[b = -\frac{1}{2}\right]$

PRACTICE EXERCISE 1

What are the slope and y-intercept of the line with equation $3x - 6y + 10 = 0$?

Answer: slope is $\frac{1}{2}$, y-intercept is $\left(0, \frac{5}{3}\right)$

Example 1 shows how to find the slope of a nonvertical line when its equation is given in general form. Solve the equation for y to obtain the slope-intercept form. The slope is always the numerical coefficient of x.

EXAMPLE 2 USING THE SLOPE-INTERCEPT FORM	**PRACTICE EXERCISE 2**

Find the general form of the equation of the line with slope -2 and y-intercept $(0, 5)$.

We first find the slope-intercept form of the equation by substituting -2 for m and 5 for b in

$$y = mx + b.$$
$$y = -2x + 5$$

By writing all terms on the left side of this equation we obtain the general form

$$2x + y - 5 = 0.$$

PRACTICE EXERCISE 2

Find the general form of the equation of the line with slope $\frac{2}{3}$ and y-intercept $(0, -3)$.

Answer: $2x - 3y - 9 = 0$

---

❸ GRAPHING USING SLOPE

In Section 12.2 we learned how to graph a line by finding the intercepts. The graph of a line in slope-intercept form can be obtained in a different way using what we know about slope. This technique is illustrated by graphing the equation $y = \frac{2}{3}x - 1$.

Since the y-intercept is $(0, -1)$, we know the graph passes through this point. But since there are infinitely many lines containing $(0, -1)$, we need to find the one with slope $\frac{2}{3}$. To do this we find a second point on the line that can be obtained from $(0, -1)$ by considering the rise and run specified by a slope of $\frac{2}{3}$. If we start at $(0, -1)$ and move up 2 units (a rise of 2) then move right 3 units (a run of 3), we are at the point $(3, 1)$. The line passes through $(0, -1)$ and $(3, 1)$, as shown in Figure 12.25.

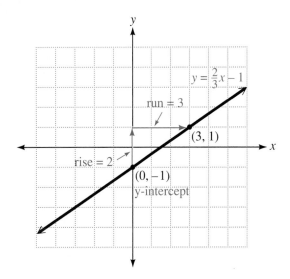

Figure 12.25   Graphing with Slope

④ POINT-SLOPE FORM

If we know that the slope of a line is m and that (x_1, y_1) is a point on the line, we can obtain the equation of the line. Suppose (x, y) is an arbitrary point on the line. Then using the formula for the slope of a line with (x, y) as (x_2, y_2), we have

$$m = \frac{y - y_1}{x - x_1}.$$

Multiplying both sides of this equation by $x - x_1$,

$$m(x - x_1) = y - y_1,$$

gives us another form of the equation of a line.

Point-Slope Form of the Equation of a Line
To find the equation of a line which has slope m and passes through the point (x_1, y_1), substitute these values into the **point-slope form** $$y - y_1 = m(x - x_1).$$

EXAMPLE 3   USING POINT-SLOPE FORM

Find the slope-intercept form of the equation of the line with slope -3 passing through the point $(1, -2)$. Graph the line.

Use the point-slope form of the equation with $m = -3$ and $(x_1, y_1) = (1, -2)$.

PRACTICE EXERCISE 3

Find the slope-intercept form of the equation of the line with slope -5 passing through the point $(2, -7)$. Graph the line.

$$y - y_1 = m(x - x_1)$$

$y - (-2) = -3(x - 1)$ $y_1 = -2, x_1 = 1, m = -3$

$y + 2 = -3x + 3$ Watch the signs

$y = -3x + 1$ Subtract 2 to obtain slope-intercept form

Since the slope $m = -3 = \frac{-3}{1} = \frac{\text{rise}}{\text{run}}$, the rise is -3 and the run is 1. Start at the y-intercept $(0, 1)$ and move down 3 units (the rise is -3), then right 1 unit (the run is 1). This locates another point on the line, $(1, -2)$. See Figure 12.26.

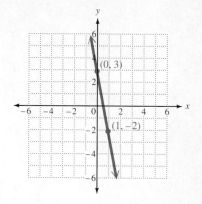

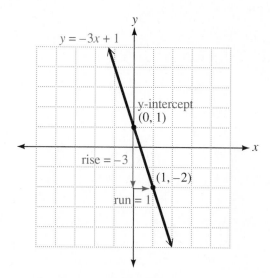

Figure 12.26

Answer: $y = -5x + 3$

To find the equation of a line passing through two points, we use both the slope formula and the point-slope form. This is shown in the next example.

EXAMPLE 4 EQUATION OF A LINE THROUGH TWO POINTS

Find the general form of the equation of the line passing through two points $(3, -2)$ and $(-1, 5)$.

   First find the slope of the line using $(3, -2) = (x_1, y_1)$ and $(-1, 5) = (x_2, y_2)$.

$$m = \frac{y_2 - y_1}{x_2 - x_1} = \frac{5 - (-2)}{-1 - 3} = \frac{7}{-4} = -\frac{7}{4}$$

Next, substitute $-\frac{7}{4}$ for m and $(3, -2)$ for (x_1, y_1) in the point-slope form. (We would get the same equation by using $(-1, 5)$ for (x_1, y_1).)

$$y - y_1 = m(x - x_1).$$

$$y - (-2) = -\frac{7}{4}(x - 3)$$

$4(y - (-2)) = -7(x - 3)$ Multiply both sides by 4

$$4y + 8 = -7x + 21$$

$$7x + 4y - 13 = 0$$

This is the general form of the equation of the desired line.

PRACTICE EXERCISE 4

Find the general form of the equation of the line passing through the points $(1, 7)$ and $(-2, 3)$.

Answer: $4x - 3y + 17 = 0$

We conclude this section with a summary of forms of the equation of a line. Note that the names *slope-intercept* and *point-slope form* help you decide which form should be used in a particular problem.

Forms of the Equation of a Line

General Form $ax + by + c = 0$

Slope-Intercept Form

$y = mx + b$ Slope m, y-intercept $(0, b)$

Point-Slope Form

$y - y_1 = m(x - x_1)$ Slope m, point (x_1, y_1) on line

12.4 EXERCISES A

Write each equation in slope-intercept form and give the slope and y-intercept.

1. $5x + y - 12 = 0$

$y = -5x + 12; -5, (0, 12)$

2. $2x - 5y + 10 = 0$

$y = \dfrac{2}{5}x + 2; \dfrac{2}{5}, (0, 2)$

3. $4y + 7 = 0$

$y = 0 \cdot x - \dfrac{7}{4}; 0, \left(0, -\dfrac{7}{4}\right)$

4. $5x + 1 = 0$

no slope-intercept form

5. $x + y = 5$

$y = -x + 5; -1, (0, 5)$

6. $2x - 3y + 1 = 0$

$y = \dfrac{2}{3}x + \dfrac{1}{3}; \dfrac{2}{3}, \left(0, \dfrac{1}{3}\right)$

Write each equation in slope-intercept form and use the y-intercept and slope to graph the equation.

7. $x + y + 2 = 0$

8. $2x - 5y + 10 = 0$

9. $4x + 3y = 15$

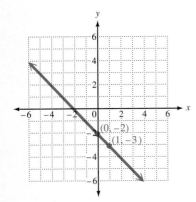

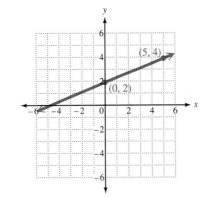

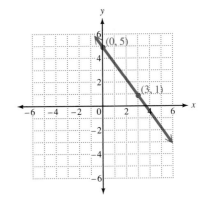

Find the general form of the equation of the line with the given slope and y-intercept.

10. $m = 2, (0, 4)$

$2x - y + 4 = 0$

11. $m = 4, (0, -3)$

$4x - y - 3 = 0$

12. $m = -\dfrac{1}{2}, (0, 6)$

$x + 2y - 12 = 0$

Find the general form of the equation of the line with the given slope passing through the given point.

13. $m = 3$, $(2, -4)$
  $3x - y - 10 = 0$

14. $m = -2$, $(1, 5)$
  $2x + y - 7 = 0$

15 $m = -\dfrac{1}{2}$, $(3, -2)$
  $x + 2y + 1 = 0$

Find the general form of the equation of the line passing through the given points.

16. $(2, 5)$ and $(4, 7)$
  $x - y + 3 = 0$

17. $(-1, 3)$ and $(2, 4)$
  $x - 3y + 10 = 0$

18 $(3, -2)$ and $(-1, -1)$
  $x + 4y + 5 = 0$

Some business problems can be described by a linear equation. For example, the cost y of producing a number of items x is given by y = mx + b, where b is the overhead cost (the cost when no items are produced) and m is the variable cost (the cost of producing a single item). Use this information to find the cost equation in 19–20.

19 Overhead cost: $25
  Variable cost: $10
  $y = 10x + 25$

20. Overhead cost: $300
  Variable cost: $10.50.
  $y = 10.50x + 300$

21. Which of the two equations, $x = 2$ or $y = -3$, is the equation of the vertical line through $(2, -3)$? $x = 2$

22. Which of the two equations, $x = 2$ or $y = -3$, is the equation of the horizontal line through $(2, -3)$? $y = -3$

23. Find the general form of the equation of the line through $(-1, 5)$ that is parallel to the line $3x + 2y - 4 = 0$.
  $3x + 2y - 7 = 0$

FOR REVIEW

The following exercises review material from Section 11.7 to help you prepare for the next section. Solve each inequality.

24. $2x + 1 < 5$
  $x < 2$

25. $-2(x - 1) \geq 8$
  $x \leq -3$

26. $3(x - 1) - (2 - x) \geq 5$
  $x \geq \dfrac{5}{2}$

ANSWERS: **1.** $y = -5x + 12$; slope is -5, y-intercept is $(0, 12)$ **2.** $y = \frac{2}{5}x + 2$; slope is $\frac{2}{5}$, y-intercept is $(0, 2)$
3. $y = -\frac{7}{4} = 0 \cdot x - \frac{7}{4}$; slope is 0, y-intercept is $\left(0, -\frac{7}{4}\right)$ **4.** The equation cannot be solved for y, hence it has no slope-intercept form. Also, it has undefined slope and no y-intercept.   **5.** $y = -x + 5$; slope is -1, y-intercept is $(0, 5)$
6. $y = \frac{2}{3}x + \frac{1}{3}$; slope is $\frac{2}{3}$, y-intercept is $\left(0, \frac{1}{3}\right)$
7. $y = -x - 2$

8. $y = \frac{2}{5}x + 2$

9. $y = -\frac{4}{3}x + 5$

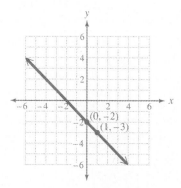

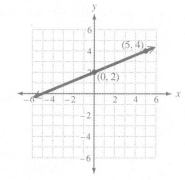

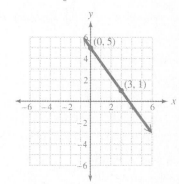

10. $2x - y + 4 = 0$ **11.** $4x - y - 3 = 0$ **12.** $x + 2y - 12 = 0$ **13.** $3x - y - 10 = 0$ **14.** $2x + y - 7 = 0$ **15.** $x + 2y + 1 = 0$ **16.** $x - y + 3 = 0$ **17.** $x - 3y + 10 = 0$ **18.** $x + 4y + 5 = 0$ **19.** $y = 10x + 25$ **20.** $y = 10.50x + 300$
21. $x = 2$ **22.** $y = -3$ **23.** $3x + 2y - 7 = 0$ **24.** $x < 2$ **25.** $x \leq -3$ **26.** $x \geq \frac{5}{2}$

12.4 EXERCISES B

Write each equation in slope-intercept form and give the slope and y-intercept.

1. $7x - y + 13 = 0$ $\quad y = 7x + 13$

2. $5x - 9y + 9 = 0$ $\quad y = \dfrac{5}{9}x + 1$

3. $2x - 1 = 0$ $\quad$ no slope-intercept form

4. $4y + 12 = 0$ $\quad y = 0x - 3$

5. $y - x = 3$ $\quad y = x + 3$

6. $3x - 5y + 2 = 0$ $\quad y = \dfrac{3}{5}x + \dfrac{2}{5}$

Write each equation in slope-intercept form and use the y-intercept and slope to graph the equation.

7. $x + y - 6 = 0$

8. $3x - 4y - 12 = 0$

9. $5x + 2y = 6$

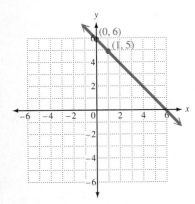
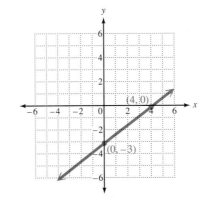
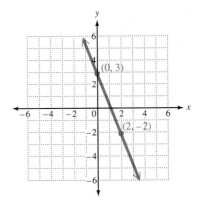

Find the general form of the equation of the line with the given slope and y-intercept.

10. $m = 3$, $(0, 1)$
$3x - y + 1 = 0$

11. $m = 5$, $(0, -2)$
$5x - y - 2 = 0$

12. $m = -\dfrac{1}{4}$, $(0, 2)$
$x + 4y - 8 = 0$

Find the general form of the equation of the line with the given slope passing through the given point.

13. $m = 5$, $(-3, 1)$
$5x - y + 16 = 0$

14. $m = -4$, $(2, 6)$
$4x + y - 14 = 0$

15. $m = -\dfrac{1}{5}$, $(-1, 4)$
$x + 5y - 19 = 0$

Find the general form of the equation of the line passing through the given points.

16. $(2, 6)$ and $(5, 9)$
$x - y + 4 = 0$

17. $(-3, 2)$ and $(2, 5)$
$3x - 5y + 19 = 0$

18. $(-4, 6)$ and $(-2, -2)$
$4x + y + 10 = 0$

Some business problems can be described by a linear equation. For example, the cost y of producing a number of items x is given by $y = mx + b$, where b is the overhead cost (the cost when no items are produced) and m is the variable cost (the cost of producing a single item). Use this information to find the cost equation in 19–20.

19. Overhead cost: $50
Variable cost: $30
$y = 30x + 50$

20. Overhead cost: $500
Variable cost: $8.50.
$y = 8.50x + 500$

21. Which of the two equations, $x = -1$ or $y = 5$, is the equation of the vertical line through $(-1, 5)$? $\quad x = -1$

22. Which of the two equations, $x = -1$ or $y = 5$, is the equation of the horizontal line through $(-1, 5)$? $y = 5$

23. Find the general form of the equation of the line through $(-1, 5)$ that is perpendicular to the line $3x + 2y - 4 = 0$.
$2x - 3y + 17 = 0$

FOR REVIEW

The following exercises review material from Section 11.7 to help you prepare for the next section. Solve each inequality.

24. $1 - 3x \leq -5$
$x \geq 2$

25. $4(2 - x) > 7 - 3x$
$x < 1$

26. $3(x + 2) - (1 - 2x) \geq 10$
$x \geq 1$

12.4 EXERCISES C

1. Find the general form of the equation of the horizontal line through the point $(6, -4)$. $y + 4 = 0$

2. Find the general form of the equation of the vertical line through the point $(5, 8)$. $x - 5 = 0$

3. The midpoint of the line segment joining (x_1, y_1) and (x_2, y_2) is $\left(\frac{x_1 + x_2}{2}, \frac{y_1 + y_2}{2}\right)$. Find the general form of the perpendicular bisector of the line segment joining $(2, -1)$ and $(-6, 3)$.   [Answer: $2x - y + 5 = 0$]

12.5 GRAPHING LINEAR INEQUALITIES IN TWO VARIABLES

STUDENT GUIDEPOSTS

1 Linear Inequalities in Two Variables

2 Graph of Solution

3 Test Point

4 Method of Graphing

In Section 11.7 we graphed linear inequalities in one variable, such as

$$3x + 2 > 5 \quad \text{and} \quad x - 1 \leq 3(x - 5) + 1,$$

on a number line. For example, if we solve

$$3x + 2 > 5$$
$$3x > 3 \qquad \text{Subtract 2}$$
$$x > 1, \qquad \text{Divide by 3}$$

the solution $x > 1$ is graphed in Figure 12.27.

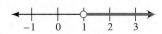

Figure 12.27

1 LINEAR INEQUALITIES IN TWO VARIABLES

We now consider **linear inequalities in two variables** in which the variables are raised only to the first power. For example,

$$2x + y < -1$$

is a linear inequality in the two variables x and y. A **solution** to such an inequality is an ordered pair of numbers which when substituted for x and y yields a true statement. Thus, $(-1, 0)$ is a solution to $2x + y < -1$ since by replacing x with -1 and y with 0 we obtain

$$2x + y < -1$$
$$2(-1) + 0 < -1$$
$$-2 < -1, \quad \text{which is true.}$$

On the other hand, $(4, -3)$ is not a solution since

$$2x + y < -1$$
$$2(4) + (-3) < -1$$
$$8 - 3 < -1$$
$$5 < -1, \quad \text{is false.}$$

EXAMPLE 1 SOLUTIONS TO AN INEQUALITY	**PRACTICE EXERCISE 1**

Tell whether $(-1, -3)$, $(1, 1)$ and $(-2, 0)$, are solutions to $3x - 2y \geq 1$.

$$3(-1) - 2(-3) \geq 1$$
$$-3 + 6 \geq 1$$
$$3 \geq 1 \qquad \text{True}$$

$(-1, -3)$ is a solution.

$$3(1) - 2(1) \geq 1$$
$$3 - 2 \geq 1$$
$$1 \geq 1 \qquad \text{True}$$

$(1, 1)$ is a solution.

$$3(-2) - 2(0) \geq 1$$
$$-6 - 0 \geq 1$$
$$-6 \geq 1 \qquad \text{False}$$

$(-2, 0)$ is not a solution.

Tell whether $(3, 0)$, $(2, 3)$, and $(1, 1)$ are solutions to $-5x + 4y < -1$.

Answer: $(3, 0)$ is a solution, $(2, 3)$ is not a solution, $(1, 1)$ is not a solution.

❷ GRAPH OF SOLUTION

The set of all solutions to a linear inequality in two variables can be displayed in a Cartesian coordinate system. Consider

$$2x + y > -3.$$

If we replace the inequality symbol with an equal sign, the resulting equation is

$$2x + y = -3.$$

To graph this equation we first plot the intercepts $\left(-\frac{3}{2}, 0\right)$ and $(0, -3)$ and draw the line as in Figure 12.28. Notice that the graph of the line divides the plane into a region above the line, the line itself, and a region below the line.

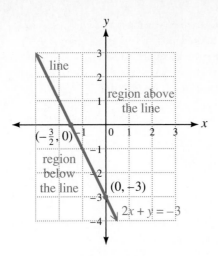

Figure 12.28

❸ TEST POINT

The graph of an inequality such as $2x + y > -3$ consists of all the points on one side of the boundary line $2x + y = -3$. The graph of $2x + y < -3$ is all the points on the other side of the line. To determine the correct side of the line to graph for an inequality such as $2x + y > -3$, we select any point not on the line as a **test point.** If we use $(0, 0)$ in this case the arithmetic is easy:

$$2x + y > -3$$
$$2(0) + 0 > -3$$
$$0 > -3. \quad \text{This is true}$$

Since this inequality is true, we graph the points on the side of the line containing the test point $(0, 0)$. If the inequality to be graphed had been $2x + y < -3$, a false inequality would have resulted using $(0, 0)$ as a test point:

$$2x + y < -3$$
$$2(0) + 0 < -3$$
$$0 < -3. \quad \text{This is false}$$

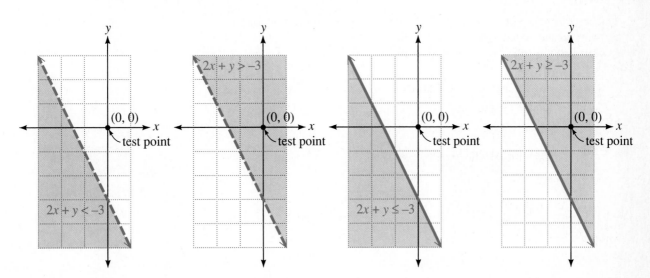

Figure 12.29

In the case when a false inequality is obtained, shade the region that does *not* contain the test point. The graphs of both $2x + y < -3$ and $2x + y > -3$ are shown in Figure 12.29 by shading the region that satisfies the inequality. The figure also has graphs of $2x + y \leq -3$ and $2x + y \geq -3$. We have used a dashed line for the boundary when the inequality is $<$ or $>$ to show that the boundary is *not* part of the graph. For the inequalities $\leq$ and $\geq$ a solid line is used to indicate that the boundary *is* part of the graph.

EXAMPLE 2 GRAPHING INEQUALITIES	PRACTICE EXERCISE 2

(a) Graph $x + 3y > 6$.

Graph the line $x + 3y = 6$ using the intercepts $(0, 2)$ and $(6, 0)$. Since the inequality is $>$, use a dashed line. Select the test point $(0, 0)$. (It is not on the line and the arithmetic is easy with $(0, 0)$.)

$$x + 3y > 6$$
$$0 + 3(0) > 6$$
$$0 > 6 \quad \text{This is false}$$

The inequality is false. Shade the region in Figure 12.30 that does not contain $(0, 0)$.

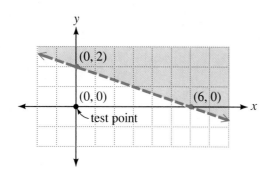

Figure 12.30

(b) Graph $2x - 3y \leq 0$.

Since both intercepts of $2x - 3y = 0$ are $(0, 0)$, we need to find another point on the boundary line. If $x = 3$ then $y = 2$, so $(3, 2)$ is a second point. Draw a solid line through $(0, 0)$ and $(3, 2)$ as in Figure 12.31.

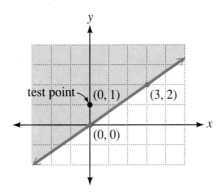

Figure 12.31

This time $(0, 0)$ cannot be used as a test point. Choosing $(0, 1)$ as a test point we obtain a true inequality:

(a) Graph $x + 3y \leq 6$.

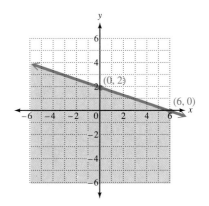

(b) Graph $2x - 3y > 0$.

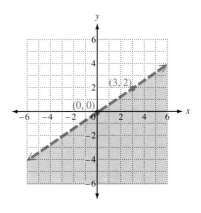

$$2x - 3y \le 0$$
$$2(0) - 3(1) \le 0$$
$$-3 \le 0.$$

Thus, we shade the region containing the test point $(0, 1)$.

Answers: (a) all points on and below the line in Figure 12.30 (b) all points below the line in Figure 12.31

Some linear inequalities in two variables have either the x-term or the y-term missing. The solution techniques are the same for these equations, but sometimes our work can be simplified by first solving the inequality for the remaining variable.

EXAMPLE 3   INEQUALITIES INVOLVING VERTICAL AND HORIZONTAL

(a) Graph $2x - 4 \ge 0$.

$$2x - 4 \ge 0$$
$$2x \ge 4$$
$$x \ge 2$$

Replacing $\ge$ with $=$ we recognize the graph of $x = 2$ as a vertical line with x-intercept $(2, 0)$. Since the inequality is $\ge$, we draw the boundary $x = 2$ as a solid line and use $(0, 0)$ as a test point.

$$x \ge 2$$
$$0 \ge 2 \qquad \text{This is false}$$

Since the inequality is false, shade the region not containing $(0, 0)$ in Figure 12.32.

(b) Graph $3y + 3 < 0$.

$$3y + 3 < 0$$
$$3y < -3$$
$$y < -1$$

Replacing $<$ with $=$ we recognize the graph of $y = -1$ as a horizontal line with y-intercept $(0, -1)$. Since the inequality is $<$, we draw the boundary $y = -1$ as a dashed line and use $(0, 0)$ as a test point.

$$y < -1$$
$$0 < -1 \qquad \text{This is false}$$

Since the inequality is false, shade the region not containing $(0, 0)$ in Figure 12.33.

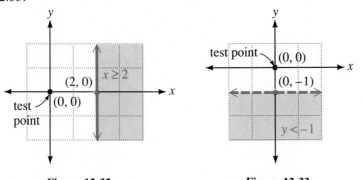

Figure 12.32   Figure 12.33

PRACTICE EXERCISE 3

(a) Graph $2x - 4 < 0$.

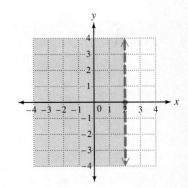

(b) Graph $3y + 3 \ge 0$.

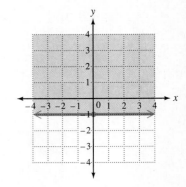

Answers: (a) all points to the left of the line in Figure 12.32   (b) all points on and above the line in Figure 12.33

④ METHOD OF GRAPHING

We now summarize the techniques that we have learned.

To Graph a Linear Inequality in Two Variables

1. Graph the boundary line using a dashed line if the inequality is $<$ or $>$ and a solid line if it is $\leq$ or $\geq$.

2. Choose a test point that is not on the boundary line and substitute it into the inequality.

3. Shade the region that includes the test point if a true inequality is obtained, and shade the region that does not contain the test point if a false inequality results.

12.5 EXERCISES A

1. In the graph of the inequality $3x - 2y \leq 5$, would the line $3x - 2y = 5$ be solid or dashed?   solid

2. In the graph of the inequality $2x + y > 9$, would the line $2x + y = 9$ be solid or dashed?   dashed

In the following exercises the boundary line for the inequality has been given. Complete each graph by shading the appropriate region.

3. $x + y \leq 5$

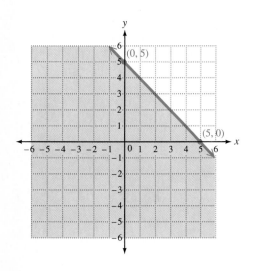

4. $2x - y < 6$

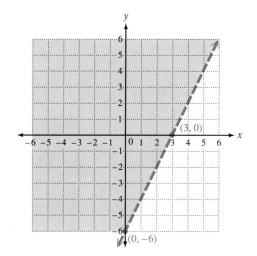

5. $3x + 12 > 0$

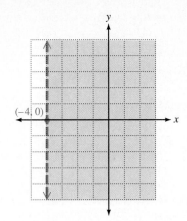

6. $1 - y \geq 0$

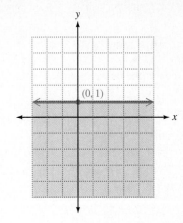

Graph the linear inequalities in two variables.

7. $x + y > 3$

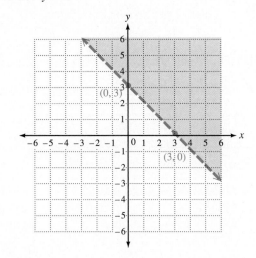

8. $x + y \leq 3$

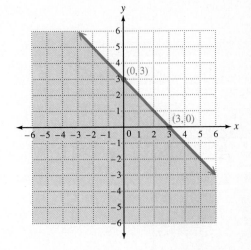

9. $x - y \geq -2$

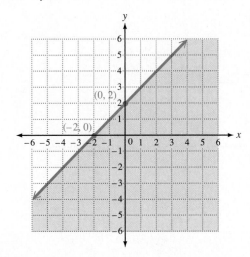

10. $x - y < -2$

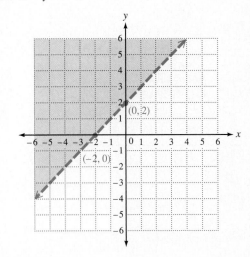

11. $x + 4y < 4$

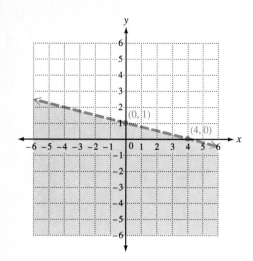

12. $x + 4y \geq 4$

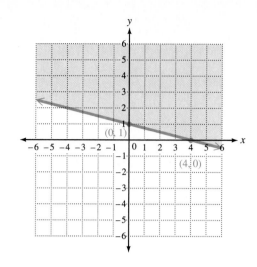

13. $4x + 12 > 0$

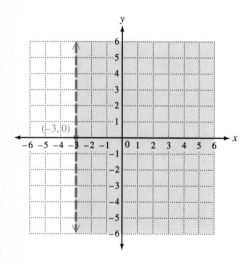

14. $4x + 12 \leq 0$

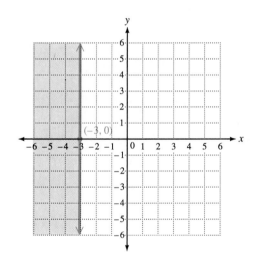

15. $2y - 8 \leq 0$

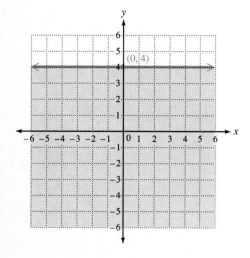

16. $2y - 8 > 0$

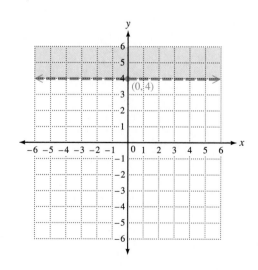

17. $x + y < 0$ **18.** $x + y \geq 0$

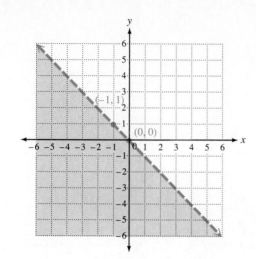   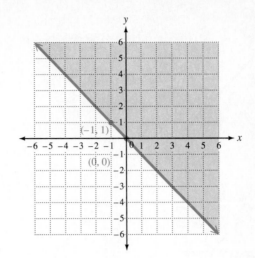

FOR REVIEW

Find the slope of the line passing through the given pair of points.

19. $(-3, 7)$ and $(10, -1)$ $-\dfrac{8}{13}$ **20.** $(6, 3)$ and $(6, -5)$ **21.** $(2, -5)$ and $(-8, -5)$ 0

undefined

Write each equation in slope-intercept form and give the slope and y-intercept.

22. $12x - 9y + 36 = 0$ $y = \dfrac{4}{3}x + 4; \dfrac{4}{3}, (0, 4)$ **23.** $5y - 15 = 0$ $y = 0 \cdot x + 3; 0, (0, 3)$

Find the general form of the equation of the line.

24. $m = \dfrac{1}{5}$, through $(-1, 4)$ $x - 5y + 21 = 0$ **25.** through $(2, 6)$ and $(-4, -3)$ $3x - 2y + 6 = 0$

26. Verify that the line l_1 between $(-9, 2)$ and $(0, 5)$ and the line l_2 between $(1, -4)$ and $(19, 2)$ are parallel.
The lines are parallel since both have slope $\frac{1}{3}$.

ANSWERS: 1. solid 2. dashed

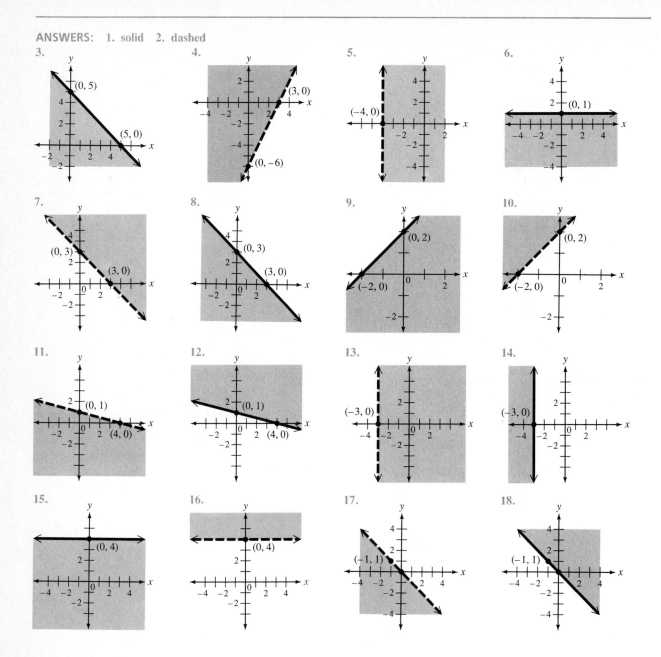

19. $-\frac{8}{13}$ 20. undefined slope 21. 0 (zero slope) 22. $y = \frac{4}{3}x + 4$; slope is $\frac{4}{3}$, y-intercept is (0, 4) 23. $y = 3 = 0 \cdot x +$ 3; slope is 0, y-intercept is (0, 3) 24. $x - 5y + 21 = 0$ 25. $3x - 2y + 6 = 0$ 26. The lines are parallel since both have slope $\frac{1}{3}$.

12.5 EXERCISES B

1. In the graph of the inequality $5x + y \geq 1$, would the line $5x + y = 1$ be solid or dashed?   solid

2. In the graph of the inequality $4x - 2y < 5$, would the line $4x - 2y = 5$ be solid or dashed?   dashed

In the following exercises the boundary line for the inequality has been given. Complete each graph by shading the appropriate region.

3. $x + y > 5$

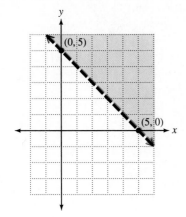

4. $2x - y \geq 6$

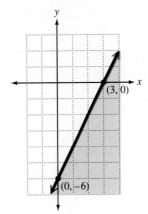

5. $3x + 12 \leq 0$

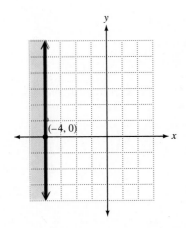

6. $1 - y < 0$

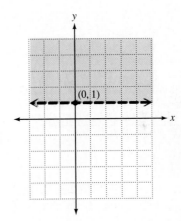

Graph the linear inequalities in two variables.

7. $x + y > 2$

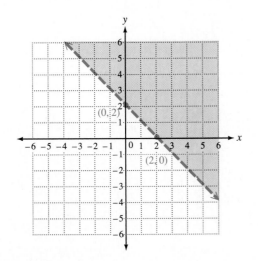

8. $x + y \leq 2$

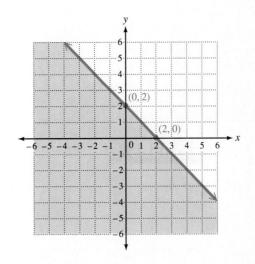

9. $x - y \geq -1$

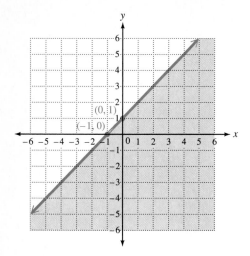

10. $x - y < -1$

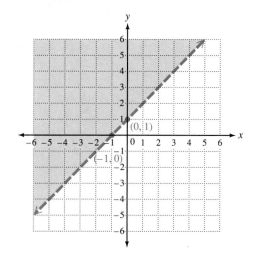

11. $3x + 4y < 12$

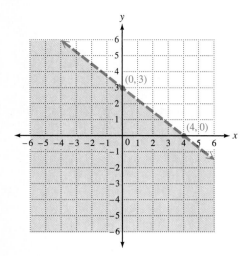

12. $3x + 4y \geq 12$

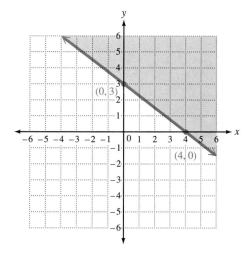

13. $4x + 8 > 0$

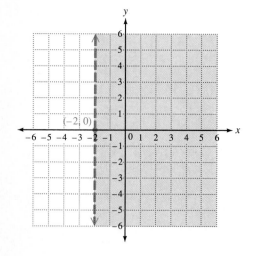

14. $4x + 8 \leq 0$

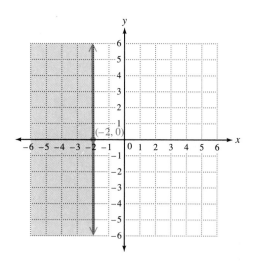

15. $2y - 6 \leq 0$

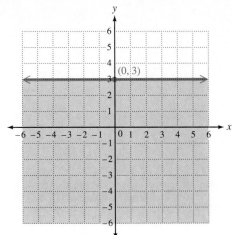

16. $2y - 6 > 0$

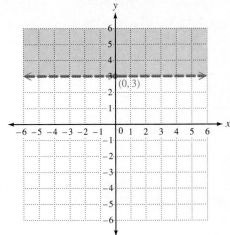

17. $2x + y < 0$

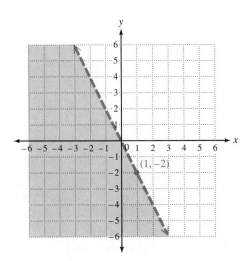

18. $2x + y \geq 0$

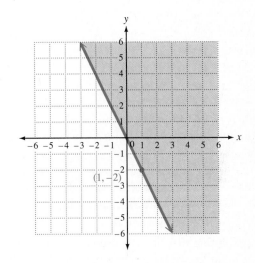

FOR REVIEW

Find the slope of the line passing through the given pair of points.

19. $(-5, 5)$ and $(3, -2)$ $-\dfrac{7}{8}$ **20.** $(-3, 10)$ and $(2, 10)$ 0 **21.** $(-9, 0)$ and $(-9, -7)$ undefined

Write each equation in slope-intercept form and give the slope and y-intercept.

22. $14x - 7y + 21 = 0$ $y = 2x + 3$; 2, $(0, 3)$ **23.** $2y + 1 = 0$ $y = 0 \cdot x - \dfrac{1}{2}$; 0, $\left(0, -\dfrac{1}{2}\right)$

Find the general form of the equation of the line.

24. $m = -3$, through $(-2, 5)$ $3x + y + 1 = 0$ **25.** through $(8, -1)$ and $(-4, 2)$ $x + 4y - 4 = 0$

26. Verify that the line l_1 between $(3, 6)$ and $(5, -4)$ and the line l_2 between $(1, -2)$ and $(-1, 8)$ are parallel.
The lines are parallel since both have slope -5.

12.5 EXERCISES C

Tell whether the given point satisfies both of the inequalities.

1. $3x - y < 4$
    $y \geq -2$
  (a) $(0, 2)$ **(b)** $(2, 0)$
    yes   no
  (c) $(-1, 5)$ **(d)** $(-3, -4)$
  [Answers: **(c)** yes; **(d)** no]

2. $x + 4y \geq 5$
    $-2x + 3y < 0$
  (a) $(0, -1)$ **(b)** $(-1, 0)$
    no   no
  (c) $(5, 2)$ **(d)** $(-3, 4)$
    yes   no

CHAPTER 12 REVIEW

KEY WORDS

12.1 A **rectangular** or **Cartesian coordinate system** is formed by using both a horizontal and a vertical number line.

The **x-axis** is the horizontal number line in a rectangular coordinate system.

The **y-axis** is the vertical number line in a rectangular coordinate system.

An **ordered pair** has x-coordinate in the first position and y-coordinate in the second position.

The **quadrants** are the four sections formed by the axes in a rectangular coordinate system.

A **linear equation in two variables** x and y is an equation of the form $ax + by = c$.

A **solution** to a linear equation in two variables is an ordered pair of numbers which when substituted for the variables results in a true equation.

12.2 The **graph** of an equation in two variables x and y is the set of points in a rectangular coordinate system that corresponds to solutions of the equation.

The **general form** of a linear equation is $ax + by + c = 0$.

The **x-intercept** is the point where the line crosses the x-axis.

The **y-intercept** is the point where the line crosses the y-axis.

12.3 The **slope** of a line gives a measure of its direction in a coordinate system.

Parallel lines never intersect.

Perpendicular lines intersect at right angles.

12.5 A **linear inequality in two variables** is an expression using $<$, $\leq$, $>$, or $\geq$ in which the variables are raised to the first power.

A **test point** is a point used to determine which side of the line is the solution of a linear inequality.

KEY CONCEPTS

12.2 1. When graphing an equation in a Cartesian coordinate system, construct a table of values.

2. Two convenient points to use for graphing a linear equation are the intercepts (if they exist).

3. An equation of the form $x = c$ is parallel to the y-axis with x-intercept $(c, 0)$.

4. An equation of the form $y = c$ is parallel to the x-axis with y-intercept $(0, c)$.

12.3 1. The slope of the line passing through points (x_1, y_1) and (x_2, y_2) is given by

$$m = \frac{y_2 - y_1}{x_2 - x_1} = \frac{\text{change in } y\text{-coordinates}}{\text{change in } x\text{-coordinates}}$$

and can be positive, negative, zero, or undefined.

2. Let m_1 and m_2 be the slopes of two distinct lines. If $m_1 = m_2$, the lines are parallel.

3. Let m_1 and m_2 be the slopes of two distinct lines. If $m_1 m_2 = -1$, the lines are perpendicular.

12.4 1. The slope-intercept form of the equation of a line is

$$y = mx + b,$$

where m is the slope and $(0, b)$ is the y-intercept.

2. The best way to find the slope of a line quickly is to solve the equation for y (if possible), putting the equation into slope-intercept form. The coefficient of x is the slope.

3. The point-slope form of the equation of a line is

$$y - y_1 = m(x - x_1),$$

where m is the slope and (x_1, y_1) is any point on the line.

12.5 To graph a linear inequality in two variables, first graph the boundary line using a dashed line (for $<$ or $>$) or a solid line (for $\leq$ or $\geq$). Choose a test point not on the boundary line and substitute it into the inequality. Shade the region containing the test point if a true inequality results, and shade the region not containing the test point if a false inequality results.

REVIEW EXERCISES

Part I

12.1 **1.** Plot the points associated with the pairs $A(-1, 3)$, $B(0, 0)$, $C(1, -2)$, $D(4, 2)$, and $E(-4, -3)$, and state which quadrant each is in.

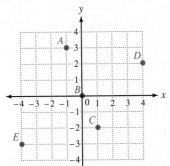

A: II
B: origin
C: IV
D: I
E: III

2. Complete the ordered pairs $(0, \ \ )$, $(\ \ , 0)$, and $(-2, \ \ )$ so that they will be solutions to the equation $2x + 3y = 6$.
$(0, 2)$, $(3, 0)$, $\left(-2, \frac{10}{3}\right)$

3. The number, y, of bacteria in a culture is approximated using time x, in days, by the equation $y = 4000x + 5000$.
Complete the ordered pair $(3, \ \ )$. After 3 days, what is the bacteria count?
$(3, 17{,}000)$; $17{,}000$

12.2 **4.** Which of the following are linear equations?
   (a) $2x + 3 = y$ **linear**   **(b)** $xy = 5$ **not linear**   **(c)** $x + y^2 = 2$ **not linear**
   (d) $x^3 + y = 3$ **not linear**   **(e)** $x = 4$ **linear**   **(f)** $3y = -1$ **linear**

Give the intercepts and graph in the Cartesian coordinate system.

5. $3x - 4y = 12$

x-intercept $= (4, 0)$

y-intercept $= (0, -3)$

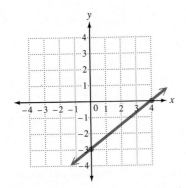

6. $2x + 3 = 0$

x-intercept $= \left(-\dfrac{3}{2}, 0\right)$

y-intercept $=$ **none**

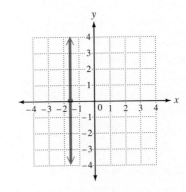

7. $x - 2y = 0$
    x-intercept =   **(0, 0)**
    y-intercept =   **(0, 0)**

8. $y - 3 = 0$
    x-intercept =   **none**
    y-intercept =   **(0, 3)**

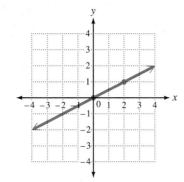
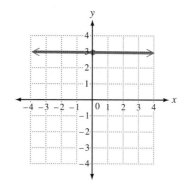

12.3 *Find the slope of the line through the given pair of points.*

9. $(-9, -2)$ and $(3, 6)$ $\dfrac{2}{3}$ **10.** $(0, 2)$ and $(-3, 0)$ $\dfrac{2}{3}$ **11.** $(-1, 8)$ and $(-1, 7)$
                                                                                                     undefined

12. What can be said about two distinct lines that both have slope $\frac{2}{3}$?   **They are parallel.**

13. What can be said about two lines with slopes $\frac{1}{3}$ and -3?   **They are perpendicular.**

12.4 14. Write $8x + y - 3 = 0$ in slope-intercept form and give the slope and y-intercept.
$y = -8x + 3$; slope is -8; y-intercept is $(0, 3)$

15. Write $3y + 4 = 0$ in slope-intercept form and give the slope and y-intercept.
$y = 0 \cdot x - \frac{4}{3}$; slope is 0; y-intercept is $\left(0, -\frac{4}{3}\right)$

16. Find the general form of the equation of the line with slope $-\frac{2}{3}$ and y-intercept $(0, 4)$.
$2x + 3y - 12 = 0$

17. Find the general form of the equation of the line passing through $(-1, -2)$ and $(5, 3)$.
$5x - 6y - 7 = 0$

12.5 *Graph the following in a Cartesian coordinate system.*

18. $x - 2y > -2$

19. $2y + 4 \le 0$

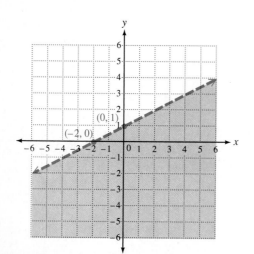
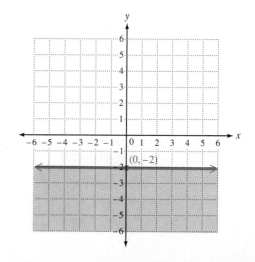

20. $3 - 3x < 0$

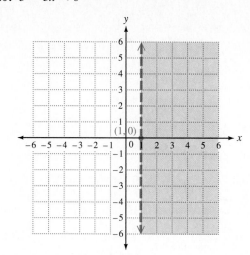

21. $3x + 2y \geq 6$

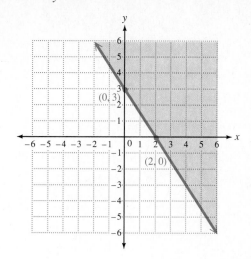

Part II

22. What can be said about two lines both with slope $\frac{1}{5}$ and *y*-intercept -5?   **They coincide.**

Find the slope of the line through the given pair of points.

23. $(-1, 2)$ and $(0, 5)$ **3**

24. $(-6, -4)$ and $(3, -4)$ **0**

Write each equation in slope-intercept form and give its slope and y-intercept.

25. $3x - 2y = 8$
$y = \frac{3}{2}x - 4$; slope is $\frac{3}{2}$; *y*-intercept is $(0, -4)$

26. $-x + 5y + 4 = 0$
$y = \frac{1}{5}x - \frac{4}{5}$; slope is $\frac{1}{5}$; *y*-intercept is $\left(0, -\frac{4}{5}\right)$

27. Find the general form of the equation of a line through $(6, -1)$ and $(4, 5)$. $3x + y - 17 = 0$

28. Find the general form of the equation of a line through $(-2, 7)$ and parallel to a line with slope $-\frac{1}{3}$. $x + 3y - 19 = 0$

29. Graph $2x + y \leq 4$.

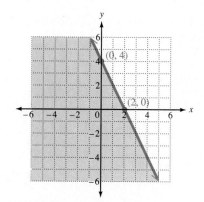

30. Use the *y*-intercept and slope to graph $3x + y + 2 = 0$.

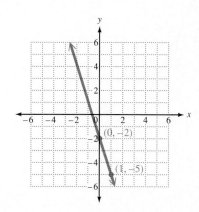

ANSWERS:

1. *A* is in II;
   B is origin;
   C is in IV;
   D is in I;
   E is in III

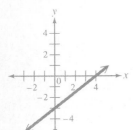

2. $(0, 2)$, $(3, 0)$, $\left(-2, \frac{10}{3}\right)$ 3. $(3, 17{,}000)$; $17{,}000$ 4. **(a)**, **(e)**, and **(f)** are linear

5. *x*-intercept $(4, 0)$;
   y-intercept $(0, -3)$

6. *x*-intercept $\left(-\frac{3}{2}, 0\right)$;
   no *y*-intercept

7. *x*-intercept $(0, 0)$;
   y-intercept $(0, 0)$;
   another point is $(2, 1)$

8. no *x*-intercept;
   y-intercept $(0, 3)$

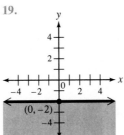

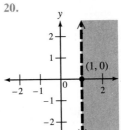

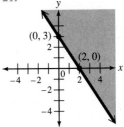

9. $\frac{2}{3}$ 10. $\frac{2}{3}$ 11. undefined slope 12. They are parallel. 13. They are perpendicular. 14. $y = -8x + 3$;
slope is -8; *y*-intercept is $(0, 3)$ 15. $y = 0 \cdot x - \frac{4}{3}$; slope is 0; *y*-intercept is $\left(0, -\frac{4}{3}\right)$ 16. $2x + 3y - 12 = 0$
17. $5x - 6y - 7 = 0$

18.
19.
20.
21.

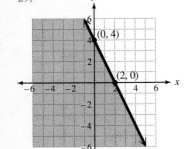

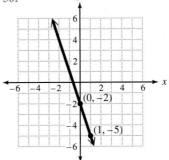

22. They coincide. 23. 3 24. 0 25. $y = \frac{3}{2}x - 4$; slope is $\frac{3}{2}$; *y*-intercept is $(0, -4)$ 26. $y = \frac{1}{5}x - \frac{4}{5}$; slope is $\frac{1}{5}$;
y-intercept is $\left(0, -\frac{4}{5}\right)$ 27. $3x + y - 17 = 0$ 28. $x + 3y - 19 = 0$

29.

30.

1. The point with coordinates $(2, -3)$ is located in which quadrant?

   1. _____ IV _____

2. Complete the ordered pair $(-3, \quad)$ so that it will be a solution to the equation $x + 5y = -1$.

   2. _____ $\left(-3, \dfrac{2}{5}\right)$ _____

3. The number, y, of items in a store during the week is approximated by using time x, in days, in the equation $y = -200x + 1000$. Complete the ordered pair $(3, \quad)$ to find the number of items in the store after three days.

   3. _____ 400 items _____

4. True or false: $x - 2y^2 = 3$ is a linear equation.

   4. _____ false _____

5. Find the slope of the line through the points $(2, -3)$ and $(-4, -5)$.

   5. _____ $\dfrac{1}{3}$ _____

6. Write $2x + 5y = 20$ in slope-intercept form and give the slope and the y-intercept.

   6. _____ $y = -\dfrac{2}{5}x + 4;\ -\dfrac{2}{5};\ (0, 4)$ _____

7. Find the general form of the equation of the line with slope -6 and passing through $(4, -5)$.

   7. _____ $6x + y - 19 = 0$ _____

8. Are lines with slopes $-\dfrac{1}{7}$ and 7 parallel, perpendicular, or neither?

   8. _____ perpendicular _____

Graph each equation in the given rectangular coordinate system.

9. $y = 2x + 4$

   9.

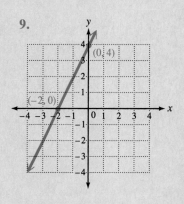

10. $3x + y = 0$

10.
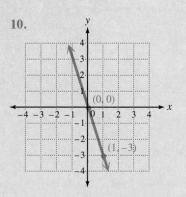

Graph each inequality in the given rectangular coordinate system.

11. $2y + x - 3 > 0$

11.
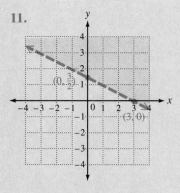

12. $y - 4 \le 0$

12.
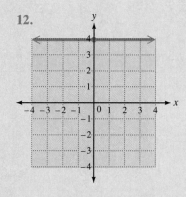

Systems of Linear Equations

13.1 PARALLEL, COINCIDING, AND INTERSECTING LINES

STUDENT GUIDEPOSTS

1 Systems of Linear Equations

2 Graphing a System

3 Relation of the Graphs of a System

1 SYSTEMS OF LINEAR EQUATIONS

In Chapter 12 we discovered that a solution to a linear equation in two variables such as

$$2x - y + 1 = 0$$

is an ordered pair of numbers that when substituted for x and y makes the equation true. Often in applied situations, two quantities are compared using two different equations. For example, a small businessman may discover that total sales and profit are related in two different ways. If the following pair of linear equations describes the relationships between sales x and profit y, it is called a **system of two linear equations in two variables,** or simply a **system of equations.**

$$3x - 2y + 1 = 0$$
$$3x + y - 5 = 0$$

A **solution** to a system of equations is an ordered pair of numbers (x, y) that is a solution to *both* equations. We can verify that $(1, 2)$ is a solution to the above system by direct substitution.

$$3(1) - 2(2) + 1 \stackrel{?}{=} 0 \quad x = 1 \text{ and } y = 2 \quad 3(1) + (2) - 5 \stackrel{?}{=} 0$$
$$3 - 4 + 1 \stackrel{?}{=} 0 \qquad\qquad\qquad 3 + 2 - 5 \stackrel{?}{=} 0$$
$$0 = 0 \qquad\qquad\qquad\qquad 0 = 0$$

2 GRAPHING A SYSTEM

Finding a solution to a system of equations is easier to understand if we first examine the graphs of the system. Remember that a linear equation in x and y has a straight line for its graph. When the linear equations in a given system are graphed together in a rectangular coordinate system, one of three possibilities occurs:

1. The lines coincide (the two equations represent the same line).

2. The lines are parallel (do not intersect).

3. The lines intersect in exactly one point.

The following three pairs of linear equations, graphed in Figure 13.1, illustrate these three possible relationships.

(a) $3x - 2y + 1 = 0$	**(c)** $3x - 2y + 1 = 0$	**(e)** $3x - 2y + 1 = 0$
(b) $6x - 4y + 2 = 0$	**(d)** $3x - 2y - 4 = 0$	**(f)** $3x + y - 5 = 0$

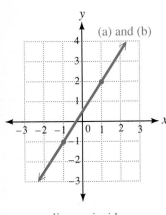

lines coincide

lines are parallel

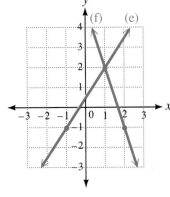

lines intersect
in exactly one point

Figure 13.1

In the first case, lines **(a)** and **(b)** coincide, in the second case, lines **(c)** and **(d)** are parallel, and in the third case, lines **(e)** and **(f)** intersect in one point.

❸ RELATION OF THE GRAPHS OF A SYSTEM

We need a way to tell quickly which case we have in a given situation. Let us solve each linear equation above for y, writing it in slope-intercept form.

Lines coincide

(a) $y = \dfrac{3}{2}x + \dfrac{1}{2}$

Slopes are the same. y-intercepts are the same.

(b) $y = \dfrac{3}{2}x + \dfrac{1}{2}$

Lines are parallel

(c) $y = \dfrac{3}{2}x + \dfrac{1}{2}$

Slopes are the same. y-intercepts are different.

(d) $y = \dfrac{3}{2}x - 2$

Lines intersect in exactly one point

(e) $y = \dfrac{3}{2}x + \dfrac{1}{2}$

↑

Slopes are
different.

↓

(f) $y = -3x + 5$

Notice that all lines with equation $y = b$ are parallel since the slopes are zero. Now we have covered all cases except when one (or both) of the equations in a system is of the form $ax = c$, since such equations cannot be solved for y. However, recalling that the graph of an equation of this form is always a line parallel to the y-axis, parallel or coinciding lines result only if both equations in the system are of this type.

EXAMPLE 1   INTERSECTING LINES

Tell how the graphs of the system are related.

$$2x + 3y + 1 = 0$$
$$x - 2y + 5 = 0$$

Solve for y.

$$3y = -2x - 1 \qquad\qquad -2y = -x - 5$$

$$\boxed{\dfrac{1}{3}} \cdot 3y = \boxed{\dfrac{1}{3}}(-2x - 1) \qquad \left(\boxed{-\dfrac{1}{2}}\right)(-2y) = \left(\boxed{-\dfrac{1}{2}}\right)(-x - 5)$$

$$y = -\dfrac{2}{3}x - \dfrac{1}{3} \qquad\qquad y = \dfrac{1}{2}x + \dfrac{5}{2}$$

Since the coefficients of x, the slopes of the lines, are $-\dfrac{2}{3}$ and $\dfrac{1}{2}$ (unequal), the lines are neither parallel nor coinciding, they intersect.

PRACTICE EXERCISE 1

Tell how the graphs of the system are related.

$$2x - 3y + 1 = 0$$
$$x + 5y - 10 = 0$$

Answer: They intersect.

EXAMPLE 2   PARALLEL LINES

Tell how the graphs of the system are related.

$$2x - 3y + 1 = 0$$
$$4x - 6y - 3 = 0$$

Solve for y.

$$-3y = -2x - 1 \qquad\qquad -6y = -4x + 3$$

$$\left(\boxed{-\dfrac{1}{3}}\right)(-3y) = \left(\boxed{-\dfrac{1}{3}}\right)(-2x - 1) \qquad \left(\boxed{-\dfrac{1}{6}}\right)(-6y) = \left(\boxed{-\dfrac{1}{6}}\right)(-4x + 3)$$

$$y = \dfrac{2}{3}x + \dfrac{1}{3} \qquad\qquad y = \dfrac{4}{6}x - \dfrac{3}{6}$$

$$y = \dfrac{2}{3}x - \dfrac{1}{2}$$

Since the coefficients of x, the slopes of the lines, are both $\dfrac{2}{3}$ (equal), the lines are either parallel or coinciding. Since the constants $\dfrac{1}{3}$ and $-\dfrac{1}{2}$ are unequal, the lines have two different y-intercepts, namely $\left(0, \dfrac{1}{3}\right)$ and $\left(0, -\dfrac{1}{2}\right)$, and the lines do not coincide. Thus, the lines are parallel.

PRACTICE EXERCISE 2

Tell how the graphs of the system are related.

$$5x - 10y + 20 = 0$$
$$-x + 2y - 4 = 0$$

Answer: They coincide.

EXAMPLE 3   COINCIDING LINES

Tell how the graphs of the system are related.

$$x - 2y + 2 = 0$$
$$-3x + 6y - 6 = 0$$

Solve for y.

$$-2y = -x - 2 \qquad\qquad 6y = 3x + 6$$

$$\left(-\frac{1}{2}\right)(-2y) = \left(-\frac{1}{2}\right)(-x - 2) \qquad \frac{1}{6}(6y) = \frac{1}{6}(3x + 6)$$

$$y = \frac{1}{2}x + 1 \qquad\qquad y = \frac{3}{6}x + \frac{6}{6}$$

$$y = \frac{1}{2}x + 1$$

Since both coefficients of x are $\frac{1}{2}$ and both constants are 1, the lines coincide.

PRACTICE EXERCISE 3

Tell how the graphs of the system are related.

$$-3x + y - 5 = 0$$
$$6x - 2y + 3 = 0$$

Answer: They are parallel.

EXAMPLE 4   ONE LINE VERTICAL

Tell how the graphs are related.

$$2x + 3y - 6 = 0$$
$$2x - 6 = 0$$

Since $2x - 6 = 0$ becomes $2x = 6$ or $x = 3$, its graph is a line parallel to the y-axis, 3 units to the right of the y-axis. Since there is a y term in $2x + 3y - 6 = 0$, the graph of this equation cannot be a line parallel to the y-axis. Thus, the two lines must intersect in one point.

PRACTICE EXERCISE 4

Tell how the graphs of the system are related.

$$7x + 3 = 0$$
$$-2x + 5 = 0$$

Answer: They are parallel.

EXAMPLE 5   ONE LINE HORIZONTAL

Tell how the graphs are related.

$$3y + 5 = 0$$
$$4x - 2y + 3 = 0$$

Solve for y.

$$3y = -5 \qquad\qquad -2y = -4x - 3$$

$$\left(\frac{1}{3}\right)(3y) = \left(\frac{1}{3}\right)(-5) \qquad \left(-\frac{1}{2}\right)(-2y) = \left(-\frac{1}{2}\right)(-4x - 3)$$

$$y = -\frac{5}{3} \qquad\qquad y = \frac{4}{2}x + \frac{3}{2}$$

$$y = 0 \cdot x - \frac{5}{3} \qquad\qquad y = 2x + \frac{3}{2}$$

Since the coefficients of x, 0 and 2, are not equal, the lines intersect in one point.

PRACTICE EXERCISE 5

Tell how the graphs of the system are related.

$$2x - 3y + 6 = 0$$
$$5y - 1 = 0$$

Answer: They intersect.

13.1 EXERCISES A

*Complete Exercises 1–4 with one of the words (**a**) parallel, (**b**) coinciding, or (**c**) intersecting.*

1. When two linear equations are both solved for y and the coefficients of x are unequal, the graphs are _____ intersecting _____ lines.

2. When two linear equations are both solved for y and the coefficients of x are equal as are the constants, the graphs are _____ coinciding _____ lines.

3. If the y term is missing in both of two linear equations, the graphs are (**a**)_____ coinciding _____ lines or lines (**b**)_____ parallel _____ to the y-axis.

4. If the y-term is missing in only one of two linear equations, the graphs are _____ intersecting _____ lines.

Tell how the graphs of the system are related.

5. $-2x + 3y + 5 = 0$
$-2x - 3y + 5 = 0$
intersecting

6. $x - 5y + 2 = 0$
$2x - 10y + 4 = 0$
coinciding

7 $2x - 7y + 1 = 0$
$-6x + 21y + 3 = 0$
parallel

8. $2y + x = 5$
$2x = 10$
intersecting

9. $x = 3y + 7$
$y = 3x + 7$
intersecting

10 $2y = 8$
$2x = 8$
intersecting

11. $2x + 1 = 3$
$4x - 3 = 1$
coinciding

12. $x + y = 5$
$3y = 15$
intersecting

13. $8y - 5 = -3x$
$-6x + 10 = 16y$
coinciding

Tell whether $(-3, 2)$ is a solution to the given system.

14 $2x + 3y = 0$
$x + 8y = 19$
no

15. $3x + y + 7 = 0$
$x + 4y - 5 = 0$
yes

16. $3y = 6$
$x - 5y = -13$
yes

Tell whether $(4, 0)$ is a solution to the given system.

17. $2x - 3y = 8$
$x - 4 = 0$
yes

18. $x + y - 2 = 0$
$2y + 2x = 4$
no

19. $2y + 1 = x$
$2x - 8 = y$
no

FOR REVIEW

To help you prepare for graphing systems in the next section, the following exercises review material from Chapter 12. Find the intercepts of each line.

20. $2x - y = 4$

$(2, 0); (0, -4)$

21. $6x + 2y = 5$

$\left(\dfrac{5}{6}, 0\right); \left(0, \dfrac{5}{2}\right)$

22. $3x - 5y + 2 = 0$

$\left(-\dfrac{2}{3}, 0\right); \left(0, \dfrac{2}{5}\right)$

ANSWERS: 1. intersecting 2. coinciding 3. (a) coinciding (b) parallel 4. intersecting 5. intersecting
6. coinciding 7. parallel 8. intersecting 9. intersecting 10. intersecting 11. coinciding 12. intersecting
13. coinciding 14. no 15. yes 16. yes 17. yes 18. no 19. no 20. x-intercept: $(2, 0)$; y-intercept: $(0, -4)$
21. x-intercept: $\left(\frac{5}{6}, 0\right)$; y-intercept: $\left(0, \frac{5}{2}\right)$ 22. x-intercept: $\left(-\frac{2}{3}, 0\right)$; y-intercept: $\left(0, \frac{2}{5}\right)$

13.1 EXERCISES B

Complete Exercises 1–4 with one of the following words: (a) parallel, (b) coinciding, or (c) intersecting.

1. When the slopes of the lines in a system of equations are equal but the y-intercepts are unequal, the graphs are ____ lines. **parallel**

2. When the slopes of the lines in a system of equations are unequal, the graphs are ____ lines. **intersecting**

3. If the x term is missing in both equations in a system of equations, the graphs are **(a)**____ lines or are lines **(b)**____ to the x-axis.
   (a) coinciding (b) parallel

4. If the x term is missing in only one of the equations in a system of equations, the graphs are ____ lines.
   intersecting

Tell how the graphs of the system are related.

5. $-5x + y - 3 = 0$
   $5x - y - 3 = 0$
   parallel

6. $x + 2y + 3 = 0$
   $-3x - 6y - 9 = 0$
   coinciding

7. $4x - y + 1 = 0$
   $x - 4y - 1 = 0$
   intersecting

8. $2x + 3y = 1$
   $y = 2$
   intersecting

9. $x = 3y - 1$
   $y = -x - 1$
   intersecting

10. $3x = 1$
    $3y = 1$
    intersecting

11. $3y + 1 = 4$
    $-y + 2 = 1$
    coinciding

12. $y - 3x = 11$
    $2x = 8$
    intersecting

13. $x + 2 = y$
    $y + 1 = x$
    parallel

Tell whether $(1, 0)$ is a solution to the given system.

14. $2x + 3y = 2$ **yes**
    $-x - 4y = -1$

15. $3x - y = 1$ **no**
    $x + y = 1$

16. $\quad\quad 3x = 3$ **yes**
    $-x + 9y = -1$

Tell whether $(-2, 1)$ is a solution to the given system.

17. $x - y = -3$ **yes**
    $y - 1 = 0$

18. $2x - 3y + 7 = 0$ **yes**
    $6y - 4x = 14$

19. $y + 2x = 3$ **no**
    $2x + 5 = y$

FOR REVIEW

To help you prepare for graphing systems in the next section, the following exercises review material from Chapter 12. Find the intercepts of each line.

20. $x - 3y = 6$

    $(6, 0); (0, -2)$

21. $4x + 5y = 8$

    $(2, 0); \left(0, \dfrac{8}{5}\right)$

22. $2x - 4y + 7 = 0$

    $\left(-\dfrac{7}{2}, 0\right); \left(0, \dfrac{7}{4}\right)$

13.1 EXERCISES C

1. Find the value of a so that the line with equation $ax - 3y + 7 = 0$ will be parallel to the line with equation $5x + y - 2 = 0$. [*Hint:* Write each equation in slope-intercept form.] $a = -15$

2. Find the value of c so that the line with equation $2x - 3y + c = 0$ will have the same y-intercept as the line with equation $4x + 2y + 3 = 0$. $c = -\dfrac{9}{2}$

13.2 SOLVING SYSTEMS OF EQUATIONS BY GRAPHING

━━━━━━━━━━ STUDENT GUIDEPOSTS ━━━━━━━━━━

| ❶ The Graphing Method | ❷ Finding the Number of Solutions |

❶ THE GRAPHING METHOD

Consider the system of equations

$$x + y = 1$$
$$2x - y = 5.$$

Solve each equation for y.

$$y = -x + 1$$
$$y = 2x - 5$$

The slopes are different, so the lines intersect in exactly one point. Since every point on each line corresponds to an ordered pair of numbers that is a solution to that equation, the point of intersection must correspond to the ordered pair that solves both equations. Hence it is the solution to the system. If we graph both equations in the same rectangular coordinate system, as in Figure 13.2, it appears that the point of intersection has coordinates $(2, -1)$. We can check to see if $(2, -1)$ is indeed a solution by substitution.

$$x + y = 1 \qquad\qquad 2x - y = 5$$
$$2 + (-1) \overset{?}{=} 1 \qquad 2(2) - (-1) \overset{?}{=} 5$$
$$1 = 1 \qquad\qquad 4 + 1 \overset{?}{=} 5$$
$$5 = 5$$

This method for solving a system is the **graphing method.**

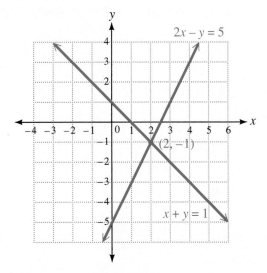

Figure 13.2

❷ FINDING THE NUMBER OF SOLUTIONS

If the graphs of the lines in a system are parallel, there can be no point of intersection, so the system has no solution. If the graphs of the lines coincide, then any solution to one equation is also a solution to the other equation, and the system has infinitely many solutions. These observations are summarized in the following statement.

Number of Solutions to a System

A system of equations has:

1. infinitely many solutions if the graphs of the equations coincide;
2. no solution if the graphs of the equations are parallel;
3. exactly one solution if the graphs of the equations intersect in one point.

EXAMPLE 1   INFINITELY MANY SOLUTIONS

Find the number of solutions to the system.

$$3x - 4y + 1 = 0$$
$$8y - 2 = 6x$$

Solve each equation for y.

$$-4y = -3x - 1 \qquad\qquad 8y = 6x + 2$$

$$\left(-\frac{1}{4}\right)(-4y) = \left(-\frac{1}{4}\right)(-3x - 1) \qquad \left(\frac{1}{8}\right)(8y) = \left(\frac{1}{8}\right)(6x + 2)$$

$$y = \frac{3}{4}x + \frac{1}{4} \qquad\qquad y = \frac{6}{8}x + \frac{2}{8}$$

$$y = \frac{3}{4}x + \frac{1}{4}$$

Since the lines coincide (why?), there are infinitely many solutions to the system.

PRACTICE EXERCISE 1

Find the number of solutions to the system.

$$x - 5y + 8 = 0$$
$$3y + 7 = -2x$$

Answer: exactly one

EXAMPLE 2   NO SOLUTION

Find the number of solutions to the system.

$$x - 5y + 1 = 0$$
$$15y - 3x = 1$$

Solve for y.

$$-5y = -x - 1 \qquad 15y = 3x + 1$$

$$y = \frac{1}{5}x + \frac{1}{5} \qquad y = \frac{3}{15}x + \frac{1}{15}$$

$$y = \frac{1}{5}x + \frac{1}{15}$$

Since the lines are parallel (why?), there is no solution to the system.

PRACTICE EXERCISE 2

Find the number of solutions to the system.

$$3x + 5y = -7$$
$$6x + 10y - 5 = 0$$

Answer: no solution

EXAMPLE 3   EXACTLY ONE SOLUTION

Find the number of solutions to the system.

$$3x - 5 = 0$$
$$3x + y = -5$$

Since the y term is missing in the first equation, it cannot be solved for y. The graph of $3x - 5 = 0$, however, is a line parallel to the y-axis, so it intersects the line $3x + y = -5$ in only one point (why?) Thus, there is exactly one solution to the system.

PRACTICE EXERCISE 3

Find the number of solutions to the system.

$$4x - 12 = 0$$
$$2x = 6$$

Answer: infinitely many

| EXAMPLE 4   USING THE GRAPHING METHOD | PRACTICE EXERCISE 4 |

Solve the system by graphing.

$$x - 2y = 1$$
$$2x + y = 7$$

We graph each equation by finding its intercepts.

$$x - 2y = 1 \qquad\qquad 2x + y = 7$$

x	y
0	$-\dfrac{1}{2}$
1	0

x	y
0	7
$\dfrac{7}{2}$	0

The point of intersection of the two lines, graphed in Figure 13.3, appears to have coordinates (3, 1). We check by substitution in both equations.

$$x - 2y = 1 \qquad\qquad 2x + y = 7$$
$$3 - 2(1) \overset{?}{=} 1 \qquad\qquad 2(3) + 1 \overset{?}{=} 7$$
$$3 - 2 \overset{?}{=} 1 \qquad\qquad 6 + 1 \overset{?}{=} 7$$
$$1 = 1 \qquad\qquad 7 = 7$$

Thus, the solution to the system is (3, 1).

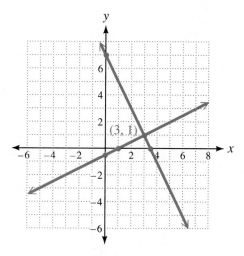

Figure 13.3

Solve the system by graphing.

$$3x - 2y = 12$$
$$2x + 3y = -5$$

Answer: $(2, -3)$

13.2 EXERCISES A

1. How many solutions will a system of equations have if the lines in the system are parallel?   **none**

2. How many solutions will a system of equations have if the lines in the system are intersecting?   **exactly one**

3. How many solutions will a system of equations have if the lines in the system are coinciding?   **infinitely many**

Find the number of solutions to the given system. For those that have exactly one solution, find that solution by graphing. Check each solution.

4. $x + 3y = 4$
$-x + y = 0$ **one: (1, 1)**

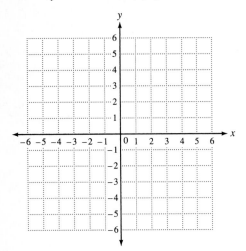

5 $x + 3y = 4$
$-2x - 6y = -8$ **infinitely many**

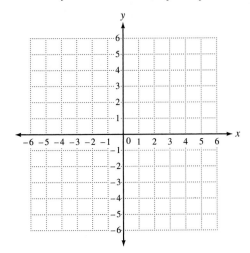

6. $x + 3y = 4$
$-2x - 6y = 0$ **no solution**

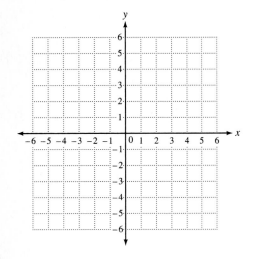

7. $2x + y = 5$
$2x - y = 3$ **one: (2, 1)**

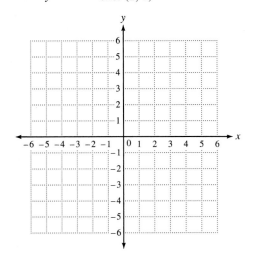

8 $x + y = 3$
$x + 3y = -1$ **one: (5, −2)**

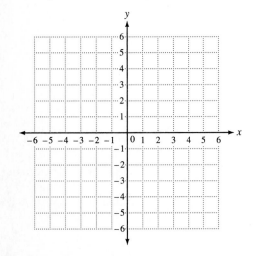

9. $2x + y = 3$
$x + y = 3$ **one: (0, 3)**

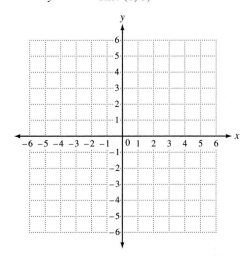

10. $3x - y = 1$
$2y = 4$ **one: (1, 2)**

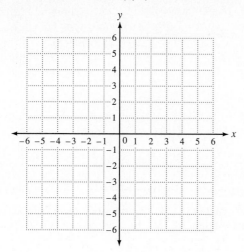

11. $2y = 3$
$3x = 2$ **one:** $\left(\dfrac{2}{3}, \dfrac{3}{2}\right)$

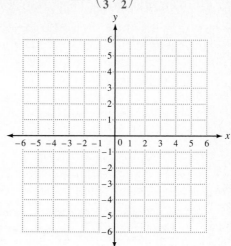

12. $x = y + 1$
$y = x - 1$ **infinitely many**

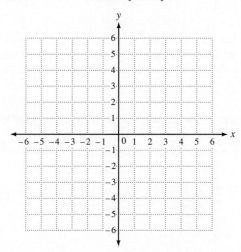

13 $x + 4y = 1$
$-x - 4y = 1$ **no solution**

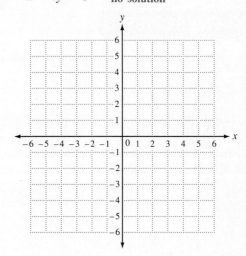

14. $4x - 3y = 2$
$-2x + y = -1$ **one:** $\left(\dfrac{1}{2}, 0\right)$

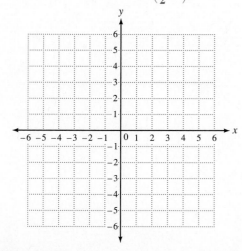

15. $2x + y = 0$
$2x - y = 0$ **one: (0, 0)**

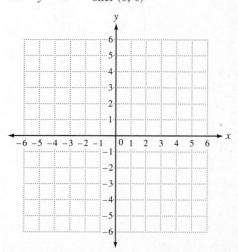

FOR REVIEW

Tell whether (0, 4) is a solution to the given system.

16. $2x + 3y - 12 = 0$
$\quad\quad x - 2y + 8 = 0$
yes

17. $x - y = -4$
$\quad 2x + y = -4$
no

18. $\quad\quad\quad\quad y = 4$
$\quad 2x - 2y + 8 = 0$
yes

ANSWERS: 1. none   2. one (exactly one)   3. infinitely many   4. one solution: (1, 1)   5. infinitely many solutions
6. no solution   7. one solution: (2, 1)   8. one solution: (5, −2)   9. one solution: (0, 3)   10. one solution: (1, 2)
11. one solution: $\left(\frac{2}{3}, \frac{3}{2}\right)$ 12. infinitely many solutions   13. no solution   14. one solution: $\left(\frac{1}{2}, 0\right)$ 15. one solution: (0, 0)   16. yes   17. no   18. yes

13.2 · EXERCISES B

1. If a system of equations has exactly one solution, how are the graphs of the system related?   **intersecting**

2. If a system of equations has no solution, how are the graphs of the system related?   **parallel**

3. If a system of equations has infinitely many solutions, how are the graphs of the system related?   **coinciding**

Find the number of the solutions to the given system. For those that have exactly one solution, find that solution by graphing. Check each solution.

4. $x + 2y = 1$
$\quad x - y = -2$ one: (−1, 1)

5. $\quad x + 2y = 1$
$\quad -x - 2y = -1$ infinitely many

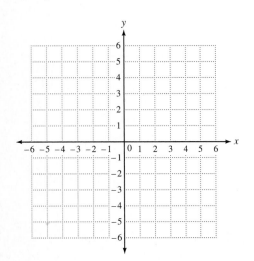

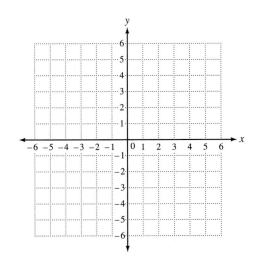

6. $x + 2y = 1$
$-x - 2y = 1$ **no solution**

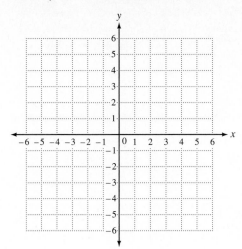

7. $x + 2y = 5$
$x - 2y = -3$ **one: (1, 2)**

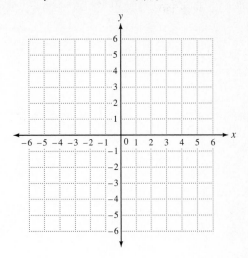

8. $x - y = -4$
$5x + y = 4$ **one: (0, 4)**

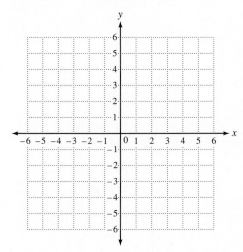

9. $x + y = 2$
$x + 2y = -1$ **one: (5, −3)**

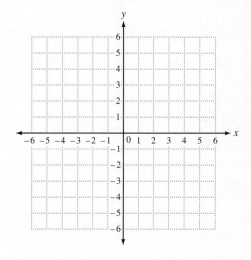

10. $2x + 3y = 6$
$3x = 3$ **one:** $\left(1, \dfrac{4}{3}\right)$

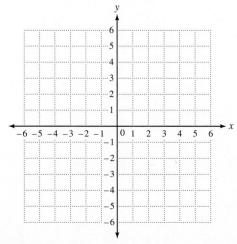

11. $3x = 4$
$4y = 3$ **one:** $\left(\dfrac{4}{3}, \dfrac{3}{4}\right)$

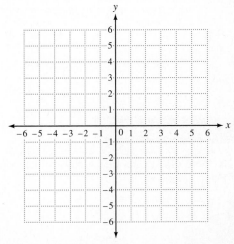

12. $x - 5y = 2$
$-x + 5y = 2$ **no solution**

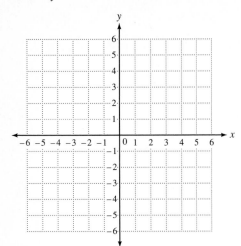

13. $y = x + 2$
$x = y - 2$ **infinitely many**

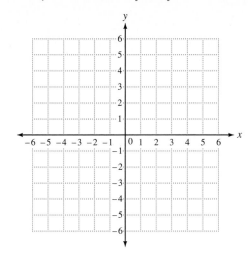

14. $3x - 3y = -1$
$x + 6y = 2$ **one:** $\left(0, \dfrac{1}{3}\right)$

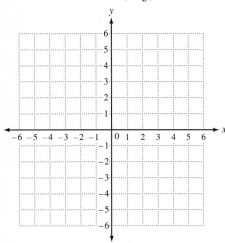

15. $x + 5y = 0$
$x - 5y = 0$ **one: (0, 0)**

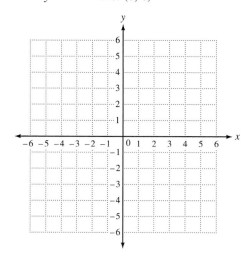

FOR REVIEW

Tell whether $(1, -5)$ *is a solution to the given system.*

16. $2x - y - 7 = 0$
$x + y = -4$
yes

17. $x - 2y = 11$
$2x - 7 = y$
yes

18. $x + 5 = 0$
$y - 1 = 0$
no

13.2 EXERCISES C

1. Find the value of c so that the system will have infinitely many solutions.

$x - 5y + c = 0$
$-3x + 15y - 7 = 0$ $\left[\text{Answer: } c = \frac{7}{3}\right]$

2. Find the value of a so that the system will have no solution.

$ax - 5y + 8 = 0$
$4x + y - 1 = 0$ $a = -20$

13.3 SOLVING SYSTEMS OF EQUATIONS BY SUBSTITUTION

═══════════════ STUDENT GUIDEPOSTS ═══════════════

| **1** The Substitution Method | **2** Contradictions and Identities |

Solving a system by graphing is time-consuming, and estimating the point of intersection depends on the accuracy of the graph. A better method for solving some systems is the **method of substitution.** Consider the system

$$x - 2y = 1$$
$$2x + y = 7.$$

We will change the system into one equation in one unknown. Solve the first equation for x,

$$x - 2y = 1$$
$$x = 2y + 1,$$

and substitute this value of x into the second equation.

$$2x + y = 7$$
$$2(2y + 1) + y = 7 \qquad x = 2y + 1$$

The result is an equation in the single variable y, which we can solve.

$$4y + 2 + y = 7$$
$$5y + 2 = 7$$
$$5y = 5$$
$$y = 1 \qquad \text{This is the } y\text{-coordinate of the solution}$$

Substitute this value of y into either of the original equations. Using the first, solve for x.

$$x - 2y = 1$$
$$x - 2(1) = 1 \qquad y = 1$$
$$x - 2 = 1$$
$$x = 3 \qquad \text{This is the } x\text{-coordinate of the solution}$$

We check the possible solution pair (3, 1) by substituting it into both original equations.

$x - 2y = 1$	$2x + y = 7$
$3 - 2(1) \stackrel{?}{=} 1$	$2(3) + 1 \stackrel{?}{=} 7$
$3 - 2 \stackrel{?}{=} 1$	$6 + 1 \stackrel{?}{=} 7$
$1 = 1$	$7 = 7$

Thus, the solution to the system is (3, 1), the same result obtained in Example 4 in Section 13.2 using the graphing method.

To Solve a System of Equations Using the Substitution Method

1. Solve one of the equations for one of the variables.
2. Substitute that value of the variable in the *remaining* equation.
3. Solve this new equation and substitute the numerical solution into either of the two *original* equations to find the numerical value of the second variable.
4. Check your solution in both original equations.

EXAMPLE 1 USING SUBSTITUTION

Solve by the substitution method.

$$5x + 3y = 17$$
$$x + 3y = 1$$

We could solve either equation for either variable. However, since the coefficient of x in the second equation is 1, we can avoid fractions if we solve for x using that equation.

$$x = 1 - 3y$$

Substitute $1 - 3y$ for x in the first equation.

$$5x + 3y = 17$$
$$5(1 - 3y) + 3y = 17$$
$$5 - 15y + 3y = 17$$
$$5 - 12y = 17$$
$$-12y = 12$$
$$y = -1 \quad \text{\textit{y-coordinate of the solution}}$$

Substitute -1 for y in the second equation.

$$x + 3y = 1$$
$$x + 3(-1) = 1$$
$$x - 3 = 1$$
$$x = 4 \quad \text{\textit{x-coordinate of the solution}}$$

The solution is $(4, -1)$.

$$\text{Check:} \quad 5x + 3y = 17 \qquad x + 3y = 1$$
$$5(4) + 3(-1) \overset{?}{=} 17 \qquad (4) + 3(-1) \overset{?}{=} 1$$
$$20 - 3 \overset{?}{=} 17 \qquad 4 - 3 \overset{?}{=} 1$$
$$17 = 17 \qquad 1 = 1$$

PRACTICE EXERCISE 1

Solve by the substitution method.

$$3x + y = 1$$
$$2x + 3y = 10$$

Start by solving the first equation for y and substituting into the second equation.

$$y = -3x + 1$$
$$2x + 3(-3x + 1) = 10$$

Answer: $(-1, 4)$

EXAMPLE 2 SYSTEM WITH NO SOLUTION

Solve by substitution.

$$4x + 2y = 1$$
$$2x + y = 8$$

If we solve the second equation for y, we obtain

$$y = 8 - 2x.$$

Substitute this value into the first equation.

$$4x + 2y = 1$$
$$4x + 2(8 - 2x) = 1$$
$$4x + 16 - 4x = 1$$
$$16 = 1$$

PRACTICE EXERCISE 2

Solve by substitution.

$$10x + 5y = 2$$
$$-2x - y = 4$$

But $16 \neq 1$. What went wrong? Return to the original system and solve both equations for y.

$$2y = 1 - 4x \qquad \text{and} \quad y = 8 - 2x$$
$$y = \frac{1}{2} - \frac{4}{2}x \qquad\qquad y = -2x + 8$$
$$y = -2x + \frac{1}{2}$$

Both coefficients of x are -2 but the constant terms are different. Thus the two lines are parallel, and there is no solution. If there is no solution (the lines are parallel), substitution results in an equation that is a contradiction (such as $16 = 1$).

Answer: no solution

| **EXAMPLE 3**   SYSTEM WITH INFINITELY MANY SOLUTIONS | **PRACTICE EXERCISE 3** |

Solve by substitution.

Solve by substitution.

$$x + 2y = 1$$
$$2y = 1 - x$$

$$3x - 6y = 9$$
$$-x + 2y = -3$$

We solve the first equation for x.

$$x = 1 - 2y$$

Substitute this value into the second equation.

$$2y = 1 - (1 - 2y)$$
$$2y = 1 - 1 + 2y \qquad \text{Watch the signs}$$
$$2y = 2y$$

Clearly, $2y = 2y$ no matter what number replaces y so that we end up with an equation that is an identity. Return to the original equations and solve them for y.

$$2y = 1 - x \qquad\qquad 2y = 1 - x$$
$$2y = -x + 1 \qquad\qquad 2y = -x + 1$$
$$y = -\frac{1}{2}x + \frac{1}{2} \qquad y = -\frac{1}{2}x + \frac{1}{2}$$

The coefficients of x are equal, as are the constants. The two lines coincide so there are infinitely many solutions to this system (any pair of numbers that is a solution to one equation is a solution to both, hence a solution to the system).

Answer: infinitely many solutions

---

❷ CONTRADICTIONS AND IDENTITIES

Examples 2 and 3 lead to the following.

When Solving a System of Equations

1. If you get a **contradiction** (an equation that is never true), the lines are parallel and there is no solution to the system.

2. If you get an **identity** (an equation that is always true for every value of the variable), the lines coincide and there are infinitely many solutions to the system.

13.3 EXERCISES A

1. Suppose while solving a system of equations using substitution we obtain the equation $3x = 3x$. What would this tell us about the system?

    There are infinitely many solutions.

2. Suppose while solving a system using substitution we obtain the equation $5 = 0$. What would this tell us about the system?

    There are no solutions.

Solve the following systems using the substitution method.

3. $y = 3x + 2$
   $5x - 2y = -7$
   (3, 11)

4. $2x - 3y = 14$
   $x = y + 10$
   (16, 6)

5. $3x + y = 35$
   $x - 2y = 7$
   (11, 2)

6. $13x - 4y = -66$
   $5x + 2y = 10$
   (−2, 10)

7 $3x - 5y = 19$
   $2x - 4y = 16$
   (−2, −5)

8. $3x - y = 7$
   $4x - 5y = 2$
   (3, 2)

9. $x - 3y = 14$
   $x - 2 = 0$
   [*Hint:* Start by solving for x in the second equation.] **(2, −4)**

10 $3x - 3y = 1$
   $x - y = -1$
   no solution

11 $2x + 2y = -6$
   $-x - y = 3$
   infinitely many solutions

12. $y = 5$
   $4x - y = 19$
   (6, 5)

13. $2x + 3y = 5$
   $4x + 7y = 11$
   (1, 1)

14 $3x + 5y = 30$
   $5x + 3y = 34$
   (5, 3)

15. Which variable in which equation is easiest to solve for in Exercise 14? (The next section explains a method for solving equations that may be better for systems such as this.)

    Neither variable is better to solve for than the other.

FOR REVIEW

16. If the lines in a system of equations are parallel, how many solutions does the system have?

   no solution

Solve by the graphing method.

17. $x + y = 5$
   $3x - y = -1$ **(1, 4)**

18. $2x + y = 4$
   $x - 2 = 0$ **(2, 0)**

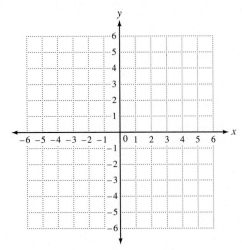

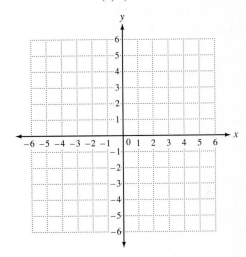

ANSWERS: **1.** There are infinitely many solutions; any solution to one equation is a solution to the other and there-fore a solution to the system.   **2.** There are no solutions.   **3.** (3, 11); first equation is already solved for y **4.** (16, 6); second equation is already solved for x **5.** (11, 2); better to solve for x in second equation   **6.** (−2, 10)   **7.** (−2, −5)   **8.** (3, 2)   **9.** (2, −4)   **10.** no solution   **11.** infinitely many solutions   **12.** (6, 5)   **13.** (1, 1); neither variable is better to solve for than the other   **14.** (5, 3)   **15.** Neither variable is better to solve for than the other.   **16.** no solution   **17.** (1, 4)   **18.** (2, 0)

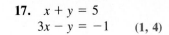

13.3 EXERCISES B

1. Suppose while solving a system of equations using substitution we obtain the equation $-3 = 0$. What would this tell us about the system?   **There are no solutions.**

2. Suppose while solving a system using substitution we obtain the equation $x + 1 = x + 1$. What would this tell us about the system?   **There are infinitely many solutions.**

Solve the following systems using the substitution method.

3. $y = -2x + 3$
   $3x - 4y = -1$
   (1, 1)

4. $5x - y = -2$
   $x = 3y - 6$
   (0, 2)

5. $x + y = 0$
   $5x + 2y = -3$
   (−1, 1)

6. $2x - 8y = -10$
   $x - 3y = -2$
   (7, 3)

7. $2x - y = 3$
   $-4x + 2y = -6$
   infinitely many

8. $4x - y = 13$
   $x + 2y = 1$
   (3, −1)

9. $3x - y = 2$
$-6x + 2y = 2$

no solution

10. $2x - y = -2$
$y - 4 = 0$

$(1, 4)$

11. $4x + y = -2$
$-2x - 3y = 1$

$\left(-\dfrac{1}{2}, 0\right)$

12. $3x - y = 14$
$x - 4 = 0$
$(4, -2)$

13. $2x + 3y = -4$
$3x - 2y = 7$
$(1, -2)$

14. $3x + 7y = 32$
$7x + 3y = 8$
$(-1, 5)$

15. What is the difficulty in using the substitution method on a system like the one in Exercise 14?
It would involve extensive work with fractions.

FOR REVIEW

16. If the lines in a system of equations are coinciding, how many solutions does the system have?
infinitely many

Solve by the graphing method.

17. $3x - y = -4$
$-x + y = 0$ $(-2, -2)$

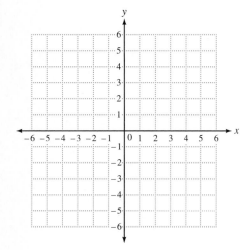

18. $x - 3y = 3$
$y + 1 = 0$ $(0, -1)$

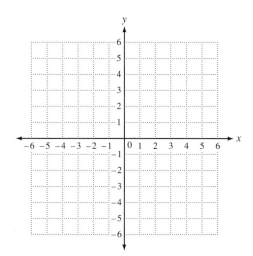

13.3 EXERCISES C

Solve using the substitution method.

1. $0.03x - 0.01y = 0.12$
$0.04x - 0.05y = 0.20$
$\left(\dfrac{40}{11}, -\dfrac{12}{11}\right)$

2. $\dfrac{x}{10} + \dfrac{2y}{5} = 1$

$\dfrac{2x}{5} + \dfrac{3y}{2} = 2$ [Answer: $(-70, 20)$]

13.4 SOLVING SYSTEMS OF EQUATIONS BY ELIMINATION

=========================== STUDENT GUIDEPOSTS ===========================

1 The Elimination Method   **3** Method to Use
2 Multiplication Rule

1 THE ELIMINATION METHOD

Consider the following system.

$$3x + 5y = 30$$
$$5x + 3y = 34$$

Solving for either variable in either equation will result in fractional coefficients. Another method, better than substitution for systems like this one, is known as the **elimination method.** It is based on the addition-subtraction rule from Chapter 11: if the same quantity is added to or subtracted from both sides of an equation, another equation with the same solutions is obtained. Suppose we illustrate the method using the following system.

$$x + y = 6$$
$$3x - y = 2.$$

In the second equation, $3x - y$ and 2 are both names for the same number, 2. Therefore we can add $3x - y$ to the left side of the first equation, and 2 to the right of it, to obtain another equation with the same solution.

$$(x + y) + (3x - y) = 6 + 2$$
$$4x = 8 \qquad y \text{ has been eliminated}$$

Generally, we add the equations vertically.

$$
\begin{array}{l}
x + y = 6 \\
\underline{3x - y = 2} \\
4x \qquad = 8 \\
\qquad x = 2 \qquad \text{Divide by 4 to obtain the } x\text{-coordinate} \\
\qquad\qquad\qquad\qquad \text{of the solution}
\end{array}
$$

By adding the two equations, we obtain one equation in one variable, x. Once we know x is 2, we substitute this value into either one of the original equations (just as in the substitution method) to find the value of y.

$$2 + y = 6 \qquad \text{Substitute in the first equation since it is simpler}$$
$$y = 4 \qquad \text{The } y\text{-coordinate of the solution}$$

Thus, the solution is (2, 4). Check this by substituting into both equations.

EXAMPLE 1 USING THE ELIMINATION METHOD	**PRACTICE EXERCISE 1**
Solve by the elimination method.	Solve by the elimination method.
$$x + y = 5$$ $$-2x + y = -4$$	$$4x + 3y = -1$$ $$2x - 3y = 13$$

If we add both equations, the resulting equation still has both variables. (Try it.) So in this case, we subtract instead of adding, to obtain an equation in one variable.

$$
\begin{array}{r}
x + y = 5 \\
+2x - y = 4 \\
\hline
3x \quad\;\; = 9 \\
x = 3
\end{array}
$$

Change all signs when subtracting and add

Substitute 3 for x in the first equation.

$$
3 + y = 5
$$
$$
y = 2
$$

The solution is the ordered pair (3, 2). Check by substitution.

Answer: $(2, -3)$

❷ MULTIPLICATION RULE

Sometimes addition or subtraction does not yield one equation in only one variable. When this occurs, we may have to multiply one (or both) of the equations by a number so that the coefficients of one variable are negatives of each other. We then add to eliminate that variable.

| EXAMPLE 2   USING MULTIPLICATION AND ELIMINATION | PRACTICE EXERCISE 2 |

Solve by the elimination method.

$$
2x - 5y = 13
$$
$$
x + 2y = 11
$$

We would not obtain one equation with one variable by simply adding or subtracting in this case. However, if we multiply both sides of the second equation by -2 the resulting equation is

$$
-2x - 4y = -22. \qquad \text{-2 times second equation}
$$

Now add this equation to the first equation.

$$
\begin{array}{r}
2x - 5y = 13 \\
-2x - 4y = -22 \\
\hline
-9y = -9 \\
y = 1
\end{array}
$$

Substitute 1 for y in the second equation.

$$
x + 2(1) = 11
$$
$$
x + 2 = 11
$$
$$
x = 9
$$

The solution is (9, 1). Check by substitution.

Solve by the elimination method.

$$
6x + 5y = 9
$$
$$
4x + y = -1
$$

Answer: $(-1, 3)$

Sometimes the multiplication rule has to be applied to both equations before adding or subtracting. In Example 3 this technique is used to solve the system at the beginning of this section.

EXAMPLE 3 USING MULTIPLICATION AND ELIMINATION	**PRACTICE EXERCISE 3**

Solve by the elimination method.

$$3x + 5y = 30$$
$$5x + 3y = 34$$

Multiply the first equation by -5 and the second equation by 3 (making the coefficients of x negatives of each other).

$$-15x - 25y = -150 \quad \text{–5 times first equation}$$
$$15x + 9y = 102 \qquad \text{3 times second equation}$$

Add to get one equation in one variable.

$$-16y = -48$$
$$y = 3$$

Substitute 3 for y in the first equation.

$$3x + 5(3) = 30$$
$$3x + 15 = 30$$
$$3x = 15$$
$$x = 5$$

The solution is $(5, 3)$. Check by substitution.

Compare this example with your work on Exercise 14-A in the last section. In this case the elimination method is faster and easier to use than the substitution method.

Practice Exercise 3 column:

Solve by the elimination method.

$$2x + 7y = -4$$
$$5x + 3y = 19$$

Answer: $(5, -2)$

To Solve a System by the Elimination Method

1. Write both equations in the form $ax + by = c$.
2. If necessary, apply the multiplication rule to each equation so that addition or subtraction will eliminate one of the variables.
3. Solve the resulting single-variable equation. Substitute this value into one of the original equations. The resulting pair of numbers is the ordered-pair solution to the system.
4. Check your answer by substitution in both original equations.

EXAMPLE 4 CONTRADICTIONS AND IDENTITIES	**PRACTICE EXERCISE 4**

Solve by the elimination method.

(a) $2x = 3y + 5$
  $-2x + 3y = -1$

Rewrite the first equation as $2x - 3y = 5$. Notice that by adding the equations, we can eliminate x.

$$2x - 3y = 5$$
$$-2x + 3y = -1$$
$$\overline{0 = 4}$$

Practice Exercise 4 column:

Solve by the elimination method.

(a) $-3x + 9y = -18$
  $x - 3y = 6$

We actually eliminated both variables and obtained a contradiction. Just as with the substitution method, this means there is no solution to the system. Verify that the lines are parallel by solving each equation for y.

(b) $2x + y = 3$
$-4x - 2y = -6$

(b) $8x = -4y + 12$
$4x + 2y = -3$

Multiply the first equation by 2 and add.

$$\begin{aligned} 4x + 2y &= 6 \\ \underline{-4x - 2y} &= \underline{-6} \\ 0 &= 0 \end{aligned}$$

Again, both variables are eliminated by this process, but this time we obtain an identity. This means there are infinitely many solutions to the system. Verify that the lines coincide by solving each equation for y.

Answers: (a) **infinitely many solutions**   (b) **no solution**

❸ METHOD TO USE

The question is often asked: "Which method should I use?" If a coefficient of one of the variables is 1 (or -1), it may be better to solve first for that variable using the substitution method. If none of the coefficients is 1 (or -1), it is probably better to use the elimination method. Keep in mind that with either method, the system has no solution if at any step a contradiction is obtained, and the system has infinitely many solutions if at any step an identity is obtained.

13.4 EXERCISES A

1. If we obtain an equation such as $0 = -3$ when using the elimination method, how many solutions does the system have?   **no solution**

2. If we obtain an equation such as $0 = 0$ when using the elimination method, how many solutions does the system have?   **infinitely many**

Solve the following systems using the elimination method. Check by substitution.

3. $x + y = 6$
$x - y = 4$

   (5, 1)

4. $3x - 2y = 12$
$2x + 2y = 13$

   $\left(5, \dfrac{3}{2}\right)$

5. $6x + 5y = 11$
$3x - 7y = -4$
(1, 1)

6. $x - \ \ y = 16$
$x - 3y = 2$
(23, 7)

7. $9x - y = 19$
$3x + 7y = -1$
(2, −1)

8. $4x + y = -3$
$-8x - 2y = 6$
infinitely many solutions

9. $-2y = 2 - x$
$3x - 6y = -4$
no solution

10 $2x + 3y = -4$
$1 + 2y = -3x$
(1, −2)

11 $3x + 7y = -21$
$7x + 3y = -9$
(0, −3)

12. $5x - 2y = 1$
$3x + 7y = -24$
(−1, −3)

Decide whether the substitution or the elimination method is more appropriate, then use it to solve the system.

13. $x + 5y = 0$
$x - 5y = 20$
(10, −2)

14. $x - 4y = -6$
$3x - 5y = -4$
(2, 2)

15. $4x - 5y = 9$
$5x + 4y = 1$
(1, −1)

16 $3x + 3y = 3$
$4x + 4y = -3$
no solution

17 $2x + 3y = -4$
$5x + 7y = -10$
(−2, 0)

18. $3x = 9$
$x - y = 7$
(3, −4)

19. $y = 3x + 1$
$4x + 5y = 24$
$(1, 4)$

20. $y - 5 = 0$
$x + 2 = 0$
$(-2, 5)$

FOR REVIEW

Exercises 21–25 review material from Section 11.4 to help you prepare for the next section. Use x for the variable and translate each sentence into symbols. Do not solve.

21. The sum of a number and 5 is 12. $x + 5 = 12$

22. Four times my age in 2 years is 60. $4(x + 2) = 60$

23. The measure of an angle is 10° less than twice its measure. $x = 2x - 10$

24. 40% of the cost of an item is $20. $0.40x = 20$

25. $60 is the sale price of an item discounted by 20%. $60 = x - 0.20x$

ANSWERS: 1. no solution   2. infinitely many   3. (5, 1)   4. $\left(5, \frac{3}{2}\right)$ 5. (1, 1)   6. (23, 7)   7. (2, −1)   8. infinitely many solutions   9. no solution   10. (1, −2)   11. (0, −3)   12. (−1, −3)   13. (10, −2)   14. (2, 2)   15. (1, −1)   16. no solution   17. (−2, 0)   18. (3, −4)   19. (1, 4)   20. (−2, 5)   21. $x + 5 = 12$ 22. $4(x + 2) = 60$ 23. $x = 2x - 10$ 24. $0.40x = 20$ 25. $60 = x - 0.20x$

13.4 EXERCISES B

1. When solving a system of equations using the elimination method, if a contradiction results, how many solutions does the system have?   **no solution**

2. When solving a system of equations using the elimination method, if an identity results, how many solutions does the system have?   **infinitely many solutions**

Solve the following systems using the elimination method. Check by substitution.

3. $x + y = 3$
$x - y = -1$
$(1, 2)$

4. $2x - 3y = -7$
$3x + 3y = 27$
$(4, 5)$

5. $x - y = 0$
$x - 5y = -12$
$(3, 3)$

6. $3x - 2y = 10$
$5x + 3y = 4$
$(2, -2)$

7. $x + 3y = 1$
$3x + y = 11$
$(4, -1)$

8. $2x + 3y = -4$
$3x - 2y = 7$
$(1, -2)$

9. $5x - 2y = 1$
$7 + 10x = 4y$
no solution

10. $2x - 2y = 5$
$4y = 3x - 7$
$\left(3, \dfrac{1}{2}\right)$

11. $4x - 2y = 2$
$-2x + y = -1$
infinitely many solutions

12. $3x + 2y = -10$
$5x - 3y = 15$
$(0, -5)$

Decide whether the substitution or the elimination method is more appropriate, then use it to solve the system.

13. $x + 3y = 4$
$x + 8y = -6$
$(10, -2)$

14. $x - 7y = 44$
$5x + 3y = -8$
$(2, -6)$

15. $5x + 7y = -2$
$7x + 5y = 2$
$(1, -1)$

16. $3x - 7y = -4$
$7x + 3y = 10$
$(1, 1)$

17. $2x = 8$
      $x - y = 1$
      (4, 3)

18. $2x + 5y = -10$
      $3x + 11y = -22$
      (0, -2)

19. $x = 4y - 3$
      $2x + 3y = 5$
      (1, 1)

20. $y + 7 = 0$
      $x - 3 = 0$
      (3, -7)

FOR REVIEW

Exercises 21–25 review material from Section 11.4 to help you prepare for the next section. Use x for the variable and translate each sentence into symbols. Do not solve.

21. A number less 2 is 14.   **$x - 2 = 14$**

22. Twice a number added to 6 is 28.   **$2x + 6 = 28$**

23. The measure of an angle is 25° less than three times its measure.   **$x = 3x - 25$**

24. 15% of the population is 3000.   **$0.15x = 3000$**

25. Her salary plus a 12% increase is $22,400.   **$x + 0.12x = 22,400$**

13.4 EXERCISES C

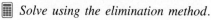 *Solve using the elimination method.*

1. $0.015x - 0.007y = 0.032$
      $-0.009x + 0.014y = -0.001$
      $\left[\text{Answer: } \left(3, \frac{13}{7}\right)\right]$

2. $-\dfrac{3x}{4} + \dfrac{7y}{10} = \dfrac{1}{5}$

   $\dfrac{5x}{2} - \dfrac{4y}{3} = -\dfrac{1}{12}$

   $\left(\dfrac{5}{18}, \dfrac{7}{12}\right)$

13.5 APPLICATIONS USING SYSTEMS

STUDENT GUIDEPOSTS

1 Problem-Solving Techniques

3 Motion Problems

2 Mixture Problems

1 PROBLEM-SOLVING TECHNIQUES

Many applied problems involving two unknown quantities can be translated into a system of linear equations in two variables. The following steps may be helpful and are illustrated in the examples.

To Solve an Applied Problem Using a System of Equations

1. Read the problem several times and try to determine what is being asked. Let x represent one of the unknowns and y the other.

2. Write two different equations using the variables x and y. Simplify the equations (if necessary) by reducing the size of the numerical coefficients and eliminating fractions or decimals.

3. Solve the system of equations.

4. Check to see that the answers actually solve the original problem.

EXAMPLE 1 NUMBER PROBLEM

The sum of two numbers is 21 and their difference is 9. Find the numbers.

Let x = the larger number,
  y = the smaller number.

Then $x + y = 21$ Represents "their sum is 21"

  $x - y = 9$ Represents "their difference is 9"

We should solve the system by the simplest method. In this case, one method is as easy as the other so we add the two equations.

$$2x = 30$$
$$x = 15$$

Now substitute 15 for x in $x + y = 21$.

$$15 + y = 21$$
$$y = 6$$

The numbers are 15 and 6. Check.

PRACTICE EXERCISE 1

If twice one number is added to a second number, the result is 25. If the first number is the larger of the two and their difference is 5, find the numbers.

Let x = the first number,
  y = the second number.

Answer: 10, 5

EXAMPLE 2 PROBLEM INVOLVING AGES

The sum of Bob's and Joe's ages is 14. In 2 years Bob will be twice as old as Joe. What are their present ages?

Let x = Bob's present age,
  y = Joe's present age.

Then both Bob and Joe will age 2 years and

$$x + 2 = \text{Bob's age in 2 years,}$$
$$y + 2 = \text{Joe's age in 2 years.}$$

In two years Bob will be twice as old as Joe

$$(x + 2) \quad = \quad 2 \quad (y + 2)$$

$$x + 2 = 2(y + 2)$$
$$x + 2 = 2y + 4$$
$$-2 = 2y - x \quad \text{or} \quad -x + 2y = -2$$

Since the sum of their ages is 14, $x + y = 14$. This gives us the following system.

$$-x + 2y = -2$$
$$x + y = 14$$

Add these equations. $3y = 12$

$$y = 4$$

Now substitute 4 for y in $x + y = 14$.

$$x + 4 = 14$$
$$x = 10$$

Bob is 10 and Joe is 4. Check.

PRACTICE EXERCISE 2

Annette is $\frac{3}{4}$ as old as Mona. How old is each if the sum of their ages is 84?

Let x = Annette's age,
  y = Mona's age,

  $\frac{3}{4}y$ = three-fourths Mona's age.

Answer: Annette is 36, Mona is 48.

<table>
<tr><td>

EXAMPLE 3 GEOMETRY PROBLEM

Two angles are **supplementary** (their sum is 180°). If one is 60° more than twice the other, find the angles.

Let x = one angle,

y = the other angle.

Then $\qquad x + y = 180.$ The angles are supplementary

If x is 60° more than $2y$, then we would add 60° to $2y$ to get x. This gives the following system.

$$x + y = 180$$
$$2y + 60 = x$$

Substitute $2y + 60$ for x in $x + y = 180$.

$$(2y + 60) + y = 180$$
$$3y + 60 = 180$$
$$3y = 120$$
$$y = 40$$

Now substitute 40 for y in $x + y = 180$.

$$x + 40 = 180$$
$$x = 140$$

The angles are 40° and 140°. Check.

</td><td>

PRACTICE EXERCISE 3

Two angles are **complementary** (their sum is 90°). If one is 12° less than twice the other, find the angles.

Answer: 34°, 56°

</td></tr>
</table>

② MIXTURE PROBLEMS

A useful application of systems is the **mixture problem.** The two equations in the system of the mixture problem are the *quantity equation* and the *value equation*.

To set up a value equation, we need to look at the idea of *total value*. For a given number of units, each of equal value, the total value can be computed by

total value = (value per unit)(number of units).

The following are examples of total value calculations.

Price Given Per Pound
Find the value of 5 lb of candy worth $1.10 per lb.

$$\text{total value} = (\text{value per lb})(\text{number of lb})$$
$$= (\$1.10)(5)$$
$$= \$5.50$$

Value of Coins
Find the value of 30 nickels.

$$\text{total value} = (\text{value per nickel})(\text{number of nickels})$$
$$= (5\cent)(30)$$
$$= 150\cent \quad \text{or} \quad \$1.50$$

The next examples are applications of the total value idea.

Percent Solutions

Find the amount of acid in 16 liters of 20% acid solution.

$$\text{total value} = \text{amount of acid in solution}$$
$$= (\text{percent acid})(\text{liters of solution})$$
$$= (0.20)(16)$$
$$= 3.2 \text{ liters of acid}$$

Interest

Find the annual interest on $8000 invested at 6.5% simple interest.

$$\text{total value} = \text{amount of interest}$$
$$= (\text{percent interest})(\text{amount invested})$$
$$= (0.065)(\$8000)$$
$$= \$520$$

These examples will be helpful in solving the following mixture problems.

EXAMPLE 4 MIXTURE OF CANDY AND NUTS	PRACTICE EXERCISE 4

A man wishes to mix candy selling for 75¢ a pound with nuts selling for 50¢ per pound to obtain a party mix to be sold for 60¢ per pound. How many pounds of each must be used to obtain 50 pounds of the mixture?

Let x = number of pounds of candy,
y = number of pounds of nuts.

The first equation we set up is the quantity equation

$$(\text{no. lb of candy}) + (\text{no. lb of nuts}) = (\text{no. lb of mixture}),$$

which translates to $x + y = 50$.

Next we find the value equation. Since the value of a quantity equals the sum of the values of its parts,

$$(\text{total value of candy}) + (\text{total value of nuts}) = \text{total value of mixture}.$$

$$75 \cdot x = \text{value of candy}$$
$$50 \cdot y = \text{value of nuts}$$
$$60 \cdot 50 = \text{value of mixture}$$

This gives the value equation

$$75x + 50y = 60 \cdot 50.$$

We need to solve the following system of equations.

$$x + \quad y = 50$$
$$75x + 50y = 3000$$

Solve the first equation for x,

$$x = 50 - y,$$

and substitute in the second equation

$$75\,(50 - y) + 50y = 3000$$
$$3750 - 75y + 50y = 3000$$
$$-25y = -750$$
$$y = 30$$

Dr. Lynn Mowen wants to mix a 10% alcohol solution and a 70% alcohol solution to obtain 30 liters of 50% alcohol solution. How much of each must she use to obtain the desired mixture?

Let x = liters of 10% solution to use,
y = liters of 70% solution to use.

$$x + y = \underline{\qquad}$$
$$0.10x + 0.70y = 0.50(30)$$

Now substitute 30 for y in $x = 50 - y$.

$$x = 50 - \boxed{30} = 20$$

The man must use 20 lb of candy and 30 lb of nuts.

Answer: 10 liters of 10% and 20 liters of 70%

EXAMPLE 5   MIXTURE OF COINS

A boy has 25 nickels and dimes. If the coins' total value is $2.10, find the number of nickels and the number of dimes in the collection.

Let x = number of nickels,
   y = number of dimes.

The quantity equation

   (no. of nickels) + (no. of dimes) = (no. coins in the collection)

translates to $x + y = 25$.

The value equation

   (value of nickels) + (value of dimes) = (value of collection)

becomes $5 \cdot x + 10 \cdot y = 210$.   Convert to cents

Solve the first equation for y,

$$y = 25 - x,$$

and substitute into the second.

$$5x + 10\,\boxed{(25 - x)} = 210$$
$$5x + 250 - 10x = 210$$
$$-5x = -40$$
$$x = 8$$

Now substitute 8 for x in $y = 25 - x$.

$$y = 25 - \boxed{8} = 17$$

There are 8 nickels and 17 dimes in the collection.

PRACTICE EXERCISE 5

Jose Garcia invests a total of $8000 in two accounts, one paying 10% simple interest and the other 12% simple interest. How much did he invest in each account if his total interest for the year was $930?

Answer: $1500 at 10% and $6500 at 12%

---

❸ MOTION PROBLEMS

Another type of problem that can be solved using systems is the **motion problem.**

EXAMPLE 6   MOTION PROBLEM

A boat travels 60 mi upstream in 4 hr and returns to the starting place in 3 hr. What is the speed of the boat in still water, and what is the speed of the stream?

Let x = speed of the boat in still water,
   y = speed of the stream,
   $x - y$ = speed of the boat going upstream (against current),
   $x + y$ = speed of the boat going downstream (with current).

PRACTICE EXERCISE 6

A boat can travel 30 mi upstream in 3 hr and return in 2 hr. What are the speed of the boat in still water and the speed of the stream?

Since $d = rt$ (distance equals rate times time), we have the following system.

$$60 = (x - y)(4)$$
$$60 = (x + y)(3)$$

Here we can simplify the system by dividing the first equation by 4 and the second equation by 3.

$$x - y = 15$$
$$\underline{x + y = 20}$$
$$2x \quad = 35 \qquad \text{Add the equations}$$
$$x = \frac{35}{2}$$
$$x = 17.5$$

Substituting 17.5 into $x + y = 20$ gives

$$17.5 + y = 20$$
$$y = 2.5.$$

Thus, the speed of the boat is 17.5 mph, and the speed of the stream is 2.5 mph.

Answer: speed of boat is 12.5 mph, speed of stream is 2.5 mph

13.5 EXERCISES A

Solve. (Some problems have been started.)

1. The sum of two numbers is 39 and their difference is 13. Find the numbers.

Let x = the first number,
y = the other number.

$x + y =$
$x - y =$
13, 26

2. Mike is 2 years younger than Burford. Four years from now the sum of their ages will be 36. How old is each now?

Let x = Burford's present age,
y = Mike's present age,

$x + 4$ = Burford's age in 4 years,
$y + 4$ = Mike's age in 4 years.
Burford is 15, Mike is 13.

3. Two angles are supplementary (their sum is 180°) and one is 5° more than six times the other. Find the angles.
25°, 155°

4. If three times the smaller of two numbers is increased by five times the larger, the result is 195. Find the two numbers if their sum is 47.

Let x = the smaller number,
y = the larger number

$3x + 5y =$
20, 27

5. Mr. Smith has two sons. If the sum of their ages is 19 and the difference between their ages is 3, how old are his sons?
11 yr, 8 yr

6. Bill is three times as old as Jim. Find the age of each if the sum of their ages is 24.
6 yr, 18 yr

7. One-third of one number is the same as twice another. The sum of the first and five times the second is 33. Find the numbers.
18, 3

8. Two angles are *complementary* (their sum is 90°) and their difference is 66°. Find the angles.
12°, 78°

9. A 30-ft rope is cut into two pieces. One piece is 8 ft longer than the other. How long is each?
11 ft, 19 ft

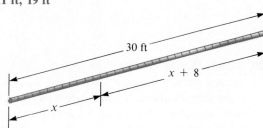

10. The starting players on a basketball team scored 40 points more than the reserves. If the team scored a total of 90 points, how many points did the starters score?
65

11. A candy mix sells for $1.10 per lb. If it is composed of two kinds of candy, one worth 90¢ per lb and the other worth $1.50 per lb, how many pounds of each would be in 30 lb of the mixture?
10 lb ($1.50), 20 lb (90¢)

12. A collection of nickels and dimes is worth $2.85. If there are 34 coins in the collection, how many of each are there?

Let x = number of nickels,
    y = number of dimes,
   $5x$ = value of nickels (in cents),
  $10y$ = value of dimes (in cents).

$5x + 10y =$
$x + \quad y =$
11 nickels, 23 dimes

13. A collection of 13 coins consists of dimes and quarters. If the value of the collection is $2.80, how many of each are there?
3 dimes, 10 quarters

14. A man wishes to mix two grades of nuts selling for 48¢ and 60¢ per lb. How many pounds of each must he use to get a 20-lb mixture that sells for 54¢ per lb? (Does your answer seem reasonable?)
10 lb (48¢), 10 lb (60¢)

15. There were 450 people at a play. If the admission price was $2.00 for adults and 75¢ for children and the receipts were $600.00, how many adults and how many children were in attendance?
210 adults, 240 children

16. A painter needs to have 24 gallons of paint that contains 30% solvent. How many gallons of paint containing 20% solvent and how many gallons containing 60% solvent should be mixed to obtain the desired paint?
18 gal of 20%, 6 gal of 60%

17 When riding his bike against the wind, Terry Larsen can go 40 mi in 4 hr. He takes 2 hr to return riding with the wind. What are Terry's average speed in still air and the average wind speed?
Terry: 15 mph, wind: 5 mph

18. Victoria Harris can ride her bike 120 mi with the wind in 5 hr and return against the wind in 10 hr. What are her speed in still air and the speed of the wind?
Victoria: 18 mph, wind: 6 mph

19 Vince McKee has $10,000 to invest, part at 10% simple interest, and the rest at 12% simple interest. If he wants to have $1160 in interest at the end of the year, how much should he invest at each rate?
$2000 at 10%, $8000 at 12%

20. Wendy Young has $20,000 invested. Part of the money earns 9% simple interest and the rest earns 11% simple interest. If her interest income for one year was $2060, how much did she have invested at each rate?
$7000 at 9%, $13,000 at 11%

21 International Car Rentals charges a daily fee plus a mileage fee. Jack Pritchard was charged $82.00 for 3 days and 400 mi, while Bill Poole was charged $120.00 for 5 days and 500 mi. What are the daily rate and the mileage rate?
$14 per day, 10¢ per mile

22. Harder-Than-Ever Car Rentals charged Terry McGinnis $76.00 for 2 days and 300 mi. For the same type of car, Linda Youngman was charged $132.00 for 6 days and 100 mi. What are the daily fee and the mileage fee?
$20 per day, 12¢ per mile

FOR REVIEW

Solve each system using either the elimination or the substitution method, whichever seems most appropriate.

23. $2x - 3y = 20$
$7x + 5y = -54$
$(-2, -8)$

24. $x - 11y = -5$
$7x + 13y = -35$
$(-5, 0)$

ANSWERS: 1. 13, 26 2. Burford is 15, Mike is 13. 3. 25°, 155° 4. 20, 27 5. 11 yr, 8 yr 6. 6 yr, 18 yr
7. 18, 3 8. 12°, 78° 9. 11 ft, 19 ft 10. 65 11. 10 lb ($1.50), 20 lb (90¢) 12. 11 nickels, 23 dimes 13. 3 dimes,
10 quarters 14. 10 lb (48¢), 10 lb (60¢) 15. 210 adults, 240 children 16. 18 gal of 20%, 6 gal of 60% 17. Terry:
15 mph, wind: 5 mph 18. Victoria: 18 mph, wind: 6 mph 19. $2000 at 10%, $8000 at 12% 20. $7000 at 9%,
$13,000 at 11% 21. $14 per day, 10¢ per mile 22. $20 per day, 12¢ per mile 23. (−2, −8) 24. (−5, 0)

13.5 EXERCISES B

Solve.

1. If three times the smaller of two numbers is increased by two times the larger, the result is 16. Find the two numbers if their sum is 7.
   2, 5

2. Julie is 4 years older than Randy. Five years from now the sum of their ages will be 24. How old is each now?
   Julie is 9, Randy is 5.

3. Two angles are supplementary (their sum is 180°) and one is 15° more than twice the other. Find the angles.
   125°, 55°

4. If four times the larger of two numbers is decreased by six times the smaller, the result is 4. Find the two numbers if they sum up to 16.
   6, 10

5. Tim Chandler has two sons. If the sum of their ages is 30 and the difference between their ages is 4, how old are his sons?
   17 yr, 13 yr

6. Sherry Maducci is five times as old as Mac Davis. Find the age of each if the sum of their ages is 42.
   Sherry is 35, Mac is 7.

7. One half of one number is the same as three times a second. The sum of twice the second and the first is 80. Find the numbers.
   60, 10

8. Two angles are complementary (their sum is 90°) and one is 15° more than four times the other. Find each angle.
   15°, 75°

9. A 45-inch wire is cut into two pieces. One piece is 11 inches longer than the other. How long is each?
   17 in, 28 in

10. The starting five on the Memphis State basketball team scored 35 more points than the reserves. If the team scored a total of 85 points, how many points did the starters score?
    60 points

11. A collection of nickels and dimes is worth $2.00. If the number of dimes minus the number of nickels is 5, how many of each are there?
    10 nickels, 15 dimes

12. A man wishes to make a party mix consisting of nuts worth 50¢ per lb and candy worth 80¢ per lb. How many pounds of each must he use to make 100 lb of the mixture that will sell for 68¢ per lb?
    40 lb of nuts, 60 lb of candy

13. Mark Bonnett has 40 coins in his piggy bank, consisting of dimes and quarters. If the value of the collection is $5.80, how many of each are in the bank?
    12 quarters, 28 dimes

14. A grocer mixes two types of candy selling for 50¢ and 70¢ per lb. How many pounds of each must he use to obtain a 30-lb mixture that sells for 60¢ per lb?
    15 lb of each

15. There were 12,000 people at a rock concert at Colorado State University. If the admission price was $7.00 for a student with an I.D. and $10.00 for a nonstudent, how many of each were in attendance if the total receipts amounted to $94,500?
    8500 students, 3500 nonstudents

16. Dr. Karen Winter needs 20 liters of 70% alcohol solution in her clinic. How many liters of 40% solution and how many liters of 90% solution must be mixed to obtain the desired solution?
    8 liters of 40%, 12 liters of 90%

17. While driving his boat upstream, Charlie Moore can go 75 mi in 5 hr. It takes him 3 hr to make the return trip downstream. What are the average speed of the boat in still water and the average stream speed?
    boat: 20 mph, stream: 5 mph

18. Maria Lopez traveled 60 mi downstream in 2 hr but needed 5 hr to go back upstream to her starting point. What are the average speed of her boat in still water and the average speed of the stream?
    boat: 21 mph, stream: 9 mph

19. Grady and Barbara wish to invest $5000, part at 8% simple interest, and the rest at 14% simple interest. If they must earn $610 by the end of one year, how much should they invest at each rate?

$1500 at 8%, $3500 at 14%

20. A student has $1200 invested. Part of this money earns 12% simple interest, and the rest earns 10% simple interest. How much is invested at each rate if the interest income for the year will be $140?

$1000 at 12%, $200 at 10%

21. Horn Rentals charges a daily fee and a mileage fee to rent a truck. Syd was charged $375 for 5 days and 1500 mi while Barry was charged $165 for 3 days and 600 mi. What are the daily rate and the mileage rate?

$15 per day, 20¢ per mile

22. Better-Than-Ever Car Rentals charges a daily fee and a mileage fee. Renee paid $110 for 10 days and 200 mi, while Pamela rented the same type of car and paid $168 for 6 days and 800 mi. What are the daily fee and the mileage fee?

$8 per day, 15¢ per mile

FOR REVIEW

Solve the system using either the elimination or the substitution method, whichever seems most appropriate.

23. $3x + 2y = 2$
    $5x - 2y = 30$
    $(4, -5)$

24. $4x - y = 6$
    $3x + 5y = -30$
    $(0, -6)$

13.5 EXERCISES C

Solve. A calculator would be helpful in these exercises.

1. On a Pacific island Angie is charged $157.92 after driving a rented car for 420 km in 5 days. Andy paid $217.56 for 7 days and 560 km. If each bill includes a 5% pleasure tax, what was the daily fee and what was the fee charged per km? [*Hint:* Find the fees before tax first.]

   $20 per day, 12¢ per km

2. John and Pat Woods invested $22,000, part at 14% simple interest and the rest at 9% simple interest. How much was invested at 14% if the total interest received in three years was $8040?

   [Answer: $14,000]

CHAPTER 13 REVIEW

KEY WORDS

13.1 A **system of equations** is a pair of linear equations.

A **solution** to a system of equations is an ordered pair of numbers (x, y) that is a solution to both equations.

Coinciding lines are the graphs of equations that represent the same line.

Parallel lines are lines that do not intersect.

Intersecting lines have exactly one point in common.

13.5 A **mixture problem** involves two quantities mixed together and is solved by finding a quantity equation and a value equation.

A **motion problem** makes use of the equation $d = rt$, distance equals rate times time.

KEY CONCEPTS

13.1, A system of equations has:
13.2 1. infinitely many solutions if the lines coincide,

2. no solution if the lines are parallel,

3. exactly one solution if the lines intersect.

13.2 The method of graphing is not usually used to solve a system since it is time-consuming and depends on our ability to graph accurately.

13.3, 1. To solve a system of equations, use either
13.4 the substitution method or the elimination
method, whichever seems more appropriate.

2. If you reach an identity (such as $0 = 0$)
when solving a system, the system has infinitely many solutions.

3. If you reach a contradiction (such as $3 = 0$)
when solving a system, the system has no solution.

4. When solving a system, be sure to find the
value of both variables, not just one of
them. For example, when solving

$$\begin{array}{r} x + y = 3 \\ \underline{x - y = 1} \\ 2x \quad\;\; = 4 \quad \text{Adding the two} \\ x = 2 \end{array}$$

do not stop here. Solve for y also.

13.5 1. The total value of a given number of units
each of equal value is found with the following equation.

(total value) = (value per unit)(no. of units)

2. The value of a quantity is equal to the sum
of the values of its parts.

REVIEW EXERCISES

Part I

13.1 *Tell how the graphs of the system are related.*

1. $2x + \;\; y = -3$
$-4x - 2y = 6$
coinciding

2. $x - \;\; 3y = 7$
$4x - 12y = -3$
parallel

3. $x + y = 3$
$2x + y = -4$
intersecting

13.2, *Solve by whichever method seems more appropriate.*
13.3,
13.4 4. $3x - \;\; y = 9$
$2x + 5y = -11$
$(2, -3)$

5. $2x - 3y = -14$
$5x + 2y = 3$
$(-1, 4)$

6. $8x - 2y = 4$
$-4x + \;\; y = -2$
infinitely many solutions

7. $x + 3 = 0$
$3x + 2y = -7$

$(-3, 1)$

8. $3x + \;\; y = -2$
$-6x - 2y = -2$

no solution

9. $x - 8 = 0$
$2y + 1 = 0$

$\left(8, -\dfrac{1}{2}\right)$

13.5 *Solve.*

10. The sum of two numbers is 37 and twice one
minus the other is 14. Find the numbers.
17, 20

11. The sum of Barb's and Cindy's ages is 17. In
4 years Barb will be twice as old as Cindy is
now. What is the present age of each?
Barb is 10, Cindy is 7.

12. A collection of nickels and dimes is worth $2.00. If the number of dimes minus the number of nickels is 5, how many of each are there?

10 nickels, 15 dimes

13. A dealer wishes to mix tea worth 80¢ per lb with tea worth $1.00 per lb to obtain a 50-lb mixture worth 94¢ per lb. How many pounds of each must he use?

15 lb (80¢), 35 lb ($1.00)

14. A boat travels 96 mi downstream in 4 hr and returns to the starting point in 6 hr. What is the speed of the boat in still water?

20 mph

15. Tina invested $7000, part at 8% simple interest and the rest at 12% simple interest. How much did she have invested at 12% if her total interest income for the year was $640?

$2000

Part II

Solve.

16. A chemist has one solution that is 50% acid and a second that is 25% acid. If she wishes to obtain 10 liters of a 40% acid solution, how many liters of each should be combined to obtain the mixture?

6 liters of 50%, 4 liters of 25%

17. Mike Brown flew his plane 600 mi with the wind in 5 hr and returned against the wind in 6 hr. What was the speed of the wind?

10 mph

18. $2x + y = 0$
$x - y = 6$

$(2, -4)$

19. $4x - y = -3$
$6x - 2y = -7$

$\left(\frac{1}{2}, 5\right)$

20. $3x + 2 = 0$
$2y - 3 = 0$

$\left(-\frac{2}{3}, \frac{3}{2}\right)$

21. Two angles are complementary (their sum is 90°) and one is 6° more than five times the other. Find each angle.

14°, 76°

22. Mrs. Carlson has two sons. If the sum of their ages is 17 and the difference between their ages is 7, how old are her sons?

12 years, 5 years

ANSWERS: 1. coinciding 2. parallel 3. intersecting 4. $(2, -3)$ 5. $(-1, 4)$ 6. infinitely many solutions 7. $(-3, 1)$ 8. no solution 9. $\left(8, -\frac{1}{2}\right)$ 10. 17, 20 11. Barb is 10, Cindy is 7. 12. 10 nickels, 15 dimes 13. 15 pounds (80¢), 35 pounds ($1.00) 14. 20 mph 15. $2000 16. 6 liters of 50%, 4 liters of 25% 17. 10 mph 18. $(2, -4)$ 19. $\left(\frac{1}{2}, 5\right)$ 20. $\left(-\frac{2}{3}, \frac{3}{2}\right)$ 21. 14°, 76° 22. 12 years, 5 years

1. Without graphing, tell how the graphs of the pair of linear equations are related (intersecting, parallel, or coinciding).

    $x + 3y = 1$
    $2x - y = 5$

1. _____ intersecting _____

2. Tell whether $(4, -2)$ is a solution to the following system.

    $3x + 4y = 4$
    $-2x - y = -10$

2. _____ no _____

Solve the system of equations by whichever method seems more appropriate.

3. $2x - 5y = 11$
    $x + 3y = 0$

3. _____ $(3, -1)$ _____

4. $3x - 2y = 1$
    $-6x + 4y = -1$

4. _____ no solution _____

5. In a system of equations, if the lines are coinciding, how many solutions will the system have?

5. _____ infinitely many _____

6. The sum of two numbers is 18 and their difference is 12. Find the two numbers.

6. _____3, 15_____

7. Two angles are supplementary and one is 12° more than six times the other. Find each angle.

7. _____24°, 156°_____

8. A collection of nickels and dimes has a value of $3.70. If there are 52 coins in the collection, how many of each are there?

8. ____30 nickels, 22 dimes____

9. An airplane can fly 1000 mi with the wind in 2 hr and return against the wind in 2.5 hr. What is the speed of the plane in still air?

9. _____450 mph_____

10. A specialty store wishes to sell a party mix of candy and nuts for $1.80 per lb. If candy is $1.20 per lb and nuts cost $2.20 per lb, how many pounds of each should be used to make 100 lb of the mixture?

10. ____40 lb candy, 60 lb nuts____

Exponents and Polynomials

14.1 INTEGER EXPONENTS

STUDENT GUIDEPOSTS

1. Exponential Notation
2. Product Rule
3. Quotient Rule
4. Power Rule
5. Powers of Products and Quotients
6. Zero Exponents
7. Negative Exponents
8. Summary of Rules for Exponents

1 EXPONENTIAL NOTATION

In Section 10.1 we introduced exponents and exponential notation.

Exponential Notation

If a is any number and n is a positive integer,

$$a^n = \underbrace{a \cdot a \cdot a \cdots a}_{n \text{ factors}},$$

where a is the base, n the exponent, and a^n the exponential expression.

EXAMPLE 1   USING EXPONENTIAL NOTATION

Write without using exponents.

(a) $4^5 = \underbrace{4 \cdot 4 \cdot 4 \cdot 4 \cdot 4}_{5 \text{ factors}}$ The product is 1024

(b) $3y^2 = \underbrace{3yy}_{2 \text{ } ys \text{ as factors}}$ 3 *is not* squared

(c) $(3y)^2 = \underbrace{(3y)(3y)}_{2 \text{ } (3y)s \text{ as factors}}$ 3 *is* squared

(d) $2^2 + 3^2 = \underbrace{2 \cdot 2}_{2 \text{ } 2s} + \underbrace{3 \cdot 3}_{2 \text{ } 3s}$ This simplifies to 13, *not*
$(2 + 3)^2 = 5^2 = 25$

PRACTICE EXERCISE 1

Write without using exponents.

(a) 9^7

(b) $6w^3$

(c) $(6w)^3$

(d) $4^2 + 8^2$

(e) 1^{27}

(f) $(-1)^5$

(g) $(-1)^{14}$

(e) $1^{32} = \underbrace{1 \cdot 1 \cdot 1 \cdots 1}_{\text{32 factors}} = 1$ 1 to any power is always 1

(f) $(-1)^3 = (-1)(-1)(-1)$
$\qquad = (+1)(-1) = -1$ −1 to an *odd* power is −1

(g) $(-1)^4 = (-1)(-1)(-1)(-1)$
$\qquad = (-1)(-1) = 1$ −1 to an *even* power is +1 or 1

Answers: (a) $9 \cdot 9 \cdot 9 \cdot 9 \cdot 9 \cdot 9 \cdot 9$
(b) $6 \cdot w \cdot w \cdot w$ (c) $(6w)(6w)(6w)$
(d) $4 \cdot 4 + 8 \cdot 8$ (e) 1 (f) -1
(g) 1

///////////// **CAUTION** //////////////

Two of the most common errors made with exponents have been shown in the above examples:

$$3y^2 \ne (3y)^2 \quad \text{and} \quad 2^2 + 3^2 \ne (2 + 3)^2.$$

///////////

Parts **(f)** and **(g)** of Example 1 illustrate the following general rule: An odd power of a negative number is negative, and an even power of a negative number is positive.

❷ PRODUCT RULE

When we combine terms containing exponential expressions by multiplication, division, or taking powers, our work can be simplified by using the basic properties of exponents. For example,

$$a^3 \cdot a^2 = \underbrace{(a \cdot a \cdot a)}_{\substack{3 \\ \text{factors}}} \underbrace{(a \cdot a)}_{\substack{2 \\ \text{factors}}} = \underbrace{a \cdot a \cdot a \cdot a \cdot a}_{\substack{5 \\ \text{factors}}} = a^5.$$

When two exponential expressions *with the same base* are multiplied, the product is that base raised to the sum of the exponents on the original expressions.

Product Rule for Exponents

If a is any number, and m and n are positive integers.

$$a^m a^n = a^{m+n}.$$

(To multiply powers with the same base, add exponents.)

EXAMPLE 2   USING THE PRODUCT RULE

Find the product.

(a) $a^3 \cdot a^4 = a^{3+4} = a^7$

(b) $5^3 \cdot 5^7 = 5^{3+7} = 5^{10}$ *Not* 25^{10}

(c) $2^3 \cdot 2^2 \cdot 2^6 = 2^{3+2+6} = 2^{11}$ The rule applies to more than two factors

(d) $3x^3 \cdot x^2 = 3x^{3+2} = 3x^5$ The 3 is not raised to the powers

PRACTICE EXERCISE 2

Find the product.

(a) $w^5 \cdot w^8$

(b) $8^2 \cdot 8^8$

(c) $3^2 \cdot 3^3 \cdot 3^5 \cdot 3^{11}$

(d) $7a^5 \cdot a^{12}$

Answers: (a) w^{13} (b) 8^{10}
(c) 3^{21} (d) $7a^{17}$

❸ QUOTIENT RULE

When two powers with the same base are divided, for example,

$$\frac{a^5}{a^2} = \frac{\overbrace{a \cdot a \cdot a \cdot a \cdot a}^{5 \text{ factors}}}{\underbrace{a \cdot a}_{2 \text{ factors}}} = \frac{a \cdot a \cdot a \cdot \cancel{a} \cdot \cancel{a}}{\cancel{a} \cdot \cancel{a}} = \underbrace{a \cdot a \cdot a}_{3 \text{ factors}} = a^3,$$

the quotient can be found by raising the base to the difference of the exponents $(5 - 2 = 3)$.

Quotient Rule for Exponents

If a is any number except zero, and m, n, and $m - n$ are positive integers, then

$$\frac{a^m}{a^n} = a^{m-n}.$$

(To divide powers with the same base, subtract exponents.)

EXAMPLE 3   USING THE QUOTIENT RULE

Find the quotient.

(a) $\dfrac{a^7}{a^3} = a^{7-3} = a^4$

(b) $\dfrac{5^8}{5^5} = 5^{8-5} = 5^3$

(c) $\dfrac{2^3}{3^4}$ Cannot be simplified using the rule of exponents since the bases are different

(d) $\dfrac{3x^3}{x^2} = 3x^{3-2} = 3x^1 = 3x$ $x^1 = x$

(e) $\dfrac{2yy^3}{y^2} = \dfrac{2y^1y^3}{y^2} = \dfrac{2y^{1+3}}{y^2}$

$= \dfrac{2y^4}{y^2} = 2y^{4-2} = 2y^2$

PRACTICE EXERCISE 3

Find the quotient.

(a) $\dfrac{x^{10}}{x^4}$

(b) $\dfrac{7^{12}}{7^9}$

(c) $\dfrac{8^3}{5^2}$

(d) $\dfrac{2a^7}{a}$

(e) $\dfrac{7z^2z^8}{z^7}$

Answers: (a) x^6 (b) 7^3
(c) cannot be simplified using the quotient rule   (d) $2a^6$ (e) $7z^3$

---

❹ POWER RULE

When a power is raised to a power, for example,

$$(a^2)^3 = \underbrace{(a^2)(a^2)(a^2)}_{3 \text{ factors}} = (a \cdot a)(a \cdot a)(a \cdot a)$$
$$= \underbrace{a \cdot a \cdot a \cdot a \cdot a \cdot a}_{6 \text{ factors}} = a^6,$$

the resulting exponential expression can be found by raising the base to the product of the exponents $(2 \cdot 3 = 6)$.

Power Rule

If a is any number, and m and n are positive integers,

$$(a^m)^n = a^{mn}.$$

(To raise a power to a power, multiply exponents.)

///////////// C A U T I O N ///////////

Do not confuse the power rule with the product rule. For example,

$$(a^5)^2 = a^{5 \cdot 2} = a^{10},$$

but

$$a^5 a^2 = a^{5+2} = a^7.$$

EXAMPLE 4   USING THE POWER RULE

Find the powers.

(a) $(a^3)^8 = a^{3 \cdot 8} = a^{24}$

(b) $2(x^3)^2 = 2x^{3 \cdot 2} = 2x^6$

PRACTICE EXERCISE 4

Find the powers.

(a) $(w^5)^7$

(b) $8(a^4)^3$

Answers: **(a)** w^{35} **(b)** $8a^{12}$

5 POWERS OF PRODUCTS AND QUOTIENTS

A product or quotient of expressions is often raised to a power. For example,

$$(3x^2)^4 = \underbrace{(3x^2) \cdot (3x^2) \cdot (3x^2) \cdot (3x^2)}_{4 \text{ factors}}$$

$$= \underbrace{3 \cdot 3 \cdot 3 \cdot 3}_{4 \text{ factors}} \cdot \underbrace{x^2 \cdot x^2 \cdot x^2 \cdot x^2}_{4 \text{ factors}} = 3^4(x^2)^4,$$

and

$$\left(\frac{2}{y^2}\right)^3 = \underbrace{\frac{2}{y^2} \cdot \frac{2}{y^2} \cdot \frac{2}{y^2}}_{3 \text{ factors}} = \frac{\overbrace{2 \cdot 2 \cdot 2}^{3 \text{ factors}}}{\underbrace{y^2 \cdot y^2 \cdot y^2}_{3 \text{ factors}}} = \frac{2^3}{(y^2)^3}.$$

These illustrate the next rule.

Powers of Products and Quotients

If a and b are any numbers, and n is a positive integer, then

$$(a \cdot b)^n = a^n \cdot b^n \qquad \text{and} \qquad \left(\frac{a}{b}\right)^n = \frac{a^n}{b^n} \quad (b \text{ not zero}).$$

| EXAMPLE 5   POWERS OF PRODUCTS AND QUOTIENTS | PRACTICE EXERCISE 5 |

Simplify.

(a) $(2y)^5 = 2^5 \cdot y^5 = 2^5 y^5 = 32y^5$

(b) $(3a^2b^3)^4 = 3^4 \cdot (a^2)^4 \cdot (b^3)^4$ Raise each factor to the fourth power

$\qquad = 3^4 a^8 b^{12}$

$\qquad = 81a^8 b^{12}$

(c) $\left(\dfrac{2}{x^3}\right)^4 = \dfrac{2^4}{(x^3)^4}$ $\qquad \left(\frac{a}{b}\right)^n = \frac{a^n}{b^n}$

$\qquad = \dfrac{16}{x^{3\cdot 4}}$ $\qquad (a^m)^n = a^{mn}$

$\qquad = \dfrac{16}{x^{12}}$

(d) $\left(\dfrac{3a^2}{b}\right)^3 = \dfrac{(3a^2)^3}{b^3}$ $\qquad \left(\frac{a}{b}\right)^n = \frac{a^n}{b^n}$

$\qquad = \dfrac{3^3(a^2)^3}{b^3}$ $\qquad (a \cdot b)^n = a^n \cdot b^n$

$\qquad = \dfrac{27a^6}{b^3}$

Simplify.

(a) $(5a)^3$

(b) $(2u^3w^5)^2$

(c) $\left(\dfrac{3}{z^2}\right)^3$

(d) $\left(\dfrac{4x}{y^3}\right)^2$

Answers: (a) $125a^3$ (b) $4u^6w^{10}$
(c) $\frac{27}{z^6}$ (d) $\frac{16x^2}{y^6}$

CAUTION

Rules similar to the product and quotient rules above do not exist for sums and differences. For example,

$$(2^2 + 3^2)^3 \quad \text{is } not \quad (2^2)^3 + (3^2)^3,$$

and $\qquad (1^3 - 4^3)^2 \quad \text{is } not \quad (1^3)^2 - (4^3)^2.$

Also, exponential expressions with different bases cannot be combined by adding exponents. For example, in general,

$$a^2 \cdot b^5 \quad \text{is } not \quad (ab)^7.$$

6 ZERO EXPONENTS

We know that if a is not zero,

$$\frac{a^m}{a^n} = a^{m-n}.$$

Suppose we extend the quotient rule to include $m = n$. Then

$$\frac{a^m}{a^m} = a^{m-m} = a^0 \quad \text{and also} \quad \frac{a^m}{a^m} = 1.$$

(Any number divided by itself is 1.) This suggests the following definition.

Zero Exponent

If a is any number except zero,

$$a^0 = 1.$$

EXAMPLE 6 USING ZERO EXPONENTS	PRACTICE EXERCISE 6

Simplify.

(a) $5^0 = 1$

(b) $(2a^2b^3)^0 = 1$ (assuming $a \neq 0$ and $b \neq 0$)

Simplify.

(a) $21^0 =$ _____.

(b) If $x \neq 0$ and $y \neq 0$,
$(7xy^5)^0 =$ _____.

Answers: (a) 1   (b) 1

7 NEGATIVE EXPONENTS

Again, consider

$$\frac{a^m}{a^n} = a^{m-n} \ (a \neq 0).$$

What happens when $n > m$? For example, if we let $n = 5$ and $m = 2$ and we extend the quotient rule to include $m - n < 0$, we have

$$\frac{a^m}{a^n} = \frac{a^2}{a^5} = a^{2-5} = a^{-3}.$$

If we look at the problem another way, we have

$$\frac{a^2}{a^5} = \frac{\not{a} \cdot \not{a}}{\not{a} \cdot \not{a} \cdot a \cdot a \cdot a} = \frac{1}{a \cdot a \cdot a} = \frac{1}{a^3}.$$

Thus, we conclude that $a^{-3} = \frac{1}{a^3}$. This suggests a way to define exponential expressions with negative integer exponents.

Negative Exponents

If $a \neq 0$ and n is a positive integer ($-n$ is a negative integer), then

$$a^{-n} = \frac{1}{a^n}.$$

EXAMPLE 7 USING NEGATIVE EXPONENTS	PRACTICE EXERCISE 7

Simplify and write without negative exponents.

(a) $5^{-3} = \frac{1}{5^3} = \frac{1}{125}$ 5^{-3} is *not* -5^3 nor $(-3)(5)$

(b) $4^{-2} = \frac{1}{4^2} = \frac{1}{16}$

(c) $\frac{1}{3^{-2}} = \frac{1}{\frac{1}{3^2}} = \frac{1}{\frac{1}{9}} = 1 \cdot \frac{9}{1} = 9 = 3^2$

(d) $(-2)^{-5} = \frac{1}{(-2)^5} = \frac{1}{-32} = -\frac{1}{32}$

Simplify and write without negative exponents.

(a) 7^{-2}

(b) 2^{-5}

(c) $\frac{1}{5^{-1}}$

(d) $(-6)^{-2}$

Answers: (a) $\frac{1}{49}$ (b) $\frac{1}{32}$ (c) 5

(d) $\frac{1}{36}$

Example 7 shows that we can "remove" negative exponents simply by moving an exponential expression with a negative exponent from numerator to denominator (or denominator to numerator) and changing the sign of the exponent.

///////////// **CAUTION** ////////////

a^{-n} is not equal to $-a^n$ nor to $(-n)a$. As shown in Example 7 (a), 5^{-3} is $\frac{1}{125}$ and not -5^3, which is -125, nor $(-3)(5)$, which is -15.

////////////

⑧ SUMMARY OF RULES FOR EXPONENTS

All the rules of exponents stated for positive integer exponents apply to all integer exponents: positive, negative, and zero. These are summarized as follows.

Rules for Exponents

Let a and b be any two numbers, m and n any two integers.

1. $a^m \cdot a^n = a^{m+n}$ 2. $\dfrac{a^m}{a^n} = a^{m-n}$ $(a \neq 0)$

3. $(a^m)^n = a^{mn}$ 4. $(a \cdot b)^n = a^n b^n$

5. $\left(\dfrac{a}{b}\right)^n = \dfrac{a^n}{b^n}$ $(b \neq 0)$ 6. $a^0 = 1$ $(a \neq 0)$

7. $a^{-n} = \dfrac{1}{a^n}$ $(a \neq 0)$ 8. $\dfrac{1}{a^{-n}} = a^n$ $(a \neq 0)$

EXAMPLE 8 USING ALL RULES

Simplify and write without negative exponents.

(a) $(2a)^{-1} = \dfrac{1}{(2a)^1} = \dfrac{1}{2a}$ $(2a)^{-1}$ is *not* $-2a$

(b) $2a^{-1} = 2\dfrac{1}{a} = \dfrac{2}{a}$ The exponent -1 is only on a, not on 2

(c) $y^3 y^{-5} = y^{3+(-5)} = y^{-2} = \dfrac{1}{y^2}$

(d) $\dfrac{a^2 b^{-3}}{a^{-1} b^5} = a^{2-(-1)} b^{-3-5} = a^3 b^{-8} = a^3 \cdot \dfrac{1}{b^8} = \dfrac{a^3}{b^8}$

(e) $\dfrac{1}{x^{-3}} = \dfrac{1}{\frac{1}{x^3}} = 1 \cdot \dfrac{x^3}{1} = x^3$

(f) $(2a^{-2}b)^{-3} = 2^{-3}(a^{-2})^{-3} b^{-3} = \dfrac{1}{2^3} a^{(-2)(-3)} \dfrac{1}{b^3}$

$= \dfrac{1}{8} \cdot a^6 \dfrac{1}{b^3} = \dfrac{a^6}{8b^3}$

PRACTICE EXERCISE 8

Simplify and write without negative exponents.

(a) $(5w)^{-2}$

(b) $5w^{-2}$

(c) $u^{-7} u^5$

(d) $\dfrac{x^4 y^{-3}}{x^{-1} y^2}$

(e) $\dfrac{1}{m^{-7}}$

(f) $(6y^{-3} z^{-1})^{-2}$

(g) $\left(\dfrac{3w^{-1}}{x^2}\right)^{-3}$

(g) $\left(\dfrac{a^2}{2y^{-3}}\right)^{-2} = \dfrac{(a^2)^{-2}}{2^{-2}(y^{-3})^{-2}} = \dfrac{a^{-4}}{\frac{1}{2^2}y^6} = \dfrac{\frac{1}{a^4}}{\frac{y^6}{2^2}}$

$$= \dfrac{1}{a^4} \cdot \dfrac{2^2}{y^6} = \dfrac{4}{a^4 y^6}$$

Answers: (a) $\frac{1}{25w^3}$ (b) $\frac{5}{w^3}$

(c) $\frac{1}{u^2}$ (d) $\frac{x^5}{y^5}$ (e) m^7 (f) $\frac{y^6 z^2}{36}$

(g) $\frac{w^3 x^6}{27}$

EXAMPLE 9 EVALUATION OF EXPONENTIAL EXPRESSIONS	**PRACTICE EXERCISE 9**

Evaluate the following when $a = -2$ and $b = 3$.

(a) $a^{-1} = (-2)^{-1} = \dfrac{1}{-2} = -\dfrac{1}{2}$ $(-2)^{-1}$ is *not* $+2$

(b) $\dfrac{a^{-2}}{b} = \dfrac{(-2)^{-2}}{3} = \dfrac{\frac{1}{(-2)^2}}{3} = \dfrac{\frac{1}{4}}{3} = \dfrac{1}{4} \cdot \dfrac{1}{3} = \dfrac{1}{12}$

(c) $(a + b)^{-1} = (-2 + 3)^{-1} = 1^{-1} = \dfrac{1}{1} = 1$

(d) $a^{-1} + b^{-1} = (-2)^{-1} + (3)^{-1}$

$$= \dfrac{1}{-2} + \dfrac{1}{3} = -\dfrac{3}{6} + \dfrac{2}{6} = -\dfrac{1}{6}$$

Evaluate the following when $x = -3$ and $y = 5$.

(a) x^{-1}

(b) $\dfrac{x^2}{y^{-1}}$

(c) $(y - x)^{-1}$

(d) $y^{-1} - x^{-1}$

Answers: (a) $\frac{1}{-3}$ (b) 45   (c) $\frac{1}{8}$

(d) $\frac{8}{15}$

14.1 EXERCISES A

Write in exponential notation.

1. $8 \cdot 8 \cdot 8 \cdot 8$ 8^4

2. $2 \cdot 2 \cdot y \cdot y \cdot y$ $2^2 y^3$

3. $(2x)(2x)(2x)(2x)$ $(2x)^4$

Write without using exponents.

4. $2y^4$ $2yyyy$

5. $(2y)^4$ $(2y)(2y)(2y)(2y)$

6. $a^2 + b^2$ $aa + bb$

Simplify and write without negative exponents.

7. $x^2 \cdot x^5$ x^7

8. $a^3 \cdot a^2 \cdot a^4$ a^9

9. $2y^2 \cdot y^8$ $2y^{10}$

10. $\dfrac{a^4}{a^3}$ a

11. $\dfrac{2y^5}{y^3}$ $2y^2$

12. $(a^3)^4$ a^{12}

13. $(2x^3)^4$ $16x^{12}$

14. $2(x^3)^4$ $2x^{12}$

15. $\left(\dfrac{2}{x^3}\right)^4$ $\dfrac{16}{x^{12}}$

16. $\dfrac{a^3}{b^5}$ cannot be simplified

17. 5^0 1

18. 0^0 undefined

19. $(2a)^0 (a \neq 0)$ **1**

20. $(2x)^{-1}$ $\dfrac{1}{2x}$

21. $2x^{-1}$ $\dfrac{2}{x}$

22. $\dfrac{2x^7}{x^9}$ $\dfrac{2}{x^2}$

23. $3y^4 y^{-7}$ $\dfrac{3}{y^3}$

24. $(3y)^{-2}$ $\dfrac{1}{9y^2}$

25. $3y^{-2}$ $\dfrac{3}{y^2}$

26. $(3y^{-2})^3$ $\dfrac{27}{y^6}$

27 $\left(\dfrac{2y}{x^3}\right)^{-2}$ $\dfrac{x^6}{4y^2}$

28. $\dfrac{b^{-2}}{a^{-4}}$ $\dfrac{a^4}{b^2}$

29. $\dfrac{a^{-4}b^2}{b^{-2}}$ $\dfrac{b^4}{a^4}$

30 $\dfrac{a^{-2}b^2}{a^4 b^{-3}}$ $\dfrac{b^5}{a^6}$

31. $(3xy^2)(4x^2 y^3)$
$12x^3 y^5$

32. $(2x^2 y^2)^2 (3x^2 y)^3$
$108x^{10} y^7$

33. $(-xy)(2x^3 y)(4xy^3)$
$-8x^5 y^5$

34. $\dfrac{x^3 y^{-5}}{x^4 y^{-6}}$ $\dfrac{y}{x}$

35. $\dfrac{3^0 a^3 b^{-8}}{ab^4}$ $\dfrac{a^2}{b^{12}}$

36. $\dfrac{3^{-1} x^{-1} y^{-5}}{x^{-6} y^2}$ $\dfrac{x^5}{3y^7}$

37. $(x^{-2} y^{-1})^{-2}$ $x^4 y^2$

38. $(x^2 y^{-3})^{-4}$ $\dfrac{y^{12}}{x^8}$

39. $(3x^{-1} y)^{-2}$ $\dfrac{x^2}{9y^2}$

40. $\left(\dfrac{a^{-5}}{b^{-1}}\right)^{-1}$ $\dfrac{a^5}{b}$

41. $\left(\dfrac{2a^{-3}}{b^3}\right)^{-2}$ $\dfrac{a^6 b^6}{4}$

42. $\left(\dfrac{2a^3 b^{-2}}{a^{-5} b}\right)^{-3}$ $\dfrac{b^9}{8a^{24}}$

Evaluate when $a = -2$ and $b = 3$.

43. $3a^2$ **12**

44. $(3a)^2$ **36**

45. $-3a^2$ **−12**

46. $(-3a)^2$ **36**

47. $-b^2$ **−9**

48. $(-b)^2$ **9**

49. $a^2 - b^2$ **−5**

50. $(a - b)^2$ **25**

51 a^{-2} $\dfrac{1}{4}$

52 $-2a$ **4**

53 $-a^2$ **−4**

54 $a^{-2} + b^{-2}$ $\dfrac{13}{36}$

55 $(a + b)^{-2}$ **1**

56. $\dfrac{a^{-1}}{b^{-2}}$ $-\dfrac{9}{2}$

57. a^{-3} $-\dfrac{1}{8}$

58. $(-a)^{-3}$ $\dfrac{1}{8}$

59 $a^{-1} b^{-1}$ $-\dfrac{1}{6}$

60 $(ab)^{-1}$ $-\dfrac{1}{6}$

ANSWERS: 1. 8^4 2. $2^2 y^3$ 3. $(2x)^4$ 4. $2yyyy$ 5. $(2y)(2y)(2y)(2y)$ 6. $aa + bb$ 7. x^7 8. a^9 9. $2y^{10}$ 10. $a^1 = a$
11. $2y^2$ 12. a^{12} 13. $16x^{12}$ 14. $2x^{12}$ 15. $\frac{16}{x^{12}}$ 16. **cannot be simplified further**   17. 1   18. **undefined**   19. 1
20. $\frac{1}{2x}$ 21. $\frac{2}{x}$ 22. $\frac{2}{x^2}$ 23. $\frac{3}{y^3}$ 24. $\frac{1}{9y^2}$ 25. $\frac{3}{y^2}$ 26. $\frac{27}{y^6}$ 27. $\frac{x^6}{4y^2}$ 28. $\frac{a^4}{b^2}$ 29. $\frac{b^4}{a^4}$ 30. $\frac{b^5}{a^6}$ 31. $12x^3 y^5$ 32. $108x^{10} y^7$
33. $-8x^5 y^5$ 34. $\frac{y}{x}$ 35. $\frac{a^2}{b^{12}}$ 36. $\frac{x^5}{3y^7}$ 37. $x^4 y^2$ 38. $\frac{y^{12}}{x^8}$ 39. $\frac{x^2}{9y^2}$ 40. $\frac{a^5}{b}$ 41. $\frac{a^6 b^6}{4}$ 42. $\frac{b^9}{8a^{24}}$ 43. 12   44. 36   45. −12
46. 36   47. −9   48. 9   49. −5   50. 25   51. $\frac{1}{4}$ 52. 4   53. −4   54. $\frac{13}{36}$ 55. 1   56. $-\frac{9}{2}$ 57. $-\frac{1}{8}$ 58. $\frac{1}{8}$ 59. $-\frac{1}{6}$
60. $-\frac{1}{6}$

14.1 EXERCISES B

Write in exponential notation.

1. $4 \cdot 4 \cdot z \cdot z \cdot z$ $16z^3$

2. $(3w)(3w)(3w)$ $(3w)^3$

3. $3www$ $3w^3$

Write without using exponents.

4. $4y^3$ $4yyy$

5. $(4y)^3$ $(4y)(4y)(4y)$

6. $x^2 + y^2$ $xx + yy$

Simplify and write without negative exponents.

7. $y^2 \cdot y^7$ y^9

8. $x^3 \cdot x^2 \cdot x^6$ x^{11}

9. $2z^2 \cdot z^5$ $2z^7$

10. $\dfrac{b^5}{b^2}$ b^3

11. $\dfrac{2y^7}{y^4}$ $2y^3$

12. $(w^3)^5$ w^{15}

13. $(2c^2)^4$ $16c^8$

14. $2(y^3)^5$ $2y^{15}$

15. $\left(\dfrac{2}{a^2}\right)^3$ $\dfrac{8}{a^6}$

16. $\dfrac{x^3}{y^4}$ $\dfrac{x^3}{y^4}$

17. 7^0 1

18. -0^0 undefined

19. $(4x)^0 (x \neq 0)$ 1

20. $(5y)^{-1}$ $\dfrac{1}{5y}$

21. $5y^{-1}$ $\dfrac{5}{y}$

22. $\dfrac{3a^3}{a^7}$ $\dfrac{3}{a^4}$

23. $4z^4z^{-9}$ $\dfrac{4}{z^5}$

24. $(5w)^{-2}$ $\dfrac{1}{25w^2}$

25. $5w^{-2}$ $\dfrac{5}{w^2}$

26. $(2a^{-2})^3$ $\dfrac{8}{a^6}$

27. $\left(\dfrac{2y}{x^3}\right)^2$ $\dfrac{4y^2}{x^6}$

28. $\dfrac{b^{-5}}{a^{-3}}$ $\dfrac{a^3}{b^5}$

29. $\dfrac{a^{-3}b^{-3}}{b^{-4}}$ $\dfrac{b}{a^3}$

30. $\dfrac{x^{-2}y^4}{x^3y^{-3}}$ $\dfrac{y^7}{x^5}$

31. $(-5x^2y^2)(3x^3y)$
    $-15x^5y^3$

32. $(x^3y^2)^3(2x^3y)^2$
    $4x^{15}y^8$

33. $(-2x)(5x^5y)(-3xy^2)$
    $30x^7y^3$

34. $\dfrac{a^4b^{-3}}{a^{-2}b^2}$ $\dfrac{a^6}{b^5}$

35. $\dfrac{5^0a^{-5}b^{-2}}{a^4b^{-1}}$ $\dfrac{1}{a^9b}$

36. $\dfrac{2^{-2}x^3y^{-4}}{x^{-2}y^{-2}}$ $\dfrac{x^5}{4y^2}$

37. $(a^{-1}b^{-5})^{-1}$ ab^5

38. $(a^{-6}b^3)^{-3}$ $\dfrac{a^{18}}{b^9}$

39. $(4a^{-1}b^{-1})^{-3}$ $\dfrac{a^3b^3}{64}$

40. $\left(\dfrac{x^{-4}}{y^{-2}}\right)^{-2}$ $\dfrac{x^8}{y^4}$

41. $\left(\dfrac{3x^4}{y^3}\right)^{-1}$ $\dfrac{y^3}{3x^4}$

42. $\left(\dfrac{3x^{-2}y^{-1}}{x^3y^{-5}}\right)^{-4}$ $\dfrac{x^{20}}{81y^{16}}$

Evaluate when $x = -3$ and $y = 2$.

43. $4x^2$ 36

44. $(4x)^2$ 144

45. $-4x^2$ -36

46. $(-4x)^2$ 144

47. $-x^2$ -9

48. $(-x)^2$ 9

49. $y^2 - x^2$ -5

50. $(y - x)^2$ 25

51. x^{-2} $\dfrac{1}{9}$

52. $-3y$ -6

53. $-y^3$ -8

54. $x^{-2} + y^{-2}$ $\dfrac{13}{36}$

55. $(x + y)^{-2}$ 1

56. $\dfrac{x^{-1}}{y^{-2}}$ $-\dfrac{4}{3}$

57. x^{-3} $-\dfrac{1}{27}$

58. $(-x)^{-3}$ $\dfrac{1}{27}$

59. $x^{-1}y^{-1}$ $-\dfrac{1}{6}$

60. $(xy)^{-1}$ $-\dfrac{1}{6}$

14.1 EXERCISES C

Simplify and write without negative exponents.

1. $\dfrac{a^{-2}b^3c^2}{a^{-3}b^{-2}c^{-2}}$ ab^5c^4

2. $\dfrac{3^0x^{-6}(y^{-1})^{-2}}{x^{-2}y^{-3}}$ $\dfrac{y^5}{x^4}$

3. $\dfrac{2^{-2}(x^2)^{-3}y^3z^{-2}}{3^{-1}x^{-1}(yz)^{-1}}$

$\left[\text{Answer: } \dfrac{3y^4}{4x^5z}\right]$

4. $\left(\dfrac{5^0a^{-6}b^2c^3}{a^2b^{-1}c^{-2}}\right)^{-1}$ $\dfrac{a^8}{b^3c^5}$

5. $\left(\dfrac{2^{-1}x^{-5}y^{-8}}{4^{-1}x^3y^{-4}}\right)^{-2}$ $\dfrac{x^{16}y^8}{4}$

6. $\left[\left(\dfrac{a^2b^{-2}}{a^{-5}b^{-1}}\right)^{-1}\right]^{-2}$

$\left[\text{Answer: } \dfrac{a^{14}}{b^2}\right]$

14.2 SCIENTIFIC NOTATION

═══════════ STUDENT GUIDEPOSTS ═══════════

1 Scientific Notation

2 Calculations Using Scientific Notation

When we compute with very large or very small numbers on a calculator, problems can arise because the display is limited (usually 8 digits). For example, if we use a calculator to multiply

$$(290{,}000)(15{,}000),$$

the product might be given as in Figure 14.1.

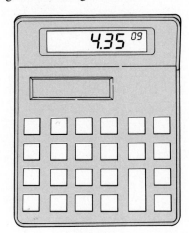

Figure 14.1

The number 4.35×10^9 is *scientific notation* for 4,350,000,000, which has ten digits, too many for the display. This shorthand notation depends on the use of integer exponents.

❶ SCIENTIFIC NOTATION

A scientist might use the number
$$235,000,000,000,000,000,000$$
but, instead of writing out all the zeros, he or she would write
$$2.35 \times 10^{20}.$$
This short form is easier to use in computations. Likewise, the number
$$0.000000000057$$
could be written as $\qquad 5.7 \times 10^{-11}.$

A number is written in **scientific notation** if it is the product of a power of 10 and a number that is greater than or equal to 1 and less than 10.

To Write a Number in Scientific Notation

1. Move the decimal point to the position immediately to the right of the first nonzero digit.
2. Multiply by a power of ten that is equal to the number of decimal places moved. The exponent on 10 is positive if the original number is greater than 10 and negative if the number is less than 1.

EXAMPLE 1 CONVERTING TO SCIENTIFIC NOTATION

Write in scientific notation.

first nonzero digit
↓
(a) $2,500,000 = 2.5 \times 10^{6}$ $\qquad$ Count 6 decimal places
6 places

first nonzero digit
↓
(b) $0.0000025 = 2.5 \times 10^{-6}$ $\qquad$ Count 6 decimal places
6 places

(c) $4,321,000,000 = 4.321 \times 10^{9}$
9 places

(d) $0.00000000001 = 1 \times 10^{-11}$
11 places

(e) $0.1 = 1 \times 10^{-1}$
1 place

(f) $4.8 = 4.8 \times 10^{0}$

PRACTICE EXERCISE 1

Write in scientific notation.

(a) 18,300,000

(b) 0.000087

(c) 65,240,000,000,000

(d) 0.0000000001

(e) 0.5

(f) 9.7

Answers: (a) 1.83×10^{7}
(b) 8.7×10^{-5} (c) 6.524×10^{13}
(d) 1×10^{-10} (e) 5×10^{-1}
(f) 9.7×10^{0}

EXAMPLE 2 CONVERTING TO STANDARD NOTATION

Write in standard notation.
(a) $5.4 \times 10^{5} = 540,000$ $\qquad$ Count 5 decimal places

(b) $5.4 \times 10^{-5} = 0.000054$ $\qquad$ Count 5 decimal places

PRACTICE EXERCISE 2

Write in standard notation.
(a) 6.1×10^{7}

(b) 6.1×10^{-7}

(c) $8.94 \times 10^{13} = 89,400,000,000,000$

(d) $2.113 \times 10^{-8} = 0.00000002113$

(c) 5.35×10^{12}

(d) 9.03×10^{-6}

Answers: **(a)** 61,000,000
(b) 0.00000061
(c) 5,350,000,000,000
(d) 0.00000903

② CALCULATIONS USING SCIENTIFIC NOTATION

Scientific notation not only shortens the notation for many numbers, but also helps in calculations.

EXAMPLE 3 CALCULATING WITH SCIENTIFIC NOTATION

Perform the indicated operations using scientific notation.

(a) $(30,000)(2,000,000) = (3 \times 10^4)(2 \times 10^6)$
$= (3 \cdot 2) \times (10^4 \times 10^6)$ Change order
$= 6 \times 10^{10}$ Add exponents

(b) $(2.4 \times 10^{-12})(4.0 \times 10^{11}) = (2.4)(4.0) \times (10^{-12} \times 10^{11})$
$= 9.6 \times 10^{-1}$

(c) $\dfrac{3.2 \times 10^{-1}}{1.6 \times 10^5} = \dfrac{3.2}{1.6} \times \dfrac{10^{-1}}{10^5} = 2 \times 10^{-6}$

PRACTICE EXERCISE 3

Perform the indicated operations using scientific notation.

(a) $(5,000,000)(200,000)$

(b) $(6.4 \times 10^{-10})(0.8 \times 10^9)$

(c) $\dfrac{8.1 \times 10^{-2}}{2.7 \times 10^{-6}}$

Answers: **(a)** 1×10^{12}
(b) 5.12×10^{-1} **(c)** 3×10^4

14.2   EXERCISES A

Write in scientific notation.

1. 370 $\mathbf{3.7 \times 10^2}$

2. 0.0037 $\mathbf{3.7 \times 10^{-3}}$

3. 98,000 $\mathbf{9.8 \times 10^4}$

4. 0.00012 $\mathbf{1.2 \times 10^{-4}}$

5. 8360 $\mathbf{8.36 \times 10^3}$

6. 0.00279 $\mathbf{2.79 \times 10^{-3}}$

7. 0.0000000000756
$\mathbf{7.56 \times 10^{-11}}$

8. 2,650,000,000
$\mathbf{2.65 \times 10^9}$

9. 0.01 $\mathbf{1 \times 10^{-2}}$

Write in standard notation.

10. 2.3×10^2 **230**

11. 2.3×10^{-2} **0.023**

12. 8.7×10^{-5}
0.000087

13. 4.58×10^6
4,580,000

14. 7.51×10^{-8}
0.0000000751

15. 6.64×10^{10}
66,400,000,000

Perform the indicated operations using scientific notation.

16. $(4 \times 10^5)(1 \times 10^6)$ 4×10^{11}

17. $(40,000,000)(20,000)$ 8×10^{11}

18. $(1 \times 10^{-10})(5 \times 10^7)$ 5×10^{-3}

(19) $(0.0000022)(300)$ 6.6×10^{-4}

20. $\dfrac{3.3 \times 10^{12}}{1.1 \times 10^{-2}}$ 3×10^{14}

21. $\dfrac{0.0000006}{0.03}$ 2×10^{-5}

22. The measure of one calorie is equal to 0.000000278 kilowatt-hours. Write this number in scientific notation. 2.78×10^{-7}

23. The distance that light will travel in 1 year is approximately 5,870,000,000,000 miles. Write the number in scientific notation. 5.87×10^{12} mi

FOR REVIEW

Evaluate when a = −2 and b = 5.

24. a^{-1} $-\dfrac{1}{2}$

25. a^{-2} $\dfrac{1}{4}$

26. $-2a$ 4

27. b^{-2} $\dfrac{1}{25}$

28. $-2b$ -10

29. $a^{-1} + b^{-1}$ $-\dfrac{3}{10}$

30. $(a + b)^{-1}$ $\dfrac{1}{3}$

31. $(a - b)^{-2}$ $\dfrac{1}{49}$

32. $a^{-2} - b^{-2}$ $\dfrac{21}{100}$

ANSWERS: 1. 3.7×10^2 2. 3.7×10^{-3} 3. 9.8×10^4 4. 1.2×10^{-4} 5. 8.36×10^3 6. 2.79×10^{-3}
7. 7.56×10^{-11} 8. 2.65×10^9 9. 1×10^{-2} 10. 230   11. 0.023   12. 0.000087   13. 4,580,000   14. 0.0000000751
15. 66,400,000,000   16. 4×10^{11} 17. 8×10^{11} 18. 5×10^{-3} 19. 6.6×10^{-4} 20. 3×10^{14} 21. 2×10^{-5}
22. 2.78×10^{-7} 23. 5.87×10^{12} mi   24. $-\frac{1}{2}$ 25. $\frac{1}{4}$ 26. 4   27. $\frac{1}{25}$ 28. −10   29. $-\frac{3}{10}$ 30. $\frac{1}{3}$ 31. $\frac{1}{49}$ 32. $\frac{21}{100}$

14.2 EXERCISES B

Write in scientific notation.

1. 5400 5.4×10^3

2. 0.0054 5.4×10^{-3}

3. 386 3.86×10^2

4. 0.000028 2.8×10^{-5}

5. 33,000 3.3×10^4

6. 0.1 1.0×10^{-1}

7. 0.000000000254
2.54×10^{-10}

8. 4,620,000,000
4.62×10^9

9. 0.001
1.0×10^{-3}

Write in standard notation.

10. 8.4×10^3
8400

11. 8.4×10^{-3}
0.0084

12. 5.42×10^{-7}
0.000000542

13. 7.25×10^7
72,500,000

14. 2.06×10^{-6}
0.00000206

15. 4.99×10^{12}
4,990,000,000,000

Perform the indicated operations using scientific notation.

16. $(5 \times 10^3)(1 \times 10^4)$ 5×10^7

17. $(60,000,000)(50,000)$ 3×10^{12}

18. $(1 \times 10^{-12})(3 \times 10^8)$ 3×10^{-4}

19. $(0.00000044)(500)$ 2.2×10^{-4}

20. $\dfrac{4.4 \times 10^{15}}{1.1 \times 10^{-3}}$ 4×10^{18}

21. $\dfrac{0.00000008}{0.04}$ 2×10^{-6}

22. The earth is approximately 93,000,000 miles from the sun. Write this number in scientific notation.
9.3×10^7 mi

23. Chemists use 602,000,000,000,000,000,000,000, known as Avogadro's number, in calculations. Write this number in scientific notation.
6.02×10^{23}

FOR REVIEW

Evaluate when $x = -3$ and $y = 2$.

24. x^{-1} $-\dfrac{1}{3}$

25. x^{-2} $\dfrac{1}{9}$

26. $-2x$ 6

27. y^{-2} $\dfrac{1}{4}$

28. $-2y$ -4

29. $x^{-1} + y^{-1}$ $\dfrac{1}{6}$

30. $(x + y)^{-1}$ -1

31. $(x - y)^{-2}$ $\dfrac{1}{25}$

32. $x^{-2} - y^{-2}$ $-\dfrac{5}{36}$

14.2 EXERCISES C

Perform the indicated operations using scientific notation.

1. $\dfrac{(2.5 \times 10^{-3})(4.2 \times 10^{-8})}{(5.0 \times 10^7)(8.4 \times 10^{-10})}$ 2.5×10^{-9}

2. $\dfrac{(0.000036)(0.0001)^{-1}}{(1,200,000)(1 \times 10^7)^{-2}}$ [Answer: 3×10^7]

14.3 BASIC CONCEPTS OF POLYNOMIALS

STUDENT GUIDEPOSTS

1. Polynomials
2. Monomials, Binomials, and Trinomials
3. Degree of a Polynomial
4. Ascending and Descending Order
5. Like Terms
6. Evaluating Polynomials

In Chapter 10 we defined a **variable** as a letter that represents a number and said that an **algebraic expression** involves sums, differences, products, quotients, or powers of numbers and variables. The **terms** in an algebraic expression are the parts that are separated by plus and minus signs. For example,

$$3x^2 - y + 2uv^3 - 9$$

is an algebraic expression with four terms, $3x^2$, $-y$, $2uv^3$, and -9. Notice that the minus sign goes with the terms $-y$ and -9.

❶ POLYNOMIALS

A **polynomial** is an algebraic expression whose terms are products of numbers and variables with whole-number exponents.

THESE ARE POLYNOMIALS:

$$3x^2 + x - 5, \quad 2m + 1, \quad 3x, \quad \frac{1}{2}z^2 - 5z^3, \quad \text{and} \quad -2x^2y + xy^2 + 3$$

The polynomial $3x^2 + x - 5$ has three terms, $3x^2$, x (or $1 \cdot x$), and -5. The **numerical coefficients** (usually just called the coefficients) of these terms are 3, 1, and -5. Similarly, the polynomial $\frac{1}{2}z^2 - 5z^3$ has terms $\frac{1}{2}z^2$ and $-5z^3$ with coefficients $\frac{1}{2}$ and -5.

THESE ARE NOT POLYNOMIALS:

$$3\sqrt{x} + 5, \quad \frac{x + 2}{x - y}, \quad x^{-2}$$

The examples above are algebraic expressions that are not polynomials. In a polynomial, a variable cannot appear under a radical, in a denominator, or with a negative exponent.

❷ MONOMIALS, BINOMIALS, AND TRINOMIALS

A polynomial with one term is a **monomial,** a **binomial** has two terms, and a **trinomial** has three terms. When a polynomial has more than three terms, no special name is used and we simply call it a polynomial. Study the polynomials in the following table.

Polynomial	Type	Terms	Coefficients
$5x - 2$	binomial	$5x, -2$	$5, -2$
6	monomial	6	6
$\frac{1}{3}y^2 + 2y - 3$	trinomial	$\frac{1}{3}y^2, 2y, -3$	$\frac{1}{3}, 2, -3$
$7a^3 + 2a^2 - 6a + 5$	polynomial	$7a^3, 2a^2, -6a, 5$	$7, 2, -6, 5$
$0.5x^3 + 0.1$	binomial	$0.5x^3, 0.1$	$0.5, 0.1$
$-4b^2a^3$	monomial	$-4b^2a^3$	-4

A **polynomial in one variable,** such as $3x^3 + 5x^2 - x + 7$, has the same variable in each of its terms. A **polynomial in several variables,** such as $-2a^2b - 3ab - b$, has two or more variables. Our primary interest is in polynomials in one variable, but some properties of polynomials with several variables will also be discussed.

③ DEGREE OF A POLYNOMIAL

For polynomials in one variable, the **degree of a term** is the exponent on the variable. The **degree of a polynomial** is the degree of the term of highest degree.

EXAMPLE 1 DEGREE OF A TERM

For the polynomial $-3y^2 + 9y^4 - 6y^7 + y + 8$, list the terms and their degree. Also give the degree of the polynomial.

Term	Degree of the term	Reason
$-3y^2$	2	Exponent on y is 2
$9y^4$	4	Exponent on y is 4
$-6y^7$	7	Exponent on y is 7
y	1	Exponent on y is 1, since $y = y^1$
8	0	Exponent on y is 0, since
		$8 = 8 \cdot 1 = 8y^0$
		(Remember that $y^0 = 1$)

Since $-6y^7$ has the highest degree, 7, the degree of the polynomial is 7.

PRACTICE EXERCISE 1

For the polynomial $-2x^3 + 9x^5 + 11$, list the terms and their degree. Also give the degree of the polynomial.

Answer: The term $-2x^3$ has degree 3, the term $9x^5$ has degree 5, and the term 11 has degree 0. The degree of the polynomial is 5.

④ ASCENDING AND DESCENDING ORDER

If the polynomial in Example 1 is written in the order

$$-6y^7 + 9y^4 - 3y^2 + y + 8,$$

we say that it is in **descending order.** Note that the highest degree term is first, followed by the next highest, and so forth. The constant term is last since 8 can be written as $8y^0$. Written in **ascending order** the same polynomial would appear as

$$8 + y - 3y^2 + 9y^4 - 6y^7.$$

Most of the time we will use descending order.

EXAMPLE 2 ASCENDING AND DESCENDING ORDER

Write $-3x + 5 - x^6 + x^3$ in both descending and ascending order.

$-x^6 + x^3 - 3x + 5$ Descending order

$5 - 3x + x^3 - x^6$ Ascending order

PRACTICE EXERCISE 2

Write $y^3 - 3y^5 + 12 - y$ in both descending and ascending order.

Answer: Descending: $-3y^5 + y^3 - y + 12$; ascending: $12 - y + y^3 - 3y^5$

⑤ LIKE TERMS

Terms that have the variable raised to the same power are called **like terms.** The like terms for the polynomial

$$7a^2 - 3a^3 + 2a - 5a^2 + 1 - 2a - 5,$$

for example, are

$$7a^2 \text{ and } -5a^2, \quad 2a \text{ and } -2a, \quad \text{and} \quad 1 \text{ and } -5.$$

A polynomial like this one can be simplified by **collecting** or **combining like terms** using the distributive law. Both words, collecting and combining, are used to describe this operation which is illustrated in the next example.

EXAMPLE 3   COLLECTING AND COMBINING LIKE TERMS

Collect like terms and write each polynomial in descending order.

(a) $7a^2 - 3a^3 + 2a - 5a^2 + 1 - 2a - 5$

$\quad = \boxed{7a^2 - 5a^2} - 3a^3 + \boxed{2a - 2a} + \boxed{1 - 5}$ 　　Commute to collect like terms

$\quad = (7 - 5)a^2 - 3a^3 + (2 - 2)a + (1 - 5)$ 　　Use distributive law to combine like terms

$\quad = 2a^2 - 3a^3 + 0 \cdot a - 4$

$\quad = 2a^2 - 3a^3 - 4$

$\quad = -3a^3 + 2a^2 - 4$ 　　Descending order

(b) $-8x^3 + x^3 - 3x + 2 + 5x - 7$

$\quad = -8x^3 + x^3 - 3x + 5x + 2 - 7$ 　　Commute

$\quad = (-8 + 1)x^3 + (-3 + 5)x + (2 - 7)$ 　　Distributive law

$\quad = -7x^3 + 2x - 5$

(c) $-3x^3 + x^7 - 7x - 4x^7 + x^3 + 1$

$\quad = x^7 - 4x^7 - 3x^3 + x^3 - 7x + 1$

$\quad = (1 - 4)x^7 + (-3 + 1)x^3 - 7x + 1$ 　　This step may be omitted

$\quad = -3x^7 - 2x^3 - 7x + 1$

PRACTICE EXERCISE 3

Collect like terms and write each polynomial in descending order.

(a) $3z^4 + z - 2z^4 - 5z + 1$

(b) $-9w^6 + 2w + 3w^6 - 5 - w^6 + 11$

(c) $y^3 - 3y + 6y^3 + 1 + 3y - 1 - 5y^3$

Answers: (a) $z^4 - 4z + 1$
(b) $-7w^6 + 2w + 6$ 　(c) $2y^3$

Like terms in polynomials with several variables must contain exactly the same variables raised to the same power. The following table lists several terms along with like and unlike terms.

Term	Like term	Unlike term
$-2xy$	$5xy$	$6y$
$8a^2b^3$	$-2a^2b^3$	$3a^3b^2$
$9xy^4z$	$4xy^4z$	$-7x^2y^4z$
2	-10	$5a^2b^2$

EXAMPLE 4   COLLECTING AND COMBINING LIKE TERMS

Collect like terms.

$3x^2y - 5xy - 2x - 4x^2y + 4x + 2xy^2$

$\quad = 3x^2y - 4x^2y - 5xy - 2x + 4x + 2xy^2$

$\quad = (3 - 4)x^2y - 5xy + (-2 + 4)x + 2xy^2$ 　　This step may be omitted

$\quad = -x^2y - 5xy + 2x + 2xy^2$

PRACTICE EXERCISE 4

Collect like terms.

$6u^2v - 3uv + 5u + 9uv - 5u^2v + uv^2$

Answer: $u^2v + uv^2 + 6uv + 5u$

⑥ EVALUATING POLYNOMIALS

Since a variable represents a real number, a polynomial in that variable also represents a real number. We can **evaluate** a polynomial when specific values for the variable(s) are given. The value of a polynomial will usually be different when different values for the variable(s) are used.

EXAMPLE 5 EVALUATING POLYNOMIALS

Evaluate the polynomials for the given values of the variables.

(a) $5x^2 - 3x + 1$ for $x = -2$

$$5x^2 - 3x + 1 = 5(-2)^2 - 3(-2) + 1 \qquad \text{Substitute } -2 \text{ for } x$$
$$= 5(4) + 6 + 1 \qquad \text{Square first, then multiply}$$
$$= 20 + 6 + 1 = 27 \qquad \text{Add}$$

(b) $5x^2 - 3x + 1$ for $x = 2$

$$5x^2 - 3x + 1 = 5(2)^2 - 3(2) + 1 \qquad \text{Substitute 2 for } x$$
$$= 5(4) - 6 + 1$$
$$= 20 - 6 + 1 = 15$$

Notice from (a) and (b) that the values of the polynomial are different for the different values of x.

(c) $3ab + 2a - 5b$ for $a = -1$ and $b = 0$

$$3ab + 2a - 5b = 3(-1)(0) + 2(-1) - 5(0) \qquad \begin{array}{l}\text{Substitute } -1 \text{ for } a \\ \text{and 0 for } b\end{array}$$
$$= 0 - 2 - 0 = -2$$

PRACTICE EXERCISE 5

Evaluate the polynomials for the given values of the variables.

(a) $6y^3 - y + 5$ for $y = -1$

(b) $6y^3 - y + 5$ for $y = 1$

(c) $-2uv + v - 7u + 1$ for $u = -2$ and $v = 0$

Answers: (a) 0   (b) 10   (c) 15

EXAMPLE 6 MANUFACTURING APPLICATION

The profit made by a machine manufacturer who sells x machines per week is given by the expression $100x^3 - 2500$. This means that fixed costs are \$2500 and the profit increases with the sale of each machine.

(a) Find the profit when 5 machines are sold.

$$100x^3 - 2500 = 100(5)^3 - 2500 \qquad \text{Substitute 5 for } x$$
$$= 100(125) - 2500$$
$$= 12{,}500 - 2500$$
$$= 10{,}000$$

The profit for the week was \$10,000.

(b) Find the profit when 2 machines are sold.

$$100x^3 - 2500 = 100(2)^3 - 2500$$
$$= 100(8) - 2500$$
$$= 800 - 2500$$
$$= -1700$$

Since the profit is negative, the company lost \$1700 that week.

PRACTICE EXERCISE 6

The cost in dollars of manufacturing a particular type of circuit board is given by $0.45n^2 + 180$, where n is the number of boards produced.

(a) Find the cost when 10 boards are made.

(b) Find the overhead cost that results when no boards are made ($n = 0$).

Answers: (a) \$225   (b) \$180

14.3 EXERCISES A

Fill in the following table.

Polynomial	Type	Terms	Coefficients
1. $-3x^2 + 2$	(a) _binomial_	(b) _$-3x^2, 2$_	(c) _$-3, 2$_
2. $2a$	(a) _monomial_	(b) _$2a$_	(c) _2_
3. $-4y^2 + 2y^3 - 7y + 1$	(a) _polynomial_	(b) _$-4y^2, 2y^3, -7y, 1$_	(c) _$-4, 2, -7, 1$_
4. $a^2 - 2a + \dfrac{1}{3}$	(a) _trinomial_	(b) _$a^2, -2a, \dfrac{1}{3}$_	(c) _$1, -2, \dfrac{1}{3}$_
5. -6	(a) _monomial_	(b) _-6_	(c) _-6_
6. $3x - 4y$	(a) _binomial_	(b) _$3x, -4y$_	(c) _$3, -4$_
7. $7a^2b^3$	(a) _monomial_	(b) _$7a^2b^3$_	(c) _7_
8. $x^2 - 2xy + y^2$	(a) _trinomial_	(b) _$x^2, -2xy, y^2$_	(c) _$1, -2, 1$_

Give the degree of each of the following terms.

9. $3x^2$ 2   **10.** $9b$ 1   **11.** -5 0   **12.** $-7b^{14}$ 14

Give the degree of each of the following polynomials.

13. $x^2 - 6x^3 + x^4 - 9x$ 4   **14.** $-3y^2 + 9y^5 + 2y - y^7$ 7   **15.** $a - 2$ 1

16. 6 0   **17.** $2x^3 - x$ 3   **18.** $-y^2 - 8y + y^4 - 3y^3 + 5$ 4

Collect like terms and write in descending order.

19. $8x - 3x$ 5x   **20.** $-5y + 2y$ -3y   **21.** $10a^2 - 7a + 5a^2 + 3a$
                                                        $15a^2 - 4a$

22. $-4b^2 - 9 + 12 - 6b^2$ **23.** $2x^4 - 2x^3 + x^3 - 3x^4 + 7 - 5$ **24.** $3x^3 - 7 + 4x^2 - 3x^3 + 6 + 2x^3$
   $-10b^2 + 3$ $-x^4 - x^3 + 2$ $2x^3 + 4x^2 - 1$

25. $7y^3 + 3y^2 - 7y^3 - 3y^2$ 0   **26.** $4y^4 - 2y^3 + 2y^3 - 4y^4 + 8$ 8

27. $6a^2 + 7a + 8a^2 - 3 - 7a - 9a^2 + 1$ **28** $-8x^{10} + x^5 - 2x^{10} - 7x^5 + 1 - x^{10} + 3$
   $5a^2 - 2$ $-11x^{10} - 6x^5 + 4$

29. $\dfrac{3}{4}y^2 - \dfrac{1}{8}y^2$ $\dfrac{5}{8}y^2$ **30.** $-0.5b^3 + 0.77b^3$ 0.27b^3

Collect like terms.

31. $2xy - 5xy$ **32.** $5x^2 + 2y^2$ **33.** $2xy - 3y^2 + 5xy + y^2$
   $-3xy$ $5x^2 + 2y^2$ (no like terms) $7xy - 2y^2$

34. $a^2b^2 - ab + a^2b^2 + ab$
$2a^2b^2$

35 $-4x^2y + 2xy^2 + x^2 - 3x^2y$
$-7x^2y + 2xy^2 + x^2$

36. $5ab^2 - 3ab + 3ab - 5ab^2$
0

Evaluate the polynomials for the given value of the variables.

37. $3x + 2$ for $x = 5$ 17

38 $7y^2 - 2y - 5$ for $y = -2$ 27

39. $8a^3 - 5$ for $a = -3$ -221

40. $2a^2 + 5b^2$ for $a = 1$ and $b = -1$ 7

41 The cost in dollars of manufacturing x bolts is given by the expression $0.05x + 15.5$. Find the cost when 440 bolts are made.   **$37.50**

42. The cost of typing a manuscript is given as two times the number of pages plus ten. Use x as the number of pages to be typed and write a polynomial to describe this cost. Find the cost of typing a one-hundred-page manuscript. $2x + 10$, $210

FOR REVIEW

Write in scientific notation.

43. 265,000
2.65 × 10⁵

44. 0.000000902
9.02 × 10⁻⁷

Write in standard notation.

45. 6.75×10^{-4}
0.000675

46. 1.06×10^8
106,000,000

Exercises 47–50 review material from Chapter 10 to help you prepare for the next section. Simplify each expression by removing the parentheses.

47. $-(-x)$
x

48. $-(x - 1)$
$-x + 1$

49. $-(-x - 1 + a)$
$x + 1 - a$

50. $-(a + x - 2)$
$-a - x + 2$

ANSWERS: 1. (a) binomial (b) $-3x^2$, 2 (c) -3, 2 2. (a) monomial (b) $2a$ (c) 2 3. (a) polynomial (b) $-4y^2$, $2y^3$, $-7y$, 1 (c) -4, 2, -7, 1 4. (a) trinomial (b) a^2, $-2a$, $\frac{1}{3}$ (c) 1, -2, $\frac{1}{3}$ 5. (a) monomial (b) -6 (c) -6
6. (a) binomial (b) $3x$, $-4y$ (c) 3, -4 7. (a) monomial (b) $7a^2b^3$ (c) 7 8. (a) trinomial (b) x^2, $-2xy$, y^2 (c) 1, -2, 1 9. 2 10. 1 11. 0 12. 14 13. 4 14. 7 15. 1 16. 0 17. 3 18. 4 19. $5x$ 20. $-3y$
21. $15a^2 - 4a$ 22. $-10b^2 + 3$ 23. $-x^4 - x^3 + 2$ 24. $2x^3 + 4x^2 - 1$ 25. 0 26. 8 27. $5a^2 - 2$
28. $-11x^{10} - 6x^5 + 4$ 29. $\frac{5}{8}y^2$ 30. $0.27b^3$ 31. $-3xy$ 32. $5x^2 + 2y^2$ (no like terms) 33. $7xy - 2y^2$ 34. $2a^2b^2$
35. $-7x^2y + 2xy^2 + x^2$ 36. 0 37. 17 38. 27 39. -221 40. 7 41. $37.50 42. $2x + 10$, $210 43. 2.65×10^5
44. 9.02×10^{-7} 45. 0.000675 46. 106,000,000 47. x 48. $-x + 1$ 49. $x + 1 - a$ 50. $-a - x + 2$

14.3 EXERCISES B

Give the type and list the terms and coefficients of the following polynomials.

1. $7x - 5$ binomial;
$7x$, -5; 7, -5

2. $3y^2 - 2y + 3$ trinomial;
$3y^2$, $-2y$, 3; 3, -2, 3

3. $-4a^3 + 6a + a^4 - 7$
polynomial;
$-4a^3$, $6a$, a^4, -7; -4, 6, 1, -7

4. 22 monomial; 22; 22

5. $-12b^{15}$
monomial; $-12b^{15}$; -12

6. $-8a + b$
binomial; $-8a$, b; -8, 1

7. $4x^2 + 4xy - y^2$
trinomial; $4x^2$, $4xy$, $-y^2$; 4, 4, -1

8. $9x^4y^6$ monomial; $9x^4y^6$; 9

Give the degree of each of the following terms.

9. $5y^7$ 7

10. $-2x^{32}$ 32

11. $4a^0$ 0

12. $44b^{12}$ 12

Give the degree of each of the following polynomials.

13. $3 + 4x^{10}$ 10

14. $6y^4 + 12y^2 - 8y + 7y^5$ 5

15. $-2 - 3a$ 1

16. -9 0

17. $-x^4 - x^2 + 9x$ 4

18. $5y^{10} + 14y^{20} - 2y^{15} + 3$ 20

Collect like terms and write in descending order.

19. $-2x + 10x$ $8x$

20. $y - 9y$ $-8y$

21. $a^3 - 3a^2 + 4a^3 - a^2$
$5a^3 - 4a^2$

22. $5 - b^4 + 3b^4 - 8$
$2b^4 - 3$

23. $3x - 8x^2 + 5x^3 - 2x^3 + 4x^2 - 2x$
$3x^3 - 4x^2 + x$

24. $3y + 2 - 7y^2 - 2y - 5$
$-7y^2 + y - 3$

25. $11a^3 - 17a^4 + 8a^2 - 5 + a^3 - 6 + a^4$
$-16a^4 + 12a^3 + 8a^2 - 11$

26. $-3b + 21b^7 - 5b^3 + 12b - b^7$
$20b^7 - 5b^3 + 9b$

27. $10 - 3y + 7y^2 - 8y^3$
$-8y^3 + 7y^2 - 3y + 10$

28. $-22x^3 + 17x^5 - 4x^3 + 2$
$17x^5 - 26x^3 + 2$

29. $0.2b - 0.8b^2 + 0.7b^2$
$-0.1b^2 + 0.2b$

30. $\dfrac{3}{2}a^3 - \dfrac{1}{3}a^4 + \dfrac{1}{4}a^3 + \dfrac{2}{9}a^4$

$-\dfrac{1}{9}a^4 + \dfrac{7}{4}a^3$

Collect like terms.

31. $-9xy + 7xy$ $-2xy$

32. $5x^2y^2 - 5$
$5x^2y^2 - 5$ (no like terms)

33. $x^2 - 3xy + 8xy - 4x^2$ $-3x^2 + 5xy$

34. $2a^2b - a^2b^2 - 2a^2b + a^2b^2$ 0

35. $2x^2y^2 - 3xy^2 + 5x^2y^2 + xy^2$
$7x^2y^2 - 2xy^2$

36. $5ab - 2a^2 - 2a^2 + 5ab + 6$
$10ab - 4a^2 + 6$

Evaluate the polynomials for the given values of the variables.

37. $-8x + 5$ for $x = 4$ -27

38. $-2y^2 + 3y - 2$ for $y = -3$ -29

39. $2a^3 + a^2$ for $a = 5$ 275

40. $5a^2b^2 - 2$ for $a = 9$ and $b = 0$ -2

41. The profit in dollars when x pairs of shoes are sold is given by the expression $2x^2 - 120$. Find the profit when 10 pairs are sold.
$80

42. The cost of making dresses is described as 10 times the number of dresses plus 8. Use x as the number of dresses and write a polynomial to describe this cost. Find the cost of making 20 dresses.
$10x + 8$; $208

FOR REVIEW

Write in scientific notation.

43. 0.0000205
2.05×10^{-5}

44. 8,600,000,000
8.6×10^9

Write in standard notation.

45. 8.11×10^5
811,000

46. 6.25×10^{-2}
0.0625

Exercises 47–50 review material from Chapter 10 to help you prepare for the next section. Simplify each expression by removing the parentheses.

47. $-(x + 1)$
$-x - 1$

48. $-(-x + 1)$
$x - 1$

49. $-(a - x + 2)$
$-a + x - 2$

50. $-(-a - x + 2)$
$a + x - 2$

14.3 EXERCISES C

The degree of a term of polynomial in several variables is found by adding the exponents on the variables. Find the degree of each term.

1. $3x^2y^3$ 5   **2.** $-8xy^6$ 7   **3.** $6xyz$ [Answer: 3]   **4.** $-9a^4b^2c^5$ 11

Write each polynomial in descending powers of x.

5. $7xy^3 - 4x^2y + 8x^4y^4$
[Answer: $8x^4y^4 - 4x^2y + 7xy^3$]

6. $-3x^2yz^4 + 4yx - 8xyz$
$-3x^2yz^4 - 8xyz + 4yz$

14.4 ADDITION AND SUBTRACTION OF POLYNOMIALS

STUDENT GUIDEPOSTS

1 Adding Polynomials   2 Subtracting Polynomials

1 ADDING POLYNOMIALS

Adding polynomials is simply a matter of collecting like terms. We work as follows.

To Add Polynomials
1. Indicate the addition with a plus sign.
2. Remove parentheses and collect like terms.

EXAMPLE 1 ADDING POLYNOMIALS

Add $2x^2 - 3x + 5$ and $-x^2 + 6x + 2$.

$(2x^2 - 3x + 5) + (-x^2 + 6x + 2)$
$= 2x^2 - 3x + 5 - x^2 + 6x + 2$ Remove parentheses
$= (2 - 1)x^2 + (-3 + 6)x + (5 + 2)$ Collect like terms using the distributive law

$= x^2 + 3x + 7$

PRACTICE EXERCISE 1

Add $3y^3 - y + 5$ and $-2y^3 + 6y - 3$.

Answer: $y^3 + 5y + 2$

When adding polynomials in one variable, we usually arrange the terms in descending order. This aids in collecting like terms and helps avoid forgetting a term.

| **EXAMPLE 2** **ARRANGING IN DESCENDING ORDER AND ADDING** | **PRACTICE EXERCISE 2** |

Add $-2x^3 - 4x^4 + 36 - 3x$ and $-14x^2 + 3x^3 - 6 + 5x$.

$(-4x^4 - 2x^3 - 3x + 36) + (3x^3 - 14x^2 + 5x - 6)$ Arrange in descending order and indicate addition

$= -4x^4 - 2x^3 - 3x + 36 + 3x^3 - 14x^2 + 5x - 6$ Remove parentheses

$= -4x^4 + (3 - 2)x^3 - 14x^2 + (5 - 3)x + (36 - 6)$ Collect like terms using the distributive law

$= -4x^4 + x^3 - 14x^2 + 2x + 30$

Add $2z - 5z^6 + z^3 - 5$ and $4 + z^6 - 8z + 3z^2$.

Answer: $-4z^6 + z^3 + 3z^2 - 6z - 1$

Another way to add polynomials is to arrange like terms in vertical columns as illustrated in Example 3. This technique will also be used later when we multiply polynomials.

| **EXAMPLE 3** **ADDING IN COLUMNS** | **PRACTICE EXERCISE 3** |

Add $2 - 3x^2$, $-x^4 + 7x - 3x^3 - 5$, and $-8x^3 + 4x^2 - 7$ using vertical columns.

$$\begin{array}{r} - 3x^2 \qquad\quad + 2 \\ -x^4 - \ 3x^3 \qquad + 7x - 5 \\ - \ 8x^3 + 4x^2 - 7x \qquad \\ \hline -x^4 - 11x^3 + \ x^2 + 0x - 3 = -x^4 - 11x^3 + x^2 - 3 \end{array}$$

Note the spaces where terms are missing

Add $y^5 + 1 - 3y$, $y - 2y^4$, and $2y - y^3 + 3y^5 - 5$ using the column method.

Answer: $4y^5 - 2y^4 - y^3 - 4$

❷ SUBTRACTING POLYNOMIALS

Since addition is commutative, the arrangement of the polynomials for addition does not matter. However, when subtracting we must be sure to write the problem in the right order. For example, to subtract

$$2x + 5 \quad \text{from} \quad x^2 - 3x + 1,$$

we need to write $\qquad (x^2 - 3x + 1) - (2x + 5).$

However, if we are given, for example,

$$(y^2 + 3) - (-2y + 1),$$

the problem is already set up for us.

Recall the definition of subtraction given in Chapter 1.

$$a - b = a + (-b)$$

That is, to subtract one expression from another, we add the negative of the second expression to the first. If a and b represent polynomials, $-b$ is found by changing all the signs in b. We can proceed as follows.

To Subtract Polynomials

1. Indicate the subtraction by putting a minus sign before the polynomial to be subtracted.

2. Remove parentheses, changing *all* signs in the polynomial being subtracted.

3. Collect like terms as in addition and simplify.

It helps to arrange terms in descending order when subtracting polynomials in one variable.

EXAMPLE 4   SUBTRACTING POLYNOMIALS

Subtract $7x - 3x^2$ from $4 - 2x - 4x^2$.

$(-4x^2 - 2x + 4) - (-3x^2 + 7x)$ Arrange in descending order and indicate the subtraction

$= -4x^2 - 2x + 4 + 3x^2 - 7x$ Remove parentheses, changing signs

$= -x^2 - 9x + 4$ Combine like terms

PRACTICE EXERCISE 4

Subtract $5z^2 + 2$ from $1 - z^2 + 8z$.

Answer: $-6z^2 + 8z - 1$

//////////////// **CAUTION** ////////////////

Be sure to change *all* signs in the polynomial being subtracted. This is shown in Example 4.

//////////////

To do the same subtraction vertically, change the signs in $7x - 3x^2$. It becomes $-7x + 3x^2$. Now arrange the terms in vertical columns.

$$\begin{array}{r} -4x^2 - 2x + 4 \\ +3x^2 - 7x \\ \hline -x^2 - 9x + 4 \end{array}$$ Change signs

Add

EXAMPLE 5   SUBTRACTING POLYNOMIALS

Subtract $-3x + 7x^4 + 5x^5 - 2$ from $2x^4 - 8x^5 - 4x + 12x^2 - x^3$.

$(-8x^5 + 2x^4 - x^3 + 12x^2 - 4x) - (5x^5 + 7x^4 - 3x - 2)$

$= -8x^5 + 2x^4 - x^3 + 12x^2 - 4x - 5x^5 - 7x^4 + 3x + 2$ Change signs

$= -8x^5 - 5x^5 + 2x^4 - 7x^4 - x^3 + 12x^2 - 4x + 3x + 2$ Commute

$= -13x^5 - 5x^4 - x^3 + 12x^2 - x + 2$

Subtracting vertically gives the same result.

$$\begin{array}{l} -8x^5 + 2x^4 - x^3 + 12x^2 - 4x \\ - 5x^5 - 7x^4 \qquad\qquad + 3x + 2 \\ \hline -13x^5 - 5x^4 - x^3 + 12x^2 - x + 2. \end{array}$$

Arrange in descending order

Change signs

Add

PRACTICE EXERCISE 5

Subtract $1 - 6w^6 + 2w - 5w^5$ from $4w + 2w^6 - 7 - 8w^5 + w^3$.

Answer: $8w^6 - 3w^5 + w^3 + 2w - 8$

EXAMPLE 6 ADDING AND SUBTRACTING

Perform the indicated operations.

$(5x^3 - 6 + x^2) + (7x^2 - 3) - (-4x^3 + 7x - 5)$

$= (5x^3 + x^2 - 6) + (7x^2 - 3) - (-4x^3 + 7x - 5)$

$= 5x^3 + x^2 - 6 + 7x^2 - 3 + 4x^3 - 7x + 5$ Change signs on polynomial being subtracted

$= (5 + 4)x^3 + (1 + 7)x^2 - 7x + (5 - 6 - 3)$

$= 9x^3 + 8x^2 - 7x - 4$

We add and subtract vertically as follows.

$$
\begin{array}{l}
5x^3 + x^2 - 6 \\
7x^2 - 3 \\
\underline{+ 4x^3 - 7x + 5} \\
9x^3 + 8x^2 - 7x - 4.
\end{array}
$$
   Change all signs

                     Add

PRACTICE EXERCISE 6

Perform the indicated operations.

$(6y^4 + 5 - y^3) - (4 - y^3)$
$ + (y^2 - 5y^4 + 2)$

Answer: $y^4 + y^2 + 3$

For polynomials in several variables, the procedures are exactly the same. Be sure that only like terms are combined.

EXAMPLE 7 ADDING IN SEVERAL VARIABLES

Add $8x^2 + 2y^2 - 3xy$ and $6x^2 - 7y^2 - 9xy + 3$.

$(8x^2 + 2y^2 - 3xy) + (6x^2 - 7y^2 - 9xy + 3)$

$= 8x^2 + 2y^2 - 3xy + 6x^2 - 7y^2 - 9xy + 3$

$= (8 + 6)x^2 + (2 - 7)y^2 + (-3 - 9)xy + 3$

$= 14x^2 - 5y^2 - 12xy + 3$

PRACTICE EXERCISE 7

Add $6a^3 + b^2 - 4ab$ and
$2b^2 - 3a^3 + 2ab - 5$.

Answer: $3a^3 + 3b^2 - 2ab - 5$

EXAMPLE 8 SUBTRACTING IN SEVERAL VARIABLES

Subtract $3ab - 5a + 9$ from $-6ab + 2a + b$.

$(-6ab + 2a + b) - (3ab - 5a + 9)$

$= -6ab + 2a + b - 3ab + 5a - 9$ Change signs

$= (-6 - 3)ab + (2 + 5)a + b - 9$ Collect like terms

$= -9ab + 7a + b - 9$

PRACTICE EXERCISE 8

Subtract $6uv - 3v^2 + 12$ from
$v^2 - 2uv + 5$.

Answer: $4v^2 - 8uv - 7$

EXAMPLE 9 SUBTRACTING SEVERAL TIMES

Perform the indicated operations.

$(7x^2y^2 - 2xy) - (5x^2y + 9xy) - (-x^2y^2 + 3xy^2 - 4xy)$

$= 7x^2y^2 - 2xy - 5x^2y - 9xy + x^2y^2 - 3xy^2 + 4xy$ Change signs

$= (7 + 1)x^2y^2 + (-2 - 9 + 4)xy - 5x^2y - 3xy^2$ Collect like terms

$= 8x^2y^2 - 7xy - 5x^2y - 3xy^2$

Note that $-5x^2y$ and $-3xy^2$ are not like terms.

PRACTICE EXERCISE 9

Perform the indicated operations.

$(9a^3b^3 + ab) - (3ab - a^3b^3)$
$ + (8a - 5a^3b^3 + 2ab)$

Answer: $5a^3b^3 + 8a$

14.4   EXERCISES A

Add.

1. $3x - 5$ and $-8x + 4$
$-5x - 1$

2. $7x^2 + 6$ and $x^2 - 2$
$8x^2 + 4$

3. $y^2 + 3y - 5$ and $-8y^2 - 5y + 9$
$-7y^2 - 2y + 4$

4. $2y - 3$ and $-4y^2 + 2$
$-4y^2 + 2y - 1$

5. $3x^5 - 2x^3 + 5x^2$ and $-5x^5 + 8x^4 - 7x^2$
$-2x^5 + 8x^4 - 2x^3 - 2x^2$

6. $-x^6 + 2x^4$ and $2x^7 - x^6 + 5x^2 + 2$
$2x^7 - 2x^6 + 2x^4 + 5x^2 + 2$

7. $8y - 6y^2 + 2$ and $-5 + 2y + 8y^2$
$2y^2 + 10y - 3$

8. $6y^5 - 2y + y^4$ and $3 - 5y^2 + 21y + 2y^5$
$8y^5 + y^4 - 5y^2 + 19y + 3$

9. $2a^2 + 3a$, $-5a^2 + 6$, and $-9a + 2$
$-3a^2 - 6a + 8$

10 $-4a^3 + 7a^4 + 3a + 2$, $5 - 3a + 7a^3$, and $17a^4 - 5 + 12a^3$
$24a^4 + 15a^3 + 2$

11. $2 - x^2 + 7x$, $9x - 7 + x^3$, $-6x$, and $12 - 3x^3$
$-2x^3 - x^2 + 10x + 7$

12. $25x^4 - 7x^5$, $12x - 17x^2 + 40x^3 - 18x^4$, and $x - 21 + 18x^5$
$11x^5 + 7x^4 + 40x^3 - 17x^2 + 13x - 21$

13. $3y^2 - 2y + 1$
  $\underline{-2y^2 + 2y + 8}$
  $y^2 \qquad\quad + 9$

14. $5x^5 \qquad\quad + \ x^3 \qquad\quad - 2x$
  $\underline{2x^5 - 8x^4 - 7x^3 + 6x^2}$
  $7x^5 - 8x^4 - 6x^3 + 6x^2 - 2x$

15. $\qquad\quad 5x^3 + 6x^2 \qquad - 7$
  $\quad -4x^4 + 3x^3 - 3x^2 - 8x$
  $\quad \underline{3x^4 - 2x^3 + 4x^2 \qquad + 1}$
  $-x^4 + 6x^3 + 7x^2 - 8x - 6$

16 $0.03y^3 - 0.75y^2 - 3y + 2$
  $-0.15y^3 \qquad\qquad\quad + 5y - 0.3$
  $\underline{0.21y^3 - 0.13y^2 \qquad + 0.6}$
  $0.09y^3 - 0.88y^2 + 2y + 2.3$

Subtract.

17. $2x + 5$ from $3x - 6$
$x - 11$

18. $6x^2 - 2$ from $4x^2 + 5$
$-2x^2 + 7$

19. $-3y^2 + 2y - 7$ from $-8y^2 + y - 9$
$-5y^2 - y - 2$

20. $-3y + 2$ from $y^2 - 4y - 5$
$y^2 - y - 7$

21. $a^4 + 5a^3 - 3a^2$ from $7a^5 - 2a^4 - a^2 + 3a$
$7a^5 - 3a^4 - 5a^3 + 2a^2 + 3a$

22. $3a - 2a^2 + 7a^3$ from $a^4 - 2a^2 + a^3 - 2$
$a^4 - 6a^3 - 3a - 2$

23 $3x^4 - 7x^5 + 15x - 32$ from $56x - 93x^3 + 21x^4 + 32x^5$
$39x^5 + 18x^4 - 93x^3 + 41x + 32$

24. $7x^{10} - 4x^5 + 1$ from $3x^5 + 1 - x^6 - 3x^2$
$-7x^{10} - x^6 + 7x^5 - 3x^2$

Perform the indicated operations.

25. $(-8y^2 + 4) - (7y^2 - 3)$
$-15y^2 + 7$

26. $(-8a^2 - 2a + 5) - (-8a^2 + a + 1)$
$-3a + 4$

27. $(6y^7 - y + y^5 + 2) - (2y + 3y^2 - 4y^5 - 2)$
$6y^7 + 5y^5 - 3y^2 - 3y + 4$

28 $(8y^{10} - y^8) - (3y^{12} + 2y^{10} - y^8)$
$-3y^{12} + 6y^{10}$

29. $(3y^2 + 2) + (-5y - 5) - (y^2 - 2y + 10)$
$2y^2 - 3y - 13$

30. $(-2a^2 + 3a) - (a^2 + a + 1) - (a^2 - 2a - 5)$
$-4a^2 + 4a + 4$

31 $(9x^4 + 3x^3 + 8x) + (3x^4 + x^3 - 7x^2) - (12x^4 - 3x^2 + x)$
$4x^3 - 4x^2 + 7x$

32. $(9y^4 + 3y^3 + 8y) - (3y^4 + y^3 - 7y^2) - (12y^4 - 3y^2 + y)$
$-6y^4 + 2y^3 + 10y^2 + 7y$

33. $(4a^2b^2 - 2ab) + (5a^2b^2 + 9ab)$
$9a^2b^2 + 7ab$

34. $(3a^2 + 2b^2 + 3) + (-6a^2 + 2b^2 - 2)$
$-3a^2 + 4b^2 + 1$

35. $(4x^2y^2 - 2xy) - (5x^2y^2 + 9xy)$
$-x^2y^2 - 11xy$

36. $(3x^2 + 2y^2 + 3) - (-6x^2 + 2y^2 - 2)$
$9x^2 + 5$

37. $(-2a^2b + ab - 4ab^2) + (6a^2b + 4ab^2)$
$4a^2b + ab$

38 $(-2a^2b + ab - 4ab^2) - (6a^2b + 4ab^2)$
$-8a^2b + ab - 8ab^2$

39. $(6x^2y - xy) + (3x^2y - 7xy^2) - (4xy - 5xy^2)$
$9x^2y - 5xy - 2xy^2$

40 $(6x^2y - xy) - (3x^2y - 7xy^2) - (4xy - 5xy^2)$
$3x^2y - 5xy + 12xy^2$

FOR REVIEW

Give the degree of each polynomial.

41. $-6y + 8y^5 + y^4 - 2$ 5

42. 14   0

43. $x^{10} - 6x^{20} + x^{30}$ 30

Evaluate the polynomial for the given value of the variable.

44. $-7a + 3$ for $a = -2$ 17

45. $3y^3 + 2y^2$ for $y = -1$ -1

46. The profit in dollars when x suits are sold is given by the expression $7x - 50$. Find the profit when 60 suits are sold.   **$370**

Exercises 47–50 review material from Chapter 10 to help you prepare for the next section. Find the product.

47. $(3x)(-2x)$ $-6x^2$ **48.** $(-3x)(-2x)$ $6x^2$ **49.** $(x^2)(4x)$ $4x^3$ **50.** $(2x^2)(-5x^2)$ $-10x^4$

ANSWERS: 1. $-5x - 1$ 2. $8x^2 + 4$ 3. $-7y^2 - 2y + 4$ 4. $-4y^2 + 2y - 1$ 5. $-2x^5 + 8x^4 - 2x^3 - 2x^2$ 6. $2x^7 -$
$2x^6 + 2x^4 + 5x^2 + 2$ 7. $2y^2 + 10y - 3$ 8. $8y^5 + y^4 - 5y^2 + 19y + 3$ 9. $-3a^2 - 6a + 8$ 10. $24a^4 + 15a^3 + 2$
11. $-2x^3 - x^2 + 10x + 7$ 12. $11x^5 + 7x^4 + 40x^3 - 17x^2 + 13x - 21$ 13. $y^2 + 9$ 14. $7x^5 - 8x^4 - 6x^3 + 6x^2 - 2x$
15. $-x^4 + 6x^3 + 7x^2 - 8x - 6$ 16. $0.09y^3 - 0.88y^2 + 2y + 2.3$ 17. $x - 11$ 18. $-2x^2 + 7$ 19. $-5y^2 - y - 2$
20. $y^2 - y - 7$ 21. $7a^5 - 3a^4 - 5a^3 + 2a^2 + 3a$ 22. $a^4 - 6a^3 - 3a - 2$ 23. $39x^5 + 18x^4 - 93x^3 + 41x + 32$
24. $-7x^{10} - x^6 + 7x^5 - 3x^2$ 25. $-15y^2 + 7$ 26. $-3a + 4$ 27. $6y^7 + 5y^5 - 3y^2 - 3y + 4$ 28. $-3y^{12} + 6y^{10}$
29. $2y^2 - 3y - 13$ 30. $-4a^2 + 4a + 4$ 31. $4x^3 - 4x^2 + 7x$ 32. $-6y^4 + 2y^3 + 10y^2 + 7y$ 33. $9a^2b^2 + 7ab$
34. $-3a^2 + 4b^2 + 1$ 35. $-x^2y^2 - 11xy$ 36. $9x^2 + 5$ 37. $4a^2b + ab$ 38. $-8a^2b + ab - 8ab^2$ 39. $9x^2y - 5xy - 2xy^2$
40. $3x^2y - 5xy + 12xy^2$ 41. 5 42. 0 43. 30 44. 17 45. -1 46. $370 47. $-6x^2$ 48. $6x^2$ 49. $4x^3$ 50. $-10x^4$

14.4 EXERCISES B

Add.

1. $9x + 3$ and $-4x - 2$
$5x + 1$
2. $-8x^2 + 5$ and $3x^2 + 2$
$-5x^2 + 7$
3. $8a^2 + a - 7$ and $-a + 3$
$8a^2 - 4$
4. $2a^4 - 5a^2 + 6$ and $3a^2$
$2a^4 - 2a^2 + 6$
5. $6x^4 + 2x^3 - 7x^2$ and $-9x^5 + 3x^4 - x^3 + 7$
$-9x^5 + 9x^4 + x^3 - 7x^2 + 7$
6. $6x^9 - 3x^6 + 2x^3$ and $2x^8 - 4x^6 - 9x^3$
$6x^9 + 2x^8 - 7x^6 - 7x^3$
7. $-8y^3 + 5 - 6y^2$ and $3 - 2y^2 + y^3$
$-7y^3 - 8y^2 + 8$
8. $3y - 5y^3 - 2y^2 + 2$ and $9 + 2y^2 - y^3$
$-6y^3 + 3y + 11$
9. $-12a^2 - 6$, $2a^2 + 5a$, and $a - 3$
$-10a^2 + 6a - 9$
10. $-9a^3 + a^4 - 6$, $2a - 5a^4 + a^3$, and
$8a^4 - 2a^3 + 5$
$4a^4 - 10a^3 + 2a - 1$
11. $x^2 - x^3 + x$, $5x - 6x^3 + 3$, 9, and $3 + 2x^3$
$-5x^3 + x^2 + 6x + 15$
12. $-3x^5 + 9x^2$, $6 - 12x^5 + 6x^2$, and
$3 - 6x + 5x^5 + x^2$
$-10x^5 + 16x^2 - 6x + 9$
13. $-9y^2 + 2y - 6$
$\underline{6y^2 + 3y + 9}$
$-3y^2 + 5y + 3$
14. $-8x^4 \qquad - 2x^2 + 6x - 8$
$\underline{7x^4 + 12x^3 + 2x^2 \qquad + 4}$
$-x^4 + 12x^3 + 6x - 4$
15. $\quad - 2x^3 + 8x^2 - 9x + 9$
$7x^4 \qquad + 2x^2 + 9x - 8$
$\underline{-6x^4 + 6x^3 - 4x^2 \qquad + 7}$
$x^4 + 4x^3 + 6x^2 + 8$
16. $0.23y^3 + 0.98y^2 - 0.6y + 0.8$
$0.54y^3 - 0.82y^2 + 0.1y$
$\underline{-0.77y^3 \qquad + 0.9y - 0.3}$
$0.16y^2 + 0.4y + 0.5$

Subtract.

17. $3x - 5$ from $2x + 8$
$-x + 13$
18. $2x^2 - 9$ from $-5x^2 + 7$
$-7x^2 + 16$
19. $6y^2 - 2y + 8$ from $-4y^2 - 2y + 1$
$-10y^2 - 7$
20. $-3y^2 + 9$ from $2y^2 - 9y + 2$
$5y^2 - 9y - 7$
21. $2a^4 - 3a^3 + 5a^2$ from $-9a^5 + 2a^4 - 3a^2 + 2a$
$-9a^5 + 3a^3 - 8a^2 + 2a$
22. $5a - 3a^2 + 5a^3$ from $a^4 + 5a^2 + 5a^3 + 3$
$a^4 + 8a^2 - 5a + 3$
23. $5x^4 + 7x^5 - 8x + 20$ from $19x + 20x^3 + 72x^4 - 10x^5$
$-17x^5 + 67x^4 + 20x^3 + 27x - 20$
24. $10x^8 + 10x^6 - 13$ from $2x^4 - 13 + 8x^8 - x^2$
$-2x^8 - 10x^6 + 2x^4 - x^2$

Perform the indicated operations.

25. $(7y^2 - 8) - (-2y^2 + 7)$
$9y^2 - 15$

26. $(9a^2 + 7a - 4) - (-2a^2 + 7a - 7)$
$11a^2 + 3$

27. $(-6y + 4y - y^5 + 8) - (5y - 4y^2 + 7y^5 - 2)$
$-6y^7 - 8y^5 + 4y^2 - y + 10$

28. $(3y^{14} - 2y^{10}) - (-y^{14} + 2y^{10} + y^6)$
$4y^{14} - 4y^{10} - y^6$

29. $(5y + 6) + (-2y^2 + 5) - (2y^2 - 3y - 8)$
$-4y^2 + 8y + 19$

30. $(-8a^2 - 2a) - (2a^2 + 3a - 1) - (a^2 - 9a + 6)$
$-11a^2 + 4a - 5$

31. $(3x^4 - 8x^3 - 4x) + (2x^4 - 8x^3 + 5x^2) - (5x^4 + 5x^2 - 6x)$
$-16x^3 + 2x$

32. $(3y^4 - 8y^3 - 4y) - (2y^4 - 8y^3 + 5y^2) - (5y^4 + 5y^2 - 6y)$
$-4y^4 - 10y^2 + 2y$

33. $(2a^2b^2 + 5ab) + (9a^2b^2 - 2ab)$
$11a^2b^2 + 3ab$

34. $(4a^2 - 3b^2 + 9) + (-4a^2 - 5b^2 + 3)$
$-8b^2 + 12$

35. $(8x^2y^2 - 9xy) - (-4x^2y^2 + 7xy)$
$12x^2y^2 - 16xy$

36. $(5x^2 + 3y^2 - 7) - (-2x^2 + 3y^2 + 7)$
$7x^2 - 14$

37. $(5a^2b - 2ab - 2ab^2) + (2a^2b - 6ab^2)$
$7a^2b - 2ab - 8ab^2$

38. $(5a^2b - 2ab - 2ab^2) - (2a^2b - 6ab^2)$
$3a^2b - 2ab + 4ab^2$

39. $(11x^2y - 2xy) + (4x^2y - 9xy^2) - (8xy - 10xy^2)$
$15x^2y - 10xy + xy^2$

40. $(11x^2y - 2xy) - (4x^2y - 9xy^2) - (8xy - 10xy^2)$
$7x^2y - 10xy + 19xy^2$

FOR REVIEW

Give the degree of each polynomial.

41. $8y^2 - 6y^{10} - y^{14}$ 14

42. $22a$ 1

43. $x^{20} + 2x^{40}$ 40

Evaluate the polynomial for the given value of the variable.

44. $6a^2 - 2a$ for $a = 3$ 48

45. $y^4 - 6y^2 + 8$ for $y = -2$ 0

46. The cost in dollars when x suits are made is given by the expression $4x^2 + 5x - 100$. Find the cost when 15 suits are made. $875

Exercises 47–50 review material from Chapter 10 to help you prepare for the next section. Find the product.

47. $(-7y)(4y)$
$-28y^2$

48. $(-7y)(-4y)$
$28y^2$

49. $(-3x^3)(6x)$
$-18x^4$

50. $(6x^3)(5x^3)$
$30x^6$

14.4 EXERCISES C

Perform the indicated operations.

1. $\left(\frac{1}{3}x^4 - \frac{1}{2}x^3 + \frac{7}{8}\right) + \left(-\frac{1}{9}x^4 + \frac{3}{4}x^3 - \frac{2}{3}x^2 + x\right) + \left(\frac{2}{9}x^4 - \frac{3}{2}x^3 + \frac{4}{3}x^2 + \frac{1}{3}x + \frac{3}{8}\right)$

$\left[\text{Answer: } \frac{4}{9}x^4 - \frac{5}{4}x^3 + \frac{2}{3}x^2 + \frac{4}{3}x + \frac{5}{4}\right]$

2. $(100a^3 - 97a^2 + 21a - 105) + (16a^3 + 45a^2 - 115a) - (88a^3 - 19a^2 - 47)$
$28a^3 - 33a^2 - 94a - 58$

3. $(-2.13a^2 + 3.25a + 7.98) - (0.32a^3 - 2.10a + 1.92)$ [Answer: $-0.32a^3 - 2.13a^2 + 5.35a + 6.06$]

4. $\left(-\frac{1}{9}x^5 + \frac{1}{3}x^4 - \frac{2}{3}x + \frac{1}{3}\right) - \left(-\frac{5}{9}x^5 - \frac{2}{9}x^4 + \frac{8}{9}x + \frac{7}{9}\right)$

$\frac{4}{9}x^5 + \frac{5}{9}x^4 - \frac{14}{9}x - \frac{4}{9}$

14.5 MULTIPLICATION OF POLYNOMIALS

STUDENT GUIDEPOSTS

1 Multiplying Monomials

2 Multiplying a Binomial by a Monomial

3 Multiplying Polynomials

4 The FOIL Method

1 MULTIPLYING MONOMIALS

In Section 14.1 we multiplied powers of a variable by adding their exponents using the product rule

$$x^m x^n = x^{m+n}.$$

This rule is used repeatedly when we multiply polynomials. We begin by reviewing multiplication of monomials.

EXAMPLE 1   MONOMIAL TIMES A MONOMIAL

Multiply.

(a) $(-3x^2)(7x^5) = (-3)(7)x^2 x^5$ The order of the factors can be changed

$\qquad = -21x^{2+5}$ Use $x^m x^n = x^{m+n}$

$\qquad = -21x^7$

(b) $(5x^4)(-x) = (5)(-1)x^4 x$ $-x = (-1) \cdot x$

$\qquad = -5x^{4+1}$ $x = x^1$

$\qquad = -5x^5$

PRACTICE EXERCISE 1

Multiply.

(a) $(4y^3)(-5y^7)$

(b) $(-8y^2)(-y)$

Answers: (a) $-20y^{10}$ (b) $8y^3$

2 MULTIPLYING A BINOMIAL BY A MONOMIAL

In Section 10.5 we introduced the distributive law. Examples in that section showed multiplication of a binomial by a monomial. Example 2 reviews this idea.

EXAMPLE 2   MONOMIAL TIMES A BINOMIAL

Multiply.

(a) $a(3a^2 + 2) = (a)(3a^2) + (a)(2)$ Distributive law

$\qquad = 3aa^2 + 2a$

$\qquad = 3a^3 + 2a$

(b) $-3a^2(8a^3 - 5a) = (-3a^2)(8a^3) + (-3a^2)(-5a)$ Distributive law

$\qquad = (-3)(8)a^2 a^3 + (-3)(-5)a^2 a$

$\qquad = -24a^{2+3} + 15a^{2+1}$

$\qquad = -24a^5 + 15a^3$

PRACTICE EXERCISE 2

Multiply.

(a) $x(5 + 3x^3)$

(b) $-4x^3(2x^2 - 7x)$

Answers: (a) $5x + 3x^4$
(b) $-8x^5 + 28x^4$

3 MULTIPLYING POLYNOMIALS

Polynomials can be multiplied by repeated use of the distributive law. This is illustrated in Example 3.

| **EXAMPLE 3   USING THE DISTRIBUTIVE LAW** | **PRACTICE EXERCISE 3** |

Multiply.

(a) $(3x + 2)(x^2 - 5x)$

$\quad = (3x + 2)\,x^2 + (3x + 2)(-5x)$ Distributive law

$\quad = (3x)(x^2) + (2)(x^2) + (3x)(-5x) + (2)(-5x)$ Distributive law

$\quad = 3x^3 + 2x^2 - 15x^2 - 10x$

$\quad = 3x^3 - 13x^2 - 10x$

(b) $(4x + 1)(3x^2 - x + 5)$

$\quad = (4x + 1)\,3x^2 + (4x + 1)(-x) + (4x + 1)\,5$

$\quad = (4x)(3x^2) + (1)(3x^2) + (4x)(-x) + (1)(-x) + (4x)(5) + (1)(5)$

$\quad = 12x^3 + 3x^2 - 4x^2 - x + 20x + 5$

$\quad = 12x^3 - x^2 + 19x + 5$

Practice Exercise 3:

Multiply.

(a) $(2z + 1)(z^3 - 3z)$

(b) $(5z - 3)(2z^2 + z - 7)$

Answers: **(a)** $2z^4 + z^3 - 6z^2 - 3z$
(b) $10z^3 - z^2 - 38z + 21$

Notice that the second step in each part of Example 3 contains all possible products of terms in the first polynomial with terms in the second. This leads to the following rule.

> **To Multiply Two Polynomials, Neither of Which Is a Monomial**
>
> 1. Multiply each term in one by each term in the other.
> 2. Collect and combine like terms.

④ THE FOIL METHOD

The distributive law guarantees that when two polynomials are multiplied, all products of all terms are found. Now suppose we use the following method to find the product in Example 3(a).

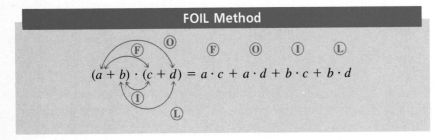

$(3x + 2)(x^2 - 5x) = (3x)(x^2) + (3x)(-5x) + (2)(x^2) + (2)(-5x)$

$\qquad\qquad\qquad\quad = 3x^3 - 15x^2 + 2x^2 - 10x$

$\qquad\qquad\qquad\quad = 3x^3 - 13x^2 - 10x$

Notice that the letters F, O, I, L spell "FOIL" and stand for the **F**irst terms, **O**utside terms, **I**nside terms, and **L**ast terms. Remember this word, and you will not omit terms when multiplying a binomial by a binomial.

> **FOIL Method**
>
> $(a + b) \cdot (c + d) = a \cdot c + a \cdot d + b \cdot c + b \cdot d$

| **EXAMPLE 4   MULTIPLYING BINOMIALS USING FOIL** | **PRACTICE EXERCISE 4** |

Use the FOIL method to multiply the binomials.

(a) $(x - 3) \cdot (x + 4) = x^2 + 4x - 3x - 12$

$= x^2 + x - 12$

(b) $(x + 7)(x + 5) = x^2 + 5x + 7x + 35$

$= x^2 + 12x + 35$

Use the FOIL method to multiply the binomials.

(a) $(a - 2)(a + 8)$

(b) $(a + 2)(a + 9)$

Answers: (a) $a^2 + 6a - 16$
(b) $a^2 + 11a + 18$

When we study factoring in Chapter 15, we will start with trinomials such as $x^2 + x - 12$ in Example 4(a) and be expected to find the factors $(x - 3)$ and $(x + 4)$. Practicing the FOIL method now will help us understand factoring later.

| **EXAMPLE 5   USING THE FOIL METHOD** | **PRACTICE EXERCISE 5** |

Use the FOIL method to multiply the binomials.

(a) $(2x - 3)(3x + 7) = 6x^2 + 14x - 9x - 21$

$= 6x^2 + 5x - 21$

(b) $(2a - 5)(2a + 5) = 4a^2 + 10a - 10a - 25$

$= 4a^2 + 0 \cdot a - 25$

$= 4a^2 - 25$

(c) $(5x + 3)(x + 8) = 5x^2 + 40x + 3x + 24$

$= 5x^2 + 43x + 24$

(d) $(7y - 2)(3y - 8) = 21y^2 - 56y - 6y + 16$

$= 21y^2 - 62y + 16$

(e) $(10a + 5)(3a - 5) = 30a^2 - 50a + 15a - 25$

$= 30a^2 - 35a - 25$

Use the FOIL method to multiply the binomials.

(a) $(3y - 1)(2y + 4)$

(b) $(6x - 7)(6x + 7)$

(c) $(8z + 3)(z - 1)$

(d) $(4a - 5)(2a - 3)$

(e) $(12x + 1)(5x - 2)$

Answers: (a) $6y^2 + 10y - 4$
(b) $36x^2 - 49$ (c) $8z^2 - 5z - 3$
(d) $8a^2 - 22a + 15$ (e) $60x^2 - 19x - 2$

A third way to multiply is to write one polynomial above the other and then arrange like terms in vertical columns.

$$
\begin{array}{r}
x^2 - 5x \\
3x + 2 \\
\hline
3x^3 - 15x^2 \\
2x^2 - 10x \\
\hline
3x^3 - 13x^2 - 10x
\end{array}
$$

3x multiplied by each term of the top polynomial
2 multiplied by each term of the top polynomial
Sum

The third method is especially good when trinomials or larger polynomials are part of the multiplication.

EXAMPLE 6   USING VERTICAL COLUMNS

Multiply.

(a)
$$5x^3 - 2x + 3$$
$$\underline{-7x^2 + 4x}$$
$$-35x^5 \qquad + 14x^3 - 21x^2 \qquad \text{—7}x^2 \text{ multiplied by top}$$
$$\text{polynomial}$$
$$\underline{20x^4 \qquad - 8x^2 + 12x} \qquad \text{4}x \text{ multiplied by top polynomial}$$
$$-35x^5 + 20x^4 + 14x^3 - 29x^2 + 12x \qquad \text{Sum}$$

We leave spaces for missing terms so we can add like terms in columns.

(b) $a^2 + 2a + 4$
$$\underline{a \; - \; 2}$$
$$a^3 + 2a^2 + 4a \qquad\qquad\qquad a \text{ multiplied by top polynomial}$$
$$\underline{- 2a^2 - 4a - 8} \qquad\qquad -2 \text{ multiplied by top polynomial}$$
$$a^3 \qquad\qquad - 8 = a^3 - 8 \qquad \text{Sum}$$

(c)
$$5x^2 - \; 3x \; + \; 7$$
$$\underline{-4x^2 + \;\; x \; + \; 8}$$
$$-20x^4 + 12x^3 - 28x^2 \qquad\qquad\qquad \text{—4}x^2 \text{ multiplied by top polynomial}$$
$$5x^3 - \;\; 3x^2 + \;\; 7x \qquad\qquad\qquad x \text{ multiplied by top polynomial}$$
$$\underline{40x^2 - 24x + 56} \qquad\qquad 8 \text{ multiplied by top polynomial}$$
$$-20x^4 + 17x^3 + \;\; 9x^2 - 17x + 56 \qquad \text{Sum}$$

All the procedures used above can be applied to the multiplication of polynomials in several variables.

EXAMPLE 7   POLYNOMIALS IN TWO VARIABLES

Use the FOIL method to multiply the binomials in two variables.

(a) $(2x + y)(x - 3y) = 2x^2 - 6xy + xy - 3y^2$
$$= 2x^2 - 5xy - 3y^2$$

(b) $(2a - 5b)(2a + 5b) = 4a^2 + 10ab - 10ab - 25b^2$
$$= 4a^2 + 0 \cdot ab - 25b^2$$
$$= 4a^2 - 25b^2$$

(c) $(5x + 3y)(x + 8y) = 5x^2 + 40xy + 3xy + 24y^2$
$$= 5x^2 + 43xy + 24y^2$$

EXAMPLE 8   USING VERTICAL COLUMNS

Use vertical columns to multiply.

$$3x^2 - 4xy + 5y^2$$
$$\underline{6x \; - \; 7y}$$
$$18x^3 - 24x^2y + 30xy^2 \qquad\qquad 6x \text{ multiplied by top polynomial}$$
$$\underline{- 21x^2y + 28xy^2 - 35y^3} \qquad\qquad \text{—7}y \text{ multiplied by top polynomial}$$
$$18x^3 - 45x^2y + 58xy^2 - 35y^3 \qquad \text{Sum}$$

We conclude this section by solving an applied problem.

PRACTICE EXERCISE 6

Multiply.

(a) $3y^2 - \; y + 2$
$$\underline{y^2 + 2y}$$

(b) $4x^2 + 6x + 9$
$$\underline{2x \; - \; 3}$$

(c) $2a^2 + 5a - 6$
$$\underline{-3a^2 + \;\; a + 5}$$

Answers: (a) $3y^4 + 5y^3 + 4y$
(b) $8x^3 - 27$ (c) $-6a^4 - 13a^3 + 33a^2 + 19a - 30$

PRACTICE EXERCISE 7

Use the FOIL method to multiply the binomials in two variables.

(a) $(5a + 2b)(a - b)$

(b) $(3x + 7y)(3x - 7y)$

(c) $(6z - 4w)(5z - 3w)$

Answers: (a) $5a^2 - 3ab - 2b^2$
(b) $9x^2 - 49y^2$ (c) $30z^2 - 38zw + 12w^2$

PRACTICE EXERCISE 8

Use vertical columns to multiply.

$$4a^2 + 3ab + b^2$$
$$\underline{5a \; - \; 2b}$$

Answer: $20a^3 + 7a^2b - ab^2 - 2b^3$

| EXAMPLE 9 BUSINESS APPLICATION | PRACTICE EXERCISE 9 |

The Valley Video Center sells blank video cassettes. From past experience it has been shown that the number n of cassettes that can be sold each week depends on the price p in dollars charged for each according to the equation $n = 1000 - 100p$. Find a formula for the weekly revenue R in terms of p. Use this formula to find the revenue when the price of a cassette is \$5, \$8, and \$10.

Find a formula for weekly revenue from a small item if $n = 300 - 20p$. Use the formula to find R when the price is \$1, \$2, and \$3.

The **revenue** R produced by selling n items for p dollars per item is given by $R = np$. In our case, the number of cassettes sold per week is related to the price of a cassette by $n = 1000 - 100p$. Notice that this relationship tells us that when p is \$9, $n = 1000 - 100(9) = 100$, and when p is reduced to \$8, $n = 1000 - 100(8) = 200$. As a result, we see that more cassettes can be sold when the price is lowered. The formula for the revenue produced weekly is

$$R = np = (1000 - 100p)p$$
$$= 1000p - 100p^2.$$

When p is \$5,

$$R = 1000(\mathbf{5}) - 100(\mathbf{5})^2$$
$$= 5000 - 2500 = 2500,$$

so the revenue is \$2500. When p is \$8,

$$R = 1000(\mathbf{8}) - 100(\mathbf{8})^2$$
$$= 8000 - 6400 = 1600,$$

so the revenue is \$1600. Finally, when p is \$10,

$$R = 1000(\mathbf{10}) - 100(\mathbf{10})^2$$
$$= 10{,}000 - 10{,}000 = 0,$$

so the revenue drops off to \$0. In other words, the consumer views \$10 as too much and refuses to purchase cassettes at this price.

Answer: $R = 300p - 20p^2$; \$280, \$520, \$720

14.5 EXERCISES A

Multiply the following monomials.

1. $(2)(6x)$
  $12x$

2. $(-8y)(4)$
  $-32y$

3. $(-7a^2)(-4a)$
  $28a^3$

4. $(4y^2)(2y^2)$
  $8y^4$

5. $(-3x^4)(2x)$
  $-6x^5$

6. $(8y^3)(-9y^2)$
  $-72y^5$

7. $(-13x^7)(10x^3)$
  $-130x^{10}$

8. $(-8y^6)(-7y^5)$
  $56y^{11}$

Multiply using the distributive law.

9. $2(x + 3)$
  $2x + 6$

10. $3x(x - 4)$
  $3x^2 - 12x$

11. $-4y(y + 8)$
  $-4y^2 - 32y$

12. $-6a^2(4a + 5)$
$-24a^3 - 30a^2$

13. $3(2x^2 - 3x + 1)$
$6x^2 - 9x + 3$

14. $4y(y^2 - 8y - 5)$
$4y^3 - 32y^2 - 20y$

15. $2y^3(-3y^2 - 2y + 5)$
$-6y^5 - 4y^4 + 10y^3$

16 $-10x^2(x^5 - 6x^3 + 7x^2)$
$-10x^7 + 60x^5 - 70x^4$

17. $-9a^3(4a^4 - 3a^2 + 2)$
$-36a^7 + 27a^5 - 18a^3$

18. $(x + 8)(x + 4)$
$x^2 + 12x + 32$

19. $(2a - 3)(3a - 2)$
$6a^2 - 13a + 6$

20. $(x - 5)(x^2 - 3x + 2)$
$x^3 - 8x^2 + 17x - 10$

21. $(2y + 7)(3y^2 + y - 1)$
$6y^3 + 23y^2 + 5y - 7$

22. $(a^2 + 1)(a^4 - a^2 + 1)$
$a^6 + 1$

23 $(x^2 + 4x - 2)(2x^2 - x + 3)$
$2x^4 + 7x^3 - 5x^2 + 14x - 6$

Use the FOIL method to multiply the following binomials.

24. $(x + 10)(x + 2)$
$x^2 + 12x + 20$

25. $(a - 6)(a + 6)$
$a^2 - 36$

26. $(y + 5)(y - 3)$
$y^2 + 2y - 15$

27. $(x - 7)(x - 3)$
$x^2 - 10x + 21$

28. $(a + 8)(a + 8)$
$a^2 + 16a + 64$

29. $(x - 12)(x + 4)$
$x^2 - 8x - 48$

30. $(a - 8)(a - 8)$
$a^2 - 16a + 64$

31. $(3x + 2)(x + 5)$
$3x^2 + 17x + 10$

32. $(5a - 3)(a - 7)$
$5a^2 - 38a + 21$

33. $(2y - 7)(y + 10)$
$2y^2 + 13y - 70$

34. $(7x + 3)(x - 5)$
$7x^2 - 32x - 15$

35. $(2a + 3)(3a + 5)$
$6a^2 + 19a + 15$

36. $(4y - 3)(2y - 5)$
$8y^2 - 26y + 15$

37 $(5x - 2)(3x + 4)$
$15x^2 + 14x - 8$

38. $(9a + 2)(2a - 7)$
$18a^2 - 59a - 14$

39. $(10y - 1)(4y + 5)$
$40y^2 + 46y - 5$

40. $(7x + 5)(3x - 8)$
$21x^2 - 41x - 40$

41 $(2z^2 + 1)(z^2 - 2)$
$2z^4 - 3z^2 - 2$

Multiply.

42. $3a^2 + 5$
$\underline{a - 4}$
$3a^3 - 12a^2 + 5a - 20$

43. $2a^3 - 3a$
$\underline{7a^2 - 5}$
$14a^5 - 31a^3 + 15a$

44. $3y^3 - 2$
$\underline{5y^2 + 4y}$
$15y^5 + 12y^4 - 10y^2 - 8y$

45. $3a^2 - 5$
$\underline{3a^2 - 5}$
$9a^4 - 30a^2 + 25$

46. $7y^3 - 2$
$\underline{7y^3 + 2}$
$49y^6 - 4$

47. $5x^4 + 3x^2$
$\underline{8x^2 - 7}$
$40x^6 - 11x^4 - 21x^2$

48. $3x^2 - 5x + 2$
$\underline{x + 4}$
$3x^3 + 7x^2 - 18x + 8$

49. $-7y^2 - 3y + 10$
$\underline{2y - 1}$
$-14y^3 + y^2 + 23y - 10$

50. $12a^3 - 3a + 8$
$\underline{4a^2 + 7a}$
$48a^5 + 84a^4 - 12a^3 + 11a^2 + 56a$

51. $12a^3 - 6a + 4$
$7a^2 - 3$
$\overline{84a^5 - 78a^3 + 28a^2 + 18a - 12}$

52. $0.3x^2 + 0.2$
$0.5x - 0.7$
$\overline{0.15x^3 - 0.21x^2 + 0.10x - 0.14}$

53. $\dfrac{2}{3}y^2 - \dfrac{3}{4}$
$\dfrac{1}{2}y + 5$
$\overline{\dfrac{1}{3}y^3 + \dfrac{10}{3}y^2 - \dfrac{3}{8}y - \dfrac{15}{4}}$

54. $a^2 - 2a + 2$
$a^2 + 2a - 2$
$\overline{a^4 - 4a^2 + 8a - 4}$

55. $3y^2 + 5y - 6$
$y^2 - 3y + 2$
$\overline{3y^4 - 4y^3 - 15y^2 + 28y - 12}$

56. $-4x^3 + 2x - 5$
$6x^2 - x + 1$
$\overline{-24x^5 + 4x^4 + 8x^3 - 32x^2 + 7x - 5}$

Multiply the polynomials in two variables.

57. $2xy(x^2 + 2xy)$
$2x^3y + 4x^2y^2$

58. $3a^2b(a^2b + 2ab - ab^2)$
$3a^4b^2 + 6a^3b^2 - 3a^3b^3$

59. $(2a - 3b)(5a + b)$
$10a^2 - 13ab - 3b^2$

60. $(x + 3y)(x + 2y)$
$x^2 + 5xy + 6y^2$

61. $(4x - y)(4x + y)$
$16x^2 - y^2$

62. $(7a - 2b)(2a - 5b)$
$14a^2 - 39ab + 10b^2$

63. $3x^2 - 2xy + 4y^2$
$2x + y$
$\overline{6x^3 - x^2y + 6xy^2 + 4y^3}$

64. $a^2b^2 + ab - 3$
$a^2 - b^2$
$\overline{a^4b^2 + a^3b - 3a^2 - a^2b^4 - ab^3 + 3b^2}$

Perform the following operations.

65. $(x - 1)(x + 1)(x + 2)$
$x^3 + 2x^2 - x - 2$

66. $(y + 3)(y - 2)(y + 2)$
$y^3 + 3y^2 - 4y - 12$

67. The length of a rectangle measures $(2x + 1)$ feet and the width is $(3x - 2)$ feet. Find the area of the rectangle in terms of x.
$(6x^2 - x - 2)$ ft^2

68. The dimensions of a box, given in centimeters, are y, $y + 1$, and $2y + 3$. Find the volume of the box in terms of y.
$(2y^3 + 5y^2 + 3y)$ cm^3

69. A wholesale distributor who sells clocks knows that the number n of clocks she can sell each month is related to the price in dollars p of each clock by the equation $n = 5000 - 100p$. Find an equation for the monthly revenue R in terms of p and use it to find R when p is \$10, \$30, and \$45.
$R = 5000p - 100p^2$; \$40,000; \$60,000; \$22,500

70. The length l of a rectangular pasture is to be 300 m longer than the width w. Find an equation that gives the area in terms of the width and use it to give the area when the width is 40 m, 100 m, and 500 m.
$A = w^2 + 300w$; 13,600 m^2; 40,000 m^2; 400,000 m^2

FOR REVIEW

Perform the indicated operations.

71. $(3x^2 + 2x - 1) + (2x + 1) - (x^2 - 2)$
$2x^2 + 4x + 2$

72. $(-4y^2 + 2y) + (y^3 - 2y) - (y^2 + 3y - 5)$
$y^3 - 5y^2 - 3y + 5$

73. $(3a^2 - 5a + 5) - (-a^5 + 4a^2 - 2) - (6a - 7)$
$a^5 - a^2 - 11a + 14$

74. $(x^2 - y^2) - (2x^2 - 3y^2) - (5x^2 - y^2)$
$-6x^2 + 3y^2$

In Exercises 75–80, find the special products using the FOIL method. This will help you prepare for the next section.

75. $(x + 5)(x + 5)$
$x^2 + 10x + 25$

76. $(x - 5)(x - 5)$
$x^2 - 10x + 25$

77. $(x + 5)(x - 5)$
$x^2 - 25$

78. $(3x + 2y)(3x - 2y)$
$9x^2 - 4y^2$

79. $(3x - 2y)(3x - 2y)$
$9x^2 - 12xy + 4y^2$

80. $(3x + 2y)(3x + 2y)$
$9x^2 + 12xy + 4y^2$

ANSWERS: 1. $12x$ 2. $-32y$ 3. $28a^3$ 4. $8y^4$ 5. $-6x^5$ 6. $-72y^5$ 7. $-130x^{10}$ 8. $56y^{11}$ 9. $2x + 6$
10. $3x^2 - 12x$ 11. $-4y^2 - 32y$ 12. $-24a^3 - 30a^2$ 13. $6x^2 - 9x + 3$ 14. $4y^3 - 32y^2 - 20y$ 15. $-6y^5 - 4y^4 + 10y^3$
16. $-10x^7 + 60x^5 - 70x^4$ 17. $-36a^7 + 27a^5 - 18a^3$ 18. $x^2 + 12x + 32$ 19. $6a^2 - 13a + 6$ 20. $x^3 - 8x^2 + 17x - 10$
21. $6y^3 + 23y^2 + 5y - 7$ 22. $a^6 + 1$ 23. $2x^4 + 7x^3 - 5x^2 + 14x - 6$ 24. $x^2 + 12x + 20$ 25. $a^2 - 36$
26. $y^2 + 2y - 15$ 27. $x^2 - 10x + 21$ 28. $a^2 + 16a + 64$ 29. $x^2 - 8x - 48$ 30. $a^2 - 16a + 64$ 31. $3x^2 + 17x + 10$
32. $5a^2 - 38a + 21$ 33. $2y^2 + 13y - 70$ 34. $7x^2 - 32x - 15$ 35. $6a^2 + 19a + 15$ 36. $8y^2 - 26y + 15$
37. $15x^2 + 14x - 8$ 38. $18a^2 - 59a - 14$ 39. $40y^2 + 46y - 5$ 40. $21x^2 - 41x - 40$ 41. $2z^4 - 3z^2 - 2$
42. $3a^3 - 12a^2 + 5a - 20$ 43. $14a^5 - 31a^3 + 15a$ 44. $15y^5 + 12y^4 - 10y^2 - 8y$ 45. $9a^4 - 30a^2 + 25$ 46. $49y^6 - 4$
47. $40x^6 - 11x^4 - 21x^2$ 48. $3x^3 + 7x^2 - 18x + 8$ 49. $-14y^3 + y^2 + 23y - 10$ 50. $48a^5 + 84a^4 - 12a^3 + 11a^2 + 56a$
51. $84a^5 - 78a^3 + 28a^2 + 18a - 12$ 52. $0.15x^3 - 0.21x^2 + 0.10x - 0.14$ 53. $\frac{1}{3}y^3 + \frac{10}{3}y^2 - \frac{3}{8}y - \frac{15}{4}$
54. $a^4 - 4a^2 + 8a - 4$ 55. $3y^4 - 4y^3 - 15y^2 + 28y - 12$ 56. $-24x^5 + 4x^4 + 8x^3 - 32x^2 + 7x - 5$ 57. $2x^3y + 4x^2y^2$
58. $3a^4b^2 + 6a^3b^2 - 3a^3b^3$ 59. $10a^2 - 13ab - 3b^2$ 60. $x^2 + 5xy + 6y^2$ 61. $16x^2 - y^2$ 62. $14a^2 - 39ab + 10b^2$
63. $6x^3 - x^2y + 6xy^2 + 4y^3$ 64. $a^4b^2 + a^3b - 3a^2 - a^2b^4 - ab^3 + 3b^2$ 65. $x^3 + 2x^2 - x - 2$ 66. $y^3 + 3y^2 - 4y - 12$
67. $(6x^2 - x - 2)$ ft^2 68. $(2y^3 + 5y^2 + 3y)$ cm^3 69. $R = 5000p - 100p^2$; \$40,000; \$60,000; \$22,500
70. $A = w^2 + 300w$; 13,600 m^2; 40,000 m^2; 400,000 m^2 71. $2x^2 + 4x + 2$ 72. $y^3 - 5y^2 - 3y + 5$
73. $a^5 - a^2 - 11a + 14$ 74. $-6x^2 + 3y^2$ 75. $x^2 + 10x + 25$ 76. $x^2 - 10x + 25$ 77. $x^2 - 25$ 78. $9x^2 - 4y^2$
79. $9x^2 - 12xy + 4y^2$ 80. $9x^2 + 12xy + 4y^2$

14.5 EXERCISES B

Multiply the following monomials.

1. $(3)(5x)$
$15x$

2. $(-4y)(8)$
$-32y$

3. $(9a)(2a^3)$
$18a^4$

4. $(-3a^3)(5a^2)$
$-15a^5$

5. $(-10y)(-8y^5)$
$80y^6$

6. $(a^7)(a^7)$
a^{14}

7. $(18x^8)(-2x^3)$
$-36x^{11}$

8. $(-10y^7)(-20y^8)$
$200y^{15}$

Multiply using the distributive law.

9. $5x(x - 6)$
$5x^2 - 30x$

10. $2x(-3x + 5)$
$-6x^2 + 10x$

11. $-8x(3x - 4)$
$-24x^2 + 32x$

12. $-9a(2a^2 - 3)$
$-18a^3 + 27a$

13. $5(3x^2 - 2x - 3)$
$15x^2 - 10x - 15$

14. $3y(y^2 - 5y + 4)$
$3y^3 - 15y^2 + 12y$

15. $6y^2(-8y^2 + 5y - 3)$
$-48y^4 + 30y^3 - 18y^2$

16. $6x^3(2x^7 - 3x^6 + 5x^4)$
$12x^{10} - 18x^9 + 30x^7$

17. $-10a^5(-2a^5 + 2a^4 - 8a^2)$
$20a^{10} - 20a^9 + 80a^7$

18. $(x + 5)(x + 10)$
$x^2 + 15x + 50$

19. $(5a - 2)(2a - 7)$
$10a^2 - 39a + 14$

20. $(x - 3)(x^2 + 5x - 4)$
$x^3 + 2x^2 - 19x + 12$

21. $(3y + 8)(2y^2 - 2y + 3)$
$6y^3 + 10y^2 - 7y + 24$

22. $(a^2 - 1)(a^4 + a^2 + 1)$
$a^6 - 1$

23. $(2x^2 + 3x - 2)(x^2 - 5x + 1)$
$2x^4 - 7x^3 - 15x^2 + 13x - 2$

Use the FOIL method to multiply the following binomials.

24. $(x + 3)(x - 9)$
$x^2 - 6x - 27$

25. $(a + 8)(a - 8)$
$a^2 - 64$

26. $(x - 4)(x - 5)$
$x^2 - 9x + 20$

27. $(y - 12)(y + 3)$
$y^2 - 9y - 36$

28. $(x + 2)(x - 8)$
$x^2 - 6x - 16$

29. $(a - 9)(a - 9)$
$a^2 - 18a + 81$

30. $(y - 9)(y + 4)$
$y^2 - 5y - 36$

31. $(5x + 3)(x + 2)$
$5x^2 + 13x + 6$

32. $(4a - 5)(a - 6)$
$4a^2 - 29a + 30$

33. $(7y - 2)(y + 5)$
$7y^2 + 33y - 10$

34. $(8x + 1)(x - 10)$
$8x^2 - 79x - 10$

35. $(4a + 5)(5a + 4)$
$20a^2 + 41a + 20$

36. $(3y - 5)(4y - 7)$
$12y^2 - 41y + 35$

37. $(2x - 9)(5x + 3)$
$10x^2 - 39x - 27$

38. $(10a + 3)(5a - 9)$
$50a^2 - 75a - 27$

39. $(7y - 3)(8y + 3)$
$56y^2 - 3y - 9$

40. $(9x + 7)(2x - 1)$
$18x^2 + 5x - 7$

41. $(3z^2 - 1)(z^2 + 1)$
$3z^4 + 2z^2 - 1$

Multiply.

42. $3x + 4$
$\underline{5x - 8}$
$15x^2 - 4x - 32$

43. $3a^3 - 5a$
$\underline{4a^2 - 7}$
$12a^5 - 41a^3 + 35a$

44. $2y^2 - 9$
$\underline{3y^2 + 5y}$
$6y^4 + 10y^3 - 27y^2 - 45y$

45. $4x^2 + 9$
$\underline{4x^2 + 9}$
$16x^4 + 72x^2 + 81$

46. $8y^3 + 5$
$\underline{8y^3 - 5}$
$64y^6 - 25$

47. $10x^4 + 7x$
$\underline{5x^2 - 2x}$
$50x^6 - 20x^5 + 35x^3 - 14x^2$

48. $2x^2 - 8x + 3$
$\underline{x + 3}$
$2x^3 - 2x^2 - 21x + 9$

49. $-6y^2 + 2y - 5$
$\underline{3y - 2}$
$-18y^3 + 18y^2 - 19y + 10$

50. $8a^4 - 5a + 1$
$\underline{3a^2 + 5}$
$24a^6 + 40a^4 - 15a^3 + 3a^2 - 25a + 5$

51. $15a^3 + 5a^2 - 3$
$\underline{6a^2 - 7}$
$90a^5 + 30a^4 - 105a^3 - 53a^2 + 21$

52. $1.2x^2 + 4.5$
$\underline{0.1x - 1.2}$
$0.12x^3 - 1.44x^2 + 0.45x - 5.40$

53. $\dfrac{1}{2}y^2 - \dfrac{5}{4}$
$\underline{\dfrac{4}{3}y + 7}$
$\frac{2}{3}y^3 + \frac{7}{2}y^2 - \frac{5}{3}y - \frac{35}{4}$

54. $a^2 + 3a - 1$
$\underline{a^2 - 3a + 1}$
$a^4 - 9a^2 + 6a - 1$

55. $4y^2 + 3y - 5$
$\underline{y^2 - 6y + 4}$
$4y^4 - 21y^3 - 7y^2 + 42y - 20$

56. $-2x^4 + 5x^2 - 3$
$\underline{8x^2 - 3x + 2}$
$-16x^6 + 6x^5 + 36x^4 - 15x^3 -$
$14x^2 + 9x - 6$

Multiply the polynomials in two variables.

57. $5xy(3xy + y^2)$
$15x^2y^2 + 5xy^3$

58. $6a^3b^2(2a^4b - 3a^2b^2 + ab)$
$12a^7b^3 - 18a^5b^4 + 6a^4b^3$

59. $(x + 4y)(x + 6y)$
$x^2 + 10xy + 24y^2$

60. $(3a - 5b)(7a + 8b)$
$21a^2 - 11ab - 40b^2$

61. $(6x - y)(6x + y)$
$36x^2 - y^2$

62. $(3a - 8b)(10a - 3b)$
$30a^2 - 89ab + 24b^2$

63. $5x^2 - 5xy + 6y^2$
$\underline{4x + 3y}$
$20x^3 - 5x^2y + 9xy^2 + 18y^3$

64. $2a^2b^2 - 3ab + 9$
$\underline{ab - 2}$
$2a^3b^3 - 7a^2b^2 + 15ab - 18$

Perform the following operations.

65. $(y - 3)(y + 3)(y - 2)$
$y^3 - 2y^2 - 9y + 18$

66. $(x + 4)(x + 2)(x - 4)$
$x^3 + 2x^2 - 16x - 32$

67. The base of a triangle measures $(3y + 2)$ inches and the height is $(2y - 1)$ inches. Find the area of the triangle in terms of y.
$\frac{1}{2}(6y^2 + y - 2)$ in^2

68. The dimensions of a box, given in feet, are x, $x + 2$, and $3x + 5$. Find the volume of the box in terms of x.
$(3x^3 + 11x^2 + 10x)$ ft^3

69. The number n of new magazine subscriptions that can be sold during each month is related to the price in dollars p of a subscription by the equation $n = 200 - 10p$. Find an equation that gives the monthly revenue R in terms of p and use it to find R when p is \$15, \$10, and \$8.

$R = 200p - 10p^2$; \$750; \$1000; \$960

70. A rectangular metal plate is to be constructed in such a way that the length is 10 cm more than twice the width. Find an equation that gives the area of the top of the plate and use it to find the area when the width is 30 cm, 70 cm, and 120 cm.

$A = 2w^2 + 10w$; 2100 cm²; 10,500 cm²; 30,000 cm²

FOR REVIEW

Perform the indicated operations.

71. $(-4x^2 + 5x - 2) + (5x - 2) - (3x^2 - 2x)$
$-7x^2 + 12x - 4$

72. $(9y^2 - 6) + (y^4 - 5y^2) - (2y^2 - 5y + 7)$
$y^4 + 2y^2 + 5y - 13$

73. $(7a^2 - 4a + 10) - (a^3 - 3a^2 + 5) - (-a + 4)$
$-a^3 + 10a^2 - 3a + 1$

74. $(x^2y^2 - xy + 2) - (5x^2y^2 - 3) - (4xy - 11)$
$-4x^2y^2 - 5xy + 16$

In Exercises 75–80, find the special products using the FOIL method. This will help you prepare for the next section.

75. $(y + 9)(y + 9)$
$y^2 + 18y + 81$

76. $(y - 9)(y - 9)$
$y^2 - 18y + 81$

77. $(y - 9)(y + 9)$
$y^2 - 81$

78. $(5x - 4y)(5x + 4y)$
$25x^2 - 16y^2$

79. $(5x - 4y)(5x - 4y)$
$25x^2 - 40xy + 16y^2$

80. $(5x + 4y)(5x + 4y)$
$25x^2 + 40xy + 16y^2$

14.5 EXERCISES C

Perform the indicated operations.

1. $(x + y)(x^2 - xy + y^2)$ [Answer: $x^3 + y^3$]

2. $(x - y)(x^2 + xy + y^2)$ $x^3 - y^3$

3. $(a - 5b)(2a + b)(2a - 3b)$
$4a^3 - 24a^2b + 17ab^2 + 15b^3$

4. $(a - b)[3a^2 - (a + b)(a - 2b)]$
[Answer: $2a^3 - a^2b + ab^2 - 2b^3$]

14.6 SPECIAL PRODUCTS

─────── STUDENT GUIDEPOSTS ───────

1 Difference of Squares   **2** Perfect Square Trinomials

1 DIFFERENCE OF SQUARES

In Section 14.5 we learned how to multiply binomials using the FOIL method. Certain products of binomials occur often enough to merit special consideration. For example, suppose we use the FOIL method to multiply the two binomials $x + 6$ and $x - 6$.

$$(x + 6)(x - 6) = x^2 - 6x + 6x - 6^2$$
$$= x^2 - 36$$

Notice that the sum of the middle terms is zero. This always happens when two binomials of the form $a + b$ and $a - b$ are multiplied.

Difference of Two Squares

The product of a sum and a difference is the square of the first minus the square of the second.

$$(a + b)(a - b) = a^2 - b^2$$

Since the order of multiplication can be changed, we also know that $(a - b)(a + b) = a^2 - b^2$.

EXAMPLE 1 DIFFERENCE OF TWO SQUARES

Find each product using both FOIL and the difference of squares formula.

(a) $(x - 7)(x + 7)$.

FOIL method

$$(x - 7)(x + 7) = x^2 + 7x - 7x - 49$$
$$= x^2 - 49$$

Using $(a - b)(a + b) = a^2 - b^2$

$$(x - 7)(x + 7) = x^2 - 7^2$$
$$= x^2 - 49$$

(b) $(3z + 4)(3z - 4)$.

FOIL method

$$(3z + 4)(3z - 4)$$
$$= 9z^2 - 12z + 12z - 16$$
$$= 9z^2 - 16$$

Using $(a + b)(a - b) = a^2 - b^2$

$$(3z + 4)(3z - 4) = (3z)^2 - 4^2$$
$$= 9z^2 - 16$$

(c) $(7x^2 - 2x)(7x^2 + 2x)$.

FOIL method

$$(7x^2 - 2x)(7x^2 + 2x)$$
$$= 49x^4 + 14x^3 - 14x^3 - 4x^2$$
$$= 49x^4 - 4x^2$$

Using $(a - b)(a + b) = a^2 - b^2$

$$(7x^2 - 2x)(7x^2 + 2x) = (7x^2)^2 - (2x)^2$$
$$= 49x^4 - 4x^2$$

PRACTICE EXERCISE 1

Find each product using both FOIL and the difference of squares formula.

(a) $(a - 5)(a + 5)$

(b) $(4x + 7)(4x - 7)$

(c) $(2y^2 - y)(2y^2 + y)$

Answers: (a) $a^2 - 25$
(b) $16x^2 - 49$ (c) $4y^4 - y^2$

❷ PERFECT SQUARE TRINOMIALS

Squaring a binomial also takes a special form. For example, suppose we square $x + 6$.

$$(x + 6)^2 = (x + 6)(x + 6) = x^2 + 6x + 6x + 36 \quad \text{FOIL method}$$
$$= x^2 + 2(6x) + 36$$
$$= x^2 + 12x + 36$$

This illustrates one of the perfect square formulas.

Perfect Square Trinomial

The square of a binomial of the form $a + b$ is the square of the first plus twice the product of first and second plus the square of the second.

$$(a + b)^2 = (a + b)(a + b) = a^2 + 2ab + b^2$$

EXAMPLE 2 SQUARING $a + b$

Find each product first using both FOIL, and the perfect square formula.

(a) $(y + 3)(y + 3)$.

FOIL method	Using $(a + b)^2 = a^2 + 2ab + b^2$
$(y + 3)(y + 3) = y^2 + 3y + 3y + 9$	$(y + 3)^2 = y^2 + 2(y)(3) + 3^2$
$= y^2 + 6y + 9$	$= y^2 + 6y + 9$

(b) $(5x + 4)(5x + 4)$.

FOIL method	Using $(a + b)^2 = a^2 + 2ab + b^2$
$(5x + 4)(5x + 4)$	$(5x + 4)^2 = (5x)^2 + 2(5x)(4) + 4^2$
$= 25x^2 + 20x + 20x + 16$	$= 25x^2 + 40x + 16$
$= 25x^2 + 40x + 16$	

PRACTICE EXERCISE 2

Find each product using both FOIL and the perfect square formula.

(a) $(x + 8)(x + 8)$

(b) $(7y + 2)(7y + 2)$

Answers: (a) $x^2 + 16x + 64$
(b) $49y^2 + 28y + 4$

To illustrate the second perfect square formula, look at the square of the binomial $x - 6$.

$$(x - 6)^2 = (x - 6)(x - 6) = x^2 - 6x - 6x + 36 \quad \text{FOIL method}$$
$$= x^2 - 2(6x) + 36 = x^2 - 12x + 36$$

Perfect Square Trinomial

The square of a binomial of the form $a - b$ is the square of the first minus twice the product of first and second plus the square of the second.

$$(a - b)^2 = (a - b)(a - b) = a^2 - 2ab + b^2$$

EXAMPLE 3 SQUARING $a - b$

Find each product using both FOIL and the perfect square formula.

(a) $(y - 2)(y - 2)$

FOIL method	Using $(a - b)^2 = a^2 - 2ab + b^2$
$(y - 2)(y - 2)$	$(y - 2)^2 = y^2 - 2(y)(2) + 2^2$
$= y^2 - 2y - 2y + 4$	$= y^2 - 4y + 4$
$= y^2 - 4y + 4$	

(b) $(5x - 4)(5x - 4)$.

FOIL method	Using $(a - b)^2 = a^2 - 2ab + b^2$
$(5x - 4)(5x - 4)$	$(5x - 4)^2 = (5x)^2 - 2(5x)(4) + 4^2$
$= 25x^2 - 20x - 20x + 16$	$= 25x^2 - 40x + 16$
$= 25x^2 - 40x + 16$	

PRACTICE EXERCISE 3

Find each product using both FOIL and the perfect square formula.

(a) $(x - 9)(x - 9)$

(b) $(7a - 1)(7a - 1)$

Answers: (a) $x^2 - 18x + 81$
(b) $49a^2 - 14a + 1$

///////////| **CAUTION** |///////////

A common mistake students make when squaring a binomial is to write

$$(a - b)^2 = a^2 - b^2 \qquad \text{THIS IS WRONG}$$

or $\qquad (a + b)^2 = a^2 + b^2. \qquad \text{THIS IS WRONG}$

To see that $(a + b)^2 \neq a^2 + b^2$, substitute 1 for a and 2 for b. Then,

$$(a + b)^2 = (1 + 2)^2 = 3^2 = 9.$$

However, $\qquad a^2 + b^2 = (1)^2 + (2)^2 = 1 + 4 = 5.$

Don't forget the middle term $2ab$ when finding $(a + b)^2$, or $-2ab$ when finding $(a - b)^2$.

///////////

EXAMPLE 4 PRODUCTS WITH TWO VARIABLES

Use the special formulas in this section to find the following products involving two variables.

(a) $(2x - 3y)(2x + 3y) = (2x)^2 - (3y)^2 \qquad (a - b)(a + b) = a^2 - b^2$
$$= 4x^2 - 9y^2$$

(b) $(2x + 3y)^2 = (2x)^2 + 2(2x)(3y) + (3y)^2 \quad (a + b)^2 = a^2 + 2ab + b^2$
$$= 4x^2 + 12xy + 9y^2$$

(c) $(2x - 3y)^2 = (2x)^2 - 2(2x)(3y) + (3y)^2 \quad (a - b)^2 = a^2 - 2ab + b^2$
$$= 4x^2 - 12xy + 9y^2$$

(d) $\left(\frac{1}{3}x - \frac{1}{2}y\right)^2 = \left(\frac{1}{3}x\right)^2 - 2\left(\frac{1}{3}x\right)\left(\frac{1}{2}y\right) + \left(\frac{1}{2}y\right)^2 \quad \begin{array}{l}(a - b)^2 = \\ a^2 - 2ab + b^2\end{array}$
$$= \frac{1}{9}x^2 - \frac{1}{3}xy + \frac{1}{4}y^2$$

PRACTICE EXERCISE 4

Use the special formulas to find each product.

(a) $(5a - b)(5a + b)$

(b) $(9a + 2b)^2$

(c) $(9a - 2b)^2$

(d) $\left(\frac{1}{3}x + \frac{1}{2}y\right)^2$

Answers: (a) $25a^2 - b^2$
(b) $81a^2 + 36ab + 4b^2$ (c) $81a^2 - 36ab + 4b^2$ (d) $\frac{1}{9}x^2 + \frac{1}{3}xy + \frac{1}{4}y^2$

14.6 EXERCISES A

Multiply using special formulas.

1. $(x - 3)(x + 3)$
$x^2 - 9$

2. $(y + 7)(y - 7)$
$y^2 - 49$

3. $(x + 5)^2$
$x^2 + 10x + 25$

4. $(x - 5)^2$
$x^2 - 10x + 25$

5. $(y - 8)(y + 8)$
$y^2 - 64$

6. $(a + 10)(a - 10)$
$a^2 - 100$

7. $(x - 12)^2$
$x^2 - 24x + 144$

8. $(x + 12)^2$
$x^2 + 24x + 144$

9. $(2y + 5)(2y - 5)$
$4y^2 - 25$

10. $(7a - 1)(7a + 1)$
$49a^2 - 1$

11. $(2x + 7)^2$
$4x^2 + 28x + 49$

12. $(2x - 7)^2$
$4x^2 - 28x + 49$

13 $(4y - 9)(4y + 9)$
$16y^2 - 81$

14. $(5a + 7)(5a - 7)$
$25a^2 - 49$

15. $(3a + 1)^2$
$9a^2 + 6a + 1$

16. $(3a - 1)^2$
$9a^2 - 6a + 1$

17. $(5x - 3)(5x + 3)$
$25x^2 - 9$

18. $(5y + 1)(5y - 1)$
$25y^2 - 1$

19. $(6a + 7)^2$
$36a^2 + 84a + 49$

20. $(6a - 7)^2$
$36a^2 - 84a + 49$

21. $(4x + 8)^2$
$16x^2 + 64x + 64$

22. $(3y - 9)^2$
$9y^2 - 54y + 81$

23. $(5a - 10)(5a + 10)$
$25a^2 - 100$

24. $(3x + 12)(3x - 12)$
$9x^2 - 144$

25. $(5y^2 + 1)^2$
$25y^4 + 10y^2 + 1$

26. $(5y^2 - 1)^2$
$25y^4 - 10y^2 + 1$

27. $(5a^2 - 7)(5a^2 + 7)$
$25a^4 - 49$

28. $(2x^2 - 3x)(2x^2 + 3x)$
$4x^4 - 9x^2$

29. $(2y^2 + 3y)^2$
$4y^4 + 12y^3 + 9y^2$

30. $(2y^2 - 3y)^2$
$4y^4 - 12y^3 + 9y^2$

31 $(0.7y - 3)^2$
$0.49y^2 - 4.2y + 9$

32. $\left(\dfrac{1}{2}a - \dfrac{1}{3}\right)\left(\dfrac{1}{2}a + \dfrac{1}{3}\right)$
$\dfrac{1}{4}a^2 - \dfrac{1}{9}$

33. $(5x - 2y)(5x + 2y)$
$25x^2 - 4y^2$

34. $(5x + 2y)^2$
$25x^2 + 20xy + 4y^2$

35. $(5x - 2y)^2$
$25x^2 - 20xy + 4y^2$

36. $(x^2 + y^2)(x^2 - y^2)$
$x^4 - y^4$

37 $(2x^2 - y)^2$
$4x^4 - 4x^2y + y^2$

38. $(8x + 3y)^2$
$64x^2 + 48xy + 9y^2$

39 $(a^2 + 2b)^2$
$a^4 + 4a^2b + 4b^2$

Perform the following operations.

40 $(x + 1)^2 - (x - 1)^2$
$4x$

41. $(a - 2b)^2 + (a + 2b)^2$
$2a^2 + 8b^2$

42. $(z - 4)(z + 4) - (z + 4)^2$
$-8z - 32$

43. $(w + 2)^2 - (w + 2)(w - 2)$
$4w + 8$

44 A field that is rectangular in shape is $(2a + 7)$ miles long and $(a - 1)$ miles wide. Find the area of the field in terms of a.
$(2a^2 + 5a - 7)$ mi^2

To multiply the polynomials in Exercises 45–50, use the formulas

$$(a + b)(a^2 - ab + b^2) = a^3 + b^3$$
$$(a - b)(a^2 + ab + b^2) = a^3 - b^3.$$

45. $(x + 2)(x^2 - 2x + 4)$
$x^3 + 8$

46. $(x - 2)(x^2 + 2x + 4)$
$x^3 - 8$

47. $(y - 5)(y^2 + 5y + 25)$
$y^3 - 125$

48. $(y + 5)(y^2 - 5y + 25)$
$y^3 + 125$

49. $(3x + y)(9x^2 - 3xy + y^2)$
$27x^3 + y^3$

50. $(3x - y)(9x^2 + 3xy + y^2)$
$27x^3 - y^3$

FOR REVIEW

Multiply.

51. $-3a^2(ab^2 - ab + 7)$
$-3a^3b^2 + 3a^3b - 21a^2$

52. $(5a + 7)(3a - 8)$
$15a^2 - 19a - 56$

53. $4x^2 - 9$
$\underline{2x\ + 3}$
$8x^3 + 12x^2 - 18x - 27$

54. $2x^2 - xy + y^2$
$\underline{x\ - y}$
$2x^3 - 3x^2y + 2xy^2 - y^3$

Exercises 55–57 review material from Section 14.1 to help you prepare for the next section. Simplify each expression using the quotient rule for exponents.

55. $\dfrac{15x^2}{3x}$ $5x$

56. $\dfrac{26x^2y}{13xy}$ $2x$

57. $\dfrac{-33x^3y^5}{33x^4y^2}$ $-\dfrac{y^3}{x}$

ANSWERS: 1. $x^2 - 9$ 2. $y^2 - 49$ 3. $x^2 + 10x + 25$ 4. $x^2 - 10x + 25$ 5. $y^2 - 64$ 6. $a^2 - 100$ 7. $x^2 - 24x + 144$
8. $x^2 + 24x + 144$ 9. $4y^2 - 25$ 10. $49a^2 - 1$ 11. $4x^2 + 28x + 49$ 12. $4x^2 - 28x + 49$ 13. $16y^2 - 81$
14. $25a^2 - 49$ 15. $9a^2 + 6a + 1$ 16. $9a^2 - 6a + 1$ 17. $25x^2 - 9$ 18. $25y^2 - 1$ 19. $36a^2 + 84a + 49$
20. $36a^2 - 84a + 49$ 21. $16x^2 + 64x + 64$ 22. $9y^2 - 54y + 81$ 23. $25a^2 - 100$ 24. $9x^2 - 144$ 25. $25y^4 + 10y^2 + 1$
26. $25y^4 - 10y^2 + 1$ 27. $25a^4 - 49$ 28. $4x^4 - 9x^2$ 29. $4y^4 + 12y^3 + 9y^2$ 30. $4y^4 - 12y^3 + 9y^2$
31. $0.49y^2 - 4.2y + 9$ 32. $\frac{1}{4}a^2 - \frac{1}{9}$ 33. $25x^2 - 4y^2$ 34. $25x^2 + 20xy + 4y^2$ 35. $25x^2 - 20xy + 4y^2$ 36. $x^4 - y^4$
37. $4x^4 - 4x^2y + y^2$ 38. $64x^2 + 48xy + 9y^2$ 39. $a^4 + 4a^2b + 4b^2$ 40. $4x$ 41. $2a^2 + 8b^2$ 42. $-8z - 32$
43. $4w + 8$ 44. $(2a^2 + 5a - 7)$ mi^2 45. $x^3 + 8$ 46. $x^3 - 8$ 47. $y^3 - 125$ 48. $y^3 + 125$ 49. $27x^3 + y^3$
50. $27x^3 - y^3$ 51. $-3a^3b^2 + 3a^3b - 21a^2$ 52. $15a^2 - 19a - 56$ 53. $8x^3 + 12x^2 - 18x - 27$
54. $2x^3 - 3x^2y + 2xy^2 - y^3$ 55. $5x$ 56. $2x$ 57. $-\frac{y^3}{x}$

14.6 EXERCISES B

Multiply using special formulas.

1. $(x - 4)(x + 4)$
$x^2 - 16$

2. $(y + 5)(y - 5)$
$y^2 - 25$

3. $(x + 10)^2$
$x^2 + 20x + 100$

4. $(x - 10)^2$
$x^2 - 20x + 100$

5. $(y - 9)(y + 9)$
$y^2 - 81$

6. $(a + 12)(a - 12)$
$a^2 - 144$

7. $(x - 4)^2$
$x^2 - 8x + 16$

8. $(x + 4)^2$
$x^2 + 8x + 16$

9. $(3y - 7)(3y + 7)$
$9y^2 - 49$

10. $(5a - 2)(5a + 2)$
$25a^2 - 4$

11. $(3x + 4)^2$
$9x^2 + 24x + 16$

12. $(3x - 4)^2$
$9x^2 - 24x + 16$

13. $(8y - 1)(8y + 1)$
$64y^2 - 1$

14. $(6a - 5)(6a + 5)$
$36a^2 - 25$

15. $(9a + 1)^2$
$81a^2 + 18a + 1$

16. $(9a - 1)^2$
$81a^2 - 18a + 1$

17. $(10x - 3)(10x + 3)$
$100x^2 - 9$

18. $(3y + 5)(3y - 5)$
$9y^2 - 25$

19. $(4a + 5)^2$
$16a^2 + 40a + 25$

20. $(4a - 5)^2$
$16a^2 - 40a + 25$

21. $(2x + 10)^2$
$4x^2 + 40x + 100$

22. $(6y - 3)^2$
$36y^2 - 36y + 9$

23. $(2a - 10)(2a + 10)$
$4a^2 - 100$

24. $(6x + 3)(6x - 3)$
$36x^2 - 9$

25. $(3y^2 + 2)^2$
$9y^4 + 12y^2 + 4$

26. $(3y^2 - 2)^2$
$9y^4 - 12y^2 + 4$

27. $(2a^2 - 5)(2a^2 + 5)$
$4a^4 - 25$

28. $(3x^2 + 4x)(3x^2 - 4x)$
$9x^4 - 16x^2$

29. $(3y^2 + 4y)^2$
$9y^4 + 24y^3 + 16y^2$

30. $(3y^2 - 4y)^2$
$9y^4 - 24y^3 + 16y^2$

31. $(0.4y - 5)^2$
$0.16y^2 - 4y + 25$

32. $\left(\dfrac{2}{3}a - \dfrac{1}{4}\right)\left(\dfrac{2}{3}a + \dfrac{1}{4}\right)$
$\frac{4}{9}a^2 - \frac{1}{16}$

33. $(7x - 3y)(7x + 3y)$
$49x^2 - 9y^2$

34. $(7x + 3y)^2$
$49x^2 + 42xy + 9y^2$

35. $(7x - 3y)^2$
$49x^2 - 42xy + 9y^2$

36. $(x^3 + y^3)(x^3 - y^3)$
$x^6 - y^6$

37. $(3x - y^2)^2$
$9x^2 - 6xy^2 + y^4$

38. $(8x - 3y)^2$
$64x^2 - 48xy + 9y^2$

39. $(a^2 - 2b)^2$
$a^4 - 4a^2b + 4b^2$

Perform the following operations.

40. $(x - 1)^2 - (x + 1)^2$
$-4x$

41. $(a + 3b)^2 + (a - 3b)^2$
$2a^2 + 18b^2$

42. $(y - 2)(y + 2) - (y - 2)^2$
$4y - 8$

43. $(z + 3)^2 - (z + 3)(z - 3)$
$6z + 18$

44. A picture frame is in the shape of a square with sides $(3z + 2)$ feet in length. Find the area of the frame in terms of z. $(9z^2 + 12z + 4)$ ft^2

Use the formulas

$$(a + b)(a^2 - ab + b^2) = a^3 + b^3$$
$$(a - b)(a^2 + ab + b^2) = a^3 - b^3$$

to multiply the polynomials in Exercises 45–50.

45. $(x + 3)(x^2 - 3x + 9)$
$x^3 + 27$

46. $(x - 3)(x^2 + 3x + 9)$
$x^3 - 27$

47. $(y - 4)(y^2 + 4y + 16)$
$y^3 - 64$

48. $(y + 4)(y^2 - 4y + 16)$
$y^3 + 64$

49. $(2x + 3y)(4x^2 - 6xy + 9y^2)$
$8x^3 + 27y^3$

50. $(2x - 3y)(4x^2 + 6xy + 9y^2)$
$8x^3 - 27y^3$

FOR REVIEW

Multiply.

51. $4ab(-2a^2b - 6a + 9b)$
$-8a^3b^2 - 24a^2b + 36ab^2$

52. $(2a - 5)(4a - 3)$
$8a^2 - 26a + 15$

53. $3x^2 + 11$
  $\underline{3x \ - 2}$
  $9x^3 - 6x^2 + 33x - 22$

54. $5x^2y^2 - 2xy + 3$
  $\underline{2xy \ \ - 1}$
  $10x^3y^3 - 9x^2y^2 + 8xy - 3$

Exercises 55–57 review material from Section 14.1 to help you prepare for the next section. Simplify each expression using the quotient rule for exponents.

55. $\dfrac{-24a^3}{8a^2}$ $-3a$

56. $\dfrac{64a^3b^2}{16ab^5}$ $\dfrac{4a^2}{b^3}$

57. $\dfrac{12a^7b}{24a^3b}$ $\dfrac{a^4}{2}$

14.6 EXERCISES C

Perform the indicated operations. Assume all exponents are positive integers.

1. $[2x^2 - (x - 2y)][2x^2 + (x - 2y)]$
[*Hint:* Multiply before removing parentheses.]
$4x^4 - x^2 + 4xy - 4y^2$

2. $(2x + y - z)^2$
$4x^2 + 4xy - 4xz + y^2 - 2yz + z^2$

3. $(a^n + b^n)(a^n - b^n)$
$a^{2n} - b^{2n}$

4. $(a^n - b^n)^2$ [Answer: $a^{2n} - 2a^nb^n + b^{2n}$]

14.7 DIVISION OF POLYNOMIALS

STUDENT GUIDEPOSTS

❶ Dividing a Polynomial by a Monomial

❷ Dividing a Polynomial by a Binomial

① DIVIDING A POLYNOMIAL BY A MONOMIAL

To illustrate the rule for dividing a polynomial by a monomial, look at the following example.

$$\frac{3x^3 - x^2}{x} = \frac{1}{x}[3x^3 - x^2] \qquad \text{Dividing by } x \text{ is the same as multiplying by } \frac{1}{x}$$

$$= \frac{1}{x}(3x^3) - \frac{1}{x}(x^2) \qquad \text{Distributive law}$$

$$= \frac{3x^3}{x} - \frac{x^2}{x} \qquad \text{Multiplication by } \frac{1}{x} \text{ is the same as dividing by } x$$

$$= 3x^{3-1} - x^{2-1} \qquad \frac{a^m}{a^n} = a^{m-n}$$

$$= 3x^2 - x$$

In general, we omit the first two steps, write

$$\frac{3x^3 - x^2}{x} = \frac{3x^3}{x} - \frac{x^2}{x},$$

and use the quotient rule for exponents.

To Divide a Polynomial by a Monomial

1. Divide each term of the polynomial by the monomial.

2. Use the rule

$$\frac{x^m}{x^n} = x^{m-n}$$

to divide the variables.

EXAMPLE 1 DIVIDING BY A MONOMIAL

Divide.

(a) $\dfrac{14x^3 - 7x^2 + 28x - 7}{7} = \dfrac{14x^3}{7} - \dfrac{7x^2}{7} + \dfrac{28x}{7} - \dfrac{7}{7}$

$$= 2x^3 - x^2 + 4x - 1$$

(b) $\dfrac{8x^4 - 6x^2 + 12x}{-2x} = \dfrac{8x^4}{-2x} - \dfrac{6x^2}{-2x} + \dfrac{12x}{-2x}$

$$= -4x^3 + 3x - 6 \qquad \frac{x^4}{x} = x^{4-1} = x^3, \frac{x^2}{x} = x^{2-1}$$
$$= x, \frac{x}{x} = x^{1-1} = x^0 = 1$$

(c) $\dfrac{7x^5 - 35x^4 + 40x}{5x^2} = \dfrac{7x^5}{5x^2} - \dfrac{35x^4}{5x^2} + \dfrac{40x}{5x^2}$

$$= \frac{7}{5}x^3 - 7x^2 + \frac{8}{x} \qquad \frac{x}{x^2} = \frac{1}{x}$$

PRACTICE EXERCISE 1

Divide.

(a) $\dfrac{6a^4 - 9a^3 + 12a^2 + 3a - 15}{3}$

(b) $\dfrac{25y^5 - 15y^3 + 30y}{-5y}$

(c) $\dfrac{18w^6 - 27w^4 - 9w^2}{9w^2}$

Answers: (a) $2a^4 - 3a^3 + 4a^2 + a - 5$ (b) $-5y^4 + 3y^2 - 6$ (c) $2w^4 - 3w^2 - 1$

//////////// **CAUTION** ///////////

Is $\frac{x+5}{5}$ equal to $x + 1$? The answer is no. Every number in the numerator must be divided by 5. The correct quotient is

$$\frac{x+5}{5} = \frac{x}{5} + \frac{5}{5} = \frac{x}{5} + 1.$$

////////////

If a division problem is expressed using the sign $\div$, change to the notation used in the examples above. To find $(3a^3b^3 - 9a^2b + 27ab) \div 9ab$ we write

$$\frac{3a^3b^3 - 9a^2b + 27ab}{9ab} = \frac{3a^3b^3}{9ab} - \frac{9a^2b}{9ab} + \frac{27ab}{9ab}$$

$$= \frac{1}{3}a^2b^2 - a + 3.$$

② DIVIDING A POLYNOMIAL BY A BINOMIAL

The procedure for dividing a polynomial by a binomial closely follows long division of one whole number by another. It might help to refer to the following numerical problem as you read through the rules given below.

```
    43 ÷ 37
    │ 61 ÷ 37
    │ │ 247 ÷ 37
    │ │ │ 252 ÷ 37
    ↓↓↓↓
    1166      Remainder 30
37)43172
    37 ←———— First digit of quotient times divisor 37
    61        Subtract and bring down next digit of dividend, 43172
    37 ←———— Second digit of quotient times 37
    247       Subtract and bring down next digit of dividend
    222 ←———— Third digit of quotient times 37
    252       Subtract and bring down next digit
    222 ←———— Fourth digit of quotient times 37
    30        Subtract; no more digits in dividend
```

The answer is 1166 with a remainder of 30, or $1166 + \frac{30}{37}$.

To Divide a Polynomial by a Binomial

1. Arrange the terms of both polynomial and binomial in descending order and set up as in long division.

2. Divide the first term of the polynomial (the dividend) by the first term of the binomial (the divisor) to obtain the first term of the quotient.

3. Multiply the first term of the quotient by the binomial and subtract the result from the dividend. Bring down the next terms to obtain a new polynomial which becomes the new dividend.

4. Divide the new dividend polynomial by the binomial. Continue the process until the variable in the first term of the remainder dividend is raised to a lower power than the variable in the first term of the divisor.

| EXAMPLE 2 DIVIDING BY A BINOMIAL | PRACTICE EXERCISE 2 |

Divide $6 - 5x + x^2$ by $x - 2$.

1. Arrange terms in descending order.

$$x - 2 \overline{)x^2 - 5x + 6}$$

2. Divide the first term of the polynomial by the first term of the binomial.

equals
$$x - 2 \overline{)x^2} - 5x + 6 \qquad x^2 \div x = x$$
divided by

3. Multiply the first term of the quotient by the binomial and subtract the results from the dividend. Bring down the next terms to obtain a new dividend.

times
$$\begin{array}{r} x \\ x - 2 \overline{)x^2 - 5x + 6} \\ \underline{x^2 - 2x} \\ -3x + 6 \end{array}$$

$x(x - 2) = x^2 - 2x$

Subtract $x^2 - 2x$ from $x^2 - 5x$ and bring down $+ 6$

4. Divide the new dividend polynomial by the binomial, using Steps 2 and 3.

$$\begin{array}{r} x - 3 \\ x - 2 \overline{)x^2 - 5x + 6} \\ \underline{x^2 - 2x} \\ -3x + 6 \\ \underline{-3x + 6} \\ 0 \end{array}$$

Divide $-3x + 6$ by $x - 2$

Multiply $x - 2$ by -3

No variable in the new dividend; the process terminates

Practice Exercise 2

Divide $8x - 9 + x^2$ by $x - 1$.

equals
$$x - 1 \overline{)x^2 + 8x - 9} \leftarrow \text{Descending order}$$
divided by
$$\begin{array}{r} x^2 - x \leftarrow x(x - 1) \\ 9x \leftarrow \text{Subtract like terms:} \end{array}$$
$x^2 - x^2 = 0$
and $8x - (-x) = 8x + x = 9x$

Bring down the -9 and continue by dividing x into $9x$.

Answer: $x + 9$

| EXAMPLE 3 DIVIDING BY A BINOMIAL | PRACTICE EXERCISE 3 |

Divide $a^3 + 27$ by $a + 3$.

The quotient of a^3 and a

$$\begin{array}{r} a^2 - 3a + 9 \\ a + 3 \overline{)a^3 \qquad\quad + 27} \\ \underline{a^3 + 3a^2} \\ -3a^2 \qquad + 27 \\ \underline{-3a^2 - 9a} \\ 9a + 27 \\ \underline{9a + 27} \\ 0 \end{array}$$

The quotient of $-3a^2$ and a
The quotient of $9a$ and a
Leave space for missing terms
The quotient a^2 times the divisor $a + 3$
Subtract $a^3 + 3a^2$ from a^3 and bring down 27
The quotient $-3a$ times $a + 3$
Subtract $-3a^2 - 9a$ from $3a^2 + 27$
The quotient 9 times $a + 3$
Subtract; no variable in new dividend

The answer is $a^2 - 3a + 9$ with a remainder of 0.

Practice Exercise 3

Divide $x^3 - 8$ by $x - 2$.

Answer: $x^2 + 2x + 4$

Note that we can check our work in any division problem by multiplying the divisor by the quotient to obtain the dividend. We can check the work in Example 3 as follows.

$$(a + 3)(a^2 - 3a + 9) = a^3 - 3a^2 + 9a + 3a^2 - 9a + 27$$
$$= a^3 + 27$$

EXAMPLE 4   DIVIDING BY A BINOMIAL

Divide $3x^4 - 19x^3 + 27x^2 - 41x + 32$ by $x - 5$.

$$
\begin{array}{r}
3x^3 - 4x^2 + 7x - 6 \\
x - 5 \overline{)3x^4 - 19x^3 + 27x^2 - 41x + 32} \\
\underline{3x^4 - 15x^3} \\
-4x^3 + 27x^2 - 41x + 32 \\
\underline{-4x^3 + 20x^2} \\
7x^2 - 41x + 32 \\
\underline{7x^2 - 35x} \\
-6x + 32 \\
\underline{-6x + 30} \\
2
\end{array}
$$

Divide $3x^4$ by x

No variable in new dividend; process ends

The answer is $3x^3 - 4x^2 + 7x - 6$ with remainder 2, or $3x^3 - 4x^2 + 7x - 6 + \frac{2}{x-5}$.

PRACTICE EXERCISE 4

Divide $2y^4 + 9y^3 - 3y^2 - 25y + 10$ by $y + 4$.

Answer: $2y^3 + y^2 - 7y + 3 - \frac{2}{y+4}$

EXAMPLE 5   DIVIDING BY A BINOMIAL

Divide $20x^3 + 18x^2 + 21x + 40$ by $5x + 7$.

$$
\begin{array}{r}
4x^2 - 2x + 7 \\
5x + 7 \overline{)20x^3 + 18x^2 + 21x + 40} \\
\underline{20x^3 + 28x^2} \\
-10x^2 + 21x + 40 \\
\underline{-10x^2 - 14x} \\
35x + 40 \\
\underline{35x + 49} \\
-9
\end{array}
$$

No variable in new dividend

The answer is $4x^2 - 2x + 7$ with remainder -9, or $4x^2 - 2x + 7 - \frac{9}{5x+7}$.

PRACTICE EXERCISE 5

Divide $16a^3 - 8a^2 + 5a - 3$ by $4a - 3$.

Answer: $4a^2 + a + 2 + \frac{3}{4a-3}$

EXAMPLE 6   DIVIDING BY A BINOMIAL

Divide $4x^4 - 8x^3 - 12x^2 + 44x - 15$ by $4x^2 - 8$.

$$
\begin{array}{r}
x^2 - 2x - 1 \\
4x^2 - 8 \overline{)4x^4 - 8x^3 - 12x^2 + 44x - 15} \\
\underline{4x^4 \qquad\quad - 8x^2} \\
-8x^3 - 4x^2 + 44x - 15 \\
\underline{-8x^3 \qquad\quad + 16x} \\
-4x^2 + 28x - 15 \\
\underline{-4x^2 \qquad\quad + 8} \\
28x - 23
\end{array}
$$

Keep like terms in same column

x is raised to first power; process ends

The answer is $x^2 - 2x - 1 + \frac{28x - 23}{4x^2 - 8}$.

PRACTICE EXERCISE 6

Divide $9x^4 + 6x^3 + 3x^2 + 7x - 1$ by $3x^2 - 1$.

Answer: $3x^2 + 2x + 2 + \frac{9x + 1}{3x^2 - 1}$

14.7   EXERCISES A

Divide.

1. $\dfrac{3x^3 - 9x^2 + 27x + 3}{3}$

$x^3 - 3x^2 + 9x + 1$

2. $\dfrac{14x^4 - 7x^2 + 28}{7}$

$2x^4 - x^2 + 4$

3. $(25a^5 - 20a^4 + 15a^3) \div (5a^3)$

$5a^2 - 4a + 3$

4. $(y^8 - y^6 + y^5) \div (y^3)$

$y^5 - y^3 + y^2$

5. $\dfrac{7x^5 - 35x^4 + 14x^2}{14x^2}$

$\dfrac{1}{2}x^3 - \dfrac{5}{2}x^2 + 1$

6. $\dfrac{8x^3 - 64x^2}{8x^2}$

$x - 8$

7 $(3a^{12} - 9a^6 + 27a^5 + 81a^4) \div (9a^2)$

$\dfrac{1}{3}a^{10} - a^4 + 3a^3 + 9a^2$

8. $(y^6 - 3y^2) \div (2y^2)$

$\dfrac{1}{2}y^4 - \dfrac{3}{2}$

9. $\dfrac{-8x^3 + 6x^2 - 4x}{-2x}$

$4x^2 - 3x + 2$

10 $\dfrac{-8x^3 + 6x^2 - 4x}{0.2x}$

$-40x^2 + 30x - 20$

11. $\dfrac{y^3 + 2y^2 + y}{y^2}$

$y + 2 + \dfrac{1}{y}$

12. $\dfrac{-8y^4 + 16y^3 - 4y}{2y^2}$

$-4y^2 + 8y - \dfrac{2}{y}$

13 $\dfrac{-5a^2b + 3ab - 2a}{-a}$

$5ab - 3b + 2$

14. $\dfrac{-6x^2y^3 + 9x^2y}{3xy}$

$-2xy^2 + 3x$

15. $\dfrac{27x^2y^4 - 18xy^6 + 36xy^3}{9xy^3}$

$3xy - 2y^3 + 4$

16. $\dfrac{12x^4y^4 - 42x^2y^2}{6x^3y^3}$

$2xy - \dfrac{7}{xy}$

17. $(x^2 - 5x + 6) \div (x - 3)$

$x - 2$

18. $(a^2 + 5a + 6) \div (a + 3)$

$a + 2$

19. $(y^2 - 2y - 3) \div (y + 1)$

$y - 3$

20. $(x^2 - 9x - 5) \div (x - 7)$

$x - 2 - \dfrac{19}{x - 7}$

21. $(3y + 2 + y^2) \div (y + 2)$
[*Hint*:   Remember descending order.]
$y + 1$

22 $(6 + 8y - y^2) \div (4 - y)$

$y - 4 - \dfrac{22}{y - 4}$

23 $(3a^2 - 23a + 40) \div (a - 5)$
$3a - 8$

24. $(14y - 6y^2 + 80) \div (3y + 8)$
$-2y + 10$

25. $4x + 6)\overline{28x^3 + 26x^2 - 44x - 36}$
$7x^2 - 4x - 5 - \dfrac{6}{4x + 6}$

26. $2x - 3)\overline{10x^4 - 15x^3 \quad\quad - 8x + 12}$
$5x^3 - 4$

27. $3a - 12)\overline{6a^3 - 18a^2 + 33a - 80}$
$2a^2 + 2a + 19 + \dfrac{148}{3a - 12}$

28 $y + 2)\overline{y^5 \quad\quad + 32}$
$y^4 - 2y^3 + 4y^2 - 8y + 16$

29. $3x^2 + 2)\overline{9x^3 + 30x^2 + x - 10}$
$3x + 10 - \dfrac{5x + 30}{3x^2 + 2}$

30 $x^2 - 1)\overline{x^4 \quad\quad - 1}$
$x^2 + 1$

FOR REVIEW

Multiply using special formulas.

31. $(4x - 1)(4x + 1)$
$16x^2 - 1$

32. $(6x - 5)^2$
$36x^2 - 60x + 25$

33. $(2a + 13)^2$
$4a^2 + 52a + 169$

34. $(a^2 - 5)(a^2 + 5)$
$a^4 - 25$

35. $(y^3 - 2)^2$
$y^6 - 4y^3 + 4$

36. $(2y^2 + 7)^2$
$4y^4 + 28y^2 + 49$

To prepare for factoring polynomials in the next chapter, factor each integer into a product of primes. (See Section 2.8.)

37. 45
$3 \cdot 3 \cdot 5$

38. 210
$2 \cdot 3 \cdot 5 \cdot 7$

39. 748
$2 \cdot 2 \cdot 11 \cdot 17$

40. 1625
$5 \cdot 5 \cdot 5 \cdot 13$

ANSWERS: 1. $x^3 - 3x^2 + 9x + 1$ 2. $2x^4 - x^2 + 4$ 3. $5a^2 - 4a + 3$ 4. $y^5 - y^3 + y^2$ 5. $\frac{1}{2}x^3 - \frac{5}{2}x^2 + 1$
6. $x - 8$ 7. $\frac{1}{3}a^{10} - a^4 + 3a^3 + 9a^2$ 8. $\frac{1}{2}y^4 - \frac{3}{2}$ 9. $4x^2 - 3x + 2$ 10. $-40x^2 + 30x - 20$ 11. $y + 2 + \frac{1}{y}$
12. $-4y^2 + 8y - \frac{2}{y}$ 13. $5ab - 3b + 2$ 14. $-2xy^2 + 3x$ 15. $3xy - 2y^3 + 4$ 16. $2xy - \frac{7}{xy}$ 17. $x - 2$ 18. $a + 2$
19. $y - 3$ 20. $x - 2 - \frac{19}{x-7}$ 21. $y + 1$ 22. $y - 4 - \frac{22}{y-4}$ 23. $3a - 8$ 24. $-2y + 10$ 25. $7x^2 - 4x - 5 - \frac{6}{4x+6}$
26. $5x^3 - 4$ 27. $2a^2 + 2a + 19 + \frac{148}{3a-12}$ 28. $y^4 - 2y^3 + 4y^2 - 8y + 16$ 29. $3x + 10 - \frac{5x+30}{3x^2+2}$ 30. $x^2 + 1$
31. $16x^2 - 1$ 32. $36x^2 - 60x + 25$ 33. $4a^2 + 52a + 169$ 34. $a^4 - 25$ 35. $y^6 - 4y^3 + 4$ 36. $4y^4 + 28y^2 + 49$
37. $3 \cdot 3 \cdot 5$ 38. $2 \cdot 3 \cdot 5 \cdot 7$ 39. $2 \cdot 2 \cdot 11 \cdot 17$ 40. $5 \cdot 5 \cdot 5 \cdot 13$

14.7 EXERCISES B

Divide.

1. $\dfrac{5x^4 - 25x^2 + 45x - 20}{5}$
$x^4 - 5x^2 + 9x - 4$

2. $\dfrac{26x^3 - 16x^2 + 10}{-2}$
$-13x^3 + 8x^2 - 5$

3. $(55a^6 - 22a^4 + 33a^2) \div (11a^2)$
$5a^4 - 2a^2 + 3$

4. $(-y^{12} + y^7 - y^5) \div (y^4)$
$-y^8 + y^3 - y$

5. $\dfrac{-100x^7 + 50x^5 - 20x^3}{10x^2}$
$-10x^5 + 5x^3 - 2x$

6. $\dfrac{15x^3 - 30x^2 + 5x}{30x}$
$\frac{1}{2}x^2 - x + \frac{1}{6}$

7. $(35a^9 - 49a^7 - 14a^6 + 56a^5) \div (7a^4)$
$5a^5 - 7a^3 - 2a^2 + 8a$

8. $(-8y^4 - 6y^3 - 12y^2) \div (-3y^2)$ $\frac{8}{3}y^2 + 2y + 4$

9. $\dfrac{54x^6 + 81x^4 - 36x^3}{9x^3}$
$6x^3 + 9x - 4$

10. $\dfrac{54x^6 + 81x^4 - 36x^3}{0.9x^3}$
$60x^3 + 90x - 40$

11. $\dfrac{y^4 - 6y^3 + y^2}{y^3}$ $y - 6 + \frac{1}{y}$

12. $\dfrac{-3y^6 + 12y^4 - 6y^2}{3y^4}$
$-y^2 + 4 - \frac{2}{y^2}$

13. $\dfrac{4a^2b^2 - 5a^2b + 7ab^2}{ab}$
$4ab - 5a + 7b$

14. $\dfrac{18x^3y^4 - 12x^4y^3}{3x^3y^3}$
$6y - 4x$

15. $\dfrac{44x^6y^6 + 20x^4y^2 - 28x^2y^4}{4x^2y}$
$11x^4y^5 + 5x^2y - 7y^3$

16. $\dfrac{-5x^6y^4 + 10x^4y^6}{5x^6y^6}$ $-\frac{1}{y^2} + \frac{2}{x^2}$

17. $(x^2 - 6x + 8) \div (x - 4)$
$x - 2$

18. $(a^2 + 6a + 8) \div (a + 2)$
$a + 4$

19. $(y^2 - 5y - 6) \div (y + 1)$
$y - 6$

20. $(x^2 + 7x - 8) \div (x + 5)$
$x + 2 - \frac{18}{x+5}$

21. $(7y + 12 + y^2) \div (y + 4)$
$y + 3$

22. $(3 - 5y - y^2) \div (3 - y)$
$y + 8 + \frac{21}{y-3}$

23. $(3a^2 + 13a - 30) \div (a + 6)$
$3a - 5$

24. $(7a - 12 + 12a^2) \div (4a - 3)$
$3a + 4$

25. $(2y^3 + y^2 - y + 1) \div (2y + 3)$
$y^2 - y + 1 - \frac{2}{2y+3}$

26. $(6x^3 + x^2 - 7x + 2) \div (3x - 1)$
$2x^2 + x - 2$

27. $(a^4 - 1) \div (a^2 + 1)$
$a^2 - 1$

28. $(x^5 - 32) \div (x - 2)$
$x^4 + 2x^3 + 4x^2 + 8x + 16$

29. $(21x^4 - 7x^3 - 6x + 2) \div (3x - 1)$
$7x^3 - 2$

30. $(2x^5 - x^3 + 16x^2 - 8) \div (2x^2 - 1)$
$x^3 + 8$

FOR REVIEW

Multiply using special formulas.

31. $(3x - 5)(3x + 5)$
$9x^2 - 25$

32. $(3x + 7)^2$
$9x^2 + 42x + 49$

33. $(2x - 11)^2$
$4x^2 - 44x + 121$

34. $(a^2 + 7)(a^2 - 7)$
$a^4 - 49$

35. $(y^3 + 5)^2$
$y^6 + 10y^3 + 25$

36. $(3y^2 - 4)^2$
$9y^4 - 24y^2 + 16$

To prepare for factoring polynomials in the next chapter, factor each integer into a product of primes. (See Section 2.8.)

37. 42
$2 \cdot 3 \cdot 7$

38. 462
$2 \cdot 3 \cdot 7 \cdot 11$

39. 910
$2 \cdot 5 \cdot 7 \cdot 13$

40. 2805
$3 \cdot 5 \cdot 11 \cdot 17$

14.7 EXERCISES C

Divide.

1. $(2x^6 - 7x^3 - 30) \div (2x^3 + 5)$
$x^3 - 6$

2. $(6x^6 + x^3 - 12) \div (3x^3 - 4)$ [Answer: $2x^3 + 3$]

3. $(5x^5 + 21 - 26x + x^2 + 10x^4 - 11x^3) \div$ $(x^2 + 2x - 3)$ $5x^3 + 4x - 7$

4. $(-34x^4 - 5x^3 + 3x^5 + 6x^6 + 54x^2 - 56 + 7x) \div$ $(2x^2 + x - 8)$ [Answer: $3x^4 - 5x^2 + 7$]

CHAPTER 14 REVIEW

KEY WORDS

14.1 An **exponential expression** is of the form a^n where a is the **base** and n is the **exponent.**

14.2 A number is written in **scientific notation** if it is the product of a power of 10 and a number that is greater than or equal to 1 and less than 10.

14.3 A **polynomial** is an algebraic expression having as terms products of numbers and variables with whole-number exponents.

The **numerical coefficient** is the numerical multiplier of the term.

A **monomial** is a polynomial with one term.

A **binomial** is a polynomial with two terms.

A **trinomial** is a polynomial with three terms.

For a polynomial in one variable, the **degree of a term** is the exponent on the variable.

The **degree of a polynomial** is the degree of the term of highest degree.

KEY CONCEPTS

14.1 **1.** If a and b are any numbers, and m and n are natural numbers, the following rules for exponents hold.

(a) $a^m \cdot a^n = a^{m+n}$

(b) $\dfrac{a^m}{a^n} = a^{m-n}$, if $a \neq 0$

(c) $(a^m)^n = a^{mn}$

(d) $(a \cdot b)^n = a^n \cdot b^n$

(e) $\left(\dfrac{a}{b}\right)^n = \dfrac{a^n}{b^n}$, if $b \neq 0$

(f) $a^0 = 1$, if $a \neq 0$ and 0^0 is undefined.

2. In an expression such as $2y^3$, only y is cubed, not $2y$. That is, $2y^3$ is not the same as $(2y)^3$, which is equal to $8y^3$.

14.3 Only like terms can be collected. For example, $5x + 3 \neq 8x$ since $5x$ and 3 are *not* like terms. Similarly, $x^2 - 2x + 1$ has no like terms to combine.

14.4 Change all signs when removing parentheses preceded by a minus sign. For example, $3x - (2x^2 - 4x - 2) = 3x - 2x^2 + 4x + 2$.

14.5 Use the FOIL method to multiply binomials, and write out all details.

14.6 **1.** $(a + b)(a - b) = a^2 - b^2$
2. $(a + b)^2 = a^2 + 2ab + b^2$ (not $a^2 + b^2$)
3. $(a - b)^2 = a^2 - 2ab + b^2$ (not $a^2 - b^2$)

14.7 Write polynomials in descending order before dividing.

REVIEW EXERCISES

Part I

14.1 *Write in exponential notation.*

1. $aaaaa$ a^5

2. $(3z)(3z)(3z)$ $(3z)^3$

3. $(a + b)(a + b)$ $(a + b)^2$

Write without using exponents.

4. b^7 $bbbbbbb$

5. $-2x^3$ $-2xxx$

6. $(-2x)^3$ $(-2x)(-2x)(-2x)$

Simplify and write without negative exponents.

7. $2y^7y^2$ $2y^9$

8. $(3x^2)^3$ $27x^6$

9. $(x^2y^3)^5$ $x^{10}y^{15}$

10. $2a^0$ $(a \neq 0)$ 2

11. $(2a)^0$ $(a \neq 0)$ 1

12. $\dfrac{2b^2b^{-5}}{b^7}$ $\dfrac{2}{b^{10}}$

13. $5a^{-1}$ $\dfrac{5}{a}$

14. $(5a)^{-1}$ $\dfrac{1}{5a}$

15. $\dfrac{a^{-1}}{b^{-1}}$ $\dfrac{b}{a}$

16. $(a^2b^{-3})^{-2}$ $\dfrac{b^6}{a^4}$

17. $\dfrac{3^0a^{-5}b^7}{a^4b^{-1}}$ $\dfrac{b^8}{a^9}$

18. $\left(\dfrac{2^{-1}x^2y^{-4}}{x^{-2}y^{-2}}\right)^{-1}$ $\dfrac{2y^2}{x^4}$

14.2 *Write in scientific notation.*

19. 0.000000411
4.11×10^{-7}

20. $549{,}000{,}000{,}000$
5.49×10^{11}

Write without using scientific notation.

21. 6.15×10^{12}
$6{,}150{,}000{,}000{,}000$

22. 4×10^{-8}
0.00000004

14.3 *Tell whether the following are monomials, binomials, or trinomials.*

23. $4x + 1$
binomial

24. $3x^2$
monomial

25. $5y^4 + y^2$
binomial

26. $-7y^3 + y - 5$
trinomial

Give the degree of each polynomial.

27. $6x^4$ 4

28. $15x^2 - 14x + 6x^7 - 4x^3$ 7

29. $-x^3 + 14x^4 - x^{10} + 8x^2$ 10

30. $2x^8 - 4x^4 - 16$ 8

Collect and combine like terms and write in descending order.

31. $5x + 2 - 6x$ $-x + 2$

32. $6y^2 - 2y + 8y^2 + 5 - 3y^2 + y$
$11y^2 - y + 5$

33. $4a^3 - 6a + 7a^2 - 4a^3 + a - a^2$
$6a^2 - 5a$

34. $-7x^2 - 30x^4 + x^5 - 3x + 5x^5 - 4x + 22x^4 - 5$
$6x^5 - 8x^4 - 7x^2 - 7x - 5$

Collect and combine like terms.

35. $8xy + 5 - 4xy - 8$
$4xy - 3$

36. $-a^2b^3 - a^2b^2 - 5a^2b^2 - ab^2 + 5a^2b^3$
$4a^2b^3 - 6a^2b^2 - ab^2$

Solve.

37. The profit in dollars when x refrigerators are sold is given by the expression $4x^3 - 140$. Find the profit when 5 refrigerators are sold. $360

14.4 **38.** Add $5x^2 - 4x + 5$ and $-7x^2 + 4x - 8$.
$-2x^2 - 3$

39. Subtract $-2x^2 - 1$ from $x^3 - 5x^2 + 14$.
$x^3 - 3x^2 + 15$

Perform the indicated operations.

40. $(4x^3 - x^4 + 6x - 2) + (3x - x^5 + 2x^3 - 5x^4 + 2)$
$-x^5 - 6x^4 + 6x^3 + 9x$

41. $(16x - 2x^4 + 3x^2 - 5) - (-6 + x^4 - 3x^3 - 3x^2 + 2x)$
$-3x^4 + 3x^3 + 6x^2 + 14x + 1$

42. $(6x^2y^2 - 5xy) + (-4xy + 3) - (3x^2y^2 - 8xy + 5)$
$3x^2y^2 - xy - 2$

Add.

43.
$$\begin{array}{r} x^4 - 2x^3 + 7x^2 \quad\quad - 20 \\ -8x^4 \quad\quad - 5x^2 + 3x + 7 \\ 36x^4 - 18x^3 \quad\quad + 5x - 31 \\ \hline 29x^4 - 20x^3 + 2x^2 + 8x - 44 \end{array}$$

44.
$$\begin{array}{r} 6.1x^3 - 0.2x^2 + 1.1x + 5 \\ -5.7x^3 \quad\quad + 2.3x - 7 \\ 1.1x^2 - 2.3x + 8 \\ \hline 0.4x^3 + 0.9x^2 + 1.1x + 6 \end{array}$$

14.5, *Multiply.*
14.6

45. $(x + 8)(x - 10)$
$x^2 - 2x - 80$

46. $(2x + 1)(3x - 4)$
$6x^2 - 5x - 4$

47. $(4a - 3)(4a + 3)$
$16a^2 - 9$

48. $(4a - 3)^2$
$16a^2 - 24a + 9$

49. $(4a + 3)^2$
$16a^2 + 24a + 9$

50. $(5y - 6)(4y - 7)$
$20y^2 - 59y + 42$

51. $\left(\dfrac{3}{4} + 2y^3\right)\left(\dfrac{3}{4} - 2y^3\right)$
$\dfrac{9}{16} - 4y^6$

52. $(4x - 9)(x + 2)$
$4x^2 - x - 18$

53. $(x + 4)(2x^2 + x - 3)$
$2x^3 + 9x^2 + x - 12$

54.
$$\begin{array}{r} 3x^2 - 2x + 1 \\ x^2 + x - 2 \\ \hline 3x^4 + x^3 - 7x^2 + 5x - 2 \end{array}$$

55.
$$\begin{array}{r} 2x^2 + xy - 3y^2 \\ x + 4y \\ \hline 2x^3 + 9x^2y + xy^2 - 12y^3 \end{array}$$

56.
$$\begin{array}{r} 6x^2y^2 - xy + 5 \\ 2xy \quad\quad - 3 \\ \hline 12x^3y^3 - 20x^2y^2 + 13xy - 15 \end{array}$$

57. $(7a - b)(4a + 3b)$
$28a^2 + 17ab - 3b^2$

58. $(5a - 7b)^2$
$25a^2 - 70ab + 49b^2$

59. $(4x - y)(4x + y)$
$16x^2 - y^2$

60. $(8x + y)^2$
$64x^2 + 16xy + y^2$

14.7 *Divide.*

61. $\dfrac{25x^4 - 50x^3 + 10x^2}{5x^2}$
$5x^2 - 10x + 2$

62. $\dfrac{x^{10} - x^8 + x^6}{-x^4}$
$-x^6 + x^4 - x^2$

63. $\dfrac{6x^4y^4 - 2x^3y^3 + 10x^2y^2}{2x^2y^2}$

$3x^2y^2 - xy + 5$

64. $\dfrac{-14a^4b^2 + 21a^2b^4 - 35ab^2}{-7a^2b^2}$

$2a^2 - 3b^2 + \dfrac{5}{a}$

65. $(x^2 + 3x - 28) \div (x - 4)$

$x + 7$

66. $(8x^2 + 10x + 3) \div (2x + 1)$

$4x + 3$

67. $2x + 3 \overline{)14x^4 + 27x^3 + 5x^2 - 16x - 15}$

$7x^3 + 3x^2 - 2x - 5$

68. $x - 5 \overline{)x^3 \qquad\qquad -100}$

$x^2 + 5x + 25 + \dfrac{25}{x - 5}$

Part II

69. The number n of records that can be sold during each week is related to the price in dollars p of each record by the equation $n = 1200 - 100p$. Find an equation that gives the weekly revenue R in terms of p and use it to find R when p is \$10.

$R = 1200p - 100p^2$; \$2000

70. Light travels at the rate of 186,000 mi per sec. Write this number in scientific notation.

1.86×10^5 mi per sec.

Perform the indicated operation.

71. $(2x + 1)(5x - 2)$

$10x^2 + x - 2$

72. $(6x^2y^2 - 3xy) - (-2x^2y^2 + 1)$

$8x^2y^2 - 3xy - 1$

73. $(x^2 + 10x + 21) \div (x + 7)$

$x + 3$

74. $(6x^5 - 4x^3) + (3x^5 + x^2) - (-x^3 + x^2)$

$9x^5 - 3x^3$

75. $(7x + 2y)(7x - 2y)$

$49x^2 - 4y^2$

76. $(6x^2y^3 - 4x^2y - 2xy^2) \div (-2xy)$

$-3xy^2 + 2x + y$

77. $(5a - b)^2$

$25a^2 - 10ab + b^2$

78. $(8x + 3)^2$

$64x^2 + 48x + 9$

79. Subtract $-x^3 - 2$ from $x^3 - 2$.

$2x^3$

80. $3a^2b(4a^3b^4 - 5a^2b - 2ab)$

$12a^5b^5 - 15a^4b^2 - 6a^3b^2$

Simplify and write without negative exponents.

81. $(-2x^2)^3$ $-8x^6$

82. $-2(x^2)^3$ $-2x^6$

83. $\left(\dfrac{x^3}{y^2}\right)^{-2}$ $\dfrac{y^4}{x^6}$

84. $(a^2b^{-4})^{-2}$ $\dfrac{b^8}{a^4}$

85. $\dfrac{2^0 x^{-4} y^3}{x^2 y^{-3}}$ $\dfrac{y^6}{x^6}$

86. $\left(\dfrac{ab^{-2}}{a^{-3}b^{-3}}\right)^{-1}$ $\dfrac{1}{a^4 b}$

Give the degree of each polynomial.

87. $6x^3 - 4x^2 + 2x^5 - 1$
5

88. $-7y + 6y^3 - 2 + y^2$
3

Tell whether the following are monomials, binomials, or trinomials.

89. $3x + 2$
binomial

90. $x^2 - 6x + 7$
trinomial

ANSWERS: 1. a^5 2. $(3z)^3$ 3. $(a+b)^2$ 4. $bbbbbb$ 5. $-2xxx$ 6. $(-2x)(-2x)(-2x)$ 7. $2y^9$ 8. $27x^6$ 9. $x^{10}y^{15}$
10. 2 11. 1 12. $\frac{2}{b^{10}}$ 13. $\frac{5}{a}$ 14. $\frac{1}{5a}$ 15. $\frac{b}{a}$ 16. $\frac{b^6}{a^4}$ 17. $\frac{b^8}{a^9}$ 18. $\frac{2y^2}{x^4}$ 19. 4.11×10^{-7} 20. 5.49×10^{11}
21. 6,150,000,000,000 22. 0.00000004 23. binomial 24. monomial 25. binomial 26. trinomial 27. 4 28. 7
29. 10 30. 8 31. $-x+2$ 32. $11y^2 - y + 5$ 33. $6a^2 - 5a$ 34. $6x^5 - 8x^4 - 7x^2 - 7x - 5$ 35. $4xy - 3$
36. $4a^2b^3 - 6a^2b^2 - ab^2$ 37. \$360 38. $-2x^2 - 3$ 39. $x^3 - 3x^2 + 15$ 40. $-x^5 - 6x^4 + 6x^3 + 9x$
41. $-3x^4 + 3x^3 + 6x^2 + 14x + 1$ 42. $3x^2y^2 - xy - 2$ 43. $29x^4 - 20x^3 + 2x^2 + 8x - 44$ 44. $0.4x^3 + 0.9x^2 + 1.1x + 6$
45. $x^2 - 2x - 80$ 46. $6x^2 - 5x - 4$ 47. $16a^2 - 9$ 48. $16a^2 - 24a + 9$ 49. $16a^2 + 24a + 9$ 50. $20y^2 - 59y + 42$
51. $\frac{9}{16} - 4y^6$ 52. $4x^2 - x - 18$ 53. $2x^3 + 9x^2 + x - 12$ 54. $3x^4 + x^3 - 7x^2 + 5x - 2$ 55. $2x^3 + 9x^2y + xy^2 - 12y^3$
56. $12x^3y^3 - 20x^2y^2 + 13xy - 15$ 57. $28a^2 + 17ab - 3b^2$ 58. $25a^2 - 70ab + 49b^2$ 59. $16x^2 - y^2$
60. $64x^2 + 16xy + y^2$ 61. $5x^2 - 10x + 2$ 62. $-x^6 + x^4 - x^2$ 63. $3x^2y^2 - xy + 5$ 64. $2a^2 - 3b^2 + \frac{5}{a}$ 65. $x + 7$
66. $4x + 3$ 67. $7x^3 + 3x^2 - 2x - 5$ 68. $x^2 + 5x + 25 + \frac{25}{x-5}$ 69. $R = 1200p - 100p^2$; \$2000
70. 1.86×10^5 mi per sec. 71. $10x^2 + x - 2$ 72. $8x^2y^2 - 3xy - 1$ 73. $x + 3$ 74. $9x^5 - 3x^3$ 75. $49x^2 - 4y^2$
76. $-3xy^2 + 2x + 2y$ 77. $25a^2 - 10ab + b^2$ 78. $64x^2 + 48x + 9$ 79. $2x^3$ 80. $12a^5b^5 - 15a^4b^2 - 6a^3b^2$
81. $-8x^6$ 82. $-2x^6$ 83. $\frac{y^4}{x^6}$ 84. $\frac{b^8}{a^4}$ 85. $\frac{y^6}{x^6}$ 86. $\frac{1}{a^4b}$ 87. 5 88. 3 89. binomial 90. trinomial

1. Write in exponential notation. $5 \cdot 5 \cdot a \cdot a \cdot a$

1. _____ $5^2 a^3$ _____

Simplify and write without negative exponents.

2. $2y^5 y^2$

2. _____ $2y^7$ _____

3. $\dfrac{a^7}{a^3}$

3. _____ a^4 _____

4. $(-2x^2)^3$

4. _____ $-8x^6$ _____

5. z^{-3}

5. _____ $\dfrac{1}{z^3}$ _____

6. $\dfrac{x^{-6}y^3}{x^2 y^{-4}}$

6. _____ $\dfrac{y^7}{x^8}$ _____

7. Evaluate y^{-3} when $y = -4$.

7. _____ $-\dfrac{1}{64}$ _____

8. Evaluate. $\dfrac{3.9 \times 10^{-4}}{1.3 \times 10^{-2}}$

8. _____ 3×10^{-2} _____

9. Is $3x^2 + 5$ a monomial, binomial, or trinomial?

9. _____ binomial _____

10. Give the degree of $x^3 - 7x^2 + 10x^4 - 8$.

10. _____ 4 _____

11. Collect and combine like terms and write in descending order.

   $3y^3 - 7y^5 + y^3 - 4y + 6 + y^5 - 2y$

11. _____ $-6y^5 + 4y^3 - 6y + 6$ _____

12. Collect and combine like terms.

   $3x^2 y^2 - 4x^2 + 3x^2 y - 7x^2 y^2 + 2x^2 y$

12. _____ $-4x^2 y^2 + 5x^2 y - 4x^2$ _____

13. The cost in dollars of making x pairs of shoes is given by the expression $20x + 45$. Find the cost when 12 pairs of shoes are made.

13. _____ \$285 _____

Add.

14. $(3x + 2) + (-5x + 8)$

14. _____ $-2x + 10$ _____

15. $(6x^3 + 5x^2 - 3x^4 + 2) + (-8x^2 + 4x^3 + 6x^4 + 5)$

15. _____ $3x^4 + 10x^3 - 3x^2 + 7$ _____

16. $(3a^2 b^2 - 2ab + 5) + (4a^2 b^2 + 8ab - 3)$

16. _____ $7a^2 b^2 + 6ab + 2$ _____

Subtract.

17. $(-5x + 4) - (2x - 3)$

17. _____ $-7x + 7$ _____

18. $(4x^2 - 6x^4 + 2) - (-2x^2 + x^3 - 2x^4 + 1)$

18. _____ $-4x^4 - x^3 + 6x^2 + 1$ _____

19. $(8a^2b - 3ab^2 - 2ab) - (4a^2b + ab^2 - ab)$

19. _____ $4a^2b - 4ab^2 - ab$ _____

Multiply.

20. $-4x^2(3x^2 - 4x + 1)$

20. _____ $-12x^4 + 16x^3 - 4x^2$ _____

21. $(x + 5)(x - 8)$

21. _____ $x^2 - 3x - 40$ _____

22. $(4a - 3)(2a + 7)$

22. _____ $8a^2 + 22a - 21$ _____

23. $(y + 8)(3y^2 - 2y + 1)$

23. _____ $3y^3 + 22y^2 - 15y + 8$ _____

24. $(2x + y)(2x - y)$

24. _____ $4x^2 - y^2$ _____

25. $(3x - 4y)^2$

25. _____ $9x^2 - 24xy + 16y^2$ _____

26. $(2x + 5)^2$

26. _____ $4x^2 + 20x + 25$ _____

27. $(3x + 2)(2x^2 - 3x + 4)$

27. _____ $6x^3 - 5x^2 + 6x + 8$ _____

28. $(x^2 - 3)(x^2 + 3)$

28. _____ $x^4 - 9$ _____

Divide.

29. $\dfrac{25x^4 - 10x^3 - 15x^2}{5x^2}$

29. _____ $5x^2 - 2x - 3$ _____

30. $(3x^3 - 11x^2 + 10x - 12) \div (x - 3)$

30. _____ $3x^2 - 2x + 4$ _____

Factoring Polynomials

15.1 COMMON FACTORS AND GROUPING

STUDENT GUIDEPOSTS

1 Greatest Common Factor 3 Factoring by Grouping
2 Factoring by Removing the GCF

Factoring is the reverse of multiplying. In Chapter 2 we factored integers such as 28 by writing

$$28 = 4 \cdot 7 = 2 \cdot 2 \cdot 7 = 2^2 \cdot 7.$$

In this form, we see that 4 and 7 are two **factors** of 28, (as are 1, 2, 14, and 28), while 2 and 7 are also **prime factors** (2 and 7 are prime numbers). Algebraic expressions can also be factored into prime factors. For example,

$$28x^3 = 2 \cdot 2 \cdot 7 \cdot x \cdot x \cdot x = 2^2 \cdot 7 \cdot x^3,$$

has 2, 7, and x as its prime factors. A **common factor** of two expressions is a factor that occurs in each of them. For instance, a common factor of 14 and 28 is 7 since

$$14 = 2 \cdot 7 \quad \text{and} \quad 28 = 2^2 \cdot 7.$$

Other common factors of 14 and 28 are 2 and 14.

1 GREATEST COMMON FACTOR

The **greatest common factor (GCF)** of two integers is the largest factor common to both integers. Thus, 14 is the greatest common factor of 14 and 28. As shown in the next example, it is easier to find the greatest common factor if integers are expressed as a product of primes.

EXAMPLE 1 GREATEST COMMON FACTOR (GCF) OF INTEGERS

Find the greatest common factor of the integers.

(a) 35 and 75

$$35 = 5 \cdot 7 \quad \text{and} \quad 75 = 3 \cdot 5 \cdot 5$$

GCF: 5

PRACTICE EXERCISE 1

Find the greatest common factor of the integers.

(a) 55 and 33

(b) 54 and 90

$$54 = 2 \cdot 3 \cdot 3 \cdot 3 \quad \text{and} \quad 90 = 2 \cdot 3 \cdot 3 \cdot 5$$

GCF: $2 \cdot 3 \cdot 3 = 18$

(b) 140 and 84

Answers: (a) 11 (b) 28

We can also find common factors of monomials such as $14x^3$ and $28x^2$. First we write the monomials in factored form.

$$14x^3 = 2 \cdot 7 \cdot x^3 = \boxed{2 \cdot 7 \cdot x^2} \cdot x$$
$$28x^2 = 2^2 \cdot 7 \cdot x^2 = 2 \cdot \boxed{2 \cdot 7 \cdot x^2}$$

— Factors common to both

Each shaded factor is a common factor of $14x^3$ and $28x^2$, and the shaded product, $2 \cdot 7 \cdot x^2 = 14x^2$, is the greatest common factor of the two monomials. Usually we are interested in finding the greatest common factor of the terms of a polynomial. For instance, $14x^2$ is the greatest common factor of the terms of the polynomial $14x^3 + 28x^2$. The following rule summarizes the technique.

To Find the Greatest Common Factor

1. Write each term as a product of prime factors, expressing repeated factors as powers.
2. Select the factors that are common to all terms and raise each to the *lowest* power that it occurs in any one term.
3. The product of these factors is the greatest common factor (GCF).

EXAMPLE 2 GCF OF A POLYNOMIAL

Find the greatest common factor of the terms of each polynomial.

(a) $12x - 9 = 2^2 \cdot 3 \cdot x - 3 \cdot 3$
Common factors: 3
Lowest power of each: 3^1
GCF: 3

(b) $18x^2 + 30x = 2 \cdot 3 \cdot 3 \cdot x \cdot x + 2 \cdot 3 \cdot 5 \cdot x$
Common factors: 2, 3, x
Lowest power of each: 2^1, 3^1, x^1
GCF: $2 \cdot 3 \cdot x = 6x$

(c) $3x^4 + 12x^2 = 3 \cdot x^2 \cdot x^2 + 2^2 \cdot 3 \cdot x^2$
Common factors: 3, x
Lowest power of each: 3^1, x^2
GCF: $3 \cdot x^2 = 3x^2$

(d) $7y + 3 = 1 \cdot 7 \cdot y + 1 \cdot 3$
Common factors: 1
Lowest power of each: 1^1
GCF: 1

(e) $14x^4 - 28x^3 + 21x^2 = 2 \cdot 7 \cdot x^2 \cdot x^2 - 2^2 \cdot 7 \cdot x^2 \cdot x + 3 \cdot 7 \cdot x^2$
Common factors: 7, x
Lowest power of each: 7^1, x^2
GCF: $7 \cdot x^2 = 7x^2$

PRACTICE EXERCISE 2

Find the greatest common factor of the terms of each polynomial.

(a) $15a - 25$

(b) $24y^2 + 20y$

(c) $7w^5 + 21w^3$

(d) $9a + 8$

(e) $18y^5 - 27y^4 + 45y^3$

(f) $22u^3v^3 - 44u^3v + 77u^2v^4$

(f) $5x^2y^3 - 15xy^2 = 5 \cdot x \cdot x \cdot y^2 \cdot y - 3 \cdot 5 \cdot x \cdot y^2$
Common factors: $5, x, y$
Lowest power of each: $5^1, x^1, y^2$
GCF: $5 \cdot x \cdot y^2 = 5xy^2$

Answers: (a) 5 (b) $4y$ (c) $7w^3$
(d) 1 (e) $9y^3$ (f) $11u^2v$

② FACTORING BY REMOVING THE GCF

A polynomial is said to be **factored by removing common factors** when the greatest common factor is removed using the distributive law. For example, since

$$14x^3 + 28x^2 = 14x^2(x + 2)$$

we have factored $14x^3 + 28x^2$ using the distributive law by removing the greatest common factor $14x^2$. Example 3 illustrates the technique summarized below.

To Factor by Removing the Greatest Common Factor

1. Write each term as a product of primes.
2. Find the greatest common factor.
3. Write each term in the form

   GCF · remaining factors.

4. Use the distributive property to remove the **GCF**. The factors left in each term are the **remaining factors.**
5. To check, multiply the GCF by the polynomial factor.

EXAMPLE 3 REMOVING THE GCF

Factor the polynomials given in Example 2.

(a) $12x - 9 = 2^2 \cdot 3 \cdot x - 3 \cdot 3$ GCF is 3
$= 3(2^2 \cdot x - 3)$ Remove the GCF using the distributive property
$= 3(4x - 3)$ Keep the GCF as a multiplier

(b) $18x^2 + 30x = 2 \cdot 3 \cdot 3 \cdot x \cdot x + 2 \cdot 3 \cdot 5 \cdot x$ GCF is $6x$
$= 2 \cdot 3 \cdot x(3x + 5)$ Distributive property
$= 6x(3x + 5)$ Keep the GCF, $6x$, as a multiplier

(c) $3x^4 + 12x^2 = 3x^2 \cdot x^2 + 3x^2 \cdot 4$ GCF is $3x^2$
$= 3x^2(x^2 + 4)$ Distributive property

(d) $7y + 3 = 1 \cdot 7y + 1 \cdot 3$ GCF is 1
$= 7y + 3$ Cannot be factored

(e) $14x^4 - 28x^3 + 21x^2 = 7x^2 \cdot 2x^2 - 7x^2 \cdot 4x + 7x^2 \cdot 3$ GCF is $7x^2$
$= 7x^2(2x^2 - 4x + 3)$

(f) $5x^2y^3 - 15xy^2 = 5 \cdot x^2 \cdot y^3 - 3 \cdot 5 \cdot x \cdot y^2$
$= 5xy^2 \cdot xy - 5xy^2 \cdot 3$ GCF is $5xy^2$
$= 5xy^2(xy - 3)$

PRACTICE EXERCISE 3

Factor by removing the greatest common factor.

(a) $15a - 25$

(b) $24y^2 + 20y$

(c) $7w^5 + 21w^3$

(d) $9a + 8$

(e) $18y^5 - 27y^4 + 45y^3$

(f) $22u^3v^3 - 44u^3v + 77u^2v^4$

Answers: (a) $5(3a - 5)$
(b) $4y(6y + 5)$ (c) $7w^3(w^2 + 3)$
(d) $9a + 8$ is a prime polynomial
(e) $9y^3(2y^2 - 3y + 5)$
(f) $11u^2v(2uv^2 - 4u + 7v^3)$

When the greatest common factor is 1, we say the polynomial cannot be factored by removing the common factor. Polynomials like $7y + 3$ in Example 3(d), which cannot be factored, are called **prime polynomials.**

///////////// **CAUTION** /////////////

When told to factor a polynomial such as $18x^2 + 30x$, a common error is to give

$$2 \cdot 3 \cdot 3 \cdot x \cdot x + 2 \cdot 3 \cdot 5 \cdot x$$

for the answer. Here the terms of the polynomial have been factored, but the polynomial itself has not. The correct factorization is given in Example 3(b).

//////////

Often when a polynomial consists of many negative terms, or when its leading coefficient is negative, we factor out the negative of the greatest common factor of its terms, as shown in the next example.

EXAMPLE 4 FACTORING OUT A NEGATIVE FACTOR	**PRACTICE EXERCISE 4**

Factor $-6x^5 - 12x^4 - 15x$.

$$-6x^5 - 12x^4 - 15x = (-3x)2x^4 + (-3x)4x^3 + (-3x)5$$
$$= -3x(2x^4 + 4x^3 + 5)$$

In this example the greatest common factor is $3x$; thus, another way to factor is $3x(-2x^4 - 4x^3 - 5)$. Either answer is considered correct, but the first is sometimes preferred.

Factor $-10x^4 - 35x^3 - 40x^2$.

Answer: $-5x^2(2x^2 + 7x + 8)$

EXAMPLE 5 FACTORING WITH 1 LEFT IN TERM	**PRACTICE EXERCISE 5**

Factor $3x^2y^3 - xy$.

$$3x^2y^3 - xy = xy \cdot 3xy^2 - xy \cdot 1 \qquad \text{GCF is } xy$$
$$= xy(3xy^2 - 1) \qquad \text{Distributive property}$$

Factor $6u^4v^4 + u^2v^3$.

Answer: $u^2v^3(6u^2v + 1)$

///////////// **CAUTION** /////////////

When we factored xy out of $xy \cdot 1$ in Example 5, we were left with a 1 in the term. Do not leave this term out and write $xy(3xy^2)$ as the factors of $3x^2y^3 - xy$. If we multiply $xy(3xy^2)$ we get $3x^2y^3$ which is just the first term of the given polynomial.

//////////

Sometimes a polynomial contains grouping symbols. For example, the polynomial

$$3x(x + 2) + 5(x + 2)$$

is really a binomial made of the two terms $3x(x + 2)$ and $5(x + 2)$. These terms have a greatest common factor $(x + 2)$.

$$3x(x + 2) + 5(x + 2) = 3x \cdot (x + 2) + 5 \cdot (x + 2)$$ GCF is $(x + 2)$

$$= (3x + 5)(x + 2)$$ Distributive property

Here the $(x + 2)$ is a single factor of each term and is removed using the distributive property.

EXAMPLE 6 POLYNOMIALS CONTAINING GROUPING SYMBOLS

Factor $3x(a + b) + y(a + b)$.

$$3x(a + b) + y(a + b) = 3x \cdot (a + b) + y \cdot (a + b)$$ GCF is $(a + b)$

$$= (3x + y)(a + b)$$ Distributive property

PRACTICE EXERCISE 6

Factor $2y(a - b) - x(a - b)$.

Answer: $(2y - x)(a - b)$

---

❸ FACTORING BY GROUPING

Notice that if we clear the parentheses in the polynomial in Example 6, we obtain

$$3x(a + b) + y(a + b) = 3xa + 3xb + ya + yb.$$

Suppose we are given the polynomial on the right to factor. We may do so by grouping the first two terms and the last two terms. Then we factor as in Example 6. This process is called **factoring by grouping** and is illustrated in the next example.

EXAMPLE 7 FACTORING BY GROUPING

Factor each polynomial by grouping.

(a) $3xa + 3xb + ya + yb$

$$= (3xa + 3xb) + (ya + yb)$$ Group the first two terms and the last two terms

$$= 3x(a + b) + y(a + b)$$ Use the distributive law on each group

$$= (3x + y)(a + b)$$ Use the distributive law to factor out $(a + b)$

(b) $5ay - 5ax + 3by - 3bx = (5ay - 5ax) + (3by - 3bx)$

$$= 5a(y - x) + 3b(y - x)$$

$$= (5a + 3b)(y - x)$$

PRACTICE EXERCISE 7

Factor each polynomial by grouping.

(a) $2ay + by + 2ax + bx$

(b) $6xu - 5yu + 6xv - 5yv$

Answers: (a) $(2a + b)(y + x)$
(b) $(6x - 5y)(u + v)$

Factoring by grouping should be tried when the polynomial has four terms. We conclude this section with an application of factoring.

EXAMPLE 8 APPLICATION OF FACTORING

If a rocket is fired straight up with a velocity of 128 feet per second, the equation $h = -16t^2 + 128t$ gives the height h of the rocket at any time t. Factor the right side of the equation and use this factored form to find the height of the rocket after 6 seconds.

Factoring, we obtain

$$h = -16t^2 + 128t = -16t^2 + 16 \cdot 8t$$

$$= -16t\,(t - 8).$$ Notice the minus sign

PRACTICE EXERCISE 8

Given a rocket as in Example 8 with $h = -16t^2 + 64t$, factor and find the height after 4 seconds.

Substitute 6 for t to find h.

$$h = -16t(t - 8)$$
$$= -16(6)(6 - 8)$$
$$= -16(6)(-2) = 192$$

The rocket will be 192 ft high after 6 sec.

Answers: $h = -16t(t - 4)$; 0

You never need to make a mistake in a factoring problem since you can always check your work by multiplying the factors and comparing the product with the given polynomial.

15.1 EXERCISES A

Find the greatest common factor.

1. 10, 15   5

2. 28, 42   14

3. 17, 23   1

4. $10x^2$, $15x$ $5x$

5. $28y^3$, $42y$ $14y$

6. $17y^2$, $23y^4$ y^2

7. 10, 15, 35   5

8. 28, 42, 12   2

9. 17, 23, 16   1

10. $10x^2$, $15x$, $35x^3$ $5x$

11. $28y^3$, $42y^2$, $12y^5$ $2y^2$

12. $17y^2$, $23y^4$, 16   1

13. $6x^3$, $12x^2$, $18x^2$ $6x^2$

14 $36y^4$, $6y^3$, $42y^5$ $6y^3$

15. $8y^2$, $9x$, $12y^5$ 1

16. $5x^2y$, $40xy^2$ $5xy$

17. $16x^2y^2$, $12x^3y^4$, $8xy^5$ $4xy^2$

18. x^4y^3, x^5y^2, x^3 x^3

Find the greatest common factor of the terms of the polynomials.

19. $4x + 8$ 4

20. $8y^2 - 16y$ $8y$

21. $3a^3 - 9a^2$ $3a^2$

22. $15x^3 - 25x^2$ $5x^2$

23. $9y^4 + 18y^2$ $9y^2$

24. $50a^{10} + 75a^8$ $25a^8$

25. $4x^3 + 2x^2 - 6x$
$2x$

26. $18y^5 - 24y^3 + 36y$
$6y$

27 $3a(a + 2) + 5(a + 2)$
$a + 2$

28. $6x^2y^2 - 4x^2y$
$2x^2y$

29 $5a^3b^3 - 15a^2b^2 + 10ab^2$
$5ab^2$

30. $x(a + b) + y(a + b)$
$a + b$

Factor.

31. $3x + 9$ $3(x + 3)$

32. $21x - 14$ $7(3x - 2)$

33. $24y - 6$ $6(4y - 1)$

34. $4a^2 + 2a$
$2a(2a + 1)$

35. $18y^2 + 11y$
$y(18y + 11)$

36. $23a^2 - 5$
cannot be factored

37. $x^{10} - x^8 + x^6$
$x^6(x^4 - x^2 + 1)$

38. $6y^4 - 24y^2 + 12y$
$6y(y^3 - 4y + 2)$

39. $16a^5 + 48a^3 - 24$
$8(2a^5 + 6a^3 - 3)$

40. $60x^3 + 50x^2 - 25x$
$5x(12x^2 + 10x - 5)$

41 $-6y^{10} - 8y^8 - 4y^5$
$-2y^5(3y^5 + 4y^3 + 2)$

42. $a^2 + 2a + 2$
cannot be factored

43. $x^2y^2 + xy$
$xy(xy + 1)$

44. $6a^2b - 2ab^2$
$2ab(3a - b)$

45. $27x^3y^2 + 45xy^2$
$9xy^2(3x^2 + 5)$

46. $15x^3y^3 + 5x^2y^2 + 10xy$
$5xy(3x^2y^2 + xy + 2)$

47 $a^2(a + 2) + 3(a + 2)$
$(a + 2)(a^2 + 3)$

48. $x^2(a + b) + y^2(a + b)$
$(a + b)(x^2 + y^2)$

Factor by grouping.

49. $a^3 + 2a^2 + 3a + 6$
$(a + 2)(a^2 + 3)$

50 $x^2a + x^2b + y^2a + y^2b$
$(a + b)(x^2 + y^2)$

51. $a^2b - a^2 + 5b - 5$
$(b - 1)(a^2 + 5)$

52. $x^3 + 2x^2 - 7x - 14$
$(x + 2)(x^2 - 7)$

53. $a^3b^2 + a^3 + 2b^2 + 2$
$(b^2 + 1)(a^3 + 2)$

54. $x^2y - 3x^2 + y - 3$
$(y - 3)(x^2 + 1)$

55. $a^4b + a^3 - ab^3 - b^2$
$(ab + 1)(a^3 - b^2)$

56 $-x^2y - x^2 - 3y - 3$
$-(x^2 + 3)(y + 1)$

57. If a principal of P dollars is invested in an account with an annual interest rate of 9% compounded annually, the amount A in the account after one year is given by the equation $A = P + 0.09P$. Factor the right side of the equation and use this factored form to find the amount in the account in one year when P is   **(a)** $100,   **(b)** $1000, and   **(c)** $4537.35.
$A = P(1.09)$; **(a)** $109   **(b)** $1090   **(c)** $4945.71

58. When asked to factor $4x^3 + 8x^2$, Burford gave $2x(2x^2 + 4x)$ for the answer. What is wrong with Burford's work? What is the correct answer?
He did not factor out the GCF. The answer should be $4x^2(x + 2)$.

59. What is wrong with the following factoring problem?

$$-3x^2y^2 + 6x^2y = -3x^2y(y + 2)$$

The answer should be $-3x^2y(y - 2)$.

60. What is the GCF of the terms of the binomial $2xy + 4x^2$? What is the GCF of the terms of the binomial $5xy - 10y$? Find the product of $2xy + 4x^2$ and $5xy - 10y$ and give the GCF of the terms of this polynomial. What can you conclude?

$2x$; $5y$; $10xy$; the GCF of the product is the product of the GCFs.

FOR REVIEW

Use the FOIL method to multiply.

61. $(x + 3)(x + 7)$
$x^2 + 10x + 21$

62. $(x + 3)(x - 7)$
$x^2 - 4x - 21$

63. $(x - 3)(x + 7)$
$x^2 + 4x - 21$

64. $(x - 3)(x - 7)$
$x^2 - 10x + 21$

65. $(x + 4y)(x + 3y)$
$x^2 + 7xy + 12y^2$

66. $(x + 4y)(x - 3y)$
$x^2 + xy - 12y^2$

ANSWERS: 1. 5 2. 14 3. 1 4. 5x 5. 14y 6. y^2 7. 5 8. 2 9. 1 10. 5x 11. $2y^2$ 12. 1 13. $6x^2$ 14. $6y^3$ 15. 1 16. 5xy 17. $4xy^2$ 18. x^3 19. 4 20. 8y 21. $3a^2$ 22. $5x^2$ 23. $9y^2$ 24. $25a^8$ 25. 2x 26. 6y 27. $a + 2$ 28. $2x^2y$ 29. $5ab^2$ 30. $a + b$ 31. $3(x + 3)$ 32. $7(3x - 2)$ 33. $6(4y - 1)$ 34. $2a(2a + 1)$ 35. $y(18y + 11)$ 36. cannot be factored 37. $x^6(x^4 - x^2 + 1)$ 38. $6y(y^3 - 4y + 2)$ 39. $8(2a^5 + 6a^3 - 3)$ 40. $5x(12x^2 + 10x - 5)$ 41. $-2y^5(3y^5 + 4y^3 + 2)$ 42. cannot be factored 43. $xy(xy + 1)$ 44. $2ab(3a - b)$ 45. $9xy^2(3x^2 + 5)$ 46. $5xy(3x^2y^2 + xy + 2)$ 47. $(a + 2)(a^2 + 3)$ 48. $(a + b)(x^2 + y^2)$ 49. $(a + 2)(a^2 + 3)$ 50. $(a + b)(x^2 + y^2)$ 51. $(b - 1)(a^2 + 5)$ 52. $(x + 2)(x^2 - 7)$ 53. $(b^2 + 1)(a^3 + 2)$ 54. $(y - 3)(x^2 + 1)$ 55. $(ab + 1)(a^3 - b^2)$ 56. $-(x^2 + 3)(y + 1)$ 57. $A = P(1 + 0.09) = P(1.09)$; (a) $109 (b) $1090 (c) $4945.71 58. He did not factor out the GCF. The answer should be $4x^2(x + 2)$ 59. The answer should be $-3x^2y(y - 2)$. 60. 2x; 5y; 10xy; the GCF of the product is the product of GCFs. 61. $x^2 + 10x + 21$ 62. $x^2 - 4x - 21$ 63. $x^2 + 4x - 21$ 64. $x^2 - 10x + 21$ 65. $x^2 + 7xy + 12y^2$ 66. $x^2 + xy - 12y^2$

15.1 EXERCISES B

Find the greatest common factor.

1. 12, 18   6

2. 32, 48   16

3. 11, 29   1

4. $12x$, $18x^3$ 6x

5. $32y^4$, $48y^2$ $16y^2$

6. $11y^3$, $29y^2$ y^2

7. 12, 18, 9   3

8. 32, 48, 64   16

9. 11, 29, 30   1

10. $12x^4$, $18x^2$, $9x^3$ $3x^2$

11. $32y^3$, $48y^5$, $64y^4$ $16y^3$

12. $11y^2$, $29y^3$, 30   1

13. $15x^2$, $20x^3$, $5x^4$ $5x^2$

14. $22y^5$, $33y^4$, $44y^3$ $11y^3$

15. $36y^4$, $15x^2$, $5xy$ 1

16. $6x^2y^3$, $24x^2y$ $6x^2y$

17. $30x^5y^5$, $2x^2y^4$, $12x^4y^3$ $2x^2y^3$

18. x^8y^7, x^4y^6, y^5 y^5

Find the greatest common factor of the terms of the polynomials.

19. $9x + 3$ 3

20. $22y^2 - 11y$ 11y

21. $15a^3 - 10a^2$ $5a^2$

22. $16x^4 - 24x^2$ $8x^2$

23. $14y^5 + 35y^4$ $7y^4$

24. $100a^{16} + 200a^{14}$ $100a^{14}$

25. $5x^3 + 15x^2 - 15x$ $5x$

26. $48y^6 - 24y^5 + 12y^2$ $12y^2$

27. $7a(a - 5) + 6(a - 5)$ $a - 5$

28. $9x^3y^3 - 12xy^4$ $3xy^3$

29. $4a^3b^3 + 6a^2b^2 - 6ab^2$ $2ab^2$

30. $2a(x + y) + b(x + y)$ $x + y$

Factor.

31. $5y - 20$
$5(y - 4)$

32. $15x - 35$
$5(3x - 7)$

33. $54y + 12$
$6(9y + 2)$

34. $22y - 11$
$11(2y - 1)$

35. $25x^4 - 15x^3$
$5x^3(5x - 3)$

36. $16a^2 - 7$
cannot be factored

37. $x^{12} - x^6 + x^4$
$x^4(x^8 - x^2 + 1)$

38. $26y^5 - 13y^3 + 39y^2$
$13y^2(2y^3 - y + 3)$

39. $8a^6 - 18a^3 + 12$
$2(4a^6 - 9a^3 + 6)$

40. $90x^4 - 45x^3 + 180x^2$
$45x^2(2x^2 - x + 4)$

41. $-5y^8 - 10y^6 - 15y^4$
$-5y^4(y^4 + 2y^2 + 3)$

42. $a^2 + a + 5$
cannot be factored

43. $x^4y^4 - x^2y$
$x^2y(x^2y^3 - 1)$

44. $11a^3b^3 - 22a^2b^2$
$11a^2b^2(ab - 2)$

45. $24x^4y^3 + 36x^3y^4$
$12x^3y^3(2x + 3y)$

46. $28x^4y^4 + 14x^3y^3 + 35x^2y^2$
$7x^2y^2(4x^2y^2 + 2xy + 5)$

47. $2a^3(a - 5) + 5(a - 5)$
$(a - 5)(2a^3 + 5)$

48. $a^2(x + y) + 2b(x + y)$
$(x + y)(a^2 + 2b)$

Factor by grouping.

49. $a^3 + 7a^2 + 2a + 14$
$(a + 7)(a^2 + 2)$

50. $a^2x + a^2y + b^2x + b^2y$
$(x + y)(a^2 + b^2)$

51. $a^3b - 2a^3 + 4b - 8$
$(b - 2)(a^3 + 4)$

52. $x^3 + 5x^2 - 2x - 10$
$(x + 5)(x^2 - 2)$

53. $a^5b^3 + 2a^5 + 6b^3 + 12$
$(b^3 + 2)(a^5 + 6)$

54. $xy - 2y + x - 2$
$(x - 2)(y + 1)$

55. $ab^2 + ab - 5b - 5$
$(b + 1)(ab - 5)$

56. $-x^3y^2 - 3x^3y - 7y - 21$
$-(y + 3)(x^3y + 7)$

57. If a bullet is fired straight up with a velocity of 256 feet per second, the equation $h = -16t^2 + 256t$ gives the height h of the bullet at any time t. Factor the right side of the equation and use this factored form to find the height of the bullet after **(a)** 6 sec, **(b)** 8 sec, **(c)** 10 sec, and **(d)** 16 sec. $h = -16t(t - 16)$; **(a)** 960 ft **(b)** 1024 ft   **(c)** 960 ft **(d)** 0 ft (the bullet has hit the ground)

58. When Burford was told to factor $12x^2 - 8x$ he gave $2 \cdot 2 \cdot 3 \cdot x \cdot x - 2 \cdot 2 \cdot 2 \cdot x$ for the answer. What is wrong with Burford's work? What is the correct answer?
Burford factored the terms of the polynomial but not the polynomial. The correct answer is $4x(3x - 2)$.

59. What is wrong with the following factoring problem?

$$4x^2y^3 - 2xy = 2xy(2xy^2)$$

The correct answer is $2xy(2xy^2 - 1)$; the -1 has been left off.

60. What is the GCF of the terms of the binomial $2x - 4$? What is the GCF of the terms of the binomial $3x + 6$? Find the product of $2x - 4$ and $3x + 6$ and give the GCF of the terms of this trinomial. What can you conclude?
2; 3; 6; the GCF of the product is the product of the GCFs.

FOR REVIEW

Use the FOIL method to multiply.

61. $(x + 2)(x + 3)$
$x^2 + 5x + 6$

62. $(x + 2)(x - 3)$
$x^2 - x - 6$

63. $(x - 2)(x + 3)$
$x^2 + x - 6$

64. $(x - 2)(x - 3)$
$x^2 - 5x + 6$

65. $(x - 2y)(x - 2y)$
$x^2 - 4xy + 4y^2$

66. $(x + 2y)(x + 2y)$
$x^2 + 4xy + 4y^2$

15.1 EXERCISES C

Factor by grouping. Do not collect like terms first.

1. $6x^2 + 21x - 10x - 35$
   $(3x - 5)(2x + 7)$

2. $10x^2 - 15xy - 2xy + 3y^2$
   $(5x - y)(2x - 3y)$

3. $a^2b^2c^2 - 5abc + 3abc - 15$
   $(abc + 3)(abc - 5)$

4. $6a^2b^2c - 15ab - 4abc^2 + 10c$
   [Answer: $(3ab - 2c)(2abc - 5)$]

15.2 FACTORING TRINOMIALS OF THE FORM $x^2 + bx + c$

STUDENT GUIDEPOSTS

1 Review of FOIL

2 Factoring $x^2 + bx + c$

3 Factoring $x^2 + bxy + cy^2$

1 REVIEW OF FOIL

To factor a trinomial into the product of two binomials, we need to recognize the pattern for multiplying binomials in reverse. Consider the following multiplication.

$$(x + 3)(x + 7) = x \cdot x + x \cdot 7 + 3 \cdot x + 3 \cdot 7$$
$$= x^2 + 7x + 3x + 21$$
$$= x^2 + 10x + 21$$

Notice that x^2 is the product of the first terms Ⓕ, 21 is the product of the last terms Ⓛ, and $10x$ is the sum of the product Ⓞ and Ⓘ.

If we are given the trinomial $x^2 + 10x + 21$ to factor, we need to reverse the multiplication steps above to find $(x + 3)(x + 7)$. That is, we must fill in the blanks in

$$x^2 + 10x + 21 = (x + \_\_)(x + \_\_).$$

The numbers in the blanks must have a product of 21 and a sum of 10.

$$x^2 + 10x + 21 = (x + \underline{3})(x + \underline{7})$$

2 FACTORING $x^2 + bx + c$

In general, to factor $x^2 + bx + c$ we look for a pair of integers whose product is c and whose sum is b.

$$x^2 + bx + c = (x + \_\_)(x + \_\_)$$

To Factor $x^2 + bx + c$

1. Write $x^2 + bx + c = (x + \underline{\quad})(x + \underline{\quad})$.
2. List all pairs of integers whose product is c.
3. Find the pair from this list whose sum is b (if there is one).
4. Fill in the blanks with this pair.

EXAMPLE 1   FACTORING TRINOMIALS

Factor $x^2 + 6x + 8$. $(b = 6$ and $c = 8)$

$$x^2 + 6x + 8 = (x + \underline{\quad})(x + \underline{\quad})$$

Factors of $c = 8$	Sum of factors
1, 8	$1 + 8 = 9$
2, 4	$2 + 4 = 6$
$-1, -8$	$-1 + (-8) = -9$
$-2, -4$	$-2 + (-4) = -6$

Since the product of 2 and 4 is 8, which equals c, and the sum of 2 and 4 is 6, which equals b, we write

$$x^2 + 6x + 8 = (x + \underline{2})(x + \underline{4}).$$

That is, the blanks are filled in with the shaded pair above. To check, we multiply using the FOIL method.

$$(x + 2)(x + 4) = x^2 + 4x + 2x + 8 = x^2 + 6x + 8$$

PRACTICE EXERCISE 1

Factor $x^2 + 9x + 18$.

Factors of $c = 18$	Sum of factors
1, 18	$1 + 18 = 19$
2, 9	$2 + 9 = 11$
3, 6	$3 + 6 = 9$

Answer: $(x + 3)(x + 6)$

EXAMPLE 2   FACTORING TRINOMIALS IN ONE VARIABLE

Factor each trinomial.

(a) $x^2 + 5x + 6$ $(b = 5$ and $c = 6)$

$$x^2 + 5x + 6 = (x + \underline{\quad})(x + \underline{\quad})$$

Factors of $c = 6$	Sum of factors
6, 1	$6 + 1 = 7$
2, 3	$2 + 3 = 5$
$-6, -1$	$-6 + (-1) = -7$
$-2, -3$	$-2 + (-3) = -5$

We fill in the blanks with the shaded pair obtaining

$$x^2 + 5x + 6 = (x + \underline{2})(x + \underline{3}).   2 \cdot 3 = 6 \text{ and } 2 + 3 = 5$$

Note in the table that since $c = 6 > 0$, the factors in each pair must have the same sign. Also, since $b = 5 > 0$, we know that the negative factors will not work.

PRACTICE EXERCISE 2

Factor each trinomial.

(a) $x^2 + 10x + 21$

(b) $x^2 - 10x + 21$

(b) $x^2 - 5x + 6$ ($b = -5$ and $c = 6$)

$$x^2 - 5x + 6 = (x + \underline{\quad})(x + \underline{\quad})$$

Factors of $c = 6$	Sum of factors
6, 1	$6 + 1 = 7$
2, 3	$2 + 3 = 5$
−6, −1	$-6 + (-1) = -7$
−2, −3	$-2 + (-3) = -5$

$$x^2 - 5x + 6 = (x - 2)(x - 3) \qquad (-2)(-3) = 6 \text{ and } -2 + (-3) = -5$$

With $c > 0$ the factors in each pair must have the same sign. Since $b = -5$, positive factors will not work.

(c) $x^2 + x - 6$ ($b = 1$ and $c = -6$)

$$x^2 + x - 6 = (x + \underline{\quad})(x + \underline{\quad})$$

Factors of $c = -6$	Sum of factors
6, −1	$6 + (-1) = 5$
−6, 1	$-6 + 1 = -5$
2, −3	$2 + (-3) = -1$
−2, 3	$-2 + 3 = 1$

$$x^2 + x - 6 = (x - 2)(x + 3) \qquad -2 \cdot 3 = -6 \text{ and } -2 + 3 = 1$$

Since $c = -6 < 0$, the factors of c must have opposite signs.

(d) $x^2 - x - 6$ ($b = -1$ and $c = -6$)

$$x^2 - x - 6 = (x + \underline{\quad})(x + \underline{\quad})$$

Factors of $c = -6$	Sum of factors
6, −1	$6 + (-1) = 5$
−6, 1	$-6 + 1 = -5$
2, −3	$2 + (-3) = -1$
−2, 3	$-2 + 3 = 1$

$$x^2 - x - 6 = (x + 2)(x - 3) \qquad 2 \cdot (-3) = -6 \text{ and } 2 + (-3) = -1$$

(c) $x^2 + 4x - 21$

(d) $x^2 - 4x - 21$

Answers: (a) $(x + 3)(x + 7)$
(b) $(x - 3)(x - 7)$
(c) $(x - 3)(x + 7)$
(d) $(x + 3)(x - 7)$

❸ FACTORING $x^2 + bxy + cy^2$

To factor trinomials in two variables such as $x^2 + bxy + cy^2$, we use the same procedures as above. But we now fill in the blanks in

$$x^2 + 5xy + 6y^2 = (x + \underline{\quad}y)(x + \underline{\quad}y).$$

The following table compares factoring trinomials in two variables with factoring similar trinomials in one variable.

One variable	Two variables
$x^2 + 5x + 6 = (x + 2)(x + 3)$	$x^2 + 5xy + 6y^2 = (x + 2y)(x + 3y)$
$x^2 - 5x + 6 = (x - 2)(x - 3)$	$x^2 - 5xy + 6y^2 = (x - 2y)(x - 3y)$
$x^2 + x - 6 = (x - 2)(x + 3)$	$x^2 + xy - 6y^2 = (x - 2y)(x + 3y)$
$x^2 - x - 6 = (x + 2)(x - 3)$	$x^2 - xy - 6y^2 = (x + 2y)(x - 3y)$

EXAMPLE 3 FACTORING TRINOMIALS IN TWO VARIABLES

Factor each trinomial.

(a) $x^2 + 7xy + 10y^2$ ($b = 7$ and $c = 10$)

$$x^2 + 7xy + 10y^2 = (x + \underline{\quad}y)(x + \underline{\quad}y)$$

Factors of $c = 10$	Sum of factors
10, 1	$10 + 1 = 11$
5, 2	$5 + 2 = 7$
$-10, -1$	$-10 + (-1) = -11$
$-5, -2$	$-5 + (-2) = -7$

$x^2 + 7xy + 10y^2 = (x + 5y)(x + 2y)$ $5 \cdot 2 = 10$ and $5 + 2 = 7$

(b) $x^2 - xy - 12y^2$ ($b = -1$ and $c = -12$)

$$x^2 - xy - 12y^2 = (x + \underline{\quad}y)(x + \underline{\quad}y)$$

Factors of $c = -12$	Sum of factors
12, -1	$12 + (-1) = 11$
-12, 1	$-12 + 1 = -11$
6, -2	$6 + (-2) = 4$
-6, 2	$-6 + 2 = -4$
4, -3	$4 + (-3) = 1$
-4, 3	$-4 + 3 = -1$

$x^2 - xy - 12y^2 = (x - 4y)(x + 3y)$ $(-4) \cdot 3 = -12$ and $-4 + 3 = -1$

With practice you will be able to find the factors of c that are more likely to sum to b without constructing the complete table.

Some trinomials are not in the form $x^2 + bxy + cy^2$ but can be put into this form by removing a common factor.

EXAMPLE 4 TRINOMIALS WITH A COMMON FACTOR

Factor $3x^2 - 12xy + 12y^2$.

$$3x^2 - 12xy + 12y^2 = 3 \cdot x^2 - 3 \cdot 4xy + 3 \cdot 4y^2$$
$$= 3(x^2 - 4xy + 4y^2)$$

PRACTICE EXERCISE 3

Factor each trinomial.

(a) $x^2 + 7xy + 12y^2$

(b) $x^2 + xy - 12y^2$

Answers: (a) $(x + 4y)(x + 3y)$
(b) $(x + 4y)(x - 3y)$

PRACTICE EXERCISE 4

Factor $7x^2 + 21xy - 70y^2$.

We can now factor $x^2 - 4xy + 4y^2$ where $b = -4$ and $c = 4$.

$$x^2 - 4xy + 4y^2 = (x + \_\_y)(x + \_\_y)$$

Factors of $c = 4$	Sum of factors
4, 1	$4 + 1 = 5$
2, 2	$2 + 2 = 4$
$-4, -1$	$-4 + (-1) = -5$
$-2, -2$	$-2 + (-2) = -4$

$$x^2 - 4xy + 4y^2 = (x - 2y)(x - 2y) \qquad (-2)(-2) = 4 \text{ and } -2 + (-2) = -4$$

Thus, $3x^2 - 12xy + 12y^2 = 3(x - 2y)(x - 2y)$.

Answer: $7(x + 5y)(x - 2y)$

Always include the common factor as part of the answer. In Example 4 we included the factor 3 in the final product.

Not all trinomials can be factored. One example is $x^2 + 5x + 2$. The only factors of $c = 2$ are 2 and 1, but $2 + 1$ is 3 and not $b = 5$.

Factoring, just like any other skill, is learned by practice. The more time spent and the more problems worked, the more proficient we become.

EXAMPLE 5 APPLICATION OF FACTORING	**PRACTICE EXERCISE 5**

A small company manufactures wood-burning stoves. The total cost c of producing n stoves is given by the equation $c = -200n^2 + 1200n + 1400$. Factor the right side of this equation and use this factored form to find the cost of producing 4 stoves.

Factoring gives the following.

$$c = -200n^2 + 1200n + 1400$$
$$= -200(n^2 - 6n - 7) \qquad \text{The GCF is } -200$$
$$= -200(n - 7)(n + 1)$$

Substitute 4 for n to find c.

$$c = -200(4 - 7)(4 + 1)$$
$$= -200(-3)(5) = 3000$$

It would cost $3000 to produce 4 stoves.

If the profit on stoves is $p = -10n^2 + 200n - 960$, factor the right side of the equation and find the profit on 10 stoves.

Answers: $p = -10(n - 12)(n - 8)$; $40

15.2 EXERCISES A

Factor and check by multiplying.

1. $x^2 + 4x + 3$

$(x + 1)(x + 3)$

2. $x^2 + 2x - 3$

$(x - 1)(x + 3)$

3. $x^2 - 2x - 3$

$(x + 1)(x - 3)$

4. $x^2 - 4x + 3$
$(x - 1)(x - 3)$

5. $u^2 - 12u + 35$
$(u - 5)(u - 7)$

6. $u^2 - 2u - 35$
$(u + 5)(u - 7)$

7. $u^2 + 12u + 35$
$(u + 5)(u + 7)$

8. $u^2 + 2u - 35$
$(u - 5)(u + 7)$

9. $y^2 + 10y + 21$
$(y + 3)(y + 7)$

10 $y^2 + 5y - 24$
$(y - 3)(y + 8)$

11. $x^2 - 12x + 27$
$(x - 3)(x - 9)$

12. $x^2 + 4x - 45$
$(x - 5)(x + 9)$

13. $y^2 - y - 56$
$(y + 7)(y - 8)$

14 $x^2 - 2x - 63$
$(x + 7)(x - 9)$

15. $x^2 - 2x - 120$
$(x + 10)(x - 12)$

16. $x^2 + 4x - 77$
$(x - 7)(x + 11)$

17. $x^2 + 4xy + 3y^2$
$(x + y)(x + 3y)$

18. $x^2 + 2xy - 3y^2$
$(x - y)(x + 3y)$

19. $x^2 - 2xy - 3y^2$
$(x + y)(x - 3y)$

20 $u^2 - 9uv + 20v^2$
$(u - 4v)(u - 5v)$

21. $x^2 - 4xy + 3y^2$
$(x - y)(x - 3y)$

22. $u^2 + uv - 20v^2$
$(u - 4v)(u + 5v)$

23. $u^2 + 9uv + 20v^2$
$(u + 4v)(u + 5v)$

24. $u^2 - uv - 20v^2$
$(u + 4v)(u - 5v)$

25. $x^2 + 13xy - 30y^2$
$(x - 2y)(x + 15y)$

26. $x^2 + 11xy + 24y^2$
$(x + 3y)(x + 8y)$

27. $u^2 - 8uv + 15v^2$
$(u - 3v)(u - 5v)$

28. $u^2 - 3uv - 40v^2$
$(u - 8v)(u + 5v)$

29. $x^2 - 10xy + 24y^2$
$(x - 4y)(x - 6y)$

30. $x^2 + 11xy + 30y^2$
$(x + 5y)(x + 6y)$

31. $u^2 + 12uv + 36v^2$
$(u + 6v)(u + 6v)$

32. $u^2 - uv - 42v^2$
$(u + 6v)(u - 7v)$

33. $x^2 + xy - 90y^2$
$(x - 9y)(x + 10y)$

34. $x^2 - 4xy - 32y^2$
$(x + 4y)(x - 8y)$

35 $u^2 - 22uv + 121v^2$
$(u - 11v)(u - 11v)$

36. $u^2 + 18uv + 77v^2$
$(u + 7v)(u + 11v)$

37. A child standing on a bridge 48 ft above the surface of the water throws a rock upward with a velocity of 32 ft per second. The equation that gives the height h in feet of the rock above the water in terms of time t in seconds is $h = -16t^2 + 32t + 48$. Factor the right side of this equation and use this factored form to find the height of the rock after **(a)** 1 sec, **(b)** 2 sec, and **(c)** 3 sec.

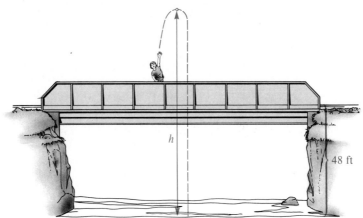

$h = -16(t - 3)(t + 1)$; (a) **64 ft** (b) **48 ft** (c) **0 ft**

38. What is wrong with the work in the following factoring problem?

$$2x^2 + 10x - 48 = 2(x^2 + 5x - 24)$$
$$= (x - 3)(x + 8)$$

The GCF 2 has been left off in the answer.

FOR REVIEW

Find the GCF and factor.

39. $35x - 70$
$35(x - 2)$

40. $88y^3 - 33y^2$
$11y^2(8y - 3)$

41. $9a^3 + 24a^2 - 15a$
$3a(3a^2 + 8a - 5)$

42. $6x^3y^2 - 12x^2y^3$
$6x^2y^2(x - 2y)$

43. $28a^4b^4 - 14a^3b^3 - 21a^2b^2$
$7a^2b^2(4a^2b^2 - 2ab - 3)$

44. $2a(x + y) - 3b(x + y)$
$(x + y)(2a - 3b)$

Factor by grouping.

45. $x^3 - 6x^2 + 5x - 30$
$(x - 6)(x^2 + 5)$

46. $5ax^2 + 2bx^2 + 5ay^2 + 2by^2$
$(5a + 2b)(x^2 + y^2)$

The following review material from Chapter 14 will help you prepare for the next section. Use the FOIL method to multiply.

47. $(2x + 1)(x + 3)$
$2x^2 + 7x + 3$

48. $(3x + 2)(x + 1)$
$3x^2 + 5x + 2$

49. $(3x - 5)(2x - 1)$
$6x^2 - 13x + 5$

50. $(5x - 2)(2x + 1)$
$10x^2 + x - 2$

51. $(2x + 5)(x - 2)$
$2x^2 + x - 10$

52. $(2x + 1)(x - 3)$
$2x^2 - 5x - 3$

ANSWERS: 1. $(x + 1)(x + 3)$ 2. $(x - 1)(x + 3)$ 3. $(x + 1)(x - 3)$ 4. $(x - 1)(x - 3)$ 5. $(u - 5)(u - 7)$
6. $(u + 5)(u - 7)$ 7. $(u + 5)(u + 7)$ 8. $(u - 5)(u + 7)$ 9. $(y + 3)(y + 7)$ 10. $(y - 3)(y + 8)$ 11. $(x - 3)(x - 9)$
12. $(x - 5)(x + 9)$ 13. $(y + 7)(y - 8)$ 14. $(x + 7)(x - 9)$ 15. $(x + 10)(x - 12)$ 16. $(x - 7)(x + 11)$
17. $(x + y)(x + 3y)$ 18. $(x - y)(x + 3y)$ 19. $(x + y)(x - 3y)$ 20. $(u - 4v)(u - 5v)$ 21. $(x - y)(x - 3y)$
22. $(u - 4v)(u + 5v)$ 23. $(u + 4v)(u + 5v)$ 24. $(u + 4v)(u - 5v)$ 25. $(x - 2y)(x + 15y)$ 26. $(x + 3y)(x + 8y)$
27. $(u - 3v)(u - 5v)$ 28. $(u - 8v)(u + 5v)$ 29. $(x - 4y)(x - 6y)$ 30. $(x + 5y)(x + 6y)$ 31. $(u + 6v)(u + 6v)$
32. $(u + 6v)(u - 7v)$ 33. $(x - 9y)(x + 10y)$ 34. $(x + 4y)(x - 8y)$ 35. $(u - 11v)(u - 11v)$ 36. $(u + 7v)(u + 11v)$
37. $h = -16(t - 3)(t + 1)$; (a) 64 ft   (b) 48 ft   (c) 0 ft (the rock has hit the water)   38. The GCF 2 has been left off
in the answer.   39. $35(x - 2)$ 40. $11y^2(8y - 3)$ 41. $3a(3a^2 + 8a - 5)$ 42. $6x^2y^2(x - 2y)$ 43. $7a^2b^2(4a^2b^2 - 2ab - 3)$
44. $(x + y)(2a - 3b)$ 45. $(x - 6)(x^2 + 5)$ 46. $(5a + 2b)(x^2 + y^2)$ 47. $2x^2 + 7x + 3$ 48. $3x^2 + 5x + 2$
49. $6x^2 - 13x + 5$ 50. $10x^2 + x - 2$ 51. $2x^2 + x - 10$ 52. $2x^2 - 5x - 3$

15.2   EXERCISES B

Factor and check by multiplying.

1. $x^2 + 6x + 5$
$(x + 1)(x + 5)$

2. $x^2 + 4x - 5$
$(x - 1)(x + 5)$

3. $x^2 - 4x - 5$
$(x + 1)(x - 5)$

4. $x^2 - 6x + 5$
$(x - 1)(x - 5)$

5. $u^2 - 8u + 15$
$(u - 3)(u - 5)$

6. $u^2 - 2u - 15$
$(u + 3)(u - 5)$

7. $u^2 + 8u + 15$
$(u + 3)(u + 5)$

8. $u^2 + 2u - 15$
$(u - 3)(u + 5)$

9. $y^2 - 10y + 21$
$(y - 3)(y - 7)$

10. $x^2 - 5x - 24$
$(x + 3)(x - 8)$

11. $u^2 + 12u + 27$
$(u + 3)(u + 9)$

12. $y^2 + y - 20$
$(y - 4)(y + 5)$

13. $x^2 - 4x - 32$
$(x + 4)(x - 8)$

14. $u^2 - 2u - 48$
$(u + 6)(u - 8)$

15. $y^2 + 16y + 60$
$(y + 6)(y + 10)$

16. $x^2 + 8x - 33$
$(x - 3)(x + 11)$

17. $x^2 + 6xy + 5y^2$
$(x + y)(x + 5y)$

18. $x^2 + 4xy - 5y^2$
$(x - y)(x + 5y)$

19. $x^2 - 4xy - 5y^2$
$(x + y)(x - 5y)$

20. $x^2 - 6xy + 5y^2$
$(x - y)(x - 5y)$

21. $u^2 - 9uv + 18v^2$
$(u - 3v)(u - 6v)$

22. $u^2 + 3uv - 18v^2$
$(u - 3v)(u + 6v)$

23. $u^2 + 9uv + 18v^2$
$(u + 3v)(u + 6v)$

24. $u^2 - 3uv - 18v^2$
$(u + 3v)(u - 6v)$

25. $x^2 + 6xy - 27y^2$
$(x - 3y)(x + 9y)$

26. $x^2 + 12xy + 20y^2$
$(x + 2y)(x + 10y)$

27. $u^2 - 11uv + 28v^2$
$(u - 4v)(u - 7v)$

28. $u^2 + uv - 30v^2$
$(u - 5v)(u + 6v)$

29. $x^2 - 13xy + 40y^2$
$(x - 5y)(x - 8y)$

30. $x^2 + 16xy + 63y^2$
$(x + 7y)(x + 9y)$

31. $u^2 + 8uv + 16v^2$
$(u + 4v)(u + 4v)$

32. $u^2 - 3uv - 40v^2$
$(u + 5v)(u - 8v)$

33. $x^2 + 6xy - 72y^2$
$(x - 6y)(x + 12y)$

34. $x^2 - 5xy - 50y^2$
$(x + 5y)(x - 10y)$

35. $u^2 - 26uv + 169v^2$
$(u - 13v)(u - 13v)$

36. $u^2 + 19uv + 88v^2$
$(u + 8v)(u + 11v)$

37. A company can produce n items at a total cost of c dollars where c is given by $c = -100n^2 + 600n + 1600$. Factor
the right side of this equation and use this factored form to find the cost of producing   **(a)** 0 items,   **(b)** 2 items,
(c) 3 items, and   **(d)** 6 items.
$c = -100(n + 2)(n - 8)$; (a) \$1600 (b) \$2400 (c) \$2500 (d) \$1600

38. What is wrong with the work in the following factoring problem?

$$-4x^2 + 12x + 40 = -4(x^2 + 3x - 10)$$
$$= -4(x + 5)(x - 2)$$

When the GCF -4 was removed, the polynomial should have been $x^2 - 3x - 10$. The correct factorization is
$-4(x - 5)(x + 2)$.

FOR REVIEW

Find the GCF and factor.

39. $18x + 36$
$18(x + 2)$

40. $35y^4 - 21y^2$
$7y^2(5y^2 - 3)$

41. $16a^5 - 24a^4 - 32a^3$
$8a^3(2a^2 - 3a - 4)$

42. $25x^4y^5 + 45x^3y^6$
$5x^3y^5(5x + 9y)$

43. $8a^5b^5 + 20a^3b^4 - 32a^2b^5$
$4a^2b^4(2a^3b + 5a - 8b)$

44. $6u(x^2 + y^2) + 7(x^2 + y^2)$
$(6u + 7)(x^2 + y^2)$

Factor by grouping.

45. $x^3 + 3x^2 - 7x - 21$ $(x + 3)(x^2 - 7)$

46. $2a^2x - 2a^2y + b^2x - b^2y$ $(x - y)(2a^2 + b^2)$

The following review material from Chapter 14 will help you prepare for the next section. Use the FOIL method to multiply.

47. $(3x + 1)(x + 7)$
$3x^2 + 22x + 7$

48. $(5x + 2)(x + 3)$
$5x^2 + 17x + 6$

49. $(3x + 5)(2x - 1)$
$6x^2 + 7x - 5$

50. $(2x - 1)(x + 10)$
$2x^2 + 19x - 10$

51. $(5x - 3)(x + 6)$
$5x^2 + 27x - 18$

52. $(2x - 5)(2x + 7)$
$4x^2 + 4x - 35$

15.2 EXERCISES C

Factor.

1. $x^2 + \dfrac{1}{3}x - \dfrac{2}{9}$ $\left(x + \dfrac{2}{3}\right)\left(x - \dfrac{1}{3}\right)$

2. $x^2 - 0.04x + 0.0003$ $(x - 0.01)(x - 0.03)$

3. $(x + 2)^2 - 10(x + 2) + 16$
[Answer: $x(x - 6)$]

4. $x^4 + 5x^2 + 6$ $(x^2 + 3)(x^2 + 2)$

15.3 FACTORING TRINOMIALS OF THE FORM $ax^2 + bx + c$

STUDENT GUIDEPOSTS

1 Factoring $ax^2 + bx + c$

2 Factoring $ax^2 + bxy + cy^2$

3 Factoring $ax^2 + bx + c$ by Grouping

❶ FACTORING $ax^2 + bx + c$

In Section 15.2 we concentrated on factoring trinomials for which the coefficient of the squared term is 1. Now we consider trinomials of the form $ax^2 + bx + c$, with $a \neq 1$. As before, it helps to look at the pattern in multiplication.

$$(5x + 2)(2x + 3) = 5x \cdot 2x + 5x \cdot 3 + 2 \cdot 2x + 2 \cdot 3$$
$$= 10x^2 + 15x + 4x + 6$$
$$= 10x^2 + \quad 19x \quad + 6$$

The term $10x^2$ is the product of the first terms Ⓕ, the 6 is the product of the last terms Ⓛ, and $19x$ is the sum of the products Ⓞ and Ⓘ. With $a \neq 1$ we can use the following trial and error method to factor. (Another method is presented later.)

To Factor $ax^2 + bx + c$

1. Write $ax^2 + bx + c = (\underline{}x + \underline{})(\underline{}x + \underline{})$.
2. List all pairs of integers whose product is a and try each of these in the first blanks in each binomial factor.
3. List all pairs of integers whose product is c and try each of these in the second blanks.
4. Use trial and error to determine which pair (if one exists) gives ⓪ + ① = b.

EXAMPLE 1 FACTORING $ax^2 + bx + c$

Factor $2x^2 + 7x + 3$. $(a = 2, b = 7, c = 3)$

$$2x^2 + 7x + 3 = (\underline{}x + \underline{})(\underline{}x + \underline{})$$

Factors of 3 (top), Factors of 2 (bottom)

Since all terms of the trinomial are positive, we need only list the positive factors of a and c. We list the factors of c in both orders as a reminder to try all possibilities.

Factors of $a = 2$	Factors of $c = 3$
2, 1	3, 1
	1, 3

$$2x^2 + 7x + 3 = (\underline{}x + \underline{})(\underline{}x + \underline{})$$
$$= (2x + \underline{})(x + \underline{}) \quad \text{The only factors of } a \text{ are 2 and 1}$$
$$\stackrel{?}{=} (2x + 3)(x + 1) \quad \text{Does not work because } 2x + 3x = 5x \neq 7x$$
$$\stackrel{?}{=} (2x + 1)(x + 3) \quad \text{This works because } 6x + x = 7x$$
$$2x^2 + 7x + 3 = (2x + 1)(x + 3)$$

To check we multiply.

$$(2x + 1)(x + 3) = 2x^2 + 6x + x + 3 = 2x^2 + 7x + 3$$

PRACTICE EXERCISE 1

Factor $3x^2 + 5x + 2$.

Factors of $a = 3$	Factors of $c = 2$
3, 1	2, 1
	1, 2

Factors of 2

$$(\underline{}x + \underline{})(\underline{}x + \underline{})$$

Factors of 3

Answer: $(3x + 2)(x + 1)$

In Example 1 we listed only the factors of c (1 and 3) in both orders and not the factors of a (1 and 2). It is not necessary to list both pairs in both orders. For example, $(2x + 1)(x + 3)$ and $(x + 3)(2x + 1)$ are two factorizations that are the same by the commutative law of multiplication.

EXAMPLE 2 FACTORING $ax^2 + bx + c$

Factor $6x^2 - 13x + 5$. $(a = 6, b = -13, c = 5)$

$$6x^2 - 13x + 5 = (\underline{}x + \underline{})(\underline{}x + \underline{})$$

Factors of 5 (top), Factors of 6 (bottom)

PRACTICE EXERCISE 2

Factor $10x^2 + x - 2$.

Here the factors of $c = 5 > 0$ must both be negative to obtain the term $-13x$. As before we list them in both orders.

Factors of $a = 6$	Factors of $c = 5$	
6, 1	−5, −1	Try 6, 1 with both −5, −1 and −1, −5
3, 2	−1, −5	Try 3, 2 with both −5, −1 and −1, −5

$6x^2 - 13x + 5 = (\underline{}x + \underline{})(\underline{}x + \underline{})$

$\overset{?}{=} (6x - 5)(x - 1)$ Does not work because
$-6x - 5x = -11x \neq -13x$

$\overset{?}{=} (6x - 1)(x - 5)$ Does not work because
$-30x - x = -31x \neq -13x$

$\overset{?}{=} (3x - 5)(2x - 1)$ This works because
$-3x - 10x = -13x$

$6x^2 - 13x + 5 = (3x - 5)(2x - 1)$

To check we multiply.

$(3x - 5)(2x - 1) = 6x^2 - 3x - 10x + 5 = 6x^2 - 13x + 5$ Answer: $(5x - 2)(2x + 1)$

EXAMPLE 3 FACTORING $ax^2 + bx + c$

Factor $8y^2 - 10y - 7$. ($a = 8$, $b = -10$, $c = -7$)
The factors of $c = -7$ will have opposite signs.

Factors of $a = 8$	Factors of $c = -7$
8, 1	7, −1
4, 2	−7, 1
	1, −7
	−1, 7

With this many cases to test, we try to minimize the number of trials. If 8 and 1 are used, the middle term will be either too big, because $8 \cdot 7y - 1 \cdot 1y = 55y$, or too small, because $8 \cdot 1y - 7 \cdot 1y = y$. Thus, we try 4 and 2 as factors of 8.

$8y^2 - 10y - 7 = (\underline{}y + \underline{})(\underline{}y + \underline{})$

$= (4y + \underline{})(2y + \underline{})$ Try 4, 2

$\overset{?}{=} (4y + 7)(2y - 1)$ Does not work

$\overset{?}{=} (4y - 7)(2y + 1)$ This works

$8y^2 - 10y - 7 = (4y - 7)(2y + 1)$

Check this by multiplying.

PRACTICE EXERCISE 3

Factor $8y^2 + 10y - 7$.

Answer: $(4y + 7)(2y - 1)$

EXAMPLE 4 FACTORING OUT A COMMON FACTOR

Factor $-4x^2 - 2x + 20$.
Unlike Examples 1, 2, and 3 this trinomial has a common factor. First we factor out the common factor including -1 to make a positive. If a is positive we only have to consider positive factors of a.

$$-4x^2 - 2x + 20 = -2(2x^2 + x - 10)$$

PRACTICE EXERCISE 4

Factor $-6x^2 + 15x + 9$.

Now we factor $2x^2 + x - 10$ where $a = 2$, $b = 1$, and $c = -10$.

Factors of $a = 2$	Factors of $c = -10$
2, 1	10, -1 and -10, 1
	-1, 10 and 1, -10
	5, -2 and -5, 2
	-2, 5 and 2, -5

$$-4x^2 - 2x + 20 = -2(2x^2 + x - 10)$$
$$= -2(2x + \underline{\quad})(x + \underline{\quad})$$
$$\overset{?}{=} -2(2x + 10)(x - 1) \qquad \text{Factors of 10 give too large}$$
$$\text{a middle term}$$
$$\overset{?}{=} -2(2x + 5)(x - 2) \qquad \text{This works}$$
$$-4x^2 - 2x + 20 = -2(2x + 5)(x - 2)$$

Answer: $-3(2x + 1)(x - 3)$

CAUTION

Do not forget to include the common factor in the answer. The common factor -2 was included in Example 4.

---

② FACTORING $ax^2 + bxy + cy^2$

Trinomials of the form $ax^2 + bxy + cy^2$ can be factored by filling in the blanks as indicated below.

$$ax^2 + bxy + cy^2 = (\underline{\quad}x + \underline{\quad}y)(\underline{\quad}x + \underline{\quad}y)$$

EXAMPLE 5 FACTORING $ax^2 + bxy + cy^2$	PRACTICE EXERCISE 5

Factor $3x^2 - 14xy + 8y^2$. ($a = 3$, $b = -14$, $c = 8$)

Both factors of $c = 8$ must be negative since $b = -14$.

Factor $3a^2 + ab - 10b^2$.

Factors of $a = 3$	Factors of $c = 8$
3, 1	-8, -1
	-1, -8
	-4, -2
	-2, -4

$$3x^2 - 14xy + 8y^2 = (3x + \underline{\quad}y)(x + \underline{\quad}y)$$
$$\overset{?}{=} (3x - 8y)(x - y) \qquad \text{Does not work}$$
$$\overset{?}{=} (3x - y)(x - 8y) \qquad \text{Does not work}$$
$$\overset{?}{=} (3x - 4y)(x - 2y) \qquad \text{Does not work}$$
$$\overset{?}{=} (3x - 2y)(x - 4y) \qquad \text{This works}$$
$$3x^2 - 14xy + 8y^2 = (3x - 2y)(x - 4y)$$

Answer: $(3a - 5b)(a + 2b)$

| EXAMPLE 6 FACTORING $ax^2 + bxy + cy^2$ | PRACTICE EXERCISE 6 |

Factor $10u^2 + 7uv - 3v^2$.   ($a = 10$, $b = 7$, $c = -3$)

Factor $10x^2 - 13xy + 3y^2$.

Factors of $a = 10$	Factors of $c = -3$
10, 1	3, −1
5, 2	−3, 1
	1, −3
	−1, 3

$10u^2 + 7uv - 3v^2 = (\_\_u + \_\_v)(\_\_u + \_\_v)$

$\overset{?}{=} (10u + 3v)(u - v)$ Does not work since
$-10uv + 3uv = -7uv$

$\overset{?}{=} (10u - 3v)(u + v)$ We know this works from the first trial

$10u^2 + 7uv - 3v^2 = (10u - 3v)(u + v)$

Answer: $(10x - 3y)(x - y)$

❸ FACTORING $ax^2 + bx + c$ BY GROUPING

There is another method for factoring trinomials. Consider the following procedure.

$$2x^2 + 13x + 15 = 2x^2 + 10x + 3x + 15 \qquad 13x = 10x + 3x$$
$$= (2x^2 + 10x) + (3x + 15) \qquad \text{Group terms}$$
$$= 2x(x + 5) + 3(x + 5) \qquad \text{Factor the groups}$$
$$= (2x + 3)(x + 5) \qquad \text{Factor out the common factor } x + 5$$

We have factored $2x^2 + 13x + 15$ by grouping the appropriate terms. But how did we decide to write $13x = 10x + 3x$? Notice that 10 and 3 are factors of $2 \cdot 15 = 30$. That is, 10 and 3 are factors of the product ac in $ax^2 + bx + c$.

> ### To Factor $ax^2 + bx + c$ by Grouping
>
> 1. Find the product ac.
> 2. List the factors of ac until a pair is found with sum b.
> 3. Write bx as a sum using these factors as coefficients of x.
> 4. Factor the result by grouping.

| EXAMPLE 7 FACTORING BY GROUPING | PRACTICE EXERCISE 7 |

Factor $3x^2 - 10x + 7$ by grouping. The product $ac = 3 \cdot 7 = 21$.

Factor $3x^2 + 10x + 7$ by grouping.

Factors of $ac = 21$	Sum of factors
21, 1	$21 + 1 = 22$
−21, −1	$-21 + (-1) = -22$
7, 3	$7 + 3 = 10$
−7, −3	$-7 + (-3) = -10$

The factors -7 and -3 will work so we write $-10x = -7x - 3x$.

$3x^2 - 10x + 7 = 3x^2 - 7x - 3x + 7$

$\qquad\qquad = x(3x - 7) - 1 \cdot (3x - 7)$ Factor out x and -1

$\qquad\qquad = (x - 1)(3x - 7)$ The common factor is $3x - 7$ Answer: $(x + 1)(3x + 7)$

| **EXAMPLE 8** FACTORING $ax^2 + bxy + cy^2$ BY GROUPING | **PRACTICE EXERCISE 8** |

Factor $5x^2 + 6xy - 8y^2$ by grouping.

   For two variables the procedure is the same. $ac = 5(-8) = -40$. We try factors that seem most likely to have a sum of 6.

Factors of $ac = -40$	Sum of factors
$8, -5$	$8 + (-5) = 3$
$-8, 5$	$-8 + 5 = -3$
$-10, 4$	$-10 + 4 = -6$
$10, -4$	$10 + (-4) = 6$

Write $6xy$ as $10xy - 4xy$.

$5x^2 + 6xy - 8y^2 = 5x^2 + 10xy - 4xy - 8y^2$

$\qquad\qquad\qquad = 5x(x + 2y) - 4y(x + 2y)$ Factor groups

$\qquad\qquad\qquad = (5x - 4y)(x + 2y)$ Common factor is $x + 2y$

Factor $5x^2 - 6xy - 8y^2$ by grouping.

Answer: $(5x + 4y)(x - 2y)$

15.3   EXERCISES A

Factor and check by multiplying.

1. $2x^2 + 7x + 5$
$(2x + 5)(x + 1)$

2. $2x^2 - 5x + 3$
$(2x - 3)(x - 1)$

3. $2u^2 - 5u - 3$
$(2u + 1)(u - 3)$

4. $2u^2 + 13u - 7$
$(2u - 1)(u + 7)$

5. $2y^2 + 13y - 24$
$(2y - 3)(y + 8)$

6. $2y^2 + 19y + 24$
$(2y + 3)(y + 8)$

7. $2y^2 - 19y + 24$
$(2y - 3)(y - 8)$

8. $2y^2 - 13y - 24$
$(2y + 3)(y - 8)$

9. $3z^2 - 14z - 5$
$(3z + 1)(z - 5)$

10 $6z^2 - 13z - 28$
$(2z - 7)(3z + 4)$

11. $5x^2 - 33x + 40$
$(5x - 8)(x - 5)$

12. $-6x^2 - 39x - 54$
[*Hint:* Don't forget the common factor.]
$(-3)(2x + 9)(x + 2)$

13. $7u^2 - 14u + 7$
$7(u-1)(u-1)$

14. $16u^2 - 16u + 4$
$4(2u-1)(2u-1)$

15. $y^2 - 9$ [*Hint:* Note that $b = 0$, $a = 1$, and $c = -9$.]
$(y+3)(y-3)$

16. $3y^3 + 11y^2 + 10y$
$y(3y+5)(y+2)$

17 $-45x^2 + 150x - 125$
$(-5)(3x-5)(3x-5)$

18. $6x^2 - 3x + 21$
$3(2x^2 - x + 7)$

19. $6u^2 - 23u + 20$
$(2u-5)(3u-4)$

20. $5u^2 + 7u - 24$
$(5u-8)(u+3)$

21. $2x^2 - 5xy + 3y^2$
$(2x-3y)(x-y)$

22 $2x^2 + 7xy + 5y^2$
$(2x+5y)(x+y)$

23. $2u^2 - 5uv - 3v^2$
$(2u+v)(u-3v)$

24. $2u^2 + 13uv - 7v^2$
$(2u-v)(u+7v)$

25. $3x^2 - 10xy + 3y^2$
$(3x-y)(x-3y)$

26. $3x^2 + 13xy + 14y^2$
$(3x+7y)(x+2y)$

27. $5u^2 + 21uv + 4v^2$
$(5u+v)(u+4v)$

28. $6u^2 + uv - v^2$
$(3u-v)(2u+v)$

29. $-3x^2 + 18xy - 24y^2$
$(-3)(x-4y)(x-2y)$

30. $x^2 + xy + y^2$
cannot be factored

31 $4u^2 - v^2$
$(2u+v)(2u-v)$

32. $3u^3v + 14u^2v^2 - 5uv^3$
$uv(3u-v)(u+5v)$

33. $6x^2 - 35xy - 6y^2$
$(6x+y)(x-6y)$

34. $6x^2 + 29xy + 30y^2$
$(3x+10y)(2x+3y)$

35. $14x^2 + 30xy + 16y^2$
$2(7x+8y)(x+y)$

36. $3x^3y - 2x^2y^2 - xy^3$
$xy(3x+y)(x-y)$

FOR REVIEW

Factor.

37. $x^2 + 17x + 72$
$(x+8)(x+9)$

38. $x^2 - 17x + 72$
$(x-8)(x-9)$

39. $y^2 - 2y - 80$
$(y+8)(y-10)$

40. $x^2 + 6xy - 27y^2$
$(x-3y)(x+9y)$

41. $x^2 - 18xy + 81y^2$
$(x-9y)(x-9y)$

42. $-3x^2 + 9x + 120$
$-3(x+5)(x-8)$

Exercises 43–48 review material from Section 14.6 to help you prepare for the next section. Find the following special products.

43. $(x + 3)(x - 3)$
$x^2 - 9$

44. $(x + 3)^2$
$x^2 + 6x + 9$

45. $(x - 3)^2$
$x^2 - 6x + 9$

46. $(4u + 5)^2$
$16u^2 + 40u + 25$

47. $(4u - 5)^2$
$16u^2 - 40u + 25$

48. $(4u + 5)(4u - 5)$
$16u^2 - 25$

ANSWERS: 1. $(2x + 5)(x + 1)$ 2. $(2x - 3)(x - 1)$ 3. $(2u + 1)(u - 3)$ 4. $(2u - 1)(u + 7)$ 5. $(2y - 3)(y + 8)$
6. $(2y + 3)(y + 8)$ 7. $(2y - 3)(y - 8)$ 8. $(2y + 3)(y - 8)$ 9. $(3z + 1)(z - 5)$ 10. $(2z - 7)(3z + 4)$ 11. $(5x - 8)(x - 5)$
12. $(-3)(2x + 9)(x + 2)$ 13. $7(u - 1)(u - 1)$ 14. $4(2u - 1)(2u - 1)$ 15. $(y + 3)(y - 3)$ 16. $y(3y + 5)(y + 2)$
17. $(-5)(3x - 5)(3x - 5)$ 18. $3(2x^2 - x + 7)$; the trinomial cannot be factored   19. $(2u - 5)(3u - 4)$
20. $(5u - 8)(u + 3)$ 21. $(2x - 3y)(x - y)$ 22. $(2x + 5y)(x + y)$ 23. $(2u + v)(u - 3v)$ 24. $(2u - v)(u + 7v)$
25. $(3x - y)(x - 3y)$ 26. $(3x + 7y)(x + 2y)$ 27. $(5u + v)(u + 4v)$ 28. $(3u - v)(2u + v)$ 29. $(-3)(x - 4y)(x - 2y)$
30. cannot be factored   31. $(2u + v)(2u - v)$ 32. $uv(3u - v)(u + 5v)$ 33. $(6x + y)(x - 6y)$ 34. $(3x + 10y)(2x + 3y)$
35. $2(7x + 8y)(x + y)$ 36. $xy(3x + y)(x - y)$ 37. $(x + 8)(x + 9)$ 38. $(x - 8)(x - 9)$ 39. $(y + 8)(y - 10)$
40. $(x - 3y)(x + 9y)$ 41. $(x - 9y)(x - 9y)$ 42. $-3(x + 5)(x - 8)$ 43. $x^2 - 9$ 44. $x^2 + 6x + 9$ 45. $x^2 - 6x + 9$
46. $16u^2 + 40u + 25$ 47. $16u^2 - 40u + 25$ 48. $16u^2 - 25$

15.3   EXERCISES B

Factor and check by multiplying.

1. $2x^2 + 5x + 3$
$(2x + 3)(x + 1)$

2. $2x^2 - 7x + 5$
$(2x - 5)(x - 1)$

3. $2u^2 + 3u - 14$
$(2u + 7)(u - 2)$

4. $2u^2 - 9u - 5$
$(2u + 1)(u - 5)$

5. $2u^2 + 9u - 5$
$(2u - 1)(u + 5)$

6. $2y^2 + 17y + 30$
$(2y + 5)(y + 6)$

7. $2x^2 - 15x + 25$
$(2x - 5)(x - 5)$

8. $2u^2 - 19u - 10$
$(2u + 1)(u - 10)$

9. $3y^2 + 10y - 8$
$(3y - 2)(y + 4)$

10. $6x^2 + x - 15$
$(2x - 3)(3x + 5)$

11. $5u^2 - 33u + 18$
$(5u - 3)(u - 6)$

12. $-4y^2 - 22y - 28$
$(-2)(2y + 7)(y + 2)$

13. $5x^2 - 20x + 20$
$5(x - 2)(x - 2)$

14. $27u^2 + 18u + 3$
$3(3u + 1)(3u + 1)$

15. $y^2 - 25$
$(y - 5)(y + 5)$

16. $2x^4 + 7x^3 + 3x^2$
$x^2(2x + 1)(x + 3)$

17. $-16u^2 + 80u - 100$
$(-4)(2u - 5)(2u - 5)$

18. $6y^2 + 36y + 36$
$6(y^2 + 6y + 6)$

19. $20x^2 - 13x + 2$
$(4x - 1)(5x - 2)$

20. $4u^2 - 4u - 35$
$(2u - 7)(2u + 5)$

21. $2x^2 + 5xy + 3y^2$
$(2x + 3y)(x + y)$

22. $2x^2 - 7xy + 5y^2$
$(2x - 5y)(x - y)$

23. $2u^2 + 3uv - 14v^2$
$(2u + 7v)(u - 2v)$

24. $2u^2 - 9uv - 5v^2$
$(2u + v)(u - 5v)$

25. $3x^2 - 7xy + 2y^2$
$(3x - y)(x - 2y)$

26. $3x^2 + 22xy + 7y^2$
$(3x + y)(x + 7y)$

27. $5u^2 + 7uv + 2v^2$
$(5u + 2v)(u + v)$

28. $6u^2 - uv - 2v^2$
$(3u - 2v)(2u + v)$

29. $-5x^2 + 25xy - 30y^2$
$(-5)(x - 2y)(x - 3y)$

30. $x^2 - 2xy + 2y^2$
cannot be factored

31. $9u^2 - v^2$
$(3u + v)(3u - v)$

32. $3u^4 - 4u^3v - 4u^2v^2$
$u^2(3u + 2v)(u - 2v)$

33. $8x^2 - 7xy - y^2$
$(8x + y)(x - y)$

34. $6x^2 + 25xy + 11y^2$
$(3x + 11y)(2x + y)$

35. $14x^2 - 30xy + 16y^2$
$2(7x - 8y)(x + y)$

36. $3x^3y + 2x^2y^2 - xy^3$
$xy(3x - y)(x + y)$

FOR REVIEW

Factor.

37. $x^2 + 17x + 70$
$(x + 10)(x + 7)$

38. $x^2 - 17x + 70$
$(x - 10)(x - 7)$

39. $y^2 - 2y - 80$
$(y - 10)(y + 8)$

40. $x^2 - 5xy - 66y^2$
$(x + 6y)(x - 11y)$

41. $x^2 + 24xy + 144y^2$
$(x + 12y)(x + 12y)$

42. $-5x^2 - 20x + 105$
$(-5)(x - 3)(x + 7)$

Exercises 43–48 review material from Section 14.6 to help you prepare for the next section. Find the following special products.

43. $(y + 4)(y - 4)$
$y^2 - 16$

44. $(y + 4)^2$
$y^2 + 8y + 16$

45. $(y - 4)^2$
$y^2 - 8y + 16$

46. $(3u + 7)^2$
$9u^2 + 42u + 49$

47. $(3u - 7)^2$
$9u^2 - 42u + 49$

48. $(3u + 7)(3u - 7)$
$9u^2 - 49$

15.3 EXERCISES C

Factor.

1. $2x^2 - \dfrac{1}{5}x - \dfrac{1}{25}$ $\left(2x + \dfrac{1}{5}\right)\left(x - \dfrac{1}{5}\right)$

2. $0.02x^2 - 0.9x + 4$
$(0.1x - 4)(0.2x - 1)$

3. $2(3x - 1)^2 - (3x - 1) - 1$
[*Hint:* Substitute u for $3x - 1$.]
$(3x - 2)(6x - 1)$

4. $3x^4 + 10x^2 - 8$
$(3x^2 - 2)(x^2 + 4)$

15.4 FACTORING PERFECT SQUARE TRINOMIALS AND DIFFERENCE OF SQUARES

STUDENT GUIDEPOSTS

1 Perfect Square Trinomials   **2** Difference of Two Squares

1 PERFECT SQUARE TRINOMIALS

We could factor trinomials such as

$$x^2 + 6x + 9 = (x + 3)(x + 3)$$
$$\text{and}\quad 4y^2 - 4y + 1 = (2y - 1)(2y - 1)$$

with the methods of the previous two sections. However, by recognizing these as *perfect square trinomials,* we can save some of the time used in trial-and-error methods. In Section 14.6 we used the formulas

$$(a + b)^2 = a^2 + 2ab + b^2$$
$$(a - b)^2 = a^2 - 2ab + b^2$$

to square binomials. Now we consider the reverse operation, factoring trinomials like those on the right into the square of a binomial. To factor $x^2 + 6x + 9$, we first observe that it fits the appropriate formula.

$$
\begin{aligned}
x^2 + 6x + 9 &= x^2 + 6x + 3^2 &&\quad x^2 \text{ and } 3^2 \text{ are perfect squares}\\
&= x^2 + 2 \cdot x \cdot 3 + 3^2 &&\quad x = a \text{ and } 3 = b \text{ in } a^2 + 2ab + b^2\\
&= (x + 3)^2 &&\quad a^2 + 2ab + b^2 = (a + b)^2
\end{aligned}
$$

From the formula and this example, it is easy to see that a **perfect square trinomial** must contain two terms that are perfect squares (x^2 and 9 in the example). The remaining term must be plus or minus twice the product of the numbers whose squares form the other terms ($2 \cdot x \cdot 3 = 6x$ in the example). Notice that $x^2 + 10x + 9 = (x + 1)(x + 9)$ is *not* a perfect square trinomial because of its middle term, even though both x^2 and 9 are perfect squares.

EXAMPLE 1 **FACTORING PERFECT SQUARES**

Factor the trinomials.

(a) $x^2 + 8x + 16 = x^2 + 8x + 4^2$ x^2 and 16 are perfect squares

$\qquad\qquad\qquad = x^2 + 2 \cdot x \cdot 4 + 4^2$ $x = a$ and $4 = b$ in $a^2 + 2ab + b^2$

$\qquad\qquad\qquad = (x + 4)^2$ $a^2 + 2ab + b^2 = (a + b)^2$

Notice that we have $(x + 4)^2$ instead of $(x - 4)^2$ since $8x$ is positive.

(b) $x^2 - 8x + 16 = x^2 - 2 \cdot x \cdot 4 + 4^2$ $x = a$ and $4 = b$

$\qquad\qquad\qquad = (x - 4)^2$ Note the minus sign in this case

(c) $16u^2 + 40u + 25 = (4u)^2 + 40u + 5^2$ $4u = a$ and $5 = b$

$\qquad\qquad\qquad = (4u)^2 + 2(4u)(5) + 5^2$ $2ab = 2(4u)(5) = 40u$

$\qquad\qquad\qquad = (4u + 5)^2$

(d) $16u^2 - 40u + 25 = (4u)^2 - 2(4u)(5) + 5^2$

$\qquad\qquad\qquad = (4u - 5)^2$

PRACTICE EXERCISE 1

Factor the trinomials.

(a) $y^2 + 22y + 121$

(b) $y^2 - 22y + 121$

(c) $9u^2 + 42u + 49$

(d) $9u^2 - 42u + 49$

Answers: (a) $(y + 11)^2$
(b) $(y - 11)^2$ (c) $(3u + 7)^2$
(d) $(3u - 7)^2$

To Factor Perfect Square Trinomials

1. Determine if two terms are perfect squares and the remaining one is twice the product of the numbers whose squares are the other terms.

2. Use the formulas

$$a^2 + 2ab + b^2 = (a + b)^2$$
$$a^2 - 2ab + b^2 = (a - b)^2.$$

EXAMPLE 2 **FACTORING PERFECT SQUARES**

Factor the trinomials in two variables.

(a) $x^2 + 18xy + 81y^2 \quad = x^2 + 18xy + (9y)^2$ x^2 and $(9y)^2$ are perfect squares

$\qquad\qquad\qquad = x^2 + 2 \cdot x \cdot 9y + (9y)^2$ $x = a$ and $9y = b$

$\qquad\qquad\qquad = (x + 9y)^2$ Use the formula

(b) $2u^2 - 40uv + 200v^2 = 2(u^2 - 20uv + 100v^2)$ Factor out common factor

$\qquad\qquad\qquad = 2(u^2 - 2 \cdot u \cdot 10v + (10v)^2)$ $u = a$ and $10v = b$

$\qquad\qquad\qquad = 2(u - 10v)^2$

PRACTICE EXERCISE 2

Factor the trinomials in two variables.

(a) $x^2 - 18xy + 81y^2$

(b) $3u^2 + 60uv + 300v^2$

Answers: (a) $(x - 9y)^2$
(b) $3(u + 10v)^2$

② DIFFERENCE OF TWO SQUARES

Another special formula can be used whenever we factor the difference of two squares.

$$a^2 - b^2 = (a + b)(a - b)$$

Observe that we are factoring a binomial in this case.

| **EXAMPLE 3** **FACTORING A DIFFERENCE OF SQUARES** | **PRACTICE EXERCISE 3** |

Factor the binomials.

(a) $x^2 - 9 = x^2 - 3^2$ x^2 and 3^2 are perfect squares

      $= (x + 3)(x - 3)$ $x = a$ and $3 = b$

(b) $y^2 - 121 = y^2 - 11^2$ y^2 and 11^2 are perfect squares

      $= (y + 11)(y - 11)$ $y = a$ and $11 = b$

(c) $16u^2 - 25 = (4u)^2 - 5^2$ $(4u)^2$ and 5^2 are perfect squares

      $= (4u + 5)(4u - 5)$ $4u = a$ and $5 = b$

(d) $5x^3 - 500x = 5x(x^2 - 100)$ Factor out common factor

      $= 5x(x^2 - 10^2)$

      $= 5x(x + 10)(x - 10)$

(e) $y^6 - 49 = (y^3)^2 - 7^2$ $(y^3)^2$ and 7^2 are perfect squares

      $= (y^3 + 7)(y^3 - 7)$ $y^3 = a$ and $7 = b$

(f) $x^2 + 9 = x^2 + 3^2$ Cannot be factored

Factor the binomials.

(a) $x^2 - 1$

(b) $y^2 - 169$

(c) $9u^2 - 49$

(d) $2y^2 - 288$

(e) $z^4 - 4$

(f) $w^2 + 25$

Answers: (a) $(x + 1)(x - 1)$
(b) $(y + 13)(y - 13)$
(c) $(3u + 7)(3u - 7)$
(d) $2(z + 12)(z - 12)$
(e) $(z^2 + 2)(z^2 - 2)$
(f) cannot be factored

⚠ CAUTION

Notice in Example 3(f) that the sum of two squares cannot be factored. Do not make the mistake of giving $(a + b)^2$ as the factors of $a^2 + b^2$.

$$(a + b)^2 = a^2 + 2ab + b^2 \neq a^2 + b^2$$

Also, $(a - b)^2 \neq a^2 - b^2.$

To Factor a Difference of Two Squares

1. Determine if the binomial is the difference of two perfect squares.
2. Use the formula

$$a^2 - b^2 = (a + b)(a - b).$$

EXAMPLE 4 **DIFFERENCE OF SQUARES IN TWO VARIABLES**	**PRACTICE EXERCISE 4**

Factor the binomials in two variables.

(a) $25x^2 - 9y^2 = (5x)^2 - (3y)^2$ $\qquad$ $(5x)^2$ and $(3y)^2$ are perfect squares

$\qquad\qquad\quad = (5x + 3y)(5x - 3y)$ $\quad$ $5x = a$ and $3y = b$

(b) $27u^3 - 48uv^2 = 3u(9u^2 - 16v^2)$ $\qquad$ Factor out common factor

$\qquad\qquad\quad = 3u[(3u)^2 - (4v)^2]$ $\qquad$ $3u = a$ and $4v = b$

$\qquad\qquad\quad = 3u(3u + 4v)(3u - 4v)$

(c) $x^4 - y^4 = (x^2)^2 - (y^2)^2$ $\qquad$ $x^2 = a$ and $y^2 = b$

$\qquad\quad = (x^2 + y^2)(x^2 - y^2)$ $\qquad$ Now $x = a$ and $y = b$ in the difference of two squares $x^2 - y^2$

$\qquad\quad = (x^2 + y^2)(x + y)(x - y)$ $\quad$ Factor again

Factor the binomials in two variables.

(a) $100a^2 - b^2$

(b) $50x^3 - 162x$

(c) $x^4 + y^4$

Answers: (a) $(10a + b)(10a - b)$
(b) $2x(5x + 9)(5x - 9)$ (c) cannot be factored

15.4   EXERCISES A

Factor using the special formulas.

1. $x^2 + 10x + 25$
$(x + 5)^2$

2. $x^2 - 10x + 25$
$(x - 5)^2$

3. $x^2 - 25$
$(x + 5)(x - 5)$

4. $x^2 + 25$
cannot be factored

5. $u^2 - 14u + 49$
$(u - 7)^2$

6. $u^2 + 14u + 49$
$(u + 7)^2$

7. $u^2 - 49$
$(u + 7)(u - 7)$

8. $u^2 + 49$
cannot be factored

9. $x^2 - 24x + 144$
$(x - 12)^2$

10. $x^2 - 121$
$(x + 11)(x - 11)$

11 $9u^2 + 6u + 1$
$(3u + 1)^2$

12. $9u^2 - 6u + 1$
$(3u - 1)^2$

13. $4y^2 - 12y + 9$
$(2y - 3)^2$

14. $4y^2 - 9$
$(2y + 3)(2y - 3)$

15. $25x^2 + 20x + 4$
$(5x + 2)^2$

16. $4x^2 + 20x + 25$
$(2x + 5)^2$

17. $9u^2 - 25$
$(3u + 5)(3u - 5)$

18. $9u^2 + 25$
cannot be factored

19 $-12y^2 + 60y - 75$
$(-3)(2y - 5)^2$

20. $4y^2 - 36y + 81$
$(2y - 9)^2$

21. $9x^2 + 30x + 25$
$(3x + 5)^2$

22. $x^2 + x + 4$
cannot be factored

23. $8u^3 - 8u$
$8u(u + 1)(u - 1)$

24. $u^6 - 16$
$(u^3 + 4)(u^3 - 4)$

25. $5y^3 - 100y^2 + 500y$
$5y(y - 10)^2$

26. $7y^4 - 63$
$7(y^2 + 3)(y^2 - 3)$

27. $x^2 + 6xy + 9y^2$
$(x + 3y)^2$

28 $25x^2 - 10xy + y^2$
$(5x - y)^2$

29. $64u^2 - 9v^2$
$(8u + 3v)(8u - 3v)$

30 $u^4 - v^4$
$(u^2 + v^2)(u + v)(u - v)$

Before we complete this exercise set we summarize the factoring techniques we have learned.

To Factor a Polynomial

1. Factor out any common factor, including -1 if a is negative in $ax^2 + bx + c$ or $ax^2 + bxy + cy^2$.

2. To factor the difference of two squares use the formula

$$a^2 - b^2 = (a + b)(a - b).$$

3. To factor a trinomial use the techniques of Section 15.2 or Section 15.3, but also consider perfect squares using the formulas

$$a^2 + 2ab + b^2 = (a + b)^2 \quad \text{and} \quad a^2 - 2ab + b^2 = (a - b)^2.$$

4. To factor a polynomial of more than three terms, try to factor by grouping.

Factor.

31. $x^2 + 9x + 18$
$(x + 3)(x + 6)$

32. $x^2 - 12x + 36$
$(x - 6)^2$

33. $u^2 - 14u + 48$
$(u - 6)(u - 8)$

34. $u^2 + 10u + 16$
$(u + 2)(u + 8)$

35. $4y^2 - 49$
$(2y + 7)(2y - 7)$

36. $4y^2 - 28y + 49$
$(2y - 7)^2$

37. $-6x^2 + 486$
$(-6)(x + 9)(x - 9)$

38. $2x^2 + 24x + 64$
$2(x + 4)(x + 8)$

39. $25u^2 - 20u + 4$
$(5u - 2)^2$

40. $16u^2 - 9$
   $(4u + 3)(4u - 3)$

41 $4y^2 - 16y + 15$
   $(2y - 3)(2y - 5)$

42. $40y^2 + 11y - 2$
   $(8y - 1)(5y + 2)$

43 $3x^9 - 147x^3$
   $3x^3(x^3 + 7)(x^3 - 7)$

44. $98x^6 - 18$
   $2(7x^3 + 3)(7x^3 - 3)$

45. $x^2 - 9y^2$
   $(x + 3y)(x - 3y)$

46. $3x^2 - 19xy - 14y^2$
   $(3x + 2y)(x - 7y)$

47. $36u^2 - 60uv + 25v^2$
   $(6u - 5v)^2$

48. $18u^2 - 98v^2$
   $2(3u + 7v)(3u - 7v)$

49 $100x^5 - 60x^4y + 9x^3y^2$
   $x^3(10x - 3y)^2$

50. $(x + y)^2 - 25$ [*Hint:* $x + y = a, 5 = b$]
   $(x + y + 5)(x + y - 5)$

FOR REVIEW

Exercises 51–56 review material from Chapter 11 to help you prepare for the next section. Solve the following equations.

51. $x - 3 = 0$ 3

52. $3x + 1 = 0$ $-\dfrac{1}{3}$

53. $2x + 5 = 0$ $-\dfrac{5}{2}$

54. $6x - 1 = 0$ $\dfrac{1}{6}$

55. $4x = 0$ 0

56. $3x + 9 = 0$ -3

ANSWERS: 1. $(x + 5)^2$ 2. $(x - 5)^2$ 3. $(x + 5)(x - 5)$ 4. cannot be factored 5. $(u - 7)^2$ 6. $(u + 7)^2$
7. $(u + 7)(u - 7)$ 8. cannot be factored 9. $(x - 12)^2$ 10. $(x + 11)(x - 11)$ 11. $(3u + 1)^2$ 12. $(3u - 1)^2$
13. $(2y - 3)^2$ 14. $(2y + 3)(2y - 3)$ 15. $(5x + 2)^2$ 16. $(2x + 5)^2$ 17. $(3u + 5)(3u - 5)$ 18. cannot be factored
19. $(-3)(2y - 5)^2$ 20. $(2y - 9)^2$ 21. $(3x + 5)^2$ 22. cannot be factored 23. $8u(u + 1)(u - 1)$ 24. $(u^3 + 4)(u^3 - 4)$
25. $5y(y - 10)^2$ 26. $7(y^2 + 3)(y^2 - 3)$ 27. $(x + 3y)^2$ 28. $(5x - y)^2$ 29. $(8u + 3v)(8u - 3v)$
30. $(u^2 + v^2)(u + v)(u - v)$ 31. $(x + 3)(x + 6)$ 32. $(x - 6)^2$ 33. $(u - 6)(u - 8)$ 34. $(u + 2)(u + 8)$
35. $(2y + 7)(2y - 7)$ 36. $(2y - 7)^2$ 37. $-6(x + 9)(x - 9)$ 38. $2(x + 4)(x + 8)$ 39. $(5u - 2)^2$
40. $(4u + 3)(4u - 3)$ 41. $(2y - 3)(2y - 5)$ 42. $(8y - 1)(5y + 2)$ 43. $3x^3(x^3 + 7)(x^3 - 7)$ 44. $2(7x^3 + 3)(7x^3 - 3)$
45. $(x + 3y)(x - 3y)$ 46. $(3x + 2y)(x - 7y)$ 47. $(6u - 5v)^2$ 48. $2(3u + 7v)(3u - 7v)$ 49. $x^3(10x - 3y)^2$
50. $(x + y + 5)(x + y - 5)$ 51. 3 52. $-\frac{1}{3}$ 53. $-\frac{5}{2}$ 54. $\frac{1}{6}$ 55. 0 56. -3

15.4 EXERCISES B

Factor using the special formulas.

1. $x^2 + 12x + 36$
   $(x + 6)^2$

2. $x^2 - 12x + 36$
   $(x - 6)^2$

3. $x^2 - 36$
   $(x + 6)(x - 6)$

4. $x^2 + 36$
   cannot be factored

5. $u^2 - 16u + 64$
   $(u - 8)^2$

6. $u^2 + 16u + 64$
   $(u + 8)^2$

7. $u^2 - 64$
   $(u + 8)(u - 8)$

8. $u^2 + 64$
   cannot be factored

9. $x^2 - 22x + 121$
   $(x - 11)^2$

10. $x^2 - 169$
   $(x + 13)(x - 13)$

11. $4u^2 + 4u + 1$
   $(2u + 1)^2$

12. $4u^2 - 4u + 1$
   $(2u - 1)^2$

13. $9y^2 - 12y + 4$
   $(3y - 2)^2$

14. $9y^2 - 4$
   $(3y + 2)(3y - 2)$

15. $36x^2 + 12x + 1$
   $(6x + 1)^2$

16. $9x^2 + 30x + 25$
   $(3x + 5)^2$

17. $4u^2 - 25$
$(2u + 5)(2u - 5)$

18. $4u^2 + 25$
cannot be factored

19. $-32y^2 + 48y - 18$
$(-2)(4y - 3)^2$

20. $9y^2 - 42y + 49$
$(3y - 7)^2$

21. $25x^2 + 30x + 9$
$(5x + 3)^2$

22. $x^2 + 2x + 4$
cannot be factored

23. $4u^3 - 16u$
$4u(u + 2)(u - 2)$

24. $u^6 - 81$
$(u^3 + 9)(u^3 - 9)$

25. $-4y^2 - 64y - 256$
$-4(y + 8)^2$

26. $y^4 - 81$
$(y^2 + 9)(y + 3)(y - 3)$

27. $x^2 + 8xy + 16y^2$
$(x + 4y)^2$

28. $36x^2 - 12xy + y^2$
$(6x - y)^2$

29. $49u^2 - 16v^2$
$(7u + 4v)(7u - 4v)$

30. $81u^4 - v^4$
$(9u^2 + v^2)(3u + v)(3u - v)$

Factor.

31. $x^2 + 12x + 35$
$(x + 5)(x + 7)$

32. $x^2 + 20x + 100$
$(x + 10)^2$

33. $u^2 - 22u + 121$
$(u - 11)^2$

34. $u^2 - 14u + 45$
$(u - 5)(u - 9)$

35. $9y^2 - 64$
$(3y + 8)(3y - 8)$

36. $-4y^2 + 400$
$(-4)(y + 10)(y - 10)$

37. $9x^2 - 60x + 100$
$(3x - 10)^2$

38. $28x^2 - 41x + 15$
$(7x - 5)(4x - 3)$

39. $-10u^2 - 21u + 10$
$-(5u - 2)(2u + 5)$

40. $18u^2 - 98$
$2(3u + 7)(3u - 7)$

41. $42y^2 - 3y - 9$
$3(7y + 3)(2y - 1)$

42. $16y^2 + 6y - 27$
$(8y - 9)(2y + 3)$

43. $25x^2 - 90x + 81$
$(5x - 9)^2$

44. $x^2 - 3x - 180$
$(x - 15)(x + 12)$

45. $x^2 - 25y^2$
$(x + 5y)(x - 5y)$

46. $2x^2 - 52xy + 338y^2$
$2(x - 13y)^2$

47. $49u^2 + 42uv + 9v^2$
$(7u + 3v)^2$

48. $200u^2 - 242v^2$
$2(10u + 11v)(10u - 11v)$

49. $9x^4y^2 - 60x^3y^3 + 100x^2y^4$
$x^2y^2(3x - 10y)^2$

50. $(x - y)^2 - 16$
$(x - y + 4)(x - y - 4)$

FOR REVIEW

Exercises 51–56 review material from Chapter 11 to help you prepare for the next section. Solve the following equations.

51. $x - 9 = 0$ 9

52. $7x + 4 = 0$ $-\dfrac{4}{7}$

53. $5x + 6 = 0$ $-\dfrac{6}{5}$

54. $4x - 2 = 0$ $\dfrac{1}{2}$

55. $\dfrac{1}{2}x = 0$ 0

56. $6x + 12 = 0$ -2

15.4 EXERCISES C

To factor the binomials in Exercises 1–6, use the formulas

$$x^3 + y^3 = (x + y)(x^2 - xy + y^2)$$
$$\text{and} \quad x^3 - y^3 = (x - y)(x^2 + xy + y^2).$$

1. $a^3 + 8$
[*Hint:* $a^3 + 8 = a^3 + 2^3$]
$(a + 2)(a^2 - 2a + 4)$

2. $a^3 - 8$
$(a - 2)(a^2 + 2a + 4)$

3. $27b^3 - 8$
$(3b - 2)(9b^2 + 6b + 4)$

4. $27b^3 + 8$
$(3b + 2)(9b^2 - 6b + 4)$

5. $8a^3 - 125b^3$ [Answer:
$(2a - 5b)(4a^2 + 10ab + 25b^2)$]

6. $8a^3 + 125b^3$
$(2a + 5b)(4a^2 - 10ab + 25b^2)$

15.5 FACTORING TO SOLVE EQUATIONS

━━━━━━━━━━━━ **STUDENT GUIDEPOSTS** ━━━━━━━━━━━━

1 Zero-Product Rule   **2** Solving Quadratic Equations

1 ZERO-PRODUCT RULE

A property of the number zero is used to solve certain types of equations. We know that any number times zero yields a zero product. For example,

$$4 \cdot 0 = 0 \quad \text{and} \quad 0 \cdot \frac{2}{3} = 0.$$

Moreover, if a product of two or more numbers is zero, then at least one of the factors must be zero. This property, called the zero-product rule, gives us a useful method for solving equations.

Solving Equations Using the Zero-Product Rule

If an equation has a product of two or more factors on one side of the equation and 0 on the other, the solutions to the equation are found by setting each factor equal to 0.

If $a \cdot b = 0$, then $a = 0$ or $b = 0$.

The zero-product rule can be used to solve for x in equations of the form

$$(x - 3)(x + 5) = 0.$$

That is, when $x - 3$ and $x + 5$ are multiplied to give zero, the rule says either $x - 3 = 0$ or $x + 5 = 0$. Thus, one solution of

$$(x - 3)(x + 5) = 0$$

is found by solving

$$x - 3 = 0$$
$$x = 3. \quad \text{One solution is 3}$$

Another solution comes by solving

$$x + 5 = 0$$
$$x = -5. \quad \text{Another solution is } -5$$

There are two solutions to the equation, 3 and -5. To check the solutions we substitute back into the original equations.

Check of 3	*Check of* -5
$(x - 3)(x + 5) = 0$	$(x - 3)(x + 5) = 0$
$(3 - 3)(3 + 5) \stackrel{?}{=} 0$	$(-5 - 3)(-5 + 5) \stackrel{?}{=} 0$
$0 \cdot 8 = 0$	$(-8) \cdot 0 = 0$
So 3 checks.	And -5 checks.

| **EXAMPLE 1** USING THE ZERO-PRODUCT RULE | **PRACTICE EXERCISE 1** |

Solve $(2x + 1)(x - 1) = 0$.

$$2x + 1 = 0 \qquad \text{or} \qquad x - 1 = 0$$
$$2x + 1 - 1 = 0 - 1 \qquad x - 1 + 1 = 0 + 1$$
$$2x = -1 \qquad\qquad x = 1$$
$$\frac{1}{2} \cdot 2x = \frac{1}{2} \cdot (-1)$$
$$x = -\frac{1}{2}$$

The solutions are $-\frac{1}{2}$ and 1.

Solve $(3x + 2)(x - 9) = 0$.

Answer: $-\frac{2}{3}$ and 9

| **EXAMPLE 2** USING THE ZERO-PRODUCT RULE | **PRACTICE EXERCISE 2** |

Solve $x(3x + 1) = 0$.

$$x = 0 \quad \text{or} \quad 3x + 1 = 0$$
$$3x + 1 - 1 = 0 - 1$$
$$3x = -1$$
$$\frac{1}{3} \cdot 3x = \frac{1}{3}(-1)$$
$$x = -\frac{1}{3}$$

The solutions are 0 and $-\frac{1}{3}$.

Solve $y(y + 5) = 0$.

Answer: 0 and -5

When one of the factors has only one term, say x, as in Example 2, one of the solutions will always be zero. Be sure that you give this solution, and do not divide both sides of the equation by x.

❷ SOLVING QUADRATIC EQUATIONS

Suppose we solve $(x - 2)(x + 3) = 0$.

$$x - 2 = 0 \quad \text{or} \quad x + 3 = 0$$
$$x = 2 \qquad\qquad x = -3$$

The solutions are 2 and -3. The equation $(x - 2)(x + 3) = 0$ is the factored form of $x^2 + x - 6 = 0$, which is a *quadratic equation*. A **quadratic equation** has the form

$$ax^2 + bx + c = 0.$$

Thus, the zero-product rule can be used to solve quadratic equations, a subject that we will study in more detail in Chapter 18.

To Solve an Equation by Factoring
1. Write all terms on the left side of the equation leaving zero on the right side.
2. Factor the left side.
3. Use the zero-product rule, set each factor equal to zero, and solve the resulting equations.
4. Check possible solutions in the original equation.

EXAMPLE 3 SOLVING A QUADRATIC EQUATION	PRACTICE EXERCISE 3

Solve $x^2 + 4x = 21$.

$$x^2 + 4x - 21 = 0 \qquad \text{Rewrite with all terms on the left}$$

$$(x + 7)(x - 3) = 0 \qquad \text{Factor the trinomial}$$

$$x + 7 = 0 \quad \text{or} \quad x - 3 = 0 \qquad \text{Use zero-product rule}$$

$$x = -7 \qquad\qquad x = 3$$

$$\text{Check of } -7 \qquad\qquad \text{Check of } 3$$

$$(-7)^2 + 4(-7) \stackrel{?}{=} 21 \qquad (3)^2 + 4(3) \stackrel{?}{=} 21$$

$$49 - 28 \stackrel{?}{=} 21 \qquad\quad 9 + 12 \stackrel{?}{=} 21$$

$$21 = 21 \qquad\qquad\quad 21 = 21$$

The solutions are -7 and 3.

Solve $2x^2 - 11x = 40$

Answer: 8 and $-\frac{5}{2}$

CAUTION

Using the zero-product rule, we can only solve equations in the form $a \cdot b = 0$. If $a \cdot b = 5$, we cannot conclude that $a = 5$ or $b = 5$. Nor can we apply this rule to $a + b = 0$ or $a - b = 0$.

EXAMPLE 4 SOLVING A QUADRATIC EQUATION	PRACTICE EXERCISE 4

Solve $x(x - 4) = 5$.

The zero-product rule does not apply immediately because this product does *not* equal zero. When this happens, try to rewrite it as a product that does equal zero. Begin by clearing the parentheses.

$$x^2 - 4x = 5$$

$$x^2 - 4x - 5 = 0 \qquad \text{Write all terms on the left}$$

$$(x - 5)(x + 1) = 0 \qquad \text{Factor}$$

$$x - 5 = 0 \quad \text{or} \quad x + 1 = 0 \qquad \text{Zero-product rule}$$

$$x = 5 \qquad\qquad x = -1$$

The solutions are 5 and -1.

Solve $x(x + 5) = 14$.

Answer: 2 and -7

| **EXAMPLE 5** LINEAR EQUATION | **PRACTICE EXERCISE 5** |

Solve $(2x + 1) - (x + 5) = 0$.

The zero-product rule does not apply because the left side of the equation is a difference, *not* a product. To solve, we remove parentheses and then combine like terms.

$$2x + 1 - x - 5 = 0$$
$$x - 4 = 0$$
$$x = 4$$

The solution is 4. Notice that this time after clearing parentheses, the result is a linear equation, not a quadratic equation which would require the zero-product rule. Also, it cannot be solved by setting $2x + 1$ and $x + 5$ equal to zero.

Solve $(x + 2) - (3x + 8) = 0$.

Answer: -3

The zero-product rule can be extended to include products of three or more factors. That is, if $a \cdot b \cdot c = 0$ then $a = 0$ or $b = 0$ or $c = 0$. We use this fact in the next example.

| **EXAMPLE 6** PRODUCT OF THREE FACTORS | **PRACTICE EXERCISE 6** |

Solve $x^3 + 2x^2 = 15x$.

$x^3 + 2x^2 - 15x = 0$	Subtract $15x$ from both sides
$x(x^2 + 2x - 15) = 0$	Factor out the GCF x
$x(x - 3)(x + 5) = 0$	Factor the trinomial
$x = 0$ or $x - 3 = 0$ or $x + 5 = 0$	Set each factor equal to 0
$x = 3$ $\qquad x = -5$	

The solutions are 0, 3, and -5. Check these in the original equation.

Solve $x^3 = 5x^2 - 6x$.

Answer: 0, 2, and 3

15.5 EXERCISES A

Solve.

1. $(x - 5)(x + 7) = 0$
   5, -7

2. $(x + 6)(x + 1) = 0$
   $-6, -1$

3. $(y - 8)(y - 9) = 0$
   8, 9

4. $(y + 10)(y - 20) = 0$
   -10, 20

5. $(u + 2)(u - 2) = 0$
   -2, 2

6. $(u - 0.5)(u + 0.2) = 0$
   0.5, -0.2

7. $\left(x - \dfrac{1}{2}\right)\left(x + \dfrac{2}{5}\right) = 0$

$\dfrac{1}{2}, -\dfrac{2}{5}$

8. $(x - 8)^2 = 0$

8

9. $(3y + 2)(y - 6) = 0$

$-\dfrac{2}{3}, 6$

10. $(y + 8)(y - 8) = 0$

$-8, 8$

11. $(5u + 1)(2u - 3) = 0$

$-\dfrac{1}{5}, \dfrac{3}{2}$

12. $(8u - 5)(3u - 8) = 0$

$\dfrac{5}{8}, \dfrac{8}{3}$

13 $x(3x + 7) = 0$

$0, -\dfrac{7}{3}$

14. $(2x - 5)(x + 12) = 0$

$\dfrac{5}{2}, -12$

15. $y^2 - 5y + 6 = 0$

2, 3

16. $y^2 - y - 6 = 0$

$-2, 3$

17. $u^2 - 8u + 7 = 0$

1, 7

18. $u^2 + u - 20 = 0$

$-5, 4$

19. $x^2 + 10x + 25 = 0$

-5

20. $x^2 - 13x + 40 = 0$

5, 8

21. $y^2 - y - 42 = 0$

$-6, 7$

22. $y^2 - 16 = 0$

$-4, 4$

23 $4u^2 - 8u = 0$

0, 2

24. $4u^2 + 12u + 9 = 0$

$-\dfrac{3}{2}$

25. $9x^2 - 49 = 0$

$\dfrac{7}{3}, -\dfrac{7}{3}$

26 $(3x - 5) - (x + 7) = 0$

6

27 $y(y + 3) = 10$

2, -5

28. $y^2 - 2y = 35$

$-5, 7$

29. $u^2 = -2u + 35$

5, -7

30. $16u^2 - 42u + 5 = 0$

$\dfrac{1}{8}, \dfrac{5}{2}$

31. $(x - 3)(x - 2)(x + 1) = 0$

3, 2, -1

32. $(2y)(y + 1)(y + 5) = 0$

0, -1, -5

33. $2x^3 - 4x^2 - 6x = 0$

0, -1, 3

FOR REVIEW

Factor.

34. $x^2 - 4x - 21$

$(x + 3)(x - 7)$

35. $x^2 + 14x + 49$

$(x + 7)^2$

36. $25y^2 - 16$

$(5y + 4)(5y - 4)$

37. $7y^2 - 70y + 175$

$7(y - 5)^2$

38. $2x^2 + 9xy - 5y^2$

$(2x - y)(x + 5y)$

39. $45x^2 - 80y^2$

$5(3x + 4y)(3x - 4y)$

Exercise 40 reviews material from Chapter 11 to help you prepare for the next section. Solve the applied problem.

40. After receiving a 9% raise Lynn Harris now makes $34,880 per year. What was her salary before the raise?
$32,000

ANSWERS: 1. 5, −7 2. −6, −1 3. 8, 9 4. −10, 20 5. −2, 2 6. 0.5, −0.2 7. $\frac{1}{2}$, −$\frac{2}{5}$ 8. 8 9. −$\frac{2}{3}$, 6
10. −8, 8 11. −$\frac{1}{5}$, $\frac{3}{2}$ 12. $\frac{5}{8}$, $\frac{8}{3}$ 13. 0, −$\frac{7}{3}$ 14. $\frac{5}{2}$, −12 15. 2, 3 16. −2, 3 17. 1, 7 18. −5, 4 19. −5
20. 5, 8 21. −6, 7 22. −4, 4 23. 0, 2 24. −$\frac{3}{2}$ 25. $\frac{7}{3}$, −$\frac{7}{3}$ 26. 6 27. 2, −5 28. −5, 7 29. 5, −7 30. $\frac{1}{8}$, $\frac{5}{2}$
31. 3, 2, −1 32. 0, −1, −5 33. 0, −1, 3 34. $(x + 3)(x − 7)$ 35. $(x + 7)^2$ 36. $(5y + 4)(5y − 4)$ 37. $7(y − 5)^2$
38. $(2x − y)(x + 5y)$ 39. $5(3x + 4y)(3x − 4y)$ 40. $32,000

15.5 EXERCISES B

Solve.

1. $(x + 2)(x − 6) = 0$
−2, 6

2. $(x + 8)(x + 3) = 0$
−8, −3

3. $(y − 5)(y − 7) = 0$
5, 7

4. $(y + 8)(y + 12) = 0$
−8, −12

5. $(u + 6)(u − 11) = 0$
−6, 11

6. $(u + 4)(u − 4) = 0$
−4, 4

7. $(x − 0.7)(x + 0.3) = 0$
0.7, −0.3

8. $\left(x + \dfrac{1}{3}\right)\left(x + \dfrac{2}{3}\right) = 0$
−$\dfrac{1}{3}$, −$\dfrac{2}{3}$

9. $(2y − 3)(y + 8) = 0$
$\dfrac{3}{2}$, −8

10. $(y − 5)^2 = 0$
5

11. $(5u − 6)(2u + 1) = 0$
$\dfrac{6}{5}$, −$\dfrac{1}{2}$

12. $(3u + 14)(2u − 5) = 0$
−$\dfrac{14}{3}$, $\dfrac{5}{2}$

13. $x(4x + 3) = 0$
0, −$\dfrac{3}{4}$

14. $(2x − 7)(2x + 9) = 0$
$\dfrac{7}{2}$, −$\dfrac{9}{2}$

15. $y^2 + 8y + 15 = 0$
−3, −5

16. $y^2 − 2y − 15 = 0$
−3, 5

17. $u^2 − 11u + 18 = 0$
2, 9

18. $u^2 − 3u − 40 = 0$
−5, 8

19. $x^2 − 18x + 81 = 0$
9

20. $x^2 + 14x + 40 = 0$
−4, −10

21. $2y^2 − 13y + 6 = 0$
$\dfrac{1}{2}$, 6

22. $49y^2 − 1 = 0$
−$\dfrac{1}{7}$, $\dfrac{1}{7}$

23. $3u^2 − 27u = 0$
0, 9

24. $9u^2 − 18u + 9 = 0$
1

25. $25y^2 − 16 = 0$
−$\dfrac{4}{5}$, $\dfrac{4}{5}$

26. $(5x + 1) − (x − 7) = 0$
−2

27. $y(y − 8) = −16$
4

28. $−63 = y^2 − 16y$
7, 9

29. $63 − u^2 = 2u$
7, −9

30. $6u^2 − 29u − 5 = 0$
−$\dfrac{1}{6}$, 5

31. $(x − 2)(x + 3)(x − 1) = 0$
2, −3, 1

32. $4y(y − 1)(y + 3) = 0$
0, 1, −3

33. $2x^3 − 4x^2 − 6x = 0$
0, 3, −1

FOR REVIEW

Factor.

34. $x^2 + 4x - 32$
$(x - 4)(x + 8)$

35. $4x^2 - 12x + 9$
$(2x - 3)^2$

36. $81y^2 - 49$
$(9y + 7)(9y - 7)$

37. $-6y^2 - 72y - 216$
$-6(y + 6)^2$

38. $15x^2 + xy - 2y^2$
$(5x + 2y)(3x - y)$

39. $40x^2 - 1210y^2$
$10(2x + 11y)(2x - 11y)$

Exercise 40 reviews material from Chapter 11 to help you prepare for the next section. Solve the applied problem.

40. The perimeter of a rectangle is 38 m. Find the dimensions if the width is 3 m less than the length.
11 m by 8 m

15.5 EXERCISES C

Solve.

1. $(x + 3)(2x - 5)(x - 4) = 0$
$-3, \dfrac{5}{2}, 4$

2. $(x - 3)^2 - 3(x - 3) = 4$
2, 7

3. $x^4 - 16 = 0$
$-2, 2$

4. $9(x + 2)^2 - 6(x + 2) = -1$
$\left[\text{Answer:} \quad -\dfrac{5}{3} \right]$

5. $16x^4 - 8x^2 + 1 = 0$
$-\dfrac{1}{2}, \dfrac{1}{2}$

6. $5(x + 4)^2 = -9(x + 4) + 2$
[*Hint:* Substitute u for $x + 4$.]
$-\dfrac{19}{5}, -6$

15.6 APPLICATIONS OF FACTORING

=== STUDENT GUIDEPOSTS ===

1 Factoring to Solve Applied Problems   **3** Business Applications
2 Geometry Applications

1 FACTORING TO SOLVE APPLIED PROBLEMS

Sometimes an applied problem can be translated into an equation that is solvable using the technique in Section 15.5.

EXAMPLE 1 NUMBER PROBLEM

If two times the square of a number minus three times the number is 27, find the number.

Let x = the number.

two times the square   three times
of the number   minus the number is 27
$$2x^2 \qquad - \qquad 3x \quad = 27$$

$$2x^2 - 3x = 27$$

$2x^2 - 3x - 27 = 0$ Subtract 27 from both sides

$(2x - 9)(x + 3) = 0$ Factor

$2x - 9 = 0$ or $x + 3 = 0$ Zero-product rule

$2x = 9$ $x = -3$

$$x = \frac{9}{2}$$

The solutions are $\frac{9}{2}$ and -3.

PRACTICE EXERCISE 1

The product of 1 less than a number and 5 more than the same number is 16. Find the number.

Let x = the number,
$x - 1 = 1$ less than the number,
$x + 5 = 5$ more than the number.
Solve the equation
$(x - 1)(x + 5) =$ ___.

Answer: 3 or -7

EXAMPLE 2 CONSECUTIVE INTEGERS

The product of two consecutive even integers is 440. Find the integers.

Let n = first integer,
$n + 2$ = next consecutive even integer,
$n(n + 2)$ = product of the two consecutive even integers.

$n(n + 2) = 440$ Product is 440

$n^2 + 2n = 440$ Remove parentheses

$n^2 + 2n - 440 = 0$ Subtract 440 from both sides

$(n + 22)(n - 20) = 0$ Factor

$n + 22 = 0$ or $n - 20 = 0$

$n = -22$ $n = 20$

$n + 2 = -20$ $n + 2 = 22$

Thus, one solution is -22 and -20 while the other is 20 and 22.

PRACTICE EXERCISE 2

The product of two consecutive positive integers is 132. Find the integers.

Let n = first integer,
$n + 1$ = next consecutive integer.

Answer: 11 and 12

---

② GEOMETRY APPLICATIONS

Some geometry problems may also require factoring. It always helps to make a sketch for a geometry problem. Also, be sure that you answer the stated question and that your answer is reasonable. Geometry formulas are found on the inside covers, and Chapter 9 gives a discussion of geometric figures.

EXAMPLE 3 GEOMETRY APPLICATION

If the area of a rectangle is 48 m^2 and the length is three times the width, find the dimensions of the rectangle.

Make a sketch, as in Figure 15.1.

PRACTICE EXERCISE 3

A triangle has area 60 ft^2, and its base is 2 ft longer than its height. Find the length of each.

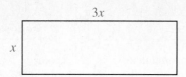

Figure 15.1

Let x = width of rectangle in meters,
   $3x$ = length of rectangle in meters.
For a rectangle.

width · length = area.

$$x \cdot 3x = 48$$
$$3x^2 = 48$$
$$3x^2 - 48 = 0$$
$$3(x^2 - 16) = 0$$
$$x^2 - 16 = 0 \quad \text{Divide both sides by 3. } \tfrac{0}{3} = 0$$
$$(x - 4)(x + 4) = 0$$
$$x - 4 = 0 \quad \text{or} \quad x + 4 = 0$$
$$x = 4 \qquad x = -4 \quad \text{Rule out } x = -4 \text{ since we want}$$
the width of a rectangle

Thus, since $x = 4$ and $3x = 12$, the rectangle is 4 m by 12 m.

Answer: base: 12 ft; height: 10 ft

❸ BUSINESS APPLICATIONS

There are numerous applications to business.

EXAMPLE 4   RETAIL APPLICATION

A clothing store owner finds that her daily profit on jeans is given by $P = n^2 - 6n - 20$, where n is the number of jeans sold. How many jeans must be sold for a profit of \$260?
   Since $P = \$260$, we need to solve the following equation.

$$P = n^2 - 6n - 20 = 260 \quad \text{Profit is to be \$260}$$
$$n^2 - 6n - 280 = 0 \quad \text{Set equal to zero}$$
$$(n + 14)(n - 20) = 0 \quad \text{Factor}$$
$$n + 14 = 0 \quad \text{or} \quad n - 20 = 0$$
$$n = -14 \qquad n = 20$$

Since n is the number of jeans sold, the -14 answer is discarded. Thus, 20 jeans must be sold to make \$260. To check we evaluate P for $n = 20$.

$$P = n^2 - 6n - 20$$
$$= (20)^2 - 6(20) - 20 \quad \text{Substitute 20 for } n$$
$$= 400 - 120 - 20$$
$$= 260 \qquad \text{This checks}$$

PRACTICE EXERCISE 4

A chemical reaction is described by the equation $C = 2n^2 - 7n + 1$, where n is always positive. Find n when C is 16.

Answer: 5

| **EXAMPLE 5** PRODUCTION APPLICATION | **PRACTICE EXERCISE 5** |

The IBX Corporation, producer of components for personal computers, has discovered that the number n of a particular type of component that it can sell each week is related to the price p of the component by the equation $n = 1100 - 100p$. Find the price that IBX should set on each component to produce a weekly revenue of \$2800.

Suppose in Example 5 $n = 500 - 50p$. Find the price if the weekly revenue is \$1250.

The revenue equation is $R = np$ where n is the number of components sold and p is the price of each component. Since $n = 1100 - 100p$, substituting we have

$$R = np$$
$$= (1100 - 100p)p$$
$$= 1100p - 100p^2.$$

We want to find p when R is 2800.

$$2800 = 1100p - 100p^2$$

Add $100p^2$ and subtract $1100p$ from both sides.

$$100p^2 - 1100p + 2800 = 0$$
$$100(p^2 - 11p + 28) = 0 \qquad \text{Factor out 100}$$
$$100(p - 4)(p - 7) = 0 \qquad \text{Factor}$$
$$p - 4 = 0 \quad \text{or} \quad p - 7 = 0 \qquad \text{Zero-product rule}$$
$$p = 4 \qquad\qquad p = 7$$

If components are sold for \$4 or for \$7, the weekly revenue will be \$2800.

Answer: \$5

15.6 EXERCISES A

Solve.

1. The product of 5 more than a number and 3 less than the number is zero. Find the number.

   Let x = the number,
      $x + 5$ = 5 more than the number,
      $x - 3$ = 3 less than the number.
   −5 or 3

2. If the square of a number plus seven times the number is 44, what is the number?
   −11 or 4

3. The product of a number and 4 less than the number is 21. Find the number.
   −3 or 7

4. The length of a rectangle is 5 cm more than the width. If the area is 24 cm^2, find the dimensions.
   3 cm, 8 cm

5 The product of two consecutive integers is 240. Find the integers.

−16, −15 or 15, 16

6. If the square of a number is increased by 3 times the number, the result is 70. Find the number.

7 or −10

7. The product of two consecutive odd integers is 143. Find the integers.

Let n = first integer,
     $n + 2$ = next odd integer.
−13, −11 or 11, 13

8 The sum of the squares of two consecutive positive integers is 100. Find the integers.

6, 8

9. The area of a rectangle is 84 ft² and the length is 5 feet more than the width. Find the dimensions of the rectangle.

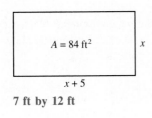

7 ft by 12 ft

10 The area of a square is numerically 4 less than the perimeter. Find the length of the side in cm.

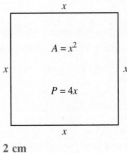

2 cm

11. The area of a triangle is 98 cm². If the base is four times the height, find the base and height.

28 cm, 7 cm

12. The area of a triangle is 56 in². If the base is 6 in less than the height, find the base and height.

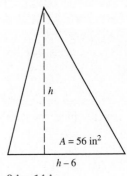

8 in, 14 in

13. A toy box is 20 in high. The length is 3 in longer than the width. If the volume is 80 in³, find the length and width.

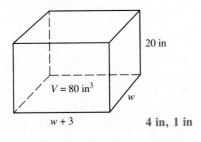

4 in, 1 in

14. The width of a box is 9 cm. The height is 6 cm less than the length. Find the height and length if the volume is 360 cm³.

4 in, 10 in

The profit on a small appliance is given by the equation $P = n^2 - 3n - 60$ where n is the number of appliances sold per day. Use this equation in Exercises 15–18.

15 What is the profit when 30 appliances are sold?
$750

16. What is the profit when 5 appliances are sold?
−$50 (loss of $50)

17 How many appliances were sold on a day when the profit was $120?   15

18. How many appliances were sold on a day when there was a $20 loss?   8

The number of ways of choosing two people from an organization to serve on a committee is given by $N = \frac{1}{2}n(n - 1)$, where n is the number of people in the organization. Use this equation to do Exercises 19–22.

19 How many committees of two can be formed from a group with 7 members?   21

20. How many committees of two can be formed from a club with 10 members?   45

21 If 10 different committees of two can be formed from the members of a club, how many members are in the club?   5

22. If 28 different committees of two can be formed from the members of a sorority, how many women are in the sorority?   8

23. A chemical reaction is described by the equation $C = 2n^2 - 7n + 1$. Find n when $C = 16$ if n is positive.   5

24. The relationship between the number n of radios a company can sell per month and the price of each radio p is given by the equation $n = 1700 - 100p$. Find the price at which a radio should be sold to produce a monthly revenue of $7000. $7 or $10

FOR REVIEW

Solve.

25. $(2x - 5)(x + 10) = 0$
$\frac{5}{2}, -10$

26. $9x^2 + 6x + 1 = 0$
$-\frac{1}{3}$

27. $50y^2 - 32 = 0$
$-\frac{4}{5}, \frac{4}{5}$

28. $3y^2 = 13y + 10$
$-\frac{2}{3}, 5$

The following exercises review material that we covered in Chapters 2 and 3. They will help you prepare for the next section. Perform the indicated operations.

29. $\frac{0}{4}$ 0

30. $\frac{4}{0}$ undefined

31. $\frac{4}{4}$ 1

32. $\frac{-4}{4}$ −1

Supply the missing numerator.

33. $\dfrac{1}{2} = \dfrac{?}{6}$ 3

34. $\dfrac{2}{3} = \dfrac{?}{15}$ 10

35. $\dfrac{-3}{7} = \dfrac{?}{42}$ -18

36. $-\dfrac{10}{21} = \dfrac{?}{105}$ -50

Reduce each fraction to lowest terms.

37. $\dfrac{20}{50}$ $\dfrac{2}{5}$

38. $\dfrac{72}{96}$ $\dfrac{3}{4}$

39. $\dfrac{30}{180}$ $\dfrac{1}{6}$

40. $\dfrac{144}{252}$ $\dfrac{4}{7}$

ANSWERS: 1. -5 or 3   2. -11 or 4   3. -3 or 7   4. 3 cm, 8 cm   5. -16, -15 or 15, 16   6. 7 or -10
7. -13, -11; 11, 13   8. 6, 8   9. 7 ft by 12 ft   10. 2 cm   11. 28 cm, 7 cm   12. 8 in, 14 in   13. 4 in, 1 in
14. 4 in, 10 in   15. $750   16. $-$50 (loss of $50)   17. 15   18. 8   19. 21   20. 45   21. 5   22. 8   23. 5
24. $7 or $10   25. $\frac{5}{2}$, -10 26. $-\frac{1}{3}$ 27. $-\frac{4}{5}$, $\frac{4}{5}$ 28. $-\frac{2}{3}$, 5   29. 0   30. undefined   31. 1   32. -1 33. 3   34. 10
35. -18 36. -50 37. $\frac{2}{5}$ 38. $\frac{3}{4}$ 39. $\frac{1}{6}$ 40. $\frac{4}{7}$

15.6   EXERCISES B

Solve.

1. The product of 4 more than a number and 6 less than the number is 0. Find the number.

-4 or 6

2. The square of a number, plus 15, is the same as 8 times the number. Find the number.

3 or 5

3. The product of 4 more than a number and 6 less than the number is 24. Find the number.

-6 or 8

4. The length of a rectangle is 10 m more than the width. If the area is 144 m^2, find the dimensions.

8 m by 18 m

5. If the length of the side of a square is increased by 3 inches, the area is 49 in^2. Find the length of the side of the original square.

4 in

6. The product of two consecutive integers is 132. Find the integers.

-12, -11 or 11, 12

7. The product of two consecutive even integers is 224. Find the integers.

-16, -14 or 14, 16

8. The sum of the squares of two consecutive odd positive integers is 202. Find the integers.

9, 11

9. The area of a rectangle is 40 cm^2 and the length is 3 cm more than the width. Find the dimensions of the rectangle.

5 cm by 8 cm

10. The area of a square is numerically 12 more than the perimeter. Find the length of a side in ft.

6 ft

11. The area of a triangle is 150 cm^2. If the base is three times the height, find the base and height.

30 cm, 10 cm

12. The area of a triangle is 30 in^2. If the base is 4 in more than the height, find the base and height.

10 in, 6 in

13. A shipping box is 6 in high. The length must be 2 in more than the width and the volume is 210 in^3. Find the length and width.

7 in, 5 in

14. The length of a box is 12 cm and the height is 4 cm less than the width. If the volume of the box is 384 cm^3, find the height and width.

4 cm, 8 cm

The profit on one type of shoe is given by the equation $P = n^2 - 5n - 200$, where n is the number of shoes sold per day. Use this equation to do Exercises 15–18.

15. What is the profit if 20 pairs of shoes are sold?

$100

16. What is the profit if 10 pairs of shoes are sold?

$-$150 ($150 loss)

17. How many pairs of shoes were sold on a day when the profit was $550?

30

18. How many pairs of shoes were sold on a day when there was a loss of $50?

15

In a basketball league containing n teams, the number of different ways that the league champion and second place team can be chosen is given by $N = n(n - 1)$. Use this equation to do Exercises 19–22.

19. How many different first and second place finishers are possible in a league with 8 members?
56

20. How many different first and second place finishers are possible in the NBA, which has 26 teams?
650

21. If it is known that there are 240 different ways for two teams to finish first and second in a league, how many teams are in the league?
16

22. If it has been determined that there are 90 different ways for two teams to finish first and second in a conference, how many teams are in the conference?
10

23. An engineer found that $E = 3u^2 - 16u + 6$ where u is positive. Find u when $E = 18$.
6

24. The relationship between the number n of tables a furniture store can sell per month and the price of each table p is given by the equation $n = 3000 - 200p$. At what price should the store sell each table to produce a monthly revenue of $10,800?
$6 or $9

FOR REVIEW

Solve.

25. $(3x + 2)(x - 8) = 0$
$-\dfrac{2}{3}, 8$

26. $x^2 - 14x + 49 = 0$
7

27. $-45y^2 + 245 = 0$
$-\dfrac{7}{3}, \dfrac{7}{3}$

28. $5y^2 = 18y + 8$
$-\dfrac{2}{5}, 4$

The following exercises review material that we covered in Chapters 2 and 3. They will help you prepare for the next section. Perform the indicated operations.

29. $\dfrac{0}{9}$ 0

30. $\dfrac{9}{0}$ undefined

31. $\dfrac{9}{9}$ 1

32. $\dfrac{9}{-9}$ -1

Supply the missing numerator.

33. $\dfrac{1}{3} = \dfrac{?}{9}$ 3

34. $\dfrac{4}{7} = \dfrac{?}{21}$ 12

35. $\dfrac{6}{-11} = \dfrac{?}{55}$ -30

36. $\dfrac{8}{25} = -\dfrac{?}{150}$ -48

Reduce each fraction to lowest terms.

37. $\dfrac{55}{22}$ $\dfrac{5}{2}$

38. $\dfrac{125}{175}$ $\dfrac{5}{7}$

39. $\dfrac{450}{750}$ $\dfrac{3}{5}$

40. $\dfrac{169}{39}$ $\dfrac{13}{3}$

15.6 EXERCISES C

Solve.

1. A chemical reaction is described by the equation $C = 6n^2 - 17n - 4$. Find n when $C = 10$ if n is a positive integer. [Answer: no solution]

2. A banker uses the equation $M = (n + 2)(n - 7)$. Find n when $M = -20$. [*Hint:* Do not set each of the given factors equal to -20.]   2 or 3

CHAPTER 15 REVIEW

KEY WORDS

15.1 A **common factor** of two expressions is a factor that occurs in both of them.

The **greatest common factor (GCF)** of two terms is the largest factor common to both terms.

A **prime polynomial** cannot be factored.

15.4 A **perfect square trinomial** is one that can be factored into the form $(a + b)^2$ or $(a - b)^2$.

KEY CONCEPTS

15.1 Always factor completely. For example, factor $20x + 30$ as $10(2x + 3)$, not $5(4x + 6)$.

15.2 To factor $x^2 + bx + c = (x + \underline{\quad})(x + \underline{\quad})$, list all pairs of integers whose product is c and fill the blanks with the pair from this list whose sum is b.

15.3 To factor $ax^2 + bx + c = (\underline{\quad}x + \underline{\quad})(\underline{\quad}x + \underline{\quad})$, list all pairs of integers whose product is a for the first blanks in

each binomial factor. Do the same for c using these in the second blanks. By trial and error, select the correct pairs that give the middle term bx. Factoring by grouping may also be used.

15.4 1. $a^2 - b^2 = (a + b)(a - b)$, not $(a - b)^2$
2. $a^2 + 2ab + b^2 = (a + b)^2$
3. $a^2 - 2ab + b^2 = (a - b)^2$

15.5 1. The zero-product rule,

if $a \cdot b = 0$,   then $a = 0$ or $b = 0$,

applies only when one side of the equation is zero. For example, if $a \cdot b = 7$, we *cannot* conclude that $a = 7$ or $b = 7$.

2. Do not use the zero-product rule on a zero-sum or zero-difference equation. For example, $(2x + 1) - (x + 5) = 0$ is *not* a zero-product equation. To solve it, clear parentheses and combine terms. *Do not* set $2x + 1$ and $x + 5$ equal to zero.

REVIEW EXERCISES

Part I

15.1 *Factor by removing the greatest common factor.*

1. $8x + 2$
$2(4x + 1)$

2. $3y^4 - 9y^2$
$3y^2(y^2 - 3)$

3. $6a^2 - 1$
cannot be factored

4. $50x^3 - 40x^2 + 20x$
$10x(5x^2 - 4x + 2)$

5. $6x^4y^2 + 12x^2y^4$
$6x^2y^2(x^2 + 2y^2)$

6. $a^2(a + b) + 5(a + b)$
$(a + b)(a^2 + 5)$

Factor by grouping.

7. $x^3 - 3x^2 + 2x - 6$
$(x - 3)(x^2 + 2)$

8. $x^2y^2 + x^2 + 6y^2 + 6$
$(y^2 + 1)(x^2 + 6)$

15.2, *Factor.*
15.3,
15.4

9. $x^2 + 3x - 10$
$(x + 5)(x - 2)$

10. $9y^2 + 6y + 1$
$(3y + 1)^2$

11. $x^2 + 7x + 10$
$(x + 5)(x + 2)$

12. $y^2 - 8y + 7$
$(y - 7)(y - 1)$

13. $4x^2 - 9$
$(2x + 3)(2x - 3)$

14. $4y^2 - 12y + 9$
$(2y - 3)^2$

15. $x^2 - 12x - 45$
$(x + 3)(x - 15)$

16. $y^2 + 12y - 45$
$(y - 3)(y + 15)$

17. $x^2 + 10x + 16$
$(x + 2)(x + 8)$

18. $2y^2 - 19y + 42$
$(2y - 7)(y - 6)$

19. $5x^2 + 12x - 9$
$(5x - 3)(x + 3)$

20. $2y^2 - 50$
$2(y + 5)(y - 5)$

21. $x^4 - 25$
$(x^2 + 5)(x^2 - 5)$

22. $y^4 - 81$
$(y^2 + 9)(y + 3)(y - 3)$

23. $6x^2 + 13x - 5$
$(3x - 1)(2x + 5)$

24. $6y^2 - 13y - 63$
$(3y + 7)(2y - 9)$

25. $x^2 - 10xy + 25y^2$
$(x - 5y)^2$

26. $x^2 - 13xy + 42y^2$
$(x - 6y)(x - 7y)$

27. $81x^2 - 64y^2$
$(9x + 8y)(9x - 8y)$

28. $81x^2 + 64y^2$
cannot be factored

29. $5x^2 - 90xy + 405y^2$
$5(x - 9y)^2$

30. $6x^2 + 13xy - 28y^2$
$(3x - 4y)(2x + 7y)$

15.5 *Solve.*

31. $(2x - 1)(12x + 5) = 0$
$\dfrac{1}{2}$ or $-\dfrac{5}{12}$

32. $81y^2 - 9 = 0$
$-\dfrac{1}{3}$ or $\dfrac{1}{3}$

33. $4x^2 + 5x + 1 = 0$
$-\dfrac{1}{4}$ or -1

34. $y^2 - 4y + 4 = 0$
2

35. $12x^2 + 6x = 0$
0 or $-\dfrac{1}{2}$

36. $y(y - 5) = 84$
-7 or 12

15.6 37. The product of a number and 5 less than the number is -6. Find the number.

2 or 3

38. The sum of the squares of two consecutive even integers is 452. Find the integers.

-16, -14 or 14, 16

39. The length of a rectangle is 5 times the width. If the area is 180 cm^2, find the dimensions.

6 cm by 30 cm

40. The profit on the sale of sport coats is given by the equation $P = n^2 - 6n - 27$, where n is the number of coats sold per day.

(a) What is the profit when 12 coats are sold?
$45

(b) How many coats were sold on a day when the profit was $160?   17

Part II

Solve.

41. $x(x - 3)(x + 8) = 0$

0, 3, or -8

42. $4x^2 + 4x - 15 = 0$

$-\dfrac{5}{2}$ or $\dfrac{3}{2}$

43. $3x^2 = 11x + 4$

$-\dfrac{1}{3}$ or 4

44. The equation $n = 2400 - 120p$ gives the relationship between the number n of balls sold and the price p of each ball. What price should be set to have a monthly revenue of $12,000? $10

Factor.

45. $9x^2 + 30x + 25$

$(3x + 5)^2$

46. $(x + 1)^2 - 25$

$(x + 6)(x - 4)$

47. $x^2 + 4$

cannot be factored

48. $2x^2 + ax - 10x - 5a$

$(2x + a)(x - 5)$

49. $8y^4 - 2$

$2(2y^2 + 1)(2y^2 - 1)$

50. $4x^2 - 29x + 7$

$(4x - 1)(x - 7)$

ANSWERS: 1. $2(4x + 1)$ 2. $3y^2(y^2 - 3)$ 3. **cannot be factored**   4. $10x(5x^2 - 4x + 2)$ 5. $6x^2y^2(x^2 + 2y^2)$
6. $(a + b)(a^2 + 5)$ 7. $(x - 3)(x^2 + 2)$ 8. $(y^2 + 1)(x^2 + 6)$ 9. $(x + 5)(x - 2)$ 10. $(3y + 1)^2$ 11. $(x + 5)(x + 2)$
12. $(y - 7)(y - 1)$ 13. $(2x + 3)(2x - 3)$ 14. $(2y - 3)^2$ 15. $(x + 3)(x - 15)$ 16. $(y - 3)(y + 15)$ 17. $(x + 2)(x + 8)$
18. $(2y - 7)(y - 6)$ 19. $(5x - 3)(x + 3)$ 20. $2(y + 5)(y - 5)$ 21. $(x^2 + 5)(x^2 - 5)$ 22. $(y^2 + 9)(y + 3)(y - 3)$
23. $(3x - 1)(2x + 5)$ 24. $(3y + 7)(2y - 9)$ 25. $(x - 5y)^2$ 26. $(x - 6y)(x - 7y)$ 27. $(9x + 8y)(9x - 8y)$ 28. **cannot be**
factored   29. $5(x - 9y)^2$ 30. $(3x - 4y)(2x + 7y)$ 31. $\frac{1}{2}$, $-\frac{5}{12}$ 32. $-\frac{1}{3}$, $\frac{1}{3}$ 33. $-\frac{1}{4}$, -1 34. 2   35. 0, $-\frac{1}{2}$
36. -7, 12   37. 2, 3   38. -16, -14; 14, 16   39. 6 cm by 30 cm   40. (a) $45 (b) 17   41. 0, 3, -8 42. $-\frac{5}{2}$, $\frac{3}{2}$
43. $-\frac{1}{3}$, 4   44. $10   45. $(3x + 5)^2$ 46. $(x + 6)(x - 4)$ 47. **cannot be factored**   48. $(2x + a)(x - 5)$
49. $2(2y^2 + 1)(2y^2 - 1)$ 50. $(4x - 1)(x - 7)$

Factor by removing the greatest common factor.

1. $20x + 12$

2. $35y - 7$

3. $24x^4 - 12x^3 + 18x^2$

4. $5x^3y^3 + 40x^3y^2$

5. Factor $7x^2y^2 - y^2 + 14x^2 - 2$ by grouping.

Factor.

6. $x^2 + 15x + 56$

7. $x^2 - 16x + 64$

8. $3y^2 - 75$

9. $x^2 - 8x - 20$

1. _____ $4(5x + 3)$

2. _____ $7(5y - 1)$

3. _____ $6x^2(4x^2 - 2x + 3)$

4. _____ $5x^3y^2(y + 8)$

5. _____ $(y^2 + 2)(7x^2 - 1)$

6. _____ $(x + 7)(x + 8)$

7. _____ $(x - 8)^2$

8. _____ $3(y + 5)(y - 5)$

9. _____ $(x + 2)(x - 10)$

10. $2x^2 + 13x - 24$

10. _____ $(2x - 3)(x + 8)$ _____

11. $3x^2 + 16xy + 5y^2$

11. _____ $(3x + y)(x + 5y)$ _____

Solve.

12. $(4y - 5)(2y + 3) = 0$

12. _____ $\dfrac{5}{4}$ or $-\dfrac{3}{2}$ _____

13. $x^2 - x - 30 = 0$

13. _____ 6 or −5 _____

14. The product of a number and 3 more than the number is 4. Find the number.

14. _____ 1 or −4 _____

15. The area of a triangle is 12 cm². If the base is 2 cm less than the height, find the base and height.

15. _____ 4 cm, 6 cm _____

16. The profit on the sale of one type of dress is $P = n^2 - 3n - 8$, where n is the number of dresses sold per day.
(a) What is the profit when 20 dresses are sold?

16. (a) _____ $332 _____

(b) How many dresses were sold on a day when the profit was $100?

16. (b) _____ 12 _____

Rational Expressions

16.1 BASIC CONCEPTS OF ALGEBRAIC FRACTIONS

STUDENT GUIDEPOSTS

1 Algebraic Fractions and Rational Expressions

2 Excluding Values that Make the Denominator Zero

3 Equivalent Fractions

4 Reducing to Lowest Terms

5 Fractions Equivalent to -1

1 ALGEBRAIC FRACTIONS AND RATIONAL EXPRESSIONS

In Chapters 3 and 4 we studied some of the properties of numerical fractions. Now we define an **algebraic fraction** as a fraction that contains a variable in the numerator or denominator or both. The following are algebraic fractions.

$$\frac{xy}{x+1}, \quad \frac{y^2+1}{y-1}, \quad \frac{\sqrt{a}+5}{2}, \quad \frac{7}{(x-2)(x+1)}, \quad \frac{x+y}{x-y}$$

If an algebraic fraction is the quotient of two polynomials it is called a **rational expression.** Of the algebraic fractions listed above only $\frac{\sqrt{a}+5}{2}$ is *not* a rational expression ($\sqrt{a}+5$ is not a polynomial because a is not raised to a whole number power). The algebraic fractions that we study in this chapter will all be rational expressions.

2 EXCLUDING VALUES THAT MAKE THE DENOMINATOR ZERO

With rational expressions, we need to avoid division by zero, which is undefined. Any value of the variable that makes the denominator zero must be excluded from consideration. The expression is defined for all other values of the variable.

To Find the Values Which Must Be Excluded in a Rational Expression

1. Set the denominator equal to zero.
2. Solve this equation; any solution must be excluded.

| **EXAMPLE 1** FINDING EXCLUDED VALUES | **PRACTICE EXERCISE 1** |

Find the value of the variable that must be excluded in each rational expression.

Find the values of the variable that must be excluded.

(a) $\dfrac{2x}{x + 7}$

Set the denominator equal to zero and solve.

$$x + 7 = 0$$
$$x = -7$$

Check: $\quad \dfrac{2(-7)}{-7 + 7} = \dfrac{-14}{0}$

Since division by zero is not defined, -7 must be excluded. The expression is defined for all x except -7.

(a) $\dfrac{a}{a - 2}$

(b) $\dfrac{x^2 + x - 3}{7}$

(b) $\dfrac{(a - 1)(a + 1)}{2}$

Set the denominator equal to zero and solve.

$$2 = 0$$

Since there is no solution to this equation, there are no values to exclude. The expression is defined for all values of a.

(c) $\dfrac{3y}{(y - 1)(y + 4)}$

We need to solve $(y - 1)(y + 4) = 0$, so we use the zero-product rule.

$$y - 1 = 0 \quad \text{or} \quad y + 4 = 0$$
$$y = 1 \qquad\qquad y = -4$$

The values to exclude are 1 and -4. The expression is defined for all values of y except 1 and -4.

(c) $\dfrac{4w}{(w + 1)(w - 7)}$

(d) $\dfrac{x^2 + 1}{x^2 + 5x + 6}$

We need to solve $x^2 + 5x + 6 = 0$. First factor and then use the zero-product rule.

$$x^2 + 5x + 6 = 0$$
$$(x + 2)(x + 3) = 0$$
$$x + 2 = 0 \quad \text{or} \quad x + 3 = 0$$
$$x = -2 \qquad\qquad x = -3$$

The values to exclude are -2 and -3. The expression is defined for all values of x except -2 and -3.

(d) $\dfrac{y^2}{3y^2 - 11y - 4}$

Answers: (a) 2   (b) There are none.   (c) -1 and 7   (d) 4 and $-\frac{1}{3}$

❸ EQUIVALENT FRACTIONS

In Chapter 4 we saw that multiplying or dividing the numerator and denominator of a fraction by the same nonzero number gives us an equivalent fraction.

$$\frac{4}{6} = \frac{4 \cdot 3}{6 \cdot 3} = \frac{12}{18} \qquad \text{$\frac{4}{6}$ is equivalent to $\frac{12}{18}$}$$

$$\frac{4}{6} = \frac{4 \div 2}{6 \div 2} = \frac{2}{3} \qquad \text{$\frac{4}{6}$ is equivalent to $\frac{2}{3}$}$$

We can generalize the idea of equivalent fractions to include algebraic fractions.

Fundamental Principle of Fractions

If the numerator and denominator of an algebraic fraction are multiplied or divided by the same nonzero expression, the resulting fraction is **equivalent** to the original. That is, if $\frac{P}{Q}$ is an algebraic fraction, $Q \neq 0$, and R is a nonzero expression, then

$$\frac{P}{Q} \text{ is equivalent to } \frac{P \cdot R}{Q \cdot R} \text{ and to } \frac{P \div R}{Q \div R}.$$

EXAMPLE 2   EQUIVALENT FRACTIONS

Are the given fractions equivalent?

(a) $\dfrac{3}{x + 2}$ and $\dfrac{3(x + 1)}{(x + 2)(x + 1)}$

$$\frac{3 \cdot (x + 1)}{(x + 2) \cdot (x + 1)} = \frac{3(x + 1)}{(x + 2)(x + 1)} \qquad \begin{array}{l}\text{Numerator and denominator}\\\text{are both multiplied by } x + 1\end{array}$$

The fractions are equivalent since the first becomes the second when both numerator and denominator are multiplied by $x + 1$.

We may also divide numerator and denominator of the second fraction by $x + 1$ to obtain the first.

$$\frac{3(x + 1) \div (x + 1)}{(x + 2)(x + 1) \div (x + 1)} = \frac{\dfrac{3(x + 1)}{(x + 1)}}{\dfrac{(x + 2)(x + 1)}{(x + 1)}} = \frac{3}{x + 2}$$

This division can be abbreviated by showing the expression is equivalent to $\frac{3}{x + 2}$ times 1.

$$\frac{3(x + 1)}{(x + 2)(x + 1)} = \frac{3}{x + 2} \cdot \frac{x + 1}{x + 1}$$

$$= \frac{3}{x + 2} \cdot 1 = \frac{3}{x + 2}$$

We usually accomplish this by dividing out the factor common to both numerator and denominator.

$$\frac{3(x + 1)}{(x + 2)(x + 1)} = \frac{3\cancel{(x + 1)}}{(x + 2)\cancel{(x + 1)}} = \frac{3}{x + 2}$$

PRACTICE EXERCISE 2

Are the given fractions equivalent?

(a) $\dfrac{a(a + 2)}{(a - 3)(a + 2)}$ and $\dfrac{a}{a - 3}$

(b) $\dfrac{5}{y + 1}$ and $\dfrac{5}{y - 1}$

The word "cancel" is often used to describe this process, but we will use "divide out."

(b) $\dfrac{3}{x+2}$ and $\dfrac{7}{x+1}$

These are not equivalent since there is no expression that one can be multiplied by, in both numerator and denominator, to give the other.

Answers: (a) yes   (b) no

---

④ REDUCING TO LOWEST TERMS

A fraction is **reduced to lowest terms** when 1 (or -1) is the only number or expression that divides both numerator and denominator. For example, we reduce $\frac{30}{12}$ to lowest terms as follows.

$$\frac{30}{12} = \frac{\cancel{2} \cdot \cancel{3} \cdot 5}{\cancel{2} \cdot 2 \cdot \cancel{3}} = \frac{5}{2}$$ Factor completely and divide out or cancel common factors

```
/////////////  CAUTION  /////////////
```

Divide out or cancel *factors* only, never cancel *terms*. The only expressions that can be divided out are those that are multiplied (never added or subtracted) by every other expression in the numerator or denominator. For example,

$$\frac{5 \cdot 6}{5} = \frac{\cancel{5} \cdot 6}{\cancel{5}} \quad \text{but} \quad \frac{5 + 6}{5} \neq \frac{\cancel{5} + 6}{\cancel{5}}.$$

```
/////////////
```

We can generalize this procedure to reduce rational expressions.

To Reduce a Rational Expression to Lowest Terms

1. Factor numerator and denominator completely.

2. Divide out all common factors.

3. Multiply the remaining factors in the numerator and multiply the remaining factors in the denominator.

EXAMPLE 3 REDUCING FRACTIONS	PRACTICE EXERCISE 3

Reduce each fraction to lowest terms.

(a) $\dfrac{12x^2}{18x^3} = \dfrac{2 \cdot 2 \cdot 3 \cdot \cancel{x} \cdot \cancel{x}}{2 \cdot 3 \cdot 3 \cdot \cancel{x} \cdot \cancel{x} \cdot x} = \dfrac{2}{3x}$ Factor completely and divide out common factors

With practice you can avoid factoring completely to prime factors. For example, the above might take the form

$$\frac{12x^2}{18x^3} = \frac{\cancel{6} \cdot 2 \cdot \cancel{x^2}}{\cancel{6} \cdot 3 \cdot \cancel{x^2} \cdot x} = \frac{2}{3x}.$$

Reduce each fraction to lowest terms.

(a) $\dfrac{15y^6}{35y^2}$

(b) $\dfrac{x}{2x^2} = \dfrac{\not{x}}{2 \cdot \not{x} \cdot x} = \dfrac{1}{2x}$

Notice that when the factors x are divided out, 1 is left in the numerator, not 0.

(c) $\dfrac{2x^2 + x}{3x^3 + 3x} = \dfrac{\not{x}(2x + 1)}{3\not{x}(x^2 + 1)} = \dfrac{2x + 1}{3(x^2 + 1)}$

(d) $\dfrac{7a + 14}{7} = \dfrac{\not{7}(a + 2)}{\not{7}} = \dfrac{a + 2}{1} = a + 2$

(e) $\dfrac{2 + 2y}{2y} = \dfrac{\not{2}(1 + y)}{\not{2}y} = \dfrac{1 + y}{y}$

We cannot divide out the ys. The y in the numerator is a term, not a factor. If we replace y with the number 2, it is easy to see that

$$\dfrac{1 + y}{y} \text{ is not equal to } \dfrac{1 + \not{y}}{\not{y}}.$$

(f) $\dfrac{x^2 + 3xy + 2y^2}{x^2 - 4y^2} = \dfrac{(x + y)(x + 2y)}{(x - 2y)(x + 2y)} = \dfrac{x + y}{x - 2y}$

Why can't we divide out the x and the y in the answer to obtain $-\frac{1}{2}$?

(b) $\dfrac{a - 3}{5(a - 3)}$

(c) $\dfrac{a^2 - 3a}{5a^3 + 10a}$

(d) $\dfrac{3}{3y - 6}$

(e) $\dfrac{-5x}{x - 5}$

(f) $\dfrac{y^2 + 2y - 15}{y^2 - 2y - 35}$

Answers: **(a)** $\frac{3y^4}{7}$ **(b)** $\frac{1}{5}$
(c) $\frac{a - 3}{5(a^2 + 2)}$ **(d)** $\frac{1}{y - 2}$ **(e)** cannot
be reduced **(f)** $\frac{y - 3}{y - 7}$

⑤ FRACTIONS EQUIVALENT TO −1

Consider the fraction $\frac{5 - x}{x - 5}$. Our first impression might be that it is reduced to lowest terms. However, if we factor -1 from the numerator, we obtain

$$\dfrac{5 - x}{x - 5} = \dfrac{(-1)(-5 + x)}{x - 5}$$
$$= \dfrac{(-1)(x - 5)}{(x - 5)} \qquad -5 + x = x - 5$$
$$= -1.$$

In general, we can show that if $a \neq b$,

$$\dfrac{a - b}{b - a} = -1.$$

For example, if $a = 5$ and $b = 2$,

$$\dfrac{a - b}{b - a} = \dfrac{5 - 2}{2 - 5} = \dfrac{3}{-3} = -1.$$

Whenever a fraction of this type results, simply replace it with -1.

⁄⁄⁄⁄⁄⁄⁄⁄⁄⁄⁄⁄ **CAUTION** ⁄⁄⁄⁄⁄⁄⁄⁄⁄⁄⁄⁄

Remember that $a + b = b + a$ so that $\frac{a + b}{b + a} = 1$, not -1.

⁄⁄⁄⁄⁄⁄⁄⁄

| EXAMPLE 4   REDUCING $\frac{a-b}{b-a}$ TO -1 | PRACTICE EXERCISE 4 |

Reduce $\dfrac{x^2 - xy}{3y - 3x}$ to lowest terms.

$$\frac{x^2 - xy}{3y - 3x} = \frac{x(x-y)}{3(y-x)} = \frac{x}{3} \cdot \frac{x-y}{y-x}$$

$$= \frac{x}{3} \cdot (-1) \qquad \frac{x-y}{y-x} = -1$$

$$= -\frac{x}{3}$$

Reduce $\dfrac{a^2b - a^3}{5a - 5b}$ to lowest terms.

Answer: $-\frac{a^2}{5}$

16.1   EXERCISES A

Find the values of the variable that must be excluded.

1. $\dfrac{3}{x+1}$ $\quad -1$

2. $\dfrac{a}{a-7}$ $\quad 7$

3. $\dfrac{y+1}{y(y+3)}$ $\quad 0, -3$

4. $\dfrac{z+2}{2(z+4)}$ $\quad -4$

5. $\dfrac{x+2}{(x-1)(x+4)}$ $\quad 1, -4$

6. $\dfrac{a-3}{(a-1)(a+1)}$ $\quad 1, -1$

7 $\dfrac{2x+7}{x^2+2x+1}$ $\quad -1$

8. $\dfrac{3y-4}{y^2-9}$ $\quad 3, -3$

9. $\dfrac{(x-1)(x+1)}{7}$ $\quad$ none

Are the given fractions equivalent? Explain.

10. $\dfrac{2}{7}, \dfrac{6}{21}$ $\quad$ yes

11. $\dfrac{3}{y}, \dfrac{-3}{-y}$ $\quad$ yes

12. $\dfrac{2x}{4}, \dfrac{x}{4}$ $\quad$ no

13. $\dfrac{a^2}{3}, \dfrac{-2a^2}{-6}$ $\quad$ yes

14. $\dfrac{x+1}{2}, \dfrac{3(x+1)}{6}$ $\quad$ yes

15. $\dfrac{x+1}{9}, \dfrac{(x+1)(x-1)}{9(x-1)}$ $\quad$ yes

16 $\dfrac{2}{x^2}, \dfrac{2+x}{x^2+x}$ $\quad$ no

17 $\dfrac{x-1}{3x-1}, \dfrac{x}{3x}$ $\quad$ no

18. $\dfrac{z-2}{z^2-4}, \dfrac{1}{z+2}$ $\quad$ yes

Reduce to lowest terms.

19. $\dfrac{21}{30}$ $\dfrac{7}{10}$

20. $\dfrac{2y^2}{10y^3}$ $\dfrac{1}{5y}$

21. $\dfrac{4x^3y}{2xy^2}$ $\dfrac{2x^2}{y}$

22. $\dfrac{22ab^5}{11ab^2}$ $2b^3$

23. $\dfrac{3x^2z}{21x^6z^2}$ $\dfrac{1}{7x^4z}$

24. $\dfrac{a(a+1)}{5(a+1)}$ $\dfrac{a}{5}$

25. $\dfrac{(x+2)(2x-3)}{(x+2)(2x+3)}$ $\dfrac{2x-3}{2x+3}$

26. $\dfrac{z+9}{z^2+9}$ $\dfrac{z+9}{z^2+9}$

27. $\dfrac{a+3}{a^2-9}$ $\dfrac{1}{a-3}$

28. $\dfrac{w^2+7w}{w^2-49}$ $\dfrac{w}{w-7}$

29 $\dfrac{x^2-3x-10}{x^2-6x+5}$ $\dfrac{x+2}{x-1}$

30. $\dfrac{2a(a+5)}{10a+2a^2}$ 1

31. $\dfrac{x-7}{7-x}$ -1

32 $\dfrac{x^2-9}{3-x}$ $-(x+3)$

33. $\dfrac{y+4}{y-4}$ $\dfrac{y+4}{y-4}$

34. $\dfrac{x+y}{x^2-y^2}$ $\dfrac{1}{x-y}$

35. $\dfrac{a^2+2ab+b^2}{a+b}$ $a+b$

36 $\dfrac{x^2-7xy+6y^2}{x^2-4xy-12y^2}$ $\dfrac{x-y}{x+2y}$

FOR REVIEW

The following exercises review material from Chapter 3 to help you prepare for the next section. Perform the indicated operations.

37. $\dfrac{5}{8} \cdot \dfrac{2}{15}$ $\dfrac{1}{12}$

38. $\dfrac{7}{5} \cdot 20$ 28

39. $\dfrac{2}{3} \div \dfrac{4}{9}$ $\dfrac{3}{2}$

40. $\dfrac{4}{5} \div 2$ $\dfrac{2}{5}$

41. Factor 300 into a product of primes.
$2 \cdot 2 \cdot 3 \cdot 5 \cdot 5$

ANSWERS: 1. -1 2. 7 3. 0, -3 4. -4 5. 1, -4 6. 1, -1 7. -1 8. 3, -3 9. none 10. yes 11. yes 12. no 13. yes 14. yes 15. yes 16. no 17. no 18. yes 19. $\frac{7}{10}$ 20. $\frac{1}{5y}$ 21. $\frac{2x^2}{y}$ 22. $2b^3$ 23. $\frac{1}{7x^4z}$ 24. $\frac{a}{5}$ 25. $\frac{2x-3}{2x+3}$ 26. $\frac{z+9}{z^2+9}$ 27. $\frac{1}{a-3}$ 28. $\frac{w}{w-7}$ 29. $\frac{x+2}{x-1}$ 30. 1 31. -1 32. $-(x+3)$ 33. $\frac{y+4}{y-4}$ 34. $\frac{1}{x-y}$ 35. $a+b$ 36. $\frac{x-y}{x+2y}$ 37. $\frac{1}{12}$ 38. 28 39. $\frac{3}{2}$ 40. $\frac{2}{5}$ 41. $2 \cdot 2 \cdot 3 \cdot 5 \cdot 5$

16.1   EXERCISES B

Find the values of the variable that must be excluded.

1. $\dfrac{5}{x-6}$ 6

2. $\dfrac{x^2+8}{x+5}$ −5

3. $\dfrac{y-3}{y(y-4)}$ 0, 4

4. $\dfrac{z-8}{2(z+10)}$ −10

5. $\dfrac{x+1}{(x-2)(x+6)}$ 2, −6

6. $\dfrac{a+5}{(a-3)(a+3)}$ 3, −3

7. $\dfrac{3x-1}{x^2-2x+1}$ 1

8. $\dfrac{3y+7}{y^2-16}$ 4, −4

9. $\dfrac{z+5}{z^2+9}$ none

Are the given fractions equivalent? Explain.

10. $\dfrac{3}{5}, \dfrac{6}{10}$ yes

11. $\dfrac{5}{y}, \dfrac{-5}{-y}$ yes

12. $\dfrac{3x}{6}, \dfrac{x}{6}$ no

13. $\dfrac{a^3}{4}, \dfrac{-2a^3}{-8}$ yes

14. $\dfrac{x+5}{5}, \dfrac{2(x+5)}{10}$ yes

15. $\dfrac{x+3}{8}, \dfrac{(x+3)(x-4)}{8(x-4)}$ yes

16. $\dfrac{3}{x^3}, \dfrac{3-x}{x^3-x}$ no

17. $\dfrac{z+3}{z^2-9}, \dfrac{1}{z-3}$ yes

18. $\dfrac{x^2+2}{3x^2+2}, \dfrac{x^2}{3x^2}$ no

Reduce to lowest terms.

19. $\dfrac{22}{55}$ $\dfrac{2}{5}$

20. $\dfrac{45x^6}{15x^2}$ $3x^4$

21. $\dfrac{9a^4b}{3ab^3}$ $\dfrac{3a^3}{b^2}$

22. $\dfrac{25xy^3}{5xy^4}$ $\dfrac{5}{y}$

23. $\dfrac{6ay}{18a^2y^3}$ $\dfrac{1}{3ay^2}$

24. $\dfrac{a^2(a-5)}{3(a-5)}$ $\dfrac{a^2}{3}$

25. $\dfrac{(2x+1)(x+5)}{(2x+1)(x-5)}$ $\dfrac{x+5}{x-5}$

26. $\dfrac{z+16}{z^2+16}$ $\dfrac{z+16}{z^2+16}$

27. $\dfrac{a+4}{a^2-16}$ $\dfrac{1}{a-4}$

28. $\dfrac{x^3+5x^2}{x(x^2-25)}$ $\dfrac{x}{x-5}$

29. $\dfrac{x^2-3x-10}{x^2-8x+15}$ $\dfrac{x+2}{x-3}$

30. $\dfrac{3a^2(a+4)}{9a+6a^2}$ $\dfrac{a(a+4)}{3+2a}$

31. $\dfrac{8-x}{x-8}$ −1

32. $\dfrac{16-x^2}{x-4}$ $(x+4)$

33. $\dfrac{2y+3}{2y-3}$ $\dfrac{2y+3}{2y-3}$

34. $\dfrac{x-y}{x^2-y^2}$ $\dfrac{1}{x+y}$

35. $\dfrac{a^2+7ab+10b^2}{a+5b}$ $a+2b$

36. $\dfrac{x^2-10xy+25y^2}{x^2-4xy-5y^2}$ $\dfrac{x-5y}{x+y}$

FOR REVIEW

The following exercises review material from Chapter 3 to help you prepare for the next section. Perform the indicated operations.

37. $\dfrac{7}{10} \cdot \dfrac{5}{14}$ $\dfrac{1}{4}$

38. $\dfrac{8}{3} \cdot 27$ 72

39. $\dfrac{3}{4} \div \dfrac{9}{16}$ $\dfrac{4}{3}$

40. $7 \div \dfrac{14}{3}$ $\dfrac{3}{2}$

41. Factor 420 into a product of primes.
$2 \cdot 2 \cdot 3 \cdot 5 \cdot 7$

16.1 EXERCISES C

Reduce to lowest terms. The factoring formulas in Exercises C of Section 15.4 are used in these exercises.

1. $\dfrac{5x^2 - 25xy + 30y^2}{10x^2 - 40y^2}$

$\dfrac{x - 3y}{2(x + 2y)}$

2. $\dfrac{27x^3 - y^3}{18x^2 + 6xy + 2y^2}$

$\left[\text{Answer:} \quad \dfrac{3x - y}{2} \right]$

3. $\dfrac{x^4 + 8xy^3}{x^4 - 16y^4}$

$\dfrac{x(x^2 + 2xy + 4y^2)}{(x - 2y)(x^2 + 4y^2)}$

16.2 MULTIPLICATION AND DIVISION OF FRACTIONS

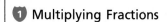 STUDENT GUIDEPOSTS

❶ Multiplying Fractions   ❷ Dividing Fractions

❶ MULTIPLYING FRACTIONS

In Chapter 3 we saw that the product of two or more fractions is equal to the product of all numerators divided by the product of all denominators. The resulting fraction should be reduced to lowest terms. For example,

$$\frac{3}{4} \cdot \frac{2}{9} = \frac{3 \cdot 2}{4 \cdot 9} = \frac{6}{36}.$$

To reduce $\frac{6}{36}$ to lowest terms, factor the numerator and denominator.

$$\frac{6}{36} = \frac{\cancel{2} \cdot \cancel{3}}{\cancel{2} \cdot \cancel{3} \cdot 2 \cdot 3} = \frac{1}{6}$$

To accomplish multiplication and reduction in one process we factor numerators and denominators first, divide out common factors, and then multiply. The resulting fraction is the product.

$$\frac{3}{4} \cdot \frac{2}{9} = \frac{3}{2 \cdot 2} \cdot \frac{2}{3 \cdot 3} = \frac{\cancel{3} \cdot \cancel{2}}{\cancel{2} \cdot 2 \cdot \cancel{3} \cdot 3} = \frac{1}{2 \cdot 3} = \frac{1}{6}$$

We can generalize this procedure to multiply algebraic fractions.

To Multiply Algebraic Fractions
1. Factor all numerators and denominators completely.
2. Place the factored product of all numerators over the factored product of all denominators.
3. Divide out common factors before multiplying the remaining numerator factors and multiplying the remaining denominator factors.

EXAMPLE 1 MULTIPLYING FRACTIONS

Multiply.

$$\frac{x}{6} \cdot \frac{3}{2x^2} = \frac{x}{2 \cdot 3} \cdot \frac{3}{2 \cdot x \cdot x} \qquad \text{Factor}$$

$$= \frac{\cancel{x} \cdot \cancel{3}}{2 \cdot \cancel{3} \cdot 2 \cdot \cancel{x} \cdot x} \qquad \text{Indicate product and divide out common factors}$$

$$= \frac{1}{2 \cdot 2 \cdot x} \qquad \text{The numerator is 1, } not \text{ 0}$$

$$= \frac{1}{4x} \qquad \text{Multiply}$$

PRACTICE EXERCISE 1

Multiply. $\dfrac{y^2}{15} \cdot \dfrac{5}{y}$

Answer: $\frac{y}{3}$

EXAMPLE 2 MULTIPLYING FRACTIONS

Multiply.

$$\frac{a^2 - 9}{a^2 - 4a + 4} \cdot \frac{a - 2}{a + 3}$$

$$= \frac{(a - 3)(a + 3)}{(a - 2)(a - 2)} \cdot \frac{(a - 2)}{(a + 3)} \qquad \text{Factor}$$

$$= \frac{(a - 3)\cancel{(a + 3)}\cancel{(a - 2)}}{(a - 2)\cancel{(a - 2)}\cancel{(a + 3)}} \qquad \text{Divide out common factors}$$

$$= \frac{a - 3}{a - 2} \qquad Do\ not \text{ divide out the term } a; \text{ it is not a factor}$$

PRACTICE EXERCISE 2

Multiply. $\dfrac{x^2 - 2x}{x^2 + 11x + 28} \cdot \dfrac{x + 7}{x - 2}$

Answer: $\frac{x}{x + 4}$

EXAMPLE 3 MULTIPLYING FRACTIONS

Multiply.

$$\frac{x^2 - y^2}{xy} \cdot \frac{xy - x^2}{x^2 - 2xy + y^2}$$

$$= \frac{(x - y)(x + y)}{xy} \cdot \frac{x(y - x)}{(x - y)(x - y)} \qquad \text{Factor}$$

$$= \frac{\cancel{(x - y)}(x + y) \cdot \cancel{x} \cdot (y - x)}{\cancel{x} \cdot y \cdot \cancel{(x - y)}(x - y)} \qquad \text{Divide out common factors}$$

$$= \frac{x + y}{y} \cdot \frac{(y - x)}{(x - y)}$$

$$= \frac{x + y}{y} \cdot (-1) \qquad \tfrac{y - x}{x - y} = -1$$

$$= -\frac{x + y}{y}$$

PRACTICE EXERCISE 3

Multiply.

$$\frac{a^2 - ab}{2b^2 - ab - a^2} \cdot \frac{2a^2 + 3ab + b^2}{2a^2 + ab}$$

Answer: $-\frac{a + b}{2b + a}$

❷ DIVIDING FRACTIONS

In Chapter 3 we divided one numerical fraction by another. In

$$\frac{2}{3} \div \frac{3}{5},$$

we divide the dividend $\frac{2}{3}$ by the divisor $\frac{3}{5}$. To find the quotient, we multiply the dividend by the **reciprocal** of the divisor. That is,

$$\frac{2}{3} \div \frac{3}{5} = \frac{2}{3} \cdot \frac{5}{3} = \frac{2 \cdot 5}{3 \cdot 3} = \frac{10}{9}. \qquad \text{$\frac{5}{3}$ is the reciprocal of $\frac{3}{5}$}$$

We use exactly the same process when dividing algebraic fractions.

To Divide Algebraic Fractions
Multiply the dividend by the reciprocal of the divisor.
$$\frac{P}{Q} \div \frac{R}{S} = \frac{P}{Q} \cdot \frac{S}{R}$$

The following table lists several algebraic fractions and their reciprocals.

Fraction	Reciprocal	
$\dfrac{2}{x}$	$\dfrac{x}{2}$	
$\dfrac{x+1}{(x-5)^2}$	$\dfrac{(x-5)^2}{x+1}$	
5	$\dfrac{1}{5}$	5 can be thought of as $\frac{5}{1}$
x	$\dfrac{1}{x}$	x is $\frac{x}{1}$
$\dfrac{1}{3}$	3	3 is $\frac{3}{1}$
$x+y$	$\dfrac{1}{x+y}$	Not $\frac{1}{x} + \frac{1}{y}$. For example, $\frac{1}{2+3} = \frac{1}{5} \neq \frac{1}{2} + \frac{1}{3} = \frac{5}{6}$

EXAMPLE 4 DIVIDING FRACTIONS

Divide.

(a) $\dfrac{x}{x+1} \div \dfrac{x}{2} = \dfrac{x}{x+1} \cdot \dfrac{2}{x}$ $\frac{2}{x}$ is the reciprocal of $\frac{x}{2}$

$\qquad = \dfrac{\cancel{x} \cdot 2}{(x+1) \cdot \cancel{x}}$ Divide out common factor

$\qquad = \dfrac{2}{x+1}$

PRACTICE EXERCISE 4

Divide.

(a) $\dfrac{y-5}{y} \div \dfrac{5}{y}$

(b) $\dfrac{y + 3}{y} \div y = \dfrac{y + 3}{y} \cdot \dfrac{1}{y}$ $\quad \frac{1}{y}$ is the reciprocal of y

$\qquad\qquad = \dfrac{y + 3}{y^2}$

(c) $\dfrac{a^2 - 4a + 4}{a^2 - 9} \div \dfrac{a - 2}{a + 3} = \dfrac{a^2 - 4a + 4}{a^2 - 9} \cdot \dfrac{a + 3}{a - 2}$

$\qquad\qquad\qquad = \dfrac{(a - 2)(a - 2)}{(a - 3)(a + 3)} \cdot \dfrac{(a + 3)}{(a - 2)}$

$\qquad\qquad\qquad = \dfrac{(a - 2)(a - 2)(a + 3)}{(a - 3)(a + 3)(a - 2)}$

$\qquad\qquad\qquad = \dfrac{a - 2}{a - 3} \quad$ Do not divide out a

(b) $\dfrac{b}{a + b} \div b$

(c) $\dfrac{y^2 + 2y - 15}{y^2 - 25} \div \dfrac{y - 3}{y - 5}$

Answers: (a) $\frac{y - 5}{5}$ (b) $\frac{1}{a + b}$
(c) 1

EXAMPLE 5 DIVIDING FRACTIONS

Divide.

$\dfrac{x^2 - 3xy + 2y^2}{x^2 - 4y^2} \div (x^2 + xy - 2y^2)$

$= \dfrac{x^2 - 3xy + 2y^2}{x^2 - 4y^2} \cdot \dfrac{1}{x^2 + xy - 2y^2} \quad$ Find reciprocal and multiply

$= \dfrac{(x - y)(x - 2y)}{(x + 2y)(x - 2y)} \cdot \dfrac{1}{(x - y)(x + 2y)} \quad$ Factor

$= \dfrac{(x - y)(x - 2y)}{(x + 2y)(x - 2y)(x - y)(x + 2y)} \quad$ Divide out common factors

$= \dfrac{1}{(x + 2y)^2}$

PRACTICE EXERCISE 5

Divide.

$\dfrac{a^2 + 8a - 33}{a^2 - 2a - 3} \div (a^2 + 12a + 11)$

Answer: $\dfrac{1}{(a + 1)^2}$

16.2 EXERCISES A

Multiply.

1. $\dfrac{1}{x^2} \cdot \dfrac{x}{3} \qquad \dfrac{1}{3x}$

2. $\dfrac{3a}{4} \cdot \dfrac{6}{9a^2} \qquad \dfrac{1}{2a}$

3. $\dfrac{2y}{5} \cdot \dfrac{25}{4y^3} \qquad \dfrac{5}{2y^2}$

4. $\dfrac{2x}{(x + 1)^2} \cdot \dfrac{x + 1}{4x^2} \qquad \dfrac{1}{2x(x + 1)}$

5. $\dfrac{x^2 - 36}{x^2 - 6x} \cdot \dfrac{x}{x + 6} \qquad 1$

6. $\dfrac{z^2}{z + 2} \cdot \dfrac{z^2 - 4}{z^2 - 2z} \qquad z$

7. $\dfrac{5x + 5}{x - 2} \cdot \dfrac{x^2 - 4x + 4}{x^2 - 1}$

$\dfrac{5(x - 2)}{x - 1}$

8. $\dfrac{y^2 - 3y - 10}{y^2 - 4y + 4} \cdot \dfrac{y - 2}{y - 5}$

$\dfrac{y + 2}{y - 2}$

9. $\dfrac{x^2 + 2x + 1}{9x^2} \cdot \dfrac{3x^3}{x^2 - 1}$

$\dfrac{x(x + 1)}{3(x - 1)}$

10 $\dfrac{z^2 - z - 20}{z^2 + 7z + 12} \cdot \dfrac{z + 3}{z^2 - 25}$

$\dfrac{1}{z + 5}$

11. $\dfrac{y^2 + 6y + 5}{7y^2 - 63} \cdot \dfrac{7y + 21}{(y + 5)^2}$

$\dfrac{y + 1}{(y - 3)(y + 5)}$

12. $\dfrac{16 - x^2}{5x - 1} \cdot \dfrac{5x^2 - x}{16 - 8x + x^2}$

$\dfrac{x(4 + x)}{4 - x}$

13. $\dfrac{a^2 - 4}{a^2 - 4a + 4} \cdot \dfrac{a^2 - 9a + 14}{a^3 + 2a^2}$

$\dfrac{a - 7}{a^2}$

14 $\dfrac{2x^2 - 5xy + 3y^2}{x^2 - y^2} \cdot (x^2 + 2xy + y^2)$

$(2x - 3y)(x + y)$

15. $\dfrac{x^2 - y^2}{(x + y)^2} \cdot \dfrac{x + y}{x - y}$ 1

16. $\dfrac{x^2 - 7xy + 6y^2}{x^2 - xy - 2y^2} \cdot \dfrac{x + y}{x - 6y}$ $\dfrac{x - y}{x - 2y}$

Find the reciprocal.

17. $\dfrac{3}{7}$ $\dfrac{7}{3}$

18. $\dfrac{1}{2}$ 2

19. 5 $\dfrac{1}{5}$

20. $\dfrac{1}{x + 1}$ $x + 1$

21. $\dfrac{2x - 3}{x + 5}$ $\dfrac{x + 5}{2x - 3}$

22 $z + 2$ $\dfrac{1}{z + 2}$

Divide.

23. $\dfrac{x}{x + 2} \div \dfrac{x}{3}$

$\dfrac{3}{x + 2}$

24. $\dfrac{2a}{a + 1} \div \dfrac{5a}{a + 1}$

$\dfrac{2}{5}$

25. $\dfrac{2(a + 3)}{7} \div \dfrac{4(a + 3)}{21}$

$\dfrac{3}{2}$

26. $\dfrac{3(x^2+5)}{4(x^2-3)} \div \dfrac{x^2+5}{4(x^2-3)}$

3

27. $\dfrac{5y^4}{y^2-1} \div \dfrac{5y^3}{y^2+2y+1}$

$\dfrac{y(y+1)}{y-1}$

28. $\dfrac{6a^2}{a^2-25} \div \dfrac{3a^3}{5a^2-25a}$

$\dfrac{10}{a+5}$

29 $(x+6) \div \dfrac{x^2-36}{x^2-6x}$

x

30. $\dfrac{z^2-4z+4}{z^2-1} \div \dfrac{z-2}{5z+5}$

$\dfrac{5(z-2)}{z-1}$

31. $\dfrac{y^2-y-20}{y^2+7y+12} \div \dfrac{y^2-25}{y+3}$

$\dfrac{1}{y+5}$

32. $\dfrac{a^2+10a+21}{a^2-2a-15} \div \dfrac{a+7}{a^2-25}$

$a+5$

33. $\dfrac{y^3-64y}{2y^2+16y} \div \dfrac{y^2-9y+8}{y^2+4y-5}$

$\dfrac{y+5}{2}$

34 $\dfrac{x^2+16x+64}{2x^2-128} \div \dfrac{3x^2+30x+48}{x^2-6x-16}$

$\dfrac{1}{6}$

35 $\dfrac{x^2-5xy+6y^2}{x^2-4y^2} \div (x^2-2xy-3y^2)$

$\dfrac{1}{(x+2y)(x+y)}$

36. $\dfrac{x+2y}{x^2-y^2} \div \dfrac{x+2y}{x+y}$

$\dfrac{1}{x-y}$

37. $\dfrac{x^2-81y^2}{x^2+18xy+81y^2} \div \dfrac{x-9y}{x+9y}$

1

38 $\dfrac{a^2-y^2}{a^2-ay} \cdot \dfrac{2a^2+ay}{a^2-4y^2} \div \dfrac{a+y}{a+2y}$

$\dfrac{2a+y}{a-2y}$

FOR REVIEW

Find the values of the variable that must be excluded.

39. $\dfrac{y-1}{y^2+2y}$ 0, −2

40. $\dfrac{a}{a^2+5}$ none

41. $\dfrac{(x+1)(x-5)}{3x}$ 0

Are the given fractions equivalent?

42. $\dfrac{x+5}{2x^2+5}, \dfrac{x}{2x^2}$ no

43. $\dfrac{x}{3}, \dfrac{x^2+2x}{3(x+2)}$ yes

44. $\dfrac{1}{x+1}, \dfrac{x+1}{1}$ no

45. Change $\frac{x}{2}$ to an equivalent fraction with the given denominator.

   (a) 10 $\dfrac{5x}{10}$

   (b) 2x $\dfrac{x^2}{2x}$

   (c) $2x^2$ $\dfrac{x^3}{2x^2}$

   (d) $2(x+1)$ $\dfrac{x(x+1)}{2(x+1)}$

   (e) $2(2x-1)$ $\dfrac{x(2x-1)}{2(2x-1)}$

   (f) $2(x^2-1)$ $\dfrac{x(x^2-1)}{2(x^2-1)}$

46. Burford reduced the fraction $\frac{1+2x}{1+5x}$ in the following way:

$$\frac{1+2x}{1+5x} = \frac{\cancel{1}+2\cancel{x}}{\cancel{1}+5\cancel{x}} = \frac{2}{5}$$

What is wrong with Burford's work?   **Burford is dividing out terms not factors.**

The following exercises review material from Chapter 4 to help you prepare for the next two sections. Perform the indicated operations.

47. $\dfrac{7}{5}+\dfrac{3}{5}$ 2

48. $\dfrac{3}{4}-\dfrac{1}{4}$ $\dfrac{1}{2}$

49. $\dfrac{3}{10}+\dfrac{2}{15}$ $\dfrac{13}{30}$

50. $\dfrac{5}{6}-\dfrac{4}{21}$ $\dfrac{9}{14}$

51. $5+\dfrac{2}{3}$ $\dfrac{17}{3}$

52. $\dfrac{13}{5}-2$ $\dfrac{3}{5}$

ANSWERS: 1. $\frac{1}{3x}$ 2. $\frac{1}{2a}$ 3. $\frac{5}{2y^2}$ 4. $\frac{1}{2x(x+1)}$ 5. 1 6. z 7. $\frac{5(x-2)}{x-1}$ 8. $\frac{y+2}{y-2}$ 9. $\frac{x(x+1)}{3(x-1)}$ 10. $\frac{1}{z+5}$ 11. $\frac{y+1}{(y-3)(y+5)}$
12. $\frac{x(4+x)}{4-x}$ 13. $\frac{a-7}{a^2}$ 14. $(2x-3y)(x+y)$ 15. 1 16. $\frac{x-y}{x-2y}$ 17. $\frac{7}{3}$ 18. 2 19. $\frac{1}{5}$ 20. $x+1$ 21. $\frac{x+5}{2x-3}$ 22. $\frac{1}{z+2}$
23. $\frac{3}{x+2}$ 24. $\frac{2}{5}$ 25. $\frac{3}{2}$ 26. 3 27. $\frac{y(y+1)}{y-1}$ 28. $\frac{10}{a+5}$ 29. x 30. $\frac{5(z-2)}{z-1}$ 31. $\frac{1}{y+5}$ 32. $a+5$ 33. $\frac{y+5}{2}$ 34. $\frac{1}{6}$
35. $\frac{1}{(x+2y)(x+y)}$ 36. $\frac{1}{x-y}$ 37. 1 38. $\frac{2a+y}{a-2y}$ 39. 0, −2 40. none 41. 0 42. no 43. yes 44. no 45. (a) $\frac{5x}{10}$ (b) $\frac{x^2}{2x}$
(c) $\frac{x^3}{2x^2}$ (d) $\frac{x(x+1)}{2(x+1)}$ (e) $\frac{x(2x-1)}{2(2x-1)}$ (f) $\frac{x(x^2-1)}{2(x^2-1)}$ 46. Burford is dividing out terms not factors. It is clear that the two fractions
are not equal when x is replaced with a number. 47. 2 48. $\frac{1}{2}$ 49. $\frac{13}{30}$ 50. $\frac{9}{14}$ 51. $\frac{17}{3}$ 52. $\frac{3}{5}$

16.2   EXERCISES B

Multiply.

1. $\dfrac{2}{x^3} \cdot \dfrac{x^2}{6}$ $\dfrac{1}{3x}$

2. $\dfrac{5a^2}{8} \cdot \dfrac{4a}{10a^3}$ $\dfrac{1}{4}$

3. $\dfrac{7y^3}{3} \cdot \dfrac{15}{5y}$ $7y^2$

4. $\dfrac{3x^2}{x+5} \cdot \dfrac{(x+5)^2}{9x}$ $\dfrac{x(x+5)}{3}$

5. $\dfrac{x^2-9}{x^2+3x} \cdot \dfrac{x}{x+9}$ $\dfrac{x-3}{x+9}$

6. $\dfrac{z^3}{z+3} \cdot \dfrac{z^2-9}{z^2-3z}$ z^2

7. $\dfrac{4x+4}{4x-4} \cdot \dfrac{x^2-2x+1}{x^2-1}$ 1

8. $\dfrac{y^2-5y+6}{y^2-9} \cdot \dfrac{y+2}{y-2}$ $\dfrac{y+2}{y+3}$

9. $\dfrac{12x^4}{x^2+8x+16} \cdot \dfrac{x^2-16}{3x^2}$ $\dfrac{4x^2(x-4)}{x+4}$

10. $\dfrac{z^2+z-20}{z^2-7x+12} \cdot \dfrac{z-3}{z^2-25}$ $\dfrac{1}{z-5}$

11. $\dfrac{y^2+8y+7}{5y^2-125} \cdot \dfrac{5y-25}{(y+1)^2}$ $\dfrac{y+7}{(y+5)(y+1)}$

12. $\dfrac{9-x^2}{3x-2} \cdot \dfrac{3x^2-2x}{9-6x+x^2}$ $\dfrac{x(3+x)}{3-x}$

13. $\dfrac{a^2-25}{a^2-10a+25} \cdot \dfrac{a^2-8a+15}{a^3-3a^2}$ $\dfrac{a+5}{a^2}$

14. $\dfrac{2x^2+xy-10y^2}{x^2-4y^2} \cdot (x^2+4xy+4y^2)$ $(2x+5y)(x+2y)$

15. $\dfrac{x^2-2xy+y^2}{x+y} \cdot \dfrac{(x+y)^2}{x-y}$ $(x-y)(x+y)$

16. $\dfrac{x^2-7xy+10y^2}{x^2-3xy-10y^2} \cdot \dfrac{x+2y}{x-2y}$ 1

Find the reciprocal.

17. 6 $\dfrac{1}{6}$

18. $\dfrac{6}{y}$ $\dfrac{y}{6}$

19. $7a$ $\dfrac{1}{7a}$

20. $\dfrac{x}{x+6}$ $\dfrac{x+6}{x}$

21. $\dfrac{3x-2}{x-2}$ $\dfrac{x-2}{3x-2}$

22. $5a-1$ $\dfrac{1}{5a-1}$

Divide.

23. $\dfrac{x+3}{x} \div \dfrac{4}{x}$ $\dfrac{x+3}{4}$

24. $\dfrac{a+5}{3a} \div \dfrac{a+5}{9a}$ 3

25. $\dfrac{6(a+4)}{5} \div \dfrac{4(a+4)}{35}$ $\dfrac{21}{2}$

26. $\dfrac{8(x^2+x+1)}{6(x^2-2)} \div \dfrac{x^2+x+1}{3(x^2-2)}$ 4

27. $\dfrac{y^2+4y+4}{8y^3} \div \dfrac{y^2-4}{2y^2}$ $\dfrac{y+2}{4y(y-2)}$

28. $\dfrac{a^2-16}{5a^3} \div \dfrac{3a^2-12a}{20a}$ $\dfrac{4(a+4)}{3a^3}$

29. $\dfrac{z^2+4z+4}{z^2-4} \div \dfrac{z+2}{3z-6}$ 3

30. $\dfrac{x^2-49}{x^2+7x} \div (x-7)$ $\dfrac{1}{x}$

31. $\dfrac{y^2+y-20}{y^2-7y+12} \div \dfrac{y^2-25}{y-3}$ $\dfrac{1}{y-5}$

32. $\dfrac{a^2-10a+21}{a^2+2a-15} \div (5a-35)$ $\dfrac{1}{5(a+5)}$

33. $\dfrac{x^2-16x+64}{3x^2-192} \div \dfrac{3x^2-30x+48}{x^2+6x-16}$ $\dfrac{1}{9}$

34. $\dfrac{5y^2+10y}{2y^3-8y} \div \dfrac{y^2+5y-6}{y^2-3y+2}$ $\dfrac{5}{2(y+6)}$

35. $\dfrac{x^2 - y^2}{x - 3y} \div \dfrac{x - y}{x - 3y}$ $x + y$

36. $\dfrac{x^2 + 8xy + 15y^2}{x^2 - 25y^2} \div (x^2 + 2xy - 3y^2)$

$$\dfrac{1}{(x - 5y)(x - y)}$$

37. $\dfrac{9x^2 - y^2}{9x^2 - 6xy + y^2} \div \dfrac{3x + y}{3x - y}$ 1

38. $\dfrac{u^2 - 4v^2}{u^2 + uv - 2v^2} \cdot \dfrac{4u^2 - 4uv - 3v^2}{2u^2 - 3uv - 2v^2} \div \dfrac{3u + v}{u - v}$

$$\dfrac{2u - 3v}{3u + v}$$

FOR REVIEW

Find the values of the variable that must be excluded.

39. $\dfrac{x - 3}{x^2 - 8x}$ $0, 8$

40. $\dfrac{a^2 + 1}{a^2 - 9}$ $-3, 3$

41. $\dfrac{y^2 + 3y}{6}$ none

Are the given fractions equivalent?

42. $\dfrac{6x^3 - 7}{7x - 7}, \dfrac{6x^3}{7x}$ no

43. $\dfrac{x^2 - 5x}{3x - 15}, \dfrac{x^2}{3x}$ yes

44. $\dfrac{2x + y}{1}, \dfrac{1}{2x + y}$ no

45. Change $\frac{x^2}{5}$ to an equivalent fraction with the given denominator.

(a) 35 $\dfrac{7x^2}{35}$

(b) $5x^3$ $\dfrac{x^5}{5x^3}$

(c) $5(x - 2)$ $\dfrac{x^2(x - 2)}{5(x - 2)}$

(d) $5(x + 3)$ $\dfrac{x^2(x + 3)}{5(x + 3)}$

(e) $5(5x + 1)$ $\dfrac{x^2(5x + 1)}{5(5x + 1)}$

(f) $5(x^3 + 2x)$ $\dfrac{x^2(x^3 + 2x)}{5(x^3 + 2x)}$

46. What is wrong with the following? $\dfrac{3 - 2y}{3 + 5y} = \dfrac{\cancel{3} - 2\cancel{y}}{\cancel{3} + 5\cancel{y}} = -\dfrac{2}{5}$ It is incorrect to divide out 3 and y since they are not factors of both numerator and denominator.

The following exercises review material from Chapter 4 to help you prepare for the next two sections. Perform the indicated operations.

47. $\dfrac{6}{7} + \dfrac{8}{7}$ 2

48. $\dfrac{7}{9} - \dfrac{4}{9}$ $\dfrac{1}{3}$

49. $\dfrac{5}{6} + \dfrac{5}{18}$ $\dfrac{10}{9}$

50. $\dfrac{7}{15} - \dfrac{17}{20}$ $-\dfrac{23}{60}$

51. $\dfrac{7}{5} + 6$ $\dfrac{37}{5}$

52. $3 - \dfrac{15}{2}$ $-\dfrac{9}{2}$

16.2 EXERCISES C

Perform the indicated operations.

1. $\dfrac{x^2 - 5x - 6}{x + 1} \div \dfrac{x^2 - 12x + 36}{x^2 - 1} \cdot \dfrac{x}{x - 1}$

$\left[\text{Answer:} \quad \dfrac{x(x + 1)}{x - 6} \right]$

2. $\dfrac{y^2 - 3y - 4}{y^2 - 1} \div \dfrac{y + 3}{y^2 - 9} \cdot \dfrac{y - 1}{y^2 - 6y + 9}$

$\dfrac{y - 4}{y - 3}$

3. $\dfrac{a^2 - b^2}{2a + b} \div \left[\dfrac{a - b}{2a^2 + 3ab + b^2} \cdot \dfrac{a - b}{a + b} \right]$

$\left[\text{Answer:} \quad \dfrac{(a + b)^3}{a - b} \right]$

4. $\dfrac{a + 2b}{a^2 - 4b^2} \div \left[\dfrac{a - b}{a^2 - 3ab + 2b^2} \cdot \dfrac{a + 2b}{a - 2b} \right]$

$\dfrac{a - 2b}{a + 2b}$

16.3 ADDITION AND SUBTRACTION OF LIKE FRACTIONS

STUDENT GUIDEPOSTS
---

1 Adding and Subtracting Like Fractions

2 Denominators Differing in Sign Only

---

1 ADDING AND SUBTRACTING LIKE FRACTIONS

Fractions that have the same denominators are called **like fractions.** For example,

$$\frac{3}{5x} \quad \text{and} \quad \frac{x+2}{5x}$$

are like fractions, while

$$\frac{1}{2x+1} \quad \text{and} \quad \frac{1}{2x}$$

are called **unlike fractions.**

To Add or Subtract Like Fractions

1. Add or subtract the numerators to find the numerator of the answer.
2. Use the common denominator as the denominator of the answer.
3. Reduce the sum or difference to lowest terms.

Symbolically, if P, Q, and R are fractions, with $R \neq 0$, then

$$\frac{P}{R} + \frac{Q}{R} = \frac{P+Q}{R} \quad \text{and} \quad \frac{P}{R} - \frac{Q}{R} = \frac{P-Q}{R}$$

EXAMPLE 1 OPERATIONS ON LIKE FRACTIONS

Perform the indicated operation.

(a) $\dfrac{3}{7} + \dfrac{2}{7} = \dfrac{3+2}{7}$ Add numerators and place sum over the common denominator 7

$\quad = \dfrac{5}{7}$

(b) $\dfrac{2+x}{x} + \dfrac{x^2+1}{x} = \dfrac{(2+x)+(x^2+1)}{x}$ Add numerators and place sum over common denominator x

$\quad = \dfrac{x^2+x+3}{x}$

(c) $\dfrac{1}{4} - \dfrac{3}{4} = \dfrac{1-3}{4} = \dfrac{-2}{4} = \dfrac{-\cancel{2}}{\cancel{2} \cdot 2} = -\dfrac{1}{2}$

Always reduce the answer to lowest terms.

(d) $\dfrac{2x+1}{x-5} - \dfrac{x-3}{x-5} = \dfrac{(2x-1)-(x-3)}{x-5}$ Use parentheses in subtraction to avoid making a sign error

$\quad = \dfrac{2x+1-x+3}{x-5}$ Be sure to change the sign to $+3$

$\quad = \dfrac{x+4}{x-5}$

PRACTICE EXERCISE 1

Perform the indicated operation.

(a) $\dfrac{1}{11} + \dfrac{5}{11}$

(b) $\dfrac{y-1}{y} + \dfrac{3+y^2}{y}$

(c) $\dfrac{5}{12} - \dfrac{1}{12}$

(d) $\dfrac{x^2}{x+y} - \dfrac{y^2}{x+y}$

Answers: (a) $\frac{6}{11}$ (b) $\frac{y^2+y+2}{y}$
(c) $\frac{1}{3}$ (d) $x-y$

///////////// **CAUTION** /////////////

When subtracting rational expressions, if you enclose the numerators in parentheses, you will eliminate sign errors when the distributive property is used. Notice how this happens in Example 1(d) when $-(x - 3)$ becomes $-x + 3$.

///////////

❷ DENOMINATORS DIFFERING IN SIGN ONLY

Some fractions have denominators that are negatives of each other. These can be made into like fractions by changing the sign of both the numerator and the denominator of *one* of the fractions. To do this, multiply both numerator and denominator by -1. Since you are multiplying the fraction by $\frac{-1}{-1} = 1$, the result is equivalent to the original fraction. This is illustrated in the next example.

| **EXAMPLE 2** **DENOMINATORS DIFFERING IN SIGN** | **PRACTICE EXERCISE 2** |

Perform the indicated operations.

(a) $\dfrac{x}{5} + \dfrac{2x + 1}{-5} = \dfrac{x}{5} + \dfrac{(-1)(2x + 1)}{(-1)(-5)}$ Multiply numerator and denominator by -1

$= \dfrac{x}{5} + \dfrac{(-2x - 1)}{5}$ The fractions are now like fractions

$= \dfrac{x + (-2x - 1)}{5}$ Add numerators

$= \dfrac{-x - 1}{5}$

(b) $\dfrac{2x}{x - 3} - \dfrac{x + 3}{3 - x} = \dfrac{2x}{x - 3} - \dfrac{(-1)(x + 3)}{(-1)(3 - x)}$ $x - 3$ and $3 - x$ are negatives of each other

$= \dfrac{2x}{x - 3} - \dfrac{(-x - 3)}{-3 + x}$

$= \dfrac{2x}{x - 3} - \dfrac{(-x - 3)}{x - 3}$

$= \dfrac{2x - (-x - 3)}{x - 3}$

$= \dfrac{2x + x + 3}{x - 3}$ Watch all signs

$= \dfrac{3x + 3}{x - 3}$

Perform the indicated operations.

(a) $\dfrac{a + 3}{2} + \dfrac{3a - 1}{-2}$

(b) $\dfrac{-3u}{u - v} - \dfrac{3u + 1}{v - u}$

Answers: (a) $2 - a$ (b) $\frac{1}{u - v}$

16.3 EXERCISES A

Perform the indicated operations.

1. $\dfrac{3}{5} + \dfrac{7}{5}$ 2

2. $\dfrac{x}{6} - \dfrac{3}{6}$ $\dfrac{x - 3}{6}$

3. $\dfrac{2}{x} + \dfrac{7}{x}$ $\dfrac{9}{x}$

4. $\dfrac{2x+1}{x-7} + \dfrac{-1-2x}{x-7}$ 0

5 $\dfrac{2y}{y+2} - \dfrac{y+1}{y+2}$ $\dfrac{y-1}{y+2}$

6. $\dfrac{z-1}{z+1} - \dfrac{2z-1}{z+1}$ $\dfrac{-z}{z+1}$

7. $\dfrac{a+1}{a+1} + \dfrac{a^2+1}{a+1}$ $\dfrac{a^2+a+2}{a+1}$

8. $\dfrac{7x^2}{3x^2-1} - \dfrac{4x^2}{3x^2-1}$ $\dfrac{3x^2}{3x^2-1}$

9. $\dfrac{a}{(a+1)^2} + \dfrac{1}{(a+1)^2}$ $\dfrac{1}{a+1}$

10 $\dfrac{z}{2} + \dfrac{3z-1}{-2}$ $\dfrac{-2z+1}{2}$

11. $\dfrac{4}{a} + \dfrac{2a-3}{-a}$ $\dfrac{7-2a}{a}$

12. $\dfrac{2x}{-3} + \dfrac{3x+7}{3}$ $\dfrac{x+7}{3}$

13. $\dfrac{7}{2z} - \dfrac{3z+1}{-2z}$ $\dfrac{3z+8}{2z}$

14 $\dfrac{2a}{a-1} + \dfrac{3a}{1-a}$ $\dfrac{-a}{a-1}$

15. $\dfrac{3x}{1-x} - \dfrac{x+1}{x-1}$ $\dfrac{4x+1}{1-x}$

16. $\dfrac{z}{(z-1)(z+1)} - \dfrac{1}{(z-1)(z+1)}$ $\dfrac{1}{z+1}$

17. $\dfrac{3}{a^2+2a+1} - \dfrac{2-a}{a^2+2a+1}$ $\dfrac{1}{a+1}$

18. $\dfrac{x}{6x-12} - \dfrac{4}{3(4-2x)}$ $\dfrac{x+4}{6(x-2)}$

19. $\dfrac{2z-3}{z^2+3z-4} - \dfrac{z-7}{z^2+3z-4}$ $\dfrac{1}{z-1}$

20 $\dfrac{2x}{x^2+x-6} + \dfrac{x-3}{6-x-x^2}$ $\dfrac{1}{x-2}$

21. $\dfrac{x^2-15}{x^2+4x+3} - \dfrac{2x}{x^2+4x+3}$ $\dfrac{x-5}{x+1}$

22. $\dfrac{y}{x} + \dfrac{2y-1}{x}$ $\dfrac{3y-1}{x}$

23. $\dfrac{3a}{a+b} - \dfrac{2a-b}{a+b}$ 1

24 $\dfrac{x+y}{x-y} - \dfrac{x+y}{y-x}$ $\dfrac{2(x+y)}{x-y}$

25 $\dfrac{5x}{x+1} + \dfrac{2x-1}{x+1} - \dfrac{3x}{x+1}$ $\dfrac{4x-1}{x+1}$

26. $\dfrac{x+2y}{2x-y} - \dfrac{3x-2y}{2x-y} - \dfrac{4y}{2x-y}$ $\dfrac{-2x}{2x-y}$

FOR REVIEW

Perform the indicated operations.

27. $\dfrac{3x-6}{2x} \cdot \dfrac{24x^2}{3(x^2-4x+4)}$ $\dfrac{12x}{x-2}$

28. $\dfrac{a^2-9}{4a+12} \div \dfrac{a-3}{6}$ $\dfrac{3}{2}$

29. $\dfrac{y^2+10y+21}{y^2-2y-15} \div (y^2+2y-35)$ $\dfrac{1}{(y-5)^2}$

30. $\dfrac{x^2-5xy+6y^2}{x^2+3xy-10y^2} \cdot \dfrac{x^2+5xy}{x^2-3xy}$ 1

ANSWERS: 1. 2 2. $\frac{x-3}{6}$ 3. $\frac{9}{x}$ 4. 0 5. $\frac{y-1}{y+2}$ 6. $\frac{-z}{z+1}$ 7. $\frac{a^2+a+2}{a+1}$ 8. $\frac{3x^2}{3x^2-1}$ 9. $\frac{1}{a+1}$ 10. $\frac{-2z+1}{2}$ 11. $\frac{7-2a}{a}$
12. $\frac{x+7}{3}$ 13. $\frac{3z+8}{2z}$ 14. $\frac{-a}{a-1}$ 15. $\frac{4x+1}{1-x}$ 16. $\frac{1}{z+1}$ 17. $\frac{1}{a+1}$ 18. $\frac{x+4}{6(x-2)}$ 19. $\frac{1}{z-1}$ 20. $\frac{1}{x-2}$ 21. $\frac{x-5}{x+1}$ 22. $\frac{3y-1}{x}$
23. 1 24. $\frac{2(x+y)}{x-y}$ 25. $\frac{4x-1}{x+1}$ 26. $\frac{-2x}{2x-y}$ 27. $\frac{12x}{x-2}$ 28. $\frac{3}{2}$ 29. $\frac{1}{(y-5)^2}$ 30. 1

16.3 EXERCISES B

Perform the indicated operations.

1. $\dfrac{x}{3} + \dfrac{7}{3}$ $\dfrac{x+7}{3}$

2. $\dfrac{4}{5} - \dfrac{9}{5}$ -1

3. $\dfrac{5}{3x} - \dfrac{8}{3x}$ $-\dfrac{1}{x}$

4. $\dfrac{6}{x+2} + \dfrac{10}{x+2}$ $\dfrac{16}{x+2}$

5. $\dfrac{3x-2}{x-4} + \dfrac{2-3x}{x-4}$ 0

6. $\dfrac{8y}{y-5} - \dfrac{4y+3}{y-5}$ $\dfrac{4y-3}{y-5}$

7. $\dfrac{a+2}{a+2} + \dfrac{3a-2}{a+2}$ $\dfrac{4a}{a+2}$

8. $\dfrac{6x^2}{5x^2+2} - \dfrac{9x^2}{5x^2+2}$ $\dfrac{-3x^2}{5x^2+2}$

9. $\dfrac{a^2}{a^2+1} - \dfrac{-1}{a^2+1}$ 1

10. $\dfrac{z}{3} + \dfrac{2z-3}{-3}$ $\dfrac{-z+3}{3}$

11. $\dfrac{9}{a} + \dfrac{5a-2}{-a}$ $\dfrac{11-5a}{a}$

12. $\dfrac{3x}{-5} + \dfrac{4x-7}{5}$ $\dfrac{x-7}{5}$

13. $\dfrac{10}{5z} - \dfrac{5z+4}{-5z}$ $\dfrac{5z+14}{5z}$

14. $\dfrac{5a}{a-4} + \dfrac{7a}{4-a}$ $\dfrac{-2a}{a-4}$

15. $\dfrac{4x}{7-x} - \dfrac{2x+5}{x-7}$ $\dfrac{6x+5}{7-x}$

16. $\dfrac{2z}{(2z+1)(2z-1)} - \dfrac{1}{(2z+1)(2z-1)}$ $\dfrac{1}{2z+1}$

17. $\dfrac{5}{a^2+4a+4} - \dfrac{3-a}{a^2+4a+4}$ $\dfrac{1}{a+2}$

18. $\dfrac{2x}{5(x-2)} - \dfrac{3}{10-5x}$ $\dfrac{2x+3}{5(x-2)}$

19. $\dfrac{3z+2}{z^2+4z-12} - \dfrac{2z-4}{z^2+4z-12}$ $\dfrac{1}{z-2}$

20. $\dfrac{2x}{x^2-2x-15} + \dfrac{x+5}{15+2x-x^2}$ $\dfrac{1}{x+3}$

21. $\dfrac{x^2-8}{x^2-4x+3} - \dfrac{-7x}{x^2-4x+3}$ $\dfrac{x+8}{x-3}$

22. $\dfrac{x+1}{y} - \dfrac{x-2}{y}$ $\dfrac{3}{y}$

23. $\dfrac{a-2b}{a-b} - \dfrac{a+2b}{a-b}$ $\dfrac{-4b}{a-b}$

24. $\dfrac{x-y}{2x-y} - \dfrac{x-y}{y-2x}$ $\dfrac{2(x-y)}{2x-y}$

25. $\dfrac{3x-1}{x+2} + \dfrac{4x-2}{x+2} - \dfrac{5x-3}{x+2}$ $\dfrac{2x}{x+2}$

26. $\dfrac{x+3y}{x-2y} - \dfrac{4x+y}{x-2y} - \dfrac{2y}{x-2y}$ $\dfrac{-3x}{x-2y}$

FOR REVIEW

Perform the indicated operations.

27. $\dfrac{5x+5}{x-2} \cdot \dfrac{x^2-4x+4}{x^2-1}$ $\dfrac{5(x-2)}{x-1}$

28. $\dfrac{a^2-16}{5a-20} \div \dfrac{a+4}{10}$ 2

29. $\dfrac{y^2-11y+28}{y^2-2y-35} \div (y^2+y-20)$ $\dfrac{1}{(y+5)^2}$

30. $\dfrac{2x^2+9xy+4y^2}{x^2+xy-12y^2} \cdot \dfrac{3xy-x^2}{2x^2+xy}$ -1

16.3 EXERCISES C

Perform the indicated operations.

1. $\dfrac{2x}{x-5} - \dfrac{2}{5-x} + \dfrac{3x}{5-x}$ $\left[\text{Answer: } \dfrac{2-x}{x-5}\right]$

2. $\dfrac{5y}{x^2-y^2} + \dfrac{5y}{y^2-x^2} - \dfrac{x-y}{y^2-x^2}$ $\quad \dfrac{1}{y+x}$

3. $\dfrac{a+5}{(a-2)(a-3)} - \dfrac{2a-1}{(2-a)(3-a)} + \dfrac{3a-2}{(a-2)(3-a)}$

$\dfrac{-4}{a-3}$

4. $\dfrac{a+b}{2a-b} - \dfrac{2b-a}{b-2a} - \dfrac{a-b}{2a-b} + \dfrac{3b-a}{b-2a}$

$\left[\text{Answer: } \dfrac{b}{2a-b}\right]$

16.4 ADDITION AND SUBTRACTION OF UNLIKE FRACTIONS

═══════════════ **STUDENT GUIDEPOSTS** ═══════════════

❶ Least Common Denominator (LCD)   **❷** Adding and Subtracting Fractions

Before we can add or subtract unlike fractions, we need to convert them to equivalent like fractions. Recall from Chapter 4 that to add $\frac{2}{15}$ and $\frac{1}{6}$ we must first find a common denominator. One such denominator is $15 \cdot 6 = 90$. Since

$$\frac{2}{15} = \frac{2 \cdot 6}{15 \cdot 6} = \frac{12}{90} \quad \text{and} \quad \frac{1}{6} = \frac{1 \cdot 15}{6 \cdot 15} = \frac{15}{90}$$

we could add as follows.

$$\frac{2}{15} + \frac{1}{6} = \frac{12}{90} + \frac{15}{90}$$

$$= \frac{12+15}{90}$$

$$= \frac{27}{90} = \frac{\cancel{9} \cdot 3}{\cancel{9} \cdot 10} = \frac{3}{10}$$

❶ LEAST COMMON DENOMINATOR (LCD)

Often it is wiser to try to find a common denominator smaller than the product of denominators. If we use the *least common denominator*, we shorten the computation needed to reduce the final sum or difference to lowest terms. In the example above, we could have used the least common denominator, 30.

$$\frac{2}{15} + \frac{1}{6} = \frac{2 \cdot 2}{15 \cdot 2} + \frac{1 \cdot 5}{6 \cdot 5} = \frac{4}{30} + \frac{5}{30} = \frac{9}{30} = \frac{3}{10}$$

In the same way, with algebraic fractions we will be concerned with finding the **least common denominator (LCD)** of all fractions.

To Find the LCD of Two or More Fractions

1. Factor the denominators completely.
2. Put each factor in the LCD as many times as it appears in the denominator where it is found the greatest number of times.

EXAMPLE 1   FINDING THE LCD

Find the LCD of the given fractions.

(a) $\dfrac{7}{90}$ and $\dfrac{5}{24}$

$$\overset{\text{one 2 two 3s one 5}}{90 = 2 \cdot \boxed{3 \cdot 3} \cdot \boxed{5}} \quad \text{and} \quad \overset{\text{three 2s one 3}}{24 = \boxed{2 \cdot 2 \cdot 2} \cdot 3.}$$

The LCD must consist of three 2s, two 3s, and one 5.

$$\text{LCD} = \boxed{2 \cdot 2 \cdot 2 \cdot 3 \cdot 3 \cdot 5} = 360$$

(b) $\dfrac{3}{x}$ and $\dfrac{2}{x + 1}$

Since x and $x + 1$ are already completely factored, and since there are no common factors in the two denominators, the LCD $= x(x + 1)$.

(c) $\dfrac{3}{2y}$ and $\dfrac{y + 1}{y^2}$

The denominators are essentially factored already. We see that the LCD has one 2 and two ys. Thus, the LCD is $2y^2$.

(d) $\dfrac{2x - 3}{x^2 - 25}$ and $\dfrac{7x}{2x - 10}$

$$x^2 - 25 = \boxed{(x - 5)}\,(x + 5) \quad \text{and} \quad 2x - 10 = \boxed{2}\,(x - 5)$$

We need one $(x - 5)$, one $(x + 5)$, and one 2 for the LCD. That is, the LCD $= 2(x - 5)(x + 5)$.

(e) $\dfrac{y}{y^2 - 4}$ and $\dfrac{4}{y^2 - 4y + 4}$

$$y^2 - 4 = (y - 2)\boxed{(y + 2)} \quad \text{and} \quad y^2 - 4y + 4 = \boxed{(y - 2)(y - 2)}$$

The LCD must consist of two $(y - 2)$s and one $(y + 2)$. Thus, the LCD $= (y - 2)^2(y + 2)$.

PRACTICE EXERCISE 1

Find the LCD of the given fractions.

(a) $\dfrac{5}{12}$ and $\dfrac{11}{126}$

(b) $\dfrac{5}{a + 1}$ and $\dfrac{1}{a - 1}$

(c) $\dfrac{7}{x^2}$ and $\dfrac{x + 11}{5x}$

(d) $\dfrac{1 - y}{7y + 21}$ and $\dfrac{4y}{y^2 - 9}$

(e) $\dfrac{w}{w^2 + 2w + 1}$ and $\dfrac{3}{w^2 - 5w - 6}$

Answers: (a) 252
(b) $(a + 1)(a - 1)$ (c) $5x^2$
(d) $7(y + 3)(y - 3)$
(e) $(w + 1)^2(w - 6)$

❷ ADDING AND SUBTRACTING FRACTIONS

To add or subtract unlike fractions, we first find their LCD and then transform each fraction into an equivalent fraction having the LCD as its denominator. We illustrate this method in the following numerical example.

$$\frac{2}{15} + \frac{3}{35}$$

$$= \frac{2}{3 \cdot 5} + \frac{3}{5 \cdot 7}$$ Since $15 = 3 \cdot 5$ and $35 = 5 \cdot 7$, the LCD is $3 \cdot 5 \cdot 7 = 105$

$$= \frac{2 \cdot 7}{3 \cdot 5 \cdot 7} + \frac{3 \cdot 3}{5 \cdot 7 \cdot 3}$$ Multiply numerator and denominator of $\frac{2}{15}$ by 7 and of $\frac{3}{35}$ by 3 so the denominators are the same

$$= \frac{2 \cdot 7 + 3 \cdot 3}{3 \cdot 5 \cdot 7}$$ Since denominators are now the same, add numerators and place the sum over the LCD

$$= \frac{14 + 9}{3 \cdot 5 \cdot 7}$$ Leave the denominator in factored form and simplify the numerator

$$= \frac{23}{3 \cdot 5 \cdot 7} = \frac{23}{105}$$ Since 23 has no factor of 3, 5, or 7, the resulting fraction is in lowest terms

We use exactly the same procedure for rational expressions except we now have polynomials for numerators and denominators.

To Add or Subtract Rational Expressions

1. Rewrite the indicated sum or difference with all denominators expressed in factored form.
2. Find the LCD.
3. Multiply the numerator and denominator of each fraction by all factors present in the LCD but missing in the denominator of the particular fraction.
4. Write out the sum or difference of all numerators, using parentheses if needed, and place the result over the LCD.
5. Simplify and factor (if possible) the resulting numerator and divide any factors common to the LCD.

EXAMPLE 2   ADDING FRACTIONS

Add.

$$\frac{5}{6x} + \frac{3}{10x^2}$$

$$= \frac{5}{2 \cdot 3 \cdot x} + \frac{3}{2 \cdot 5 \cdot x \cdot x}$$ Factor denominators; LCD = $2 \cdot 3 \cdot 5 \cdot x \cdot x$

$$= \frac{5 \cdot 5 \cdot x}{2 \cdot 3 \cdot x \cdot 5 \cdot x} + \frac{3 \cdot 3}{2 \cdot 5 \cdot x \cdot x \cdot 3}$$ Supply missing factors

$$= \frac{5 \cdot 5 \cdot x + 3 \cdot 3}{2 \cdot 3 \cdot 5 \cdot x \cdot x}$$ Add numerators over LCD

$$= \frac{25x + 9}{30x^2}$$ No common factors exist, so this is in lowest terms

PRACTICE EXERCISE 2

Add.

$$\frac{1}{12y^3} + \frac{7}{4y}$$

Answer: $\frac{1 + 21y^2}{12y^3}$

EXAMPLE 3 SUBTRACTING FRACTIONS

Subtract.

$$\frac{2}{y+2} - \frac{2}{y+3}$$

Denominators already factored; LCD $= (y+2)(y+3)$

$$= \frac{2(y+3)}{(y+2)(y+3)} - \frac{2(y+2)}{(y+2)(y+3)}$$

Supply missing factors

$$= \frac{2(y+3) - 2(y+2)}{(y+2)(y+3)}$$

Subtract numerators over LCD

$$= \frac{2y+6-2y-4}{(y+2)(y+3)}$$

Watch signs when using the distributive law

$$= \frac{2}{(y+2)(y+3)}$$

No common factors exist, so this is in lowest terms

PRACTICE EXERCISE 3

Subtract.

$$\frac{3}{a-1} - \frac{1}{a+3}$$

Answer: $\dfrac{2(a+5)}{(a-1)(a+3)}$

EXAMPLE 4 ADDING FRACTIONS

Add.

(a) $$\frac{5}{x^2-4} + \frac{7}{x^2+2x}$$

$$= \frac{5}{(x-2)(x+2)} + \frac{7}{x(x+2)}$$

Factor denominators; LCD $= x(x-2)(x+2)$

$$= \frac{5(x)}{(x-2)(x+2)(x)} + \frac{7(x-2)}{x(x+2)(x-2)}$$

Supply missing factors

$$= \frac{5x + 7(x-2)}{x(x+2)(x-2)}$$

Add numerators

$$= \frac{5x + 7x - 14}{x(x+2)(x-2)}$$

Clear parentheses using the distributive law

$$= \frac{12x - 14}{x(x+2)(x-2)}$$

$$= \frac{2(6x-7)}{x(x+2)(x-2)}$$

No common factors exist, so this is in lowest terms

(b) $$\frac{5}{x^2+x-6} + \frac{3x}{x^2-4x+4}$$

$$= \frac{5}{(x-2)(x+3)} + \frac{3x}{(x-2)(x-2)}$$

LCD $= (x+3)(x-2)^2$

$$= \frac{5(x-2)}{(x-2)(x+3)(x-2)} + \frac{3x(x+3)}{(x-2)(x-2)(x+3)}$$

$$= \frac{5(x-2) + 3x(x+3)}{(x+3)(x-2)^2}$$

$$= \frac{5x - 10 + 3x^2 + 9x}{(x+3)(x-2)^2}$$

$$= \frac{3x^2 + 14x - 10}{(x+3)(x-2)^2}$$

This is in lowest terms

PRACTICE EXERCISE 4

Add.

(a) $$\frac{3}{y^2-5y} + \frac{-2}{y^2-3y-10}$$

(b) $$\frac{y}{y^2+2y-15} + \frac{1-y}{y^2-6y+9}$$

Answers: (a) $\dfrac{y+6}{y(y-5)(y+2)}$

(b) $\dfrac{5-7y}{(y-3)^2(y+5)}$

We use the same procedures even if there are two variables and more than two terms.

| **EXAMPLE 5** OPERATIONS WITH TWO VARIABLES | **PRACTICE EXERCISE 5** |

Perform the indicated operations.

$$\frac{2x}{x+y} - \frac{y}{x-y} - \frac{-4xy}{x^2-y^2}$$

$$= \frac{2x}{x+y} - \frac{y}{x-y} - \frac{-4xy}{(x+y)(x-y)} \qquad \text{LCD} = (x+y)(x-y)$$

$$= \frac{2x(x-y)}{(x+y)(x-y)} - \frac{y(x+y)}{(x-y)(x+y)} - \frac{-4xy}{(x+y)(x-y)} \qquad \text{Supply missing factors}$$

$$= \frac{2x(x-y) - y(x+y) + 4xy}{(x+y)(x-y)}$$

$$= \frac{2x^2 - 2xy - xy - y^2 + 4xy}{(x+y)(x-y)}$$

$$= \frac{2x^2 + xy - y^2}{(x+y)(x-y)} \qquad \text{This will simplify}$$

$$= \frac{(2x-y)(x+y)}{(x+y)(x-y)} \qquad \text{Divide out } (x+y)$$

$$= \frac{2x-y}{x-y}$$

Perform the indicated operations.

$$\frac{2a}{a+2b} - \frac{b}{a-2b} + \frac{2b^2+ab}{a^2-4b^2}$$

Answer: $\frac{2a}{a+2b}$

16.4 EXERCISES A

Find the LCD of the given fractions.

1. $\dfrac{1}{20}$ and $\dfrac{7}{30}$ **60**

2. $\dfrac{2}{39}$ and $\dfrac{4}{35}$ **1365**

3. $\dfrac{y+1}{y^2}$ and $\dfrac{1}{3y}$ $3y^2$

4. $\dfrac{a+1}{9a^2}$ and $\dfrac{5}{12a^3}$
$36a^3$

5. $\dfrac{2}{3x}$ and $\dfrac{x+2}{3x+3}$
$3x(x+1)$

6. $\dfrac{7}{y+1}$ and $\dfrac{y}{y-1}$
$(y+1)(y-1)$

7. $\dfrac{3}{2z+4}$ and $\dfrac{-5}{3z+6}$
$6(z+2)$

8. $\dfrac{2a+1}{a^2-25}$ and $\dfrac{3}{a+5}$
$(a+5)(a-5)$

9. $\dfrac{3x+2}{3x+6}$ and $\dfrac{x}{x^2-4}$
$3(x+2)(x-2)$

10 $\dfrac{1}{x^2+2x+1}$ and $\dfrac{2}{x^2-1}$
$(x-1)(x+1)^2$

11. $\dfrac{3}{z^2-9}$ and $\dfrac{2}{z^2+z-6}$
$(z-3)(z+3)(z-2)$

12. $\dfrac{2}{a^2-9}$ and $\dfrac{2}{a^2+2a-3}$
$(a-3)(a+3)(a-1)$

13. $\dfrac{1}{x^2 - y^2}$ and $\dfrac{3xy}{x - y}$

$(x + y)(x - y)$

14 $\dfrac{5x}{x + y}$, $\dfrac{7y}{2x + 2y}$, and $\dfrac{2xy}{3x + 3y}$

$6(x + y)$

15. $\dfrac{16}{5x}$, $\dfrac{5}{3y}$, and $\dfrac{xy}{2x - y}$

$15xy(2x - y)$

Perform the indicated operations.

16. $\dfrac{4}{21} + \dfrac{5}{14}$ $\quad \dfrac{23}{42}$

17. $\dfrac{5}{12} - \dfrac{11}{30}$ $\quad \dfrac{1}{20}$

18. $\dfrac{2}{9x} + \dfrac{4}{15x^2}$ $\quad \dfrac{2(5x + 6)}{45x^2}$

19. $\dfrac{a - 5}{a} - \dfrac{3a - 1}{4a}$

$\dfrac{a - 19}{4a}$

20. $\dfrac{5}{y + 5} - \dfrac{3}{y - 5}$

$\dfrac{2(y - 20)}{(y + 5)(y - 5)}$

21. $\dfrac{3a}{a + 2} + \dfrac{a}{a - 2}$

$\dfrac{4a(a - 1)}{(a + 2)(a - 2)}$

22. $\dfrac{2}{3x + 21} - \dfrac{3}{5x + 35}$

$\dfrac{1}{15(x + 7)}$

23 $\dfrac{4y}{y^2 - 36} - \dfrac{4}{y + 6}$

$\dfrac{24}{(y - 6)(y + 6)}$

24. $\dfrac{8}{7 - a} - \dfrac{8a}{49 - a^2}$

$\dfrac{56}{(7 - a)(7 + a)}$

25. $\dfrac{3}{2x^2 - 2x} - \dfrac{5}{2x - 2}$

$\dfrac{3 - 5x}{2x(x - 1)}$

26. $\dfrac{-8}{y^2 - 4} - \dfrac{4}{y + 2}$

$\dfrac{-4y}{(y - 2)(y + 2)}$

27. $\dfrac{8}{a^2 - 16} + \dfrac{1}{a + 4}$

$\dfrac{1}{a - 4}$

28. $\dfrac{3z + 2}{3z + 6} + \dfrac{z - 2}{z^2 - 4}$

$\dfrac{3z + 5}{3(z + 2)}$

29. $\dfrac{2}{y^2 - 1} + \dfrac{1}{y^2 + 2y - 3}$

$\dfrac{3y + 7}{(y - 1)(y + 1)(y + 3)}$

30. $\dfrac{2}{z^2 - 9} - \dfrac{2}{z^2 + 2z - 3}$

$\dfrac{4}{(z - 3)(z + 3)(z - 1)}$

31 $\dfrac{3}{a^2 - a - 12} - \dfrac{2}{a^2 - 9}$

$\dfrac{a - 1}{(a - 4)(a + 3)(a - 3)}$

32. $\dfrac{5}{x^2 - 4x + 3} + \dfrac{7}{x^2 + x - 2}$

$\dfrac{12x - 11}{(x - 3)(x - 1)(x + 2)}$

33. $\dfrac{y + 1}{y + 4} + \dfrac{4 - y^2}{y^2 - 16}$

$\dfrac{-3y}{(y - 4)(y + 4)}$

34. $\dfrac{2}{a - 1} + \dfrac{a}{a^2 - 1}$

$\dfrac{3a + 2}{(a - 1)(a + 1)}$

35 $\dfrac{3}{x + 1} + \dfrac{5}{x - 1} - \dfrac{10}{x^2 - 1}$

$\dfrac{8}{x + 1}$

36. $\dfrac{5x}{x + y} + \dfrac{6y - 2x}{x - y}$

$\dfrac{3x^2 - xy + 6y^2}{(x + y)(x - y)}$

37 $\dfrac{2y}{y+5} - \dfrac{3y}{y-2} - \dfrac{2y^2}{y^2+3y-10}$

$\dfrac{-y(3y+19)}{(y+5)(y-2)}$

38. $\dfrac{1}{x(x-y)} - \dfrac{2}{y(x+y)}$

$\dfrac{y^2+3xy-2x^2}{xy(x+y)(x-y)}$

FOR REVIEW

Perform the indicated operations.

39. $\dfrac{3}{z^2+2z+1} - \dfrac{2-z}{z^2+2z+1}$

$\dfrac{1}{z+1}$

40. $\dfrac{2}{2x-1} - \dfrac{3}{1-2x}$

$\dfrac{5}{2x-1}$

41. $\dfrac{2a-1}{a-5} - \dfrac{6-a}{5-a}$

$\dfrac{a+5}{a-5}$

42. $\dfrac{3-y}{y-7} + \dfrac{2y-5}{7-y}$

$\dfrac{8-3y}{y-7}$

43. $\dfrac{2x}{3x-2} - \dfrac{5x}{3x-2} - \dfrac{-2}{3x-2}$

-1

44. $\dfrac{x^2}{x-y} + \dfrac{x^2}{x-y} - \dfrac{y^2}{y-x}$

$\dfrac{2x^2+y^2}{x-y}$

The following exercises review material that we covered in Chapters 11 and 15. They will help you prepare for the next section. Solve.

45. $2 - 3(x+1) = 5$ -2

46. $y - 4(2y+1) = y + 4$ -1

47. $x^2 - x - 6 = 0$ 3 or -2

48. $x^2 + 5x = 0$ 0 or -5

49. Eric Wade receives 11% simple interest on his savings. How much interest will he earn at the end of one year if he deposits $875?
$96.25

50. Two families leave their homes at the same time planning to meet for a reunion at a point between them. If one travels at a rate of 55 mph, the other travels at 50 mph, and they live 273 miles apart, how long will it take for them to meet?
2.6 hr

ANSWERS: 1. 60 2. 1365 3. $3y^2$ 4. $36a^3$ 5. $3x(x + 1)$ 6. $(y + 1)(y - 1)$ 7. $6(z + 2)$ 8. $(a + 5)(a - 5)$
9. $3(x + 2)(x - 2)$ 10. $(x - 1)(x + 1)^2$ 11. $(z - 3)(z + 3)(z - 2)$ 12. $(a - 3)(a + 3)(a - 1)$ 13. $(x + y)(x - y)$
14. $6(x + y)$ 15. $15xy(2x - y)$ 16. $\frac{23}{42}$ 17. $\frac{1}{20}$ 18. $\frac{2(5x + 6)}{45x^2}$ 19. $\frac{a - 19}{4a}$ 20. $\frac{2(y - 20)}{(y + 5)(y - 5)}$ 21. $\frac{4a(a - 1)}{(a + 2)(a - 2)}$ 22. $\frac{1}{15(x + 7)}$
23. $\frac{24}{(y - 6)(y + 6)}$ 24. $\frac{56}{(7 - a)(7 + a)}$ 25. $\frac{3 - 5x}{2x(x - 1)}$ 26. $\frac{-4y}{(y - 2)(y + 2)}$ 27. $\frac{1}{a - 4}$ 28. $\frac{3z + 5}{3(z + 2)}$ 29. $\frac{3y + 7}{(y - 1)(y + 1)(y + 3)}$
30. $\frac{4}{(z - 3)(z + 3)(z - 1)}$ 31. $\frac{a - 1}{(a - 4)(a + 3)(a - 3)}$ 32. $\frac{12x - 11}{(x - 3)(x - 1)(x + 2)}$ 33. $\frac{-3y}{(y - 4)(y + 4)}$ 34. $\frac{3a + 2}{(a - 1)(a + 1)}$ 35. $\frac{8}{x + 1}$ 36. $\frac{3x^2 - xy + 6y^2}{(x + y)(x - y)}$
37. $\frac{-y(3y + 19)}{(y + 5)(y - 2)}$ 38. $\frac{y^2 + 3xy - 2x^2}{xy(x + y)(x - y)}$ 39. $\frac{1}{z + 1}$ 40. $\frac{5}{2x - 1}$ 41. $\frac{a + 5}{a - 5}$ 42. $\frac{8 - 3y}{y - 7}$ 43. -1 44. $\frac{2x^2 + y^2}{x - y}$ 45. -2
46. -1 47. 3 or -2 48. 0 or -5 49. \$96.25 50. 2.6 hr

16.4 EXERCISES B

Find the LCD of the given fractions.

1. $\dfrac{1}{40}$ and $\dfrac{1}{30}$ 120

2. $\dfrac{4}{39}$ and $\dfrac{2}{15}$ 195

3. $\dfrac{3}{x}$ and $\dfrac{8}{7x}$ $7x$

4. $\dfrac{a + 1}{6a^2}$ and $\dfrac{7}{9a^3}$ $18a^3$

5. $\dfrac{3}{5x}$ and $\dfrac{x + 5}{5x + 5}$ $5x(x + 1)$

6. $\dfrac{3}{x + 1}$ and $\dfrac{x}{x - 1}$
$(x + 1)(x - 1)$

7. $\dfrac{-8}{3z - 6}$ and $\dfrac{2}{7z - 14}$
$21(z - 2)$

8. $\dfrac{2z + 3}{z^2 - 9}$ and $\dfrac{12}{z + 3}$
$(z + 3)(z - 3)$

9. $\dfrac{4x - 1}{2x + 6}$ and $\dfrac{5x + 1}{x^2 - 9}$
$2(x + 3)(x - 3)$

10. $\dfrac{4}{y^2 - 1}$ and $\dfrac{y}{y^2 - 2y + 1}$
$(y + 1)(y - 1)$

11. $\dfrac{a}{a^2 - 4}$ and $\dfrac{2a + 5}{a^2 - a - 6}$
$(a + 2)(a - 2)(a - 3)$

12. $\dfrac{6y^2}{y^2 - 25}$ and $\dfrac{3y}{y^2 + y - 20}$
$(y + 5)(y - 5)(y - 4)$

13. $\dfrac{1}{4x^2 - y^2}$ and $\dfrac{x^2}{2x + y}$
$(2x + y)(2x - y)$

14. $\dfrac{2}{4x - 4y}$, $\dfrac{3x}{2x - 2y}$, and
$\dfrac{3xy}{5x - 5y}$ $20(x - y)$

15. $\dfrac{21y}{8x}$, $\dfrac{15x}{7y}$, and $\dfrac{x^2y^2}{x + 3y}$
$56xy(x + 3y)$

Perform the indicated operations.

16. $\dfrac{2}{15} + \dfrac{3}{25}$ $\dfrac{19}{75}$

17. $\dfrac{11}{20} - \dfrac{8}{35}$ $\dfrac{9}{28}$

18. $\dfrac{3}{4y} + \dfrac{5}{8y^2}$ $\dfrac{6y + 5}{8y^2}$

19. $\dfrac{a + 8}{2a} - \dfrac{5a - 3}{3a}$
$\dfrac{-7a + 30}{6a}$

20. $\dfrac{6}{x - 6} - \dfrac{4}{x + 6}$
$\dfrac{2(x + 30)}{(x + 6)(x - 6)}$

21. $\dfrac{5y}{y + 4} + \dfrac{2y}{y - 4}$
$\dfrac{y(7y - 12)}{(y + 4)(y - 4)}$

22. $\dfrac{7}{4x + 8} - \dfrac{5}{7x + 14}$
$\dfrac{29}{28(x + 2)}$

23. $\dfrac{4}{z + 2} + \dfrac{8}{z^2 - 4}$
$\dfrac{4z}{(z + 2)(z - 2)}$

24. $\dfrac{14}{49 - a^2} - \dfrac{7}{7 - a}$
$\dfrac{-7(a + 5)}{(7 + a)(7 - a)}$

25. $\dfrac{5}{3x^2 - 3x} - \dfrac{7}{3x - 3}$
$\dfrac{5 - 7x}{3x(x - 1)}$

26. $\dfrac{8}{y^2 - 16} + \dfrac{1}{y + 4}$
$\dfrac{1}{y - 4}$

27. $\dfrac{10}{x^2 - 25} - \dfrac{3}{x - 5}$
$\dfrac{-3x - 5}{(x + 5)(x - 5)}$

28. $\dfrac{2z-3}{2z-4} + \dfrac{z+2}{z^2-4}$

$\dfrac{2z-1}{2(z-2)}$

29. $\dfrac{2}{a^2-a-2} + \dfrac{a}{a^2-1}$

$\dfrac{a^2-2}{(a-2)(a+1)(a-1)}$

30. $\dfrac{y}{y^2-1} - \dfrac{y+2}{y^2+y-2}$

$\dfrac{-1}{(y+1)(y-1)}$

31. $\dfrac{4}{x^2-9} - \dfrac{4}{x^2-2x-3}$

$\dfrac{-8}{(x+1)(x+3)(x-3)}$

32. $\dfrac{8}{x^2+4x+3} + \dfrac{3}{x^2-x-2}$

$\dfrac{11x-7}{(x+1)(x+3)(x-2)}$

33. $\dfrac{y-3}{y+5} + \dfrac{9-y^2}{y^2-25}$

$\dfrac{-8(y-3)}{(y+5)(y-5)}$

34. $\dfrac{4}{a-2} + \dfrac{2a}{a^2-4}$

$\dfrac{2(3a+4)}{(a+2)(a-2)}$

35. $\dfrac{2}{x+5} + \dfrac{3}{x-5} - \dfrac{7}{x^2-25}$

$\dfrac{5x-2}{(x+5)(x-5)}$

36. $\dfrac{2y}{x-y} + \dfrac{3x-y}{x+y}$

$\dfrac{3x^2-2xy+3y^2}{(x+y)(x-y)}$

37. $\dfrac{3y}{y+4} - \dfrac{2y}{y-3} - \dfrac{y^2}{y^2+y-12}$

$\dfrac{-17y}{(y+4)(y-3)}$

38. $\dfrac{3}{x(x+y)} - \dfrac{2}{y(x-y)}$

$\dfrac{-(3y^2-xy+2x^2)}{xy(x+y)(x-y)}$

FOR REVIEW

Perform the indicated operations.

39. $\dfrac{z}{z^2+2z+1} + \dfrac{2+z}{z^2+2z+1}$ $\quad \dfrac{2}{z+1}$

40. $\dfrac{3x}{x-1} - \dfrac{-2x}{1-x}$ $\quad \dfrac{x}{x-1}$

41. $\dfrac{y+2}{y-2} - \dfrac{y^2-2}{2-y}$ $\quad \dfrac{y(y+1)}{y-2}$

42. $\dfrac{2x^2}{2x^2-1} + \dfrac{x^2}{1-2x^2}$ $\quad \dfrac{x^2}{2x^2-1}$

43. $\dfrac{6x}{2x-3} - \dfrac{8x}{2x-3} - \dfrac{5x}{2x-3}$ $\quad \dfrac{-7x}{2x-3}$

44. $\dfrac{2x^2}{2x-y} + \dfrac{3y^2}{2x-y} - \dfrac{x^2+y^2}{y-2x}$ $\quad \dfrac{3x^2+4y^2}{2x-y}$

The following exercises review material that we covered in Chapters 11 and 15. They will help you prepare for the next section. Solve.

45. $3 - 4(-x+2) = 7$ $\quad$ 3

46. $2y - 3(3y-2) = -y-6$ $\quad$ 2

47. $x^2 - 5x + 6 = 0$ $\quad$ 2 or 3

48. $2x^2 - 6x = x^2$ $\quad$ 0 or 6

49. If \$16.80 is charged as tax on the sale of a washing machine priced at \$420, what is the tax rate?
4%

50. Clem hiked to a waterfall at a rate of 3 mph and returned at a rate of 4 mph. If the total time of the trip was 7 hours, how long did it take to reach the waterfall? What was the total distance hiked?
4 hr, 24 mi

16.4 EXERCISES C

Perform the indicated operations.

1. $\dfrac{2}{x^2 - y^2} - \dfrac{3}{x^2 + 2xy + y^2} - \dfrac{3}{x^2 - 2xy + y^2}$

$\left[\text{Answer: } \dfrac{-4(x^2 + 2y^2)}{(x + y)^2(x - y)^2}\right]$

2. $\dfrac{3}{x^2 - 7xy + 10y^2} + \dfrac{2}{x^2 + 2xy - 8y^2} -$
$\dfrac{2}{x^2 - xy - 20y^2} \qquad \dfrac{3(x + 2y)}{(x - 2y)(x + 4y)(x - 5y)}$

3. $\dfrac{1}{x + 5y} \div \dfrac{1}{x - 5y} - \dfrac{1}{x + 5y} \cdot \dfrac{-6xy}{x - y}$

$\dfrac{x^2 + 5y^2}{(x - y)(x + 5y)}$

4. $\dfrac{2x - y}{x^2 - y^2} \div \dfrac{2x - y}{x + y} - \dfrac{x + y}{(x - y)^2} \div \dfrac{x + y}{2y}$

$\left[\text{Answer: } \dfrac{x - 3y}{(x - y)^2}\right]$

16.5 SOLVING FRACTIONAL EQUATIONS

STUDENT GUIDEPOSTS

1 Fractional Equations

2 Applications of Fractional Equations

3 Work Problems

1 FRACTIONAL EQUATIONS

An equation that has one or more algebraic fractions or rational expressions is called a **fractional equation.** Once all the fractions have been eliminated (called **clearing the fractions**) by multiplying both sides by the LCD of the denominators in the equation, the resulting equation can often be solved by the techniques of Chapter 11 or Section 15.5. Every solution to this new equation must be checked in the original equation to be sure that it does not make one of the denominators equal to zero. If this occurs it must be discarded.

To Solve a Fractional Equation

1. Find the LCD of all fractions in the equation.

2. Multiply both sides of the equation by the LCD, making sure that *all* terms are multiplied. Once simplified, the resulting equation will be free of fractions.

3. Solve this equation.

4. Check your solutions in the original equation to be certain your answers do not make one of the denominators zero.

EXAMPLE 1   SOLVING A FRACTIONAL EQUATION

Solve $\dfrac{2}{5} + \dfrac{x}{15} = \dfrac{1}{3}$.

$15\left(\dfrac{2}{5} + \dfrac{x}{15}\right) = 15 \cdot \dfrac{1}{3}$ Multiply both sides by the LCD, 15

$15 \cdot \dfrac{2}{5} + 15 \cdot \dfrac{x}{15} = 5$ Clear the parentheses using the distributive law

$6 + x = 5$

$6 - 6 + x = 5 - 6$ Subtract 6 from both sides

$x = -1$

Check: $\dfrac{2}{5} + \dfrac{-1}{15} \overset{?}{=} \dfrac{1}{3}$

$\dfrac{2 \cdot 3}{5 \cdot 3} + \dfrac{-1}{15} \overset{?}{=} \dfrac{1 \cdot 5}{3 \cdot 5}$

$\dfrac{6}{15} + \dfrac{-1}{15} \overset{?}{=} \dfrac{5}{15}$

$\dfrac{5}{15} = \dfrac{5}{15}$

The solution is -1.

PRACTICE EXERCISE 1

Solve $\dfrac{x}{24} + \dfrac{3}{8} = \dfrac{2}{3}$.

Answer: 7

Notice that we are using the LCD in this section to clear the fractions and not to write them as equivalent fractions with the same denominator.

EXAMPLE 2   SOLVING A FRACTIONAL EQUATION

Solve $\dfrac{1}{y} = \dfrac{1}{8 - y}$.

$y(8 - y)\left(\dfrac{1}{y}\right) = y(8 - y)\left(\dfrac{1}{8 - y}\right)$ Multiply both sides by the LCD, $y(8 - y)$

$8 - y = y$

$8 = 2y$ Add y to both sides

$4 = y$

Check: $\dfrac{1}{4} \overset{?}{=} \dfrac{1}{8 - 4}$

$\dfrac{1}{4} = \dfrac{1}{4}$

The solution is 4.

PRACTICE EXERCISE 2

Solve $\dfrac{2}{x + 3} = \dfrac{5}{x - 1}$.

Answer: $-\frac{17}{3}$

EXAMPLE 3 Solving a Fractional Equation

Solve $\dfrac{3x}{x-3} - 3 = \dfrac{1}{x}$.

$x(x-3)\left[\dfrac{3x}{x-3} - 3\right] = x(x-3)\left[\dfrac{1}{x}\right]$ The LCD = $x(x-3)$

$x(3x) - 3\,x(x-3) = x - 3$ Be sure to multiply 3 by LCD

$3x^2 - 3x^2 + 9x = x - 3$

$8x = -3$

$x = -\dfrac{3}{8}$

Substituting $-\frac{3}{8}$ into the original equation to check would require much calculation. Instead, we might note that the only two numbers that make a denominator zero are 0 and 3, neither of which is our answer, $-\frac{3}{8}$. After checking our work for mistakes, we conclude that the solution is $-\frac{3}{8}$.

PRACTICE EXERCISE 3

Solve $\dfrac{2x}{x-4} - 2 = \dfrac{-24}{x}$.

Answer: 3

EXAMPLE 4 Equation with No Solution

Solve $\dfrac{x}{x+4} - \dfrac{4}{x-4} = \dfrac{x^2 + 16}{x^2 - 16}$.

The LCD = $(x-4)(x+4)$.

$(x-4)(x+4)\left(\dfrac{x}{x+4} - \dfrac{4}{x-4}\right) = (x-4)(x+4)\left(\dfrac{x^2+16}{x^2-16}\right)$

$x(x-4) - 4(x+4) = x^2 + 16$

$x^2 - 4x - 4x - 16 = x^2 + 16$

$-8x - 16 = 16$

$-8x = 32$

$x = -4$

Check: $\dfrac{-4}{-4+4} - \dfrac{4}{-4-4} \overset{?}{=} \dfrac{(-4)^2 + 16}{(-4)^2 - 16}$.

But $\frac{-4}{-4+4} = \frac{-4}{0}$, which is undefined. Thus, -4 *does not check*. There are no solutions.

PRACTICE EXERCISE 4

Solve $\dfrac{2}{y-3} - \dfrac{12}{y^2-9} = \dfrac{3}{y+3}$.

Answer: no solution (notice that 3 makes the denominators of the fractions on the left side equal to zero)

////////// **CAUTION** //////////

Be sure to check your answers in the *original* equation. Example 4 shows that sometimes a possible solution may have to be excluded because it makes one or more denominators equal to zero.

//////////

In some fractional equations, when the fractions are cleared the resulting equation must be solved by the factoring method that we studied in Chapter 15.

EXAMPLE 5 USING THE ZERO-PRODUCT RULE

Solve $\dfrac{2}{x + 2} + \dfrac{3}{x - 2} = 1$.

$$(x + 2)(x - 2)\left[\dfrac{2}{x + 2} + \dfrac{3}{x - 2}\right] = (x + 2)(x - 2)(1) \qquad \text{LCD} = (x + 2)(x - 2)$$

$$2(x - 2) + 3(x + 2) = x^2 - 4$$

$$2x - 4 + 3x + 6 = x^2 - 4$$

$$5x + 2 = x^2 - 4$$

$$-x^2 + 5x + 6 = 0$$

$$x^2 - 5x - 6 = 0 \qquad \text{Multiply by } -1$$

$$(x - 6)(x + 1) = 0 \qquad \text{Factor}$$

$$x - 6 = 0 \quad \text{or} \quad x + 1 = 0 \qquad \text{Zero-product Rule}$$

$$x = 6 \qquad\qquad x = -1$$

Check: $\dfrac{2}{6 + 2} + \dfrac{3}{6 - 2} \stackrel{?}{=} 1 \qquad \dfrac{2}{-1 + 2} + \dfrac{3}{-1 - 2} \stackrel{?}{=} 1$

$\qquad\qquad \dfrac{2}{8} + \dfrac{3}{4} \stackrel{?}{=} 1 \qquad\qquad \dfrac{2}{1} + \dfrac{3}{-3} \stackrel{?}{=} 1$

$\qquad\qquad \dfrac{1}{4} + \dfrac{3}{4} = 1 \qquad\qquad\qquad 2 - 1 = 1$

The solutions are 6 and -1.

PRACTICE EXERCISE 5

Solve $\dfrac{y}{y - 1} + \dfrac{2}{y + 1} = \dfrac{8}{y^2 - 1}$.

Answer: 2 and -5

---

② APPLICATIONS OF FRACTIONAL EQUATIONS

In Chapter 11 we looked at applications that resulted in simple equations. We now consider similar problems that result in fractional equations.

EXAMPLE 6 AGE PROBLEM

The sum of Bill's and Bob's ages is 45 years, and Bill's age divided by Bob's age is $\frac{5}{4}$. How old is each?

Let $x = $ Bill's age,
$45 - x = $ Bob's age.　　The sum of x and $45 - x$ is 45

We need to solve the following equation.

$$\dfrac{x}{45 - x} = \dfrac{5}{4} \qquad \text{The LCD} = 4(45 - x)$$

$$4(45 - x)\left(\dfrac{x}{45 - x}\right) = 4(45 - x)\left(\dfrac{5}{4}\right) \qquad \text{Multiply by the LCD}$$

$$4x = (45 - x)5$$

$$4x = 225 - 5x$$

$$9x = 225$$

$$x = 25 \qquad \text{Bill's age is 25}$$

Since $45 - x = 45 - 25 = 20$, Bill is 25 years old and Bob is 20.

PRACTICE EXERCISE 6

Terry's age divided by Pam's age is $\frac{7}{5}$, and the sum of their ages is 24. How old is each?
Let $x = $ Terry's age,
$24 - x = $ Pam's age.

Answer: Terry is 14 and Pam is 10.

| EXAMPLE 7 NUMBER PROBLEM | PRACTICE EXERCISE 7 |

The reciprocal of 2 more than a number is three times the reciprocal of the number. Find the number.

Let x = the number,

   $x + 2$ = 2 more than the number,

   $\dfrac{1}{x + 2}$ = the reciprocal of 2 more than the number,

   $\dfrac{1}{x}$ = the reciprocal of the number,

$3 \cdot \dfrac{1}{x} = \dfrac{3}{x}$ = three times the reciprocal of the number.

The equation we need to solve is $\dfrac{1}{x + 2} = \dfrac{3}{x}$.

$$x(x + 2)\left(\dfrac{1}{x + 2}\right) = x(x + 2)\left(\dfrac{3}{x}\right) \qquad \text{The LCD} = x(x + 2)$$

$$x = (x + 2)3$$
$$x = 3x + 6$$
$$-2x = 6$$
$$x = -3$$

The number is -3.

The numerator of a fraction is 3 more than the denominator, and the value of the fraction is $\frac{4}{3}$. Find the fraction.

Let x = denominator of fraction,

   $x + 3$ = numerator of fraction,

   $\dfrac{x + 3}{x}$ = the fraction.

The equation to solve is

$$\dfrac{x + 3}{x} = \underline{\hspace{1.5cm}}.$$

Answer: $\frac{12}{9}$

---

❸ WORK PROBLEMS

Consider the following problem: Bob can do a job in 2 hours and Pete can do the same job in 4 hours. How long would it take them to do the job if they worked together?

To solve this type of problem, usually called a *work problem*, three principles must be kept in mind:

1. The time it takes to do a job when the individuals work together must be less than the time it takes for the faster worker to complete the job alone. Thus, the time together is *not* the average of the two times (which would be 3 hours in this case). Since Bob can do the job alone in 2 hr, with help the time must clearly be less than 2 hr.

2. If a job can be done in t hr, in 1 hour $\frac{1}{t}$ of the job would be completed. For example, since Bob can do the job in 2 hours, in 1 hour he would do $\frac{1}{2}$ the job. Similarly, Pete would do $\frac{1}{4}$ the job in 1 hr since he does it all in 4 hr.

3. The amount of work done by Bob in 1 hr added to the amount of work done by Pete in 1 hr equals the amount of work done together in 1 hr. So if t is the time it takes to do the job working together,

$$\text{(amount Bob does in 1 hr)} + \text{(amount Pete does in 1 hr)}$$
$$= \text{(amount done together in 1 hr)}$$

translates to the equation

$$\dfrac{1}{2} + \dfrac{1}{4} = \dfrac{1}{t}.$$

We see that a work problem translates to a fractional equation that may result in either a first-degree equation or a second-degree (quadratic) equation. In our example, we obtain a first-degree equation after we multiply both sides by the LCD, $4t$. We will work only with first-degree equations in this section.

$$4t\left(\frac{1}{2}+\frac{1}{4}\right) = 4t\left(\frac{1}{t}\right)$$

$$4t\cdot\frac{1}{2}+4t\cdot\frac{1}{4} = 4$$

$$2t+t = 4$$

$$3t = 4$$

$$t = \frac{4}{3}$$

Thus, it would take $\frac{4}{3}$ hr (1 hr 20 min) for Bob and Pete to do the job working together.

EXAMPLE 8 WORK PROBLEM	**PRACTICE EXERCISE 8**

Jan can wash the dishes in 20 minutes and her mother can wash them in worked together?

    20 = number of minutes for Jan to do the job,

    $\dfrac{1}{20}$ = amount done by Jan in 1 minute,

    15 = number of minutes for Jan's mother to do the job,

    $\dfrac{1}{15}$ = amount done by Jan's mother in 1 minute.

    Let t = time it takes to do the job if Jan and her mother work together.

Then $\dfrac{1}{t}$ = amount done in 1 minute when they work together.

Thus, we have the following equation.

    (amount by Jan) + (amount by mother) = (amount together)

$$\frac{1}{20}+\frac{1}{15} = \frac{1}{t}$$

$$60t\left(\frac{1}{20}+\frac{1}{15}\right) = 60t\left(\frac{1}{t}\right) \qquad \text{Multiply by the LCD, } 60t$$

$$60t\cdot\frac{1}{20}+60t\cdot\frac{1}{15} = 60t\cdot\frac{1}{t}$$

$$3t+4t = 60$$

$$7t = 60$$

$$t = \frac{60}{7} = 8\frac{4}{7}$$

Working together, it would take Jan and her mother $8\frac{4}{7}$ minutes to wash the dishes. Is this reasonable in view of principle (1)?

Barbara can do a job in 3 days and Paula can do the same job in 8 days. How long will it take them to do the job if they work together?

Let t = number of days to do the job together

    $\dfrac{1}{t}$ = amount both do in 1 day

    $\dfrac{1}{3}$ = amount Barbara does in 1 day

    $\dfrac{1}{8}$ = _____

The equation to solve is

$$\frac{1}{3}+\frac{1}{8} = \underline{\hspace{2cm}}.$$

Answer: $\frac{24}{11}$ days

The next example is a variation of a work problem. It asks for the time it takes for one working unit to do a job when the time for the other and the time together are given.

EXAMPLE 9   FINDING THE TIME OF ONE WORKING UNIT

When the valves on a 6-inch pipe and a 2-inch pipe are opened together, it takes 5 hr to fill an oil storage tank. If the 6-inch pipe can fill the tank in 7 hr, how long would it take the 2-inch pipe to fill the tank by itself?

5 = number of hr to fill the tank together,

$\dfrac{1}{5}$ = amount filled together in 1 hr,

7 = number of hr for 6-inch pipe to fill the tank,

$\dfrac{1}{7}$ = amount filled by the 6-inch pipe in 1 hr.

Let t = number of hr for 2-inch pipe to fill the tank,

$\dfrac{1}{t}$ = amount filled by the 2-inch pipe in 1 hr.

We have the following equation.

(amount by 6-inch pipe) + (amount by 2-inch pipe) = (amount together)

$$\frac{1}{7} + \frac{1}{t} = \frac{1}{5}$$

$$35t\left(\frac{1}{7} + \frac{1}{t}\right) = 35t\left(\frac{1}{5}\right) \qquad \text{The LCD is } 35t$$

$$35t\left(\frac{1}{7}\right) + 35t\left(\frac{1}{t}\right) = 35t \cdot \frac{1}{5}$$

$$5t + 35 = 7t$$

$$35 = 2t$$

$$\frac{35}{2} = t$$

The 2-inch pipe would take $\frac{35}{2}$ hours or $17\frac{1}{2}$ hours to fill the tank alone. (Does this make sense in view of principle (1) of work problems?) Notice that the 6 inches and 2 inches only describe the size of the pipes and do not enter into the calculations.

Working together, Phil Mortensen and his son Brad can build a barn in 45 days. Working by himself, Phil can build the barn in 60 days. How long would it take Brad working alone?

Answer: 180 days

16.5   EXERCISES A

Solve.

1. $\dfrac{3x + 5}{6} = \dfrac{4 + 3x}{5}$ $\dfrac{1}{3}$

2. $\dfrac{1}{x} - \dfrac{2}{x} = 6$ $-\dfrac{1}{6}$

3. $\dfrac{2}{y + 1} = \dfrac{3}{y}$ -3

4 $\dfrac{1}{z-3} + \dfrac{3}{z-5} = 0$ $\dfrac{7}{2}$

5. $\dfrac{1}{2x-3} - \dfrac{3}{4x-15} = 0$ -3

6. $\dfrac{a-1}{a+1} = \dfrac{a-3}{a+2}$ $-\dfrac{1}{3}$

7. $\dfrac{2y+1}{3y-2} = \dfrac{2y-3}{3y+2}$ $\dfrac{1}{5}$

8 $\dfrac{z}{z+1} - 1 = \dfrac{1}{z}$ $-\dfrac{1}{2}$

9. $\dfrac{2x}{x+2} - 2 = \dfrac{1}{x}$ $-\dfrac{2}{5}$

10. $\dfrac{1}{y+2} + \dfrac{1}{y-2} = \dfrac{1}{y^2-4}$ $\dfrac{1}{2}$

11 $\dfrac{4}{z-3} + \dfrac{2z}{z^2-9} = \dfrac{1}{z+3}$
no solution

12. $\dfrac{3}{x+3} + \dfrac{1}{x-3} = \dfrac{10}{x^2-9}$ 4

13. $\dfrac{y}{y+3} - \dfrac{3}{y-3} = \dfrac{y^2+9}{y^2-9}$
no solution

14. $\dfrac{10}{x^2-25} + 1 = \dfrac{x}{x+5}$ 3

15. $\dfrac{y}{y-4} - 1 = \dfrac{8}{y^2-16}$ -2

16 $\dfrac{3}{x-3} + \dfrac{2}{x+3} = -1$ $1, -6$

17. $\dfrac{4}{y-4} + \dfrac{6}{y+4} = -1$ $2, -12$

18. $\dfrac{2z}{z-2} - 1 = \dfrac{1}{z^2-4}$ $-1, -3$

Solve. (Some problems have been started.)

19. Find two numbers whose sum is 42 and whose quotient is $\frac{3}{4}$.

    Let $x =$ one number,
  $42 - x =$ second number. [The sum of the two is
      $x + (42 - x) = 42$.]

18, 24

20. Two-thirds of a certain number is one more than five-eighths of the number. Find the number.
24

21. The denominator of a certain fraction is 2 greater than the numerator, and the value of the fraction is $\frac{4}{5}$. Find the fraction.

$\frac{8}{10}$

22. If Joe is three-fourths as old as Jeff and the difference in their ages is 6 years, how old is each?

Let x = Jeff's age,

$\frac{3}{4}x$ = Joe's age.

Joe is 18, Jeff is 24.

23 Find the number which when added to both the numerator and denominator of $\frac{5}{7}$ will produce a fraction equivalent to $\frac{4}{5}$.

3

24. The reciprocal of 3 less than a number is four times the reciprocal of the number. Find the number.

Let x = the desired number,

$x - 3$ = 3 less than the number,

$\dfrac{1}{x - 3}$ = the reciprocal of 3 less than the number.

4

25 The numerator of a fraction is 4 less than its denominator. If both the numerator and denominator are increased by 1, the value of the fraction is $\frac{2}{3}$. Find the fraction.

$\frac{7}{11}$

26. Mary's age is 4 years more than Ruth's age, and the quotient of their ages is $\frac{7}{5}$. How old is each?
Mary is 14, Ruth is 10.

27. The denominator of a fraction exceeds the numerator by 6, and the value of the fraction is $\frac{5}{6}$. Find the fraction.

Let x = numerator of fraction,

$x + 6$ = denominator of fraction.

$\frac{30}{36}$

28 The reciprocal of 4 less than a number is three times the reciprocal of the number. Find the number.

6

29. If Burford can mow a lawn in 7 hours and Hilda can mow the same lawn in 3 hours, how long would it take them to mow it if they worked together?

7 = number of hr for Burford to do the job

$\frac{1}{7}$ = amount done by Burford in 1 hr

3 = number of hr for Hilda to do the job

$\frac{1}{3}$ = amount done by Hilda in 1 hr

Let t = time to do the job working together

$\frac{1}{t}$ = amount done in 1 hr working together

$\frac{1}{7} + \frac{1}{3} = \frac{1}{t}$

$\frac{21}{10}$ **hr**

30 Bob can paint a house alone in 12 days. When he and his father work together it takes only 9 days. How long would it take his father to paint the house if he worked alone?
36 days

31. A 6-inch pipe can fill a reservoir in 3 weeks. When this pipe and a second pipe are both turned on, it takes only 2 weeks to fill the reservoir. How long would it take the second pipe to fill it by itself?
6 weeks

32 If Irv takes 4 times longer to repair a car than Max, and together they can repair it in 8 hours, how long does it take each working alone?
10 hr, 40 hr

33. If a 1-inch pipe takes 40 minutes to drain a pond and a 2-inch pipe takes 10 minutes to drain the same pond, how long would it take them to drain it together?
8 min

34. If Art can frame a shed in 4 days and together Art and Clyde can frame the same shed in 3 days, how long would it take Clyde to frame it working by himself?
12 days

35. An older machine requires twice as much time to do a job as a modern one. Together they accomplish the job in 8 minutes. How long would it take each to do the job alone?
12 min, 24 min

36. Susan can make 100 widgets in 3 hours. It takes Harry 5 hours to produce 100 widgets. How long would it take them to produce 100 widgets if they worked together?
$\frac{15}{8}$ **hr**

FOR REVIEW

Perform the indicated operations.

37. $\dfrac{3}{5 - x} - \dfrac{2 - x}{x - 5}$

1

38. $\dfrac{2}{x - 3} + \dfrac{5}{x + 2}$

$\dfrac{7x - 11}{(x - 3)(x + 2)}$

39. $\dfrac{-5y}{y^2 - 25} - \dfrac{y}{y + 5}$

$\dfrac{-y^2}{y^2 - 25}$

40. $\dfrac{5}{y^2 - 25} + \dfrac{5}{y^2 - 3y - 10}$

$\dfrac{5(2y + 7)}{(y + 5)(y - 5)(y + 2)}$

41. $\dfrac{3}{x^2 - x - 2} - \dfrac{2}{x^2 + 6x + 5}$

$\dfrac{x + 19}{(x - 2)(x + 1)(x + 5)}$

42. $\dfrac{x + y}{x - y} - \dfrac{x - y}{x + y}$

$\dfrac{4xy}{(x + y)(x - y)}$

ANSWERS: 1. $\frac{1}{3}$ 2. $-\frac{1}{6}$ 3. -3 4. $\frac{7}{2}$ 5. -3 6. $-\frac{1}{3}$ 7. $\frac{1}{5}$ 8. $-\frac{1}{2}$ 9. $-\frac{2}{5}$ 10. $\frac{1}{2}$ 11. no solution 12. 4
13. no solution 14. 3 15. -2 16. 1, -6 17. 2, -12 18. $-1, -3$ 19. 18, 24 20. 24 21. $\frac{8}{10}$ 22. Joe is 18,
Jeff is 24 23. 3 24. 4 25. $\frac{7}{11}$ 26. Mary is 14, Ruth is 10 27. $\frac{30}{36}$ 28. 6 29. $\frac{21}{10}$ hr 30. 36 days 31. 6 weeks
32. 10 hr, 40 hr 33. 8 min 34. 12 days 35. 12 min, 24 min 36. $\frac{15}{8}$ hr 37. 1 38. $\frac{7x - 11}{(x - 3)(x + 2)}$ 39. $\frac{-y^2}{y^2 - 25}$
40. $\frac{5(2y + 7)}{(y + 5)(y - 5)(y + 2)}$ 41. $\frac{x + 19}{(x - 2)(x + 1)(x + 5)}$ 42. $\frac{4xy}{(x + y)(x - y)}$

16.5 EXERCISES B

Solve.

1. $\dfrac{5x-2}{3} = \dfrac{2+7x}{4}$ -14

2. $\dfrac{3}{x} - \dfrac{4}{x} = 3$ $-\dfrac{1}{3}$

3. $\dfrac{8}{a+4} = \dfrac{7}{a-5}$ 68

4. $\dfrac{2}{z-4} + \dfrac{3}{z-6} = 0$ $\dfrac{24}{5}$

5. $\dfrac{2}{3x-1} - \dfrac{1}{4x-5} = 0$ $\dfrac{9}{5}$

6. $\dfrac{x-3}{x-2} = \dfrac{x-2}{x+3}$ $\dfrac{13}{4}$

7. $\dfrac{2y-3}{4y-1} = \dfrac{y-6}{2y-1}$ $\dfrac{3}{17}$

8. $\dfrac{2z}{z+3} - 2 = \dfrac{3}{z}$ -1

9. $\dfrac{x+1}{x} = 1 + \dfrac{1}{x+1}$

    no solution

10. $\dfrac{1}{y+3} + \dfrac{2}{y-3} = \dfrac{3}{y^2-9}$ 0

11. $\dfrac{5}{z+2} + \dfrac{3z}{z^2-4} = \dfrac{2}{z-2}$ $\dfrac{7}{3}$

12. $\dfrac{-2}{x+5} + \dfrac{1}{x-5} = \dfrac{4}{x^2-25}$

    11

13. $\dfrac{y}{y+5} - \dfrac{5}{y-5} = \dfrac{y^2+25}{y^2-25}$

    no solution

14. $\dfrac{64}{x^2-16} + 2 = \dfrac{2x}{x-4}$

    no solution

15. $\dfrac{y}{y+6} - 1 = \dfrac{6}{y^2-36}$

    5

16. $\dfrac{6}{x+5} + \dfrac{1}{x-5} = 1$ 0 or 7

17. $\dfrac{1}{y-6} - \dfrac{6}{y+6} = -1$ 2 or 3

18. $\dfrac{3z}{z-3} - 2 = \dfrac{10}{z^2-9}$

    -1 or -8

Solve.

19. Find two numbers whose sum is 32 and whose quotient is $\frac{3}{5}$.   **12, 20**

20. Three-fourths of a number is three more than three-eighths of the number. Find the number.   **8**

21. The denominator of a fraction is 6 greater than the numerator, and the value of the fraction is $\frac{7}{9}$. Find the fraction. $\frac{21}{27}$

22. Art is five-eighths as old as Toni and the difference in their ages is 21 years. How old is each?

    Art is 35, Toni is 56.

23. Find the number which when added to both the numerator and denominator of $\frac{3}{11}$ will produce a fraction equivalent to $\frac{1}{2}$.   **5**

24. The reciprocal of 6 less than a number is four times the reciprocal of the number. Find the number.

    8

25. The numerator of a fraction is 5 less than the denominator. If both the numerator and the denominator are decreased by 2, the value of the fraction is $\frac{1}{6}$. Find the fraction. $\frac{3}{8}$

26. Cindy is 4 years younger than Becky and the quotient of their ages is $\frac{7}{9}$. How old is each?

    Cindy is 14, Becky is 18.

27. The numerator of a fraction exceeds the denominator by 10, and the value of the fraction is $\frac{7}{5}$. Find the fraction. $\frac{35}{25}$

28. The reciprocal of 6 more than a number is four times the reciprocal of the number. Find the number.

    -8

29. Alfonse can paint a garage in 20 hours and Heidi can do the same job in 10 hours. How long would it take them if they worked together? $\frac{20}{3}$ **hr**

30. Walter can dig a well in 8 days. When he and Sean work together they can dig the same size well in 6 days. How long would it take Sean to do the job working alone?   **24 days**

31. If Andy takes 3 times as long to put new brakes on a car as Hortense does, and together they can do the job in 4 hours, how long does it take each working alone? $\frac{16}{3}$ **hr, 16 hr**

32. There are two drains to a water tank. If one is used, the tank is drained in 24 minutes. The other requires 32 minutes. How long will it take if both are opened? $\frac{96}{7}$ **min**

33. If Nancy can process an order of 1000 appliances in 20 minutes and together she and Hans can process the order in 12 minutes, how long will it take Hans to process the order by himself?   **30 min**

34. Using old equipment it requires 5 times as long to make 100 sportcoats as with new equipment. If together the equipment can accomplish the job in $7\frac{1}{2}$ hours, how long would it take each working alone?
9 hr, 45 hr

35. Charles can make 200 dolls in 3 weeks and Wilma can make 200 dolls in $2\frac{1}{2}$ weeks. How long will it take to make 200 dolls if they work together?
$\frac{15}{11}$ **weeks**

36. Bertha can clean the stable in 4 hours. If Bertha and Jose work together they can do the same job in 1 hour. How long would Jose take to do the job working alone? $\frac{4}{3}$ **hr**

FOR REVIEW

Perform the indicated operations.

37. $\dfrac{2}{9 - x} - \dfrac{3 - x}{x - 9}$ $\dfrac{x - 5}{x - 9}$

38. $\dfrac{8}{z - 7} + \dfrac{2}{z + 1}$ $\dfrac{2(5z - 3)}{(z - 7)(z + 1)}$

39. $\dfrac{-9y}{y^2 - 16} - \dfrac{5y}{y - 4}$ $\dfrac{-(5y^2 + 29y)}{(y + 4)(y - 4)}$

40. $\dfrac{3}{y^2 - 6y + 8} + \dfrac{7}{y^2 - 16}$ $\dfrac{2(5y - 1)}{(y - 2)(y - 4)(y + 4)}$

41. $\dfrac{5}{x^2 + x - 12} - \dfrac{3}{x^2 + 10x + 24}$ $\dfrac{2x + 39}{(x - 3)(x + 4)(x + 6)}$

42. $\dfrac{x - y}{x + y} - \dfrac{x + y}{x - y}$ $\dfrac{-4xy}{(x + y)(x - y)}$

16.5 EXERCISES C

Solve.

1. $\dfrac{3}{x^2 - 6x + 5} - \dfrac{2}{x^2 - 25} = \dfrac{4}{x^2 + 4x - 5}$ $\dfrac{37}{3}$

2. $\dfrac{2x}{x^2 - 3x - 4} - \dfrac{x - 1}{x^2 + 2x + 1} = \dfrac{x + 4}{x^2 + 5x + 4}$
[Answer: 0]

3. A man walks a distance of 12 mi at a rate 8 mph slower than the rate he rides a bicycle for a distance of 24 mi. If the total time of the trip is 5 hr, how fast does he walk? How fast does he ride?
He walks 4 mph, he rides 12 mph.

4. A swimming pool can be filled by an inlet pipe in 6 hr. It can be drained by an outlet in 9 hr. If the inlet and outlet are both opened, how long will it take to fill the empty pool?   [Answer: 18 hr]

16.6 RATIO, PROPORTION, AND VARIATION

STUDENT GUIDEPOSTS

1 Ratios
2 Proportions
3 Means-Extremes Property
4 Applying Proportions
5 Direct Variation
6 Inverse Variation

1 RATIOS

The **ratio** of one number x to another number y is the quotient

$$x \div y \quad \text{or} \quad \frac{x}{y}.$$

We sometimes express the ratio $\frac{x}{y}$ with the notation $x:y$, which is read "the ratio of x to y." Ratios occur in applications such as percent, rate of speed, gas mileage, unit cost, and number comparisons. Consider the following examples.

Applications	Ratio
5% sales tax	$\frac{\$5}{\$100} = \$5$ per $\$100$
200 miles in 4 hours	$\frac{200 \text{ mi}}{4 \text{ hr}} = 50\frac{\text{mi}}{\text{hr}} = 50$ mi per hr
200 miles on 10 gallons of gas	$\frac{200 \text{ mi}}{10 \text{ gal}} = 20\frac{\text{mi}}{\text{gal}} = 20$ mi per gal
$12 for 2 kg of meat	$\frac{\$12}{2 \text{ kg}} = \$6$ per kg
30 children and 6 adults	$\frac{30 \text{ children}}{6 \text{ adults}} = 5$ children per adult

② PROPORTIONS

Ratios are used to define *proportions* which can describe many practical problems.

Proportions

An equation which states that two ratios are equal is called a **proportion.** That is, if $\frac{a}{b}$ and $\frac{c}{d}$ are two ratios, the equation

$$\frac{a}{b} = \frac{c}{d}$$

is a proportion. It can be read "*a* is to *b* as *c* is to *d*."

③ MEANS-EXTREMES PROPERTY

For example,

$$\frac{2}{3} = \frac{4}{6}$$

is a proportion. The numbers 2 and 6 are the **extremes** of the proportion, and the numbers 3 and 4 are the **means.**

$$2 \text{ is to } 3 \text{ as } 4 \text{ is to } 6$$

means

extremes

Consider the proportion

$$\frac{a}{b} = \frac{c}{d},$$

with extremes a and d and means b and c. If we multiply both sides by the LCD of the two fractions, bd, we obtain

$$bd\frac{a}{b} = \frac{c}{d}bd$$

or $\quad ad = bc.$

We have just verified the following property of proportions.

Means-Extremes Property of Proportions
In any proportion, the product of the extremes is equal to the product of the means. That is, if $$\frac{a}{b} = \frac{c}{d} \quad \text{then} \quad ad = bc.$$

This property of proportions allows us to solve for an unknown term. We call this **solving the proportion.** For example, suppose we solve the proportion

$$\frac{15}{25} = \frac{x}{5}.$$

$15 \cdot 5 = 25 \cdot x$	The product of the extremes equals the product of the means
$75 = 25x$	Simplify
$\dfrac{75}{25} = x$	Divide both sides by 25
$3 = x$	Reduce the fraction

❹ APPLYING PROPORTIONS

We now show how proportions can be used to solve applied problems. Consider the following: If it takes 3 hours to travel 150 miles, how many hours would it take to travel 250 miles? The simple proportion

$$\text{first time} \rightarrow \ \frac{3}{150} = \frac{x}{250} \ \leftarrow \text{second time}$$
$$\text{first distance} \rightarrow \qquad\qquad \leftarrow \text{second distance}$$

where x is the number of hours it takes to travel 250 miles, completely describes this problem. That is, the ratio of the first time to the first distance must equal the ratio of the second time to the second distance. Setting the product of the extremes equal to the product of the means, we have

$$\left(\begin{array}{c}\text{first}\\\text{time}\end{array}\right) \cdot \left(\begin{array}{c}\text{second}\\\text{distance}\end{array}\right) = \left(\begin{array}{c}\text{first}\\\text{distance}\end{array}\right) \cdot \left(\begin{array}{c}\text{second}\\\text{time}\end{array}\right)$$

$$3 \quad\cdot\quad 250 \quad = \quad 150 \quad\cdot\quad x$$

or
$$5 = x.$$

It would take 5 hours to go 250 miles.

EXAMPLE 1 AN APPLICATION OF PROPORTION	**PRACTICE EXERCISE 1**

If a family drinks 5 gallons of milk in two weeks, how many gallons of milk will be consumed in a year (52 weeks)?

    Let x = number of gallons of milk consumed in 52 weeks. The proportion is

$$\begin{array}{c}\text{gallons consumed} \rightarrow\\\text{in 2 weeks}\end{array} \quad \frac{5}{2} = \frac{x}{52}. \quad \begin{array}{c}\leftarrow \text{gallons consumed}\\\text{in 52 weeks}\end{array}$$

$5 \cdot 52 = 2x$	Product of the extremes equals product of the means
$130 = x$	

The family drinks 130 gallons of milk in a year.

A sample of 235 television sets contained 4 that were defective. How many defective sets would you expect to find in a sample of 1645?

Answer: 28

| EXAMPLE 2   AN APPLICATION OF PROPORTION | PRACTICE EXERCISE 2 |

If a secretary spends 37 minutes typing 3 pages of a report, how long will it take her to type the remaining 10 pages?

Let x be the number of minutes it takes to complete the typing. We set up the following proportion.

minutes to   minutes to
type 3 pages → $\dfrac{3}{37} = \dfrac{10}{x}$.   ← type 10 pages

$$3x = 370 \qquad \text{Product of the extremes}$$
$$\text{equals product of the means}$$

$$x = 123\frac{1}{3}.$$

It will take another $123\frac{1}{3}$ minutes or 2 hours, 3 minutes, and 20 seconds.

PRACTICE EXERCISE 2

If it costs $18.00 to operate a freezer for 3 months, how much would it cost to operate the freezer for one year?

Answer: $72.00

⑤ DIRECT VARIATION

Another way of looking at proportion problems is in terms of **variation.** For example, the distance that a car travels varies (or changes) as we vary or change the speed. Also, the total cost of a roast will vary as the price per pound varies. The equation in each of these examples can be written as a ratio equal to a constant. This type of variation is called **direct variation.**

Direct Variation

If two variables have a constant ratio k,

$$\frac{y}{x} = k \quad \text{or} \quad y = kx,$$

we say that y **varies directly** as x and call k the **constant of variation.**

One common example of this is the fact that the ratio of the circumference to the diameter of any circle is π.

$$\frac{c}{d} = \pi \quad \text{or} \quad c = \pi d \quad k = \pi$$

In a variation problem if we know the value of y for one value of x, we can find k. With k known we can find y for any x.

| EXAMPLE 3   USING DIRECT VARIATION | PRACTICE EXERCISE 3 |

Find the equation of variation if y varies directly as x and we know that $y = 30$ when $x = 6$. Also, find y when $x = 10$.

First, write an equation stating that y *varies directly as x*. Then substitute $y = 30$ and $x = 6$ to find k.

$$y = kx \qquad \text{y varies directly as x}$$
$$30 = k \cdot 6 \qquad \text{$y = 30$ and $x = 6$}$$
$$\frac{30}{6} = k \qquad \text{k is a ratio}$$
$$5 = k$$

PRACTICE EXERCISE 3

Find the equation of variation if u varies directly as v and we know that $u = 4$ when $v = 3$. Use this equation to find u when $v = 45$.

The equation of variation is

$$y = 5x.$$

To find y when $x = 10$, substitute 10 for x in the equation of variation.

$y = 5x$	Equation of variation
$y = 5(10)$	Substitute 10 for x
$y = 50$	

Thus y is 50 when x is 10.

Answer: $u = \frac{4}{3}v$; 60

EXAMPLE 4 APPLYING DIRECT VARIATION	PRACTICE EXERCISE 4

The cost c of steak varies directly as the weight w of the steak purchased. If 8 kg of steak cost \$36, how much will 30 kg of steak cost?

First write the equation for c varies directly as w. Then substitute $c = 36 and $w = 8$ kg to find k.

$c = kw$	c varies directly as w
$36 = k \cdot 8$	$c = 36 and $w = 8$ kg
$4.5 = k$	$\frac{36}{8} = 4.5$

The equation of variation that we can use for any weight of meat is

$$c = 4.5\,w.$$

To find the cost of 30 kg of steak substitute 30 for w.

$$c = 4.5w$$
$$c = 4.5(30)$$
$$= 135$$

Thus 30 kg (about 66 lb) of steak would cost \$135.

The weight m of an object on the moon varies directly as its weight e on earth. If a man weighing 150 lb on earth weighs 24 lb on the moon, how much will a module that weighs 9500 lb on earth weigh on the moon?

Answer: 1520 lb

---

⑥ INVERSE VARIATION

In many applications the product of the variables is a constant. This is called **inverse variation.**

Inverse Variation

If two variables have a constant product k,

$$xy = k \quad \text{or} \quad y = \frac{k}{x}$$

where k is the **constant of variation,** we say that y **varies inversely as** x.

EXAMPLE 5 APPLYING INVERSE VARIATION	**PRACTICE EXERCISE 5**

The time t it takes to drive a fixed distance d varies inversely as the rate r (or speed) of the car. If it takes 3 hours when the speed is 40 mph, how long will it take when the speed is 50 mph?

The equation of variation in this case is

$$t = \frac{d}{r}.$$

Since the distance is fixed, d is the constant of variation. We are given that $t = 3$ hr when $r = 40$ mph.

$$t = \frac{d}{r}$$

$$3 = \frac{d}{40}$$

$$120 = d$$

Thus

$$t = \frac{120}{r} \quad \text{Equation of variation}$$

$$t = \frac{120}{50} \quad r = 50 \text{ mph}$$

$$t = 2.4.$$

It will take 2.4 hours to go the distance at 50 mph.

The rate r of a car when traveling a fixed distance d varies inversely as the time t of travel. If the rate is 55 mph when the time is 4 hr, what is the rate when the time is 5.5 hr?

Answer: 40 mph

16.6 EXERCISES A

Indicate as a ratio and simplify.

1. 320 mi in 8 hr
   40 mph

2. $10 for 4 lb of meat
   $2.50 per lb

3. 20 children for 2 adults
   10 children per adult

4. 420 mi on 12 gal of gas
   35 mpg

5. 200 trees on 4 acres
   50 trees per acre

6. 45°F in 30 min
   1.5°F per min

Solve the proportions.

7. $\frac{x}{2} = \frac{9}{27}$ $\frac{2}{3}$

8. $\frac{y+8}{y-2} = \frac{7}{3}$ $\frac{19}{2}$

9. $\frac{w-5}{6} = \frac{w+6}{2}$ $-\frac{23}{2}$

10. $\dfrac{5}{x-7} = \dfrac{4}{x+4}$ -48

11 $\dfrac{y-3}{y+2} = \dfrac{y-2}{y+3}$

no solution

12. $\dfrac{z+5}{z-2} = \dfrac{z-7}{z-3}$ $\dfrac{29}{11}$

Solve. (Some problems have been started.)

13. In an election the winning candidate won by a 5 to 3 margin. If she received 1025 votes, how many votes did the losing candidate receive?

Let x = no. of votes for losing candidate

$\dfrac{1025}{x} = \dfrac{5}{3}$

615 votes

14. If a boat uses 7 gal of gas to go 51 mi, how many gallons would be needed to go 255 mi?

Let x = no. of gallons needed to go 255 mi

$\dfrac{7}{51} = \dfrac{x}{255}$

35 gal

15. If 50 feet of wire weigh 15 lb, what will 80 feet of the same wire weigh?

Let x = weight of 80 feet of wire

24 lb

16. If $\frac{3}{4}$ inch on a map represents 10 mi, how many miles will be represented by 9 in?

Let x = no. of miles represented by 9 in

120 mi

17 A sample of 184 tires contained 6 that were defective. How many defective tires would you expect in a sample of 1288?

42

18. If 2 bricks weigh 9 pounds, how much will 558 bricks weigh?

2511 lb

19. In an election, the successful candidate won by a 3 to 2 margin. If he received 324 votes, how many did the losing candidate receive?

216 votes

20. If a car uses 8 gallons of gas for a trip of 132 miles, how much gas will be used for a trip of 550 miles?

$33\frac{1}{3}$ **gal**

21. The reciprocal of 4 more than a number is equal to five times the reciprocal of the number. Find the number.

-5

22. If Sal is 7 years older than Ev and the ratio of their ages is equal to $\frac{2}{3}$, how old is each?

Ev is 14, Sal is 21.

23. On a map 2 cm represent 14 km. How many centimeters are used to represent 35 km?

5 cm

24. In a sample of 212 light bulbs, 3 were defective. How many defective bulbs would you expect in a shipment of 3240?

approximately 46

25. Pat was charged a commission of $65.00 on the purchase of 400 shares of stock. What commission would be charged on 700 shares of the same stock?
$113.75

26 To determine the approximate number of fish in a lake, the State Game and Fish Department caught 100 fish, tagged their fins, and returned them to the water. After a period of time they caught 70 fish and discovered that 14 were tagged. Approximately how many fish are in the lake?
500 fish

Solve using direct variation.

27. If y varies directly as x, and $y = 10$ when $x = 15$, find y when $x = 42$.
28

28. If u varies directly as w, and $u = 24$ when $w = 6$, find u when $w = 20$.
80

29. If y varies directly as x, and $y = 12$ when $x = 9$, find y when $x = 6$.
8

30 The cost c of peaches varies directly as the total weight w of the peaches. If 12 pounds of peaches cost $4.80, what would 66 pounds cost?
$26.40

31. The number d of defective tires in a shipment varies directly as the number n of tires in the shipment. If there were 10 defective tires in a shipment of 2200, how many defective tires would you expect in a shipment of 8800 tires?
40

32. The number g of gallons of gas needed varies directly as the number n of miles traveled. If 25 gallons are needed to travel 650 miles, how many gallons will be needed to travel 3900 miles?
150 gal

33. The size of Maria's paycheck varies directly as the number of hours she works. If she is paid $334.00 for working 40 hours each week, how much would she be paid for working 2080 hours (1 year)?
$17,368

34 The amount of money a man spends on recreation varies directly as his total income. If he spends $1920 on recreation when earning $16,000, how much would he spend if his income is increased by $5,000?
$2520

Solve using inverse variation.

35. If y varies inversely as x, and $y = 12$ when $x = 9$, find y when $x = 6$.
18

36. If p varies inversely as q, and $p = 520$ when $q = 80$, find p when $q = 200$.
208

37. The time t needed to travel a fixed distance varies inversely as the rate r of travel. If 6 hr are needed when the rate is 500 mph, how long would it take at 1200 mph?

2.5 hours

38. The rate r of travel for a fixed distance varies inversely as the time t. If the rate is 120 km per hour when the time is 4 hr, find the rate when the time is 6 hr.

80 km/hr

39. The volume V of a given amount of gas varies inversely as the pressure P. If the volume is 24 in^3 when the pressure is 8 lb/in^2, what is the volume when the pressure is 6 lb/in^2?

32 in^3

40. The current I in an electrical circuit with constant voltage varies inversely as the resistance R of the circuit. If the current is 3 amps when the resistance is 4 ohms, find I when $R = 6$ ohms.

2 amps

FOR REVIEW

Solve.

41. $\dfrac{3}{x+1} - \dfrac{2}{x+1} = 5$

$-\dfrac{4}{5}$

42. $\dfrac{3}{2y+1} - \dfrac{1}{2y-1} = 0$

1

43. $\dfrac{x+1}{x-1} - \dfrac{x-1}{x+1} = 0$

0

44. $\dfrac{3x}{x-3} - 3 = \dfrac{1}{x}$

$-\dfrac{3}{8}$

45. $\dfrac{1}{y-5} + \dfrac{1}{y+5} = \dfrac{1}{y^2-25}$

$\dfrac{1}{2}$

46. $\dfrac{2x}{x-3} - 1 = \dfrac{16}{x^2-9}$

1 or -7

47. Phillip can paint a wall in 3 hr and Elizabeth can paint the same size wall in 2 hr. How long would it take to paint the wall if they worked together?

$\frac{6}{5}$ **hr**

48. Steve can type a manuscript 3 times as fast as Randy. Working together they can do the job in 5 hr. How long would it take each working alone?

Steve $\frac{20}{3}$ hr; Randy 20 hr

The following exercises review material from Chapter 3 to help you prepare for the next section. Find the reciprocal of each fraction.

49. $\dfrac{7}{11}$ $\dfrac{11}{7}$

50. 4 $\dfrac{1}{4}$

Divide the fractions.

51. $\dfrac{4}{9} \div \dfrac{8}{27}$ $\dfrac{3}{2}$

52. $8 \div \dfrac{10}{9}$ $\dfrac{36}{5}$

ANSWERS: 1. 40 mph 2. $2.50 per lb 3. 10 children per adult 4. 35 mpg 5. 50 trees per acre 6. 1.5°F per min 7. $\frac{2}{3}$ 8. $\frac{19}{2}$ 9. $-\frac{23}{2}$ 10. -48 11. no solution 12. $\frac{29}{11}$ 13. 615 votes 14. 35 gal 15. 24 lb 16. 120 mi 17. 42 18. 2511 lb 19. 216 votes 20. $33\frac{1}{3}$ gal 21. -5 22. Ev is 14, Sal is 21 23. 5 cm 24. approximately 46 25. $113.75 26. 500 fish 27. 28 28. 80 29. 8 30. $26.40 31. 40 32. 150 gal 33. $17,368 34. $2520 35. 18 36. 208 37. 2.5 hours 38. 80 km/hr 39. 32 in^3 40. 2 amps 41. $-\frac{4}{5}$ 42. 1 43. 0 44. $-\frac{3}{8}$ 45. $\frac{1}{2}$ 46. 1, -7 47. $\frac{6}{5}$ hr 48. Steve $\frac{20}{3}$ hr; Randy 20 hr 49. $\frac{11}{7}$ 50. $\frac{1}{4}$ 51. $\frac{3}{2}$ 52. $\frac{36}{5}$

16.6 EXERCISES B

Give as a ratio and simplify.

1. 480 mi in 6 hr
   80 mph

2. $6 for 15 lb of apples
   $0.40 per lb

3. 60 children for 4 adults
   15 children per adult

4. 2232 mi on 72 gal of gas
   31 mpg

5. 600 bushels from 12 acres
   50 bushels per acre

6. 320 gal in 10 min
   32 gal per min

Solve the proportions.

7. $\dfrac{x}{18} = \dfrac{12}{20}$ **10.8**

8. $\dfrac{y-3}{y-7} = \dfrac{5}{8}$ $-\dfrac{11}{3}$

9. $\dfrac{w+8}{5} = \dfrac{w-4}{3}$ **22**

10. $\dfrac{6}{x+10} = \dfrac{5}{x-10}$ **110**

11. $\dfrac{y+8}{y-4} = \dfrac{y+1}{y-6}$ $\dfrac{44}{5}$

12. $\dfrac{z-1}{z-2} = \dfrac{z-3}{z-4}$ **no solution**

Solve.

13. In an election the winning candidate won by a 6 to 5 margin. If the loser received 1520 votes, how many did the winner receive?
   1824 votes

14. If a boat uses 12 gallons of gas to go 82 miles, how many gallons would be needed to go 123 miles?
   18 gal

15. If 20 m of hose weighs 6 kg, what will 320 m of the same hose weigh?
   96 kg

16. If 3 cm on a map represent 14 km, how many km are represented by 36 cm?
   168 km

17. A sample of 350 bolts contained 5 that were defective. How many defective bolts should be expected in a shipment of 9870 bolts?
   141 bolts

18. If 5 blocks weigh 14 pounds, how much will 255 blocks weigh?
   714 lb

19. In an election the losing candidate lost by a 3 to 4 margin. If he received 5280 votes, how many votes did the winner receive?
   7040 votes

20. A boat uses 12 gal of gas in 4 hr. Under the same conditions how many gallons will be used in 10 hr?
   30 gal

21. The reciprocal of 5 less than a number is equal to the reciprocal of twice the number. Find the number.
   -5

22. If Walter is 3 years older than Molly and the ratio of their ages is equal to $\frac{5}{6}$, how old is each?
   Walter is 18, Molly is 15.

23. On a map 5 inches represent 24 mi. How many miles are represented by 2 in?
   9.6 mi

24. In a sample of 280 toasters, 8 were defective. How many defective toasters should be expected in a warehouse containing 42,700 toasters?
   1220 toasters

25. Marvin received a commission of $1200 on the sale of $150,000 worth of life insurance. If the same person has bought $250,000 worth of insurance, what commission would Marvin have earned?
   $2000

26. To determine the number of antelope on a game preserve, a ranger caught 50 antelope and tagged their ears. Some time later he captured 18 antelope and discovered that 5 were tagged. Approximately how many antelope are on the preserve?
   180 antelope

Solve using direct variation.

27. If y varies directly as x, and $y = 5$ when $x = 35$, find y when $x = 10$. $\frac{10}{7}$

28. If u varies directly as w, and $u = 52$ when $w = 13$, find u when $w = 22$.   **88**

29. If y varies directly as x, and $y = 7$ when $x = 3$, find y when $x = 39$.   **91**

30. The cost c of carrots varies directly as the total weight w. If 5 kg of carrots cost $3.20, what would 80 kg of carrots cost?   **$51.20**

31. The number d of defective bolts in a shipment varies directly as the number n of bolts in the shipment. If there were 16 defective bolts in a shipment of 4800, how many would be expected in a shipment of 8400?   **28 bolts**

32. The number g of gallons of gas needed varies directly as the number n of miles traveled. If 16 gallons are needed to travel 840 miles, how many gallons will be needed to travel 2800 miles? $53\frac{1}{3}$ **gal**

33. The amount of tax paid on real estate varies directly as its assessed value. If the Austins paid $320.00 in taxes on a home valued at $90,000, how much tax will the Wilsons pay on their home valued at $65,000?   **$231.11**

34. The amount of money a family spends on food varies directly as its income. If the Carpenters spend $3780 on food and have an income of $18,000, how much will the Woods spend if their income is $31,000?   **$6510**

Solve using inverse variation.

35. If y varies inversely as x, and $y = 7$ when $x = 3$, find y when $x = 39$. $\frac{7}{13}$

36. If p varies inversely as q, and $p = 580$ when $q = 20$, find p when $q = 50$.   **232**

37. The time t it takes to travel a fixed distance varies inversely as the rate r of travel. If it takes 12 hr when the rate is 24 mph, how long would it take at 36 mph?   **8 hr**

38. The rate r of travel for a fixed distance varies inversely as the time t. If the rate is 580 km per hour when the time is 8 hr, find the rate when the time is 20 hr.   **232 km/hr**

39. The volume V of a given amount of gas varies inversely as the pressure P. If the volume is 120 ft^3 when the pressure is 15 lb per sq ft, what is the volume when the pressure is 50 lb per sq ft?   **36 ft^3**

40. The resistance R of an electrical circuit with constant voltage varies inversely as the current I of the circuit. If the resistance is 10 ohms when the current is 5 amps, find the resistance when the current is 25 amps.   **2 ohms**

FOR REVIEW

Solve.

41. $\dfrac{5}{x-2} - \dfrac{1}{x-2} = 8$ $\dfrac{5}{2}$

42. $\dfrac{10}{3y-2} + \dfrac{3}{3y+1} = 0$ $-\dfrac{4}{39}$

43. $\dfrac{x-2}{x+2} - \dfrac{x+2}{x-2} = 0$ **0**

44. $\dfrac{2x}{x+5} - 2 = \dfrac{1}{x}$ $-\dfrac{5}{11}$

45. $\dfrac{1}{y+6} + \dfrac{1}{y-6} = \dfrac{4}{y^2-36}$ **2**

46. $\dfrac{3x}{x+4} - 2 = \dfrac{12}{x^2-16}$ **2 or 10**

47. Grady can build a log cabin in 5 weeks and Smitty can build the same type of cabin in 4 weeks. How long would it take to build a cabin if they worked together? $\frac{20}{9}$ **weeks**

48. Lynda can process an order twice as fast as Maria, and working together they can do the job in 3 hours. How long would it take each working alone?   **Lynda:** $\frac{9}{2}$ **hr; Maria: 9 hr**

The following exercises review material from Chapter 3 to help you prepare for the next section. Find the reciprocal of each fraction.

49. $-\dfrac{17}{8}$ $-\dfrac{8}{17}$

50. $\dfrac{1}{99}$ 99

Divide the fractions.

51. $-\dfrac{24}{25} \div \dfrac{16}{15}$ $-\dfrac{9}{10}$

52. $\dfrac{49}{36} \div 14$ $\dfrac{7}{72}$

16.6 EXERCISES C

Solve.

1. In a sample of 120 tires, 4 were defective. Because of manufacturing problems it is estimated that the defective rate will increase by 50% on the next shipment of 480 tires. How many of the new shipment would you predict to be defective?   [Answer: 24]

2. A light-bulb manufacturer detected 12 defective bulbs in a sample of 840. After buying new equipment he was able to decrease the number of defective bulbs by 40 percent. How many defective bulbs should now be expected in a shipment of 1400?
12

3. If z varies directly as x and inversely as y, show that x varies directly as y and z.
$x = \frac{1}{k}yz = k_1 yz$

4. The volume V of a gas varies directly as the temperature T and inversely as the pressure P. If the volume is 50 ft^3 when the temperature is 300° and the pressure is 30 lb per sq ft, find V when $T = 400°$ and $P = 20$ lb/ft^2.   [Answer: 100 ft^3]

16.7 SIMPLIFYING COMPLEX FRACTIONS

STUDENT GUIDEPOSTS

1 Complex Fractions   **2** Simplifying Complex Fractions

1 COMPLEX FRACTIONS

A **complex fraction** is a fraction that contains at least one other fraction within it. The following are complex fractions.

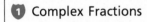

$$\frac{\frac{2}{3}}{5}, \quad \frac{\frac{1}{2}}{\frac{4}{5}}, \quad \frac{\frac{1}{x}}{3}, \quad \frac{1 + \frac{1}{x}}{2}, \quad \frac{1 + \frac{1}{a}}{1 - \frac{1}{a}}$$

2 SIMPLIFYING COMPLEX FRACTIONS

A complex fraction is **simplified** when all its component fractions have been eliminated. There are two basic methods for simplifying a complex fraction. We will present both methods since with some complex fractions one method might be more easily applied than the other. The first method involves writing the numerator and denominator as single fractions and then dividing.

To Simplify a Complex Fraction (Method 1)
1. Change the numerator and denominator to single fractions.
2. Divide the two fractions.
3. Reduce the result to lowest terms.

EXAMPLE 1 SIMPLIFYING COMPLEX FRACTIONS (METHOD 1)

Simplify the complex fractions.

(a) $\dfrac{\dfrac{x}{4}}{\dfrac{x^2}{2}}$

Since the numerator and denominator are already single fractions, all we need to do is divide.

$$\frac{\dfrac{x}{4}}{\dfrac{x^2}{2}} = \frac{x}{4} \div \frac{x^2}{2} = \frac{x}{4} \cdot \frac{2}{x^2} = \frac{\cancel{x} \cdot \cancel{2}}{\cancel{2} \cdot 2 \cdot \cancel{x} \cdot x} = \frac{1}{2x}$$

(b) $\dfrac{1 + \dfrac{1}{2}}{1 - \dfrac{1}{3}}$

$$1 + \frac{1}{2} = \frac{2}{2} + \frac{1}{2} = \frac{3}{2} \qquad \text{Add the numerator fractions}$$

$$1 - \frac{1}{3} = \frac{3}{3} - \frac{1}{3} = \frac{2}{3} \qquad \text{Subtract the denominator fractions}$$

Thus, $\dfrac{1 + \dfrac{1}{2}}{1 - \dfrac{1}{3}} = \dfrac{\dfrac{3}{2}}{\dfrac{2}{3}} = \dfrac{3}{2} \div \dfrac{2}{3} = \dfrac{3}{2} \cdot \dfrac{3}{2} = \dfrac{9}{4}$.

(c) $\dfrac{1 + \dfrac{1}{x}}{2}$

$$1 + \frac{1}{x} = \frac{x}{x} + \frac{1}{x} = \frac{x+1}{x} \qquad \text{Add the numerator fractions}$$

Thus, $\dfrac{1 + \dfrac{1}{x}}{2} = \dfrac{\dfrac{x+1}{x}}{2} = \dfrac{x+1}{x} \div 2 = \dfrac{x+1}{x} \cdot \dfrac{1}{2} = \dfrac{x+1}{2x}$.

PRACTICE EXERCISE 1

Simplify the complex fractions.

(a) $\dfrac{\dfrac{y^2}{10}}{\dfrac{y^4}{15}}$

(b) $\dfrac{2 + \dfrac{3}{4}}{3 - \dfrac{1}{6}}$

(c) $\dfrac{2 + \dfrac{2}{y}}{2y}$

Answers: (a) $\frac{3}{2y^2}$ (b) $\frac{33}{34}$ (c) $\frac{y+1}{y^2}$

The second method involves clearing all fractions within the complex fraction by multiplying numerator and denominator by the LCD of all the internal fractions.

To Simplify a Complex Fraction (Method 2)
1. Find the LCD of all fractions within the complex fraction.
2. Multiply numerator and denominator of the complex fraction by the LCD to obtain an equivalent fraction.
3. Reduce the result to lowest terms.

EXAMPLE 2 SIMPLIFYING A COMPLEX FRACTION (METHOD 2)	**PRACTICE EXERCISE 2**

Simplify $\dfrac{1 + \dfrac{2}{a}}{1 - \dfrac{4}{a^2}}$.

Simplify $\dfrac{\dfrac{1}{x} + 3}{\dfrac{1}{x^2 - 9}}$.

The denominator in $1 + \frac{2}{a}$ is a, and in $1 - \frac{4}{a^2}$ is a^2. Thus, the LCD $= a^2$.

$$\frac{1 + \dfrac{2}{a}}{1 - \dfrac{4}{a^2}} = \frac{\left(1 + \dfrac{2}{a}\right) \cdot a^2}{\left(1 - \dfrac{4}{a^2}\right) \cdot a^2} \qquad \text{Multiply by LCD} = a^2$$

$$= \frac{1 \cdot a^2 + \dfrac{2}{a} \cdot a^2}{1 \cdot a^2 - \dfrac{4}{a^2} \cdot a^2} \qquad \text{Multiply all terms by } a^2$$

$$= \frac{a^2 + 2a}{a^2 - 4}$$

$$= \frac{a\cancel{(a + 2)}}{\cancel{(a + 2)}(a - 2)} \qquad \text{Factor and reduce to lowest terms}$$

$$= \frac{a}{a - 2}$$

Answer: $\frac{x}{1 - 3x}$

EXAMPLE 3 SIMPLIFYING A COMPLEX FRACTION (METHOD 2)	**PRACTICE EXERCISE 3**

Simplify $\dfrac{1 + \dfrac{2}{x - 2}}{\dfrac{2}{x + 2} - 1}$.

Simplify $\dfrac{\dfrac{1}{y + 2} - 4}{4 + \dfrac{15}{y - 2}}$.

The denominator in $1 + \frac{2}{x - 2}$ is $x - 2$, and in $\frac{2}{x + 2} - 1$ is $x + 2$. Thus, the LCD $= (x - 2)(x + 2)$.

$$\frac{1 + \dfrac{2}{x - 2}}{\dfrac{2}{x + 2} - 1} = \frac{\left(1 + \dfrac{2}{x - 2}\right)(x - 2)(x + 2)}{\left(\dfrac{2}{x + 2} - 1\right)(x - 2)(x + 2)} \qquad \text{Multiply by LCD}$$

$$= \frac{1 \cdot (x - 2)(x + 2) + \dfrac{2}{\cancel{x - 2}}\cancel{(x - 2)}(x + 2)}{\dfrac{2}{x + 2}(x - 2)(x + 2) - 1 \cdot (x - 2)(x + 2)}$$

$$= \frac{(x^2 - 4) + 2(x + 2)}{2(x - 2) - (x^2 - 4)}$$

$$= \frac{x^2 - 4 + 2x + 4}{2x - 4 - x^2 + 4} \qquad \begin{array}{l}\text{Watch signs when clearing} \\ \quad \text{parentheses using the distributive law}\end{array}$$

$$= \frac{x^2 + 2x}{-x^2 + 2x}$$

$$= \frac{\cancel{x}(x + 2)}{\cancel{x}(-x + 2)} = \frac{x + 2}{2 - x} \qquad \text{Factor and reduce to lowest terms}$$

Answer: $\frac{2 - y}{y + 2}$

16.7 EXERCISES A

Simplify.

1. $\dfrac{\dfrac{2}{3}}{\dfrac{1}{3}}$

2

2. $\dfrac{3 + \dfrac{2}{3}}{1 - \dfrac{1}{3}}$

$\dfrac{11}{2}$

3. $\dfrac{\dfrac{1}{3} + \dfrac{1}{5}}{\dfrac{2}{3} - \dfrac{3}{5}}$

8

4. $\dfrac{\dfrac{2}{x}}{\dfrac{3}{x}}$

$\dfrac{2}{3}$

5. $\dfrac{6}{1 + \dfrac{1}{y}}$

$\dfrac{6y}{y + 1}$

6. $\dfrac{1 - \dfrac{1}{a}}{3}$

$\dfrac{a - 1}{3a}$

7. $\dfrac{\dfrac{1}{x} + 3}{\dfrac{1}{x} - 3}$

$\dfrac{1 + 3x}{1 - 3x}$

8. $\dfrac{y - 1}{y - \dfrac{1}{y}}$

$\dfrac{y}{y + 1}$

9. $\dfrac{1 - \dfrac{4}{y}}{\dfrac{4 - y}{y}}$

-1

10. $\dfrac{\dfrac{2}{a} + a}{\dfrac{a}{2} + a}$

$\dfrac{4 + 2a^2}{3a^2}$

11 $\dfrac{\dfrac{1}{2x} - 1}{\dfrac{1}{x} - 2}$

$\dfrac{1}{2}$

12. $\dfrac{\dfrac{2}{y} + 3}{2 - \dfrac{3}{y}}$

$\dfrac{2 + 3y}{2y - 3}$

13. $\dfrac{\dfrac{1}{x} + 1}{\dfrac{1}{x} - 1}$

$\dfrac{1 + x}{1 - x}$

14. $\dfrac{4 - \dfrac{1}{a^2}}{2 - \dfrac{1}{a}}$

$\dfrac{2a + 1}{a}$

15 $\dfrac{z - 2 - \dfrac{3}{z}}{1 + \dfrac{1}{z}}$

$z - 3$

16. $\dfrac{1 - \dfrac{2}{y} - \dfrac{3}{y^2}}{1 + \dfrac{1}{y}}$

$\dfrac{y - 3}{y}$

17 $\dfrac{x - 3 + \dfrac{2}{x}}{x - 4 + \dfrac{3}{x}}$

$\dfrac{x - 2}{x - 3}$

18. $\dfrac{\dfrac{3}{a - 3} + 1}{\dfrac{3}{a + 3} - 1}$

$\dfrac{a + 3}{3 - a}$

19 An important formula in physics gives V in terms of S_1 and S_2 by the formula

$$V = \frac{3}{\dfrac{1}{S_1} + \dfrac{1}{S_2}}.$$

Express V as a simple fraction and use the result to find V when S_1 is 2 and S_2 is 5.

$$V = \frac{3S_1S_2}{S_1 + S_2}; \ \frac{30}{7}$$

20. A transportation engineer uses the formula

$$R = \frac{2}{\dfrac{1}{g} + \dfrac{1}{r}}$$

to calculate average rate of speed on a round trip. Simplify this complex fraction.

$$R = \frac{2rg}{r + g}$$

FOR REVIEW

Solve.

21. $\dfrac{x + 10}{x - 8} = \dfrac{2}{3}$

-46

22. $\dfrac{y + 3}{5} = \dfrac{y - 7}{3}$

22

23. $\dfrac{z + 1}{z - 5} = \dfrac{z + 2}{z - 6}$

2

24. A boat can go 38 mi on 6 gal of gas. How far can it go on 21 gal of gas?

133 mi

25. If y varies directly as x, and $y = 42$ when $x = 18$, find y when $x = 3$.

7

26. The cost c of fish varies directly as the total weight w of the fish. If 8 lb of fish cost $18.60, what would 60 lb of fish cost?

$139.50

27. A recipe calls for 2.5 cups of flour to feed 6 people. How many cups would be needed to feed 18 people?

7.5 cups

The following exercises review material from Chapter 10 to help you prepare for the first section of the next chapter. Simplify each expression.

28. 3^2 9

29. $(-3)^2$ 9

30. -3^2 -9

31. $2^2 \cdot 5^2$ 100

32. $(2 \cdot 5)^2$ 100

33. $|4|$ 4

34. $|-4|$ 4

35. $-(4 - 5)^2$ -1

ANSWERS: 1. 2 2. $\frac{11}{2}$ 3. 8 4. $\frac{2}{3}$ 5. $\frac{6y}{y+1}$ 6. $\frac{a-1}{3a}$ 7. $\frac{1+3x}{1-3x}$ 8. $\frac{y}{y+1}$ 9. -1 10. $\frac{4+2a^2}{3a^2}$ 11. $\frac{1}{2}$ 12. $\frac{2+3y}{2y-3}$
13. $\frac{1+x}{1-x}$ 14. $\frac{2a+1}{a}$ 15. $z-3$ 16. $\frac{y-3}{y}$ 17. $\frac{x-2}{x-3}$ 18. $\frac{a+3}{3-a}$ 19. $V=\frac{3S_1S_2}{S_1+S_2}$; $\frac{30}{7}$ 20. $R=\frac{2rg}{r+g}$ 21. -46 22. 22
23. 2 24. 133 mi 25. 7 26. \$139.50 27. 7.5 cups 28. 9 29. 9 30. -9 31. 100 32. 100 33. 4 34. 4
35. -1

16.7 EXERCISES B

Simplify.

1. $\dfrac{\frac{3}{5}}{\frac{2}{5}}$ $\frac{3}{2}$

2. $\dfrac{7+\frac{1}{2}}{2-\frac{1}{2}}$ 5

3. $\dfrac{\frac{1}{6}+\frac{1}{8}}{\frac{5}{6}-\frac{1}{8}}$ $\frac{7}{17}$

4. $\dfrac{\frac{4}{x}}{\frac{2}{x}}$ 2

5. $\dfrac{10}{1-\frac{1}{y}}$ $\frac{10y}{y-1}$

6. $\dfrac{2+\frac{1}{a}}{7}$ $\frac{2a+1}{7a}$

7. $\dfrac{\frac{1}{x}+1}{\frac{1}{x}-1}$ $\frac{1+x}{1-x}$

8. $\dfrac{2-\frac{1}{y}}{\frac{2}{y}}$ $\frac{2y-1}{2}$

9. $\dfrac{a-\frac{1}{a}}{1-\frac{1}{a}}$ $a+1$

10. $\dfrac{\frac{3}{a}+a}{\frac{a}{3}+a}$ $\frac{3(3+a^2)}{4a^2}$

11. $\dfrac{\frac{1}{3y}-2}{\frac{1}{y}+2}$ $\frac{1-6y}{3(1+2y)}$

12. $\dfrac{\frac{1}{a}+\frac{2}{a}}{\frac{3}{a}+\frac{4}{a}}$ $\frac{3}{7}$

13. $\dfrac{\frac{2}{x-2}+1}{\frac{2}{x+2}-1}$ $-\frac{x+2}{x-2}$

14. $\dfrac{\frac{1}{y^2}-4}{\frac{1}{y}-2}$ $\frac{2y+1}{y}$

15. $\dfrac{\frac{3}{z}+z+4}{1+\frac{3}{z}}$ $z+1$

16. $\dfrac{1-\frac{2}{x}-\frac{3}{x^2}}{\frac{1}{x}+1}$ $\frac{x-3}{x}$

17. $\dfrac{\frac{5}{y+5}-1}{\frac{5}{y-5}+1}$ $\frac{5-y}{y+5}$

18. $\dfrac{z-5+\frac{6}{z}}{z-1-\frac{2}{z}}$ $\frac{z-3}{z+1}$

19. An environmental chemist might use the formula
$$w=\frac{\frac{1}{c}}{1+\frac{1}{d}}.$$
Express w as a simple fraction and use the result to find w when c is 3 and d is 14.
$w=\frac{d}{cd+c}$; $\frac{14}{45}$

20. A navigation expert uses the formula
$$V=\frac{v_1+v_2}{1+\frac{v_1v_2}{c^2}}$$
to calculate the velocity of an airplane. Simplify this complex fraction.
$V=\frac{c^2(v_1+v_2)}{c^2+v_1v_2}$

FOR REVIEW

Solve.

21. $\dfrac{x-3}{x+4}=\dfrac{7}{8}$ 52

22. $\dfrac{2y+8}{6}=\dfrac{y-9}{6}$ -10.5

23. $\dfrac{z-3}{z-4}=\dfrac{z-6}{z-5}$ $\frac{9}{2}$

24. In a shipment of 200 tires, 6 were defective. How many defective tires would you expect in a shipment of 9000 tires?   **270**

25. If y varies directly as x, and $y = 14$ when $x = 5$, find y when $x = 20$.   **56**

26. The distance d traveled varies directly as the rate r of travel. If 280 miles are traveled at a rate of 40 mph, what distance could be traveled at 60 mph?   **420 mi**

27. If 1 quart contains 0.95 liters, how many quarts are in 190 liters?
200 quarts

The following exercises review material from Chapter 10 to help you prepare for the first section of the next chapter. Simplify each expression.

28. 5^2 25

29. $(-5)^2$ 25

30. -5^2 -25

31. $4^2 \cdot 3^2$ 144

32. $(4 \cdot 3)^2$ 144

33. $|9|$ 9

34. $|-9|$ 9

35. $-(6-8)^2$ -4

16.7 EXERCISES C

Simplify.

1. $\dfrac{\dfrac{a+b}{a-b} - \dfrac{a-b}{a+b}}{\dfrac{a}{a-b} + \dfrac{b}{a+b}}$

$\dfrac{4ab}{a^2 + 2ab - b^2}$

2. $\dfrac{\dfrac{1}{xy} + \dfrac{1}{yz} + \dfrac{1}{xz}}{\dfrac{x+y+z}{xyz}}$

[Answer: 1]

3. $a - \dfrac{a}{1 - \dfrac{a}{1-a}}$

$\left[\text{Answer: } \dfrac{-a^2}{1-2a}\right]$

CHAPTER 16 REVIEW

KEY WORDS

16.1 An **algebraic fraction** is a fraction that contains a variable in the numerator or denominator or both.

A **rational expression** is the quotient of two polynomials.

A fraction is **reduced to lowest terms** when 1 (or -1) is the only number or expression that divides both numerator and denominator.

16.2 The **reciprocal** of the fraction $\frac{a}{b}$ is $\frac{b}{a}$.

16.3 **Like fractions** have the same denominator.

Unlike fractions have different denominators.

16.4 The **least common denominator (LCD)** of two or more fractions is the smallest of all common denominators.

16.5 A **fractional equation** is one that contains one or more algebraic fractions.

16.6 The **ratio** of one number x to another number y is the quotient $x \div y$.

A **proportion** is an equation which states that two ratios are equal.

The **extremes** of the proportion $\frac{a}{b} = \frac{c}{d}$ are a and d.

The **means** of the proportion $\frac{a}{b} = \frac{c}{d}$ are b and c.

If $y = kx$ (**direct variation**) or $y = \frac{k}{x}$ (**inverse variation**), k is called the **constant of variation**.

16.7 A **complex fraction** is a fraction that contains at least one other fraction within it.

KEY CONCEPTS

16.1 **1.** Always be aware of values that must be excluded when working with algebraic fractions. For example, 4 cannot replace x in $\frac{3}{x-4}$.

2. When reducing fractions, divide out factors only, never terms.

3. When reducing fractions by dividing out common factors, do not make the numerator 0 when it is actually 1. For example,

$$\frac{\cancel{(x+1)}}{3\cancel{(x+1)}} = \frac{1}{3}, \quad not \;\; \frac{0}{3}.$$

16.2 When multiplying or dividing fractions, factor numerators and denominators and divide out common factors. Do not find the LCD.

16.3, **1.** Find the LCD of all fractions when adding
16.4 or subtracting.

2. When subtracting fractions, use parentheses to avoid sign errors. For example,

$$\frac{2x+1}{x-5} - \frac{x-3}{x-5}$$

$$= \frac{2x+1-(x-3)}{x-5}$$

$$= \frac{2x+1-x+3}{x-5} = \frac{x+4}{x-5}.$$

3. Do not divide out terms when adding or subtracting algebraic fractions. For example,

$$\frac{3(x-2)}{(x+2)(x-2)} - \frac{7(x+2)}{(x-2)(x+2)}$$

$$= \frac{3(x-2)-7(x+2)}{(x+2)(x-2)}$$

$$= \frac{3x-6-7x-14}{(x+2)(x-2)},$$

$$not \quad \frac{3\cancel{(x-2)}-7\cancel{(x+2)}}{\cancel{(x+2)}\cancel{(x-2)}}$$

16.5 **1.** To solve a fractional equation, multiply both sides by the LCD of all fractions.

2. Check all answers in the *original* fractional equation. Watch for division by zero.

16.6 **1.** If y varies directly as x, then

$$y = kx$$

where k is the constant of variation.

2. If y varies inversely as x, then

$$y = \frac{k}{x}$$

where k is the constant of variation.

16.7 One way to simplify a complex fraction is to write the numerator as a single fraction, the denominator as a single fraction, and divide. Another way is to multiply each term of a complex fraction by the LCD of all fractions. For example,

$$\frac{1 + \dfrac{1}{x}}{1 - \dfrac{1}{x}} = \frac{\left(1 + \dfrac{1}{x}\right) \cdot x}{\left(1 - \dfrac{1}{x}\right) \cdot x}$$

$$= \frac{1 \cdot x + \dfrac{1}{x} \cdot x}{1 \cdot x - \dfrac{1}{x} \cdot x} = \frac{x+1}{x-1}.$$

REVIEW EXERCISES

Part I

16.1 *Find the values of the variable that must be excluded.*

1. $\dfrac{x+2}{x(x-1)}$
0, 1

2. $\dfrac{a}{a^2+1}$
none

3. $\dfrac{y^2}{(y+1)(y-5)}$
5, −1

Are the given fractions equivalent?

4. $\dfrac{x-2}{5x-2}, \dfrac{x}{5x}$
no

5. $\dfrac{a-3}{a^2-9}, \dfrac{1}{a+3}$
yes

6. $\dfrac{x+y}{x}, \dfrac{x}{x+y}$
no

Reduce to lowest terms.

7. $\dfrac{x^2 + 2x}{x}$

$x + 2$

8. $\dfrac{x^2 - 9}{x + 3}$

$x - 3$

9. $\dfrac{x^2 + 2xy + y^2}{x + y}$

$x + y$

16.2 *Perform the indicated operations.*

10. $\dfrac{x}{x + 6} \cdot \dfrac{x^2 - 36}{x^2 - 6x}$

1

11. $\dfrac{y^2 - y - 2}{y^2 - 2y - 3} \cdot \dfrac{y^2 - 3y}{y + 2}$

$\dfrac{y(y - 2)}{y + 2}$

12. $\dfrac{x^2 - 4y^2}{x + 2y} \cdot \dfrac{x + y}{x - 2y}$

$x + y$

13. $\dfrac{a + 7}{a - 7} \div \dfrac{a^2 + 7}{a^2 - 49}$

$\dfrac{(a + 7)^2}{a^2 + 7}$

14. $\dfrac{4y^4}{y^2 - 1} \div \dfrac{2y^3}{y^2 - 2y + 1}$

$\dfrac{2y(y - 1)}{y + 1}$

15. $\dfrac{9x^2 - y^2}{x + y} \div \dfrac{3x + y}{x^2 - y^2}$

$(3x - y)(x - y)$

16.3, *Perform the indicated operations.*
16.4

16. $\dfrac{3x}{x - 2} + \dfrac{2x - 1}{2 - x}$

$\dfrac{x + 1}{x - 2}$

17. $\dfrac{2x}{x^2 - x} + \dfrac{x}{x^2 - 1}$

$\dfrac{3x + 2}{(x - 1)(x + 1)}$

18. $\dfrac{2}{x^2 - 4x + 3} + \dfrac{3}{x^2 + x - 2}$

$\dfrac{5}{(x - 3)(x + 2)}$

19. $\dfrac{x + 1}{x - 3} - \dfrac{x - 1}{3 - x}$

$\dfrac{2x}{x - 3}$

20. $\dfrac{2}{y^2 - y - 12} - \dfrac{1}{y^2 - 9}$

$\dfrac{y - 2}{(y - 4)(y + 3)(y - 3)}$

21. $\dfrac{a}{a^2 - 1} - \dfrac{a + 2}{a^2 + a - 2}$

$\dfrac{-1}{(a + 1)(a - 1)}$

22. $\dfrac{3}{x - 2} + \dfrac{x + 2}{x - 2} - \dfrac{2x}{2 - x}$

$\dfrac{3x + 5}{x - 2}$

23. $\dfrac{6x}{x^2 - 9} - \dfrac{2}{x - 3} - \dfrac{5}{x + 3}$

$\dfrac{9 - x}{(x + 3)(x - 3)}$

24. $\dfrac{x}{x + y} + \dfrac{y}{x - y}$

$\dfrac{x^2 + y^2}{x^2 - y^2}$

16.5 *Solve.*

25. $\dfrac{2}{z+5} = \dfrac{3}{z}$

-15

26. $\dfrac{3}{x+2} + \dfrac{2}{x-2} = \dfrac{3}{x^2-4}$

1

27. $\dfrac{x}{x-1} + 1 = \dfrac{x^2+1}{x^2-1}$

-2

28. The denominator of a fraction is 2 less than the numerator. If 2 is added to both the numerator and denominator, the result has value $\frac{4}{3}$. Find the fraction.

$\frac{6}{4}$

29. Al is 9 years older than Jim and the quotient of their ages is $\frac{5}{6}$. How old is each?

Al is 54, Jim is 45

30. If Tim can do a job in 3 days and Bob can do the same job in 11 days, how long would it take them to do the job if they worked together?

$\frac{33}{14}$ **days**

31. If it takes pipe A three times longer to fill a tank than pipe B and together they can fill it in 6 hours, how long would it take pipe B to fill the tank?

8 hr

16.6 *Give as a ratio and simplify.*

32. 450 mi on 15 gal of gas

30 mpg

33. 1200 trees on 6 acres

200 trees per acre

34. $12 for 8 lb of meat

$1.50 per lb

Solve.

35. $\dfrac{x-2}{x+2} = \dfrac{3}{2}$

-10

36. $\dfrac{y+5}{y-4} = \dfrac{y+4}{y-5}$

no solution

37. $\dfrac{z-8}{z+3} = \dfrac{z+9}{z-2}$

$-\dfrac{1}{2}$

38. In an election, the winner won by a 6 to 5 margin. If the winner had 270 votes, how many votes did the loser receive?

225 votes

39. If 70 ft of wire weighs 15 lb, how much will 210 ft of the same wire weigh?

45 lb

40. If y varies directly as x, and $y = 20$ when $x = 4$, find y when $x = 14$.

70

41. The cost c of candy varies directly as the total weight w of the candy. If 4 lb of candy cost $10.40, what would 6 lb cost?

$15.60

16.7 *Simplify.*

42. $\dfrac{2 - \dfrac{1}{8}}{3 + \dfrac{3}{4}}$ $\dfrac{1}{2}$

43. $\dfrac{4x - \dfrac{1}{x}}{2 - \dfrac{1}{x}}$ $2x + 1$

44. $\dfrac{y - 2 + \dfrac{1}{y}}{1 - \dfrac{1}{y}}$ $y - 1$

Part II

Solve.

45. If Sandy can frame a shed in 4 days and together Sandy and Clyde can frame the same shed in 3 days, how long would it take Clyde to frame it working by himself?

12 days

46. If y varies inversely as x, and $y = 45$ when $x = 5$, find y when $x = 15$.

15

47. $\dfrac{3}{x - 4} + \dfrac{4}{x + 4} = \dfrac{3}{x^2 - 16}$ 1

48. $\dfrac{x + 1}{x} = \dfrac{5}{7}$ $-\dfrac{7}{2}$

49. $x + \dfrac{6}{x} = 5$ 2 or 3

50. $\dfrac{x}{x - 2} - 1 = \dfrac{3}{x^2 - 4}$ $-\dfrac{1}{2}$

Find the values of the variable that must be excluded.

51. $\dfrac{x + 5}{(x - 4)(x + 1)}$ 4, −1

52. $\dfrac{y^2}{x^2 - 9}$ 3, −3

Perform the indicated operation and simplify.

53. $\dfrac{2}{x-5} + \dfrac{x}{x+5}$

$\dfrac{x^2 - 3x + 10}{(x-5)(x+5)}$

54. $\dfrac{x+1}{x-2} - \dfrac{x-1}{x+3}$

$\dfrac{7x+1}{(x-2)(x+3)}$

55. $\dfrac{x^2 - 4x + 3}{x^2 + 7x + 10} \cdot \dfrac{x^2 - 25}{x^2 - 9}$

$\dfrac{(x-1)(x-5)}{(x+2)(x+3)}$

56. $\dfrac{y^2 - 3y}{y^2 - y - 6} \div \dfrac{y}{y+2}$

1

57. $\dfrac{1 + \dfrac{3}{x} - \dfrac{10}{x^2}}{1 - \dfrac{5}{x} + \dfrac{6}{x^2}}$

$\dfrac{x+5}{x-3}$

58. $\dfrac{5}{y-1} - \dfrac{6}{y+1} - \dfrac{1}{2}$

$\dfrac{-y^2 - 2y + 23}{2(y-1)(y+1)}$

ANSWERS: 1. 0, 1   2. none   3. 5, −1   4. no   5. yes   6. no   7. $x+2$ 8. $x-3$ 9. $x+y$ 10. 1   11. $\frac{y(y-2)}{y+2}$
12. $x+y$ 13. $\frac{(a+7)^2}{a^2+7}$ 14. $\frac{2y(y-1)}{y+1}$ 15. $(3x-y)(x-y)$ 16. $\frac{x+1}{x-2}$ 17. $\frac{3x+2}{(x-1)(x+1)}$ 18. $\frac{5}{(x-3)(x+2)}$ 19. $\frac{2x}{x-3}$
20. $\frac{y-2}{(y-4)(y+3)(y-3)}$ 21. $\frac{-1}{(a+1)(a-1)}$ 22. $\frac{3x+5}{x-2}$ 23. $\frac{9-x}{(x+3)(x-3)}$ 24. $\frac{x^2+y^2}{x^2-y^2}$ 25. −15   26. 1   27. −2   28. $\frac{6}{4}$ 29. Al is
54, Jim is 45   30. $\frac{33}{14}$ days   31. 8 hr   32. 30 mpg   33. 200 trees per acre   34. $1.50 per lb   35. −10   36. no solu-
tion   37. $-\frac{1}{2}$ 38. 225 votes   39. 45 lb   40. 70   41. $15.60   42. $\frac{1}{2}$ 43. $2x+1$ 44. $y-1$ 45. 12 days   46. 15
47. 1   48. $-\frac{7}{2}$ 49. 2 or 3   50. $-\frac{1}{2}$ 51. 4, −1   52. 3, −3   53. $\frac{x^2-3x+10}{(x-5)(x+5)}$ 54. $\frac{7x+1}{(x-2)(x+3)}$ 55. $\frac{(x-1)(x-5)}{(x+2)(x+3)}$ 56. 1
57. $\frac{x+5}{x-3}$ 58. $\frac{-y^2-2y+23}{2(y-1)(y+1)}$

1. What are the values of the variable that must be excluded in

   $$\frac{5}{x(x + 5)}?$$

   1. _____ $0, -5$ _____

2. Are the fractions $\dfrac{y}{5}$ and $\dfrac{6y}{30}$ equivalent?

   2. _____ yes _____

3. Reduce to lowest terms. $\dfrac{a + 2}{a^2 + 2a}$

   3. _____ $\dfrac{1}{a}$ _____

4. Multiply. $\dfrac{x^2 - 16}{7x^2} \cdot \dfrac{35x}{x - 4}$

   4. _____ $\dfrac{5(x + 4)}{x}$ _____

5. Divide. $\dfrac{y}{y^2 + y - 12} \div \dfrac{y^2 + 3y}{y^2 + 7y + 12}$

   5. _____ $\dfrac{1}{y - 3}$ _____

6. Add. $\dfrac{3}{a - b} + \dfrac{2}{b - a}$

   6. _____ $\dfrac{1}{a - b}$ _____

7. Add. $\dfrac{x^2 - 6x}{x^2 - 4} + \dfrac{x}{x - 2}$

   7. _____ $\dfrac{2x}{x + 2}$ _____

8. Subtract. $\dfrac{4}{y^2 - 1} - \dfrac{3}{y^2 - y - 2}$

   8. _____ $\dfrac{y - 5}{(y - 1)(y + 1)(y - 2)}$ _____

9. Solve. $\dfrac{x}{x+1} = \dfrac{x^2+1}{x^2-1} + \dfrac{3}{x-1}$

9. _____no solution_____

10. The denominator of a fraction is 3 more than the numerator. If 1 is subtracted from both the numerator and denominator, the result has value $\frac{3}{4}$. Find the fraction.

10. _____$\dfrac{10}{13}$_____

11. A car needed 16 gal of gas to go 352 mi. How many gallons would be needed to go 814 mi?

11. _____37 gallons_____

12. The cost c of ground beef varies directly as the total weight w of the meat purchased. If 5 lb of ground beef cost $5.75, what would 12 lb cost?

12. _____$13.80_____

13. Ralph can do a job in 3 days and Harry can do the same job in 5 days. How long would it take them to do the same job if they work together?

13. _____$\dfrac{15}{8}$ days_____

14. Simplify. $\dfrac{\dfrac{1}{y}-1}{\dfrac{1}{y}-y}$

14. _____$\dfrac{1}{1+y}$_____

15. If y varies inversely as x, and $y = 20$ when $x = 4$, find y when $x = 16$.

15. _____5_____

Radicals

17.1 ROOTS AND RADICALS

STUDENT GUIDEPOSTS

1. Perfect Squares and Square Roots
2. Evaluating $\sqrt{x^2}$
3. Using a Calculator
4. Applications of Radicals

1 PERFECT SQUARES AND SQUARE ROOTS

To introduce our study of radicals we review material from Section 10.6. Consider the following.

$$2^2 = 2 \cdot 2 = 4$$
$$(-2)^2 = (-2)(-2) = 4$$

When an integer is squared, the result is a **perfect square.** Thus 4 is a perfect square, since $4 = 2^2$. Either of the identical factors of a perfect square is a *square root* of the number. From above, 2 is a square root of 4 and since $(-2)^2 = 4$, -2 is also a square root of 4.

In addition to whole number perfect squares, there are also fractional perfect squares. For example,

$$\frac{4}{9} \text{ is a perfect square since } \frac{4}{9} = \left(\frac{2}{3}\right)^2,$$

$$\text{and } \frac{2}{3} \text{ is a square root of } \frac{4}{9}.$$

There are fractional perfect squares that do not have perfect square numerators and perfect square denominators. For example, $\frac{8}{18}$ is a perfect square since

$$\frac{8}{18} = \frac{4 \cdot 2}{9 \cdot 2} = \frac{4}{9},$$

which is a perfect square.

We saw in Chapter 10 that some square roots are irrational numbers. As we will see later, these numbers are included in the following definition. If a and x are real numbers such that

$$a^2 = x$$

then a is a **square root** of x. In general, every *positive* real number has two square roots. If a is a square root of x, then both a and $-a$ are square roots of x since

$$a^2 = x \quad \text{and} \quad (-a)^2 = x.$$

The nonnegative square root of a positive real number is called its **principal square root,** and is denoted by a **radical,** $\sqrt{}$. For example, if we write $\sqrt{4}$, we call 4 the **radicand** and 2 its principal square root. That is,

$$\sqrt{4} = 2. \qquad \text{2 is the principal (nonnegative) square root of 4}$$

To indicate the other square root of 4, we place a minus sign in front of the radical. Thus,

$$-\sqrt{4} = -2.$$

The positive and negative square roots of a positive real number can be represented together with the symbol $\pm$. For example,

$$\pm\sqrt{4} = \pm2$$

represents both 2 and -2, and is read "plus or minus 2." Of course 0 has only one square root, namely 0, since $-0 = 0$.

We have seen that every positive real number has two square roots (one positive and one negative), that 0 has only one square root (namely 0 itself), but what about negative numbers? Does -9 have a square root, for example? The obvious choices would be -3 or 3, but

$$(-3)^2 = 9 \quad \text{and} \quad 3^2 = 9.$$

In fact, since $a^2 \geq 0$ for any real number a, a^2 could never be equal to the *negative* number -9. Thus, since no real number squared can be negative, negative numbers have no real square roots. We say that $\sqrt{-9}$, for example, is not a real number.

EXAMPLE 1 EVALUATING SQUARE ROOTS	PRACTICE EXERCISE 1

Evaluate the following radicals.

(a) $\sqrt{100} = 10 \qquad 10^2 = 100$

(b) $-\sqrt{100} = -10$

(c) $\pm\sqrt{100} = \pm10$ (two numbers, 10 and -10)

(d) $\sqrt{-100}$ is not a real number

(e) $\sqrt{\dfrac{25}{81}} = \dfrac{5}{9}$

(f) $\sqrt{\dfrac{12}{75}} = \sqrt{\dfrac{4 \cdot 3}{25 \cdot 3}} = \sqrt{\dfrac{4}{25}} = \dfrac{2}{5}$

(g) $\sqrt{1} = 1$

(h) $\sqrt{5^2} = \sqrt{25} = 5$

(i) $\sqrt{(-5)^2} = \sqrt{25} = 5$

(j) $\sqrt{-5^2} = \sqrt{-25}$ is not a real number.
Note the difference between $(-5)^2$ and -5^2.

Evaluate the following radicals.

(a) $\sqrt{169}$

(b) $-\sqrt{169}$

(c) $\pm\sqrt{169}$

(d) $\sqrt{-169}$

(e) $\sqrt{\dfrac{144}{25}}$

(f) $\sqrt{\dfrac{50}{32}}$

(g) $-\sqrt{1}$

(h) $\sqrt{9^2}$

(i) $\sqrt{(-9)^2}$

(j) $\sqrt{-9^2}$

Answers: (a) 13   (b) -13
(c) ±13 (d) not a real number
(e) $\frac{12}{5}$ (f) $\frac{5}{4}$ (g) -1 (h) 9
(i) 9   (j) not a real number

❷ EVALUATING $\sqrt{x^2}$

In Example 1 we saw that

$$\sqrt{5^2} = 5 \quad \text{and} \quad \sqrt{(-5)^2} = 5.$$

This is also true in general. Thus, if x is a nonnegative number, $\sqrt{x^2} = x$. For example, $\sqrt{5^2} = 5$. However, if x is negative then $\sqrt{x^2} = -x$. For example, $\sqrt{(-5)^2} = -(-5) = 5$. To make it clear that the principal square root of a number is positive, we write

$$\sqrt{x^2} = |x|,$$

where $|x|$ is the absolute value of x. To avoid problems of this nature, we will assume that *all variables represent nonnegative real values*. Then for $x \geq 0$,

$$\sqrt{x^2} = x. \quad \text{Since } |x| = x \text{ in this case}$$

We will also assume that *all algebraic expressions under radicals are nonnegative real numbers*. Thus, for example,

$$\sqrt{(x-7)^2} = x - 7,$$

since we are assuming that $x - 7 \geq 0$.

EXAMPLE 2 EVALUATING RADICALS

Evaluate the following radicals assuming all variables and algebraic expressions are nonnegative.

(a) $\sqrt{49} = \sqrt{7^2} = 7$

(b) $\sqrt{y^2} = y$

(c) $\sqrt{25x^2} = \sqrt{5^2 x^2} = \sqrt{(5x)^2} = 5x \qquad a^n b^n = (ab)^n$

(d) $\sqrt{x^2 + 6x + 9} = \sqrt{(x+3)^2} = x + 3$

(e) $-\sqrt{81} = -\sqrt{9^2} = -9$

(f) $-\sqrt{u^2} = -u$

PRACTICE EXERCISE 2

Evaluate the following radicals assuming all variables and algebraic expressions are nonnegative.

(a) $\sqrt{36}$

(b) $\sqrt{a^2}$

(c) $\sqrt{36a^2}$

(d) $\sqrt{x^2 - 6x + 9}$

(e) $-\sqrt{196}$

(f) $-\sqrt{w^2}$

Answers: (a) 6 (b) a (c) $6a$ (d) $x-3$ (e) -14 (f) $-w$

---

③ USING A CALCULATOR

The square roots we have looked at so far have all resulted in rational numbers. For example, $\sqrt{25}$ is the rational number 5. Many square roots, such as $\sqrt{5}$, do not have this property. Since 5 is not a perfect square, that is, since there is no integer a such that $a^2 = 5$, $\sqrt{5}$ cannot be simplified further. In fact, $\sqrt{5}$ is an irrational number for which an approximate rational number must be used in a practical situation. There are two ways to find such an approximation, with a table of square roots (like the one at the back of this text), or with a calculator. Whenever we need to find the approximate value of a square root, we will use a calculator with a $\boxed{\sqrt{}}$ key.

To find a rational-number approximation for $\sqrt{5}$, use the following steps on your calculator.

$$5 \; \boxed{\sqrt{}} \rightarrow \boxed{2.2360680}$$

To see that 2.2360680 is indeed an approximation of $\sqrt{5}$, enter 2.2360680 on your calculator and square it using the $\boxed{x^2}$ key.

$$2.2360680 \; \boxed{x^2} \rightarrow \boxed{5.0000001}$$

④ APPLICATIONS OF RADICALS

In applied problems such as the next example, the approximate value of a square root is needed.

EXAMPLE 3 APPLYING RADICALS	PRACTICE EXERCISE 3

If P dollars are invested in an account for two years and at the end of this period the value of the account is A dollars, then the annual rate of return r on the investment can be found by the formula

$$r = \sqrt{\frac{A}{P}} - 1.$$

Suppose Diane purchased a rare book for $1000 and two years later sold it for $1500. What was the annual rate of return on Diane's investment?

Since $A = 1500$ and $P = 1000$, we substitute into the formula to find r.

$$r = \sqrt{\frac{A}{P}} - 1 = \sqrt{\frac{1500}{1000}} - 1$$
$$\approx 0.2247449$$

The calculator steps used to find r are:

$$1500 \boxed{\div} \ 1000 \ \boxed{=} \ \boxed{\sqrt{}} \ \boxed{-} \ 1 \ \boxed{=} \rightarrow \boxed{0.2247449}^{[1]}$$

Rounded to three decimal places, $r \approx 0.225$, which translates to 22.5%. Thus, Diane realized an approximate annual return of 22.5% on her investment.

If Pamela invested $5000 in bonds and two years later sold the bonds for $6200, what was the annual rate of return?

Answer: 11.4%

///////////// **CAUTION** //////////////

Do not get in the habit of reaching immediately for your calculator when given a radical expression. Using a calculator, for example, to evaluate $\sqrt{4}$ would be as ridiculous as using it to find $2 + 3$. Throughout this chapter you should use a calculator only for approximate values of radicals whose square roots are not integers.

17.1 EXERCISES A

Tell whether each number has 0, 1, or 2 square roots.

1. 25   2

2. −25   0

3. 0   1

4. 175   2

Evaluate each square root assuming that all variables and algebraic expressions under the radical are nonnegative.

5. $\sqrt{6^2}$ 6

6. $\sqrt{(-6)^2}$ 6

7. $\sqrt{-6^2}$ meaningless

8. $\sqrt{121}$ 11

---

[1]Calculator steps are given using algebraic logic. If you have a calculator that uses Reverse Polish Notation (RPN), consult your operator's manual.

9. $-\sqrt{121}$ -11 **10.** $\pm\sqrt{121}$ ± 11 **11.** $\sqrt{\dfrac{9}{4}}$ $\dfrac{3}{2}$ **12.** $\pm\sqrt{\dfrac{9}{4}}$ $\pm\dfrac{3}{2}$

13. $\sqrt{\dfrac{144}{25}}$ $\dfrac{12}{5}$ **14** $\sqrt{\dfrac{1000}{10}}$ 10 **15.** $\sqrt{\dfrac{50}{2}}$ 5 **16.** $\sqrt{-\dfrac{49}{9}}$ meaningless

17. $\sqrt{(x+1)^2}$ $x+1$ **18.** $\sqrt{(x-1)^2}$ $x-1$ **19** $\sqrt{49a^2}$ $7a$ **20.** $\sqrt{x^2y^2}$ xy

21. $\sqrt{-x^2y^2}$ meaningless **22.** $-\sqrt{x^2y^2}$ $-xy$ **23** $\sqrt{x^2+10x+25}$ $x+5$ **24.** $\sqrt{y^2-8y+16}$ $y-4$

In a particular manufacturing plant, the productivity, p, is related to the work force, w, by the equation $p = \sqrt{w}$.

25 Find p when $w = 25$. 5 **26.** Find p when $w = 8^2$. 8 **27.** Find p when $w = 0$. 0

28 Find w when $p = 4$. 16 **29.** Find w when $p = 12$. 144 **30.** Find p when $w = -4$.
no value for p when w is negative

In Exercises 31–38, use a calculator to find each square root correct to three decimal places.

31. $\sqrt{3}$ 1.732 **32.** $\sqrt{8}$ 2.828 **33.** $\sqrt{41}$ 6.403 **34.** $\sqrt{73}$ 8.544

35. $\sqrt{26.5}$ 5.148 **36.** $\sqrt{34.8}$ 5.899 **37.** $\sqrt{627.84}$ 25.057 **38.** $\sqrt{543.09}$ 23.304

Use the formula given in Example 3 to solve each investment problem given in Exercises 39–40.

39. Suppose Janet Condon invested $300 in the stock market and two years later the stocks were worth $525. Find the approximate annual rate of return on Janet's investment.
32.3%

40. If Peter Horn purchased an antique car for $4500 and sold it two years later for $6700, what was the approximate annual rate of return on his investment?
22.0%

FOR REVIEW

Exercises 41–44 review material from Section 14.1 to help you prepare for the next section. Simplify each exponential expression.

41. $9x^5x^2$ $9x^7$ **42.** $(2x^3)(18x^7)$ $36x^{10}$ **43.** $\dfrac{8x^6}{2x^2}$ $4x^4$ **44.** $\dfrac{50x^2y^3}{2x^6y}$ $\dfrac{25y^2}{x^4}$

ANSWERS: 1. 2 2. 0 3. 1 4. 2 5. 6 6. 6 7. meaningless 8. 11 9. -11 10. ±11 (11 and -11) 11. $\frac{3}{2}$
12. $\pm\frac{3}{2}$ $\left(\frac{3}{2}\text{ and }-\frac{3}{2}\right)$ 13. $\frac{12}{5}$ 14. 10 15. 5 16. meaningless 17. $x+1$ 18. $x-1$ 19. $7a$ 20. xy
21. meaningless 22. $-xy$ 23. $x+5$ 24. $y-4$ 25. 5 26. 8 27. 0 28. 16 29. 144 30. no value for p when w
is negative 31. 1.732 32. 2.828 33. 6.403 34. 8.544 35. 5.148 36. 5.899 37. 25.057 38. 23.304 39. 32.3%
40. 22.0% 41. $9x^7$ 42. $36x^{10}$ 43. $4x^4$ 44. $\frac{25y^2}{x^4}$

17.1 EXERCISES B

Tell whether each number has 0, 1, or 2 square roots.

1. 81   2
2. −81   0
3. 19   2
4. −0   1

Evaluate each square root assuming that all variables and algebraic expressions under the radical are nonnegative.

5. $\sqrt{49}$ 7
6. $\sqrt{(-7)^2}$ 7
7. $\sqrt{-7^2}$ **meaningless**
8. $\sqrt{144}$ 12

9. $-\sqrt{144}$ −12
10. $\pm\sqrt{144}$ ± 12
11. $\pm\sqrt{\dfrac{16}{9}}$ $\pm\dfrac{4}{3}$
12. $\sqrt{\dfrac{121}{16}}$ $\dfrac{11}{4}$

13. $\sqrt{\dfrac{75}{3}}$ 5
14. $\pm\sqrt{\dfrac{64}{25}}$ $\pm\dfrac{8}{5}$
15. $\sqrt{-0}$ 0
16. $\sqrt{-\dfrac{9}{16}}$ **meaningless**

17. $\sqrt{(y+2)^2}$ y + 2
18. $\sqrt{(y-2)^2}$ y − 2
19. $\sqrt{36x^2}$ 6x
20. $\sqrt{u^2v^2}$ uv

21. $\sqrt{-u^2v^2}$ **meaningless**
22. $-\sqrt{u^2v^2}$ −uv
23. $\sqrt{x^2+12x+36}$ x + 6
24. $\sqrt{y^2-14y+49}$ y − 7

In a particular retail operation, the cost of production c, is related to the number of items sold, n, by the equation $c = \sqrt{n}$.

25. Find c when n = 49.   7
26. Find c when n = 0.   0
27. Find c when n = 3².   3

28. Find n when c = 6.   36
29. Find n when c = 11.   121
30. Find c when n = −1.
no value for c when n is negative

In Exercises 31–38, use a calculator to find each square root correct to three decimal places.

31. $\sqrt{7}$ 2.646
32. $\sqrt{11}$ 3.317
33. $\sqrt{57}$ 7.550
34. $\sqrt{91}$ 9.539

35. $\sqrt{14.8}$ 3.847
36. $\sqrt{63.7}$ 7.981
37. $\sqrt{340.22}$ 18.445
38. $\sqrt{992.13}$ 31.498

Use the formula given in Example 3 to solve each investment problem given in Exercises 39–40.

39. Robert Weaver invested $2150 in savings. At the end of two years the account was worth $2490. Find the approximate annual rate of return.
7.6%

40. Michele Furr bought a painting for $3200. Two years later it was valued at $4100. What was the approximate annual rate of return?
13.2%

FOR REVIEW

Exercises 41–44 review material from Section 14.1 to help you prepare for the next section. Simplify each exponential expression.

41. $21y^8y^4$ $21y^{12}$
42. $(3y^5)(15y^3)$ $45y^8$
43. $\dfrac{30y^6}{10y^5}$ 3y
44. $\dfrac{99x^3y^7}{11x^7y}$ $\dfrac{9y^6}{x^4}$

17.1 EXERCISES C

Use the definition of cube root, $\sqrt[3]{a^3} = a$, to do the following exercises.

1. $\sqrt[3]{5^3}$ 5
2. $\sqrt[3]{27}$ 3
3. $\sqrt[3]{8x^3}$ [Answer: 2x]
4. $\sqrt[3]{(2x+1)^6}$ $(2x+1)^2$

17.2 SIMPLIFYING RADICALS

1 SQUARE ROOTS OF PRODUCTS

In the last section we saw that $\sqrt{9} = \sqrt{3^2} = 3$, but what about square roots such as $\sqrt{18}$? We might use a calculator but this would result in only an approximation of $\sqrt{18}$. However, since $18 = 9 \cdot 2$, we can write

$$\sqrt{18} = \sqrt{9 \cdot 2} = \sqrt{9} \cdot \sqrt{2}$$
$$= 3\sqrt{2}.$$

This is an example of the following rule.

Simplifying Rule for Products

If $a \geq 0$ and $b \geq 0$, then

$$\sqrt{ab} = \sqrt{a}\,\sqrt{b}.$$

The square root of a product is equal to the product of square roots.

What do we look for to take advantage of this rule? In the example

$$\sqrt{18} = \sqrt{9 \cdot 2} = \sqrt{9}\,\sqrt{2} = 3\sqrt{2},$$

9 is a perfect square. Thus we look for perfect square factors of the radicand (18 in this case). Notice that when there are no perfect square factors the rule does not help us. For example, it does no good in trying to simplify $\sqrt{6} = \sqrt{2 \cdot 3}$ since neither 2 nor 3 is a perfect square. In fact $\sqrt{6}$ is in its simplest form.

EXAMPLE 1 SIMPLIFYING SQUARE ROOTS OF PRODUCTS	**PRACTICE EXERCISE 1**

Simplify the radicals.

(a) $\sqrt{27} = \sqrt{9 \cdot 3}$ 9 is a perfect square

      $= \sqrt{9}\,\sqrt{3}$ $\sqrt{ab} = \sqrt{a}\,\sqrt{b}$

      $= 3\sqrt{3}$ $\sqrt{9} = \sqrt{3^2} = 3$

(b) $\sqrt{52} = \sqrt{4 \cdot 13}$ 4 is a perfect square

      $= \sqrt{4}\,\sqrt{13}$ $\sqrt{ab} = \sqrt{a}\,\sqrt{b}$

      $= 2\sqrt{13}$ $\sqrt{4} = \sqrt{2^2} = 2$

(c) $\sqrt{15} = \sqrt{3 \cdot 5}$ Cannot be simplified since neither 3 nor 5 is a perfect square

      $= \sqrt{15}$

(d) $\sqrt{3^4} = \sqrt{(3^2)^2}$ $3^4 = 3^{2 \cdot 2} = (3^2)^2$, which is a perfect square

      $= 3^2 = 9$ $\sqrt{(3^2)^2} = 3^2$

Simplify the radicals.

(a) $\sqrt{45}$

(b) $\sqrt{44}$

(c) $\sqrt{55}$

(d) $\sqrt{2^4}$

(e) $2\sqrt{3^6} = 2\sqrt{(3^3)^2}$ $3^6 = 3^{3\cdot2} = (3^3)^2$

$\qquad = 2 \cdot 3^3 = 2 \cdot 27 = 54$

(f) $\sqrt{3^7} = \sqrt{3^6 \cdot 3}$ $3^7 = 3^{6+1} = 3^6 \cdot 3^1 = 3^6 \cdot 3$

$\qquad = \sqrt{(3^3)^2}\,\sqrt{3}$

$\qquad = 3^3\sqrt{3} = 27\sqrt{3}$

(e) $3\sqrt{2^6}$

(f) $\sqrt{2^9}$

Answers: (a) $3\sqrt{5}$ **(b)** $2\sqrt{11}$
(c) cannot be simplified **(d) 4**
(e) 24 **(f)** $16\sqrt{2}$

As Example 1 illustrates, to simplify radicals we need to recall the rules of exponents such as

$$a^{m+n} = a^m a^n \quad \text{and} \quad a^{m \cdot n} = (a^m)^n.$$

Notice in Example 1(d) and (e) when the exponent under the radical is even, the radicand is a perfect square. When the exponent under the radical is odd, we rewrite the radicand as a product with a perfect square factor. For example, in Example 1(f), we rewrite $3^7 = 3^6 \cdot 3$ and then simplify further. This process also works with variables. For example,

$$\sqrt{x^7} = \sqrt{x^6 \cdot x} \qquad x^6 \text{ is a perfect square}$$

$$= \sqrt{x^6}\,\sqrt{x} \qquad \sqrt{ab} = \sqrt{a}\,\sqrt{b}$$

$$= \sqrt{(x^3)^2}\,\sqrt{x} \qquad x^6 = x^{3\cdot2} = (x^3)^2$$

$$= x^3\sqrt{x}.$$

Remember that all our variables are positive.

| **EXAMPLE 2** SQUARE ROOTS INVOLVING VARIABLES | **PRACTICE EXERCISE 2** |

Simplify the radicals.

(a) $\sqrt{x^4} = \sqrt{(x^2)^2}$ $(x^2)^2$ is a perfect square

$\qquad = x^2$

(b) $\sqrt{x^5} = \sqrt{x^4 \cdot x}$ $x^5 = x^{4+1} = x^4 \cdot x$

$\qquad = \sqrt{x^4}\,\sqrt{x}$ $\sqrt{ab} = \sqrt{a}\,\sqrt{b}$

$\qquad = \sqrt{(x^2)^2}\,\sqrt{x}$

$\qquad = x^2\sqrt{x}$

(c) $\sqrt{27a^3} = \sqrt{9 \cdot 3 \cdot a^2 \cdot a}$ 9 and a^2 are perfect squares

$\qquad = \sqrt{9 \cdot a^2 \cdot 3a}$ $3a$ will be left under the radical

$\qquad = \sqrt{9} \cdot \sqrt{a^2} \cdot \sqrt{3a}$ The simplifying rule expanded to three factors, $\sqrt{abc} = \sqrt{a}\,\sqrt{b}\,\sqrt{c}$

$\qquad = 3 \cdot a \cdot \sqrt{3a}$

$\qquad = 3a\sqrt{3a}$

(d) $\sqrt{x^4y^7} = \sqrt{x^4y^6 \cdot y}$ x^4 and y^6 are perfect squares

$\qquad = \sqrt{x^4}\,\sqrt{y^6}\,\sqrt{y}$ $\sqrt{abc} = \sqrt{a}\,\sqrt{b}\,\sqrt{c}$

$\qquad = \sqrt{(x^2)^2}\,\sqrt{(y^3)^2}\,\sqrt{y}$

$\qquad = x^2y^3\sqrt{y}$

Simplify the radicals.

(a) $\sqrt{y^6}$

(b) $\sqrt{y^7}$

(c) $\sqrt{45x^2}$

(d) $\sqrt{a^5b^8}$

(e) $\sqrt{288ab^2} = \sqrt{2 \cdot 144 \cdot a \cdot b^2}$ 144 and b^2 are perfect squares

$\qquad = \sqrt{144 \cdot b^2 \cdot 2a}$

$\qquad = \sqrt{144}\ \sqrt{b^2}\ \sqrt{2a}$

$\qquad = 12b\sqrt{2a}$

(e) $\sqrt{162x^3b^4}$

Answers: (a) y^3 (b) $y^3\sqrt{y}$
(c) $3x\sqrt{5}$ (d) $a^2b^4\sqrt{a}$
(e) $9xb^2\sqrt{2x}$

② SQUARE ROOTS OF QUOTIENTS

We can use the simplifying rule above to simplify

$$\sqrt{\frac{9}{4}} = \sqrt{\left(\frac{3}{2}\right)^2} = \frac{3}{2}.$$

This can also be simplified as

$$\sqrt{\frac{9}{4}} = \frac{\sqrt{9}}{\sqrt{4}} = \frac{3}{2}.$$

This is an example of the following rule which is used for radical expressions involving fractions.

Simplifying Rule for Quotients

If $a \geq 0$ and $b > 0$, then

$$\sqrt{\frac{a}{b}} = \frac{\sqrt{a}}{\sqrt{b}}.$$

The square root of a quotient is equal to the quotient of the square roots.

EXAMPLE 3 SIMPLIFYING SQUARE ROOTS OF QUOTIENTS

Simplify the radicals.

(a) $\sqrt{\frac{27}{4}} = \frac{\sqrt{27}}{\sqrt{4}}$ $\qquad \sqrt{\frac{a}{b}} = \frac{\sqrt{a}}{\sqrt{b}}$

$\qquad = \frac{\sqrt{9 \cdot 3}}{2}$ $\qquad \sqrt{4} = 2$

$\qquad = \frac{\sqrt{9}\ \sqrt{3}}{2} = \frac{3\sqrt{3}}{2}$ $\quad \sqrt{ab} = \sqrt{a}\ \sqrt{b}$

(b) $\sqrt{\frac{4x^3}{y^2}} = \frac{\sqrt{4x^3}}{\sqrt{y^2}}$ $\qquad \sqrt{\frac{a}{b}} = \frac{\sqrt{a}}{\sqrt{b}}$

$\qquad = \frac{\sqrt{4 \cdot x^2 \cdot x}}{y}$ $\qquad \sqrt{y^2} = y$

$\qquad = \frac{\sqrt{4}\ \sqrt{x^2}\ \sqrt{x}}{y}$ $\quad \sqrt{abc} = \sqrt{a}\ \sqrt{b}\ \sqrt{c}$

$\qquad = \frac{2x\sqrt{x}}{y}$

PRACTICE EXERCISE 3

Simplify the radicals.

(a) $\sqrt{\frac{8}{9}}$

(b) $\sqrt{\frac{25a^5}{w^2}}$

(c) $\sqrt{\dfrac{48x^3y^2}{3xy}} = \sqrt{\dfrac{3 \cdot 16 \cdot x^3 \cdot y^2}{3xy}}$ Simplify under the radical first

$= \sqrt{16x^2y}$ $\dfrac{x^3}{x} = x^{3-1} = x^2,\ \dfrac{y^2}{y} = y^{2-1} = y$

$= \sqrt{16}\ \sqrt{x^2}\ \sqrt{y}$

$= 4x\sqrt{y}$

(c) $\sqrt{\dfrac{32a^4b^3}{2a^2b^2}}$

Answers: (a) $\dfrac{2\sqrt{2}}{3}$ (b) $\dfrac{5a^2\sqrt{a}}{w}$

(c) $4a\sqrt{b}$

Notice that in Example 3(c) we used

$$\frac{a^m}{a^n} = a^{m-n}.$$

❸ RATIONALIZING A DENOMINATOR

Consider the following problem.

$$\sqrt{\frac{9}{7}} = \frac{\sqrt{9}}{\sqrt{7}} = \frac{3}{\sqrt{7}} \qquad \sqrt{7} \text{ is in the denominator}$$

When a square root is left in the denominator, the radical expression is not considered completely simplified. The process of removing radicals from the denominator (making the denominator a rational number) is called **rationalizing the denominator.** We do this by making the expression under the radical in the denominator a perfect square.

To rationalize denominators we need the following property. Since $\sqrt{ab} = \sqrt{a}\ \sqrt{b}$, if we let $a = b$, then

$$\sqrt{aa} = \sqrt{a}\ \sqrt{a}$$
$$\sqrt{a^2} = \sqrt{a}\ \sqrt{a}$$
$$a = \sqrt{a}\ \sqrt{a}.$$

Thus, $\sqrt{7}\ \sqrt{7} = 7$ and $\sqrt{x^3}\ \sqrt{x^3} = x^3$.

EXAMPLE 4 RATIONALIZING DENOMINATORS	PRACTICE EXERCISE 4

Rationalize the denominators.

(a) $\dfrac{3}{\sqrt{7}} = \dfrac{3\sqrt{7}}{\sqrt{7}\ \sqrt{7}}$ Multiply numerator and denominator by $\sqrt{7}$

$= \dfrac{3\sqrt{7}}{7}$ Since $\sqrt{7}\ \sqrt{7} = 7$, the denominator is now rational

(b) $\sqrt{\dfrac{25x^2}{y}} = \dfrac{\sqrt{25x^2}}{\sqrt{y}}$ $\sqrt{\dfrac{a}{b}} = \dfrac{\sqrt{a}}{\sqrt{b}}$

$= \dfrac{\sqrt{25}\ \sqrt{x^2}}{\sqrt{y}}$

$= \dfrac{5x}{\sqrt{y}}$

$= \dfrac{5x\sqrt{y}}{\sqrt{y}\ \sqrt{y}} = \dfrac{5x\sqrt{y}}{y}$ The denominator is rationalized

Rationalize the denominators.

(a) $\dfrac{5}{\sqrt{11}}$

(b) $\sqrt{\dfrac{49u^4}{w}}$

(c) $\sqrt{\dfrac{25x^2y}{yz^2}} = \sqrt{\dfrac{25x^2\cancel{y}}{\cancel{y}z^2}}$ Simplify under the radical first

$= \dfrac{\sqrt{25x^2}}{\sqrt{z^2}}$ $\sqrt{\dfrac{a}{b}} = \dfrac{\sqrt{a}}{\sqrt{b}}$

$= \dfrac{5x}{z}$ Do not need to rationalize

(c) $\sqrt{\dfrac{32x^7y^9}{3y^4}}$

Answers: (a) $\dfrac{5\sqrt{11}}{11}$ (b) $\dfrac{7u^2\sqrt{w}}{w}$

(c) $\dfrac{4x^3y^2\sqrt{6xy}}{3}$

④ SIMPLIFIED RADICALS

A Radical Is Considered Simplified

1. When there are no perfect square factors of the radicand.
2. When there are no fractions under the radical.
3. When there are no radicals in the denominator.

EXAMPLE 5 APPLICATION TO PHYSICS

If an object is dropped from a height of h feet, the time t in seconds it takes the object to reach the ground is given by the formula

$$t = \sqrt{\dfrac{h}{16}}.$$

Simplify this radical expression and use it to find the time it would take for a rock to reach the ground if dropped from a hot air balloon 144 feet above the ground. See Figure 17.1.

Figure 17.1

Simplify t.

$$t = \sqrt{\dfrac{h}{16}} = \dfrac{\sqrt{h}}{\sqrt{16}} = \dfrac{\sqrt{h}}{4}$$

PRACTICE EXERCISE 5

Simplify the expression

$$t = \sqrt{\dfrac{25h}{169}}$$

and use it to find t when $h = 49$.

Substitute 144 for h.

$$t = \frac{\sqrt{h}}{4} = \frac{\sqrt{144}}{4} = \frac{12}{4} = 3$$

It would take 3 seconds for the rock to reach the ground.

Answer: $\frac{35}{13}$

17.2 EXERCISES A

Simplify each of the radicals. Assume that all variables are positive.

1. $\sqrt{75}$ $5\sqrt{3}$
2. $\sqrt{10}$ $\sqrt{10}$
3. $\sqrt{32}$ $4\sqrt{2}$
4. $3\sqrt{200}$ $30\sqrt{2}$

5. $\sqrt{147}$ $7\sqrt{3}$
6. $5\sqrt{36}$ 30
7. $\sqrt{5^4}$ 25
8. $\sqrt{625}$ 25

9. $\sqrt{7^3}$ $7\sqrt{7}$
10. $\sqrt{5^5}$ $25\sqrt{5}$
11. $\sqrt{75y^2}$ $5y\sqrt{3}$
12. $\sqrt{25x^3}$ $5x\sqrt{x}$

13. $\sqrt{75y^3}$ $5y\sqrt{3y}$
14. $\sqrt{3^2x^3}$ $3x\sqrt{x}$
15. $\sqrt{3^3y^3}$ $3y\sqrt{3y}$
16. $\sqrt{147x^5}$ $7x^2\sqrt{3x}$

17. $\sqrt{9x^2y^2}$ $3xy$
18. $\sqrt{27x^3y^2}$ $3xy\sqrt{3x}$
19. $\sqrt{48x^2y^3}$ $4xy\sqrt{3y}$
20. $\sqrt{75x^9y^5}$ $5x^4y^2\sqrt{3xy}$

21. $\sqrt{\frac{75}{49}}$ $\frac{5\sqrt{3}}{7}$
22. $\sqrt{\frac{8}{121}}$ $\frac{2\sqrt{2}}{11}$
23. $\sqrt{\frac{50}{32}}$ $\frac{5}{4}$
24. $\sqrt{\frac{25x^2}{y^2}}$ $\frac{2x}{y}$

25. $\sqrt{\frac{16x^4}{y^4}}$ $\frac{4x^2}{y^2}$
26. $\sqrt{\frac{75x^2}{y^2}}$ $\frac{5x\sqrt{3}}{y}$
27. $\sqrt{\frac{75x^2y^4}{3}}$ $5xy^2$
28. $\sqrt{\frac{x^3y^3}{49}}$ $\frac{xy\sqrt{xy}}{7}$

29. $\frac{5}{\sqrt{3}}$ $\frac{5\sqrt{3}}{3}$
30. $\frac{8}{\sqrt{7}}$ $\frac{8\sqrt{7}}{7}$
31. $\sqrt{\frac{9}{5}}$ $\frac{3\sqrt{5}}{5}$
32. $\sqrt{\frac{45}{5}}$ 3

33. $\sqrt{\frac{75x^2y^3}{yz^2}}$ $\frac{5xy\sqrt{3}}{z}$
34. $\sqrt{\frac{36y^2}{x}}$ $\frac{6y\sqrt{x}}{x}$
35. $\sqrt{\frac{25x^3}{y^3z^4}}$ $\frac{5x\sqrt{xy}}{y^2z^2}$
36. $\sqrt{\frac{48x^2y^2}{3x^4y^4}}$ $\frac{4}{xy}$

In a wholesale operation, it has been estimated that the cost, c, is related to the number of items sold, n, by the equation
$c = \sqrt{n}$.

37. Find c if $n = 81$.   9

38. Find c if $n = 8$. $2\sqrt{2}$

39. Find c if $n = 0$.   0

40. Find c if $n = 75$. $5\sqrt{3}$

41. Find c if $n = 2^4$.   4

42. Find c if $n = 2^5$. $4\sqrt{2}$

A child drops a coin from the top of a building. Use the formula given in Example 5 to find the time it will take for the coin to reach the ground for each height h given in Exercises 43–46. Use a calculator to give the answer correct to the nearest tenth of a second if appropriate.

43. 256 ft   4 sec

44. 100 ft   2.5 sec

45. 150 ft   3.1 sec

46. 740 ft   6.8 sec

FOR REVIEW

Evaluate.

47. $-\sqrt{121}$ -11

48. $\sqrt{-121}$ meaningless

49. $\sqrt{-0}$ 0

50. $\sqrt{\dfrac{49}{25}}$ $\dfrac{7}{5}$

51. $\sqrt{(x-7)^2}$ $x - 7$

52. $\sqrt{x^2 - 10x + 25}$ $x - 5$

ANSWERS: 1. $5\sqrt{3}$ 2. $\sqrt{10}$ 3. $4\sqrt{2}$ 4. $30\sqrt{2}$ 5. $7\sqrt{3}$ 6. 30 7. 25 8. 25 9. $7\sqrt{7}$ 10. $25\sqrt{5}$ 11. $5y\sqrt{3}$
12. $5x\sqrt{x}$ 13. $5y\sqrt{3y}$ 14. $3x\sqrt{x}$ 15. $3y\sqrt{3y}$ 16. $7x^2\sqrt{3x}$ 17. $3xy$ 18. $3xy\sqrt{3x}$ 19. $4xy\sqrt{3y}$ 20. $5x^4y^2\sqrt{3xy}$
21. $\dfrac{5\sqrt{3}}{7}$ 22. $\dfrac{2\sqrt{2}}{11}$ 23. $\dfrac{5}{4}$ 24. $\dfrac{5x}{y}$ 25. $\dfrac{4x^2}{y^2}$ 26. $\dfrac{5x\sqrt{3}}{y}$ 27. $5xy^2$ 28. $\dfrac{xy\sqrt{xy}}{7}$ 29. $\dfrac{5\sqrt{3}}{3}$ 30. $\dfrac{8\sqrt{7}}{7}$ 31. $\dfrac{3\sqrt{5}}{5}$
32. 3 33. $\dfrac{5xy\sqrt{3}}{z}$ 34. $\dfrac{6y\sqrt{x}}{x}$ 35. $\dfrac{5x\sqrt{xy}}{y^2z^2}$ 36. $\dfrac{4}{xy}$ 37. 9 38. $2\sqrt{2}$ 39. 0 40. $5\sqrt{3}$ 41. 4 42. $4\sqrt{2}$
43. 4 sec 44. 2.5 sec 45. 3.1 sec 46. 6.8 sec 47. -11 48. meaningless 49. 0 50. $\dfrac{7}{5}$ 51. $x - 7$ 52. $x - 5$

17.2 EXERCISES B

Simplify each of the radicals. Assume that all variables are positive.

1. $\sqrt{98}$ $7\sqrt{2}$

2. $\sqrt{35}$ $\sqrt{35}$

3. $\sqrt{50}$ $5\sqrt{2}$

4. $2\sqrt{300}$ $20\sqrt{3}$

5. $\sqrt{108}$ $6\sqrt{3}$

6. $4\sqrt{64}$ 32

7. $\sqrt{3^4}$ 9

8. $\sqrt{81}$ 9

9. $\sqrt{5^3}$ $5\sqrt{5}$

10. $\sqrt{7^5}$ $49\sqrt{7}$

11. $\sqrt{98y^2}$ $7y\sqrt{2}$

12. $\sqrt{49x^3}$ $7x\sqrt{x}$

13. $\sqrt{98y^3}$ $7y\sqrt{2y}$

14. $\sqrt{5^2x^3}$ $5x\sqrt{x}$

15. $\sqrt{5^3x^3}$ $5x\sqrt{5x}$

16. $\sqrt{125x^5}$ $5x^2\sqrt{5x}$

17. $\sqrt{16x^2y^2}$ $4xy$

18. $\sqrt{32x^3y^2}$ $4xy\sqrt{2x}$

19. $\sqrt{27x^2y^3}$ $3xy\sqrt{3y}$

20. $\sqrt{12x^7y^9}$ $2x^3y^4\sqrt{3xy}$

21. $\sqrt{\dfrac{32}{81}}$ $\dfrac{4\sqrt{2}}{9}$

22. $\sqrt{\dfrac{75}{144}}$ $\dfrac{5\sqrt{3}}{12}$

23. $\sqrt{\dfrac{45}{20}}$ $\dfrac{3}{2}$

24. $\sqrt{\dfrac{49x^2}{y^2}}$ $\dfrac{7x}{y}$

25. $\sqrt{\dfrac{81x^4}{y^4}}$ $\dfrac{9x^2}{y^2}$ **26.** $\sqrt{\dfrac{147x^2}{y^2}}$ $\dfrac{7x\sqrt{3}}{y}$ **27.** $\sqrt{\dfrac{x^5y^5}{25}}$ $\dfrac{x^2y^2\sqrt{xy}}{5}$ **28.** $\sqrt{\dfrac{147x^4y^7}{3}}$ $7x^2y^3\sqrt{y}$

29. $\dfrac{3}{\sqrt{5}}$ $\dfrac{3\sqrt{5}}{5}$ **30.** $\dfrac{12}{\sqrt{11}}$ $\dfrac{12\sqrt{11}}{11}$ **31.** $\sqrt{\dfrac{16}{3}}$ $\dfrac{4\sqrt{3}}{3}$ **32.** $\sqrt{\dfrac{63}{7}}$ 3

33. $\sqrt{\dfrac{4x^4}{y}}$ $\dfrac{2x^2\sqrt{y}}{y}$ **34.** $\sqrt{\dfrac{147x^3y^2}{xz^2}}$ $\dfrac{7xy\sqrt{3}}{z}$ **35.** $\sqrt{\dfrac{50x^3y^3}{2x^5y^5}}$ $\dfrac{5}{xy}$ **36.** $\sqrt{\dfrac{16x^5}{y^5z^2}}$ $\dfrac{4x^2\sqrt{xy}}{y^3z}$

In a manufacturing process, two quantities p and r are related by the equation $p = \sqrt{r}$.

37. Find p when $r = 64$.   8   **38.** Find p when $r = 27$. $3\sqrt{3}$ **39.** Find p when $r = 0$.   0

40. Find p when $r = 32$. $4\sqrt{2}$ **41.** Find p when $r = 3^4$.   9   **42.** Find p when $r = 3^5$. $9\sqrt{3}$

A child drops a coin from the top of a building. Use the formula given in Example 5 to find the time it will take for the coin to reach the ground for each height h given in Exercises 43–46. Use a calculator to give the answer correct to the nearest tenth of a second if appropriate.

43. 64 ft   2 sec   **44.** 196 ft   3.5 sec   **45.** 270 ft   4.1 sec   **46.** 580 ft   6.0 sec

FOR REVIEW

Evaluate.

47. $-\sqrt{144}$ -12 **48.** $\sqrt{-144}$ meaningless   **49.** $\sqrt{0}$ 0

50. $\sqrt{\dfrac{81}{4}}$ $\dfrac{9}{2}$ **51.** $\sqrt{(x-3)^2}$ $x-3$ **52.** $\sqrt{x^2+10x+25}$ $x+5$

17.2 EXERCISES C

Simplify each of the radicals. Assume all variables are positive.

1. $\sqrt{\dfrac{27x^9y^7}{2x^6}}$ $\dfrac{3xy^3\sqrt{6xy}}{2}$ **2.** $\sqrt{\dfrac{243x^5y^4}{8xyz^3}}$ $\dfrac{9x^2y\sqrt{6yz}}{4z^2}$ **3.** $\sqrt{\dfrac{75xy}{9x^5y^6z^7}}$

$\left[\text{Answer: } \dfrac{5\sqrt{3yz}}{3x^2y^3z^4}\right]$

Use $\sqrt[3]{a^3} = a$ in the following exercises.

4. $\sqrt[3]{y^5}$ $y\sqrt[3]{y^2}$ **5.** $\sqrt[3]{\dfrac{81x^5y^4}{8z^3}}$ $\dfrac{3xy\sqrt[3]{3x^2y}}{2z}$ **6.** $\sqrt[3]{\dfrac{24x^4}{y}}$

$\left[\text{Answer: } \dfrac{2x\sqrt[3]{3xy^2}}{y}\right]$

17.3 MULTIPLICATION AND DIVISION OF RADICALS

1 PRODUCT OF SQUARE ROOTS

The simplifying rule for products, introduced in Section 17.2, allows us to simplify radicals like the following.

$$\sqrt{4 \cdot 9} = \sqrt{4}\,\sqrt{9} = 2 \cdot 3 = 6$$

The rule can also be used in the reverse order to multiply radicals. For example,

$$
\begin{aligned}
\sqrt{2}\,\sqrt{18} &= \sqrt{2 \cdot 18} && \text{Reverse of the simplifying rule} \\
&= \sqrt{2 \cdot 2 \cdot 9} && \text{Factor 18 into } 2 \cdot 9 \\
&= \sqrt{4 \cdot 9} && \text{4 and 9 are perfect squares} \\
&= \sqrt{4}\,\sqrt{9} = 2 \cdot 3 = 6.
\end{aligned}
$$

Here we used the simplifying rule for products from Section 17.2 in reverse order. This gives us the rule for multiplying radicals.

> ### Multiplication Rule
>
> If $a \geq 0$ and $b \geq 0$, then
>
> $$\sqrt{a}\,\sqrt{b} = \sqrt{ab}.$$
>
> The product of square roots is equal to the square root of the product.

EXAMPLE 1   MULTIPLYING RADICALS

Multiply and then simplify.

(a) $\sqrt{7}\,\sqrt{7} = \sqrt{7 \cdot 7}$ $\sqrt{a}\,\sqrt{b} = \sqrt{ab}$

$\qquad\quad = \sqrt{7^2} = 7$ We did this in Section 17.2

(b) $\sqrt{5}\,\sqrt{20} = \sqrt{5 \cdot 20}$ $\sqrt{a}\,\sqrt{b} = \sqrt{ab}$

$\qquad\quad = \sqrt{5 \cdot 5 \cdot 4}$

$\qquad\quad = \sqrt{25 \cdot 4}$ 25 and 4 are perfect squares

$\qquad\quad = \sqrt{25}\,\sqrt{4}$ $\sqrt{ab} = \sqrt{a}\,\sqrt{b}$

$\qquad\quad = 5 \cdot 2 = 10$

(c) $\sqrt{6}\,\sqrt{75} = \sqrt{6 \cdot 75}$ $\sqrt{a}\,\sqrt{b} = \sqrt{ab}$

$\qquad\quad = \sqrt{2 \cdot 3 \cdot 3 \cdot 25}$ Factor to find perfect squares

$\qquad\quad = \sqrt{9 \cdot 25 \cdot 2}$ 2 will be left under the radical

$\qquad\quad = \sqrt{9}\,\sqrt{25}\,\sqrt{2}$

$\qquad\quad = 3 \cdot 5 \cdot \sqrt{2} = 15\sqrt{2}$

(d) $\sqrt{5}\,\sqrt{7y} = \sqrt{5 \cdot 7 \cdot y}$ $\sqrt{a}\,\sqrt{b} = \sqrt{ab}$

$\qquad\quad = \sqrt{35y}$ This is considered a simpler form than $\sqrt{5}\,\sqrt{7y}$

PRACTICE EXERCISE 1

Multiply and then simplify.

(a) $\sqrt{11}\,\sqrt{11}$

(b) $\sqrt{2}\,\sqrt{18}$

(c) $\sqrt{48}\,\sqrt{6}$

(d) $\sqrt{2x}\,\sqrt{13}$

(e) $\sqrt{3x}\,\sqrt{75x} = \sqrt{3x \cdot 75x}$ $\sqrt{a}\,\sqrt{b} = \sqrt{ab}$

$\qquad\qquad\quad = \sqrt{3 \cdot 3 \cdot 25 \cdot x^2}$ Look for perfect squares

$\qquad\qquad\quad = \sqrt{9}\,\sqrt{25}\,\sqrt{x^2}$ 9, 25, and x^2 are perfect squares

$\qquad\qquad\quad = 3 \cdot 5 \cdot x = 15x$

(e) $\sqrt{24a}\,\sqrt{6a}$

Answers: (a) 11   **(b)** 6
(c) $12\sqrt{2}$ (d) $\sqrt{26x}$ (e) $12a$

| EXAMPLE 2   MULTIPLYING RADICALS | PRACTICE EXERCISE 2 |

Multiply and then simplify.

(a) $\sqrt{5x^3}\,\sqrt{3x^3}\,\sqrt{5x} = \sqrt{5x^3 \cdot 3x^3 \cdot 5x}$ $\sqrt{a}\,\sqrt{b}\,\sqrt{c} = \sqrt{abc}$

$\qquad\qquad\qquad\quad = \sqrt{5 \cdot 5 \cdot x^3 \cdot x^3 \cdot 3x}$ Look for perfect squares

$\qquad\qquad\qquad\quad = \sqrt{5^2 \cdot (x^3)^2 \cdot 3x}$ 5^2 and $(x^3)^2$ are perfect squares

$\qquad\qquad\qquad\quad = \sqrt{5^2}\,\sqrt{(x^3)^2}\,\sqrt{3x}$

$\qquad\qquad\qquad\quad = 5x^3\sqrt{3x}$

(b) $\sqrt{2xy}\,\sqrt{2x^3y^3} = \sqrt{2xy \cdot 2x^3y^3}$ $\sqrt{a}\,\sqrt{b} = \sqrt{ab}$

$\qquad\qquad\qquad = \sqrt{2^2x^4y^4}$

$\qquad\qquad\qquad = \sqrt{2^2}\,\sqrt{(x^2)^2}\,\sqrt{(y^2)^2}$ $x^4 = (x^2)^2$ and $y^4 = (y^2)^2$

$\qquad\qquad\qquad = 2x^2y^2$

(c) $7\sqrt{6ab}\,\sqrt{3a} = 7\sqrt{6ab \cdot 3a}$ Multiplication rule

$\qquad\qquad\quad = 7\sqrt{2 \cdot 3 \cdot 3 \cdot a^2 \cdot b}$ 3^2 and a^2 are perfect squares

$\qquad\qquad\quad = 7\sqrt{3^2a^2 \cdot 2b}$

$\qquad\qquad\quad = 7\sqrt{3^2}\,\sqrt{a^2}\,\sqrt{2b}$

$\qquad\qquad\quad = 7 \cdot 3 \cdot a \cdot \sqrt{2b}$

$\qquad\qquad\quad = 21a\sqrt{2b}$

Multiply and then simplify.

(a) $\sqrt{7w^3}\,\sqrt{2w}\,\sqrt{7w^5}$

(b) $\sqrt{3ab^3}\,\sqrt{3ab}$

(c) $5\sqrt{14xy}\,\sqrt{7y}$

Answers: (a) $7w^4\sqrt{2w}$ (b) $3ab^2$
(c) $35y\sqrt{2x}$

---

② QUOTIENT OF SQUARE ROOTS

The simplifying rule for quotients, presented in Section 17.2, allows us to simplify problems such as

$$\sqrt{\frac{9}{4}} = \frac{\sqrt{9}}{\sqrt{4}} = \frac{3}{2}.$$

Again, we can reverse this rule to simplify quotients of radicals. For example,

$$\frac{\sqrt{45}}{\sqrt{20}} = \sqrt{\frac{45}{20}} = \sqrt{\frac{5 \cdot 9}{5 \cdot 4}} = \sqrt{\frac{9}{4}} = \frac{\sqrt{9}}{\sqrt{4}} = \frac{3}{2}.$$

We see that a quotient of radicals may be simplified by first considering the radical of the quotient, then reducing the quotient. This technique is summarized in the next rule.

| Division Rule |

If $a \geq 0$ and $b > 0$, then

$$\frac{\sqrt{a}}{\sqrt{b}} = \sqrt{\frac{a}{b}}.$$

The quotient of square roots is equal to the square root of the quotient.

EXAMPLE 3 DIVIDING RADICALS

Divide and then simplify.

(a) $\dfrac{\sqrt{75}}{\sqrt{12}} = \sqrt{\dfrac{75}{12}}$ $\dfrac{\sqrt{a}}{\sqrt{b}} = \sqrt{\dfrac{a}{b}}$

$= \sqrt{\dfrac{3 \cdot 25}{3 \cdot 4}}$ Look for perfect squares and common factors

$= \sqrt{\dfrac{25}{4}}$ 25 and 4 are perfect squares

$= \dfrac{\sqrt{25}}{\sqrt{4}} = \dfrac{5}{2}$ $\sqrt{\dfrac{a}{b}} = \dfrac{\sqrt{a}}{\sqrt{b}}$

(b) $\dfrac{\sqrt{15x^3}}{\sqrt{3x}} = \sqrt{\dfrac{15x^3}{3x}}$ $\dfrac{\sqrt{a}}{\sqrt{b}} = \sqrt{\dfrac{a}{b}}$

$= \sqrt{\dfrac{3 \cdot 5 \cdot x \cdot x^2}{3x}}$ 3 and x are common factors and x^2 is a perfect square

$= \sqrt{5x^2} = x\sqrt{5}$

(c) $\dfrac{\sqrt{12y^5}}{\sqrt{75y^9}} = \sqrt{\dfrac{12y^5}{75y^9}}$ Division rule

$= \sqrt{\dfrac{3 \cdot 4 \cdot y^5}{3 \cdot 25 \cdot y^5 \cdot y^4}}$ 3 and y^5 are common factors since $y^9 = y^{5+4} = y^5 \cdot y^4$

$= \sqrt{\dfrac{4}{25y^4}}$

$= \dfrac{\sqrt{4}}{\sqrt{25}\,\sqrt{y^4}} = \dfrac{2}{5y^2}$

PRACTICE EXERCISE 3

Divide and then simplify.

(a) $\dfrac{\sqrt{72}}{\sqrt{50}}$

(b) $\dfrac{\sqrt{21a^5}}{\sqrt{7a}}$

(c) $\dfrac{\sqrt{20a}}{\sqrt{405a^3}}$

Answers: (a) $\frac{6}{5}$ (b) $a^2\sqrt{3}$
(c) $\frac{2}{9a}$

EXAMPLE 4 DIVIDING RADICALS

Divide and then simplify.

(a) $\dfrac{\sqrt{3xy^3}}{\sqrt{4x^3y}} = \sqrt{\dfrac{3xy^3}{4x^3y}}$ Division rule

$= \sqrt{\dfrac{3 \cdot x \cdot y^2 \cdot y}{4 \cdot x \cdot x^2 \cdot y}}$ x and y are common factors

$= \sqrt{\dfrac{3y^2}{4x^2}}$

$= \dfrac{\sqrt{y^2}\,\sqrt{3}}{\sqrt{4}\,\sqrt{x^2}} = \dfrac{y\sqrt{3}}{2x}$

PRACTICE EXERCISE 4

Divide and then simplify.

(a) $\dfrac{\sqrt{7ab^3}}{\sqrt{9a^5b}}$

(b) $\dfrac{2\sqrt{3x}}{\sqrt{12x^3}} = 2\sqrt{\dfrac{3x}{12x^3}}$ Division rule

$= 2\sqrt{\dfrac{\cancel{3}\cdot\cancel{x}}{\cancel{3}\cdot 4\cdot x^2\cdot\cancel{x}}}$ 3 and x are common factors

$= 2\sqrt{\dfrac{1}{4x^2}}$ 1 is left in the numerator

$= \dfrac{2\sqrt{1}}{\sqrt{4}\sqrt{x^2}}$

$= \dfrac{\cancel{2}\cdot 1}{\cancel{2}x} = \dfrac{1}{x}$ 2 is a common factor and $\sqrt{1} = 1$

(c) $\dfrac{\sqrt{2x^2}}{\sqrt{6y}} = \sqrt{\dfrac{2x^2}{6y}}$ Division rule

$= \sqrt{\dfrac{x^2}{3y}}$ 2 is a common factor

$= \dfrac{\sqrt{x^2}}{\sqrt{3y}}$

$= \dfrac{x}{\sqrt{3y}}$

$= \dfrac{x\sqrt{3y}}{\sqrt{3y}\sqrt{3y}}$ Rationalize the denominator

$= \dfrac{x\sqrt{3y}}{3y}$

(b) $\dfrac{5\sqrt{2w^4}}{\sqrt{50w^6}}$

(c) $\dfrac{\sqrt{3u^4}}{\sqrt{15v}}$

Answers: (a) $\dfrac{b\sqrt{7}}{3a^2}$ (b) $\dfrac{1}{w}$

(c) $\dfrac{u^2\sqrt{5v}}{5v}$

17.3 EXERCISES A

Multiply and simplify.

1. $\sqrt{2}\,\sqrt{50}$ 10

2. $\sqrt{3}\,\sqrt{75}$ 15

3. $\sqrt{15}\,\sqrt{5}$ $5\sqrt{3}$

4. $\sqrt{18}\,\sqrt{98}$ 42

5. $\sqrt{6}\,\sqrt{7}$ $\sqrt{42}$

6. $\sqrt{6}\,\sqrt{30}$ $6\sqrt{5}$

7. $\sqrt{20}\,\sqrt{48}$ $8\sqrt{15}$

8. $2\sqrt{24}\,\sqrt{42}$ $24\sqrt{7}$

9. $\sqrt{98}\,\sqrt{75}$ $35\sqrt{6}$

10. $\sqrt{2}\,\sqrt{8x}$ $4\sqrt{x}$

11. $\sqrt{27x}\,\sqrt{3x}$ $9x$

12. $\sqrt{15y}\,\sqrt{5}$ $5\sqrt{3y}$

13. $\sqrt{15y}\ \sqrt{20y}$
$10y\sqrt{3}$

14 $\sqrt{3a^2}\ \sqrt{12a}$
$6a\sqrt{a}$

15. $\sqrt{3x^3}\ \sqrt{3x}$
$3x^2$

16. $\sqrt{2xy}\ \sqrt{8xy}$
$4xy$

17 $\sqrt{6x^2y}\ \sqrt{2xy^2}$
$2xy\sqrt{3xy}$

18. $\sqrt{7x^3y^3}\ \sqrt{42xy^3}$
$7x^2y^3\sqrt{6}$

19 $\sqrt{5x}\ \sqrt{15xy}\ \sqrt{3y}$
$15xy$

20. $\sqrt{2xy}\ \sqrt{6xy}\ \sqrt{9xy}$
$6xy\sqrt{3xy}$

Divide and simplify.

21. $\dfrac{\sqrt{50}}{\sqrt{2}}$ 5

22. $\dfrac{\sqrt{18}}{\sqrt{8}}$ $\dfrac{3}{2}$

23 $\dfrac{\sqrt{98}}{\sqrt{18}}$ $\dfrac{7}{3}$

24. $\dfrac{\sqrt{6}}{\sqrt{50}}$ $\dfrac{\sqrt{3}}{5}$

25. $\dfrac{\sqrt{54}}{\sqrt{50}}$ $\dfrac{3\sqrt{3}}{5}$

26 $\dfrac{5\sqrt{15}}{\sqrt{75}}$ $\sqrt{5}$

27. $\dfrac{\sqrt{4}}{\sqrt{3}}$ $\dfrac{2\sqrt{3}}{3}$

28. $\dfrac{\sqrt{20}}{\sqrt{35}}$ $\dfrac{2\sqrt{7}}{7}$

29. $\dfrac{\sqrt{10}}{\sqrt{12}}$ $\dfrac{\sqrt{30}}{6}$

30. $\dfrac{\sqrt{50x^2}}{\sqrt{2x^2}}$ 5

31 $\dfrac{\sqrt{3y}}{\sqrt{4y}}$ $\dfrac{\sqrt{3}}{2}$

32 $\dfrac{\sqrt{27a^3}}{\sqrt{3}}$ $3a\sqrt{a}$

33. $\dfrac{\sqrt{27x^2}}{\sqrt{3}}$ $3x$

34. $\dfrac{\sqrt{4y}}{\sqrt{y^3}}$ $\dfrac{2}{y}$

35. $\dfrac{\sqrt{9x^2y}}{\sqrt{4y}}$ $\dfrac{3x}{2}$

36. $\dfrac{\sqrt{50x^3y^3}}{\sqrt{2xy}}$ $5xy$

37 $\dfrac{5\sqrt{2xy}}{\sqrt{25x}}$ $\sqrt{2y}$

38 $\dfrac{\sqrt{18x}}{\sqrt{2y}}$ $\dfrac{3\sqrt{xy}}{y}$

39. $\dfrac{\sqrt{25}}{\sqrt{x}}$ $\dfrac{5\sqrt{x}}{x}$

40. $\dfrac{\sqrt{2y^2}}{\sqrt{5x}}$ $\dfrac{y\sqrt{10x}}{5x}$

In an environmental study, three quantities m, p, and q have been found to be related by the equation $m = \sqrt{p}\sqrt{q}$.

41. Find m if $p = 4$ and $q = 9$.
6

42. Find m if $p = 25$ and $q = 8$.
$10\sqrt{2}$

43. Find m if $p = 8$ and $q = 18$.
12

44. Find m if $p = 0$ and $q = 5$.
0

45. Find m if $p = 4$ and $q = -4$.
no value of m if q is negative

46. Find m if $p = 6$ and $q = 15$.
$3\sqrt{10}$

FOR REVIEW

Simplify.

47. $\sqrt{288x^2}$
$12x\sqrt{2}$

48. $\sqrt{147x^2y}$
$7x\sqrt{3y}$

49. $\sqrt{\dfrac{98x^2}{25}}$
$\dfrac{7x\sqrt{2}}{5}$

50. $\sqrt{\dfrac{36x^2}{5y}}$
$\dfrac{6x\sqrt{5y}}{5y}$

Exercises 51–56 review material from Section 10.5 to prepare for the next section. Use the distributive law to collect like terms.

51. $2x + 5x$ $7x$

52. $8y - 3y$ $5y$

53. $7x + 9y$
no like terms to collect

54. $4x - x + 6x$ $9x$

55. $\dfrac{1}{2}y - \dfrac{3}{4}y + \dfrac{1}{8}y$ $-\dfrac{1}{8}y$

56. $4x + 3y - 2x - 9y$ $2x - 6y$

ANSWERS: 1. 10 2. 15 3. $5\sqrt{3}$ 4. 42 5. $\sqrt{42}$ 6. $6\sqrt{5}$ 7. $8\sqrt{15}$ 8. $24\sqrt{7}$ 9. $35\sqrt{6}$ 10. $4\sqrt{x}$ 11. $9x$ 12. $5\sqrt{3y}$ 13. $10y\sqrt{3}$ 14. $6a\sqrt{a}$ 15. $3x^2$ 16. $4xy$ 17. $2xy\sqrt{3xy}$ 18. $7x^2y^3\sqrt{6}$ 19. $15xy$ 20. $6xy\sqrt{3xy}$ 21. 5 22. $\dfrac{3}{2}$ 23. $\dfrac{7}{3}$ 24. $\dfrac{\sqrt{3}}{5}$ 25. $\dfrac{3\sqrt{3}}{5}$ 26. $\sqrt{5}$ 27. $\dfrac{2\sqrt{3}}{3}$ 28. $\dfrac{2\sqrt{7}}{7}$ 29. $\dfrac{\sqrt{30}}{6}$ 30. 5 31. $\dfrac{\sqrt{3}}{2}$ 32. $3a\sqrt{a}$ 33. $3x$ 34. $\dfrac{2}{y}$ 35. $\dfrac{3x}{2}$ 36. $5xy$ 37. $\sqrt{2y}$ 38. $\dfrac{3\sqrt{xy}}{y}$ 39. $\dfrac{5\sqrt{x}}{x}$ 40. $\dfrac{y\sqrt{10x}}{5x}$ 41. 6 42. $10\sqrt{2}$ 43. 12 44. 0 45. no value for m if q is negative 46. $3\sqrt{10}$ 47. $12x\sqrt{2}$ 48. $7x\sqrt{3y}$ 49. $\dfrac{7x\sqrt{2}}{5}$ 50. $\dfrac{6x\sqrt{5y}}{5y}$ 51. $7x$ 52. $5y$ 53. no like terms to collect 54. $9x$ 55. $-\dfrac{1}{8}y$ 56. $2x - 6y$

17.3 EXERCISES B

Multiply and simplify.

1. $\sqrt{5}\sqrt{45}$ 15

2. $\sqrt{7}\sqrt{28}$ 14

3. $\sqrt{6}\sqrt{2}$ $2\sqrt{3}$

4. $\sqrt{20}\sqrt{45}$ 30

5. $\sqrt{7}\sqrt{11}$ $\sqrt{77}$

6. $\sqrt{10}\sqrt{18}$ $6\sqrt{5}$

7. $\sqrt{18}\sqrt{28}$ $6\sqrt{14}$

8. $3\sqrt{35}\sqrt{45}$ $45\sqrt{7}$

9. $\sqrt{72}\sqrt{48}$ $24\sqrt{6}$

10. $\sqrt{5}\sqrt{20x}$ $10\sqrt{x}$

11. $\sqrt{8x}\sqrt{2x}$ $4x$

12. $\sqrt{12y}\sqrt{4}$ $4\sqrt{3y}$

13. $\sqrt{14y}\sqrt{50y}$ $10y\sqrt{7}$

14. $\sqrt{5a^2}\sqrt{20a}$ $10a\sqrt{a}$

15. $\sqrt{5x}\sqrt{5x^3}$ $5x^2$

16. $\sqrt{3xy}\sqrt{27xy}$ $9xy$

17. $\sqrt{10x^2y}\sqrt{2xy^2}$
$2xy\sqrt{5xy}$

18. $\sqrt{5x^3y^3}\sqrt{35xy^3}$
$5x^2y^3\sqrt{7}$

19. $\sqrt{2x}\sqrt{14xy}\sqrt{7y}$
$14xy$

20. $\sqrt{3xy}\sqrt{15xy}\sqrt{xy}$
$3xy\sqrt{5xy}$

Divide and simplify.

21. $\dfrac{\sqrt{3}}{\sqrt{48}}$ $\dfrac{1}{4}$

22. $\dfrac{\sqrt{27}}{\sqrt{12}}$ $\dfrac{3}{2}$

23. $\dfrac{\sqrt{125}}{\sqrt{45}}$ $\dfrac{5}{3}$

24. $\dfrac{\sqrt{14}}{\sqrt{72}}$ $\dfrac{\sqrt{7}}{6}$

25. $\dfrac{\sqrt{150}}{\sqrt{8}}$ $\dfrac{5\sqrt{3}}{2}$

26. $\dfrac{6\sqrt{10}}{\sqrt{72}}$ $\sqrt{5}$

27. $\dfrac{\sqrt{9}}{\sqrt{2}}$ $\dfrac{3\sqrt{2}}{2}$

28. $\dfrac{\sqrt{50}}{\sqrt{14}}$ $\dfrac{5\sqrt{7}}{7}$

29. $\dfrac{\sqrt{5}}{\sqrt{28}}$ $\dfrac{\sqrt{35}}{14}$

30. $\dfrac{\sqrt{98x^2}}{\sqrt{2x^2}}$ 7

31. $\dfrac{\sqrt{5y}}{\sqrt{9y}}$ $\dfrac{\sqrt{5}}{3}$

32. $\dfrac{\sqrt{45x^2}}{\sqrt{5}}$ $3x$

33. $\dfrac{\sqrt{8a^3}}{\sqrt{2}}$ $2a\sqrt{a}$

34. $\dfrac{\sqrt{9y}}{\sqrt{y^3}}$ $\dfrac{3}{y}$

35. $\dfrac{\sqrt{25xy^2}}{\sqrt{9x}}$ $\dfrac{5y}{3}$

36. $\dfrac{\sqrt{98x^3y^3}}{\sqrt{2xy}}$ $7xy$

37. $\dfrac{7\sqrt{3xy}}{\sqrt{49y}}$ $\sqrt{3x}$

38. $\dfrac{\sqrt{16}}{\sqrt{x}}$ $\dfrac{4\sqrt{x}}{x}$

39. $\dfrac{\sqrt{75y}}{\sqrt{3x}}$ $\dfrac{5\sqrt{xy}}{x}$

40. $\dfrac{\sqrt{3x^2}}{\sqrt{7y}}$ $\dfrac{x\sqrt{21y}}{7y}$

During a scientific experiment, it was found that three quantities v, w, and d are related by the equation $v = \dfrac{\sqrt{w}}{\sqrt{d}}$.

41. Find v if $w = 4$ and $d = 25$.

$\dfrac{2}{5}$

42. Find v if $w = 8$ and $d = 16$.

$\dfrac{\sqrt{2}}{2}$

43. Find v if $w = 8$ and $d = 50$.

$\dfrac{2}{5}$

44. Find v if $w = 0$ and $d = 5$.

0

45. Find v if $w = -9$ and $d = 7$.

no value for v when w is negative

46. Find v if $w = 21$ and $d = 14$.

$\dfrac{\sqrt{6}}{2}$

FOR REVIEW

Simplify.

47. $\sqrt{242x^2}$ $11x\sqrt{2}$

48. $\sqrt{75xy^2}$ $5y\sqrt{3x}$

49. $\sqrt{\dfrac{147x^2}{16}}$ $\dfrac{7x\sqrt{3}}{4}$

50. $\sqrt{\dfrac{25x^2}{6y}}$ $\dfrac{5x\sqrt{6y}}{6y}$

Exercises 51–56 review material from Section 10.5 to prepare for the next section. Use the distributive law to collect like terms.

51. $4x + 9x$ $13x$

52. $7y - 2y$ $5y$

53. $-2x + 3y$

no like terms to collect

54. $3x + 7x - 6x$ $4x$

55. $\dfrac{2}{3}y - \dfrac{1}{9}y + \dfrac{1}{18}y$ $\dfrac{11}{18}y$

56. $-5x + 4y + 2x - 6y$

$-3x - 2y$

17.3 EXERCISES C

Multiply or divide and simplify.

1. $\sqrt{3x^3y^{-3}}\,\sqrt{6xy^{-1}}$

$\dfrac{3x^2\sqrt{2}}{y^2}$

2. $\dfrac{\sqrt{125x^{-6}y^{-3}}}{\sqrt{45x^4y^{-6}}}$

$\dfrac{5y\sqrt{y}}{3x^5}$

3. $\dfrac{\sqrt{16x^2y^5z^{-1}}}{\sqrt{6x^3y^{-4}z^{-3}}}$

$\left[\text{Answer: } \dfrac{2y^4z\sqrt{6xy}}{3x}\right]$

Use $\sqrt[3]{a^3} = a$ in the following exercises.

4. $\sqrt[3]{81x^4y^6} \; \sqrt[3]{18x^5y^5}$

$9x^3y^3\sqrt[3]{2y^2}$

5. $\dfrac{\sqrt[3]{9x^4y^7}}{\sqrt[3]{24xy^2}}$

$\dfrac{xy\sqrt[3]{3y^2}}{2}$

6. $\dfrac{\sqrt[3]{40x^6y}}{\sqrt[3]{25xy^5}}$

$\left[\text{Answer:} \quad \dfrac{2x\sqrt[3]{25x^2y^2}}{5y^2}\right]$

17.4 ADDITION AND SUBTRACTION OF RADICALS

STUDENT GUIDEPOSTS
---

1 Like Radicals and Unlike Radicals   **2** Adding and Subtracting Radicals

The rules for multiplication and division of radicals,

$$\sqrt{a}\,\sqrt{b} = \sqrt{ab} \quad \text{and} \quad \frac{\sqrt{a}}{\sqrt{b}} = \sqrt{\frac{a}{b}},$$

do not have counterparts relative to addition and subtraction. For example,

$$5 = \sqrt{25} = \sqrt{9 + 16} \neq \sqrt{9} + \sqrt{16} = 3 + 4 = 7, \qquad 5 \neq 7$$

so that in general,

$$\sqrt{a + b} \neq \sqrt{a} + \sqrt{b}.$$

Also, since

$$4 = \sqrt{16} = \sqrt{25 - 9} \neq \sqrt{25} - \sqrt{9} = 5 - 3 = 2, \qquad 4 \neq 2$$

in general,

$$\sqrt{a - b} \neq \sqrt{a} - \sqrt{b}.$$

1 LIKE RADICALS AND UNLIKE RADICALS

We may, however, use the distributive law to add or subtract *like radicals*. **Like radicals** have the same radicand. Thus, the terms

$$\sqrt{11}, \quad 3\sqrt{11}, \quad -5\sqrt{11}$$

have like radicals. Also, terms such as

$$\sqrt{x}, \quad -7\sqrt{x}, \quad 100\sqrt{x}$$

contain like radicals. However,

$$\sqrt{11}, \quad \sqrt{x}, \quad 3\sqrt{y}$$

do not have like radicals; they are **unlike radicals.**

2 ADDING AND SUBTRACTING RADICALS

We add or subtract like radicals just as we collect like terms of polynomials. Remember,

$$3x + 5x = (3 + 5)x = 8x.$$

We collect like radicals in the same way.

$$3\sqrt{x} + 5\sqrt{x} = (3 + 5)\sqrt{x} = 8\sqrt{x}$$

| EXAMPLE 1   ADDING AND SUBTRACTING LIKE RADICALS | PRACTICE EXERCISE 1 |

Add or subtract.

(a) $9\sqrt{5} + 4\sqrt{5} = (9 + 4)\sqrt{5}$ Distributive law, $ac + bc = (a + b)c$

$\qquad\qquad = 13\sqrt{5}$

(b) $9\sqrt{5} - 4\sqrt{5} = (9 - 4)\sqrt{5}$ Distributive law, $ac - bc = (a - b)c$

$\qquad\qquad = 5\sqrt{5}$

(c) $3\sqrt{xy} - 10\sqrt{xy} = (3 - 10)\sqrt{xy}$ Distributive law

$\qquad\qquad\quad = -7\sqrt{xy}$

(d) $7\sqrt{x} + 5\sqrt{y}$ cannot be simplified since $\sqrt{x}$ and $\sqrt{y}$ are not like radicals.

Add or subtract.

(a) $2\sqrt{11} + 5\sqrt{11}$

(b) $2\sqrt{11} - 5\sqrt{11}$

(c) $6\sqrt{u} - 5\sqrt{u}$

(d) $8\sqrt{a} - 2\sqrt{w}$

Answers: (a) $7\sqrt{11}$ (b) $-3\sqrt{11}$ (c) $\sqrt{u}$ (d) cannot be simplified since $\sqrt{a}$ and $\sqrt{w}$ are not like radicals

Some unlike radicals can be changed into like radicals if we simplify. For example, $\sqrt{2}$ and $\sqrt{8}$ are not like radicals since they have different radicands. However,

$$\sqrt{8} = \sqrt{4 \cdot 2} = \sqrt{4}\,\sqrt{2} = 2\sqrt{2}.$$

Now the terms $\sqrt{2}$ and $2\sqrt{2}$ do contain like radicals.

To Add or Subtract Radicals

1. Simplify each radical as much as possible.
2. Use the distributive law to collect any like radicals.

| EXAMPLE 2   SIMPLIFYING BEFORE ADDING OR SUBTRACTING | PRACTICE EXERCISE 2 |

Add or subtract.

(a) $3\sqrt{2} + 5\sqrt{8} = 3\sqrt{2} + 5\sqrt{4 \cdot 2}$ 4 is a perfect square

$\qquad\qquad\quad = 3\sqrt{2} + 5\sqrt{4}\,\sqrt{2}$ $\sqrt{ab} = \sqrt{a}\,\sqrt{b}$

$\qquad\qquad\quad = 3\sqrt{2} + 5 \cdot 2\sqrt{2}$ $\sqrt{4} = 2$

$\qquad\qquad\quad = 3\,\boxed{\sqrt{2}} + 10\,\boxed{\sqrt{2}}$ Terms now contain like radicals

$\qquad\qquad\quad = (3 + 10)\,\boxed{\sqrt{2}} = 13\sqrt{2}$ Distributive law

(b) $\sqrt{12} + \sqrt{75} = \sqrt{4 \cdot 3} + \sqrt{25 \cdot 3}$ 4 and 25 are perfect squares

$\qquad\qquad\quad = \sqrt{4}\,\sqrt{3} + \sqrt{25}\,\sqrt{3}$ $\sqrt{ab} = \sqrt{a}\,\sqrt{b}$

$\qquad\qquad\quad = 2\,\boxed{\sqrt{3}} + 5\,\boxed{\sqrt{3}}$ $\sqrt{4} = 2$ and $\sqrt{25} = 5$

$\qquad\qquad\quad = (2 + 5)\,\boxed{\sqrt{3}} = 7\sqrt{3}$ Distributive law

(c) $5\sqrt{12} - 7\sqrt{27} = 5\sqrt{4 \cdot 3} - 7\sqrt{9 \cdot 3}$ 4 and 9 are perfect squares

$\qquad\qquad\quad = 5\sqrt{4}\,\sqrt{3} - 7\sqrt{9}\,\sqrt{3}$ $\sqrt{ab} = \sqrt{a}\,\sqrt{b}$

$\qquad\qquad\quad = 5 \cdot 2 \cdot \sqrt{3} - 7 \cdot 3 \cdot \sqrt{3}$

$\qquad\qquad\quad = 10\sqrt{3} - 21\sqrt{3}$ Terms contain like radicals

$\qquad\qquad\quad = (10 - 21)\sqrt{3} = -11\sqrt{3}$ Distributive law

Add or subtract.

(a) $5\sqrt{5} + 3\sqrt{20}$

(b) $7\sqrt{32} + 3\sqrt{18}$

(c) $6\sqrt{8} - \sqrt{50}$

(d) $2\sqrt{25y} - 4\sqrt{36y} = 2\sqrt{25}\ \sqrt{y} - 4\sqrt{36}\ \sqrt{y}$ 25 and 36 are
perfect squares

$$= 2 \cdot 5\sqrt{y} - 4 \cdot 6\sqrt{y}$$

$$= 10\sqrt{y} - 24\sqrt{y}$$

$$= (10 - 24)\sqrt{y} = -14\sqrt{y}$$ Distributive law

(d) $8\sqrt{4x} - 3\sqrt{49x}$

Answers: (a) $11\sqrt{5}$ (b) $37\sqrt{2}$
(c) $7\sqrt{2}$ (d) $-5\sqrt{x}$

| **EXAMPLE 3   ADDING SEVERAL RADICALS** | **PRACTICE EXERCISE 3** |

Perform the indicated operations.

(a) $4\sqrt{18} + 3\sqrt{27} - 6\sqrt{12}$

$$= 4\sqrt{9 \cdot 2} + 3\sqrt{9 \cdot 3} - 6\sqrt{4 \cdot 3}$$

$$= 4\sqrt{9}\ \sqrt{2} + 3\sqrt{9}\ \sqrt{3} - 6\sqrt{4}\ \sqrt{3}$$

$$= 4 \cdot 3\sqrt{2} + 3 \cdot 3\sqrt{3} - 6 \cdot 2\sqrt{3}$$

$$= 12\sqrt{2} + 9\ \sqrt{3} - 12\ \sqrt{3}$$ Only the last two terms are like terms

$$= 12\sqrt{2} + (9 - 12)\ \sqrt{3}$$ Distributive law

$$= 12\sqrt{2} - 3\sqrt{3}$$ As simple as possible

(b) $3\sqrt{x} - 2\sqrt{4x} - 5\sqrt{9x} = 3\sqrt{x} - 2\sqrt{4}\ \sqrt{x} - 5\sqrt{9}\ \sqrt{x}$

$$= 3\sqrt{x} - 2 \cdot 2\sqrt{x} - 5 \cdot 3\sqrt{x}$$

$$= 3\ \sqrt{x} - 4\ \sqrt{x} - 15\ \sqrt{x}$$

$$= (3 - 4 - 15)\ \sqrt{x} = -16\sqrt{x}$$

Perform the indicated operations.

(a) $3\sqrt{75} - \sqrt{18} + 2\sqrt{27}$

(b) $8\sqrt{w} - 3\sqrt{25w} + 5\sqrt{4w}$

Answers: (a) $21\sqrt{3} - 3\sqrt{2}$
(b) $3\sqrt{w}$

To obtain like terms, it may be necessary to rationalize denominators.

| **EXAMPLE 4   RATIONALIZING BEFORE ADDING** | **PRACTICE EXERCISE 4** |

Add or subtract.

(a) $3\sqrt{2} + \dfrac{5}{\sqrt{2}} = 3\sqrt{2} + \dfrac{5\ \sqrt{2}}{\sqrt{2}\ \sqrt{2}}$ Rationalize denominator

$$= 3\sqrt{2} + \dfrac{5\sqrt{2}}{2}$$ $\sqrt{2}\ \sqrt{2} = 2$

$$= 3\ \sqrt{2} + \dfrac{5}{2}\ \sqrt{2}$$ $\dfrac{5\sqrt{2}}{2} = \dfrac{5 \cdot \sqrt{2}}{2 \cdot 1} = \dfrac{5}{2} \cdot \dfrac{\sqrt{2}}{1} = \dfrac{5}{2}\sqrt{2}$

$$= \left(3 + \dfrac{5}{2}\right)\sqrt{2}$$ Distributive law

$$= \left(\dfrac{6}{2} + \dfrac{5}{2}\right)\sqrt{2} = \dfrac{11}{2}\sqrt{2} = \dfrac{11\sqrt{2}}{2}$$

Add or subtract.

(a) $5\sqrt{3} + \dfrac{1}{\sqrt{3}}$

(b) $\dfrac{6}{\sqrt{5}} - 4\sqrt{5} = \dfrac{6}{\sqrt{5}}\dfrac{\sqrt{5}}{\sqrt{5}} - 4\sqrt{5}$ Rationalize the denominator   **(b)** $\dfrac{9}{\sqrt{7}} - 6\sqrt{7}$

$\qquad\qquad = \dfrac{6\sqrt{5}}{5} - 4\sqrt{5}$

$\qquad\qquad = \dfrac{6}{5}\sqrt{5} - 4\sqrt{5}$ $\dfrac{6\sqrt{5}}{5} = \dfrac{6 \cdot \sqrt{5}}{5 \cdot 1} = \dfrac{6}{5} \cdot \dfrac{\sqrt{5}}{1} = \dfrac{6}{5}\sqrt{5}$

$\qquad\qquad = \left(\dfrac{6}{5} - 4\right)\sqrt{5}$ Distributive law

$\qquad\qquad = \left(\dfrac{6}{5} - \dfrac{20}{5}\right)\sqrt{5}$ Answers: (a) $\dfrac{16\sqrt{3}}{3}$

$\qquad\qquad = -\dfrac{14}{5}\sqrt{5} = -\dfrac{14\sqrt{5}}{5}$ (b) $-\dfrac{33\sqrt{7}}{7}$

17.4 EXERCISES A

Add or subtract.

1. $9\sqrt{2} + \sqrt{2}$ $10\sqrt{2}$ **2.** $-4\sqrt{5} + 3\sqrt{5}$ $-\sqrt{5}$ **3.** $-18\sqrt{xy} - 7\sqrt{xy}$ $-25\sqrt{xy}$

4. $3\sqrt{2} + 2\sqrt{3}$ $3\sqrt{2} + 2\sqrt{3}$ **⑤** $5\sqrt{12} - 3\sqrt{12}$ $4\sqrt{3}$ **6.** $\sqrt{50} + \sqrt{98}$ $12\sqrt{2}$

7. $\sqrt{18} - \sqrt{8}$ $\sqrt{2}$ **8.** $6\sqrt{2} + 3\sqrt{8}$ $12\sqrt{2}$ **9.** $5\sqrt{3} - \sqrt{27}$ $2\sqrt{3}$

⑩ $4\sqrt{50} + 7\sqrt{18}$ $41\sqrt{2}$ **11.** $3\sqrt{18} - 5\sqrt{12}$ $9\sqrt{2} - 10\sqrt{3}$ **12.** $-18\sqrt{7} - 2\sqrt{28}$ $-22\sqrt{7}$

13. $15\sqrt{45} + 4\sqrt{20}$ $53\sqrt{5}$ **⑭** $\sqrt{147} - 2\sqrt{75}$ $-3\sqrt{3}$ **15.** $6\sqrt{44} + 2\sqrt{99}$ $18\sqrt{11}$

16. $\sqrt{121} - \sqrt{144}$ -1 **17.** $2\sqrt{75} - 3\sqrt{125}$ **⑱** $2\sqrt{4x} + 7\sqrt{9x}$ $25\sqrt{x}$
$\qquad\qquad\qquad\qquad\qquad\qquad 10\sqrt{3} - 15\sqrt{5}$

19. $-3\sqrt{25y} - 2\sqrt{9y}$ $-21\sqrt{y}$ **20.** $8\sqrt{50y} + 2\sqrt{18y}$ $46\sqrt{2y}$ **㉑** $x\sqrt{y^3} + y\sqrt{x^2y}$ $2xy\sqrt{y}$

Perform the indicated operations.

22 $\sqrt{18} + \sqrt{50} + \sqrt{72}$ $14\sqrt{2}$

23. $\sqrt{27} - \sqrt{75} - \sqrt{108}$ $-8\sqrt{3}$

24. $5\sqrt{20} + 2\sqrt{45} - 9\sqrt{80}$ $-20\sqrt{5}$

25 $3\sqrt{75} + 2\sqrt{12} - 5\sqrt{48}$ $-\sqrt{3}$

26. $-3\sqrt{24} - 5\sqrt{150} - 4\sqrt{54}$ $-43\sqrt{6}$

27. $4\sqrt{99} - 7\sqrt{44} + 5\sqrt{52}$ $-2\sqrt{11} + 10\sqrt{13}$

28. $4\sqrt{98} - 8\sqrt{72} + 5\sqrt{32}$ 0

29 $4\sqrt{300} - 2\sqrt{500} + 4\sqrt{125}$ $40\sqrt{3}$

30 $\sqrt{4x} + \sqrt{16x} - \sqrt{25x}$ $\sqrt{x}$

31. $3\sqrt{5x} + 2\sqrt{20x} - 8\sqrt{45x}$ $-17\sqrt{5x}$

32 $9\sqrt{100a^2b^2} - 4\sqrt{36a^2b^2} - 10\sqrt{a^2b^2}$ $56ab$

33. $2\sqrt{25ab^3} - b\sqrt{36ab} - 5\sqrt{49ab^3}$ $-31b\sqrt{ab}$

Rationalize the denominator and then add or subtract.

34 $\sqrt{3} + \dfrac{1}{\sqrt{3}}$ $\dfrac{4\sqrt{3}}{3}$

35. $2\sqrt{5} - \dfrac{3}{\sqrt{5}}$ $\dfrac{7\sqrt{5}}{5}$

36. $\dfrac{7}{\sqrt{2}} + 6\sqrt{2}$ $\dfrac{19\sqrt{2}}{2}$

37. $3\sqrt{7} - \dfrac{1}{\sqrt{7}}$ $\dfrac{20\sqrt{7}}{7}$

38 $3\sqrt{7} - \sqrt{\dfrac{1}{7}}$ $\dfrac{20\sqrt{7}}{7}$

39 $-3\sqrt{5} - \dfrac{9}{\sqrt{5}}$ $\dfrac{-24\sqrt{5}}{5}$

40. $\dfrac{\sqrt{2}}{\sqrt{3}} + 2\sqrt{6}$ $\dfrac{7\sqrt{6}}{3}$

41 $\dfrac{\sqrt{2}}{\sqrt{10}} - \sqrt{5}$ $\dfrac{-4\sqrt{5}}{5}$

42. $\sqrt{\dfrac{2}{3}} - 2\sqrt{6}$ $\dfrac{-5\sqrt{6}}{3}$

Suppose we are given the formula $F = \sqrt{f} + 2\sqrt{g}$, relating the three quantities, F, f, and g.

43. Find F when $f = 8$ and $g = 18$.
 $8\sqrt{2}$

44. Find F when $f = 60$ and $g = 15$.
 $4\sqrt{15}$

45. Find F when $f = 0$ and $g = 7$.
 $2\sqrt{7}$

46. Find F when $f = -4$ and $g = 9$.
 no value for F when f is negative

FOR REVIEW

Simplify.

47. $3\sqrt{10}\,\sqrt{40}$ 60

48. $\sqrt{3a^2}\,\sqrt{15a}$ $3a\sqrt{5a}$

49. $\dfrac{\sqrt{150x}}{\sqrt{3x}}$ $5\sqrt{2}$

50. $\dfrac{\sqrt{12x^2}}{\sqrt{5y}}$ $\dfrac{2x\sqrt{15y}}{5y}$

Exercises 51–54 review material from Sections 14.3 and 14.4. They will help you prepare for the next section. Multiply the polynomials.

51. $(x - y)(x + 7y)$
$x^2 + 6xy - 7y^2$

52. $(x - 3y)^2$
$x^2 - 6xy + 9y^2$

53. $(x + 4y)(x - 4y)$
$x^2 - 16y^2$

54. $(3x - 7y)(3x + 7y)$
$9x^2 - 49y^2$

ANSWERS: 1. $10\sqrt{2}$ 2. $-\sqrt{5}$ 3. $-25\sqrt{xy}$ 4. $3\sqrt{2} + 2\sqrt{3}$ (cannot be simplified) 5. $4\sqrt{3}$ 6. $12\sqrt{2}$ 7. $\sqrt{2}$
8. $12\sqrt{2}$ 9. $2\sqrt{3}$ 10. $41\sqrt{2}$ 11. $9\sqrt{2} - 10\sqrt{3}$ 12. $-22\sqrt{7}$ 13. $53\sqrt{5}$ 14. $-3\sqrt{3}$ 15. $18\sqrt{11}$ 16. -1
17. $10\sqrt{3} - 15\sqrt{5}$ 18. $25\sqrt{x}$ 19. $-21\sqrt{y}$ 20. $46\sqrt{2y}$ 21. $2xy\sqrt{y}$ 22. $14\sqrt{2}$ 23. $-8\sqrt{3}$ 24. $-20\sqrt{5}$
25. $-\sqrt{3}$ 26. $-43\sqrt{6}$ 27. $-2\sqrt{11} + 10\sqrt{13}$ 28. 0 29. $40\sqrt{3}$ 30. $\sqrt{x}$ 31. $-17\sqrt{5x}$ 32. $56ab$ 33. $-31b\sqrt{ab}$
34. $\dfrac{4\sqrt{3}}{3}$ 35. $\dfrac{7\sqrt{5}}{5}$ 36. $\dfrac{19\sqrt{2}}{2}$ 37. $\dfrac{20\sqrt{7}}{7}$ 38. $\dfrac{20\sqrt{7}}{7}$ 39. $\dfrac{-24\sqrt{5}}{5}$ 40. $\dfrac{7\sqrt{6}}{3}$ 41. $\dfrac{-4\sqrt{5}}{5}$ 42. $\dfrac{-5\sqrt{6}}{3}$
43. $8\sqrt{2}$ 44. $4\sqrt{15}$ 45. $2\sqrt{7}$ 46. no value for F when f is negative 47. 60 48. $3a\sqrt{5a}$ 49. $5\sqrt{2}$ 50. $\dfrac{2x\sqrt{15y}}{5y}$
51. $x^2 + 6xy - 7y^2$ 52. $x^2 - 6xy + 9y^2$ 53. $x^2 - 16y^2$ 54. $9x^2 - 49y^2$

17.4 EXERCISES B

Add or subtract.

1. $5\sqrt{3} - 9\sqrt{3}$ $-4\sqrt{3}$

2. $-5\sqrt{7} - 9\sqrt{7}$ $-14\sqrt{7}$

3. $-9\sqrt{xy} - 3\sqrt{xy}$ $-12\sqrt{xy}$

4. $5\sqrt{5} + 3\sqrt{3}$ $5\sqrt{5} + 3\sqrt{3}$

5. $2\sqrt{27} - 4\sqrt{27}$ $-6\sqrt{3}$

6. $\sqrt{48} + \sqrt{75}$ $9\sqrt{3}$

7. $\sqrt{45} - \sqrt{20}$ $\sqrt{5}$

8. $7\sqrt{3} + 2\sqrt{12}$ $11\sqrt{3}$

9. $4\sqrt{2} - \sqrt{50}$ $-\sqrt{2}$

10. $8\sqrt{32} + 3\sqrt{50}$ $47\sqrt{2}$

11. $6\sqrt{27} - 4\sqrt{18}$ $18\sqrt{3} - 12\sqrt{2}$

12. $-8\sqrt{5} + 2\sqrt{45}$ $-2\sqrt{5}$

13. $4\sqrt{28} + 8\sqrt{63}$ $32\sqrt{7}$

14. $3\sqrt{75} - 5\sqrt{147}$ $-20\sqrt{3}$

15. $-8\sqrt{99} + 8\sqrt{44}$ $-8\sqrt{11}$

16. $\sqrt{169} - \sqrt{121}$ 2

17. $5\sqrt{45} - 2\sqrt{48}$ $15\sqrt{5} - 8\sqrt{3}$

18. $4\sqrt{25x} + 2\sqrt{16x}$ $28\sqrt{x}$

19. $-4\sqrt{4y} - 9\sqrt{25y}$ $-53\sqrt{y}$

20. $6\sqrt{8y} + 5\sqrt{32y}$ $32\sqrt{2y}$

21. $y\sqrt{x^3} + x\sqrt{xy^2}$ $2xy\sqrt{x}$

22. $\sqrt{8} + \sqrt{72} + \sqrt{98}$ $15\sqrt{2}$

23. $\sqrt{12} - \sqrt{27} - \sqrt{147}$ $-8\sqrt{3}$

24. $7\sqrt{48} - 4\sqrt{75} - 6\sqrt{12}$
$-4\sqrt{3}$

25. $3\sqrt{125} + 4\sqrt{80} - 6\sqrt{5}$
$25\sqrt{5}$

26. $-2\sqrt{6} - 2\sqrt{54} - 5\sqrt{150}$
$-33\sqrt{6}$

27. $5\sqrt{28} - 2\sqrt{63} + 4\sqrt{99}$
$4\sqrt{7} + 12\sqrt{11}$

28. $5\sqrt{147} - 2\sqrt{108} + 7\sqrt{48}$ $51\sqrt{3}$

29. $7\sqrt{700} - 4\sqrt{175} + 8\sqrt{44}$ $50\sqrt{7} + 16\sqrt{11}$

30. $\sqrt{25x} + \sqrt{36x} - \sqrt{49x}$ $4\sqrt{x}$

31. $5\sqrt{3x} + 6\sqrt{27x} - 7\sqrt{75x}$ $-12\sqrt{3x}$

32. $5\sqrt{25a^2b^2} - 3\sqrt{16a^2b^2} - 20\sqrt{a^2b^2}$ $-7ab$

33. $2a\sqrt{16ab} + \sqrt{4a^3b} - 3\sqrt{25a^3b}$ $-5a\sqrt{ab}$

Rationalize the denominator and then add or subtract.

34. $\sqrt{2} + \dfrac{1}{\sqrt{2}}$ $\dfrac{3\sqrt{2}}{2}$

35. $3\sqrt{5} - \dfrac{4}{\sqrt{5}}$ $\dfrac{11\sqrt{5}}{5}$

36. $\dfrac{8}{\sqrt{3}} + 2\sqrt{3}$ $\dfrac{14\sqrt{3}}{3}$

37. $4\sqrt{5} - \dfrac{1}{\sqrt{5}}$ $\dfrac{19\sqrt{5}}{5}$

38. $4\sqrt{5} - \sqrt{\dfrac{1}{5}}$ $\dfrac{19\sqrt{5}}{5}$

39. $-4\sqrt{7} - \dfrac{3}{\sqrt{7}}$ $\dfrac{-31\sqrt{7}}{7}$

40. $\dfrac{\sqrt{3}}{\sqrt{2}} + 5\sqrt{6}$ $\dfrac{11\sqrt{6}}{2}$

41. $\sqrt{\dfrac{3}{2}} - 5\sqrt{6}$ $\dfrac{-9\sqrt{6}}{2}$

42. $\dfrac{\sqrt{5}}{\sqrt{10}} - \sqrt{2}$ $-\dfrac{\sqrt{2}}{2}$

Suppose we have the formula $P = \sqrt{n} - 3\sqrt{p}$, relating the three quantities P, n, and p.

43. Find P when $n = 49$ and $p = 45$. $7 - 9\sqrt{5}$

44. Find P when $n = 52$ and $p = 117$. $-7\sqrt{13}$

45. Find P when $n = -9$ and $p = 14$.
   no value for P when n is negative

46. Find P when $n = 11$ and $p = 0$. $\sqrt{11}$

FOR REVIEW

Simplify.

47. $5\sqrt{48}\,\sqrt{3}$ 60

48. $\sqrt{8a^2}\,\sqrt{10a}$ $4a\sqrt{5a}$

49. $\dfrac{\sqrt{150x}}{\sqrt{2x}}$ $5\sqrt{3}$

50. $\dfrac{\sqrt{18y^2}}{\sqrt{7x}}$ $\dfrac{3y\sqrt{14x}}{7x}$

Exercises 51–54 review material from Sections 14.3 and 14.4. They will help you prepare for the next section. Multiply the polynomials.

51. $(2x - y)(x + 4y)$
   $2x^2 + 7xy - 4y^2$

52. $(4x - y)^2$
   $16x^2 - 8xy + y^2$

53. $(2x - y)(2x + y)$
   $4x^2 - y^2$

54. $(5x + 2y)(5x - 2y)$
   $25x^2 - 4y^2$

17.4 EXERCISES C

Add or subtract.

1. $\sqrt{12x^4 - 4x^2y^2} - \sqrt{27x^4 - 9x^2y^2}$
   $-x\sqrt{3x^2 - y^2}$

2. $\sqrt{2x^2 - 20xy + 50y^2} + \sqrt{8x^2 - 80xy + 200y^2}$
   [Answer: $3(x - 5y)\sqrt{2}$]

3. $3\sqrt[3]{40x^4y^4} - 2x\sqrt[3]{5xy^4}$ [*Hint:* $\sqrt[3]{a^3} = a$]
   $4xy\sqrt[3]{5xy}$

4. $-6ab\sqrt[3]{27a^5b^3} + 10\sqrt[3]{27a^8b^6}$
   [Answer: $12a^2b^2\sqrt[3]{a^2}$]

17.5 SUMMARY OF TECHNIQUES AND RATIONALIZING DENOMINATORS

STUDENT GUIDEPOSTS

1 Simplifying Radical Expressions

2 Multiplying Binomial Radical Expressions

3 Rationalizing Binomial Denominators

1 SIMPLIFYING RADICAL EXPRESSIONS

Below is a summary of the simplifying techniques from the previous sections.

To Simplify a Radical Expression

1. Combine all like radicals using the distributive laws.
2. When needed, use the rules of multiplication and division,

$$\sqrt{a}\,\sqrt{b} = \sqrt{ab} \quad \text{and} \quad \frac{\sqrt{a}}{\sqrt{b}} = \sqrt{\frac{a}{b}}.$$

3. Remove all perfect squares from under the radicals.
4. Rationalize all denominators.

The following table illustrates the various techniques.

Simplification	*Type*
$3\sqrt{2} - \sqrt{2} = 2\sqrt{2}$	Combining like radicals
$\sqrt{2}\,\sqrt{3} = \sqrt{6}$	Using multiplication rule
$\dfrac{\sqrt{12}}{\sqrt{3}} = \sqrt{\dfrac{12}{3}} = \sqrt{4} = 2$	Using division rule and removing perfect squares
$\sqrt{9x^2y} = 3x\sqrt{y}$	Removing perfect squares
$\dfrac{3}{\sqrt{2}} = \dfrac{3\sqrt{2}}{\sqrt{2}\,\sqrt{2}} = \dfrac{3\sqrt{2}}{2}$	Rationalizing the denominator

To multiply radical expressions, we often use one or more of the simplifying techniques, as in the next example.

EXAMPLE 1 MULTIPLYING USING THE DISTRIBUTIVE LAW

Multiply.

(a) $\sqrt{5}(\sqrt{15} + \sqrt{5}) = \sqrt{5}\,\sqrt{15} + \sqrt{5}\,\sqrt{5}$ Distributive law

$\qquad = \sqrt{5 \cdot 15} + \sqrt{5 \cdot 5}$ $\sqrt{a}\,\sqrt{b} = \sqrt{ab}$

$\qquad = \sqrt{5 \cdot 5 \cdot 3} + \sqrt{5 \cdot 5}$ Factor

$\qquad = \sqrt{5^2}\,\sqrt{3} + \sqrt{5^2}$ 5^2 is a perfect square

$\qquad = 5\sqrt{3} + 5$

PRACTICE EXERCISE 1

Multiply.

(a) $\sqrt{2}(\sqrt{6} + 2\sqrt{2})$

(b) $\sqrt{3}(\sqrt{27} - \sqrt{12}) = \sqrt{3}\,\sqrt{27} - \sqrt{3}\,\sqrt{12}$ Distributive law
$$= \sqrt{3 \cdot 3^3} - \sqrt{3 \cdot 3 \cdot 4}\qquad \sqrt{a}\,\sqrt{b} = \sqrt{ab}$$
$$= \sqrt{3^4} - \sqrt{3^2 \cdot 2^2}\qquad 3^4,\ 3^2,\ \text{and } 2^2 \text{ are}$$
$$\phantom{= \sqrt{3^4} - \sqrt{3^2 \cdot 2^2}\qquad}\text{perfect squares}$$
$$= 3^2 - 3 \cdot 2$$
$$= 9 - 6 = 3$$

(b) $\sqrt{5}(\sqrt{125} - \sqrt{20})$

(c) $\sqrt{7}\left(5\sqrt{7} - \dfrac{8}{\sqrt{7}}\right) = \sqrt{7}(5\sqrt{7}) - \sqrt{7}\left(\dfrac{8}{\sqrt{7}}\right)$ Distributive law

$$= 5(\sqrt{7})^2 - \dfrac{\sqrt{7}}{\sqrt{7}} \cdot 8\qquad \begin{array}{l}\text{Commutative and}\\\text{associative laws}\end{array}$$

$$= 5 \cdot 7 - 1 \cdot 8\qquad (\sqrt{7})^2 = \sqrt{7}\,\sqrt{7} = 7$$

$$= 35 - 8 = 27$$

(c) $\sqrt{3}\left(2\sqrt{3} - \dfrac{1}{\sqrt{3}}\right)$

Answers: (a) $2\sqrt{3} + 4$ (b) 15
(c) 5

---

② MULTIPLYING BINOMIAL RADICAL EXPRESSIONS

In Chapter 14 we used the FOIL method (*F*—First terms, *O*—Outside terms, *I*—Inside terms, and *L*—Last terms) to multiply binomials. Recall that, for example, to multiply $x + 2y$ and $2x - y$ we proceed as follows.

$$(x + 2y)(2x - y) = 2x \cdot x - x \cdot y + 2x \cdot 2y - 2y \cdot y$$

The same rule can be used to multiply binomial radical expressions.

$$(\sqrt{3} + 2\sqrt{2})(2\sqrt{3} - \sqrt{2}) = 2\sqrt{3} \cdot \sqrt{3} - \sqrt{3} \cdot \sqrt{2} + 2\sqrt{2} \cdot 2\sqrt{3} - 2\sqrt{2} \cdot \sqrt{2}$$
$$= 2 \cdot 3 - \sqrt{6} + 4\sqrt{6} - 2 \cdot 2$$
$$= 6 - \sqrt{6} + 4\sqrt{6} - 4$$
$$= 2 + 3\sqrt{6}$$

We multiplied the first terms Ⓕ, the outside terms Ⓞ, the inside terms Ⓘ, and the last terms Ⓛ, and then used the product rule and collected like terms.

| **EXAMPLE 2   USING FOIL** | **PRACTICE EXERCISE 2** |

Multiply.

(a) $(\sqrt{3} - \sqrt{2})(\sqrt{3} + 5\sqrt{2})$
$$= \sqrt{3}\,\sqrt{3} + \sqrt{3}(5\sqrt{2}) - \sqrt{2}\,\sqrt{3} - \sqrt{2}(5\sqrt{2})$$
$$= 3 + 5\sqrt{6} - \sqrt{6} - 5(2)$$
$$= 3 - 10 + (5 - 1)\sqrt{6}$$
$$= -7 + 4\sqrt{6}$$

Multiply.
(a) $(\sqrt{7} - 2\sqrt{3})(\sqrt{7} + \sqrt{3})$

(b) $(\sqrt{3} - \sqrt{2})(\sqrt{3} + \sqrt{2})$

$\quad = \sqrt{3}\,\sqrt{3} + \sqrt{3}\,\sqrt{2} - \sqrt{2}\,\sqrt{3} - \sqrt{2}\,\sqrt{2}$

$\quad = 3 + \sqrt{6} - \sqrt{6} - 2$

$\quad = 3 - 2$

$\quad = 1$

(b) $(\sqrt{11} - 2\sqrt{2})(\sqrt{11} + 2\sqrt{2})$

Answers: (a) $1 - \sqrt{21}$ (b) 3

❸ RATIONALIZING BINOMIAL DENOMINATORS

The product in Example 2(**b**) is the rational number 1. This is a special case of a more general result involving products of expressions of the form

$$(a - b)(a + b) = a^2 - b^2.$$

Notice that

$$(\sqrt{3} - \sqrt{2})(\sqrt{3} + \sqrt{2}) = (\sqrt{3})^2 - (\sqrt{2})^2 \quad \sqrt{3} = a \text{ and } \sqrt{2} = b$$
$$= 3 - 2 = 1.$$

Any time that a sum and a difference like those above are multiplied, the result will be a rational number free of radicals. This observation leads to a way to rationalize the denominator of an expression with a binomial in the denominator. For example, we can rationalize the denominator of $\dfrac{1}{\sqrt{3} - \sqrt{2}}$ by multiplying the numerator and denominator by $\sqrt{3} + \sqrt{2}$.

$$\frac{1}{\sqrt{3} - \sqrt{2}} = \frac{1(\sqrt{3} + \sqrt{2})}{(\sqrt{3} - \sqrt{2})(\sqrt{3} + \sqrt{2})}$$

$$= \frac{\sqrt{3} + \sqrt{2}}{(\sqrt{3})^2 - (\sqrt{2})^2} = \frac{\sqrt{3} + \sqrt{2}}{3 - 2}$$

$$= \frac{\sqrt{3} + \sqrt{2}}{1} = \sqrt{3} + \sqrt{2}$$

Had the denominator been $\sqrt{3} + \sqrt{2}$, we would have rationalized by multiplying both numerator and denominator by $\sqrt{3} - \sqrt{2}$. The binomials $\sqrt{3} + \sqrt{2}$ and $\sqrt{3} - \sqrt{2}$ are called **conjugates** as are binomials of the form $\sqrt{3} + 2$ and $\sqrt{3} - 2$.

To Rationalize a Binomial Denominator

Multiply both numerator and denominator of the fraction by the conjugate of the denominator and simplify.

| EXAMPLE 3 RATIONALIZING DENOMINATORS | PRACTICE EXERCISE 3 |

EXAMPLE 3 RATIONALIZING DENOMINATORS

Rationalize the denominators.

(a) $\dfrac{2}{\sqrt{7}+\sqrt{5}}$

$= \dfrac{2(\sqrt{7}-\sqrt{5})}{(\sqrt{7}+\sqrt{5})(\sqrt{7}-\sqrt{5})}$ Multiply numerator and denominator by $\sqrt{7}-\sqrt{5}$, the conjugate of $\sqrt{7}+\sqrt{5}$

$= \dfrac{2(\sqrt{7}-\sqrt{5})}{(\sqrt{7})^2-(\sqrt{5})^2}$ $(a+b)(a-b)=a^2-b^2$

$= \dfrac{2(\sqrt{7}-\sqrt{5})}{7-5}$ $(\sqrt{7})^2=7$ and $(\sqrt{5})^2=5$

$= \dfrac{2(\sqrt{7}-\sqrt{5})}{2}$

$= \sqrt{7}-\sqrt{5}$

(b) $\dfrac{\sqrt{3}}{\sqrt{3}-2} = \dfrac{\sqrt{3}(\sqrt{3}+2)}{(\sqrt{3}-2)(\sqrt{3}+2)}$ Multiply numerator and denominator by $\sqrt{3}+2$, the conjugate of $\sqrt{3}-2$

$= \dfrac{\sqrt{3}(\sqrt{3}+2)}{(\sqrt{3})^2-(2)^2}$ $(a-b)(a+b)=a^2-b^2$

$= \dfrac{\sqrt{3}(\sqrt{3}+2)}{3-4}$ $(\sqrt{3})^2=3$ and $(2)^2=4$

$= \dfrac{\sqrt{3}(\sqrt{3}+2)}{-1}$

$= -\sqrt{3}(\sqrt{3}+2)$ $\dfrac{a}{-1}=-a$

$= -\sqrt{3}\cdot\sqrt{3}-\sqrt{3}\cdot2$ Distributive law

$= -3-2\sqrt{3}$

PRACTICE EXERCISE 3

Rationalize the denominators.

(a) $\dfrac{4}{\sqrt{7}-\sqrt{3}}$

(b) $\dfrac{\sqrt{5}}{\sqrt{5}+2}$

Answers: (a) $\sqrt{7}+\sqrt{3}$
(b) $5-2\sqrt{5}$

17.5 EXERCISES A

Refer to the summary of techniques and simplify.

1. $\sqrt{20}-\sqrt{45}$ $-\sqrt{5}$

2. $\sqrt{2}\,\sqrt{32}$ 8

3. $\sqrt{3x}\,\sqrt{12x}$ $6x$

4. $\dfrac{\sqrt{18xy^2}}{\sqrt{2x}}$ $3y$

5 $\dfrac{\sqrt{25x^2}}{\sqrt{3y}}$ $\dfrac{5x\sqrt{3y}}{3y}$

6. $\dfrac{3}{\sqrt{2}}+\dfrac{1}{\sqrt{2}}$ $2\sqrt{2}$

7. $\sqrt{147} - \sqrt{108}$ $\sqrt{3}$ **8.** $\sqrt{3} \sqrt{12} + \sqrt{5} \sqrt{20}$ 16   **9.** $\sqrt{2} \sqrt{10} - \sqrt{15} \sqrt{3}$ $-\sqrt{5}$

10 $\dfrac{\sqrt{2}}{\sqrt{50}} - \dfrac{\sqrt{3}}{\sqrt{75}}$ 0   **11.** $\dfrac{\sqrt{98}}{\sqrt{2}} + \dfrac{\sqrt{125}}{\sqrt{5}}$ 12   **12.** $\dfrac{\sqrt{3}}{\sqrt{2}} - \dfrac{\sqrt{5}}{\sqrt{2}}$ $\dfrac{\sqrt{6} - \sqrt{10}}{2}$

13. $\sqrt{3x} \sqrt{6x} + \sqrt{5x} \sqrt{10x}$ $8x\sqrt{2}$ **14** $\sqrt{x} \sqrt{y} + \dfrac{\sqrt{xy^2}}{\sqrt{y}}$ $2\sqrt{xy}$ **15.** $\sqrt{5x} - \dfrac{\sqrt{x}}{\sqrt{5}}$ $\dfrac{4\sqrt{5x}}{5}$

16. $6\sqrt{2} \sqrt{6} + 2\sqrt{5} \sqrt{15} - 3\sqrt{12} \sqrt{25}$ $-8\sqrt{3}$ **17** $\sqrt{x^3} \sqrt{y} - 3x\sqrt{x} \sqrt{y} + \dfrac{4x\sqrt{xy^2}}{\sqrt{y}}$ $2x\sqrt{xy}$

Multiply and simplify.

18. $\sqrt{3}(\sqrt{3} + \sqrt{2})$ $3 + \sqrt{6}$ **19.** $\sqrt{5}(\sqrt{125} - \sqrt{45})$ 10

20 $\sqrt{2}(\sqrt{27} - \sqrt{12})$ $\sqrt{6}$ **21.** $\sqrt{7}(\sqrt{5} - \sqrt{125})$ $-4\sqrt{35}$

22. $\sqrt{5}(\sqrt{5} + 1)$ $5 + \sqrt{5}$ **23.** $\sqrt{6}(\sqrt{2} - 1)$ $2\sqrt{3} - \sqrt{6}$

24. $\sqrt{3}\left(\sqrt{3} - \dfrac{1}{\sqrt{3}}\right)$ 2   **25.** $\sqrt{7}\left(\dfrac{1}{\sqrt{7}} + \sqrt{7}\right)$ 8

26 $\sqrt{15}\left(\dfrac{\sqrt{3}}{\sqrt{5}} + \dfrac{1}{\sqrt{15}}\right)$ 4

27. $\sqrt{8}\left(\sqrt{2} - \dfrac{1}{\sqrt{8}}\right)$ 3

28 $\sqrt{2}\left(\sqrt{6} + \dfrac{\sqrt{2}}{\sqrt{3}}\right)$ $\dfrac{8\sqrt{3}}{3}$

29. $\sqrt{5}\left(\sqrt{10} - \dfrac{\sqrt{5}}{\sqrt{2}}\right)$ $\dfrac{5\sqrt{2}}{2}$

30. $(\sqrt{5} + 1)(\sqrt{5} + 1)$ $6 + 2\sqrt{5}$

31. $(\sqrt{5} + 1)(\sqrt{5} - 1)$ 4

32. $(\sqrt{2} - \sqrt{5})(\sqrt{2} - \sqrt{5})$ $7 - 2\sqrt{10}$

33. $(\sqrt{2} - \sqrt{5})(\sqrt{2} + \sqrt{5})$ -3

34. $(2\sqrt{3} + 1)(\sqrt{3} + 1)$ $7 + 3\sqrt{3}$

35 $(\sqrt{2} - 2)(\sqrt{2} + 1)$ $-\sqrt{2}$

36. $(2\sqrt{5} + \sqrt{7})(2\sqrt{5} - \sqrt{7})$ 13

37 $(3\sqrt{3} - \sqrt{2})(2\sqrt{3} + \sqrt{2})$ $16 + \sqrt{6}$

Rationalize the denominator.

38. $\dfrac{1}{\sqrt{2} - 1}$ $\sqrt{2} + 1$

39. $\dfrac{2}{\sqrt{5} - \sqrt{3}}$ $\sqrt{5} + \sqrt{3}$

40. $\dfrac{2}{\sqrt{7} + \sqrt{5}}$ $\sqrt{7} - \sqrt{5}$

41. $\dfrac{3}{\sqrt{5} - 2}$ $3\sqrt{5} + 6$

42. $\dfrac{6}{\sqrt{5} + 2}$ $6\sqrt{5} - 12$

43 $\dfrac{-8}{\sqrt{3} - \sqrt{7}}$ $2\sqrt{3} + 2\sqrt{7}$

44 $\dfrac{2\sqrt{3}}{\sqrt{3}-1}$ $3 + \sqrt{3}$

45. $\dfrac{\sqrt{2}}{\sqrt{2}+1}$ $2 - \sqrt{2}$

46. $\dfrac{2\sqrt{5}}{\sqrt{5}-\sqrt{3}}$ $5 + \sqrt{15}$

47 $\dfrac{x-1}{\sqrt{x}-1}$ $\sqrt{x}+1$

FOR REVIEW

Add or subtract.

48. $2\sqrt{75} - 6\sqrt{48} + 9\sqrt{12}$ $4\sqrt{3}$

49. $6\sqrt{20x} - 2\sqrt{45x} - \sqrt{5x}$ $5\sqrt{5x}$

50. An economist might use the formula $E = 3\sqrt{c} + \sqrt{e}$, relating the three quantities E, c, and e. Find E when $c = 98$ and $e = 50$. $26\sqrt{2}$

Exercises 51–56 review material from Chapter 11 and Section 15.5 to help you prepare for the next section. Solve each equation.

51. $7x = 42$ 6

52. $x + 8 = 31$ 23

53. $2(x - 7) = x + 4$ 18

54. $x^2 - 1 = x(x + 1) + 2$ -3

55. $x^2 - 3x - 4 = 0$ $4 \text{ or } -1$

56. $x^2 + 2x = 0$ $0 \text{ or } -2$

ANSWERS: 1. $-\sqrt{5}$ 2. 8 3. $6x$ 4. $3y$ 5. $\dfrac{5x\sqrt{3y}}{3y}$ 6. $2\sqrt{2}$ 7. $\sqrt{3}$ 8. 16 9. $-\sqrt{5}$ 10. 0 11. 12

12. $\dfrac{\sqrt{6}-\sqrt{10}}{2}$ 13. $8x\sqrt{2}$ 14. $2\sqrt{xy}$ 15. $\dfrac{4\sqrt{5x}}{5}$ 16. $-8\sqrt{3}$ 17. $2x\sqrt{xy}$ 18. $3 + \sqrt{6}$ 19. 10 20. $\sqrt{6}$

21. $-4\sqrt{35}$ 22. $5 + \sqrt{5}$ 23. $2\sqrt{3} - \sqrt{6}$ 24. 2 25. 8 26. 4 27. 3 28. $\dfrac{8\sqrt{3}}{3}$ 29. $\dfrac{5\sqrt{2}}{2}$ 30. $6 + 2\sqrt{5}$ 31. 4

32. $7 - 2\sqrt{10}$ 33. -3 34. $7 + 3\sqrt{3}$ 35. $-\sqrt{2}$ 36. 13 37. $16 + \sqrt{6}$ 38. $\sqrt{2} + 1$ 39. $\sqrt{5} + \sqrt{3}$ 40. $\sqrt{7} - \sqrt{5}$
41. $3\sqrt{5} + 6$ 42. $6\sqrt{5} - 12$ 43. $2\sqrt{3} + 2\sqrt{7}$ 44. $3 + \sqrt{3}$ 45. $2 - \sqrt{2}$ 46. $5 + \sqrt{15}$ 47. $\sqrt{x} + 1$ 48. $4\sqrt{3}$
49. $5\sqrt{5x}$ 50. $26\sqrt{2}$ 51. 6 52. 23 53. 18 54. -3 55. $4 \text{ or } -1$ 56. $0 \text{ or } -2$

17.5 EXERCISES B

Refer to the summary of techniques and simplify.

1. $\sqrt{98} + \sqrt{50}$ $12\sqrt{2}$

2. $\sqrt{5y}\,\sqrt{75y}$ $5y\sqrt{15}$

3. $\dfrac{\sqrt{108}}{\sqrt{3}}$ 6

4. $\dfrac{\sqrt{50x^2y}}{\sqrt{2y}}$ $5x$

5. $\dfrac{\sqrt{4y^2}}{\sqrt{3x}}$ $\dfrac{2y\sqrt{3x}}{3x}$

6. $\dfrac{5}{\sqrt{3}} + \dfrac{1}{\sqrt{3}}$ $2\sqrt{3}$

7. $2\sqrt{175} - 3\sqrt{28}$ $4\sqrt{7}$

8. $\sqrt{2}\,\sqrt{8} + \sqrt{3}\,\sqrt{27}$ 13

9. $\sqrt{5}\,\sqrt{10} - \sqrt{6}\,\sqrt{3}$ $2\sqrt{2}$

10. $\dfrac{\sqrt{3}}{\sqrt{75}} + \dfrac{\sqrt{5}}{\sqrt{125}}$ $\dfrac{2}{5}$

11. $\dfrac{\sqrt{108}}{\sqrt{3}} + \dfrac{\sqrt{147}}{\sqrt{3}}$ 13

12. $\dfrac{\sqrt{2}}{\sqrt{5}} - \dfrac{\sqrt{3}}{\sqrt{5}}$ $\dfrac{\sqrt{10} - \sqrt{15}}{5}$

13. $\sqrt{2x}\,\sqrt{6x} + \sqrt{15x}\,\sqrt{5x}$ $7x\sqrt{3}$

14. $\dfrac{\sqrt{x^2y}}{\sqrt{x}} - \sqrt{x}\,\sqrt{y}$ 0

15. $\dfrac{\sqrt{x}}{\sqrt{7}} + \sqrt{7x}$ $\dfrac{8\sqrt{7x}}{7}$

16. $3\sqrt{5}\,\sqrt{15} + 5\sqrt{27} - 6\sqrt{6}\,\sqrt{8}$ $6\sqrt{3}$

17. $\dfrac{3y\sqrt{x^2y}}{\sqrt{y}} + 3x\sqrt{y^2} - \dfrac{5x\sqrt{y^3}}{\sqrt{y}}$ xy

Multiply and simplify.

18. $\sqrt{2}(\sqrt{3} + \sqrt{2})$ $\sqrt{6} + 2$

19. $\sqrt{3}(\sqrt{27} - \sqrt{12})$ 3

20. $\sqrt{5}(\sqrt{7} - \sqrt{28})$ $-\sqrt{35}$

21. $\sqrt{5}(\sqrt{12} - \sqrt{48})$ $-2\sqrt{15}$

22. $\sqrt{7}(1 - \sqrt{7})$ $\sqrt{7} - 7$

23. $\sqrt{10}(\sqrt{5} + 1)$ $5\sqrt{2} + \sqrt{10}$

24. $\sqrt{5}\left(\sqrt{5} - \dfrac{1}{\sqrt{5}}\right)$ 4

25. $\sqrt{11}\left(\sqrt{11} + \dfrac{1}{\sqrt{11}}\right)$ 12

26. $\sqrt{18}\left(\sqrt{2} - \dfrac{1}{\sqrt{18}}\right)$ 5

27. $\sqrt{21}\left(\dfrac{\sqrt{3}}{\sqrt{7}} + \dfrac{1}{\sqrt{21}}\right)$ 4

28. $\sqrt{3}\left(\sqrt{6} + \dfrac{\sqrt{3}}{\sqrt{2}}\right)$ $\dfrac{9\sqrt{2}}{2}$

29. $\sqrt{2}\left(\sqrt{10} - \dfrac{\sqrt{2}}{\sqrt{5}}\right)$ $\dfrac{8\sqrt{5}}{5}$

30. $(\sqrt{7} + 1)(\sqrt{7} + 1)$
$8 + 2\sqrt{7}$

31. $(\sqrt{7} + 1)(\sqrt{7} - 1)$
6

32. $(\sqrt{3} - \sqrt{7})(\sqrt{3} - \sqrt{7})$
$10 - 2\sqrt{21}$

33. $(\sqrt{3} - \sqrt{7})(\sqrt{3} + \sqrt{7})$
-4

34. $(3\sqrt{5} + 1)(\sqrt{5} + 1)$
$16 + 4\sqrt{5}$

35. $(\sqrt{3} - 3)(\sqrt{3} + 1)$
$-2\sqrt{3}$

36. $(2\sqrt{2} - \sqrt{3})(3\sqrt{2} + \sqrt{3})$
$9 - \sqrt{6}$

37. $(2\sqrt{3} + \sqrt{5})(2\sqrt{3} - \sqrt{5})$
7

Rationalize the denominator.

38. $\dfrac{2}{\sqrt{3} - 1}$
$\sqrt{3} + 1$

39. $\dfrac{3}{\sqrt{5} - \sqrt{2}}$
$\sqrt{5} + \sqrt{2}$

40. $\dfrac{4}{\sqrt{7} + \sqrt{3}}$
$\sqrt{7} - \sqrt{3}$

41. $\dfrac{-2}{\sqrt{7} - 3}$
$\sqrt{7} + 3$

42. $\dfrac{6}{\sqrt{5} - \sqrt{7}}$
$-3\sqrt{5} - 3\sqrt{7}$

43. $\dfrac{9}{\sqrt{7} + 2}$
$3\sqrt{7} - 6$

44. $\dfrac{2\sqrt{3}}{\sqrt{3} + 1}$
$3 - \sqrt{3}$

45. $\dfrac{4\sqrt{5}}{\sqrt{5} - 1}$
$5 + \sqrt{5}$

46. $\dfrac{3\sqrt{5}}{\sqrt{5} - \sqrt{2}}$
$5 + \sqrt{10}$

47. $\dfrac{1 - x}{1 - \sqrt{x}}$
$1 + \sqrt{x}$

FOR REVIEW

Add or subtract.

48. $5\sqrt{125} + 3\sqrt{5} - 8\sqrt{45}$ $4\sqrt{5}$

49. $7\sqrt{12x} - 3\sqrt{48x} - 2\sqrt{75x}$ $-8\sqrt{3x}$

50. A meteorologist may use the formula $W = 5\sqrt{f} - \sqrt{g}$, relating the three quantities W, f, and g. Find W when $f = 72$ and $g = 8$. $28\sqrt{2}$

Exercises 51–56 review material from Chapter 11 and Section 15.5 to help you prepare for the next section. Solve each equation.

51. $6x = 54$ 9

52. $4 + x = 15$ 11

53. $5(x - 3) = 2x - 6$ 3

54. $4 - x^2 = x(3 - x) - 2$ 2

55. $x^2 + x - 6 = 0$ 2 or -3

56. $2x^2 - x = 0$ 0 or $\dfrac{1}{2}$

17.5 EXERCISES C

Multiply and simplify.

1. $(\sqrt{ax} - \sqrt{by})(\sqrt{ax} + \sqrt{by})$ $ax - by$

2. $(\sqrt{x} - \sqrt{y})(x + \sqrt{xy} + y)$
[Answer: $x\sqrt{x} - y\sqrt{y}$]

Rationalize the denominator and simplify.

3. 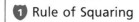 $\dfrac{x\sqrt{y} - y\sqrt{x}}{\sqrt{x} - \sqrt{y}}$ $\sqrt{xy}$

4. $\dfrac{x^2}{\sqrt{4 - x^2}} + \sqrt{4 - x^2}$ $\left[\text{Answer: } \dfrac{4\sqrt{4 - x^2}}{4 - x^2}\right]$

17.6 SOLVING RADICAL EQUATIONS

STUDENT GUIDEPOSTS

❶ Rule of Squaring

❷ Solving Equations Involving Radicals

❶ RULE OF SQUARING

Sometimes an equation contains a radical expression with a variable in the radicand, such as $\sqrt{x} = 5$. We use the following rule to solve such equations.

Rule of Squaring

If $a = b$,   then $a^2 = b^2$.

❷ SOLVING EQUATIONS INVOLVING RADICALS

When $\sqrt{x} = 5$, then the rule of squaring gives $(\sqrt{x})^2 = 5^2$ or $x = 25$. When this rule is applied to an equation with a variable, the resulting equation may have more solutions than the original. For instance, 2 is the only solution to the equation $x = 2$, but the equation formed by squaring both sides, $x^2 = 4$, has two solutions, 2 and -2. We call -2 an **extraneous root** in this case. Thus, when the rule of squaring is used, any possible solutions *must be* checked in the original equation, and extraneous roots must be discarded.

To Solve an Equation Involving Radicals

1. If only one radical is present, isolate this radical on one side of the equation and proceed to 3.
2. If two radicals are present, isolate one of the radicals on one side of the equation.
3. Square both sides to obtain an equation without the isolated radical expression(s).
4. Solve the resulting equation.
5. Check all possible solutions in the original equation.

EXAMPLE 1   SOLVING A RADICAL EQUATION

Solve.

$\sqrt{3x} - 5 = 7$

$\qquad \sqrt{3x} = 12$ Add 5 to both sides to isolate radical on the left

$\quad (\sqrt{3x})^2 = (12)^2$ Use the rule of squaring

$\qquad\quad 3x = 144$ Squaring

$\qquad\quad\ x = 48$ Divide both sides by 3

Check: $\sqrt{3(48)} - 5 \stackrel{?}{=} 7$

$\qquad\qquad \sqrt{144} - 5 \stackrel{?}{=} 7$

$\qquad\qquad\quad 12 - 5 \stackrel{?}{=} 7$

$\qquad\qquad\qquad\quad 7 = 7$

The solution is 48.

PRACTICE EXERCISE 1

Solve. $\sqrt{5y} + 2 = 12$

Answer: 20

EXAMPLE 2   SOLVING A RADICAL EQUATION

Solve.

$\sqrt{x + 3} + 4 = 11$

$\quad \sqrt{x + 3} = 11 - 4$ Isolate the radical

$\quad \sqrt{x + 3} = 7$

$\ (\sqrt{x + 3})^2 = 7^2$ Rule of squaring

$\qquad\ x + 3 = 49$

$\qquad\qquad x = 46$

Check: $\sqrt{46 + 3} + 4 \stackrel{?}{=} 11$

$\qquad\qquad \sqrt{49} + 4 \stackrel{?}{=} 11$

$\qquad\qquad\quad 7 + 4 \stackrel{?}{=} 11$

$\qquad\qquad\qquad 11 = 11$

The solution is 46.

PRACTICE EXERCISE 2

Solve. $\sqrt{z - 2} - 1 = 3$

Answer: 18

| **EXAMPLE 3** Equation with Two Radicals | **PRACTICE EXERCISE 3** |

Solve.

$$3\sqrt{2x-5} - \sqrt{x+23} = 0$$

$$3\sqrt{2x-5} = \sqrt{x+23} \qquad \text{Isolate the radicals}$$

$$(3\sqrt{2x-5})^2 = (\sqrt{x+23})^2$$

$$9(2x-5) = x+23 \qquad \text{Be sure to square the 3 on the left}$$

$$18x - 45 = x + 23$$

$$17x = 68$$

$$x = 4$$

Check: $\quad 3\sqrt{2 \cdot 4 - 5} - \sqrt{4+23} \stackrel{?}{=} 0$

$$3\sqrt{3} - \sqrt{27} \stackrel{?}{=} 0$$

$$3\sqrt{3} - 3\sqrt{3} = 0$$

The solution is 4.

Solve. $2\sqrt{3y+1} - \sqrt{y+15} = 0$

Answer: 1

| **EXAMPLE 4** Equation with No Solution | **PRACTICE EXERCISE 4** |

Solve.

$$\sqrt{x^2 - 5} - x + 5 = 0$$

$$\sqrt{x^2 - 5} = x - 5 \qquad \text{Isolate the radical}$$

$$(\sqrt{x^2 - 5})^2 = (x-5)^2 \qquad \text{Do \textit{not} square the right side}$$
$$\qquad\qquad\qquad\qquad\quad \text{as } x^2 - 25$$

$$x^2 - 5 = x^2 - 10x + 25$$

$$-5 = -10x + 25 \qquad \text{Subtract } x^2 \text{ from both sides}$$

$$-30 = -10x$$

$$x = \frac{-30}{-10} = 3$$

Check: $\quad \sqrt{3^2 - 5} - 3 + 5 \stackrel{?}{=} 0$

$$\sqrt{9 - 5} - 3 + 5 \stackrel{?}{=} 0$$

$$\sqrt{4} - 3 + 5 \stackrel{?}{=} 0$$

$$2 - 3 + 5 \stackrel{?}{=} 0 \qquad \sqrt{4} \text{ is 2 \textit{not} } -2$$

$$4 \neq 0 \qquad\qquad 3 \text{ does not check}$$

The equation has no solution.

Solve. $\sqrt{y^2 + 9} - y + 1 = 0$

Answer: no solution (-4 does not check)

CAUTION

Remember that the radical represents the principal (nonnegative) root. Thus, in Example 4, we cannot replace $\sqrt{4}$ with -2 in the check.

Sometimes the equation that results when the radical is eliminated must be solved by factoring and using the zero-product rule. We see this in Example 5.

| EXAMPLE 5 USING THE ZERO-PRODUCT RULE | PRACTICE EXERCISE 5 |

Solve.

$$\sqrt{2x + 2} - x + 3 = 0$$

$\sqrt{2x + 2} = x - 3$	Isolate the radical
$(\sqrt{2x + 2})^2 = (x - 3)^2$	
$2x + 2 = x^2 - 6x + 9$	$(a - b)^2 = a^2 - 2ab + b^2$
$0 = x^2 - 8x + 7$	Subtract 2x and 2
$0 = (x - 1)(x - 7)$	Factor
$x - 1 = 0 \quad \text{or} \quad x - 7 = 0$	Zero-product rule
$x = 1 \qquad\qquad x = 7$	

Check: $\quad \sqrt{2(1) + 2} - 1 + 3 \stackrel{?}{=} 0 \qquad \sqrt{2(7) + 2} - 7 + 3 \stackrel{?}{=} 0$

$$\sqrt{4} - 1 + 3 \stackrel{?}{=} 0 \qquad\qquad \sqrt{16} - 7 + 3 \stackrel{?}{=} 0$$

$$2 - 1 + 3 \stackrel{?}{=} 0 \qquad\qquad 4 - 7 + 3 \stackrel{?}{=} 0$$

$$4 \neq 0 \qquad\qquad\qquad 0 = 0$$

1 does not check. $\qquad$ 7 does check.

The only solution is 7.

Solve. $\sqrt{17 - 4x} - x - 1 = 0$

Answer: 2

17.6 EXERCISES A

Solve the following equations.

1. $\sqrt{x} = 4 \qquad$ 16

2 $3\sqrt{2x} - 9 = 0 \qquad \dfrac{9}{2}$

3. $\sqrt{4y} - 6 = 2 \qquad$ 16

4. $\sqrt{y + 2} = 6 \qquad$ 34

5. $\sqrt{4x - 3} = 5 \qquad$ 7

6 $3\sqrt{a - 3} + 6 = 0$
no solution

7. $4\sqrt{2a - 1} - 2 = 0 \qquad \dfrac{5}{8}$

8. $3\sqrt{x} = \sqrt{x + 16} \qquad$ 2

9. $\sqrt{y + 3} = \sqrt{y + 5}$
no solution

10. $3\sqrt{y + 3} = \sqrt{y + 35} \qquad$ 1

11. $\sqrt{3a + 2} - \sqrt{a + 8} = 0 \qquad$ 3

12. $\sqrt{5a - 3} - \sqrt{2a + 1} = 0$
$\dfrac{4}{3}$

13 $3\sqrt{x-1} - \sqrt{x+31} = 0$ 5

14. $\sqrt{2x+1} = \sqrt{x+10}$ 9

15. $3\sqrt{y+7} = 4\sqrt{y}$ 9

16. $2\sqrt{2x+5} - 3\sqrt{3x-2} = 0$ 2

17 $\sqrt{a^2-5} + a - 5 = 0$ 3

18. $\sqrt{a^2+2} - a - 2 = 0$ $-\dfrac{1}{2}$

19. $-\sqrt{x^2-12} + x + 6 = 0$ −4

20 $\sqrt{x^2+3} + x = 6$ $\dfrac{11}{4}$

21. $\sqrt{x^2+16} - x + 8 = 0$
no solution

22. $\sqrt{x^2-8} + 3 = x$ $\dfrac{17}{6}$

23. $\sqrt{x+2} = x$ 2

24. $\sqrt{x+1} = 1 - x$ 0

25 $\sqrt{x-3} - x + 5 = 0$ 7

26. $\sqrt{x+4} - x + 2 = 0$ 5

27. $\sqrt{x + 15} - x - 3 = 0$ $\quad$ 1 $\qquad$ **28** $\sqrt{x + 10} + x - 2 = 0$ $\quad$ −1

FOR REVIEW

Simplify.

29. $\sqrt{5}(\sqrt{15} - \sqrt{12})$ $\quad$ $5\sqrt{3} - 2\sqrt{15}$ $\qquad$ **30.** $(\sqrt{3} + \sqrt{5})(2\sqrt{3} - \sqrt{5})$ $\quad$ $1 + \sqrt{15}$

31. $\dfrac{4}{\sqrt{11} - \sqrt{7}}$ $\quad$ $\sqrt{11} + \sqrt{7}$ $\qquad$ **32.** $\dfrac{3\sqrt{5}}{\sqrt{5} - \sqrt{2}}$ $\quad$ $5 + \sqrt{10}$

33. If $c = \sqrt{a^2 + b^2}$, find c when $a = 4$ and $b = 3$. $\qquad$ **34.** If $b = \sqrt{c^2 - a^2}$, find b when $a = 5$ and $c = 13$.
5 $\qquad\qquad\qquad\qquad\qquad\qquad\qquad\qquad\qquad\qquad\qquad\qquad$ 12

ANSWERS: 1. 16 2. $\frac{9}{2}$ 3. 16 4. 34 5. 7 6. no solution 7. $\frac{5}{8}$ 8. 2 9. no solution 10. 1 11. 3 12. $\frac{4}{3}$
13. 5 14. 9 15. 9 16. 2 17. 3 18. $-\frac{1}{2}$ 19. −4 20. $\frac{11}{4}$ 21. no solution 22. $\frac{17}{6}$ 23. 2 24. 0 25. 7 26. 5
27. 1 28. −1 29. $5\sqrt{3} - 2\sqrt{15}$ 30. $1 + \sqrt{15}$ 31. $\sqrt{11} + \sqrt{7}$ 32. $5 + \sqrt{10}$ 33. 5 34. 12

17.6 EXERCISES B

Solve the following equations.

1. $\sqrt{x} = 9$ $\quad$ 81 $\qquad$ **2.** $5\sqrt{3x} - 10 = 0$ $\quad$ $\dfrac{4}{3}$ $\qquad$ **3.** $\sqrt{9y} - 6 = 9$ $\quad$ 25

4. $\sqrt{y + 6} = 2$ $\quad$ −2 $\qquad$ **5.** $\sqrt{8x + 2} = 2$ $\quad$ $\dfrac{1}{4}$ $\qquad$ **6.** $5\sqrt{a - 2} + 10 = 0$
$\qquad\qquad\qquad\qquad\qquad\qquad\qquad\qquad\qquad\qquad\qquad\qquad\qquad\qquad\qquad$ no solution

7. $3\sqrt{3a + 1} - 6 = 0$ $\quad$ 1 $\qquad$ **8.** $5\sqrt{x} = \sqrt{x + 3}$ $\quad$ $\dfrac{1}{8}$ $\qquad$ **9.** $\sqrt{2y + 1} = \sqrt{2y + 3}$
$\qquad\qquad\qquad\qquad\qquad\qquad\qquad\qquad\qquad\qquad\qquad\qquad\qquad\qquad\qquad$ no solution

10. $3\sqrt{y - 2} = \sqrt{y + 12}$ $\quad$ $\dfrac{15}{4}$ $\qquad$ **11.** $\sqrt{5a + 3} - \sqrt{a + 4} = 0$ $\quad$ $\dfrac{1}{4}$ $\qquad$ **12.** $\sqrt{7a - 10} - \sqrt{4a - 5} = 0$ $\quad$ $\dfrac{5}{3}$

13. $2\sqrt{x - 5} - \sqrt{x + 22} = 0$ $\quad$ 14 $\qquad$ **14.** $\sqrt{3x + 5} = \sqrt{x + 15}$ $\quad$ 5 $\qquad$ **15.** $3\sqrt{y + 2} = 5\sqrt{y}$ $\quad$ $\dfrac{9}{8}$

16. $3\sqrt{2x - 4} - 2\sqrt{3x + 3} = 0$ $\quad$ 8 $\qquad$ **17.** $\sqrt{a^2 + 5} - a + 5 = 0$ $\qquad$ **18.** $\sqrt{a^2 - 8} - a - 4 = 0$ $\quad$ −3
$\qquad\qquad\qquad\qquad\qquad\qquad\qquad\qquad\qquad$ no solution

19. $-\sqrt{x^2 - 20} + x + 10 = 0$ $\quad$ −6 $\qquad$ **20.** $\sqrt{x^2 + 7} - x + 7 = 0$ $\qquad$ **21.** $\sqrt{x^2 - 15} + x = 5$ $\quad$ 4
$\qquad\qquad\qquad\qquad\qquad\qquad\qquad\qquad\qquad\qquad\qquad\qquad$ no solution

22. $\sqrt{x^2 + 3} - 3 = x$ $\quad$ −1 $\qquad$ **23.** $\sqrt{x + 6} = x$ $\quad$ 3 $\qquad$ **24.** $\sqrt{x + 1} = x - 1$ $\quad$ 3

25. $\sqrt{x + 3} - x + 3 = 0$ $\quad$ 6 $\qquad$ **26.** $\sqrt{x + 6} - x - 4 = 0$ $\quad$ −2 $\qquad$ **27.** $\sqrt{x + 21} + x + 1 = 0$ $\quad$ −5

28. $\sqrt{x + 5} + x - 1 = 0$ $\quad$ −1

FOR REVIEW

Simplify.

29. $\sqrt{7}(\sqrt{14} - \sqrt{21})$ $7\sqrt{2} - 7\sqrt{3}$

30. $(\sqrt{5} - \sqrt{3})(\sqrt{5} - 2\sqrt{3})$ $11 - 3\sqrt{15}$

31. $\dfrac{4}{\sqrt{11} + \sqrt{7}}$ $\sqrt{11} - \sqrt{7}$

32. $\dfrac{4\sqrt{7}}{\sqrt{7} - \sqrt{5}}$ $14 + 2\sqrt{35}$

33. If $c = \sqrt{a^2 + b^2}$, find c if $a = 6$ and $b = 8$.
   10

34. If $a = \sqrt{c^2 - b^2}$, find a when $c = 7$ and $b = 5$.
   $2\sqrt{6}$

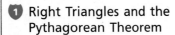

17.6   EXERCISES C

Solve.

1. $\sqrt{x + 5} - \sqrt{x - 3} = 2$ 4
   [*Hint:*   You will need to square twice.]

2. $\sqrt{x - 3} + \sqrt{x + 2} = 5$ 7

17.7   APPLICATIONS USING RADICALS

STUDENT GUIDEPOSTS

❶ Right Triangles and the Pythagorean Theorem

❷ 30°–60° and 45°–45° Right Triangles

❶ RIGHT TRIANGLES AND THE PYTHAGOREAN THEOREM

Many applied problems use radicals. For example, finding one side of a right triangle when the other two sides are given requires taking a square root. Remember that a **right triangle** is a triangle with a 90° (right) angle. The side opposite the 90° angle is the **hypotenuse** of the right triangle and the remaining two sides are its **legs.** We will agree to label the legs a and b and the hypotenuse c as in Figure 17.2. The next well-known theorem is named after the Greek mathematician Pythagoras.

Pythagorean Theorem
In a right triangle with legs a and b and hypotenuse c. $$a^2 + b^2 = c^2.$$

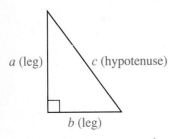

Figure 17.2 $a^2 + b^2 = c^2$

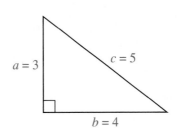

Figure 17.3

| **EXAMPLE 1** FINDING THE HYPOTENUSE | **PRACTICE EXERCISE 1** |

If the legs of a right triangle are 3 and 4, find the hypotenuse. See Figure 17.3.

By the Pythagorean theorem,

$$c^2 = a^2 + b^2$$
$$= 3^2 + 4^2 \quad \text{$a = 3$ and $b = 4$}$$
$$= 9 + 16 = 25.$$

We need to solve $c^2 = 25$.

$$c^2 = 25$$
$$c^2 - 25 = 0 \quad \text{Subtract 25}$$
$$c^2 - 5^2 = 0 \quad \text{$25 = 5^2$}$$
$$(c + 5)(c - 5) = 0 \quad \text{Factor using $a^2 - b^2 = (a + b)(a - b)$}$$
$$c + 5 = 0 \quad \text{or} \quad c - 5 = 0$$
$$c = -5 \qquad c = 5$$

Since we want the length of the hypotenuse, we discard -5; $c = 5$. Notice that

$$c^2 = 25$$
$$c = \sqrt{25} = 5.$$

That is, if we take the principal square root of both sides of the equation, the result is 5. This technique is useful when the equation cannot be factored.

If the legs of a right triangle are 9 and 12, find the hypotenuse.

Answer: 15

| **EXAMPLE 2** FINDING A LEG OF THE TRIANGLE | **PRACTICE EXERCISE 2** |

If $c = 7$ and $a = 3$, find b. See Figure 17.4.

If $c = 9$ and $b = 3$, find a.

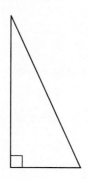

Figure 17.4

$$a^2 + b^2 = c^2 \qquad \text{Pythagorean theorem}$$
$$3^2 + b^2 = 7^2 \qquad \text{$a = 3$ and $c = 7$}$$
$$9 + b^2 = 49$$
$$b^2 = 40$$
$$b = \sqrt{40} \qquad \text{Take the principal square root on both sides}$$
$$b = \sqrt{4 \cdot 10} = 2\sqrt{10}$$

Answer: $6\sqrt{2}$

EXAMPLE 3 Application of Radicals	**PRACTICE EXERCISE 3**

The base of a 20-ft ladder used at a construction site is 8 ft from the base of a wall. How far is the top of the ladder from the ground? See Figure 17.5.

$$a^2 + b^2 = c^2 \qquad \text{Pythagorean theorem}$$
$$a^2 + 8^2 = 20^2 \qquad b = 8 \text{ and } c = 20$$
$$a^2 + 64 = 400$$
$$a^2 = 336$$
$$a = \sqrt{336} \qquad \text{Take the principal square}$$
$$ = \sqrt{16 \cdot 21} \qquad \text{root on both sides}$$
$$ = 4\sqrt{21}$$

A guy wire from the top of a pole to the ground measures 45 ft. If the pole is 30 ft in length, how far is the wire anchored from the base of the pole?

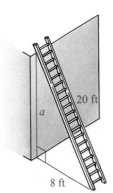

20 ft
a

8 ft

Figure 17.5

The top of the ladder is $4\sqrt{21}$ ft from the ground. It is usually practical to give an approximate solution. Using a calculator, we find that $\sqrt{21} \approx 4.58$, so that $4\sqrt{21} \approx 18.3$. Thus, the ladder reaches about 18.3 ft up the wall.

Answer: $15\sqrt{5} \approx 33.5$ ft

Applications from the business world, such as the next one, use square roots.

EXAMPLE 4 Application to Business	**PRACTICE EXERCISE 4**

A manufacturer of novelty items has found that his total daily expenses for production are given by the equation

$$e = 10\sqrt{n} + 25,$$

where e represents total expenses and n is the number of items produced.

(a) Find the total expenses when no items are produced. This is called **overhead.**

$$e = 10\sqrt{n} + 25 \qquad \text{Expense equation}$$
$$ = 10\sqrt{0} + 25 \qquad \text{Substitute 0 for } n$$
$$ = 10\sqrt{25}$$
$$ = 10 \cdot 5 = 50$$

Overhead is $50 per day.

A manufacturer of puzzles has found that his daily expenses for production can be estimated by $e = 6\sqrt{n} + 49$, where e represents total expenses and n is the number of puzzles produced.

(a) Find the total expenses when no puzzles are produced.

(b) Find the total expenses when 600 items are produced.

$$e = 10\sqrt{n + 25} \qquad \text{Expense equation}$$
$$= 10\sqrt{600 + 25} \qquad \text{Substitute 600 for } n$$
$$= 10\sqrt{625}$$
$$= 10(25) = 250 \qquad (25)^2 = 625$$

Total expenses are $250 per day when 600 items are produced.

(c) How many items are made when the cost is $120 per day?

$$e = 10\sqrt{n + 25} \qquad \text{Expense equation}$$
$$120 = 10\sqrt{n + 25} \qquad \text{We solve for } n \text{ when } e \text{ is } 120$$
$$12 = \sqrt{n + 25} \qquad \text{To simplify, divide both sides by 10}$$
$$(12)^2 = (\sqrt{n + 25})^2 \qquad \text{Square both sides}$$
$$144 = n + 25$$
$$119 = n$$

Thus, 119 items are made on the day when expenses total $120.

(b) Find the total expense of producing 95 puzzles.

(c) How many puzzles were made on a day when the expenses were $78.

Answers: (a) $42   (b) $72
(c) 120

EXAMPLE 5   APPLICATION TO AERONAUTICS

If an airplane is flying above the earth, an equation that gives the approximate distance to the horizon in terms of the altitude of the plane is

$$d = \sqrt{8000h},$$

where d is the distance to the horizon in miles and h is the altitude (distance above the earth) in miles. See Figure 17.6.

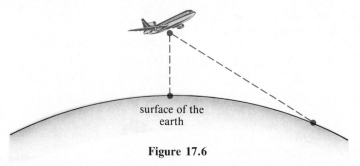

surface of the earth

Figure 17.6

(a) Give the distance to the horizon when viewed from a plane at an altitude of 5 miles.

$$d = \sqrt{8000h} \qquad \text{Horizon equation}$$
$$d = \sqrt{(8000)(5)} \qquad h = 5 \text{ miles}$$
$$d = \sqrt{40000}$$
$$= \sqrt{(200)^2} \qquad (200)^2 = 40{,}000$$
$$= 200$$

The horizon is approximately 200 miles away.

PRACTICE EXERCISE 5

Use the equation $d = \sqrt{8000h}$ in the following:

(a) Give the distance to the horizon when viewed from a balloon at an altitude of 1 mile.

(b) Find the altitude of a plane when the horizon is 50 miles away.

$$d = \sqrt{8000h} \qquad \text{Horizon equation}$$
$$50 = \sqrt{8000h} \qquad d = 50 \text{ miles, solve for } h$$
$$(50)^2 = (\sqrt{8000h})^2 \qquad \text{Square both sides}$$
$$2500 = 8000h$$
$$\frac{2500}{8000} = h$$
$$\frac{5}{16} = h$$

The plane is about $\frac{5}{16}$ of a mile (about 1650 ft) above the earth.

(b) Find the altitude of a balloon when the horizon is 150 miles away.

Answers: (a) $40\sqrt{5} \approx 89.4$ mi
(b) approximately 2.8 mi

2 30°–60° AND 45°–45° RIGHT TRIANGLES

To conclude this section we look at two special right triangles which have several applications in mathematics. A right triangle in which the two acute angles have measure 30° and 60° is called a **30°–60° right triangle.** The sides in such a triangle satisfy the following property.

> ### Property of 30°–60° Right Triangle
>
> In a 30°–60° right triangle, the length of the leg opposite the 30° angle is one-half the length of the hypotenuse.

Consider the right triangle in Figure 17.7. Since the hypotenuse c is 2, a must be 1 (one-half of c). We can find b using one of the equivalent forms of the Pythagorean theorem.

$$b^2 = c^2 - a^2$$
$$= 2^2 - 1^2$$
$$= 4 - 1$$
$$= 3$$
$$b = \sqrt{3} \qquad \text{Take the principal square root of both sides}$$

In general, if a is the side opposite the 30° angle in a right triangle, $c = 2a$ and $b = \sqrt{3}a$.

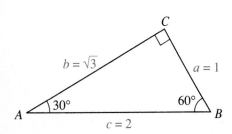

Figure 17.7 30°–60° Right Triangle

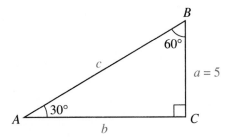

Figure 17.8

| EXAMPLE 6   SOLVING A 30°–60° RIGHT TRIANGLE | PRACTICE EXERCISE 6 |

In right $\triangle ABC$, $\angle C$ is the right angle, $\angle A$ has measure 30°, and $a = 5$. Find c and b.

First sketch the triangle as in Figure 17.8. Since $a = 5$ is one-half hypotenuse, c must be 10. Also $b = \sqrt{3}a = \sqrt{3}(5) = 5\sqrt{3}$. We reach the same result using the Pythagorean theorem.

$$b^2 = c^2 - a^2$$
$$= 10^2 - 5^2$$
$$= 100 - 25 = 75$$
$$b = \sqrt{75} = \sqrt{25 \cdot 3} = 5\sqrt{3}$$

In right $\triangle ABC$, $\angle C$ is the right angle, $\angle B$ has measure 60°, and $c = 6$. Find a and b.

Answer: $a = 3$, $b = 3\sqrt{3}$

A right triangle with both acute angles measuring 45° is called a **45°–45° right triangle.** Since a 45°–45° right triangle is isosceles, the legs have equal length, and the Pythagorean theorem can be used to find the length of the hypotenuse. In Figure 17.9, since a and b are both 1, and

$$c^2 = a^2 + b^2$$
$$= 1^2 + 1^2 = 1 + 1 = 2,$$

we have that $c = \sqrt{2}$.

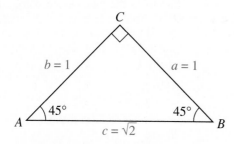

Figure 17.9   45°–45° Right Triangle

In general, if the legs are of length a then the hypotenuse is $\sqrt{2}a$.

17.7   EXERCISES A

Find the unknown leg or hypotenuse.

1. $a = 12$, $b = 5$ 13

2. $a = 7$, $c = 10$ $\sqrt{51}$

3. $b = 5$, $c = 9$ $2\sqrt{14}$

4. $a = 6$, $b = 8$ 10

5. $a = 1$, $c = 5$ $2\sqrt{6}$

6. $b = 10$, $c = 20$ $10\sqrt{3}$

Solve.

7. A ladder 8 m long is placed on a building. If the base is 3 m from the building, how high up the side of the building is the top of the ladder?

$\sqrt{55}$ m $\approx$ 7.4 m

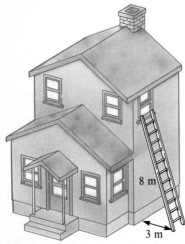

8 m

3 m

8. The diagonal of a square is 12 centimeters. Find the length of the sides.   [*Hint:* $a^2 + a^2 = 12^2$.]

$6\sqrt{2}$ m $\approx$ 8.5 cm

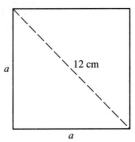

a

12 cm

a

9. Find the length of the diagonal of a rectangle if the sides are 20 inches and 30 inches.

$10\sqrt{13}$ in $\approx$ 36.1 in

10 How long must a wire be to reach from the top of a 40-foot telephone pole to a point on the ground 30 feet from the base of the pole?   **50 ft**

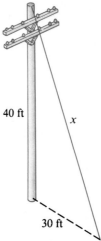

40 ft

x

30 ft

11. Three times the square root of a number is 6. Find the number.   **4**

12. Twice the square root of a number is 6 less than 12. Find the number.   **9**

13 The square root of 5 more than a number is twice the square root of the number. Find the number.

$\dfrac{5}{3}$

14. Twice the square root of one more than a number is the same as the square root of 13 more than the number. Find the number.   **3**

A manufacturer finds that the total cost per day of producing appliances is given by the equation $c = 100\sqrt{n} + 36$ where c is the total cost and n is the number produced. Use this information for Exercises 15–18.

15. What is c when no appliances are produced (overhead cost)?   **$600**

16 What is the cost when 64 appliances are produced?   **$1000**

17. What is the cost when 39 appliances are produced?   **500\sqrt{3}$ ≈ $866**

18. How many appliances were made on a day when the cost was $2000?   **364**

A retailer finds that the total cost per day of selling sportcoats is given by the equation $c = 10\sqrt{n} + 100$, where c is the total cost and n is the number sold. Use this information in Exercises 19–22.

19. Find c when no sportcoats are sold (overhead cost).   **$100**

20. Find the cost when 64 sportcoats are sold.   **$180**

21. Find the number of coats sold on a day when the cost was $240.   **196**

22 Find the number of coats sold on a day when the cost was $50.   **no solution**

The distance in miles to the horizon from a point h miles above the earth is given by $d = \sqrt{8000h}$. Use this information in Exercises 23–26.

23. Find d when the altitude is 20 miles.   **400 mi**

24. Find the distance to the horizon when the altitude is $\frac{1}{2}$ mile.   **20$\sqrt{10}$ mi ≈ 63.2 mi**

25. Find the altitude when the distance to the horizon is 120 miles.   **1.8 mi**

26. Find the altitude when the distance to the horizon is 20 miles.   **0.05 mi (264 ft)**

Use the Pythagorean theorem to find the length of the missing legs or hypotenuse in each triangle.

27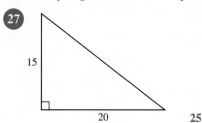
25

28.

29. 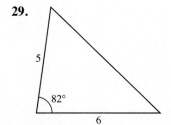   **not a right triangle**

30.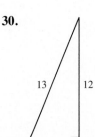

Use the properties of 30°–60° or 45°–45° right triangles to find the missing legs or hypotenuse.

31.

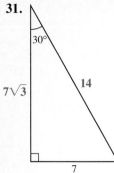

32.

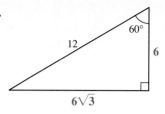

33.

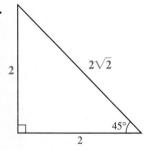

34.

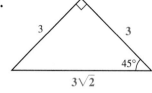

FOR REVIEW

Solve.

35. $\sqrt{3x + 1} = 5$ 8

36. $\sqrt{4x - 5} - \sqrt{2x + 9} = 0$ 7

37. $2\sqrt{y^2 + 1} - 2y - 1 = 0$ $\dfrac{3}{4}$

38. $2\sqrt{3y + 1} - \sqrt{15y - 11} = 0$ 5

ANSWERS: 1. 13 2. $\sqrt{51}$ 3. $2\sqrt{14}$ 4. 10 5. $2\sqrt{6}$ 6. $10\sqrt{3}$ 7. $\sqrt{55}$ m (approximately 7.4 m) 8. $6\sqrt{2}$ cm (approximately 8.5 cm) 9. $10\sqrt{13}$ in (approximately 36.1 in) 10. 50 ft 11. 4 12. 9 13. $\frac{5}{3}$ 14. 3 15. $600 16. $1000 17. $500\sqrt{3}$ (approximately $866) 18. 364 19. $100 20. $180 21. 196 22. no solution 23. 400 mi 24. $20\sqrt{10}$ mi (approximately 63.2 mi) 25. 1.8 mi 26. 0.05 mi (264 ft) 27. 25 28. 24 29. The Pythagorean theorem does not apply since the triangle is not a right triangle. 30. 5 31. hypotenuse 14; leg $7\sqrt{3}$ 32. 6; $6\sqrt{3}$ 33. hypotenuse $2\sqrt{2}$; leg 2 34. hypotenuse $3\sqrt{2}$; leg 3 35. 8 36. 7 37. $\frac{3}{4}$ 38. 5

17.7 EXERCISES B

Find the unknown leg or hypotenuse.

1. $a = 4$, $b = 4$ $4\sqrt{2}$

2. $a = 5$, $c = 8$ $\sqrt{39}$

3. $b = 14$, $c = 20$ $2\sqrt{51}$

4. $a = 9$, $b = 12$ 15

5. $a = 3$, $c = 7$ $2\sqrt{10}$

6. $b = 20$, $c = 30$ $10\sqrt{5}$

Solve.

7. How long would a ladder need to be to reach the top of a 10-m building if the base of the ladder is 4 m from the building?
$2\sqrt{29}$ m $\approx$ 10.8 m

8. The diagonal of a square is 10 cm. Find the length of the sides.
$5\sqrt{2}$ cm $\approx$ 7.07 cm

9. Find the length of the diagonal of a rectangular pasture if the sides are 40 yards and 60 yards.
$20\sqrt{13}$ yd $\approx$ 72.1 yd

10. A wire 50 feet long is attached to the top of a tower and to the ground 20 feet from the base of the tower. How tall is the tower?
$10\sqrt{21}$ ft $\approx$ 45.8 ft

11. Four times the square root of a number is 20. Find the number.   **25**

12. Twice the square root of a number is 5 more than 15. Find the number.   **100**

13. The square root of 8 more than a number is 3 times the square root of the number. Find the number.
1

14. Twice the square root of 4 less than a number is the same as the square root of 1 less than the number. Find the number.   **5**

A manufacturer finds that the total cost per day of producing dresses is given by the equation $c = 20\sqrt{n} + 16$, where c is the total cost and n is the number produced. Use this information in Exercises 15–18.

15. What is c when no dresses are produced (overhead cost)?   **$80**

16. What is the cost when 84 dresses are produced?
$200

17. What is the cost when 24 dresses are produced?
$40\sqrt{10} \approx$ $126.49

18. How many dresses were made on a day when the cost was $360?   **308**

A retailer finds that the total cost per day of selling widgets is given by the equation $c = 50\sqrt{n} + 200$, where c is the total cost and n is the number sold. Use this information for Exercises 19–22.

19. Find c when no widgets are sold (overhead cost).
$200

20. Find the cost when 81 widgets are sold.
$650

21. Find the number of widgets sold on a day when the cost was $950.   **225**

22. Find the number of widgets sold on a day when the cost was $150.   **no solution**

The distance d in miles to the horizon from a point h miles above the earth is given by $d = 10\sqrt{80h}$. Use this information in Exercises 23–26.

23. Find d when the altitude is $\frac{1}{5}$ mile.   **40 mi**

24. Find the distance to the horizon when the altitude is 10 miles.   **$200\sqrt{2}$ mi $\approx$ 283 mi**

25. Find the altitude when the distance to the horizon is 320 miles.   **12.8 mi**

26. Find the altitude when the distance to the horizon is 8 miles.   **0.008 mi $\approx$ 42 ft**

Use the Pythagorean theorem to find the length of the missing legs or hypotenuse in each triangle.

27.
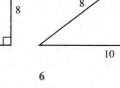
15

28.
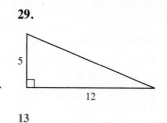
6

29.

5

12

13

30.
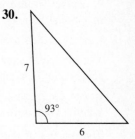

not a right triangle

Use the properties of 30°–60° or 45°–45° right triangles to find the missing legs or hypotenuse.

31. 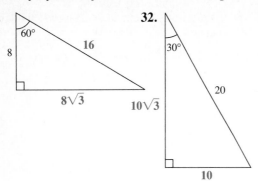 **32.**

33. 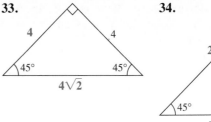 **34.**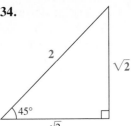

FOR REVIEW

Solve.

35. $\sqrt{2x - 3} = 3$ 6

36. $\sqrt{3x + 2} - \sqrt{2x + 3} = 0$ 1

37. $\sqrt{y^2 + 2} - y - 1 = 0$ $\frac{1}{2}$

38. $\sqrt{2y + 1} + \sqrt{y + 1} = 0$ no solution

17.7 EXERCISES C

Solve.

1. An airplane flew 120 km north from an airport and then 200 km west. How far was the plane from the airport? $40\sqrt{34}$ **km ≈ 233 km**

2. A scientist uses the formula $a = \sqrt{1 + \frac{b}{x}}$. Solve this equation for b and find b when $a = 3$ and $x = \frac{1}{2}$. [Answer: $b = x(a^2 - 1)$; 4]

CHAPTER 17 REVIEW

KEY WORDS

17.1 A **perfect square** is the square of an integer or rational number.

If $a^2 = x$, then a is a **square root** of x.

The **principal square root** is the nonnegative square root of a positive real number.

In $\sqrt{x}$ the symbol $\sqrt{\ }$ is called a **radical** and x is the **radicand.**

17.2 The process of removing radicals from the denominator is called **rationalizing the denominator.**

17.4 **Like radicals** have the same radicand.

Unlike radicals have different radicands.

17.5 The expressions $\sqrt{a} + \sqrt{b}$ and $\sqrt{a} - \sqrt{b}$ are **conjugates** of each other.

17.6 An **extraneous root** is a "solution" obtained which is not a solution to the original problem.

17.7 A **right triangle** is a triangle with a 90° angle.

The **hypotenuse** is the side opposite the right angle and the **legs** are the other sides of a right triangle.

KEY CONCEPTS

17.1 The radical alone indicates the principal (non-negative) square root of a number. Thus $\sqrt{9} = 3$, not -3.

17.2, 17.3 **1.** If $a \geq 0$ and $b \geq 0$, then

$$\sqrt{ab} = \sqrt{a}\,\sqrt{b}. \qquad \text{Multiplication rule}$$

2. If $a \geq 0$ and $b > 0$, then

$$\sqrt{\frac{a}{b}} = \frac{\sqrt{a}}{\sqrt{b}}. \qquad \text{Division rule}$$

3. When removing a perfect square from under a radical, multiply all factors; do not add. For example,

$$3\sqrt{4a} = 3\sqrt{4}\,\sqrt{a} = 3 \cdot 2\sqrt{a} = 6\sqrt{a}, \textit{ not } (3 + 2)\sqrt{a}, \text{ which is } 5\sqrt{a}.$$

17.4 $\sqrt{a + b} \neq \sqrt{a} + \sqrt{b}$ and $\sqrt{a - b} \neq \sqrt{a} - \sqrt{b}$.

17.5 To rationalize a binomial denominator, multiply both numerator and denominator of the fraction by the conjugate of the denominator and simplify.

17.6 **1.** To solve a radical equation, isolate a radical on one side of the equation and square both sides.

2. Check all solutions to a radical equation in the *original* equation.

17.7 **1.** Pythagorean theorem: In a right triangle with legs a and b and hypotenuse c, $a^2 + b^2 = c^2$.

2. In the 30°–60° triangle below,

$$c = 2a \text{ and } b = \sqrt{3}a.$$

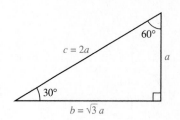

3. In the 45°–45° triangle below,

$$c = \sqrt{2}a \text{ and } b = a.$$

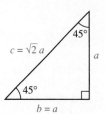

REVIEW EXERCISES

Part I

17.1 *Give the number of square roots of each number.*

1. 81   2

2. -81 0

3. 0   1

Evaluate each square root. Assume all variables and expressions are nonnegative.

4. $\sqrt{16}$ 4

5. $\sqrt{9^2}$ 9

6. $\sqrt{(-9)^2}$ 9

7. $\pm\sqrt{144}$ ± 12

8. $\sqrt{\dfrac{25}{16}}$ $\dfrac{5}{4}$

9. $\sqrt{-16}$
    meaningless

10. $\sqrt{0}$ 0

11. $\sqrt{4a^2}$ $2a$

12. $\sqrt{x^2 + 2x + 1}$ $x + 1$

17.2 *Simplify each of the radicals.*

13. $\sqrt{50}$ $\quad 5\sqrt{2}$

14. $5\sqrt{8}$ $\quad 10\sqrt{2}$

15. $\sqrt{3^4}$ $\quad 9$

16. $\sqrt{16x^3}$ $\quad 4x\sqrt{x}$

17. $\sqrt{8y^2}$ $\quad 2y\sqrt{2}$

18. $\sqrt{50x^2y^3}$ $\quad 5xy\sqrt{2y}$

19. $\sqrt{\dfrac{32}{81}}$ $\quad \dfrac{4\sqrt{2}}{9}$

20. $\sqrt{\dfrac{50}{8}}$ $\quad \dfrac{5}{2}$

21. $\sqrt{\dfrac{16y^2}{x^2}}$ $\quad \dfrac{4y}{x}$

22. $\sqrt{\dfrac{x^2y^3}{81}}$ $\quad \dfrac{xy\sqrt{y}}{9}$

23. $\dfrac{6}{\sqrt{7}}$ $\quad \dfrac{6\sqrt{7}}{7}$

24. $\sqrt{\dfrac{50x^2}{y}}$ $\quad \dfrac{5x\sqrt{2y}}{y}$

17.3 *Multiply and simplify.*

25. $\sqrt{8}\ \sqrt{18}$ $\quad 12$

26. $\sqrt{7}\ \sqrt{7}$ $\quad 7$

27. $3\sqrt{5}\ \sqrt{7}$ $\quad 3\sqrt{35}$

28. $\sqrt{12y}\ \sqrt{3y}$ $\quad 6y$

29. $\sqrt{3xy}\ \sqrt{2xy^2}$ $\quad xy\sqrt{6y}$

30. $\sqrt{5x}\ \sqrt{5y}\ \sqrt{5xy}$ $\quad 5xy\sqrt{5}$

Divide and simplify.

31. $\dfrac{\sqrt{99}}{\sqrt{11}}$ $\quad 3$

32. $\dfrac{\sqrt{50}}{\sqrt{8}}$ $\quad \dfrac{5}{2}$

33. $\dfrac{3\sqrt{35}}{\sqrt{20}}$ $\quad \dfrac{3\sqrt{7}}{2}$

34. $\dfrac{\sqrt{25x^2}}{\sqrt{2x^2}}$ $\dfrac{5\sqrt{2}}{2}$ **35.** $\dfrac{\sqrt{18x^2y}}{\sqrt{8y}}$ $\dfrac{3x}{2}$ **36.** $\dfrac{\sqrt{25x^2}}{\sqrt{3y}}$ $\dfrac{5x\sqrt{3y}}{3y}$

17.4 *Add or subtract.*

37. $5\sqrt{7} + 12\sqrt{7}$ $17\sqrt{7}$ **38.** $2\sqrt{5} - 7\sqrt{5}$ $-5\sqrt{5}$ **39.** $7\sqrt{5} - 3\sqrt{3}$ $7\sqrt{5} - 3\sqrt{3}$

40. $3\sqrt{8} + 7\sqrt{8}$ $20\sqrt{2}$ **41.** $\sqrt{49} - \sqrt{144}$ -5 **42.** $3\sqrt{16x} + 4\sqrt{36x}$
                                                                                            $36\sqrt{x}$

43. $\sqrt{50} - \sqrt{98} + \sqrt{32}$ $2\sqrt{2}$ **44.** $2\sqrt{28} - 5\sqrt{112} + 3\sqrt{63}$ **45.** $3\sqrt{12x} + 2\sqrt{27x} - \sqrt{48x}$
                                                                $-7\sqrt{7}$ $8\sqrt{3x}$

46. $5\sqrt{48x^2} - \sqrt{27x^2} - \sqrt{3x^2}$ **47.** $\sqrt{5} + \dfrac{1}{\sqrt{5}}$ $\dfrac{6\sqrt{5}}{5}$ **48.** $\sqrt{\dfrac{3}{2}} - \sqrt{6}$ $-\dfrac{\sqrt{6}}{2}$
   $16x\sqrt{3}$

17.5 *Perform the indicated operations and simplify.*

49. $\sqrt{5}\,\sqrt{125} + \sqrt{3}\,\sqrt{27}$ 34 **50.** $\dfrac{\sqrt{5}}{\sqrt{2}} - \dfrac{\sqrt{3}}{\sqrt{2}}$ $\dfrac{\sqrt{10} - \sqrt{6}}{2}$ **51.** $\sqrt{5}(\sqrt{3} - \sqrt{5})$ $\sqrt{15} - 5$

52. $\sqrt{7}\left(\sqrt{7} - \dfrac{1}{\sqrt{7}}\right)$ 6

53. $(\sqrt{7} + 1)(\sqrt{7} + 1)$

$8 + 2\sqrt{7}$

54. $(\sqrt{7} + 1)(\sqrt{7} - 1)$ 6

55. $(\sqrt{2} + \sqrt{3})(2\sqrt{2} - \sqrt{3})$ $1 + \sqrt{6}$

56. $(\sqrt{5} + 3)(\sqrt{5} - 1)$ $2 + 2\sqrt{5}$

Rationalize the denominator.

57. $\dfrac{2}{\sqrt{3} + 1}$

$\sqrt{3} - 1$

58. $\dfrac{-4}{\sqrt{7} - \sqrt{5}}$

$-2\sqrt{7} - 2\sqrt{5}$

59. $\dfrac{\sqrt{3}}{\sqrt{3} - \sqrt{2}}$

$3 + \sqrt{6}$

60. $\dfrac{\sqrt{7}}{\sqrt{7} - 1}$

$\dfrac{7 + \sqrt{7}}{6}$

17.6 *Solve.*

61. $2\sqrt{x} = \sqrt{x + 2}$ $\dfrac{2}{3}$

62. $\sqrt{7a - 1} - \sqrt{5a + 3} = 0$ 2

63. $\sqrt{x^2 - 13} - x + 1 = 0$ 7

64. $\sqrt{y + 5} + y - 1 = 0$ -1

17.7 *Find the unknown leg or hypotenuse.*

65. $a = 6$, $b = 12$ $6\sqrt{5}$

66. $a = 5$, $c = 13$ 12

Solve.

67. Find the length of a diagonal of a rectangular table with sides 8 feet and 12 feet.
$4\sqrt{13}$ ft ≈ 14.4 ft

68. Twice the square root of 4 less than a number is equal to the square root of 8 more than the number. Find the number. 8

A manufacturer finds that the total cost per day of producing wabbles is given by the equation $c = 500\sqrt{n} + 144$, where c is the total cost and n is the number produced. Use this information in Exercises 69–70.

69. Find the cost when 25 wabbles are produced.
$6500

70. How many wabbles were produced on a day when the cost was $10,000?   256

🖩 *The distance d in miles to the horizon from a point h miles above the earth is given by $d = 10\sqrt{80h}$. Use this information in Exercises 71–72.*

71. Find the distance to the horizon when the altitude is 0.05 miles.   **20 mi**

72. Find the altitude when the distance to the horizon is 80 miles.   **0.8 mi**

73. In a 30°–60° right triangle, a is the shorter leg and $a = 10$. Find b and c. $b = 10\sqrt{3}; c = 20$

74. In a 45°–45° right triangle, $a = 7$. Find b and c. $b = 7; c = 7\sqrt{2}$

Part II

In Exercises 75–86, perform the indicated operations and simplify. Rationalize all denominators.

75. $(\sqrt{6} - \sqrt{2})(\sqrt{6} + \sqrt{2})$ 4

76. $\sqrt{x} + \dfrac{x}{\sqrt{x}}$ $2\sqrt{x}$

77. $\sqrt{5xy}\ \sqrt{10x^2y}$ $5xy\sqrt{2x}$

78. $\sqrt{y^7}$ $y^3\sqrt{y}$

79. $\sqrt{\dfrac{14x^2y^3}{7xy}}$ $y\sqrt{2x}$

80. $\dfrac{3}{\sqrt{5}}$ $\dfrac{3\sqrt{5}}{5}$

81. $\dfrac{5}{\sqrt{7} - \sqrt{2}}$ $\sqrt{7} + \sqrt{2}$

82. $6x\sqrt{8y^2} - 4xy\sqrt{2}$ $8xy\sqrt{2}$

83. $(\sqrt{2} - \sqrt{5})(3\sqrt{2} + \sqrt{5})$
$1 - 2\sqrt{10}$

84. $\sqrt{11}\left(\sqrt{11} + \dfrac{1}{\sqrt{11}}\right)$ 12

85. $\dfrac{\sqrt{32x^2y^5}}{\sqrt{2xy^3}}$ $4y\sqrt{x}$

86. $\sqrt{169} - \sqrt{196}$ -1

The pressure *p* in pounds per square foot of a wind blowing against a flat surface is related to the velocity of the wind by the formula

$$v = \sqrt{\frac{p}{0.003}}$$

where v is in miles per hour. Use this information in Exercises 87–90.

87. A pressure gauge on a building registers a wind pressure of 10.8 lb/ft². What is the velocity of the wind?   **60 mph**

88. During a storm, the wind pressure against the side of a mobile home measured 14.7 lb/ft². What was the velocity of the wind?   **70 mph**

89. What is the pressure of the wind blowing 50 mph against the side of a bridge?   **7.5 lb/ft²**

90. If the wind is blowing 15 mph, what pressure does it exert against the side of an apartment building?   **0.675 lb/ft²**

91. $\sqrt{3x + 1} - 5 = 2$ **16**

92. $\sqrt{y - 1} - y + 3 = 0$ **5**

93. If *a* is the shorter of the two legs of a 30°–60° right triangle and *a* = 12 cm, find *b* and *c*.
$b = 12\sqrt{3}$ **cm;** *c* = **24 cm**

94. If the hypotenuse of a 45°–45° right triangle is $7\sqrt{2}$ ft, find the length of each leg.   **7 ft**

ANSWERS: 1. 2   2. 0   3. 1   4. 4   5. 9   6. 9   7. ±12 (12 and −12)   8. $\frac{5}{4}$ 9. meaningless   10. 0   11. 2*a*

12. *x* + 1   13. 5$\sqrt{2}$ 14. 10$\sqrt{2}$ 15. 9   16. 4*x*$\sqrt{x}$ 17. 2*y*$\sqrt{2}$ 18. 5*xy*$\sqrt{2y}$ 19. $\frac{4\sqrt{2}}{9}$ 20. $\frac{5}{2}$ 21. $\frac{4y}{x}$ 22. $\frac{xy\sqrt{y}}{9}$

23. $\frac{6\sqrt{7}}{7}$ 24. $\frac{5x\sqrt{2y}}{y}$ 25. 12   26. 7   27. 3$\sqrt{35}$ 28. 6*y*   29. *xy*$\sqrt{6y}$ 30. 5*xy*$\sqrt{5}$ 31. 3   32. $\frac{5}{2}$ 33. $\frac{3\sqrt{7}}{2}$

34. $\frac{5\sqrt{2}}{2}$ 35. $\frac{3x}{2}$ 36. $\frac{5x\sqrt{3y}}{3y}$ 37. 17$\sqrt{7}$ 38. −5$\sqrt{5}$ 39. 7$\sqrt{5}$ − 3$\sqrt{3}$ 40. 20$\sqrt{2}$ 41. −5   42. 36$\sqrt{x}$ 43. 2$\sqrt{2}$

44. −7$\sqrt{7}$ 45. 8$\sqrt{3x}$ 46. 16*x*$\sqrt{3}$ 47. $\frac{6\sqrt{5}}{5}$ 48. −$\frac{\sqrt{6}}{2}$ 49. 34   50. $\frac{\sqrt{10} - \sqrt{6}}{2}$ 51. $\sqrt{15}$ − 5   52. 6

53. 8 + 2$\sqrt{7}$ 54. 6   55. 1 + $\sqrt{6}$ 56. 2 + 2$\sqrt{5}$ 57. $\sqrt{3}$ − 1   58. −2$\sqrt{7}$ − 2$\sqrt{5}$ 59. 3 + $\sqrt{6}$ 60. $\frac{7 + \sqrt{7}}{6}$ 61. $\frac{2}{3}$

62. 2   63. 7   64. −1   65. 6$\sqrt{5}$ 66. 12   67. 4$\sqrt{13}$ ft (approximately 14.4 ft)   68. 8   69. $6500   70. 256

71. 20 mi   72. 0.8 mi   73. *b* = 10$\sqrt{3}$; *c* = 20   74. *b* = 7; *c* = 7$\sqrt{2}$ 75. 4   76. 2$\sqrt{x}$ 77. 5*xy*$\sqrt{2x}$ 78. $y^3\sqrt{y}$

79. *y*$\sqrt{2x}$ 80. $\frac{3\sqrt{5}}{5}$ 81. $\sqrt{7}$ + $\sqrt{2}$ 82. 8*xy*$\sqrt{2}$ 83. 1 − 2$\sqrt{10}$ 84. 12   85. 4*y*$\sqrt{x}$ 86. −1   87. 60 mph

88. 70 mph   89. 7.5 lb/ft²   90. 0.675 lb/ft²   91. 16   92. 5   93. *b* = 12$\sqrt{3}$ cm; *c* = 24 cm   94. 7 ft

1. How many square roots does 36 have?

1. _____ 2 _____

Evaluate each square root. Assume all variables and expressions are nonnegative.

2. $\sqrt{(-5)^2}$

2. _____ 5 _____

3. $\pm\sqrt{121}$

3. _____ ± 11 _____

4. $\sqrt{\dfrac{81}{16}}$

4. _____ $\dfrac{9}{4}$ _____

5. $\sqrt{16x^2}$

5. _____ $4x$ _____

6. $3\sqrt{48}$

6. _____ $12\sqrt{3}$ _____

7. $\sqrt{\dfrac{8x}{y^2}}$

7. _____ $\dfrac{2\sqrt{2x}}{y}$ _____

8. $\sqrt{\dfrac{98x^2}{y}}$

8. _____ $\dfrac{7x\sqrt{2y}}{y}$ _____

Multiply or divide and simplify.

9. $2\sqrt{6}\ \sqrt{24}$

9. _____ 24 _____

10. $\dfrac{\sqrt{27x^3}}{\sqrt{3x}}$

10. _____ $3x$ _____

Perform the indicated operations and simplify.

11. $\sqrt{20} + \sqrt{80}$

11. _____ $6\sqrt{5}$ _____

12. $3\sqrt{18} - 5\sqrt{50}$

12. _____ $-16\sqrt{2}$ _____

13. $6\sqrt{12x} + 3\sqrt{3x} - \sqrt{75x}$

13. _____ $10\sqrt{3x}$ _____

14. $\sqrt{11} + \dfrac{1}{\sqrt{11}}$

14. _____ $\dfrac{12\sqrt{11}}{11}$ _____

15. $\sqrt{2}\,\sqrt{8} + \sqrt{3}\,\sqrt{12}$

15. _____ 10 _____

16. $(\sqrt{5} + \sqrt{3})(\sqrt{5} - \sqrt{3})$

16. _____ 2 _____

17. $(\sqrt{2} + \sqrt{7})(2\sqrt{2} - \sqrt{7})$

17. _____ $-3 + \sqrt{14}$ _____

Rationalize the denominator.

18. $\dfrac{\sqrt{5}}{\sqrt{5} - 1}$

18. _____ $\dfrac{5 + \sqrt{5}}{4}$ _____

Solve.

19. $\sqrt{x + 6} = \sqrt{3x - 1}$

19. _____ $\dfrac{7}{2}$ _____

20. $\sqrt{x^2 + 7} + x - 1 = 0$

20. _____ -3 _____

21. Find the length of the diagonal of a rectangle with sides 6 cm and 8 cm.

21. _____ 10 cm _____

22. Given $d = 10\sqrt{80h}$.
   (a) Find d when h is $\frac{1}{16}$.

22(a) _____ $10\sqrt{5}$ _____

   (b) Find h when d is 400.

22(b) _____ 20 _____

23. In a right triangle, $\angle A = 30°$ and $a = 20$. Find b and c.

23. _____ $b = 20\sqrt{3},\ c = 40$ _____

24. In a right triangle, $\angle B = 45°$ and the hypotenuse is 8. Find the length of each leg.

24. _____ $4\sqrt{2}$ _____

Quadratic Equations

18.1 FACTORING AND TAKING ROOTS

1 QUADRATIC EQUATIONS

Most of the equations we have solved have had the variable raised only to the first power. The only exception to this was in Section 13.5 where we briefly introduced the idea of a *quadratic equation*. An equation that can be expressed in the form

$$ax^2 + bx + c = 0,$$

where a, b, and c are constant real numbers and $a \neq 0$, is called a **quadratic equation.** (If a were 0, the term in which the variable is squared would be missing and the equation would not be quadratic.) We call $ax^2 + bx + c = 0$ the **general form** of a quadratic equation.

EXAMPLE 1 IDENTIFYING CONSTANTS IN GENERAL FORM

Write each equation in general form and identify each constant a, b, and c.

(a) $3x^2 + 2x - 6 = 0$ $(3x^2 + 2x + (-6) = 0)$
This equation is already in general form with $a = 3$, $b = 2$ and $c = -6$. Note that $c = -6$ and not 6. In the general form of an equation, all signs are + so any negative sign goes with the constant a, b, or c.

(b) $2y^2 = 3y + 1$
Collect all terms on one side of the equation to obtain

$$2y^2 - 3y - 1 = 0$$

so that $a = 2$, $b = -3$, and $c = 1$.

PRACTICE EXERCISE 1

Write each equation in general form and identify each constant a, b, and c.

(a) $5x^2 - 4x + 8 = 0$

(b) $6y^2 = 1 - y$

(c) $y^2 = 3y + 1$
Collect all terms on the left side.

$$y^2 - 3y - 1 = 0$$

Then $a = 1$, $b = -3$, and $c = -1$. If we collect all terms on the right side we obtain

$$0 = -y^2 + 3y + 1$$

so that $a = -1$, $b = 3$, and $c = 1$. Both of these are correct since the equations are equivalent. However, it is common practice to write the equation so that the coefficient of the squared term is positive.

(d) $2x + 1 = 3 - 4x + 2$
Since there is no term in which the variable is squared, this is not a quadratic equation.

(e) $2y - 5y^2 = 3y^2 + 2y + 1$
Collect all terms to obtain

$$8y^2 + 1 = 0 \quad \text{or} \quad 8y^2 + 0 \cdot y + 1 = 0$$

so that $a = 8$, $b = 0$, and $c = 1$.

(f) $4 = 6x + 12x^2$
Collect all terms on one side.

$$12x^2 + 6x - 4 = 0$$

This is a general form of the equation. However, we should simplify by factoring out 2 and multiplying both sides by $\frac{1}{2}$. That is,

$$2(6x^2 + 3x - 2) = 0 \quad \text{or} \quad 6x^2 + 3x - 2 = 0.$$

After removing common factors, we have $a = 6$, $b = 3$, and $c = -2$.

(c) $-4y + 3 = -2y^2$

(d) $6 - 3x = 2 - (x - 5)$

(e) $3y + 7 + 7y^2 = 8y^2 - 3y + 2$

(f) $5 + 20x = 10x^2$

Answers:
(a) $5x^2 - 4x + 8 = 0$; $a = 5$, $b = -4$, $c = 8$
(b) $6y^2 + y - 1 = 0$; $a = 6$, $b = 1$, $c = -1$
(c) $2y^2 - 4y + 3 = 0$; $a = 2$, $b = -4$, $c = 3$
(d) not a quadratic equation
(e) $y^2 - 6y - 5 = 0$; $a = 1$, $b = -6$, $c = -5$
(f) $2x^2 - 4x - 1 = 0$; $a = 2$, $b = -4$, $c = -1$

❷ SOLVING BY FACTORING

Recall that an equation of the type

$$(x - 2)(x + 1) = 0$$

is solved by using the zero-product rule, which states that if the product of two numbers is zero then one or both of the numbers must be zero. Using this rule we set each factor equal to zero and solve.

$$x - 2 = 0 \quad \text{or} \quad x + 1 = 0$$
$$x = 2 \quad \text{or} \quad x = -1$$

Since $(x - 2)(x + 1) = x^2 - x - 2$, in effect we have solved the quadratic equation

$$x^2 - x - 2 = 0.$$

To Solve a Quadratic Equation by Factoring

1. Write the equation in general form.
2. Divide out any common numerical factor from each term.
3. Factor the left side into a product of two factors.
4. Use the zero-product rule to obtain two equations whose solutions are the solutions to the original quadratic equation.

EXAMPLE 2 Solving a Quadratic Equation by Factoring	**Practice Exercise 2**

Solve $2x^2 - x - 1 = 0$.

$$(2x + 1)(x - 1) = 0 \qquad \text{Factor}$$

$$2x + 1 = 0 \quad \text{or} \quad x - 1 = 0 \qquad \text{Zero-product rule}$$

$$2x = -1 \qquad\qquad x = 1$$

$$x = -\frac{1}{2}$$

Check: $\quad 2\left(-\frac{1}{2}\right)^2 - \left(-\frac{1}{2}\right) - 1 \stackrel{?}{=} 0 \qquad 2\,(1)^2 - (1) - 1 \stackrel{?}{=} 0$

$$2\left(\frac{1}{4}\right) + \frac{1}{2} - 1 \stackrel{?}{=} 0 \qquad\qquad 2 - 1 - 1 \stackrel{?}{=} 0$$

$$\frac{2}{4} + \frac{1}{2} - 1 \stackrel{?}{=} 0 \qquad\qquad 2 - 2 = 0$$

$$1 - 1 = 0$$

The solutions are 1 and $-\frac{1}{2}$.

Solve $3x^2 + x - 2 = 0$.

Answer: $\frac{2}{3}$, -1

EXAMPLE 3 Removing Common Factor First	**Practice Exercise 3**

Solve $2x^2 - 2x = 12$.

$$2x^2 - 2x - 12 = 0 \qquad \text{Write in general form}$$

$$2(x^2 - x - 6) = 0 \qquad \text{2 is a common factor}$$

$$x^2 - x - 6 = 0 \qquad \text{Divide both sides by 2}$$

$$(x - 3)(x + 2) = 0 \qquad \text{Factor}$$

$$x - 3 = 0 \quad \text{or} \quad x + 2 = 0 \qquad \text{Zero-product rule}$$

$$x = 3 \qquad\qquad x = -2$$

The solutions are 3 and -2. Check.

Solve $4x^2 + 18x = 10$.

Answer: $\frac{1}{2}$, -5

EXAMPLE 4 One Solution Zero	**Practice Exercise 4**

Solve $5x^2 = 3x$.

$$5x^2 - 3x = 0 \qquad \text{Write in general form}$$

$$x(5x - 3) = 0 \qquad \text{Factor}$$

$$x = 0 \quad \text{or} \quad 5x - 3 = 0 \qquad \text{Zero-product rule}$$

$$5x = 3$$

$$x = \frac{3}{5}$$

The solutions are 0 and $\frac{3}{5}$. Check.

Solve $7x = 14x^2$.

Answer: 0, $\frac{1}{2}$

Whenever we have a quadratic equation in general form with $c = 0$, that is, $ax^2 + bx = 0$, one factor is always x so zero is one solution to the equation.

///////// **CAUTION** ,/////////

Never divide both sides of an equation by an expression involving the variable. In Example 4 if we had divided by x we would have lost the solution 0.

///////

| EXAMPLE 5   A MORE COMPLEX EQUATION | PRACTICE EXERCISE 5 |

Solve $5z(z - 6) - 8z = 2z$.

$$5z^2 - 30z - 8z = 2z$$
$$5z^2 - 38z = 2z$$
$$5z^2 - 40z = 0 \quad \text{Write in general form}$$
$$5(z^2 - 8z) = 0$$
$$z^2 - 8z = 0 \quad \text{Divide out the common factor of 5}$$
$$z(z - 8) = 0 \quad \text{Factor}$$
$$z = 0 \quad \text{or} \quad z - 8 = 0$$
$$z = 8$$

The solutions are 0 and 8. Check.

Solve $y(y - 3) - 7 = 2y + 7$.

Answer: $7, -2$

③ SOLVING BY TAKING ROOTS

A quadratic equation with $b = 0$, that is, $ax^2 + c = 0$, can be solved using a different technique. Consider the equation

$$x^2 - 4 = 0$$
or
$$x^2 = 4.$$

What number squared is 4? Clearly $2^2 = 4$ and $(-2)^2 = 4$ so that x is either 2 or -2. By taking the square root of both sides of the equation and using the symbol $\pm$ to indicate both the positive (principal) square root and the negative square root, we have

$$x = \pm\sqrt{4}$$
$$x = \pm 2.$$

Remember that $x = \pm 2$ is simply a shorthand way of writing $x = +2$ *or* $x = -2$.

To Solve a Quadratic Equation of the Form $ax^2 + c = 0$

1. Subtract c from both sides, obtaining $ax^2 = -c$.

2. Divide both sides by a, obtaining $x^2 = -\dfrac{c}{a}$, where $-\dfrac{c}{a} > 0$ if a and c have different signs.

3. Take the square root of both sides, obtaining $x = \pm\sqrt{-\dfrac{c}{a}}$.

//////////// **CAUTION** ////////////

This method gives real-number solutions only when the radicand $-\frac{c}{a}$ is positive. For this to happen a and c must have opposite signs.

////////////

| **EXAMPLE 6 SOLVING A QUADRATIC EQUATION BY TAKING ROOTS** | **PRACTICE EXERCISE 6** |

Solve $3x^2 - 75 = 0$.

$$3x^2 = 75 \qquad \text{Add 75 to both sides}$$

$$x^2 = \frac{75}{3} \qquad \text{Divide both sides by 3}$$

$$x^2 = 25$$

$$x = \pm\sqrt{25} \qquad \text{Take square root of both sides}$$

$$= \pm 5$$

Check: $3(+5)^2 - 75 \stackrel{?}{=} 0 \qquad 3(-5)^2 - 75 \stackrel{?}{=} 0$

$ \; 3(25) - 75 \stackrel{?}{=} 0 \qquad 3(25) - 75 \stackrel{?}{=} 0$

$ \qquad 75 - 75 = 0 \qquad 75 - 75 = 0$

The solutions are 5 and -5.

Solve $64 - 4x^2 = 0$.

Answer: ± 4

| **EXAMPLE 7 AN EQUATION WITH IRRATIONAL SOLUTIONS** | **PRACTICE EXERCISE 7** |

Solve $5x^2 - 40 = 0$.

$$5x^2 = 40 \qquad \text{Add 40}$$

$$x^2 = \frac{40}{5} \qquad \text{Divide by 5}$$

$$x = \pm\sqrt{8} \qquad \text{Take square root of both sides}$$

$$x = \pm\sqrt{4 \cdot 2} = \pm 2\sqrt{2}$$

The solutions are $2\sqrt{2}$ and $-2\sqrt{2}$. Check.

Solve $8x^2 - 96 = 0$.

Answer: $\pm 2\sqrt{3}$

| **EXAMPLE 8 AN EQUATION WITH NO SOLUTION** | **PRACTICE EXERCISE 8** |

Solve $x^2 + 5 = 0$.

$$x^2 = -5 \qquad \text{Subtract 5}$$

Recall from Section 17.1 that no real number squared can be negative. Hence, there are no real-number solutions to the equation $x^2 = -5$, nor to any equation that has x^2 equal to a negative number.

Solve $x^2 + 17 = 0$.

Answer: no real number solutions

| **EXAMPLE 9**   EQUATION IN FORM FOR TAKING ROOTS | **PRACTICE EXERCISE 9** |

Solve $(y + 3)^2 = 16$.

Although not in the form $ax^2 + c = 0$, we can use the same basic technique of taking the square root of both sides to solve.

$$y + 3 = \pm\sqrt{16}$$
$$y + 3 = \pm 4$$
$$y = -3 \pm 4 \qquad \text{Subtract 3 from both sides}$$
$$y = \begin{cases} -3 + 4 = 1 & \text{Use + to obtain solution 1} \\ -3 - 4 = -7 & \text{Use − to obtain solution −7} \end{cases}$$

It is easy to substitute 1 and −7 into the original equation to see that they check and are the solutions.

Solve $(x + 7)^2 = 49$.

Answer: 0, −14

18.1 EXERCISES A

Write each equation in general quadratic form and identify the constants a, b, and c.

1. $2x^2 + x = 5$
$2x^2 + x - 5 = 0$;
$a = 2, b = 1, c = -5$

2. $y - y^2 = 5$
$y^2 - y + 5 = 0$;
$a = 1, b = -1, c = 5$

3. $x^2 = 3x$
$x^2 - 3x = 0$;
$a = 1, b = 3, c = 0$

4. $y^2 = -7y - 12$
$y^2 + 7y + 12 = 0$;
$a = 1, b = 7, c = 12$

5. $2y^2 - y = 3 + y - y^2$
$3y^2 - 2y - 3 = 0$;
$a = 3, b = -2, c = -3$

6. $2(x + x^2) = 3x - 1$
$2x^2 - x + 1 = 0$;
$a = 2, b = -1, c = 1$

7. $x^2 + 1 = 0$

$x^2 + 1 = 0$;
$a = 1, b = 0, c = 1$

8. $\frac{1}{2}y^2 - y = 3 - y$

$y^2 - 6 = 0$;
$a = 1, b = 0, c = -6$

9. $35y - 2 = 0$

not a quadratic
equation

10. $2(x^2 + x) + x^3 = x^3 - 5$
$2x^2 + 2x + 5 = 0$;
$a = 2, b = 2, c = 5$

11. $2y^3 + y^2 + 1 = 0$
not a quadratic
equation

12. $3x = 3(x + 1)$
not a quadratic
equation

13. Given the quadratic equation $-x^2 + 2x - 3 = 0$, we might conclude that $a = -1$, $b = 2$, and $c = -3$. Would $a = 1$, $b = -2$, and $c = 3$ be incorrect? Explain.
   No; multiplying the original equation on both sides by −1 yields a general form with the given constants.

Solve.

14. $x^2 - 3x - 28 = 0$
7, −4

15. $y^2 + 5y + 6 = 0$
−2, −3

16. $x^2 = 2x + 35$
7, −5

17. $2z^2 - 5x = 3$
$-\frac{1}{2}, 3$

18. $6x^2 - 2x - 8 = 0$
$\frac{4}{3}, -1$

19 $4z^2 = 7z$
$0, \frac{7}{4}$

20. $12x^2 + 6x = 0$
$0, -\frac{1}{2}$

21. $y^2 - 10y = -25$
5

22. $z^2 - 6 = z$
$3, -2$

23. $3(y^2 + 8) = 24 - 15y$
$0, -5$

24. $z(8 - 2z) = 6$
$1, 3$

25. $x(3x + 20) = 7$
$\frac{1}{3}, -7$

26 $(3z - 1)(2z + 1) = 3(2z + 1)$
$\frac{4}{3}, -\frac{1}{2}$

27. $(2y - 3)(y + 1) = 4(2y - 3)$
$3, \frac{3}{2}$

28 $4x^2 - 9 = 3$
$\sqrt{3}, -\sqrt{3}$

29. $x^2 = 16$
± 4

30. $z^2 - 81 = 0$
± 9

31 $3x^2 - 48 = -48$
0

32. $2y^2 - 128 = 0$
± 8

33. $y^2 - \frac{9}{4} = 0$
$\pm \frac{3}{2}$

34. $-3z^2 = -27$
± 3

35 $2x^2 - 150 = 0$
$\pm 5\sqrt{3}$

36. $4y^2 = 72$
$\pm 3\sqrt{2}$

37 $2x^2 + 50 = 0$
no solution

38. $-3y^2 = -6$
$\pm \sqrt{2}$

39. $3 - y^2 = y^2 - 3$
$\pm \sqrt{3}$

40. $z^2 + 2z = 2z - 4$
no solution

41 $(x - 2)^2 = 9$
5, −1

42. $(y + 1)^2 = 49$
6, −8

43. $(z - 1)^2 = 5$
$1 \pm \sqrt{5}$

44. The square of a number, decreased by 2, is equal to the negative of the number. Find the number.
1, −2

45. The square of 2 more than a number is 64. Find the number.
6, −10

The amount of money A that will result if a principal P is invested at r percent interest compounded annually for 2 years is given by $A = P(1 + r)^2$. Use this formula to find the interest rate for the given values of A and P in Exercises 46–47. Give the answer to the nearest tenth of a percent if appropriate.

46. $A = \$2645$, $P = \$2000$
15%

47. $A = \$1765$, $P = \$1500$
8.5%

Tell what is wrong with each of the "solutions" in Exercises 48–49.

48. $x^2 - x = 3(x - 1)$
$x(x - 1) = 3(x - 1)$
$x = 3$
The solution 1 has been lost by dividing both sides by $x − 1$.

49. $x^2 - x = 2$
$x(x - 1) = 2$
$x = 2$ or $x - 1 = 2$
$x = 3$
The product is 2, not zero. Cannot use zero-product rule.

FOR REVIEW

The following exercises review material from Sections 14.4 and 15.4. They will help you prepare for the next section. Find the products in Exercises 50–52.

50. $(x + 6)^2$

$x^2 + 12x + 36$

51. $(x - 10)^2$

$x^2 - 20x + 100$

52. $\left(x + \dfrac{1}{9}\right)^2$

$x^2 + \frac{2}{9}x + \frac{1}{81}$

Factor the trinomials in Exercises 53–55.

53. $x^2 - 4x + 4$

$(x - 2)^2$

54. $x^2 + 10x + 25$

$(x + 5)^2$

55. $x^2 + x + \dfrac{1}{4}$

$\left(x + \frac{1}{2}\right)^2$

Simplify the square roots in Exercises 56–58.

56. $\sqrt{8}$

$2\sqrt{2}$

57. $\sqrt{\dfrac{49}{4}}$

$\frac{7}{2}$

58. $\sqrt{\dfrac{7}{4}}$

$\frac{\sqrt{7}}{2}$

ANSWERS: 1. $2x^2 + x - 5 = 0$; $a = 2$, $b = 1$, $c = -5$ 2. $y^2 - y + 5 = 0$; $a = 1$, $b = -1$, $c = 5$
3. $x^2 - 3x = 0$; $a = 1$, $b = -3$, $c = 0$ 4. $y^2 + 7y + 12 = 0$; $a = 1$, $b = 7$, $c = 12$ 5. $3y^2 - 2y - 3 = 0$; $a = 3$,
$b = -2$, $c = -3$ 6. $2x^2 - x + 1 = 0$; $a = 2$, $b = -1$, $c = 1$ 7. $x^2 + 1 = 0$; $a = 1$, $b = 0$, $c = 1$
8. $\frac{1}{2}y^2 - 3 = 0$; $a = \frac{1}{2}$, $b = 0$, $c = -3$ or $y^2 - 6 = 0$ (multiply by 2 to clear fractions and make a, b, and c integers);
$a = 1$, $b = 0$, $c = -6$ 9. not a quadratic equation   10. $2x^2 + 2x + 5 = 0$; $a = 2$, $b = 2$, $c = 5$ 11. not a quadratic
equation   12. not a quadratic equation   13. No; multiplying the original equation on both sides by -1 yields a general
form with the given constants.   14. 7, -4 15. -2, -3 16. 7, -5 17. $-\frac{1}{2}$, 3 18. $\frac{4}{3}$, -1 19. 0, $\frac{7}{4}$ 20. 0, $-\frac{1}{2}$
21. 5 22. 3, -2 23. 0, -5 24. 1, 3 25. $\frac{1}{3}$, -7 26. $\frac{4}{3}$, $-\frac{1}{2}$ 27. 3, $\frac{3}{2}$ 28. $\sqrt{3}$, $-\sqrt{3}$ 29. ± 4 30. ± 9 31. 0
32. ± 8 33. $\pm \frac{3}{2}$ 34. ± 3 35. $\pm 5\sqrt{3}$ 36. $\pm 3\sqrt{2}$ 37. no solution   38. $\pm \sqrt{2}$ 39. $\pm \sqrt{3}$ 40. no solution   41. 5,
-1 42. 6, -8 43. $1 \pm \sqrt{5}$ 44. 1, -2 45. 6, -10 46. 15%   47. 8.5%   48. The solution 1 has been lost by
dividing both sides by an expression containing the variable, $x - 1$.   49. The product is 2, *not* zero. Do not try to
use the zero-product rule at this point.   50. $x^2 + 12x + 36$ 51. $x^2 - 20x + 100$ 52. $x^2 + \frac{2}{9}x + \frac{1}{81}$ 53. $(x - 2)^2$
54. $(x + 5)^2$ 55. $\left(x + \frac{1}{2}\right)^2$ 56. $2\sqrt{2}$ 57. $\frac{7}{2}$ 58. $\frac{\sqrt{7}}{2}$

18.1   EXERCISES B

Write each equation in general form and identify the constants a, b, *and* c.

1. $3x^2 + x = 7$
$3x^2 + x - 7 = 0$;
$a = 3$, $b = 1$, $c = -7$

2. $y - y^2 = 4$
$y^2 - y + 4 = 0$;
$a = 1$, $b = -1$, $c = 4$

3. $z^2 = 9z$
$z^2 - 9z = 0$;
$a = 1$, $b = -9$, $c = 0$

4. $y^2 = -10y - 18$
$y^2 + 10y + 18 = 0$;
$a = 1$, $b = 10$, $c = 18$

5. $3z^2 - z = 5 + z - z^2$
$4z^2 - 2z - 5 = 0$;
$a = 4$, $b = -2$, $c = -5$

6. $4(z + z^2) = 2z + 1$
$4z^2 + 2z - 1 = 0$;
$a = 4$, $b = 2$, $c = -1$

7. $w^2 + 5 = 0$
$w^2 + 5 = 0$;
$a = 1$, $b = 0$, $c = 5$

8. $\frac{1}{3}x^2 - x = 5 - x$
$x^2 - 15 = 0$;
$a = 1$, $b = 0$, $c = -15$

9. $4x - 8 = 0$
not a quadratic equation

10. $3(y^2 + y) - y^3 = 6 - y^3$
$y^2 + y - 2 = 0$;
$a = 1$, $b = 1$, $c = -2$

11. $5x^3 = x^2 + 7$
not a quadratic equation

12. $3(w - 3) = 4w$
not a quadratic equation

13. Given the quadratic equation $-y^2 + y - 7 = 0$, we might conclude that $a = -1$, $b = 1$, and $c = -7$. Would
$a = 1$, $b = -1$, and $c = 7$ be incorrect? Explain.
No; multiplying the original equation on both sides by -1 yields a general form with the given constants.

Solve.

14. $x^2 + x - 6 = 0$ 2, -3

15. $y^2 + 3y + 2 = 0$ -1, -2

16. $x^2 = 2x + 24$ 6, -4

17. $2z^2 + 3 = 7z$ $\frac{1}{2}$, 3

18. $3x^2 - 2x - 5 = 0$ $\frac{5}{3}$, -1

19. $2z^2 = 3z$ 0, $\frac{3}{2}$

20. $3x^2 - 18x = 0$ 0, 6

21. $y^2 - 8y + 16 = 0$ 4

22. $z^2 - 20 = z$ 5, -4

23. $3(y^2 + 4y) = -6y$ 0, -6

24. $z(z - 3) = -2$ 1, 2

25. $x(2x + 9) = 5$ $\frac{1}{2}$, -5

26. $(z - 2)(z + 1) = 3(z + 1)$
-1, 5

27. $5x^2 - 10 = 5$ $\sqrt{3}$, $-\sqrt{3}$

28. $y(2y - 1) = 7(1 - 2y)$ $\frac{1}{2}$, -7

29. $x^2 = 49$ ± 7

30. $z^2 - 36 = 0$ ± 6

31. $5x^2 - 7 = -7$ 0

32. $2y^2 - 98 = 0$ ± 7

33. $y^2 - \dfrac{25}{4} = 0$ $\pm \frac{5}{2}$

34. $-8z^2 = -32$ ± 2

35. $2x^2 - 100 = 0$ $\pm 5\sqrt{2}$ **36.** $3y^2 = 60$ $\pm 2\sqrt{5}$ **37.** $3x^2 + 75 = 0$ no solution

38. $-6y^2 = -18$ $\pm \sqrt{3}$ **39.** $7 - x^2 = x^2 - 7$ $\pm \sqrt{7}$ **40.** $z^2 + 5z = 5z - 1$
                                                                                                          no solution

41. $(x - 5)^2 = 4$ 7, 3   **42.** $(y + 2)^2 = 1$ $-1, -3$ **43.** $(z - 3)^2 = 3$ $3 \pm \sqrt{3}$

44. The square of a number, less 10 is equal to three times the number. Find the number.   5, -2 **45.** The square of 5 more than a number is 9. Find the number. $-2, -8$

The amount of money A that will result if a principal P is invested at r percent interest compounded annually for 2 years is given by $A = P(1 + r)^2$. Use this formula to find the interest rate for the given values of A and P in Exercises 46–47. Give the answer to the nearest tenth of a percent if appropriate.

46. $A = \$1210$, $P = \$1000$ 10%   **47.** $A = \$2785$, $P = \$2400$ 7.7%

Tell what is wrong with each of the ''solutions'' in Exercises 48–49.

48. $x^2 - 2x = 0$
          $x(x - 2) = 0$
          $\dfrac{x(x - 2)}{x} = \dfrac{0}{x}$
          $x - 2 = 0$ The solution 0 is lost by divid-
          $x = 2$ ing both sides by x

49. $x(x - 5) = 6$
        $x = 6$ or $x - 5 = 6$
                    $x = 11$

The product is 6, not zero. Cannot use the zero-product rule.

FOR REVIEW

The following exercises review material from Sections 14.4 and 15.4. They will help you prepare for the next section. Find the products in Exercises 50–52.

50. $(x + 3)^2$
        $x^2 + 6x + 9$

51. $(x - 7)^2$
        $x^2 - 14x + 49$

52. $\left(x - \dfrac{1}{6}\right)^2$
        $x^2 - \frac{1}{3}x + \frac{1}{36}$

Factor the trinomials in Exercises 53–55.

53. $x^2 + 22x + 121$
        $(x + 11)^2$

54. $x^2 - 20x + 100$
        $(x - 10)^2$

55. $x^2 + \dfrac{2}{3}x + \dfrac{1}{9}$
        $\left(x + \frac{1}{3}\right)^2$

Simplify the square roots in Exercises 56–58.

56. $\sqrt{32}$ $4\sqrt{2}$

57. $\sqrt{\dfrac{144}{25}}$ $\frac{12}{5}$

58. $\sqrt{\dfrac{11}{9}}$ $\frac{\sqrt{11}}{3}$

18.1 EXERCISES C

Solve.

1. $6x^2 + 47x - 8 = 0$
       $\frac{1}{6}$, -8

2. $(2x - 1)(x + 5) =$
       $(4x + 3)(x + 5)$ [Answer: 5, -2]

3. $(x^2 - 5)^2 = 8$
       $\pm \sqrt{5 \pm 2\sqrt{2}}$

18.2 SOLVING BY COMPLETING THE SQUARE

1 Completing the Square   **2** Solving by Completing the Square

1 COMPLETING THE SQUARE

A trinomial $x^2 + bx + c$ is a perfect square trinomial if it factors into the square of a binomial. Notice that the trinomials in the left column below are perfect square trinomials. Also notice in the right column that for each perfect square trinomial, the constant term, c, is the square of one half the coefficient of the x term, b.

$$x^2 + 6x + 9 = (x + 3)^2 \qquad 9 = \left(\tfrac{1}{2} \cdot 6\right)^2$$
$$x^2 - 10x + 25 = (x - 5)^2 \qquad 25 = \left(\tfrac{1}{2} \cdot (-10)\right)^2$$
$$x^2 - x + \frac{1}{4} = \left(x - \frac{1}{2}\right)^2 \qquad \tfrac{1}{4} = \left(\tfrac{1}{2} \cdot (-1)\right)^2$$

In general, whenever we are given the first two terms of a trinomial, $x^2 + bx$, we can add $\left(\tfrac{1}{2} \cdot b\right)^2$ to it to get a perfect square trinomial. This process, called **completing the square,** can be used only when the coefficient in the squared term is 1.

EXAMPLE 1 COMPLETING THE SQUARE	**PRACTICE EXERCISES 1**

What must be added to complete the square?

(a) $x^2 + 12x +$ _____

Add $\left(\dfrac{1}{2} \cdot 12\right)^2 = 6^2 = 36$.

(b) $x^2 - 20x +$ _____

Add $\left(\dfrac{1}{2} \cdot (-20)\right)^2 = (-10)^2 = 100$.

(c) $x^2 + \dfrac{2}{3}x +$ _____

Add $\left(\dfrac{1}{2} \cdot \dfrac{2}{3}\right)^2 = \left(\dfrac{1}{3}\right)^2 = \dfrac{1}{9}$.

What must be added to complete the square?

(a) $x^2 + 14x +$ _____

(b) $x^2 - 22x +$ _____

(c) $x^2 - \dfrac{2}{3}x +$ _____

Answers: (a) 49   (b) 121   (c) $\frac{1}{9}$

---

2 SOLVING BY COMPLETING THE SQUARE

We can solve quadratic equations by completing the square (a method that will have applications in the next algebra course). For example, the equation

$$x^2 + 3x - 10 = 0$$

which can be solved by factoring,

$$(x + 5)(x - 2) = 0$$
$$x = -5, 2$$

can also be solved by completing the square. To do so, first isolate the constant term on the right side. (Leave space as indicated.)

$$x^2 + 3x \qquad = 10$$

To complete the square on the left side, add the square of half the coefficient of x.

$$\left(\frac{1}{2} \cdot 3\right)^2 = \left(\frac{3}{2}\right)^2 = \frac{9}{4}$$

But if $\frac{9}{4}$ is added on the left, to keep equality we must add $\frac{9}{4}$ on the right.

$$x^2 + 3x + \frac{9}{4} = 10 + \frac{9}{4}$$

$$\left(x + \frac{3}{2}\right)^2 = \frac{49}{4}$$

Take the square root of each side, remembering to write both the positive and negative root.

$$x + \frac{3}{2} = \pm \sqrt{\frac{49}{4}} = \pm \frac{7}{2}$$

$$x = -\frac{3}{2} \pm \frac{7}{2} \qquad \text{Subtract } \tfrac{3}{2} \text{ from both sides}$$

$$x = -\frac{3}{2} + \frac{7}{2} \quad \text{or} \quad x = -\frac{3}{2} - \frac{7}{2}$$

$$x = \frac{4}{2} = 2 \qquad\qquad x = \frac{-10}{2} = -5$$

The solutions are the same as before, 2 and -5.

To Solve a Quadratic Equation by Completing the Square

1. Isolate the constant term on the right side of the equation.
2. If the coefficient of the squared item is 1, proceed to step 4.
3. If the coefficient of the squared term is not 1, divide each term by that coefficient.
4. Complete the square on the left side and add the same number to the right side.
5. Factor (check by multiplying the factors) and take the square root of both sides (positive and negative roots).
6. Use both the positive root and the negative root to obtain the solutions to the original equation.

EXAMPLE 2 SOLVING BY COMPLETING THE SQUARE

Solve $y^2 - 6y + 8 = 0$ by completing the square.

$$y^2 - 6y \qquad = -8 \qquad \text{Isolate the constant, } b = -6$$

$$y^2 - 6y + 9 = -8 + 9 \qquad \text{Add 9 to both sides; } \left(\tfrac{-6}{2}\right)^2 = (-3)^2 = 9$$

$$(y - 3)^2 = 1$$

$$y - 3 = \pm\sqrt{1} \qquad \text{Take the square root of both sides}$$

$$y - 3 = \pm 1 \qquad \sqrt{1} = 1$$

$$y = 3 \pm 1$$

$$y = 3 + 1 \quad \text{or} \quad y = 3 - 1$$

$$y = 4 \qquad\qquad y = 2$$

The solutions are 2 and 4.

PRACTICE EXERCISE 2

Solve $y^2 - 8y + 12 = 0$ by completing the square.

Answer: 2, 6

As mentioned earlier, we can use the technique above to complete the square when the coefficient of the squared item is 1. If this is not the case, we need to divide each term of the equation by the coefficient of the squared term before we can use the method.

EXAMPLE 3 MAKING THE COEFFICIENT OF x^2 ONE

Solve $2x^2 + 2x - 3 = 0$ by completing the square.

$$2x^2 + 2x = 3 \qquad \text{Isolate the constant}$$

$$x^2 + x = \frac{3}{2} \qquad \text{Divide through by 2; } b = 1$$

$$x^2 + x + \frac{1}{4} = \frac{3}{2} + \frac{1}{4} \qquad \text{Complete the square by adding } \tfrac{1}{4} \text{ to both sides}$$

$$\left(x + \frac{1}{2}\right)^2 = \frac{7}{4} \qquad \tfrac{3}{2} + \tfrac{1}{4} = \tfrac{6}{4} + \tfrac{1}{4} = \tfrac{7}{4}$$

$$x + \frac{1}{2} = \pm\sqrt{\frac{7}{4}} = \pm\frac{\sqrt{7}}{\sqrt{4}} = \pm\frac{\sqrt{7}}{2}$$

$$x + \frac{1}{2} = \frac{\sqrt{7}}{2} \qquad \text{or} \qquad x + \frac{1}{2} = -\frac{\sqrt{7}}{2}$$

$$x = -\frac{1}{2} + \frac{\sqrt{7}}{2} \qquad\qquad x = -\frac{1}{2} - \frac{\sqrt{7}}{2}$$

$$x = \frac{-1 + \sqrt{7}}{2} \qquad\qquad x = \frac{-1 - \sqrt{7}}{2}$$

The solutions are $\dfrac{-1 \pm \sqrt{7}}{2}$.

PRACTICE EXERCISE 3

Solve $3x^2 - 2x - 6 = 0$ by completing the square.

Answer: $\dfrac{1 \pm \sqrt{19}}{3}$

18.2 EXERCISES A

What must be added to complete the square?

1. $x^2 - 8x +$ ___16___

2. $y^2 - 18y +$ ___81___

3. $z^2 + 7z +$ ___$\frac{49}{4}$___

4. $x^2 - x +$ ___$\frac{1}{4}$___

5. $y^2 + 9y +$ ___$\frac{81}{4}$___

6 $z^2 + \dfrac{1}{3}z +$ ___$\frac{1}{36}$___

7. $x^2 - \dfrac{1}{2}x +$ ___$\frac{1}{16}$___

8. $y^2 - \dfrac{4}{3}y +$ ___$\frac{4}{9}$___

9. $z^2 + z +$ ___$\frac{1}{4}$___

Solve by completing the square.

10. $x^2 + 8x + 15 = 0$
$-3, -5$

11. $y^2 + 10y + 21 = 0$
$-3, -7$

12. $z^2 - 2z - 8 = 0$
$4, -2$

13 $x^2 - 2x - 1 = 0$
$1 \pm \sqrt{2}$

14. $y^2 + 4y = 96$
$-12, 8$

15 $3z^2 + 12z = 135$
$5, -9$

16 $2x^2 + x - 1 = 0$
$\frac{1}{2}, -1$

17. $2y^2 - 5y = -2$
$2, \frac{1}{2}$

18. $z^2 + 16z + 50 = 0$
$-8 \pm \sqrt{14}$

19. $x^2 - 10x + 22 = 0$
$5 \pm \sqrt{3}$

20 $y^2 + 2y + 5 = 0$
no solution

21. $2z^2 - z - 3 = 0$
$\frac{3}{2}, -1$

Solve by any method.

22. $x^2 - 5x = -4$
4, 1 (best to factor)

23 $5y^2 - 4y - 33 = 0$
$3, -\frac{11}{5}$ **(best to factor)**

24. $z^2 + 6z = 16$
2, −8 (best to factor)

25. $2x^2 - 5x - 3 = 0$
$3, -\frac{1}{2}$ **(factor)**

26. $2y^2 - 3y = 1$
$\dfrac{3 \pm \sqrt{17}}{4}$ **(complete the square)**

27. $z^2 + z = z + 10$
$\pm\sqrt{10}$ **(take roots)**

28. A free-falling object, starting from rest, will fall d feet in t seconds according to the formula $d = 16t^2$.
  (a) How far will the object fall in 1 second?   **16 feet**
  (b) How far will the object fall in 5 seconds?   **400 feet**
  (c) How many seconds does it take for the object to fall 64 feet?   **2 seconds**

FOR REVIEW

The following exercises review material from Sections 10.1, 16.4, 17.2, 17.4, and 17.7. They will help prepare for the next section. Evaluate each expression in Exercises 29–34 if $a = 2$, $b = -4$, and $c = 1$.

29. $-b$ **4**

30. b^2 **16**

31. $4ac$ **8**

32. $b^2 - 4ac$ **8**

33. $\sqrt{b^2 - 4ac}$ $2\sqrt{2}$

34. $-b + \sqrt{b^2 - 4ac}$
$4 + 2\sqrt{2}$

Solve each equation in Exercises 35–36.

35. $\dfrac{1}{x-1} + \dfrac{2}{x+1} = \dfrac{5}{x^2-1}$ 2

36. $\sqrt{x^2-3} - x = 1$ no solution

The Pythagorean theorem states that $a^2 + b^2 = c^2$ where a and b are legs of a right triangle and c is the hypotenuse. In Exercises 37–38, find the missing leg or hypotenuse.

37. $a = 6$ and $b = 4$ $2\sqrt{13}$

38. $b = \dfrac{1}{2}$ and $c = \dfrac{3}{2}$ $\sqrt{2}$

ANSWERS: 1. 16   2. 81   3. $\frac{49}{4}$ 4. $\frac{1}{4}$ 5. $\frac{81}{4}$ 6. $\frac{1}{36}$ 7. $\frac{1}{16}$ 8. $\frac{4}{9}$ 9. $\frac{1}{4}$ 10. $-3, -5$ 11. $-3, -7$ 12. $4, -2$ 13. $1 \pm \sqrt{2}$ 14. $-12, 8$ 15. $5, -9$ 16. $\frac{1}{2}, -1$ 17. $2, \frac{1}{2}$ 18. $-8 \pm \sqrt{14}$ 19. $5 \pm \sqrt{3}$ 20. no solution (square root of a negative number does not exist)   21. $\frac{3}{2}, -1$ 22. 4, 1 (best to factor)   23. $3, -\frac{11}{5}$ (factor)   24. 2, -8 (factor)   25. $3, -\frac{1}{2}$ (factor)   26. $\frac{3 \pm \sqrt{17}}{4}$ (complete the square)   27. $\pm\sqrt{10}$ (take roots)   28. (a) 16 feet (b) 400 feet (c) 2 seconds   29. 4   30. 16   31. 8   32. 8   33. $2\sqrt{2}$ 34. $4 + 2\sqrt{2}$ 35. 2   36. no solution (-2 does not check)   37. $2\sqrt{13}$ 38. $\sqrt{2}$

18.2   EXERCISES B

What must be added to complete the square?

1. $x^2 + 24x + \underline{\quad 144 \quad}$

2. $y^2 - 4y + \underline{\quad 4 \quad}$

3. $z^2 + 3z + \underline{\quad \frac{9}{4} \quad}$

4. $x^2 + x + \underline{\quad \frac{1}{4} \quad}$

5. $y^2 - 11y + \underline{\quad \frac{121}{4} \quad}$

6. $z^2 + \dfrac{1}{2}z + \underline{\quad \frac{1}{16} \quad}$

7. $x^2 - \dfrac{1}{3}x + \underline{\quad \frac{1}{36} \quad}$

8. $y^2 + \dfrac{2}{3}y + \underline{\quad \frac{1}{9} \quad}$

9. $z^2 - z + \underline{\quad \frac{1}{4} \quad}$

Solve by completing the square.

10. $x^2 + 6x + 5 = 0$
   $-1, -5$

11. $y^2 + 3y - 18 = 0$
   $3, -6$

12. $z^2 - 3z - 10 = 0$
   $5, -2$

13. $x^2 - 4x - 1 = 0$ $2 \pm \sqrt{5}$

14. $y^2 = 8y + 20$ $-2, 10$

15. $z^2 + 5z = 24$ $3, -8$

16. $2x^2 - 3x + 1 = 0$ $\frac{1}{2}, 1$

17. $2y^2 - 5y = 3$ $3, -\frac{1}{2}$

18. $z^2 - z = 3$ $\dfrac{1 \pm \sqrt{13}}{2}$

19. $2x^2 + 6x + 1 = 0$ $\dfrac{-3 \pm \sqrt{7}}{2}$

20. $y^2 + 2y = -9$ no solution

21. $z^2 + 15 = 8z$ $3, 5$

Solve by any method.

22. $x^2 + 8x - 9 = 0$ $1, -9$

23. $3y^2 + 11y - 4 = 0$ $-4, \frac{1}{3}$

24. $z^2 - 14z + 49 = 0$ 7

25. $2x^2 - 3x - 2 = 0$ $2, -\frac{1}{2}$

26. $3y^2 - 6 = 2y$ $\dfrac{1 \pm \sqrt{19}}{3}$

27. $z^2 - z = 8 - z$ $\pm 2\sqrt{2}$

28. A free falling object, starting at rest, will fall d feet in t seconds according to the formula $d = 16t^2$.

  (a) How far will the object fall in 2 seconds?   **64 feet**
  (b) How far will the object fall in 10 seconds?   **1600 feet**
  (c) How many seconds does it take for the object to fall 144 feet?   **3 seconds**

FOR REVIEW

The following exercises review material from Sections 10.1, 16.4, 17.2, 17.4, and 17.7. They will help prepare for the next section. Evaluate each expression in Exercises 29–34 if $a = 3$, $b = -6$ and $c = 2$.

29. $-b$ 6

30. b^2 36

31. $4ac$ 24

32. $b^2 - 4ac$ 12

33. $\sqrt{b^2 - 4ac}$ $2\sqrt{3}$

34. $-b + \sqrt{b^2 - 4ac}$ $6 + 2\sqrt{3}$

Solve each equation in Exercises 35–36.

35. $\dfrac{3}{x - 2} + \dfrac{4}{x + 2} = \dfrac{-1}{x^2 - 4}$ $\frac{1}{7}$

36. $\sqrt{x^2 + 5} - x = 1$ 2

The Pythagorean theorem states that $a^2 + b^2 = c^2$ where a and b are legs of a right triangle and c is the hypotenuse. In Exercises 37–38, find the missing leg or hypotenuse.

37. $a = 9$ and $b = 2$ $\sqrt{85}$

38. $a = \dfrac{1}{3}$ and $c = \dfrac{4}{3}$ $\dfrac{\sqrt{15}}{3}$

18.2 EXERCISES C

Solve by completing the square.

1. $5x^2 - 2x - 10 = 0$
$\dfrac{1 \pm \sqrt{51}}{5}$

2. $x^2 + bx + c = 0$
$\left[\text{Answer:} \quad x = \dfrac{-b \pm \sqrt{b^2 - 4ac}}{2} \right]$

18.3 SOLVING BY THE QUADRATIC FORMULA

========= STUDENT GUIDEPOSTS =========

1 Quadratic Formula

3 The Discriminant

2 Using the Quadratic Formula

4 Applications Requiring the Quadratic Formula

1 QUADRATIC FORMULA

Factoring and taking roots are usually faster and easier methods than the method of completing the square. Completing the square becomes tedious since it repeats the same process time and time again. To avoid such repetition, we can work through the process once in a general case and memorize the result. This is how the *quadratic formula* is developed. Starting with a quadratic equation in general form

$$ax^2 + bx + c = 0 \quad (a \neq 0)$$

we complete the square.

$ax^2 + bx = -c$	Isolate the constant term
$\dfrac{\cancel{a}x^2}{\cancel{a}} + \dfrac{b}{a}x = -\dfrac{c}{a}$	Divide by the coefficient of x^2
$x^2 + \dfrac{b}{a}x + \left(\dfrac{b}{2a}\right)^2 = -\dfrac{c}{a} + \left(\dfrac{b}{2a}\right)^2$	Add $\left(\dfrac{1}{2} \cdot \dfrac{b}{a}\right)^2 = \left(\dfrac{b}{2a}\right)^2$
$\left(x + \dfrac{b}{2a}\right)^2 = -\dfrac{c}{a} + \dfrac{b^2}{4a^2}$	Factor
$\left(x + \dfrac{b}{2a}\right)^2 = -\dfrac{4ac}{4a^2} + \dfrac{b^2}{4a^2}$	Find a common denominator
$\qquad\qquad = \dfrac{b^2 - 4ac}{4a^2}$	Subtract the fractions
$x + \dfrac{b}{2a} = \pm\sqrt{\dfrac{b^2 - 4ac}{4a^2}}$	Take square roots
$\qquad\qquad = \dfrac{\pm\sqrt{b^2 - 4ac}}{2a}$	
$x = -\dfrac{b}{2a} \pm \dfrac{\sqrt{b^2 - 4ac}}{2a}$	Subtract $\dfrac{b}{2a}$
$x = \dfrac{-b \pm \sqrt{b^2 - 4ac}}{2a}$	Note that $2a$ is the denominator of the entire expression

② USING THE QUADRATIC FORMULA

This last formula is called the **quadratic formula** and it must be memorized. To use it to solve a quadratic equation, we identify the constants, a, b, and c and substitute into the quadratic formula.

To Solve a Quadratic Equation Using the Quadratic Formula

1. Write the equation in general form ($ax^2 + bx + c = 0$).
2. Identify the constants a, b, and c.
3. Substitute the values for a, b, and c into the quadratic formula,

$$x = \frac{-b \pm \sqrt{b^2 - 4ac}}{2a}.$$

4. Simplify the numerical expression to obtain the solutions.

EXAMPLE 1   USING THE QUADRATIC FORMULA

Solve $x^2 - 5x + 6 = 0$ using the quadratic formula.
   We have $a = 1$, $b = -5$ (not 5), and $c = 6$.

$$x = \frac{-b \pm \sqrt{b^2 - 4ac}}{2a} = \frac{-(-5) \pm \sqrt{(-5)^2 - 4(1)(6)}}{2(1)} \quad \text{Substitute}$$

$$= \frac{5 \pm \sqrt{25 - 24}}{2} \quad \text{Watch all signs}$$

$$= \frac{5 \pm \sqrt{1}}{2} = \frac{5 \pm 1}{2} = \begin{cases} \dfrac{5+1}{2} = \dfrac{6}{2} = 3 \\ \dfrac{5-1}{2} = \dfrac{4}{2} = 2 \end{cases}$$

The solutions are 2 and 3.

PRACTICE EXERCISE 1

Solve $x^2 + 6x + 9 = 0$ using the quadratic formula.

Answer: -3

EXAMPLE 2   IRRATIONAL SOLUTIONS

Solve $3x^2 - 5 = -4x$ using the quadratic formula.
   First we write the equation in general form, $3x^2 + 4x - 5 = 0$, and identify $a = 3$, $b = 4$, and $c = -5$ (not 5), then substitute.

$$x = \frac{-b \pm \sqrt{b^2 - 4ac}}{2a} = \frac{-(4) \pm \sqrt{(4)^2 - 4(3)(-5)}}{2(3)}$$

$$= \frac{-4 \pm \sqrt{16 + 60}}{6} \quad \text{Watch the signs}$$

$$= \frac{-4 \pm \sqrt{76}}{6} = \frac{-4 \pm \sqrt{4 \cdot 19}}{6}$$

$$= \frac{-4 \pm 2\sqrt{19}}{6} = \frac{2(-2 \pm \sqrt{19})}{2 \cdot 3}$$

$$= \frac{-2 \pm \sqrt{19}}{3}$$

The solutions are $\dfrac{-2 \pm \sqrt{19}}{3}$.

PRACTICE EXERCISE 2

Solve $7x - 2 = 2x^2$ using the quadratic formula.

Answer: $\dfrac{7 \pm \sqrt{33}}{4}$

To Solve a Quadratic Equation

1. First try factoring or taking roots.
2. If the first two methods do not work, go directly to the quadratic formula.

   Notice that the quadratic equation in Example 1 could be solved much more quickly by factoring than by using the quadratic formula.

$$x^2 - 5x + 6 = 0$$
$$(x - 2)(x - 3) = 0$$
$$x - 2 = 0 \quad \text{or} \quad x - 3 = 0$$
$$x = 2 \quad \text{or} \quad x = 3$$

❸ THE DISCRIMINANT

In a quadratic equation $ax^2 + bx + c = 0$, the number $b^2 - 4ac$, called the **discriminant,** can be used to find out whether the equation has solutions. Notice that the discriminant is the number under the radical sign in the quadratic formula $x = \dfrac{-b \pm \sqrt{b^2 - 4ac}}{2a}$. Since the square root of a negative number is not a real number, a quadratic equation has no real-number solutions when the discriminant is a negative number. Consider

$$x^2 + x + 1 = 0.$$

Here $a = 1$, $b = 1$, and $c = 1$, so $b^2 - 4ac = (1) - 4(1)(1) = 1 - 4 = -3$. Since -3 is a negative number, $\sqrt{-3}$ is not a real number and $x^2 + x + 1 = 0$ has no real-number solutions.

It is helpful to clear all fractions first when solving quadratic equations with fractional coefficients. To solve

$$\frac{1}{3}x^2 - x + \frac{1}{3} = 0,$$

it would be difficult to use $a = \frac{1}{3}$, $b = -1$, and $c = \frac{1}{3}$ in the quadratic formula. However, if we multiply through by the LCD 3, we obtain the equivalent equation

$$x^2 - 3x + 1 = 0,$$

in which $a = 1$, $b = -3$, and $c = 1$. Avoiding fractions makes the arithmetic simpler.

❹ APPLICATIONS REQUIRING THE QUADRATIC FORMULA

In many applications of quadratic equations, solutions containing radicals must be estimated by using an approximate value for the radical.

EXAMPLE 3 **APPLICATION OF THE QUADRATIC FORMULA**	**PRACTICE EXERCISE 3**

A large wheat field is in the shape of a right triangle with hypotenuse 5 km long and one leg 4 km longer than the other. Find the measure of each leg (side) of the field.

The triangular field is shown in Figure 18.1. Remember the Pythagorean theorem: the sum of the squares of the legs of a right triangle is equal to the square of the hypotenuse.

Meg and Joe need a water line the length of their garden. If the garden is a rectangle with diagonal 8 meters and the length is 2 meters longer than the width, approximately how much pipe will be needed?

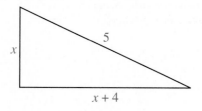

Figure 18.1

Let $x = $ the measure of one leg,
$x + 4 = $ the measure of the other leg.
Use the Pythagorean theorem.

$$x^2 + (x + 4)^2 = 5^2$$
$$x^2 + x^2 + 8x + 16 = 25 \qquad \text{Don't forget the } 8x$$
$$2x^2 + 8x - 9 = 0$$

Since we cannot factor, use the quadratic formula.

$$x = \frac{-b \pm \sqrt{b^2 - 4ac}}{2a}$$

$$= \frac{-8 \pm \sqrt{64 - 4(2)(-9)}}{2(2)} \qquad a = 2, \, b = 8, \text{ and } c = -9$$

$$= \frac{-8 \pm \sqrt{64 + 72}}{4}$$

$$= \frac{-8 \pm \sqrt{136}}{4}$$

$$= \frac{-8 \pm \sqrt{4 \cdot 34}}{4}$$

$$= \frac{-8 \pm 2\sqrt{34}}{4} \qquad \sqrt{4 \cdot 34} = \sqrt{4}\sqrt{34} = 2\sqrt{34}$$

$$= \frac{2[-4 \pm \sqrt{34}]}{2 \cdot 2} \qquad \text{Factor and divide out 2}$$

$$= \frac{-4 \pm \sqrt{34}}{2}$$

Since $\dfrac{-4 - \sqrt{34}}{2}$ is a negative number, we can discard it (a length must be positive or zero). Thus, the length of one leg is exactly $\dfrac{-4 + \sqrt{34}}{2}$ km. Using a calculator to get a decimal approximation, we find $\sqrt{34} \approx 5.8$. Thus,

$$\frac{-4 + \sqrt{34}}{2} \approx \frac{-4 + 5.8}{2} = \frac{1.8}{2} = 0.9,$$

and the lengths of the sides of the field are approximately 0.9 km and 4.9 km ($x = 0.9$, $x + 4 = 4.9$). As a check,

$$(0.9)^2 + (4.9)^2 = 0.81 + 24.01 = 24.82 \approx 25 = 5^2.$$

Answer: approximately 6.6 m

Unless you are working an applied problem in which an approximate value would make more sense, or unless instructed otherwise, always leave answers in radical form.

18.3 EXERCISES A

Solve using the quadratic formula.

1. $x^2 + x - 20 = 0$

4, −5

2. $y^2 - 3y = 10$

5, −2

3. $z^2 + 8 = 9z$

1, 8

4. $x^2 - 9 = 0$

  3, −3

5 $0 = 7y - 2 + 15y^2$

  $\frac{1}{5}$, $-\frac{2}{3}$

6. $5z - 2 = 3z^2$

  1, $\frac{2}{3}$

7 $5x^2 - x = 1$

  $\dfrac{1 \pm \sqrt{21}}{10}$

8. $2y^2 - 4y + 1 = 0$

  $\dfrac{2 \pm \sqrt{2}}{2}$

9. $4z^2 + 9z + 3 = 0$

  $\dfrac{-9 \pm \sqrt{33}}{8}$

10. $x^2 + 2x = 5$

  $-1 \pm \sqrt{6}$

11 $z^2 = z - 5$

  no solution

12. $y^2 - 6y = 1$

  $3 \pm \sqrt{10}$

13. $x^2 - 3x - 9 = 0$

  $\dfrac{3 \pm 3\sqrt{5}}{2}$

14 $y(y + 1) + 2y = 4$

  1, −4

15. $(z + 1)^2 = 5 + 3z$

  $\dfrac{1 \pm \sqrt{17}}{2}$

16. $x^2 - 2x + 5 = 0$

  no solution

17. $4y^2 - 1 = 0$

  $\frac{1}{2}$, $-\frac{1}{2}$

18 $(2z - 1)(z - 2) - 11 =$
  $2(z + 4) - 8$

  $\frac{9}{2}$, −1

Solve using any method.

19. $2x^2 - 7x + 3 = 0$

  3, $\frac{1}{2}$

20 $3z^2 + 10z - 1 = 0$

  $\dfrac{-5 \pm 2\sqrt{7}}{3}$

21. $3y^2 - 12 = 0$

  2, −2

22. $3x^2 - 2 = -5x$

  $\frac{1}{3}$, −2

23. $x^2 + x - 1 = 0$

  $\dfrac{-1 \pm \sqrt{5}}{2}$

24. $-x^2 - x + 1 = 0$

  $\dfrac{1 \pm \sqrt{5}}{-2}$

25. How do the solutions to Exercises 23 and 24 compare? Explain.
  Same. The equations are equivalent.

26 $\dfrac{2}{3}x^2 - \dfrac{1}{3}x - 1 = 0$

  $\frac{3}{2}$, −1

27. $\dfrac{1}{2}x^2 + x - 1 = 0$

  $-1 \pm \sqrt{3}$

28. $\dfrac{1}{4}x^2 + \dfrac{1}{2}x - \dfrac{3}{4} = 0$

  1, −3

29. Write the solution to Exercise 7 using a decimal approximation.   **0.56, −0.36**

30. Write the solution to Exercise 8 using a decimal approximation.   **1.7, 0.3**

Solve.

31. The hypotenuse of a right triangle is 3 cm long, and one leg is 1 cm more than the other. Find the approximate measure of each leg.   **1.6 cm, 2.6 cm**

32 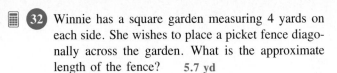 Winnie has a square garden measuring 4 yards on each side. She wishes to place a picket fence diagonally across the garden. What is the approximate length of the fence?   **5.7 yd**

4 yd   4 yd

33. If an object is thrown upward with initial velocity of 128 feet per second, its height, h, above the ground in t seconds is given by $h = 128t - 16t^2$. How long will it take for the object to reach a height of 240 feet?   **3 sec and 5 sec**

34. Show that the sum of the solutions to the equation $y^2 - 6y - 1 = 0$ is equal to $-\frac{b}{a}$. Note that $a = 1$ and $b = -6$.
$(3 + \sqrt{10}) + (3 - \sqrt{10}) = 6 = -\frac{-6}{1} = -\frac{b}{a}$

FOR REVIEW

35. Solve $x^2 + x = 13 + x$ by taking roots.

$\pm\sqrt{13}$

36. Solve $2y^2 + 2y = 3$ by completing the square.
$\dfrac{-1 \pm \sqrt{7}}{2}$

Exercises 37–38 review material from Sections 16.5 and 17.6 to prepare for the next section. Solve each equation.

37. $\dfrac{1}{x} + \dfrac{2}{x - 3} = \dfrac{4}{x^2 - 3x}$

$\dfrac{7}{3}$

38. $\sqrt{x^2 - 11} - x + 1 = 0$

6

ANSWERS: 1. $4, -5$ 2. $5, -2$ 3. $1, 8$ 4. $3, -3$ 5. $\frac{1}{5}, -\frac{2}{3}$ 6. $1, \frac{2}{3}$ 7. $\dfrac{1 \pm \sqrt{21}}{10}$ 8. $\dfrac{2 \pm \sqrt{2}}{2}$ 9. $\dfrac{-9 \pm \sqrt{33}}{8}$

10. $-1 \pm \sqrt{6}$ 11. no solution 12. $3 \pm \sqrt{10}$ 13. $\dfrac{3 \pm 3\sqrt{5}}{2}$ 14. $1, -4$ 15. $\dfrac{1 \pm \sqrt{17}}{2}$ 16. no solution 17. $\frac{1}{2}, -\frac{1}{2}$

18. $\frac{9}{2}, -1$ 19. $3, \frac{1}{2}$ 20. $\dfrac{-5 \pm 2\sqrt{7}}{3}$ 21. $2, -2$ 22. $\frac{1}{3}, -2$ 23. $\dfrac{-1 \pm \sqrt{5}}{2}$ 24. $\dfrac{1 \pm \sqrt{5}}{-2}$ 25. Same (Multiply

numerator and denominator of $\dfrac{-1 \pm \sqrt{5}}{2}$ by -1 to obtain $\dfrac{1 \pm \sqrt{5}}{-2}$). The equations are equivalent (multiply both sides

of 23 by -1 to get 24).   26. $\frac{3}{2}, -1$ 27. $-1 \pm \sqrt{3}$ 28. $1, -3$ 29. $0.56, -0.36$ (using $\sqrt{21} \approx 4.6$) 30. $1.7, 0.3$

(using $\sqrt{2} \approx 1.4$) 31. 1.6 cm, 2.6 cm 32. 5.7 yards 33. 3 seconds (on the way up) then again at 5 seconds (on the

way down) 34. The solutions are $3 + \sqrt{10}$ and $3 - \sqrt{10}$. Their sum is $(3 + \sqrt{10}) + (3 - \sqrt{10}) = 6 = -\frac{-6}{1} = -\frac{b}{a}$.

35. $\pm\sqrt{13}$ 36. $\dfrac{-1 \pm \sqrt{7}}{2}$ 37. $\frac{7}{3}$ 38. 6

18.3 EXERCISES B

Solve using the quadratic formula.

1. $x^2 - 4x - 21 = 0$ $-3, 7$

2. $y^2 - 10y + 9 = 0$ $1, 9$

3. $z^2 + 5 = 6z$ $1, 5$

4. $x^2 - 49 = 0$ $-7, 7$

5. $0 = 20y^2 - y - 1$ $\frac{1}{4}, -\frac{1}{5}$

6. $7z - 3 = 4z^2$ $1, \frac{3}{4}$

7. $2x^2 - x = 2$ $\dfrac{1 \pm \sqrt{17}}{4}$

8. $y^2 + 5y + 5 = 0$ $\dfrac{-5 \pm \sqrt{5}}{2}$

9. $3z^2 + 3 = 7z$ $\dfrac{7 \pm \sqrt{13}}{6}$

10. $x^2 - 4x = 1$ $2 \pm \sqrt{5}$

11. $3y^2 + 2y = 7$ $\dfrac{-1 \pm \sqrt{22}}{3}$

12. $z^2 = z - 9$ no solution

13. $x^2 - 5x + 3 = 0$ $\dfrac{5 \pm \sqrt{13}}{2}$

14. $y(y + 3) + 2y = 6$ $1, -6$

15. $3(z - 1)^2 = 3z$ $\dfrac{3 \pm \sqrt{5}}{2}$

16. $x^2 - 3x + 7 = 0$ no solution

17. $6y^2 + 7y = 3$ $\frac{1}{3}, -\frac{3}{2}$

18. $(z + 1)^2 = 5(1 + z)$ $4, -1$

Solve using any method.

19. $2x^2 - x - 1 = 0$ $1, -\frac{1}{2}$

20. $7y^2 - 49 = 0$ $\pm \sqrt{7}$

21. $z^2 - 5z - 2 = 0$ $\dfrac{5 \pm \sqrt{33}}{2}$

22. $2x^2 = \frac{1}{2}(27x + 7)$ $7, -\frac{1}{4}$

23. $x^2 + 6x - 1 = 0$ $-3 \pm \sqrt{10}$

24. $-x^2 - 6x + 1 = 0$ $-3 \pm \sqrt{10}$

25. How do the solutions to Exercises 23 and 24 compare? Explain.   Same. The equations are equivalent.

26. $x^2 - \frac{1}{2}x - 5 = 0$ $\frac{5}{2}, -2$

27. $\frac{1}{3}x^2 + x + \frac{1}{6} = 0$ $\dfrac{-3 \pm \sqrt{7}}{2}$

28. $\frac{1}{11}x^2 + \frac{9}{11}x - 2 = 0$ $2, -11$

29. Write the solution to Exercise 7 using a decimal approximation. $1.28, -0.78$

30. Write the solution to Exercise 8 using a decimal approximation. $-1.4, -3.6$

31. The hypotenuse of a right triangle is 7 cm long, and one leg is 3 cm more than the other. Find the approximate measure of each leg. 3.2 cm, 6.2 cm

32. A baseball diamond is a square 90 ft on each side. How far is it from home plate directly across to second base?   approximately 127 ft

33. If an object is thrown upward with initial velocity of 64 feet per second its height, h, above the ground in t seconds is given by $h = 64t - 16t^2$. How long will it take for the object to reach a height of 48 feet?   1 sec and 3 sec

34. Show that the product of the solutions to the equation $y^2 - 6y - 1 = 0$ is equal to $\frac{c}{a}$. Note that $a = 1$ and $c = -1$.
$(3 + \sqrt{10})(3 - \sqrt{10}) = 9 - 10 = -1 = \frac{-1}{1} = \frac{c}{a}$

FOR REVIEW

35. Solve $2y^2 - 50 = 0$ by taking roots. ± 5

36. Solve $2x^2 - 6x + 3 = 0$ by the method of completing the square. $\dfrac{3 \pm \sqrt{3}}{2}$

Exercises 37–38 review material from Sections 16.5 and 17.6 to prepare for the next section. Solve each equation.

37. $\dfrac{3}{x + 5} - \dfrac{2}{x} = \dfrac{5}{x^2 + 5x}$ 15

38. $\sqrt{x^2 + 7} - x = 1$ 3

18.3 EXERCISES C

Solve for x using the quadratic formula.

1. $0.3x^2 - 0.1x - 1.6 = 0$ [*Hint:* Multiply by 10.]
$\dfrac{1 \pm \sqrt{193}}{6}$

2. $x^2 + 3xy + y^2 = 0$ $\left[\text{Answer: } \dfrac{-y(3 \pm \sqrt{5})}{2}\right]$

Use the quadratic formula and your calculator to solve each quadratic equation. Give answers to two decimal places.

3. $\sqrt{2}x^2 - 3x - 7.1 = 0$
[Answer: 3.54, −1.42]

4. $3x^2 + \sqrt{5}x - \dfrac{1}{2} = 0$
0.18, −0.93

18.4 SOLVING FRACTIONAL AND RADICAL EQUATIONS

STUDENT GUIDEPOSTS

1 Fractional Equations   **2** Radical Equations

1 FRACTIONAL EQUATIONS

In Chapter 16 we learned that to solve fractional equations we multiply both sides by the LCD. The fractional equations we look at in this section will result in quadratic equations when both sides are multiplied by the LCD. Remember always to check for invalid answers when solving fractional equations.

EXAMPLE 1 SOLVING A FRACTIONAL EQUATION	PRACTICE EXERCISE 1

Solve $\dfrac{6}{x} - x = 5$.

The LCD is x.

$$x\left[\dfrac{6}{x} - x\right] = x \cdot 5 \quad \text{Multiply by LCD}$$

$$\cancel{x} \cdot \dfrac{6}{\cancel{x}} - x \cdot x = 5x$$

$$6 - x^2 = 5x$$

$$x^2 + 5x - 6 = 0$$

$$(x + 6)(x - 1) = 0$$

$$x + 6 = 0 \quad \text{or} \quad x - 1 = 0$$

$$x = -6 \qquad\qquad x = 1$$

Check: $\dfrac{6}{(-6)} - (-6) \overset{?}{=} 5 \qquad \dfrac{6}{(1)} - (1) \overset{?}{=} 5$

$$-1 + 6 = 5 \qquad\qquad 6 - 1 = 5$$

The solutions are −6 and 1.

Solve $x - \dfrac{11}{x} = \dfrac{10}{x} + 4$.

Answer: 7, −3

| EXAMPLE 2 SOLVING A FRACTIONAL EQUATION | PRACTICE EXERCISE 2 |

Solve $\dfrac{12}{x^2 - 4} - \dfrac{3}{x - 2} = -1$.

The LCD $= (x - 2)(x + 2)$.

$$(x - 2)(x + 2)\left[\dfrac{12}{x^2 - 4} - \dfrac{3}{x - 2}\right] = (x - 2)(x + 2)(-1) \quad \text{Multiply both sides by LCD}$$

$$\cancel{(x - 2)(x + 2)} \cdot \dfrac{12}{\cancel{(x - 2)(x + 2)}} = \cancel{(x - 2)}(x + 2) \cdot \dfrac{3}{\cancel{(x - 2)}} \quad \text{Distribute}$$

$$= (x - 2)(x + 2)(-1)$$

$$12 - 3(x + 2) = (x^2 - 4)(-1) \quad \text{Watch parentheses}$$

$$12 - 3x - 6 = -x^2 + 4 \quad \text{Watch signs}$$

$$x^2 - 3x + 2 = 0 \quad \text{Collect terms}$$

$$(x - 1)(x - 2) = 0 \quad \text{Factor}$$

$$x - 1 = 0 \quad \text{or} \quad x - 2 = 0$$

$$x = 1 \qquad\qquad x = 2$$

Check: $\dfrac{12}{(1)^2 - 4} - \dfrac{3}{(1) - 2} \overset{?}{=} -1 \qquad \dfrac{12}{(2)^2 - 4} - \dfrac{3}{(2) - 2} \overset{?}{=} -1$

$$\dfrac{12}{-3} - \dfrac{3}{-1} \overset{?}{=} -1 \qquad\qquad \dfrac{2}{4 - 4} - \dfrac{3}{2 - 2} \overset{?}{=} -1$$

$$-4 + 3 = -1 \qquad\qquad \dfrac{12}{0} - \dfrac{3}{0} \neq -1$$

Cannot divide by 0

The only solution is 1. Notice that 2 makes a denominator zero in the original equation. In general, any number for which a fraction is undefined must be discarded as a solution.

PRACTICE EXERCISE 2

Solve $\dfrac{3}{x + 3} + \dfrac{4}{x^2 - 9} = 1$.

Answer: **4, −1**

| EXAMPLE 3 SOLVING A FRACTIONAL EQUATION | PRACTICE EXERCISE 3 |

Solve $\dfrac{y + 1}{y} = \dfrac{-2}{y - 2}$.

The LCD $= y(y - 2)$.

$$\cancel{y}(y - 2)\left[\dfrac{y + 1}{\cancel{y}}\right] = y\cancel{(y - 2)}\left[\dfrac{-2}{\cancel{y - 2}}\right] \quad \text{Multiply by LCD}$$

$$(y - 2)(y + 1) = -2y$$

$$y^2 - y - 2 = -2y$$

$$y^2 + y - 2 = 0$$

$$(y + 2)(y - 1) = 0$$

$$y + 2 = 0 \quad \text{or} \quad y - 1 = 0$$

$$y = -2 \qquad\qquad y = 1$$

PRACTICE EXERCISE 3

Solve $\dfrac{x^2}{x - 10} = \dfrac{3}{x - 10}$. Start by multiplying both sides by the LCD, $x - 10$.

Check: $\dfrac{(-2) + 1}{(-2)} \overset{?}{=} \dfrac{-2}{(-2) - 2}$ $\dfrac{(1) + 1}{(1)} \overset{?}{=} \dfrac{-2}{(1) - 2}$

$\dfrac{-1}{-2} \overset{?}{=} \dfrac{-2}{-4}$ $\dfrac{2}{1} \overset{?}{=} \dfrac{-2}{-1}$

$\dfrac{1}{2} = \dfrac{1}{2}$ $2 = 2$

The solutions are -2 and 1.

Answer: $\pm\sqrt{3}$

---

② RADICAL EQUATIONS

With radical equations in Chapter 17, we squared both sides of the equation to eliminate a radical. Now we consider other radical equations that become quadratic equations when both sides are squared. We always need to check our answers when solving radical equations, since answers that do not satisfy the *original* equation may be introduced by the process of squaring.

| EXAMPLE 4 SOLVING A RADICAL EQUATION | PRACTICE EXERCISE 4 |

Solve $\sqrt{12 - x} = x$.

$(\sqrt{12 - x})^2 = x^2$ Indicate the square of both sides

$12 - x = x^2$ Square both sides

$x^2 + x - 12 = 0$ Collect terms

$(x - 3)(x + 4) = 0$ Factor

$x - 3 = 0$ or $x + 4 = 0$

$x = 3$ $x = -4$

Check: $\sqrt{12 - (3)} \overset{?}{=} 3$ $\sqrt{12 - (-4)} \overset{?}{=} (-4)$

$\sqrt{9} \overset{?}{=} 3$ $\sqrt{16} \neq -4$ ($\sqrt{16} = 4$,

$3 = 3$ *not* -4.)

The only solution is 3.

Solve $x = \sqrt{8x + 20}$.

Answer: 10 (-2 does not check)

| EXAMPLE 5 SOLVING A RADICAL EQUATION | PRACTICE EXERCISE 5 |

Solve $x + 2 = \sqrt{x + 8}$.

$(x + 2)^2 = (\sqrt{x + 8})^2$ Indicate the square of both sides

$x^2 + 4x + 4 = x + 8$ Square both sides, $(x + 2)^2 \neq (x^2 + 4)$

$x^2 + 3x - 4 = 0$ Collect terms

$(x + 4)(x - 1) = 0$ Factor

$x + 4 = 0$ or $x - 1 = 0$

$x = -4$ $x = 1$

Check: $(-4) + 2 \overset{?}{=} \sqrt{(-4) + 8}$ $(1) + 2 \overset{?}{=} \sqrt{(1) + 8}$

$-2 \neq \sqrt{4}$ $3 \overset{?}{=} \sqrt{9}$

$3 = 3$

The only solution is 1.

Solve $\sqrt{13 - 6x} = x - 3$.

Answer: no solution

| **EXAMPLE 6**   SOLVING A RADICAL EQUATION | **PRACTICE EXERCISE 6** |

Solve $x = 4\sqrt{x + 1} - 4$.

$x + 4 = 4\sqrt{x + 1}$	Add 4 to both sides to isolate the radical
$(x + 4)^2 = (4\sqrt{x + 1})^2$	Indicate the square of both sides
$x^2 + 8x + 16 = 16(x + 1)$	Do not forget the middle term on left and to square 4 on right
$x^2 + 8x + 16 = 16x + 16$	Distribute the 16
$x^2 - 8x = 0$	Collect terms
$x(x - 8) = 0$	Factor out x

$x = 0$ or $x - 8 = 0$

$\qquad\qquad\quad x = 8$

Check: $(0) \overset{?}{=} 4\sqrt{(0) + 1} - 4$ $\qquad$ $(8) \overset{?}{=} 4\sqrt{(8) + 1} - 4$

$\qquad\qquad 0 \overset{?}{=} 4\sqrt{1} - 4$ $\qquad\qquad\quad 8 \overset{?}{=} 4\sqrt{9} - 4$

$\qquad\qquad 0 \overset{?}{=} 4 - 4$ $\qquad\qquad\qquad 8 \overset{?}{=} 4 \cdot 3 - 4$

$\qquad\qquad 0 = 0$ $\qquad\qquad\qquad\quad 8 \overset{?}{=} 12 - 4$

$\qquad\qquad\qquad\qquad\qquad\qquad\qquad 8 = 8$

The solutions are 0 and 8.

Solve $\sqrt{x + 5} - x + 1 = 0$.

Answer: 4

18.4   EXERCISES A

Solve.

1. $\dfrac{3}{x - 4} = 1 + \dfrac{5}{x + 4}$

6, −8

2. $\dfrac{y - 3}{y} = \dfrac{-4}{y + 1}$

1, −3

3. $\dfrac{3}{1 + z} + \dfrac{2}{1 - z} = -1$

−3, 2

4. $\dfrac{a^2}{a + 1} - \dfrac{9}{a + 1} = 0$

3, −3

5. $\dfrac{1}{x + 2} + \dfrac{x}{x - 2} = \dfrac{1}{2}$

0, −6

6. $\dfrac{60}{y + 3} = \dfrac{60}{y} - 1$

12, −15

7 $\dfrac{5}{z + 3} - \dfrac{1}{z + 3} = \dfrac{z + 3}{z + 2}$

−1

8 $\dfrac{x - 1}{x} = \dfrac{x}{x + 1}$

no solution

9. $\dfrac{16}{a + 2} = 1 + \dfrac{2}{a - 4}$

6, 10

10. $\dfrac{1}{x-3} + \dfrac{1}{x+3} = \dfrac{1}{x^2-9}$

$\frac{1}{2}$

11. $\dfrac{x^2}{2x+1} - \dfrac{5}{2x+1} = 0$

[*Hint:* Multiply through by $2x+1$.]

$\pm\sqrt{5}$

12 $\dfrac{x+2}{2} = \dfrac{x+5}{x+2}$

$-1 \pm \sqrt{7}$

13. $x = \sqrt{3x+10}$

5

14. $\sqrt{5y+6} - y = 0$

6

15. $z - 7 = \sqrt{z-5}$

9

16. $\sqrt{15-a} = a - 3$

6

17 $1 + 2\sqrt{x-1} = x$

5, 1

18. $y = 1 + 6\sqrt{y-9}$

25, 13

19. $\sqrt{1-2a} = a - 1$

no solution

20. $\sqrt{2x+7} - 4 = x$

-3

21. $\sqrt{z^2+2} = z + 1$

$\frac{1}{2}$

22 $3\sqrt{y+1} - y - 1 = 0$

$-1, 8$

23. $\sqrt{5-4x} = 2 - x$

1, -1

24 $\sqrt{2x^2-5} = x$

$\sqrt{5}$

25 A number increased by its reciprocal is the same as $\frac{5}{2}$. Find the number.

$2, \frac{1}{2}$

26. The principal square root of 2 more than a number is equal to the number itself. Find the number.

2

FOR REVIEW

Solve.

27. $z^2 - 8z + 8 = 0$
$4 \pm 2\sqrt{2}$

28. $x^2 - \dfrac{2}{3}x - 1 = 0$
$\dfrac{1 \pm \sqrt{10}}{3}$

29. $2y^2 + 3y - 2 = 0$
$\frac{1}{2}, -2$

30. $5z^2 + 2z + 1 = 0$
no solution

31. It can be shown that a polygon (many-sided geometric figure) with x sides has a total of n diagonals where $n = \frac{1}{2}x^2 - \frac{3}{2}x$. A hexagon has 6 sides; how many diagonals does it have? Draw a hexagon and verify your answer by actual count.
9

32. Use the formula in Exercise 31 to find the number of sides that a polygon has if it is known to have 20 diagonals. Draw a polygon with the number of sides that you determine and count its diagonals.
8

ANSWERS: 1. 6, −8 2. 1, −3 3. −3, 2 4. 3, −3 5. 0, −6 6. 12, −15 7. −1 8. no solution 9. 6, 10
10. $\frac{1}{2}$ 11. $\pm\sqrt{5}$ 12. $-1 \pm \sqrt{7}$ 13. 5 14. 6 15. 9 16. 6 17. 5, 1 18. 25, 13 19. no solution 20. −3 21. $\frac{1}{2}$
22. −1, 8 23. 1, −1 24. $\sqrt{5}$ 25. 2, $\frac{1}{2}$ 26. 2 27. $4 \pm 2\sqrt{2}$ 28. $\dfrac{1 \pm \sqrt{10}}{3}$ 29. $\frac{1}{2}$, −2 30. no solution 31. 9
32. 8

18.4 EXERCISES B

Solve.

1. $\dfrac{3}{x-1} = 1 + \dfrac{2}{x+1}$
3, −2

2. $\dfrac{y}{1+y} = \dfrac{3-y}{4}$
−3, 1

3. $\dfrac{1}{x+2} + \dfrac{1}{x-2} = \dfrac{3}{8}$
$-\frac{2}{3}$, 6

4. $\dfrac{a^2}{a+5} - \dfrac{4}{a+5} = 0$
2, −2

5. $\dfrac{6}{x+3} + \dfrac{x}{x-3} = 1$ 1

6. $\dfrac{32}{y+5} = \dfrac{6}{y} + 2$ 3, 5

7. $\dfrac{4}{z+1} - \dfrac{1}{z+1} = \dfrac{z-1}{z+3}$
5, −2

8. $x + \dfrac{2}{x-2} = \dfrac{1}{x-2}$ 1

9. $\dfrac{z+2}{z} = \dfrac{1}{z+2}$
no solution

10. $\dfrac{3}{a-1} + 1 = \dfrac{6}{a^2-1}$
−4

11. $\dfrac{x^2}{1+3x} - \dfrac{11}{1+3x} = 0$
$\pm\sqrt{11}$

12. $\dfrac{x+3}{x+1} = \dfrac{x+1}{3}$
$\dfrac{1 \pm \sqrt{33}}{2}$

13. $x = \sqrt{2x + 24}$ 6

14. $\sqrt{4y + 5} - y = 0$ 5

15. $z + 3 = \sqrt{12z + 9}$ 0, 6

16. $\sqrt{5y + 21} = y + 3$ 3

17. $\sqrt{2x - 1} = x - 2$ 5

18. $z = \sqrt{z + 7} - 1$ 2

19. $\sqrt{7 - 3a} = 1 + a$ 1

20. $2\sqrt{3 - x} - 5 = x$ −1

21. $\sqrt{y^2 + 8} = y - 2$
no solution

22. $2\sqrt{y + 15} + y - 9 = 0$ 1

23. $\sqrt{4x + 13} = x + 2$ 3

24. $\sqrt{5x^2 - 11} = 2x$ $\sqrt{11}$

25. A number decreased by its reciprocal is the same as $\frac{15}{4}$. Find the number.
$4, -\frac{1}{4}$

26. The principal square root of 8 more than a number is three times the number. Find the number.
1

FOR REVIEW

Solve.

27. $x^2 - 6x + 6 = 0$ $3 \pm \sqrt{3}$

28. $y^2 - \frac{1}{3}y - 1 = 0$ $\dfrac{1 \pm \sqrt{37}}{6}$

29. $5z^2 + 14z - 3 = 0$ $\frac{1}{5}, -3$

30. $4x^2 + x + 2 = 0$ no solution

31. It can be shown that a polygon with x sides has a total of n diagonals where $n = \frac{1}{2}x^2 - \frac{3}{2}x$. An octagon has 8 sides; how many diagonals does it have? Draw an octagon and verify your answer by actual count.
20

32. Use the formula in Exercise 31 to find the number of sides that a polygon has if it is known to have 5 diagonals. Draw a polygon with the number of sides that you determine and count its diagonals.
5

18.4 EXERCISES C

Solve.

1. $\dfrac{3}{x^2 - 5x + 6} - \dfrac{2}{x^2 + 3x - 10} = \dfrac{x + 2}{x^2 + 2x - 15}$

$\left[\text{Answer:} \quad \dfrac{1 \pm \sqrt{101}}{2} \right]$

2. $\sqrt{2x - 2} - \sqrt{x + 6} = -1$
3

18.5 MORE APPLICATIONS OF QUADRATIC EQUATIONS

STUDENT GUIDEPOSTS

1 Basic Applications

3 Work Applications

2 Geometry Applications

4 Motion Applications

1 BASIC APPLICATIONS

Many applied problems can be solved using quadratic equations. We illustrated this in the preceding sections using several types of problems, and now we consider a wider variety of these applications.

EXAMPLE 1   AGE PROBLEM

Murphy's age in 3 years will be four times the square of his age now. How old is Murphy?

Let x = Murphy's present age,

$x + 3$ = Murphy's age in 3 years,

$4x^2$ = four times the square of his present age.

The last two expressions are equal, so we solve the following equation.

$$4x^2 = x + 3$$
$$4x^2 - x - 3 = 0$$
$$(4x + 3)(x - 1) = 0$$
$$4x + 3 = 0 \quad \text{or} \quad x - 1 = 0$$
$$4x = -3 \qquad\qquad x = 1$$
$$x = -\frac{3}{4}$$

Since $-\frac{3}{4}$ could not be a person's age, $x = 1$ is the only possible solution.

Check: $4(\mathbf{1})^2 \overset{?}{=} \mathbf{1} + 3$

$4 = 4$

Murphy is 1 year old.

PRACTICE EXERCISE 1

The square of Roberta's age plus the square of her age in 5 years is 97. How old is Roberta?

Answer: 4 years old

EXAMPLE 2   CONSECUTIVE INTEGERS

Twice the product of two consecutive positive even integers is 160. Find the integers.

Let x = the first positive even integer,

$x + 2$ = the next consecutive even integer.

Twice the product of these two integers is 160, so we should solve the following equation.

$$2x(x + 2) = 160$$
$$x(x + 2) = 80 \qquad \text{Multiply both sides by } \tfrac{1}{2}$$
$$x^2 + 2x = 80$$
$$x^2 + 2x - 80 = 0$$
$$(x - 8)(x + 10) = 0$$
$$x - 8 = 0 \quad \text{or} \quad x + 10 = 0$$
$$x = 8 \qquad\qquad x = -10$$

Since x must be positive, 8 is the only possible solution.

Check: $2 \cdot \mathbf{8}\,(\mathbf{8} + 2) \overset{?}{=} 160$

$2 \cdot 8(10) \overset{?}{=} 160$

$160 = 160$

The numbers are 8 and 10. ($x + 2 = 10$.)

PRACTICE EXERCISE 2

Four times the product of two consecutive negative odd integers is 780. Find the integers.

Answer: $-15, -13$

❷ GEOMETRY APPLICATIONS

When working a geometry problem remember to sketch a figure as shown in the next two examples.

EXAMPLE 3 RECTANGLE PROBLEM	PRACTICE EXERCISE 3

Find the length and width of a rectangle if the length is 3 cm more than the width and the area is 180 cm².

Make a sketch as in Figure 18.2.

Let $x =$ width of the rectangle,
    $x + 3 =$ length of the rectangle.

Use the formula for the area of a rectangle, $A = l \cdot w$, to get the following equation.

$$x(x + 3) = 180$$
$$x^2 + 3x = 180$$
$$x^2 + 3x - 180 = 0$$
$$(x + 15)(x - 12) = 0$$

$$x + 15 = 0 \quad \text{or} \quad x - 12 = 0$$
$$x = -15 \qquad\qquad x = 12$$
$$(x + 3 = 15)$$

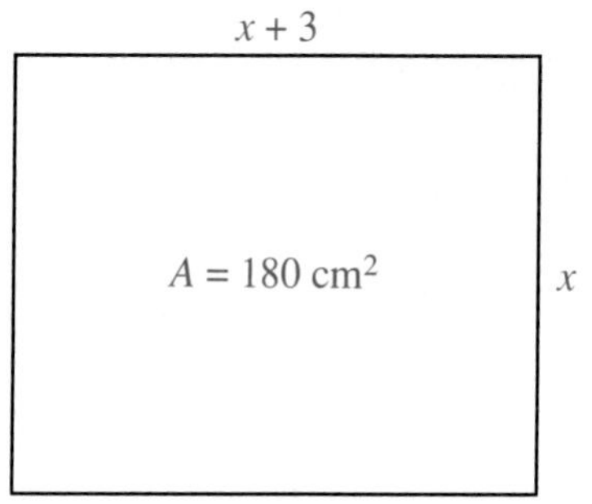

Figure 18.2

Since -15 could not be the width of a rectangle, 12 is the only possible solution.

Check: $12\,(12 + 3) \stackrel{?}{=} 180$
           $12 \cdot 15 \stackrel{?}{=} 180$
           $180 = 180$

The width is 12 cm and the length is 15 cm.

Practice Exercise 3: Find the base and height of a triangle if the base is 2 cm more than the height and the area is 24 cm².

Answer: 8 cm, 6 cm

EXAMPLE 4   AREA AND PERIMETER

The number of square inches in the area of a square is 12 more than the number of inches in its perimeter. Find the length of a side.

A sketch of the square is shown in Figure 18.3.

Let x = length of a side (all sides are the same length, x),

x^2 = area of the square ($A = l \cdot w = x \cdot x$),

$4x$ = perimeter of the square ($P = 2 \cdot l + 2 \cdot w = 2 \cdot x + 2 \cdot x = 4x$).

Since the area is 12 more than the perimeter, adding 12 to the perimeter, $4x$, will give us the area, x^2.

$$x^2 = 4x + 12$$
$$x^2 - 4x - 12 = 0$$
$$(x + 2)(x - 6) = 6$$
$$x + 2 = 0 \quad \text{or} \quad x - 6 = 0$$
$$x = -2 \qquad\qquad x = 6$$

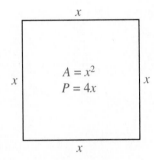

Figure 18.3

Since the length of a side cannot be negative, 6 is the only possible solution. Each side is 6 in.

PRACTICE EXERCISE 4

A box is 12 inches high and the volume is 420 in³. Find the length and width if the length is 2 in more than the width.

Let x = width of box,
$x + 2$ = length of box.
$12x(x + 2) = \underline{\qquad}$

Answer: 7 inches, 5 inches

---

③ WORK APPLICATIONS

Another variation of the work problems introduced in Section 16.5 results in a quadratic equation.

EXAMPLE 5   WORK PROBLEM

When each works alone, Ralph can do a job in 3 hours less time than Bert. When they work together, it takes 2 hours. How long does it take each to do the job by himself?

Since 2 = number of hrs to do the job together,

then $\dfrac{1}{2}$ = amount done together in 1 hr.

PRACTICE EXERCISE 5

Gloria can do a job in 2 hours less time than Evelyn. How long would it take each working alone, if together they can do the job in 5 hours?

Let x = number of hrs for Bert to do the job,

$\dfrac{1}{x}$ = amount done by Bert in 1 hr.

Then $x - 3$ = number of hrs for Ralph to do the job,

$\dfrac{1}{(x-3)}$ = amount done by Ralph in 1 hr.

(Note: $\dfrac{1}{(x-3)} \neq \dfrac{1}{x} - \dfrac{1}{3}$. To see this, substitute $x = 2$.)

We need to solve the following equation.

$$\dfrac{1}{x} + \dfrac{1}{(x-3)} = \dfrac{1}{2} \qquad \begin{array}{l}\text{(amount by Bert)} + \\ \text{(amount by Ralph)} = \\ \text{(amount together)}\end{array}$$

$$2x(x-3)\left(\dfrac{1}{x} + \dfrac{1}{(x-3)}\right) = 2x(x-3) \cdot \dfrac{1}{2} \qquad \text{The LCD is } 2x(x-3)$$

$$2x(x-3)\dfrac{1}{x} + 2x(x-3)\dfrac{1}{(x-3)} = x(x-3)$$

$$2(x-3) + 2x = x^2 - 3x$$

$$2x - 6 + 2x = x^2 - 3x$$

$$0 = x^2 - 7x + 6$$

$$0 = (x-6)(x-1)$$

$$x - 6 = 0 \quad \text{or} \quad x - 1 = 0$$

$$x = 6 \qquad\qquad x = 1$$

If Bert did the job in 1 hr, Ralph would do it in $1 - 3 = -2$ hr, which makes no sense. Thus, Bert would take 6 hr and Ralph 3 hr.

Answer: Gloria approximately 9.1 hr; Evelyn approximately 11.1 hr

④ MOTION APPLICATIONS

The last example in this section can be classified as a motion or rate problem.

EXAMPLE 6 MOTION PROBLEM

A backpacker can hike 10 miles up a mountain and then return in a total time of 7 hours. Her hiking rate uphill is 3 mph slower than her rate downhill. At what rate does she hike up the mountain?

Let x = the rate hiking up the mountain.

Then $x + 3$ = the rate hiking down the mountain.

We use the distance formula $d = rt$, solved for $t = \dfrac{d}{r}$, to express the two times. Since the distance hiked uphill and the return distance downhill are both 10, we have

$$\dfrac{10}{x} = \text{the time spent hiking up the mountain,}$$

$$\dfrac{10}{x+3} = \text{the time spent hiking down the mountain.}$$

Since the total time of the hike is 7 hours, we need to solve the following equation.

PRACTICE EXERCISE 6

A small plane flies 720 mi with the wind and returns against the wind in a total time of 15 hours. If the speed of the wind is 20 mph, what is the speed of the plane in still air?

$$\frac{10}{x} + \frac{10}{x+3} = 7 \qquad \text{LCD} = x(x+3)$$

$$x(x+3)\left[\frac{10}{x} + \frac{10}{x+3}\right] = 7x(x+3) \qquad \begin{array}{l}\text{Multiply both sides}\\ \text{by the LCD}\end{array}$$

$$x(x+3)\frac{10}{x} + x(x+3)\frac{10}{x+3} = 7x(x+3) \qquad \begin{array}{l}\text{Use the distributive}\\ \text{law}\end{array}$$

$$10(x+3) + 10x = 7x^2 + 21x$$

$$10x + 30 + 10x = 7x^2 + 21x$$

$$20x + 30 = 7x^2 + 21x$$

$$0 = 7x^2 + x - 30 \qquad \begin{array}{l}\text{Collect like terms}\\ \text{in the quadratic}\\ \text{equation}\end{array}$$

$$0 = (7x + 15)(x - 2) \qquad \text{Factor}$$

$$7x + 15 = 0 \quad \text{or} \quad x - 2 = 0$$

$$x = \frac{-15}{7} \quad \text{or} \quad x = 2$$

Since $\frac{-15}{7}$ cannot represent a rate in this problem, we can discard it. Thus, her rate up the mountain is 2 mph and down the mountain is 5 mph.

Answer: 100 mph

//////////// **C A U T I O N** ////////////

It is important when working any word problem to be neat and complete. Do not try to take shortcuts, especially when writing down the pertinent information. Writing complete descriptions of the variables can eliminate time-consuming errors.

//////////

18.5 EXERCISES A

Solve. (Some problems have been started.)

1. If five is added to the square of a number the result is 41. Find the number.

Let $\quad x =$ the desired number,
$\quad x^2 + 5 =$ five added to the square of the number.

6, −6

2. If the square of Mary's age is decreased by 44, the result is 100. How old is Mary?

Let $\quad x =$ Mary's age,
$\quad x^2 - 44 =$ the square of Mary's age decreased by 44.

12 years old

3. If the square of a number is decreased by 10 the result is three times the number. Find the number.

Let $x =$ the number.
5, −2

4. Twice the square of Ernie's age, less 9, is the same as seventeen times his age. How old is Ernie?

Let $x =$ Ernie's age.
9 years old

5. The product of 1 more than a number and 1 less than the number is 99. Find the number.

Let x = the desired number,
   $x + 1$ = 1 more than the number,
   $x - 1$ = 1 less than the number.

10, −10

6. Sam's present age times his age in five years is 84. How old is Sam?

7 years old

7. The product of two positive consecutive even integers is 120. Find the integers.

Let x = the first even integer,
   $x + 2$ = the next even integer.

10, 12

8 Three times the product of two positive consecutive odd integers is 297. Find the integers.

9, 11

9. Adding 4 to the square of Marvin's age is the same as subtracting 3 from eight times his age. How old is Marvin?

1 year old or 7 years old

10. The sum of the squares of two consecutive even positive integers is 100. Find the integers.

6, 8

11. One number is 6 larger than another. The square of the larger is 96 more than the square of the smaller. Find the numbers.

5, 11

12. Find the length and width of a rectangle if the length is 4 cm longer than the width and the area is 140 cm².

Let x = width of rectangle,
   $x + 4$ = length of rectangle.

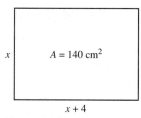

14 cm, 10 cm

13. The number of square inches in the area of a square is 21 more than the number of inches in its perimeter. Find the length of a side.

Let x = length of sides of square.

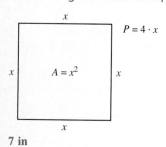

7 in

14. If the hypotenuse of a right triangle is 13 feet long and one leg is 7 feet longer than the other, find the measure of each leg.

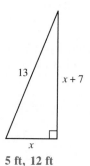

5 ft, 12 ft

15. If the sides of a square are lengthened by 2 cm, its area becomes 169 cm². Find the length of a side.

Let x = length of sides of square,
$\quad x + 2$ = length of sides increased by 2 cm.
11 cm

16 The area of a triangle is 24 ft² and the base is $\frac{1}{3}$ as long as the height. Find the base and height.
12 ft, 4 ft

17. A box is 6 inches high. The length is 8 inches longer than the width and the volume is 1080 in³. Find the width and length.
10 in, 18 in

18. The area of a circle is 1256 cm². Find the radius. (Use $\pi \approx 3.14$.)
20 cm

19 The area of a parallelogram is 55 ft². If the base is 1 ft greater than twice the height, find the base and height.
11 ft, 5 ft

20. The perimeter of a rectangle is 62 inches and the length is 3 inches more than the width. Find the dimensions.
17 in, 14 in

21. It takes one pipe 6 hours longer to fill a tank than it takes a second pipe. Working together they fill the tank in 4 hours. How long would it take each working alone to fill the tank?
6 hr, 12 hr

22 It takes Joe 9 hours longer to grade a set of tests than Dan. If they work together they can grade them in 20 hours. How long would it take each to grade the tests if they worked alone?
36 hr, 45 hr

23. A picture frame is 24 in by 18 in as in the sketch below. If the area of the picture itself is 216 in², what is the width of the frame?

Let x = the width of the frame.
$(24 - 2x)(18 - 2x) = 216$

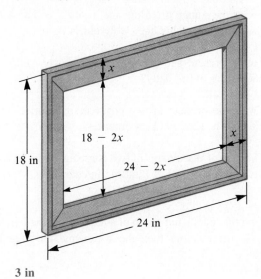

18 − 2x

18 in

24 − 2x

x

24 in

3 in

24. The pressure p, in lb per sq ft, of a wind blowing v mph can be approximated by the equation $p = 0.003v^2$. What is the approximate velocity of the wind when it creates a pressure of 1.875 lb per sq ft against the side of a building?
25 mph

25 The amount of money A that will result if a principal P is invested at r percent integer compounded annually for 2 years is given by $A = P(1 + r)^2$. If $1000 grows to $1210 in 2 years using this formula, what is the interest rate?
10%

26 A boat travels 48 miles upstream and then returns in a total time of 10 hours. If the speed of the stream is 2 mph, what is the speed of the boat in still water?
10 mph

27. Mary Conners can walk 8 miles up a mountain and then return in a total time of 6 hours. Her speed downhill is 2 mph faster than her speed uphill. What is her speed uphill?   **2 mph**

FOR REVIEW

Solve.

28. $\dfrac{x^2}{x-3} = \dfrac{9}{x-3} + 10$ **7**

29. $x = \sqrt{7x+18} - 4$ **1, −2**

30. The square root of Neil's age in 6 years is the same as one-fifth of his present age. How old is Neil?
30 years old

ANSWERS: 1. 6, −6   2. 12 years old   3. 5, −2   4. 9 years old   5. 10, −10   6. 7 years old   7. 10, 12   8. 9, 11
9. 1 year old or 7 years old   10. 6, 8   11. 5, 11   12. 14 cm, 10 cm   13. 7 in   14. 5 ft, 12 ft   15. 11 cm   16. 12 ft,
4 ft   17. 10 in, 18 in   18. 20 cm   19. 11 ft, 5 ft   20. 17 in, 14 in   21. 6 hr, 12 hr   22. 36 hr, 45 hr   23. 3 in
24. 25 mph   25. 10%   26. 10 mph   27. 2 mph   28. 7   29. 1, −2   30. 30 years old

18.5 EXERCISES B

Solve.

1. If 4 is added to the square of a number, the result is 85. Find the number.
9, −9

2. If the square of Rosemary's age is decreased by 24, the result is 300. How old is Rosemary?
18 years old

3. If the square of a number is decreased by 32, the result is four times the number. Find the number.
8, −4

4. Twice the square of Arlo's age, less 22, is the same as twenty times his age. How old is Arlo?
11 years old

5. The product of 1 more than a number and 1 less than the number is 224. Find the number.
15, −15

6. Troy's present age times his age in seven years is 260. How old is Troy?
13 years old

7. The product of two positive consecutive even integers is 168. Find the integers.
12, 14

8. Twice the product of two positive consecutive odd integers is 126. Find the integers.
7, 9

9. Subtracting 100 from the square of Raul's age is the same as adding 60 to twelve times his age. How old is Raul?
20 years old

10. The sum of the squares of two consecutive even positive integers is 164. Find the integers.
8, 10

11. One number is 5 more than another. The square of the larger exceeds the square of the smaller by 95. Find the number.
7, 12

12. Find the length and width of a rectangle if the length is 7 meters more than the width and the area is 198 m².
11 m, 18 m

13. The number of square inches in the area of a square is 5 more than the number of inches in its perimeter. Find the length of a side.

5 in

15. If the sides of a square are lengthened by 2 feet, its area becomes 100 ft². Find the length of a side.

8 ft

17. A box is 8 yards high. The length is 4 yards longer than the width and the volume is 360 yd³. Find the width and length.

5 yd, 9 yd

19. The area of a parallelogram is 175 ft². If the base is 4 feet more than three times the height, find the base and height.

7 ft, 25 ft

21. It takes Jeff 16 hours longer to repair his car than it takes his father, who is a mechanic, to do the same job. If they could do the job together in 6 hours, how long would it take each if they worked alone?

Jeff: 24 hr; Father: 8 hr

23. A rectangular backyard is to have a sidewalk placed entirely around its perimeter in such a way that 875 ft² of lawn area are enclosed inside the walk. If the dimensions of the yard are 40 ft by 30 ft, what is the width of the sidewalk?

$\frac{5}{2}$ **ft**

25. The amount of money A that will result if a principal P is invested at r percent interest compounded annually for 2 years is given by $A = P(1 + r)^2$. If $2000 grows to $2645 in 2 years using this formula, what is the interest rate?

15%

27. Chuck Little rode a bicycle with the wind for 18 miles. He then returned against the wind and the total time of his trip was 5 hours. If his speed with the wind was 3 mph faster than against the wind, what was his speed against the wind?

6 mph

14. If the hypotenuse of a right triangle is 26 cm long and one leg is 14 cm longer than the other, find the measure of each leg.

10 cm, 24 cm

16. The area of a triangle is 18 m² and the base is $\frac{1}{4}$ the height. Find the base and height.

3 m, 12 m

18. The area of a circle is 200.96 cm². Find the radius. (Use $\pi \approx 3.14$.)

8 cm

20. The perimeter of a rectangular garden is 28 meters and the length is 4 meters more than the width. Find the dimensions of the garden.

5 m, 9 m

22. Graydon, the registered cheese cutter at Perko's Delicatessen, is training a new assistant, Burford. It takes Burford 24 hours longer to process the Tillamook cheese shipment than it takes Graydon. If together they could process the cheese in 5 hours, how long would it take each, working alone, to cut and display the cheese?

Graydon: 6 hr; Burford: 30 hr

24. The pressure p, in lb per sq ft, of a wind blowing v miles per hour can be approximated by the equation $p = 0.003v^2$. What is the approximate wind velocity when a pressure of 2.7 lb per sq ft is exerted against the side of a camper?

30 mph

26. Pat Marx swims 4 miles downstream and then returns in a total time of 3 hours. If the speed of the stream is 1 mph, what is Pat's speed in still water?

3 mph

FOR REVIEW

Solve.

28. $\dfrac{z^2}{z - 2} = \dfrac{4}{z - 2} + 5$ **3**

29. $\sqrt{8y - 7} = y$ **1, 7**

30. The square root of 6 more than a number is the same as 6 less than the number. Find the number.   **10**

18.5 EXERCISES C

Solve.

1. Two boats leave an island with one heading south and the other west. After 4 hours they are 100 miles apart. What is the speed of each boat if one travels 5 mph faster than the other? [Answer: 15 mph, 20 mph]

2. A boat requires one hour longer to go 80 miles upstream than to make the return trip downstream. What is the speed of the boat in still water if the speed of the stream is 2 mph? **18 mph**

18.6 SOLVING FORMULAS

STUDENT GUIDEPOSTS

1 Linear Equations   **2** Quadratic and Radical Equations

Many times, equations involve a variable and constants represented by letters instead of numerical constants. Formulas are excellent examples of these types of equations. For instance, we have used such formulas as

$$A = lw \qquad P = 2l + 2w \qquad c^2 = a^2 + b^2 \qquad c = \pi d.$$

To solve a given formula for a particular letter, remember that the letters play the same role as numbers and all of our equation-solving rules apply. If you have trouble solving for a letter in a formula, it may be helpful to make up a similar equation with numbers instead of letters, and pattern your solution steps after the procedure you use to solve the new equation. We will demonstrate this technique in the examples.

 LINEAR EQUATIONS

Recall that solving a formula for a particular variable is a process of isolating the variable on one side of the equation.

| **EXAMPLE 1** ADDITION-SUBTRACTION RULE | **PRACTICE EXERCISE 1** |

Solve $a + x = b$ for x.

Solve $2y + z = x$ for z.

| *Formula* | *Similar numerical equation* |

$a + x = b$ $\qquad\qquad$ $3 + x = 7$

$a - a + x = b - a$ Subtract a $\quad$ $3 - 3 + x = 7 - 3$ Subtract 3

$0 + x = b - a$ $\qquad\qquad$ $0 + x = 7 - 3$

$x = b - a$ $\qquad\qquad\qquad$ $x = 4$

The solution is $x = b - a$.

Answer: $z = x - 2y$

EXAMPLE 2 MULTIPLICATION-DIVISION RULE

Solve $cx = d$ for x.

Formula	*Similar numerical equation*

$$cx = d \qquad\qquad 5x = 8$$

$$\boxed{\frac{1}{c}} \cdot cx = \boxed{\frac{1}{c}} \cdot d \quad \text{Multiply by } \tfrac{1}{c} \qquad \boxed{\frac{1}{5}} \cdot 5x = \boxed{\frac{1}{5}} \cdot 8 \quad \text{Multiply by } \tfrac{1}{5}$$

$$1 \cdot x = \frac{d}{c} \qquad\qquad 1 \cdot x = \frac{8}{5}$$

$$x = \frac{d}{c} \qquad\qquad x = \frac{8}{5}$$

The solution is $x = \dfrac{d}{c}$.

PRACTICE EXERCISE 2

Solve $5uv = w$ for u.

Answer: $u = \dfrac{w}{5v}$

EXAMPLE 3 COMBINATION OF RULES

Solve $ax + b = c$ for x.

Formula	*Similar numerical equation*

$$ax + b = c \qquad\qquad 3x + 5 = 20$$

$$ax + b \boxed{-\,b} = c \boxed{-\,b} \quad \text{Subtract } b \qquad 3x + 5 \boxed{-\,5} = 20 \boxed{-\,5} \quad \text{Subtract 5}$$

$$ax = c - b \qquad\qquad 3x = 15$$

$$\boxed{\frac{1}{a}} \cdot ax = \boxed{\frac{1}{a}}(c - b) \quad \text{Multiply by } \tfrac{1}{a} \qquad \boxed{\frac{1}{3}} \cdot 3x = \boxed{\frac{1}{3}} \cdot 15 \quad \text{Multiply by } \tfrac{1}{3}$$

$$x = \frac{c - b}{a} \qquad\qquad x = \frac{15}{3} = 5$$

The solution is $x = \dfrac{c - b}{a}$.

PRACTICE EXERCISE 3

Solve $mn - k = 6$ for n.

Answer: $n = \dfrac{k + 6}{m}$

EXAMPLE 4 FORMULA FOR THE AREA OF RECTANGLE

Solve $A = lw$ for l.

Formula	*Similar numerical equation*

$$A = lw \qquad\qquad 12 = l \cdot 4$$

$$A \cdot \boxed{\frac{1}{w}} = lw \cdot \boxed{\frac{1}{w}} \quad \text{Multiply by } \tfrac{1}{w} \qquad 12 \cdot \boxed{\frac{1}{4}} = l \cdot 4 \cdot \boxed{\frac{1}{4}} \quad \text{Multiply by } \tfrac{1}{4}$$

$$\frac{A}{w} = l \cdot 1 \qquad\qquad \frac{12}{4} = l \cdot 1$$

$$\frac{A}{w} = l \qquad\qquad 3 = l$$

The solution is $l = \dfrac{A}{w}$.

PRACTICE EXERCISE 4

Solve $PV = RT$ for R.

Answer: $R = \dfrac{PV}{T}$

EXAMPLE 5 DIVIDING BY TWO VARIABLES

Solve $a = bcx$ for x.

$$a = bcx$$

$$\frac{1}{bc} \cdot a = \frac{1}{bc} \cdot bcx \qquad \text{Multiply by } \frac{1}{bc}, \text{ the reciprocal of the coefficient of } x$$

$$\frac{a}{bc} = 1 \cdot x$$

$$\frac{a}{bc} = x$$

The solution is $x = \dfrac{a}{bc}$.

PRACTICE EXERCISE 5

Solve $3xyz = uv$ for x.

Answer: $x = \dfrac{uv}{3yz}$

---

② QUADRATIC AND RADICAL EQUATIONS

Some formulas involve quadratic equations or radical equations. We shall assume that the variables and constants are chosen so that division by zero and negative numbers under a radical sign are avoided.

EXAMPLE 6 QUADRATIC EQUATION

Solve $a = bx^2$ for x.

Formula   *Similar numerical equation*

$$a = bx^2 \qquad\qquad\qquad 50 = 2x^2$$

$$\frac{1}{b} \cdot a = \frac{1}{b} \cdot bx^2 \quad \text{Multiply by } \frac{1}{b} \qquad \frac{1}{2} \cdot 50 = \frac{1}{2} \cdot 2x^2 \quad \text{Multiply by } \frac{1}{2}$$

$$\frac{a}{b} = x^2 \qquad\qquad\qquad 25 = x^2$$

$$\pm\sqrt{\frac{a}{b}} = x \quad \begin{array}{l}\text{Take square root}\\ \text{of both sides}\end{array} \qquad \pm 5 = \pm\sqrt{25} = x$$

The solutions are $x = \pm\sqrt{\dfrac{a}{b}}$.

PRACTICE EXERCISE 6

Solve $uy^2 - v = 0$ for y.

Answer: $y = \pm\sqrt{\dfrac{v}{u}}$

EXAMPLE 7 RADICAL EQUATION

Solve $a = \sqrt{\dfrac{x}{b}}$ for x.

PRACTICE EXERCISE 7

Solve $m = \sqrt{\dfrac{2k}{n}}$ for n.

Formula		Similar numerical equation

$$a = \sqrt{\dfrac{x}{b}}$$

$$5 = \sqrt{\dfrac{x}{2}}$$

$$(a)^2 = \left(\sqrt{\dfrac{x}{b}}\right)^2 \quad \text{Square both sides} \quad (5)^2 = \left(\sqrt{\dfrac{x}{2}}\right)^2$$

$$a^2 = \dfrac{x}{b}$$

$$25 = \dfrac{x}{2}$$

$$b \cdot a^2 = b \cdot \dfrac{x}{b} \quad \text{Multiply by } b \quad 2 \cdot 25 = 2 \cdot \dfrac{x}{2}$$

$$ba^2 = x$$

$$50 = x$$

The solution is $x = ba^2$.

Answer: $n = \dfrac{2k}{m^2}$

18.6 EXERCISES A

1. Solve $g = x + h$ for x.
$g - h$

2. Solve $t - x = k$ for x.
$t - k$

3. Solve $bx + d = e$ for x.
$\dfrac{e - d}{b}$

4 Solve $g - ax = m$ for x.
$\dfrac{g - m}{a}$

5. Solve $d = rt$ for t.
$\dfrac{d}{r}$

6. Solve $d = rt$ for r.
$\dfrac{d}{t}$

7. Solve $E = IR$ for I.
$\dfrac{E}{R}$

8. Solve $I = prt$ for t.
$\dfrac{I}{pr}$

9. Solve $I = prt$ for r.
$\dfrac{I}{pt}$

10. Solve $I = prt$ for p.

$\dfrac{I}{rt}$

11 Solve $A = \dfrac{1}{2}bh$ for b.

$\dfrac{2A}{h}$

12. Solve $A = \dfrac{1}{2}bh$ for h.

$\dfrac{2A}{b}$

13. Solve $r = \dfrac{d}{t}$ for d.

rt

14 Solve $r = \dfrac{d}{t}$ for t.

$\dfrac{d}{r}$

15. Solve $A = \pi r^2$ for r.

$\pm\sqrt{\dfrac{A}{\pi}}$

16 Solve $S = \dfrac{1}{2}gt^2$ for t.

$\pm\sqrt{\dfrac{2S}{g}}$

17. Solve $E = mc^2$ for m.

$\dfrac{E}{c^2}$

18. Solve $E = mc^2$ for c.

$\pm\sqrt{\dfrac{E}{m}}$

19. Solve $A = \dfrac{4}{3}\pi r^2$ for r.

$\pm\sqrt{\dfrac{3A}{4\pi}}$

20 Solve $U = \sqrt{\dfrac{a}{c}}$ for c.

$\dfrac{a}{U^2}$

21. Solve $U = \sqrt{\dfrac{a}{c}}$ for a.

cU^2

22. Solve $\sqrt{x + a} = b$ for x.

$b^2 - a$

FOR REVIEW

Solve.

23. The product of 2 less than a number and 3 more than the number is 176. Find the number.

−14, 13

24. The hypotenuse of a right triangle is 5 cm and one leg is 1 cm longer than the other. Find the measure of each leg.

3 cm, 4 cm

25. The area of a triangle is 24 ft², and the base is 3 times as long as the height. Find the base and height.

12 ft, 4 ft

26. Laura takes 2 days to type a manuscript and Holly takes 8 days. How long would it take if they worked together?

$\frac{8}{5}$ days

Exercises 27–28 review material from Chapter 12 to help you prepare for the next section. Graph each equation.

27. $y + 2x - 4 = 0$

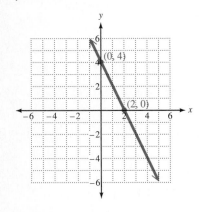

28. $y = 3x + 6$

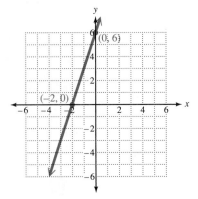

ANSWERS: 1. $g - h$ 2. $t - k$ 3. $\dfrac{e - d}{b}$ 4. $\dfrac{g - m}{a}$ 5. $\dfrac{d}{r}$ 6. $\dfrac{d}{t}$ 7. $\dfrac{E}{R}$ 8. $\dfrac{I}{pr}$ 9. $\dfrac{I}{pt}$ 10. $\dfrac{I}{rt}$ 11. $\dfrac{2A}{h}$

12. $\dfrac{2A}{b}$ 13. rt 14. $\dfrac{d}{r}$ 15. $\pm\sqrt{\dfrac{A}{\pi}}$ 16. $\pm\sqrt{\dfrac{2S}{g}}$ 17. $\dfrac{E}{c^2}$ 18. $\pm\sqrt{\dfrac{E}{m}}$ 19. $\pm\sqrt{\dfrac{3A}{4\pi}}$ 20. $\dfrac{a}{U^2}$ 21. cU^2 22. $b^2 - a$

23. −14, 13   24. 3 cm, 4 cm   25. 12 ft, 4 ft   26. $\frac{8}{5}$ days

27.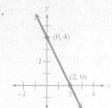

28.

18.6 EXERCISES B

1. Solve $x + k = w$ for x.
$w - k$

2. Solve $t = m - x$ for x.
$m - t$

3. Solve $ax + w = m$ for x.
$\dfrac{m - w}{a}$

4. Solve $p - bx = w$ for x.
$\dfrac{p - w}{b}$

5. Solve $A = bh$ for h.
$\dfrac{A}{b}$

6. Solve $A = bh$ for b.
$\dfrac{A}{h}$

7. Solve $T = cn$ for n.
$\dfrac{T}{c}$

8. Solve $V = lwh$ for h.
$\dfrac{V}{lw}$

9. Solve $V = lwh$ for w.
$\dfrac{V}{lh}$

10. Solve $V = lwh$ for l.
$\dfrac{V}{wh}$

11. Solve $W = \dfrac{1}{3}pq$ for p.
$\dfrac{3W}{q}$

12. Solve $W = \dfrac{1}{3}pq$ for q.
$\dfrac{3W}{p}$

13. Solve $a = \dfrac{b}{m}$ for b.
am

14. Solve $a = \dfrac{b}{m}$ for m.
$\dfrac{b}{a}$

15. Solve $V = \pi r^2 h$ for h.
$\dfrac{V}{\pi r^2}$

16. Solve $V = \pi r^2 h$ for r.
$\pm \sqrt{\dfrac{V}{\pi h}}$

17. Solve $S = a\pi r^2$ for r.
$\pm \sqrt{\dfrac{S}{a\pi}}$

18. Solve $S = a\pi r^2$ for a.
$\dfrac{S}{\pi r^2}$

19. Solve $W = \dfrac{5}{4}am^2$ for m.
$\pm \sqrt{\dfrac{4W}{5a}}$

20. Solve $T = \sqrt{\dfrac{a}{w}}$ for a.
wT^2

21. Solve $T = \sqrt{\dfrac{a}{w}}$ for w.
$\dfrac{a}{T^2}$

22. Solve $\sqrt{x + t} = p$ for x. $p^2 - t$

Solve.

23. Of two positive numbers, one is 4 larger than the other and the sum of the squares of the two is 136. Find the numbers.
6, 10

24. The height of a triangle is 8 inches more than the base. If the area is 90 in², find the height and base.
10 in, 18 in

25. The length of a rectangle is 4 cm more than the width and the area is 96 cm². Find the dimensions.
8 cm, 12 cm

26. The amount of money A that will result if a principal P is invested at r percent interest compounded annually for 2 years is given by $A = P(1 + r)^2$. If $5000 grows to $6498 in 2 years using this formula, what is the interest rate?
14%

Exercises 27–28 review material from Chapter 12 to help you prepare for the next section. Graph each equation.

27. $2y - x - 6 = 0$

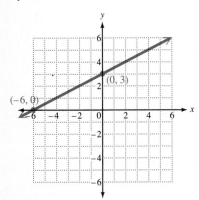

28. $y = -3x - 3$

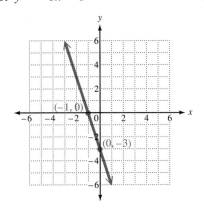

18.6 EXERCISES C

1. Solve $\dfrac{u^2}{1 + v^2} = 2$ for positive v.

$$v = \sqrt{\dfrac{u^2 - 2}{2}}$$

2. Solve $\dfrac{x}{\sqrt{2x + y}} = 1$ for x.

[Answer: $x = 1 \pm \sqrt{1 + y}$]

18.7 GRAPHING QUADRATIC EQUATIONS

═══════ **STUDENT GUIDEPOSTS** ═══════

1 Parabolas
2 Graphing Quadratic Equations
3 Graphs for $a > 0$ and $a < 0$
4 Vertex of a Parabola

1 PARABOLAS

In Chapter 12 we learned how to graph a linear equation

$$ax + by + c = 0 \qquad \text{General form}$$

$$\text{or} \quad y = mx + b \qquad \text{Slope-intercept form}$$

by plotting the intercepts (or the intercept and one additional point if the intercepts are both $(0, 0)$) and drawing the straight line through them. The graph of every **quadratic equation in two variables** of the form

$$y = ax^2 + bx + c \quad (a \neq 0)$$

is a **parabola,** a U-shaped curve similar to the one shown in Figure 18.4. Applications of parabolas or surfaces in the shape of a parabola are numerous in science, engineering, business, and architecture.

❷ GRAPHING QUADRATIC EQUATIONS

To graph a quadratic equation, we choose several values for x, calculate the corresponding y-values, and plot the resulting points. A table of values helps us keep a record of the points.

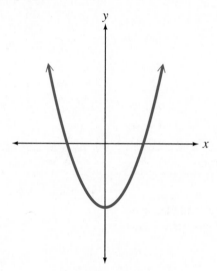

Figure 18.4

EXAMPLE 1 GRAPHING A PARABOLA

Graph $y = x^2$.

　　Find y-values for the following values of x.

If $x = 0$, $y = x^2 = 0^2 = 0$.　　　If $x = -1$, $y = x^2 = (-1)^2 = 1$.

If $x = 1$, $y = x^2 = 1^2 = 1$.　　　If $x = -2$, $y = x^2 = (-2)^2 = 4$.

If $x = 2$, $y = x^2 = 2^2 = 4$.

PRACTICE EXERCISE 1

Graph $y = x^2 - 1$.

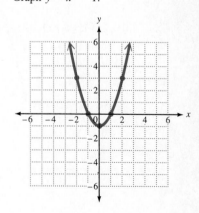

The completed table is shown below. Plotting the ordered pairs, we obtain the parabola in Figure 18.5.

x	y
0	0
1	1
2	4
−1	1
−2	4

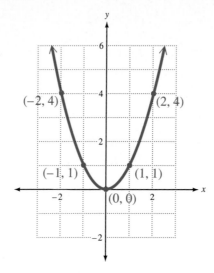

Figure 18.5

Answer: same as Figure 18.5 except all points down one unit

EXAMPLE 2 GRAPHING A PARABOLA

Graph $y = -x^2$.

Be careful when finding values for y in this case. For example, if $x = 1$, $y = -x^2 = -(1)^2 = -1$. The completed table appears beside the graph given in Figure 18.6.

x	y
0	0
1	−1
2	−4
−1	−1
−2	−4

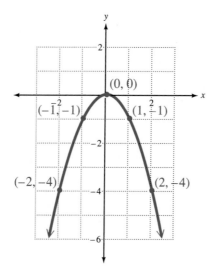

Figure 18.6

PRACTICE EXERCISE 2

Graph $y = -x^2 + 2$.

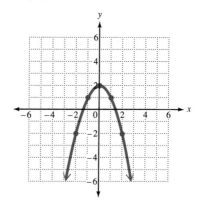

Answer: same as Figure 18.6 except all points up two units

③ GRAPHS FOR $a > 0$ AND $a < 0$

Notice that the parabola in Example 1 opens up while the one in Example 2 opens down. In general, the direction in which a parabola opens depends on the sign of the coefficient of the x^2-term, a. If $a > 0$ (as in Example 1) the parabola opens up, and if $a < 0$ (as in Example 2), the parabola opens down.

Direction a Parabola Opens

The graph of $y = ax^2 + bx + c$ $(a \neq 0)$ is a parabola which

1. opens up when the coefficient of x^2 is positive,

2. opens down when the coefficient of x^2 is negative.

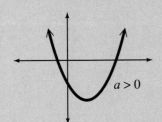
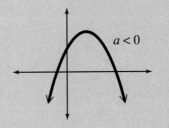

EXAMPLE 3   GRAPH OPENING UP

Graph $y = x^2 - 2x - 3$.

We know that the parabola opens up since the coefficient of x^2 is positive. From the values computed in the table, the graph is plotted in Figure 18.7.

x	y
0	-3
1	-4
2	-3
3	0
-1	0

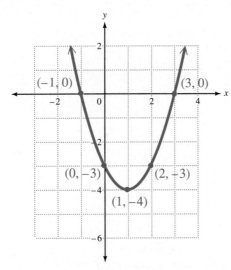

Figure 18.7

PRACTICE EXERCISE 3

Graph $y = -x^2 + 2x + 3$.

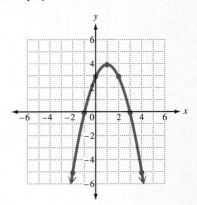

Answer: like Figure 18.7 except opening down

| EXAMPLE 4 GRAPH OPENING DOWN | PRACTICE EXERCISE 4 |

Graph $y = -x^2 + 2x$.

Use care when substituting for x. The graph is in Figure 18.8.

x	y
0	0
1	1
2	0
3	−3
−1	−3

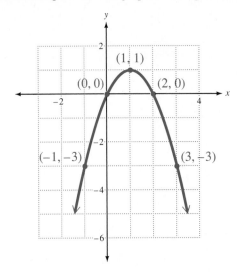

Figure 18.8

Graph $y = x^2 - 2x$.

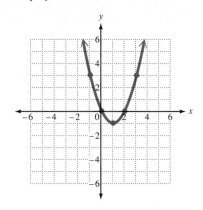

Answer: like Figure 18.8 except opening up

---

❹ VERTEX OF A PARABOLA

Consider the following equation.

$$y = (x - 1)^2 - 4$$
$$= x^2 - 2x + 1 - 4$$
$$= x^2 - 2x - 3$$

Thus, $y = (x - 1)^2 - 4$ and $y = x^2 - 2x - 3$ have the same graph, given in Figure 18.7. Notice that $(1, -4)$ is the low point or **vertex** of the parabola. equation of a parabola is written in the form

$$y = (x - h)^2 + k,$$

then (h, k) is the vertex and the graph is a parabola opening up.

A similar rule holds for parabolas opening down. Consider

$$y = -(x - 1)^2 + 1$$
$$= -(x^2 - 2x + 1) + 1$$
$$= -x^2 + 2x - 1 + 1$$
$$= -x^2 + 2x.$$

Thus, $y = -(x - 1)^2 + 1$ and $y = -x^2 + 2x$ have the same graph, given in Figure 18.8, with vertex $(1, 1)$. If the equation of a parabola is written in the form

$$y = -(x - h)^2 + k,$$

then (h, k) is the vertex, and the graph is a parabola opening down.

| **EXAMPLE 5   FINDING THE VERTEX** | **PRACTICE EXERCISE 5** |

Find the vertex of each parabola.

(a) $y = (x + 3)^2 - 5$

We must have the form $y = (x - h)^2 + k$.

$$y = (x + 3)^2 - 5$$
$$= [x - (-3)]^2 + (-5)$$

Thus, $h = -3$ and $k = -5$. The vertex is $(-3, -5)$.

(b) $y = -(x + 1)^2 + 7$

Write in the form $y = -(x - h)^2 + k$.

$$y = -(x + 1)^2 + 7$$
$$= -[x - (-1)]^2 + 7$$

Thus, $h = -1$ and $k = 7$. The vertex is $(-1, 7)$.

(c) $y = (x + 5)^2$

Write in the form $y = (x - h)^2 + k$.

$$y = (x + 5)^2$$
$$= [x - (-5)]^2 + 0$$

Thus, $h = -5$ and $k = 0$. The vertex is $(-5, 0)$.

Find the vertex of each parabola.

(a) $y = (x - 4)^2 - 5$

(b) $y = -(x + 2)^2 + 4$

(c) $y = -(x + 8)^2$

Answers: (a) $(4, -5)$
(b) $(-2, 4)$ (c) $(-8, 0)$

18.7   EXERCISES A

Before graphing, tell whether the parabola opens up or down. Then graph the equation.

1. $y = x^2 + 1$

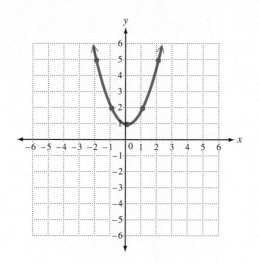

2. $y = -x^2 + 1$

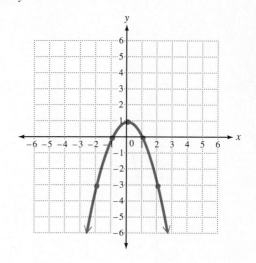

3. $y = x^2 - 2$

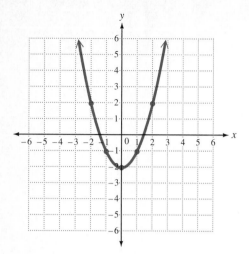

4. $y = -x^2 + 2$

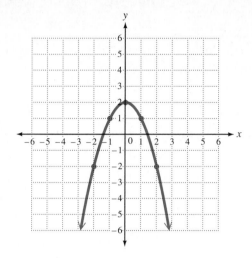

5. $y = x^2 + 2x + 1$

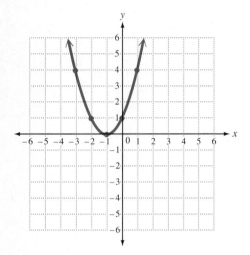

6. $y = -x^2 - 2x - 1$

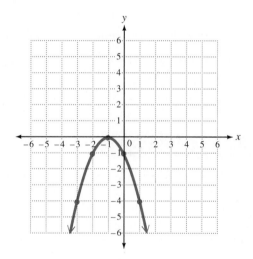

7. $y = x^2 + 2x$

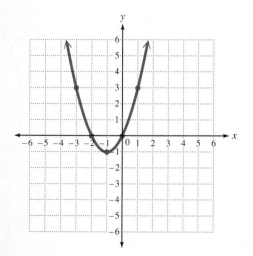

8. $y = -x^2 - 2x$

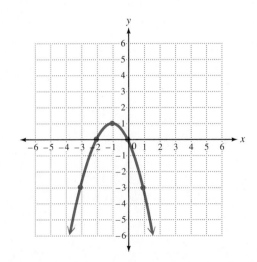

9. $y = x^2 - 3x + 2$

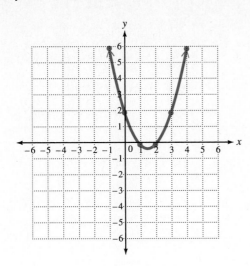

10. $y = -x^2 + 3x - 2$

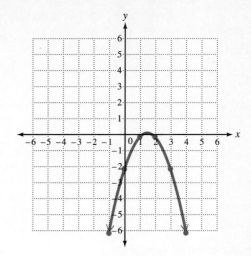

Give the vertex of the parabola.

11. $y = (x - 4)^2 + 2$
(4, 2)

12. $y = -(x - 1)^2 + 5$
(1, 5)

13. $y = (x + 3)^2 - 2$
(−3, −2)

14. $y = -(x + 10)^2 - 9$
(−10, −9)

15 $y = x^2 - 3$
(0, −3)

16. $y = -x^2 - 3$
(0, −3)

FOR REVIEW

17. Solve $p = ax + t$ for x.

$$\frac{p - t}{a}$$

18. Solve $M = \frac{1}{3}wt^2$ for t.

$$\pm\sqrt{\frac{3M}{w}}$$

19. Solve $A = \frac{v}{2w}$ for v.

$$2Aw$$

20. Solve $A = \frac{v}{2w}$ for w.

$$\frac{v}{2A}$$

21. Solve $S = \sqrt{\frac{m}{r}}$ for m.

$$S^2 r$$

22. Solve $\sqrt{3x + u} = a$ for x.

$$\frac{a^2 - u}{3}$$

ANSWERS:

1.
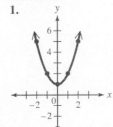

2.
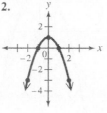

3.
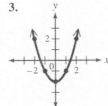

4.
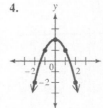

5.

6. 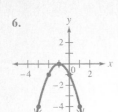 **7.** 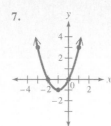 **8.** 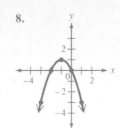 **9.** 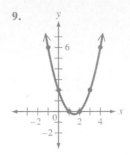 **10.**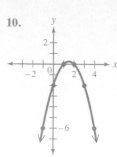

11. (4, 2) **12.** (1, 5) **13.** (−3, −2) **14.** (−10, −9) **15.** (0, −3) **16.** (0, −3) **17.** $\dfrac{p - t}{a}$ **18.** $\pm\sqrt{\dfrac{3M}{w}}$ **19.** $2Aw$

20. $\dfrac{v}{2A}$ **21.** $S^2 r$ **22.** $\dfrac{a^2 - u}{3}$

18.7 EXERCISES B

Before graphing, tell whether the parabola opens up or down. Then graph the equation.

1. $y = x^2 + 2$

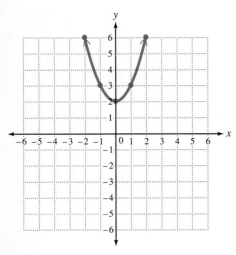

2. $y = -x^2 - 1$

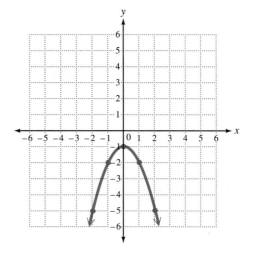

3. $y = x^2 - 1$

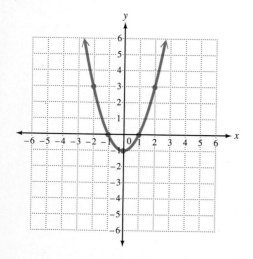

4. $y = -x^2 + 3$

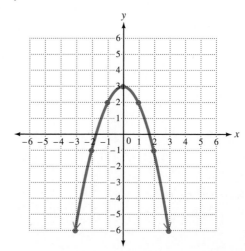

5. $y = x^2 - 2x + 1$

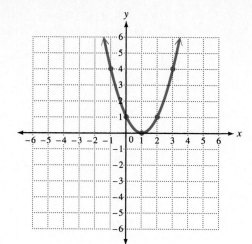

6. $y = -x^2 + 2x - 1$

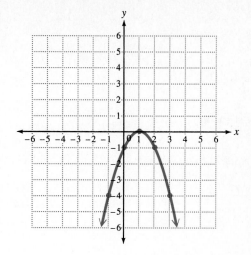

7. $y = x^2 + 4x$

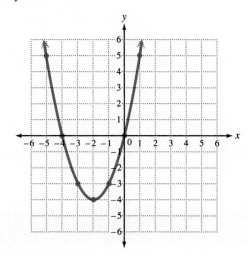

8. $y = -x^2 - 4x$

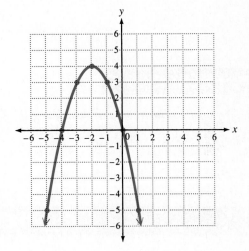

9. $y = x^2 + 3x + 2$

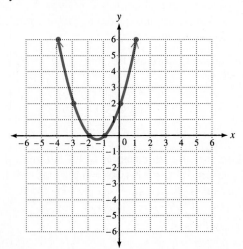

10. $y = -x^2 - 3x - 2$

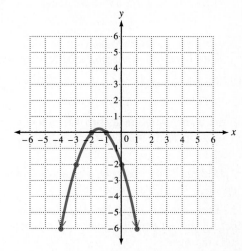

Give the vertex of the parabola.

11. $y = (x - 7)^2 + 1$
(7, 1)

12. $y = -(x - 4)^2 + 8$
(4, 8)

13. $y = (x + 12)^2 - 8$
(−12, −8)

14. $y = -(x + 9)^2 - 14$
(−9, −14)

15. $y = (x + 1)^2$
(−1, 0)

16. $y = -x^2 + 8$
(0, 8)

FOR REVIEW

17. Solve $q = bx - w$ for x.

$\dfrac{q + w}{b}$

18. Solve $V = \dfrac{4}{3}am^2$ for m.

$\pm\sqrt{\dfrac{3V}{4a}}$

19. Solve $R = \dfrac{I}{2E}$ for I.

$2ER$

20. Solve $R = \dfrac{I}{2E}$ for E.

$\dfrac{I}{2R}$

21. Solve $W = \sqrt{\dfrac{a}{n}}$ for n.

$\dfrac{a}{W^2}$

22. Solve $\sqrt{a + 2x} = z$ for x.

$\dfrac{z^2 - a}{2}$

18.7 EXERCISES C

Find the vertex of each parabola by completing the square.

1. $y = x^2 + 10x + 20$ [Answer: (−5, −5)]

2. $y = -x^2 + 6x - 4$ (3, 5)

CHAPTER 18 REVIEW

KEY WORDS

18.1 A **quadratic equation** is an equation that can be written in the form $ax^2 + bx + c = 0$, $a \neq 0$.

18.2 To **complete the square** on $x^2 + bx$ add $\left(\frac{1}{2}b\right)^2$ to make the expression a perfect square.

18.3 The **discriminant** of $ax^2 + bx + c = 0$ is $b^2 - 4ac$.

18.7 A **quadratic equation in two variables** is an equation of the form $y = ax^2 + bx + c$, $a \neq 0$.

A **parabola** is the graph of a quadratic equation in two variables.

The **vertex** is the low point of a parabola opening up and the high point of a parabola opening down.

KEY CONCEPTS

18.1, The best ways to solve a quadratic equation are
18.2 by factoring and taking roots. If these methods do not work, use the quadratic formula instead of completing the square.

18.3 The quadratic formula for solving $ax^2 + bx + c = 0$ is

$$x = \frac{-b \pm \sqrt{b^2 - 4ac}}{2a}.$$

Make sure that an equation is put into the above general form before trying to identify the constants a, b, and c. Also, the entire numerator $-b \pm \sqrt{b^2 - 4ac}$, is divided by $2a$, *not* just the radical term.

18.4 1. To solve a fractional equation, multiply both sides by the LCD of all fractions. Be sure to check all possible solutions in the original equations and exclude those that make any denominator zero.

2. To solve a radical equation, isolate a radical and square both sides of the equation. If a binomial occurs on one side, don't forget the middle term when squaring. For example,

$$x + 4 = 3\sqrt{x + 1} \quad \text{becomes}$$
$$x^2 + 8x + 16 = 9(x + 1).$$

Be sure to square 3 also. Finally, check all possible answers in the *original* equa-

tion and remember that the radical only represents the positive (principal) square root.

18.5 Be precise and write out all details (including complete descriptions of the variable) when solving word problems.

18.6 Solving a similar numerical equation can often help when solving a formula for a particular variable.

18.7 The graph of a quadratic equation $y = ax^2 + bx + c$ $(a \neq 0)$ is a parabola that opens up if $a > 0$ and down if $a < 0$.

REVIEW EXERCISES

Part I

18.1 *Solve the following quadratic equations.*

1. $2x^2 - 32 = 0$
± 4

2. $-3z^2 = -15$
$\pm\sqrt{5}$

3. $x^2 + 6x - 72 = 0$
$6, -12$

4. $2y^2 + y = 21$
$3, -\frac{7}{2}$

5. $(z - 5)^2 = 25$
$0, 10$

6. $21x^2 - 4x - 32 = 0$
$\frac{4}{3}, -\frac{8}{7}$

18.2 *What must be added to complete the square?*

7. $x^2 + x + \underline{\quad\frac{1}{4}\quad}$

8. $y^2 + \frac{1}{4}y + \underline{\quad\frac{1}{64}\quad}$

9. Solve $x^2 - 3x + 1 = 0$ by completing the square. $\dfrac{3 \pm \sqrt{5}}{2}$

18.3 *Solve.*

10. $2y^2 + y = 5$ $\dfrac{-1 \pm \sqrt{41}}{4}$

11. $2z^2 + 11z = -(10 + z)$ $-1, -5$

12. $x^2 + 2x - 5 = 0$ $-1 \pm \sqrt{6}$

13. $2y^2 + y + 5 = 0$ no solution

18.4 *Solve.*

14. $\dfrac{z + 2}{-z} = \dfrac{1}{z + 2}$
$-1, -4$

15. $\dfrac{3}{1 + x} = -1 - \dfrac{2}{1 - x}$
$2, -3$

16. $x = 1 + \sqrt{4 - 4x}$
1

17. $4 = 4\sqrt{x + 1} - x$
$0, 8$

18.5 **18.** The product of two positive consecutive integers is 420. Find the integers.
20, 21

19. Mike's present age times his age in 7 years is 30. Find his present age.
3 years old

20. The number of square inches in a square is 3 less than the number of inches in its perimeter. Find the length of its sides.
1 in or 3 in

21. Find the number whose square is 18 more than three times the number.
6, −3

22. The pressure p, in lb per sq ft, of a wind blowing v mph can be approximated by the equation $p = 0.003v^2$. What is the approximate wind velocity when a pressure of 3.675 lb per sq ft is exerted against the side of a skyscraper?
35 mph

23. Use $A = P(1 + r)^2$ to find the interest rate if $3000 grows to $3630 in 2 years.
10%

18.6 **24.** Solve $c = \dfrac{1}{3}dh$ for h.

$\dfrac{3c}{d}$

25. Solve $a - x = 2b$ for x.

$a - 2b$

26. Solve $D = au^2$ for u.

$\pm\sqrt{\dfrac{D}{a}}$

27. Solve $D = au^2$ for a.

$\dfrac{D}{u^2}$

28. Solve $g = \sqrt{\dfrac{a}{d}}$ for a.

dg^2

29. Solve $g = \sqrt{\dfrac{a}{d}}$ for d.

$\dfrac{a}{g^2}$

18.7 **30.** Tell whether the parabola with the given equation opens up or down.
   (a) $y = 3x^2 + x - 1$ **(b)** $y = -2x^2 + 5$
   up   **down**

Graph the given equations.

31. $y = x^2 - 3$

32. $y = x^2 - 6x + 8$

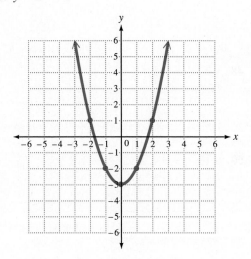

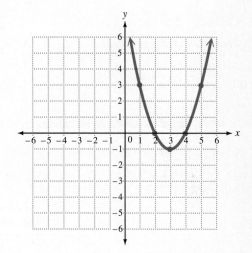

Give the vertex of the parabola.

33. $y = -(x + 4)^2 - 3$ **(−4, −3)**

34. $y = x^2 + 6$ **(0, 6)**

Part II

Solve.

35. Solve $a = cu^2$ for u.

$\pm\sqrt{\dfrac{a}{c}}$

36. Solve $2hr^2 = 3V$ for r.

$\pm\sqrt{\dfrac{3V}{2h}}$

37. The square of a number, less 7, is equal to 9. Find the number.

4, −4

38. Twice the square root of 2 more than a number is the same as 1 less than the number. Find the number.

7

39. Lennie has a garden in the shape of a square 12 yards on a side. He wishes to lay a water pipe from one corner diagonally across to the other. What is the approximate length of this pipe?

16.8 yd

40. If it takes Bill 12 hours longer to paint a room than Peter and together they can complete the job in 8 hours, how long would it take each to do it alone?

Peter: 12 hr, Bill: 24 hr

41. $\dfrac{5}{x-2} + 1 = \dfrac{2}{x-3}$

$1 \pm \sqrt{6}$

42. $x^2 - 6x + 2 = 0$

$3 \pm \sqrt{7}$

43. $1 - x^2 = 7x - 5$

$\dfrac{-7 \pm \sqrt{73}}{2}$

44. $\sqrt{x+1} + 1 = x$

3

45. $9x^2 = 1$

$\pm\frac{1}{3}$

46. $2x^2 = 9x + 5$

$-\frac{1}{2},\, 5$

Give the vertex of the parabola.

47. $y = -x^2 - 8$

$(0, -8)$

48. $y = -(x-2)^2 + 5$

$(2, 5)$

What must be added to complete the square?

49. $x^2 - 14x + \underline{}$ 49

50. $y^2 + \dfrac{1}{3}x + \underline{\phantom{\frac{1}{36}}}$ $\frac{1}{36}$

ANSWERS: 1. ± 4 2. $\pm\sqrt{5}$ 3. $6, -12$ 4. $3, -\frac{7}{2}$ 5. $0, 10$ 6. $\frac{4}{3}, -\frac{8}{7}$ 7. $\frac{1}{4}$ 8. $\frac{1}{64}$ 9. $\dfrac{3 \pm \sqrt{5}}{2}$

10. $\dfrac{-1 \pm \sqrt{41}}{4}$ 11. $-1, -5$ 12. $-1 \pm \sqrt{6}$ 13. no solution 14. $-1, -4$ 15. $2, -3$ 16. 1 17. $0, 8$

18. $20, 21$ 19. 3 years old 20. 1 in or 3 in 21. $6, -3$ 22. 35 mph 23. 10% 24. $\dfrac{3c}{d}$ 25. $a - 2b$ 26. $\pm\sqrt{\dfrac{D}{a}}$

27. $\dfrac{D}{u^2}$ 28. dg^2 29. $\dfrac{a}{g^2}$ 30. (a) up (b) down

31.

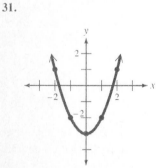

32.

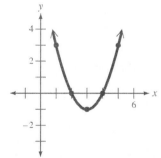

33. $(-4, -3)$ 34. $(0, 6)$ 35. $\pm\sqrt{\dfrac{a}{c}}$ 36. $\pm\sqrt{\dfrac{3V}{2h}}$ 37. $4, -4$ 38. 7 39. 16.8 yd (using $\sqrt{2} \approx 1.4$) 40. Peter: 12 hr,

Bill: 24 hr 41. $1 \pm \sqrt{6}$ 42. $3 \pm \sqrt{7}$ 43. $\dfrac{-7 \pm \sqrt{73}}{2}$ 44. 3 45. $\pm\frac{1}{3}$ 46. $-\frac{1}{2}, 5$ 47. $(0, -8)$ 48. $(2, 5)$ 49. 49

50. $\frac{1}{36}$

Solve the following quadratic equations.

1. $x^2 + 13x + 36 = 0$

1. _____ $-4, -9$ _____

2. $3z^2 - z = 27 - z$

2. _____ $3, -3$ _____

3. $y(y - 2) = 3y$

3. _____ $0, 5$ _____

4. $x^2 - 4x + 1 = 0$

4. _____ $2 \pm \sqrt{3}$ _____

5. What must be added to complete the square?

$y^2 + \dfrac{1}{2}y +$ _____

5. _____ $\dfrac{1}{16}$ _____

Solve.

6. $\dfrac{2}{x - 1} + 2 = \dfrac{3}{x - 2}$

6. _____ $3, \dfrac{1}{2}$ _____

7. $4 = x + \sqrt{2 + x}$

7. _____ 2 _____

Solve.

8. The product of two positive consecutive even integers is 120. Find the integers.

8. _____ 10, 12 _____

9. It takes 8 hours less time to fill a pond using a large pipe than it takes using a smaller pipe. If when used together they fill the pond in 3 hours, how long would it take each to fill it alone?

9. ____ large: 4 hr, small: 12 hr ____

10. A boat travels 70 miles upstream and then returns in a total time of 12 hours. If the speed of the stream is 2 mph, what is the speed of the boat in still water?

10. _____ 12 mph _____

11. Solve $w = \sqrt{\dfrac{b}{m}}$ for m.

11. _____ $m = \dfrac{b}{w^2}$ _____

12. You are given the equation $y = x^2 + 2$.
   (a) Does the graph of this parabola open up or down?
   (b) What is the vertex of the parabola?
   (c) Graph the equation.

12(a) _____ up _____
12(b) _____ (0, 2) _____
12(c)

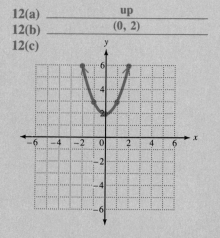

CHAPTER 1

1. Consider the number line.

    0   1 *a* 3   *b* 8 *c*

    (a) What number is paired with the point a?   2

    (b) What number is paired with the point b?   7

    Place the correct symbol ($<$ or $>$) between the numbers in each pair.

    (c) $1 < a$ **(d)** $a < 5$ **(e)** $a < b$ **(f)** $c > 4$

2. Write 27,143,206 in expanded notation. $20{,}000{,}000 + 7{,}000{,}000 + 100{,}000 + 40{,}000 + 3000 + 200 + 6$

3. Write $2{,}000{,}000 + 40{,}000 + 300 + 20 + 5$ in standard notation.   2,040,325

4. Write a word name for 86,247,159.   eighty-six million, two hundred forty-seven thousand, one hundred fifty-nine

Find the following sums and check your work.

5.
```
  279
+  48
  327
```

6.
```
  305
+ 299
  604
```

7.
```
  2187
  6532
+   27
  8746
```

8.
```
  4020
    37
   408
+   90
  4555
```

9. Round 27,485 to the nearest:

    (a) ten;   27,490   **(b)** hundred;   27,500   **(c)** thousand.   27,000

10. Estimate the given sum by rounding to the nearest **(a)** ten **(b)** hundred.
```
  1953      (a)      4940           (b)      4900
  2247
+  738
```

Solve.

11. An electronics shop has in stock 38 radios, 127 televisions, 72 calculators, and 12 watches. How many items does the owner have in his inventory?
249

12. On Monday, Sue jogged 4 mi, on Tuesday she jogged 7 mi, on Wednesday 6 mi, on Thursday 8 mi, and on Friday 3 mi. How many miles did she jog during the week?   28 mi

Find the following differences and check your work.

13.   54
   − 28
     26

14.   607
   − 421
    186

15.   3208
   − 1157
    2051

16.   2007
   − 1348
    659

Solve.

17. On the first of November, Mr. Schwacker had a balance of $427 in his checking account. During the month he made deposits of $370 and $1845. He wrote checks for $238, $475, $87, $5, and $614, and paid a service charge of $6. What was his balance at the end of the month? $1217

18. Pete bought an antique car for $750. He spent $1230 restoring the car, then sold it for $4750. How much money did he make? $2770

CHAPTER 2

19. $3 \div 1 =$ ___3___.

20. $3 \div 3 =$ ___1___.

21. In the expression 7^3, 3 is called what?
exponent

22. (a) What is the square of 4?   16
(b) What is the square root of 4?   2

Find the following products and check your work by estimating the product.

23.   79
   × 41
   3239

24.   819
   × 78
  63,882

25.   3015
   × 602
 1,815,030

26.   6499
   × 4000
 25,996,000

Find the following quotients and check your work.

27. $7\overline{)3367}$
Q: 481, R: 0

28. $34\overline{)2341}$
Q: 68, R: 29

29. $615\overline{)140006}$
Q: 227, R: 401

Perform the indicated operations.

30. $3 - 4 \div 2 + 6$ 7

31. $(2 + 3) \cdot (2 - 1) - \sqrt{25}$ 0

32. $(1 + 3) \div 2 + 3 \cdot 4$ 14

33. Find all divisors of 150.   1, 2, 3, 5, 6, 10, 15, 25, 30, 50, 75, 150

34. Express 546 as a product of primes. $546 = 2 \cdot 3 \cdot 7 \cdot 13$

35. Is 90 divisible by
(a) 2?   yes   (b) 3?   yes   (c) 4?   no   (d) 5?   yes   (e) 10?   yes

36. Simplify each radical expression.
(a) $\sqrt{9}$ 3   (b) $\sqrt{225}$ 15   (c) $\sqrt{4}$ 2   (d) $\sqrt{144}$ 12   (e) $\sqrt{1}$ 1

37. Sharon wants to carpet her living room with carpet selling for $21 per sq yd. A sketch showing the floor plan of the room is shown. How much will the carpet cost? $777

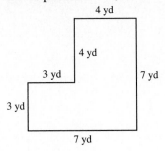

38. Assuming equal quality, which is the better buy, 32 ounces of soda for $1.60 (160¢) or 28 ounces for $1.12 (112¢)?

28 ounces of soda at 4¢ an ounce is the better buy than 32 ounces at 5¢ an ounce

CHAPTER 3

39. What fractional part of the figure has been shaded?

$\frac{2}{3}$

40. Let the figure be one unit.

$\frac{1}{4}$	$\frac{1}{4}$	$\frac{1}{4}$	$\frac{1}{4}$

What fraction of the unit is shaded below? [Consider the whole figure. There is one answer, and it is an improper fraction.] $\frac{5}{4}$

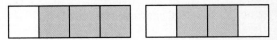

Express as a whole number.

41. $\dfrac{11}{11}$ 1

42. $\dfrac{0}{6}$ 0

43. Tell whether $\frac{7}{4}$ and $\frac{19}{11}$ are equal by using the cross-product rule. $\frac{7}{4} \neq \frac{19}{11}$

44. Find the missing term of the fraction. $\dfrac{6}{27} = \dfrac{2}{?}$ 9

45. Reduce $\frac{70}{385}$ to lowest terms. $\frac{2}{11}$

Find the product.

46. $\dfrac{27}{6} \cdot \dfrac{2}{9}$ 1

47. $\dfrac{42}{15} \cdot \dfrac{3}{7}$ $\dfrac{6}{5}$

Find the value of each power or root.

48. $\left(\dfrac{3}{4}\right)^2$ $\dfrac{9}{16}$

49. $\sqrt{\dfrac{4}{81}}$ $\dfrac{2}{9}$

Find the quotient.

50. $\dfrac{5}{8} \div \dfrac{3}{16}$ $\dfrac{10}{3}$

51. $\dfrac{15}{4} \div \dfrac{20}{6}$ $\dfrac{9}{8}$

Find the missing number.

52. $\frac{2}{3} \cdot \square = 12$ 18

53. $\frac{11}{4}$ of $\square$ is 22   8

Change to a mixed number.

54. $\frac{41}{9}$ $4\frac{5}{9}$

55. $\frac{122}{11}$ $11\frac{1}{11}$

Change to an improper fraction.

56. $13\frac{2}{5}$ $\frac{67}{5}$

57. $32\frac{1}{4}$ $\frac{129}{4}$

Perform the indicated operations.

58. $1\frac{3}{4} \cdot 3\frac{1}{8}$ $5\frac{15}{32}$

59. $6\frac{2}{3} \div 2\frac{2}{5}$ $2\frac{7}{9}$

Solve.

60. Laura got $\frac{3}{5}$ of the votes for president of her club. If the club has 75 members, how many members voted for her?
45

61. A container holds 162 gallons of water when it is $\frac{2}{3}$ full. What is the capacity of the container?
243 gallons

62. A container holds 162 gallons of water when it is full. If $\frac{1}{3}$ of a full container is drained off, how much water remains?   108 gallons

63. A child's ride makes $7\frac{2}{3}$ revolutions per minute. How many revolutions will be made during a $3\frac{3}{5}$ minute ride? $27\frac{3}{5}$ revolutions

CHAPTER 4

64. Use the listing method to find the LCM of 6 and 9.   18

65. Use prime factors to find the LCM of 220 and 390.   8580

66. Use the special algorithm to find the LCM of 840 and 1260.   2520

Perform the indicated operations.

67. $\frac{7}{12} + \frac{1}{12}$ $\frac{2}{3}$

68. $\frac{8}{15} - \frac{2}{15}$ $\frac{2}{5}$

69. $\frac{7}{12} + \frac{1}{18}$ $\frac{23}{36}$

70. $\frac{4}{15} + \frac{11}{35}$ $\frac{61}{105}$

71. $\frac{3}{5} - \frac{1}{7}$ $\frac{16}{35}$

72. $\frac{9}{20} - \frac{5}{12}$ $\frac{1}{30}$

73. $4\frac{3}{5} + 2\frac{7}{8}$ $7\frac{19}{40}$

74. $9\frac{1}{4} - 3\frac{5}{6}$ $5\frac{5}{12}$

75. $\frac{7}{4} - 1 + \frac{1}{5}$ $\frac{19}{20}$

Evaluate each expression.

76. $\left(\dfrac{3}{8} - \dfrac{1}{8}\right)^2 + \sqrt{\dfrac{1}{16}}$ $\dfrac{5}{16}$

77. $\dfrac{2}{3} \div \dfrac{4}{9} - \dfrac{3}{11} \cdot \dfrac{22}{9}$ $\dfrac{5}{6}$

Answer true or false.

78. $\dfrac{11}{32} < \dfrac{21}{61}$ true

79. $8\dfrac{15}{19} > 8\dfrac{7}{11}$ true

Solve.

80. Ann worked $7\frac{1}{2}$ hours, $6\frac{3}{4}$ hours, and $4\frac{3}{8}$ hours during a three-day period. How many hours did she work? $18\frac{5}{8}$ hours

81. It takes Graydon $60\frac{1}{2}$ minutes to run a particular course. If he has been running for $38\frac{3}{4}$ minutes, how much longer will it be before he completes the course? $21\frac{3}{4}$ minutes

CHAPTER 5

Give a formal word name for each decimal.

82. 42.5

     forty-two and five tenths

Write each number in decimal notation.

83. 3.0007

     three and seven ten thousandths

84. two and nine thousandths

     2.009

85. six hundred twenty-seven and nine hundredths

     627.09

Change each decimal to a fraction (or mixed number).

86. 0.47 $\frac{47}{100}$

87. 8.131 $8\frac{131}{1000}$

Change each fraction to a decimal.

88. $\dfrac{4}{3}$ $1.\overline{3}$

89. $\dfrac{5}{8}$ 0.625

90. $\dfrac{4}{9}$ $0.\overline{4}$

Find the sum.

91. $2.7 + 8.32 + 105.006$ **116.026**

92. $5.2 + 5.02 + 0.502 + 50.02$ **60.742**

93. Find the total deposit on the given portion of a deposit slip.   **$7206.16**

	Dollars	Cents
Cash	$ 493	07
Checks	647	51
	18	30
	6047	28
Total	$	

94. Find the perimeter of the figure.   **23.13 in**

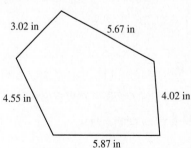

3.02 in   5.67 in

4.55 in   4.02 in

5.87 in

Find the difference.

95. $37.2 - 5.09$ **32.11**

96. $7 - 1.006$ **5.994**

97. Cindy received a check for $185.70. She spent $72.43 on a coat, $24.17 on a pair of shoes, and $13.98 for a pair of pajamas. How much did she have left?   **$75.12**

98. Round 4.355 to two decimal places.   **4.36**

99. Round \$47.51 to the nearest dollar.   **\$48**

100. Round $0.9\overline{2}$ to the nearest hundredth.   **0.93**

101. Round $0.9\overline{2}$ to three decimal places.   **0.929**

Find the product.

102. $\begin{array}{r} 24.07 \\ \times\ 2.8 \\ \hline 67.396 \end{array}$

103. $\begin{array}{r} 6.1109 \\ \times\ 4.02 \\ \hline 24.565818 \end{array}$

104. $\begin{array}{r} 0.90001 \\ \times 1.9 \\ \hline 1.710019 \end{array}$

105. Find the area of a rectangle which is 2.31 ft long and 6.5 ft wide.   **15.015 sq ft**

Find the quotient.

106. $3.7\overline{)22.385}$ **6.05**

107. $0.025\overline{)0.1175}$ **4.7**

108. $52\overline{)607.36}$ **11.68**

109. What is the approximate cost of 11 pounds of nuts selling for \$3.05 per pound? The exact cost?
\$30 (10 × \$3), \$33.55

Find the sum.

110. $4.35 + \dfrac{4}{5}$ **5.15**

111. $5\dfrac{3}{4} + 7.84$ **13.59**

112. If Robert earns \$17.13 an hour, how much will he be paid for working $6\frac{1}{2}$ hours?   **\$111.35**

113. If $16\frac{3}{4}$oz of breakfast drink sell for \$1.89, to the nearest tenth of a cent, what is the price per ounce (the unit price)?   **11.3¢**

114. Arrange 0.26, 2.6, 2.06, 0.026, and 20.06 in order of size with the smallest on the left.
0.026, 0.26, 2.06, 2.6, 20.06

Perform the indicated operations.

115. $5.2 \div \dfrac{1}{2} - (1.2)^2$ **8.96**

116. $\dfrac{5}{3} \cdot \dfrac{9}{20} - (0.5)^2$ $\dfrac{1}{2}$

CHAPTER 6

Are the given pairs of numbers proportional?

117. 6, 5 and 7, 6   **no**

118. 9, 12 and 30, 40   **yes**

Solve the proportion.

119. $\dfrac{a}{12} = \dfrac{5}{3}$ **20**

120. $\dfrac{14}{a} = \dfrac{35}{8}$ $\dfrac{16}{5}$

121. If 48 m of wire weighs 16 kg, what will 360 m of the same wire weigh?   **120 kg**

122. A boat can travel 90 miles on 40 gallons of gas. How far can it travel on 56 gallons of gas?
126 mi

Change from percents to fractions.

123. 37% $\dfrac{37}{100}$

124. 0.06% $\dfrac{3}{5000}$

125. $9\dfrac{3}{8}\%$ $\dfrac{3}{32}$

Change from percents to decimals.

126. 58%   0.58

127. 220%   2.2

128. $10\dfrac{1}{2}\%$ 0.105

Change to percent notation.

129. 0.81   81%

130. 0.005   0.5%

131. 20   2000%

132. $\dfrac{1}{6}$ $16.\overline{6}\%$ or $16\dfrac{2}{3}\%$

133. $\dfrac{9}{25}$ 36%

134. $\dfrac{4}{9}$ $44.\overline{4}\%$

Solve the following percent problems.

135. What is 38% of 40?   15.2

136. What percent of 96 is 12?   12.5%

137. 72 is 120% of what number?   60

138. 2.7 is what percent of 900?   0.3%

139. There were 202,000 votes cast for the two people in an election. If Helen received 52% of the votes, how many votes did John receive?   96,960

140. If there are 25 g of acid in a solution of total weight 40 g, what is the percent acid in the solution?   62.5%

141. What is the sales tax on a refrigerator priced at $720 if the tax rate is 5%? What is the total price, including tax? $36, $756

142. If a tax of $1.86 is charged on a dress selling for $62.00, what is the sales-tax rate?   3%

143. If the tax rate for Social Security was 7.51% in 1989, how much tax did Walter pay on an income of $20,800? $1562.08

144. Birgreta receives a 1.5% commission each month on all sales of $30,000 or less. For sales over $30,000, she is paid a 2% commission. What is her commission on sales of $58,400 during the month of June? $1018

145. Considering the total commission that Birgreta received in Exercise 144, what is her overall commission rate to the nearest tenth of a percent?   1.7%

146. Shirts are advertised for $4.50 off the original price. If the discount rate is 30%, what was the original price? $15

147. Manuel borrowed $1500 from his mother and agreed to pay her 8% per year simple interest. How much must he pay at the end of 18 months (a year and a half)?

$1680

148. Karen has $5200 to invest. One account pays 10% compounded annually and another pays 10.5% simple interest. Which account would give her the most interest over a 2-year period?

each account pays $1092 interest over 2 years

149. The price of steak increased from $1.95 per pound to $2.35 per pound. To the nearest tenth of a percent, what was the percent increase?   **20.5%**

150. A dealer bought a toaster for $23.50. The markup was 30%. She was later forced to sell the toaster at a markdown of 28%. What was the percent of profit or loss based on the original cost?

6.4% loss

CHAPTER 7

Exercises 151–54 refer to the bar graph showing the number of students in each class who applied for a scholarship.

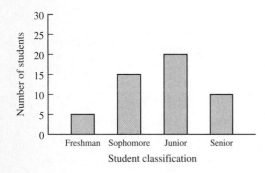

151. How many students applied?   **50**

152. What is the ratio of juniors to sophomores who applied? $\dfrac{4}{3}$

153. What percent of those applying were freshmen?   **10%**

154. What is the ratio of freshmen and sophomores to juniors and seniors? $\dfrac{2}{3}$

Exercises 155–59 refer to the broken-line graph showing the number of vehicles passing through an intersection (in hundreds) during an 8-hour interval.

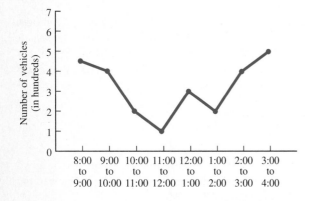

155. How many vehicles passed through the intersection from 8:00 to 9:00?   **450**

156. How many vehicles passed through the intersection from 11:00 to 12:00?   **100**

157. How many vehicles passed through the intersection from 8:00 to 12:00?   1150

158. How many vehicles passed through the intersection during the period from 8:00 to 4:00?   2550

159. What percent (to the nearest tenth) of the total number of vehicles passed through the intersection from 8:00 to 12:00?   45.1%

Find the mean and median. Give answers to nearest tenth.

160. 9, 10, 17, 6   10.5, 9.5

161. 6.2, 8.5, 4.7, 3.3, 9.4   6.4, 6.2

162. A truck was driven 826 miles on 203 gallons of fuel. To the nearest tenth, what was the average miles per gallon?   4.1 mpg

163. Find the mode. 42, 35, 42, 50, 35, 70, 60, 35   35

CHAPTER 8

Complete.

164. 42 in = _3.5_ ft

165. 5400 cm = _0.054_ km

166. 2.5 g = _2500_ mg

167. 5 lb = _2270_ g

168. $100 \dfrac{\text{km}}{\text{hr}}$ = _62.1_ $\dfrac{\text{mi}}{\text{hr}}$

169. 60°C = _140_ °F

170. If sugar costs 36¢ per lb, how much will the sugar cost for a cake which requires 24 oz of sugar?
54¢

171. If a car's gasoline tank holds 25 gal, how many liters does it hold?
94.6

Perform the indicated operations and simplify.

172.   21 hr 15 min
    + 6 hr 48 min   28 hr 3 min

173.   13 yd 1 ft 7 in
    − 2 yd 2 ft 4 in   10 yd 2 ft 3 in

174.   25.2 cm
    + 7.9 cm   33.1 cm

175.   421.6 ml
    − 382.7 ml   38.9 ml

176. 4 × (2 gal 2 qt 1 pt)   10 gal 2 qt

177. 62 km ÷ 8 km   7.75

CHAPTER 9

178. Use inductive reasoning to determine the next element in the list. 1, 3, 9, 27, 81   243

In Exercises 179–80, determine if the conclusion follows logically from the premises, and state whether the reasoning is inductive or deductive.

179. Premise: If one angle in a triangle is 90°, then the triangle is a right triangle.
Premise: One angle of the triangle is 90°.
Conclusion: The triangle is a right triangle.
follows logically; deductive reasoning

180. Premise: The answer to problem one is correct.
Premise: The answer to problem two is correct.
Premise: The answer to problem three is correct.
Conclusion: The answer to problem four is correct.
does not follow logically; inductive reasoning

Use the figure and deductive reasoning to answer true or false in Exercises 181–86. If the answer is false, tell why.

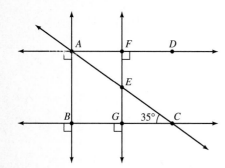

181. △ABC is a right triangle.
true

182. ∠CAD = 35°.
true

183. $\overleftrightarrow{AD}$ is parallel to $\overleftrightarrow{AB}$.
false

184. $\overleftrightarrow{AD}$ is perpendicular to $\overleftrightarrow{FG}$.
true

185. △ABC is similar to △EGC.
true

186. If AF = CG, then △AFE is congruent to △CGE.
true

Use the figure and deductive reasoning in Exercises 187–88.

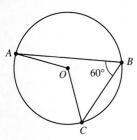

187. What is the measure of ∠AOC?
120°

188. What is the measure of $\overset{\frown}{ABC}$?
240°

Find the perimeter and area of the rectangle or square.

189. 5 ft by 9 ft   **28 ft, 45 ft²**

190. $12\frac{1}{2}$ m by 6.2 m   **37.4 m, 77.5 m²**

191. The area shown is to be carpeted. How much will it cost if carpet is $22.50 per square yard?   **$945**

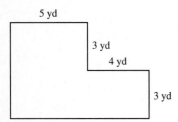

5 yd

3 yd

4 yd

3 yd

Find the perimeter and area of each figure.

192.

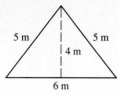

16 m, 12 m²

193.

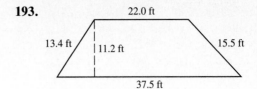

88.4 ft, 333.2 ft²

Find the circumference and area of each circle. Use 3.14 *for* π *and give answers to nearest tenth.*

194. $r = 2.6$ ft   16.3 ft, 21.2 ft²

195. $d = 9.2$ m   28.9 m, 66.4 m²

196. A 14-in diameter pizza costs $7.50 and a 16-in pizza costs $8.50. Which costs less per square inch?
14 in: 0.049 per sq in
16 in: 0.042 per sq in
16 in costs less

Find the volume and surface area of each solid.

197. Rectangular solid: 5 m by 2 m by 8 m
80 m³, 132 m²

198. Cube: 3.6 ft edge
46.7 ft³, 77.8 ft²

199. Cylinder: $d = 5.2$ dm, $h = 7.8$ dm
165.6 dm³, 169.8 dm²

200. Sphere: $r = 15$ in
14,130 in³, 2826 in²

201. A spherical tank has radius 42 cm. How many liters of gasoline will it hold (to the nearest liter)?
310 L

CHAPTER 10

202. Write *yyyyyy* in exponential notation. y^6

203. Evaluate $x^2 + 4(x - y)$ for $x = 5$ and $y = 3$.
33

204. If $A = P(1 + r)^t$, find A when P is $2000, the interest is 8% compounded annually, and t is 2 years.
$2332.80

205. Find the reciprocal of $-\frac{3}{4}$. $-\frac{4}{3}$

206. Find $|-8|$.   8

Place the appropriate symbol =, <, or > between the pairs of numbers.

207. $-8 > -10$

208. $\dfrac{8}{31} > \dfrac{4}{17}$

Perform the indicated operations.

209. $-4 - (-16)$ 12   **210.** $14 + (-24)$ -10 **211.** $\left(-\dfrac{5}{9}\right) \cdot \left(-\dfrac{3}{20}\right)$ $\dfrac{1}{12}$ **212.** $(7.5) \div (-2.5)$ -3

Evaluate.

213. $2 \cdot 4 + (8 - 12) - 2(3 - 1)$ 0   **214.** $-8 - [-(-8)]$ -16

Factor.

215. $2y + 12$ $2(y + 6)$ **216.** $-14x - 21$ $-7(2x + 3)$

Collect like terms.

217. $-8a + 3b - 4a - 5 - b$ $-12a + 2b - 5$ **218.** $7y - (2y - 5) + 3$ $5y + 8$

219. Evaluate $2u^2 - v^2$ when $u = -2$ and $v = 4$. -8 **220.** Write $-5x^4$ without exponents. $-5xxxx$

221. Evaluate. $-\sqrt{144}$ -12 **222.** Evaluate. $\sqrt{\dfrac{12}{75}}$ $\dfrac{2}{5}$

CHAPTER 11

Solve.

223. $x + 10 = 19$ 9   **224.** $16x = 4$ $\dfrac{1}{4}$ **225.** $x - \dfrac{5}{3} = \dfrac{7}{3}$ 4

226. $-5.5x = 33$ -6 **227.** $\dfrac{2}{3}x = 26$ 39   **228.** $\dfrac{x}{\frac{3}{4}} = 12$ 9

229. $4x + 8 = 7x - 4$ 4   **230.** $-3x - 2 = -5x + 9$ $\dfrac{11}{2}$

231. $5(x + 2) - 2(x - 10) = 0$ -10 **232.** $9x - (2x + 3) = 4$ 1

233. The sum of two consecutive even integers is 134. What are the integers?
66, 68

234. A suit is put on sale at a 20% discount. If the original price was $160, what is the sale price?
$128

235. The area of a triangle is 85 in^2 and the base is 10 inches. What is the altitude?
17 in

236. Two trains leave Portland, one traveling north and the other south. If one is traveling 30 mph faster than the other, how fast is each traveling if they are 360 miles apart after 3 hours?
45 mph, 75 mph

Graph on the number line.

237. $3x + 8 = 5$

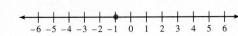

238. $4(x - 2) \geq 4 - (x - 3)$

Solve the inequality.

239. $4x + 9 \geq 7x - 3$ $x \leq 4$

240. $2x - (3x - 2) < 9x - 8$ $x > 1$

CHAPTER 12

241. The point with coordinates $(-3, -5)$ is located in which quadrant?   **third**

242. Complete the ordered pair (5,) so that it is a solution to $2x + 3y = 10$.   **(5, 0)**

243. Cost, c, is related to number of items produced, n, by $c = 100n + 50$. Find c when n is 18.
$1850

244. True or False: $5 + 3y = 2x$ is a linear equation.
true

245. Find the slope of the line through the points $(7, -1)$ and $(2, -5)$. Determine the general form of the equation of the line.
$\frac{4}{5}$, $4x - 5y - 33 = 0$

246. Write $3x - 4y = 12$ in slope-intercept form and give the slope and the y-intercept.
$y = \frac{3}{4}x - 3$, $\frac{3}{4}$, $(0, -3)$

247. Find the general form of the equation of the line with slope $-\frac{4}{5}$ and y-intercept $(0, 7)$.
$4x + 5y - 35 = 0$

248. If l_1 passes through $(1, 2)$ and $(-1, 6)$ and l_2 passes through $(0, -3)$ and $(6, 0)$, are the lines parallel, perpendicular, or neither?
perpendicular

Graph in the given Cartesian coordinate system.

249. $y = -3x + 6$

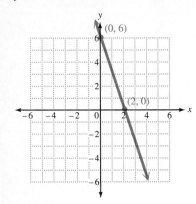

250. $2x - 3y + 6 \leq 0$

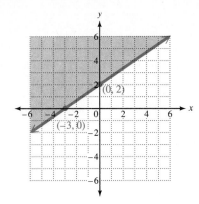

CHAPTER 13

251. Without graphing, determine the nature of the graphs (intersecting, parallel, or coinciding).
$3x + 4y = 7$
$9x + 12y = 5$ **parallel**

252. If the lines in a system of equations are intersecting, the system has _____ solution(s).
exactly one

Solve the system.

253. $3x - y = -2$
$-5x + 2y = 5$ **(1, 5)**

254. $2x + 3y = 8$
$4x + 7y = 16$ **(4, 0)**

255. $2x - 3y = 12$
$3x + 9 = 0$ **(−3, −6)**

256. $4x - 1 = 0$
$3y + 2 = 0$ $\left(\dfrac{1}{4}, -\dfrac{2}{3}\right)$

257. A boat travels 40 miles upstream in 4 hours and returns to the starting point in 2 hours. What is the speed of the stream?   **5 mph**

258. Pam Johnson earned $1180 in one year on an investment of $12,000. Part of the money was invested at 9% and part at 11% simple interest. How much was invested at 11%?   **$5000**

259. Twice one number plus a second is 13. The first minus the second is 5. Find the numbers.   **(6, 1)**

260. A candy shop has one candy that sells for 95¢ per pound and a second that sells for 65¢ per pound. How much of each should be used to make 60 pounds of candy selling for 75¢ per pound?
20 lb of 95¢ candy
40 lb of 65¢ candy

CHAPTER 14

Simplify and write without negative exponents.

261. $3x^2x^3$ $3x^5$ **262.** $\dfrac{y^6}{y^2}$ y^4 **263.** $(-3a^3)^2$ $9a^6$ **264.** a^{-2} $\dfrac{1}{a^2}$

265. Write 36,200,000 in scientific notation.
3.62×10^7

266. Write 7.2×10^{-4} without using scientific notation.
0.00072

Evaluate using scientific notation.

267. $(0.0000021)(3,000,000)$ $6.3 \times 10^0 = 6.3$

268. $\dfrac{0.000055}{11,000,000}$ 5×10^{-12}

269. Is $-8x^3y$ a monomial, binomial, or trinomial?
monomial

270. Give the degree of $3x^4 - 6x^5 + 5x - 2$.
5

271. Write in descending order.
$3x - 4x^3 - 6x^5 + 4x^2 + x^7$
$x^7 - 6x^5 - 4x^3 + 4x^2 + 3x$

272. Profit in dollars when x coats are sold is given by $2x^2 - 5$. What is the profit when 6 coats are sold?
$67

Add.

273. $(-5a + 2) + (6a - 1)$
$a + 1$

274. $(2x^2y^2 - 3xy + 2) + (-6x^2y^2 - xy - 5)$
$-4x^2y^2 - 4xy - 3$

Subtract.

275. $(9y - 3) - (-2y + 5)$
$11y - 8$

276. $(3x^2 - 2x + 5) - (4x^2 + 9x - 6)$
$-x^2 - 11x + 11$

277. $(7a - 6) - (-a + 1) - (3a - 5)$
$5a - 2$

278. $(6x^2y - 4xy) - (4xy^2 + 2xy) - (2x^2y + xy^2)$
$4x^2y - 5xy^2 - 6xy$

Multiply.

279. $-3a^2(4a - 2)$
$-12a^3 + 6a^2$

280. $(3x + 4y)(9x - 2y)$
$27x^2 + 30xy - 8y^2$

281. $(a - 2)(3a^2 + 2a - 5)$
$3a^3 - 4a^2 - 9a + 10$

282. $(5x + 7)^2$
$25x^2 + 70x + 49$

283. $(2x + 5y)(2x - 5y)$
$4x^2 - 25y^2$

284. $(x^2 - 2)^2$
$x^4 - 4x^2 + 4$

Divide.

285. $\dfrac{6y^9 - 2y^6 + 10y^3}{2y^2}$
$3y^7 - y^4 + 5y$

286. $(2x^3 - 27x + 20) \div (x + 4)$
$2x^2 - 8x + 5$

CHAPTER 15

Factor by removing the greatest common factor.

287. $24x - 16$
$8(3x - 2)$

288. $6x^3y^2 - 3x^2y^3 + 9x^2y^2$
$3x^2y^2(2x - y + 3)$

Factor by grouping.

289. $3x^3 + 2x^2 + 6x + 4$
$(x^2 + 2)(3x + 2)$

290. $a^2b^2 - a^2 + 3b^2 - 3$
$(a^2 + 3)(b + 1)(b - 1)$

Factor.

291. $x^2 + 13x + 40$
$(x + 8)(x + 5)$

292. $y^2 - 18y + 81$
$(y - 9)^2$

293. $6x^2 - 24y^2$
$6(x + 2y)(x - 2y)$

294. $x^2 + xy - 20y^2$
$(x - 4y)(x + 5y)$

295. $2x^2 - 9x - 5$
$(2x + 1)(x - 5)$

296. $9x^2 + 12xy + 4y^2$
$(3x + 2y)^2$

Solve.

297. $(2x - 1)(x + 8) = 0$
$\dfrac{1}{2}, -8$

298. $x^2 - 8x - 20 = 0$
$-2, 10$

299. The length of a rectangle is four times the width. If the area is 144 cm^2, find the dimensions.
6 cm, 24 cm

300. Profit is given by $P = n^2 - 2n + 3$, where n is the number of items sold. How many items were sold when the profit was \$38?   **7**

CHAPTER 16

301. What values of the variable must be omitted in $\dfrac{3x}{(x + 1)(x - 8)}$?   **−1, 8**

302. Reduce $\dfrac{a^2 - b^2}{(a - b)^2}$ to lowest terms.
$\dfrac{a + b}{a - b}$

Perform the indicated operations.

303. $\dfrac{x^2 - x - 2}{x^2 - 2x - 3} \cdot \dfrac{x^2 - 3x}{x + 2}$ $\dfrac{x(x - 2)}{x + 2}$

304. $\dfrac{y^2 - 1}{y^2 + 2y} \div \dfrac{y - 1}{y + 2}$ $\dfrac{y + 1}{y}$

305. $\dfrac{5}{y^2 - 4y + 3} + \dfrac{7}{y^2 + y - 2}$ $\dfrac{12y - 11}{(y + 2)(y - 3)(y - 1)}$

306. $\dfrac{x + 6}{x^2 - 4} - \dfrac{x}{x - 2}$ $\dfrac{-x - 3}{x + 2}$

307. Simplify. $\dfrac{x - \dfrac{1}{x}}{1 - \dfrac{1}{x}}$ $x + 1$

308. Solve. $\dfrac{1}{y + 2} + \dfrac{1}{y - 2} = \dfrac{1}{y^2 - 4}$ $\dfrac{1}{2}$

309. A car goes 205 miles on 15 gallons of gasoline. How many miles can the car go on 75 gallons of gasoline?   1025 mi

310. The denominator of a fraction is 5 more than the numerator. If 2 is added to both the numerator and the denominator the result is $\frac{1}{2}$. Find the fraction.
$\frac{3}{8}$

311. The cost c of nuts varies directly as the weight w of the nuts. If 12 lb of nuts cost \$30, how much would 80 lb cost?   \$200

312. The volume V of a gas varies inversely as the pressure P. If V is 16 in^3 when the pressure is 3 pounds per square inch (psi), what is the volume when $P = 12$ psi?   4 in^3

CHAPTER 17

Evaluate. Assume all variables and expressions are nonnegative.

313. $\sqrt{(-16)^2}$ 16

314. $\pm\sqrt{169}$ ± 13

315. $\sqrt{81x^2}$ $9x$

316. $5\sqrt{8x^2y^2}$ $10xy\sqrt{2}$

317. $\sqrt{\dfrac{27x^2}{4y^2}}$ $\dfrac{3x\sqrt{3}}{2y}$

318. $\sqrt{\dfrac{49y^2}{x}}$ $\dfrac{7y\sqrt{x}}{x}$

Perform the indicated operations and simplify.

319. $5\sqrt{15}\ \sqrt{45}$ $75\sqrt{3}$

320. $\dfrac{\sqrt{24x^3y}}{\sqrt{3xy}}$ $2x\sqrt{2}$

321. $\sqrt{98} + \sqrt{50}$ $12\sqrt{2}$

322. $3\sqrt{45} - 2\sqrt{80} + \sqrt{125}$ $6\sqrt{5}$

323. $\sqrt{28x} + 3\sqrt{7x} - 2\sqrt{63x}$ $-\sqrt{7x}$

324. $\sqrt{3} - \dfrac{2}{\sqrt{3}}$ $\dfrac{\sqrt{3}}{3}$

325. $(\sqrt{7} - 2\sqrt{2})(\sqrt{7} + \sqrt{2})$ $3 - \sqrt{14}$

326. $\dfrac{\sqrt{6}}{\sqrt{3} - \sqrt{2}}$ $3\sqrt{2} + 2\sqrt{3}$

Solve.

327. $\sqrt{2x - 1} - \sqrt{3x - 7} = 0$ 6

328. $\sqrt{x^2 + 3} + 1 - x = 0$ no solution

329. One side of a rectangle is 12 m long and the diagonal is 13 m. Find the length of the other side.
5 m

330. Three times the square root of 5 more than a number is equal to 15. Find the number.   20

331. In a 30°–60° right triangle a is the side opposite the 30° angle. If $a = 22$, find b and c.

$b = 22\sqrt{3},\ c = 44$

332. In a 45°–45° right triangle $b = 45$ cm. Find a and c.

$a = 45$ cm, $c = 45\sqrt{2}$ cm

CHAPTER 18

Solve.

333. $x^2 - 12x + 35 = 0$ 5, 7

334. $x^2 + 14x + 49 = 0$ −7

335. $2x^2 - 11x + 5 = 0$ $\dfrac{1}{2},\ 5$

336. $x^2 - 3x - 1 = 0$ $\dfrac{3 \pm \sqrt{13}}{2}$

337. $3y^2 - 5 = y(y + 3)$ $\dfrac{5}{2},\ -1$

338. $(z - 2)(z + 2) = 5z$ $\dfrac{5 \pm \sqrt{41}}{2}$

339. $\sqrt{7 - 2x} = x - 2$ 3

340. $\dfrac{x + 2}{x} = \dfrac{2}{x - 1}$ 2, −1

341. The number of square inches in a square is 3 less than the perimeter of the square. Find the length of a side.

3 in, 1 in

342. Al can do a job in 3 hours less time than Joe. Together they can do the job in 2 hours. How long would it take each to do the job?

Al needs 3 hr and Joe 6 hr

343. Solve $a = \sqrt{\dfrac{2b}{c}}$ for b. $b = \dfrac{a^2 c}{2}$

344. Graph $y = -x^2 + 1$.

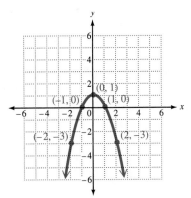

ANSWERS: 1. (a) 2   (b) 7   (c) <   (d) <   (e) <   (f) >   2. $20{,}000{,}000 + 7{,}000{,}000 + 100{,}000 + 40{,}000 + 3000 + 200 + 6$ 3. 2,040,325   4. eighty-six million, two hundred forty-seven thousand, one hundred fifty-nine   5. 327
6. 604   7. 8746   8. 4555   9. (a) 27,490   (b) 27,500   (c) 27,000   10. (a) 4940   (b) 4900   11. 249   12. 28 mi
13. 26   14. 186   15. 2051   16. 659   17. $1217   18. $2770   19. 3   20. 1   21. exponent   22. (a) 16   (b) 2
23. 3239   24. 63,882   25. 1,815,030   26. 25,996,000   27. Q: 481, R: 0   28. Q: 68, R: 29   29. Q: 227, R: 401

30. 7   31. 0   32. 14   33. 1, 2, 3, 5, 6, 10, 15, 25, 30, 50, 75, 150   34. $546 = 2 \cdot 3 \cdot 7 \cdot 13$ 35. (a) yes   (b) yes
(c) no   (d) yes   (e) yes   36. (a) 3   (b) 15   (c) 2   (d) 12   (e) 1   37. \$777   38. 28 ounces of soda at 4¢ an ounce is
the better buy than the 32 ounces at 5¢ an ounce   39. $\frac{2}{3}$ 40. $\frac{5}{4}$ 41. 1   42. 0   43. $\frac{7}{4} \neq \frac{19}{11}$ 44. 9   45. $\frac{2}{11}$ 46. 1
47. $\frac{6}{5}$ 48. $\frac{9}{16}$ 49. $\frac{2}{9}$ 50. $\frac{10}{3}$ 51. $\frac{9}{8}$ 52. 18   53. 8   54. $4\frac{5}{9}$ 55. $11\frac{1}{11}$ 56. $\frac{67}{5}$ 57. $\frac{129}{32}$ 58. $5\frac{15}{32}$ 59. $2\frac{7}{9}$ 60. 45
61. 243 gallons   62. 108 gallons   63. $27\frac{3}{5}$ revolutions   64. 18   65. 8580   66. 2520   67. $\frac{2}{3}$ 68. $\frac{2}{5}$ 69. $\frac{23}{36}$ 70. $\frac{61}{105}$
71. $\frac{16}{35}$ 72. $\frac{1}{30}$ 73. $7\frac{19}{40}$ 74. $5\frac{5}{12}$ 75. $\frac{19}{20}$ 76. $\frac{5}{16}$ 77. $\frac{5}{6}$ 78. true   79. true   80. $18\frac{5}{8}$ hours   81. $21\frac{3}{4}$ minutes
82. forty-two and five tenths   83. three and seven ten thousandths   84. 2.009   85. 627.09   86. $\frac{47}{100}$ 87. $8\frac{131}{1000}$
88. $1.\overline{3}$ 89. 0.625   90. $0.\overline{4}$ 91. 116.026   92. 60.742   93. \$7206.16   94. 23.13 in   95. 32.11   96. 5.994   97. \$75.12
98. 4.36   99. \$48   100. 0.93   101. 0.929   102. 67.396   103. 24.565818   104. 1.710019   105. 15.015 sq ft
106. 6.05   107. 4.7   108. 11.68   109. \$30 ($10 \times \$3$), \$33.55   110. 5.15   111. 13.59   112. \$111.35   113. 11.3¢
114. 0.026, 0.26, 2.06, 2.6, 20.06   115. 8.96   116. $\frac{1}{2}$ 117. no   118. yes   119. 20   120. $\frac{16}{5}$ 121. 120 kg   122. 126
miles   123. $\frac{37}{100}$ 124. $\frac{3}{5000}$ 125. $\frac{3}{32}$ 126. 0.58   127. 2.2   128. 0.105   129. 81%   130. 0.5%   131. 2000%
132. $16.\overline{6}$% or $16\frac{2}{3}$%   133. 36%   134. $44.\overline{4}$%   135. 15.2   136. 12.5%   137. 60   138. 0.3%   139. 96,960
140. 62.5%   141. \$36, \$756   142. 3%   143. \$1562.08   144. \$1018   145. 1.7%   146. \$15   147. \$1680   148. each
account pays \$1092 interest over 2 years   149. 20.5%   150. 6.4% loss   151. 50   152. $\frac{4}{3}$ 153. 10%   154. $\frac{2}{3}$
155. 450   156. 100   157. 1150   158. 2550   159. 45.1%   160. 10.5, 9.5   161. 6.4, 6.2   162. 4.1 mpg   163. 35
164. 3.5   165. 0.054   166. 2500   167. 2270   168. 62.1   169. 140   170. 54¢   171. 94.6   172. 28 hr 3 min
173. 10 yd 2 ft 3 in   174. 33.1 cm   175. 38.9 ml   176. 10 gal 2 qt   177. 7.75   178. 243   179. follows logically; deductive reasoning   180. does not follow logically; inductive reasoning   181. true   182. true   183. false (perpendicular)
184. true   185. true   186. true   187. 120°   188. 240°   189. 28 ft, 45 ft²   190. 37.4 m, 77.5 m²   191. \$945
192. 16 m, 12 m²   193. 88.4 ft, 333.2 ft²   194. 16.3 ft, 21.2 ft²   195. 28.9 m, 66.4 m²   196. 14 in: \$0.049 per square
inch, 16 in: \$0.042 per square inch, 16 in costs less   197. 80 m³, 132 m²   198. 46.7 ft³, 77.8 ft²   199. 165.6 dm³,
169.8 dm²   200. 14,130 in, 2826 in²   201. 310 L   202. y^6 203. 33   204. \$2332.80   205. $-\frac{4}{3}$ 206. 8   207. >
208. >   209. 12   210. -10 211. $\frac{1}{12}$ 212. -3 213. 0   214. -16 215. $2(y + 6)$ 216. $-7(2x + 3)$ 217. $-12a +$
$2b - 5$ 218. $5y + 8$ 219. -8 220. $-5xxxx$ 221. -12 222. $\frac{2}{5}$ 223. 9   224. $\frac{1}{4}$ 225. 4   226. -6 227. 39
228. 9   229. 4   230. $\frac{11}{2}$ 231. -10 232. 1   233. 66, 68   234. \$128   235. 17 in   236. 45 mph, 75 mph
237.   238.   239. $x \leq 4$ 240. $x > 1$ 241. third
242. (5, 0)   243. \$1850   244. true   245. $\frac{4}{5}$, $4x - 5y - 33 = 0$ 246. $y = \frac{3}{4}x - 3$, $\frac{3}{4}$, $(0, -3)$ 247. $4x + 5y - 35 = 0$
248. perpendicular   249.   250.   251. parallel   252. exactly one

253. (1, 5)   254. (4, 0)   255. $(-3, -6)$ 256. $(\frac{1}{4}, -\frac{2}{3})$ 257. 5 mph   258. \$5000   259. (6, 1)   260. 20 lbs of 95¢
candy and 40 lbs of 65¢ candy   261. $3x^5$ 262. y^4 263. $9a^6$ 264. $\frac{1}{a^2}$ 265. 3.62×10^7 266. 0.00072
267. $6.3 \times 10^0 = 6.3$ 268. 5×10^{-12} 269. monomial   270. 5   271. $x^7 - 6x^5 - 4x^3 + 4x^2 + 3x$ 272. \$67
273. $a + 1$ 274. $-4x^2y^2 - 4xy - 3$ 275. $11y - 8$ 276. $-x^2 - 11x + 11$ 277. $5a - 2$ 278. $4x^2y - 5xy^2 - 6xy$
279. $-12a^3 + 6a^2$ 280. $27x^2 + 30xy - 8y^2$ 281. $3a^3 - 4a^2 - 9a + 10$ 282. $25x^2 + 70x + 49$ 283. $4x^2 - 25y^2$
284. $x^4 - 4x^2 + 4$ 285. $3y^7 - y^4 + 5y$ 286. $2x^2 - 8x + 5$ 287. $8(3x - 2)$ 288. $3x^2y^2(2x - y + 3)$
289. $(x^2 + 2)(3x + 2)$ 290. $(a^2 + 3)(b + 1)(b - 1)$ 291. $(x + 8)(x + 5)$ 292. $(y - 9)^2$ 293. $6(x + 2y)(x - 2y)$
294. $(x - 4y)(x + 5y)$ 295. $(2x + 1)(x - 5)$ 296. $(3x + 2y)^2$ 297. $\frac{1}{2}$, -8 298. -2, 10   299. 6 cm, 24 cm   300. 7
301. -1, 8   302. $\frac{a+b}{a-b}$ 303. $\frac{x(x-2)}{x+2}$ 304. $\frac{y+1}{y}$ 305. $\frac{12y-11}{(y+2)(y-3)(y-1)}$ 306. $\frac{-x-3}{x+2}$ 307. $x + 1$ 308. $\frac{1}{2}$ 309. 1025 mi
310. $\frac{3}{8}$ 311. \$200   312. 4 in³   313. 16   314. ± 13 315. $9x$ 316. $10xy\sqrt{2}$ 317. $\frac{3x\sqrt{3}}{2y}$ 318. $\frac{7y\sqrt{x}}{x}$ 319. $75\sqrt{3}$
320. $2x\sqrt{2}$ 321. $12\sqrt{2}$ 322. $6\sqrt{5}$ 323. $-\sqrt{7x}$ 324. $\frac{\sqrt{3}}{3}$ 325. $3 - \sqrt{14}$ 326. $3\sqrt{2} + 2\sqrt{3}$ 327. 6   328. no solution   329. 5 m   330. 20   331. $b = 22\sqrt{3}$, $c = 44$ 332. $a = 45$ cm, $c = 45\sqrt{2}$ cm   333. 5, 7   334. -7 335. $\frac{1}{2}$, 5
336. $\frac{3 \pm \sqrt{13}}{2}$ 337. $\frac{5}{2}$, -1 338. $\frac{5 \pm \sqrt{41}}{2}$ 339. 3   340. 2, -1 341. 3 in, 1 in   342. Al needs 3 hr and Joe 6 hr
343. $b = \frac{a^2c}{2}$ 344.

TABLE OF SQUARES AND SQUARE ROOTS

n	n^2	$\sqrt{n}$	$\sqrt{10n}$	n	n^2	$\sqrt{n}$	$\sqrt{10n}$
1	1	1.000	3.162	51	2601	7.141	22.583
2	4	1.414	4.472	52	2704	7.211	22.804
3	9	1.732	5.477	53	2809	7.280	23.022
4	16	2.000	6.325	54	2916	7.348	23.238
5	25	2.236	7.071	55	3025	7.416	23.452
6	36	2.449	7.746	56	3136	7.483	23.664
7	49	2.646	8.367	57	3249	7.550	23.875
8	64	2.828	8.944	58	3364	7.616	24.083
9	81	3.000	9.487	59	3481	7.681	24.290
10	100	3.162	10.000	60	3600	7.746	24.495
11	121	3.317	10.488	61	3721	7.810	24.698
12	144	3.464	10.954	62	3844	7.874	24.900
13	169	3.606	11.402	63	3969	7.937	25.100
14	196	3.742	11.832	64	4096	8.000	25.298
15	225	3.873	12.247	65	4225	8.062	25.495
16	256	4.000	12.649	66	4356	8.124	25.690
17	289	4.123	13.038	67	4489	8.185	25.884
18	324	4.243	13.416	68	4624	8.246	26.077
19	361	4.359	13.784	69	4761	8.307	26.268
20	400	4.472	14.142	70	4900	8.367	26.458
21	441	4.583	14.491	71	5041	8.426	26.646
22	484	4.690	14.832	72	5184	8.485	26.833
23	529	4.796	15.166	73	5329	8.544	27.019
24	576	4.899	15.492	74	5476	8.602	27.203
25	625	5.000	15.811	75	5625	8.660	27.386
26	676	5.099	16.125	76	5776	8.718	27.568
27	729	5.196	16.432	77	5929	8.775	27.749
28	784	5.292	16.733	78	6084	8.832	27.928
29	841	5.385	17.029	79	6241	8.888	28.107
30	900	5.477	17.321	80	6400	8.944	28.284
31	961	5.568	17.607	81	6561	9.000	28.460
32	1024	5.657	17.889	82	6724	9.055	28.636
33	1089	5.745	18.166	83	6889	9.110	28.810
34	1156	5.831	18.439	84	7056	9.165	28.983
35	1225	5.916	18.708	85	7225	9.220	29.155
36	1296	6.000	18.974	86	7396	9.274	29.326
37	1369	6.083	19.235	87	7569	9.327	29.496
38	1444	6.164	19.494	88	7744	9.381	29.665
39	1521	6.245	19.748	89	7921	9.434	29.833
40	1600	6.325	20.000	90	8100	9.487	30.000
41	1681	6.403	20.248	91	8281	9.539	30.166
42	1764	6.481	20.494	92	8464	9.592	30.332
43	1849	6.557	20.736	93	8649	9.644	30.496
44	1936	6.633	20.976	94	8836	9.695	30.659
45	2025	6.708	21.213	95	9025	9.747	30.822
46	2116	6.782	21.448	96	9216	9.798	30.984
47	2209	6.856	21.679	97	9409	9.849	31.145
48	2304	6.928	21.909	98	9604	9.899	31.305
49	2401	7.000	22.136	99	9801	9.950	31.464
50	2500	7.071	22.361	100	10000	10.000	31.623

Chapter 1

1. 1, 2, 3, 4   2. "is less than"   3. associative law   4. (a) 2   (b) 8   (c) $a < b$ 5. 30,000,000 + 5,000,000 + 400,000 + 8000 + 400 + 30 + 8   6. 380,407   7. one million, six hundred fifty-nine thousand, two hundred eighty-eight
8. 120   9. 515   10. 9247   11. 13,217   12. (a) 8890   (b) 8900   13. 37   14. 349   15. 207   16. 6069   17. $72
18. estimated savings: $500; exact savings: $484   19. 1255 gallons   20. 46 feet   21. $458

Chapter 2

1. commutative law   2. composite   3. when its ones digit is 0, 2, 4, 6, or 8   4. 1316   5. 843,733   6. 3,878,000
7. 1,800,000   8. 1   9. 0   10. 15   11. undefined   12. Q:236; R:0   13. Q:154; R:14   14. Q:458; R:12
15. $350 = 2 \cdot 5 \cdot 5 \cdot 7$ or $2 \cdot 5^2 \cdot 7$ 16. $622   17. $600   18. 5   19. 625   20. 8   21. 5   22. 5   23. 9   24. 54
25. 216   26. 3   27. $550,000

Chapter 3

1. proper   2. associative law   3. $\frac{1}{6}$ 4. $\frac{9}{2} = \frac{36}{8}$ 5. 21   6. 33   7. $\frac{7}{3}$ 8. $\frac{12}{13}$ 9. $\frac{1}{28}$ 10. $\frac{3}{11}$ 11. 15   12. $\frac{9}{64}$ 13. $\frac{7}{12}$
14. $\frac{55}{2}$ or $27\frac{1}{2}$ 15. $2\frac{2}{5}$ 16. $\frac{79}{11}$ 17. $22\frac{1}{10}$ 18. $\frac{66}{83}$ 19. 315   20. $2\frac{1}{7}$ 21. $20\frac{1}{4}$ gallons   22. 550 revolutions

Chapter 4

1. 210   2. 120   3. $\frac{1}{3}$ 4. $\frac{2}{3}$ 5. $\frac{23}{24}$ 6. $\frac{17}{60}$ 7. $\frac{9}{7}$ or $1\frac{2}{7}$ 8. $\frac{7}{24}$ 9. $9\frac{8}{9}$ 10. $8\frac{13}{15}$ 11. $17\frac{2}{3}$ 12. $8\frac{5}{8}$ 13. $8\frac{1}{6}$ 14. $\frac{1}{90}$
15. true   16. false   17. $13\frac{5}{8}$ yd   18. $3\frac{11}{12}$ hours

Chapter 5

1. seventy-three and five thousandths   2. 92.06   3. $\frac{4}{25}$ 4. 0.4375   5. 6.36   6. 45.1   7. 0.75   8. 110.02   9. 5.997
10. $24.60   11. 50¢; 48¢   12. 127.89   13. 10.08   14. 635.7   15. 0.6357   16. 0.006357   17. 63.57   18. $24.04
19. 24.8 mpg   20. $36.00 [using 400×9¢]; $38.60   21. 14.1   22. $8.03 \div 2.5 = 3.212$

Chapter 6

1. $\frac{13}{7}$ 2. 3   3. 279 pounds   4. $\frac{19}{20}$ 5. 0.008   6. 506%   7. 180%   8. 20%   9. 21.6   10. 40   11. $1156.54
12. $20.65   13. $610.20   14. 18.3%   15. 2% profit   16. $1.60; $2.00   17. $20.93   18. 12%

Chapter 7

1. 180   2. 150   3. $\frac{11}{3}$ or 11 to 3   4. 40   5. 90   6. $\frac{2}{1}$ or 2 to 1   7. $26.\overline{6}$%   8. 60%   9. 25   10. 80   11. $\frac{7}{16}$ or 7 to
16   12. 25%   13. 20   14. 18   15. 8   16. (a) 254.5 pounds   (b) 255 pounds   (c) no mode

Chapter 8

1. 2.5   2. 0.028   3. 14.3   4. 149   5. 138   6. 396   7. 110.2   8. 610,000   9. 8 yd 2 ft 7 in   10. 4 hr 43 min 50 sec
11. 23 yd 2 ft 2 in   12. 162 m^2 13. 4   14. 16.3 L   15. no; 59°F = 15°C

Chapter 9

1. $\frac{1}{5}$ 2. no; inductive reasoning   3. true   4. false (supplementary)   5. false ($\overleftrightarrow{AB}$ and $\overrightarrow{FG}$ intersect at B)   6. true
7. yes   8. 110°   9. 250°   10. 55°   11. false   12. true   13. true   14. true   15. true   16. true   17. false   18. true
19. true   20. true   21. $945   22. 16.3 ft   23. 66.4 m^2 24. (a) 46.7 ft^3 (b) 77.8 ft^2 25. 16-in (14-in: $0.049 per in^2; 16-in: $0.042 per in^2)   26. 12 ft^2 27. 31 m   28. 2009.6 cm^3 29. 5024 in^2 30. $78.75

Chapter 10

1. a^3 2. 1   3. 75   4. 225   5. 125   6. 35   7. 5   8. $594.05   9. false   10. 13   11. true
12. <   13. 141°   14. −22   15. 8   16. $-\frac{1}{8}$ 17. 3   18. 5   19. −18   20. −19   21. −6   22. $-3(y - 5)$
23. $5 - b - 2y$ 24. 3   25. 15   26. $-12y + 6$ 27. 11   28. $-\frac{5}{4}$ 29. true   30. false

Chapter 11

1. 15   2. 8   3. $\frac{1}{2}$ 4. -2 5. 36   6. 4   7. -3 8. 13   9. 3   10. 83, 85   11. $62.10   12. 40 cm   13. 48 mph, 57 mph   14. $x \geq -3$ 15. $x < \frac{1}{2}$ 16. 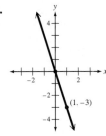   17.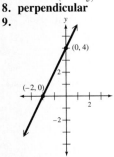

Chapter 12

1. IV   2. $\left(-3, \frac{2}{5}\right)$ 3. 400 items   4. false   5. $\frac{1}{3}$ 6. $y = -\frac{2}{5}x + 4$; $-\frac{2}{5}$; $(0, 4)$ 7. $6x + y - 19 = 0$
8. perpendicular
9.   10.   11.   12.

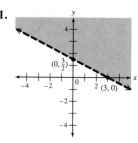
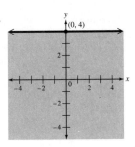

Chapter 13

1. intersecting lines   2. no   3. $(3, -1)$ 4. no solution   5. infinitely many   6. 3, 15   7. 24°, 156°   8. 30 nickels, 22 dimes   9. 450 mph   10. 40 lb candy; 60 lb nuts

Chapter 14

1. $5^2 a^3$ 2. $2y^7$ 3. a^4 4. $-8x^6$ 5. $\frac{1}{3}$ 6. $\frac{y^7}{x^8}$ 7. $-\frac{1}{64}$ 8. 3×10^{-2} 9. binomial   10. 4   11. $-6y^5 + 4y^3 - 6y + 6$
12. $-4x^2y^2 + 5x^2y - 4x^2$ 13. $285   14. $-2x + 10$ 15. $3x^4 + 10x^3 - 3x^2 + 7$ 16. $7a^2b^2 + 6ab + 2$ 17. $-7x + 7$
18. $-4x^4 - x^3 + 6x^2 + 1$ 19. $4a^2b - 4ab^2 - ab$ 20. $-12x^4 + 16x^3 - 4x^2$ 21. $x^2 - 3x - 40$ 22. $8a^2 + 22a - 21$
23. $3y^3 + 22y^2 - 15y + 8$ 24. $4x^2 - y^2$ 25. $9x^2 - 24xy + 16y^2$ 26. $4x^2 + 20x + 25$ 27. $6x^3 - 5x^2 + 6x + 8$
28. $x^4 - 9$ 29. $5x^2 - 2x - 3$ 30. $3x^2 - 2x + 4$

Chapter 15

1. $4(5x + 3)$ 2. $7(5y - 1)$ 3. $6x^2(4x^2 - 2x + 3)$ 4. $5x^3y^2(y + 8)$ 5. $(y^2 + 2)(7x^2 - 1)$ 6. $(x + 7)(x + 8)$
7. $(x - 8)^2$ 8. $3(y + 5)(y - 5)$ 9. $(x + 2)(x - 10)$ 10. $(2x - 3)(x + 8)$ 11. $(3x + y)(x + 5y)$ 12. $\frac{5}{4}, -\frac{3}{2}$
13. 6, -5 14. 1 or -4 15. 4 cm, 6 cm   16. (a) $332   (b) 12

Chapter 16

1. 0, -5 2. yes   3. $\frac{1}{a}$ 4. $\frac{5(x + 4)}{x}$ 5. $\frac{1}{y - 3}$ 6. $\frac{1}{a - b}$ 7. $\frac{2x}{x + 2}$ 8. $\frac{y - 5}{(y - 1)(y + 1)(y - 2)}$ 9. no solution   10. $\frac{10}{13}$ 11. 37 gallons   12. $13.80   13. $\frac{15}{8}$ days   14. $\frac{1}{1 + y}$ 15. 5

Chapter 17

1. 2   2. 5   3. ± 11 4. $\frac{9}{4}$ 5. $4x$ 6. $12\sqrt{3}$ 7. $\frac{2\sqrt{2x}}{y}$ 8. $\frac{7x\sqrt{2y}}{y}$ 9. 24   10. $3x$ 11. $6\sqrt{5}$ 12. $-16\sqrt{2}$ 13. $10\sqrt{3x}$
14. $\frac{12\sqrt{11}}{11}$ 15. 10   16. 2   17. $-3 + \sqrt{14}$ 18. $\frac{5 + \sqrt{5}}{4}$ 19. $\frac{7}{2}$ 20. -3 21. 10 cm   22. (a) $10\sqrt{5}$ (b) 20   23. $b = 20\sqrt{3}$; $c = 40$ 24. $4\sqrt{2}$

Chapter 18

1. $-4, -9$ 2. 3, -3 3. 0, 5   4. $2 \pm \sqrt{3}$ 5. $\frac{1}{16}$ 6. 3, $\frac{1}{2}$ 7. 2   8. 10, 12   9. large pipe: 4 hours, small pipe: 12 hours   10. 12 mph   11. $m = \frac{b}{w^2}$ 12. (a) up   (b) (0, 2)   (c)

All exercises are selected from the A *sets of exercises.*

1.3

32. To find the total miles driven during two days, add the miles driven each day.

$$
\begin{array}{r}
135 \\
+467 \\
\hline
602
\end{array}
\quad
\begin{array}{l}
\textbf{Miles driven first day} \\
\textbf{Miles driven second day} \\
\textbf{Total miles driven}
\end{array}
$$

Thus, Todd drove 602 miles on the trip.

1.4

16.
$$
\begin{array}{rcr}
23 & \longrightarrow & 20 \\
16 & \longrightarrow & 20 \\
54 & \longrightarrow & 50 \\
+79 & \longrightarrow & +80 \\
\hline
 & & 170
\end{array}
\quad
\begin{array}{l}
\textbf{Round 23 to 20} \\
\textbf{Round 16 to 20} \\
\textbf{Round 54 to 50} \\
\textbf{Round 79 to 80} \\
\textbf{The estimated sum}
\end{array}
$$

20.
$$
\begin{array}{rcr}
329 & \longrightarrow & 300 \\
556 & \longrightarrow & 600 \\
113 & \longrightarrow & 100 \\
+925 & \longrightarrow & +900 \\
\hline
 & & 1900
\end{array}
\quad
\begin{array}{l}
\textbf{Round 329 to 300} \\
\textbf{Round 556 to 600} \\
\textbf{Round 113 to 100} \\
\textbf{Round 925 to 900} \\
\textbf{The estimated sum}
\end{array}
$$

26.
$$
\begin{array}{rcr}
19501 & \longrightarrow & 20000 \\
19499 & \longrightarrow & 19000 \\
38201 & \longrightarrow & 38000 \\
+21005 & \longrightarrow & +21000 \\
\hline
 & & 98000
\end{array}
\quad
\begin{array}{l}
\textbf{Round 19501 to 20000} \\
\textbf{Round 19499 to 19000} \\
\textbf{Round 38201 to 38000} \\
\textbf{Round 21005 to 21000} \\
\textbf{The estimated sum}
\end{array}
$$

34. Round 7105 to 7100 and round 6898 to 6900. Then add to estimate the total number of books.

$$
\begin{array}{r}
7100 \\
+6900 \\
\hline
14000
\end{array}
$$

There are about 14,000 books on the two floors.

1.6

29. Round 389 to 400 and 605 to 600. Subtract to obtain an estimate for the increase in payment.

$$
\begin{array}{r}
600 \\
-400 \\
\hline
200
\end{array}
\quad \textbf{Estimated increase}
$$

Subtract to find the exact increase.

$$
\begin{array}{r}
605 \\
-389 \\
\hline
216
\end{array}
\quad
\begin{array}{l}
\textbf{New payment} \\
\textbf{Former payment} \\
\textbf{Exact increase in payment}
\end{array}
$$

The estimated increase is $200 and the exact increase is $216.

1.7

13. First add the two deposits to the starting balance.

$$
\begin{array}{r}
625 \\
300 \\
+235 \\
\hline
1160
\end{array}
\quad
\begin{array}{l}
\textbf{Starting balance} \\
\textbf{One deposit} \\
\textbf{Other deposit}
\end{array}
$$

Next add the five amounts of the checks written.

$$
\begin{array}{r}
23 \\
18 \\
113 \\
410 \\
+\ 32 \\
\hline
596
\end{array}
$$

To find the balance at the end of the month, subtract the value of the checks written from the amount in the account.

$$
\begin{array}{r}
1160 \\
-\ 596 \\
\hline
564
\end{array}
$$

Thus, the balance at the end of the month was $564.

15. First find the three distances from Salt Lake City to Seattle.

$$
\begin{array}{r}
455 \\
735 \\
+597 \\
\hline
1787
\end{array}
\quad
\begin{array}{l}
\textbf{Salt Lake City to Cheyenne} \\
\textbf{Cheyenne to Butte} \\
\textbf{Butte to Seattle}
\end{array}
$$

$$
\begin{array}{r}
419 \\
+597 \\
\hline
1016
\end{array}
\quad
\begin{array}{l}
\textbf{Salt Lake City to Butte} \\
\textbf{Butte to Seattle}
\end{array}
$$

$$
\begin{array}{r}
528 \\
564 \\
+175 \\
\hline
1267
\end{array}
\quad
\begin{array}{l}
\textbf{Salt Lake City to Reno} \\
\textbf{Reno to Portland} \\
\textbf{Portland to Seattle}
\end{array}
$$

Thus, the shortest distance is 1016 miles (through Butte) which is 251 miles shorter than the next shortest distance (through Reno and Portland).

18. To find the sales on Thursday, subtract $305 from the sales on Friday.

$$
\begin{array}{r}
1245 \quad \textbf{Friday sales} \\
-\ 305 \\
\hline
940 \quad \textbf{Thursday sales}
\end{array}
$$

To find the sales on Saturday, add $785 to the sales on Friday.

$$
\begin{array}{r}
1245 \quad \textbf{Friday sales} \\
+\ 785 \\
\hline
2030 \quad \textbf{Saturday sales}
\end{array}
$$

22. Since Larry had $12 more than Jim, add $12 to Jim's total to find Larry's total.

$$
\begin{array}{r}
43 \quad \textbf{Jim's total} \\
+12 \\
\hline
55 \quad \textbf{Larry's total}
\end{array}
$$

Mike and Jim together have:

$$
\begin{array}{r}
25 \\
+43 \\
\hline
68
\end{array}
$$

Since Henri has $9 less than this total, subtract.

$$
\begin{array}{r}
68 \\
-\ 9 \\
\hline
59 \quad \textbf{Henri's total}
\end{array}
$$

Add the four amounts to find the total together.

$$
\begin{array}{r}
25 \quad \textbf{Mike's total} \\
43 \quad \textbf{Jim's total} \\
55 \quad \textbf{Larry's total} \\
+59 \quad \textbf{Henri's total} \\
\hline
182
\end{array}
$$

Thus, the four together have $182.

2.2

34. Since 230 calories are burned in 1 hour, in 4 hours there will be 4 times 230 calories.

$$
\begin{array}{r}
230 \\
\times\ \ 4 \\
\hline
920
\end{array}
$$

Thus, 920 calories are burned in 4 hours.

2.3

25. Round 103 to 100 and 289 to 300.

$$
\begin{array}{r}
300 \\
\times\ 100 \\
\hline
30000 \quad \textbf{Estimated total}
\end{array}
$$

Thus, about $30,000 was taken in on Saturday.

2.5

16.

```
          ┌── Don't forget this zero
      3027
   2)6054
      6
      ─
      05
       4
      ──
      14
      14
      ──
       0
```

21.

```
         ┌── 223 ÷ 23
        97 ← 161 ÷ 23
   23)2231
      207
      ───
      161
      161
      ───
        0
```

23.

```
          ┌── 1324 ÷ 473
         28 ← 3784 ÷ 473
   473)13244
       946
       ────
       3784
       3784
       ────
          0
```

29.

```
         ┌── 755 ÷ 83
        91 ← 83 ÷ 83                Check:      83
   83)7553                                    × 91
      747                                     ────
      ───                                       83
       83                                      747
       83                                     ─────
       ──                                     7553
        0
```

Thus, the quotient is 91 and the remainder is 0.

32. To find the approximate number of calories in each bar, divide 4000 by 20, which is 200. Thus, there are about 200 calories in each bar. To find the exact number, divide 4011 by 21.

```
           191
   21)4011
       21
       ───
       191
       189
       ───
        21
        21
        ──
         0
```

Thus, there are exactly 191 calories in each bar.

2.6

4. **(a)** Multiply 35 by the value of each.

$$
\begin{array}{r}
329 \quad \textbf{Price of 1 snow blower} \\
\times 35 \quad \textbf{Number of snow blowers} \\
\hline
1645 \\
987 \\
\hline
11515
\end{array}
$$

Thus, $11,515 was taken in on the sale of snow blowers.

(b) Multiply 40 by the value of each.

$$
\begin{array}{rl}
289 & \textbf{Price of 1 chain saw} \\
\underline{\times 40} & \textbf{Number of chain saws} \\
11560 &
\end{array}
$$

Thus, $11,560 was taken in on the sale of chain saws.

(c) Add to find the total taken in on both items.

$$
\begin{array}{r}
11515 \\
\underline{+11560} \\
23075
\end{array}
$$

Thus, $23,075 was taken in on both.

8. First find the number of square feet of mirror.

$$
\begin{array}{rl}
8 & \textbf{Length of mirror} \\
\underline{\times 5} & \textbf{Width of mirror} \\
40 & \textbf{Area of mirror in square feet}
\end{array}
$$

Since each square foot costs $3, multiply to find the total cost.

$$
\begin{array}{r}
40 \\
\underline{\times 3} \\
120
\end{array}
$$

It will cost $120 to put a mirror on the wall.

13. (a) 5 sections, each seating 400, results in

$$5 \times 400 = 2000 \text{ seats.}$$

8 sections, each seating 1200, results in

$$8 \times 1200 = 9600 \text{ seats.}$$

Since there are 20 sections in total, and 13 have been considered $(5 + 8 = 13)$, there are 7 sections left $(20 - 13 = 7)$, each seating 950.

$$7 \times 950 = 6650$$

Add to find the total number of seats.

$$
\begin{array}{rl}
2000 & \\
9600 & \\
\underline{+6650} & \\
18250 & \textbf{Total seats in the dome}
\end{array}
$$

(b) Multiply 18,250 by the value of each seat, $5, to find the total amount of money taken in.

$$
\begin{array}{r}
18250 \\
\underline{\times 5} \\
91250
\end{array}
$$

Thus, $91,250 was taken in.

17. $\dfrac{372¢}{12} = 31¢$ **Price per pound of the 12-pound bag**

$\dfrac{288¢}{9} = 32¢$ **Price per pound of the 9-pound bag**

Thus, the 12-pound bag is the better buy.

20. To find the student-teacher ratio, divide the number of students (12,100) by the number of faculty members (550).

$$
\begin{array}{r}
22 \\
550\overline{)12100} \\
\underline{1100} \\
1100 \\
\underline{1100} \\
0
\end{array}
$$

Thus, the student-teacher ratio is 22 students per teacher.

2.7

13.

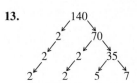

Thus, $140 = 2 \cdot 2 \cdot 5 \cdot 7$.

2.8

12.

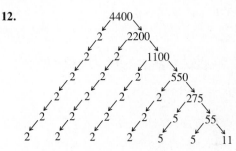

Thus, $4400 = 2 \cdot 2 \cdot 2 \cdot 2 \cdot 5 \cdot 5 \cdot 11 = 2^4 \cdot 5^2 \cdot 11$.

32. $5^2 - \sqrt{49} = 25 - 7$ **Square and take root first**

$\qquad\qquad\quad = 18$ **Then subtract**

34. $\sqrt{25} + 15 - 2^2 \cdot 5$

$\quad = 5 + 15 - 4 \cdot 5$ **Take root and square first**

$\quad = 5 + 15 - 20$ **Multiply next**

$\quad = 20 - 20$ **Add**

$\quad = 0$ **Then subtract**

41. $\sqrt{4}(9 \div 3) + 2^3$

$\quad = \sqrt{4}(3) + 2^3$ **Divide inside parentheses first**

$\quad = 2(3) + 8$ **Take root and cube next**

$\quad = 6 + 8$ **Multiply next**

$\quad = 14$ **Then add**

43. $(3 \cdot 7)^2 = (21)^2$ **Multiply first inside parentheses**

$\qquad\qquad = 441$ **Then square**

44. $3 \cdot 7^2 = 3 \cdot 49$ **Square first**

$\qquad\quad = 147$ **Then multiply**

Note the difference between Exercises 43 and 44.

46. $(3 + 7)^2 = 10^2$ **Add first inside parentheses**

$\qquad\qquad = 100$ **Then square**

47. $3^2 + 7^2 = 9 + 49$ **Square first**

$\qquad\qquad = 58$ **Then add**

Note the difference between Exercises 46 and 47.

3.1

14. Since each shaded part of the figure corresponds to $\frac{1}{4}$ of a unit, there are five $\frac{1}{4}$'s shaded, or $\frac{5}{4}$ of the unit is shaded.

3.2

8. $\dfrac{8}{9}$ and $\dfrac{96}{108}$

$\qquad\qquad 8 \cdot 108 = 864$ and $9 \cdot 96 = 864$

Since the cross products are equal, the fractions are equal.

13. $\dfrac{9}{18} = \dfrac{?}{6}$

To obtain 6 from 18, divide 18 by 3. Thus 9 is divided by 3, giving 3.

$\qquad \dfrac{9}{18} = \dfrac{9 \div 3}{18 \div 3} = \dfrac{3}{6}$ $\longleftarrow$ **The desired number**

23. $\dfrac{36}{14} = \dfrac{18}{?}$

To get 18 from 36, divide by 2.

$\qquad \dfrac{36}{14} = \dfrac{36 \div 2}{14 \div 2} = \dfrac{18}{7}$ $\longleftarrow$ **The desired number**

31. $\dfrac{210}{105} = \dfrac{5 \cdot 42}{5 \cdot 21} = \dfrac{\cancel{5} \cdot 2 \cdot \cancel{21}}{\cancel{5} \cdot \cancel{21}} = \dfrac{2}{1} = 2$

3.3

20. $\dfrac{18}{84} \cdot \dfrac{36}{27} = \dfrac{\overset{2}{\cancel{18}}}{\underset{7}{\cancel{84}}} \cdot \dfrac{\overset{3}{\cancel{36}}}{\underset{3}{\cancel{27}}} = \dfrac{2}{7} \cdot \dfrac{3}{3}$

$\qquad\qquad = \dfrac{2}{7} \cdot 1 = \dfrac{2}{7}$

31. To find the area multiply the length, $\frac{7}{8}$, by the width, $\frac{4}{7}$.

$\qquad \dfrac{7}{8} \cdot \dfrac{4}{7} = \dfrac{\overset{1}{\cancel{7}}}{\underset{2}{\cancel{8}}} \cdot \dfrac{\overset{1}{\cancel{4}}}{\underset{1}{\cancel{7}}} = \dfrac{1}{2}$

The area is $\frac{1}{2}$ sq km.

3.4

14. $\dfrac{13}{15} \div \dfrac{39}{5} = \dfrac{13}{15} \cdot \dfrac{5}{39}$ **Multiply by reciprocal**

$\qquad = \dfrac{\overset{1}{\cancel{13}}}{\underset{3}{\cancel{15}}} \cdot \dfrac{\overset{1}{\cancel{5}}}{\underset{3}{\cancel{39}}} = \dfrac{1}{9}$

22. $\dfrac{10}{9}$ of $\square$ is 25

$\qquad \downarrow \ \downarrow \ \downarrow \ \ \downarrow \ \downarrow$

$\qquad \dfrac{10}{9} \cdot \square = 25$

$\qquad \square = 25 \div \dfrac{10}{9}$ **Related division sentence**

$\qquad = \dfrac{25}{1} \cdot \dfrac{9}{10} = \dfrac{5 \cdot \cancel{5} \cdot 9}{2 \cdot \cancel{5}} = \dfrac{45}{2}$

28. The problem translates to:

$\qquad \dfrac{5}{6}$ of $\boxed{\text{trip}}$ is 365

$\qquad \downarrow \downarrow \ \ \downarrow \ \ \downarrow \ \downarrow$

$\qquad \dfrac{5}{6} \cdot \ \square = 365$

$\qquad\qquad \square = 365 \div \dfrac{5}{6}$ **Related division sentence**

$\qquad\qquad = \dfrac{365}{1} \cdot \dfrac{6}{5}$

$\qquad\qquad = \dfrac{\cancel{5} \cdot 73 \cdot 6}{1 \cdot \cancel{5}} = 438$

The total length of the trip was 438 miles.

3.5

8. $\dfrac{257}{9}$

$\qquad \begin{array}{r} 28 \\ 9\overline{)257} \\ \underline{18} \\ 77 \\ \underline{72} \\ 5 \end{array}$

$\qquad$ Thus, $\dfrac{257}{9} = 28\dfrac{5}{9}$.

17. $3\overset{\curvearrowright}{2}\overset{7}{\underset{\curvearrowleft}{10}}$ $\qquad \dfrac{10 \cdot 32 + 7}{10} = \dfrac{320 + 7}{10}$

$\qquad\qquad\qquad = \dfrac{327}{10}$

3.6

8. $2\dfrac{3}{5} \cdot 6\dfrac{1}{4} \cdot 12$

$\qquad = \dfrac{13}{5} \cdot \dfrac{25}{4} \cdot \dfrac{12}{1}$ **Change to improper fractions**

$\qquad = \dfrac{13 \cdot \cancel{5} \cdot 5 \cdot \cancel{4} \cdot 3}{\cancel{5} \cdot \cancel{4}}$

$\qquad = \dfrac{13 \cdot 5 \cdot 3}{1} = 195$

22. If we knew how many uniforms could be made, $\square$, we would multiply that number by $3\frac{1}{8}$ to obtain 25. That is,

$\qquad 3\dfrac{1}{8} \cdot \square = 25.$

$$\square = 25 \div 3\tfrac{1}{8} \qquad \textbf{Related division sentence}$$

$$= \frac{25}{1} \div \frac{25}{8}$$

$$= \frac{25}{1} \cdot \frac{8}{25}$$

$$= \frac{25 \cdot 8}{1 \cdot 25} = \frac{8}{1} = 8$$

Thus, 8 uniforms could be made.

3.7

10. Amount spent for food $= \dfrac{1}{5}$ of 18000

$$= \frac{1}{5} \cdot 18000$$

$$= \frac{18000}{5} = \$3600$$

Amount spent for housing $= \dfrac{1}{4}$ of 18000

$$= \frac{1}{4} \cdot 18000$$

$$= \frac{18000}{4} = \$4500$$

Amount spent for clothing $= \dfrac{1}{10}$ of 18000

$$= \frac{1}{10} \cdot 18000$$

$$= \frac{18000}{10} = \$1800$$

Amount spent for taxes $= \dfrac{3}{8}$ of 18000

$$= \frac{3}{8} \cdot 18000$$

$$= \frac{3 \cdot 18000}{8} = \$6750$$

13. If we knew the number of minutes, $\square$, we would multiply by 45 to obtain $321\tfrac{1}{2}$.

$$45 \cdot \square = 321\tfrac{1}{2}$$

$$\square = 321\tfrac{1}{2} \div 45 \qquad \textbf{Related division sentence}$$

$$= \frac{643}{2} \cdot \frac{1}{45}$$

$$= \frac{643}{90} = 7\frac{13}{90}$$

It will take $7\frac{13}{90}$ minutes to revolve $321\tfrac{1}{2}$ times.

4.1

8. $\dfrac{2}{35} + \dfrac{18}{35} + \dfrac{25}{35} = \dfrac{2 + 18 + 25}{35}$ **Add numerators**

$$= \frac{45}{35}$$

$$= \frac{5 \cdot 9}{5 \cdot 7} = \frac{9}{7} \qquad \textbf{Reduce to lowest terms}$$

14. $\dfrac{28}{35} - \dfrac{18}{35} = \dfrac{28 - 18}{35}$ **Subtract numerators**

$$= \frac{10}{35}$$

$$= \frac{5 \cdot 2}{5 \cdot 7} = \frac{2}{7} \qquad \textbf{Reduce to lowest terms}$$

20. $\dfrac{7}{10} - \dfrac{1}{10} + \dfrac{3}{10} = \dfrac{7 - 1 + 3}{10}$

$$= \frac{6 + 3}{10}$$

$$= \frac{9}{10}$$

23. First add the amounts done the first two days.

$$\frac{5}{12} + \frac{3}{12} = \frac{8}{12} = \frac{4 \cdot 2}{4 \cdot 3} = \frac{2}{3}$$

Subtract this total from 1 to find the amount to be done the third day.

$$1 - \frac{2}{3} = \frac{3}{3} - \frac{2}{3} = \frac{3 - 2}{3} = \frac{1}{3}$$

Thus, $\tfrac{1}{3}$ of the job must be done the third day.

4.2

14. Factor each number into a product of primes

$$18 = 2 \cdot 3 \cdot 3$$
$$24 = 2 \cdot 2 \cdot 2 \cdot 3$$
$$30 = 2 \cdot 3 \cdot 5$$

The LCM must consist of three 2's, two 3's, and one 5. Thus, the LCM $= 2 \cdot 2 \cdot 2 \cdot 3 \cdot 3 \cdot 5 = 360$.

29. The shortest length of wall will be the LCM of 8, 9, and 14. Factor each number into a product of primes.

$$8 = 2 \cdot 2 \cdot 2$$
$$9 = 3 \cdot 3$$
$$14 = 2 \cdot 7$$

The LCM must consist of three 2's, two 3's, and one 7. The LCM $= 2 \cdot 2 \cdot 2 \cdot 3 \cdot 3 \cdot 7 = 504$. Thus, the shortest length is 504 inches.

4.3

8. $4 + \dfrac{4}{5} = \dfrac{4 \cdot 5}{5} + \dfrac{4}{5}$ **The LCD is 5**

$\quad = \dfrac{20}{5} + \dfrac{4}{5}$

$\quad = \dfrac{20 + 4}{5} = \dfrac{24}{5}$

11. $\dfrac{1}{6} = \dfrac{5}{5 \cdot 6} = \dfrac{5}{30}$ **The LCD is 30**

$+\dfrac{3}{5} = +\dfrac{3 \cdot 6}{5 \cdot 6} = +\dfrac{18}{30}$

$\qquad\qquad\qquad \dfrac{23}{30}$

17. $4 - \dfrac{4}{5} = \dfrac{4 \cdot 5}{5} - \dfrac{4}{5}$ **The LCD is 5**

$\quad = \dfrac{20}{5} - \dfrac{4}{5}$

$\quad = \dfrac{20 - 4}{5} = \dfrac{16}{5}$

23. $\dfrac{8}{15} = \dfrac{8}{3 \cdot 5} = \dfrac{8 \cdot 4}{3 \cdot 4 \cdot 5} = \dfrac{32}{60}$

$-\dfrac{3}{20} = -\dfrac{3}{4 \cdot 5} = -\dfrac{3 \cdot 3}{3 \cdot 4 \cdot 5} = -\dfrac{9}{60}$

$\qquad\qquad\qquad\qquad\qquad\qquad \dfrac{23}{60}$

29. $\dfrac{8}{15} + \dfrac{1}{12} - \dfrac{5}{20} = \dfrac{8}{3 \cdot 5} + \dfrac{1}{2 \cdot 2 \cdot 3} - \dfrac{5}{2 \cdot 2 \cdot 5}$

$\quad = \dfrac{8 \cdot 2 \cdot 2}{2 \cdot 2 \cdot 3 \cdot 5} + \dfrac{1 \cdot 5}{2 \cdot 2 \cdot 3 \cdot 5} - \dfrac{5 \cdot 3}{2 \cdot 2 \cdot 3 \cdot 5}$

$\quad = \dfrac{32}{60} + \dfrac{5}{60} - \dfrac{15}{60}$

$\quad = \dfrac{32 + 5 - 15}{60}$

$\quad = \dfrac{22}{60} = \dfrac{2 \cdot 11}{2 \cdot 30} = \dfrac{11}{30}$

4.4

8. $\quad 15\dfrac{7}{8}$

$\quad +\ 4$

$\quad \overline{19\dfrac{7}{8}}$ **Add the whole numbers**

10. $215\dfrac{7}{8} = 215\dfrac{7}{8}$ **The LCD is 8**

$+147\dfrac{1}{2} = +147\dfrac{4}{8}$

$\quad\quad = 362\dfrac{11}{8} = 362 + \dfrac{8}{8} + \dfrac{3}{8}$

$\quad\quad = 362 + 1 + \dfrac{3}{8}$

$\quad\quad = 363\dfrac{3}{8}$

20. $\quad 17\dfrac{6}{11}$

$\quad -\ 8$

$\quad \overline{9\dfrac{6}{11}}$ **Subtract the whole numbers**

22. $485\dfrac{9}{10} = 485\dfrac{9}{10}$

$-316\dfrac{2}{5} = -316\dfrac{4}{10}$ **The LCD is 10**

$\quad\quad\overline{169\dfrac{5}{10}} = 169\dfrac{1}{2}$

26. $7\dfrac{1}{5} = 7\dfrac{1}{5} = 7\dfrac{3}{15}$

$3\dfrac{2}{3} = 3\dfrac{2}{3} = 3\dfrac{10}{15}$

$+1\dfrac{1}{15} = 1\dfrac{1}{3 \cdot 5} = 1\dfrac{1}{15}$

$\quad\quad = 11\dfrac{14}{15}$

$\quad$ **The LCD $= 3 \cdot 5 = 15$**

30. Subtract $1\dfrac{2}{3}$ from $7\dfrac{1}{8}$.

$7\dfrac{1}{8} = 7\dfrac{3}{24} = 6\dfrac{27}{24}$

$-1\dfrac{2}{3} = 1\dfrac{16}{24} = 1\dfrac{16}{24}$

$\quad\quad = 5\dfrac{11}{24}$

The LCD is 24 and we had to borrow $\frac{24}{24}$ from 7 to be able to subtract $\frac{16}{24}$.

4.5

8. $\left(\dfrac{3}{5}\right)^2 - \dfrac{1}{5} + \dfrac{6}{25}$

$= \dfrac{9}{25} - \dfrac{1}{5} + \dfrac{6}{25}$ **Square first**

$= \dfrac{9}{25} - \dfrac{5}{25} + \dfrac{6}{25}$ **LCD $= 25$**

$= \dfrac{9 - 5 + 6}{25} = \dfrac{10}{25} = \dfrac{2}{5}$

10. $\sqrt{\dfrac{1}{9}} + \dfrac{6}{7} \div \dfrac{3}{14} - 3 \cdot \dfrac{1}{9}$

$= \dfrac{1}{3} + \dfrac{6}{7} \div \dfrac{3}{14} - 3 \cdot \dfrac{1}{9}$ **Take root first**

$$= \frac{1}{3} + \frac{6}{7} \cdot \frac{14}{3} - 3 \cdot \frac{1}{9}$$ **Divide by multiplying by the reciprocal**

$$= \frac{1}{3} + 4 - 3 \cdot \frac{1}{9} \qquad \frac{6}{7} \cdot \frac{14}{3} = \frac{2 \cdot \cancel{3} \cdot 2 \cdot \cancel{7}}{\cancel{7} \cdot \cancel{3}} = 4$$

$$= \frac{1}{3} + 4 - \frac{1}{3} \qquad 3 \cdot \frac{1}{9} = \frac{3 \cdot 1}{9} = \frac{1}{3}$$

$$= 4 \qquad \frac{1}{3} - \frac{1}{3} = 0$$

17. $\dfrac{17}{30} = \dfrac{17 \cdot 4}{30 \cdot 4} = \dfrac{68}{120}$

$\dfrac{21}{40} = \dfrac{21 \cdot 3}{40 \cdot 3} = \dfrac{63}{120}$

Since $68 > 63$, $\dfrac{17}{30} > \dfrac{21}{40}$.

26. $\dfrac{1}{4} + \dfrac{7}{40}$ is the amount done Monday.

$\dfrac{1}{4} + \dfrac{1}{5}$ is the amount done Tuesday.

Add these pairs.

$$\frac{1}{4} + \frac{7}{40} = \frac{1 \cdot 10}{4 \cdot 10} + \frac{7}{4 \cdot 10} = \frac{10 + 7}{40} = \frac{17}{40}$$

$$\frac{1}{4} + \frac{1}{5} = \frac{1 \cdot 5}{4 \cdot 5} + \frac{1 \cdot 4}{4 \cdot 5} = \frac{5 + 4}{20} = \frac{9}{20}$$

To compare $\frac{17}{40}$ and $\frac{9}{20}$, write each with the LCD of 40. Since $\frac{9}{20} = \frac{9 \cdot 2}{20 \cdot 2} = \frac{18}{40}$ and $\frac{18}{40} > \frac{17}{40}$, Carl did more on Tuesday. The amount is more by $\frac{18}{40} - \frac{17}{40} = \frac{1}{40}$.

4.6

9. Add the number of hours worked each day.

$$7\frac{3}{8} + 6\frac{2}{5} + 5\frac{4}{5} + 8\frac{1}{4} + 7\frac{1}{2}$$

$$(7 + 6 + 5 + 8 + 7) + \left(\frac{3}{8} + \frac{2}{5} + \frac{4}{5} + \frac{1}{4} + \frac{1}{2}\right)$$

$$= 33 + \left(\frac{15}{40} + \frac{16}{40} + \frac{32}{40} + \frac{10}{40} + \frac{20}{40}\right) \quad \textbf{LCD = 40}$$

$$= 33 + \frac{15 + 16 + 32 + 10 + 20}{40}$$

$$= 33 + \frac{93}{40}$$

$$= 33 + 2\frac{13}{40} = 35\frac{13}{40}$$

Thus, Lon worked $35\frac{13}{40}$ hr.

10. To find the amount Lon was paid, multiply the number of hours worked, $35\frac{13}{40}$, by the pay per hour, $\$6\frac{1}{4}$.

$$35\frac{13}{40} \cdot 6\frac{1}{4} = \frac{1413}{40} \cdot \frac{25}{4}$$

$$= \frac{1413 \cdot \cancel{5} \cdot 5}{\cancel{5} \cdot 8 \cdot 4}$$

$$= \frac{7065}{32} = 220\frac{25}{32}$$

Thus, Lon was paid $\$220\frac{25}{32}$.

12. The closing value was

$$121\frac{3}{8} + 1\frac{3}{4} - 3\frac{1}{8} + 2\frac{7}{8}.$$

First add $121\frac{3}{8} + 1\frac{3}{4} + 2\frac{7}{8}$.

$$121\frac{3}{8} + 1\frac{3}{4} + 2\frac{7}{8}$$

$$= (121 + 1 + 2) + \left(\frac{3}{8} + \frac{3}{4} + \frac{7}{8}\right)$$

$$= 124 + \left(\frac{3}{8} + \frac{6}{8} + \frac{7}{8}\right) \quad \textbf{LCD = 8}$$

$$= 124 + \frac{16}{8} = 124 + 2 = 126.$$

Now subtract $3\frac{1}{8}$ from the sum.

$$\begin{array}{rcl} 126 & = & 125\frac{8}{8} \\[4pt] -\ 3\frac{1}{8} & = & -\ 3\frac{1}{8} \\ \hline & & 122\frac{7}{8} \end{array}$$

Thus, the closing value was $\$122\frac{7}{8}$.

5.2

7. To change 0.1302 to a fraction, note that 0.1302 is "one thousand three hundred two ten thousandths." Thus,

$$0.1302 = \frac{1302}{10,000} = \frac{\cancel{2} \cdot 651}{\cancel{2} \cdot 5000} = \frac{651}{5000}$$

22. To change $\frac{19}{16}$ to a decimal, divide 19 by 16.

$$\begin{array}{r} 1.1875 \\ 16\overline{)19.0000} \\ \underline{16} \\ 3\ 0 \\ \underline{1\ 6} \\ 1\ 40 \\ \underline{1\ 28} \\ 120 \\ \underline{112} \\ 80 \\ \underline{80} \\ 0 \end{array}$$

Thus, $\frac{19}{16} = 1.1875$.

5.3

34. (a) Round $0.\overline{35}$ to the nearest tenth.

Since this digit is 5,
change 3 to 4 and
$0.\overline{35} = 0.3535\ldots$ drop remaining digits

↑
Round to here

Thus, $0.\overline{35} \approx 0.4$.   Rounded to nearest tenth

(b) Round $0.\overline{35}$ to the nearest hundredth.

Since this digit is 3,
$0.\overline{35} = 0.3535\ldots$ drop remaining digits

↑
Round to here

Thus, $0.\overline{35} \approx 0.35$.   Rounded to nearest hundredth

(c) Round $0.\overline{35}$ to the nearest thousandth.

Since this digit is 5, change
3 to 4 and drop remaining
$0.\overline{35} = 0.3535\ldots$ digits

↑
Round to here

Thus, $0.\overline{35} \approx 0.354$.   Rounded to nearest thousandth

47. To approximate the total cost, use 10¢ for the price of a brick and 300 for the number of bricks. Then the approximate cost is

$$300 \cdot 10¢ = 3000¢ = \$30.00.$$

5.4

29. First find the total of all deductions.

$1237.40
725.36
2347.01
+ 444.83
$4754.60   Total deductions

Now subtract the total deductions from the total income.

$31255.75
− 4754.60
$26501.15

The taxable income was $26,501.15.

36. (a) To find the total of the four injections add

$$2.65 + 2.75 + 3.5 + 4.0 = 12.9 \text{ milligrams.}$$

(b) The amount left in the bottle was

$$30.5 - 12.9 = 17.6 \text{ milligrams.}$$

37. From the figure,

$$x = 2.3 + 4.75 + 2.3 = 9.35 \text{ ft.}$$

Also,

$$y = 11.15 - 3.25 = 7.9 \text{ ft.}$$

5.5

31. First find the area of the wall.

$$\text{Area} = \text{length} \cdot \text{width}$$
$$= (13.5) \cdot (7.6) = 102.6 \text{ square feet}$$

Then multiply the area by the cost per square foot.

$$(102.6) \cdot (0.67) = 68.742$$

To the nearest cent, the cost of the project is $68.74.

35. Since each tire costs $38.75, the approximate cost of a tire is $40.00. Thus, the approximate cost of 4 tires is

$$(4)(\$40.00) = \$160.00.$$

To find the exact cost, multiply $38.75 by 4.

$$(4)(\$38.75) = \$155.00$$

Four tires will cost $155.00.

5.6

8.

```
          7.2
0.31.)2.23.2
      2 17
        6 2
        6 2
          0
```

17.

```
         6.562  ⟵  2 < 5 so drop remaining digits
48)315.000
   288
    27 0
    24 0
     3 00
     2 88
       120
        96
        24
```

The quotient, to the nearest hundredth, is 6.56.

32.

```
           11.77
1.8.)21.2.00
     18
      3 2
      1 8
      1 4 0
      1 2 6
        1 40
        1 26
          14
```

Since it is clear that 7 will continue to repeat, the quotient is $11.\overline{7}$.

37. Divide $382.40 by 0.147.

$$
\begin{array}{r}
2\ 601.360 \\
0.147.\overline{)382.400.000} \\
294 \\
\hline
88\ 4 \\
88\ 2 \\
\hline
200 \\
147 \\
\hline
53\ 0 \\
44\ 1 \\
\hline
8\ 90 \\
8\ 82 \\
\hline
80
\end{array}
$$

To the nearest cent, the assessed value is $2601.36.

5.7

2. $3.25 - \dfrac{3}{8} = 3\dfrac{1}{4} - \dfrac{3}{8}$

$\qquad = \dfrac{13}{4} - \dfrac{3}{8}$

$\qquad = \dfrac{26}{8} - \dfrac{3}{8}$ **LCD = 8**

$\qquad = \dfrac{23}{8} = 2\dfrac{7}{8}$

8. $3.25 - \dfrac{3}{8} = 3.25 - 0.375$ **Change $\frac{3}{8}$ to a decimal**

$$
\begin{array}{r}
3.250 \\
-0.375 \\
\hline
2.875
\end{array}
$$

Thus, $3.25 - \frac{3}{8} \approx 2.88$ to the nearest hundredth.

20. $(6.2)^2 - 6.6 \div 1.1$

$\qquad = 38.44 - 6.6 \div 1.1$ **Square first**

$\qquad = 38.44 - 6$ **Divide next**

$\qquad = 32.44$ **Then subtract**

28. Note that $\frac{2}{3} = 0.666 \ldots = 0.\overline{6}$ and $\frac{7}{10} = 0.7$. Thus, we are arranging

$$0.\overline{6},\ 0.65,\ 6.05,\ 0.7,\ 6.5.$$

Write each number correct to two decimal places; round $0.\overline{6}$ to 0.67.

$$0.67,\ 0.65,\ 6.05,\ 0.70,\ 6.50$$

Ignoring decimal points, the order is

$$65,\ 67,\ 70,\ 605,\ 650,$$

or

$$0.65,\ 0.67,\ 0.70,\ 6.05,\ 6.50.$$

Thus, the order is $0.65,\ \frac{2}{3},\ \frac{7}{10},\ 6.05,\ 6.5.$

5.8

34. $5 \div 1.2 + 6.5$

$\qquad = 4.1\overline{6} + 6.5$ **Divide first**

$\qquad = 10.\overline{6}$ **Then add**

Don't round after dividing 5 by 1.2; let your calculator hold the quotient. Then add to that value 6.5. The result, $10.\overline{6}$, will be displayed.

6.1

33. Let a be the weight of 110 ft of the wire. The proportion described by the situation is

$$\frac{70}{84} = \frac{110}{a}.$$

$70a = 84 \cdot 110$ **Cross-product equation**

$\qquad a = \dfrac{84 \cdot 110}{70}$

$\qquad = 132$

Thus, 110 ft of wire weighs 132 pounds.

36. Let a be the number of antelope in the preserve. The proportion described by the situation is

$$\frac{58}{a} = \frac{7}{29}.$$

$58 \cdot 29 = 7a$ **Cross-product equation**

$\dfrac{58 \cdot 29}{7} = a$

$\qquad 240 \approx a$

There are about 240 antelope in the preserve.

38. Let a be the number of pounds of garbage produced by 7 families of four. The proportion is

$$\frac{1}{115} = \frac{7}{a}.$$

$1 \cdot a = 7 \cdot 115$

$\qquad a = 805$

Thus, 7 families will produce 805 pounds of garbage. Notice that the number "four" in "family of four" is not used in the calculation.

6.2

46. $\dfrac{100}{3} = 33.\overline{3}$ **Divide**

$\qquad = 33.33\overline{3}$

$\qquad = 3333.3\%$ **Move decimal point right two places and attach %**

$\qquad = 3333\dfrac{1}{3}\%$

49. $\frac{4}{7} = 0.\overline{571428} \approx 0.571$ **To nearest thousandth**
Move decimal point two places to the right and attach %.

$$\frac{4}{7} \approx 57.1\%$$

6.3

15. The question is: What is 62% of 50?

A is the unknown, $P = 62$, and $B = 50$.

$$A = P\% \cdot B$$
$$A = (62)(0.01)(50)$$
$$= 31$$

Alternatively, we could use the percent proportion.

$$\frac{A}{50} = \frac{62}{100}$$
$$100(A) = (50)(62)$$
$$A = \frac{(50)(62)}{100} = 31$$

The team won 31 games.

18. The question is: 22 is what percent of 30?

P is the unknown, $A = 22$, and $B = 30$.

$$A = P\% \cdot B$$
$$22 = P(0.01)(30)$$
$$22 = P(0.3)$$
$$\frac{22}{0.3} = P$$
$$73.\overline{3} = P$$

Alternatively, using the percent proportion we have:

$$\frac{22}{30} = \frac{P}{100}.$$
$$(22)(100) = 30(P)$$
$$\frac{(22)(100)}{30} = P$$
$$73.\overline{3} = P$$

Thus, to the nearest percent, 73% of her answers were correct.

22. The actual decrease in weight was $220 - 187 = 33$ pounds. The question is:

33 is what percent of 220?

P is the unknown. $A = 33$, and $B = 220$.

$$A = P\% \cdot B$$
$$33 = P(0.01)(220)$$
$$33 = P(2.2)$$

$$\frac{33}{2.2} = P$$
$$15 = P$$

Alternatively, using the percent proportion we have:

$$\frac{33}{220} = \frac{P}{100}.$$
$$(33)(100) = 220(P)$$
$$\frac{(33)(100)}{220} = P$$
$$15 = P$$

The percent decrease was 15%.

6.4

5. The question is: $3.05 is 2.5% of what number?

The unknown is B, $A = 3.05$, and $P = 2.5$.

$$A = P\% \cdot B$$
$$3.05 = (2.5)(0.01)B$$
$$3.05 = (0.025)B$$
$$\frac{3.05}{0.025} = B$$
$$122 = B$$

Alternatively, using the percent proportion we have:

$$\frac{3.05}{B} = \frac{2.5}{100}.$$
$$(3.05)(100) = 2.5(B)$$
$$\frac{(3.05)(100)}{2.5} = B$$
$$122 = B$$

The price before tax is $122.

10. The Social Security tax rate in 1989 was 7.51%. Since the maximum amount on which the tax must be paid was $48,000, and since Toni made more than this, she paid 7.51% of $48,000.

$$(0.0751)(48,000) = 3604.80$$

Thus, Toni paid $3604.80 in Social Security tax.

6.5

4. The question is: $639.12 is 12% of what number?

The unknown is B, $A = 639.12$, and $P = 12$.

$$639.12 = (12)(0.01)B$$
$$639.12 = (0.12)B$$
$$\frac{639.12}{0.12} = B$$
$$5326 = B$$

Using the percent proportion we would have:

$$\frac{639.12}{B} = \frac{12}{100}.$$

$$(639.12)(100) = 12(B)$$

$$\frac{(639.12)(100)}{12} = B$$

$$5326 = B$$

Thus, Angela's total sales were $5326.

10. The discount is $40 − $34 = $6. Thus, the question is: $6 is what percent of $40? The unknown is P, $A = 6$, and $B = 40$.

$$A = P\% \cdot B$$

$$6 = P(0.01)(40)$$

$$6 = P(0.4)$$

$$\frac{6}{0.4} = P$$

$$15 = P$$

Alternatively, using the percent proportion we have:

$$\frac{6}{40} = \frac{P}{100}.$$

$$(6)(100) = 40(P)$$

$$\frac{(6)(100)}{40} = P$$

$$15 = P$$

Thus, the discount rate is 15%.

14. First take 25% of $142.80.

$$(0.25)(142.80) = 35.70$$

The first discount was $35.70 and the first sale price was $142.80 − $35.70 = $107.10.

Next take 20% of $107.10.

$$(0.20)(107.10) = 21.42$$

The second discount was $21.42 making the second sale price $107.10 − $21.42 = $85.68.

6.6

4. Since 9 months is $\frac{9}{12}$ or $\frac{3}{4}$ of a year, we have $P = 800$, $R = 12\% = 0.12$, and $T = \frac{3}{4}$.

$$I = P \cdot R \cdot T$$

$$= (800)(0.12)\left(\frac{3}{4}\right)$$

$$= 72$$

To pay off the loan, Lisa must pay back the principal plus the interest which is

$$A = P + I = \$800 + \$72 = \$872.$$

8. Since 18 months is $\frac{18}{12}$ or $\frac{3}{2}$ of a year, we have $P = 1213.40$, $R = 16\% = 0.16$, and $T = \frac{3}{2}$.

$$I = P \cdot R \cdot T$$

$$= (1213.40)(0.16)\left(\frac{3}{2}\right)$$

$$= 291.22 \quad \textbf{Rounded to nearest cent}$$

To pay off the debt, Robert must pay $1213.40 + $291.22 = $1504.62. To find the amount paid each month, divide by 18.

$$\frac{\$1504.62}{18} = \$83.59$$

Thus, Robert pays $83.59 each month.

11. Since the interest is 12% compounded semiannually, the rate is 6% (one half of 12%). Since the time is 18 months, the number of periods is 3.

$$(\$625.50)(0.06) = \$37.53$$

The interest the first period is $37.53 and the principal for the second period is $625.50 + $37.53 = $663.03.

$$(\$663.03)(0.06) = \$39.78 \quad \textbf{Nearest cent}$$

The interest the second period is $39.78 and the principal for the third period is $663.03 + $39.78 = $702.81.

$$(\$702.81)(0.06) = \$42.17 \quad \textbf{Nearest cent}$$

The amount to be paid back is $702.81 + $42.17 = $744.98.

6.7

5. The actual increase in population was $415 − 329 = 86$. The question is:

86 is what percent of 329?

The unknown is P, $A = 86$, and $B = 329$.

$$A = P\% \cdot B$$

$$86 = P(0.01)(329)$$

$$86 = P(3.29)$$

$$\frac{86}{3.29} = P$$

$$26 \approx P \quad \textbf{To the nearest unit}$$

Alternatively, using the percent proportion we have:

$$\frac{86}{329} = \frac{P}{100}$$

$$(100)(86) = 329(P)$$

$$\frac{(100)(86)}{329} = P$$

$$26 \approx P \quad \textbf{To the nearest unit}$$

Thus, to the nearest percent, the percent increase was 26%.

8. The question is: $1824 is 15% of what number?

B is the unknown, $A = 1824$, and $P = 15$.

$$A = P\% \cdot B$$
$$1824 = 15(0.01)B$$
$$1824 = (0.15)B$$
$$\frac{1824}{0.15} = B$$
$$12{,}160 = B$$

Alternatively, using the percent proportion we have:

$$\frac{1824}{B} = \frac{15}{100}.$$
$$(1824)(100) = 15(B)$$
$$12{,}160 = B$$

Thus, Forrest's original salary was $12,160.

16. The total percent increase is $13\% + 9\% = 22\%$. The question is: 22% of 30 is what number? The unknown is A, $P = 22$, and $B = 30$.

$$A = P\% \cdot B$$
$$A = (22)(0.01)(30) = 6.6$$

Alternatively,

$$\frac{A}{30} = \frac{22}{100}.$$
$$(100)A = (30)(22)$$
$$A = \frac{(30)(22)}{100} = 6.6$$

Thus, the new mileage per gallon would be $30 + 6.6 = 36.6$ mpg.

6.8

8. The actual markdown is $125 - $95 = $30. The question is: $30 is what percent of $125? The unknown is P, $A = 30$, and $B = 125$.

$$A = P\% \cdot B$$
$$30 = P(0.01)(125)$$
$$30 = P(1.25)$$
$$\frac{30}{1.25} = P$$
$$24 = P$$

Alternatively,

$$\frac{30}{125} = \frac{P}{100}.$$
$$(30)(100) = 125(P)$$
$$\frac{(30)(100)}{125} = P$$
$$24 = P$$

The markdown rate is 24%.

10. The markup was

$$20\% \text{ of } \$120$$
$$\downarrow \quad \downarrow \quad \downarrow$$
$$0.20 \cdot \$120 = \$24.$$

The first selling price of the tools was $120 + $24 = $144. The markdown was

$$15\% \text{ of } \$144$$
$$\downarrow \quad \downarrow \quad \downarrow$$
$$0.15 \cdot \$144 = \$21.60.$$

The second selling price was $144 - $21.60 = $122.40.

11. Since the tools were bought for $120.00 and sold for $122.40, the profit was

$$\$122.40 - \$120.00 = \$2.40.$$

To find the percent profit, ask:

$$\$2.40 \text{ is what percent of } \$120?$$

The unknown is P, $A = 2.40$, and $B = 120$.

$$A = P\% \cdot B$$
$$2.40 = P(0.01)(120)$$
$$2.40 = P(1.2)$$
$$\frac{2.40}{1.2} = P$$
$$2 = P$$

Alternatively,

$$\frac{2.40}{120} = \frac{P}{100}.$$
$$(2.40)(100) = 120(P)$$
$$\frac{(2.40)(100)}{120} = P$$
$$2 = P$$

Thus, the percent profit was 2%.

7.1

15. With 54% of the students in K–8 and 22% of the students in high school, 76% of the students are in K–8 and high school combined. Find

$$76\% \text{ of } 42{,}500.$$
$$\downarrow \quad \downarrow \quad \downarrow$$
$$0.76 \cdot 42{,}500 = 32{,}300$$

There are 32,300 students in these two groups.

22. The ratio of math majors to the total number of students is

$$\frac{300}{1300} = \frac{3}{13}.$$

25. The ratio of science majors to the total number of students is

$$\frac{400}{1300} = \frac{4}{13}.$$

Change $\frac{4}{13}$ to a percent to obtain 30.8%.

7.2

22. The number of employees who earn $9 or less an hour is the total of the numbers who earn $3 to $5, $5 to $7, and $7 to $9. This total is

$$5 + 10 + 18 = 33.$$

25. The number of employees who earn more than $9 an hour is 18. Thus, the ratio of these employees to the total number of employees is $\frac{18}{51} = \frac{6}{17}$. Changing $\frac{6}{17}$ to a percent gives 35.3%.

7.3

5. To find the mean, add the seven numbers and divide the result by 7.

$$\frac{7.2 + 6.1 + 8.6 + 9.2 + 5.5 + 7.2 + 8.8}{7} = \frac{52.6}{7} \approx 7.5$$

11. To find the median, first arrange the seven numbers by size.

$$5.5, \ 6.1, \ 7.2, \ 7.2, \ 8.6, \ 8.8, \ 9.2$$

The median is the middle number, 7.2, which is the fourth number from the left.

23. The right column in the table is:

$$
\begin{aligned}
(16,200)(8) &= 129,600 \\
(20,800)(10) &= 208,000 \\
(28,600)(7) &= 200,200 \\
(32,000)(2) &= 64,000 \\
(46,000)(1) &= \underline{46,000} \\
& 647,800
\end{aligned}
$$

The total is 647,800. To find the average salary divide 647,800 by $8 + 10 + 7 + 2 + 1 = 28$.

$$\frac{647,800}{28} = 23,135.714$$

Thus, to the nearest cent, the average salary is $23,135.71.

Since there are 28 salaries, when ranked by size, the median is the average of the middle two, both of which are $20,800. Thus, the median salary is $20,800.

The mode of the salaries is the one that occurs most frequently. Since $20,800 occurs 10 times, more frequently than any other salary, the mode is $20,800.

8.1

14. Use two unit fractions, $\dfrac{1 \text{ min}}{60 \text{ sec}}$ and $\dfrac{1 \text{ hr}}{60 \text{ min}}$.

$$5400 \text{ sec} = \frac{5400 \text{ sec}}{1} \times \frac{1 \text{ min}}{60 \text{ sec}} \times \frac{1 \text{ hr}}{60 \text{ min}}$$

$$= 5400 \times \frac{1}{60} \times \frac{1}{60} \text{ hr} = 1.5 \text{ hr}$$

23. $88 \dfrac{\text{ft}}{\text{sec}} = \dfrac{88 \text{ ft}}{1 \text{ sec}} \times \dfrac{60 \text{ sec}}{1 \text{ min}} \times \dfrac{60 \text{ min}}{1 \text{ hr}} \times \dfrac{1 \text{ mi}}{5280 \text{ ft}}$

$$= \frac{88 \times 60 \times 60}{1 \times 5280} \frac{\text{mi}}{\text{hr}} = 60 \frac{\text{mi}}{\text{hr}}$$

31. Since the price is given in cents per foot, change 3 mi to feet.

$$3 \text{ mi} = \frac{3 \text{ mi}}{1} \times \frac{5280 \text{ ft}}{1 \text{ mi}}$$

$$= 15,840 \text{ ft}$$

$$\frac{15,840 \text{ ft}}{1} \times \frac{20 \text{ cents}}{1 \text{ ft}} = 316800\cent = \$3168.00$$

8.2

10.
$$
\begin{array}{r}
6 \text{ yd } 2 \text{ ft } 11 \text{ in} \\
+7 \text{ yd } 1 \text{ ft } 9 \text{ in} \\
\hline
13 \text{ yd } 3 \text{ ft } 20 \text{ in}
\end{array}
$$
$= 13 \text{ yd} + 3 \text{ ft} + 1 \text{ ft} + 8 \text{ in}$
$= 13 \text{ yd} + 1 \text{ yd} + 1 \text{ ft} + 8 \text{ in}$
$= 14 \text{ yd } 1 \text{ ft } 8 \text{ in}$

14. $5 \text{ da } 3 \text{ hr } 8 \text{ min} = 4 \text{ da} + 1 \text{ da} + 2 \text{ hr} + 1 \text{ hr} + 8 \text{ min}$
$= 4 \text{ da} + 24 \text{ hr} + 2 \text{ hr} + 60 \text{ min} + 8 \text{ min}$
$= 4 \text{ da } 26 \text{ hr } 68 \text{ min}$

$$
\begin{array}{r}
4 \text{ da } 26 \text{ hr } 68 \text{ min} \\
-4 \text{ da } 6 \text{ hr } 40 \text{ min} \\
\hline
0 \text{ da } 20 \text{ hr } 28 \text{ min}
\end{array} = 20 \text{ hr } 28 \text{ min}
$$

17. $5 \times (4 \text{ gal } 3 \text{ qt}) = 20 \text{ gal } 15 \text{ qt}$
$= 20 \text{ gal} + 12 \text{ qt} + 3 \text{ qt}$
$= 20 \text{ gal} + 3 \text{ gal} + 3 \text{ qt}$
$= 23 \text{ gal } 3 \text{ qt}$

23. $(7 \text{ hr}) \div 3 = 2 \text{ hr}$ with remainder 1 hr

$1 \text{ hr} = 60 \text{ min}$

$60 \text{ min} + 33 \text{ min} = 93 \text{ min}$

$93 \text{ min} \div 3 = 31 \text{ min}$

Thus, $(7 \text{ hr } 33 \text{ min}) \div 3 = 2 \text{ hr } 31 \text{ min}.$

28. To make 7 loaves it would take 7 times the amount of flour necessary for one loaf.

$$(1 \text{ lb } 12 \text{ oz}) \times 7 = 7 \text{ lb} + 84 \text{ oz}$$
$$= 7 \text{ lb} + 80 \text{ oz} + 4 \text{ oz}$$
$$= 7 \text{ lb} + 5 \text{ lb} + 4 \text{ oz}$$
$$= 12 \text{ lb } 4 \text{ oz}$$

8.3

17. 1000 g   100 g   10 g
    (kg)   (hg)   (dag)

$$0.00721 \text{ kg} = 000.721 \text{ dag}$$
$$= 0.721 \text{ dag}$$

23. 100 g   10 g   1 g   0.1 g
  (hg)   (dag)   (g)   (dg)

$$4728 \text{ dg} = 4.728 \text{ hg}$$
$$= 4.728 \text{ hg}$$

32. $20 \dfrac{\text{cg}}{\text{da}} = \dfrac{20 \text{ cg}}{1 \text{ da}} \times \dfrac{7 \text{ da}}{1 \text{ wk}} \times \dfrac{1 \text{ g}}{100 \text{ cg}}$

$$= 1.4 \dfrac{\text{g}}{\text{wk}}$$

41. 600 gm = 0.6 kg

$$\dfrac{0.6 \text{ kg}}{1} \times \dfrac{6 \text{ dollars}}{1 \text{ kg}} = \$3.60$$

8.4

14. $40 \text{ L} = \dfrac{40 \text{ L}}{1} \times \dfrac{1.06 \text{ qt}}{1 \text{ L}} \times \dfrac{1 \text{ gal}}{4 \text{ qt}}$

$$= \dfrac{(40)(1.06)}{4} \text{ gal}$$

$$= 10.6 \text{ gal}$$

19. $F = \dfrac{9}{5} C + 32$

$$= \dfrac{9}{5}(40) + 32$$

$$= 72 + 32 + 104$$

Thus 40°C = 104°F.

22. $C = \dfrac{5}{9}(F - 32)$

$$= \dfrac{5}{9}(200 - 32)$$

$$= \dfrac{5}{9}(168) \approx 93$$

Thus, 200°F ≈ 93°C.

9.1

4. Since $(1)(2) = 2$, $(2)(2) = 4$, $4(2) = 8$, and $(8)(2) = 16$, each element after the first one in the list is obtained by multiplying the preceding element by 2. Thus, the next element should be $(16)(2) = 32$.

14. The problem is *not* that you have 8 apples, take away 5, and have 3 left. The problem states that you *take* 5 apples, and thus you will *have* 5 apples.

9.2

16. Since $\angle EDF = 0°$ and $\angle DFE = 180°$, the sum of their measures is 180°, making them supplementary.

9.3

9. The third angle in the first triangle is $180° - (80° + 40°) = 60°$. Thus no angle in the first triangle is 70°, and the two triangles cannot be similar.

9.4

9. $C = 2\pi r$

$$= 2\pi(7.7 \text{ mi})$$

$$= 15.4\pi \text{ mi}$$

12. $C = 2\pi r$

$$= 2\pi(0.05 \text{ cm})$$

$$= 2(3.14)(0.05) \text{ cm}$$

$$= 0.314 \text{ cm}$$

9.5

16. Divide the area into two rectangles. One is 7 yd long and 3 yd wide and the other is 3 yd long and 2 yd wide.

7 yd × 3 yd = 21 yd²

3 yd × 2 yd = 6 yd²

21 yd² + 6 yd² = 27 yd²

$$\dfrac{\$16.50}{1 \text{ yd}^2} \times \dfrac{27 \text{ yd}^2}{1} = (\$16.50)(27)$$

$$= \$445.50$$

27. $A = bh$

$$= (22.2 \text{ yd})(22.2 \text{ yd})$$

$$\approx 492.8 \text{ yd}^2$$

32. First find the area of the rectangle which is 28 cm long and 14 cm wide.

$$A = (28 \text{ cm})(14 \text{ cm}) = 392 \text{ cm}^2$$

Now find the area of the two triangles and subtract from the area of the rectangle. The base of each triangle is 10 cm and the height is 14 cm − 7 cm = 7 cm.

$$A = \dfrac{1}{2}(10 \text{ cm})(7 \text{ cm}) = 35 \text{ cm}^2$$

$$2A = 70 \text{ cm}^2$$

Thus, the area is 392 cm² − 70 cm² = 322 cm².

37. $A = \pi r^2$

$$= \pi(7.7 \text{ mi})^2$$

$$= 59.29\pi \text{ mi}^2$$

40. $A = \pi r^2$

$= \pi(0.05 \text{ cm})^2$

$\approx (3.14)(0.0025) \text{ cm}^2$

$= 0.00785 \text{ cm}^2$

41. Calculate the area for each hose and subtract. The radius of the first hose is $\frac{3}{8}$ in and the radius of the second hose is $\frac{1}{4}$ in.

$$A = \pi r^2$$

$$\approx 3.14\left(\frac{3}{8} \text{ in}\right)^2$$

$$\approx 0.442 \text{ in}^2$$

$$A = \pi r^2$$

$$\approx 3.14\left(\frac{1}{4} \text{ in}\right)^2$$

$$\approx 0.196 \text{ in}^2$$

$$0.442 \text{ in}^2 - 0.196 \text{ in}^2 = 0.246 \text{ in}^2$$

9.6

3. Convert each number to decimals:
9.5 in by 3.4 in by 8.75 in

$$V = (9.5 \text{ in})(3.4 \text{ in})(8.75 \text{ in})$$

$$= 282.625 \text{ in}^3$$

10. $A = 2lh + 2wh + 2lw$

$= 2(15.5 \text{ cm})(20.2 \text{ cm}) + 2(6.4 \text{ cm})(20.2 \text{ cm}) +$
$2(15.5 \text{ cm})(6.4 \text{ cm})$

$= 1083.16 \text{ cm}^2$

15. $V = e^3$

$= (22 \text{ cm})^3$

$= 10,648 \text{ cm}^3$

Since water weighs $1 \frac{\text{gm}}{\text{cm}^3}$ the answer is 10,648 g.

16. Area of bottom $= (22 \text{ cm})^2 = 484 \text{ cm}^2$

Area of 4 sides $= 4(22 \text{ cm})^2 = 4(484)\text{cm}^2$

$= 1936 \text{ cm}^2$

Total required $= 484 \text{ cm}^2 + 1936 \text{ cm}^2$

$= 2420 \text{ cm}^2$

19. $V = \pi r^2 h$

$= \pi(12.2 \text{ in})^2(30 \text{ in})$ **Radius is $\frac{1}{2}$ the diameter**

$\approx (3.14)(12.2)^2(30)\text{in}^2$

$\approx 14,020.7 \text{ in}^3$

22. $A = 2\pi rh + 2\pi r^2$

$\approx 2(3.14)(12.2 \text{ in})(30 \text{ in}) + 2(3.14)(12.2 \text{ in})^2$

$\approx 3233.2 \text{ in}^2$

25. $V = \frac{4}{3}\pi r^3$

$= \frac{4}{3}\pi\left(\frac{3}{4} \text{ in}\right)^3$

$\approx \left(\frac{4}{3}\right)(3.14)\left(\frac{3}{4}\right)^3 \text{ in}^3$

$\approx 1.766 \text{ in}^3$

29. $A = 4\pi r^2$

$= 4\pi(2.8 \text{ yd})^2$

$\approx 4(3.14)(2.8)^2 \text{ yd}^2$

$\approx 98.47 \text{ yd}^2$

30. Calculate the volume of the can and the volume of the pie pan and compare the two.

$$V = \pi r^2 h$$

$$= \pi(4 \text{ cm})^2(12 \text{ cm})$$

$$\approx (3.14)(4)^2(12) \text{ cm}^3$$

$$= 602.88 \text{ cm}^3$$

$$V = \pi r^2 h$$

$$= \pi(10 \text{ cm})^2(3 \text{ cm})$$

$$\approx (3.14)(10)^2(3) \text{ cm}^3$$

$$= 942 \text{ cm}^3$$

$$\frac{942 \text{ cm}^3}{1} \times \frac{1 \text{ can}}{602.88 \text{ cm}^3} = 1.6 \text{ cans}$$

35. Find the area of the triangle and subtract 3 times the area of one hole.

$$A = \frac{1}{2} bh$$

$$= \frac{1}{2}(9 \text{ in})(7.8 \text{ in})$$

$$= 35.1 \text{ in}^2$$

$$A = \pi r^2$$

$$= \pi(1 \text{ in})^2 \quad \textbf{Radius is } \frac{1}{2} \textbf{ (2 in)} = \textbf{1 in}$$

$$\approx 3.14 \text{ in}^2$$

The area of the three circles is

$$3(3.14 \text{ in}^2) = 9.42 \text{ in}^2$$

$$35.1 \text{ in}^2 - 9.42 \text{ in}^2 = 25.68 \text{ in}^2$$

10.1

51. $x - [a(b + 1) - c] = 12 - [2(3 + 1) - 5] =$
$12 - [2 \cdot 4 - 5] = 12 - [8 - 5] = 12 - 3 = 9$

60. $A = 2\pi rh + 2\pi r^2 = 2(3.14)(2)(10) + 2(3.14)(2)^2 =$
$125.60 + 25.12 = 150.72 \text{ cm}^2$

10.2

26. Given $\frac{17}{43}$ and $\frac{28}{79}$. The first cross product is $17 \cdot 79 =$ 1343, and the second cross product is $43 \cdot 28 = 1204$. Since $1343 > 1204$, we have $\frac{17}{43} > \frac{28}{79}$.

10.3

51. The difference in temperature was from $+35°$ to $-13°$. That is, we must find $35 - (-13) = 35 + 13 = 48°$. Thus, the total change in temperature was a decrease of $48°$.

10.4

44. Since 20 penalty points must be deducted from the total 3 times, we have $3(-20) = -60$ represents the deduction. Thus, $423 + (-60) = 363$ points is her total after losing the penalty points.

50. $3 - \dfrac{2 + (-4)}{1 - 5} = 3 - \dfrac{-2}{-4}$

$$= 3 - \dfrac{1}{2}$$

$$= \dfrac{6}{2} - \dfrac{1}{2} = \dfrac{5}{2}$$

63. When $a = -1$, $b = 2$, and $c = -4$,

$2[3(a + 1) + 2(4 + c)]$

$= 2[3((-1) + 1) + 2(4 + (-4))]$

$= 2[3(0) + 2(0)] = 2[0 + 0] = 2[0] = 0$.

75. -3000 represents the amount owed, $+1200$ represents the amount repaid, and -650 is the amount again borrowed. We must compute $(-3000) + (1200) + (-650) = -2450$. Thus, the Brisebois family owes $2450, so that -2450 represents the status of their account.

10.5

37. $\dfrac{1}{2}x - \dfrac{2}{3}y - \dfrac{5}{2}x - \dfrac{1}{3}y$

$= \dfrac{1}{2}x - \dfrac{5}{2}x - \dfrac{2}{3}y - \dfrac{1}{3}y$

$= \left(\dfrac{1}{2} - \dfrac{5}{2}\right)x - \left(\dfrac{2}{3} + \dfrac{1}{3}\right)y$

$= -2x - y$

43. $2x - (-x + 1) + 3$

$= 2x + x - 1 + 3$

$= 3x + 2$

47. $a - [3a - (1 - 2a)] = a - [3a - 1 + 2a] =$
$a - [5a - 1] = a - 5a + 1 = -4a + 1$

55. When $x = -3$, $-2x^2 = -2(-3)^2 = -2 \cdot 9 = -18$. Note that we square -3 before multiplying the result by -2.

56. When $x = -3$, $(-2x)^2 = ((-2)(-3))^2 = (6)^2 = 36$. Note that we first multiply -2 by -3, obtaining 6, then square the result. Compare Exercise 55 with Exercise 56.

68. When $x = -3$ and $y = -1$, $|x - y| = |(-3) - (-1)| = |(-3) + 1| = |-2| = 2$. Notice that we first simplify inside the absolute value bars.

10.6

17. Always reduce fractions when they appear under a radical. Thus,

$$\sqrt{\dfrac{8}{18}} = \sqrt{\dfrac{2 \cdot 4}{2 \cdot 9}} = \sqrt{\dfrac{4}{9}} = \dfrac{2}{3}.$$

29. Since

$$3^2 = 9 < 15 < 16 = 4^2,$$

we have

$$\sqrt{3^2} = 3 < \sqrt{15} < 4 = \sqrt{4^2}.$$

Thus, $\sqrt{15}$ is between 3 and 4.

11.1

28. Change $2\frac{1}{2}$ to $\frac{5}{2}$ and subtract $\frac{5}{2}$ from both sides of the equation.

$$x + 2\dfrac{1}{2} = 5$$

$$x + \dfrac{5}{2} = 5$$

$$x + \dfrac{5}{2} - \dfrac{5}{2} = 5 - \dfrac{5}{2}$$

$$x + 0 = \dfrac{10}{2} - \dfrac{5}{2}$$

$$x = \dfrac{5}{2} = 2\dfrac{1}{2}$$

11.2

15. Since $-\frac{1}{8}$ is the divisor of z we must multiply both sides by $-\frac{1}{8}$.

$$\dfrac{z}{-\frac{1}{8}} = 16$$

$$\left(-\dfrac{1}{8}\right)\dfrac{z}{-\frac{1}{8}} = 16\left(-\dfrac{1}{8}\right)$$

$$\dfrac{\left(-\frac{1}{8}\right)}{\left(-\frac{1}{8}\right)}z = -\dfrac{16}{8} \qquad \dfrac{-\frac{1}{8}}{-\frac{1}{8}} = 1$$

$$1 \cdot z = -2$$

$$z = -2$$

16. First multiply both sides by 3.

$$\dfrac{2x}{3} = 4$$

$$(3)\dfrac{2x}{3} = 4(3)$$

$$\dfrac{3}{3} \cdot 2x = 12$$

$$2x = 12$$

$$\frac{2x}{2} = \frac{12}{2} \quad \textbf{Now divide by 2}$$

$$x = 6$$

22. First change $2\frac{1}{2}$ to $\frac{5}{2}$.

$$2\frac{1}{2}x = 10$$

$$\frac{5}{2}x = 10$$

$$\frac{2}{5} \cdot \frac{5}{2}x = 10 \cdot \frac{2}{5} \quad \textbf{The product } \frac{2}{5} \cdot \frac{5}{2} \textbf{ is 1}$$

$$1 \cdot x = \frac{10}{1} \cdot \frac{2}{5}$$

$$x = 4 \qquad \textbf{10 ÷ 5 = 2 and 2 · 2 = 4}$$

11.3

11. Collect like terms on the left side.

$$-\frac{1}{7}y + y = 18$$

$$\left(-\frac{1}{7} + 1\right)y = 18$$

$$\left(-\frac{1}{7} + \frac{7}{7}\right)y = 18$$

$$\frac{6}{7}y = 18$$

$$\frac{7}{6} \cdot \frac{6}{7}y = 18 \cdot \frac{7}{6} \quad \frac{7}{6} \textbf{ is the reciprocal of } \frac{6}{7}$$

$$1 \cdot y = \frac{18}{1} \cdot \frac{7}{6}$$

$$y = 21 \qquad \begin{array}{l}\textbf{18 ÷ 6 = 3}\\ \textbf{and 3 · 7 = 21}\end{array}$$

14. $-7 + 21y + 23 = 16$

$$21y + 16 = 16 \quad \textbf{Collect like terms}$$

$$21y = 0 \quad \textbf{Subtract 16 from both sides}$$

$$y = 0 \quad \textbf{Divide both sides by 21}$$

17. $6y - 4y + 1 = 12 + 2y - 11$

$$2y + 1 = 1 + 2y \quad \textbf{Collect like terms}$$

Since this is an identity, every real number is a solution.

19. Clear fractions by multiplying through by 2 and collect like terms on both sides.

$$3y + \frac{5}{2}y + \frac{3}{2} = \frac{1}{2}y + \frac{5}{2}y$$

$$2\left(3y + \frac{5}{2}y + \frac{3}{2}\right) = 2\left(\frac{1}{2}y + \frac{5}{2}y\right)$$

$$6y + 5y + 3 = y + 5y \quad \begin{array}{l}\textbf{Multiply each term}\\ \textbf{by 2}\end{array}$$

$$11y + 3 = 6y \quad \textbf{Collect like terms}$$

$$11y = 6y - 3 \quad \begin{array}{l}\textbf{Subtract 3 from}\\ \textbf{both sides}\end{array}$$

$$11y - 6y = -3 \quad \begin{array}{l}\textbf{Subtract 6y from}\\ \textbf{both sides}\end{array}$$

$$5y = -3$$

$$y = -\frac{3}{5} \quad \begin{array}{l}\textbf{Divide both sides}\\ \textbf{by 5}\end{array}$$

29. $5(2 - 3x) = 15 - (x + 7)$

$$10 - 15x = 15 - x - 7 \quad \textbf{Clear parentheses}$$

$$10 - 15x = 8 - x \quad \textbf{Collect like terms}$$

$$10 = 8 + 14x \quad \textbf{Add 15x to both sides}$$

$$2 = 14x \quad \textbf{Subtract 8 from both sides}$$

$$\frac{2}{14} = x \quad \textbf{Divide both sides by 14}$$

$$\frac{1}{7} = x$$

36. The fractions could be cleared by multiplying both sides by the LCD of 2 and 3, which is 6. However, in this exercise the fractions are cleared when we use the distributive law.

$$\frac{1}{3}(6x - 9) = \frac{1}{2}(8x - 4)$$

$$\frac{1}{3} \cdot 6x - \frac{1}{3} \cdot 9 = \frac{1}{2} \cdot 8x - \frac{1}{2} \cdot 4$$

$$2x - 3 = 4x - 2$$

$$2x = 4x - 2 + 3$$

$$2x - 4x = 1$$

$$-2x = 1$$

$$\frac{-2x}{-2} = \frac{1}{-2}$$

$$x = -\frac{1}{2}$$

37. $5(x + 1) - 4x = x - 5$

$$5x + 5 - 4x = x - 5 \quad \textbf{Clear parentheses}$$

$$x + 5 = x - 5 \quad \textbf{Collect like terms}$$

$$5 = -5 \quad \textbf{Subtract x from both sides}$$

Since we obtain a contradiction, the equation has no solution.

38. Clear all parentheses first.

$$3(2 - 4x) = 4(2x - 1) - 2(1 + x)$$

$$6 - 12x = 8x - 4 - 2 - 2x$$

$$6 - 12x = 8x - 2x - 4 - 2 \quad \begin{array}{l}\textbf{Collect like}\\ \textbf{terms}\end{array}$$

$$-12x = 6x - 6 - 6$$

$$-12x - 6x = -12$$

$$-18x = -12$$

$$\frac{-18x}{18} = \frac{-12}{-18}$$

$$x = \frac{2}{3}$$

41. $-4(-x + 1) + 3(2x + 3) - 7x = -10$

$4x - 4 + 6x + 9 - 7x = -10$	**Clear parentheses**
$3x + 5 = -10$	**Collect like terms**
$3x = -15$	**Subtract 5 from both sides**
$x = -5$	**Divide both sides by 3**

11.5

17. Write out complete details when solving a word problem.

Let $x =$ length of one piece of rope,

$x + 7 =$ length of other piece of rope (it is 7 feet longer than the first).

$x + (x + 7) = 17$	**The sum of the lengths must be 17**

$$2x + 7 = 17$$
$$2x = 10$$
$$x = 5$$
$$x + 7 = 12$$

Thus, one length is 5 ft and the other is 12 ft.

18. Let $x =$ Tony's age,

$\frac{1}{4}x =$ Angela's age (If Tony is 4 times as old as Angela, then Angela is $\frac{1}{4}$ as old as Tony.),

$2x =$ Theresa's age (If Tony is $\frac{1}{2}$ as old as Theresa, then Theresa is 2 times as old as Tony.).

$x + \frac{1}{4}x + 2x = 39$	**The sum of their ages is 39**
$4\left(x + \frac{1}{4}x + 2x\right) = 4(39)$	**Clear fractions**

$$4x + x + 8x = 156$$
$$13x = 156$$
$$\frac{13}{13}x = \frac{156}{13}$$

$x = 12$	**Tony's age**
$\frac{1}{4}x = 3$	**Angela's age**
$2x = 24$	**Theresa's age**

32. Write out complete details. Let $x =$ amount that Percy had invested. We are given that the amount of interest was \$31.50 and the percent interest was 6%.

(% interest)·(amount invested) = (amount of interest)

$$0.06 \qquad x \qquad = \qquad 31.50$$

$$0.06x = 31.50$$

$$\frac{0.06}{0.06}x = \frac{31.50}{0.06}$$

$$x = 525$$

The amount invested was \$525.

34. Let $x =$ the original price of the dress,
$0.30x =$ amount of discount.

(original price) − (amount of discount) = (new price)

$$x \qquad - \qquad 0.30x \qquad = \qquad 51.45$$

$$(1 - 0.30)x = 51.45$$

$$0.7x = 51.45$$

$$x = \frac{51.45}{0.7} = 73.50$$

The original price of the dress was \$73.50.

38. $x =$ sum invested. $0.045x =$ amount of interest.

(sum invested) + (amount of interest) = (amt. in one year)

$$x \qquad + \qquad 0.045x \qquad = \qquad 1254$$

$$(1 + 0.045)x = 1254$$

$$1.045x = 1254$$

$$x = \frac{1254}{1.045} = 1200$$

The sum invested was \$1200.

11.6

14. The volume of a rectangular solid is $V = lwh$. Since we have $V = 4725$ m^3, $w = 15$ m, and $h = 15$ m, substitute and solve for l.

$$4725 = l(15)(15)$$
$$4725 = l(225)$$
$$\frac{4725}{225} = l$$
$$21 = l$$

Thus, the length is 21 m.

17. Let $x =$ number of feet that the water rises. The volume of the cube is $10 \cdot 10 \cdot 10 = 1000$ ft^3. Thus, 1000 ft^3 of water will be the increase in the level of the tank. But this volume is represented by $40 \cdot 50 \cdot x$.

Thus,

$$(40)(50)x = 1000$$
$$2000x = 1000$$
$$x = \frac{1000}{2000} = \frac{1}{2} = 0.5.$$

The level of the tank will increase 0.5 ft.

19. Let x = number of meters that the level rises. From Exercise 18 the volume increase is 33.5 m^3 (the volume of the sphere). Thus,

$$(10)(8)x = 33.5$$
$$80x = 33.5$$
$$x = \frac{33.5}{80} \approx 0.4.$$

Thus, the level increased about 0.4 m.

20. Let x = cost in dollars to enclose the pasture. The perimeter of the pasture is $P = 2l + 2w = 2(3) + 2(2) = 6 + 4 = 10$ mi. The length in feet is $10(5280) = 52,800$. Now multiply the number of feet (52,800 ft) by the cost per foot in dollars, \$0.35. $x = (52,800)(0.35) = 18,480$.
The total cost is \$18,480.

25. The distance one travels in 8 hours is $(8)(22) = 176$ nautical miles. The distance the second travels in 8 hours is $(8)(17) = 136$ nautical miles. Thus, in 8 hours they will be $176 - 136 = 40$ nautical miles apart.

Supplementary Applied Problems

5. Let x = the length of the pole. We can obtain an equation to use by setting the length of the pole equal to the sum of the lengths.

$$\text{(total length)} = \begin{pmatrix}\text{length in}\\\text{sand}\end{pmatrix} + \begin{pmatrix}\text{length in}\\\text{water}\end{pmatrix} + \begin{pmatrix}\text{length in}\\\text{air}\end{pmatrix}$$

$$x = \frac{1}{4}x + 9 + \frac{3}{8}x$$

$$x = \frac{1}{4}x + 9 + \frac{3}{8}x$$
$$8x = 2x + 72 + 3x \quad \textbf{Multiply by the LCD 8}$$
$$8x = 5x + 72$$
$$3x = 72$$
$$x = \frac{72}{3} = 24$$

The pole is 24 ft long.

10.   Let x = measure of first angle of the triangle,

$x + 30$ = measure of second angle of the triangle,

$4(x + 30)$ = measure of third angle of the triangle (If the second is $\frac{1}{4}$ the third, then the third is 4 times the second).

$$x + (x + 30) + 4(x + 30) = 180 \quad \textbf{Sum of the angles is 180}$$

$$x + x + 30 + 4x + 120 = 180$$
$$6x + 150 = 180$$
$$6x = 30$$
$$x = 5$$
$$x + 30 = 35$$
$$4(x + 30) = 140$$

The first angle is 5°, the second is 35°, and the third is 140°.

13.   Let x = width of rectangle,

$2x$ = length of rectangle,

$x + 8$ = width increased by 8,

$2x - 8$ = length decreased by 8.

For a square the length and width are equal.

$$x + 8 = 2x - 8$$
$$x = 2x - 16$$
$$-x = -16$$
$$x = 16$$
$$2x = 32$$

Thus, the width is 16 ft and the length is 32 ft.

19. x = population in 1980,

$0.20x$ = amount of decrease.

(population in 1980) − (amt. of decrease) = (pop. in 1990)

$$x \quad - \quad 0.20x \quad = \quad 740$$
$$(1 - 0.20)x = 740$$
$$0.8x = 740$$
$$x = \frac{740}{0.8} = 925$$

There were 925 people in Deserted in 1980.

20. x = amount invested at 14% interest,

$10,000 - x$ = amount invested at 11% interest (If x dollars are invested at 14%, the amount left to invest at 11% is $10,000 - x$),

$0.14x$ = amount of interest earned on 14% investment,

$0.11(10,000 - x)$ = amount of interest earned on 11% investment.

$$0.14x + 0.11(10,000 - x) = 1310$$
$$14x + 11(10,000 - x) = 131,000 \quad \textbf{Multiply by 100 to clear decimals}$$
$$14x + 110,000 - 11x = 131,000$$
$$3x = 21,000$$

$$x = 7000$$
$$10,000 - x = 3000$$

Thus, $7000 was invested at 14% and $3000 at 11%.

24. $x =$ score Becky must make to have 90 average. The average of her 4 scores must be 90.

$$\frac{96 + 78 + 91 + x}{4} = 90$$

$$\frac{265 + x}{4} = 90$$

$$4\left(\frac{265 + x}{4}\right) = 90(4)$$

$$265 + x = 360$$

$$x = 95$$

Becky must make at least 95 on her fourth test to get an A.

11.7

27. $2a + 1 > a - 3$

$2a > a - 4$

$a > -4$

33. Multiply both sides by -4 to make the coefficient of y 1. When multiplying by a negative number we must reverse the inequality.

$$-\frac{1}{4}y \geq 2$$

$$-4\left(-\frac{1}{4}\right)y \leq 2(-4)$$

$$y \leq -8$$

45. $3(2x + 3) - (3x + 2) < 12$

$6x + 9 - 3x - 2 < 12$

$3x + 7 < 12$

$3x < 5$

$x < \dfrac{5}{3}$

50. Multiply by the LCD, 8, to clear the fractions. Since 8 is positive the inequality stays the same.

$$\frac{3}{4}z - \frac{3}{8} < \frac{3}{2} + \frac{1}{8}z$$

$$8\left(\frac{3}{4}z - \frac{3}{8}\right) < 8\left(\frac{3}{2} + \frac{1}{8}z\right)$$

$$6z - 3 < 12 + z$$

$$5z < 15$$

$$z < \frac{15}{5} \qquad \textbf{5 is positive}$$

$$z < 3$$

57. $x =$ the number,

$3x =$ product of the number and 3,

$x - 8 =$ the number less 8.

$3x \geq x - 8$

$2x \geq -8$

$x \geq \dfrac{-8}{2}$

$x \geq -4$

60. Let $x =$ number of cars he must sell in August. The average number of cars in the three month period is given by $\frac{47 + 62 + x}{3}$.

To win the trip

$$\frac{47 + 62 + x}{3} \geq 50.$$

$$47 + 62 + x \geq 150 \quad \textbf{Multiply both sides by 3}$$

$$109 + x \geq 150$$

$$x \geq 41$$

Thus, Darrell must sell at least 41 cars in August.

12.1

17. (a) Substitute 0 for x in $x - 5 = 0$.

$$0 - 5 = 0$$

$$-5 = 0$$

This is a contradiction so x cannot be 0. In fact, x must always be 5.

(b) Substitute 0 for y in $x - 5 = 0$. But y does not occur in the equation. Thus, if y is 0, x is still 5, so that (5, 0) is the completed ordered pair.

(c) Substitute 5 for x in $x - 5 = 0$.

$$5 - 5 = 0$$

$$0 = 0$$

Thus, x is 5 and y can be any number. That is, (5, any number) is a solution to the equation.

(d) As in part (b), y can be -10 but x is still 5. Thus, $(5, -10)$ is the completed ordered pair.

22. Using the equation $y = 55x$, we must complete each pair.

(a) For (1,): Substitute 1 for x to obtain $y = 55 \cdot 1 = 55$. Thus, (1, 55) is the completed pair and the car travels 55 miles in 1 hour.

(b) For (5,): Substitute 5 for x to obtain $y = 55 \cdot 5 = 275$. Thus, (5, 275) is the completed pair and the car travels 275 miles in 5 hours.

(c) For (10,): Substitute 10 for x to obtain $y = 55 \cdot 10 = 550$. Thus, the completed pair is (10, 550) and the car travels 550 miles in 10 hours.

12.3

14. The three points will all lie on the same straight line if the slopes of the lines through (2, 3) and (0, 2) and through (0, 2) and (−2, 1) are equal. The slope of the line through (2, 3) and (0, 2) is

$$\frac{3 - 2}{2 - 0} = \frac{1}{2},$$

and the slope of the line through (0, 2) and (−2, 1) is

$$\frac{2 - 1}{0 - (-2)} = \frac{1}{2}.$$

Thus, the three points all lie on the same straight line.

16. First choose two points on $5x + 1 = 0$. Since all points on this line have x-coordinate $-\frac{1}{5}$ ($5x + 1 = 0$ means $5x = -1$ or $x = -\frac{1}{5}$), two such points are $\left(-\frac{1}{5}, 1\right)$ and $\left(-\frac{1}{5}, 2\right)$. Substitute into the slope formula.

$$m = \frac{y_2 - y_1}{x_2 - x_1} = \frac{2 - 1}{\left(-\frac{1}{5}\right) - \left(-\frac{1}{5}\right)} = \frac{1}{0}$$

Since m is undefined, the slope is undefined (the line is parallel to the y-axis).

12.4

15. Substitute $-\frac{1}{2}$ for m and (3, −2) for (x_1, y_1) in

$$y - y_1 = m(x - x_1).$$
$$y - (-2) = -\tfrac{1}{2}(x - 3)$$

$2(y + 2) = -(x - 3)$	**Clear fractions**
$2y + 4 = -x + 3$	
$x + 2y + 1 = 0$	**General form**

18. First find the slope.

$$m = \frac{y_2 - y_1}{x_2 - x_1} = \frac{-2 - (-1)}{3 - (-1)}$$

$$= \frac{-2 + 1}{3 + 1} = \frac{-1}{4}$$

Then substitute into the slope-intercept form using (3, −2) for (x_1, y_1) and $-\frac{1}{4}$ for m.

$$y - y_1 = m(x - x_1)$$
$$y - (-2) = -\tfrac{1}{4}(x - 3)$$

$4(y + 2) = -(x - 3)$	**Multiply both sides by 4**
$4y + 8 = -x + 3$	
$x + 4y + 5 = 0$	**General form**

19. We are given that
$$y = mx + b$$
where $b = $ overhead cost and $m = $ variable cost. In this problem overhead cost $= b = \$25$ and variable cost $= m = \$10$. Thus,

$$y = mx + b$$
$$= 10x + 25.$$

13.1

7. Solve for y.

$$2x - 7y + 1 = 0 \qquad\qquad -6x + 21y + 3 = 0$$
$$-7y = -2x - 1 \qquad\qquad 21y = 6x - 3$$
$$y = \frac{2}{7}x + \frac{1}{7} \qquad\qquad y = \frac{6}{21}x - \frac{3}{21}$$
$$y = \frac{2}{7}x - \frac{1}{7}$$

Since the coefficients of x are both $\frac{2}{7}$, the lines are either parallel or coinciding (they have the same slope). Since the constants are unequal, the y-intercepts are different so the lines do not coincide. Thus, the lines are parallel.

10. Since the line $2y = 8$ is the same as $y = 4$, its graph is a horizontal line with y-intercept (0, 4). Also, the line $2x = 8$ is the same as $x = 4$, and its graph is a vertical line with x-intercept (4, 0). Since a horizontal line and a vertical line are perpendicular, they are intersecting.

14. To determine whether (−3, 2) is a solution to the system, substitute into both equations.

$$2x + 3y = 0 \qquad\qquad x + 8y = 19$$
$$2(-3) + 3(2) \overset{?}{=} 0 \qquad (-3) + 8(2) \overset{?}{=} 19$$
$$-6 + 6 \overset{?}{=} 0 \qquad\qquad -3 + 16 \overset{?}{=} 19$$
$$0 = 0 \qquad\qquad 13 \neq 19$$

Thus, although (−3, 2) is a solution to the first equation, it is not a solution to the system since it does not solve the second equation.

13.2

5. First determine the number of solutions by writing each equation in slope-intercept form.

$$x + 3y = 4 \qquad\qquad -2x - 6y = -8$$
$$3y = -x + 4 \qquad\qquad -6y = 2x - 8$$
$$y = -\frac{1}{3}x + \frac{4}{3} \qquad\qquad y = -\frac{2}{6}x + \frac{8}{6}$$
$$= -\frac{1}{3}x + \frac{4}{3}$$

Since the equations are the same, the lines are coinciding so there will be infinitely many solutions. We might also have observed sooner that if the second equation is multiplied by $-\frac{1}{2}$ on both sides the result is the first equation. Thus, obviously the two equations are the same.

8. The system will have exactly one solution since the two lines have slope −1 and $-\frac{1}{3}$. Graph each equation by finding its intercepts.

$x + y = 3$ $x + 3y = -1$

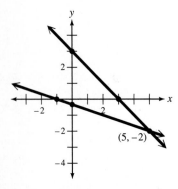

x	y
0	3
3	0

x	y
0	$-\frac{1}{3}$
-1	0

The point of intersection appears to have coordinates $(5, -2)$. Check by substitution.

$x + y = 3$ $x + 3y = -1$

$5 + (-2) \stackrel{?}{=} 3$ $5 + 3(-2) \stackrel{?}{=} -1$

$3 = 3$ $-1 = -1$

Thus, the solution is the ordered pair $(5, -2)$.

13. Write in slope-intercept form.

$x + 4y = 1$ $-x - 4y = 1$

$4y = -x + 1$ $-4y = x + 1$

$y = -\frac{1}{4}x + \frac{1}{4}$ $y = -\frac{1}{4}x - \frac{1}{4}$

Although the slopes are the same $\left(\text{both are } -\frac{1}{4}\right)$, the y-intercepts are different. Thus, the lines are parallel, and the system has no solution.

13.3

7. $3x - 5y = 19$

$2x - 4y = 16$

Simplify by multiplying the second equation by $\frac{1}{2}$.

$x - 2y = 8$

Solve for x.

$x = 2y + 8$

Substitute into the first equation.

$3(2y + 8) - 5y = 19$

$6y + 24 - 5y = 19$

$y + 24 = 19$

$y = -5$

Substitute -5 for y in $x - 2y = 8$.

$x - 2(-5) = 9$

$x + 10 = 8$

$x = -2$

Thus, the solution is the ordered pair $(-2, -5)$.

10. $3x - 3y = 1$

$x - y = -1$

Solve the second equation for x, $x = y - 1$, and substitute into the first.

$3(y - 1) - 3y = 1$

$3y - 3 - 3y = 1$

$-3 = 1$

Since a contradiction is obtained, the system has no solution.

11. $2x + 2y = -6$

$-x - y = 3$

Solve the second equation for y.

$-y = x + 3$

$y = -x - 4$

Substitute into the first equation.

$2x + 2(-x - 3) = -6$

$2x - 2x - 6 = -6$

$-6 = -6$

Since an identity is obtained, the system has infinitely many solutions.

14. $3x + 5y = 30$

$5x + 3y = 34$

Solving for either variable in either equation will result in fractions. Perhaps it is best to solve for x in the first.

$3x = -5y + 30$

$x = -\frac{5}{3}y + 10$

Substitute into the second.

$5\left(-\frac{5}{3}y + 10\right) + 3y = 34$

$-\frac{25}{3}y + 50 + 3y = 34$

$-25y + 150 + 9y = 102$ **Multiply through by 3 to clear the fraction**

$-16y = -48$

$y = 3$

Substitute 3 for y in the second equation.

$$5x + 3(3) = 34$$
$$5x = 25$$
$$x = 5$$

Thus, the solution pair is $(5, 3)$.

13.4

5. $6x + 5y = 11$

$3x - 7y = -4$

Multiply the second equation by -2 and add.

$$\begin{array}{r} 6x + 5y = 11 \\ -6x + 14y = 8 \\ \hline 19y = 19 \\ y = 1 \end{array}$$

Substitute 1 for y in the first equation and solve for x.

$$6x + 5(1) = 11$$
$$6x + 5 = 11$$
$$6x = 6$$
$$x = 1$$

Thus, the solution is the pair $(1, 1)$.

10. $2x + 3y = -4$

$1 + 2y = -3x$

First rewrite the system.

$$2x + 3y = -4$$
$$3x + 2y = -1$$

Multiply the first equation by 3 and the second equation by -2 and add.

$$\begin{array}{r} 6x + 9y = -12 \\ -6x - 4y = 2 \\ \hline 5y = -10 \\ y = -2 \end{array}$$

Now substitute -2 for y in the first equation.

$$2x + 3(-2) = -4$$
$$2x - 6 = -4$$
$$2x = 2$$
$$x = 1$$

Thus, the solution is $(1, -2)$.

11. $3x + 7y = -21$

$7x + 3y = -9$

If we multiply the first equation by -7 and the second by 3, the coefficients of x will be -21 and 21. Then by adding, we eliminate x and obtain an equation in the one variable y.

$$\begin{array}{r} -21x - 49y = 147 \\ 21x + 9y = -27 \\ \hline -40y = 120 \\ y = -3 \end{array}$$

Substitute -3 for y in the first equation.

$$3x + 7(-3) = -21$$
$$3x = 0$$
$$x = 0$$

Thus, the solution is the pair $(0, -3)$.

16. $3x + 3y = 3$

$4x + 4y = -3$

Multiply the first equation by -4, the second by 3, and add.

$$\begin{array}{r} -12x - 12y = -12 \\ 12x + 12y = -9 \\ \hline 0 = -21 \end{array}$$

Since a contradiction is obtained, the system has no solution.

17. $2x + 3y = -4$

$5x + 7y = -10$

Multiply the first equation by -5, the second by 2, and add to eliminate x.

$$\begin{array}{r} -10x - 15y = 20 \\ 10x + 14y = -20 \\ \hline -y = 0 \\ y = 0 \end{array}$$

Substitute 0 for y in the first equation.

$$2x + 3(0) = -4$$
$$2x = -4$$
$$x = -2$$

Thus, the solution is $(-2, 0)$.

13.5

11. Let x = number of pounds of 90¢ candy,

y = number of pounds of $1.50 candy.

Since there are 30 pounds in the mixture,

$$x + y = 30.$$

The value of the mixture is

$$90x + 150y = 110(30).$$

Notice that the units were changed to cents, and don't forget to multiply 110¢ by 30 pounds on the right side. Solve the first equation for x and substitute.

$$x = 30 - y$$
$$90(30 - y) + 150y = 110(30)$$
$$9(30 - y) + 15y = 11(30) \quad \textbf{Divide through by}$$
$$\textbf{10}$$
$$270 - 9y + 15y = 330$$
$$6y = 60$$
$$y = 10$$

Then $x = 30 - y = 30 - 10 = 20$. Thus, there are 20 lb of the 90¢ candy and 10 lb of the $1.50 candy.

17. Let x = Terry's average speed riding in still air,
$\quad y$ = average wind speed,
$x + y$ = Terry's average speed riding with the wind,
$x - y$ = Terry's average speed riding against the wind.

Use the distance formula, $d = rt$, twice noting that both distances are 40 miles.

$$40 = (x - y)4 \quad \textbf{Distance against the wind}$$
$$40 = (x + y)2 \quad \textbf{Distance with the wind}$$

Thus, dividing the first equation by 4 and the second by 2, we have the system:

$$x - y = 10$$
$$\underline{x + y = 20}$$
$$2x = 30$$
$$x = 15.$$

Substitute 15 for x in the first equation.

$$15 - y = 10$$
$$-y = -5$$
$$y = 5$$

Thus, Terry's rate riding is 15 mph, and the wind speed is 5 mph.

19. Let x = amount to be invested at 10% interest,
$\quad y$ = amount to be invested at 12% interest.

Then,
$\quad x + y = 10,000.$ The total to invest is $10,000

The total interest is the sum of the interests earned on each part.
$$(0.10)x + (0.12)y = 1160$$

Multiply through by 100 to clear all decimals,
$$10x + 12y = 116,000,$$

then divide through by 2 to decrease the coefficients.
$$5x + 6y = 58,000$$

Solve the first equation for x and substitute.

$$5(10,000 - y) + 6y = 58,000$$
$$50,000 - 5y + 6y = 58,000$$
$$y = 8000$$

Then $x = 10,000 - y = 10,000 - 8000 = 2000$. Thus, $2000 is invested at 10% and $8000 is invested at 12%.

21. Let x = daily rate charged,
$\quad y$ = mileage rate charged.

The system of equations to solve is:

$$3x + 400y = 82$$
$$5x + 500y = 120.$$

Divide the second equation by 5 and solve for x.

$$x + 100y = 24$$
$$x = 24 - 100y$$

Substitute into the first equation.

$$3(24 - 100y) + 400y = 82$$
$$72 - 300y + 400y = 82$$
$$100y = 10$$
$$y = 0.10$$

Then $x = 24 - 100y = 24 - 100(0.10) = 24 - 10 = 14$. Thus, the charges are $14 per day and 10¢ per mile.

14.1

27. $\left(\dfrac{2y}{x^3}\right)^{-2} = \dfrac{(2y)^{-2}}{(x^3)^{-2}} \qquad \left(\dfrac{a}{b}\right)^n = \dfrac{a^n}{b^n}$

$\qquad = \dfrac{2^{-2}y^{-2}}{(x^3)^{-2}} \quad (ab)^n = a^n b^n$

$\qquad = \dfrac{2^{-2}y^{-2}}{x^{-6}} \quad (a^m)^n = a^{mn}$

$\qquad = \dfrac{x^6}{2^2 y^2}$

$\qquad = \dfrac{x^6}{4y^2}$

30. $\dfrac{a^{-2}b^2}{a^4 b^{-3}} = a^{-2-4}b^{2-(-3)} \quad \dfrac{a^m}{a^n} = a^{m-n}$

$\qquad = a^{-6}b^5$

$\qquad = \dfrac{1}{a^6} \cdot b^5 \qquad a^{-n} = \dfrac{1}{a^n}$

$\qquad = \dfrac{b^5}{a^6}$

51. When $a = -2$, $a^{-2} = (-2)^{-2} = \dfrac{1}{(-2)^2} = \dfrac{1}{4}$.

52. When $a = -2$, $-2a = -2(-2) = 4$.

53. When $a = -2$, $-a^2 = -(-2)^2 = -(4) = -4$.
Note the difference between Exercises 51, 52, and 53.

54. $a^{-2} + b^{-2} = (-2)^{-2} + (3)^{-2}$ **Substitute −2 for a and 3 for b**

$\qquad = \dfrac{1}{(-2)^2} + \dfrac{1}{3^2} \qquad a^{-n} = \dfrac{1}{a^n}$

$\qquad = \dfrac{1}{4} + \dfrac{1}{9}$

$\qquad = \dfrac{13}{36}$

55. $(a + b)^{-2} = ((-2) + 3)^{-2}$ **Substitute −2 for a and 3 for b**

$$= (1)^{-2}$$

$$= \frac{1}{1^2} = 1$$

Note the difference between Exercises 54 and 55.

59. $a^{-1}b^{-1} = (-2)^{-1}(3)^{-1}$

$$= \left(\frac{1}{-2}\right)\left(\frac{1}{3}\right) \quad a^{-1} = \frac{1}{a^1} = \frac{1}{a}$$

$$= -\frac{1}{6}$$

60. $(ab)^{-1} = ((-2)(3))^{-1}$

$$= (-6)^{-1}$$

$$= \frac{1}{(-6)^1}$$

$$= -\frac{1}{6}$$

14.2

19. $(0.0000022)(300) = (2.2 \times 10^{-6})(3.0 \times 10^2) = (2.2)(3.0) \times (10^{-6})(10^2) = 6.6 \times 10^{-4}$

14.3

28. $-8x^{10} + x^5 - 2x^{10} - 7x^5 + 1 - x^{10} + 3$

$$= (-8x^{10} - 2x^{10} - x^{10}) + (x^5 - 7x^5) + (1 + 3)$$

$$= (-8 - 2 - 1)x^{10} + (1 - 7)x^5 + (1 + 3)$$

$$= -11x^{10} - 6x^5 + 4$$

35. $-4x^2y + 2xy^2 + x^2 - 3x^2y$

$$= (-4x^2y - 3x^2y) + 2xy^2 + x^2$$

$$= (-4 - 3)x^2y + 2xy^2 + x^2$$

$$= -7x^2y + 2xy^2 + x^2$$

38. $7y^2 - 2y - 5 = 7(-2)^2 - 2(-2) - 5$

$$= 7 \cdot 4 + 4 - 5$$

$$= 28 + 4 - 5 = 27$$

41. We must evaluate $0.05x + 15.5$ when $x = 440$.

$$0.05(440) + 15.5$$

$$= 22 + 15.5$$

$$= 37.5$$

Thus, the cost is $37.50 to manufacture 440 bolts.

14.4

10. $(-4a^3 + 7a^4 + 3a + 2) + (5 - 3a + 7a^3) + (17a^4 - 5 + 12a^3)$

$$= (7a^4 - 4a^3 + 3a + 2) + (7a^3 - 3a + 5) + (17a^4 + 12a^3 - 5)$$

$$= 7a^4 + 17a^4 - 4a^3 + 7a^3 + 12a^3 + 3a - 3a + 2 + 5 - 5$$

$$= (7 + 17)a^4 + (-4 + 7 + 12)a^3 + (3 - 3)a + (2 + 5 - 5)$$

$$= 24a^4 + 15a^3 + 0 \cdot a + 2$$

$$= 24a^4 + 15a^3 + 2$$

16.
$$\begin{array}{r} 0.03y^3 - 0.75y^2 - 3y + 2 \\ -0.15y^3 \qquad\quad + 5y - 0.3 \\ \underline{0.21y^3 - 0.13y^2 \qquad\quad + 0.6} \\ 0.09y^3 - 0.88y^2 + 2y + 2.3 \end{array}$$

Simply add the coefficients of like terms (down the columns) to obtain the coefficients in the sum.

23. $(56x - 93x^3 + 21x^4 + 32x^5) - (3x^4 - 7x^5 + 15x - 32)$

$$= (32x^5 + 21x^4 - 93x^3 + 56x) - (-7x^5 + 3x^4 + 15x - 32)$$

$$= 32x^5 + 21x^4 - 93x^3 + 56x + 7x^5 - 3x^4 - 15x + 32$$

$$= 32x^5 + 7x^5 + 21x^4 - 3x^4 - 93x^3 + 56x - 15x + 32$$

$$= (32 + 7)x^5 + (21 - 3)x^4 - 93x^3 + (56 - 15)x + 32$$

$$= 39x^5 + 18x^4 - 93x^3 + 41x + 32$$

28. $(8y^{10} - y^8) - (3y^{12} + 2y^{10} - y^8)$

$$= 8y^{10} - y^8 - 3y^{12} - 2y^{10} + y^8 \quad \text{Clear parentheses}$$

$$= -3y^{12} + 8y^{10} - 2y^{10} - y^8 + y^8$$

$$= -3y^{12} + (8 - 2)y^{10} + (-1 + 1)y^8 \quad \text{Distributive laws}$$

$$= -3y^{12} + 6y^{10}$$

31. $(9x^4 + 3x^3 + 8x) + (3x^4 + x^3 - 7x^2) - (12x^4 - 3x^2 + x)$

$$= 9x^4 + 3x^3 + 8x + 3x^4 + x^3 - 7x^2 - 12x^4 + 3x^2 - x$$

$$= (9 + 3 - 12)x^4 + (3 + 1)x^3 + (-7 + 3)x^2 + (8 - 1)x$$

$$= 0 \cdot x^4 + 4x^3 - 4x^2 + 7x$$

$$= 4x^3 - 4x^2 + 7x$$

38. $(-2a^2b + ab - 4ab^2) - (6a^2b + 4ab^2)$

$$= -2a^2b + ab - 4ab^2 - 6a^2b - 4ab^2$$

$$= (-2 - 6)a^2b + ab + (-4 - 4)ab^2$$

$$= -8a^2b + ab - 8ab^2$$

40. $(6x^2y - xy) - (3x^2y - 7xy^2) - (4xy - 5xy^2)$

$$= 6x^2y - xy - 3x^2y + 7xy^2 - 4xy + 5xy^2$$

Clear parentheses

$$= 6x^2y - 3x^2y - xy - 4xy + 7xy^2 + 5xy^2$$

$$= (6 - 3)x^2y + (-1 - 4)xy + (7 + 5)xy^2$$

Distributive laws

$$= 3x^2y - 5xy + 12xy^2$$

14.5

16. $-10x^2(x^5 - 6x^3 + 7x^2)$

$= (-10x^2)(x^5) - (-10x^2)(6x^3) +$ **Distributive**
$(-10x^2)(7x^2)$ **laws**

$= -10x^2x^5 + 60x^2x^3 - 70x^2x^2$ **Multiply**
                                                              coefficients

$= -10x^7 + 60x^5 - 70x^4$ $a^m a^n = a^{m+n}$

23. $(x^2 + 4x - 2)(2x^2 - x + 3)$

$= (x^2 + 4x - 2)(2x^2) + (x^2 + 4x - 2)(-x) +$
$(x^2 + 4x - 2)(3)$

$= (x^2)(2x^2) + (4x)(2x^2) + (-2)(2x^2) + (x^2)(-x) +$
$(4x)(-x) + (-2)(-x) + (x^2)(3) + (4x)(3) +$
$(-2)(3)$

$= 2x^4 + 8x^3 - 4x^2 - x^3 - 4x^2 + 2x + 3x^2 +$
$12x - 6$

$= 2x^4 + (8 - 1)x^3 + (-4 + 3)x^2 + (2 + 12)x - 6$

$= 2x^4 + 7x^3 - x^2 + 14x - 6$

37. $(5x - 2)(3x + 4)$

$= (5x)(3x) + (5x)(4) + (-2)(3x) + (-2)(4)$

$= 15x^2 + 20x - 6x - 8$

$= 15x^2 + 14x - 8$

41. $(2z^2 + 1)(z^2 - 2)$

$= (2z^2)(z^2) + (2z^2)(-2) + (1)(z^2) + (1)(-2)$

$= 2z^4 - 4z^2 + z^2 - 2$

$= 2z^4 - 3z^2 - 2$ **Collect like terms**

52. $0.3x^2 + 0.2$
$\underline{0.5x - 0.7}$
$0.15x^3 + 0.10x$
$\underline{- 0.21x^2 - 0.14}$
$0.15x^3 - 0.21x^2 + 0.10x - 0.14$

55. $3y^2 + 5y - 6$
$\underline{y^2 - 3y + 2}$
$3y^4 + 5y^3 - 6y^2$
$- 9y^3 - 15y^2 + 18y$
$\underline{6y^2 + 10y - 12}$
$3y^4 - 4y^3 - 15y^2 + 28y - 12$

59. $(2a - 3b)(5a + b)$

$= (2a)(5a) + (2a)(b) + (-3b)(5a) + (-3b)(b)$

$= 10a^2 + 2ab - 15ab - 3b^2$

$= 10a^2 - 13ab - 3b^2$

65. To find the product $(x - 1)(x + 1)(x + 2)$, we multiply the first two factors then the result times the third.

$(x - 1)(x + 1) = x^2 + x - x - 1$ **FOIL**

$= x^2 - 1$

Now multiply by $(x + 2)$.

$(x^2 - 1)(x + 2) = x^3 + 2x^2 - x - 2$ **FOIL**

Thus,

$(x - 1)(x + 1)(x + 2) = x^3 + 2x^2 - x - 2.$

67. Use the formula $A = lw$ and substitute $(2x + 1)$ for l and $(3x - 2)$ for w.

$A = (2x + 1)(3x - 2)$

$= (2x)(3x) + (2x)(-2) + (1)(3x) + (1)(-2)$ **FOIL**

$= 6x^2 - 4x + 3x - 2$

$= 6x^2 - x - 2$ **Collect like terms**

Thus, the area is $(6x^2 - x - 2)$ ft^2.

14.6

13. $(4y - 9)(4y + 9)$

$= (4y)^2 - (9)^2$ $(a - b)(a + b) = a^2 - b^2$

$= 4^2y^2 - 81$

$= 16y^2 - 81$

31. $(0.7y - 3)^2 = (0.7y)^2 - 2(0.7y)(3) + (3)^2$

$= (0.7)^2y^2 - 6(0.7)y + 9 = 0.49y^2 - 4.2y + 9$

37. $(2x^2 - y)^2 = (2x^2)^2 - 2(2x^2)(y) + (y)^2$

$= 2^2(x^2)^2 - 4x^2y + y^2 = 4x^4 - 4x^2y + y^2$

39. $(a^2 + 2b)^2$

$= (a^2)^2 + 2(a^2)(2b) + (2b)^2$ **Use a^2 for a and $2b$ for**
                                                          b in $(a + b)^2 =$
                                                          $a^2 + 2ab + b^2$

$= a^4 + 4a^2b + 4b^2$

40. $(x + 1)^2 - (x - 1)^2$

$= (x^2 + (2)(x)(1) + 1^2) - (x^2 - (2)(x)(1) + 1^2)$

Use both perfect square formulas and be sure to enclose the products in parentheses.

$= (x^2 + 2x + 1) - (x^2 - 2x + 1)$

$= x^2 + 2x + 1 - x^2 + 2x - 1$ **Clear parentheses**

$= 4x$

Remember to use parentheses when subtracting in order to avoid a common sign error.

44. Substitute $(2a + 7)$ for l and $(a - 1)$ for w in $A = lw$.

$A = (2a + 7)(a - 1)$

$= 2a^2 - 2a + 7a - 7$ **FOIL**

$= 2a^2 + 5a - 7$ **Collect like terms**

Thus, the area is $(2a^2 + 5a - 7)$ mi^2.

14.7

7. $(3a^{12} - 9a^6 + 27a^5 + 81a^4) \div 9a^2$

$= \dfrac{3a^{12} - 9a^6 + 27a^5 + 81a^4}{9a^2}$

$= \dfrac{3a^{12}}{9a^2} - \dfrac{9a^6}{9a^2} + \dfrac{27a^5}{9a^2} + \dfrac{81a^4}{9a^2}$

$= \dfrac{1}{3}a^{12-2} - 1 \cdot a^{6-2} + 3a^{5-2} + 9a^{4-2}$

$= \dfrac{1}{3}a^{10} - a^4 + 3a^3 + 9a^2$

10. $\dfrac{-8x^3 + 6x^2 - 4x}{0.2x} = \dfrac{-8x^3}{0.2x} + \dfrac{6x^2}{0.2x} - \dfrac{4x}{0.2x}$

$= -40x^{3-1} + 30x^{2-1} - 20x^{1-1}$ $\quad \dfrac{8}{0.2} = \dfrac{80}{2} = \mathbf{40}$

$= -40x^2 + 30x - 20x^0$

$\quad -40x^2 + 30x - 20 \quad \boldsymbol{x^0 = 1}$

13. $\dfrac{-5a^2b + 3ab - 2a}{-a} = \dfrac{-5a^2b}{-a} + \dfrac{3ab}{-a} - \dfrac{2a}{-a} =$

$5a^{2-1}b - 3a^{1-1} + 2a^{1-1} = 5ab - 3a^0b + 2a^0 =$

$5ab - 3b + 2$

22. $(6 + 8y - y^2) \div (4 - y)$

$= \dfrac{6 + 8y - y^2}{4 - y}$

$= \dfrac{(-1)(6 + 8y - y^2)}{(-1)(4 - y)}$ To make y in denominator positive

$= \dfrac{-6 - 8y + y^2}{-4 + y}$

$= \dfrac{y^2 - 8y - 6}{y - 4}$ Descending order

$$
\begin{array}{r}
y - 4 \\
y - 4 \overline{)\, y^2 - 8y - 6\,} \\
\underline{y^2 - 4y} \\
-4y - 6 \\
\underline{-4y + 16} \\
-22
\end{array}
$$

The answer is $y - 4 - \dfrac{22}{y - 4}$.

23.
$$
\begin{array}{r}
3a - 8 \\
a - 5 \overline{)\, 3a^2 - 23a + 40\,} \\
\underline{3a^2 - 15a } \\
-8a + 40 \\
\underline{-8a + 40} \\
0
\end{array}
$$
Remember to subtract at this step then bring down the 40
Remember again to subtract

Thus, the quotient is $3a - 8$.

28.
$$
\begin{array}{r}
y^4 - 2y^3 + 4y^2 - 8y + 16 \\
y + 2 \overline{)\, y^5 + 32\,} \\
\underline{y^5 + 2y^4 } \\
-2y^4 \\
\underline{-2y^4 - 4y^3 } \\
4y^3 \\
\underline{4y^3 + 8y^2 } \\
-8y^2 \\
\underline{-8y^2 - 16y } \\
16y + 32 \\
\underline{16y + 32} \\
0
\end{array}
$$

30.
$$
\begin{array}{r}
x^2 + 1 \\
x^2 - 1 \overline{)\, x^4 - 1\,} \\
\underline{x^4 - x^2 } \\
x^2 - 1 \\
\underline{x^2 - 1} \\
0
\end{array}
$$
Subtract, don't add

15.1

14. $36y^4 = 2^2 \cdot 3^2 \cdot y^4$

$6y^3 = 2 \cdot 3 \cdot y^3$

$42y^5 = 2 \cdot 3 \cdot 7 \cdot y^5$

The factors 2, 3, and y are all three monomials. The lowest power of 2 and 3 is 1 and the lowest power of y is 3. Thus, the GCF is $2 \cdot 3 \cdot y^3 = 6y^3$.

27. The only common factor of $3a(a + 2)$ and $5(a + 2)$ is $(a + 2)$. Thus, the GCF is $(a + 2)$.

29. Since 5 is the GCF of 5, -15, and 10, and since ab^2 is the GCF of a^3b^3, a^2b^2, and ab^2, the GCF of $5a^3b^3 - 15a^2b^2 + 10ab^2$ is $5ab^2$.

41. $-6y^{10} - 8y^8 - 4y^5$

$= (-2y^5) \cdot 3y^5 + (-2y^5) \cdot 4y^3 + (-2y^5) \cdot 2$

$= -2y^5(3y^5 + 4y^3 + 2)$

47. $a^2(a + 2) + 3(a + 2)$

$= (a + 2) \cdot a^2 + (a + 2) \cdot 3 \quad$ GCF is $(a + 2)$

$= (a + 2)(a^2 + 3)$

50. $x^2a + x^2b + y^2a + y^2b = (x^2a + x^2b) + (y^2a + y^2b)$

$\qquad\qquad\qquad\qquad = x^2(a + b) + y^2(a + b)$

$\qquad\qquad\qquad\qquad = (x^2 + y^2)(a + b)$

56. $-x^2y - x^2 - 3y - 3$

$= (-1)(x^2y + x^2 + 3y + 3) \qquad$ First factor -1 from all terms

$= (-1)[x^2(y + 1) + 3(y + 1)] \quad$ Factor x^2 from first two and 3 from second two

$= (-1)[(x^2 + 3)(y + 1)] \qquad (y + 1)$ is common to the two terms in brackets

$= -(x^2 + 3)(y + 1)$

15.2

10. $y^2 + 5y - 24 = (y + \underline{\quad})(y + \underline{\quad})$
$b = 5$ and $c = -24$

Factors of $c = -24$	Sum of factors
12, -2	10
-2, 12	-10
8, -3	5

With c negative, we know that the factors must have opposite signs. Since we have found factors of c whose sum is $b = 5$, we can stop the table.
$y^2 + 5y - 24 = (y + 8)(y - 3)$

14. $x^2 - 2x - 63 = (x + \underline{\quad})(x + \underline{\quad})$
$b = -2$ and $c = -63$

Factors of $c = -63$	Sum of factors
21, -3	18
-21, 3	-18
9, -7	2
-9, 7	-2

With c negative, we only try factors with opposite signs. Since $b = -2$ the factors are -9 and 7.
$$x^2 - 2x - 63 = (x - 9)(x + 7)$$

20. $u^2 - 9uv + 20v^2 = (u + \underline{\quad}v)(u + \underline{\quad}v)$
$b = -9$ and $c = 20$

Factors of $c = 20$	Sum of factors
$-10, -2$	-12
$-5, -4$	-9

With c positive and b negative, we only try factors that are both negative. Thus, $u^2 - 9uv + 20v^2 = (u - 5v)(u - 4v)$.

35. $u^2 - 22uv + 121v^2 = (u + \underline{\quad}v)(u + \underline{\quad}v)$
$b = -22$ and $c = 121$

Factors of $c = 121$	Sum of factors
$-121, -1$	-122
$-11, -11$	-22

With c positive and b negative, we only try factors that are both negative. Thus, $u^2 - 22uv + 121v^2 = (u - 11v)(u - 11v)$.

15.3

10. For $6z^2 - 13z - 28$ we have $a = 6$, $b = -13$, and $c = -28$. The factors of $c = -28$ will have opposite signs.

Factors of $a = 6$	Factors of $c = -28$
6, 1	28, -1 and -28, 1
3, 2	14, -2 and -14, 2
	7, -4 and -7, 4

$6z^2 - 13z - 28 = (\underline{\quad}z + \underline{\quad})(\underline{\quad}z + \underline{\quad})$

$\overset{?}{=} (6z - 1)(z + 28)$ **Does not work**

$\overset{?}{=} (3z - 2)(2z + 14)$ **There would have to be a common factor of 2 for the 14, 2 factors to work**

$\overset{?}{=} (3z - 4)(2z + 7)$ **Does not work**

$\overset{?}{=} (3z + 4)(2z - 7)$ **This works**

$6z^2 - 13z - 28 = (3z + 4)(2z - 7)$

17. $-45x^2 + 150x - 125$
$= (-5)9x^2 + (-5)(-30x) + (-5)(25)$
$= -5(9x^2 - 30x + 25)$
Now factor $9x^2 - 30x + 25$.

Factors of $a = 9$	Factors of $c = 25$
9, 1	$-25, -1$
3, 3	$-5, -5$

$9x^2 - 30x + 25 = (\underline{\quad}x + \underline{\quad})(\underline{\quad}x + \underline{\quad})$
$\overset{?}{=} (9x - 1)(x - 25)$ **Does not work**

$\overset{?}{=} (3x - 5)(3x - 5)$ **This works**

$9x^2 - 30x + 25 = (3x - 5)(3x - 5)$

We must include the common factor in the final answer.
$-45x^2 + 150x - 125 = -5(3x - 5)(3x - 5)$

22. $2x^2 + 7xy + 5y^2$ has $a = 2$, $b = 7$, and $c = 5$.

Factors of $a = 2$	Factors of $c = 5$
2, 1	5, 1

$2x^2 + 7xy + 5y^2 = (\underline{\quad}x + \underline{\quad}y)(\underline{\quad}x + \underline{\quad}y)$
$\overset{?}{=} (2x + y)(x + 5y)$ **Does not work**

$\overset{?}{=} (2x + 5y)(x + y)$ **This works**

$2x^2 + 7xy + 5y^2 = (2x + 5y)(x + y)$

31. $4u^2 - v^2$ has $a = 4$, $b = 0$, and $c = -1$.

Factors of $a = 4$	Factors of $c = -1$
4, 1	1, -1
2, 2	

$4u^2 - v^2 = (\underline{\quad}u + \underline{\quad}v)(\underline{\quad}u + \underline{\quad}v)$

$\overset{?}{=} (4u + v)(u - v)$ **Does not work**

$\overset{?}{=} (2u + v)(2u - v)$ **This works**

$4u^2 - v^2 = (2u + v)(2u - v)$

15.4

11. $9u^2 + 6u + 1$
$= (3u)^2 + 6u + 1^2$ **$(3u)^2$ and 1^2 are perfect squares**

$= (3u)^2 + 2 \cdot 3u \cdot 1 + 1^2$ **$3u = a$ and $1 = b$**

$= (3u + 1)^2$ **$a^2 + 2ab + b^2 = (a + b)^2$**

19. $-12y^2 + 60y - 75$
$= (-3)4y^2 + (-3)(-20y) + (-3)(25)$ **Common factor first**

$= -3(4y^2 - 20y + 25)$

$= -3[(2y)^2 - 2 \cdot 2y \cdot 5 + (5)^2]$ **$2y = a$ and $5 = b$**

$= -3(2y - 5)^2$

28. $25x^2 - 10xy + y^2 = (5x)^2 - 2 \cdot 5x \cdot y + (y)^2$
$= (5x - y)^2$

30. $u^4 - v^4 = (u^2)^2 - (v^2)^2$
$= (u^2 + v^2)(u^2 - v^2)$
$= (u^2 + v^2)(u + v)(u - v)$

41. In $4y^2 - 16y + 15$ the constants are $a = 4$, $b = -16$, and $c = 15$.

Factors of $a = 4$	Factors of $c = 15$
4, 1	$-15, -1$
2, 2	$-5, -3$

Since $b = -16$, the most likely combinations are 4, 1 with -5, -3 or 2, 2 with -5, -3.
$4y^2 - 16y + 15 = (\underline{\quad}y + \underline{\quad})(\underline{\quad}y + \underline{\quad})$
$\overset{?}{=} (4y - 5)(y - 3)$ **Does not work**

$\overset{?}{=} (2y - 5)(2y - 3)$ **This works**

$4y^2 - 16y + 15 = (2y - 5)(2y - 3)$

43. $3x^9 - 147x^3 = (3x^3)x^6 - (3x^3)49$ **Common factor is $3x^3$**

$= 3x^3(x^6 - 49)$

$$= 3x^3[(x^3)^2 - (7)^2] \qquad x^3 = a \text{ and}$$
$$ 7 = b$$
$$= 3x^3(x^3 + 7)(x^3 - 7)$$

49. $100x^5 - 60x^4y + 9x^3y^2$
$$= (x^3)100x^2 - (x^3)60xy + x^3(9y^2)$$
$$= x^3(100x^2 - 60xy + 9y^2)$$
$$= x^3[(10x)^2 - 2(10x)(3y) + (3y)^2]$$
$$= x^3(10x - 3y)^2$$

15.5

13. $x(3x + 7) = 0$

$\quad x = 0 \qquad$ or $\qquad 3x + 7 = 0 \qquad$ **Zero-product rule;**
$$\phantom{x = 0 \qquad \text{or} \qquad} \textbf{do not divide both}$$
$$\phantom{x = 0 \qquad \text{or} \qquad} 3x = -7 \qquad \textbf{sides by } x$$

$\quad x = 0 \qquad$ or $\qquad x = -\dfrac{7}{3}$

Thus, the solutions are 0 and $-\frac{7}{3}$. Remember that whenever one factor is x, 0 will always be one solution.

23. $\qquad\qquad 4u^2 - 8u = 0$
$$(4u) \cdot u - (4u) \cdot 2 = 0$$
$$4u(u - 2) = 0$$

$u = 0 \qquad$ or $\qquad u - 2 = 0$
$$\phantom{u = 0 \qquad \text{or} \qquad} u = 2$$

The solutions are 0 and 2. Do not forget the zero solution.

26. $(3x - 5) - (x + 7) = 0$

Do not try to use the zero-product rule on a problem like this. It involves a zero-difference, not a zero-product. Clear parentheses and solve.

$3x - 5 - x - 7 = 0 \quad$ **Watch the signs**
$$2x - 12 = 0 \quad \textbf{Collect like terms}$$
$$2x = 12$$
$$x = 6$$

Thus, the solution is 6.

27. $\qquad\qquad\qquad y(y + 3) = 10$
$$y^2 + 3y = 10$$
$$y^2 + 3y - 10 = 0$$
$$(y - 2)(y + 5) = 0$$

$y - 2 = 0 \qquad$ or $\qquad y + 5 = 0$
$$y = 2 \phantom{\qquad \text{or} \qquad} y = -5$$

The solutions are 2 and -5.

15.6

5. $\qquad\qquad$ Let $n = $ first integer,

$\qquad n + 1 = $ next consecutive integer,

$\qquad n(n + 1) = $ product of consecutive integers.

$$n(n + 1) = 240$$
$$n^2 + n = 240$$
$$n^2 + n - 240 = 0$$
$$(n - 15)(n + 16) = 0$$

$n - 15 = 0 \qquad$ or $\qquad n + 16 = 0$
$$n = 15 \phantom{\qquad \text{or} \qquad} n = -16$$
$$n + 1 = 16 \phantom{\qquad \text{or} \qquad} n + 1 = -15$$

Thus, 15 and 16 form one solution and -16 and -15 the other.

8. Let $n = $ the first even positive integer,

$\quad n + 2 = $ the next consecutive even positive integer.

The equation to solve is

$$n^2 + (n + 2)^2 = 100.$$
$$n^2 + n^2 + 4n + 4 = 100$$
$$2n^2 + 4n - 96 = 0$$
$$n^2 + 2n - 48 = 0 \quad \textbf{Divide out common factor 2}$$
$$(n + 8)(n - 6) = 0 \quad \textbf{Factor}$$

$n + 8 = 0 \qquad$ or $\qquad n - 6 = 0 \quad$ **Zero-product rule**
$$n = -8 \phantom{\qquad \text{or} \qquad} n = 6$$

Since n must be positive, we discard -8. Thus, the integers are 6 and 8 (since $n + 2 = 8$).

10. Let $x = $ length of one side,

$\quad x^2 = $ area of square,

$\quad 4x = $ perimeter of square.

If we add 4 to the area, the value is the same as the perimeter.

$$x^2 + 4 = 4x$$
$$x^2 - 4x + 4 = 0$$
$$(x - 2)^2 = 0$$

$x - 2 = 0 \qquad$ or $\qquad x - 2 = 0$
$$x = 2 \phantom{\qquad \text{or} \qquad} x = 2$$

The side is 2 cm.

15. We must find P when n is 30.
$$P = n^2 - 3n - 60$$
$$= (30)^2 - 3(30) - 60 \quad \textbf{Substitute 30 for } n$$
$$= 900 - 90 - 60$$
$$= 750$$

Thus, the profit is \$750 when 30 appliances are sold.

17. We must find n when P is 120.

$$P = n^2 - 3n - 60$$
$$120 = n^2 - 3n - 60 \quad \textbf{Substitute 120 for } P$$
$$0 = n^2 - 3n - 180$$
$$0 = (n + 12)(n - 15)$$

$n + 12 = 0 \qquad$ or $\qquad n - 15 = 0 \quad$ **Zero-product rule**
$$n = 12 \phantom{\qquad \text{or} \qquad} n = 15$$

Since the number of appliances cannot be negative, we discard -12. Thus, 15 were sold when the profit was $120.

19. We must find N when n is 7.

$$N = \frac{1}{2}n(n-1)$$

$$= \frac{1}{2}(7)(7-1) \quad \text{Substitute 7 for } n$$

$$= \frac{1}{2}(7)(6) = 21$$

Thus, 21 committees of two can be formed.

21. We must find n when N is 10.

$$N = \frac{1}{2}n(n-1)$$

$$10 = \frac{1}{2}n(n-1)$$

$$20 = n(n-1) \quad \text{Clear fraction}$$

$$20 = n^2 - n$$

$$0 = n^2 - n - 20$$

$$0 = (n+4)(n-5)$$

$$n + 4 = 0 \quad \text{or} \quad n - 5 = 0$$

$$n = -4 \qquad\qquad n = 5$$

Since n cannot be negative, there are 5 members of the organization.

16.1

7. $\dfrac{2x+7}{x^2+2x+1}$ We must solve $x^2 + 2x + 1 = 0$.

First factor, then use the zero-product rule.

$$x^2 + 2x + 1 = 0$$

$$(x+1)(x+1) = 0$$

$$x + 1 = 0 \quad \text{or} \quad x + 1 = 0$$

$$x = -1 \quad \text{or} \quad x = -1$$

Thus, -1 is the only number to be excluded.

16. $\frac{2}{x^2}$ and $\frac{2+x}{x^2+x}$ are not equivalent since one cannot be obtained from the other by multiplying both numerator and denominator by the same expression. Notice that the second fraction can be formed from the first by adding x to both numerator and denominator, but this process does not make the fraction equivalent. To see this, substitute 1 for x in both fractions.

17. $\frac{x-1}{3x-1}$ and $\frac{x}{3x}$ are not equivalent.
To obtain one from the other we would have to add or subtract the same number in both the numerator and denominator, and this process (unlike multiplying or dividing) does not yield equivalent fractions.

29. $\dfrac{x^2 - 3x - 10}{x^2 - 6x + 5} = \dfrac{(x-5)(x+2)}{(x-5)(x-1)} = \dfrac{x+2}{x-1}$

32. $\dfrac{x^2 - 9}{3 - x} = \dfrac{(x-3)(x+3)}{3-x} = \dfrac{x-3}{3-x} \cdot (x+3) =$

$$(-1)(x+3) = -(x+3)$$

36. $\dfrac{x^2 - 7xy + 6y^2}{x^2 - 4xy - 12y^2}$

$$= \dfrac{(x-y)(x-6y)}{(x+2y)(x-6y)} \quad \begin{array}{l}\text{Factor numerator and}\\ \text{denominator}\end{array}$$

$$= \dfrac{x-y}{x+2y} \quad \text{Divide out factors } (x-6y)$$

Do not divide out the x's since they are terms, not factors, of the numerator and denominator.

16.2

10. $\dfrac{z^2 - z - 20}{z^2 + 7z + 12} \cdot \dfrac{z+3}{z^2 - 25}$

$$= \dfrac{(z-5)(z+4)}{(z+3)(z+4)} \cdot \dfrac{(z+3)}{(z-5)(z+5)}$$

$$= \dfrac{(z-5)(z+4)(z+3)}{(z+3)(z+4)(z-5)(z+5)} = \dfrac{1}{z+5}$$

14. $\dfrac{2x^2 - 5xy + 3y^2}{x^2 - y^2} \cdot (x^2 + 2xy + y^2)$

$$= \dfrac{(2x-3y)(x-y)}{(x-y)(x+y)} \cdot \dfrac{(x+y)(x+y)}{1}$$

$$= \dfrac{(2x-3y)(x-y)(x+y)(x+y)}{(x-y)(x+y)}$$

$$= (2x - 3y)(x + y)$$

22. The reciprocal of $z + 2$ is $\frac{1}{z+2}$ and *not* $\frac{1}{z} + \frac{1}{2}$; a common error.

29. $(x+6) \div \dfrac{x^2 - 36}{x^2 - 6x}$

$$= \dfrac{(x+6)}{1} \cdot \dfrac{x^2 - 6x}{x^2 - 36} \quad \begin{array}{l}\text{Multiply by the}\\ \text{reciprocal}\end{array}$$

$$= \dfrac{(x+6)}{1} \cdot \dfrac{x(x-6)}{(x+6)(x-6)} \quad \text{Factor}$$

$$= \dfrac{(x+6)x(x-6)}{(x+6)(x-6)} \quad \begin{array}{l}\text{Divide out common}\\ \text{factors}\end{array}$$

$$= x$$

34. $\dfrac{x^2 + 16x + 64}{2x^2 - 128} \div \dfrac{3x^2 + 30x + 48}{x^2 - 6x - 16}$

$$= \dfrac{x^2 + 16x + 64}{2x^2 - 128} \cdot \dfrac{x^2 - 6x - 16}{3x^2 + 30x + 48}$$

$$= \dfrac{(x+8)(x+8)}{2(x-8)(x+8)} \cdot \dfrac{(x-8)(x+2)}{3(x+8)(x+2)}$$

$$= \dfrac{(x+8)(x+8)(x-8)(x+2)}{2(x-8)(x+8) \cdot 3 \cdot (x+8)(x+2)}$$

$$= \dfrac{1}{2 \cdot 3} = \dfrac{1}{6}$$

35. $\dfrac{x^2 - 5xy + 6y^2}{x^2 - 4y^2} \div (x^2 - 2xy - 3y^2)$

$= \dfrac{x^2 - 5xy + 6y^2}{x^2 - 4y^2} \cdot \dfrac{1}{x^2 - 2xy - 3y^2}$

$= \dfrac{(x - 2y)(x - 3y)}{(x - 2y)(x + 2y)} \cdot \dfrac{1}{(x - 3y)(x + y)}$

$= \dfrac{1}{(x + 2)(x + y)}$

38. $\dfrac{a^2 - y^2}{a^2 - ay} \cdot \dfrac{2a^2 + ay}{a^2 - 4y^2} \div \dfrac{a + y}{a + 2y}$

$= \dfrac{(a - y)(a + y)}{a(a - y)} \cdot \dfrac{a(2a + y)}{(a - 2y)(a + 2y)} \cdot \dfrac{(a + 2y)}{a + y}$

$= \dfrac{(a - y)(a + y)a(2a + y)(a + 2y)}{a(a - y)(a - 2y)(a + 2y)(a + y)}$

$= \dfrac{2a + y}{a - 2y}$

16.3

5. $\dfrac{2y}{y + 2} - \dfrac{y + 1}{y + 2} = \dfrac{2y - (y + 1)}{y + 2}$ **Use parentheses**

$= \dfrac{2y - y - 1}{y + 2} = \dfrac{y - 1}{y + 2}$

10. $\dfrac{z}{2} + \dfrac{3z - 1}{-2}$

$= \dfrac{z}{2} + \dfrac{(-1)(3z - 1)}{(-1)(-2)}$

$= \dfrac{z}{2} + \dfrac{1 - 3z}{2}$

$= \dfrac{z + (1 - 3z)}{2}$

$= \dfrac{z + 1 - 3z}{2} = \dfrac{-2z + 1}{2}$

14. $\dfrac{2a}{a - 1} + \dfrac{3a}{1 - a} = \dfrac{2a}{a - 1} + \dfrac{3a}{(-1)(a - 1)} =$

$\dfrac{2a}{a - 1} + \dfrac{-3a}{a - 1} = \dfrac{2a - 3a}{a - 1} = \dfrac{-a}{a - 1}$

20. $\dfrac{2x}{x^2 + x - 6} + \dfrac{x - 3}{6 - x - x^2}$

$= \dfrac{2x}{x^2 + x - 6} + \dfrac{x - 3}{(-1)(-6 + x + x^2)}$

$= \dfrac{2x}{x^2 + x - 6} + \dfrac{(-1)(x - 3)}{x^2 + x - 6}$

$= \dfrac{2x + (-1)(x - 3)}{x^2 + x - 6}$

$= \dfrac{2x - x + 3}{x^2 + x - 6} = \dfrac{(x + 3)}{(x + 3)(x - 2)} = \dfrac{1}{x - 2}$

24. $\dfrac{x + y}{x - y} - \dfrac{x + y}{y - x}$

$= \dfrac{x + y}{x - y} - \dfrac{(-1)(x + y)}{(-1)(y - x)}$

$= \dfrac{x + y}{x - y} + \dfrac{x + y}{x - y}$

$= \dfrac{(x + y) + (x + y)}{x - y}$

$= \dfrac{2x + 2y}{x - y} = \dfrac{2(x + y)}{x - y}$

25. $\dfrac{5x}{x + 1} + \dfrac{2x - 1}{x + 1} - \dfrac{3x}{x + 1} = \dfrac{5x + 2x - 1 - 3x}{x + 1}$

$= \dfrac{4x - 1}{x + 1}$

16.4

10. $\dfrac{1}{x^2 + 2x + 1}$ and $\dfrac{2}{x^2 - 1}$

$x^2 + 2x + 1 = (x + 1)(x + 1)$ **The LCD consists of one $(x - 1)$ and two $(x + 1)$'s**

$x^2 - 1 = (x + 1)(x - 1)$
Thus, LCD $= (x - 1)(x + 1)^2$.

14. $\dfrac{5x}{x + y}, \dfrac{7y}{2x + 2y}$, and $\dfrac{2xy}{3x + 3y}$

The factored denominators are $x + y$, $2(x + y)$, and $3(x + y)$. The LCD must consist of one $(x + y)$, one 2, and one 3. Thus, LCD $= 2 \cdot 3 \cdot (x + y) = 6(x + y)$.

23. $\dfrac{4y}{y^2 - 36} - \dfrac{4}{y + 6}$

$= \dfrac{4y}{(y - 6)(y + 6)} - \dfrac{4(y - 6)}{(y + 6)(y - 6)}$

$= \dfrac{4y - 4(y - 6)}{(y - 6)(y + 6)}$

$= \dfrac{4y - 4y + 24}{(y - 6)(y + 6)} = \dfrac{24}{(y - 6)(y + 6)}$

31. $\dfrac{3}{a^2 - a - 12} - \dfrac{2}{a^2 - 9}$

$= \dfrac{3}{(a - 4)(a + 3)} - \dfrac{2}{(a - 3)(a + 3)}$

$= \dfrac{3(a - 3)}{(a - 4)(a + 3)(a - 3)} - \dfrac{2(a - 4)}{(a - 3)(a + 3)(a - 4)}$

$= \dfrac{3(a - 3) - 2(a - 4)}{(a - 4)(a + 3)(a - 3)}$

$= \dfrac{3a - 9 - 2a + 8}{(a - 4)(a + 3)(a - 3)} = \dfrac{a - 1}{(a - 4)(a + 3)(a - 3)}$

35. $\dfrac{3}{x+1} + \dfrac{5}{x-1} - \dfrac{10}{x^2-1}$

$= \dfrac{3}{x+1} + \dfrac{5}{x-1} - \dfrac{10}{(x-1)(x+1)}$

    Factor denominators, the LCD $= (x-1)(x+1)$

$= \dfrac{3(x-1)}{(x-1)(x+1)} + \dfrac{5(x+1)}{(x-1)(x+1)} - \dfrac{10}{(x-1)(x+1)}$

    Supply missing factors

$= \dfrac{3(x-1) + 5(x+1) - 10}{(x-1)(x+1)}$ Add and subtract numerators over LCD

$= \dfrac{3x - 3 + 5x + 5 - 10}{(x-1)(x+1)}$

$= \dfrac{8x - 8}{(x-1)(x+1)}$ Collect like terms

$= \dfrac{8(x-1)}{(x-1)(x+1)} = \dfrac{8}{x+1}$ Divide out common factor

37. $\dfrac{2y}{y+5} - \dfrac{3y}{y-2} - \dfrac{2y^2}{y^2+3y-10}$

$= \dfrac{2y}{y+5} - \dfrac{3y}{y-2} - \dfrac{2y^2}{(y+5)(y-2)}$

$= \dfrac{2y(y-2)}{(y+5)(y-2)} - \dfrac{3y(y+5)}{(y-2)(y+5)} - \dfrac{2y^2}{(y+5)(y-2)}$

$= \dfrac{2y(y-2) - 3y(y+5) - 2y^2}{(y+5)(y-2)}$

$= \dfrac{2y^2 - 4y - 3y^2 - 15y - 2y^2}{(y+5)(y-2)}$

$= \dfrac{-3y^2 - 19y}{(y+5)(y-2)} = \dfrac{-y(3y+19)}{(y+5)(y-2)}$

16.5

4. $\dfrac{1}{z-3} + \dfrac{3}{z-5} = 0$

$\dfrac{1}{z-3} = -\dfrac{3}{z-5}$ Subtract $\dfrac{3}{z-5}$ from both sides

$(1)(z-5) = (-3)(z-3)$ Multiply by LCD

$z - 5 = -3z + 9$ Clear parentheses

$4z = 14$

$z = \dfrac{14}{4} = \dfrac{7}{2}$

Since $\frac{7}{2}$ will check in the original equation, it is the solution.

8. $\dfrac{z}{z+1} - 1 = \dfrac{1}{z}$

Multiply through by the LCD, $z(z+1)$.

$z(z+1)\dfrac{z}{z+1} - z(z+1)(1) = z(z+1)\dfrac{1}{z}$

$z^2 - (z^2 + z) = z + 1$ Use parentheses in this step to avoid a common sign error

$z^2 - z^2 - z = z + 1$

$-2z = 1$

$z = -\dfrac{1}{2}$

The solution is $-\frac{1}{2}$, which does indeed check in the original equation.

11. $\dfrac{4}{z-3} + \dfrac{2z}{z^2-9} = \dfrac{1}{z+3}$

The LCD $= (z-3)(z+3)$; so multiply both sides by it.

$\left[\dfrac{4}{z-3} + \dfrac{2z}{(z-3)(z+3)} \right](z-3)(z+3)$

$= \dfrac{1}{z+3}(z-3)(z+3)$

$\dfrac{4}{(z-3)}(z-3)(z+3) +$

$\dfrac{2z}{(z-3)(z+3)}(z-3)(z+3)$

$= \dfrac{1}{z+3}(z-3)(z+3)$

$4(z+3) + 2z = z - 3$

$4z + 12 + 2z = z - 3$

$6z + 12 = z - 3$

$5z = -15$

$z = -3$

However, if -3 is substituted into the original equation, we obtain zero in two denominators. That is, -3 must be excluded as a replacement for z. Thus, the equation has no solution.

16. $\dfrac{3}{x-3} + \dfrac{2}{x+3} = -1$

The LCD is $(x-3)(x+3)$. Multiply both sides by it, and do not forget to multiply on the right side

$\left[\dfrac{3}{x-3} + \dfrac{2}{x+3} \right](x-3)(x+3)$

$= (-1)(x-3)(x+3)$

$\dfrac{3}{(x-3)}(x-3)(x+3) + \dfrac{2}{(x+3)}(x-3)(x+3) =$

$(-1)(x^2 - 9)$

$3(x+3) + 2(x-3) = -x^2 + 9$

$3x + 9 + 2x - 6 = -x^2 + 9$

$5x + 3 = -x^2 + 9$

$$x^2 + 5x - 6 = 0$$

$$(x + 6)(x - 1) = 0$$

$x + 6 = 0$ or $x - 1 = 0$ **Zero-product rule**

$x = -6$ or $x = 1$

Both -6 and 1 are solutions since both check in the original equation.

23. Let x = the desired number.
The equation to solve is
$$\frac{5 + x}{7 + x} = \frac{4}{5}.$$

$5(5 + x) = 4(7 + x)$ **Multiply by LCD**

$25 + 5x = 28 + 4x$ **Clear parentheses**

$x = 3$

25. Let x = the numerator of the fraction,

$x + 4$ = the denominator of the fraction.
Then the fraction is $\frac{x}{x+4}$.
Now we increase *both* the numerator and the denominator by 1, and set the result equal to $\frac{2}{3}$.

$$\frac{x + 1}{x + 4 + 1} = \frac{2}{3}$$

$$\frac{x + 1}{x + 5} = \frac{2}{3}$$ **Multiply both sides by the LCD = 3(x + 5)**

$$3(x + 5) \cdot \frac{(x + 1)}{(x + 5)} = 3(x + 5) \cdot \frac{2}{3}$$

$$3(x + 1) = (x + 5)2$$

$$3x + 3 = 2x + 10$$

$$x = 7$$

Thus, the fraction (which was what we were asked to find) is $\frac{7}{11}$.

28. Let n = the desired number,

$n - 4$ = 4 less than the number,

$\dfrac{1}{n - 4}$ = the reciprocal of 4 less than the number,

$\dfrac{1}{n}$ = the reciprocal of the desired number.

The equation we must solve is

$$\frac{1}{n - 4} = 3 \cdot \frac{1}{n}$$

$$\frac{1}{n - 4} = \frac{3}{n}.$$

Multiplying both sides by $n(n - 4)$ results in the equation

$$n = 3(n - 4)$$

$$n = 3n - 12$$

$$-2n = -12$$

$$n = 6.$$

Thus, the desired number is 6, and 6 does satisfy the description of the problem.

30. Let t = time required for Bob's father to paint house,

12 = time required for Bob to paint house,

9 = time required for Bob and his father to paint the house working together.

Then,

$$\frac{1}{12} + \frac{1}{t} = \frac{1}{9}.$$

That is, the amount Bob does in 1 day plus the amount his father does in 1 day equals the amount they do together in 1 day.
Multiply through by $36t$, the LCD.

$$36t \cdot \frac{1}{12} + 36t \cdot \frac{1}{t} = 36t \cdot \frac{1}{9}$$

$$3t + 36 = 4t$$

$$36 = t$$

Thus, it would take Bob's father 36 days to paint the house.

32. Let x = time for Max to repair the car working alone,

$4x$ = time for Irv to repair the car working alone,

$\dfrac{1}{x}$ = amount done by Max in 1 hr,

$\dfrac{1}{4x}$ = amount done by Irv in 1 hr,

$\dfrac{1}{8}$ = amount done together in 1 hr.

The equation to solve is

$$\frac{1}{x} + \frac{1}{4x} = \frac{1}{8}.$$

Multiply through by the LCD, $8x$.

$$8x\left(\frac{1}{x}\right) + 8x\left(\frac{1}{4x}\right) = 8x\left(\frac{1}{8}\right)$$

$$8 + 2 = x$$

$$10 = x$$

Thus, it will take Max 10 hr and Irv 40 hr to repair the car working alone.

16.6

11. $\dfrac{y - 3}{y + 2} = \dfrac{y - 2}{y + 3}$

$(y - 3)(y + 3) = (y + 2)(y - 2)$ **Product of means equals product of extremes**

$$y^2 - 9 = y^2 - 4$$

$$-9 = -4$$

Since we obtain a contradiction, the proportion has no solutions.

17. Let x = the number of defective tires expected in a sample of 1288 tires. Then

$$\frac{6}{184} = \frac{x}{1288}.$$

$$6 \cdot 1288 = 184x$$

$$7728 = 184x$$

$$42 = x$$

Thus, you would expect 42 tires to be defective.

26. Let x = approximate number of fish in the lake. The ratio of total fish to the number tagged should be about the same as the ratio of the number of fish caught in the sample to the number of fish in the sample that were tagged.

That is, we must solve

$$\frac{x}{100} = \frac{70}{14}.$$

$14x = (70)(100)$ Product of means equals product of extremes

$x = 500$

Thus, there are about 500 fish in the lake.

30.

$c = kw$	This is the direct variation equation
$480 = k \cdot 12$	Substitute values for c and w, and let c be in cents to avoid decimals
$40 = k$	The value of the constant k is 40
$c = 40w$	This is the variation equation using the known value for the constant
$c = 40 \cdot 66 = 2640$	Substitute the new value of w

Thus, the cost of 66 lb of peaches would be $26.40.

34. The variation equation is

$$a = kt,$$

where a is the amount spent on recreation and t is total income. Then

$$1920 = k(16,000)$$

$$\frac{1920}{16,000} = k$$

$$0.12 = k.$$

Thus, the variation equation is

$$a = 0.12t.$$

Next we find a when his total income is increased by $5000; that is, when $t = 21,000$.

$$a = 0.12(21,000)$$

$$a = 2520$$

Hence, the man would spend about $2520 on recreation under these conditions.

37. The variation equation is

$$t = \frac{d}{r}$$

where t is the time, d is the fixed distance (constant), and r is the rate.

Then

$$6 = \frac{d}{500}$$

$$6(500) = d$$

$$3000 = d.$$

Thus, the variation equation is

$$t = \frac{3000}{r}.$$

When $r = 1200$ mph we have

$$t = \frac{3000}{1200}$$

$$= 2.5 \text{ hours}.$$

16.7

11. $\dfrac{\dfrac{1}{2x} - 1}{\dfrac{1}{x} - 2}$ The LCD of all fractions is $2x$. Multiply both numerator and denominator by $2x$.

$$\frac{\left[\dfrac{1}{2x} - 1\right]2x}{\left[\dfrac{1}{x} - 2\right]2x} = \frac{\dfrac{1}{2x} \cdot 2x - 1 \cdot 2x}{\dfrac{1}{x} \cdot 2x - 2 \cdot 2x} = \frac{1 - 2x}{2 - 4x} =$$

$$\frac{(1 - 2x)}{2(1 - 2x)} = \frac{1}{2}$$

15.

$$\frac{z - 2 - \dfrac{3}{z}}{1 + \dfrac{1}{z}} = \frac{\left[z - 2 - \dfrac{3}{z}\right]z}{\left[1 + \dfrac{1}{z}\right]z} = \frac{z^2 - 2z - 3}{z + 1}$$

$$= \frac{(z - 3)(z + 1)}{(z + 1)} = z - 3$$

17.

$$\frac{x - 3 + \dfrac{2}{x}}{x - 4 + \dfrac{3}{x}} = \frac{x\left(x - 3 + \dfrac{2}{x}\right)}{x\left(x - 4 + \dfrac{3}{x}\right)}$$ The LCD is x

$$= \frac{x^2 - 3x + 2}{x^2 - 4x + 3}$$

$$= \frac{(x - 1)(x - 2)}{(x - 1)(x - 3)}$$

$$= \frac{x - 2}{x - 3}$$

19. $V = \dfrac{3}{\dfrac{1}{S_1} + \dfrac{1}{S_2}}$

$$= \frac{3(S_1 S_2)}{\left(\dfrac{1}{S_1} + \dfrac{1}{S_2}\right)(S_1 S_2)}$$ $S_1 S_2$ is the LCD

$$= \frac{3S_1S_2}{\dfrac{S_1S_2}{S_1} + \dfrac{S_1S_2}{S_2}}$$

$$= \frac{3S_1S_2}{S_2 + S_1} = \frac{3S_1S_2}{S_1 + S_2}$$

If $S_1 = 2$ and $S_2 = 5$, then

$$V = \frac{3S_1S_2}{S_1 + S_2} = \frac{3(2)(5)}{2 + 5} = \frac{30}{7}.$$

17.1

14. $\sqrt{\dfrac{1000}{10}} = \sqrt{100}$ Reduce the fraction

$$= 10$$

19. $\sqrt{49a^2} = \sqrt{7^2a^2}$ Factor 49

$$= \sqrt{(7a)^2} \quad a^n b^n = (ab)^n$$

$$= 7a$$

23. $\sqrt{x^2 + 10x + 25} = \sqrt{(x + 5)^2}$ Factor the radicand

$$= x + 5$$

Do not make the mistake of trying to take the square roots term-by-term.

25. We are given that $p = \sqrt{w}$ and that w is 25. Thus,

$$p = \sqrt{w} = \sqrt{25} = 5.$$

The productivity is 5.

28. With $p = \sqrt{w}$ and $p = 4$ we have

$$p = \sqrt{w}$$

$$4 = \sqrt{w}$$

$$(4)^2 = (\sqrt{w})^2$$

$$16 = w.$$

17.2

28. $\sqrt{\dfrac{x^3y^3}{49}} = \dfrac{\sqrt{x^3y^3}}{\sqrt{49}} = \dfrac{\sqrt{x^2 \cdot x \cdot y^2 \cdot y}}{7} = \dfrac{\sqrt{x^2}\sqrt{y^2}\sqrt{xy}}{7}$

$$= \frac{xy\sqrt{xy}}{7}$$

31. $\sqrt{\dfrac{9}{5}} = \dfrac{\sqrt{9}}{\sqrt{5}} = \dfrac{3}{\sqrt{5}}$

$$= \frac{3\sqrt{5}}{\sqrt{5}\sqrt{5}} \quad \text{Rationalize the denominator}$$

$$= \frac{3\sqrt{5}}{5}$$

34. $\sqrt{\dfrac{36y^2}{x}} = \dfrac{\sqrt{36y^2}}{\sqrt{x}} = \dfrac{\sqrt{36}\sqrt{y^2}}{\sqrt{x}} = \dfrac{6y}{\sqrt{x}} =$

$$\frac{6y\sqrt{x}}{\sqrt{x}\,\sqrt{x}} = \frac{6y\sqrt{x}}{x}$$

35. $\sqrt{\dfrac{25x^3}{y^3z^4}} = \sqrt{\dfrac{25x^2x}{y^2y(z^2)^2}}$ Identify perfect squares

$$= \frac{5x}{yz^2}\sqrt{\frac{x}{y}} \quad \text{Remove perfect squares}$$

$$= \frac{5x}{yz^2}\sqrt{\frac{xy}{y^2}} \quad \begin{array}{l}\text{Multiply numerator and denomi-}\\\text{nator of radicand by } y \text{ to obtain}\\\text{a perfect square denominator}\end{array}$$

$$= \frac{5x}{yz^2} \cdot \frac{\sqrt{xy}}{y}$$

$$= \frac{5x\sqrt{xy}}{y^2z^2}$$

17.3

14. $\sqrt{3a^2}\sqrt{12a} = \sqrt{3a^2 \cdot 12a} = \sqrt{3^2 \cdot 2^2 \cdot a^2 \cdot a} =$
$\sqrt{3^2}\sqrt{2^2}\sqrt{a^2}\sqrt{a} = 3 \cdot 2 \cdot a\sqrt{a} = 6a\sqrt{a}$

17. $\sqrt{6x^2y}\sqrt{2xy^2} = \sqrt{6x^2y \cdot 2xy^2} =$
$\sqrt{2^2 \cdot 3 \cdot x^2 \cdot y \cdot x \cdot y^2} = \sqrt{2^2x^2y^2 \cdot 3xy} =$
$\sqrt{2^2}\sqrt{x^2}\sqrt{y^2}\sqrt{3xy} = 2xy\sqrt{3xy}$

19. $\sqrt{5x}\sqrt{15xy}\sqrt{3y} = \sqrt{5x \cdot 15xy \cdot 3y} = \sqrt{5^2 \cdot 3^2 \cdot x^2y^2}$
$= \sqrt{5^2}\sqrt{3^2} \cdot \sqrt{x^2} \cdot \sqrt{y^2} = 5 \cdot 3 \cdot xy = 15xy$

23. $\dfrac{\sqrt{98}}{\sqrt{18}} = \sqrt{\dfrac{98}{18}} \qquad \dfrac{\sqrt{a}}{\sqrt{b}} = \sqrt{\dfrac{a}{b}}$

$$= \sqrt{\frac{\cancel{2} \cdot 49}{\cancel{2} \cdot 9}} \quad \begin{array}{l}\text{Factor, look for perfect squares,}\\\text{divide out the 2's}\end{array}$$

$$= \frac{\sqrt{49}}{\sqrt{9}} = \frac{7}{3}$$

26. $\dfrac{5\sqrt{15}}{\sqrt{75}} = 5\sqrt{\dfrac{15}{75}} = 5\sqrt{\dfrac{\cancel{3} \cdot 5}{\cancel{3} \cdot 25}}$

$$= 5\sqrt{\frac{5}{25}} \quad \begin{array}{l}\text{Do not cancel 5's since 25 is a}\\\text{perfect square}\end{array}$$

$$= 5\frac{\sqrt{5}}{\sqrt{25}} = \frac{\cancel{5}\sqrt{5}}{\cancel{5}} = \sqrt{5}$$

31. $\dfrac{\sqrt{3y}}{\sqrt{4y}} = \sqrt{\dfrac{3\cancel{y}}{4\cancel{y}}} = \sqrt{\dfrac{3}{4}} = \dfrac{\sqrt{3}}{\sqrt{4}} = \dfrac{\sqrt{3}}{2}$

32. $\dfrac{\sqrt{27a^3}}{\sqrt{3}} = \sqrt{\dfrac{27a^3}{3}} \qquad \dfrac{\sqrt{a}}{\sqrt{b}} = \sqrt{\dfrac{a}{b}}$

$$= \sqrt{9a^2a}$$

$$= 3a\sqrt{a}$$

37. $\dfrac{5\sqrt{2xy}}{\sqrt{25x}} = 5\sqrt{\dfrac{2xy}{25x}} = 5\sqrt{\dfrac{2y}{25}} = 5\dfrac{\sqrt{2y}}{\sqrt{25}} =$

$$\frac{\cancel{5}\sqrt{2y}}{\cancel{5}} = \sqrt{2y}$$

38. $\dfrac{\sqrt{18x}}{\sqrt{2y}} = \sqrt{\dfrac{18x}{2y}} = \sqrt{\dfrac{9x}{y}} = \dfrac{\sqrt{9}\sqrt{x}}{\sqrt{y}} =$

$$\frac{3\sqrt{x}\sqrt{y}}{\sqrt{y}\sqrt{y}} = \frac{3\sqrt{xy}}{y}$$

17.4

5. $5\sqrt{12} - 3\sqrt{12} = (5-3)\sqrt{12}$ **Distributive law**

$\qquad = 2\sqrt{12}$

$\qquad = 2\sqrt{4 \cdot 3}$ **Look for perfect squares**

$\qquad = 2 \cdot 2\sqrt{3}$

$\qquad = 4\sqrt{3}$

10. $4\sqrt{50} + 7\sqrt{18} = 4\sqrt{25 \cdot 2} + 7\sqrt{9 \cdot 2} =$

$4\sqrt{25}\sqrt{2} + 7\sqrt{9}\sqrt{2} = 4 \cdot 5\sqrt{2} + 7 \cdot 3\sqrt{2} =$

$20\sqrt{2} + 21\sqrt{2} = (20 + 21)\sqrt{2} = 41\sqrt{2}$

14. $\sqrt{147} - 2\sqrt{75} = \sqrt{49 \cdot 3} - 2\sqrt{25 \cdot 3}$

$\qquad = 7\sqrt{3} - 2 \cdot 5\sqrt{3}$

$\qquad = 7\sqrt{3} - 10\sqrt{3}$

$\qquad = (7 - 10)\sqrt{3}$

$\qquad = -3\sqrt{3}$

18. $2\sqrt{4x} + 7\sqrt{9x} = 2\sqrt{4}\sqrt{x} + 7\sqrt{9}\sqrt{x} =$

$2 \cdot 2\sqrt{x} + 7 \cdot 3\sqrt{x} = 4\sqrt{x} + 21\sqrt{x} =$

$(4 + 21)\sqrt{x} = 25\sqrt{x}$

21. $x\sqrt{y^3} + y\sqrt{x^2y} = x\sqrt{y^2 \cdot y} + y\sqrt{x^2y} =$

$x\sqrt{y^2}\sqrt{y} + y\sqrt{x^2}\sqrt{y} = xy\sqrt{y} + yx\sqrt{y} =$

$(xy + xy)\sqrt{y} = 2xy\sqrt{y}$

22. $\sqrt{18} + \sqrt{50} + \sqrt{72} =$

$\sqrt{9 \cdot 2} + \sqrt{25 \cdot 2} + \sqrt{36 \cdot 2} =$

$\sqrt{9}\sqrt{2} + \sqrt{25}\sqrt{2} + \sqrt{36}\sqrt{2} =$

$3\sqrt{2} + 5\sqrt{2} + 6\sqrt{2} = (3 + 5 + 6)\sqrt{2} =$

$14\sqrt{2}$

25. $3\sqrt{75} + 2\sqrt{12} - 5\sqrt{48} =$

$3\sqrt{25 \cdot 3} + 2\sqrt{4 \cdot 3} - 5\sqrt{16 \cdot 3} =$

$3\sqrt{25}\sqrt{3} + 2\sqrt{4}\sqrt{3} - 5\sqrt{16}\sqrt{3} =$

$3 \cdot 5\sqrt{3} + 2 \cdot 2\sqrt{3} - 5 \cdot 4\sqrt{3} =$

$15\sqrt{3} + 4\sqrt{3} - 20\sqrt{3} = -\sqrt{3}$

29. $4\sqrt{300} - 2\sqrt{500} + 4\sqrt{125}$

$= 4\sqrt{100 \cdot 3} - 2\sqrt{100 \cdot 5} + 4\sqrt{25 \cdot 5}$ **Look for perfect squares**

$= 4 \cdot 10\sqrt{3} - 2 \cdot 10\sqrt{5} + 4 \cdot 5\sqrt{5}$

$= 40\sqrt{3} - 20\sqrt{5} + 20\sqrt{5}$

$= 40\sqrt{3}$

30. $\sqrt{4x} + \sqrt{16x} - \sqrt{25x} =$

$\sqrt{4}\sqrt{x} + \sqrt{16}\sqrt{x} - \sqrt{25}\sqrt{x} =$

$2\sqrt{x} + 4\sqrt{x} - 5\sqrt{x} = \sqrt{x}$

32. $9\sqrt{100a^2b^2} - 4\sqrt{36a^2b^2} - 10\sqrt{a^2b^2}$

$= 9\sqrt{100}\sqrt{a^2}\sqrt{b^2} - 4\sqrt{36}\sqrt{a^2}\sqrt{b^2} -$

$10\sqrt{a^2}\sqrt{b^2}$

$= 9 \cdot 10 \cdot ab - 4 \cdot 6 \cdot ab - 10 \cdot ab$

$= 90ab - 24ab - 10ab = 56ab$

34. $\sqrt{3} + \dfrac{1}{\sqrt{3}} = \sqrt{3} + \dfrac{\sqrt{3}}{\sqrt{3}\sqrt{3}} =$

$\sqrt{3} + \dfrac{\sqrt{3}}{3} = \left(1 + \dfrac{1}{3}\right)\sqrt{3} = \left(\dfrac{3}{3} + \dfrac{1}{3}\right)\sqrt{3} =$

$\dfrac{4}{3}\sqrt{3} = \dfrac{4\sqrt{3}}{3}$

38. $3\sqrt{7} - \sqrt{\dfrac{1}{7}} = 3\sqrt{7} - \dfrac{1}{\sqrt{7}}$

$\qquad = 3\sqrt{7} - \dfrac{1 \cdot \sqrt{7}}{\sqrt{7}\sqrt{7}}$ **Rationalize**

$\qquad = 3\sqrt{7} - \dfrac{\sqrt{7}}{7}$

$\qquad = \left(3 - \dfrac{1}{7}\right)\sqrt{7}$ **Distributive law**

$\qquad = \left(\dfrac{21}{7} - \dfrac{1}{7}\right)\sqrt{7}$

$\qquad = \dfrac{20\sqrt{7}}{7}$

39. $-3\sqrt{5} - \dfrac{9}{\sqrt{5}} = -3\sqrt{5} - \dfrac{9\sqrt{5}}{\sqrt{5}\sqrt{5}} =$

$-3\sqrt{5} - \dfrac{9\sqrt{5}}{5} = \left(-3 - \dfrac{9}{5}\right)\sqrt{5} =$

$\left(-\dfrac{15}{5} - \dfrac{9}{5}\right)\sqrt{5} = -\dfrac{24\sqrt{5}}{5}$

41. $\dfrac{\sqrt{2}}{\sqrt{10}} - \sqrt{5} = \sqrt{\dfrac{2}{10}} - \sqrt{5} = \sqrt{\dfrac{1}{5}} - \sqrt{5} =$

$\dfrac{1}{\sqrt{5}} - \sqrt{5} = \dfrac{\sqrt{5}}{\sqrt{5}\sqrt{5}} - \sqrt{5} = \dfrac{\sqrt{5}}{5} - \sqrt{5} =$

$\left(\dfrac{1}{5} - 1\right)\sqrt{5} = -\dfrac{4\sqrt{5}}{5}$

17.5

5. $\dfrac{\sqrt{25x^2}}{\sqrt{3y}} = \dfrac{\sqrt{25}\sqrt{x^2}}{\sqrt{3y}} = \dfrac{5x}{\sqrt{3y}} \cdot \dfrac{\sqrt{3y}}{\sqrt{3y}} = \dfrac{5x\sqrt{3y}}{3y}$

10. $\dfrac{\sqrt{2}}{\sqrt{50}} - \dfrac{\sqrt{3}}{\sqrt{75}} = \sqrt{\dfrac{2}{50}} - \sqrt{\dfrac{3}{75}} =$

$\sqrt{\dfrac{\cancel{2}}{25 \cdot \cancel{2}}} - \sqrt{\dfrac{\cancel{3}}{25 \cdot \cancel{3}}} = \sqrt{\dfrac{1}{25}} - \sqrt{\dfrac{1}{25}} = 0$

14. $\sqrt{x}\,\sqrt{y} + \dfrac{\sqrt{xy^2}}{\sqrt{y}} = \sqrt{xy} + \sqrt{\dfrac{xy^2}{y}} =$

$\sqrt{xy} + \sqrt{xy} = 2\sqrt{xy}$

17. $\sqrt{x^3}\sqrt{y} - 3x\sqrt{x}\sqrt{y} + \dfrac{4x\sqrt{xy^2}}{\sqrt{y}}$

$= \sqrt{x^3y} - 3x\sqrt{xy} + 4x\sqrt{\dfrac{xy^2}{y}}$

$= \sqrt{x^2xy} - 3x\sqrt{xy} + 4x\sqrt{xy}$

$= x\sqrt{xy} - 3x\sqrt{xy} + 4x\sqrt{xy}$

$= (x - 3x + 4x)\sqrt{xy} = 2x\sqrt{xy}$

20. $\sqrt{2}(\sqrt{27} - \sqrt{12}) = \sqrt{2}\,\sqrt{27} - \sqrt{2}\,\sqrt{12}$
$$= \sqrt{2 \cdot 27} - \sqrt{2 \cdot 12}$$
$$= \sqrt{9 \cdot 6} - \sqrt{4 \cdot 6}$$
$$= 3\sqrt{6} - 2\sqrt{6}$$
$$= \sqrt{6}$$

26. $\sqrt{15}\left(\dfrac{\sqrt{3}}{\sqrt{5}} + \dfrac{1}{\sqrt{15}}\right) = \dfrac{\sqrt{15}\sqrt{3}}{\sqrt{5}} + \dfrac{\sqrt{15}}{\sqrt{15}} =$

$\dfrac{\sqrt{5 \cdot 3 \cdot 3}}{\sqrt{5}} + 1 = \dfrac{\sqrt{5}\sqrt{9}}{\sqrt{5}} + 1 = 3 + 1 = 4$

28. $\sqrt{2}\left(\sqrt{6} + \dfrac{\sqrt{2}}{\sqrt{3}}\right) = \sqrt{2}\,\sqrt{6} + \dfrac{\sqrt{2}\,\sqrt{2}}{\sqrt{3}} =$

$\sqrt{2 \cdot 2 \cdot 3} + \dfrac{2}{\sqrt{3}} = \sqrt{4}\,\sqrt{3} + \dfrac{2}{\sqrt{3}} =$

$2\sqrt{3} + \dfrac{2}{\sqrt{3}} \cdot \dfrac{\sqrt{3}}{\sqrt{3}} = 2\sqrt{3} + \dfrac{2\sqrt{3}}{3} =$

$\left(2 + \dfrac{2}{3}\right)\sqrt{3} = \dfrac{8\sqrt{3}}{3}$

35. $(\sqrt{2} - 2)(\sqrt{2} + 1)$
$$= \sqrt{2} \cdot \sqrt{2} + \sqrt{2} \cdot 1 + (-2)\sqrt{2} + (-2) \cdot 1$$
$$= 2 + \sqrt{2} - 2\sqrt{2} - 2$$
$$= 2 - 2 + \sqrt{2} - 2\sqrt{2} = -\sqrt{2}$$

37. $(3\sqrt{3} - \sqrt{2})(2\sqrt{3} + \sqrt{2})$
$$= (3\sqrt{3})(2\sqrt{3}) + (3\sqrt{3})(\sqrt{2}) + (-\sqrt{2})(2\sqrt{3}) +$$
$$(-\sqrt{2})(\sqrt{2}) = 6 \cdot 3 + 3\sqrt{6} - 2\sqrt{6} - 2$$
$$= 18 - 2 + 3\sqrt{6} - 2\sqrt{6} = 16 + \sqrt{6}$$

43. $\dfrac{-8}{\sqrt{3} - \sqrt{7}} = \dfrac{-8(\sqrt{3} + \sqrt{7})}{(\sqrt{3} - \sqrt{7})(\sqrt{3} + \sqrt{7})} =$

$\dfrac{-8(\sqrt{3} + \sqrt{7})}{(\sqrt{3})^2 - (\sqrt{7})^2} = \dfrac{-8(\sqrt{3} + \sqrt{7})}{3 - 7} =$

$\dfrac{-8(\sqrt{3} + \sqrt{7})}{-4} = 2(\sqrt{3} + \sqrt{7}) = 2\sqrt{3} + 2\sqrt{7}$

44. $\dfrac{2\sqrt{3}}{\sqrt{3} - 1} = \dfrac{2\sqrt{3}(\sqrt{3} + 1)}{(\sqrt{3} - 1)(\sqrt{3} + 1)}$ **Rationalize**

$= \dfrac{2\sqrt{3}(\sqrt{3} + 1)}{3 - 1}$

$= \dfrac{\cancel{2}\sqrt{3}(\sqrt{3} + 1)}{\cancel{2}}$

$= 3 + \sqrt{3}$

47. $\dfrac{x - 1}{\sqrt{x} - 1} = \dfrac{(x - 1)(\sqrt{x} + 1)}{(\sqrt{x} - 1)(\sqrt{x} + 1)}$

$= \dfrac{(x - 1)(\sqrt{x} + 1)}{(\sqrt{x})^2 - (1)^2}$

$= \dfrac{\cancel{(x - 1)}(\sqrt{x} + 1)}{\cancel{x - 1}}$

$= \sqrt{x} + 1$

17.6

2. $3\sqrt{2x} - 9 = 0$

     $3\sqrt{2x} = 9$ **Isolate the radical**

     $\sqrt{2x} = 3$ **Simplify**

     $(\sqrt{2x})^2 = (3)^2$ **Square both sides**

     $2x = 9$

     $x = \dfrac{9}{2}$

Check: $3\sqrt{2 \cdot \dfrac{9}{2}} - 9 = 3\sqrt{9} - 9 = 3 \cdot 3 - 9 = 0$

6. $3\sqrt{a - 3} + 6 = 0$

     $3\sqrt{a - 3} = -6$ **Isolate the radical**

     $\sqrt{a - 3} = -2$ **Divide by 3**

     $(\sqrt{a - 3})^2 = (-2)^2$ **Square both sides**

     $a - 3 = 4$

     $a = 7$

Check: $3\sqrt{a - 3} + 6 = 3\sqrt{7 - 3} + 6$
$$= 3\sqrt{4} + 6 = 3 \cdot 2 + 6 \neq 0$$
Since 7 does not check, there is no solution.

13. $3\sqrt{x - 1} - \sqrt{x + 31} = 0$

     $3\sqrt{x - 1} = \sqrt{x + 31}$ **Isolate the radicals**

     $(3\sqrt{x - 1})^2 = (\sqrt{x + 31})^2$ **Square both sides**

     $9(x - 1) = x + 31$ **Square the 3 also**

     $9x - 9 = x + 31$

     $8x = 40$

     $x = 5$

Check: $3\sqrt{5 - 1} - \sqrt{5 + 31} = 3\sqrt{4} - \sqrt{36} =$
$$3 \cdot 2 - 6 = 0$$

17. $\sqrt{a^2 - 5} + a - 5 = 0$

     $\sqrt{a^2 - 5} = 5 - a$ **Subtract a and add 5**

     $(\sqrt{a^2 - 5})^2 = (5 - a)^2$

     $a^2 - 5 = 25 - 10a + a^2$ **Do not forget the $-10a$**

     $-5 = 25 - 10a$ **Subtract a^2**

     $10a = 30$

     $a = 3$

Check: $\sqrt{3^2 - 5} + 3 - 5 = \sqrt{9 - 5} - 2 =$
$$\sqrt{4} - 2 = 2 - 2 = 0$$

20. $\sqrt{x^2 + 3} + x = 6$

     $\sqrt{x^2 + 3} = 6 - x$

     $(\sqrt{x^2 + 3})^2 = (6 - x)^2$

     $x^2 + 3 = 36 - 12x + x^2$

     $3 = 36 - 12x$

     $12x = 33$

     $x = \dfrac{33}{12} = \dfrac{11}{4}$

Check: $\sqrt{\left(\dfrac{11}{4}\right)^2 + 3} + \dfrac{11}{4} = \sqrt{\dfrac{121}{16} + \dfrac{48}{16}} + \dfrac{11}{4}$

$= \sqrt{\dfrac{169}{16}} + \dfrac{11}{4} = \dfrac{13}{4} + \dfrac{11}{4} = \dfrac{24}{4} = 6$

25. $\sqrt{x - 3} - x + 5 = 0$

$\sqrt{x - 3} = x - 5$

$(\sqrt{x - 3})^2 = (x - 5)^2$

$x - 3 = x^2 - 10x + 25$

$0 = x^2 - 11x + 28$

$0 = (x - 7)(x - 4)$

$x - 7 = 0$ or $x - 4 = 0$

$x = 7$ or $x = 4$

Check: $\sqrt{7 - 3} - 7 + 5 = \sqrt{4} - 2 =$

$2 - 2 = 0$

$\sqrt{4 - 3} - 4 + 5 = \sqrt{1} + 1 =$

$1 + 1 \neq 0$

Since 7 checks and 4 does not, the only solution is 7.

28. $\sqrt{x + 10} + x - 2 = 0$

$\sqrt{x + 10} = 2 - x$ **Isolate radical**

$(\sqrt{x + 10})^2 = (2 - x)^2$ **Square both sides**

$x + 10 = 4 - 4x + x^2$ **Don't forget the middle term** $-4x$

$0 = x^2 - 5x - 6$

$0 = (x - 6)(x + 1)$

$x - 6 = 0$ or $x + 1 = 0$

$x = 6$ $x = -1$

Check: $x = 6$ Check: $x = -1$

$\sqrt{6 + 10} + 6 - 2 \overset{?}{=} 0$ $\sqrt{-1 + 10} + (-1) - 2 \overset{?}{=} 0$

$\sqrt{16} + 6 - 2 \overset{?}{=} 0$ $\sqrt{9} - 1 - 2 \overset{?}{=} 0$

$4 + 6 - 2 \neq 0.$ $3 - 1 - 2 = 0$

Since 6 does not check, the only solution is -1.

17.7

10. Let x = length of wire needed to reach top. Use the Pythagorean theorem.

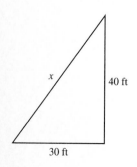

$x^2 = (40)^2 + (30)^2$

$x^2 = 1600 + 900$

$x^2 = 2500$

$x = \sqrt{2500} = 50$

The wire must be 50 ft long.

13.   Let x = the number,

$x + 5$ = five more than the number,

$\sqrt{x + 5}$ = square root of 5 more than the number,

$2\sqrt{x}$ = twice the square root of the number.

$\sqrt{x + 5} = 2\sqrt{x}$

$(\sqrt{x + 5})^2 = (2\sqrt{x})^2$

$x + 5 = 4x$

$5 = 3x$

$\tfrac{5}{3} = x$

Check: $\sqrt{\dfrac{5}{3} + 5} = \sqrt{\dfrac{5}{3} + \dfrac{15}{3}} = \sqrt{\dfrac{20}{3}} = \sqrt{\dfrac{4 \cdot 5}{3}} =$

$\sqrt{4}\sqrt{\dfrac{5}{3}} = 2\sqrt{\dfrac{5}{3}}$

16. We are given that $n = 64$ and asked to find c.

$c = 100\sqrt{n + 36} = 100\sqrt{64 + 36} = 100\sqrt{100} =$

$100(10) = 1000$

The cost is \$1000 when 64 appliances are produced.

22. We must find n when c is 50.

$c = 10\sqrt{n} + 100$

$50 = 10\sqrt{n} + 100$ **Substitute 50 for c**

$-50 = 10\sqrt{n}$

$-5 = \sqrt{n}$

At this point we can stop if we realize that $\sqrt{n}$ must be nonnegative and hence cannot be -5. Thus, there is no value of n for $c = 50$.

27. We must find c, the hypotenuse, given the two legs $a = 15$ and $b = 20$. By the Pythagorean theorem,

$$c^2 = a^2 + b^2$$

$$= 15^2 + 20^2$$

$$= 225 + 400 = 625$$

$$c = \sqrt{625} = 25$$

18.1

19. $4z^2 = 7z$

$4z^2 - 7z = 0$

$z(4z - 7) = 0$

$z = 0$ or $4z - 7 = 0$

$4z = 7$

$z = \dfrac{7}{4}$

Thus, the solutions are 0 and $\tfrac{7}{4}$.

26. $(3z - 1)(2z + 1) = 3(2z + 1)$

$$6z^2 + z - 1 = 6z + 3$$

$$6z^2 - 5z - 4 = 0$$

$$(3z - 4)(2z + 1) = 0$$

$$3z - 4 = 0 \quad \text{or} \quad 2z + 1 = 0$$

$$3z = 4 \qquad\qquad 2z = -1$$

$$z = \frac{4}{3} \qquad\qquad z = -\frac{1}{2}$$

Thus, the solutions are $\frac{4}{3}$ and $-\frac{1}{2}$.

28. $4x^2 - 9 = 3$

$$4x^2 = 12$$

$$x^2 = \frac{12}{4}$$

$$x^2 = 3$$

$$x = \pm\sqrt{3}$$

Thus, the solutions are $\sqrt{3}$ and $-\sqrt{3}$.

31. $3x^2 - 48 = -48$

$$3x^2 = 0 \qquad \text{Add 48 to both sides}$$

$$x^2 = 0 \qquad \text{Divide both sides by 3}$$

$$x = \pm\sqrt{0} \qquad \text{Take square root of both sides}$$

$$x = \pm 0 = 0 \quad \text{+0 and −0 are both 0}$$

Thus, 0 is the only solution.

35. $2x^2 - 150 = 0$

$$2x^2 = 150$$

$$x^2 = 75$$

$$x = \pm\sqrt{75} = \pm\sqrt{25 \cdot 3} = \pm 5\sqrt{3}$$

Thus, the two solutions are $5\sqrt{3}$ and $-5\sqrt{3}$.

37. $2x^2 + 50 = 0$

$$2x^2 = -50$$

$$x^2 = -25$$

Since $\sqrt{-25}$ is not a real number, there are no real solutions to this equation.

41. $(x - 2)^2 = 9$

$$(x - 2) = \pm\sqrt{9} \quad \text{Take square root of both sides}$$

$$x - 2 = \pm 3$$

$$x = 2 \pm 3 \quad \text{Add 2 to both sides}$$

$$x = 2 + 3 = 5 \quad \text{or} \quad x = 2 - 3 = -1$$

Thus, the solutions are 5 and −1.

18.2

6. $z^2 + \dfrac{1}{3}z + \underline{\hspace{2cm}}$

$$\left[\frac{1}{2} \text{ of } \frac{1}{3}\right]^2 = \left[\frac{1}{2} \cdot \frac{1}{3}\right]^2 = \left[\frac{1}{6}\right]^2 = \frac{1}{36}$$

Thus, we must add $\frac{1}{36}$ to complete the square.

---

13. $x^2 - 2x - 1 = 0$

$$x^2 - 2x \quad = 1$$

$$x^2 - 2x + 1 = 1 + 1 \qquad 1 = \left[\frac{1}{2} \text{ coefficient of } x\right]^2$$

$$(x - 1)^2 = 2$$

$$x - 1 = \pm\sqrt{2} \quad \text{Take square root of both sides}$$

$$x = 1 \pm \sqrt{2} \quad \text{Add 1 to both sides}$$

15. $\qquad 3z^2 + 12z = 135$

$$3z^2 + 12z - 135 = 0$$

$$z^2 + 4z - 45 = 0 \qquad \text{Divide through by 3}$$

$$z^2 + 4z \quad = 45$$

$$z^2 + 4z + 4 = 45 + 4 \quad \left[\frac{1}{2} \text{ of } 4\right]^2 = [2]^2 = 4$$

$$(z + 2)^2 = 49$$

$$z + 2 = \pm\sqrt{49} = \pm 7$$

$$z = -2 \pm 7$$

$$z = -2 + 7 = 5 \quad \text{or} \quad z = -2 - 7 = -9$$

Thus, the solutions are 5 and −9.

16. $2x^2 + x - 1 = 0$

$$x^2 + \frac{1}{2}x - \frac{1}{2} = 0 \qquad \text{Divide through by 2 to make coefficient of } x^2\text{-term equal to 1}$$

$$x^2 + \frac{1}{2}x + \frac{1}{16} = \frac{1}{2} + \frac{1}{16} \quad \left[\frac{1}{2} \text{ of } \frac{1}{2}\right]^2 = \left[\frac{1}{4}\right]^2 = \frac{1}{16}$$

$$\left(x + \frac{1}{4}\right)^2 = \frac{9}{16}$$

$$x + \frac{1}{4} = \pm\sqrt{\frac{9}{16}} = \pm\frac{3}{4}$$

$$x = -\frac{1}{4} \pm \frac{3}{4}$$

$$x = -\frac{1}{4} + \frac{3}{4} = \frac{2}{4} = \frac{1}{2} \quad \text{or}$$

$$x = -\frac{1}{4} - \frac{3}{4} = -1$$

Thus, the solutions are $\frac{1}{2}$ and −1.

20. $y^2 + 2y + 5 = 0$

$$y^2 + 2y \quad = -5$$

$$y^2 + 2y + 1 = -5 + 1$$

$$(y + 1)^2 = -4$$

But since $\sqrt{-4}$ is not a real number, there are no real solutions.

23. $5y^2 - 4y - 33 = 0$

We first try to factor.

$$(5y + 11)(y - 3) = 0$$

$$5y + 11 = 0 \quad \text{or} \quad y - 3 = 0 \quad \text{Zero-product rule}$$

$$5y = -11 \qquad\qquad y = 3$$

$$y = -\frac{11}{5} \quad \text{Thus, the solutions are } -\frac{11}{5} \text{ and 3.}$$

18.3

5. $0 = 7y - 2 + 15y^2$

$15y^2 + 7y - 2 = 0$ **Write in general form**

$y = \dfrac{-7 \pm \sqrt{49 - 4(15)(-2)}}{2 \cdot 15}$ $a = 15, b = 7,$ $c = -2$

$= \dfrac{-7 \pm \sqrt{49 + 120}}{30} = \dfrac{-7 \pm \sqrt{169}}{30} = \dfrac{-7 \pm 13}{30}$

$y = \dfrac{-7 + 13}{30} = \dfrac{6}{30} = \dfrac{1}{5}$ or

$y = \dfrac{-7 - 13}{30} = \dfrac{-20}{30} = -\dfrac{2}{3}$

Thus, the solutions are $\frac{1}{5}$ and $-\frac{2}{3}$.

7. $5x^2 - x = 1$

$5x^2 - x - 1 = 0$ **General form, $a = 5, b = -1,$** $c = -1$

$x = \dfrac{-b \pm \sqrt{b^2 - 4ac}}{2a}$

$= \dfrac{-(-1) \pm \sqrt{(-1)^2 - 4(5)(-1)}}{2(5)}$

$= \dfrac{1 \pm \sqrt{1 - (-20)}}{10}$

$= \dfrac{1 \pm \sqrt{21}}{10}$

11. $z^2 = z - 5$

$z^2 - z + 5 = 0$ **General form, $a = 1, b = -1,$** $c = 5$

$z = \dfrac{-b \pm \sqrt{b^2 - 4ac}}{2a}$

$= \dfrac{-(-1) \pm \sqrt{(-1)^2 - 4(1)(5)}}{2(1)}$

$= \dfrac{1 \pm \sqrt{1 - 20}}{2} = \dfrac{1 \pm \sqrt{-19}}{2}$

Since $\sqrt{-19}$ is not a real number, the equation has no real solution.

14. $y(y + 1) + 2y = 4$

$y^2 + y + 2y = 4$

$y^2 + 3y - 4 = 0$

$y = \dfrac{-3 \pm \sqrt{9 - 4(1)(-4)}}{2 \cdot 1}$ $a = 1, b = 3,$ $c = -4$

$= \dfrac{-3 \pm \sqrt{9 + 16}}{2} = \dfrac{-3 \pm \sqrt{25}}{2} = \dfrac{-3 \pm 5}{2}$

$y = \dfrac{-3 + 5}{2} = \dfrac{2}{2} = 1$ or

$y = \dfrac{-3 - 5}{2} = \dfrac{-8}{2} = -4$

Thus, the solutions are 1 and -4.

18. $(2z - 1)(z - 2) - 11 = 2(z + 4) - 8$

$2z^2 - 5z + 2 - 11 = 2z + 8 - 8$

$2z^2 - 7z - 9 = 0$

$z = \dfrac{7 \pm \sqrt{49 - 4(2)(-9)}}{2 \cdot 2}$ $a = 2, b = -7,$ $c = -9$

$= \dfrac{7 \pm \sqrt{49 + 72}}{4} = \dfrac{7 \pm \sqrt{121}}{4} = \dfrac{7 \pm 11}{4}$

$z = \dfrac{7 + 11}{4} = \dfrac{18}{4} = \dfrac{9}{2}$ or

$z = \dfrac{7 - 11}{4} = \dfrac{-4}{4} = -1$

Thus, the solutions are $\frac{9}{2}$ and -1.

20. $3z^2 + 10z - 1 = 0$

$z = \dfrac{-10 \pm \sqrt{100 - 4(3)(-1)}}{2 \cdot 3}$ $a = 3, b = 10,$ $c = -1$

$= \dfrac{-10 \pm \sqrt{100 + 12}}{6} = \dfrac{-10 \pm \sqrt{112}}{6}$

$= \dfrac{-10 \pm \sqrt{16 \cdot 7}}{6}$

$= \dfrac{-10 \pm 4\sqrt{7}}{6} = \dfrac{\cancel{2}(-5 \pm 2\sqrt{7})}{\cancel{2} \cdot 3}$

$= \dfrac{-5 \pm 2\sqrt{7}}{3}$

Thus, the solutions are $\frac{-5 + 2\sqrt{7}}{3}$ and $\frac{-5 - 2\sqrt{7}}{3}$.

26. $\dfrac{2}{3}x^2 - \dfrac{1}{3}x - 1 = 0$

$2x^2 - x - 3 = 0$ **Multiply by 3 to clear fractions**

$(2x - 3)(x + 1) = 0$

$2x - 3 = 0$ or $x + 1 = 0$

$2x = 3$ $x = -1$

$x = \dfrac{3}{2}$

Thus, the solutions are $\frac{3}{2}$ and -1.

32. Let $x =$ length of the fence.
By the Pythagorean theorem,

$x^2 = 4^2 + 4^2$

$x^2 = 16 + 16$

$x^2 = 16 \cdot 2$

$x = \pm\sqrt{16 \cdot 2} = \pm 4\sqrt{2}.$

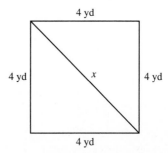

Since we are interested in the length of the fence, we discard the negative solution, $-4\sqrt{2}$. Thus, the length of the fence is $4\sqrt{2}$ yd, which is approximately 5.7 yards (using $\sqrt{2} \approx 1.414$).

18.4

7. $\dfrac{5}{z+3} - \dfrac{1}{z+3} = \dfrac{z+3}{z+2}$ **LCD = (z + 3)(z + 2)**

$\left[\dfrac{5}{z+3} - \dfrac{1}{z+3}\right](z+3)(z+2)$

$= \dfrac{z+3}{z+2}(z+3)(z+2)$ **Multiply by LCD**

$\dfrac{5}{z+3}(z+3)(z+2) - \dfrac{1}{z+3}(z+3)(z+2)$

$= (z+3)(z+3)$

$5(z+2) - (z+2) = z^2 + 6z + 9$

$5z + 10 - z - 2 = z^2 + 6z + 9$

$0 = z^2 + 2z + 1$

$0 = (z+1)(z+1)$

Thus, $z + 1 = 0$ or $z = -1$, which does indeed check in the original equation.

8. $\dfrac{x-1}{x} = \dfrac{x}{x+1}$ **LCD = x(x + 1)**

$\dfrac{x-1}{x}x(x+1) = \dfrac{x}{x+1}x(x+1)$ **Multiply by LCD**

$x^2 - 1 = x^2$

$-1 = 0$

Since we obtain a contradiction, there is no solution.

12. $\dfrac{x+2}{2} = \dfrac{x+5}{x+2}$

$(x+2)(x+2) = 2(x+5)$ **Cross-product equation**

$x^2 + 4x + 4 = 2x + 10$

$x^2 + 2x - 6 = 0$

Since this will not factor, we use the quadratic formula with $a = 1$, $b = 2$, and $c = -6$.

$x = \dfrac{-b \pm \sqrt{b^2 - 4ac}}{2a}$

$= \dfrac{-2 \pm \sqrt{2^2 - 4(1)(-6)}}{2(1)}$

$= \dfrac{-2 \pm \sqrt{4 + 24}}{2}$

$= \dfrac{-2 \pm \sqrt{28}}{2}$

$= \dfrac{-2 \pm 2\sqrt{7}}{2}$ $\sqrt{28} = \sqrt{4 \cdot 7} = 2\sqrt{7}$

$= \dfrac{2(-1 \pm \sqrt{7})}{2} = -1 \pm \sqrt{7}$

Thus, the solutions are $-1 \pm \sqrt{7}$.

17. $1 + 2\sqrt{x-1} = x$

$2\sqrt{x-1} = x - 1$ **Isolate the radical**

$(2\sqrt{x-1})^2 = (x-1)^2$ **Square both sides**

$4(x-1) = x^2 - 2x + 1$ **Remember to square 2 on the left**

$4x - 4 = x^2 - 2x + 1$

$0 = x^2 - 6x + 5$

$0 = (x-1)(x-5)$

$x - 1 = 0$ or $x - 5 = 0$

$x = 1$ or $x = 5$

Thus, since both 1 and 5 check in the original equation, they are the solutions.

22. $3\sqrt{y+1} - y - 1 = 0$

$3\sqrt{y+1} = y + 1$ **Isolate the radical**

$(3\sqrt{y+1})^2 = (y+1)^2$ **Square both sides**

$9(y+1) = y^2 + 2y + 1$

$9y + 9 = y^2 + 2y + 1$

$0 = y^2 - 7y - 8$

$0 = (y-8)(y+1)$

$y - 8 = 0$ or $y + 1 = 0$

$y = 8$ $y = -1$

Thus, the solutions are 8 and -1, which do check.

24. $\sqrt{2x^2 - 5} = x$

$(\sqrt{2x^2 - 5})^2 = x^2$ **Square both sides**

$2x^2 - 5 = x^2$

$x^2 = 5$

$x = \pm\sqrt{5}$

But if we check $-\sqrt{5}$, we have

$\sqrt{2(-\sqrt{5})^2 - 5} \stackrel{?}{=} -\sqrt{5}$

$\sqrt{5} \neq -\sqrt{5}.$

Since $-\sqrt{5}$ does not check, it must be discarded. Thus, $\sqrt{5}$ is the only solution that checks.

25.   *Let n* = the desired number,

$\dfrac{1}{n}$ = reciprocal of the number,

$n + \dfrac{1}{n}$ = the number increased by its reciprocal.

$n + \dfrac{1}{n} = \dfrac{5}{2}$ **LCD = 2n**

$2n\left[n + \dfrac{1}{n}\right] = 2n \cdot \dfrac{5}{2}$

$2n^2 + 2 = 5n$

$2n^2 - 5n + 2 = 0$

$(2n-1)(n-2) = 0$

$2n - 1 = 0$ or $n - 2 = 0$

$2n = 1$ or $n = 2$

$$n = \frac{1}{2}$$

Thus, the solutions are $\frac{1}{2}$ and 2.

18.5

8. Let x = a positive odd integer,

$x + 2$ = the next consecutive positive odd integer.

$3x(x + 2) = 297$

$x(x + 2) = 99$ **Divide both sides by 3**

$x^2 + 2x - 99 = 0$

$(x - 9)(x + 11) = 0$

$x - 9 = 0$ or $x + 11 = 0$

$x = 9$ or $x = -11$

Since x must be positive, we discard -11. Thus, the two integers are 9 and 11.

16. Let x = height of a triangle,

$\frac{1}{3}x$ = base of the triangle.

Since the formula for the area of a triangle is

$A = \frac{1}{2}bh = \frac{1}{2}$(base)(height), we must solve:

$$24 = \frac{1}{2}\left(\frac{1}{3}x\right)(x)$$

$$24 = \frac{1}{6}x^2$$

$144 = x^2$ **Multiply both sides by 6**

$\pm\sqrt{144} = x$

$\pm 12 = x.$

But since the height of a triangle must be positive we discard -12. Thus, the height is 12 ft and the base is 4 ft.

19. The area of a parallelogram is given by $A = bh$.

Let x = height of parallelogram,

$2x + 1$ = base of parallelogram.

The equation to solve is

$55 = x(2x + 1).$

$55 = 2x^2 + x$

$0 = 2x^2 + x - 55$

$0 = (2x + 11)(x - 5)$

$2x + 11 = 0$ or $x - 5 = 0$

$x = -\dfrac{11}{2}$ $x = 5$

Since the height cannot be negative, we discard $-\frac{11}{2}$. Thus, the height is 5 ft and the base is 11 ft (since $2x + 1 = 2(5) + 1 = 11$).

22. Let x = number of hrs required for Dan to grade tests, $x + 9$ = number of hrs required for Joe to grade tests.

$\dfrac{1}{x} + \dfrac{1}{x + 9} = \dfrac{1}{20}$ **Number each grades in 1 hr is $\frac{1}{20}$ of the total**

$20x(x + 9)\dfrac{1}{x} + 20x(x + 9)\dfrac{1}{(x + 9)} = 20x(x + 9)\dfrac{1}{20}$

$20x + 180 + 20x = x^2 + 9x$

$0 = x^2 - 31x - 180$

$0 = (x + 5)(x - 36)$

$x + 5 = 0$ or $x - 36 = 0$

$x = -5$ or $x = 36$

Discard -5. Thus, Dan takes 36 hrs and Joe takes 45 hrs to grade the tests.

25. $A = P(1 + r)^2$

$1210 = 1000(1 + r)^2$

$$\frac{1210}{1000} = (1 + r)^2$$

$\pm\sqrt{1.21} = 1 + r$

$\pm 1.1 = 1 + r$

Thus, $r = -1 \pm 1.1$, or $r = -2.1$ and $r = 0.1$. Since the negative value for r must be discarded, $r = 0.1$ or $r = 10\%$.

26. Let x = speed of boat in still water.

$x + 2$ = speed of boat downstream,

$x - 2$ = speed of boat upstream.

We use the distance formula $d = rt$, solved for t, to obtain:

$\dfrac{48}{x - 2}$ = time of travel upstream,

$\dfrac{48}{x + 2}$ = time of travel downstream.

Since the total time is 10 hr, we must solve

$$\frac{48}{x - 2} + \frac{48}{x + 2} = 10.$$

Multiply both sides by the LCD, $(x - 2)(x + 2)$ to obtain

$48(x + 2) + 48(x - 2) = 10(x - 2)(x + 2).$

$48x + 96 + 48x - 96 = 10x^2 - 40$

$0 = 10x^2 - 96x - 40$

$0 = 5x^2 - 48x - 20$

$0 = (5x + 2)(x - 10)$

$5x + 2 = 0$ or $x - 10 = 0$

$$x = -\frac{2}{5} \qquad\qquad x = 10$$

We can discard $-\frac{2}{5}$, so the speed of the boat in still water is 10 mph.

18.6

4. $g - ax = m$

     $-ax = m - g$ Subtract g from both sides

     $x = \dfrac{m - g}{-a}$ Divide both sides by $-a$

     $x = \dfrac{g - m}{a}$ Multiplying numerator and denominator by -1

11. Solve $A = \dfrac{1}{2}bh$ for b.

     $2A = bh$ Multiply both sides by 2

     $\dfrac{2A}{h} = b$ Divide both sides by h

14. $r = \dfrac{d}{t}$

     $rt = d$ Multiply both sides by t

     $t = \dfrac{d}{r}$ Divide both sides by r

16. Solve $S = \dfrac{1}{2}gt^2$ for t.

     $2S = gt^2$ Multiply by 2

     $\dfrac{2S}{g} = t^2$ Divide by g

     $\pm\sqrt{\dfrac{2S}{g}} = t$ Take square root of both sides

20. Solve $U = \sqrt{\dfrac{a}{c}}$ for c.

     $U^2 = \dfrac{a}{c}$ Square both sides

     $U^2c = a$ Multiply both sides by c

     $c = \dfrac{a}{U^2}$ Divide both sides by U^2

18.7

15. To determine the vertex of the parabola with equation

$$y = x^2 - 3$$

we must observe that the equation can be written in the form

$$y = (x - 0)^2 - 3.$$

Then it is clear that $h = 0$ and $k = -3$, so the vertex is $(0, -3)$.

GLOSSARY

Absolute value of a number The number of units that the number is from zero on a number line. **(10.2)**

Acute angle An angle with measure between 0° and 90°. **(9.2)**

Addends Two or more numbers to be added. **(1.2)**

Additive identity The number 0 is the additive identity. **(10.3)**

Additive inverse *See* Negatives.

Algebraic expression An expression containing variables and numbers combined by the basic operations of addition, subtraction, multiplication, and division. **(10.1)**

Algebraic fraction A fraction that contains a variable in the numerator or denominator or both. **(16.1)**

Algorithm A procedure or method used to perform a certain task. **(2.5)**

Angle A geometric figure consisting of two rays that share a common endpoint called the *vertex* of the angle. **(9.2)**

Area The number of square units in a two-dimensional figure. **(9.5)**

Associative law The law which states that regrouping sums or products does not change the results. **(1.3)**

Average (arithmetic mean) The sum of the quantities divided by the number of quantities. **(7.3)**

Bar graph A method of presenting numerical data by using a bar to show quantity. **(7.1)**

Base The number or variable in an exponential expression that is raised to a power. **(10.1)**

Binary operation An operation on two numbers at a time. **(1.3)**

Binomial A polynomial with two terms. **(14.3)**

Borrowing in a subtraction problem The reverse of carrying in an addition problem. **(1.6)**

Canceling factors The process of dividing out common factors by crossing them out. **(3.2)**

Carry number A number which must be added to the next column to the left in an addition problem. **(1.3)**

Cartesian coordinate system *See* Rectangular coordinate system.

Circle A plane figure with the property that all of its points are the same distance from a point called the *center*. **(9.4)**

Circle graph A method of presenting numerical data in pie-shaped sectors of a circle. **(7.1)**

Circumference The distance around a circle. **(9.4)**

Coefficient (numerical coefficient) The numerical factor of a term of an algebraic expression. **(10.5)**

Coinciding lines The graphs of equations that represent the same line. **(13.1)**

Commission Income that is a percent of sales. **(6.5)** *and* **(11.5)**

Common factor A factor of two expressions that occurs in both of them. **(15.1)**

Common multiple A number that is a multiple of two or more numbers. **(4.2)**

Commutative law The law which states that changing the order of sums or products does not change the results. **(1.2)**

Complementary angles Two angles whose measures total 90°. **(9.2)**

Complete the square To complete the square on $x^2 + bx$, add $[(1/2)b]^2$ to make the expression a perfect square. **(18.2)**

Complex fraction A fraction that contains at least one other fraction within it. **(16.7)**

Composite number A natural number greater than 1 that is not prime. **(2.7)**

Compound interest A type of interest that is paid on interest as well as principle. **(6.6)**

Conditional equation An equation that is true for some replacements of the variable and false for others. **(11.1)**

Congruent triangles Triangles that can be made to coincide by placing one on top of the other. **(9.3)**

Conjugates Expressions like $\sqrt{a} + \sqrt{b}$ and $\sqrt{a} - \sqrt{b}$ are conjugates of each other. **(17.5)**

Consecutive integers Integers with the property that one immediately follows the other in the regular counting order. **(11.5)**

Contradiction An equation that has no solutions. **(11.1)**

Counting number *See* Natural number. **(1.1)**

Cross products Products formed by multiplying the numerator of one fraction by the denominator of another, and vice-versa. **(3.2)**

Cube A rectangular solid which has $l = w = h$. **(9.6)**

Cube of a number A number raised to the third power. **(2.8)**

Cubic unit The volume of a cube that has measure one unit on each edge. **(9.6)**

Cylinder A solid with circular ends of the same radius. **(9.6)**

Decimal fractions (decimals) A fraction which has denominator equal to a power of 10. **(5.1)**

Decimal point The period which separates the ones digit from the tenths digit in a decimal fraction. **(5.1)**

Deductive reasoning Reasoning used to reach a specific conclusion based on a collection of generally accepted assumptions. **(9.1)**

Degree (of a polynomial) The degree of the term of highest degree in the polynomial. **(14.3)**

Degree (of a term) The exponent on the variable in the term of the polynomial. **(14.3)**

Denominator The bottom number in a fraction. **(3.1)**

Diameter The number which is twice the radius of a circle. **(9.4)**

Difference The result of a subtraction problem. **(1.5)**

Digit One of the first ten whole numbers, 0, 1, 2, 3, 4, 5, 6, 7, 8, or 9. **(1.1)**

Direct variation y varies directly as x means that $y = kx$. where k is the *constant of variation*. **(16.6)**

Discount A reduction in the regular price of an item when it is put on sale. **(6.5)**

Discriminant The discriminant of $ax^2 + bx + c = 0$ is $b^2 - 4ac$. **(18.3)**

Distributive law The law which states that quantities like the following are equal.

$$5 \times (7 + 2) = (5 \times 7) + (5 \times 2)$$
$$5 \times (7 - 2) = (5 \times 7) - (5 \times 2) \quad \textbf{(2.3)}$$

Dividend The number being divided in a division problem. **(2.4)**

Divisor The dividing number in a division problem. **(2.4)**

English system of measurement The measurement system used in the United States. **(8.1)**

Equation A statement that two quantities or expressions are equal. **(11.1)**

Equivalent equations (inequalities) Equations (inequalities) that have the same solutions. **(11.1)**

Exponent The number which tells how many times a number (or variable) is multiplied by itself. **(2.8)** *and* **(10.1)**

Exponential expression An expression of the form a^n where a is the *base* and n is the *exponent*. **(14.1)**

Extraneous root A ''solution'' obtained which is not a solution to the original equation. **(17.6)**

Extremes of a proportion The numbers a and d in the proportion $a/b = c/d$. **(16.6)**

Factor Numbers or variables (letters) which are multiplied in a term. **(10.5)**

Fraction A quotient of two whole numbers. **(3.1)**

Fractional equation An equation that contains one or more algebraic fractions. **(16.5)**

General form of a linear equation A linear equation written as $ax + by + c = 0$ is in general form. **(12.2)**

Gram The unit for weight in the metric system. **(8.3)**

Graph The set of points in a rectangular coordinate system that corresponds to solutions of an equation. **(12.2)**

Greatest common factor (GCF) of two terms The largest factor common to both terms. **(15.1)**

Histogram A type of bar graph on which the width of the bar corresponds to the class interval and the length of the bar to the class frequency. **(7.2)**

Hypotenuse The side opposite the right angle in a right triangle. **(17.7)**

Identities for addition and multiplication *See* Additive identity *and* Multiplicative identity.

Identity An equation that has every number as a solution. **(11.1)**

Improper fraction A fraction that has numerator greater than or equal to the denominator. **(3.1)**

Inductive reasoning Reasoning used to reach a general conclusion based on a limited collection of specific observations. **(9.1)**

Integers The numbers which consist of the natural numbers and their negatives together with zero. **(10.2)**

Interest Money paid to borrow money or money paid on an investment. **(6.6)** *and* **(11.5)**

Intersecting lines Lines having exactly one point in common. **(13.1)**

Inverse variation y varies inversely as x means that $y = k/x$, where k is the *constant of variation*. **(16.6)**

Irrational numbers All numbers on the number line that are not rational numbers, or numbers that have a decimal representation that does not terminate nor repeat. **(10.6)**

Least common denominator (LCD) The least common multiple (LCM) of all denominators in an addition or subtraction problem. **(4.3)** *and* **(16.4)**

Least common multiple (LCM) The smallest of all common multiples of two or more counting numbers. **(4.2)**

Legs The two sides of a right triangle that are not the hypotenuse. **(17.7)**

Like fractions Fractions having the same denominator. **(4.1)** *and* **(16.3)**

Like radicals Radical expressions with the same radicand. **(17.4)**

Like terms *See* Similar terms.

Line All points on a segment together with all the points beyond the endpoints of the segment. **(9.2)**

Linear equation (one variable) An equation in which the variable occurs to the first power only and which can be written in the form $ax + b = 0$. **(11.1)**

Linear equation (two variables) An equation of the form $ax + by = c$, where x and y are variables and a, b, and c are constants. **(12.1)**

Linear inequality (one variable) A statement involving the inequality symbols $<$, $>$, $\leq$, or $\geq$, such as $x + 1 < 7$ or $2x - 1 \geq 3$. **(11.7)**

Linear inequality (two variables) A statement involving two variables and the inequality symbols $<$, $>$, $\leq$, or $\geq$, in which the variables are raised to the first power. **(12.5)**

Liter The unit of volume in the metric system. **(8.3)**

Markdown The amount of a price reduction. **(6.8)**

Markup The amount by which the selling price is increased over the cost. **(6.8)**

Mean (average) The sum of values divided by the number of values. **(7.3)**

Means of a proportion The numbers b and c in the proportion $a/b = c/d$. **(16.6)**

Median The middle value, or average of the two middle quantities, of several values. **(7.3)**

Meter The unit for length in the metric system. **(8.3)**

Metric system of measurement The measurement system used in most countries of the world. **(8.3)**

Minuend The number before the minus sign or the top number in a subtraction problem. **(1.5)**

Mixed number The sum of a whole number and a proper fraction. **(3.5)**

Mixture problem A problem involving two quantities mixed together and is solved by finding a quantity equation and a value equation. **(13.5)**

Mode The value occuring most frequently in a set of values. **(7.3)**

Monomial A polynomial with one term. **(14.3)**

Motion problem A problem that makes use of the equation $d = rt$, distance equals rate times time. **(13.5)**

Multiple of a number The product of the number and a whole number. **(2.1)**

Multiplicative identity The number 1 is the multiplicative identity. **(10.4)**

Multiplicative inverse The multiplicative inverse of a ($a \neq 0$) is $1/a$. **(10.4)**

Natural number The numbers used in counting: 1, 2, 3, 4, 5, 6, . . . **(1.1)**

Negatives The negative of a is $-a$. **(10.3)**

Number line A line used to "picture" numbers. **(1.1)**

Numeral A symbol used to represent a number. **(1.1)**

Numerator The top number in a fraction **(3.1)**

Numerical coefficient The numerical multiplier of a term. **(14.3)**

Obtuse angle An angle with measure between 90° and 180°. **(9.2)**

Ordered pair A pair of numbers (x, y) with x-coordinate in the first position and y-coordinate in the second position. **(12.1)**

Parabola The U-shaped curve that is the graph of a quadratic equation in two variables. **(18.7)**

Parallel lines Distinct lines in a plane that do not intersect. **(9.2)**, **(12.3)**, *and* **(13.1)**

Parallelogram A four-sided figure whose opposite sides are parallel and equal. **(9.5)**

Percent The number of parts in one hundred parts. **(6.1)**

Perfect square (number) The result when a whole number, an integer, or a rational number is squared. **(2.8)**, **(10.6)**, *and* **(17.1)**

Perfect square (trinomial) A trinomial that can be factored into one of the forms $(a + b)^2$ or $(a - b)^2$. **(15.4)**

Perimeter The distance around a figure. **(9.5)**

Perpendicular lines Two lines that intersect in right angles. **(9.2)** *and* **(12.3)**

Plane A surface with the property that any two points in it can be joined by a line, all points of which are also contained in the surface. **(9.2)**

Polynomial An algebraic expression having, as terms, products of numbers and variables with whole-number exponents. **(14.3)**

Prime number A natural number having exactly two different divisors, 1 and itself. **(2.7)**

Prime polynomial A polynomial that cannot be factored. **(15.1)**

Principal Money borrowed or invested. **(6.6)**

Principal square root The nonnegative square root of a positive real number. **(17.1)**

Product The result of a multiplication problem. **(2.1)**

Proper fraction A fraction which has numerator less than the denominator. **(3.1)**

Proportion A statement that two ratios are equal. **(6.1)** *and* **(16.6)**

Quadrants The four sections formed by the axes in a rectangular coordinate system. **(12.1)**

Quadratic equation (one variable) An equation that can be written in the form $ax^2 + bx + c = 0$, $a \neq 0$. **(18.1)**

Quadratic equation (two variables) An equation of the form $y = ax^2 + bx + c$, $a \neq 0$. **(18.7)**

Quotient The result of a division problem. **(2.4)**

Radical The symbol $\sqrt{\ }$ used to indicate the square root. **(2.8)** *and* **(17.1)**

Radicand The number or expression written under the radical. **(17.1)**

Radius of a circle The distance from the center to each point on a circle. **(9.4)**

Rate of interest The percent interest. **(6.6)**

Ratio A comparison of two numbers using division. **(6.1)** *and* **(16.6)**

Rational expression The quotient of two polynomials. **(16.1)**

Rational numbers The fractions formed by dividing pairs of integers. **(10.2)**

Rationalizing a denominator The process of removing radicals from the denominator of an expression. **(17.2)**

Ray That part of a line lying on one side of point P on the line together with the point P. **(9.2)**

Real numbers The rational numbers together with the irrational numbers. **(10.6)**

Reciprocal The fraction formed by interchanging numerator and denominator. **(3.4)** *and* **(16.2)**

Rectangle A four–sided figure with opposite sides parallel and adjacent sides perpendicular. **(9.5)**

Rectangular coordinate system Used to graph equations and inequalities in two variables and formed by a horizontal and a vertical number line. **(12.1)**

Reduced to lowest terms A fraction has been reduced to lowest terms when 1 (or −1) is the only divisor of both the numerator and denominator. **(16.1)**

Remainder The final difference in a division problem. **(2.5)**

Right angle An angle with measure 90°. **(9.2)**

Right triangle A triangle with a 90° angle. **(17.7)**

Sale price Regular price minus the discount. **(6.5)**

Scientific notation A number written as the product of a power of 10 and a number that is greater than or equal to 1 and less than 10. **(14.2)**

Secant A line that intersects a circle in two points. **(9.4)**

Segment A geometric figure made up of two endpoints and all points between them. **(9.2)**

Similar terms Terms containing the same variables. **(10.5)**

Similar triangles Two triangles with the property that the angles of one are equal to the angles of the other. **(9.3)**

Simple interest Interest charged only on the principal. **(6.6)**

Slope of a line The measure of the direction of a line in a coordinate system. **(12.3)**

Solution to an equation (or inequality) in one variable A number which makes the equation (or inequality) true. **(11.1)** *and* **(11.7)**

Solution to an equation (or inequality) in two variables An ordered pair of numbers which, when substituted for the variables, results in a true equation (or inequality). **(12.1)** *and* **(12.5)**

Solution to a system of equations An ordered pair of numbers that is a solution to both equations in the system. **(13.1)**

Sphere A solid in the shape of a ball. **(9.6)**

Square A rectangle with all sides equal. **(9.5)**

Square of a number A number raised to the second power. **(2.8)**

Square root of a number If $a^2 = x$, then a is a square root of x. **(17.1)**

Square root of a perfect square The number that is squared to give the perfect square. **(2.8)** *and* **(10.6)**

Square unit The area of a square that measures one unit on each side. **(9.5)**

Statistics An area of mathematics dealing with the collection, presentation, and interpretation of numerical data. **(7.1)**

Straight angle An angle with measure 180°. **(9.2)**

Subtrahend The number following the minus sign or the bottom number in a subtraction problem. **(1.5)**

Sum The result of an addition problem. **(1.2)**

Supplementary angles Two angles whose measures total 180°. **(9.2)**

Surface area The area of all the surfaces of an object. **(9.6)**

System of equations A pair of linear equations. **(13.1)**

Tangent A line intersecting a circle in exactly one point. **(9.4)**

Term A part of an algebraic expression which is a product of numbers and variables and which is separated from the rest of the expression by plus (or minus) signs. **(10.5)**

Test point A point used to determine which side of the line is the solution of a linear inequality in two variables. **(12.5)**

Transversal A line intersecting each of two parallel lines. **(9.2)**

Trapezoid A four–sided figure with two parallel sides. **(9.5)**

Triangle A three–sided figure. **(9.3)**

Trinomial A polynomial with three terms. **(14.3)**

Unit fraction A fraction which has value 1 and is used in measurement conversions. **(8.1)**

Unit price The total price divided by the number of units. **(5.7)**

Unlike fractions Fractions with different denominators. **(4.1)** *and* **(16.3)**

Unlike radicals Radical expressions with different radicands. **(17.4)**

Variable A letter that can be replaced by various numbers. **(10.1)**

Vertex of an angle The common endpoint of the rays that form the angle. **(9.2)**

Vertex of a parabola The low point of a parabola

opening up and the high point of a parabola opening down. **(18.7)**

Vertical angles Two non-adjacent angles formed by intersecting lines. **(9.2)**

Volume The number of cubic units in a solid. **(9.6)**

Whole number The natural numbers together with zero: 0, 1, 2, 3, 4, . . . **(1.1)**

x-**axis** The horizontal number line in a rectangular coordinate system. **(12.1)**

x-**intercept** The point where the line crosses the *x*-axis. **(12.2)**

y-**axis** The vertical number line in a rectangular coordinate system. **(12.1)**

y-**intercept** The point where the line crosses the *y*-axis. **(12.2)**

Second cross product, 537
Segment, 466
Selling price, 351–52
Semicircles, 486
Similar (like) terms, 563, 592
Simplifying an expression, 564
Simplifying radicals, 929–34, 952, 953
Slope, 669–73
Slope-intercept form, 678
Social Security tax, 354–55
Solution(s), 582, 648
  to proportions, 899–900
  to a system of equations, 705, 711, 712
Solving equations, 582, 711
  addition-subtraction rule, 583, 627, 1024
  formulas, 1024–27
  fractional, 887–90
  graphing, 711
  multiplication-division rule, 587
  quadratic, 838–40, 986–1035
  by the quadratic formula, 1000–1004
  radical, 960–63, 1010, 1011, 1026, 1027
  with parentheses, 565
Sphere, 511, 512
  surface area of, 512
  volume of, 512
Square, 492
  area of, 494
  perimeter of, 492
Square of number, 114, 151, 525
Square root(s), 114, 151, 570, 923
  negative, 924
  positive, 924
  principal, 571, 924
  table of, 115, 571
Standard notation, 4
Statistics, 395
  using ratio in, 396–98
Substitution method
  solving systems of equations by, 719–21
Subtraction
  basic facts, 34, 546
  on the calculator, 306
  checking, 42–44
  and the column method, 769
  of decimals, 260–61
  difference, 34
  distributive law of, 562
  of fractions, 188–89, 201–5, 209–13
  horizontal display, 34
  of integers, 546, 547
  of like algebraic fractions, 875
  of measurement numbers, 437
  of mixed numbers, 210–12
  of polynomials, 768
  of radicals, 944–47
  of rational numbers, 546, 547
  symbols used, 34
  take-away method, 34–36
  of unlike algebraic fractions, 880–82
  vertical display, 34
  of whole numbers, 34–44
Subtrahend, 34

Sum, 11
Sum of two cubes, 836
Supplementary angles, 469
Surface area, 508
  of a cube, 509
  of a cylinder, 511
  of a rectangular solid, 508
  of a sphere, 512
Symbols
  addition (+), 11
  approximately equal to (≈), 251
  division (÷), 80
  equal to (=), 535
  greater than (>), 2, 534
  greater than or equal to (≥), 535, 536
  grouping, 116–17, 526
  of inequality, 2
  less than (<), 2, 534
  less than or equal to (≤), 535, 536
  multiplication (·), 61–62
  percent (%), 329
  radical, 115, 571
  subtraction (−), 34
  unequal (≠), 42, 527, 535
Systems of equations, 705
  solutions to 705, 711, 712
    techniques, 731
  solving by elimination, 725–28
  solving by graphing, 705, 711, 712
  solving by substitution, 719–21
  work problems using, 731–33

Tables
  addition, 12
  decimals, fractions, and percents, 247, 334
  of household units, 430
  multiplication, 63
  units of measure in the English system, 427, 430, 449
  units of measure in the metric system, 443–46, 449
Take-away method, 34–36
Taking roots, 988–90
Tangent, 486, 487
Tax
  income, 262, 355–56
  sales, 351–54
    rate, 351–52
  Social Security, 354–55
Temperature, 453–55
Terminating decimals, 243–44
Term(s), 563, 760
  coefficient of, 760
  collecting like (similar), 762
  degree of, 761
  factors of, 563
Test point, 687
Times sign, 61–62
Total value, 94, 733
Transversal, 470, 471
Trapezoid, 497, 498
Triangle(s), 475–79
  area, 496
  classifying, 475, 476
  congruent, 477, 478

right, 476, 966
  similar, 478, 479
Trinomials, 760
  factoring, 814–34
  perfect square, 785, 786, 830, 831
    and factoring, 830, 831

Unequal to (≠), 42, 527, 535
Unit fraction, 428
Unit price, 300
Units of measure, 427–32, 442–46, 492
  English system, 427, 430, 449
  metric system, 442–46
  table, 427, 430, 443–46, 449
Unlike fractions, 873
Unlike radicals, 944
Unlike terms, 762

Variable(s), 524, 592, 760
Variation, 901
  constants of, 902
Varies directly, 901
Varies inversely, 902
Vertex
  of angle, 468
  of parabola, 1034, 1035
  of triangle, 475
Vertical axis, 646
Volume, 506
  of a cube, 507
  of a cylinder, 510
  of a rectangular solid, 506
  of a sphere, 512

Whole numbers, 1
  addition of, 11–24
  division of, 80–94
  fractions, 132
  multiplication of, 66–71
  order of, 2, 3
  quotients of two, 289
  subtraction of, 34–44
  word names for, 5
Word names
  for decimals, 233–36
  for whole numbers, 5
Word problems. See Application.

x-axis, 646
x-coordinate, 646
x-intercept, 658, 659

y-axis, 646
y-coordinate, 646
y-intercept, 658, 659

Zero
  in addition, 13
  in division, 82, 83, 856
  as exponent, 749
  in multiplication, 62–63
  as place-holder, 4, 89
  in subtraction, 41–42
Zero-product rule, 837
  and quadratic equations, 837–40, 986–88

NOTES

NOTES

NOTES

NOTES

NOTES